Linear Functions

$f(x) = mx + b$
$Slope = m > 0$

$f(x) = mx + b$
$Slope = m < 0$

Constant Function $f(x) = b$

Identity Function $f(x) = x$

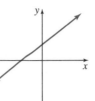 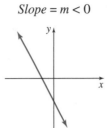 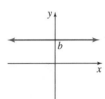 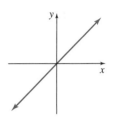

Square Function $f(x) = x^2$

Cube Function $f(x) = x^3$

Square Root Function $f(x) = \sqrt{x}$

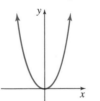 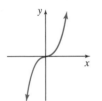 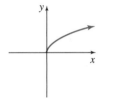

Greatest Integer Function $f(x) = [x]$

Absolute Value Function $f(x) = |x|$

 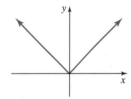

Power Functions

$f(x) = x^n$ (n even)

$f(x) = x^n$ (n odd)

Reciprocal Functions

$f(x) = \dfrac{1}{x}$

$f(x) = \dfrac{1}{x^2}$

 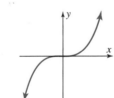 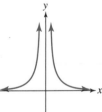

Exponential Functions

$f(x) = b^x$ $(b > 1)$

$f(x) = b^x$ $(0 < b < 1)$

Logarithmic Functions

$f(x) = \log x$

$f(x) = \ln x$

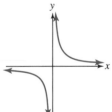

Continues→

Catalog of Basic Functions   (*continued*)

Trigonometric Functions

$f(t) = \sin t$

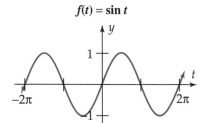

$f(t) = \cos t$

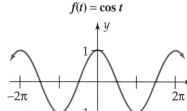

$f(t) = \tan t$

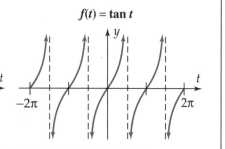

Rectangular and Parametric Equations for Conic Sections

Circles
Center (h, k), radius r

$$(x - h)^2 + (y - k)^2 = r^2$$

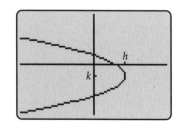

$x = r \cos t + h$
$y = r \sin t + k$ $\quad (0 \le t \le 2\pi)$

Ellipse
Center (h, k)

$$\frac{(x - h)^2}{a^2} + \frac{(y - k)^2}{b^2} = 1$$

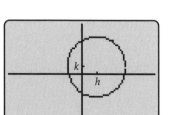

$x = a \cos t + h$
$y = b \sin t + k$ $\quad (0 \le t \le 2\pi)$

Parabola
Vertex (h, k)

$$(x - h)^2 = 4p(y - k)$$

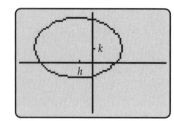

$x = t$ $\quad (t$ any real$)$
$y = \dfrac{(t - h)^2}{4p} + k$

Parabola
Vertex (h, k)

$$(y - k)^2 = 4p(x - h)$$

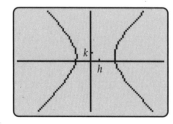

$x = \dfrac{(t - k)^2}{4p} + h$
$y = t$ $\quad (t$ any real$)$

Hyperbola
Center (h, k)

$$\frac{(x - h)^2}{a^2} - \frac{(y - k)^2}{b^2} = 1$$

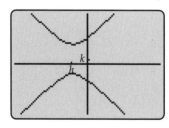

$x = \dfrac{a}{\cos t} + h$
$y = b \tan t + k$ $\quad (0 \le t \le 2\pi)$

Hyperbola
Center (h, k)

$$\frac{(y - k)^2}{a^2} - \frac{(x - h)^2}{b^2} = 1$$

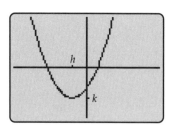

$x = b \tan t + h$
$y = \dfrac{a}{\cos t} + k$ $\quad (0 \le t \le 2\pi)$

Contemporary Precalcul

A Graphing Approach

5e

Contemporary Precalculus

A Graphing Approach

THOMAS W. HUNGERFORD
Saint Louis University

DOUGLAS J. SHAW
University of Northern Iowa

THOMSON
™
BROOKS/COLE

Australia • Brazil • Canada • Mexico • Singapore • Spain
United Kingdom • United States

Contemporary Precalculus: A Graphing Approach, **Fifth Edition**

Thomas W. Hungerford, Douglas J. Shaw

Mathematics Editor: *Gary Whalen*
Development Editors: *Leslie Lahr, Kari Hopperstead*
Assistant Editor: *Natasha Coats*
Editorial Assistant: *Rebecca Dashiell*
Technology Project Manager: *Lynh Pham*
Marketing Manager: *Joe Rogove*
Marketing Assistant: *Ashley Pickering*
Marketing Communications Manager: *Darlene Amidon-Brent*
Project Manager, Editorial Production: *Hal Humphrey*
Art Director: *Vernon Boes*
Print Buyer: *Karen Hunt*
Production Service: *Laura Horowitz / Hearthside Publishing Services*

Text Designer: *Terri Wright / Terri Wright Design*
Photo Researcher: *Terri Wright / Terri Wright Design*
Copy Editor: *Grace Lefrancois*
Illustrator: *Hearthside Publishing Services*
Cover Designer: *Terri Wright / Terri Wright Design*
Cover Image: (front: Grand Canyon Skywalk, Hualapai Reservation, Arizona) *ART FOXALL / UPI / Landov*; (back) *AP / Wide World Photos*
Cover Printer: *R. R. Donnelley / Willard*
Compositor: *ICC Macmillan Inc.*
Printer: *R. R. Donnelley / Willard*

Library of Congress Control Number: 2007939806

STUDENT EDITION:
ISBN-13: 978-0-495-10833-7
ISBN-10: 0-495-10833-2

For more information about our products, contact us at:
Thomson Learning Academic Resource Center
1-800-423-0563
For permission to use material from this text or product, submit a request online at
http://www.thomsonrights.com.
Any additional questions about permissions can be submitted by e-mail to thomsonrights@thomson.com.

Thomson Higher Education
10 Davis Drive
Belmont, CA 94002-3098
USA

To the memory of my mother and my aunts,
Whose presence in my life has greatly enriched it:
Grace Parks Hungerford
Irene Parks Mills
Florence M. Parks
Ellen McGillicuddy

For my daughter Francebelle Shaw, whose response to
"Stop being so cute, I seriously can't take it"
was to give me a big smile and kiss me on the tip of my nose.

Contents

Preface   xi
To the Instructor   xiv
Ancillaries   xvii
To the Student   xix

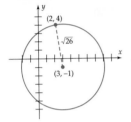
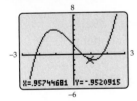
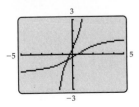

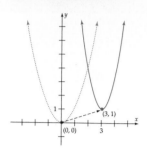
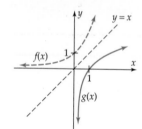
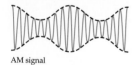

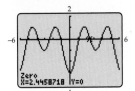
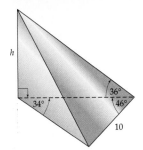
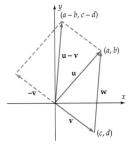
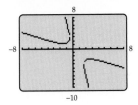

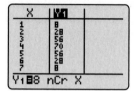
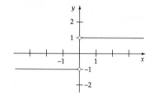

Preface

This book provides the mathematical background needed for calculus for students who have had two or three years of high school mathematics. Topics that are essential for success in calculus are thoroughly covered, including functional notation, graph reading, the natural exponential and logarithmic functions, average rates of change, trigonometric functions of a real variable, and limits.

The mathematics is presented in an informal manner that stresses meaningful motivation, careful explanations, and numerous examples, with an ongoing focus on real-world problem solving. Technology is integrated into the presentation and students are expected to use it to participate actively in exploring topics from algebraic, graphical, and numerical perspectives.

Content Changes in the Fifth Edition

The major changes in this edition include the following.

Chapter Tests   In addition to the Chapter Review exercises, each chapter now has a chapter test (in two parts for longer chapters), with answers for both odd and even questions in the answer section.

Section Objectives   The goals for student learning are summarized at the beginning of each section.

Discovery Projects   A new Discovery Project has been added to Chapter 1.

Exercises   Every exercise in the previous edition was examined and a significant number of them were updated, revised, or replaced. All in all, more than 20% of the exercises in this edition are new.

A number of smaller changes and additions—ranging from a few lines to a page or so—have been made throughout the book to improve coverage and clarity.

Organizational Changes

Several changes in the order and arrangement of topics have been made to provide more flexibility for the instructor.

Parametric Graphing   The brief introduction in Chapter 3 is now in a section by itself and graph reading is covered in Section 3.3 (Graphs of Functions). Parametric equations for conic sections are now in a separate (optional) section of Chapter 10, rather than being included as parts of other sections.

Arc Length and Angular Speed   These topics are now in a separate (optional) section instead of being included in Section 6.1.

Calculator Investigations   These formerly appeared at the end of selected sections at the beginning of the book and are now incorporated into the exercise sets for these sections.

Ongoing Features

All the helpful pedagogical features of earlier editions are retained here, such as:

A Flexible Approach to Trigonometry that allows instructors to use the order of topics they prefer: Trig functions of a real variable can be covered before or after triangle trigonometry or both approaches can be combined (as explained in the chart on page xiv);

Graphing Explorations, in which students discover and develop useful mathematics on their own;

Cautions that alert students to common misconceptions and mistakes;

Technology Tips that provide assistance in carrying out various procedures on specific calculators;

Exercises that proceed from routine drill to those requiring some thought, including graph interpretation and applied problems, as well as *Thinkers* that challenge students to "think outside the box" (most of these are not difficult—just different);

Chapter Reviews that include a list of important concepts (referenced by section and page number), a summary of important facts and formulas, and a set of review questions;

Algebra Review Appendix for students who need a review of basic algebra;

Geometry Review Appendix that summarizes frequently used facts from plane geometry, with examples and exercises;

Program Appendix that provides a small number of programs that are useful for updating older calculator models or for more easily carrying out procedures discussed in the text.

Acknowledgments

First of all, the senior author is happy to welcome Doug Shaw as a coauthor. His insights and enthusiasm have had a very positive impact on this revision and will, I trust, continue to do so in the future.

We are particularly grateful to three people who have been associated with several editions of this book:

Leslie Lahr, our Developmental Editor, who has been involved in almost all aspects of the book, and from whose sage advice we have greatly benefited;

Phil Embree of William Woods University, one of our accuracy checkers and the author of several Discovery Projects, whose helpful suggestions have definitely improved the final product;

Laura Horowitz of Hearthside Publication Services, whose calm coordination of all aspects of production has kept us sane and happy.

Thanks are also due to Heidi A. Howard, who pitched in on a variety of tasks whenever needed.

It is a pleasure to acknowledge the invaluable assistance of the Brooks/Cole staff:

Gary Whalen, Acquisitions Editor
Joe Rogove, Marketing Manager
Natasha Coats, Assistant Editor
Lynh Pham, Editorial Assistant
Hal Humphrey, Senior Content Project Manager
Vernon Boes, Senior Art Director

We also want to thank the outside production staff:

Grace Lefrancois, Copy Editor
Jade Myers, Art Rendering
Terri Wright of Terri Wright Design

Special thanks go to

John Samons, Florida Community College,

who did most of the accuracy checking of the manuscript, and to

Jerret Dumouchel, Florida Community College
Arkady Hanjiev, North Hills College,

who prepared the Solution Manuals.

The students who have assisted in manuscript preparation also have our thanks: Varun Khanna (St. Louis University), and Jordan Meyer, Ren Waddell and Kevan Irvine (University of Northern Iowa).

Finally, we want to thank the reviewers whose constructive comments have played a crucial role in revising this text.

Jared Abwawo, Tacoma Community College
Tony Akhlaghi, Bellevue Community College
Donna Bernardy, Lane Community College
Patrick DeFazio, Onondaga Community College
Linda Horner, Columbia State Community College
Douglas Nelson, Central Oregon Community College
Stephen Nicoloff, Paradise Valley Community College
Bogdan Nita, Montclair State University
Dennis Reissig, Suffolk County Community College
Arnavaz Taraporevala, New York City College of Technology

Thanks also go the reviewers of previous editions:

Deborah Adams, Jacksonville University
Kelly Bach, University of Kansas
David Blankenbaker, University of New Mexico
Kathleen A. Cantone, Onondaga Community College
Bettyann Daley, University of Delaware
Margaret Donlan, University of Delaware
Patricia Dueck, Arizona State University
Betsy Farber, Bucks County Community College
Alex Feldman, Boise State University

Robert Fliess, West Liberty State College
Betty Givan, Eastern Kentucky University
William Grimes, Central Missouri State University
Frances Gulick, University of Maryland
John Hamm, University of New Mexico
Lonnie Hass, North Dakota State University
Larry Howe, Rowan University
Conrad D. Krueger, San Antonio College
Ann Lawrance, Wake Technical Community College
Charles Laws, Cleveland State Community College
Anatoly S. Libgober, University of Illinois at Chicago
Martha Lisle, Prince George's Community College
Matthew Liu, University of Wisconsin-Stevens Point
Sergey Lvin, University of Maine
George Matthews, Onondaga Community College
Nancy Matthews, University of Oklahoma
Ruth Meyering, Grand Valley State University
William Miller, Central Michigan University
Philip Montgomery, University of Kansas
Roger Nelsen, Lewis and Clark College
Jack Porter, University of Kansas
Robert Rogers, University of New Mexico
Barbara Sausen, Fresno City College
Hugo Sun, California State University at Fresno
Stuart Thomas, University of Oregon
Trung G. Tran, Tacoma Community College
Bettie Truitt, Black Hawk College
Jan Vandever, South Dakota State University
Judith Wolbert, Kettering University
Cathleen M. Zucco-Teveloff, Trinity College

The last word, as always, goes to our wives, Mary Alice Hungerford and Laurel Shaw, who have, as always, provided understanding and support when it was most needed.

Thomas W. Hungerford
Douglas J. Shaw

This book contains more than enough material for two semesters. By using the chart on the facing page (and the similar ones at the beginning of each chapter that show the interdependence of sections within the chapter), you can easily design a course to fit the needs of your students and the constraints of time. When planning your syllabus, two things are worth noting.

Special Topics   Sections with this label are usually related to the immediately preceding section and are not prerequisites for other sections of the text. Judgments vary as to which of these topics are essential and which can be omitted. So feel free to include as many or as few Special Topics sections as you wish.

Trigonometry   We believe that a precalculus course should emphasize trigonometric functions of a real variable from the beginning. We also think that the advent of technology has made the cotangent, secant, and cosecant functions less necessary than was the case when calculations were done by hand. So they are not introduced until the end of Chapter 6. This is the approach given in the first column of the chart below.

However, we are well aware that some instructors feel strongly that triangle trigonometry should be introduced first, or that both approaches should be presented together, or that all six trig functions should be introduced at the same time, etc. The following chart (and its footnotes) show how to arrange things to suit your personal preferences.

OPTIONS FOR COVERING TRIGONOMETRY			
Real Variable First	**Triangles First**	**Mixed** [Trig functions of a real variable *and* basic triangle trig introduced as soon as possible]	
		Real Variable First	*Triangles First*
Chapter 6*	Chapter 8 (using Alternate 8.1 in place of 8.1) Chapter 6 (using Alternate 6.2 in place of 6.2)*	Sections 6.1–6.4* 8.1–8.2 6.5–6.6	Section 6.1 Alternate 8.1 8.2 6.2–6.6*
Chapters 7, 8, 9 in any order†	Chapters 7 & 9 in either order†	Chapter 7, Sections 8.3–8.4 and Chapter 9 in any order†	Chapter 7, Sections 8.3–8.4 and Chapter 9 in any order†

*For early introduction of cotangent, secant, and cosecant, incorporate Section 6.6 into Sections 6.2–6.4 by covering

Part I of 6.6 at the end of 6.2;
Part II of 6.6 at the end of 6.3;
Part III of 6.6 at the end of 6.4.

†Within each of these chapters, sections can be covered in several different orders. See the interdependence-of-sections chart at the beginning of the chapter.

Interdependence of Chapters

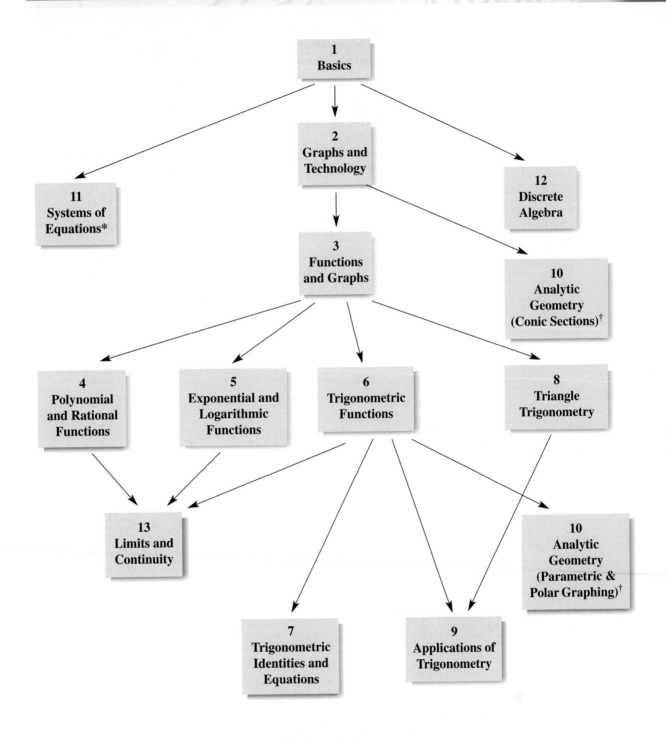

1 Basics

2 Graphs and Technology

11 Systems of Equations*

12 Discrete Algebra

3 Functions and Graphs

10 Analytic Geometry (Conic Sections)†

4 Polynomial and Rational Functions

5 Exponential and Logarithmic Functions

6 Trigonometric Functions

8 Triangle Trigonometry

13 Limits and Continuity

10 Analytic Geometry (Parametric & Polar Graphing)†

7 Trigonometric Identities and Equations

9 Applications of Trigonometry

*Section 2.1 (graphs) is a prerequisite for Special Topics 11.1.A (Systems of Nonlinear Equations).
†Sections 10.1–10.4 (Conic Sections) depend only on Chapter 2. Trigonometry is an additional prerequisite for Special Topics 10.3.A and 10.4.A and Sections 10.5–10.7.

This text assumes the use of technology (a graphing calculator or computer with appropriate software). Discussions of calculators in the text apply (with obvious modifications) to computer software.

To address the fact that many students are unaware of the power of their graphing calculators, there are Technology Tips in the margins throughout the text. They provide general information and advice, as well as listing the proper menus or keys needed to care out procedures on specific calculators.* Unless noted otherwise,

Tips for	**also apply to**
TI-84+	TI-82, TI-83, TI-83+
TI-86	TI-85
TI-89	TI-92
Casio 9850	Casio 9750, Casio 9860, Casio 9970
HP-39gs	HP-38, HP-39, HP-39+

*In addition there is a Program Appendix that provides users of older calculators with several helpful programs that are built in to newer calculators.

Ancillaries

Supplements for Instructors

COMPLETE SOLUTIONS MANUAL

ISBN-10: 0-495-55398-0 | ISBN-13: 978-0-495-55398-4
The complete solutions manual provides worked out solutions to all of the problems in the text.

TEST BANK

ISBN-10: 0-495-55400-6 | ISBN-13: 978-0-495-55400-4
The Test Bank includes 8 tests per chapter as well as 3 final exams, each combining multiple-choice, free-response, and fill-in-the-blank questions.

EXAMVIEW® WITH ALGORITHMIC EQUATIONS

ISBN-10: 0-495-55417-0 | ISBN-13: 978-0-495-55417-2
Create, deliver, and customize tests and study guides (both print and online) in minutes with this easy-to-use assessment and tutorial system, which includes the Test Bank questions in electronic format. ExamView offers both a Quick Test Wizard and an Online Test Wizard that guide you step-by-step through the process of creating tests—you can even see the test you are creating on the screen exactly as it will print or display online.

TEXT-SPECIFIC DVDs

ISBN-10: 0-495-55401-4 | ISBN-13: 978-0-495-55401-1
This set of video segments, available upon adoption of the text, offers a 10- to 20-minute problem-solving lesson for each section of each chapter.

JOININ™ STUDENT RESPONSE SYSTEM

ISBN-10: 0-495-55415-4 | ISBN-13: 978-0-495-55415-8
JoinIn™ content tailored to this text (on Microsoft® PowerPoint® slides) allows you to pose book-specific questions and display answers seamlessly within the Microsoft PowerPoint slides of your own lecture. Use the slides with your favorite response system software.

ENHANCED WEBASSIGN—AN ASSIGNABLE HOMEWORK AND TUTORIAL TOOL THAT SAVES YOU TIME

www.webassign.net/brookscole

Proven and reliable, Enhanced WebAssign allows you to easily assign, collect, grade, and homework assignments via the web. You save time, and students get interactive assistance plus relevant, immediate feedback. Features include:

- Thousands of algorithmically generated homework problems based on up to 1,500 end-of-section exercises
- Read It links to PDFs of relevant text sections
- Watch It links to videos that provide further instruction on problems
- Master It tutorials for step-by-step support
- Chat About It links to live, online tutoring
- Windows® and Apple® Macintosh compatible; works with most web browsers (Firefox, Internet Explorer, Mozilla, Safari); does not require proprietary plug-ins

Supplements for Students

STUDENT SOLUTIONS MANUAL

ISBN-10: 0-495-55399-9 I ISBN-13: 978-0-495-55399-1

The student solutions manual provides worked out solutions to the odd-numbered problems in the text.

This text assumes the use of technology (a graphing calculator or computer with appropriate software). Discussions of calculators in the text apply (with obvious modifications) to computer software.

To help you get the most from your graphing calculator, there are Technology Tips in the margins throughout the text. They provide general information and advice, as well as listing the proper menus or keys needed to care out procedures on specific calculators.* Unless noted otherwise,

Tips for	also apply to
TI-84+	TI-82, TI-83, TI-83+
TI-86	TI-85
TI-89	TI-92
Casio 9850	Casio 9750, Casio 9860, Casio 9970
HP-39gs	HP-38, HP-39, HP-39+

Getting the Most Out of This Course

With all this talk about calculators, don't lose sight of this crucial fact:

Technology is only a *tool* for doing mathematics.

You can't build a house if you only use a hammer. A hammer is great for pounding nails, but useless for sawing boards. Similarly, a calculator is great for computations and graphing, but it is not the right tool for every mathematical task. To succeed in this course, you must develop and use your algebraic and geometric skills, your reasoning power and common sense, and you must be willing to work.

The key to success is to use all of the resources at your disposal: your instructor, your fellow students, your calculator (and its instruction manual), and this book. Here are some tips for making the most of these resources.

Ask Questions   Remember the words of Hillel:

The bashful do not learn.

There is no such thing as a "dumb question" (assuming, of course, that you have attended class and read the text). Your instructor will welcome questions that arise from a serious effort on your part.

Read the Book   Not just the homework exercises, but the rest of the text as well. There is no way your instructor can possibly cover the essential topics, clarify ambiguities, explain the fine points, and answer all your questions during class time. You simply will not develop the level of understanding you need to succeed in this course and in calculus unless you read the text fully and carefully.

---

*In addition there is a Program Appendix that provides users of older calculators with several helpful programs that are built in to newer calculators.

Be an Interactive Reader   You can't read a math book the way you read a novel or history book. You need pencil, paper, and your calculator at hand to work out the statements you don't understand and to make notes of things to ask your fellow students and/or your instructor.

Do the Graphing Explorations   When you come to a box labeled "Graphing Exploration," use your calculator as directed to complete the discussion. Typically, this will involve graphing one or more equations and answering some questions about the graphs. Doing these explorations as they arise will improve your understanding and clarify issues that might otherwise cause difficulties.

Do Your Homework   Remember that

Mathematics is not a spectator sport.

You can't expect to learn mathematics without doing mathematics, any more than you could learn to swim without getting wet. Like swimming or dancing or reading or any other skill, mathematics takes practice. Homework assignments are where you get the practice that is essential for passing this course and succeeding in calculus.

BASICS

On a clear day, can you see forever?

If you are at the top of the Smith Tower in Seattle, how far can you see? In earlier centuries, the lookout on a sailing ship was posted atop the highest mast because he could see farther from there than from the deck. How much farther? These questions, and similar ones, can be answered (at least approximately) by using basic algebra and geometry. See Example 4 on page 9 and Exercise 94 on page 15.

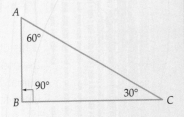

Chapter Outline

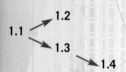
This chapter reviews the essential facts about real numbers, equations, the coordinate plane, and lines that are needed in this course and in calculus. The Algebra Review Appendix at the end of the book is a prerequisite for this material.

1.1 The Real Number System

Section Objectives

- Identify important types of real numbers.
- Simplify mathematical expressions.
- Represent sets of real numbers with interval notation.
- Graph intervals on a number line.
- Use scientific notation.
- Understand and apply the properties of square roots.
- Understand and apply the properties of absolute value.
- Compute the distance between two points on the number line.

Most of this book deals with the real number system, so it may be helpful to review the types of real numbers.

Name	Definition/Description
Natural numbers	1, 2, 3, 4, 5, . . . Natural numbers are also called **counting numbers** or **positive integers.**
Integers	. . . , −5, −4, −3, −2, −1, 0, 1, 2, 3, 4, 5, . . . The integers consist of the natural numbers, their negatives, and zero.

Continued

Name	Definition/Description
Rational numbers	A rational number is a number that can be expressed as a fraction $\frac{r}{s}$, with r and s integers and $s \neq 0$, such as $$\frac{1}{2}, \quad -9.83 = \frac{-983}{100}, \quad 47 = \frac{47}{1}, \quad 8\frac{3}{5} = \frac{43}{5}.$$
	Alternatively, rational numbers are numbers that can be expressed as terminating decimals, such as $.25 = \frac{1}{4}$, or as nonterminating repeating decimals in which a single digit or block of digits eventually repeats forever, such as $$\frac{5}{3} = 1.66666 \cdots \quad \text{or} \quad \frac{362}{1665} = .2174174174 \cdots.$$
Irrational numbers	An irrational number is a number that cannot be expressed as a fraction with an integer numerator and denominator, such as the number π, which is used to calculate the area of a circle.*
	Alternatively, irrational numbers are numbers that can be expressed as nonterminating, nonrepeating decimals (no block of digits repeats forever).

More information about decimal expansions of real numbers is given in Special Topics 1.1.A.

The relationships among the types of numbers in the preceding table are summarized in Figure 1–1, in which each set of numbers is contained in the set to its right. So, for example, integers are also rational numbers and real numbers, but not irrational numbers.

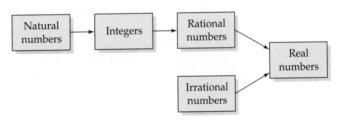

Figure 1–1

ARITHMETIC

To avoid ambiguity when dealing with expressions such as $6 + 3 \times 5$, mathematicians have made the following agreement, which is also followed by your calculator.

Order of Operations

> In an expression without parentheses, multiplication and division are performed first (from left to right). Addition and subtraction are performed last (from left to right).

In light of this convention, there is only one correct way to interpret $6 + 3 \times 5$:

$$6 + 3 \times 5 = 6 + 15 = 21. \qquad \text{[Multiplication first, addition last]}$$

---

*The proof that π is irrational is beyond the scope of this book. In the past you may have used 22/7 or 3.1416 as π and a calculator may display π as 3.141592654. However, these numbers are just *approximations* of π.

On the other hand, if you want to "add 6 + 3 and then multiply by 5," you must use parentheses:

$$(6 + 3) \cdot 5 = 9 \cdot 5 = 45.$$

This is an illustration of the first of two basic rules for dealing with parentheses.

Rules for Parentheses

1. Do all computations inside the parentheses before doing any computations outside the parentheses.

2. When dealing with parentheses within parentheses, begin with the innermost pair and work outward.

For example,

$$8 + [11 - (6 \times 3)] = 8 + (11 - 18) = 8 + (-7) = 1.$$

Inside parentheses first

We assume that you are familiar with the basic properties of real number arithmetic, particularly the following fact.

Distributive Law

For all real numbers a, b, c,

$$a(b + c) = ab + ac \qquad \text{and} \qquad (b + c)a = ba + ca.$$

The distributive law doesn't usually play a direct role in easy computations, such as $4(3 + 5)$. Most people don't say $4 \cdot 3 + 4 \cdot 5 = 12 + 20 = 32$. Instead, they mentally add the numbers in parentheses and say 4 times 8 is 32. But when symbols are involved, you can't do that, and the distributive law is essential. For example,

$$4(3 + x) = 4 \cdot 3 + 4x = 12 + 4x.$$

THE NUMBER LINE AND ORDER

The real numbers are often represented geometrically as points on a **number line,** as in Figure 1–2. We shall assume that there is exactly one point on the line for every real number (and vice versa) and use phrases such as "the point 3.6" or "a number on the line." This mental identification of real numbers and points on the line is often helpful.

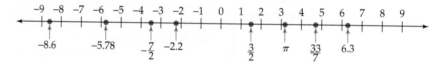

Figure 1–2

The statement $c < d$, which is read "**c is less than d**," and the statement $d > c$ (read "**d is greater than c**") mean exactly the same thing:

c lies to the *left* of d on the number line.

For example, Figure 1–2 shows that $-5.78 < -2.2 \quad$ and $\quad 4 > \pi$.

The statement $c \leq d$, which is read "**c is less than or equal to d**," means

Either c is less than d or c is equal to d.

Only one part of an "either . . . or" statement needs to be true for the entire statement to be true. So the statement $5 \leq 10$ is true because $5 < 10$, and the statement $5 \leq 5$ is true because $5 = 5$. The statement $d \geq c$ (read "**d is greater than or equal to c**") means exactly the same thing as $c \leq d$.

The statement $b < c < d$ means

$$b < c \qquad \text{and simultaneously} \qquad c < d.$$

For example, $3 < x < 7$ means that x is a number that is strictly between 3 and 7 on the number line (greater than 3 and less than 7). Similarly, $b \leq c < d$ means

$$b \leq c \qquad \text{and simultaneously} \qquad c < d,$$

and so on.

Certain sets of numbers, defined in terms of the order relation, appear frequently enough to merit special notation. Let c and d be real numbers with $c < d$. Then

Interval Notation

$[c, d]$ denotes the set of all real numbers x such that $c \leq x \leq d$.

(c, d) denotes the set of all real numbers x such that $c < x < d$.

$[c, d)$ denotes the set of all real numbers x such that $c \leq x < d$.

$(c, d]$ denotes the set of all real numbers x such that $c < x \leq d$.

All four of these sets are called **intervals** from c to d. The numbers c and d are the **endpoints** of the interval. $[c, d]$ is called the **closed interval** from c to d (both endpoints included and *square* brackets), and (c, d) is called the **open interval** from c to d (neither endpoint included and *round* brackets). Some examples are shown in Figure 1–3.*

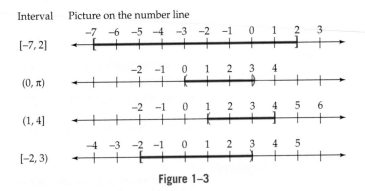

Figure 1–3

---

*In Figures 1–3 and 1–4, a round bracket such as) or (indicates that the endpoint is *not* included, whereas a square bracket such as] or [indicates that the endpoint *is* included.

If b is a real number, then the half-line extending to the right or left of b is also called an **interval.** Depending on whether or not b is included, there are four possibilities.

Interval Notation

[b, ∞) denotes the set of all real numbers x such that $x \geq b$.

(b, ∞) denotes the set of all real numbers x such that $x > b$.

($-\infty$, b] denotes the set of all real numbers x such that $x \leq b$.

($-\infty$, b) denotes the set of all real numbers x such that $x < b$.

Some examples are shown in Figure 1–4.

NOTE

The symbol ∞ is read "infinity," and we call the set [b, ∞) "the interval from b to infinity." The symbol ∞ does *not* denote a real number; it is simply part of the notation used to label the first two sets of numbers defined in the previous box. Analogous remarks apply to the symbol $-\infty$, which is read "negative infinity."

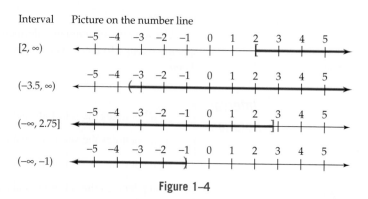

Figure 1–4

In a similar vein, $(-\infty, \infty)$ **denotes the set of all real numbers.**

NEGATIVE NUMBERS AND NEGATIVES OF NUMBERS

The **positive numbers** are those to the right of 0 on the number line, that is,

All numbers c with $c > 0$.

The **negative numbers** are those to the left of 0, that is,

All numbers c with $c < 0$.

The **nonnegative** numbers are the numbers c with $c \geq 0$.

The word "negative" has a second meaning in mathematics. The **negative *of* a number** c is the number $-c$. For example, the negative of 5 is -5, and the negative of -3 is $-(-3) = 3$. Thus the negative of a negative number is a positive number. Zero is its own negative, since $-0 = 0$. In summary,

Negatives

The negative of the number c is $-c$.

If c is a positive number, then $-c$ is a negative number.

If c is a negative number, then $-c$ is a positive number.

SCIENTIFIC NOTATION

In mid-2006 the U.S. national debt was 8.4 trillion dollars. Since one trillion is 10^{12} (that is, 1 followed by 12 zeros), the national debt is the number

$$8.4 \times 10^{12}.$$

This is an example of *scientific notation*. A positive number is said to be in **scientific notation** when it is written in the form

$$a \times 10^n \qquad \text{where } 1 \le a < 10 \text{ and } n \text{ is an integer.}$$

You should be able to translate between scientific notation and ordinary notation and vice versa.

EXAMPLE 1

Express these numbers in ordinary notation:

(a) 1.55×10^9 (b) 2.3×10^{-8}

SOLUTION

(a) $1.55 \times 10^9 = 1.55 \times 1{,}000{,}000{,}000 = 1{,}550{,}000{,}000$

You can do this multiplication in your head by using the fact that multiplying by 10^9 is equivalent to moving the decimal point 9 places to the right.

(b) $2.3 \times 10^{-8} = \dfrac{2.3}{10^8} = .000000023*$

This computation can also be done mentally by using the fact that dividing by 10^8 is equivalent to moving the decimal point 8 places to the left. ∎

EXAMPLE 2

Write each of these numbers in scientific notation:

(a) 356   (b) 1,564,000   (c) .072   (d) .00000087

SOLUTION   In each case, move the decimal point to the left or right to obtain a number between 1 and 10; then write the original number in scientific notation as follows.

(a) Move the decimal point in 356 two places to the left to obtain the number 3.56, which is between 1 and 10. You can get the original number back by multiplying by 100: $356 = 3.56 \times 100$. So the scientific notation form is

$$356 = 3.56 \times 10^2$$

Note that the original decimal point is moved 2 places to the *left* and 10 is raised to the power 2.

(b) $1{,}564{,}000 = 1.564 \times 1{,}000{,}000 = 1.564 \times 10^6$ [Decimal point is moved 6 places to the *left*, and 10 is raised to the 6th power.]

(c) $.072 = 7.2 \times \dfrac{1}{100} = 7.2 \times 10^{-2}$ [Decimal point is moved 2 places to the *right*, and 10 is raised to -2.]

---

*Negative exponents are explained in the first section of the Algebra Review Appendix.

(d) $.00000087 = 8.7 \times \dfrac{1}{10,000,000} = 8.7 \times 10^{-7}$ [Decimal point is moved 7 places to the *right*, and 10 is raised to the -7.]   ∎

Scientific notation is useful for computations with very large or very small numbers.

EXAMPLE 3

$$(.00000002)(4,300,000,000) = (2 \times 10^{-8})(4.3 \times 10^{9})$$

$$= 2(4.3)10^{-8+9} = (8.6)10^{1} = 86.$$ ∎

Calculators automatically switch to scientific notation whenever a number is too large or too small to be displayed in the standard way. If you try to enter a number with more digits than the calculator can handle, such as 45,000,000,333,222,111, a typical calculator will approximate it using scientific notation as 4.500000033 E 16, that is, as 45,000,000,330,000,000.

SQUARE ROOTS

A *square root* of a nonnegative real number d is any number whose square is d. For instance both 5 and -5 are square roots of 25 because

$$5^{2} = 25 \qquad \text{and} \qquad (-5)^{2} = 25.$$

The nonnegative square root (in this case, 5) is given a special name and notation.

> If d is a nonnegative real number, the **principal square root** of d is the *nonnegative* number whose square is d. It is denoted $\sqrt{d}$.* Thus,
>
> $$\sqrt{d} \geq 0 \qquad \text{and} \qquad \left(\sqrt{d}\right)^{2} = d.$$

For example,

$$\sqrt{25} = 5 \qquad \text{because} \qquad 5 \geq 0 \text{ and } 5^{2} = 25.$$

The radical symbol always denotes a nonnegative number. To express the negative square root of 25 in terms of radicals, we write $-5 = -\sqrt{25}$.

Although $-\sqrt{25}$ is a real number, the expression $\sqrt{-25}$ is *not defined* in the real numbers because there is no real number whose square is -25. In fact, since the square of every real number is nonnegative,

No negative number has a square root in the real numbers.

Some square roots can be found (or verified) by hand, such as

$$\sqrt{225} = 15 \qquad \text{and} \qquad \sqrt{1.21} = 1.1.$$

Usually, however, a calculator is needed to obtain rational *approximations* of roots. For instance, we know that $\sqrt{87}$ is between 9 and 10 because $9^{2} = 81$ and $10^{2} = 100$. A calculator shows that $\sqrt{87} \approx 9.327379$.[†]

TECHNOLOGY TIP

To compute $\sqrt{7^{2} + 51} + 3$ on a calculator, you must use a pair of parentheses:

$$\sqrt{(7^{2} + 51)} + 3.$$

Otherwise the calculator will not compute the correct answer, which is:

$$\sqrt{7^{2} + 51} + 3 = \sqrt{49 + 51} + 3$$

$$= \sqrt{100} + 3$$

$$= 13.$$

Try it!

---

*The symbol $\sqrt{}$ is called a **radical.**
[†] $\approx$ means "approximately equal."

If c and d are positive real numbers, then

$$\sqrt{c+d} \neq \sqrt{c} + \sqrt{d}.$$

For example,

$$\sqrt{9+16} = \sqrt{25} = 5,$$

but

$$\sqrt{9} + \sqrt{16} = 3 + 4 = 7.$$

We shall often use the following property of square roots:

$$\sqrt{cd} = \sqrt{c}\sqrt{d} \text{ for any nonnegative real numbers } c \text{ and } d.$$

For example, $\sqrt{9 \cdot 16} = \sqrt{9}\sqrt{16} = 3 \cdot 4 = 12$. But be careful—*there is no similar property for sums*, as the Caution in the margin demonstrates.

EXAMPLE 4

Suppose you are located h feet above the ground. Because of the curvature of the earth, the maximum distance you can see is approximately d miles, where

$$d = \sqrt{1.5h + (3.587 \times 10^{-8})h^2}.$$

How far can you see from the 500-foot-high Smith Tower in Seattle and from the 1454-foot-high Sears Tower in Chicago?

SOLUTION   For the Smith Tower, substitute 500 for h in the formula, and use your calculator:

$$d = \sqrt{1.5(500) + (3.587 \times 10^{-8})500^2} \approx 27.4 \text{ miles.}$$

For the Sears Tower, you can see almost 47 miles because

$$d = \sqrt{1.5(1454) + (3.587 \times 10^{-8})1454^2} \approx 46.7 \text{ miles.} \qquad \blacksquare$$

 ABSOLUTE VALUE

On an informal level, most students think of absolute value like this:

The absolute value of a positive number is the number itself.

The absolute value of a negative number is found by "erasing the minus sign."

If $|c|$ denotes the absolute value of c, then, for example, $|5| = 5$ and $|-4| = 4$.
This informal approach is inadequate, however, for finding the absolute value of a number such as $\pi - 6$. It doesn't make sense to "erase the minus sign" here. So we must develop a more precise definition. The statement $|5| = 5$ suggests that the absolute value of a nonnegative number ought to be the number itself. For negative numbers, such as -4, note that $|-4| = 4 = -(-4)$, that is, the absolute value of the negative number -4 is the *negative* of -4. These facts are the basis of the formal definition.

Absolute Value

> The **absolute value** of a real number c is denoted $|c|$ and is defined as follows.
>
> If $c \geq 0$, then $|c| = c$.
>
> If $c < 0$, then $|c| = -c$.

EXAMPLE 5

(a) $|3.5| = 3.5$ and $|-7/2| = -(-7/2) = 7/2$.

(b) To find $|\pi - 6|$, note that $\pi \approx 3.14$, so $\pi - 6 < 0$. Hence, $|\pi - 6|$ is defined to be the *negative* of $\pi - 6$, that is,

$$|\pi - 6| = -(\pi - 6) = -\pi + 6.$$

(c) $|5 - \sqrt{2}| = 5 - \sqrt{2}$ because $5 - \sqrt{2} \geq 0$. $\qquad \blacksquare$

Here are the important facts about absolute value.

Properties of Absolute Value

Property	Description
1. $\lvert c \rvert \geq 0$	The absolute value of a number is nonnegative.
2. If $c \neq 0$, then $\lvert c \rvert > 0$.	The absolute value of a nonzero number is positive.
3. $\lvert c \rvert = \lvert -c \rvert$	A number and its negative have the same absolute value.
4. $\lvert cd \rvert = \lvert c \rvert \cdot \lvert d \rvert$	The absolute value of the product of two numbers is the product of their absolute values.
5. $\left\lvert \dfrac{c}{d} \right\rvert = \dfrac{\lvert c \rvert}{\lvert d \rvert}$ $(d \neq 0)$	The absolute value of the quotient of two numbers is the quotient of their absolute values.

EXAMPLE 6

TECHNOLOGY TIP

To find $\lvert 9 - 3\pi \rvert$ on a calculator, key in

$$Abs\,(9 - 3\pi).$$

The *Abs* key is located in this menu/submenu:

    TI: MATH/NUM
    Casio: OPTN/NUM
    HP-39gs: Keyboard

Here are examples of the last three properties in the box.

3. $\lvert 3 \rvert = 3$ and $\lvert -3 \rvert = 3$, so $\lvert 3 \rvert = \lvert -3 \rvert$.

4. If $c = 6$ and $d = -2$, then

$$\lvert cd \rvert = \lvert 6(-2) \rvert = \lvert -12 \rvert = 12$$

and

$$\lvert c \rvert \cdot \lvert d \rvert = \lvert 6 \rvert \cdot \lvert -2 \rvert = 6 \cdot 2 = 12,$$

so $\lvert cd \rvert = \lvert c \rvert \cdot \lvert d \rvert$.

5. If $c = -5$ and $d = 4$, then

$$\left\lvert \frac{c}{d} \right\rvert = \left\lvert \frac{-5}{4} \right\rvert = \left\lvert -\frac{5}{4} \right\rvert = \frac{5}{4} \qquad \text{and} \qquad \frac{\lvert c \rvert}{\lvert d \rvert} = \frac{\lvert -5 \rvert}{\lvert 4 \rvert} = \frac{5}{4},$$

so $\left\lvert \dfrac{c}{d} \right\rvert = \dfrac{\lvert c \rvert}{\lvert d \rvert}$. ∎

When c is a positive number, then $\sqrt{c^2} = c$, but when c is negative, this is *false*. For example, if $c = -3$, then

$$\sqrt{c^2} = \sqrt{(-3)^2} = \sqrt{9} = 3 \qquad (not\ -3),$$

so $\sqrt{c^2} \neq c$. In this case, however, $\lvert c \rvert = \lvert -3 \rvert = 3$, so $\sqrt{c^2} = \lvert c \rvert$. The same thing is true for any negative number c. It is also true for positive numbers (since $\lvert c \rvert = c$ when c is positive). In other words,

Square Roots of Squares

For every real number c,

$$\sqrt{c^2} = \lvert c \rvert.$$

When dealing with long expressions inside absolute value bars, do the computations inside first, and then take the absolute value.

EXAMPLE 7

(a) $|5(2-4)+7| = |5(-2)+7| = |-10+7| = |-3| = 3.$

(b) $4 - |3-9| = 4 - |-6| = 4 - 6 = -2.$ ∎

CAUTION

When c and d have opposite signs, $|c+d|$ is *not equal* to $|c| + |d|$. For example, when $c = -3$ and $d = 5$, then

$$|c+d| = |-3+5| = 2,$$

but

$$|c| + |d| = |-3| + |5| = 3 + 5 = 8.$$

The caution shows that $|c+d| < |c| + |d|$ when $c = -3$ and $d = 5$. In the general case, we have the following fact.

The Triangle Inequality

For any real numbers c and d,

$$|c+d| \le |c| + |d|.$$

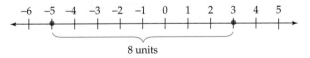 **DISTANCE ON THE NUMBER LINE**

Observe that the distance from -5 to 3 on the number line is 8 units:

Figure 1–5

This distance can be expressed in terms of absolute value by noting that

$$|(-5) - 3| = 8.$$

That is, the distance is the *absolute value of the difference* of the two numbers. Furthermore, the order in which you take the difference doesn't matter; $|3 - (-5)|$ is also 8. This reflects the geometric fact that the distance from -5 to 3 is the same as the distance from 3 to -5. The same thing is true in the general case.

Distance on the Number Line

The distance between c and d on the number line is the number

$$|c - d| = |d - c|.$$

EXAMPLE 8

The distance from 4.2 to 9 is $|4.2 - 9| = |-4.8| = 4.8$, and the distance from 6 to $\sqrt{2}$ is $|6 - \sqrt{2}|$. ∎

When $d = 0$, the distance formula shows that $|c - 0| = |c|$. Hence,

Distance to Zero

> $|c|$ is the distance between c and 0 on the number line.

Algebraic problems can sometimes be solved by translating them into equivalent geometric problems. The key is to interpret statements involving absolute value as statements about distance on the number line.

EXAMPLE 9

Solve the equation $|x + 5| = 3$ geometrically.

SOLUTION   We rewrite it as $|x - (-5)| = 3$. In this form it states that

*The distance between x and −5 is 3 units.**

Figure 1–6 shows that −8 and −2 are the only two numbers whose distance to −5 is 3 units:

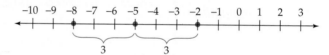

Figure 1–6

Thus $x = -8$ and $x = -2$ are the solutions of $|x + 5| = 3$. ∎

EXAMPLE 10

The solutions of $|x - 1| \geq 2$ are all numbers x such that

The distance between x and 1 is greater than or equal to 2.

Figure 1–7 shows that the numbers 2 or more units away from 1 are the numbers x such that

$$x \leq -1 \qquad \text{or} \qquad x \geq 3.$$

So these numbers are the solutions of the inequality. ∎

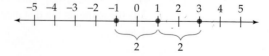

Figure 1–7

---

*It's necessary to rewrite the equation first because the distance formula involves the *difference* of two numbers, not their sum.

EXAMPLE 11

The solutions of $|x - 7| < 2.5$ are all numbers x such that

The distance between x and 7 is less than 2.5.

Figure 1–8 shows that the solutions of the inequality, that is, the numbers within 2.5 units of 7, are the numbers x such that $4.5 < x < 9.5$, that is, the interval $(4.5, 9.5)$. ∎

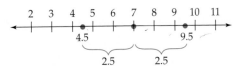

Figure 1–8

EXERCISES 1.1

1. Draw a number line and mark the location of each of these numbers: $0, -7, 8/3, 10, -1, -4.75, 1/2, -5$, and 2.25.

2. Use your calculator to determine which of the following rational numbers is the best approximation of the irrational number π.

$$\frac{22}{7}, \quad \frac{355}{113}, \quad \frac{103,993}{33,102}, \quad \frac{2,508,429,787}{798,458,000}.$$

If your calculator says that one of these numbers equals π, it's lying. All you can conclude is that the number agrees with π for as many decimal places as your calculator can handle (usually 12–14).

In Exercises 3–8, b, c, and d are real numbers such that $b < 0$, $c > 0$, and $d < 0$. Determine whether the given number is positive or negative.

3. $-b$ **4.** $-c$ **5.** bcd

6. $b - c$ **7.** $bc - bd$ **8.** $b^2c - c^2d$

In Exercises 9 and 10, use a calculator and list the given numbers in order from smallest to largest.

9. $\dfrac{189}{37}, \quad \dfrac{4587}{691}, \quad \sqrt{47}, \quad 6.735, \quad \sqrt{27}, \quad \dfrac{2040}{523}$

10. $\dfrac{385}{177}, \quad \sqrt{10}, \quad \dfrac{187}{63}, \quad \pi, \quad \sqrt{\sqrt{85}}, \quad 2.9884$

In Exercises 11–19, express the given statement in symbols.

11. -4 is greater than -8.

12. -17 is less than 6.

13. π is less than 100.

14. x is nonnegative.

15. z is greater than or equal to -4.

16. t is negative.

17. d is not greater than 7.

18. c is at most 3.

19. z is at least -17.

In Exercises 20–24, fill the blank with $<$, $=$, or $>$ so that the resulting statement is true.

20. -6 _____ -2 **21.** 5 _____ -3

22. $3/4$ _____ $.75$ **23.** 3.1416 _____ π

24. $1/3$ _____ $.33$

*The consumer price index for urban consumers (CPI-U) measures the cost of consumer goods and services such as food, housing, transportation, medical costs, etc. The table shows the yearly percentage increase in the CPI-U over a decade.**

Year	Percentage change
1996	3.0
1997	2.3
1998	1.6
1999	2.2
2000	3.4
2001	2.8
2002	1.6
2003	2.3
2004	2.7
2005	2.5

*U.S. Bureau of Labor Statistics; data for 2005 is for the first half.

In Exercises 25–29, let p denote the yearly percentage increase in the CPI-U. Find the number of years in this period which satisfied the given inequality.

25. $p \geq 2.8$ **26.** $p < 2.6$ **27.** $p > 2.3$

28. $p \leq 3.0$ **29.** $p > 3.4$

In Exercises 30–36, fill the blank so as to produce two equivalent statements. For example, the arithmetic statement "a is negative" is equivalent to the geometric statement "the point a lies to the left of the point 0."

Arithmetic Statement	Geometric Statement
30. $a \geq b$	_____
31. _____	a lies c units to the right of b
32. _____	a lies between b and c
33. $a - b > 0$	_____
34. a is positive	_____
35. _____	a lies to the left of b
36. $a + b < c$ $(b > 0)$	_____

In Exercises 37–42, draw a picture on the number line of the given interval.

37. $(0, 8]$ **38.** $(0, \infty)$ **39.** $[-2, 1]$

40. $(-1, 1)$ **41.** $(-\infty, 0]$ **42.** $[-2, 7)$

In Exercises 43–48, use interval notation to denote the set of all real numbers x that satisfy the given inequality.

43. $5 \leq x \leq 10$ **44.** $-2 \leq x \leq 7$

45. $-3 < x < 14$ **46.** $7 < x < 77$

47. $x \geq -9$ **48.** $x \geq 12$

In Exercises 49–53, express the given numbers (based on 2006 estimates) in scientific notation.

49. Population of the world: 6,506,000,000

50. Population of the United States: 298,400,000

51. Average distance from Earth to Pluto: 5,910,000,000,000 meters

52. Radius of a hydrogen atom: .00000000001 meter

53. Width of a DNA double helix: .000000002 meter

In Exercises 54–57, express the given number in normal decimal notation.

54. Speed of light in a vacuum: 2.9979×10^8 miles per second

55. Average distance from the earth to the sun: 1.50×10^{11} meters

56. Electron charge: 1.602×10^{-27} coulomb

57. Proton mass: 1.6726×10^{-19} kilogram

58. One light-year is the distance light travels in a 365-day year. The speed of light is about 186,282.4 miles per second.

(a) How long is 1 light-year (in miles)? Express your answer in scientific notation.

(b) Light from the North Star takes 680 years to reach the earth. How many miles is the North Star from the earth?

59. The gross federal debt was about 8365 billion dollars in 2006, when the U.S. population was approximately 298.4 million people.

(a) Express the debt and the population in scientific notation.

(b) At that time, what was each person's share of the federal debt?

60. Apple reported that it had sold 28 million iPods through the end of 2005 and that 14 million iPods were sold in the first quarter of 2006. If the rate in the first quarter of 2006 continues through the end of 2008, how many iPods will be sold? Express your answer in scientific notation.

In Exercises 61–68, simplify the expression without using a calculator. Your answer should not have any radicals in it.

61. $\sqrt{2}\sqrt{8}$ **62.** $\sqrt{12}\sqrt{3}$

63. $\sqrt{\dfrac{3}{5}}\sqrt{\dfrac{12}{5}}$ **64.** $\sqrt{\dfrac{1}{2}}\sqrt{\dfrac{1}{6}}\sqrt{\dfrac{1}{12}}$

65. $\sqrt{6} + \sqrt{2}(\sqrt{2} - \sqrt{3})$ **66.** $\sqrt{12}(\sqrt{3} - \sqrt{27})$

67. $\sqrt{u^4}$ (u any real number)   **68.** $\sqrt{3x}\sqrt{75x^3}$ ($x \geq 0$)

In Exercises 69–78, simplify, and write the given number without using absolute values.

69. $|3 - 14|$ **70.** $|(-2)3|$ **71.** $3 - |2 - 5|$

72. $-2 - |-2|$ **73.** $|(-13)^2|$ **74.** $-|-5|^2$

75. $|\pi - \sqrt{2}|$ **76.** $|\sqrt{2} - 2|$ **77.** $|3 - \pi| + 3$

78. $|4 - \sqrt{2}| - 5$

In Exercises 79–84, fill the blank with <, =, or > so that the resulting statement is true.

79. $|-2|$ ____ $|-5|$ **80.** 5 ____ $|-2|$

81. $|3|$ ____ $-|4|$ **82.** $|-3|$ ____ 0

83. -7 ____ $|-1|$ **84.** $-|-4|$ ____ 0

In Exercises 85–92, find the distance between the given numbers.

85. -3 and 4 **86.** 7 and 107

87. -7 and $15/2$ **88.** $-3/4$ and -10

89. π and 3 **90.** π and -3

91. $\sqrt{2}$ and $\sqrt{3}$ **92.** π and $\sqrt{2}$

93. Galileo discovered that the period of a pendulum depends only on the length of the pendulum and the acceleration of gravity. The period T of a pendulum (in seconds) is

$$T = 2\pi\sqrt{\dfrac{l}{g}},$$

where l is the length of the pendulum in feet and $g \approx 32.2$ ft/sec^2 is the acceleration due to gravity. Find the period of a pendulum whose length is 4 feet.

94. Suppose you are k miles (not feet) above the ground. The radius of the earth is approximately 3960 miles. At the point where your line of sight meets the earth, it is perpendicular to the radius of the earth, as shown in the figure.

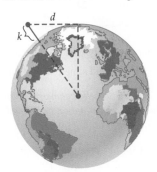

(a) Use the Pythagorean Theorem (see the Geometry Review Appendix) to show that
$$d = \sqrt{(3960 + k)^2 - 3960^2}.$$

(b) Show that the equation in part (a) simplifies to
$$d = \sqrt{7920k + k^2}.$$

(c) If you are h feet above the ground, then you are $h/5280$ miles high (why?). Use this fact and the equation in part (b) to obtain the formula used in Example 4.

95. According to data from the Center for Science in the Public Interest, the healthy weight range for a person depends on the person's height. For example,

Height	**Healthy Weight Range (lb)**
5 ft 8 in.	143 ± 21
6 ft 0 in.	163 ± 26

Express each of these ranges as an absolute value inequality in which x is the weight of the person.

96. At Statewide Insurance, each department's expenses are reviewed monthly. A department can fail to pass the budget variance test in a category if either (i) the absolute value of the difference between actual expenses and the budget is more than \$500 or (ii) the absolute value of the difference between the actual expenses and the budget is more than 5% of the budgeted amount. Which of the following items fail the budget variance test? Explain your answers.

Item	Budgeted Expense ($)	Actual Expense ($)
Wages	220,750	221,239
Overtime	10,500	11,018
Shipping and Postage	530	589

The wind-chill factor, shown in the table, calculates how a given temperature feels to a person's skin when the wind is taken into account. For example, the table shows that a temperature of $20°$ in a 40 mph wind feels like $-1°$. *

			Wind	(mph)					
	Calm	5	10	15	20	25	30	35	40
Temperature (°F)	40	36	34	32	30	29	28	28	27
	30	25	21	19	17	16	15	14	13
	20	13	9	6	4	3	1	0	−1
	10	1	−4	−7	−9	−11	−12	−14	−15
	0	−11	−16	−19	−22	−24	−26	−27	−29
	−10	−22	−28	−32	−35	−37	−39	−41	−43
	−20	−34	−41	−45	−48	−51	−53	−55	−57
	−30	−46	−53	−58	−61	−64	−67	−69	−71
	−40	−57	−66	−71	−74	−78	−80	−82	−84

In Exercises 97–99, find the absolute value of the difference of the two given wind-chill factors. For example, the difference between the wind-chill at $30°$ with a 15 mph wind and one at $-10°$ with a 10 mph wind is $|19-(-28)| = 47°$ or, equivalently, $|-28 - 19| = 47°$.

97. $10°$ with a 25 mph wind and $20°$ with a 20 mph wind

98. $30°$ with a 10 mph wind and $10°$ with a 30 mph wind

99. $-30°$ with a 5 mph wind and $0°$ with a 10 mph wind

100. The graph shows the number of Schedule C and C-EZ forms (in millions) that were filed with the IRS over a six-year period.[†]

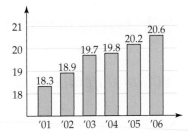

In what years was the following statement true:
$$|x - 19,500,000| \geq 600,000,$$
where x is the number of Schedule C and C-EZ forms in that year?

In Exercises 101–108, write the given expression without using absolute values.

101. $|t^2|$

102. $|-2 - y^2|$

103. $|b - 3|$ if $b \geq 3$

104. $|a - 5|$ if $a < 5$

---

*Table from the Joint Action Group for Temperature Indices, 2001.
†Internal Revenue Service. These schedules are for self-employed individuals.

105. $|c - d|$ if $c < d$ **106.** $|c - d|$ if $c \geq d$

107. $|u - v| - |v - u|$ **108.** $\dfrac{|u - v|}{|v - u|}$ if $u \neq v,\, u \neq 0,\, v \neq 0$

In Exercises 109 and 110, explain why the given statement is true for any numbers c and d. [Hint: Look at the properties of absolute value on page 10.]

109. $|(c - d)^2| = c^2 - 2cd + d^2$

110. $\sqrt{9c^2 - 18cd + 9d^2} = 3|c - d|$

In Exercises 111–116, express the given geometric statement about numbers on the number line algebraically, using absolute values.

111. The distance from x to 5 is less than 4.

112. x is more than 6 units from c.

113. x is at most 17 units from -4.

114. x is within 3 units of 7.

115. c is closer to 0 than b is.

116. x is closer to 1 than to 4.

In Exercises 117–120, translate the given algebraic statement into a geometric statement about numbers on the number line.

117. $|x - 3| < 2$ **118.** $|x - c| > 6$

119. $|x + 7| \leq 3$ **120.** $|u + v| \geq 2$

121. Match each of the following graphs with the appropriate absolute value equation or inequality.

(a) ⊢———⊣ (10, 24)   i. $|x - 17| = 7$

(b) •———• (10, 24)   ii. $|x - 17| = -7$

(c) ⟷   iii. $|x - 17| \leq 7$

(d))———((10, 24)   iv. $|x - 17| \geq -7$

(e) ⟷   v. $|x - 17| > 7$

122. Explain geometrically why this statement is always false:
$$|c - 1| < 2 \text{ and simultaneously } |c - 12| < 3.$$

In Exercises 123–134, use the geometric approach explained in the text to solve the given equation or inequality.

123. $|x| = 1$ **124.** $|x| = 3/2$

125. $|x - 2| = 1$ **126.** $|x + 3| = 2$

127. $|x + \pi| = 4$ **128.** $\left|x - \dfrac{3}{2}\right| = 5$

129. $|x| < 7$ **130.** $|x| \geq 5$

131. $|x - 5| < 2$ **132.** $|x - 6| > 2$

133. $|x + 2| \geq 3$ **134.** $|x + 4| \leq 2$

THINKERS

135. Explain why the statement $|a| + |b| + |c| > 0$ is algebraic shorthand for "at least one of the numbers a, b, c, is different from zero."

136. Find an algebraic shorthand version of the statement "none of the numbers a, b, c, is zero."

1.1.A *SPECIAL TOPICS* Decimal Representation of Real Numbers

Section Objectives ■ Convert a repeating decimal to a rational number, and vice versa.
■ Distinguish between rational and irrational numbers.

Every rational number can be expressed as a terminating or repeating decimal. For instance, $3/4 = .75$. To express $15/11$ as a decimal, divide the numerator by the denominator:

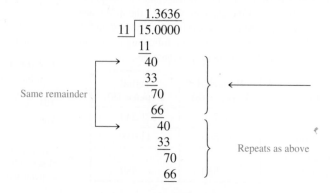

Since the remainder at the first step (namely, 4) occurs again at the third step, it is clear that the division process goes on forever with the two-digit block "36" repeating over and over in the quotient $15/11 = 1.3636363636 \cdots$.

The method used in the preceding example can be used to express any rational number as a decimal. During the division process, some remainder *necessarily repeats*. If the remainder at which this repetition starts is 0, the result is a repeating decimal ending in zeros—that is, a terminating decimal (for instance, $.75000 \cdots = .75$). If the remainder at which the repetition starts is nonzero, then the result is a nonterminating repeating decimal, as in the example above.

Conversely, there is a simple method for converting any repeating decimal into a rational number.

TECHNOLOGY TIP

To convert repeating decimals to fractions on TI, use *Frac* in this menu/submenu:

TI-84+: MATH

TI-86: MATH/MISC

On HP-39gs, select *Fraction* number format in the MODE menu; then enter the decimal.

On Casio, use the FRAC program in the Program Appendix.

EXAMPLE 1

Write $d = .272727 \cdots$ as a rational number.

SOLUTION   Assuming that the usual rules of arithmetic hold, we see that

$$100d = 27.272727 \cdots \qquad \text{and} \qquad d = .272727 \cdots.$$

Now subtract d from $100d$:

$$
\begin{aligned}
100d &= 27.272727 \cdots \\
-d &= -.272727 \cdots \\
\hline
99d &= 27
\end{aligned}
$$

Dividing both sides of this last equation by 99 shows that $d = 27/99 = 3/11$. ∎

 IRRATIONAL NUMBERS

Many nonterminating decimals are *nonrepeating* (that is, no block of digits repeats forever), such as $.202002000200002 \cdots$ (where after each 2 there is one more zero than before). Although the proof is too long to give here, it is true that every nonterminating and nonrepeating decimal represents an *irrational* real number. Conversely every irrational number can be expressed as a nonterminating and nonrepeating decimal (no proof to be given here).

A typical calculator can hold only the first 10–14 digits of a number in decimal form. Consequently, a calculator can contain the *exact* value only of those rational numbers whose decimal expansion terminates after 10–14 places. It must *approximate* all other real numbers.

Since every real number is either a rational number or an irrational one, the preceding discussion can be summarized as follows.

Decimal Representation

1. Every real number can be expressed as a decimal.

2. Every decimal represents a real number.

3. The terminating decimals and the nonterminating repeating decimals are the rational numbers.

4. The nonterminating, nonrepeating decimals are the irrational numbers.

EXERCISES 1.1.A

In Exercises 1–6, express the given rational number as a repeating decimal.

1. 7/9   **2.** 19/88   **3.** 9/11

4. 2/13   **5.** 22/7   **6.** 1/19 (long)

In Exercises 7–13, express the given repeating decimal as a fraction.

7. .373737 · · ·   **8.** .929292 · · ·

9. 76.63424242 · · · [*Hint:* Consider 10,000d − 100d, where d = 76.63424242 · · ·.]

10. 13.513513 · · · [*Hint:* Consider 1000d − d, where d = 13.513513 · · ·.]

11. .135135135 · · · [*Hint:* See Exercise 10.]

12. .33030303 · · ·   **13.** 52.31272727 · · ·

14. If two real numbers have the same decimal expansion through three decimal places, how far apart can they be on the number line?

In Exercises 15–22, state whether a calculator can express the given number exactly.

15. 2/3   **16.** 7/16   **17.** 1/64   **18.** 1/22

19. 3π/2   **20.** π − 3   **21.** 1/.625   **22.** 1/.16

23. Use the methods in Exercises 7–13 to show that both .74999 · · · and .75000 · · · are decimal expansions of 3/4. [Every terminating decimal can also be expressed as a decimal ending in repeated 9's. It can be proved that these are the only real numbers with more than one decimal expansion.]

Finding remainders with a calculator

24. If you use long division to divide 369 by 7, you obtain:

$$\begin{array}{r} 52 \leftarrow \text{Quotient} \\ \text{Divisor} \rightarrow \ 7\overline{)369} \leftarrow \text{Dividend} \\ \underline{35} \\ 19 \\ \underline{14} \\ 5 \leftarrow \text{Remainder} \end{array}$$

If you use a calculator to find 369 ÷ 7, the answer is displayed as 52.71428571. Observe that the integer part of this calculator answer, 52, is the quotient when you do the problem by long division. The usual "checking procedure" for long division shows that

$$7 \cdot 52 + 5 = 369 \quad \text{or, equivalently} \quad 369 - 7 \cdot 52 = 5.$$

Thus, the remainder is

$$\text{Dividend} - (\text{divisor}) \left(\begin{array}{c} \text{integer part of} \\ \text{calculator answer} \end{array} \right).$$

Use this method to find the quotient and remainder in these problems:

(a) 5683 ÷ 9   (b) 1,000,000 ÷ 19

(c) 53,000,000 ÷ 37

In Exercises 25–30, find the decimal expansion of the given rational number. All these expansions are too long to fit in a calculator but can be readily found by using the hint in Exercise 25.

25. 1/17 [*Hint:* The first part of dividing 1 by 17 involves working this division problem: 1,000,000 ÷ 17. The method of Exercise 24 shows that the quotient is 58,823 and the remainder is 9. Thus the decimal expansion of 1/17 begins .058823, and the next block of digits in the expansion will be the quotient in the problem 9,000,000 ÷ 17. The remainder when 9,000,000 is divided by 17 is 13, so the next block of digits in the expansion of 1/17 is the quotient in the problem 13,000,000 ÷ 17. Continue in this way until the decimal expansion repeats.]

26. 3/19   **27.** 1/29   **28.** 3/43   **29.** 283/47

30. 768/59

THINKERS

31. If your calculator has a *Frac* key or program (see the Program Appendix), test its limitations by entering each of the following numbers and then pressing the *Frac* key.

(a) .058823529411   (b) .0588235294117

(c) .058823529411724   (d) .0588235294117985

Which of your answers are correct? [*Hint:* Exercise 25 may be helpful.]

32. (a) Show that there are at least as many irrational numbers (nonrepeating decimals) as there are terminating decimals. [*Hint:* With each terminating decimal associate a nonrepeating decimal.]

(b) Show that there are at least as many irrational numbers as there are repeating decimals. [*Hint:* With each repeating decimal, associate a nonrepeating decimal by inserting longer and longer strings of zeros: for instance, with .11111111 · · · associate the number .101001000100001 · · ·.]

1.2 Solving Equations Algebraically

Section Objectives
- Understand the basic principles for solving equations.
- Solve linear equations.
- Solve quadratic equations by factoring, completing the square, or using the quadratic formula.
- Use the discriminant to determine the number of real solutions of a quadratic equation.
- Solve some types of higher-degree equations.
- Solve fractional equations.

This section deals with equations such as

$$3x - 6 = 7x + 4, \qquad x^2 - 5x + 6 = 0, \qquad 2x^4 - 13x^2 = 3.$$

A **solution** of an equation is a number that, when substituted for the variable x, produces a true statement.* For example, 5 is a solution of $3x + 2 = 17$ because $3 \cdot 5 + 2 = 17$ is a true statement. To **solve** an equation means to find all its solutions. Throughout this chapter, we shall deal only with **real solutions,** that is, solutions that are real numbers.

Two equations are said to be **equivalent** if they have the same solutions. For example, $3x + 2 = 17$ and $x - 2 = 3$ are equivalent because 5 is the only solution of each one.

Basic Principles for Solving Equations

Performing any of the following operations on an equation produces an equivalent equation:

1. Add or subtract the same quantity from both sides of the equation.

2. Multiply or divide both sides of the equation by the same *nonzero* quantity.

The usual strategy in equation solving is to use these basic principles to transform a given equation into an equivalent one whose solutions are known.

A **first-degree,** or **linear, equation** is one that can be written in the form

$$ax + b = 0$$

for some constants a, b, with $a \neq 0$. Every first-degree equation has exactly one solution, which is easily found.

EXAMPLE 1

To solve $3x - 6 = 7x + 4$, we use the basic principles to transform this equation into an equivalent one whose solution is obvious:

$$3x - 6 = 7x + 4$$

Add 6 to both sides: $\qquad 3x = 7x + 10$

Subtract $7x$ from both sides: $\qquad -4x = 10$

Divide both sides by -4: $\qquad x = \dfrac{10}{-4} = -\dfrac{5}{2}.$

*Any letter may be used for the variable.

Since $-5/2$ is the only solution of this last equation, $-5/2$ is the only solution of the original equation, $3x - 6 = 7x + 4$. ∎

EXAMPLE 2

Solve the equation $a^2x + b^2y = 2$ for y.

SOLUTION   Since we are to solve for y, we treat y as the variable and treat all other letters as constants. We begin by getting the y-term on one side and all other terms on the other side of the equation.

$$a^2x + b^2y = 2$$

Subtract a^2x from both sides: $$b^2y = 2 - a^2x$$

Divide both sides by b^2: $$y = \frac{2 - a^2x}{b^2}$$ ∎

EXAMPLE 3

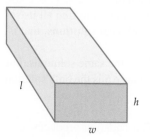

Figure 1–9

The surface area S of the rectangular box in Figure 1–9 is given by

$$2lh + 2lw + 2wh = S.$$

Solve this equation for h.

SOLUTION   Treat h as the variable and all the other letters as constants. First, get all the terms involving the variable h on one side of the equation and everything else on the other side.

$$2lh + 2lw + 2wh = S$$

Subtract $2lw$ from both sides: $$2lh + 2wh = S - 2lw$$

Factor out h on the left side: $$(2l + 2w)h = S - 2lw$$

Divide both sides by $(2l + 2w)$: $$h = \frac{S - 2lw}{2l + 2w}.$$ ∎

QUADRATIC EQUATIONS

A **second-degree,** or **quadratic, equation** is an equation that can be written in the form

$$ax^2 + bx + c = 0$$

for some constants a, b, c, with $a \neq 0$. There are several techniques for solving such equations. We begin with the **factoring method,** which makes use of this property of the real numbers:

Zero Products

If a product of real numbers is zero, then at least one of the factors is zero; in other words,

If $cd = 0$, then $c = 0$ or $d = 0$ (or both).

EXAMPLE 4

To solve $3x^2 - x = 10$, we first rearrange the terms to make one side 0 and then factor:

Subtract 10 from each side: $\qquad 3x^2 - x - 10 = 0$

Factor left side: $\qquad (3x + 5)(x - 2) = 0.$

If a product of real numbers is 0, then at least one of the factors must be 0. So this equation is equivalent to

$$3x + 5 = 0 \qquad \text{or} \qquad x - 2 = 0$$
$$3x = -5 \qquad\qquad\qquad x = 2$$
$$x = -5/3$$

Therefore the solutions are $-5/3$ and 2. ∎

CAUTION

You cannot use the factoring method unless one side of the equation is 0. Otherwise, you'll get the wrong answer, as is the case here:

$$x^2 + 3x + 2 = 1$$
$$(x + 2)(x + 1) = 1$$
$$x + 2 = 1 \quad \text{or} \quad x + 1 = 1 \qquad \text{Mistake here!}$$
$$x = -1 \quad \text{or} \qquad x = 0$$

These are NOT solutions of the original equation, as you can easily verify.

There are two numbers whose square is 7, namely, $\sqrt{7}$ and $-\sqrt{7}$. So the solutions of $x^2 = 7$ are $\sqrt{7}$ and $-\sqrt{7}$, or in abbreviated form, $\pm\sqrt{7}$. The same argument works for any positive real number d:

The solutions of $x^2 = d$ are $\sqrt{d}$ and $-\sqrt{d}$.

The same reasoning enables us to solve other equations.

EXAMPLE 5

Solve: $(z - 2)^2 = 5$.

SOLUTION The equation says that $z - 2$ is a number whose square is 5. Since there are only two numbers whose square is 5, namely, $\sqrt{5}$ and $-\sqrt{5}$, we must have

$$z - 2 = \sqrt{5} \qquad \text{or} \qquad z - 2 = -\sqrt{5}$$
$$z = \sqrt{5} + 2 \qquad\qquad\qquad z = -\sqrt{5} + 2.$$

In compact notation, the solutions of the equation are $\pm\sqrt{5} + 2$. ∎

We now use a slight variation of Example 5 to develop a method for solving quadratic equations that don't readily factor. Consider, for example, the expression $x^2 + 6x$. If you add 9, the result is a perfect square:

$$x^2 + 6x + 9 = (x + 3)^2.$$

The process of adding a constant to produce a perfect square is called **completing the square.** Note that one-half the coefficient of x in $x^2 + 6x$ is $\frac{6}{2} = 3$. We added 9, which is 3^2, and the resulting perfect square was $(x + 3)^2$. The same idea works in the general case, as is proved in Exercise 94.

Completing
the Square

> To change $x^2 + bx$ into a perfect square, add $\left(\dfrac{b}{2}\right)^2$. The resulting polynomial
> $x^2 + bx + \left(\dfrac{b}{2}\right)^2$ factors as $\left(x + \dfrac{b}{2}\right)^2$.

The following example shows how completing the square can be used to solve quadratic equations.

EXAMPLE 6

To solve $x^2 + 6x + 1 = 0$, we first rewrite the equation as $x^2 + 6x = -1$. Next we complete the square on the left side by adding the square of half the coefficient of x, namely, $(6/2)^2 = 9$. To have an equivalent equation, we must add 9 to *both* sides:

$$x^2 + 6x + 9 = -1 + 9.$$

Factor left side: $(x + 3)^2 = 8.$

Thus $x + 3$ is a number whose square is 8. The only numbers whose squares equal 8 are $\sqrt{8}$ and $-\sqrt{8}$. So we must have

$$x + 3 = \sqrt{8} \qquad \text{or} \qquad x + 3 = -\sqrt{8}$$
$$x = \sqrt{8} - 3 \qquad \text{or} \qquad x = -\sqrt{8} - 3.$$

Therefore the solutions of the original equation are $\sqrt{8} - 3$ and $-\sqrt{8} - 3$ or, in more compact notation, $\pm\sqrt{8} - 3$. ∎

We can use the completing-the-square method to solve *any* quadratic equation:*

$$ax^2 + bx + c = 0$$

Divide both sides by a: $x^2 + \dfrac{b}{a}x + \dfrac{c}{a} = 0$

Subtract $\dfrac{c}{a}$ from both sides: $x^2 + \dfrac{b}{a}x = -\dfrac{c}{a}$

CAUTION

Completing the square only works when the coefficient of x^2 is 1. In an equation such as

$$5x^2 - x + 2 = 0,$$

you must first divide every term on both sides by 5 and *then* complete the square.

---

*If you have trouble following any step here, do it for a numerical example, such as the case when $a = 3, b = 11, c = 5$.

Add $\left(\dfrac{b}{2a}\right)^2$ to both sides:* $x^2 + \dfrac{b}{a}x + \left(\dfrac{b}{2a}\right)^2 = \left(\dfrac{b}{2a}\right)^2 - \dfrac{c}{a}$

Factor left side: $\left(x + \dfrac{b}{2a}\right)^2 = \left(\dfrac{b}{2a}\right)^2 - \dfrac{c}{a}$

Find common denominator
for right side: $\left(x + \dfrac{b}{2a}\right)^2 = \dfrac{b^2}{4a^2} - \dfrac{c}{a} = \dfrac{b^2 - 4ac}{4a^2}.$

Since the square of $x + \dfrac{b}{2a}$ equals $\dfrac{b^2 - 4ac}{4a^2}$, we must have

$$x + \frac{b}{2a} = \pm\sqrt{\frac{b^2 - 4ac}{4a^2}} = \pm\frac{\sqrt{b^2 - 4ac}}{2a}$$

Subtract $\dfrac{b}{2a}$ from
both sides: $x = \dfrac{-b}{2a} \pm \dfrac{\sqrt{b^2 - 4ac}}{2a} = \dfrac{-b \pm \sqrt{b^2 - 4ac}}{2a}.$

We have proved

**The Quadratic
Formula**

The solutions of the quadratic equation $ax^2 + bx + c = 0$ are

$$x = \frac{-b \pm \sqrt{b^2 - 4ac}}{2a}.$$

You should memorize the quadratic formula.

EXAMPLE 7

Solve $x^2 + 3 = -8x$.

SOLUTION   Rewrite the equation as $x^2 + 8x + 3 = 0$, and apply the quadratic formula with $a = 1$, $b = 8$, and $c = 3$:

$$x = \frac{-b \pm \sqrt{b^2 - 4ac}}{2a} = \frac{-8 \pm \sqrt{8^2 - 4 \cdot 1 \cdot 3}}{2 \cdot 1}$$

$$= \frac{-8 \pm \sqrt{52}}{2} = \frac{-8 \pm \sqrt{4 \cdot 13}}{2} = \frac{-8 \pm \sqrt{4}\sqrt{13}}{2}$$

$$= \frac{-8 \pm 2\sqrt{13}}{2} = -4 \pm \sqrt{13}.$$

The equation has two distinct real solutions, $-4 + \sqrt{13}$ and $-4 - \sqrt{13}$.   ∎

---

*This is the square of half the coefficient of x.

EXAMPLE 8

Solve $x^2 - 194x + 9409 = 0$.

SOLUTION   Use a calculator and the quadratic formula with $a = 1, b = -194$, and $c = 9409$:

$$x = \frac{-b \pm \sqrt{b^2 - 4ac}}{2a} = \frac{-(-194) \pm \sqrt{(-194)^2 - 4 \cdot 1 \cdot 9409}}{2 \cdot 1}$$

$$= \frac{194 \pm \sqrt{37636 - 37636}}{2} = \frac{194 \pm 0}{2} = 97.$$

Thus, 97 is the only solution of the equation.   ■

EXAMPLE 9

Solve $2x^2 + x + 3 = 0$.

SOLUTION   Use the quadratic formula with $a = 2, b = 1$, and $c = 3$:

$$x = \frac{-b \pm \sqrt{b^2 - 4ac}}{2a} = \frac{-1 \pm \sqrt{1^2 - 4 \cdot 2 \cdot 3}}{2 \cdot 2} = \frac{-1 \pm \sqrt{1 - 24}}{4}$$

$$= \frac{-1 \pm \sqrt{-23}}{4}.$$

Since $\sqrt{-23}$ is not a real number, this equation has *no real solutions* (that is, no solutions in the real number system).   ■

The expression $b^2 - 4ac$ in the quadratic formula is called the **discriminant.** As the last three examples illustrate, the discriminant determines the *number* of real solutions of the equation $ax^2 + bx + c = 0$.

Real Solutions of a Quadratic Equation

Discriminant $b^2 - 4ac$	Number of Real Solutions of $ax^2 + bx + c = 0$	Example
> 0	Two distinct real solutions	Example 7
$= 0$	One real solution	Example 8
< 0	No real solutions	Example 9

EXAMPLE 10

Use the discriminant to determine the number of real solutions of each of these equations.

(a) $4x^2 - 20x + 25 = 0$ (b) $7x^2 + 3 = 5x$ (c) $.5x^2 + 6x - 2 = 0$.

SOLUTION

(a) Here $a = 4$, $b = -20$, and $c = 25$. So the discriminant is

$$b^2 - 4ac = (-20)^2 - 4 \cdot 4 \cdot 25 = 400 - 400 = 0.$$

The equation has one real solution.

(b) First rewrite the equation as $7x^2 - 5x + 3 = 0$. The discriminant is

$$b^2 - 4ac = (-5)^2 - 4 \cdot 7 \cdot 3 = 25 - 84 = -59,$$

so the equation has no real solutions.

(c) The discriminant is $b^2 - 4ac = 6^2 - 4 \cdot (.5) \cdot (-2) = 36 + 4 = 40$. There are two real solutions. ∎

The quadratic formula and a calculator can be used to solve any quadratic equation with nonnegative discriminant.

EXAMPLE 11

The number of identity theft complaints (in thousands) in year x is approximated by

$$-5.5x^2 + 77.1x + 26.3 \qquad (0 \le x \le 9),$$

where x is the number of years since 2000.* Use the quadratic formula and a calculator to find the year in which there were 247,000 complaints.

SOLUTION   Complaints are measured in thousands, so we must solve

$$-5.5x^2 + 77.1x + 26.3 = 247.$$

Subtracting 247 from both sides produces the equivalent equation

$$-5.5x^2 + 77.1x - 220.7 = 0.$$

To solve this equation, we compute the radical part of the quadratic formula

$$\sqrt{b^2 - 4ac} = \sqrt{77.1^2 - 4(-5.5)(-220.7)}$$

and store the result in memory D. By the quadratic formula, the solutions of the equation are

$$x = \frac{-b \pm \sqrt{b^2 - 4ac}}{2a} = \frac{-b \pm D}{2a} = \frac{-77.1 \pm D}{2(-5.5)}.$$

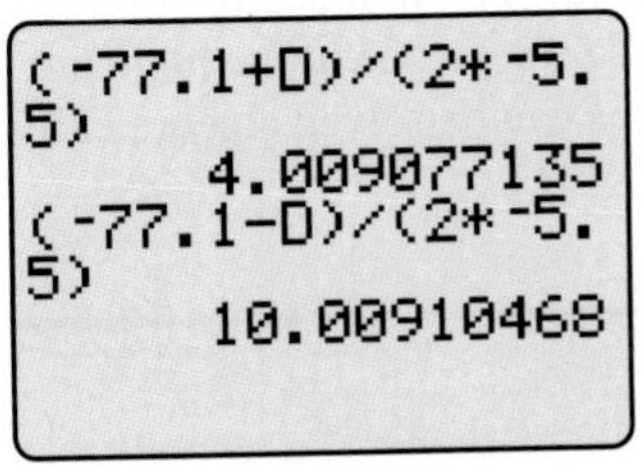

```
(-77.1+D)/(2*-5.
5)
        4.009077135
(-77.1-D)/(2*-5.
5)
        10.00910468
```

Figure 1–10

So the approximate solutions are

$$x = \frac{-77.1 + D}{2(-5.5)} \approx 4.009 \qquad \text{and} \qquad x = \frac{-77.1 - D}{2(-5.5)} \approx 10.009,$$

as shown in Figure 1–10. Since we are given that $0 \le x \le 9$, the only applicable solution here is $x \approx 4.009$, which corresponds to early 2004. ∎

---

*Based on data from the Identity Theft Clearinghouse of the Federal Trade Commission.

EXAMPLE 12

If an object is thrown upward, dropped, or thrown downward and travels in a straight line subject only to gravity (with wind resistance ignored), the height h of the object above the ground (in feet) after t seconds is given by

$$h = -16t^2 + v_0 t + h_0,$$

where h_0 is the height of the object when $t = 0$ and v_0 is the initial velocity at time $t = 0$. The value of v_0 is taken as positive if the object moves upward and negative if it moves downward. If a baseball is thrown down from the top of a 640-foot-high building with an initial velocity of 52 feet per second, how long does it take to reach the ground?

SOLUTION   In this case, v_0 is -52 and h_0 is 640, so that the height equation is

$$h = -16t^2 - 52t + 640.$$

The object is on the ground when $h = 0$, so we must solve the equation

$$0 = -16t^2 - 52t + 640.$$

Using the quadratic formula and a calculator, we see that

$$t = \frac{-(-52) \pm \sqrt{(-52)^2 - 4(-16)(640)}}{2(-16)} = \frac{52 \pm \sqrt{43{,}664}}{-32} = \begin{cases} -8.15 \\ \text{or} \\ 4.90. \end{cases}$$

Only the positive answer makes sense in this case. So it takes about 4.9 seconds for the baseball to reach the ground.   ∎

> **TECHNOLOGY TIP**
>
> Most calculators have built-in polynomial equation solvers that will solve quadratic and other polynomial equations. See Exercise 105.

HIGHER-DEGREE EQUATIONS

A **polynomial equation of degree n** is one that can be written in the form

$$a_n x^n + \cdots + a_3 x^3 + a_2 x^2 + a_1 x + a_0 = 0,$$

where n is a positive integer, each a_i is a constant, and $a_n \neq 0$. For instance,

$$4x^6 - 3x^5 + x^4 + 7x^3 - 8x^2 + 4x + 9 = 0$$

is a polynomial equation of degree 6. As a general rule, polynomial equations of degree 3 and above are best solved by the numerical or graphical methods presented in Section 2.2. However, some such equations can be solved algebraically by making a suitable substitution, as we now see.

EXAMPLE 13

To solve $4x^4 - 13x^2 + 3 = 0$, substitute u for x^2 and solve the resulting quadratic equation:

$$4x^4 - 13x^2 + 3 = 0$$

$$4(x^2)^2 - 13x^2 + 3 = 0$$

$$\text{Let } u = x^2: \qquad 4u^2 - 13u + 3 = 0$$

$$(u - 3)(4u - 1) = 0$$

$$u - 3 = 0 \qquad \text{or} \qquad 4u - 1 = 0$$

$$u = 3 \qquad\qquad\qquad 4u = 1$$

$$u = \frac{1}{4}.$$

Since $u = x^2$, we see that

$$x^2 = 3 \qquad \text{or} \qquad x^2 = \frac{1}{4}$$

$$x = \pm\sqrt{3} \qquad\qquad x = \pm\frac{1}{2}.$$

Hence, the original equation has four solutions: $-\sqrt{3}, \sqrt{3}, -1/2, 1/2$.   ∎

EXAMPLE 14

To solve $x^4 - 4x^2 + 1 = 0$, let $u = x^2$:

$$x^4 - 4x^2 + 1 = 0$$

$$u^2 - 4u + 1 = 0.$$

The quadratic formula shows that

$$u = \frac{-(-4) \pm \sqrt{(-4)^2 - 4 \cdot 1 \cdot 1}}{2 \cdot 1} = \frac{4 \pm \sqrt{12}}{2}$$

$$= \frac{4 \pm \sqrt{4 \cdot 3}}{2} = \frac{4 \pm \sqrt{4}\sqrt{3}}{2}$$

$$= \frac{4 \pm 2\sqrt{3}}{2} = 2 \pm \sqrt{3}.$$

Since $u = x^2$, we have the equivalent statements:

$$x^2 = 2 + \sqrt{3} \qquad \text{or} \qquad x^2 = 2 - \sqrt{3}$$

$$x = \pm\sqrt{2 + \sqrt{3}} \qquad\qquad x = \pm\sqrt{2 - \sqrt{3}}.$$

Therefore the original equation has four solutions.   ∎

FRACTIONAL EQUATIONS

The first step in solving an equation that involves fractional expressions is to multiply both sides by a common denominator to eliminate the fractions. Then use the methods presented earlier in this chapter to solve the resulting equation. However, care must be used.

Multiplying both sides of an equation by a quantity involving the variable (which may be zero for some values) may lead to an **extraneous solution,** a number that does not satisfy the original equation.* To avoid errors in such situations, always check your solutions in the *original* equation.

---

*The second basic principle for solving equations (page 19) applies only to nonzero quantities.

EXAMPLE 15

Solve $\dfrac{x+2}{x-3} - \dfrac{7}{x+3} = \dfrac{30}{x^2-9}$.

SOLUTION   Note that $(x-3)(x+3) = x^2 - 9$. Multiply both sides of the equation by $(x-3)(x+3)$ to eliminate the fractions.

$$\frac{x+2}{x-3}(x-3)(x+3) - \frac{7}{x+3}(x-3)(x+3) = \frac{30}{x^2-9}(x-3)(x+3)$$

Cancel common factors: $(x+2)(x+3) - 7(x-3) = 30$

Multiply out left side: $x^2 + 5x + 6 - 7x + 21 = 30$

Simplify: $x^2 - 2x + 27 = 30$

Subtract 30 from both sides: $x^2 - 2x - 3 = 0$

Factor: $(x-3)(x+1) = 0$

Zero Product Property: $x - 3 = 0 \quad\text{or}\quad x + 1 = 0$

$$x = 3 \qquad\qquad x = -1$$

So the *possible* solutions are 3 and -1. When 3 is substituted for x in the original equation, its first term becomes $\dfrac{3+3}{3-3} = \dfrac{6}{0}$, which is not defined. So 3 is *not* a solution of the original equation. Next we substitute -1 for x in the original equation and obtain:

Left side: $\dfrac{-1+2}{-1-3} - \dfrac{7}{-1+3} = \dfrac{1}{-4} - \dfrac{7}{2} = -\dfrac{1}{4} - \dfrac{14}{4} = -\dfrac{15}{4}$

Right side: $\dfrac{30}{(-1)^2-9} = \dfrac{30}{-8} = -\dfrac{15}{4}$.

So the left and right sides are equal. Therefore, -1 is the only solution of the original equation.   ∎

EXAMPLE 16

The relationship between the focal length F of a digital camera lens, the distance u from the object being photographed to the lens, and the distance v from the lens to the camera's sensor is given by

$$\frac{1}{F} = \frac{1}{v} + \frac{1}{u}.$$

Express u in terms of F and v.

SOLUTION   We must solve the equation for u. First we eliminate fractions by multiplying both sides by Fvu.

$$Fvu \cdot \frac{1}{F} = Fvu\left(\frac{1}{v} + \frac{1}{u}\right)$$

Distributive Law: $Fvu \cdot \dfrac{1}{F} = Fvu \cdot \dfrac{1}{v} + Fvu \cdot \dfrac{1}{u}$

Cancel: $\qquad vu = Fu + Fv$

Subtract Fu from both sides: $\qquad vu - Fu = Fv$

Distributive Law: $\qquad (v - F)u = Fv$

Divide both sides by $v - F$: $\qquad u = \dfrac{Fv}{v - F}.$ ∎

EXERCISES 1.2

In Exercises 1–6, solve the equation.

1. $3x + 2 = 26$

2. $\dfrac{y}{5} - 3 = 14$

3. $3x + 2 = 9x + 7$

4. $-7(t + 2) = 3(4t + 1)$

5. $\dfrac{3y}{4} - 6 = y + 2$

6. $2(1 + x) = 3x + 5$

In Exercises 7–12, solve the equation for the indicated variable.

7. $x = 3y - 5$ for y

8. $5x - 2y = 1$ for x

9. $A = \dfrac{h}{2}(b + c)$ for b

10. $V = \pi b^2 c$ for c

11. $V = \dfrac{\pi d^2 h}{4}$ for h

12. $\dfrac{1}{r} = \dfrac{1}{s} + \dfrac{1}{t}$ for r

In Exercises 13–22, solve the equation by factoring.

13. $x^2 - 8x + 15 = 0$

14. $x^2 - 5x + 6 = 0$

15. $x^2 - 5x = 14$

16. $x^2 + x = 20$

17. $2y^2 + 5y - 3 = 0$

18. $3t^2 - t - 2 = 0$

19. $4t^2 + 9t + 2 = 0$

20. $9t^2 + 2 = 11t$

21. $3u^2 - 4u = 4$

22. $5x^2 + 26x = -5$

In Exercises 23–26, solve the equation by completing the square.

23. $x^2 - 2x = 12$

24. $x^2 - 4x - 30 = 0$

25. $x^2 - x - 1 = 0$

26. $x^2 + 3x - 2 = 0$

In Exercises 27–32, find the number of real solutions of the equation by computing the discriminant.

27. $x^2 + 4x + 1 = 0$

28. $4x^2 - 4x - 3 = 0$

29. $9x^2 = 12x + 1$

30. $9t^2 + 15 = 30t$

31. $25t^2 + 49 = 70t$

32. $49t^2 + 5 = 42t$

In Exercises 33–42, use the quadratic formula to solve the equation.

33. $x^2 - 4x + 1 = 0$

34. $x^2 + 2x - 1 = 0$

35. $x^2 + 6x + 7 = 0$

36. $x^2 + 4x - 3 = 0$

37. $x^2 + 6 = 2x$

38. $x^2 + 11 = 6x$

39. $4x^2 + 4x = 7$

40. $4x^2 - 4x = 11$

41. $4x^2 - 8x + 1 = 0$

42. $2t^2 + 4t + 1 = 0$

In Exercises 43–52, solve the equation by any method.

43. $x^2 + 9x + 18 = 0$

44. $3t^2 - 11t - 20 = 0$

45. $4x(x + 1) = 1$

46. $25y^2 = 20y + 1$

47. $2x^2 = 7x + 15$

48. $2x^2 = 6x + 3$

49. $t^2 + 4t + 13 = 0$

50. $5x^2 + 2x = 2$

51. $\dfrac{7x^2}{3} = \dfrac{2x}{3} - 1$

52. $x^2 + \sqrt{2}x - 3 = 0$

In Exercises 53–56, use a calculator to find approximate solutions of the equation.

53. $4.42x^2 - 10.14x + 3.79 = 0$

54. $8.06x^2 + 25.8726x - 25.047256 = 0$

55. $3x^2 - 82.74x + 570.4923 = 0$

56. $7.63x^2 + 2.79x = 5.32$

In Exercises 57–64, find all real solutions of the equation exactly.

57. $y^4 - 7y^2 + 6 = 0$

58. $x^4 - 2x^2 + 1 = 0$

59. $x^4 - 2x^2 - 35 = 0$

60. $x^4 - 2x^2 - 24 = 0$

61. $2y^4 - 9y^2 + 4 = 0$

62. $6z^4 - 7z^2 + 2 = 0$

63. $10x^4 + 3x^2 = 1$

64. $6x^4 - 7x^2 = 3$

In Exercises 65–74, solve the equation and check your answers.

65. $\dfrac{1}{2t} - \dfrac{2}{5t} = \dfrac{1}{10t} - 1$

66. $\dfrac{1}{2} + \dfrac{2}{y} = \dfrac{1}{3} + \dfrac{3}{y}$

67. $\dfrac{2x - 7}{x + 4} = \dfrac{5}{x + 4} - 2$

68. $\dfrac{z + 4}{z + 5} = \dfrac{-1}{z + 5}$

69. $25x + \dfrac{4}{x} = 20$

70. $1 - \dfrac{3}{x} = \dfrac{40}{x^2}$

71. $\dfrac{2}{x^2} - \dfrac{5}{x} = 4$

72. $\dfrac{x}{x-1} + \dfrac{x+2}{x} = 3$

73. $\dfrac{4x^2+5}{3x^2+5x-2} = \dfrac{4}{3x-1} - \dfrac{3}{x+2}$

74. $\dfrac{x+3}{x-2} - \dfrac{3}{x+2} = \dfrac{20}{x^2-4}$

The gross federal debt y (in trillions of dollars) in year x is approximated by

$$y = .79x + 3.93 \quad (x \geq 3),$$

where x is the number of years after 2000.* In Exercises 75–76, find the year in which the approximate federal debt is:

75. \$12.62 billion

76. \$14.2 billion

The total health care expenditures E in the United States (in billions of dollars) can be approximated by

$$E = 73.04x + 625.6,$$

where x is the number of years since 1990.† Determine the year in which health care expenditures are at the given level.

77. \$1794.25 billion

78. \$1940.3 billion

In a simple model of the economy (by John Maynard Keynes), equilibrium between national output and national expenditures is given by the equilibrium equation

$$Y = C + I + G + (X - M),$$

where Y is the national income, C is consumption (which depends on national income), I is the amount of investment, G is government spending, X is exports, and M is imports (which also depend on national income). In Exercises 79 and 80, solve the equilibrium equation for Y under the given conditions.

79. $C = 120 + .9Y$, $M = 20 + .2Y$, $I = 140$, $G = 150$, and $X = 60$

80. $C = 60 + .85Y$, $M = 35 + .2Y$, $I = 95$, $G = 145$, and $X = 50$

In Exercises 81–84, use the height equation in Example 12. Note that an object that is dropped (rather than thrown downward) has initial velocity $v_0 = 0$.

81. How long does it take a baseball to reach the ground if it is dropped from the top of a 640-foot-high building? Compare with Example 12.

82. You are standing on a cliff that is 200 feet high. How long will it take a rock to reach the ground if

(a) you drop it?

(b) you throw it downward at an initial velocity of 40 feet per second?

(c) How far does the rock fall in 2 seconds if you throw it downward with an initial velocity of 40 feet per second?

83. A rocket is fired straight up from ground level with an initial velocity of 800 feet per second.

(a) How long does it take the rocket to rise 3200 feet?

(b) When will the rocket hit the ground?

84. A rocket loaded with fireworks is to be shot vertically upward from ground level with an initial velocity of 200 feet per second. When the rocket reaches a height of 400 feet on its upward trip the fireworks will be detonated. How many seconds after liftoff will this take place?

85. The atmospheric pressure a (in pounds per square foot) at height h thousand feet above sea level is approximately

$$a = .8315h^2 - 73.93h + 2116.1.$$

(a) Find the atmospheric pressure at sea level and at the top of Mount Everest, the tallest mountain in the world (29,035 feet*). [Remember that h is measured in thousands.]

(b) The atmospheric pressure at the top of Mount Rainier is 1223.43 pounds per square foot. How high is Mount Rainier?

86. Data from the U.S. Department of Health and Human Services indicates that the cumulative number N of reported cases of AIDS in the United States in year x can be approximated by the equation

$$N = 3362.1x^2 - 17{,}270.3x + 24{,}043,$$

where $x = 0$ corresponds to 1980. In what year did the total reach 550,000?

87. According to data from the U.S. Census Bureau, the population P of Cleveland, Ohio (in thousands) in year x can be approximated by $P = .08x^2 - 13.08x + 927$, where $x = 0$ corresponds to 1950. In what year in the past was the population about 804,200?

88. The number N of AIDS cases diagnosed to date (in thousands) is approximated by $N = -.37x^2 + 59.5x + 247.26$, where x is the number of years since 1990.† Assuming that this equation remains valid through 2011, determine when the number of diagnosed cases of AIDS was or will be

(a) 825,000

(b) 1.2 million

---

*Based on data and projections from the Congressional Budget Office in 2005.

†Based on data from the U.S. Centers for Medicare and Medicaid Services.

*Based on measurements in 1999 by climbers sponsored by the Boston Museum of Science and the National Geographic Society, using satellite-based technology.

†Based on data from the U.S. Department of Health and Human Services.

89. The number N of Walgreens drugstores in year x can be approximated by $N = 6.82x^2 - 1.55x + 666.8$, where $x = 0$ corresponds to 1980.* Determine when the number of stores was or will be

(a) 4240   (b) 5600   (c) 7000

90. The total resources T (in billions of dollars) of the Pension Benefit Guaranty Corporation, the government agency that insures pensions, can be approximated by the equation $T = -.26x^2 + 3.62x + 30.18$, where x is the number of years after 2000.† Determine when the total resources are at the given level.

(a) $42.5 billion
(b) $30 billion
(c) When will the Corporation be out of money ($T = 0$)?

91. According to data from the National Highway Traffic Safety Administration, the driver fatality rate D per 1000 licensed drivers every 100 million miles can be approximated by the equation $D = .0031x^2 - .291x + 7.1$, where x is the age of the driver.

(a) For what ages is the driver fatality rate about 1 death per 1000?
(b) For what ages is the rate three times greater than in part (a)?

92. The combined resistance of two resistors, with resistances R_1 and R_2 respectively, connected in parallel is given by $\dfrac{R_1 R_2}{R_1 + R_2}$. If the first resistor has 8 ohms resistance and the combined resistance is 4.8 ohms, what is the resistance of the second resistor?

93. The *cost-benefit equation* $\dfrac{18x}{100 - x} = D$ relates the cost D (in thousands of dollars) needed to remove x percent of a pollutant from the emissions of a factory. Find the percent of the pollutant removed when the following amounts are spent.

(a) $50,000 [Here $D = 50$]
(b) $100,000
(c) $200,000

94. (a) Let b be a real number. Multiply out the expression $\left(x + \dfrac{b}{2}\right)^2$.

(b) Explain why your computation in part (a) shows that this statement is true: If you add $\left(\dfrac{b}{2}\right)^2$ to the expression $x^2 + bx$, the resulting polynomial is a perfect square.

THINKERS

In Exercises 95–98, find a number k such that the given equation has exactly one real solution.

95. $x^2 + kx + 25 = 0$ **96.** $x^2 - kx + 49 = 0$

97. $kx^2 + 8x + 1 = 0$ **98.** $kx^2 + 24x + 16 = 0$

In Exercises 99–101, the discriminant of the equation $ax^2 + bx + c = 0$ (with a, b, c integers) is given. Use it to determine whether or not the solutions of the equation are rational numbers.

99. $b^2 - 4ac = 25$

100. $b^2 - 4ac = 0$

101. $b^2 - 4ac = 72$

102. Find the error in the following "proof" that $6 = 3$.

	$x = 3$
Multiply both sides by x:	$x^2 = 3x$
Subtract 9 from both sides:	$x^2 - 9 = 3x - 9$
Factor each side:	$(x - 3)(x + 3) = 3(x - 3)$
Divide both sides by $x - 3$:	$x + 3 = 3$
Since $x = 3$:	$3 + 3 = 3$
	$6 = 3$

103. Find a number k such that 4 and 1 are the solutions of $x^2 - 5x + k = 0$.

104. Suppose a, b, c are fixed real numbers such that $b^2 - 4ac \geq 0$. Let r and s be the solutions of

$$ax^2 + bx + c = 0.$$

(a) Use the quadratic formula to show that $r + s = -b/a$ and $rs = c/a$.
(b) Use part (a) to verify that $ax^2 + bx + c = a(x - r)(x - s)$.
(c) Use part (b) to factor $x^2 - 2x - 1$ and $5x^2 + 8x + 2$.

105. (a) Solve $x^2 + 5x + 2 = 0$ (exact answer required).

(b) If you have one of the calculators listed below, use its polynomial solver to solve the equation in part (a). Does your answer agree with the one in part (a)?

Calculator	Use this menu/choice
TI-84+	APPS/*PolySmlt**
TI-86	POLY†
TI-89	ALGEBRA/*Solve*‡
Casio 9850	EQUATION (Main Menu)
HP-39gs	MATH/POLYNOM/*Polyroot*§

(c) Use the solver to solve $3x^4 - 2x^3 - 5x^2 + 2x + 1 = 0$.

*If *PolySmlt* is not in the APPS menu, you can download it from TI.
†When asked for "order", enter the degree of the polynomial.
‡Syntax: Solve($x^2 + 5x + 2 = 0, x$)
§Syntax: Polyroot([1, 5, 2])

*Based on data from the Walgreen Company.
†Based on data from the Center on Federal Financial Institutions.

1.2.A *SPECIAL TOPICS*   Absolute Value Equations

Section Objective ■ Solve absolute value equations algebraically.

If c is a real number, then by the definition of absolute value, $|c|$ is either c or $-c$ (whichever one is positive). This fact can be used to solve absolute value equations algebraically.

EXAMPLE 1

To solve $|3x - 4| = 8$, apply the fact stated above with $c = 3x - 4$. Then $|3x - 4|$ is either $3x - 4$ or $-(3x - 4)$, so

$$3x - 4 = 8 \qquad \text{or} \qquad -(3x - 4) = 8$$
$$3x = 12 \qquad\qquad\qquad -3x + 4 = 8$$
$$x = 4 \qquad\qquad\qquad\qquad -3x = 4$$
$$x = -4/3.$$

So there are two possible solutions of the original equation $|3x - 4| = 8$. You can readily verify that both 4 and $-4/3$ actually are solutions.   ■

EXAMPLE 2

Solve $|x + 4| = 5x - 2$.

SOLUTION   The left side of the equation is either $x + 4$ or $-(x + 4)$ (why?). Hence,

$$x + 4 = 5x - 2 \qquad \text{or} \qquad -(x + 4) = 5x - 2$$
$$-4x + 4 = -2 \qquad\qquad\qquad -x - 4 = 5x - 2$$
$$-4x = -6 \qquad\qquad\qquad\qquad -6x = 2$$
$$x = \frac{-6}{-4} = \frac{3}{2} \qquad\qquad\qquad x = \frac{2}{-6} = -\frac{1}{3}.$$

We must check each of these possible solutions in the original equation,

$$|x + 4| = 5x - 2.$$

We see that $x = 3/2$ is a solution because

$$\left| \frac{3}{2} + 4 \right| = \frac{11}{2} \qquad \text{and} \qquad 5\left(\frac{3}{2}\right) - 2 = \frac{11}{2}.$$

However, $x = -1/3$ is not a solution, since

$$\left| -\frac{1}{3} + 4 \right| = \frac{11}{3} \qquad \text{but} \qquad 5\left(-\frac{1}{3}\right) - 2 = -\frac{11}{3}.$$ ■

EXAMPLE 3

Solve the equation $|x^2 + 4x - 3| = 2$.

SOLUTION   The equation is equivalent to

$$x^2 + 4x - 3 = 2 \qquad \text{or} \qquad -(x^2 + 4x - 3) = 2$$
$$x^2 + 4x - 5 = 0 \qquad\qquad\qquad -x^2 - 4x + 3 = 2$$
$$-x^2 - 4x + 1 = 0$$
$$x^2 + 4x - 1 = 0.$$

The first of these equations can be solved by factoring and the second by the quadratic formula:

$$(x + 5)(x - 1) = 0 \qquad \text{or} \qquad x = \frac{-4 \pm \sqrt{4^2 - 4 \cdot 1 \cdot (-1)}}{2 \cdot 1}$$

$$x = -5 \qquad \text{or} \qquad x = 1 \qquad\qquad x = \frac{-4 \pm \sqrt{20}}{2} = \frac{-4 \pm 2\sqrt{5}}{2}$$

$$x = -2 + \sqrt{5} \quad \text{or} \quad x = -2 - \sqrt{5}.$$

Verify that all four of these numbers are solutions of the original equation. ∎

EXERCISES 1.2.A

In Exercises 1–12, find all real solutions of each equation.

1. $|2x + 3| = 9$

2. $|3x - 5| = 7$

3. $|6x - 9| = 0$

4. $|4x - 5| = -9$

5. $|2x + 3| = 4x - 1$

6. $|3x - 2| = 5x + 4$

7. $|x - 3| = x$

8. $|2x - 1| = 2x + 1$

9. $|x^2 + 4x - 1| = 4$

10. $|x^2 + 2x - 9| = 6$

11. $|x^2 - 5x + 1| = 3$

12. $|12x^2 + 5x - 7| = 4$

13. In statistical quality control, one needs to find the proportion of the product that is not acceptable. The upper and lower control limits are found by solving the following equation (in which $\bar{p}$ is the mean percent defective and n is the sample size) for CL.

$$|CL - \bar{p}| = 3\sqrt{\frac{\bar{p}(1 - \bar{p})}{n}}$$

Find the control limits when $\bar{p} = .02$ and $n = 200$.

1.2.B *SPECIAL TOPICS*   Variation

Section Objectives
- Set up and solve a direct variation equation.
- Set up and solve an indirect variation equation.
- Set up and solve combined variation equations.

If a plane flies 400 mph for t hours, then it travels a distance of $400t$ miles. So the distance d is related to the time t by the equation $d = 400t$. When time t increases, so does the distance d. We say that d *varies directly as t* and that the *constant of variation* is 400. Direct variation also occurs in other settings and is formally defined as follows.

Direct Variation

> Suppose the quantities u and v are related by the equation
>
> $$v = ku,$$
>
> where k is a nonzero constant. We say that v **varies directly as u** or that v **is directly proportional to u.** In this case, k is called the **constant of variation** or the **constant of proportionality.**

EXAMPLE 1

When you swim underwater, the pressure p varies directly with the depth d at which you swim. At a depth of 20 feet, the pressure is 8.6 pounds per square inch. What is the pressure at 65 feet deep?

SOLUTION   The variation equation is $p = kd$ for some constant k. We first use the given information to find k. Since $p = 8.6$ when $d = 20$, we have

$$p = kd$$

$$8.6 = k(20)$$

$$k = \frac{8.6}{20} = .43.$$

Therefore, the variation equation is $p = .43d$. To find the pressure at 65 feet, substitute 65 for d in the equation:

$$p = .43d = .43(65) = 27.95 \text{ pounds per square inch.} \qquad \blacksquare$$

The basic idea in direct variation is that the two quantities grow or shrink together. For instance, when $v = 3u$, then as u gets large, so does v. But two quantities can also be related in such a way that one grows as the other shrinks or vice versa. For example if $v = 5/u$, then when u is large (say, $u = 500$), v is small: $v = 5/500 = .01$. This is an example of *inverse variation*.

Inverse Variation

> Suppose the quantities u and v are related by the equation
>
> $$v = \frac{k}{u},$$
>
> where k is a nonzero constant. We say that v **varies inversely as u** or that v **is inversely proportional to u.** Once again, k is called the **constant of variation** or the **constant of proportionality.**

EXAMPLE 2

According to one of Parkinson's laws, the amount of time the Math Department spends discussing an item in its budget is inversely proportional to the cost of the item. If the department spent 40 minutes discussing the purchase of a copying machine for $2100, how much time will it spend discussing a $280 appropriation for the annual picnic?

SOLUTION   Let t denote time (in minutes) and c the cost of an item. Then the variation equation is

$$t = \frac{k}{c}$$

for some constant k. We know that $t = 40$ when $c = 2100$. Substituting these numbers in the equation, we see that

$$40 = \frac{k}{2100}$$

$$k = 40 \cdot 2100 = 84{,}000.$$

So the variation equation is

$$t = \frac{84{,}000}{c}.$$

To find the time spent discussing the picnic, let $c = 280$. Hence,

$$t = \frac{84{,}000}{280} = 300 \text{ minutes } (= 5 \text{ hours!}).$$ ∎

The terminology of variation also applies to situations involving powers of variables, as illustrated in the following table.

Example	Terminology	General Case*
$y = 7x^4$	y varies directly as the fourth power of x. y is directly proportional to the fourth power of x. (Constant of variation is 7.)	$v = ku^n$
$A = \pi r^2$	A varies directly as the square of r. A is directly proportional to the square of r. (Constant of variation is π).	
$y = \frac{10}{x^3}$	y varies inversely as the cube of x. y is inversely proportional to the cube of x. (Constant of variation is 10.)	$v = \frac{k}{u^n}$
$W = \frac{3.48 \times 10^9}{d^2}$	W varies inversely as the square of d. W is inversely proportional to the square of d. (Constant of variation is 3.48×10^9.)	

EXAMPLE 3

The density of light I falling on an object (measured in lumens per square foot) is inversely proportional to the square of the distance d from the light source. If an object 2 feet from a standard 100-watt light bulb receives 34.82 lumens per square foot, how much light falls on an object 10 feet from this bulb?

SOLUTION   The variation equation is

$$I = \frac{k}{d^2}$$

*k is a nonzero constant; u and v are variables.

for some constant k. Since $I = 34.82$, when $d = 2$, we have

$$34.82 = \frac{k}{2^2}$$

$$k = 34.82(4) = 139.28.$$

Therefore, the variation equation is

$$I = \frac{139.28}{d^2}.$$

So the light density at 10 feet is

$$I = \frac{139.28}{10^2} = 1.3928 \text{ lumens per square foot.} \qquad \blacksquare$$

In many situations, variation involves more than two variables. Typical terminology in such cases is illustrated in the following table.

Example	Terminology
$I = 100rt$	I varies jointly as r and t. I is jointly proportional to r and t.
$H = .06sd^3$	H varies jointly as s and the cube of d. H is jointly proportional to s and the cube of d.
$P = \dfrac{3T}{V}$	P varies directly as T and inversely as V. P is directly proportional to T and inversely proportional to V.
$S = \dfrac{2.5wt^2}{\ell}$	S varies directly as w and the square of t and inversely as ℓ. S is directly proportional to w and the square of t and inversely proportional to ℓ.

EXAMPLE 4

The electrical resistance R of wire (of uniform material) varies directly as the length L and inversely as the square of the diameter d. A 2-meter-long piece of wire with a diameter of 4.4 millimeters has a resistance of 500 ohms. What diameter wire should be used if a 10-meter piece is to have a resistance of 1300 ohms?

SOLUTION   The variation equation is

$$R = \frac{kL}{d^2}$$

for some constant k. So we substitute the known information ($R = 500$ when $L = 2$ and $d = 4.4$) and solve for k:

$$R = \frac{kL}{d^2}$$

$$500 = \frac{k \cdot 2}{(4.4)^2}$$

$$k = \frac{500(4.4)^2}{2} = 4840.$$

Hence, the equation is

$$R = \frac{4840L}{d^2}.$$

We must find d when $L = 10$ and $R = 1300$:

$$1300 = \frac{4840 \cdot 10}{d^2}$$

$$1300d^2 = 48{,}400$$

$$d^2 = \frac{48{,}400}{1300} \approx 37.2308.$$

Since the diameter is positive, we must have $d \approx \sqrt{37.2308} \approx 6.1$ mm. ∎

EXERCISES 1.2.B

In Exercises 1–6, express each geometric formula as a statement about variation by filling in the blanks in this sentence:

_____ varies _____ as _____;
the constant of variation is _____.

1. Area of a circle of radius r: $A = \pi r^2$.

2. Volume of a sphere of radius r: $V = \frac{4}{3}\pi r^3$.

3. Area of a rectangle of length l and width w: $A = lw$.

4. Area of a triangle of base b and height h: $A = \frac{bh}{2}$.

5. Volume of a right circular cone of base radius r and height h: $V = \frac{\pi r^2 h}{3}$.

6. Volume of a triangular cylinder of base b, height h, and length l: $V = \frac{1}{2}bhl$.

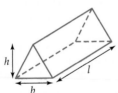

In Exercises 7–12, find the variation equation. Use k for the constant of variation.

7. a varies inversely as b.

8. r is proportional to t.

9. z varies jointly as x, y, and w.

10. The weight w of an object varies inversely as the square of the distance d from the object to the center of the earth.

11. The distance d one can see to the horizon varies directly as the square root of the height h above sea level.

12. The pressure p exerted on the floor by a person's shoe heel is directly proportional to the weight w of the person and inversely proportional to the square of the width r of the heel.

In Exercises 13–22, express the given statement as an equation and find the constant of variation.

13. v varies directly as u; $v = 8$ when $u = 2$.

14. v is directly proportional to u; $v = .4$ when $u = .8$.

15. v varies inversely as u; $v = 8$ when $u = 2$.

16. v is inversely proportional to u; $v = .12$ when $u = .1$.

17. t varies jointly as r and s; $t = 24$ when $r = 2$ and $s = 3$.

18. B varies inversely as u and v; $B = 4$ when $u = 1$ and $v = 3$.

19. w varies jointly as x and y^2; $w = 96$ when $x = 3$ and $y = 4$.

20. p varies directly as the square of z and inversely as r; $p = 32/5$ when $z = 4$ and $r = 10$.

21. T varies jointly as p and the cube of v and inversely as the square of u; $T = 24$ when $p = 3$, $v = 2$, and $u = 4$.

22. D varies jointly as the square of r and the square of s and inversely as the cube of t; $D = 18$ when $r = 4$, $s = 3$, and $t = 2$.

In Exercises 23–30, use an appropriate variation equation to find the required quantity.

23. If r varies directly as t, and $r = 6$ when $t = 3$, find r when $t = 2$.

24. If r is directly proportional to t, and $r = 4$ when $t = 2$, find t when $r = 2$.

25. If b varies inversely as x, and $b = 9$ when $x = 3$, find b when $x = 12$.

26. If b is inversely proportional to x, and $b = 10$ when $x = 4$, find x when $b = 12$.

27. Suppose w is directly proportional to the sum of u and the square of v. If $w = 200$ when $u = 1$ and $v = 7$, then find u when $w = 300$ and $v = 5$.

28. Suppose z varies jointly as x and y. If $z = 30$ when $x = 5$ and $y = 2$, then find x when $z = 45$ and $y = 3$.

29. Suppose r varies inversely as s and t. If $r = 12$ when $s = 3$ and $t = 1$, then find r when $s = 6$ and $t = 2$.

30. Suppose u varies jointly as r and s and inversely as t. If $u = 1.5$ when $r = 2$, $s = 3$, and $t = 4$, then find r when $u = 27$, $s = 9$, and $t = 5$.

31. A resident of Michigan whose taxable income was $24,000 dollars paid state income tax of $936 in 2005.* Use the fact that state income tax varies directly with the taxable income to determine the state tax paid by someone with a taxable income of $39,000.

32. By experiment, you discover that the amount of water that comes from your garden hose varies directly with the water pressure. A pressure of 10 pounds per square inch is needed to produce a flow of 3 gallons per minute.

    (a) What pressure is needed to produce a flow of 4.2 gallons per minute?

    (b) If the pressure is 5 pounds per square inch, what is the flow rate?

33. According to Hooke's law, the distance d that a spring stretches when an object is attached to it is directly proportional to the weight of the object. Suppose that the string stretches 15.75 inches when a 7-pound weight is attached. What weight is required to stretch the spring 27 inches?

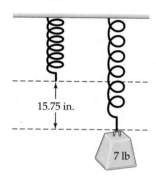

15.75 in.

7 lb

34. In a thunderstorm, there is a gap between the time you see the lightning and the time you hear the thunder. Your distance from the storm is directly proportional to the time interval between the lightning and the thunder. If you hear thunder from a storm that is 1.36 miles away 6.6 seconds after you see lightning, how far away is a storm for which the time gap between lightning and thunder is 12 seconds?

35. At a fixed temperature, the pressure of an enclosed gas is inversely proportional to its volume. The pressure is 50 kilograms per square centimeter when the volume is 200 cubic centimeters. If the gas is compressed to 125 cubic centimeters, what is the pressure?

36. The electrical resistance in a piece of wire of a given length and material varies inversely as the square of the diameter of the wire. If a wire of diameter .01 cm has a resistance of .4 ohm, what is the resistance of a wire of the same length and material but with diameter .025 cm?

37. The distance traveled by a falling object (subject only to gravity, with wind resistance ignored) is directly proportional to the square of the time it takes to fall that far. If an object falls 100 feet in 2.5 seconds, how far does it fall in 5 seconds?

38. The weight of an object varies inversely with the square of its distance from the center of the earth. Assume that the radius of the earth is 3960 miles. If an astronaut weighs 180 pounds on the surface of the earth, what does that astronaut weigh when traveling 300 miles above the surface of the earth?

39. The weight of a cylindrical can of Glop varies jointly as the height and the square of the base radius. The weight is 250 ounces when the height is 20 inches and the base radius is 5 inches. What is the height when the weight is 960 ounces and the base radius is 8 inches?

40. The force of the wind blowing directly on a flat surface varies directly with the area of the surface and the square of the velocity of the wind. A 10 mile an hour (mph) wind blowing on a wooden gate that measures 3 by 6 feet exerts a force of 15 pounds. What is the force on a 1.5 by 2 foot traffic sign from a 50 mph wind?

41. The force needed to keep a car from skidding on a circular curve varies inversely as the radius of the curve and jointly as the weight of the car and the square of the speed. It takes 1500 kilograms of force to keep a 1000-kilogram car from skidding on a curve of radius 200 meters at a speed of 50 kilometers per hour. What force is needed to keep the same car from skidding on a curve of radius 320 meters at 100 kilometers per hour?

42. Boyle's law states that the volume of a given mass of gas varies directly as the temperature and inversely as the pressure. If a gas has a volume of 14 cubic inches when the temperature is 40° and the pressure is 280 pounds per square inch, what is the volume at a temperature of 50° and pressure of 175 pounds per square inch?

43. The maximum safe load that a rectangular beam can support varies directly with the width w and the square of the height h, and inversely with the length l of the beam. A 6-foot-long beam that is 2 inches high and 4 inches wide has a maximum safe load of 1000 pounds.

    (a) What is the maximum safe load for a 10-foot-long beam that is 4 inches high and 4 inches wide?

    (b) How long should a 4 inch by 4 inch beam be to safely support a 6000-pound load?

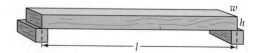

w

h

l

---

*The exercise assumes that there are no nonrefundable credits.

44. The period of a pendulum is the time it takes for the pendulum to make one complete swing (going both left and right) and return to its starting point. The period P varies directly with the *square root* of the length l of the pendulum.

(a) Write the variation equation that describes this situation, using k for the constant of variation.

(b) What happens to the period if the length of the pendulum is quadrupled?

1.3 The Coordinate Plane

Section Objectives

■ Locate points in the coordinate plane.
■ Create a scatter plot and line graph from a data set.
■ Determine the distance between two points in the plane.
■ Find the midpoint of a line segment.
■ Understand the relationship between equations and their graphs.
■ Find the intercepts of a graph.
■ Find the equation of a circle.
■ Identify equations whose graphs are circles.

Just as real numbers are identified with points on the number line, ordered *pairs* of real numbers can be identified with points in the plane. To do this, draw two number lines in the plane, one vertical and one horizontal, as in Figure 1–11. The horizontal line is usually called the **x-axis,** and the vertical line the **y-axis,** but other letters may be used if desired. The point where the axes intersect is the **origin.** The axes divide the plane into four regions, called **quadrants,** that are numbered as in Figure 1–11.

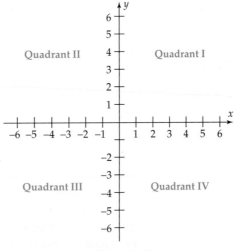

Figure 1–11

If P is a point in the plane, draw vertical and horizontal lines through P to the coordinate axes, as shown in Figure 1–12, on the next page. These lines intersect the x-axis at some number c and the y-axis at d. We say that P has **coordinates** (c, d).

The number c is the **x-coordinate** of P, and d is the **y-coordinate** of P. The plane is said to have a **rectangular** (or **Cartesian**) **coordinate system.**

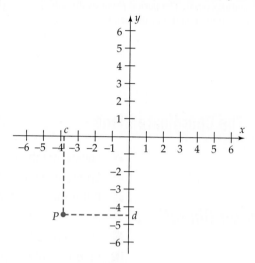

Figure 1–12

You can think of the coordinates of a point as directions for locating it. For instance, to find $(4, -3)$, start at the origin and move 4 units to the right along the x-axis, then move 3 units downward, as shown in Figure 1–13, which also shows other points and their coordinates.

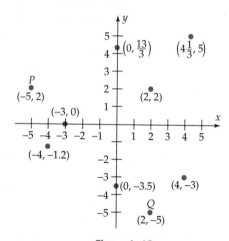

Figure 1–13

EXAMPLE 1

The following table, from the U.S. Department of Education shows the maximum Pell Grant for college students in selected years.

Year	1990	1992	1994	1996	1998	2000	2002	2004
Amount	2300	2400	2300	2470	3000	3300	4000	4050

One way to represent this data graphically is to represent each year's maximum by a point; for instance (1990, 2300) and (2004, 4050). Alternatively, to avoid using large numbers, we can let x be the number of years since 1990, so that $x = 0$ is 1990 and $x = 14$ is 2004. We can also list the dollar amounts in *hundreds,* so the points for 1990 and 2004 are (0, 23) and (14, 40.5). Plotting all the data in this way leads to the **scatter plot** in Figure 1–14. Connecting these data points with line segments produces the **line graph** in Figure 1–15. ∎

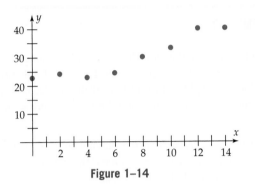

Figure 1–14

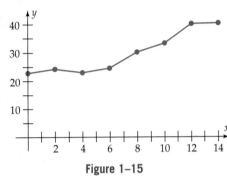

Figure 1–15

THE DISTANCE FORMULA

We shall often identify a point with its coordinates and refer, for example, to the point (2, 3). When dealing with several points simultaneously, it is customary to label the coordinates of the first point (x_1, y_1), the second point (x_2, y_2), the third point (x_3, y_3), and so on.* Once the plane is coordinatized, it's easy to compute the distance between any two points:

The Distance Formula

> The distance between points (x_1, y_1) and (x_2, y_2) is
> $$\sqrt{(x_1 - x_2)^2 + (y_1 - y_2)^2}.$$

Before proving the distance formula, we shall see how it is used.

EXAMPLE 2

To find the distance between the points $(-1, -3)$ and $(2, -4)$ in Figure 1–16, substitute $(-1, -3)$ for (x_1, y_1) and $(2, -4)$ for (x_2, y_2) in the distance formula:

Distance formula: $\text{Distance} = \sqrt{(x_1 - x_2)^2 + (y_1 - y_2)^2}$

Substitute: $= \sqrt{(-1 - 2)^2 + (-3 - (-4))^2}$

Simplify: $= \sqrt{(-3)^2 + (-3 + 4)^2}$

$= \sqrt{9 + 1} = \sqrt{10}.$

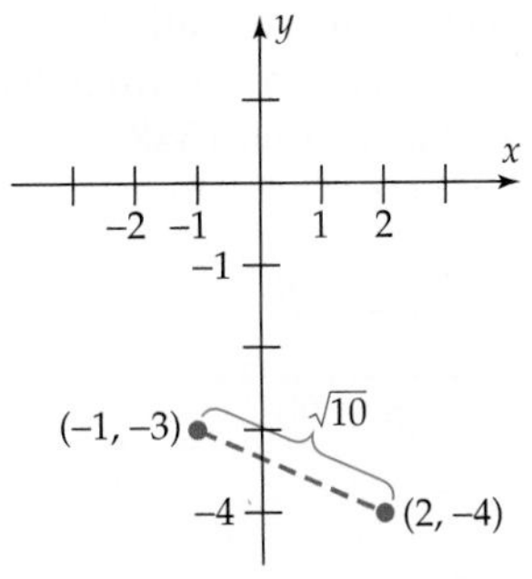

Figure 1–16

*"x_1" is read "x-one" or "x-sub-one"; it is a *single symbol* denoting the first coordinate of the first point, just as c denotes the first coordinate of (c, d). Analogous remarks apply to y_1, x_2, and so on.

The order in which the points are used in the distance formula doesn't make a difference. If we substitute $(2, -4)$ for (x_1, y_1) and $(-1, -3)$ for (x_2, y_2), we get the same answer:

$$\sqrt{[2 - (-1)]^2 + [-4 - (-3)]^2} = \sqrt{3^2 + (-1)^2} = \sqrt{10}. \quad \blacksquare$$

EXAMPLE 3

In a Cubs game at Wrigley Field, a fielder catches the ball near the right-field corner and throws it to second base. The right-field corner is 353 feet from home plate along the right-field foul line. If the fielder is 5 feet from the outfield wall and 5 feet from the foul line, how far must he throw the ball?

SOLUTION   Imagine that the playing field is placed on the coordinate plane, with home plate at the origin and the right-field foul line along the positive *x*-axis, as shown in Figure 1–17 (not to scale).

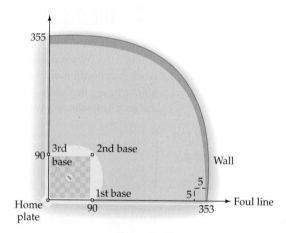

Figure 1–17

Since the four bases form a square whose sides measure 90 ft each, second base has coordinates $(90, 90)$. The fielder is located 5 feet from the wall, so his *x*-coordinate is $353 - 5 = 348$. His *y*-coordinate is 5, since he is 5 feet from the foul line. Therefore the distance he throws is the distance from $(348, 5)$ to $(90, 90)$, which can be found as follows.

Distance formula: $\text{Distance} = \sqrt{(x_1 - x_2)^2 + (y_1 - y_2)^2}$

Substitute: $= \sqrt{(348 - 90)^2 + (5 - 90)^2}$

Simplify: $= \sqrt{258^2 + (-85)^2}$

$= \sqrt{73{,}789} \approx 271.6 \text{ feet.}$

Therefore, he must throw about 272 feet.   ■

EXAMPLE 4

To find the distance from (a, b) to $(2a, -b)$, where a and b are fixed real numbers, substitute a for x_1, b for y_1, $2a$ for x_2, and $-b$ for y_2 in the distance formula:

$$\sqrt{(x_1 - x_2)^2 + (y_1 - y_2)^2} = \sqrt{(a - 2a)^2 + (b - (-b))^2}$$
$$= \sqrt{(-a)^2 + (b + b)^2} = \sqrt{a^2 + (2b)^2}$$
$$= \sqrt{a^2 + 4b^2}. \qquad \blacksquare$$

PROOF OF THE DISTANCE FORMULA

Figure 1–18 shows typical points P and Q in the plane. We must find length d of line segment PQ.

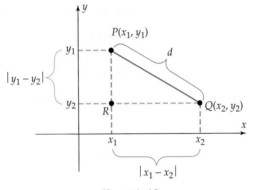

Figure 1–18

As shown in Figure 1–18, the length of RQ is the same as the distance from x_1 to x_2 on the x-axis (number line), namely, $|x_1 - x_2|$. Similarly, the length of PR is the same as the distance from y_1 to y_2 on the y-axis, namely, $|y_1 - y_2|$. According to the Pythagorean Theorem* the length d of PQ is given by

$$(\text{Length } PQ)^2 = (\text{length } RQ)^2 + (\text{length } PR)^2$$
$$d^2 = |x_1 - x_2|^2 + |y_1 - y_2|^2.$$

Since $|c|^2 = |c| \cdot |c| = |c^2| = c^2$ (because $c^2 \geq 0$), this equation becomes

$$d^2 = (x_1 - x_2)^2 + (y_1 - y_2)^2.$$

Since the length d is nonnegative, we must have

$$d = \sqrt{(x_1 - x_2)^2 + (y_1 - y_2)^2}. \qquad \blacksquare$$

The distance formula can be used to prove the following useful fact (see Exercise 96).

The Midpoint Formula

The midpoint of the line segment from (x_1, y_1) to (x_2, y_2) is

$$\left(\frac{x_1 + x_2}{2}, \frac{y_1 + y_2}{2} \right).$$

*See the Geometry Review Appendix.

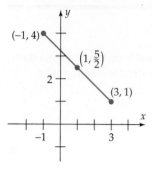

Figure 1–19

To find the midpoint of the segment joining $(-1, 4)$ and $(3, 1)$, use the formula in the box with $x_1 = -1$, $y_1 = 4$, $x_2 = 3$, and $y_2 = 1$. The midpoint is

$$\left(\frac{x_1 + x_2}{2}, \frac{y_1 + y_2}{2}\right) = \left(\frac{-1 + 3}{2}, \frac{4 + 1}{2}\right) = \left(1, \frac{5}{2}\right)$$

as shown in Figure 1–19. ∎

EXAMPLE 6

The annual revenues of the Dell Computer company were \$31.2 billion in 2002 and \$55.9 billion in 2006.* Assume that revenues are growing approximately linearly and estimate the revenues in 2004.

SOLUTION   Let the point (x, y) denote the revenues y (in billions of dollars) in year x. Then the points $(2002, 31.2)$ and $(2006, 55.9)$ represent the given data. The midpoint of the line segment joining these points is

$$\left(\frac{2002 + 2006}{2}, \frac{31.2 + 55.9}{2}\right) = (2004, 43.55).$$

Since the data is growing linearly, this suggests that 2004 revenues were approximately \$43.55 billion. ∎

GRAPHS

A **graph** is a set of points in the plane. Some graphs are based on data points, such as Figures 1–14 and 1–15. Other graphs arise from equations, as follows. A **solution** of an equation in variables x and y is a pair of numbers such that the substitution of the first number for x and the second for y produces a true statement. For instance, $(3, -2)$ is a solution of $5x + 7y = 1$ because

$$5 \cdot 3 + 7(-2) = 1,$$

and $(-2, 3)$ is *not* a solution because $5(-2) + 7 \cdot 3 \neq 1$. The **graph of an equation** in two variables is the set of points in the plane whose coordinates are solutions of the equation. Thus the graph is a *geometric picture of the solutions.*

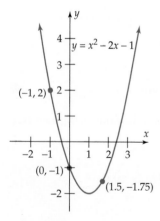

Figure 1–20

EXAMPLE 7

The graph of $y = x^2 - 2x - 1$ is shown in Figure 1–20. You can readily verify that each of the points whose coordinates are labeled is a solution of the equation. For instance, $(0, -1)$ is a solution because $-1 = 0^2 - 2(0) - 1$. ∎

A graph may intersect the x- or y-axis at one or more points. The x-coordinate of a point where the graph intersects the x-axis is called an **x-intercept** of the graph. Similarly, the y-coordinate of a point where the graph intersects the y-axis is called a **y-intercept** of the graph. Figure 1–21 shows some examples.

*Dell, Inc.

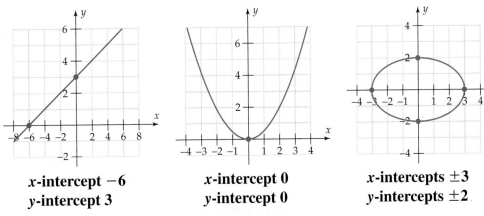

x-intercept −6
y-intercept 3

x-intercept 0
y-intercept 0

x-intercepts ±3
y-intercepts ±2

Figure 1–21

When the x- and y-intercepts cannot easily be read from the graph, they can often be found algebraically.

EXAMPLE 8

Find the x- and y-intercepts of the graph of $y = x^2 - 2x - 1$ in Figure 1–20.

SOLUTION   The points where the graph intersects the x-axis have 0 as their y-coordinate (see Figure 1–20). We can find their x-coordinates by setting $y = 0$ and solving the resulting equation,

$$x^2 - 2x - 1 = 0.$$

By the quadratic formula

$$x = \frac{-(-2) \pm \sqrt{(-2)^2 - 4 \cdot 1 \cdot (-1)}}{2 \cdot 1} = \frac{2 \pm \sqrt{8}}{2} \approx \begin{cases} -.4142 \\ 2.4142. \end{cases}$$

So the x-intercepts are approximately −.4142 and 2.4142.
In this case, you can read the y-intercept from the graph; it is −1. Because points on the y-axis have 0 as their x-coordinate, the y-intercept can be found algebraically by setting $x = 0$ in the equation and solving for y. ∎

The process in Example 8 can be summarized as follows.

x- and y-Intercepts

> To find the x-intercepts of the graph of an equation, set $y = 0$ and solve for x.
>
> To find the y-intercepts, set $x = 0$ and solve for y.

Reading and interpreting information from graphs is an essential skill if you want to succeed in this course and calculus.

EXAMPLE 9

Leslie Lahr takes out a 30-year mortgage on which her monthly payment is $850. One of the graphs in Figure 1–22 shows the portion of each payment that goes to interest and the other shows the portion that goes to paying off the principal.

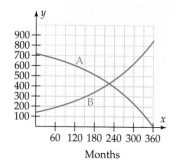

900
800
700
600
500
400
300
200
100

A

B

60 120 180 240 300 360

Months

Figure 1–22

(a) Which graph is the interest portion and which is the principal portion?

(b) At the end of ten years (120 months), about how much of each payment goes for interest and how much for the principal?

SOLUTION

(a) The interest portion of the payment is the monthly interest due on the unpaid balance. This balance (and hence, the interest) is large at the beginning but slowly decreases as more payments are made. So the interest graph begins high and ends low—it must be graph A. Consequently, graph B shows the portion of each payment that goes to reducing the principal.

(b) The point (120, 600) on graph A shows that about $600 of the $850 payment was for interest. Hence, $250 was for principal, as the point (120, 250) on graph B indicates.   ∎

CIRCLES

If (c, d) is a point in the plane and r a positive number, then the **circle with center (c, d) and radius r** consists of all points (x, y) that lie r units from (c, d), as shown in Figure 1–23. According to the distance formula, the statement that "the distance from (x, y) to (c, d) is r units" is equivalent to:

$$\sqrt{(x - c)^2 + (y - d)^2} = r$$

Squaring both sides shows that (x, y) satisfies this equation:

$$(x - c)^2 + (y - d)^2 = r^2$$

Reversing the procedure shows that any solution (x, y) of this equation is a point on the circle. Therefore

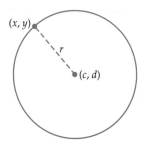

(x, y)

r

(c, d)

Figure 1–23

Circle Equation

The circle with center (c, d) and radius r is the graph of
$$(x - c)^2 + (y - d)^2 = r^2.$$

We say that $(x - c)^2 + (y - d)^2 = r^2$ is the **equation of the circle** with center (c, d) and radius r.

EXAMPLE 10

Identify the graph of the equation $(x - 4)^2 + (y - 2)^2 = 9$.

SOLUTION   Since $9 = 3^2$, we can write the equation as

$$(x - 4)^2 + (y - 2)^2 = 3^2.$$

Now the equation is of the form shown in the box above, with $c = 4$, $d = 2$ and $r = 3$. So the graph is a circle with center $(4, 2)$ and radius 3, as shown in Figure 1–24.   ∎

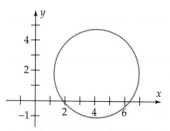

Figure 1–24

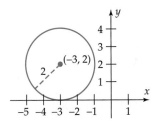

EXAMPLE 11

Find the equation of the circle with center $(-3, 2)$ and radius 2 and sketch its graph.

SOLUTION Here the center is $(c, d) = (-3, 2)$ and the radius is $r = 2$, so the equation of the circle is

$$(x - c)^2 + (y - d)^2 = r^2$$
$$[x - (-3)]^2 + (y - 2)^2 = 2^2$$
$$(x + 3)^2 + (y - 2)^2 = 4.$$

Its graph is shown in Figure 1–25. ■

Figure 1–25

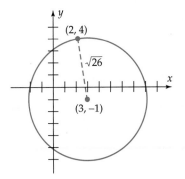

EXAMPLE 12

Find the equation of the circle with center $(3, -1)$ that passes through $(2, 4)$.

SOLUTION We must first find the radius. Since $(2, 4)$ is on the circle, the radius is the distance from $(2, 4)$ to $(3, -1)$ as shown in Figure 1–26, namely,

$$\sqrt{(2 - 3)^2 + (4 - (-1))^2} = \sqrt{1 + 25} = \sqrt{26}.$$

The equation of the circle with center at $(3, -1)$ and radius $\sqrt{26}$ is

$$(x - 3)^2 + (y - (-1))^2 = (\sqrt{26})^2$$
$$(x - 3)^2 + (y + 1)^2 = 26$$
$$x^2 - 6x + 9 + y^2 + 2y + 1 = 26$$
$$x^2 + y^2 - 6x + 2y - 16 = 0. \qquad ■$$

Figure 1–26

The equation of any circle can always be written in the form

$$x^2 + y^2 + Bx + Cy + D = 0$$

for some constants B, C, D, as in Example 12 (where $B = -6$, $C = 2$, $D = -16$). Conversely, the graph of such an equation can always be determined.

EXAMPLE 13

Show that the graph of

$$3x^2 + 3y^2 - 12x - 30y + 45 = 0$$

is a circle and find its center and radius.

SOLUTION We will be completing the square, which requires that x^2 and y^2 each have coefficient 1. So we begin by dividing both sides of the equation by 3 and regrouping the terms.

$$x^2 + y^2 - 4x - 10y + 15 = 0$$
$$(x^2 - 4x) + (y^2 - 10y) = -15.$$

Next we complete the square in both expressions in parentheses (see page 22). To complete the square in $x^2 - 4x$, we add 4 (the square of half the coefficient of x), and to complete the square in $y^2 - 10y$, we add 25 (why?). To have an equivalent equation, we must add these numbers to *both* sides:

$$(x^2 - 4x + 4) + (y^2 - 10y + 25) = -15 + 4 + 25$$

$$(x - 2)^2 + (y - 5)^2 = 14$$

Since $14 = (\sqrt{14})^2$, this is the equation of the circle with center $(2, 5)$ and radius $\sqrt{14}$. ∎

When the center of a circle of radius r is at the origin $(0, 0)$, its equation takes a simpler form.

Circle at the Origin

> The circle with center $(0, 0)$ and radius r is the graph of
>
> $$x^2 + y^2 = r^2.$$

Proof   Substitute $c = 0$ and $d = 0$ in the equation for the circle with center (c, d) and radius r.

$$(x - c)^2 + (y - d)^2 = r^2$$

$$(x - 0)^2 + (y - 0)^2 = r^2$$

$$x^2 + y^2 = r^2.$$ ∎

EXAMPLE 14

Letting $r = 1$ shows that the graph of $x^2 + y^2 = 1$ is the circle of radius 1 centered at the origin, as shown in Figure 1–27. This circle is called the **unit circle**. ∎

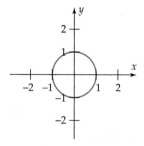

Figure 1–27

EXERCISES 1.3

1. Find the coordinates of points *A–I*.

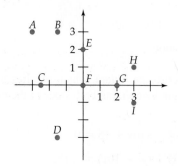

In Exercises 2–5, find the coordinates of the point P.

2. *P* lies 4 units to the left of the *y*-axis and 5 units below the *x*-axis.

3. *P* lies 3 units above the *x*-axis and on the same vertical line as $(-6, 7)$.

4. *P* lies 2 units below the *x*-axis, and its *x*-coordinate is three times its *y*-coordinate.

5. *P* lies 4 units to the right of the *y*-axis, and its *y*-coordinate is half its *x*-coordinate.

In Exercises 6–8, sketch a scatter plot and a line graph of the given data.

6. Tuition and fees at four-year public colleges in the fall of each year are shown in the table (*Source:* The College Board). Let $x = 0$ correspond to 2000.

Year	2000	2001	2002	2003	2004	2005
Tuition and Fees	3487	3725	4081	4694	5127	5491

7. The table shows sales of personal digital video recorders.* Let $x = 0$ correspond to 2000, and measure y in thousands.

Year	Number Sold
2000	257,000
2001	129,000
2002	143,000
2003	214,000
2004	315,000
2005	485,000

8. The maximum yearly contribution to an individual retirement account (IRA) was $3000 in 2003. It changed to $4000 in 2005 and will change to $5000 in 2008. Assuming 3% inflation, however, the picture is somewhat different. The table shows the maximum IRA contribution in fixed 2003 dollars. Let $x = 0$ correspond to 2000.

Year	Maximum Contribution
2003	3000
2004	2910
2005	3764
2006	3651
2007	3541
2008	4294

9. (a) If the first coordinate of a point is greater than 3 and its second coordinate is negative, in what quadrant does it lie?

(b) What is the answer in part (a) if the first coordinate is less than 3?

10. In what quadrant(s) does a point lie if the product of its coordinates is
(a) positive?   (b) negative?

11. (a) Plot the points $(3, 2)$, $(4, -1)$, $(-2, 3)$, and $(-5, -4)$.

(b) Change the sign of the y-coordinate in each of the points in part (a), and plot these new points.

(c) Explain how the points (a, b) and $(a, -b)$ are related graphically. [*Hint:* What are their relative positions with respect to the x-axis?]

12. (a) Plot the points $(5, 3)$, $(4, -2)$, $(-1, 4)$, and $(-3, -5)$.

(b) Change the sign of the x-coordinate in each of the points in part (a), and plot these new points.

(c) Explain how the points (a, b) and $(-a, b)$ are related graphically. [*Hint:* What are their relative positions with respect to the y-axis?]

In Exercises 13–20, find the distance between the two points and the midpoint of the segment joining them.

13. $(-3, 5), (2, -7)$ **14.** $(2, 4), (3, 6)$

15. $(-2, 5), (-1, 2)$ **16.** $(-2, 3), (-3, 2)$

17. $(\sqrt{2}, 1), (\sqrt{3}, 2)$ **18.** $(-1, \sqrt{5}), (\sqrt{2}, -\sqrt{3})$

19. $(a, b), (b, a)$ **20.** $(s, t), (0, 0)$

21. Which of the following points is closest to the origin?

$$(4, 4.2), (-3.5, 4.6), (-3, -5), (2, -5.5)$$

22. Which of the following points is closest to $(3, 2)$?

$$(0, 0), (4, 5.3), (-.6, 1.5), (1, -1)$$

23. Find the perimeter of the shaded region in the figure.

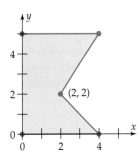

24. What is the perimeter of the triangle with vertices $(1, 1)$, $(5, 4)$, and $(-2, 5)$?

25. Find the area of the shaded region in Exercise 23. [*Hint:* What is the area of the triangle with vertices $(4, 0)$, $(2, 2)$, and $(4, 5)$?]

26. Find the area of the triangle with vertices $(1, 4)$, $(4, 3)$, and $(-2, -5)$. You may assume that there is a right angle at vertex $(1, 4)$.

In Exercises 27–29, show that the three points are the vertices of a right triangle, and state the length of the hypotenuse. [You may assume that a triangle with sides of lengths a, b, c is a right triangle with hypotenuse c provided that $a^2 + b^2 = c^2$.]

27. $(0, 0), (1, 1), (2, -2)$

28. $(3, -2), (0, 4), (-2, 3)$

29. $(1, 4), (5, 2), (3, -2)$

30. Suppose a baseball playing field is placed on the coordinate plane, as in Example 3.

(a) Find the coordinates of first and third base.
(b) If the left fielder is at the point (50, 325), how far is he from first base?
(c) How far is the left fielder in part (b) from the right fielder, who is at the point (280, 20)?

31. A standard football field is 100 yards long and $53\frac{1}{3}$ yards wide. The quarterback, who is standing on the 10-yard line, 20 yards from the left sideline, throws the ball to a receiver who is on the 45-yard line, 5 yards from the right sideline, as shown in the figure.

(a) How long was the pass? [*Hint:* Place the field in the first quadrant of the coordinate plane, with the left sideline on the y-axis and the goal line on the x-axis. What are the coordinates of the quarterback and the receiver?]
(b) A player is standing halfway between the quarterback and the receiver. What are his coordinates?

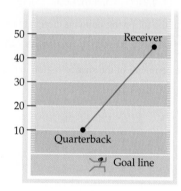

32. How far is the quarterback in Exercise 31 from a player who is on the 50-yard line, halfway between the sidelines?

33. The number of passengers annually on U.S. commercial airlines was 650 million in 2002 and is expected to be 1.05 billion in 2016.*

(a) Represent this data graphically by two points.
(b) Find the midpoint of the line segment joining these two points.
(c) How might this midpoint be interpreted? What assumptions, if any, are needed to make this interpretation?

34. The net revenues of Pepsico were $26,971 million in 2003 and $32,562 million in 2005.† Estimate the net revenue in 2004.

In Exercises 35–40, determine whether the point is on the graph of the given equation.

35. $(2, -1)$; $3x - y - 5 = 0$

36. $(2, -1)$; $x^2 + y^2 - 6x + 8y = -15$

---

*Federal Aviation Agency.
†Pepsico annual reports.

37. $(6, 2)$; $3y + x = 12$

38. $(1, -2)$; $3x + y = 12$

39. $(1, -4)$; $(x - 2)^2 + (y + 5)^2 = 4$

40. $(1, -1)$; $\dfrac{x^2}{2} + \dfrac{y^2}{3} = 1$

In Exercises 41–46, find the x- and y-intercepts of the graph of the equation.

41. $x^2 - 6x + y + 5 = 0$

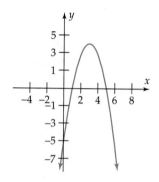

42. $x^2 - 2xy + 3y^2 = 1$

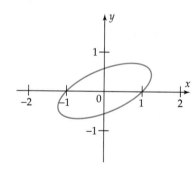

43. $(x - 2)^2 + y^2 = 9$

44. $(x + 1)^2 + (y - 2)^2 = 4$

45. $9x^2 + 24xy + 16y^2 + 90x - 128y = 0$

46. $2x^2 - 4xy + 2y^2 + 3x + 5y = 10$

47. The graph on the next page, which is based on data from the Actuarial Society of South Africa and assumes no changes in current behavior, shows the projected new cases of AIDS in South Africa (in millions) in coming years ($x = 0$ corresponds to 2000).

(a) Estimate the number of new cases in 2010.
(b) Estimate the year in which the largest number of new cases will occur. About how many new cases will there be in that year?
(c) In what years will the number of new cases be below 7,000,000?

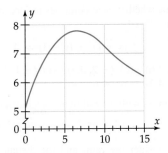

48. The graph shows the total number of alcohol-related car crashes in Ohio at a particular time of the day for the years 1991–2000.* Time is measured in hours after midnight. During what periods is the number of crashes

(a) below 5000?
(b) above 15,000?

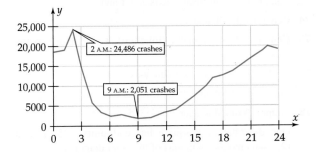

49. Many companies are changing their traditional employee pension plans to so-called cash balance plans. The graph shows pension accrual by age for two hypothetical plans.†

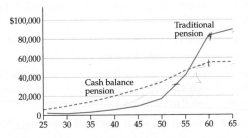

(a) Assuming that you can take your accrued pension benefits in cash when you leave the company before retirement, for what age group is the cash balance plan better?
(b) At what age is the accrued amount the same for either type of pension plan?
(c) If you remain with the company until retirement, how much better off are you with a traditional instead of a cash balance plan?

*The Cleveland *Plain Dealer.*
†Data from Steve J. Kopp and Lawrence W. Sher, *The Pension Forum,* Vol 11, No. 1. Graph from "What if a Pension Shift Hit Lawmakers Too?" by M. W. Walsh, *New York Times,* March 9, 2003. Copyright © 2003 The New York Times Co.

50. In an ongoing consumer confidence survey, respondents are asked two questions: Are jobs plentiful? Are jobs hard to get? The graph shows the percentage of people answering "yes" to each question over the years.*

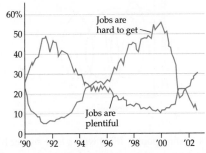

(a) In what year did the most people feel that jobs were plentiful? In that year, approximately what percentage of people felt that jobs were hard to get?
(b) In what year did the most people feel that jobs were hard to get? In that year approximately what percentage of people felt that jobs were plentiful?
(c) In what years was the percentage of those who thought jobs were plentiful the same as the percentage of those that thought jobs were hard to get?

In Exercises 51–54, determine which of graphs A, B, C best describes the given situation.

51. You have a job that pays a fixed salary for the week. The graph shows your salary.

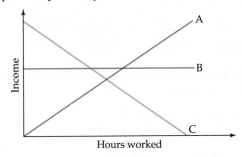

52. You have a job that pays an hourly wage. The graph shows your salary.

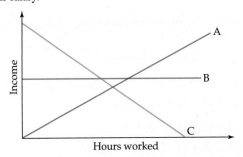

*Data for The Conference Board. Graph from "Tight U.S. Job Market Adds to Jitters Among Consumers." by A. Berenson, *The New York Times,* March 1, 2003. Copyright © 2003 The New York Times Co.

53. You take a ride on a Ferris wheel. The graph shows your distance from the ground.*

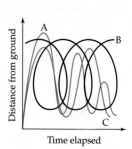

54. Alison's wading pool is filled with a hose by her big sister Emily, and Alison plays in the pool. When they are finished, Emily empties the pool. The graph shows the water level of the pool.

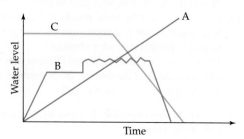

In Exercises 55–58, find the equation of the circle with given center and radius r.

55. $(-3, 4);$ $r = 2$ **56.** $(-3, -5);$ $r = 3$

57. $(0, 0);$ $r = \sqrt{3}$ **58.** $(5, -2);$ $r = 1$

In Exercises 59–62, sketch the graph of the equation. Label the x- and y-intercepts.

59. $(x - 5)^2 + (y + 2)^2 = 5$

60. $(x + 6)^2 + y^2 = 4$

61. $(x + 1)^2 + (y - 3)^2 = 9$

62. $(x - 2)^2 + (y - 4)^2 = 1$

In Exercises 63–68, find the center and radius of the circle whose equation is given.

63. $x^2 + y^2 + 8x - 6y - 15 = 0$

64. $15x^2 + 15y^2 = 10$

65. $x^2 + y^2 + 6x - 4y - 15 = 0$

66. $x^2 + y^2 + 10x - 75 = 0$

67. $x^2 + y^2 + 25x + 10y = -12$

68. $3x^2 + 3y^2 + 12x + 12 = 18y$

69. Determine whether each point lies inside, or outside, or on the circle

$$(x - 1)^2 + (y - 3)^2 = 4.$$

(a) $(2.2, 4.6)$ (b) $(-.2, 4.7)$ (c) $(-.1, 1.4)$
(d) $(2.6, 4.3)$ (e) $(-.6, 1.8)$

70. Do the circles with the following equations intersect?

$$(x - 3)^2 + (y + 2)^2 = 25 \quad \text{and} \quad (x + 3)^2 + (y - 2)^2 = 4$$

[*Hint:* Consider the radii and the distance between the centers.]

In Exercises 71–78, find the equation of the circle.

71. Center $(3, 3)$; passes through the origin.

72. Center $(-1, -3)$; passes through $(-4, -2)$.

73. Center $(1, 2)$; intersects x-axis at -1 and 3.

74. Center $(3, 1)$; diameter 2.

75. Center $(-5, 4)$; tangent (touching at one point) to the x-axis.

76. Center $(2, -6)$; tangent to the y-axis.

77. Endpoints of a diameter are $(3, 3)$ and $(1, -1)$.

78. Endpoints of a diameter are $(-3, 5)$ and $(7, -5)$.

79. One diagonal of a square has endpoints $(-3, 1)$ and $(2, -4)$. Find the endpoints of the other diagonal.

80. Find the vertices of all possible squares with this property: Two of the vertices are $(2, 1)$ and $(2, 5)$. [*Hint:* There are three such squares.]

81. Do Exercise 80 with (c, d) and (c, k) in place of $(2, 1)$ and $(2, 5)$.

82. Find the three points that divide the line segment from $(-4, 7)$ to $(10, -9)$ into four parts of equal length.

83. Find all points P on the x-axis that are 5 units from $(3, 4)$. [*Hint:* P must have coordinates $(x, 0)$ for some x and the distance from P to $(3, 4)$ is 5.]

84. Find all points on the y-axis that are 8 units from $(-2, 4)$.

85. Find all points with first coordinate 3 that are 6 units from $(-2, -5)$.

86. Find all points with second coordinate -1 that are 4 units from $(2, 3)$.

87. Find a number x such that $(0, 0)$, $(3, 2)$, and $(x, 0)$ are the vertices of an isosceles triangle, neither of whose two equal sides lie on the x-axis.

88. Do Exercise 87 if one of the two equal sides lies on the positive x-axis.

89. Show that the midpoint M of the hypotenuse of a right triangle is equidistant from the vertices of the triangle. [*Hint:* Place the triangle in the first quadrant of the plane, with right angle at the origin so that the situation looks like the figure.]

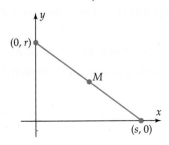

90. Show that the diagonals of a parallelogram bisect each other. [*Hint:* Place the parallelogram in the first quadrant with a vertex at the origin and one side along the x-axis so that the situation looks like the figure.]

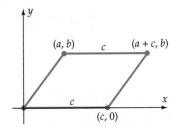

91. Show that the diagonals of a rectangle have the same length. [*Hint:* Place the rectangle in the first quadrant of the plane and label its vertices appropriately, as in Exercises 89–90.]

92. If the diagonals of a parallelogram have the same length, show that the parallelogram is actually a rectangle. [*Hint:* See Exercise 90.]

THINKERS

93. For each nonzero real number k, the graph of $(x - k)^2 + y^2 = k^2$ is a circle. Describe all possible such circles.

94. Suppose every point in the coordinate plane is moved 5 units straight up.
 (a) To what point does each of these points go:$(0, -5)$, $(2, 2)$, $(5, 0)$, $(5, 5)$, $(4, 1)$?
 (b) Which points go to each of the points in part (a)?
 (c) To what point does (a, b) go?
 (d) To what point does $(a, b - 5)$ go?
 (e) What point goes to $(-4a, b)$?
 (f) What points go to themselves?

95. Let (c, d) be any point in the plane with $c \neq 0$. Prove that (c, d) and $(-c, -d)$ lie on the same straight line through the origin, on opposite sides of the origin, the same distance from the origin. [*Hint:* Find the midpoint of the line segment joining (c, d) and $(-c, -d)$.]

96. *Proof of the Midpoint Formula* Let P and Q be the points (x_1, y_1) and (x_2, y_2), respectively, and let M be the point with coordinates

$$\left(\frac{x_1 + x_2}{2}, \frac{y_1 + y_2}{2} \right).$$

Use the distance formula to compute the following:
 (a) The distance d from P to Q;
 (b) The distance d_1 from M to P;
 (c) The distance d_2 from M to Q.
 (d) Verify that $d_1 = d_2$.
 (e) Show that $d_1 + d_2 = d$. [*Hint:* Verify that $d_1 = \frac{1}{2}d$ and $d_2 = \frac{1}{2}d$.]
 (f) Explain why parts (d) and (e) show that M is the midpoint of PQ.

1.4 Lines

Section Objectives
- Find the slope of a line.
- Understand what its slope tells you about a line.
- Construct and interpret the slope-intercept form of the equation of a line.
- Identify the equations of horizontal and vertical lines.
- Use point-slope form of the equation of a line.
- Recognize the general form of the equation of a line.
- Understand the relationship between parallel lines and their equations.
- Understand the relationship between perpendicular lines and their equations.
- Interpret slope as a rate of change.

When you move from a point P to a point Q on a line,* two numbers are involved, as illustrated in Figure 1–28:

(i) The vertical distance you move (the **change in y,** denoted Δy);

(ii) The horizontal distance you move (the **change in x,** denoted Δx).†

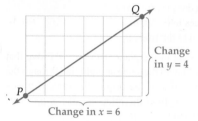

$$\frac{\Delta y}{\Delta x} = \frac{\text{Change in } y}{\text{Change in } x} = \frac{4}{6} = \frac{2}{3}$$

(a)

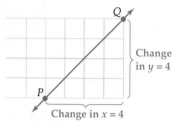

$$\frac{\Delta y}{\Delta x} = \frac{\text{Change in } y}{\text{Change in } x} = \frac{4}{4} = 1$$

(b)

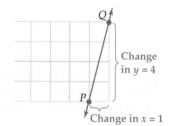

$$\frac{\Delta y}{\Delta x} = \frac{\text{Change in } y}{\text{Change in } x} = \frac{4}{1} = 4$$

(c)

Figure 1–28

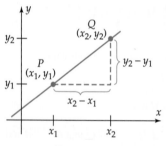

Figure 1–29

The number $\dfrac{\text{change in } y}{\text{change in } x}$ measures the steepness of the line: the steeper the line, the larger the number. In Figure 1–28, the grid allowed us to measure the change in y and the change in x. When the coordinates of P and Q are given, as in Figure 1–29, then:

The change in y is the difference of the y-coordinates of P and Q;

The change in x is the difference of the x-coordinates of P and Q.

Consequently, we have the following definition.

Slope of a Line

> If (x_1, y_1) and (x_2, y_2) are points with $x_1 \neq x_2$, then the **slope** of the line through these points is the number
> $$\frac{\Delta y}{\Delta x} = \frac{\text{change in } y}{\text{change in } x} = \frac{y_2 - y_1}{x_2 - x_1}.$$

EXAMPLE 1

Find the slope of the line through the two points.

(a) $(0, -1)$ and $(4, 1)$ (b) $(-2, 3)$ and $(2, -1)$

(c) $(1, 1)$ and $(3, 1)$ (d) $(3, -1)$ and $(3, 2)$

SOLUTION

(a) We apply the formula in the preceding box, with $x_1 = 0$, $y_1 = -1$ and $x_2 = 4$, $y_2 = 1$:

$$\text{Slope} = \frac{\Delta y}{\Delta x} = \frac{y_2 - y_1}{x_2 - x_1} = \frac{1 - (-1)}{4 - 0} = \frac{2}{4} = \frac{1}{2}.$$

---

*In this section, "line" means "straight line" and movement is from left to right.
†Δ (pronounced "delta") is the Greek letter D.

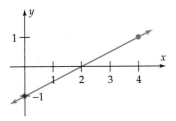

Figure 1–30

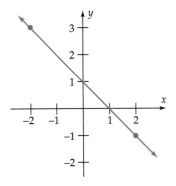

Figure 1–31

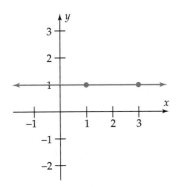

Figure 1–32

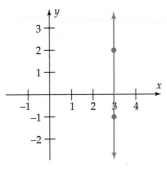

Figure 1–33

The order of the points makes no difference; if you use $(4, 1)$ for (x_1, y_1) and $(0, -1)$ for (x_2, y_2), you obtain the same number:

$$\text{Slope} = \frac{\Delta y}{\Delta x} = \frac{y_2 - y_1}{x_2 - x_1} = \frac{-1 - 1}{0 - 4} = \frac{-2}{-4} = \frac{1}{2}.$$

The slope is positive and the line through the two points rises from left to right, as shown in Figure 1–30.

(b) We have $(x_1, y_1) = (-2, 3)$ and $(x_2, y_2) = (2, -1)$, as shown in Figure 1–31. Hence,

$$\text{Slope} = \frac{\Delta y}{\Delta x} = \frac{y_2 - y_1}{x_2 - x_1} = \frac{-1 - 3}{2 - (-2)} = \frac{-4}{4} = -1.$$

The slope is negative and the line through the points falls from left to right.

(c) The points $(1, 1)$ and $(3, 1)$ lie on a horizontal line, as shown in Figure 1–32, and

$$\text{Slope} = \frac{\Delta y}{\Delta x} = \frac{y_2 - y_1}{x_2 - x_1} = \frac{1 - 1}{3 - 1} = \frac{0}{2} = 0.$$

A similar argument shows that *every horizontal line has slope* 0.

(d) Figure 1–33 shows that the points lie on a vertical line. Applying the slope formula to $(x_1, y_1) = (3, -1)$ and $(x_2, y_2) = (3, 2)$ yields

$$\text{Slope} = \frac{y_2 - y_1}{x_2 - x_1} = \frac{2 - (-1)}{3 - 3} = \frac{3}{0} \text{ not defined!}$$

The same argument works for any vertical line: *the slope of a vertical line is not defined.* ∎

CAUTION

When finding slopes, you must subtract the y-coordinates and x-coordinates in the same order. With the points $(3, 4)$ and $(1, 8)$, for instance, if you use $8 - 4$ in the numerator, you must use $1 - 3$ in the denominator (*not* $3 - 1$).

EXAMPLE 2

The lines shown in Figure 1–34 on the next page are determined by these points:

L_1: $(-1, -1)$ and $(0, 2)$ L_2: $(0, 2)$ and $(2, 4)$ L_3: $(-6, 2)$ and $(3, 2)$

L_4: $(-3, 5)$ and $(3, -1)$ L_5: $(1, 0)$ and $(2, -2)$.

Their slopes are as follows:

Points	Slopes
$(-1, -1)$ and $(0, 2)$	$L_1: \dfrac{2 - (-1)}{0 - (-1)} = \dfrac{3}{1} = 3$
$(0, 2)$ and $(2, 4)$	$L_2: \dfrac{4 - 2}{2 - 0} = \dfrac{2}{2} = 1$
$(-6, 2)$ and $(3, 2)$	$L_3: \dfrac{2 - 2}{3 - (-6)} = \dfrac{0}{9} = 0$
$(-3, 5)$ and $(3, -1)$	$L_4: \dfrac{-1 - 5}{3 - (-3)} = \dfrac{-6}{6} = -1$
$(1, 0)$ and $(2, -2)$	$L_5: \dfrac{-2 - 0}{2 - 1} = \dfrac{-2}{1} = -2$

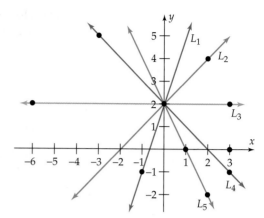

Figure 1–34

As Examples 1 and 2 illustrate, the slope is a number m that measures how steeply a line rises or falls, as summarized below.

Properties of Slope

There are four possibilities for a line L with slope m.

$m > 0$	$m < 0$	$m = 0$	m is not defined
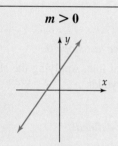		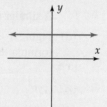	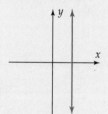
The line rises from left to right.	The line falls from left to right.	The line is horizontal.	The line is vertical.
The larger m is, the more steeply the line rises.	The larger $|m|$ is, the more steeply the line falls.		
[*See Example 1(a) and lines L_1 and L_2 in Example 2.*]	[*See Example 1(b) and lines L_4 and L_5 in Example 2.*]	[*See Example 1(c) and line L_3 in Example 2.*]	[*See Example 1(d)*]

SLOPE-INTERCEPT FORM

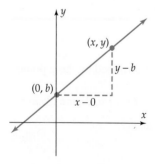

Figure 1–35

Let L be a nonvertical line with slope m and y-intercept b. Then $(0, b)$ is a point on L. Let (x, y) be any other point on L. Using the points $(0, b)$ and (x, y) to compute the slope of L (see Figure 1–35), we have

$$\text{Slope of } L = \frac{y - b}{x - 0}.$$

Since the slope of L is m, this equation becomes

$$m = \frac{y - b}{x}$$

Multiply both sides by x: $mx = y - b$

Rearrange terms: $y = mx + b$

Thus the coordinates of any point on L satisfy the equation $y = mx + b$. So we have the following fact.

Slope-Intercept
Form

> The line with slope m and y-intercept b is the graph of the equation
>
> $$y = mx + b.$$

EXAMPLE 3

List the slope and y-intercept of the line whose equation is given and describe its graph.

(a) $y = 3x - 5$ (b) $2x + y = 7$

SOLUTION

(a) The equation has the form in the preceding box, with $m = 3$ and $b = -5$, so the line has slope 3 and y-intercept -5. This shows that the line rises from left to right and passes through $(0, -5)$.

(b) First, solve the equation for y: $y = -2x + 7$. Here the slope is $m = -2$ and the y-intercept is $b = 7$. The line falls from left to right and passes through $(0, 7)$. ∎

EXAMPLE 4

Show that the graph of $2y - 5x = 2$ is a straight line. Find its slope, and graph the line.

SOLUTION   We begin by solving the equation for y:

Add $5x$ to both sides: $2y = 5x + 2$

Divide both sides by 2: $y = 2.5x + 1.$

The equation now has the form in the preceding box, with $m = 2.5$ and $b = 1$. Therefore, its graph is the line with slope 2.5 and y-intercept 1.

Since the y-intercept is 1, the point $(0, 1)$ is on the graph. To find another point on the line, choose a value for x, say, $x = 2$, and compute the corresponding value of y:

$$y = 2.5x + 1 = 2.5(2) + 1 = 6.$$

Hence, $(2, 6)$ is on the line. Plotting the line through $(0, 1)$ and $(2, 6)$ produces Figure 1–36. ∎

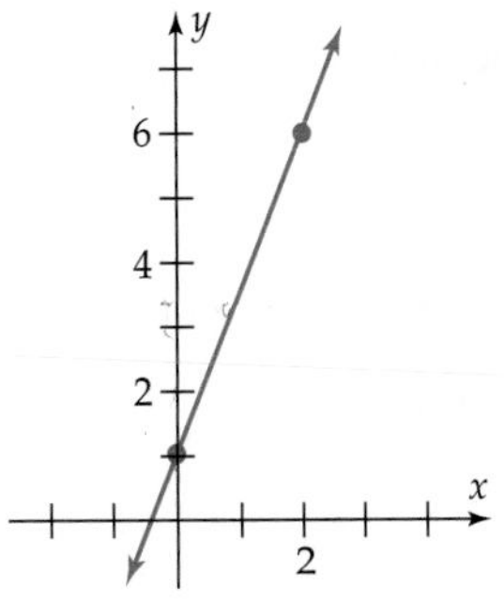

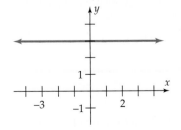

Figure 1–36

EXAMPLE 5

Describe and sketch the graph of the equation $y = 3$.

SOLUTION   We can write $y = 3$ as $y = 0x + 3$. So its graph is a line with slope 0, which means that the line is horizontal, and y-intercept 3, which means that the line crosses the y-axis at 3. This is sufficient information to obtain the graph in Figure 1–37.

Figure 1–37

Example 5 is an illustration of this fact.

Horizontal Lines

The horizontal line with y-intercept b is the graph of the equation

$$y = b.$$

Because slope is not defined for vertical lines, the equations of such lines have a different form from those examined above.

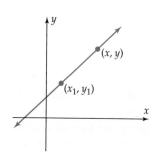

Figure 1–38

EXAMPLE 6

We can easily list some points on the line in Figure 1–38: $(2, 0)$, $(2, 1)$, $(2, -1)$, $(2, -1.5)$, and so on. Every point on this line has first coordinate 2. So every point satisfies $x + 0y = 2$. Hence, the line is the graph of $x = 2$. ∎

Example 6 illustrates these facts:

Vertical Lines

The vertical line with x-intercept c is the graph of the equation

$$x = c.$$

The slope of this line is undefined.

POINT-SLOPE FORM

Suppose the line L passes through the point (x_1, y_1) and has slope m. Let (x, y) be any other point on L. Using the points (x_1, y_1) and (x, y) to compute the slope m of L (see Figure 1–39), we have

Figure 1–39

$$\frac{y - y_1}{x - x_1} = \text{slope of } L$$

$$\frac{y - y_1}{x - x_1} = m$$

Multiply both sides by $x - x_1$: $\qquad y - y_1 = m(x - x_1).$

Thus, the coordinates of every point on L satisfy the equation

$$y - y_1 = m(x - x_1),$$

and we have this fact:

Point-Slope Form

The line with slope m through the point (x_1, y_1) is the graph of the equation

$$y - y_1 = m(x - x_1).$$

EXAMPLE 7

The tangent line to the graph of $y = 2x^2 - 8$ at the point $(1, -6)$ is shown in Figure 1–40. In calculus, it is shown that this line passes through $(1, -6)$ and has slope 4. Find the equation of the tangent line.

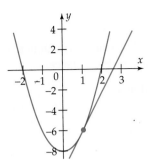

Figure 1–40

SOLUTION Substitute 4 for m and $(1, -6)$ for (x_1, y_1) in the point-slope equation:

$$y - y_1 = m(x - x_1)$$

$$y - (-6) = 4(x - 1) \qquad \text{[point-slope form]}$$

$$y + 6 = 4x - 4$$

$$y = 4x - 10 \qquad \text{[slope-intercept form]} \qquad \blacksquare$$

EXAMPLE 8

The annual out-of-pocket spending (per person) on doctors and clinical services was approximately $105 in 1997. According to projections from the Health Care Financing Committee, this cost is expected to rise linearly to $196 in 2010, as indicated in Figure 1–41.

(a) Find an equation that gives the out-of-pocket cost y in year x.

(b) Use this equation to estimate the out-of-pocket costs in 2006 and 2009.

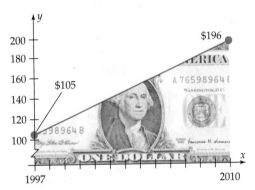

Figure 1–41

SOLUTION

(a) Let $x = 0$ correspond to 1997, so that $x = 13$ corresponds to 2010. Then the given information can be represented by the points $(0, 105)$ and $(13, 196)$. We must find the equation of the line through these points. Its slope is

$$\frac{196 - 105}{13 - 0} = \frac{91}{13} = 7.$$

Now we use the slope 7 and one of the points $(0, 105)$ or $(13, 196)$ to find the equation of the line. It doesn't matter which point, since both lead to the same equation.

$$y - y_1 = m(x - x_1) \qquad\qquad y - y_1 = m(x - x_1)$$

$$y - 105 = 7(x - 0) \qquad\qquad y - 196 = 7(x - 13)$$

$$y = 7x + 105 \qquad\qquad y - 196 = 7x - 91$$

$$y = 7x + 105.$$

(b) Since 2006 corresponds to $x = 9$, the projected out-of-pocket costs in 2006 are

$$y = 7x + 105 = 7 \cdot 9 + 105 = \$168.$$

The costs in 2009 ($x = 12$) are

$$y = 7x + 105 = 7 \cdot 12 + 105 = \$189. \qquad \blacksquare$$

GENERAL FORM

By rearranging terms, if necessary, the equation of any line can be written in the **general form** $Ax + By = C$ for some constants A, B, and C. For instance,

This equation	can be written as
$y = 4x - 10$	$4x - 1y = 10$
$y - 196 = 7(x - 13)$	$-7x + 1y = 105$
$y = 3$	$0x + 1y = 3$
$x = 2$	$1x + 0y = 2$

In summary:

*General
Form*

> Every line is the graph of an equation of the form
>
> $$Ax + By = C,$$
>
> where A and B are not both zero.

EXAMPLE 9

Graph $3x + 2y = 6$.

SOLUTION   From the preceding box, we know that the graph is a line. It is easily graphed by finding its intercepts.

y-intercept	*x*-intercept
Set $x = 0$ and solve for y.	Set $y = 0$ and solve for x.
$3x + 2y = 6$	$3x + 2y = 6$
$3 \cdot 0 + 2y = 6$	$3x + 2 \cdot 0 = 6$
$2y = 6$	$3x = 6$
$y = 3$	$x = 2$

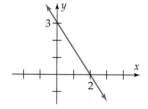

Figure 1–42

The y-intercept is 3 and the x-intercept is 2, which leads to the graph in Figure 1–42.   ∎

 PARALLEL AND PERPENDICULAR LINES

The slope of a line measures how steeply it rises or falls. Since parallel lines rise or fall equally steeply, the following fact should be plausible (see Exercises 93–94 for a proof).

*Parallel
Lines*

> Two nonvertical lines are parallel exactly when they have the same slope.

EXAMPLE 10

Find the equation of the line L through $(2, -1)$ that is parallel to the line M whose equation is $3x - 2y + 6 = 0$.

SOLUTION   First find the slope of M by rewriting its equation in slope-intercept form:

$$3x - 2y + 6 = 0$$
$$-2y = -3x - 6$$
$$y = \frac{3}{2}x + 3.$$

Therefore M has slope $3/2$. The parallel line L must have the same slope, $3/2$. Since $(2, -1)$ is on L, we can use the point-slope form to find its equation:

$$y - y_1 = m(x - x_1)$$

$$y - (-1) = \frac{3}{2}(x - 2) \qquad \text{[point-slope form]}$$

$$y + 1 = \frac{3}{2}x - 3$$

$$y = \frac{3}{2}x - 4 \qquad \text{[slope-intercept form]} \qquad \blacksquare$$

Two lines that meet in a right angle (90° angle) are said to be **perpendicular.** As you might suspect, there is a close relationship between the slopes of two perpendicular lines.

Perpendicular Lines

> Two nonvertical lines, with slopes m_1 and m_2, are perpendicular exactly when the product of their slopes is -1, that is,
>
> $$m_1 m_2 = -1, \quad \text{or equivalently,} \quad m_1 = -\frac{1}{m_2}.$$

A proof of this fact is outlined in Exercise 96.

EXAMPLE 11

In Figure 1–43, the line L through $(0, 2)$ and $(1, 5)$ appears to be perpendicular to the line M through $(-3, -2)$ and $(3, -4)$. Verify algebraically that the lines are perpendicular.

SOLUTION   Compute the slope of the lines:

$$\text{Slope } L = \frac{5 - 2}{1 - 0} = 3 \quad \text{and} \quad \text{slope } M = \frac{-4 - (-2)}{3 - (-3)} = \frac{-2}{6} = -\frac{1}{3}.$$

Since $3(-1/3) = -1$, the lines L and M are perpendicular. $\blacksquare$

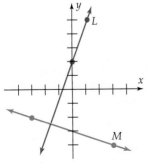

Figure 1–43

EXAMPLE 12

Find the equation of the perpendicular bisector of the line segment with endpoints $(-5, -4)$ and $(7, 2)$.

SOLUTION   The perpendicular bisector M goes through the midpoint of the line segment from $(-5, -4)$ and $(7, 2)$. The midpoint formula (page 43) shows that this midpoint is

$$\left(\frac{x_1 + x_2}{2}, \frac{y_1 + y_2}{2}\right) = \left(\frac{-5 + 7}{2}, \frac{-4 + 2}{2}\right) = (1, -1).$$

The line L through $(-5, -4)$ and $(7, 2)$ has slope

$$\frac{y_2 - y_1}{x_2 - x_1} = \frac{2 - (-4)}{7 - (-5)} = \frac{6}{12} = \frac{1}{2}.$$

Since M is perpendicular to L, we have (slope M)(slope L) $= -1$, so that

$$\text{Slope } M = \frac{-1}{\text{slope } L} = \frac{-1}{1/2} = -2.$$

Thus M is the line through $(1, -1)$ with slope -2, and its equation is

$$y - (-1) = -2(x - 1) \qquad \text{[point-slope form]}$$

$$y = -2x + 1. \qquad \text{[slope-intercept form]} \qquad \blacksquare$$

 ## RATES OF CHANGE

We have seen that the geometric interpretation of slope is that it measures the *steepness* or direction of a line. The next examples show that slope can also be interpreted as a *rate of change*.

EXAMPLE 13

According to the *Kelley Blue Book,* a Ford Focus ZX5 hatchback that is worth $14,632 today will be worth $10,120 in 3 years (if it is in good condition with average mileage).

(a) Assuming linear depreciation, find the equation that gives the value y of the car in year x.

(b) At what rate is the car depreciating?

(c) What will the car be worth in 6 years?

SOLUTION

(a) Linear depreciation means that the value equation is linear. So the equation is of the form $y = mx + b$ for some constants m and b. Since the car is worth $14,632 now (that is, $y = 14,632$ when $x = 0$), we have

$$y = mx + b$$

Let $x = 0$ and $y = 14,632$: $$14,632 = m \cdot 0 + b$$

$$b = 14,632.$$

So the equation is $y = mx + 14{,}632$. Since the car is worth \$10,120 in 3 years (that is, $y = 10{,}120$ when $x = 3$), we have

$$y = mx + 14{,}632$$

Let $x = 3$ and $y = 10{,}120$: $\quad 10{,}120 = m \cdot 3 + 14{,}632$

Subtract 14,632 from both sides: $\quad -4512 = 3m$

Divide both sides by 3: $\quad m = -1504$

Therefore, the value equation is $y = -1504x + 14{,}632$.

(b) Consider this table:

Year x	0	1	2	3	4
$y = -1504x + 14{,}632$	14,632	13,128	11,624	10,120	8616

You can easily verify that the car depreciates (decreases in value) \$1504 each year (that is, each time x changes by 1). In other words, the value changes at the *rate* of -1504 per year. This *rate* is the *slope* of the line $y = -1504x + 14{,}632$.

(c) The value of the car after 6 years ($x = 6$) is given by

$$y = -1504x + 14{,}632$$

Let $x = 6$: $\quad y = -1504(6) + 14{,}632 = 5608.$

The car is worth \$5608 in 6 years. ∎

EXAMPLE 14

A factory that makes can openers has fixed costs (for building, fixtures, machinery, etc.) of \$26,000. The variable cost (materials and labor) for making one can opener is \$2.75.

(a) Find the cost equation that gives the total cost y of producing x can openers and sketch its graph.

(b) At what rate does the total cost increase as more can openers are made?

(c) What is the total cost of making 1000 can openers? 20,000? 40,000?

(d) In part (c), what is the average cost per can opener in each case?

SOLUTION

(a) Since each can opener costs \$2.75, the variable cost of making x can openers is 2.75x. The total cost y of making x can openers is

$$y = \text{variable costs} + \text{fixed costs}$$
$$y = 2.75x + 26{,}000.$$

The graph of this equation is the line in Figure 1–44.

(b) The cost equation shows that y increases by 2.75 each time x increases by 1. That is, total cost is increasing at the rate of \$2.75 per can opener. This rate of change is the slope the cost equation line $y = 2.75x + 26{,}000$.

(c) The cost of making 1000 can openers is

$$y = 2.75x + 26{,}000 = 2.75(1000) + 26{,}000 = \$28{,}750.$$

Similarly, the cost of making 20,000 can openers is

$$y = 2.75(20{,}000) + 26{,}000 = \$81{,}000,$$

and the cost of 40,000 is

$$y = 2.75(40{,}000) + 26{,}000 = \$136{,}000.$$

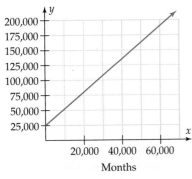

Months

Figure 1–44

(d) The average cost per can opener in each case is the total cost divided by the number of can openers. So the average cost per can opener is as follows.

For 1000: $28,750/1000 = $28.75 per can opener;
For 20,000: $81,000/20,000 = $4.05 per can opener;
For 40,000: $136,000/40,000 = $3.40 per can opener.   ∎

Examples 13 and 14 illustrate this fact.

Linear Rate of Change

The slope m of the line with equation

$$y = mx + b$$

is the rate of change of y with respect to x.

EXERCISES 1.4

1. For which of the line segments in the figure is the slope
   (a) largest?   (b) smallest?
   (c) largest in absolute value? (d) closest to zero?

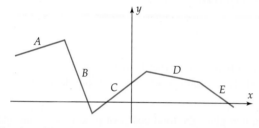

2. The doorsill of a campus building is 5 feet above ground level. To allow wheelchair access, the steps in front of the door are to be replaced by a straight ramp with constant slope $1/12$, as shown in the figure. How long must the ramp be? [The answer is *not* 60 feet.]

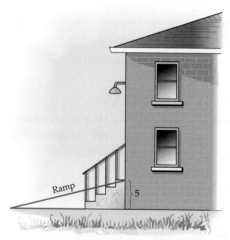

In Exercises 3–6, find the slope of the line through the given points.

3. $(1, 2); (3, 7)$ 4. $(-1, -2); (2, -1)$

5. $(1/4, 0); (3/4, 2)$ 6. $(\sqrt{2}, -1); (2, -9)$

In Exercises 7–10, find a number t such that the line passing through the two given points has slope -2.

7. $(0, t); (9, 4)$ 8. $(1, t); (-2, 4)$

9. $(t + 1, 5); (6, -3t + 7)$ 10. $(t, t); (5, 9)$

11. Let L be a nonvertical straight line through the origin. L intersects the vertical line through $(1, 0)$ at a point P. Show that the second coordinate of P is the slope of L.

12. On one graph, sketch five line segments, not all meeting at a single point, whose slopes are five different positive numbers. Do this in such a way that the left-hand line has the largest slope, the second line from the left has the next largest slope, and so on.

In Exercises 13–16, match the given equation with the line shown below that most closely resembles its graph.

(a) 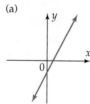   (b)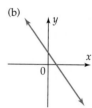

(c) 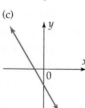   (d)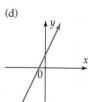

13. $y = 3x + 2$ 14. $y = -3x + 2$

15. $y = 3x - 2$ 16. $y = -3x - 2$

In Exercises 17–20, find the equation of the line with y-intercept b and slope m.

17. $b = 5, m = 4$ 18. $b = -3, m = -7$

19. $b = 1.5, m = -2.3$ **20.** $b = -4.5, m = 2.5$

In Exercises 21–24, find the equation of the line.

21.

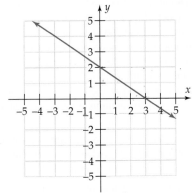

22.

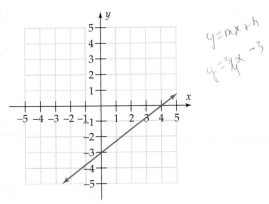

23.

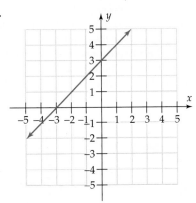

$y = mx + b$

$y = \frac{3}{4}x - 3$

24.

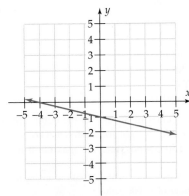

In Exercises 25–28, find the slope and y-intercept of the line whose equation is given.

25. $2x - y + 5 = 0$ **26.** $4x + 3y = 5$

27. $3(x - 2) + y = 7 - 6(y + 4)$

28. $2(y - 3) + (x - 6) = 4(x + 1) - 2$

In Exercises 29–32, find the equation of the line with slope m that passes through the given point.

29. $m = 1; (4, 7)$ **30.** $m = 2; (-2, 1)$

31. $m = -1; (6, 2)$ **32.** $m = 0; (-4, -5)$

In Exercises 33–36, find the equation of the line through the given points.

33. $(0, -5)$ and $(-3, -2)$ **34.** $(4, 3)$ and $(2, -1)$

35. $(6/5, 3/5)$ and $(1/5, 3)$ **36.** $(6, 7)$ and $(6, 15)$

In Exercises 37–42, graph the equation. Label all intercepts.

37. $3x + 5y = 15$ **38.** $2x - 3y = 12$

39. $2y - x = 2$ **40.** $4x + 5y = -10$

41. $3x - 2y = 0$ **42.** $2x + 6y = 0$

In Exercises 43–46, determine whether the line through P and Q is parallel or perpendicular to the line through R and S or neither.

43. $P = (2, 5), Q = (-1, -1)$ and $R = (4, 2), S = (6, 1)$.

44. $P = (0, 3/2), Q = (1, 1)$ and $R = (2, 7), S = (3, 9)$.

45. $P = (-3, 1/3), Q = (1, -1)$ and $R = (2, 0), S = (4, -2/3)$.

46. $P = (3, 3), Q = (-3, -1)$ and $R = (2, -2), S = (4, -5)$.

In Exercises 47–49, determine whether the lines whose equations are given are parallel, perpendicular, or neither.

47. $2x + y - 2 = 0$ and $4x + 2y + 18 = 0$.

48. $3x + y - 3 = 0$ and $6x + 2y + 17 = 0$.

49. $y = 2x + 4$ and $.5x + y = -3$.

50. Do the points $(-4, 6), (-1, 12)$, and $(-7, 0)$ all lie on the same straight line? [*Hint:* Use slopes.]

51. Are $(9, 6), (-1, 2)$, and $(1, -3)$ the vertices of a right triangle? [*Hint:* Use slopes.]

52. Are the points $(-5, -2) (-3, 1), (3, 0)$, and $(5, 3)$ the vertices of a parallelogram?

In Exercises 53–56, find the equation of the perpendicular bisector of the line segment joining the two given points.

53. $(1, 3), (3, 7)$ **54.** $(-3, 6), (7, 2)$

55. $(2, -3), (4, 7)$ **56.** $(-6, 2), (6, -7)$

In Exercises 57–64, find an equation for the line satisfying the given conditions.

57. Through $(-2, 1)$ with slope 3.

58. y-intercept -7 and slope 1.

59. Through $(2, 3)$ and parallel to $3x - 2y = 5$.

60. Through $(1, -2)$ and perpendicular to $y = 2x - 3$.

61. x-intercept 5 and y-intercept -5.

62. Through $(-5, 2)$ and parallel to the line through $(1, 2)$ and $(4, 3)$.

63. Through $(-1, 3)$ and perpendicular to the line through $(0, 1)$ and $(2, 3)$.

64. y-intercept 3 and perpendicular to $2x - y + 6 = 0$.

65. Find a real number k such that $(3, -2)$ is on the line $kx - 2y + 7 = 0$.

66. Find a real number k such that the line $3x - ky + 2 = 0$ has y-intercept -3.

If P is a point on a circle with center C, then the tangent line to the circle at P is the straight line through P that is perpendicular to the radius CP. In Exercises 67–70, find the equation of the tangent line to the circle at the given point.

67. $x^2 + y^2 = 25$ at $(3, 4)$ [*Hint:* Here C is $(0, 0)$ and P is $(3, 4)$; what is the slope of radius CP?]

68. $x^2 + y^2 = 169$ at $(-5, 12)$

69. $(x - 1)^2 + (y - 3)^2 = 5$ at $(2, 5)$

70. $x^2 + y^2 + 6x - 8y + 15 = 0$ at $(-2, 1)$

71. Let A, B, C, D be nonzero real numbers. Show that the lines $Ax + By + C = 0$ and $Ax + By + D = 0$ are parallel.

72. Let L be a line that is neither vertical nor horizontal and that does not pass through the origin. Show that L is the graph of $\dfrac{x}{a} + \dfrac{y}{b} = 1$, where a is the x-intercept and b is the y-intercept of L.

73. Worldwide motor vehicle production was about 60 million in 2000 and about 66 million in 2005.
    (a) Let the x-axis denote time and the y-axis the number of vehicles (in millions). Let $x = 0$ correspond to 2000. Fill in the blanks: the given data is represented by the points (___, 60) and (5, ___).
    (b) Find the linear equation determined by the two points in part (a).
    (c) Use the equation in part (b) to estimate the number of vehicles produced in 2004.
    (d) If this model remains accurate, when will vehicle production reach 72 million?

74. Carbon dioxide (CO_2) concentration is measured regularly at the Mauna Loa observatory in Hawaii. The mean annual concentration in parts per million in various years is given in the table.*

Year	Concentration (ppm)
1984	344.4
1989	352.9
1994	358.9
1999	368.3
2004	377.4

(a) Let $x = 0$ correspond to 1980. List the five data points given by the table. Do these points all lie on a single line? How can you tell?
(b) Use the data points from 1984 and 2004 to write a linear equation to model CO_2 concentration over time.
(c) Do part (b), using the data points from 1994 and 2004.
(d) Use the two models to estimate the CO_2 concentration in 1989 and 1999. Do the models overestimate or underestimate the concentration?
(e) What do the two models say about the concentration in 2008? Which model do you think is the more accurate? Why?

75. Suppose you drive along the Ohio Turnpike in an area where the grade of the road is 3% (which means that the line representing the road in the figure has slope .03.)
    (a) Find the equation of the line representing the road. [*Hint:* Your trip begins at the origin.]
    (b) If you drive on the road for one mile, how many feet higher are you at the end of the mile than you were at the beginning? [*Hint:* Express one mile as 5280 feet. Use the equation from part (a) to express x in terms of y, and then use the Pythagorean Theorem to find y.]

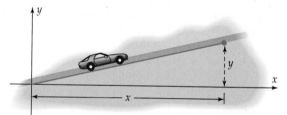

76. The Missouri American Water Company charges residents of St. Louis County $6.15 per month plus $2.0337 per thousand gallons used.*
    (a) Find the monthly bill when 3000 gallons of water are used. What is the bill when no water is used?
    (b) Write a linear equation that gives the monthly bill y when x thousand gallons are used.
    (c) If the monthly bill is $22.42, how much water was used?

77. At sea level, water boils at 212°F. At a height of 1100 feet, water boils at 210°F. The relationship between boiling point and height is linear.
    (a) Find an equation that gives the boiling point y of water at a height of x feet.

    Find the boiling point of water in each of the following cities (whose altitudes are given).
    (b) Cincinnati, OH (550 feet)
    (c) Springfield, MO (1300 feet)
    (d) Billings, MT (3120 feet)
    (e) Flagstaff, AZ (6900 feet)

78. According to the Center of Science in the Public Interest, the maximum healthy weight for a person who is 5 feet 5 inches tall is 150 pounds, and the maximum healthy

---

*C. D. Keeling and T. P. Whorf, Scripps Institution of Oceanography

*Residential rates for a 5/8 inch meter in March 2006, assuming monthly billing and maximum usage of 16,000 gallons.

weight for someone 6 feet 3 inches tall is 200 pounds. The relationship between weight and height here is linear.

(a) Find a linear equation that gives the maximum healthy weight y for a person whose height is x inches over 4 feet 10 inches. (Thus $x = 0$ corresponds to 4 feet 10 inches, $x = 2$ to 5 feet, etc.)

(b) What is the maximum healthy weight for a person whose height is 5 feet? 6 feet?

(c) How tall is a person who is at a maximum healthy weight of 220 pounds?

79. The number of unmarried couples in the United States who live together was 3.2 million in 1990 and grew in a linear fashion to 5.5 million in 2000.*

(a) Let $x = 0$ correspond to 1990. Write a linear equation expressing the number y of unmarried couples living together (in millions) in year x.

(b) Assuming the equation remains accurate, estimate the number of unmarried couples living together in 2010.

(c) When will the number of unmarried couples living together reach 10,100,000?

80. The percentage of people 25 years old and older who have a Bachelor's degree or higher was about 25.6 in 2000 and 27.7 in 2004.*

(a) Find a linear equation that gives the percentage of people 25 and over who have a Bachelor's degree or higher in terms of time t, where t is the number of years since 2000. Assume that this equations remains valid in the future.

(b) What will the percentage be in 2010?

(c) When will 34% of those 25 and over have a Bachelor's degree or higher?

81. At the Factory in Example 14, the cost of producing x can openers is given by $y = 2.75x + 26,000$.

(a) Write an equation that gives the average cost per can opener when x can openers are produced.

(b) How many can openers should be made to have an average cost of $3 per can opener?

82. Suppose the cost of making x TV sets is given by $y = 145x + 120,000$.

(a) Write an equation that gives the average cost per set when x sets are made.

(b) How many sets should be made in order to have an average cost per set of $175?

83. The profit p (in thousands of dollars) on x thousand units of a specialty item is $p = .6x - 14.5$. The cost c of manufacturing x thousand items is given by $c = .8x + 14.5$.

(a) Find an equation that gives the revenue r from selling x thousand items.

(b) How many items must be sold for the company to break even (i.e., for revenue to equal cost)?

84. A publisher has fixed costs of $110,000 for a mathematics text. The variable costs are $50 per book. The book sells for $72. Find equations that give

(a) The cost c of making x books

(b) The revenue r from selling x books

(c) The profit p from selling x books

(d) What is the publisher's break-even point (see Exercise 83(b))?

Use the graph and the following information for Exercises 85–86. Rocky is an "independent" ticket dealer who markets choice tickets for Los Angeles Lakers home games. (California currently has no laws against ticket scalping.) Each graph shows how many tickets will be demanded by buyers at a particular price. For instance, when the Lakers play the Chicago Bulls, the graph shows that at a price of $160, no tickets are demanded. As the price (y-coordinate) gets lower, the number of tickets demanded (x-coordinate) increases.

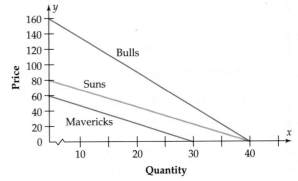

85. Write a linear equation that relates the quantity x of tickets demanded at price y when the Lakers play the

(a) Dallas Mavericks   (b) Phoenix Suns

(c) Chicago Bulls

[*Hint:* In each case, use the x- and y-intercepts to determine its slope.]

86. Use the equations from Exercise 85 to find the number of tickets Rocky would sell at a price of $40 for a game against the

(a) Mavericks   (b) Bulls

87. The Fahrenheit and Celsius scales for measuring temperatures are linearly related. They are calibrated using the freezing and boiling points of water at sea level.

Temperature Scale	Fahrenheit Scale	Celsius Scale
Water Freezes	32°	0°
Water Boils	212°	100°

(a) Use the data in the table to write a formula that relates the Fahrenheit temperature F to the Celsius temperature C. Your answer should be in the form $F = mC + b$.

(b) Solve the equation in part (a) for C to find a formula that relates the Celsius temperature to the Fahrenheit temperature.

(c) When is the temperature in degrees Fahrenheit the same as the temperature in degrees Celsius?

88. (a) If the temperature changes 1° Fahrenheit, how many degrees does the Celsius temperature change? [*Hint:* See Exercise 87.]

(b) What happens to the Fahrenheit temperature when the Celsius temperature changes 1°?

(c) How are your answers in parts (a) and (b) related to the formulas in Exercises 87?

89. A 75-gallon water tank is being emptied. The graph shows the amount of water in the tank after x minutes.

(a) At what rate is the tank emptying during the first 2 minutes? During the next 3 minutes? During the last minute?

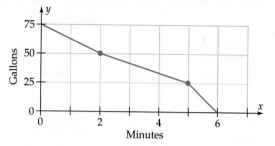

(b) Suppose the tank is emptied at a constant rate of 10 gallons per minute. Draw the graph that shows the amount of water after x minutes. What is the equation of the graph?

90. The poverty level income for a family of four was $13,359 in 1990. Because of inflation and other factors, the poverty level rose approximately linearly to $19,307 in 2004.*

(a) At what rate is the poverty level increasing?

(b) Estimate the poverty level in 2000 and 2009.

91. A Honda Civic LX sedan is worth $15,350 now and will be worth $9910 in four years.

(a) Assuming linear depreciation, find the equation that gives the value y of the car in year x.

(b) At what rate is the car depreciating?

(c) Estimate the value of the car six years from now.

92. A house in Shaker Heights, Ohio was bought for $160,000 in 1980. It increased in value in an approximately linear fashion and sold for $359,750 in 1997.

(a) At what rate did the house appreciate (increase in value) during this period?

(b) If this appreciation rate remained accurate what would the house be worth in 2010?

THINKERS

93. Show that two nonvertical lines with the same slope are parallel. [*Hint:* The equations of distinct lines with the same slope must be of the form $y = mx + b$ and $y = mx + c$ with $b \neq c$ (why?). If (x_1, y_1) were a point on both lines, its coordinates would satisfy both equations. Show that this leads to a contradiction, and conclude that the lines have no point in common.]

94. Prove that nonvertical parallel lines L and M have the same slope, as follows. Suppose M lies above L, and choose two points (x_1, y_1) and (x_2, y_2) on L.

(a) Let P be the point on M with first coordinate x_1. Let b denote the vertical distance from P to (x_1, y_1). Show that the second coordinate of P is $y_1 + b$.

(b) Let Q be the point on M with first coordinate x_2. Use the fact that L and M are parallel to show that the second coordinate of Q is $y_2 + b$.

(c) Compute the slope of L using (x_1, y_1) and (x_2, y_2). Compute the slope of M using the points P and Q. Verify that the two slopes are the same.

95. Show that the diagonals of a square are perpendicular. [*Hint:* Place the square in the first quadrant of the plane, with one vertex at the origin and sides on the positive axes. Label the coordinates of the vertices appropriately.]

96. This exercise provides a proof of the statement about slopes of perpendicular lines in the box on page 61. First, assume that L and M are nonvertical perpendicular lines that both pass through the origin. L and M intersect the vertical line $x = 1$ at the points $(1, k)$ and $(1, m)$, respectively, as shown in the figure.

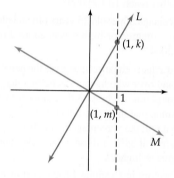

(a) Use $(0, 0)$ and $(1, k)$ to show that L has slope k. Use $(0, 0)$ and $(1, m)$ to show that M has slope m.

(b) Use the distance formula to compute the length of each side of the right triangle with vertices $(0, 0)$, $(1, k)$, and $(1, m)$.

(c) Use part (b) and the Pythagorean Theorem to find an equation involving k, m, and various constants. Show that this equation simplifies to $km = -1$. This proves half of the statement.

(d) To prove the other half, assume that $km = -1$, and show that L and M are perpendicular as follows. You may assume that a triangle whose sides a, b, c satisfy $a^2 + b^2 = c^2$ is a right triangle with hypotenuse c. Use this fact, and do the computation in part (b) in reverse (starting with $km = -1$) to show that the triangle with vertices $(0, 0)$, $(1, k)$, and $(1, m)$ is a right triangle, so that L and M are perpendicular.

(e) Finally, to prove the general case when L and M do not intersect at the origin, let L_1 be a line through the origin that is parallel to L, and let M_1 be a line through the origin that is parallel to M. Then L and L_1 have the same slope, and M and M_1 have the same slope (why?). Use this fact and parts (a)–(d) to prove that L is perpendicular to M exactly when $km = -1$.

Chapter 1 Review

IMPORTANT CONCEPTS

Section 1.1

Real numbers, integers, rationals,
   irrationals   2–3
Order of operations   3
Distributive law   4
Number line   4
Order ($<, \leq, >, \geq$)   4–5
Intervals, open intervals, closed
   intervals   5
Negatives   6
Scientific notation   7
Square roots   8
Absolute value   9
Distance on the number line   11

Special Topics 1.1.A

Repeating and nonrepeating
   decimals   17

Section 1.2

Basic principles for solving
   equations   19
First-degree equations   19–20

Quadratic equations   20
Factoring   20–21
Completing the square   22
Quadratic formula   23
Discriminant   24
Higher-degree equations   26
Fractional equations   27

Special Topics 1.2.A

Absolute value equations   32

Special Topics 1.2.B

Direct variation   34
Inverse variation   34
Constant of variation   34

Section 1.3

Coordinate plane, x-axis, y-axis,
   quadrants   39–40
Scatter plots and line graphs   41
Distance formula   41
Midpoint formula   43
Graph of an equation   44

x- and y-intercepts   44–45
Circle, center, radius   46
Equation of the circle   46
Unit circle   48

Section 1.4

Change in x, change in y 54
Slope   54
Properties of slope   56
Slope-intercept form of the equation
   of a line   57
Horizontal lines   58
Vertical lines   58
Point-slope form of the equation
   of a line   58
General form of the equation
   of a line   60
Slopes of parallel lines   60
Slopes of perpendicular lines   61
Fixed and variable costs   63
Linear rate of change   64

IMPORTANT FACTS & FORMULAS

- $|c - d|$ = distance from c to d on the number line.
- *Quadratic Formula*: If $a \neq 0$, then the solutions of $ax^2 + bx + c = 0$ are

$$x = \frac{-b \pm \sqrt{b^2 - 4ac}}{2a}.$$

- If $a \neq 0$, then the number of real solutions of $ax^2 + bx + c = 0$ is 0, 1, or 2, depending on whether the discriminant $b^2 - 4ac$ is negative, zero, or positive.
- *Distance Formula*: The distance from (x_1, y_1) to (x_2, y_2) is

$$\sqrt{(x_1 - x_2)^2 + (y_1 - y_2)^2}.$$

- *Midpoint Formula*: The midpoint of the line segment from (x_1, y_1) to (x_2, y_2) is

$$\left(\frac{x_1 + x_2}{2}, \frac{y_1 + y_2}{2}\right).$$

- Equation of the circle with center (c, d) and radius r is

$$(x - c)^2 + (y - d)^2 = r^2.$$

- The slope of the line through (x_1, y_1) and (x_2, y_2) (where $x_1 \neq x_2$) is

$$\frac{y_2 - y_1}{x_2 - x_1}.$$

- The equation of the line with slope m and y-intercept b is

$$y = mx + b.$$

- The equation of the line through (x_1, y_1) with slope m is

$$y - y_1 = m(x - x_1).$$

- Two nonvertical lines are parallel exactly when they have the same slope.
- Two nonvertical lines are perpendicular exactly when the product of their slopes is -1.

REVIEW QUESTIONS

1. Fill the blanks with one of the symbols $<$, $=$, or $>$ so that the resulting statement is true.

   (a) 142 ____ $|-51|$ (b) $\sqrt{2}$ ____ $|-2|$

   (c) -1000 ____ $\dfrac{1}{10}$ (d) $|-2|$ ____ $-|6|$

   (e) $|u - v|$ ____ $|v - u|$, where u and v are fixed real numbers.

2. List two real numbers that are *not* rational numbers.

3. Express in symbols:

   (a) y is negative, but greater than -10.
   (b) x is nonnegative and not greater than 10.

4. Express in symbols:

   (a) $c - 7$ is nonnegative.
   (b) $.6$ is greater than $|5x - 2|$.

5. Express in interval notation:

   (a) The set of all real numbers that are strictly greater than -8;
   (b) The set of all real numbers that are less than or equal to 5.

6. Express in interval notation:

   (a) The set of all real numbers that are strictly between -6 and 9;
   (b) The set of all real numbers that are greater than or equal to 5, but strictly less than 14.

7. Express in scientific notation:

   (a) 12,320,000,000,000,000   (b) .0000000000789

8. Express in decimal notation:

   (a) 4.78×10^8 (b) 6.53×10^{-9}

9. Express in symbols:

   (a) x is less than 3 units from -7 on the number line.
   (b) y is farther from 0 than x is from 3 on the number line.

10. Simplify: $|b^2 - 2b + 1|$ 11. Solve: $|x - 5| = 3$

12. Solve: $|x + 2| = 4$

13. Solve: $|x + 3| = \dfrac{5}{2}$

14. Solve: $|x - 5| \leq 2$ 15. Solve: $|x + 2| > 2$

16. When John-Paul carves pumpkins, he will not use any that weigh less than 2 pounds or more than 10 pounds. If x represents the weight of a pumpkin in pounds that he will *not* use, which of the following statements is always true?

   (a) $|x - 2| > 10$
   (b) $|x - 4| > 6$
   (c) $|x - 5| > 5$
   (d) $|x - 6| > 4$
   (e) $|x - 10| > 4$

17. (a) $|\pi - 7| =$ ____   (b) $|\sqrt{23} - \sqrt{3}| =$ ____

18. If c and d are real numbers with $c \neq d$ what are the possible values of $\dfrac{c - d}{|c - d|}$?

19. Express $.282828 \cdots$ as a fraction.

20. Express $.362362362 \cdots$ as a fraction.

21. Solve for x:

$$2\left(\frac{x}{5} + 7\right) - 3x = \frac{x + 2}{5} - 4.$$

22. Solve for x in terms of y: $xy + 3 = x - 2y$

23. Solve for x: $3x^2 - 2x + 5 = 0$

24. Solve for y: $3y^2 - 2y = 5$

25. Solve for z: $5z^2 + 6z = 7$

26. The population P (in thousands) of St. Louis, Missouri, can be approximated by

$$P = .11x^2 - 15.95x + 864,$$

   where $x = 0$ corresponds to 1950. When did the population first drop below 450,000?

27. How many times larger is the viewing area on a 21-inch computer monitor than on a 14-inch monitor? Remember

that the size of a monitor is its diagonal measurement, as shown in the figure, and assume that the height of the screen is three-fourths of its width.

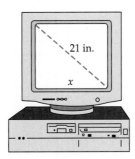

21 in.

x

28. The median sales price S (in thousands of dollars) for a single-family home in the midwestern United States can be approximated by $S = .1x^2 + 3.9x + 74.5$, where $x = 0$ corresponds to 1990.* When did the median price reach $136,000?

29. Find the *number* of real solutions of the equation $20x^2 + 12 = 31x$.

30. For what value of k does the equation
$$kt^2 + 5t + 2 = 0$$
have exactly one real solution for t?

31. Solve for x: $\dfrac{3}{x} + \dfrac{5}{x + 2} = 2$.

32. Solve for z: $2 - \dfrac{1}{z - 2} = \dfrac{z - 3}{z - 2}$.

33. If x and y are directly proportional and $x = 12$ when $y = 36$, then what is y when $x = 2$?

34. Driving time varies inversely as speed. If it takes 3 hours to drive to the beach at an average speed of 48 mph, how long will it take to drive home at an average speed of 54 mph?

35. Find the constant of variation when T varies directly as the square of R and inversely as S if $T = .6$ when $R = 3$ and $S = 15$.

36. The statement "r varies directly as s and the square root of t and inversely as the cube of x" means that for some constant k,

(a) $r = kstx^3$ (b) $rx^3 = ks\sqrt{t}$

(c) $r = \dfrac{kx^3\sqrt{t}}{s}$ (d) $k = \dfrac{rs\sqrt{t}}{x^3}$

(e) $kx^3s = k\sqrt{t}$

In Questions 37–40, find all real solutions of the equation. Do not approximate.

37. $x^4 - 11x^2 + 18 = 0$ **38.** $x^6 - 4x^3 + 4 = 0$

39. $|3x - 1| = 4$ **40.** $|2x - 1| = x + 4$

---

*Department of Housing and Urban Development

41. Find the distance from $(1, -2)$ to $(4, 5)$.

42. Find the distance from $(3/2, 4)$ to $(3, 5/2)$.

43. Find the distance from (c, d) to $(c - d, c + d)$.

44. Find the midpoint of the line segment from $(-4, 7)$ to $(9, 5)$.

45. Find the midpoint of the line segment from (c, d) to $(2d - c, c + d)$.

46. Find the equation of the circle with center $(-3, 4)$ that passes through the origin.

47. (a) If $(1, 1)$ is on a circle with center $(2, -3)$, what is the radius of the circle?
(b) Find the equation of the circle in part (a).

48. Sketch the graph of $3x^2 + 3y^2 = 12$.

49. Sketch the graph of $(x - 5)^2 + y^2 - 9 = 0$.

50. Find the center and radius of the circle whose equation is
$$x^2 + y^2 - 2x + 6y + 1 = 0.$$

51. Which of statements (a)–(d) are descriptions of the circle with center $(0, -2)$ and radius 5?

(a) The set of all points (x, y) that satisfy
$$|x| + |y + 2| = 5.$$
(b) The set of all points whose distance from $(0, -2)$ is 5.
(c) The set of all points (x, y) such that
$$x^2 + (y + 2)^2 = 5.$$
(d) The set of all points (x, y) such that
$$\sqrt{x^2 + (y + 2)^5} = 5.$$

52. If the equation of a circle is $3x^2 + 3(y - 2)^2 = 12$, which of the following statements is true?

(a) The circle has diameter 3.
(b) The center of the circle is $(2, 0)$.
(c) The point $(0, 0)$ is on the circle.
(d) The circle has radius $\sqrt{12}$.
(e) The point $(1, 1)$ is on the circle.

53. The graph of one of the equations below is *not* a circle. Which one?

(a) $x^2 + (y + 5)^2 = \pi$
(b) $7x^2 + 4y^2 - 14x + 3y^2 - 2 = 0$
(c) $3x^2 + 6x + 3 = 3y^2 + 15$
(d) $2(x - 1)^2 - 8 = -2(y + 3)^2$
(e) $\dfrac{x^2}{4} + \dfrac{y^2}{4} = 1$

54. The point $(7, -2)$ is on the circle whose center is on the midpoint of the segment joining $(3, 5)$ and $(-5, -1)$. Find the equation of this circle.

55. The table on the next page shows fatal crash involvements per 100 million miles traveled by drivers of selected ages.* Sketch a scatter plot and a line graph for this data.

---

*Insurance Institute for Highway Safety

Age of Driver	16	17	18	19	23	28	33
Fatal crashes	9.3	8.3	6.5	7.2	4.3	2.3	1.6

56. The table shows the average speed (mph) of the winning car in the Indianapolis 500 race in selected years.* Sketch a scatter plot and a line graph for this data, letting $x = 0$ correspond to 1992.

Year	1992	1994	1996	1998	2000	2002	2004
Speed	134	161	148	145	168	166	139

57. (a) What is the y-intercept of the graph of the line

$$y = x - \frac{x-2}{5} + \frac{3}{5}?$$

(b) What is the slope of the line?

58. Find the equation of the line passing through $(1, 3)$ and $(2, 5)$.

59. Find the equation of the line passing through $(2, -1)$ with slope 3.

60. Find all points on the graph of $y = 3x$ whose distance to the origin is 2.

61. Find the equation of the line that crosses the y-axis at $y = 1$ and is perpendicular to the line $2y - x = 5$.

62. (a) Find the y-intercept of the line $2x + 3y - 4 = 0$.
(b) Find the equation of the line through $(1, 3)$ that has the same y-intercept as the line in part (a).

63. Find the equation of the line through $(-4, 5)$ that is parallel to the line through $(1, 3)$ and $(-4, 2)$.

64. Sketch the graph of the line $3x + y - 1 = 0$.

65. As a balloon is launched from the ground, the wind is blowing it due east. The conditions are such that the balloon is ascending along a straight line with slope $1/5$. After 1 hour the balloon is 5000 feet vertically above the ground. How far east has the balloon blown?

66. The point (u, v) lies on the line $y = 5x - 10$. What is the slope of the line passing through (u, v) and the point $(0, -10)$?

In Questions 67–73, determine whether the statement is true or false.

67. The graph of $x = 5y + 6$ has y-intercept 6.

68. The graph of $2y - 8 = 3x$ has y-intercept 4.

69. The lines $3x + 4y = 12$ and $4x + 3y = 12$ are perpendicular.

70. Slope is not defined for horizontal lines.

71. The line in the figure has positive slope.

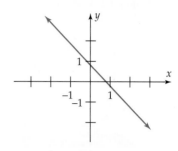

72. The line in the figure above does not pass through the third quadrant.

73. The y-intercept of the line in the figure above is negative.

74. Consider the *slopes* of the lines shown in the figure below. Which line has the slope with the largest *absolute value*?

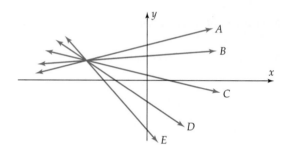

75. Which of the following lines rises most steeply from left to right?
(a) $y = -4x - 10$ (b) $y = 3x + 4$
(c) $20x + 2y - 20 = 0$ (d) $4x = y - 1$
(e) $4x = 1 - y$

76. Which of the following lines is *not* perpendicular to the line $y = x + 5$?
(a) $y = 4 - x$ (b) $y + x = -5$
(c) $4 - 2x - 2y = 0$ (d) $x = 1 - y$
(e) $y - x = \frac{1}{5}$

77. Which of the following does *not* pass through the third quadrant?
(a) $y = x$ (b) $y = 4x - 7$
(c) $y = -2x - 5$ (d) $y = 4x + 7$
(e) $y = -2x + 5$

78. Let a, b be fixed real numbers. Where do the lines $x = a$ and $y = b$ intersect?
(a) Only at (b, a).   (b) Only at (a, b).
(c) These lines are parallel, so they don't intersect.
(d) If $a = b$, then these are the same line, so they have infinitely many points of intersection.
(e) Since these equations are not of the form $y = mx + b$, the graphs are not lines.

79. Which of the following is an equation of a line with y-intercept 2 and x-intercept 3?

(a) $-2x + 3y = 6$ (b) $-2x + 3y = 4$
(c) $2x + 3y = 6$ (d) $2x + 3y = 4$
(e) $3x + 2y = 6$

80. For what values of k will the graphs of $2y + x + 3 = 0$ and $3y + kx + 2 = 0$ be perpendicular lines?

81. Average life expectancy increased linearly from 74.7 years for a person born in 1985 to 77.8 years for a person born in 2005.

(a) Find a linear equation that gives the average life expectancy y of a person born in year x, with $x = 0$ corresponding to 1985.
(b) Use the equation in part (a) to estimate the average life expectancy of a person born in 1990.
(c) Assuming the equation remains valid, in what year will the average life expectancy be 80 years for people born in that year?

82. The population of San Diego, California, grew in an approximately linear fashion from 1,110,600 in 1990 to 1,263,700 in 2004.

(a) Find a linear equation that gives the population y of San Diego (in thousands) in year x, with $x = 0$ corresponding to 1990.
(b) Estimate the population of San Diego in 2010.
(c) Assuming the equation remains accurate, when will San Diego's population reach 1.5 million people?

In Exercises 83–86, match the given information with one of the graphs (a)–(d), and determine the slope of the graph.

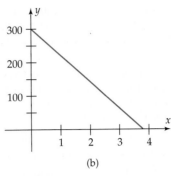

(b)

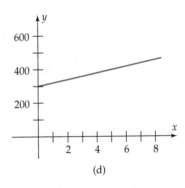

(c)

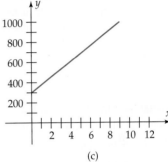

(d)

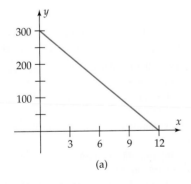

(a)

83. A salesman is paid $300 per week plus $75 for each unit sold.

84. A person is paying $25 per week to repay a $300 loan.

85. A gold coin that was purchased for $300 appreciates $20 per year.

86. A CD player that was purchased for $300 depreciates $80 per year.

Chapter 1 Test

Sections 1.1 and 1.2; Special Topics 1.1.A and 1.2.A.

1. (a) Draw a picture on the number line of the interval $[-3, 2)$.
(b) Use interval notation to denote the set of all real numbers x that satisfy $x \leq -2$.

2. The cost c of manufacturing x items is given by $c = .7x + 19.6$ and the revenue r from selling these items is given by $r = 1.1x$. How many items must be sold for the company to break even (that is, for revenue to equal cost)?

3. The atmospheric pressure a (in pounds per square foot) at height h thousand feet above sea level is approximately $a = .8315h^2 - 73.93h + 2116.1$ $(0 \le h \le 40)$.

   (a) Find the atmospheric pressure (rounded to two decimal places) at the top of Mount Annapurna (26,504 feet).
   (b) The atmospheric pressure at the top of Mount Rainier is 1238.41 pounds per square foot. How high (to the nearest foot) is Mount Rainier?

4. The gross federal debt was about $5622 billion in 2000, when the U.S. population was approximately 281.1 million people.

   (a) Express the debt in scientific notation.
   (b) Express the population in scientific notation.
   (c) In 2000, what was each person's share of the federal debt?

5. Solve for t: $\dfrac{1}{3t} - \dfrac{3}{4t} = \dfrac{1}{12t} + 1$.

6. Solve for x: $2x^2 + 13x - 7 = 0$.

7. Express the infinite decimal $.14141414\cdots$ as a fraction, without using a calculator. Show your work.

8. Solve for b: $E = \dfrac{h}{2}(b + c)$.

9. Solve for x: $4x^2 = 6x + 5$.

10. (a) Describe in words the solutions of $|x - 16| < 6$.
    (b) Use absolute value to describe all real numbers c that are more than 5 units from 16 on the number line.

11. Find a real number k such that the equation $x^2 - kx + 16 = 0$ has exactly one real solution.

12. Express each of the following without using absolute value bars. Assume $x \ge 1$.

    (a) $|y^2 - 2y + 1|$
    (b) $|x^2 - 1|$

13. Solve for x: $|4x - 5| = 12$.

Sections 1.3 and 1.4; Special Topics 1.2.B

14. Find the equation of the circle that passes through $(-7, -5)$ and has center $(-1, -6)$.

15. Two lines have equations

    $$4x + y - 4 = 0 \quad \text{and} \quad 6x + 3y + 13 = 0.$$

    Are the lines parallel, perpendicular, or neither? Given reasons for your answer.

16. (a) Find the midpoint of the line segment joining $(-1, 3)$ and $(2, -1)$.
    (b) Find the length of the line segment in part (a).

17. Find the slope of the line through $(\sqrt{5}, -6)$ and $(5, -8)$.

18. Find the x-intercepts and y-intercepts of the graph of $3x^2 + x - y - 2 = 0$. You need not graph the equation.

19. The poverty level income for a family of four was $13,359 in 1990. It grew approximately linearly to $19,309 in 2004.

    (a) At what rate was the poverty level increasing during this period?
    (b) Estimate the poverty level in 2000.
    (c) Assuming the growth rate remains the same, estimate the poverty level in 2010.

20. If P is a point on a circle with center C, then the *tangent line to the circle at P* is the straight line through P that is perpendicular to the radius CP.

    (a) Find the center C of the circle .
    $$(x - 1)^2 + (y - 1)^2 = 5.$$
    (b) Find the equation of the tangent line to this circle at the point $P = (2, 3)$.

21. Find the center and the radius of the circle whose equation is

    $$3x^2 + 3y^2 + 6x + 24 = 24y.$$

22. Graph the equation $2x - 5y = 10$. Label all intercepts.

23. Find the perimeter of the shaded area in the figure.

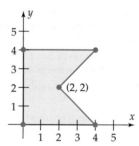

24. The age-adjusted death rate from heart disease was 412.1 per 100,000 population in 1980 and 232.3 in 2003.

    (a) Assuming that the death rate declined linearly, find an equation of the form $y = mx + b$ that gives the number y of deaths per 100,000 in year x, with $x = 0$ corresponding to 1980. Round m to two decimal places.
    (b) Use the equation in part (a) to estimate the death rate in 2001.
    (c) Assuming that the equation remains accurate, estimate the death rate in 2009.
    (d) Is the equation likely to remain accurate over the next three decades? Give reasons for your answer.

25. If r varies inversely as t and $r = 9$ when $t = 3$, find r when $t = 12$.

26. Determine whether the line through P and Q is parallel, or perpendicular to the line through R and S, or neither, when

    $$P = \left(0, \tfrac{3}{2}\right), \quad Q = (1, 1), \quad R = (6, 4), \quad \text{and} \quad S = (7, 5).$$

DISCOVERY PROJECT 1   Taxicab Geometry

A man was arrested in New York City in 2002 for selling drugs within 1000 feet of a school. This carries a heavier penalty than does a sale more than 1000 feet from the school. In a case that went to the state's highest court, the man argued that his distance from the school should not be measured "as the crow flies" but as a person would have to walk along the street (being unable to cut through buildings that the crow would fly over). In his case, the walking distance was more than 1000 feet. The court rejected his argument and he is now serving a 6 to 12 year sentence.*

Measuring distance as one walks—or as a taxi drives—from one point in the city to another is an example of *taxicab geometry*. The **taxicab distance** from point A to point B in Figure 1 is defined to be the sum $u + v$, whereas the ordinary (crow flying) distance is the length of the red line ($\sqrt{u^2 + v^2}$ according to the Pythagorean Theorem).

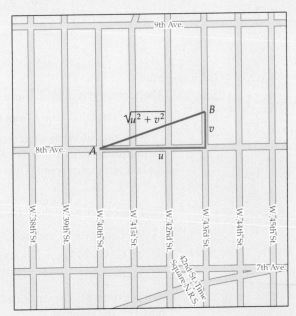

Figure 1

Find the taxicab distance and the ordinary distance between each pair of points.

1. $(1, 1)$ and $(5, 4)$
2. $(0, 0)$ and $(500, 500)$
3. $(0, 0)$ and $(1000, 0)$

---

*The same issue (with a different outcome) arose in a federal court. In a case involving the Family and Medical Leave Act, the 10th Circuit Court of Appeals upheld a Labor Department rule requiring that distance be measured in "surface miles, using surface transportation" rather than the distance as the crow flies. See *The New York Times* of November 23, 2005 and May 28, 2007.

More generally, suppose A and B have coordinates (x_1, y_1) and (x_2, y_2) respectively. Then Figure 2 shows that $u = |x_2 - x_1|$ and $v = |y_2 - y_1|$ (why?).

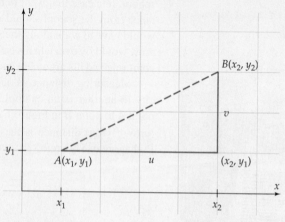

Figure 2

Therefore, the taxicab distance formula is

Taxicab distance from (x_1, y_1) to $(x_2, y_2) = |x_2 - x_1| + |y_2 - y_1|$.

The **taxicab circle** with center (a, b) and radius r consists of all points (x, y) whose taxicab distance from (a, b) is r. Substituting (a, b) and (x, y) for (x_1, y_1) and (x_2, y_2) in the taxicab distance formula, we see that the equation of this taxicab circle is

$$r = |x - a| + |y - b|, \quad \text{or equivalently,} \quad |x - a| + |y - b| = r.$$

Find the equations of these taxicab circles:

4. Center $(0, 0)$, radius 4

5. Center $(0, 0)$, radius 1000

6. Center $(3, 4)$, radius 5

According to the equations above, the taxicab unit circle (center at $(0, 0)$, radius 1) is the graph of

$$|x - 0| + |y - 0| = 1, \quad \text{that is,} \quad |x| + |y| = 1.$$

7. Graph the taxicab unit circle. [*Hint:* You can plot points, but it's better to find the graph one quadrant at a time. Use the definition of absolute value to rewrite the equation for each of the four quadrants. In quadrant I, $x \geq 0$ and $y \geq 0$, so the equation is $x + y = 1$. In quadrant II, $x \leq 0$ and $y \geq 0$ so the equation is $(-x) + y = 1$, and so on.]

Suppose the drug dealer in the first paragraph of this project is standing at the origin.

8. Graph the taxicab circle of radius 1000 with the drug dealer as center.

9. On the same coordinate plane as in Exercise 8, graph the ordinary circle of radius 1000 with the drug dealer as center.

Chapter 2

GRAPHS AND TECHNOLOGY

Are the financial rewards worth it?

Data from past years shows that the typical college graduate earned more than the typical high school graduate (with no college). Technology can be used to construct a linear model to estimate the median earnings of both groups of graduates in future years and to determine whether the gap between the two groups is increasing or decreasing. See Exercises 20 and 21 on page 130.

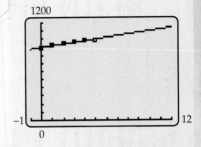

© Barry Austin Photography/Getty Images

Chapter Outline

Technology for graphing, solving equations, and building mathematical models is introduced in this chapter. It is a powerful tool for dealing with complicated mathematical situations and real-world problems that might otherwise be intractable. Keep in mind, however, that technology is not a substitute for mathematical knowledge or common sense. To use it effectively, you need a sound grounding in algebra and geometry.

2.1   Graphs

Section Objectives
- Graph an equation by plotting points.
- Graph an equation using technology.
- Use the graphing tools available with graphing technology.
- Use technology to create a scatter plot and line graph.

The traditional method of graphing an equation "by hand" is as follows: Construct a table of values with a reasonable number of entries, plot the corresponding points, and use whatever algebraic or other information is available to make an educated guess about the rest.

EXAMPLE 1

The graph of $y = x^2$ consists of all points (x, x^2), where x is a real number. You can easily construct a table of values and plot the corresponding points, as in Figure 2–1.

x	$y = x^2$
-2.5	6.25
-2	4
-1.5	2.25
-1	1
$-.5$	.25
0	0
.5	.25
1	1
1.5	2.25
2	4
2.5	6.25

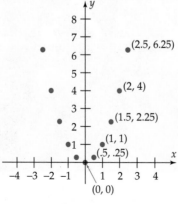

Figure 2–1

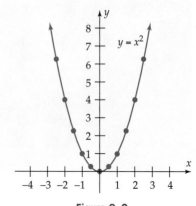

Figure 2–2

78

These points suggest that the graph looks like the one in Figure 2–2, which is obtained by connecting the plotted points and extending the graph upward. ∎

GRAPHING WITH TECHNOLOGY

Graphing calculators and computer graphing programs use no algebraic reasoning when sketching graphs. They plot 95 or more points and simultaneously connect them with line segments. These graphs are generally quite accurate, but no technology is perfect. Algebra and geometry may be needed to interpret misleading or incorrect screen images. The basic procedure for graphing with technology is summarized in the following box and explained in Example 2 below.

Graphing Equations with Technology

1. Solve the equation for y and enter it in the equation memory.

2. Set the viewing window—the portion of the coordinate plane that will appear on the screen.

3. Graph the equation.

4. If necessary, adjust the viewing window for a better view.

EXAMPLE 2

Graph the equation

$$2x^3 - 8x - 2y + 4 = 0$$

(a) using a calculator;

(b) using a computer graphing program.

SOLUTION

(a) For calculator graphing, use the four steps in the preceding box.

Step 1   Solve the equation for y:

$$2x^3 - 8x - 2y + 4 = 0$$

Rearrange terms: $$-2y = -2x^3 + 8x - 4$$

Divide by -2: $$y = x^3 - 4x + 2.$$

Now call up the **equation memory** by pressing Y= on TI (or SYMB on HP-39gs or GRAPH (main menu) on Casio).* Use the Technology Tip in the margin to enter the equation, as shown in Figure 2–3.

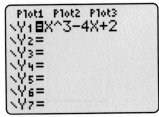

Figure 2–3

TECHNOLOGY TIP

When entering an equation for graphing, use the "variable" key rather than the ALPHA X key. It has a label such as X, T, θ or X, T, θ, n or X, θ, T or x-VAR. On TI-89, however, use the x key on the keyboard.

---

*On TI-86, press GRAPH first, then y(x)= will appear as a menu choice.

Step 2 Since we don't know where the graph lies, we'll use the *standard window*—the one with $-10 \leq x \leq 10$ and $-10 \leq y \leq 10$. We can change it later if necessary. Press WINDOW or WIND on TI (or V-WINDOW on Casio or PLOT SETUP on HP), and enter the appropriate numbers, as in Figure 2–4. Figure 2–5 shows how these entries determine the portion of the plane to be shown and the placement of axis tick marks.*

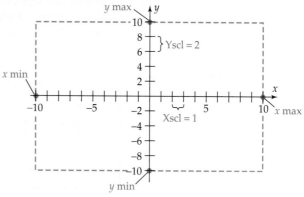

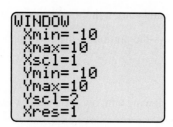

Figure 2–4

Figure 2–5

TECHNOLOGY TIP

On TI-83/86 you may get an error message when you press GRAPH if one of Plot 1, Plot 2, or Plot 3, at the top of the Y= menu (see Figure 2–3) is shaded. In this case, move the cursor over the shading and press ENTER to remove it.

Step 3 Press GRAPH on TI (or DRAW on Casio or PLOT on HP) to obtain Figure 2–6. Because of the limited resolution of a calculator screen, the graph appears to consist of short adjacent line segments rather than a smooth unbroken curve.

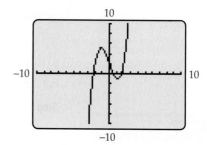

Figure 2–6

Step 4 The graph in Figure 2–6 is squeezed into the middle of the screen. So we change the viewing window (Figure 2–7) and press GRAPH again to obtain Figure 2–8.

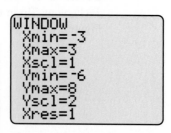

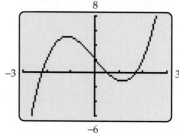

Figure 2–7

Figure 2–8

---

*Xscl is labeled "X scale" on Casio and "Xtick" on HP-39gs. Xres should normally be set at 1 on TI and "res" should normally be set at "detail" on HP. Casio has no resolution setting.

(b) The procedure for a typical computer graphing program is basically the same as for calculators, but various details may be a bit different. So check your instruction manual or help key index. On Maple, for example, the following command produces Figure 2–9, in which the range of y-values was chosen automatically by Maple.

$$\text{plot}(x\text{\textasciicircum}3 - 4*x + 2, x = -3..3);$$

To duplicate Figure 2–8, we specify the range of y-values and the number of tick marks on each axis:

$$\text{plot}(x\text{\textasciicircum}3 - 4*x + 2, x = -3..3, y = -6..8,$$
$$\text{xtickmarks} = 6, \text{ytickmarks} = 6);$$

The result is Figure 2–10.
   As is typical of computer-generated graphs, this one appears smooth and connected, as it should. ∎

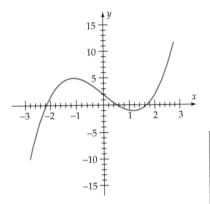

Figure 2–9

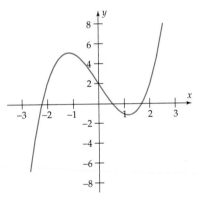

Figure 2–10

NOTE

An equation stays in a calculator's equation memory until you delete it. When several equations are in the memory, you must turn "on" those you want graphed and turn "off" those you don't want graphed. An equation is "on" if its equal sign is shaded or if there is a check mark next to it. Only equations that are "on" are graphed when you press GRAPH. To turn an equation "on" or "off" on TI-84+, move the cursor over the equal sign and press ENTER. On other calculators, move the cursor to the equation and press SELECT or CHECK.

GRAPHING TOOLS: THE TRACE FEATURE

A calculator obtains a graph by plotting points and simultaneously connecting them. To see which points were actually plotted, press TRACE, and a flashing cursor appears on the graph. Use the left and right arrow keys to move the cursor along the graph. The coordinates of the point the cursor is on appear at the bottom of the screen. Figure 2–11 illustrates this for the graph of $y = x^3 - 4x + 2$.

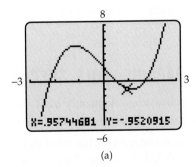

(a)

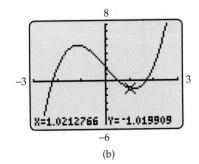

(b)

Figure 2–11

The trace cursor displays only the points that the calculator actually plotted. For instance, $(1, -1)$ is on the graph of $y = x^3 - 4x + 2$, as you can easily verify, but was not one of the points the calculator plotted to produce Figure 2–11. So the trace lands on two nearby points that *were* plotted, but skips $(1, -1)$.

GRAPHING TOOLS: ZOOM IN/OUT

The ZOOM menu makes it easy to change the size of the viewing window.

EXAMPLE 3

How many x-intercepts does the graph of $y = x^3 - 1.8x + .97$ have between -5 and 5?

SOLUTION   The graph of this equation in Figure 2–12 suggests that there is one x-intercept near -1.5 and another one between 0 and 1. But appearances can be deceiving. Use the Technology Tip in the margin to set the zoom factors at 10, then move the cursor to the apparent positive x-intercept. Choose *Zoom In* in the ZOOM menu and press ENTER. The result is Figure 2–13, which shows clearly that the graph does not touch the x-axis there. So there is only one x-intercept. ∎

> **TECHNOLOGY TIP**
>
> If ZOOM is not on the keyboard, it will appear on screen after you press GRAPH (or PLOT or DRAW).
>
> To set the zoom factors, look for *Fact, ZFact,* or *(Set) Factors* in the ZOOM menu (or its MEMORY submenu).

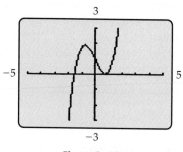

Figure 2–12

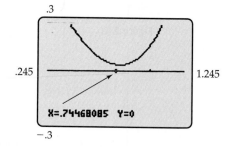

X=.74468085   Y=0

Figure 2–13

The calculator automatically changes the range of x- and y-values when zooming, but does not change the Xscl or Yscl settings. This may cause occasional viewing problems.

> **NOTE**
> The heading GRAPHING EXPLORATION indicates that you are to use your calculator or computer as directed to complete the discussion.

GRAPHING EXPLORATION

Graph $y = x^3 - 1.8x + .97$ in the same window as Figure 2–12. Then zoom *out* from the origin by a factor of 10. Can you read the tick marks on the axes? Change the settings to Xscl = 5 and Yscl = 5, and regraph. Can you read them now?

GRAPHING TOOLS: SPECIAL VIEWING WINDOWS

On most calculators, frequently used viewing windows can be obtained with a single keystroke. These include the following.

1. The **standard viewing window** has $-10 \le x \le 10$ and $-10 \le y \le 10$. It is labeled *ZStandard, ZStd,* or *ZoomStd* in the ZOOM menu on TI, and *Std* in the V-WINDOW menu on Casio.

TECHNOLOGY TIP

For a decimal window in which both the horizontal and vertical distance between adjacent pixels is .1, use this menu/choice

TI-84+: ZOOM/*Zdecimal*
TI-86: ZOOM/*Zdecm*
TI-89: ZOOM/*ZoomDec*
Casio 9850: V-WINDOW/*Init*
HP-39gs: ZOOM/*Decimal*

On these windows you may need to change the Ymin and Ymax settings to get the full picture.

TECHNOLOGY TIP

For an approximately square window, the y-axis should be 2/3 as long as the x-axis on TI-84+ and 3/5 as long on TI-86. It should be half as long on TI-89, Casio, and HP-39gs.
    On calculators other than TI-86, the decimal window is a square window.

2. A **decimal window** is one in which the *horizontal* distance between two adjacent pixels is .1. The width of a decimal window depends on the width of your calculator screen. To find this width, call up the preset decimal window (see the Technology Tip) and look in WINDOW. For example, the preset decimal window in the TI-84+ has $-4.7 \le x \le 4.7$, so its width is $4.7 - (-4.7) = 9.4$. Consequently, any TI-84+ window with Xmax− Xmin = 9.4 (such as $0 \le x \le 9.4$ or $5 \le x \le 14.4$) is a decimal window.

GRAPHING EXPLORATION

Graph $y = x^4 - 2x^2 - 2$ in the standard window. Use the TRACE key, and watch the values of the x-coordinates. Now regraph in a decimal window (use the Technology Tip). Then use the TRACE key again. How do the x-coordinates change at each step? Finally, look in WINDOW to find the width of your decimal window.

3. In a **square window,** a one-unit segment on the x-axis has the same length on the screen as a one-unit segment on the y-axis. Because calculator screens are wider than they are high, the y-axis in a square window must be shorter than the x-axis (see the Technology Tip in the margin).

EXAMPLE 4

Graph the circle $x^2 + y^2 = 9$ on a calculator.

SOLUTION   First, we solve the equation for y:

$$y^2 = 9 - x^2$$
$$y = \sqrt{9 - x^2} \qquad \text{or} \qquad y = -\sqrt{9 - x^2}.$$

Graphing both of these equations on the same screen will produce the graph of the circle (Figure 2–14). However, the graph does not look like a circle because the standard window in Figure 2–14 is *not* square (a one-unit segment on the x-axis is a tad longer than a one-unit segment on the y-axis).

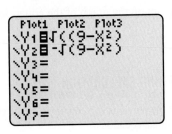 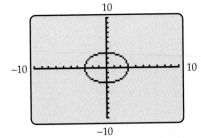

Figure 2–14

If we select *Square* (or *ZSquare, ZSqr, ZoomSqr,* or *Sqr*) in the ZOOM menu, the length of the x-axis is adjusted to produce a square window in which the circle looks round (Figure 2–15 on the next page).* Alternatively, we can change the WINDOW settings by hand to obtain a square window such as Figure 2–16.† ∎

---

*On HP-39gs, SQUARE adjusts the length of the y-axis to produce a square window.
†The gap between the top and bottom of the circle (more apparent here than in Figure 2–15) is caused by the low resolution of the calculator screen.

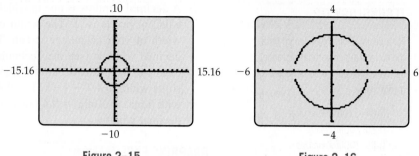

Figure 2–15   Figure 2–16

Although any convenient viewing window is usually OK, you should use a square window when you want circles to look round and perpendicular lines to look perpendicular.

GRAPHING EXPLORATION

The lines $y = .5x$ and $y = -2x + 2$ are perpendicular (why?). Graph them in the standard viewing window. Do they look perpendicular? Now graph them in a square window. Do they look perpendicular?

The method used to graph the circle $x^2 + y^2 = 9$ in Example 4 can be used to graph any equation that can be solved for y.

EXAMPLE 5

To graph $12x^2 - 4y^2 + 16x + 12 = 0$, solve the equation for y:

$$4y^2 = 12x^2 + 16x + 12$$
$$y^2 = 3x^2 + 4x + 3$$
$$y = \pm\sqrt{3x^2 + 4x + 3}.$$

Every point on the graph of the equation is on the graph of either

$$y = \sqrt{3x^2 + 4x + 3} \qquad \text{or} \qquad y = -\sqrt{3x^2 + 4x + 3}.$$

GRAPHING EXPLORATION

Graph the previous two equations on the same screen. The result will be the graph of the original equation.

■

GRAPHING TOOLS: THE MAXIMUM/MINIMUM FINDER

Many graphs have peaks and valleys (for instance, see Figure 2–12 on page 82 or Figure 2–17 on the next page). A calculator's maximum finder or minimum finder can locate these points with a high degree of accuracy, as illustrated in the next example.

EXAMPLE 6

The Cortopassi Computer Company can produce a maximum of 100,000 computers a year. Their annual profit is given by

$$y = -.003x^4 + .3x^3 - x^2 + 5x - 4000,$$

where y is the profit (in thousands of dollars) from selling x thousand computers. Use graphical methods to estimate how many computers should be sold to make the largest possible profit.

SOLUTION   We first choose a viewing window. The number x of computers is nonnegative, and no more than 100,000 can be produced, so that $0 \le x \le 100$ (because x is measured in thousands). The profit y may be positive or negative (the company could lose money). So we try a window with $-25,000 \le y \le 25,000$ and obtain Figure 2–17. For each point on the graph,

The x-coordinate is the number of thousands of computers produced;

The y-coordinate is the profit (in thousands) on that number of computers.

The largest possible profit occurs at the point with the largest y-coordinate, that is, the highest point in the window. The maximum finder on a TI-84+ (see the Technology Tip in the margin) produced Figure 2–18. Since x and y are measured in thousands, we see that making about 72,789 computers results in a maximum profit of about $22,547,757. ∎

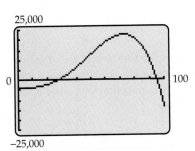

25,000

0 ┤————————— 100

−25,000

Figure 2–17

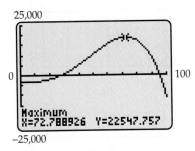

25,000

0 ┤————————— 100

Maximum
X=72.788926 Y=22547.757

−25,000

Figure 2–18

GRAPHING EXPLORATION

Graph $y = .3x^3 + .8x^2 - 2x - 1$ in the window with $-5 \le x \le 5$ and $-5 \le y \le 5$. Use your maximum finder to approximate the coordinates of the highest point to the left of the y-axis. Then use your minimum finder (in the same menu) to approximate the coordinates of the *lowest* point to the right of the y-axis. How do these answers compare with the ones you get by using the trace feature?

COMPLETE GRAPHS

A viewing window is said to display a **complete graph** if it shows all the important features of the graph (peaks, valleys, intercepts, etc.) and suggests the general shape of the portions of the graph that aren't in the window. Many different windows may show a complete graph. It's usually best to use a window that is small enough to show as much detail as possible.

    In later chapters we shall develop algebraic facts that will enable us to know when certain graphs are complete. For the present, however, the best you may be able to do is try several different windows to see which, if any, appear to display a complete graph.

EXAMPLE 7

Sketch a complete graph of

$$y = .007x^5 - .2x^4 + 1.332x^3 - .004x^2 + 10.$$

SOLUTION Four different viewing windows for this graph are shown in Figure 2–19. Graph (a) (the standard window) is certainly not complete, since it shows no points to the right of the *y*-axis. Graph (b) is not complete, since it indicates that parts of the graph lie outside the window.

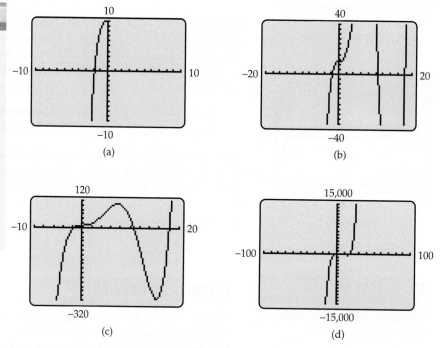

Figure 2–19

Graph (d) tends to confirm what Graph (c) suggests: that the graph keeps climbing sharply forever as you move to the right and that it keeps falling sharply forever as you move to the left. Because of its large scale, however, graph (d) doesn't show the features of the graph near the origin. So we conclude that graph (c) is probably a complete graph, since it shows important features (twists and turns) near the origin, as well as suggesting the shape of the graph farther out. ∎

EXAMPLE 8

If you graph $y = -2x^3 + 26x^2 + 18x + 50$ in the standard viewing window, you get a blank screen (try it!). In such cases, you can usually find at least one point on the graph by setting $x = 0$ and determining the corresponding value of y (the y-intercept of the graph). If $x = 0$ here, then $y = 50$, so the point $(0, 50)$ is on

the graph. Consequently, the y-axis of our viewing window should extend well
beyond 50. An alternative method of finding some points on the graph is in the
Technology Tip in the margin.

GRAPHING EXPLORATION

Find a complete graph of this equation. [*Hint:* The graph crosses the x-axis once and
has one "peak" and one "valley."]

∎

SCATTER PLOTS AND LINE GRAPHS

A somewhat different procedure must be used to graph scatter plots and construct
line graphs on a calculator.

EXAMPLE 9

The winning speeds for the Indianapolis 500 race (rounded to the nearest mph) are
shown in the table.

Year	2000	2001	2002	2003	2004	2005	2006
Speed	168	142	166	156	139	158	157

For this data, construct a

(a) scatter plot

(b) line graph.

SOLUTION   Consult the Technology Tip after the example for directions on
how to carry out the steps listed below.

(a) Let $x = 0$ correspond to 2000, so that the data points are (0, 168), (1, 142),
(2, 166), and so on. Enter these points in the statistics editor as two lists
(x-coordinates in the first list and the corresponding y-coordinates in the
second), as shown in Figure 2–20. In the stat plot setup screen (Figure 2–21),
select "scatter plot" as the type. Enter an appropriate viewing window and
press GRAPH to obtain the scatter plot in Figure 2–22.

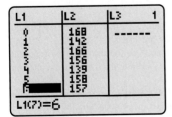

Figure 2–20

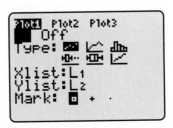

Figure 2–21

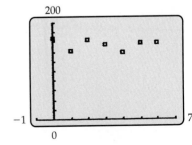

Figure 2–22

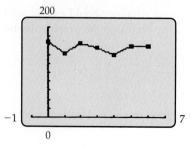

Figure 2–23

(b) For a line graph of the data, change the type to "line graph" in the stat plot setup screen. Then press GRAPH to obtain Figure 2–23. ∎

TECHNOLOGY TIP

Most calculators allow you to store three or more statistics graphs (identified by number); the following directions assume that the first one is used.

To call up the statistics editor, use these commands:

    TI-84+: STAT   EDIT   (lists are L_1, L_2, . . .);

    TI-86: STAT   EDIT   (built-in lists are *x*-stat, *y*-stat; it is usually better to create your own lists; we use L1 and L2 as list names here);*

    TI-89: APPS   DATA/MATRIX EDITOR   NEW; then choose DATA as the TYPE, enter a VARIABLE name (we use L here), and key in ENTER (lists are C_1, C_2, . . .);

    Casio 9850: STAT   (lists are List 1, List 2, . . .);

    HP-39gs: APLET   STATISTICS   (lists are C_1, C_2, . . .).

Enter the data points as two lists (*x*-coordinates in the first, and the corresponding *y*-coordinates in the second).

    To create a scatter plot or line graph of the data points (after the lists have been entered), call up the stat plot setup screen with these commands:

    TI-84+: STAT PLOT (on keyboard)   PLOT 1;

    TI-86: PLOT (on STAT menu)   PLOT 1;

    TI-89: PLOT-SETUP and DEFINE (in the Data Editor);

    Casio 9850: GRPH   SET (on the STAT list screen);

    HP-39gs: PLOT-SETUP (on the keyboard)

Enter the request information, such as graph number, lists to be used, the mark used for the data points, and the plot type.[†] Finally, select a viewing window and press GRAPH on TI (or GPH 1 on Casio or PLOT on HP).[‡]

---

*When statistical computations are run on TI-86, the lists used are automatically copied into the *x*-stat and *y*-stat lists, replacing whatever was there before. So use the *x*-stat and *y*-stat lists only if you don't want to save them. See your instruction manual to find out how to create new lists in the statistics editor.

[†]Types are shown schematically on TI-84+. The shaded one in Figure 2–21 indicates a scatter plot and the one to its right a line graph. Scatter plots and line graphs are listed as "scat" and "*xy* line" respectively on TI-86 and Casio 9850. On HP-39gs, check "connect" on the second page of the stat plot setup screen for a line graph and uncheck it for a scatter plot. To enter the lists to be graphed on HP, press SYMB (ignore the "fit" setting).

[‡]On Casio, the viewing window will be chosen automatically unless "stat wind" in the main setup menu (SHIFT MENU) is set to "manual" (recommended). On HP, the viewing window is selected in the plot setup screen. If a line or curve appears on the scatter plot, press MENU and FIT to remove it.

EXERCISES 2.1

In Exercises 1–6, graph the equation by hand by plotting no more than six points and filling in the rest of the graph as best you can. Then use the calculator to graph the equation and compare the results.

1. $y = |x - 2|$ 2. $y = \sqrt{x + 5}$

3. $y = x^2 - x$ 4. $y = x^2 + x + 1$

5. $y = x^3 + 1$ 6. $y = \dfrac{1}{x}$

In Exercises 7–12, find the graph of the equation in the standard window.

7. $3 + y = .5x$ 8. $y - 2x = 4$

9. $y = x^2 - 5x + 2$ 10. $y = .3x^2 + x - 4$

11. $y = .2x^3 + .1x^2 - 4x + 1$

12. $y = .2x^4 - .2x^3 - 2x^2 - 2x + 5$

13. (a) Graph

$$y = \frac{1}{x^2 + 1}$$

   in the standard window.

   (b) Does the graph appear to stop abruptly partway along the x-axis? Use the trace feature to explain why this happens. [*Hint:* In this viewing window, each pixel represents a rectangle that is approximately .32 unit high.]

   (c) Find a viewing window with $-10 \le x \le 10$ that shows a complete graph that does not fade into the x-axis.

14. (a) Graph $y = x^3 - 2x^2 + x - 2$ in the standard window.

   (b) Use the trace feature to show that the portion of the graph with $0 \le x \le 1.5$ is not actually horizontal. [*Hint:* All the points on a horizontal segment must have the same y-coordinate (why?).]

   (c) Find a viewing window that clearly shows that the graph is not horizontal when $0 \le x \le 1.5$.

In Exercises 15–24, use the techniques of Examples 4 and 5 to graph the equation in a suitable square viewing window.

15. $x^2 + y^2 = 16$ 16. $y^2 = x - 2$

17. $3x^2 + 2y^2 = 48$

18. $25(x - 5)^2 + 36(y + 4)^2 = 900$

19. $(x - 4)^2 + (y + 2)^2 = 25$

20. $9x^2 + 4y^2 = 36$ 21. $4x^2 - 9y^2 = 36$

22. $9y^2 - x^2 = 9$ 23. $9x^2 + 5y^2 = 45$

24. $x = y^2 - 2$

25. Use your minimum finder to approximate the x-coordinates of the lowest point on the graph of $y = x^3 - 2x + 5$ in the

window with $0 \le x \le 5$ and $-3 \le y \le 8$. The correct answer is

$$x = \sqrt{\frac{2}{3}} \approx .816496580928.$$

How good is your approximation?

In Exercises 26 and 27, use zoom-in or a maximum/minimum finder to determine the highest and lowest point on the graph in the given window.

26. $y = .07x^5 - .3x^3 + 1.5x^2 - 2$ $(-3 \le x \le 2$ and $-6 \le y \le 6)$

27. $y = .04x^4 + .01x^3 - 1.7x^2 + 2.5x + 3$ $(0 \le x \le 5$ and $-5 \le y \le 5)$

28. According to United Nations projections, the population of China (in millions) between now and 2050 is given by

$$y = -.00096x^3 - .1x^2 + 11.3x + 1274 \quad (0 \le x \le 50),$$

   where $x = 0$ corresponds to 2000. In what year will China reach its maximum population, and what will that population be?

29. The national average interest rate for a 30-year fixed rate mortgage is approximated by

$$y = .015x^3 + .112x^2 - 1.562x + 8.67 \quad (0 \le x \le 6),$$

   where $x = 0$ corresponds to 2000.* According to this model, when was the rate the lowest? What was the lowest rate?

30. The following equation gives the approximate number of thefts at Cleveland businesses each hour of the day:

$$y = .00357x^4 - .3135x^3 + 6.87x^2 - 38.3x + 118.4$$
$$(0 \le x \le 23),$$

   where x is measured in hours after midnight.[†]

   (a) About how many thefts occurred around 5 A.M.?
   (b) When did the largest number of thefts occur?

In Exercises 31–36, determine which of the following viewing windows gives the best view of the graph of the given equation.

   (a) $-10 \le x \le 10;$ $-10 \le y \le 10$
   (b) $-5 \le x \le 25;$ $0 \le y \le 20$
   (c) $-10 \le x \le 10;$ $-100 \le y \le 100$
   (d) $-20 \le x \le 15;$ $-60 \le y \le 250$
   (e) None of a, b, c, d gives a complete graph.

31. $y = 18x - 3x^2$ 32. $y = 4x^2 + 80x + 350$

33. $y = \dfrac{1}{3}x^3 - 25x + 100$ 34. $y = x^4 + x - 5$

35. $y = x^2 + 50x + 625$ 36. $y = .01(x - 15)^4$

*http://mortgage-x.com
†Cleveland Police Department

In Exercises 37–42, obtain a complete graph of the equation by trying various viewing windows. List a viewing window that produces this complete graph. (Many correct answers are possible; consider your answer to be correct if your window shows all the features in the window given in the answer section.)

37. $y = 7x^3 + 35x + 10$ **38.** $y = x^3 - 5x^2 + 5x - 6$

39. $y = \sqrt{x^2} - x$ **40.** $y = 1/x^2$

41. $y = -.1x^4 + x^3 + x^2 + x + 50$

42. $y = .002x^5 + .06x^4 - .001x^3 + .04x^2 - .2x + 15$

In Exercises 43–46, use technology to construct a scatter plot and a line graph of the data.

43. Last year's electric bills for one of the authors are shown in the table. Let $x = 3$ correspond to March, $x = 4$ to April, etc.

Month	Bill ($)
March	61
April	50
May	116
June	187
July	149
August	182
September	77

44. The table shows the population of Kansas City, Missouri in various years.* Let $x = 0$ correspond to 1950.

Year	1950	1960	1970	1980	1990	2004
Population	457	476	507	448	435	444

45. The table gives the average SAT math score in 2005 for students in states with a population of 5 to 6 million people.[†] Let $x = 1$ be the first state on the list, $x = 2$ the second, etc.

State	Score
Arizona	530
Indiana	508
Maryland	515
Minnesota	597
Missouri	588
Tennessee	563
Washington	534
Wisconsin	599

46. The average monthly rainfall (in inches) in Cleveland, Ohio (based on a thirty year average) is shown in the table.* Let $x = 1$ correspond to January, $x = 3$ to March, etc.

Month	Rainfall
January	2.5
March	2.9
May	3.5
July	3.5
September	3.8
November	3.4

47. (a) Graph $y = -.3(x - 3)^4 + 8$ in the standard window.
  (b) To see exactly which points the calculator actually graphed, change the graphing mode to *Dot* (or *Draw Dot* or *Plot*) in the menu/submenu list below, and graph again.

  TI-84+: MODE
  TI-86: GRAPH/FORMAT
  TI-89: Y=/STYLE
  Casio: SETUP/DRAW TYPE
  HP-39gs: uncheck "connect" on the second page of the PLOT SETUP menu.

  (c) Why does the graph in part (b) look "solid" at the top, but consists of isolated points elsewhere?

48. (a) Graph $y = .3x^3 - 2x^2 + 6$ in the standard window.
  (b) Use trace to move to a point whose x-coordinate is close to 1.
  (c) Set the zoom factors of your calculator to 10. Zoom-in once or twice. Does the graph appear to be a straight line near the point?
  (d) Repeat parts (a)—(c) at lowest point to the right of the y-axis. Is the result the same? If not, keep zooming in until it is (at each stage move the flashing cursor up or down, so it is on the graph).
  (e) What do parts (a)—(d) suggest about the graph?

In Exercises 49–54, use your algebraic *knowledge to state whether or not the two equations have the same graph. Confirm your answer by graphing the equations in the standard window.*

49. $y = |x + 3|$ and $y = |x| + 3$

50. $y = |x| - 4$ and $y = |x - 4|$

51. $y = \sqrt{x^2}$ and $y = |x|$

52. $y = \sqrt{x^2 + 6x + 9}$ and $y = |x + 3|$

53. $y = \sqrt{x^2 + 9}$ and $y = x + 3$

*U.S. Census Bureau
[†]The College Board

*National Climatic Data Center

54. $y = \dfrac{1}{x^2 + 2}$ and $y = \dfrac{1}{x^2} + \dfrac{1}{2}$

55. (a) Confirm the accuracy of the factorization $x^2 - 5x + 6 = (x - 2)(x - 3)$ graphically. [*Hint:* Graph $y = x^2 - 5x + 6$ and $y = (x - 2)(x - 3)$ on the same screen. If the factorization is correct, the graphs will be identical (which means that you will see only a single graph on the screen).]

  (b) Show graphically that $(x + 5)^2 \neq x^2 + 5^2$. [*Hint:* Graph $y = (x + 5)^2$ and $y = x^2 + 5^2$ on the same screen. If the graphs are different, then the two expressions cannot be equal.]

True or False. *In Exercises 56–58, use the technique of Exercise 55 to determine graphically whether the given statement is possibly true or definitely false. (We say "possibly true" because two graphs that appear identical on a calculator screen may actually differ by small amounts or at places not shown in the window.)*

56. $x^3 - 7x - 6 = (x + 1)(x + 2)(x - 3)$

57. $(1 - x)^6 = 1 - 6x + 15x^2 - 20x^3 + 15x^4 - 6x^5 + x^6$

58. $x^5 - 8x^4 + 16x^3 - 5x^2 + 4x - 20$
$$= (x - 2)^2(x - 5)(x^2 + x + 1)$$

59. A toy rocket is shot straight up from ground level and then falls back to earth; wind resistance is negligible. Use your calculator to determine which of the following equations has a graph whose portion above the x-axis provides the most plausible model of the path of the rocket.

  (a) $y = .1(x - 3)^3 - .1x^2 + 5$
  (b) $y = -x^4 + 16x^3 - 88x^2 + 192x$
  (c) $y = -16x^2 + 117x$
  (d) $y = .16x^2 - 3.2x + 16$
  (e) $y = -(.1x - 3)^6 + 600$

60. Monthly profits at DayGlo Tee Shirt Company appear to be given by the equation

$$y = -.00027(x - 15,000)^2 + 60,000,$$

where x is the number of shirts sold that month and y is the profit. DayGlo's maximum production capacity is 15,000 shirts per month.

  (a) If you plan to graph the profit equation, what range of x values should you use? [*Hint:* You can't make a negative number of shirts.]
  (b) The president of DayGlo wants to motivate the sales force (who are all in the profit-sharing plan), so he asks you to prepare a graph that shows DayGlo's profits increasing *dramatically* as sales increase. Using the profit equation and the x range from part (a), what viewing window would be suitable?
  (c) The City Council is talking about imposing more taxes. The president asks you to prepare a graph showing that DayGlo's profits are essentially flat. Using the profit equation and the x range from part (a), what viewing window would be suitable?

In each of the applied situations in Exercises 61–64, find an appropriate viewing window for the equation (that is, a window that includes all the points relevant to the problem but does not include large regions that are not relevant to the problem, and has easily readable tick marks on the axes). Explain why you chose this window. See the Hint in Exercise 60(a).

61. Beginning in 1905 the deer population in a region of Arizona rapidly increased because of a lack of natural predators. Eventually food resources were depleted to such a degree that the deer population completely died out. In the equation $y = -.125x^5 + 3.125x^4 + 4000$, y is the number of deer in year x, where $x = 0$ corresponds to 1905.

62. A cardiac test measures the concentration y of a dye x seconds after a known amount is injected into a vein near the heart. In a normal heart

$$y = -.006x^4 + .14x^3 - .053x^2 + 179x.$$

63. The concentration of a certain medication in the bloodstream at time x hours is approximated by the equation

$$y = \frac{375x}{.1x^3 + 50},$$

where y is measured in milligrams per liter. After two days the medication has no effect.

64. A winery can produce x barrels of red wine and y barrels of white wine, where

$$y = \frac{200,000 - 50x}{2000 + x}.$$

In Exercises 65–67, use the viewing windows found in Exercises 61–63.

65. (a) Use the trace feature to estimate the year when the deer population in Exercise 61 first reached 40,000. In what later year was it also 40,000?
  (b) When was the deer population at its maximum? Approximately how many deer were there at that time?

66. (a) In the cardiac test of Exercise 62, use the trace feature to estimate the time at which all dye is gone from the body.
  (b) At what time during the test was the concentration of the dye the greatest?

67. (a) When was the concentration of the medication in Exercise 63 at its peak?
  (b) Use the trace feature to determine approximately when the concentration dropped below 10 milligrams per liter and stayed below that level.

68. The total resources (in billions of dollars) of the Pension Benefit Guaranty Corporation, the government agency that insures pensions, is approximated by

$$y = -.279x^2 + 4.006x + 28.412 \quad (4 \leq x \leq 20),$$

where $x = 4$ corresponds to 2004.*

(continued)

---

*Center on Federal Financial Institutions

(a) When are resources the greatest?

(b) Use the trace feature to find the approximate time when the Corporation will run out of money.

In Exercises 69–72, graph all four equations on the same screen, using a sufficiently large square viewing window, and answer this question: What is the geometric relationship of graphs (b), (c), and (d) to graph (a)?

69. (a) $y = x^2$ (b) $y = x^2 + 5$
     (c) $y = x^2 - 5$ (d) $y = x^2 - 2$

70. (a) $y = \sqrt{x}$ (b) $y = \sqrt{x - 3}$
     (c) $y = \sqrt{x + 3}$ (d) $y = \sqrt{x - 6}$

71. (a) $y = \sqrt{x}$ (b) $y = 2\sqrt{x}$
     (c) $y = 3\sqrt{x}$ (d) $y = \frac{1}{2}\sqrt{x}$

72. (a) $y = x^2$ (b) $y = -x^2$
     (c) $y = -\frac{1}{2}x^2$ (d) $y = -2x^2$

In Exercises 73–75, graph the two given equations and the equation y = x on the same screen, using a sufficiently large square viewing window, and answer this question: What is the geometric relationship between graphs (a) and (b)?

73. (a) $y = x^3$ (b) $y = \sqrt[3]{x}$

74. (a) $y = \frac{1}{2}x^3 - 4$ (b) $y = \sqrt[3]{2x + 8}$

75. (a) $y = 5x - 15$ (b) $y = .2x + 3$

76. Put your calculator in *radian* mode, and use the viewing window given by $0 \le x \le 6.28$ and $-2 \le y \le 2$.*

(a) Graph $y = \sin x$

(b) Graph $y = \sin (2x)$

(c) Graph $y = \sin (3x)$

(d) On the basis of parts (a)–(c), what do you think the graphs of $y = \sin (4x)$, $y = \sin (5x)$, $y = \sin (6x)$, and so on, will look like? Use the calculator to verify your answer.

(e) On the basis of part (d), what do you think the graphs of $y = \sin (50x)$ and $y = \sin (100x)$ will look like? What does a calculator display instead? What might explain the graphs of the calculator?

---

*You don't need to know what radian mode is or what "sin" means. Just use the calculator key with this label.

2.2   Solving Equations Graphically and Numerically

Section Objectives
- ■ Use technology to solve equations graphically and numerically.
- ■ Solve equations by the intercept method.
- ■ Solve equations the intersection method.
- ■ Use an equation solver to solve equations numerically.
- ■ Learn strategies for solving radical and fractional equations with technology.

Algebraic techniques can be used to solve linear, quadratic, and some higher-degree equations, as we saw in Section 1.2. For many other equations, however, graphical and numerical approximation methods are the only practical alternatives. These methods are based on a connection between equations and graphs that we now examine.

Recall that the *x*-intercepts of the graph of

$$y = x^4 - 4x^3 + 3x^2 - x - 2$$

are the *x*-coordinates of the points where the graph intersects the *x*-axis (see Figure 2–24). Since points on the *x*-axis have 0 second coordinates, the *x*-intercepts are found algebraically by setting $y = 0$ and solving for *x*, that is, by solving the equation

$$x^4 - 4x^3 + 3x^2 - x - 2 = 0.$$

Graphical equation solving amounts to running this in the opposite direction: An equation is solved by finding the *x*-intercepts, as illustrated in the next example.

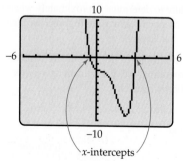

Figure 2–24

EXAMPLE 1

Solve $x^4 - 4x^3 + 3x^2 - x - 2 = 0$ graphically.

SOLUTION   As we have just seen, the solutions are the x-intercepts of the graph of

$$y = x^4 - 4x^3 + 3x^2 - x - 2.$$

These x-intercepts may be found by graphing the equation and using the graphical root finder, as in Figure 2–25. (See the Technology Tip in the margin.)

TECHNOLOGY TIP

The graphical root finder is labeled *Root* or *Zero* and is in this menu/submenu:

TI-84+: CALC

TI-86/89: GRAPH/MATH

Casio: DRAW/G-SOLV

HP-39gs: PLOT/FCN

On most TI calculators, you must select a left and a right bound (x-values to the left and right of the intercept) and make an initial guess.

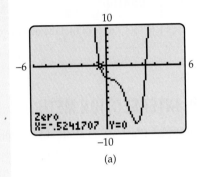

(a)

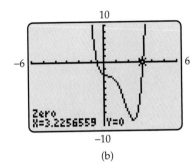

(b)

Figure 2–25

Therefore, the approximate solutions of the original equation are $x \approx -.5242$ and $x \approx 3.2257$. ∎

The process in Example 1 works in the general case as well.

The Intercept Method

To solve a one-variable equation of the form,

$$\text{Expression in } x = 0,$$

1. Graph the two-variable equation

$$y = \text{expression in } x.$$

2. Use a graphical root finder to determine the x-intercepts of the graph.

The x-intercepts are the real solutions of the equation.

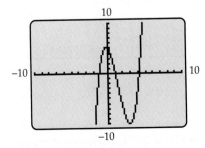

Figure 2–26

EXAMPLE 2

Solve $x^3 - 4x^2 - 2x + 5 = 0$.

SOLUTION   We graph $y = x^3 - 4x^2 - 2x + 5$ (Figure 2–26) and see that it has three x-intercepts. So the equation has three real solutions. A root finder shows that one solution is $x \approx -1.1926$ (Figure 2–27 on the next page). ∎

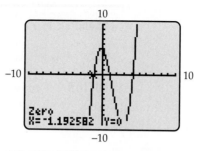

Figure 2–27

Use your graphical root finder to find the other two solutions (*x*-intercepts) of the equation in Example 2.

THE INTERSECTION METHOD

The next examples illustrate an alternative graphical method, which is sometimes more convenient for solving equations.

EXAMPLE 3

Solve $|x^2 - 4x - 3| = x^3 + x - 6$.

SOLUTION   Let $y_1 = |x^2 - 4x - 3|$ and $y_2 = x^3 + x - 6$, and graph both equations on the same screen (Figure 2–28). Consider the point where the two graphs intersect. Since it is on the graph of y_1, its second coordinate is $|x^2 - 4x - 3|$, and since it is also on the graph of y_2, its second coordinate is $x^3 + x - 6$. So for this number x, we must have $|x^2 - 4x - 3| = x^3 + x - 6$. In other words, *the x-coordinate of the intersection point is the solution of the equation.*

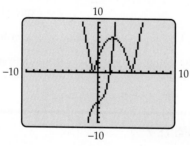

Figure 2–28

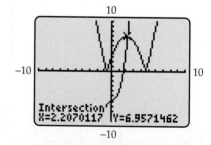

Figure 2–29

TECHNOLOGY TIP

The graphical intersection finder is in the same menu/submenu as the root finder. It is labeled *Intersection, Intersect, Isect,* or *ISCT.*

This coordinate can be approximated by using a graphical intersection finder (see the Technology Tip in the margin), as shown in Figure 2–29. Therefore, the solution of the original equation is $x \approx 2.207$. ∎

GRAPHING EXPLORATION

Show that the equation $x^2 - 2x - 6 = \sqrt{2x + 7}$ has two real solutions by graphing the left and right sides in the standard window and counting the number of intersection points. The positive solution is $x \approx 4.3094$. Find the negative solution.

The technique used in Example 3 and the preceding exploration works in the general case

The Intersection Method

To solve an equation of the form

$$\text{First expression in } x = \text{Second expression in } x,$$

1. Set y_1 equal to the left side and y_2 equal to the right side.

2. Graph y_1 and y_2 on the same screen.

3. Use a graphical intersection finder to find the x-coordinate of each point where the graphs intersect.

These x-coordinates are the real solutions of the equation.

EXAMPLE 4

According to data from the U.S. Census Bureau, the approximate population y (in millions) of Detroit is given by

$$y = (-6.985 \times 10^{-7})x^4 + (9.169 \times 10^{-5})x^3 - .00373x^2 + .0268x + 1.85,$$

where $x = 0$ corresponds to 1950. The approximate population of San Diego during the same period is given by

$$y = (-6.216 \times 10^{-6})x^3 + (4.569 \times 10^{-4})x^2 + .01035x + .346.$$

(a) Approximately when did the two cities have the same population?

(b) In what year was the population of Detroit about 1,500,000?

SOLUTION

(a) The populations of the two cities are equal when

$$(-6.985 \times 10^{-7})x^4 + (9.169 \times 10^{-5})x^3 - .00373x^2 + .0268x + 1.85$$
$$= (-6.216 \times 10^{-6})x^3 + (4.569 \times 10^{-4})x^2 + .01035x + .346.$$

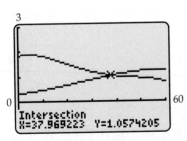

Figure 2–30

We solve this equation by graphing the left side as y_1 and the right side as y_2 and finding the intersection point, as in Figure 2–30. Rounding the x-coordinate to the nearest year ($x \approx 38$), we see that the populations were equal around 1988.

(b) To find when the population of Detroit was 1,500,000 (that is, 1.5 million), we must solve the equation

$$(-6.985 \times 10^{-7})x^4 + (9.169 \times 10^{-5})x^3 - .00373x^2 + .0268x + 1.85 = 1.5.$$

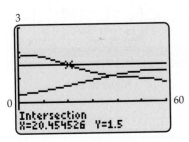

Figure 2–31

Since the left side of this equation has already been graphed as y_1, we need only graph $y_3 = 1.5$ on the same screen and find the intersection point of y_1 and y_3, as in Figure 2–31. The solution $x \approx 20.45$ shows that Detroit had a population of 1.5 million in 1970. ∎

NUMERICAL METHODS

In addition to graphical tools for solving equations, calculators and computer algebra systems have **equation solvers** that can find or approximate the solutions of most equations. On most calculators, the solver finds one solution at a time. You must enter the equation and an initial guess and possibly an interval in which to search. A few calculators (such as TI-89) and most computer systems have **one-step equation solvers** that will find or approximate all the solutions in a single step. Check you calculator instruction manual or your computer's help menu for directions on using these solvers.

EXAMPLE 5

Use an equation solver to solve

$$\frac{5}{\sqrt{x^2 + 1}} = \frac{x^4}{x + 5}.$$

SOLUTION   On the TI-84+ solver, the equation must be put in the form $\frac{5}{\sqrt{x^2 + 1}} - \frac{x^4}{x + 5} = 0.$ When asked to find a solution in the interval $-10 \le x \le 10$, with an initial guess of 1, the solver found the one in Figure 2–32. We changed the initial guess to -1 to produce the solution in Figure 2–33.*

The one-step solver on a TI-89 produced both solutions of the equation (Figure 2–34).

The first solution in Figure 2–34 was found on Maple with the command

fsolve(5/sqrt(x^2 + 1) − x^4/(x + 5) = 0, x);

and the second was found by changing the search interval to $-10 \le x \le 0$, with the command

fsolve(5/sqrt(x^2 + 1) − x^4/(x + 5) = 0, x, x = −10. . 0).   ∎

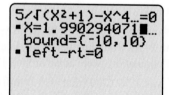

Figure 2–32

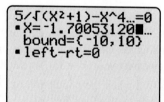

Figure 2–33

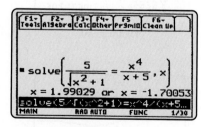

Figure 2–34

---

*In some cases, you may also have to change the search interval to find additional solutions.

Several calculators and many computer algebra systems also have **polynomial solvers,** one-step solvers designed specifically for polynomial equations; you need only enter the degree and coefficients of the polynomial. A few of these (such as Casio 9850) are limited to equations of degree 2 and 3. Directions for using polynomial solvers on calculators are in Exercise 105 on page 31. Computer users should check the help menu for the proper syntax.

EXAMPLE 6

Use a polynomial solver to solve

$$4x^5 - 12x^3 + 8x - 1 = 0.$$

SOLUTION   We enter the degree and the coefficients of the equation in the polynomial solver (Figure 2–35) and press SOLVE to obtain the five solutions (Figure 2–36).* The fsolve command on Maple also produced all five solutions. ∎

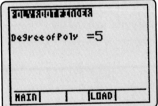

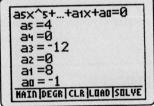

Figure 2–35

Figure 2–36

STRATEGIES FOR SPECIAL CASES

Because of various technological shortcomings, some equations are easier to solve if an indirect approach is used.

EXAMPLE 7

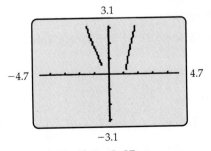

Figure 2–37

Solve $\sqrt{x^4 + x^2 - 2x - 1} = 0$.

SOLUTION   The graph of $y = \sqrt{x^4 + x^2 - 2x - 1}$ in Figure 2–37 does not even appear to touch the x-axis (although it actually does touch at two places). Consequently, some solvers and graphical root finders will return an error message (check yours). Even if your root finder or solver can handle this equation, it may fail in other similar situations. So the best approach is to use the following fact:

The only number whose square root is 0 is 0 itself.

*This illustrates the procedure for TI-84+ and 86. For HP-39gs, see Exercise 105 on page 31.

Thus, $\sqrt{x^4 + x^2 - 2x - 1} = 0$ exactly when $x^4 + x^2 - 2x - 1 = 0$. So you need only solve the polynomial equation

$$x^4 + x^2 - 2x - 1 = 0,$$

which is easily done on any calculator or computer. In fact, if you have a polynomial or one-step solver, you can find all the solutions at once, which is faster than solving the original equation with a root finder or other solvers.

GRAPHING EXPLORATION

If you have a polynomial or one-step solver, use it to find all the real solutions of $x^4 + x^2 - 2x - 1 = 0$ at once. Otherwise, graph $y = x^4 + x^2 - 2x - 1$ and use the graphical root finder to obtain the solutions one at a time. Verify that the real solutions of this equation, and hence, of the original one, are $x \approx -.4046978$ and $x \approx 1.1841347$.

The technique used in Example 7 is recommended for all similar situations.

Radical Equations

> To solve an equation of the form
>
> $$\sqrt{\text{Expression in } x} = 0,$$
>
> set the expression under the radical equal to 0 and solve the resulting equation.

EXAMPLE 8

Solve

$$\frac{2x^2 + x - 1}{9x^2 - 9x + 2} = 0.$$

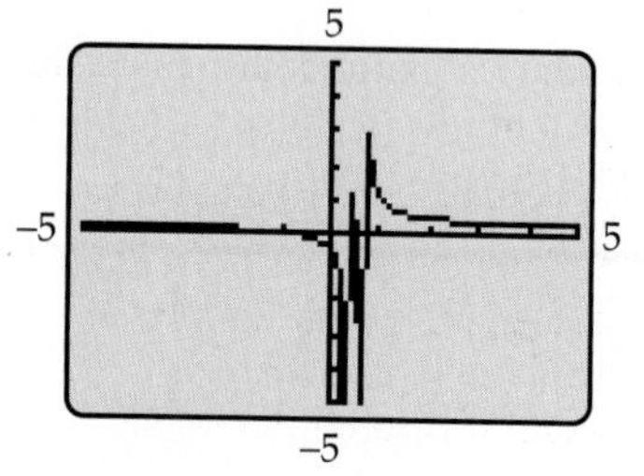

SOLUTION   If you graph

$$y = \frac{2x^2 + x - 1}{9x^2 - 9x + 2},$$

Figure 2–38

you may get "garbage," as in Figure 2–38. You could experiment with other viewing windows, but it's easier to use this fact:

A fraction is 0 exactly when its numerator is 0 and its denominator is nonzero.

To find where the numerator is 0, we need only solve $2x^2 + x - 1 = 0$. This is easily done algebraically.*

$$2x^2 + x - 1 = 0$$

$$(2x - 1)(x + 1) = 0$$

$$x = 1/2 \quad \text{or} \quad x = -1$$

You can readily verify that neither of these numbers make the denominator 0. Hence, the solutions of the original equation are $1/2$ and -1.   ■

---

*In other cases, you may need technology to solve the numerator equation.

Example 8 illustrates a useful technique.

Fractional Equations

To solve an equation of the form

$$\text{Fraction} = 0,$$

set the numerator equal to 0 and solve the resulting equation. The solutions that do *not* make the denominator of the fraction 0 are the solutions of the original equation.*

 CHOOSING A SOLUTION METHOD

We have seen that equations can be solved by algebraic, graphical, and numerical methods. Each method has both advantages and disadvantages, as summarized in the table.

Solution Method	Advantages	Possible Disadvantages
Algebraic	Produces exact solutions. Easiest method for most linear and quadratic equations.	May be difficult or impossible to use with complicated equations.
Graphical **Root Finder or** **Intersection Finder**	Works well for a large variety of equations. Gives visual picture of the location of the solutions.	Solutions may be approximations. Finding a useable viewing window may take a lot of time.
Numerical **Equation Solvers**		Solutions may be approximations.
Polynomial solver	Fast and easy.	Works only for polynomial equations.
One-step solver	Fast and easy.	May miss some solutions or be unable to solve certain equations.
Other solvers		May require a considerable amount of work to find the particular solution you want.

The choice of solution method is up to you. In the rest of this book, we normally use algebraic means for solving linear and quadratic equations because this is often the fastest and most accurate method. Naturally, any such equation can also be solved graphically or numerically (and you may want to do that as a check against errors). Except in special cases, graphical and numerical methods will normally be used for more complicated equations.

---

*A number that makes *both* numerator and denominator 0 is *not* a solution of the original equation because 0/0 is not defined.

EXERCISES 2.2

In Exercises 1–6, determine graphically the number of solutions of the equation, but don't solve the equation. You may need a viewing window other than the standard one to find all the x-intercepts.

1. $x^5 + 5 = 3x^4 + x$

2. $x^3 + 5 = 3x^2 + 24x$

3. $x^7 - 10x^5 + 15x + 10 = 0$

4. $x^5 + 36x + 25 = 13x^3$

5. $x^4 + 500x^2 - 8000x = 16x^3 - 32,000$

6. $6x^5 + 80x^3 + 45x^2 + 30 = 45x^4 + 86x$

In Exercises 7–20, use graphical approximation (a root finder or an intersection finder) to find a solution of the equation in the given open interval.

7. $x^3 + 4x^2 + 10x + 15 = 0;$ $(-3, -2)$

8. $x^3 + 9 = 3x^2 + 6x;$ $(1, 2)$

9. $x^4 + x - 3 = 0;$ $(-\infty, 0)$

10. $x^5 + 5 = 3x^4 + x;$ $(2, \infty)$

11. $\sqrt{x^4 + x^3 - x - 3} = 0;$ $(-\infty, 0)$

12. $\sqrt{8x^4 - 14x^3 - 9x^2 + 11x - 1} = 0;$ $(-\infty, 0)$

13. $\sqrt{\dfrac{2}{5}x^5 + x^2 - 2x} = 0;$ $(0, \infty)$

14. $\sqrt{x^4 + x^2 - 3x + 1} = 0;$ $(0, 1)$

15. $x^2 = \sqrt{x + 5};$ $(-2, -1)$

16. $\sqrt{x^2 - 1} - \sqrt{x + 9} = 0;$ $(3, 4)$

17. $\dfrac{2x^5 - 10x + 5}{x^3 + x^2 - 12x} = 0;$ $(1, \infty)$

18. $\dfrac{3x^5 - 15x + 5}{x^7 - 8x^5 + 2x^2 - 5} = 0;$ $(1, \infty)$

19. $\dfrac{x^3 - 4x + 1}{x^2 + x - 6} = 0;$ $(-\infty, 0)$

20. $\dfrac{4}{x + 2} - \dfrac{3}{x + 1} = 0;$ $(0, \infty)$ [*Hint:* Write the left side as a single fraction.]

In Exercises 21–34, use algebraic, graphical, or numerical methods to find all real solutions of the equation, approximating when necessary.

21. $2x^3 - 4x^2 + x - 3 = 0$

22. $6x^3 - 5x^2 + 3x - 2 = 0$

23. $x^5 - 6x + 6 = 0$

24. $x^3 - 3x^2 + x - 1 = 0$

25. $10x^5 - 3x^2 + x - 6 = 0$

26. $\dfrac{1}{4}x^4 - x - 4 = 0$

27. $2x - \dfrac{1}{2}x^2 - \dfrac{1}{12}x^4 = 0$

28. $\dfrac{1}{4}x^4 + \dfrac{1}{3}x^2 + 3x - 1 = 0$

29. $\dfrac{5x}{x^2 + 1} - 2x + 3 = 0$

30. $\dfrac{2x}{x + 5} = 1$

31. $|x^2 - 4| = 3x^2 - 2x + 1$

32. $|x^3 + 2| = 5 + x - x^2$

33. $\sqrt{x^2 + 3} = \sqrt{x - 2} + 5$

34. $\sqrt{x^3 + 2} = \sqrt{x + 5} + 4$

In Exercises 35–40, find an exact solution of the equation in the given open interval. (For example, if the graphical approximation of a solution begins .3333, check to see whether 1/3 is the exact solution. Similarly, $\sqrt{2} \approx 1.414$; so if your approximation begins 1.414, check to see whether $\sqrt{2}$ is a solution.)

35. $3x^3 - 2x^2 + 3x - 2 = 0;$ $(0, 1)$

36. $4x^3 - 3x^2 - 3x - 7 = 0;$ $(1, 2)$

37. $12x^4 - x^3 - 12x^2 + 25x - 2 = 0;$ $(0, 1)$

38. $8x^5 + 7x^4 - x^3 + 16x - 2 = 0;$ $(0, 1)$

39. $4x^4 - 13x^2 + 3 = 0;$ $(1, 2)$

40. $x^3 + x^2 - 2x - 2 = 0;$ $(1, 2)$

Exercises 41–46 deal with exponential, logarithmic, and trigonometric equations, which will be dealt with in later chapters. If you are familiar with these concepts, solve each equation graphically or numerically.

41. $10^x - \dfrac{1}{4}x = 28$

42. $e^x - 6x = 5$

43. $x + \sin\left(\dfrac{x}{2}\right) = 4$

44. $x^3 + \cos\left(\dfrac{x}{3}\right) = 5$

45. $5 \ln x + x^3 - x^2 = 5$

46. $\ln x - x^2 + 3 = 0$

47. According to data from the U.S. Department of Education, the average cost y of tuition and fees at four-year public colleges and universities in year x is approximated by

$$y = \sqrt{180,115x^2 + 2,863,851x + 11,383,876}$$

where $x = 0$ corresponds to 2000. If this model continues to be accurate, in what year will tuition and fees reach $7000? Round your answer to the nearest year.

48. Use the information in Example 4 to determine the year in which the population of San Diego reached 1.1 million people.

49. According to data from the U.S. Department of Health and Human Services, the cumulative number y of AIDS cases (in thousands) as of year x is approximated by

$$y = .004x^3 - 1.367x^2 + 54.35x + 569.72 \quad (0 \le x < 11),$$

where $x = 0$ corresponds to 1995.

(a) When did the cumulative number of cases reach 944,000?

(b) If this model remains accurate after 2006, in what year will the cumulative number of cases reach 1.1 million?

50. The enrollment in public high schools (in millions of students) in year x is approximated by

$$y = -.000035606x^4 + .0021x^3 - .02714x^2 - .12059x + 14.2996 \qquad (0 \le x < 35),$$

where $x = 0$ corresponds to 1975.* During the current century, when was enrollment 13.9 million students?

51. In Example 4 of Section 1.1 (page 9), a formula is given for determining how far you can see from a given height. Suppose you are on a cruise ship and that you can see the top of a lighthouse 12 miles away. About how high above water level are you?

52. According to the U.S. Centers for Medicare and Medicaid Services, total medical expenditures (in billions of dollars) in the United States in year x are expected to be given by

10

$$y = -.035x^4 + 1.01x^3 - 4.91x^2 + 126.94x + 1309.6,$$

where $x = 0$ corresponds to 2000. When will expenditures be $2.6 trillion?   *260 billion*

53. When is the population of China expected to reach 1.4 billion people? [*Hint:* The equation that estimates the population of China is in Exercise 28 of Section 2.1 (page 89).]

54. (a) How many real solutions does the equation

$$.2x^5 - 2x^3 + 1.8x + k = 0$$

have when $k = 0$?
(b) How many real solutions does it have when $k = 1$?
(c) Is there a value of k for which the equation has just one real solution?
(d) Is there a value of k for which the equation has no real solutions?

---

*Based on data from the National Center for Educational Statistics.

2.3 Applications of Equations

Section Objectives

■ Set up applied problems.
■ Translate verbal statements into mathematical language.
■ Solve applied problems.

Actual problem situations are usually described verbally. To solve such problems, you must interpret this verbal information and express it as an equivalent mathematical problem. The following guidelines may be helpful.

Setting up Applied Problems

1. *Read* the problem carefully, and determine what is asked for.

2. *Label* the unknown quantities by letters (variables), and, if appropriate, draw a picture of the situation.

3. *Translate* the verbal statements in the problem and the relationships between the known and unknown quantities into mathematical language.

4. *Consolidate* the mathematical information into an equation in one variable that can be solved or an equation in two variables that can be graphed to determine at least one of the unknown quantities.

Here are some examples of how these guidelines are applied.

EXAMPLE 1

Set up the following problem: The average of two real numbers is 41.125, and their product is 1683. What are the numbers?

SOLUTION   *Read:* We are asked for two numbers. *Label:* Call the numbers x and y. *Translate:*

English Language	Mathematical Language
Two numbers	x and y
Their average is 41.125.	$\dfrac{x + y}{2} = 41.125$
Their product is 1683.	$xy = 1683$

Consolidate: One technique to use when you have two unknowns is to express one in terms of the other and use this to obtain an equation in one variable. In this case, we can do that by solving the second equation for y:

$$xy = 1683$$

$$y = \frac{1683}{x}$$

and substituting the result in the first equation:

$$\frac{x + y}{2} = 41.125$$

$$\frac{x + \dfrac{1683}{x}}{2} = 41.125$$

Multiply both sides by 2: $x + \dfrac{1683}{x} = 82.25.$

The solution of this equation is one of the numbers, and $1683/x$ is the other.   ∎

EXAMPLE 2

Set up the following problem: A rectangle is twice as long as it is wide. If it has an area of 24.5 square inches, what are its dimensions?

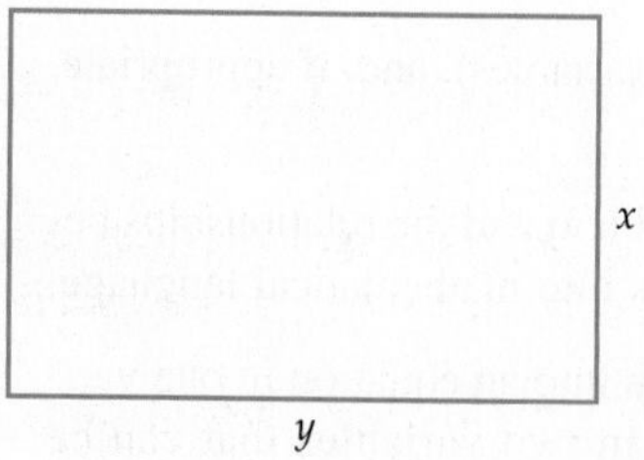

SOLUTION   *Read:* We are asked to find the length and width. *Label:* Let x denote the width and y the length, and draw a picture of the situation, as in Figure 2–39. *Translate:* Use the fact that the area of a rectangle is length × width.

x

y

Figure 2–39

English Language	Mathematical Language
The width and length of the rectangle	x and y
The length is twice the width.	$y = 2x$
The area is 24.5 square inches.	$xy = 24.5$

Consolidate: Substitute $y = 2x$ in the area equation:

$$xy = 24.5$$

$$x(2x) = 24.5.$$

So the equation to be solved is $2x^2 = 24.5$.   ∎

EXAMPLE 3

Set up this problem: A rectangular box with a square base and no top is to have a volume of 20,000 cubic centimeters. If the surface area of the box is 4000 square centimeters, what are its dimensions?

SOLUTION   *Read:* We must find the length, width, and height of the box. *Label:* Let x denote the length. Since the base is square, the length and width are the same. Let h denote the height, as in Figure 2–40. *Translate:* Recall that the volume of a box is given by the product length $\times$ width $\times$ height and that the surface area is the sum of the area of the base and the area of the four sides of the box. Then we have these translations:

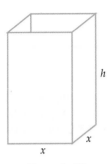

Figure 2–40

English Language	Mathematical Language
The length, width, and height	x, x, and h
The volume is 20,000 cm^3.	$x^2 h = 20{,}000$
The surface area is 4000 cm^2.	$x^2 + 4xh = 4000$

Consolidate: We have two equations in two variables, so we solve the first equation for h

$$h = \frac{20{,}000}{x^2}$$

and substitute this result in the second equation:

$$x^2 + 4x\left(\frac{20{,}000}{x^2}\right) = 4000$$

$$x^2 + \frac{80{,}000}{x} = 4000.$$

The solution of this last equation will provide the solution of the problem. ■

Once you are comfortable with the process for setting up problems, you can often do much of it mentally. In the rest of this section, the setup process will be simplified or shortened.

Setting up a problem is only half the job. You must then solve the equation you have obtained. Whenever possible, linear and quadratic equations should be solved algebraically, giving exact answers. Other equations may require graphical or numerical methods to find approximate solutions. Finally, you must check your answers in the original problem:

Interpret your answers in terms of the original problem.
Do they make sense? Do they satisfy the required conditions?

In particular, an equation may have several solutions, some of which might r make sense in the context of the problem. For instance, distance can't be nega' the number of people in a room cannot be a proper fraction, etc.

APPLICATIONS

We begin with some problems involving interest. Recall that 8% means .08 and that "8% *of* 227" means ".08 *times* 227," that is, .08(227) = 18.16. The basic rule of annual simple interest is

$$\text{Interest} = \text{rate} \times \text{amount.}$$

EXAMPLE 4

A high-risk stock pays dividends at a rate of 12% per year, and a savings account pays 6% interest per year. How much of a $9000 investment should be put in the stock and how much in the savings account to obtain a return of 8% per year on the total investment?

SOLUTION   *Label:* Let x be the amount invested in stock. Then the rest of the $9000, namely, $(9000 - x)$ dollars, goes in the savings account. *Translate:* We want the total return on $9000 to be 8%, so we have

$$\left(\begin{array}{c}\text{Return on } x \text{ dollars}\\ \text{of stock at 12\%}\end{array}\right) + \left[\begin{array}{c}\text{Return on } (9000 - x)\\ \text{dollars of savings at 6\%}\end{array}\right] = 8\% \text{ of } \$9000$$

$$(12\% \text{ of } x \text{ dollars}) + [6\% \text{ of } (9000 - x) \text{ dollars}] = 8\% \text{ of } \$9000$$

$$.12x + .06(9000 - x) = .08(9000)$$
$$.12x + .06(9000) - .06x = .08(9000)$$
$$.12x + 540 - .06x = 720$$
$$.12x - .06x = 720 - 540$$
$$.06x = 180$$
$$x = \frac{180}{.06} = 3000.$$

Therefore, $3000 should be invested in stock and $(9000 - 3000) = \$6000$ in the savings account. If this is done, the total return will be 12% of $3000 ($360) plus 6% of $6000 ($360), a total of $720, which is precisely 8% of $9000. ∎

EXAMPLE 5

A car radiator contains 12 quarts of fluid, 20% of which is antifreeze. How much fluid should be drained and replaced with pure antifreeze in order that the resulting mixture be 50% antifreeze?

SOLUTION   Let x be the number of quarts of fluid to be replaced by pure antifreeze.* When x quarts are drained, there are $12 - x$ quarts of fluid left in the radiator, 20% of which is antifreeze. So we have

$$\begin{pmatrix} \text{Amount of} \\ \text{antifreeze in} \\ \text{radiator after} \\ \text{draining } x \text{ quarts} \\ \text{of fluid} \end{pmatrix} + \begin{pmatrix} x \text{ quarts of} \\ \text{antifreeze} \end{pmatrix} = \begin{pmatrix} \text{Amount of} \\ \text{antifreeze in} \\ \text{final mixture} \end{pmatrix}$$

$$20\% \text{ of } (12 - x) \quad + \quad x \quad = \quad 50\% \text{ of } 12$$

$$.20(12 - x) + x = .50(12)$$

$$2.4 - .2x + x = 6$$

$$-.2x + x = 6 - 2.4$$

$$.8x = 3.6$$

$$x = \frac{3.6}{.8} = 4.5.$$

Therefore, 4.5 quarts should be drained and replaced with pure antifreeze. ∎

The two preceding examples used only algebraic models (equations in one variable). Sometimes a diagram (geometrical model) is helpful in visualizing the situation and setting up an appropriate equation.

EXAMPLE 6

A landscaper wants to put a cement walk of uniform width around a rectangular garden that measures 24 by 40 feet. She has enough cement to cover 660 square feet. How wide should the walk be to use all the cement?

SOLUTION   Let x denote the width of the walk (in feet), and draw a picture of the situation (Figure 2–41).

The length of the outer rectangle is $40 + 2x$ (the garden length plus walks on each end), and its width is $24 + 2x$.

$$\begin{pmatrix} \text{Area of outer} \\ \text{rectangle} \end{pmatrix} - \begin{pmatrix} \text{Area of} \\ \text{garden} \end{pmatrix} = \text{Area of walk}$$

$$\text{Length} \cdot \text{Width} - \text{Length} \cdot \text{Width} = 660$$

$$(40 + 2x)(24 + 2x) - 40 \cdot 24 = 660$$

$$960 + 128x + 4x^2 - 960 = 660$$

$$4x^2 + 128x - 660 = 0.$$

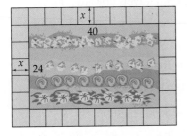

Figure 2–41

---

*Hereafter, we omit the headings *Label*, *Translate*, etc.

Dividing both sides by 4 and applying the quadratic formula yields

$$x^2 + 32x - 165 = 0$$

$$x = \frac{-32 \pm \sqrt{(32)^2 - 4 \cdot 1 \cdot (-165)}}{2 \cdot 1}$$

$$x = \frac{-32 \pm \sqrt{1684}}{2} \approx \begin{cases} 4.5183 \\ \text{or} \\ -36.5183. \end{cases}$$

Only the positive solution makes sense in the context of this problem. The walk should be approximately 4.5 feet wide. ∎

The basic formula for problems involving distance and a uniform rate of speed is

Distance = rate × time.

For instance, if you drive at a rate of 55 mph for 2 hours, you travel a distance of 55 · 2 = 110 miles.

EXAMPLE 7

A stone is dropped from the top of a cliff. Five seconds later, the sound of the stone hitting the ground is heard. Use the fact that the speed of sound is 1100 feet per second and that the distance traveled by a falling object in t seconds is $16t^2$ feet to determine the height of the cliff.

SOLUTION   If t is the time it takes the rock to fall to the ground, then the height of the cliff is $16t^2$ feet. The total time for the rock to fall and the sound of its hitting ground to return to the top of the cliff is 5 seconds. So the sound must take $5 - t$ seconds to return to the top, as indicated in Figure 2–42. Therefore,

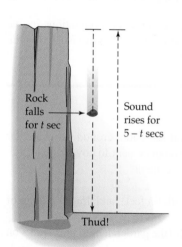

Rock falls for t sec

Sound rises for $5 - t$ secs

Thud!

Figure 2–42

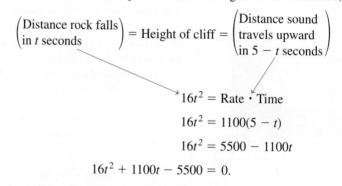

$$\begin{pmatrix} \text{Distance rock falls} \\ \text{in } t \text{ seconds} \end{pmatrix} = \text{Height of cliff} = \begin{pmatrix} \text{Distance sound} \\ \text{travels upward} \\ \text{in } 5 - t \text{ seconds} \end{pmatrix}$$

$$16t^2 = \text{Rate} \cdot \text{Time}$$

$$16t^2 = 1100(5 - t)$$

$$16t^2 = 5500 - 1100t$$

$$16t^2 + 1100t - 5500 = 0.$$

The quadratic formula and a calculator show that

$$t = \frac{-1100 \pm \sqrt{(1100)^2 - 4 \cdot 16 \cdot (-5500)}}{2 \cdot 16}$$

$$= \frac{-1100 \pm \sqrt{1,562,000}}{32} \approx \begin{cases} 4.6812 \\ \text{or} \\ -73.4312. \end{cases}$$

Since the time must be a number between 0 and 5, we see that $t \approx 4.6812$ seconds. Therefore the height of the cliff is

$$16t^2 = 16(4.6812)^2 \approx 350.6 \text{ feet.} \quad \blacksquare$$

EXAMPLE 8

A pilot wants to make the 840-mile round trip from Cleveland to Peoria and back in 5 hours flying time. Going to Peoria, there will be a headwind of 30 mph, that is, a wind opposite to the direction the plane is flying. It is estimated that on the return trip to Cleveland, there will be a 40-mph tailwind (in the direction the plane is flying). At what constant speed should the plane be flown?

SOLUTION   Let x be the engine speed of the plane. On the trip to Peoria, the actual speed will be $x - 30$ mph because of the headwind.

The distance from Cleveland to Peoria is 420 miles (half the round trip). So we have

$$\text{Rate} \cdot \text{Time} = \text{Distance}$$

$$\text{Time} = \frac{\text{Distance}}{\text{Rate}} = \frac{420}{x - 30}.$$

On the return trip, the actual speed will be $x + 40$ because of the tailwind.

So we have

$$\text{Rate} \cdot \text{Time} = \text{Distance}$$

$$\text{Time} = \frac{\text{Distance}}{\text{Rate}} = \frac{420}{x + 40}.$$

Therefore,

$$5 = \left(\begin{array}{l}\text{Time from Cleveland} \\ \text{to Peoria}\end{array}\right) + \left(\begin{array}{l}\text{Time from Peoria} \\ \text{to Cleveland}\end{array}\right)$$

$$5 = \frac{420}{x - 30} + \frac{420}{x + 40}.$$

Multiplying both sides by the common denominator $(x - 30)(x + 40)$ and simplifying, we have

$$5(x - 30)(x + 40) = \frac{420}{x - 30} \cdot (x - 30)(x + 40) + \frac{420}{x + 40} \cdot (x - 30)(x + 40)$$

$$5(x - 30)(x + 40) = 420(x + 40) + 420(x - 30)$$

$$(x - 30)(x + 40) = 84(x + 40) + 84(x - 30)$$

$$x^2 + 10x - 1200 = 84x + 3360 + 84x - 2520$$

$$x^2 - 158x - 2040 = 0$$

$$(x - 170)(x + 12) = 0$$

$$x - 170 = 0 \qquad \text{or} \qquad x + 12 = 0$$

$$x = 170 \qquad\qquad x = -12.$$

Obviously, the negative solution doesn't apply. Since we multiplied both sides by a quantity involving the variable, we must check that 170 actually is a solution of the original equation. It is, so the plane should be flown at a speed of 170 mph.   ∎

EXAMPLE 9

A rectangular box with a square base and no top is to have a volume of 20,000 cubic centimeters. If the surface area of the box is 4000 square centimeters and the box is required to be higher than it is wide, what are its dimensions?

SOLUTION   This is essentially Example 3 with an extra condition (the box must be higher than it is wide). If the length, width, and height of the box are x, x, and h as shown in Figure 2–43, then as we saw in Example 3,

$$x^2 + \frac{80,000}{x} = 4000 \qquad \text{and} \qquad h = \frac{20,000}{x^2}.$$

Multiplying both sides of the first equation by x (which is nonzero because it is a length), we obtain

$$x^3 + 80,000 = 4000x$$

$$x^3 - 4000x + 80,000 = 0.$$

This equation can be solved in more than one way.

Graphical: The graph of $y = x^3 - 4000x + 80,000$ in Figure 2–44 shows that the equation has three solutions. The negative solution does not apply here, nor does the one near 50 (because a value of x in that range makes the corresponding height $h = 20,000/x^2$ smaller than the width x, as indicated in Figure 2–45). The third solution, $x \approx 23.069$, is easily found (Figure 2–46).

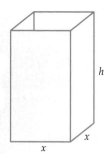

Figure 2–43

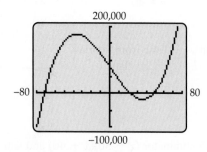

Figure 2–44

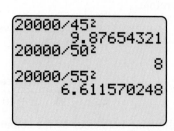

Figure 2–45

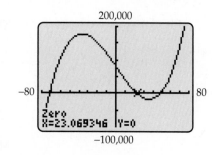

Figure 2–46

Numerical: The polynomial solver on a TI-86 or the "fsolve" command on Maple produces the three solutions (Figure 2–47). As above, only one of them is applicable here, namely, $x \approx 23.069$.

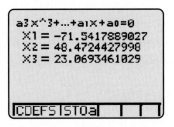

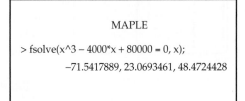

Figure 2–47

Therefore, the height is

$$h = \frac{20{,}000}{x^2} \approx \frac{20{,}000}{23.069^2} \approx 37.581.$$

So the approximate dimensions of the box are 23.069 by 23.069 by 37.581 cm. ∎

EXAMPLE 10

A box (with no top) of volume 1000 cubic inches is to be made from a 22 × 30 inch sheet of cardboard by cutting squares of equal size from each corner and folding up the flaps, as shown in Figure 2–48. If the box must be at least 4 inches high, what size square should be cut from each corner?

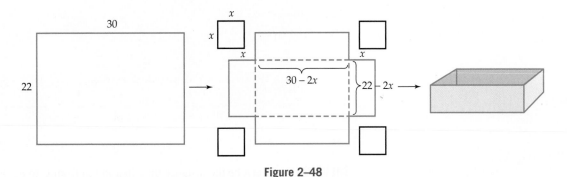

Figure 2–48

SOLUTION   Let x denote the length of the side of the square to be cut from each corner. The dashed rectangle in Figure 2–48 is the bottom of the box. Its length is $30 - 2x$ as shown in the figure. Similarly, the width of the box will be $22 - 2x$, and its height will be x inches. Therefore,

$$\text{Length} \times \text{Width} \times \text{Height} = \text{Volume of box}$$

$$(30 - 2x) \cdot (22 - 2x) \cdot x = 1000$$

$$(660 - 104x + 4x^2)x = 1000$$

$$4x^3 - 104x^2 + 660x - 1000 = 0.$$

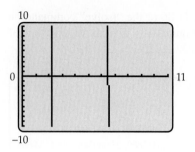

Figure 2–49

Since the cardboard is 22 inches wide, x must be less than 11 (otherwise, you can't cut out two squares of length x). Since x is a length, it is positive. So we need only find solutions of the equation between 0 and 11. We graph

$$y = 4x^3 - 104x^2 + 660x - 1000$$

in a window with $0 \leq x \leq 11$ (Figure 2–49). A complete graph isn't needed here, only the x-intercepts (solutions). The one between 2 and 3 is not relevant here because x is the height of the box, which must be at least 4 inches.

GRAPHING EXPLORATION

Use a root finder or a polynomial solver on a calculator or computer to find the solution of the equation between 6 and 7. This is the side x of the square that should be cut from each corner. Round the value of x to two decimal places, and find the dimensions of the resulting box.

EXAMPLE 11

Sharon Mahoney was at point A on the bank of a 2.5-kilometer wide river and traveled to point B, 15 kilometers downstream on the opposite bank. She rowed to point C and then ran to B, as shown in Figure 2–50. She rowed at a rate of 4 kilometers per hour and ran at 8 kilometers per hour. Her trip took 3 hours. How far did she run?

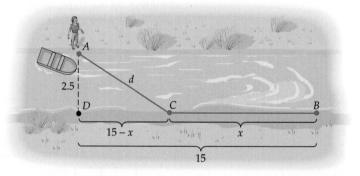

Figure 2–50

SOLUTION Let x be the distance that Sharon ran from C to B. Using the basic formula for distance, we have

$$\text{Rate} \times \text{Time} = \text{Distance}$$

$$\text{Time} = \frac{\text{Distance}}{\text{Rate}} = \frac{x}{8}.$$

Similarly, the time required to row distance d is

$$\text{Time} = \frac{\text{Distance}}{\text{Rate}} = \frac{d}{4}.$$

Since $15 - x$ is the distance from D to C, the Pythagorean Theorem applied to right triangle ADC shows that

$$d^2 = (15 - x)^2 + 2.5^2 \qquad \text{or, equivalently,} \qquad d = \sqrt{(15 - x)^2 + 6.25}.$$

Therefore, the total time for the trip is given by

$$\text{Rowing time} + \text{Running time} = \frac{d}{4} + \frac{x}{8} = \frac{\sqrt{(x-15)^2 + 6.25}}{4} + \frac{x}{8}.$$

If the trip took 3 hours, then we must solve the equation

$$\frac{\sqrt{(x-15)^2 + 6.25}}{4} + \frac{x}{8} = 3.$$

GRAPHING EXPLORATION

Using the viewing window with $0 \le x \le 15$ and $-2 \le y \le 2$, graph

$$y = \frac{\sqrt{(x-15)^2 + 6.25}}{4} + \frac{x}{8} - 3$$

and use a root finder to find its x-intercept (the solution of the equation).

This Graphing Exploration shows that Sharon ran approximately 6.74 kilometers from C to B. ∎

EXERCISES 2.3

In Exercises 1–4, a problem situation is given.
(a) Decide what is being asked for, and label the unknown quantities.
(b) Translate the verbal statements in the problem and the relationships between the known and unknown quantities into mathematical language, using a table as in Examples 1–3 (pages 101–103). The table is provided in Exercises 1 and 2. You need not find an equation to be solved.

1. A student has exam scores of 88, 62, and 79. What score does he need on the fourth exam to have an average of 80?

English Language	Mathematical Language
Score on fourth exam	
Sum of scores on four exams	
Average of scores on four exams	

2. How many gallons of a 12% salt solution should be combined with 10 gallons of an 18% salt solution to obtain a 16% solution?

English Language	Mathematical Language
Gallons of 12% solution	
Total gallons of mixture	
Amount of salt in 10 gallons of the 18% solution	
Amount of salt in the 12% solution	
Amount of salt in the mixture	

3. A rectangle has perimeter of 45 centimeters and an area of 112.5 square centimeters. What are its dimensions?

4. A triangle has area 96 square inches, and its height is two-thirds of its base. What are the base and height of the triangle?

In Exercises 5–8, set up the problem by labeling the unknowns, translating the given information into mathematical language, and finding an equation that will produce the solution to the problem. You need not solve this equation.

5. A worker gets an 8% pay raise and now makes $2619 per month. What was the worker's old salary?

6. A merchant has 5 pounds of mixed nuts that cost $30. He wants to add peanuts that cost $1.50 per pound and cashews that cost $4.50 per pound to obtain 50 pounds of a mixture that costs $2.90 per pound. How many pounds of peanuts are needed?

7. The diameter of a circle is 16 cm. By what amount must the radius be decreased to decrease the area by 48π square centimeters?

8. A corner lot has dimensions 25 by 40 yards. The city plans to take a strip of uniform width along the two sides bordering the streets to widen these roads. How wide should the strip be if the remainder of the lot is to have an area of 844 square yards?

In the remaining exercises, solve the applied problems.

9. You have already invested $550 in a stock with an annual return of 11%. How much of an additional $1100 should be

invested at 12% and how much at 6% so that the total return on the entire $1650 is 9%?

10. If you borrow $500 from a credit union at 12% annual interest and $250 from a bank at 18% annual interest, what is the *effective annual interest rate* (that is, what single rate of interest on $750 would result in the same total amount of interest)?

11. A radiator contains 8 quarts of fluid, 40% of which is antifreeze. How much fluid should be drained and replaced with pure antifreeze so that the new mixture is 60% antifreeze?

12. A radiator contains 10 quarts of fluid, 30% of which is antifreeze. How much fluid should be drained and replaced with pure antifreeze so that the new mixture is 40% antifreeze?

13. An airplane flew with the wind for 2.5 hours and returned the same distance against the wind in 3.5 hours. If the cruising speed of the plane was a constant 360 mph in air, how fast was the wind blowing?

14. A train leaves New York for Boston, 200 miles away, at 3:00 P.M. and averages 75 mph. Another train leaves Boston for New York on an adjacent set of tracks at 5:00 P.M. and averages 45 mph. At what time will the trains meet?

15. The average of two real numbers is 41.125, and their product is 1683. What are the numbers? [*Hint:* See Example 1.]

16. A rectangle is twice as long as it is wide. If it has an area of 24.5 square inches, what are its dimensions? [*Hint:* See Example 2.]

17. Two cars leave a gas station at the same time, one traveling north and the other south. The northbound car travels at 50 mph. After 3 hours, the cars are 345 miles apart. How fast is the southbound car traveling?

18. A student leaves the university at noon, bicycling south at a constant rate. At 12:30 P.M., a second student leaves the same point and heads west, bicycling 7 mph faster than the first student. At 2:00 P.M., they are 30 miles apart. How fast is each one going?

19. A 13-foot-long ladder leans on a wall, as shown in the figure. The bottom of the ladder is 5 feet from the wall. If the bottom is pulled out 3 feet farther from the wall, how far does the top of the ladder move down the wall? [*Hint:* Draw pictures of the right triangle formed by the ladder, the ground, and the wall before and after the ladder is moved. In each case, use the Pythagorean Theorem to find the distance from the top of the ladder to the ground.]

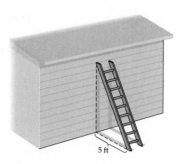

5 ft

20. A 15-foot-long pole leans against a wall. The bottom is 9 feet from the wall. How much farther should the bottom be pulled away from the wall so that the top moves the same amount down the wall?

21. A concrete walk of uniform width is to be built around a circular pool, as shown in the figure. The radius of the pool is 12 meters, and enough concrete is available to cover 52π square meters. If all the concrete is to be used, how wide should the walk be?

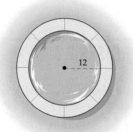

12

22. In Example 14 of Section 1.4 (page 63), how many can openers must be produced to have an average cost per can opener of $3?

23. Tom drops a rock into a well and 3 seconds later hears the sound of its splash. How deep is the well? [Make the same assumptions about falling rocks and sound as in Example 7 (page 106).]

Rock falls — Sound rises

24. A group of homeowners are to share equally in the $210 cost of repairing a bus-stop shelter near their homes. At the last moment, two members of the group decide not to participate, and this raises the share of each remaining person by $28. How many people were in the group at the beginning?

25. Red Riding Hood drives the 432 miles to Grandmother's house in 1 hour less than it takes the Wolf to drive the same route. Her average speed is 6 mph faster than the Wolf's average speed. How fast does each drive?

26. To get to work, Sam jogs 3 kilometers to the train and then rides the remaining 5 kilometers. If the train goes 40 kilometers per hour faster than Sam's constant rate of jogging and the entire trip takes 30 minutes, how fast does Sam jog?

27. Suppose that the open-top box being made from a sheet of cardboard in Example 10 (page 109) may have any height, but is required to have at least one of its dimensions *greater* than 18 inches. What size square should be cut from each corner?

28. The dimensions of a rectangular box are consecutive integers. If the box has volume of 13,800 cubic centimeters, what are its dimensions?

29. The surface area S of the right circular cone in the figure below is given by $S = \pi r \sqrt{r^2 + h^2}$. What radius should be used to produce a cone of height 5 inches and surface area 100 square inches?

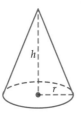

30. What is the radius of the base of a cone whose surface area is 18π square centimeters and whose height is 4 cm?

31. Find the radius of the base of a conical container whose height is $1/3$ of the radius and whose volume is 180 cubic inches. [*Note:* The volume of a cone of radius r and height h is $\pi r^2 h / 3$.]

32. The surface area of the right square pyramid in the figure below is given by $S = b\sqrt{b^2 + 4h^2}$. If the pyramid has height 10 feet and surface area 100 square feet, what is the length of a side b of its base?

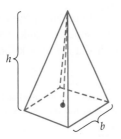

33. A rope is to be stretched at uniform height from a tree to a 35-foot-long fence, which is 20 feet from the tree, and then to the side of a building at a point 30 feet from the fence, as shown in the figure. If 63 feet of rope is to be used, how far from the building wall should the rope meet the fence?

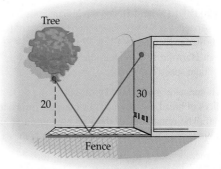

34. Anne is standing on a straight road and wants to reach her helicopter, which is located 2 miles down the road from her, a mile from the road in a field (see the figure). She can run 5 miles per hour on the road and 3 miles per hour in the field. She plans to run down the road, then cut diagonally across the field to reach the helicopter. If she runs less than a mile on the road and arrives at the helicopter in 42 minutes (.7 hour), exactly how far did she run on the road?

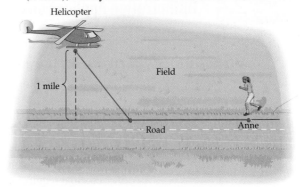

35. A power plant is located on the bank of a river that is $\frac{1}{2}$ mile wide. Wiring is to be laid across the river and then along the shore to a substation 8 miles downstream, as shown in the figure. It costs $12,000 per mile for underwater wiring and $8000 per mile for wiring on land. If $72,000 is to be spent on the project, how far from the substation should the wiring come to shore?

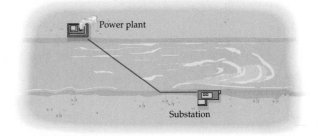

36. A spotlight is to be placed on the side of a 28-foot tall building to illuminate a bench that is 32 feet from the base of the building. The intensity I of the light at the bench is known to be x/d^3, where x is the height of the spotlight above the ground and d is the distance from the bench to the spotlight. If the intensity is to be .00035, how high should the spotlight be?

37. In 2005 Dan Wheldon won the Indianapolis 500 (mile) race. His speed was 83 mph faster than the speed of Ray Harroun, who won the race in 1911. Wheldon took 3.53 hours less than Harroun to complete the race. What was Wheldon's average speed?

38. A homemade loaf of bread turns out to be a perfect cube. Five slices of bread, each .6 inch thick, are cut from one end of the loaf. The remainder of the loaf now has a volume of 235 cubic inches. What were the dimensions of the original loaf?

39. A rectangular bin with an open top and volume of 38.72 cubic feet is to be built. The length of its base must be twice the width, and the bin must be at least 3 feet high. Material for the base of the bin costs $12 per square foot, and material for the sides costs $8 per square foot. If it costs $538.56 to build the bin, what are its dimensions?

THINKER

40. One corner of an 8.5 × 11 inch piece of paper is folded over to the opposite side, as shown in the figure. The area of the darkly shaded triangle at the lower left of the figure is 6 square inches and we want to find the length x.

(a) Take a piece of paper this size and experiment. Approximately, what is the largest value x could have (and still have the paper look like the figure)? With this value of x, what is the approximate area of the triangle? Try some other possibilities.

(b) Now find an exact answer by constructing and solving a suitable equation. Explain why one of the solutions to the equation is not an answer to this problem.

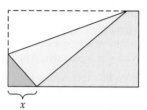

2.4   Optimization Applications*

Section Objectives   ■ Set up maximum and minimum problems.
■ Use graphing technology to find approximate solutions for such problems.

Many real-life situations require you to find the largest or smallest quantity that satisfies certain conditions. For instance, automotive engineers want to design engines with maximum fuel efficiency. Similarly, a cereal manufacturer who needs a box of volume 300 cubic inches might want to know the dimensions of the box that requires the least amount of cardboard (and hence is cheapest). The exact solutions of such minimum/maximum problems require calculus. However, graphing technology can provide very accurate approximate solutions, as we shall see. Before reading the following examples, you should review the guidelines for setting up applied problems (page 101).

EXAMPLE 1

Sharon was at point A on the bank of a 2.5 kilometer wide river and traveled to point B, 15 kilometers downstream on the opposite bank. She rowed at 4 kilometers per hour to point C, and ran from C to B at 8 kilometers per hour. If

---

*This section may be omitted or postponed until Chapter 4 (Polynomial and Rational Functions). It will be used only in occasional examples and in clearly identifiable exercises.

she made the trip in the shortest possible time, what was that time and how far did she run?

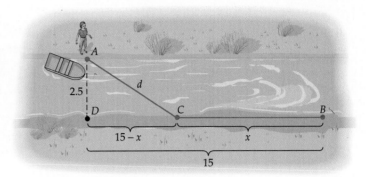

Figure 2–50

SOLUTION   In Example 11 of Section 2.3, we showed that the total time y for Sharon's trip is given by

$$y = \frac{\sqrt{(x - 15)^2 + 6.25}}{4} + \frac{x}{8},$$

where x is the distance she ran. In the graph of this equation in Figure 2–51, each point (x, y) on the graph represents one of the possibilities:

   x represents the distance Sharon runs;

   y represents her total time for the journey.

To find the shortest possible time, we must find the point on the graph with the smallest y-coordinate, that is, the lowest point on the graph.

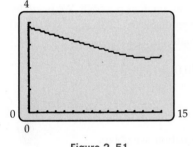

Figure 2–51

GRAPHING EXPLORATION

Graph the equation and use a minimum finder to verify that the lowest point is approximately (13.56, 2.42).

Therefore, the shortest time for the trip is 2.42 hours and occurs when Sharon runs 13.56 miles (that is, she lands 13.56 miles from B). ∎

EXAMPLE 2

A landscaper wants to build a 10,000 square foot garden along the side of a river. The rectangular garden will be fenced on three sides (no fence on the river side). She plans to use ornamental steel fencing at $18 per foot on the side opposite the river and chain link fence at $4 per foot on the other two sides. What dimensions for the garden will minimize her cost? What is the minimum cost?

SOLUTION   First draw a sketch of the situation and labels the sides of the garden as in Figure 2–52.

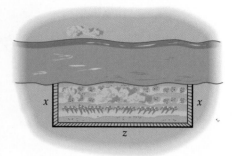

Figure 2–52

For the parallel sides, she needs $2x$ feet of fencing at \$4 per foot. For the side opposite the river, she needs z feet of fencing at \$18 per foot. So her cost is

$$C = 2x(4) + z(18) = 8x + 18z.$$

We must find the values of x and z that make C as small as possible. We begin by expressing C in terms of a single variable. The garden is to have an area of 10,000 square feet and the area of a rectangle is its length times its width. So

$$xz = \text{area} = 10{,}000$$

$$z = \frac{10{,}000}{x}.$$

Substituting this expression for z in the cost equation, we obtain

$$C = 8x + 18z$$

$$C = 8x + 18\left(\frac{10{,}000}{x}\right)$$

$$C = 8x + \frac{180{,}000}{x}.$$

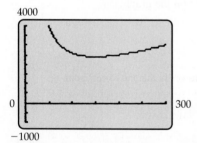

Figure 2–53

Now graph $y = 8x + \dfrac{180{,}000}{x}$ (Figure 2–53). If (x, y) is a point on the graph, then

x represents the side length of the garden;

y represents the cost of fencing a garden with this side length.

To find the minimum cost we use a minimum finder to determine the lowest point on the graph (that is, the one with the smallest y-coordinate), as in Figure 2–54. Therefore, the minimum cost is \$2400 and the garden with this minimum cost has dimensions

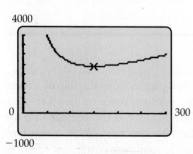

Figure 2–54

$$x = 150 \text{ ft} \quad \text{and} \quad z = \frac{10{,}000}{x} = \frac{10{,}000}{150} = 66\tfrac{2}{3} \text{ ft.} \qquad \blacksquare$$

EXAMPLE 3

A box with no top is to be made from a 22 × 30 inch sheet of cardboard by cutting squares of equal size from each corner and bending up the flaps, as shown in Figure 2–55. To the nearest hundredth of an inch, what size square should be cut from each corner in order to obtain a box with the largest possible volume, and what is the volume of this box?

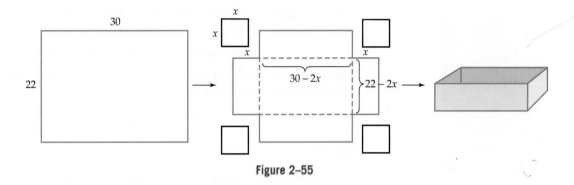

Figure 2–55

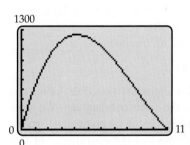

Figure 2–56

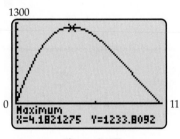

Figure 2–57

SOLUTION Let x denote the length of the side of the square to be cut from each corner. Then, as we saw in Example 10 of Section 2.3 (page 109),

$$\text{Volume of box} = \text{Length} \times \text{Width} \times \text{Height}$$

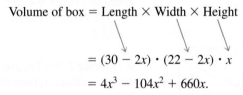

$$= (30 - 2x) \cdot (22 - 2x) \cdot x$$
$$= 4x^3 - 104x^2 + 660x.$$

Thus, the equation $y = 4x^3 - 104x^2 + 660x$ gives the volume y of the box that results from cutting a square of side x from each corner. Since the shortest side of the cardboard is 22 inches, the length x of the side of the cut-out square must be less than 11 (why?).

Each point on the graph of $y = 4x^3 - 104x^2 + 660x$ $(0 < x < 11)$ in Figure 2–56 represents one of the possibilities:

The x-coordinate is the size of the square to be cut from each corner;

The y-coordinate is the volume of the resulting box.

The box with the largest volume corresponds to the point with the largest y-coordinate, that is, the highest point in the viewing window. A maximum finder (Figure 2–57) shows that this point is approximately (4.182, 1233.809). Therefore, a square measuring approximately 4.18 × 4.18 inches should be cut from each corner, producing a box of volume approximately 1233.81 cubic inches. ∎

EXAMPLE 4

A cylindrical can of volume 58 cubic inches (approximately 1 quart) is to be designed. For convenient handling, it must be at least 1 inch high and 2 inches in diameter. What dimensions will use the least amount of material?

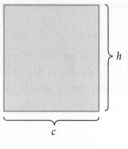

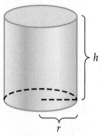

Figure 2–58

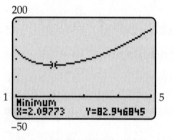

Figure 2–59

SOLUTION   We can construct a can by rolling a rectangular sheet of metal into a tube and then attaching the top and bottom, as shown in Figure 2–58. The surface area of the can (which determines the amount of material) is

$$\underset{\text{sheet}}{\text{Area of rectangular}} + \underset{\text{top}}{\text{Area of}} + \underset{\text{bottom}}{\text{Area of}}$$

$$ch + \pi r^2 + \pi r^2 = ch + 2\pi r^2.$$

When the sheet is rolled into a tube, the width c of the sheet is the circumference of the end of the can, so that $c = 2\pi r$, and hence,

$$\text{Surface area} = ch + 2\pi r^2 = 2\pi rh + 2\pi r^2.$$

The volume of a cylinder of radius r and height h is $\pi r^2 h$. Since the can is to have volume 58 cubic inches, we have

$$\pi r^2 h = 58, \quad \text{or equivalently,} \quad h = \frac{58}{\pi r^2}.$$

Therefore,

$$\text{Surface area} = 2\pi rh + 2\pi r^2 = 2\pi r\left(\frac{58}{\pi r^2}\right) + 2\pi r^2 = \frac{116}{r} + 2\pi r^2.$$

Note that r must be greater than 1 (since the diameter $2r$ must be at least 2). Furthermore, r cannot be more than 5 (if $r > 5$ and $h \geq 1$, then the volume $\pi r^2 h$ would be at least $\pi \cdot 25 \cdot 1$, which is greater than 58). The situation can be represented by the graph of the equation

$$y = 116/x + 2\pi x^2,$$

as in Figure 2–59. The x-coordinate of each point represents a possible radius, and the y-coordinate represents the surface area of the corresponding can. We must find the point with the smallest y-coordinate, that is, the lowest point on the graph. A graphical minimum finder (Figure 2–59) shows that the coordinates of this point are approximately (2.09773, 82.946845). If the radius is 2.09773, then the height is $58/(\pi 2.09773^2) \approx 4.1955$. As a practical matter, it would probably be best to round to one decimal place and construct a can of radius 2.1 and height 4.2 inches. ∎

EXERCISES 2.4

In Exercises 1–6, find the coordinates of the highest or lowest point on the part of the graph of the equation in the given viewing window. Only the range of x-coordinates for the window are given; you must choose an appropriate range of y-coordinates.

1. $y = 2x^3 - 3x^2 - 12x + 1$;   highest point when $-3 \leq x \leq 3$

2. $y = 2x^6 + 3x^5 + 3x^3 - 2x^2$;   lowest point when $-3 \leq x \leq 3$

3. $y = \dfrac{4}{x^2} - \dfrac{7}{x} + 1$;   lowest point when $-10 \leq x \leq 10$

4. $y = \dfrac{1}{x^2 + 2x + 2}$;   highest point when $-5 \leq x \leq 5$

5. $y = \dfrac{x^2(x + 1)^3}{(x - 2)^2(x - 4)^2}$;   highest point when $-1 \leq x \leq 0$

   [*Hint:* Think small.]

6. $y = \dfrac{x^2(x + 1)^3}{(x - 2)(x - 4)^2}$;   lowest point when $-10 \leq x \leq -1$

7. Find the highest point on the part of the graph of $y = x^3 - 3x + 2$ that is shown in the given window. The answers are not all the same.
   (a) $-2 \leq x \leq 0$ (b) $-2 \leq x \leq 2$
   (c) $-2 \leq x \leq 3$

8. Find the lowest point on the part of the graph of $y = x^3 - 3x + 2$ that is shown in the given window.

(a) $0 \le x \le 2$ (b) $-2 \le x \le 2$
(c) $-3 \le x \le 2$

9. The fuel economy y of a representative car (in miles per gallon) can be approximated by

$$y = -.00000636x^4 + .001032x^3 - .067x^2 + 2.19x + 8.6,$$

where x is the speed of the car (in miles per hour).* At what speed does this car get the most miles per gallon?[†]

10. Between 1997 and 2005, the number y of unemployed (in thousands) was approximated by

$$y = -53.4x^3 + 1772.33x^2 - 18,681.32x + 69,188.1$$
$$(7 \le x \le 15),$$

where $x = 7$ corresponds to 1997.[‡] In what year was unemployment the highest?

11. In the situation described in Exercise 33 of Section 2.3 (page 113), how far from the building wall should the rope meet the fence, if as little rope as possible is to be used.

12. In the situation described in Exercise 34 of Section 2.3 (page 113), how far should Anne run along the road if she wants her trip to the helicopter to take the least possible amount of time?

13. A farmer has 1800 feet of fencing. He plans to enclose a rectangular region bordering a river (with no fencing needed along the river side). What dimensions should he use to have an enclosure of largest possible area?

14. A rectangular field will be fenced on all four sides. Fencing for the north and south sides costs $5 per foot and fencing for the other two sides costs $10 per foot. What is the maximum area that can be enclosed for $5000?

15. A rectangular area of 24,200 square feet is to be fenced on all four sides. Fencing for the east and west sides costs $10 per foot and fencing for the other two sides costs $20 per foot. What is the cost of the least expensive fence?

16. A fence is needed to enclose an area of 30,246 square feet. One side of the area is bounded by an existing fence, so no new fencing is needed there. Fencing for the side opposite the existing fence costs $18 per foot. Fencing for the other two sides costs $6 per foot. What is the cost of the least expensive fence?

17. Find the dimensions of the rectangular box with a square base and no top that has volume 20,000 cubic centimeters and the smallest possible surface area. [*Hint:* See Example 3 in Section 2.3 (page 103).]

18. An open-top box with a square base is to be constructed from 120 square centimeters of material. What dimensions will produce a box

(a) of volume 100 cubic centimeters?
(b) with largest possible volume?

19. A 20-inch square piece of metal is to be used to make an open-top box by cutting equal-sized squares from each corner and folding up the sides (as in Example 3 on page 117). The length, width, and height of the box are each to be less than 14 inches. What size squares should be cut out to produce a box with

(a) volume 550 cubic inches?
(b) largest possible volume?

20. A cylindrical waste container with no top, a diameter of at least 2 feet, and a volume of 25 cubic feet is to be constructed. What should its radius be if

(a) 65 square feet of material are to be used to construct it?
(b) the smallest possible amount of material is to be used to construct it? In this case, how much material is needed?

21. If $c(x)$ is the cost of producing x units, then $c(x)/x$ is the *average cost* per unit.* Suppose the cost of producing x units is given by $c(x) = .13x^3 - 70x^2 + 10,000x$ and that no more than 300 units can be produced per week.

(a) If the average cost is $1100 per unit, how many units are being produced?
(b) What production level should be used in order to minimize the average cost per unit? What is the minimum average cost?

22. A manufacturer's revenue (in cents) from selling x items per week is given by $200x - .02x^2$. It costs $60x + 30,000$ cents to make x items.

(a) Approximately how many items should be made each week to make a profit of $1100? (Don't confuse cents and dollars.)
(b) How many items should be made each week to have the largest possible profit? What is that profit?

23. (a) A company makes novelty bookmarks that sell for $142 per hundred. The cost (in dollars) of making x hundred bookmarks is $x^3 - 8x^2 + 20x + 40$. Because of other projects, a maximum of 600 bookmarks per day can be manufactured. Assuming that the company can sell all the bookmarks it makes, how many should it make each day to maximize profits?
(b) Owing to a change in other orders, as many as 1600 bookmarks can now be manufactured each day. How many should be made to maximize profits?

24. If the cost of material to make the can in Example 4 on pages 117–118 is 5 cents per square inch for the top and bottom and 3 cents per square inch for the sides, what dimensions should be used to minimize the cost of making the can? [The answer is not the same as in Example 4.]

25. A certain type of fencing comes in rigid 10-foot-long segments. Four uncut segments are used to fence in a garden on the side of a building, as shown in the figure on the next page.

---

*Based on data from the U.S. Department of Energy.
[†]The most fuel efficient speed for a particular car may differ by 3 or 4 mph from the answer here.
[‡]Based on data from the U.S. Department of Labor Statistics.

---

*Depending on the situation, a unit of production might consist of a single item or several thousand items. Similarly, the cost of x units might be measured in thousands of dollars.

What value of x will result in a garden of the largest possible area, and what is that area?

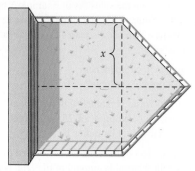

26. A rectangle is to be inscribed in a semicircle of radius 2, as shown in the figure. What is the largest possible area of such a rectangle? [*Hint:* The width of the rectangle is the second coordinate of the point P (why?), and P is on the top half of the circle $x^2 + y^2 = 4$.]

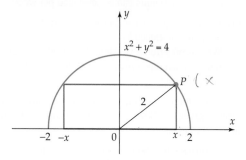

27. Find the point on the graph of $y = 5 - x^2$ that is closest to the point $(0, 1)$ and has positive coordinates. [*Hint:* The distance from the point (x, y) on the graph to $(0, 1)$ is $\sqrt{(x - 0)^2 + (y - 1)^2}$; express y in terms of x.]

28. A manufacturer's cost (in thousands of dollars) of producing x thousand units is $x^3 - 6x^2 + 15x$ dollars, and the revenue (in thousands) from x thousand units is $9x$ dollars. What production level(s) will result in the largest possible profit?

29. A hardware store sells ladders throughout the year. It costs \$20 every time an order for ladders is placed and \$10 to store a ladder until it is sold. When ladders are ordered x times per year, then an average of $300/x$ ladders are in storage at any given time. How often should the company order ladders each year to minimize its total ordering and storage costs? [*Be careful:* The answer must be an integer.]

30. A mathematics book has 36 square inches of print per page. Each page has a left side margin of 1.5 inches and top, bottom, and right side margins of .5 inch. If a page cannot be wider than 7.5 inches, what should its length and width be to use the least amount of paper?

2.5   Linear Models*

Section Objectives
- Construct a linear model.
- Gauge the accuracy of a linear model using residuals.
- Use linear regression to find the least squares regression line.
- Interpret the correlation coefficient r.

People working in business, medicine, agriculture, and other fields often need to make judgments based on past data. For instance, a stock analyst might use the past profits of a company to estimate next year's profits, or a doctor might use data on previous patients to determine the ideal dosage of a drug for a new patient. In such situations, the available data can sometimes be used to construct a **mathematical model,** such as an equation or graph, that approximates the likely outcome in cases that are not included in the data. In this section, we consider applications in which the data can be modeled by a linear equation.

The simplest way to construct a linear model is to use the line determined by two of the data points, as illustrated in the following example.

---

*This section is optional. It will be used only in clearly identifiable exercises or (sub)sections of the text that can be omitted by those not interested.

EXAMPLE 1

The profits of the General Electric Company (in billions of dollars) during the first part of this decade are shown in the table.*

Year	2000	2001	2002	2003	2004	2005
Profit	13	14	14	15	17	16

Let $x = 0$ correspond to 2000, so that the data points are $(0, 13)$, $(1, 14)$, $(2, 14)$, $(3, 15)$, $(4, 17)$ and $(5, 16)$.

(a) Make a scatter plot of the data points.

(b) Use the points $(0, 13)$ and $(5, 16)$ to find a line that models the data and graph this line.

(c) Use the points $(1, 14)$ and $(5, 16)$ to find another line that models the data and graph this line.

SOLUTION

(a)

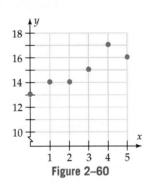

Figure 2–60

Figure 2–60 shows that the data is approximately linear, so a line would be a reasonable model.

(b) *Points:* $(0, 13)$ and $(5, 16)$

  Slope of line: $\dfrac{16 - 13}{5 - 0} = \dfrac{3}{5} = .6$

  Equation: $y - y_1 = m(x - x_1)$

     $y - 13 = .6(x - 0)$

       $y = .6x + 13$

(c) *Points:* $(1, 14)$ and $(5, 16)$

  Slope: $\dfrac{16 - 14}{5 - 1} = \dfrac{2}{4} = .5$

  Equation: $y - y_1 = m(x - x_1)$

     $y - 14 = .5(x - 1)$

     $y - 14 = .5x - .5$

       $y = .5x + 13.5$

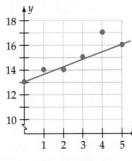

Figure 2–61

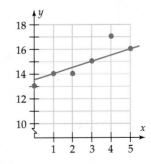

Figure 2–62

---

*GE Annual Report 2005. The profit figures (net earnings) are rounded to the nearest billion.

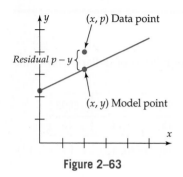

Figure 2–63

Both models in Example 1 seem reasonable. To determine the better one, we first measure the errors in each model by computing the difference between the actual profit p and the amount y given by the model. If the data point is (x, p) and the corresponding point on the line is (x, y), then the difference $p - y$ measures the error in the model for that particular value of x. The number $p - y$ is called a **residual.** As shown in Figure 2–63, the residual $p - y$ is the vertical distance from the data point to the line (positive when the data point is above the line, negative when it is below the line, and 0 when it is on the line).

The universally accepted measure of the overall accuracy of a linear model is the sum of the *squares* of its residuals—the smaller this number is, the better the model fits the data points.* Using this number has the effect of emphasizing large errors (those with absolute value greater than 1) because the square is greater than the residual and minimizing small errors (those with absolute value less than 1) because the square is less than the residual. So a smaller sum means that the line has less overall error and fits the data better.

EXAMPLE 2

Two models for GE's profits were constructed in Example 1:

$$y = .6x + 13 \quad \text{and} \quad y = .5x + 13.5.$$

For each model, determine the residuals, the squares of the residuals, and the sum of these squares. Decide which model is a better fit.

SOLUTION   The following tables give the required information.

| $y = .6x + 13$ | | | |
Data Point (x, p)	Model Point (x, y)	Residual $p - y$	Squared Residual $(p - y)^2$
(0, 13)	(0, 13)	0	0
(1, 14)	(1, 13.6)	.4	.16
(2, 14)	(2, 14.2)	−.2	.04
(3, 15)	(3, 14.8)	.2	.04
(4, 17)	(4, 15.4)	1.6	2.56
(5, 16)	(5, 16)	0	0
			Sum: 2.8

| $y = .5x + 13.5$ | | | |
Data Point (x, p)	Model Point (x, y)	Residual $p - y$	Squared Residual $(p - y)^2$
(0, 13)	(0, 13.5)	−.5	.25
(1, 14)	(1, 14)	0	0
(2, 14)	(2, 14.5)	−.5	.25
(3, 15)	(3, 15)	0	0
(4, 17)	(4, 15.5)	1.5	2.25
(5, 16)	(5, 16)	0	0
			Sum: 2.75

Since the sum of the squares of the residuals of $y = .5x + 13.5$ is less than the sum of the squares of the residuals of $y = .6x + 13$, we conclude that $y = .5x + 13.5$ fits the data better.   ∎

---

*The sum of the residuals themselves is less useful. As Exercise 1 shows, the residuals for two different models of the data may each sum to 0 (which doesn't mean there is no error, but only that the positive and negative errors cancel each other out). In contrast, the sum of the *squares* of the residuals is 0 only when every residual is 0, that is, when all the data points lie on the model line.

LINEAR REGRESSION

The following fact, which requires multivariate calculus for its proof, shows that there is always a best possible model for linear data.

Linear Regression Theorem

> For any set of data points, there is one and only one line for which the sum of the squares of the residuals is as small as possible. This line is called the **least squares regression line.**

The computational process for finding the least squares regression line is called **linear regression.** The linear regression formulas are quite complicated and can be tiresome to use with a large data set. Fortunately, however, linear regression routines are built into most calculators and are also available in spreadsheet and other computer programs.

EXAMPLE 3

As we saw in Example 1, GE's profits (in billions of dollars) during the first part of this decade were as follows.

Year	2000	2001	2002	2003	2004	2005
Profit	13	14	14	15	17	16

Use technology to find the least squares regression line that models this data. Graph the data points and the regression line.

SOLUTION   Let $x = 0$ correspond to 2000, so that the data points are (0, 13), (1, 14), and so on. Enter the data points as two lists in the calculator's statistics editor (Figure 2–64) and make a scatter plot of the data (Figure 2–65).*

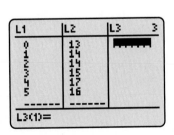

Figure 2–64

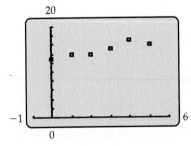

Figure 2–65

In the statistics calculation menu, choose linear regression (Figure 2–66 on the next page) and enter the list names and the variable where the regression line graph should be stored (Figure 2–67).† Press ENTER to obtain the equation of the regression line (Figure 2–68).

---

*Directions for doing this are in the Technology Tip on page 88.
†See the Technology Tip at the end of this section for directions on how to do this.

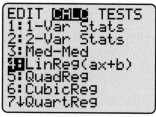

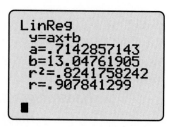

Figure 2–66   **Figure 2–67**   **Figure 2–68**

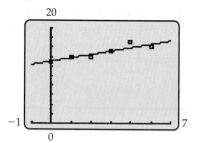

Figure 2–69

So the equation of the least squares regression line, which best fits the data, is

$$y = .7142857143x + 13.04761905.$$

Finally, press GRAPH to see the graph of the regression line (Figure 2–69).   ■

In addition to the coefficients of the regression line, Figure 2–68 contains a number r (and its square). The number r, which is called the **correlation coefficient,** is always between -1 and 1. It is a statistical measure of how well the regression line fits the data. The closer the absolute value of r is to 1, the better the fit. For instance, the regression line in Example 3 is a good fit because $r \approx .91$, as shown in Figure 2–68. When $r = 1$ or $r = -1$, the fit is perfect: All the data points are on the regression line. Conversely, a regression coefficient near 0 indicates a poor fit, as shown in the summary below.

Correlation Coefficient r	
$0 < r \leq 1$	$-1 \leq r < 0$
The regression line has positive slope and moves upward from left to right. As x increases, y also increases. We say that the data has **positive correlation.** *Examples:*	The regression line has negative slope and moves downward from left to right. As x increases, y decreases. We say that the data has **negative correlation.** *Examples:*

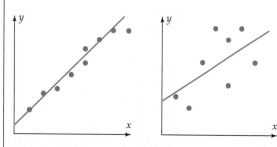

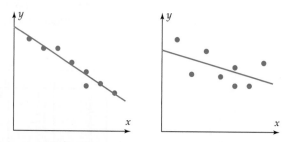

r is very close to 0 (regardless of sign)
The regression line is a very poor fit for the data. We say that the data has **no correlation.** *Examples:*

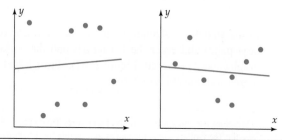

APPLICATIONS

The various possibilities in the preceding box are illustrated in the following applied examples.

EXAMPLE 4

The table shows the poverty level for a family of four in selected years (families whose income is below this level are considered to be living in poverty).*

Year	1996	1998	2000	2002	2004
Income	$15,141	$16,660	$17,603	$18,392	$19,307

(a) Use linear regression to find an equation that models this data.

(b) If the model fits the data well, use the equation to estimate the poverty level in 2003 and in 2008.

SOLUTION

(a) Let $x = 0$ correspond to 1990 and write the income in thousands (for instance, 15.141 in place of 15,141). Then the data points are (6, 15.141), . . . , (14, 19.307). We proceed as in Example 3 to find the least squares regression line (Figure 2–70) and its graph (Figure 2–71).

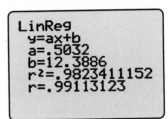

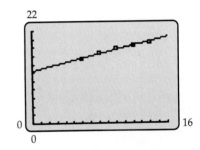

Figure 2–70   Figure 2–71

The correlation coefficient $r \approx .99$ shows that the data is positively correlated and that the model is an excellent fit. The graph confirms this.

(b) The year 2003 corresponds to $x = 13$. To estimate the poverty level in 2003, substitute $x = 13$ in the regression equation:

$$y = .5032x + 12.3886$$
$$= .5032(13) + 12.3886$$
$$= 18.9302.$$

So the regression model estimates the poverty level in 2003 to be $18,930. The actual level in that year was $18,811, a difference of only $119. Similarly, for 2008 ($x = 18$) the model estimates the poverty level to be

$$y = .5032(18) + 12.3886 = 21.4462$$

*U.S. Census Bureau.

or $21,446. Because 2008 is outside the range of the data points, this figure might not be as accurate as the estimate for 2003, which lies within the data range.   ■

EXAMPLE 5

The death rates from heart disease (per 100,000 population) are given in the table.*

Year	Death Rate	Year	Death Rate
1980	412.1	2001	247.8
1985	375.0	2002	240.8
1990	321.8	2003	232.3
1995	296.3	2004	217.5
2000	257.6		

Find the linear regression model for this data. If the model fits the data well, use it to estimate the death rates in 1998 and 2008.

SOLUTION   We let $x = 0$ correspond to 1980, enter the data points in the statistical editor of a calculator, and find the equation of the regression line (Figure 2–72). Since $r \approx -.98$, the data is negatively correlated and the model fits it very well, as the graph confirms (Figure 2–73).

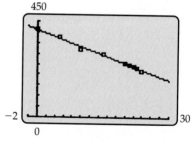

Figure 2–72   Figure 2–73

To find the death rates in 1998 and 2008, we evaluate the equation when $x = 18$ and $x = 28$.[†]

$$y = -7.8199x + 410.6649 \qquad\qquad y = -7.8199x + 410.6649$$

$$= -7.8199(18) + 410.6649 \qquad = -7.8199(28) + 410.6649$$

$$\approx 269.9 \qquad\qquad\qquad \approx 191.7$$

So the death rates are estimated to be 269.9 in 1998 and 191.7 in 2008.   ■

---

*U.S. National Center for Health Statistics.
[†]The coefficients are rounded for easier reading; this doesn't affect the answer here.

EXAMPLE 6

The number of unemployed people in the labor force (in millions) for 1988–2005 is shown in the table.* Determine whether a linear equation is a good model for this data.

Year	Unemployed	Year	Unemployed	Year	Unemployed
1988	6.701	1994	7.996	2000	5.692
1989	6.528	1995	7.404	2001	6.801
1990	7.047	1996	7.236	2002	8.378
1991	8.628	1997	6.739	2003	8.774
1992	9.613	1998	6.210	2004	8.149
1993	8.940	1999	5.880	2005	7.591

SOLUTION   Let $x = 0$ correspond to 1980. After entering the data points as two lists in the statistics editor, you can test it graphically or analytically.

Graphical: The scatter plot of the data points in Figure 2–74 does not have a linear pattern (unemployment tends to rise and fall).

Analytical: Linear regression (Figure 2–75) produces an equations whose correlation coefficient that is almost 0, namely, $r \approx -.017$. This indicates that there is no correlation and that the regression line is a very poor fit for the data.

Therefore, a linear equation is not a good model for this data. ∎

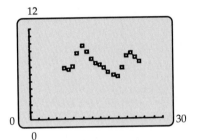

Figure 2–74

```
LinReg
 y=ax+b
 a=-.0036707946
 b=7.522068111
 r²=3.058433ᴇ-4
 r=-.0174883773
 ■
```

Figure 2–75

GRAPHING EXPLORATION

Enter the data from Example 6 in the statistics editor of your calculator. Graph the data points. Graph the least squares regression line on the same screen to see how poorly it fits the data.

The following Technology Tip may be helpful for learning how to use the regression feature of your calculator.

TECHNOLOGY TIP

If the data points have been entered in the statistics editor as lists L_1 and L_2, use these TI commands to find the least squares regression line and store its equation as y_1 in the equation memory:

TI-84+: STAT CALC *LinReg* L_1, L_2, Y_1
TI-86: STAT CALC *Lin R* L1, L2, y_1
TI-89: From the Data Editor, choose CALC (F5); enter LIN REG as the CALCULATION TYPE, list C_1 as x, list C_2 as y, and choose $y_1(x)$ in STORE REGEQ.

To graph the equation and the data points (assuming Plot 1 is ON and set for L_1 and L_2), press GRAPH or DRAW.
    On HP-39gs, plot the data points, press MENU FIT to graph the regression line and SYMB to see its equation.
    On Casio 9850, plot the data points, press X for the equation of the regression line and DRAW for its graph.

*U.S. Bureau of Labor Statistics.

EXERCISES 2.5

In Exercises 1–4, two linear models are given for the data. For each model,
(a) *Find the residuals and their sum;*
(b) *Find the sum of the squares of the residuals;*
(c) *Determine which model is the better fit.*

1. The weekly amount spent on advertising and the weekly sales revenue of a small store over a five-week period are shown in the table. Two models are $y = x$ and $y = .5x + 1.5$.

Advertising Expenditures x (in hundreds of dollars)	1	2	3	4	5
Sales Revenue y (in thousands of dollars)	2	2	3	3	5

2. Advertising expenditures in the United States (in billions of dollars) in selected years are shown in the table.* Two models are $y = 12x + 215$ and $y = 11x + 218$, where $x = 0$ corresponds to 2000.

Year	2001	2002	2003	2004
Amount	231	237	245	264

3. The table gives the consumer price index (CPI) in April of selected years.[†] Two models are $y = 4.9x + 170$ and $y = 5x + 171$, where $x = 0$ corresponds to 2000.

Year	2000	2002	2004	2006
CPI	171.3	179.8	188	201.5

4. Revenues for U.S. public elementary and secondary schools (in billions of dollars) in the fall of selected years are shown in the table.[‡] Two models are

$$y = 21x + 370 \quad \text{and} \quad y = 21.6x + 372,$$

where $x = 0$ corresponds to 2000.

Year	2000	2002	2003	2004
Revenues	370	416	444	453

5. The regression model for GE's profits in Example 3 on page 123 (with coefficients rounded) was $y = .71x + 13.05$. Compute the sum of the squares of its residuals and verify that this model is better than either of those in Example 2.

Advertising Age.
[†]*U.S. Bureau of Labor Statistics.*
[‡]*Statistical Abstract of the United States: 2006.*

In Exercises 6–9, determine whether the given scatter plot of the data indicates that there is a positive correlation, a negative correlation, or very little correlation.

6.
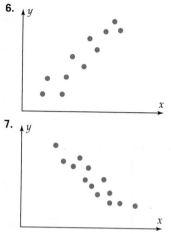

7.
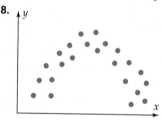

8.
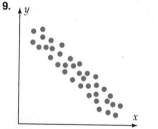

9.

In Exercises 10–15, construct a scatter plot for the data and answer these questions: (a) Does the data appear to be linear? (b) If so, is there a positive or negative correlation?

10. Consumer debt (in trillions of dollars) is shown in the table.* Let $x = 0$ correspond to 1990.

Year	1995	2000	2002	2003	2004
Debt	1.1	1.7	1.9	2	2.1

11. The U.S. gross domestic product (GDP) is the total value of all goods and services produced in the United States. The table shows the GDP in billions of 2000 dollars.[†] Let $x = 0$ correspond to 1990.

Federal Reserve Bulletin. The amounts are primarily credit card balances; home mortgages are not included.
[†]U.S. Bureau of Economic Analysis.

Year	1990	1995	1998	2000	2002	2004
GDP	7113	8032	9067	9817	10,075	10,842

12. In mid-2002, Grange Life Insurance advertised the following monthly premium rates for a $50,000 term policy for a female nonsmoker. Let x represent age and y the monthly premium.

Age	30	35	40	45	50
Premium	6.48	6.52	6.78	7.88	9.93

Age	55	60	65	70
Premium	12.69	16.71	23.67	36.79

13. The table shows the average monthly temperature (in degrees Fahrenheit) in Cleveland, Ohio.* Let $x = 2$ correspond to February, $x = 4$ to April, etc.

Month	Feb	April	June	Aug	Oct	Dec
Temperature	27.3	47.5	67.5	70.3	52.7	30.9

14. The vapor pressure y of water depends on the temperature x, as given in the table.

Temperature (°C)	Pressure (mm Hg)
0	4.6
10	9.2
20	17.5
30	31.8
40	55.3
50	92.5
60	149.4
70	233.7
80	355.1
90	525.8
100	760

15. The table shows the U.S. Bureau of Census population data for St. Louis, Missouri. Let $x = 0$ correspond to 1950.

Year	Population
1950	856,796
1970	622,236
1980	452,801
1990	396,685
2000	348,189
2005	352,572

*National climatic Data Center.

In Exercises 16–26, use linear regression to find the requested linear model.

16. The table shows the median time (in months) for the U.S. Food and Drug Administration to approve a generic drug.

Year	Median Approval Time
1997	19.3
1998	18
1999	18.6
2000	18.2
2001	18.1
2002	18.2
2003	17
2004	15.7

(a) Make a scatter plot of the data, with $x = 0$ corresponding to 1990.
(b) Find a linear model for the data.
(c) Assume that the model remains accurate and find the median approval time in 2010.

17. The table shows the size of a room air conditioner (in BTUs) needed to cool a room of the given area (in square feet).

Room Size	BTUs
150	5000
175	5500
215	6000
250	6500
280	7000
310	7500
350	8000
370	8500
420	9000
450	9500

(a) Find a linear model for the data.
(b) Use the model to find the number of BTUs required to cool a rooms of size 150 sq ft, 280 sq ft, and 420 sq ft. How well do the model estimates agree with the actual data values?
(c) Use the model to estimate how many BTUs are needed to cool a 235 sq ft room. If air conditioners are available only with the BTU choices in the table, which size should be chosen?

18. Enrollment in public colleges (in thousands) in selected years is shown in the table on next page.*

*Data and projections from the U.S. National Center for Education Statistics.

Year	2000	2001	2002	2004	2006	2008
Enrollment	15,313	15,928	16,612	17,095	17,664	18,350

(a) Find a linear model for this data, with $x = 0$ corresponding to 2000.

(b) Use the model to estimate public college enrollment in 2005 and 2010.

(c) According to this model, when will public college enrollment reach 21 million?

19. The graph shows Intel's expenditures on research and development (in billions of dollars) over a ten-year period.*

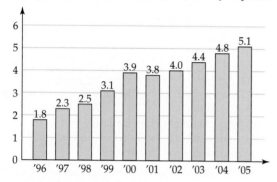

(a) List the data points, with $x = 6$ corresponding to 1996.

(b) Find a linear model for the data.

(c) If the trend shown continues, what will Intel spend on research and development in 2010?

In Exercises 20 and 21, use the following table, which gives the median weekly earnings of full-time workers 25 years and older by their educational level.[†]

Year	No High School Diploma	High School Graduate	Some College	College Graduate
2000	$360	$506	$598	$896
2001	$378	$520	$621	$924
2002	$388	$536	$617	$941
2003	$396	$554	$622	$963
2004	$401	$574	$642	$986
2005	$412	$584	$654	$996

20. (a) Find linear models for the median weekly earnings of full-time workers 25 years and older who did not graduate from high school and for those who graduated from high school, but did not attend college. Let $x = 0$ correspond to 2000.

(b) Estimate the median weekly income for each group in part (a) in 2009.

21. (a) Find linear models for the median weekly earnings of full-time workers 25 years and older who attended college, but did not graduate, and those who graduated from college. Let $x = 0$ correspond to 2000.

(b) Use the models in Exercises 20(a) and 21(a) to determine the approximate yearly rate at which the median weekly earnings of each of the four groups in the table are increasing. What does this say about the value of education?

22. The table shows the relationship between the temperature (in degree Fahrenheit) and the rate at which the striped ground cricket chirps.*

Temperature	Chirps per second
88.6	20.0
71.6	16.0
93.3	19.8
84.3	18.4
80.6	17.1
75.2	15.5
69.7	14.7
82.0	17.1
69.4	15.4
83.3	16.2
79.6	15.0
82.6	17.2
80.6	16.0
83.5	17.0
76.3	14.4

(a) Find a linear model for this data and lists its correlation coefficient.

(b) Estimate the number of chirps per second at a temperature of 73°.

(c) If the chirping rate is 18 times per second, estimate the temperature.

23. The table on the next page gives the life expectancy at birth (in years) in selected years in the United States.[†]

(a) Find a linear model for men's life expectancy, with $x = 70$ corresponding to 1970.

(b) Do part (a) for women's life expectancy.

(c) Suppose the models in parts (a) and (b) remain accurate in the future. Will men's life expectancy ever be the same as women's? If so, in what birth year will this occur?

---

*Source: *The Songs of Insects* by George W. Pierce. Cambridge, MA: Harvard University Press, copyright © 1948 by the President and Fellows of Harvard College.
[†]U.S. Center for Health Statistics.

*Intel Corporation Annual Report, 2005.
[†]U.S. Bureau of Labor Statistics.

Birth Year	Life Expectancy	
	Men	Women
1970	67.1	74.7
1975	68.8	76.6
1980	70	77.4
1985	71.1	78.2
1990	71.8	78.8
1995	72.5	78.9
1998	73.8	79.5
2000	74.3	79.7
2003	74.8	80.1
2005	74.9	80.7

24. The projected number of new cases of Alzheimer's disease (in thousands) in the United States in selected years is shown in the table.*

Year	2000	2010	2020	2030	2040	2050
New Cases	400	467	489	600	800	956

(a) Find a linear model for this data, with $x = 0$ corresponding to 2000, and use it to estimate the number of new cases in the years 2005, 2015, and 2025.

*U.S. Center for Health Statistics.

(b) Use the model to predict the years in which the number of new cases will be 750,000 and 1,000,000.

25. The projected number of scheduled passengers on U.S. commercial airlines (in billions) is given in the table.*

Year	2002	2003	2004	2005	2006	2007	2008	2009
Passengers	.63	.64	.69	.72	.77	.79	.8	.84

(a) Find a linear model for this data, with $x = 2$ corresponding to 2002.
(b) Estimate the number of passengers in 2012 and 1016.

26. During the current decade, the amount of money each American spends annually on prescription drugs (in addition to the amounts paid by insurance) is projected to increase, as shown in the table.†

Year	2000	2002	2004	2006	2008	2010
Amount	$143	$175	$207	$231	$287	$351

(a) Find a linear model for this data, with $x = 0$ corresponding to 2000.
(b) Use the model to estimate the amount each person will spend on prescription drugs in 2005, 2007, and 2011.
(c) According to this model, in what year will each American spend $400 on prescription drugs?

*Data and projections from the Federal Aviation Administration.
†Centers for Medicare and Medicaid Services.

Chapter 2 Review

IMPORTANT CONCEPTS

REVIEW QUESTIONS

In Questions 1–6,
(a) Determine which of the viewing windows a–e shows a complete graph of the equation.
(b) For each viewing window that does not show a complete graph, explain why.
(c) Find a viewing window that gives a "better" complete graph than windows a–e (meaning that the window is small enough to show as much detail as possible, yet large enough to show a complete graph).

  (a) Standard viewing window
  (b) $-10 \le x \le 10, -200 \le y \le 200$
  (c) $-20 \le x \le 20, -500 \le y \le 500$
  (d) $-50 \le x \le 50, -50 \le y \le 50$
  (e) $-1000 \le x \le 1000, -1000 \le y \le 1000$

1. $y = .2x^3 - .8x^2 - 2.2x + 6$

2. $y = x^3 - 11x^2 - 25x + 275$

3. $y = x^4 - 7x^3 - 48x^2 + 180x + 200$

4. $y = x^3 - 6x^2 - 4x + 24$

5. $y = .03x^5 - 3x^3 + 69.12x$

6. $y = .00000002x^6 - .0000014x^5 - .00017x^4 +$
    $.0107x^3 + .2568x^2 - 12.096x$

In Questions 7–10, sketch a complete graph of the equation, and give reasons why it is complete.

7. $y = x^2 - 10$ **8.** $y = x^3 + x + 4$

9. $y = \sqrt{x - 5}$ **10.** $y = x^4 + x^2 - 6$

In Questions 11–14, sketch a complete graph of the equation.

11. $y = x^2 - 13x + 43$ **12.** $y = |x|$

13. $y = |x + 5|$ **14.** $y = 1/x$

In Questions 15–22, solve the equation graphically. You need only find solutions in the given interval.

15. $x^3 + 2x^2 = 11x + 6$; $[0, \infty)$

16. $x^3 + 2x^2 = 11x + 6$; $(-\infty, 0)$

17. $x^4 + x^3 - 10x^2 = 8x + 16$; $[0, \infty)$

18. $2x^4 + x^3 - 2x^2 + 6x + 2 = 0$; $(-\infty, -1)$

19. $\dfrac{x^3 + 2x^2 - 3x + 4}{x^2 + 2x - 15} = 0$; $(-10, \infty)$

20. $\dfrac{3x^4 + x^3 - 6x^2 - 2x}{x^5 + x^3 + 2} = 0$; $[0, \infty)$

21. $\sqrt{x^3 + 2x^2 - 3x - 5} = 0$; $[0, \infty)$

22. $\sqrt{1 + 2x - 3x^2 + 4x^3 - x^4} = 0$; $(-5, 5)$

23. A jeweler wants to make a 1-ounce ring consisting of gold and silver, using $200 worth of metal. If gold costs $600 per ounce and silver costs $50 per ounce, how much of each metal should she use?

24. A calculator is on sale for 15% less than the list price. The sale price, plus a 5% shipping charge, totals $210. What is the list price?

25. Karen can do a job in 5 hours, and Claire can do the same job in 4 hours. How long will it take them to do the job together?

26. A car leaves the city traveling at 54 mph. One-half hour later, a second car leaves from the same place and travels at 63 mph along the same road. How long will it take for the second car to catch up with the first?

27. A 12-foot-long rectangular board is cut in two pieces so that one piece is four times as long as the other. How long is the bigger piece?

28. George owns 200 shares of stock, 40% of which are in the computer industry. How many more shares must he buy to have 50% of his total shares in computers?

29. A square region is changed into a rectangular one by making it 2 feet wider and twice as long. If the area of the rectangular region is three times larger than the area of the original square region, what was the length of a side of the square before it was changed?

30. The radius of a circle is 10 inches. By how many inches should the radius be increased so that the area increases by 5π square inches?

31. The cost of manufacturing x caseloads of ballpoint pens is

$$\frac{600x^2 + 600x}{x^2 + 1}$$

dollars. How many caseloads should be manufactured to have an *average cost* of $25? [Average cost was defined in Exercise 21 of Section 2.4.]

32. An open-top box with a rectangular base is to be constructed. The box is to be at least 2 inches wide and twice as long as it is wide and is to have a volume of 150 cubic inches. What should the dimensions of the box be if the surface area is to be

  (a) 90 square inches?   (b) as small as possible?

33. A farmer has 120 yards of fencing and wants to construct a rectangular pen, divided in two parts by an interior fence, as shown in the figure. What should the dimensions of the pen be to enclose the maximum possible area?

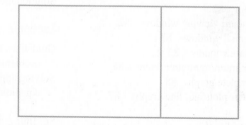

34. The top and bottom margins of a rectangular poster are each 5 inches, and each side margin is 3 inches. The printed material on the poster occupies an area of 400 square inches. Find the dimensions that will use the least possible amount of posterboard.

35. A rectangle has one side on the *x*-axis, and its other two corners sit on the graph of $y = 9 - x^2$, as shown in the figure. What value of *x* gives a rectangle of maximum area?

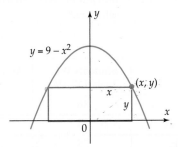

36. The window in the figure has a rectangular bottom, with a semicircle of radius *r* lying on top of it, and a perimeter of 40 feet. In order that the window have the maximum possible area, what should *r* and *h* be?

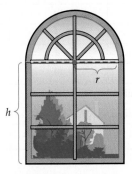

37. The table shows monthly premiums (as of 2002) for a $100,000 term life insurance policy from Grange Life Insurance for a male smoker.

Age	Premium
25	$20.30
30	$20.39
35	$20.39
40	$22.05
45	$28.61
50	$41.65
55	$60.90
60	$85.58
65	$132.91

(a) Make a scatter plot of the data, using *x* for age and *y* for premiums.
(b) Does the data appear to be linear?

38. For which of the following scatter plots would a linear model be reasonable? Which sets of data show positive correlation, and which show negative correlation?

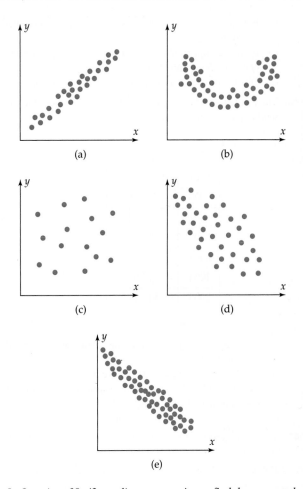

(a)

(b)

(c)

(d)

(e)

In Questions 39–42, use linear regression to find the requested linear model.

39. The table shows how the circumference (in inches) of a white oak tree (the Illinois state tree) is related to the approximate age of the tree (in years).*

Circumference (inches)	Approximate Age (years)
5	8
10	16
20	32
30	48
40	64
50	80
60	95
80	127
100	159

---

*The table assumes that the circumference is measured 4.5 ft above ground level.

(a) Find a linear model that gives the age y in terms of the circumference x.

(b) Find the approximate age of trees whose circumferences are 56 in and 68 in.

(c) What is the *diameter* of a tree that is 151 years old?

40. The table shows the average hourly earnings of production workers in manufacturing.*

Year	Hourly Earnings
2000	$14.00
2001	$14.53
2002	$14.95
2003	$15.35
2004	$15.67
2005	$16.11

(a) Find a linear model for the data, with $x = 0$ corresponding to 2000.

(b) Use the model to estimate the average hourly earnings in 2001 and 2004. How do the estimates of the model compare with the actual figures?

(c) Estimate the average hourly earnings in 2008.

41. The table shows the total amount of charitable giving (in billions of dollars) in the United States in recent years.[†]

Year	Charitable Giving
1994	119.2
1996	138.6
1998	177.4
2000	227.7
2001	229.0
2002	234.1
2003	240.7

(a) Find a linear model for the data, with $x = 0$ corresponding to 1990.

(b) Estimate the amount of charitable giving in 1999 and 2006.

(c) According to the model, in what year will charitable giving reach $372 billion?

42. The table shows, for selected states, the percent of high school students in the class of 2005 who took the SAT and the average SAT math score.*

State	Students Who Took SAT (%)	Average Math Score
Connecticut	86	517
Delaware	74	502
Georgia	75	496
Idaho	21	542
Indiana	66	508
Iowa	5	608
Montana	31	540
Nevada	39	513
New Mexico	13	547
North Dakota	4	605
Ohio	29	543
Pennsylvania	75	503
South Carolina	64	499
Washington	55	534

(a) Make a scatter plot of average SAT math score y and percent x of students who took the SAT, with the data points arranged in order of increasing values of x.

(b) Find a linear model for the data.

(c) What is the slope of your linear model? What does this mean in the context of the problem?

(d) Here are the data on four additional states. How well does the model match the actual figures for these states?

State	Students Who Took SAT (%)	Average Math Score
Alaska	52	519
Arizona	33	530
Hawaii	61	516
Oklahoma	7	563

*U.S. Bureau of Labor Statistics.
[†]*Statistical Abstract of the United States: 2006.*

*The College Board.

Chapter 2 Test

Sections 2.1 and 2.2

1. Find a viewing window (or windows) that shows a complete graph of
$$y = .02x^5 - .32x^4 + .78x^3 + 2.48x^2 - 3.44x - 4.8.$$
Your graph should clearly show all peaks, valleys, and intercepts.

2. Solve $\dfrac{7x^2}{x^2 + 24x} = x.$

3. Find the highest point on the graph of
$$y = -.04x^4 + .4x^3 + .4x^2 + .4x + 26.$$
Round the coordinates of the point to four decimal places.

4. (a) How many real solutions does the equation
$$.3x^5 - 2x^3 + x + k = 0$$
have when $k = 0$?
   (b) Find a value of k for which the equation has just one real solution.

5. Find a square window that shows a complete graph of
$$x^2 + 4y^2 = 144.$$

6. Solve $\dfrac{x^3 - 2x^2 - 4x + 8}{x^2 + x - 6} = 0.$

7. The concentration of a certain medication y in the bloodstream at time x hours is approximated by
$$y = \frac{500x}{.1x^3 + 100},$$
where y is measured in milligrams per liter. After two days the medication has no effect.
   (a) Find a viewing window that contains only those points on the graph of the equation that are relevant to the situation.
   (b) At what time is the concentration of the medicine the highest and what is the concentration at that time? Round your answers to three decimal places.

8. Solve: $\sqrt{x^4 - x^2 + 2x + 1} = 0$. Round your answers to four decimal places.

Sections 2.3 and 2.4

9. Find an equation whose solution provides the answer to the following problem. You need not solve the equation.

   A corner lot has dimensions 35 by 40 yards. The city plans to take a strip of uniform width along the two sides of the lot that border the streets to widen these roads. How wide should the strip be if the lot is to have an area of 785 square yards?

10. A physics book has 40 square inches of print per page. Each page has a left-side margin of 1.7 inches, and top, bottom, and right-side margins of .4 inch. If a page cannot be wider than 7.4 inches, what should its dimensions be to use the least amount of paper? Round your answer to three decimal places.

11. A radiator contains 8 quarts of fluid, 30% of which is antifreeze. How much fluid should be drained and replaced with pure antifreeze so that the new mixture is 40% antifreeze? Round your answer to two decimal places.

12. Find the lowest point on the graph of $y = x^3 - 3x + 3.1$ shown in the given viewing windows.
   (a) $0 \le x \le 3$ and $-3 \le y \le 3$
   (b) $-2 \le x \le .99$ and $-3 \le y \le 6$

13. The dimensions of a rectangular box are consecutive integers. If the box has volume 21,924 cubic centimeters, what are its dimensions?

14. The population P of Cleveland, Ohio (in thousands) in year x is approximated by
$$P = .0000352x^4 - .0049x^3 - .08x^2 + 22.706x + 375.2,$$
where $x = 0$ corresponds to 1990. According to this model, in what year was the population of Cleveland largest?

15. A rectangular bin with an open top and a volume of 42.32 cubic feet is to be built. The length of its base must be twice the width, and the bin must be at least 3 feet high. Material for the base of the bin costs $13 per square foot and material for the sides costs $9 per square foot. If it coasts $634.34 to build the bin, what are its dimensions?

16. A 10-inch square piece of metal is to be used to make an open-top box by cutting equal-sized squares from each corner and folding up the sides. The length, width, and height of the box are each to be less than 6 inches. Round your answers to the following questions to three decimal places.
   (a) If the box is to have a volume of 50 cubic inches, what size squares should be cut from each corner?
   (b) What size squares should be cut to produce a box with the largest possible volume?

Section 2.5

17. The table shows the population of Nowhere, Missouri in selected years.

Year	Population
1950	930,568
1970	681,185
1980	528,162
1990	401,901
2000	354,474
2007	250,040

(a) Construct a scatter plot of the data, with $x = 0$ corresponding to 1950 and P measured in thousands.
(b) Does the data appear to be approximately linear? If so, is there a positive or negative correlation?
(c) Use regression to find a model for the data. Round the coefficients in your model to three decimal places.

18. The table shows the consumer price index (CPI) in April of selected years.

Year	1994	1998	2000	2002
CPI	147.4	162.5	171.3	179.8

For each of the following two models, in which $x = 0$ corresponds to 1990, compute the required information for each blank.

(a) Model: $y = 4x + 131$

Data Point	Model Point	Residual	Residual Squared
(4, 147.4)	_____	_____	_____
(8, 162.5)	_____	_____	_____
(10, 171.3)	_____	_____	_____
(12, 179.8)	_____	_____	_____
		Sum: _____	Sum: _____

(b) Model: $y = 4.1x + 131.5$

Data Point	Model Point	Residual	Residual Squared
(4, 147.4)	_____	_____	_____
(8, 162.5)	_____	_____	_____
(10, 171.3)	_____	_____	_____
(12, 179.8)	_____	_____	_____
		Sum: _____	Sum: _____

(c) Which of the preceding models is the better fit for the data?

19. China's oil consumption (in millions of barrels per day) in selected years is shown in the table.*

Year	2000	2005	2010	2020	2030
Oil Consumption	4.8	6.8	8.5	11.5	14.8

(a) Let $x = 0$ correspond to 2000 and use linear regression to find a model for this data. Round the coefficients to four decimal places.
(b) Estimate Chinese oil consumption in 2008 and 2018.

20. The approximate sales of Lexus automobiles are shown in the table.[†]

Year	Vehicle Sold
2000	207,000
2001	229,000
2002	236,000
2003	264,000
2004	289,000
2005	300,000

(a) Let $x = 0$ correspond to 2000 and use linear regression to find a model for this data (in which the number of vehicles sold is in thousands). Round the coefficients to four decimal places.
(b) Use your model to estimate sales in 2007.
(c) According to your model, at what rate are sales increasing?

---

*Data and projections by the Energy Information Administration.
[†]Based on data from Autodata Corporation.

DISCOVERY PROJECT 2   Supply and Demand

Economists study the forces at play as buyers and sellers interact in what we call the **market** for a product. Because their money is limited, consumers tend to buy less of a particular product if the price is higher and more if the price is lower. This is the essence of the **demand curve,** which is just a depiction of the relationship between the price of an item and the maximum quantity that people will buy at that price. The demand curve represents the aggregate demand of all consumers in a market.

Manufacturers, however, have other considerations. Production costs and limited resources (materials, labor, technology, and capital, for instance) factor into the decision to produce at different levels. Ultimately, as the ones supplying (selling) the product, they want to sell more of them if the price is high and less of them if the price is low. This is the essence of the **supply curve,** a depiction of the relationship between the price of an item and the number of those items the manufacturers are willing to sell at that price. Like the demand curve, the supply curve represents the aggregate of all suppliers of a given product.

Here is a simple example of these ideas. The table below shows the quantity demanded for a hypothetical product at several different prices. You can easily verify that the equation $p = 5.85 - .005q$ describes this relationship.

Price p (in dollars)	Quantity q (in thousands)
3.25	520
3.00	570
2.75	620
2.50	670
2.25	720

Suppose the supply curve for this product is given by the equation $p = .01q - 3.75$ and that the producer would like to set the price of the product at $2.80. To find the supply and demand at this price we must solve the appropriate equation for q when $p = 2.80$.*

Supply

$$.01q - 3.75 = p$$

$$.01q - 3.75 = 2.80$$

$$.01q = 6.55$$

$$q = \frac{6.55}{.01} = 655$$

Demand

$$5.85 - .005q = p$$

$$5.85 - .005q = 2.80$$

$$-.005q = -3.05$$

$$q = \frac{-3.05}{-.005} = 610.$$

---

*We use algebra to solve the equation, but you could also use graphical means or an equation solver. The technological methods are often the best choice for nonlinear supply and demand equations.

Claus Meyer/Black Star Publishing/PictureQuest

These results pose a problem: At a price of $2.80 the producer would be willing to sell 655,000 items, but the demand curve says that only 610,000 items would be demanded by consumers at this price. In technical terms, there is a **surplus** because supply exceeds demand.

You can readily verify that the situation is reversed when the price is $2.40. Solving the supply and demand equations when $p = 2.40$ shows that consumer demand is 690,000 items, but producers are only willing to sell 615,000. In this case, there is a **shortage** because demand exceeds supply.

When there is a surplus, sellers are inclined to lower the price because storing unsold items (or having to discard perishable ones) is expensive. When there is a shortage, consumers are willing to pay more in order to get a product they want. These tendencies push the market toward the **equilibrium point,** at which buyers and sellers agree on both price and quantity. This occurs when

$$\text{quantity demanded} = \text{quantity supplied.}$$

The price p at this point is called the **equilibrium price,** the price at which supply equals demand.

The equilibrium point can be found graphically by finding the intersection point of the supply curve and the demand curve. Because economists normally put price (p) on the vertical axis and quantity (q) on the horizontal axis, we follow the same convention here (using X in place of q and Y in place of p when graphing on a calculator). Figure 1 shows that the equilibrium point for our example is (640, 2.65), which means that the equilibrium price is $2.65 and that 640,000 items will be demanded and supplied at that price.

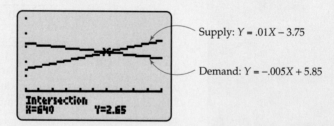

Figure 1

Supply and demand curves provide quantities that correspond to a limited collection of feasible prices at a given point in time. So the corresponding equations are valid only in this range and may not be valid outside of it. In our example, the equations are valid when $520 \le q \le 720$, which determines the viewing window in Figure 1.

When these ideas are used to make business decisions, the first step is to determine the supply and demand equations. This is often done by using statistical data from a trial market, as in the following exercises.

1. An economist obtained the following data about the demand for oranges.

Price per pound (dollars)	0.93	0.90	0.84	0.70	0.65	0.63	0.60
Quantity demanded (in 100,000 tons)	10	10.6	11.5	13.8	14.7	15.1	15.5

Find a linear demand equation of the form $p = mq + b$ either by using linear regression *or* by using the first and last data points [(10, .93) and (15.5, .60)]. If you use regression, round the coefficients to four decimal places.

2. Use the model you found in Exercise 1 to find the quantity demanded at a price of 75¢ per pound.

3. The economist in Exercise 1 also obtained this data about the supply of oranges.

Price per pound (dollars)	0.20	0.36	0.45	0.55	0.65	0.75	1.00
Quantity supplied (in 100,000 tons)	10	11.4	12.4	13.5	14.6	15.5	18

Find a linear supply equation of the form $p = mq + b$ either by using linear regression *or* by using the first and last data points, as in Exercise 1.

4. Use the model you found in Exercise 3 to find the price at which producers are willing to supply 1,300,000 tons of oranges.

5. Find the equilibrium point for the situation in Exercises 1 and 3. What is the equilibrium price? How many oranges will be supplied/demanded at this price?

6. Here are the equations of the curves for the sale of apples:

$$p = .06q - .2 \quad \text{and} \quad p = -.04q + .7,$$

where p is in dollars and q is in 100,000 tons.

(a) Which curve is supply and which one is demand? How can you tell?

(b) Find the equilibrium point. What is the equilibrium price? How many apples will be supplied/demanded at this price?

7. Graph $p = .28$ together with the demand and supply curves of Exercise 6. Is this price above or below the equilibrium price? Would this price lead to a surplus, a shortage, or neither? Explain.

Chapter 3

FUNCTIONS AND GRAPHS

Looking for a house?

If you buy a house, you'll probably need a mortgage. If you can get a low interest rate, your monthly payments will be lower (or, alternatively, you can afford a more expensive house). The timing of your purchase can make a difference because mortgage interest rates constantly fluctuate. In mathematical terms, rates are a function of time. The graph of this function provides a picture of how interest rates change. See Exercise 47 on page 174.

© Bob Perzel/Mira.com/drr.net

Chapter Outline

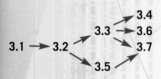

Interdependence of Sections

The concepts of functions and functional notation are central to modern mathematics and its applications. In this chapter, you will be introduced to functions and operations on functions, learn how to use functional notation, and develop skill in constructing and interpreting graphs of functions.

3.1   Functions

Section Objectives
- Understand the definition of a function.
- Recognize functions in various formats: table, graph, verbal description.
- Define a function using an equation or a graph.
- Create a table of inputs and outputs.

While it is possible to think of functions as completely abstract mathematical objects, it is usually simpler to picture a function as a description of how one quantity determines another.

EXAMPLE 1

The amount of state income tax Louisiana residents pay depends on their income. The way that the income determines the tax is given by the following tax law.*

Income		Tax
At Least	But Less Than	
0	$12,500	2.1%
$12,500	$25,000	$262.50 + 3.45% of amount over $12,500
$25,000		$693.75 + 4.8% of amount over $25,000

*2006 rates for a single person with one exemption and no deductions; actual tax amount may vary slightly from this formula when tax tables are used.

142

So if a single person's income were \$30,000 the income tax would be 693.75 + .048(5,000) = \$933.75. ∎

EXAMPLE 2

The graph in Figure 3–1 shows the temperatures in Cleveland, Ohio, on April 11, 2001, as recorded by the U.S. Weather Bureau at Hopkins Airport. The graph indicates the temperature that corresponds to each given time. For example, at 8 A.M. on April 11, 2001, the temperature was 47°F. ∎

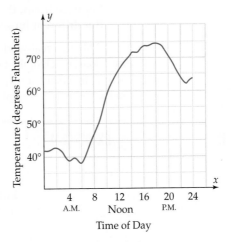

Figure 3–1

EXAMPLE 3

Suppose a rock is dropped straight down from a high place. Physics tells us that the distance traveled by the rock in t seconds is $16t^2$ feet. Therefore, after 5 seconds the rock has fallen $16(5^2) = 400$ feet. ∎

These examples share several common features. Each involves two sets of numbers, which we can think of as inputs and outputs. In each case, there is a rule by which each input determines an output, as summarized here.

	Set of Inputs	Set of Outputs	Rule
Example 1	All incomes	All tax amounts	The tax law
Example 2	Hours since midnight	Temperatures during the day	Time/temperature graph
Example 3	Seconds elapsed after dropping the rock	Distance rock travels	Distance = $16t^2$

Each of these examples may be mentally represented by an idealized calculator that has a single operation key: A number is entered [*input*], the rule key is pushed [*rule*], and an answer is displayed [*output*]. The formal definition of function incorporates these common features (input/rule/output), with a slight change in terminology.

Functions

> A **function** consists of
>
> A set of inputs (called the **domain**);
>
> A **rule** by which each input determines exactly one output;
>
> A set of outputs (called the **range**).

Think about the phrase "exactly one output." In Example 2, for each time of day, there is exactly one temperature. But it is quite possible to have the same temperature (output) occur at different times (inputs). In general,

For each input, the rule of a function determines exactly one output. But different inputs may produce the same output.

Although real-world situations, such as Examples 1–3, are the motivation for functions, much of the emphasis in mathematics courses is on the functions themselves, independent of possible interpretations in specific situations, as illustrated in the following examples.

EXAMPLE 4

The graph in Figure 3–2 defines a function whose rule is as follows:

For input x, the output is the unique number y such that (x, y) is on the graph.

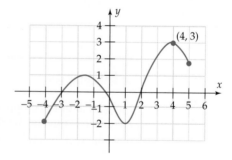

Figure 3–2

Input 4, for example, produces output 3 because (4, 3) is on the graph. Similarly, $(-3, 0)$ is on the graph, which means that input -3 produces output 0. Since the first coordinates of all points on the graph (the inputs) lie between -4 and 5, the domain of this function is the interval $[-4, 5]$. The range is the interval $[-2, 3]$ because all the second coordinates of points on the graph (the outputs) lie between -2 and 3.   ∎

EXAMPLE 5

Could either of the following be the table of values of a function?

(a)

Input	-4	-2	0	2	4
Output	21	7	1	3	7

(b)

Input	3	2	1	3	5
Output	4	0	2	6	9

SOLUTION

(a) Two different inputs (-2 and 4) produce the same output, but that's okay because each input produces exactly one output. So this table could represent a function.

(b) The input 3 produces two different outputs (4 and 6), so this table cannot possibly represent a function. ∎

EXAMPLE 6

Using the procedure of Example 4, does this graph define a function?

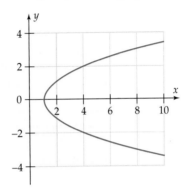

Figure 3–3

SOLUTION   No, because the input 4, for example, produces two outputs 2 and -2. (Both $(4, 2)$ and $(4, -2)$ are on the graph.) ∎

EXAMPLE 7

The **greatest integer function** is the function whose domain is the set of all real numbers, whose range is the set of integers, and whose rule is:

> For each input x, the output is the largest integer that is less than or equal to x. We denote the output by $[\![x]\!]$. For example:

$$[\![5]\!] = 5 \quad [\![4.124]\!] = 4 \quad \left[\!\left[\frac{5}{3}\right]\!\right] = 1 \quad [\![\pi]\!] = 3$$

$$[\![-3]\!] = -3 \quad [\![-1.5]\!] = -2 \quad [\![-0.01]\!] = -1 \quad [\![-\pi]\!] = -4 \quad ∎$$

FUNCTIONS DEFINED BY EQUATIONS AND GRAPHS

Equations in two variables are *not* the same things as functions. However, many equations can be used to define functions.

EXAMPLE 8

The equation $4x - 2y^3 + 5 = 0$ can be solved uniquely for y:

$$2y^3 = 4x + 5$$

$$y^3 = 2x + \frac{5}{2}$$

$$y = \sqrt[3]{2x + \frac{5}{2}}.$$

If a number is substituted for x in this equation, then exactly one value of y is produced. In other words, for every real number x there exists exactly one y such that the equation $4x - 2y^3 + 5 = 0$ is true.

So we can define a function whose domain is the set of all real numbers and whose rule is

The input x produces the output $\sqrt[3]{2x + 5/2}$.

In this situation, we say that the equation defines **y as a function of x.**

The original equation can also be solved for x:

$$4x = 2y^3 - 5$$
$$x = \frac{2y^3 - 5}{4}.$$

Now if a number is substituted for y, exactly one value of x is produced. So we can think of y as the input and the corresponding x as the output and say that the equation defines **x as a function of y.** ∎

EXAMPLE 9

Does the equation $x^2 - y + 1 = 0$ define y as a function of x, or x as a function of y, or both?

SOLUTION   Solving for y, we have

$$x^2 + 1 = y$$
$$y = x^2 + 1.$$

This equation defines y as a function of x, since each value of x produces exactly one value of y.

Solving for x, we obtain

$$x^2 = y - 1$$
$$x = \pm\sqrt{y - 1}.$$

This equation does *not* define x as a function of y because, for example, the input $y = 5$ produces two outputs: $x = \pm 2$. ∎

EXAMPLE 10

A group of students drives from Cleveland to Seattle, a distance of 2350 miles, at an average speed of 52 mph.

(a) Express their distance from Cleveland as a function of time.

(b) Express their distance from Seattle as a function of time.

SOLUTION

(a) Let t denote the time traveled in hours after leaving Cleveland, and let D be the distance from Cleveland at time t. Then the equation that expresses D as a function of t is

$$D = \text{Distance traveled in } t \text{ hours at 52 mph} = 52t.$$

(b) At time t, the car has traveled $52t$ miles of the 2350-mile journey, so the distance K remaining to Seattle is given by $K = 2350 - 52t$. This equation expresses K as a function of t. ∎

Graphing calculators are designed to deal with equations that define y as a function of x. Calculators can evaluate such functions (that is, produce the outputs from various inputs). One method is illustrated in the next example.

EXAMPLE 11

The equation $y = x^3 - 2x + 3$ defines y as a function of x. Use the table feature of a calculator to find the outputs for each of the following inputs:

(a) $-4, -3, -2, -1, 0, 1, 2, 3, 4$ (b) $-5, -11, 8, 7.2, -.44$

TECHNOLOGY TIP

To find the table setup screen, look for TBLSET (or RANG or NUM SETUP) on the keyboard or in the TABLE menu.

The increment is called ∆TBL on TI (and PITCH or NUMSTEP on others).

The table type is called INDPNT on TI, and NUMTYPE on HP-39gs.

SOLUTION

(a) To use the table feature, we first enter $y = x^3 - 2x + 3$ in the equation memory, say, as y_1. Then we call up the setup screen (see the Technology Tip in the margin and Figure 3–4) and enter the *starting number* (-4), the *increment* (the amount the input changes for each subsequent entry, which is 1 here), and the *table type* (AUTO, which means the calculator will compute all the outputs at once).* Then press TABLE to obtain the table in Figure 3–5. To find values that don't appear on the screen in Figure 3–5, use the up and down arrow keys to scroll through the table.

(b) With an apparently random list of inputs, as here, we change the table type to ASK (or USER or BUILD YOUR OWN).† Then key in each value of x, and hit ENTER. This produces the table one line at a time, as in Figure 3–6. ∎

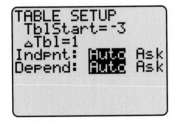

Figure 3–4

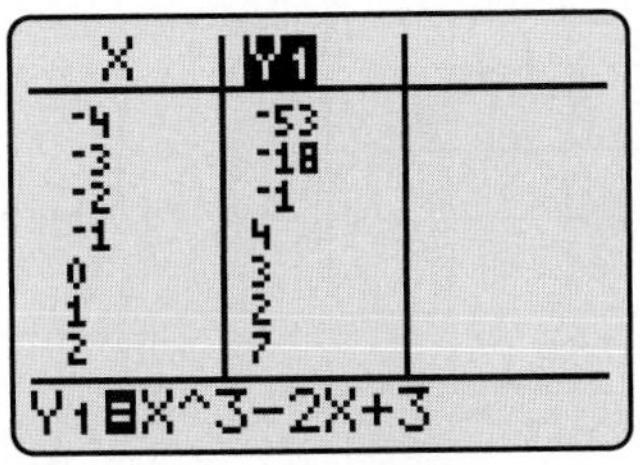

Figure 3–5

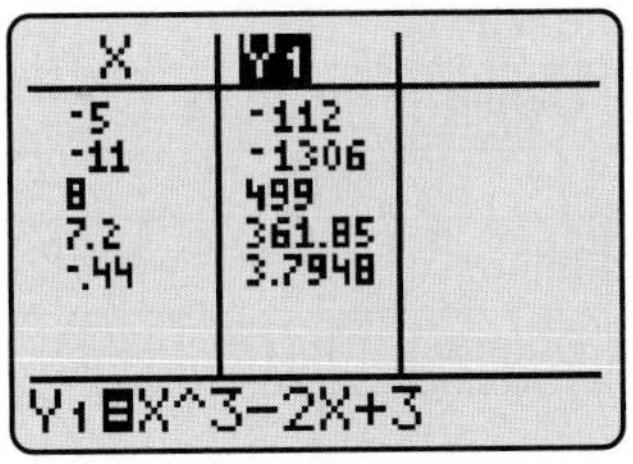

Figure 3–6

CALCULATOR EXPLORATION

Construct a table of values for the function in Example 11 that shows the outputs for these inputs: 2, 2.4, 2.8, 3.2, 3.6, and 4. What is the increment here?

EXAMPLE 12

The revenues y of MTV in year x can be approximated by the equation

$$y = .257x^3 + 11.6x^2 + 4.5x + 177.5 \qquad (0 \le x \le 14),$$

---

*There is no table type selection on Casio, but you must enter a maximum value for x.
†Casio users should see Exercise 51.

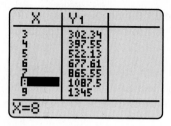

Figure 3–7

where $x = 0$ corresponds to 1987 and y is in millions of dollars.* So the revenue y is a function of the year x. In what year did revenues first exceed one billion dollars?

SOLUTION   Since y is in millions, one billion dollars corresponds to $y = 1000$. Make a table of values for the function (Figure 3–7). It shows that the revenue was approximately \$865,550,000 in 1994 ($x = 7$) and \$1,087,500,000 in 1995 ($x = 8$). ∎

*Based on data from MTV Networks.

EXERCISES 3.1

In Exercises 1–4, determine whether or not the given table could possibly be a table of values of a function. Give reasons for your answer.

1.

Input	1	0	3	1	−5
Output	2	3	−2.5	2	14

2.

Input	−5	3	0	−3	5
Output	0	3	0	5	−3

3.

Input	−5	1	3	−5	7
Output	0	2	4	6	8

4.

Input	1	−1	2	−2	3
Output	1	−2	±5	−6	8

Exercises 5–10 deal with the greatest integer function of Example 7, which is given by the equation $y = [\![x]\!]$. Compute the following values of the function:

5. $[\![6.75]\!]$

6. $[\![.75]\!]$

7. $[\![-4/3]\!]$

8. $[\![5/3]\!]$

9. $[\![-16.0001]\!]$

10. Does the equation $y = [\![x]\!]$ define x as a function of y? Give reasons for your answer.

In Exercises 11–18, determine whether the equation defines y as a function of x or defines x as a function of y.

11. $y = 3x^2 - 12$

12. $y = 2x^4 + 3x^2 - 2$

13. $y^2 = 4x + 1$

14. $5x - 4y^4 + 64 = 0$

15. $3x + 2y = 12$

16. $y - 4x^3 - 14 = 0$

17. $x^2 + y^2 = 9$

18. $x^2 + 2xy + y^2 = 0$

In Exercises 19–22, each equation defines y as a function of x. Create a table that shows the values of the function for the given values of x.

19. $y = x^2 + x - 4$; $x = -2, -1.5, -1, \ldots, 3, 3.5, 4$.

20. $y = \sqrt{4 - x^2}$; $x = -2, -1.2, -.4, .4, 1.2, 2$

21. $y = |x^2 - 5|$; $x = -8, -6, \ldots, 8, 10, 12$

22. $y = x^{10} - 100x$ $x = -1, 0, 1, 2, 3, 4$

23. Consider the Louisiana Tax law in Example 1. Find the output (tax amount) that is produced by each of the following inputs (incomes):

$400 $1509 $25000

$20,000 $12,500 $55,342

24. One proposed tax code looks like this:

Income		Tax
At Least	But Less Than	
0	$12,500	0
$12,500	$100,000	5% of amount over $12,500
$100,000		$5,000 + 10% of amount over $100,000

Find four different numbers in the domain of this function that produce the same output (number in the range).

25. Explain why your answer in Exercise 24 does *not* contradict the definition of a function (in the box on page 144).

26. Is it possible to do Exercise 24 if all four numbers in the domain are required to be greater than 12,500? Why or why not?

27. The amount of postage required to mail a first-class letter is determined by its weight. In this situation, is weight a function of postage? Or vice versa? Or both?

28. Chinese philosopher Laotze (600 BC) said, "the farther one travels, the less one knows." Let x be the distance one travels, and y be the amount one knows. If Laotze is right, is y a function of x? Is x a function of y? Why or Why not?

29. Could the following statement ever be the rule of a function?

> For input x, the output is the number whose square is x.

Why or why not? If there is a function with this rule, what is its domain and range?

30. (a) Use the following chart to make two tables of values (one for an average man and one for an average woman) in which the inputs are the number of drinks per hour and the outputs are the corresponding blood alcohol contents.*

Blood alcohol content

A look at the number of drinks consumed and blood alcohol content in one hour under optimum conditions:

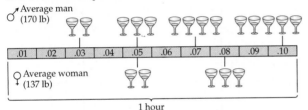

(b) Does each of these tables define a function? If so, what are the domain and range of each function? [Remember that you can have part of a drink.]

31. The prime rate is the rate that large banks charge their best corporate customers for loans. The graph shows how the prime rate charged by a particular bank has varied in recent years.[†] Answer the following questions by reading the graph as best you can.

(a) What was prime rate in January, 2000? In January 2001? In mid-2005?

(b) In what time period was the prime rate below 5%?

(c) On the basis of the data provided by this graph, can the prime rate be considered a function of time? Can time be considered a function of the prime rate?

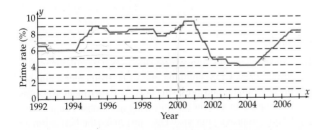

32. Find an equation that expresses the area A of a circle as a function of its

    (a) radius r (b) diameter d

33. Find an equation that expresses the area of a square as a function of its

    (a) side x (b) diagonal d

34. A box with a square base of side x is four times higher than it is wide. Express the volume V of the box as a function of x.

35. The surface area of a cylindrical can of radius r and height h is $2\pi r^2 + 2\pi rh$. If the can is twice as high as the diameter of its top, express its surface area S as a function of r.

36. Suppose you drop a rock from the top of a 400-foot-high building. Express the distance D from the rock to the ground as a function of time t. What is the range of this function? [*Hint:* See Example 3.]

37. A bicycle factory has weekly fixed costs of $26,000. In addition, the material and labor costs for each bicycle are $125. Express the total weekly cost C as a function of the number x of bicycles that are made.

38. The table below shows the percentage of single-parent families in various years.*

Year	1960	1970	1980	1990	2000	2003
Percent	12.8	13.2	17.5	20.8	23.2	27.5

(a) The equation

$$y = (2.4542 \times 10^{-6})x^5 - (2.2459 \times 10^{-4})x^4 + (.0065232)x^3 - (.055795)x^2 + .14568x + 12.8$$

in which $x = 0$ corresponds to 1960, defines y as a function of x. Make a table of values that includes the x-values corresponding to the years in the Census Bureau table.

(b) How do the values of y in your table compare with the percentages in the Census Bureau table? Does this equation seem to provide a reasonable model of the Census Bureau data?

(c) Use the equation to estimate the percentage of single-parent families in 1995 and 2005.

(d) Assuming that this model remains reasonably accurate, in what year will 50% of families be single-parent families?

39. The table shows the amount spent on student scholarships (in millions of dollars) by Oberlin College in recent years. [1995 indicates the school year 1995–96, and so on.]

Year	1995	1996	1997	1998	1999	2000	2001
Scholarships	19.8	22.0	25.7	27.5	28.7	31.1	34.3

---

*National Highway Traffic Safety Administration. Art by AP/Amy Kranz.
[†]Federal Reserve Board.

---

*U.S. Census Bureau.

(a) Use linear regression to find an equation that expresses the amount of scholarships y as a function of the year x, with $x = 0$ corresponding to 1995.

(b) Assuming that the function in part (a) remains accurate, estimate the amount spent on scholarships in 2004.

Use the following figures for Exercises 40–45. Each of the graphs in the figure defines a function, as in Example 4.

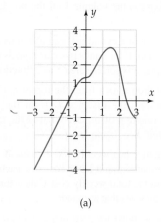

(a)

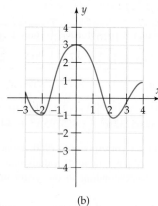

(b)

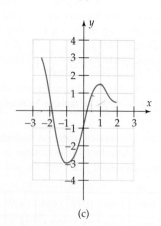

(c)

40. State the domain and range of the function defined by graph (a).

41. State the output (number in the range) that the function of Exercise 40 produces from the following inputs (numbers in the domain): $-2, -1, 0, 1$.

42. State the domain and range of the function defined by graph (b).

43. State the output (number in the range) that the function of Exercise 42 produces from the following inputs (numbers in the domain): $-2, 0, 1, 2.5, -1.5$.

44. State the domain and range of the function defined by graph (c).

45. State the output (number in the range) that the function of Exercise 44 produces from the following inputs (numbers in the domain): $-2, -1, 0, 1/2, 1$.

46. Explain why none of the graphs in the figure below defines a function according to the procedure in Example 6. What goes wrong?

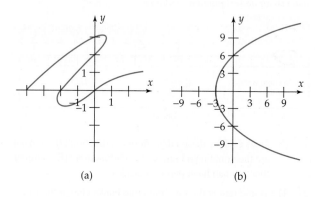

(a)   (b)

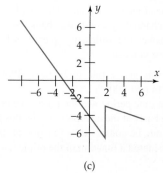

(c)

THINKERS

47. Consider the function whose rule uses a calculator as follows: "Press COS, and then press LN; then enter a number in the domain, and press ENTER."* Experiment with this function, then answer the following questions. You may not be able to prove your answers—just make the best estimate you can based on the evidence from your experiments.

---

*You don't need to know what these keys mean to do this exercise.

(a) What is the largest set of real numbers that could be used for the domain of this function? [If applying the rule to a number produces an error message or a complex number, that number cannot be in the domain.]

(b) Using the domain in part (a), what is the range of this function?

48. Do Exercise 47 for the function whose rule is "Press 10^x, and then press TAN; then enter a number in the domain, and press ENTER."

49. The *integer part* function has the set of all real numbers (written as decimals) as its domain. The rule is "For each input number, the output is the part of the number to the left of the decimal point." For instance, the input 37.986 produces the output 37, and the input -1.5 produces the output -1. On most calculators, the integer part function is denoted "iPart." On calculators that use "Intg" or "Floor" for the greatest integer function, the integer part function is denoted by "INT."

(a) For each nonnegative real number input, explain why both the integer part function and the greatest integer function [Example 7] produce the same output.

(b) For which negative numbers do the two functions produce the same output?

(c) For which negative numbers do the two functions produce different outputs?

50. It is possible to write every even natural number uniquely as the product of two natural numbers, one odd and one a power of two. For example:

$$46 = 23 \times 2 \qquad 36 = 9 \times 2^2 \qquad 8 = 1 \times 2^3.$$

Consider the function whose input is the set of even integers and whose output is the odd number you get in the above process. So if the input is 36, the output is 9. If the input is 46, the output is 23.

(a) Write a table of values for inputs 2, 4, 6, 8, 10, 12 and 14.

(b) Find five different inputs that give an output of 3.

51. Example 11(b) showed how we create a table of values for a function when you get to choose all the values of the inputs. The technique presented does not work for Casio calculators. This exercise is designed for users of Casio calculators.

- Enter an equation such as $y = x^3 - 2x + 3$ in the equation memory. This can be done by selecting TABLE in the MAIN menu.
- Return to the MAIN menu and select LIST. Enter the numbers at which you want to evaluate the function as List 1.
- Return to the MAIN menu and select TABLE. Then press SET-UP [that is, 2nd MENU] and select LIST as the Variable; on the LIST menu, choose List 1. Press EXIT and then press TABL to produce the table.
- Use the up/down arrow key to scroll through the table. If you change an entry in the X column, the corresponding y_1 value will automatically change.

(a) Use this technique to duplicate the table in Example 11(b).

(b) Change the number -11 to 10, and confirm that you've obtained $10^3 - 2(10) + 3$.

3.2 Functional Notation

Section Objectives
- Use functional notation.
- Compute the difference quotient of a function.
- Identify common mistakes made with functional notation.
- Determine the domain of a function.
- Use a piecewise-defined function.

Functional notation is a convenient shorthand language that facilitates the analysis of mathematical problems involving functions. It arises from real-life situations, such as the following.

EXAMPLE 1

In Section 3.1, we saw that the 2006 Louisiana state income tax rates (for a single person with one exemption and no deductions) were as follows:

Income		Tax
At Least	But Less Than	
0	$12,500	2.1%
$12,500	$25,000	$262.50 + 3.45% of amount over $12,500
$25,000		$693.75 + 4.8% of amount over $25,000

Let I denote income, and write $T(I)$ (read "T of I") to denote the amount of tax on income I. In this shorthand language, $T(7500)$ denotes "the tax on an income of $7500." The sentence "The tax on an income of $7500 is $157.50" is abbreviated as $T(7500) = 157.5$. Similarly, $T(25,000) = 693.75$ says that the tax on an income of $25,000 is $693.75. There is nothing that forces us to use the letters T and I here:

Any choice of letters will do, provided that we make clear what is meant by these letters. ∎

EXAMPLE 2

Recall that a falling rock travels $16t^2$ feet after t seconds. Let $d(t)$ stand for the phrase "the distance the rock has traveled after t seconds." Then the sentence "The distance the rock has traveled after t seconds is $16t^2$ feet" can be abbreviated as $d(t) = 16t^2$. For instance,

$$d(1) = 16 \cdot 1^2 = 16$$

means "the distance the rock has traveled after 1 second is 16 feet," and

$$d(4) = 16 \cdot 4^2 = 256$$

means "the distance the rock has traveled after 4 seconds is 256 feet." ∎

Functional notation is easily adapted to mathematical settings, in which the particulars of time, distance, etc., are not mentioned. Suppose a function is given. Denote the function by f and let x denote a number in the domain. Then

$f(x)$ denotes the output produced by input x.

For example, $f(6)$ is the output produced by the input 6. The sentence

"y is the output produced by input x according to the rule of the function f"

is abbreviated

$$y = f(x),$$

which is read "y equals f of x." The output $f(x)$ is sometimes called the **value** of the function f at x.

In actual practice, functions are seldom presented in the style of domain, rule, range, as they have been here. Usually, you will be given a phrase such

> **CAUTION**
>
> The parentheses in $d(t)$ do *not* denote multiplication as in the algebraic equation $3(a + b) = 3a + 3b$. The entire symbol $d(t)$ is part of a *shorthand language.* In particular
>
> $d(1 + 4)$ is *not* equal to $d(1) + d(4)$.
>
> We saw above that $d(1) = 16$ and $d(4) = 256$, so $d(1) + d(4) = 16 + 256 = 272$. But $d(1 + 4)$ is "the distance traveled after $1 + 4$ seconds," that is, the distance after 5 seconds, namely, $16 \cdot 5^2 = 400$. In general,
>
> **Functional notation is a convenient shorthand for phrases and sentences in the English language. It is *not* the same as ordinary algebraic notation.**

as "the function $f(x) = \sqrt{x^2 + 1}$." This should be understood as a set of directions:

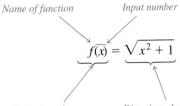

Name of function

Input number

$$f(x) = \sqrt{x^2 + 1}$$

Output number

Directions that tell you what to do with input x to produce the corresponding output f(x), namely, "square it, add 1, and take the square root of the result."

For example, to find $f(3)$, the output of the function f for input 3, simply replace x by 3 in the formula:

$$f(x) = \sqrt{x^2 + 1}$$
$$f(3) = \sqrt{3^2 + 1} = \sqrt{10}.$$

Similarly, replacing x by -5 and 0 shows that

$$f(-5) = \sqrt{(-5)^2 + 1} = \sqrt{26} \qquad \text{and} \qquad f(0) = \sqrt{0^2 + 1} = 1.$$

EXAMPLE 3

The expression

$$h(x) = \frac{x^2 + 5}{x - 1}$$

defines the function h whose rule is

For input x, the output is the number $\dfrac{x^2 + 5}{x - 1}$.

Find each of the following:

$$h(\sqrt{3}), \qquad h(-2), \qquad h(-a), \qquad h(r^2 + 3), \qquad h(\sqrt{c + 2}).$$

SOLUTION   To find $h(\sqrt{3})$ and $h(-2)$, replace x by $\sqrt{3}$ and -2, respectively, in the rule of h:

$$h(\sqrt{3}) = \frac{(\sqrt{3})^2 + 5}{\sqrt{3} - 1} = \frac{8}{\sqrt{3} - 1} \qquad \text{and} \qquad h(-2) = \frac{(-2)^2 + 5}{-2 - 1} = -3.$$

The value of the function h at any quantity, such as $-a, r^2 + 3$, etc., can be found by using the same procedure: *Replace x in the formula for $h(x)$ by that quantity:*

$$h(-a) = \frac{(-a)^2 + 5}{-a - 1} = \frac{a^2 + 5}{-a - 1}$$

$$h(r^2 + 3) = \frac{(r^2 + 3)^2 + 5}{(r^2 + 3) - 1} = \frac{r^4 + 6r^2 + 9 + 5}{r^2 + 2} = \frac{r^4 + 6r^2 + 14}{r^2 + 2}$$

$$h(\sqrt{c + 2}) = \frac{(\sqrt{c + 2})^2 + 5}{\sqrt{c + 2} - 1} = \frac{c + 2 + 5}{\sqrt{c + 2} - 1} = \frac{c + 7}{\sqrt{c + 2} - 1}.$$

TECHNOLOGY TIP

One way to evaluate a function $f(x)$ is to enter its rule as an equation $y = f(x)$ in the equation memory and use TABLE or (on TI-86) EVAL; see Example 11 in Section 3.1.

When functional notation is used in expressions such as $f(-x)$ or $f(x + h)$, the same basic rule applies: Replace x in the formula by the *entire* expression in parentheses.

EXAMPLE 4

If $f(x) = x^2 + x - 2$, then

$$f(-3) = (-3)^2 + (-3) - 2 = 4$$

$$-f(3) = -(3^2 + 3 - 2) = -10$$

$$f(-x) = (-x)^2 + (-x) - 2 = x^2 - x - 2.$$

Note that for this function, $f(-x)$ is *not* the same as $-f(x)$, because $-f(x)$ is the negative of the number $f(x)$, that is,

$$-f(x) = -(x^2 + x - 2) = -x^2 - x + 2.$$ ∎

EXAMPLE 5

If $f(x) = x^2 - x + 2$ and $h \neq 0$, find

(a) $f(x + h)$ (b) $f(x + h) - f(x)$ (c) $\dfrac{f(x + h) - f(x)}{h}$.

SOLUTION

(a) Replace x by $x + h$ in the rule of the function:

$$f(x + h) = (x + h)^2 - (x + h) + 2$$

$$= x^2 + 2xh + h^2 - x - h + 2.$$

(b) By part (a),

$$f(x + h) - f(x) = [(x + h)^2 - (x + h) + 2] - [x^2 - x + 2]$$

$$= [x^2 + 2xh + h^2 - x - h + 2] - [x^2 - x + 2]$$

$$= x^2 + 2xh + h^2 - x - h + 2 - x^2 + x - 2$$

$$= 2xh + h^2 - h.$$

(c) By part (b), we have

$$\frac{f(x + h) - f(x)}{h} = \frac{2xh + h^2 - h}{h}$$

$$= \frac{h(2x + h - 1)}{h}$$

$$= 2x + h - 1.$$ ∎

If f is a function, then the quantity $\dfrac{f(x + h) - f(x)}{h}$, as in Example 5(c), is called the **difference quotient** of f. Difference quotients, whose significance is explained in Section 3.6, play an important role in calculus.

EXAMPLE 6

Compute and simplify the difference quotient for the function $f(x) = x^3 + 2x$.

SOLUTION

$$\frac{f(x+h) - f(x)}{h} = \frac{\overbrace{[(x+h)^3 + 2(x+h)]}^{f(x+h)} - \overbrace{[x^3 + 2x]}^{f(x)}}{h}$$

$$= \frac{[x^3 + 3x^2h + 3xh^2 + h^3 + 2x + 2h] - [x^3 + 2x]}{h}$$

$$= \frac{x^3 + 3x^2h + 3xh^2 + h^3 + 2x + 2h - x^3 - 2x}{h}$$

$$= \frac{3x^2h + 3xh^2 + h^3 + 2h}{h}$$

$$= \frac{h(3x^2 + 3xh + h^2 + 2)}{h}$$

$$= 3x^2 + 3xh + h^2 + 2. \qquad \blacksquare$$

As the preceding examples illustrate, functional notation is a specialized shorthand language. Treating it as ordinary algebraic notation may lead to mistakes.

EXAMPLE 7

> **CAUTION**
>
> It is common to make mistakes with functional notation.
> Remember that, in general:
>
> $f(a + b) \neq f(a) + f(b)$
>
> $f(a - b) \neq f(a) - f(b)$
>
> $f(ab) \neq f(a)f(b)$
>
> $f(ab) \neq af(b)$
>
> $f(ab) \neq f(a)b$

Many students make untrue assumptions when working with functional notation. Here are examples of three of the items listed in the Caution box.

If $f(x) = x^2$, then

$$f(3 + 2) = f(5) = 5^2 = 25.$$

But

$$f(3) + f(2) = 3^2 + 2^2 = 9 + 4 = 13.$$

So $f(3 + 2) \neq f(3) + f(2)$.

If $f(x) = x + 7$, then

$$f(3 \cdot 4) = f(12) = 12 + 7 = 19.$$

But

$$f(3)f(4) = (3 + 7)(4 + 7) = 10 \cdot 11 = 110.$$

So $f(3 \cdot 4) \neq f(3)f(4)$.

If $f(x) = x^2 + 1$, then

$$f(2 \cdot 3) = (2 \cdot 3)^2 + 1 = 36 + 1 = 37.$$

But

$$f(2) \cdot 3 = (2^2 + 1)3 = 5 \cdot 3 = 15.$$

So $f(2 \cdot 3) \neq f(2) \cdot 3.$ $\qquad \blacksquare$

DOMAINS

When the rule of a function is given by a formula, as in Examples 3–7, its domain (set of inputs) is determined by the following convention.

Domain Convention

> Unless specific information to the contrary is given, the domain of a function f includes every real number (input) for which the rule of the function produces a real number as output.

Thus, the domain of a polynomial function such as $f(x) = x^3 - 4x + 1$ is the set of all real numbers, since $f(x)$ is defined for every value of x. In cases in which applying the rule of a function leads to division by zero or to the square root of a negative number, however, the domain may not consist of all real numbers, as illustrated in the next example.

EXAMPLE 8

Find the domain of the function given by

(a) $k(x) = \dfrac{x^2 + 5}{x - 1}$

(b) $f(u) = \sqrt{u + 2}$

SOLUTION

(a) When $x = 1$, the denominator of $\dfrac{x^2 + 5}{x - 1}$ is 0, and the fraction is not defined. When $x \neq 1$, however, the denominator is nonzero and the fraction *is* defined. Therefore, the domain of the function k consists of all real numbers *except* 1.

(b) Since negative numbers do not have real square roots, $\sqrt{u + 2}$ is a real number only when $u + 2 \geq 0$, that is, when $u \geq -2$. Therefore, the domain of f consists of all real numbers greater than or equal to -2, that is, the interval $[-2, \infty)$. ■

EXAMPLE 9

A **piecewise-defined** function is one whose rule includes several formulas, such as

$$f(x) = \begin{cases} 2x + 3 & \text{if } x < 4 \\ x^2 - 1 & \text{if } 4 \leq x \leq 10. \end{cases}$$

Find each of the following.

(a) $f(-5)$ (b) $f(8)$ (c) $f(k)$ (d) The domain of f.

SOLUTION

(a) Since $-5 < 4$, the first part of the rule applies:

$$f(-5) = 2(-5) + 3 = -7.$$

(b) Since 8 is between 4 and 10, the second part of the rule applies:

$$f(8) = 8^2 - 1 = 63.$$

(c) We cannot find $f(k)$ unless we know whether $k < 4$ or $4 \le k \le 10$.

(d) The rule of f gives no directions when $x > 10$, so the domain of f consists of all real numbers x with $x \le 10$, that is, $(-\infty, 10]$. ■

EXAMPLE 10

Use Example 1 to write the rule of the piecewise-defined function T that gives the Louisiana state income tax $T(x)$ on an income of x dollars.

SOLUTION   By translating the information in the table in Example 1 into functional notation, we obtain

$$T(x) = \begin{cases} .021x & \text{if} \quad 0 \le x < 12{,}500 \\ 262.50 + .0345(x - 12{,}500) & \text{if} \quad 12{,}500 \le x < 25{,}000 \\ 693.75 + .048(x - 25{,}000) & \text{if} \quad x \ge 25{,}000. \end{cases}$$ ■

 APPLICATIONS

The domain convention does not always apply when dealing with applications. Consider, for example, the distance function for falling objects, $d(t) = 16t^2$ (see Example 2). Since t represents time, only nonnegative values of t make sense here, even though the rule of the function is defined for all values of t.

A real-life situation may lead to a function whose domain is smaller than the one dictated by the domain convention.

EXAMPLE 11

A glassware factory has fixed expenses (mortgage, taxes, machinery, etc.) of $12,000 per week. It costs 80 cents to make one cup (labor, materials, shipping). A cup sells for $1.95. At most 18,000 cups can be manufactured and sold each week.

(a) Express the weekly revenue as a function of the number x of cups made.

(b) Express the weekly costs as a function of x.

(c) Find the domain and the rule of the weekly profit function.

SOLUTION

(a) If $R(x)$ is the weekly revenue from selling x cups, then

$$R(x) = (\text{price per cup}) \times (\text{number sold})$$

$$R(x) = 1.95x.$$

(b) If $C(x)$ is the weekly cost of manufacturing x cups, then

$$C(x) = (\text{cost per cup}) \times (\text{number sold}) + (\text{fixed expenses})$$

$$C(x) = .80x + 12{,}000.$$

(c) If $P(x)$ is the weekly profit from selling x cups, then

$$P(x) = \text{Revenue} - \text{Cost}$$
$$P(x) = R(x) - C(x)$$
$$P(x) = 1.95x - (.80x + 12{,}000) = 1.95x - .80x - 12{,}000$$
$$P(x) = 1.15x - 12{,}000.$$

Although this rule is defined for all real numbers x, the domain of the function P consists of the possible number of cups that can be made each week. Since you can make only whole cups and the maximum production is 18,000, the domain of P consists of all integers from 0 to 18,000. ∎

EXAMPLE 12

Let P be the profit function in Example 11.

(a) What is the profit from selling 5000 cups? From 14,000 cups?

(b) What is the break-even point?

SOLUTION

(a) We evaluate the function $P(x) = 1.15x - 12{,}000$ at the required values of x:

$$P(5000) = 1.15(5000) - 12{,}000 = -\$6250$$

$$P(14{,}000) = 1.15(14{,}000) - 12{,}000 = \$4100.$$

Thus, sales of 5000 cups produce a loss of $6250, while sales of 14,000 produce a profit of $4100.

(b) The break-even point occurs when revenue equals costs (that is, when profit is 0). So we set $P(x) = 0$ and solve for x:

$$1.15x - 12{,}000 = 0$$

$$1.15x = 12{,}000$$

$$x = \frac{12{,}000}{1.15} \approx 10{,}434.78.$$

Thus, the break-even point occurs between 10,434 and 10,435 cups. There is a slight loss from selling 10,434 cups and a slight profit from selling 10,435. ∎

EXERCISES 3.2

In Exercises 1 and 2, find the indicated values of the function by hand and by using the table feature of a calculator (or the EVAL key on TI-85/86). If your answers do not agree with each other or with those at the back of the book, you are either making algebraic mistakes or incorrectly entering the function in the equation memory.

1. $f(x) = \dfrac{x - 3}{x^2 + 4}$

    (a) $f(-1)$ (b) $f(0)$ (c) $f(1)$ (d) $f(2)$ (e) $f(3)$

2. $g(x) = \sqrt{x + 4} - 2$

    (a) $g(-2)$ (b) $g(0)$ (c) $g(4)$ (d) $g(5)$ (e) $g(12)$

Exercises 3–24 refer to these three functions:

$$f(x) = \sqrt{x + 3} - x + 1$$

$$g(t) = t^2 - 1$$

$$h(x) = x^2 + \frac{1}{x} + 2.$$

In each case, find the indicated value of the function.

3. $f(0)$

4. $f(1)$

5. $f(\sqrt{2})$

6. $f(\sqrt{2} - 1)$

7. $f(-2)$

8. $f(-3/2)$

9. $h(-4)$

10. $h(3/2)$

11. $h(\pi + 1)$

12. $h(m)$

13. $h(a + k)$

14. $f(a)$

15. $h(-x)$

16. $h(2 - x)$

17. $h(x - 3)$

18. $g(3)$

19. $g(s + 1)$

20. $g(1 - r)$

21. $g(-t)$

22. $g(t + h)$

23. $f(g(3))$

24. $f(g(t))$

In Exercises 25–34, assume $h \neq 0$. Compute and simplify the difference quotient

$$\frac{f(x + h) - f(x)}{h}$$

25. $f(x) = x + 1$

26. $f(x) = -10x$

27. $f(x) = 3x + 7$

28. $f(x) = x^2$

29. $f(x) = x - x^2$

30. $f(x) = x^3$

31. $f(x) = \sqrt{x}$

32. $f(x) = 1/x$

33. $f(x) = x^2 + 3$

34. $f(x) = 3$

35. In each part, compute $f(a)$, $f(b)$, and $f(a + b)$, and determine whether the statement "$f(a + b) = f(a) + f(b)$" is true or false for the given function.

(a) $f(x) = x^2$ (b) $f(x) = 3x$ (c) $f(x) = 5$

36. In each part, compute $g(a)$, $g(b)$, and $g(ab)$, and determine whether the satement "$g(ab) = g(a) \cdot g(b)$" is true or false for the given function.

(a) $g(x) = x^3$ (b) $g(x) = 5x$ (c) $g(x) = -2$

37. If $f(x) = x^3 + cx^2 + 4x - 1$ for some constant c and $f(1) = 2$, find c. [*Hint:* Use the rule of f to compute $f(1)$.]

38. If $f(x) = \dfrac{dx - 5}{x - 3}$ and $f(4) = 3$, find d.

39. The rule of the function f is given by the graph, as in Example 4 of Section 3.1. Find

(a) The domain of f

(b) The range of f

(c) $f(-3)$

(d) $f(-1)$

(e) $f(1)$

(f) $f(2)$

40. The rule of the function g is given by the graph. Find

(a) The domain of g $[-3, 4]$

(b) The range of g $[-2, 3]$

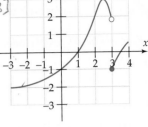

(c) $g(-3)$

(d) $g(-1)$

(e) $g(1)$

(f) $g(4)$

41. Let $f(x) = \begin{cases} -x & \text{if } x < 2 \\ x & \text{if } x \geq 0 \end{cases}$

This function is identical to a function you already know. What is that function?

42. Let $f(x) = \begin{cases} 1/x & \text{if } x \neq 0 \\ 2 & \text{if } x = 0 \end{cases}$

Find the domain of f.

43. If $f(x) = \begin{cases} x^2 + 2x & \text{if } x < 2 \\ 3x - 5 & \text{if } 2 \leq x \leq 20 \end{cases}$ find

(a) The domain of f

(b) $f(-3)$ (c) $f(-1)$ (d) $f(2)$ (e) $f(7/3)$

44. If $g(x) = \begin{cases} 2x - 3 & \text{if } x < -1 \\ |x| - 5 & \text{if } -1 \leq x \leq 2 \\ x^2 & \text{if } x > 2 \end{cases}$ find

(a) The domain of g

(b) $g(-2.5)$ (c) $g(-1)$ (d) $g(2)$ (e) $g(4)$

In Exercises 45–58, determine the domain of the function according to the usual convention.

45. $f(x) = x^2$

46. $g(x) = \dfrac{1}{x^2} + 2$

47. $h(t) = |t| - 1$

48. $k(u) = \sqrt{u}$

49. $k(x) = |x| + \sqrt{x} - 1$

50. $h(x) = \sqrt{(x + 1)^2}$

51. $g(u) = \dfrac{|u|}{u}$

52. $h(x) = \dfrac{\sqrt{x - 1}}{x^2 - 1}$

53. $g(y) = [\![-y]\!]$

54. $f(t) = \sqrt{-t}$

55. $g(u) = \dfrac{u^2 + 1}{u^2 - u - 6}$

56. $f(t) = \sqrt{4 - t^2}$

57. $f(x) = -\sqrt{9 - (x - 9)^2}$

58. $f(x) = \sqrt{-x} + \dfrac{2}{x + 1}$

59. Give an example of two different functions f and g that have all of the following properties:

$$f(-1) = 1 = g(-1) \quad \text{and} \quad f(0) = 0 = g(0)$$
$$\text{and} \quad f(1) = 1 = g(1).$$

60. Give an example of a function g with the property that $g(x) = g(-x)$ for every real number x.

61. Give an example of a function g with the property that $g(-x) = -g(x)$ for every real number x.

In Exercises 62–65, find the values of x for which $f(x) = g(x)$.

62. $f(x) = 2x^2 + 4x - 4$; $g(x) = x^2 + 12x + 6$

63. $f(x) = 2x^2 + 13x - 14$; $g(x) = 8x - 2$

64. $f(x) = 3x^2 - x + 5$; $g(x) = x^2 - 2x + 26$

65. $f(x) = 2x^2 - x + 1$; $g(x) = x^2 - 4x + 4$

In Exercises 66–68, the rule of a function f is given. Write an algebraic formula for $f(x)$.

66. Triple the input, subtract 8, and take the square root of the result.

67. Square the input, multiply by 3, and subtract the result from 8.

68. Cube the input, add 6, and divide the result by 5.

69. A potato chip factory has a daily overhead from salaries and building costs of $1800. The cost of ingredients and packaging to produce a pound of potato chips is 50¢. A pound of potato chips sells for $1.20. Show that the factory's daily profit is a function of the number of pounds of potato chips sold, and find the rule of this function. (Assume that the factory sells all the potato chips it produces each day.)

70. Jack and Jill are salespersons in the suit department of a clothing store. Jack is paid $200 per week plus $5 for each suit he sells, whereas Jill is paid $10 for every suit she sells.

(a) Let $f(x)$ denote Jack's weekly income, and let $g(x)$ denote Jill's weekly income from selling x suits. Find the rules of the functions f and g.

(b) Use algebra or a table to find $f(20)$ and $g(20), f(35)$ and $g(35), f(50)$ and $g(50)$.

(c) If Jack sells 50 suits a week, how many must Jill sell to have the same income as Jack?

71. A person who needs crutches can determine the correct length as follows: a 50-inch-tall person needs a 38-inch-long crutch. For each additional inch in the person's height, add .72 inch to the crutch length.

(a) If a person is y inches taller than 50 inches, write an expression for the proper crutch length.

(b) Write the rule of a function f such that $f(x)$ is the proper crutch length (in inches) for a person who is x inches tall. [*Hint:* Replace y in your answer to part (a) with an expression in x. How are x and y related?]

72. The table shows the 2006 federal income tax rates for a single person.

Taxable Income		Tax
Over	But Not Over	
0	$7,300	10% of income
$7,300	$29,700	$730.00 + 15% of amount over $7,300
$29,700	$71,950	$4090.00 + 25% of amount over $29,700
$71,950	$150,150	$14,652.50 + 28% of amount over $71,950
$150,150	$326,450	$36,548.50 + 33% of amount over $150,150
$326,450		$94,727.50 + 35% of amount over $326,450

(a) Write the rule of a piecewise-defined function T such that $T(x)$ is the tax due on a taxable income of x dollars.

(b) Find $T(24,000)$, $T(35,000)$, and $T(200,000)$.

73. Suppose a car travels at a constant rate of 55 mph for 2 hours and travels at 45 mph thereafter. Show that distance traveled is a function of time, and find the rule of the function.

74. A man walks for 45 minutes at a rate of 3 mph, then jogs for 75 minutes at a rate of 5 mph, then sits and rests for 30 minutes, and finally walks for $1\frac{1}{2}$ hours. Find the rule of the function that expresses his distance traveled as a function of time. [*Caution:* Don't mix up the units of time; use either minutes or hours, not both.]

75. Suppose that the width and height of the box in the figure are equal and that the sum of the length and the girth is 108 (the maximum size allowed by the post office).

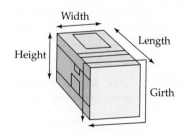

(a) Express the length y as a function of the width x. [*Hint:* Use the girth.]

(b) Express the volume V of the box as a function of the width x. [*Hint:* Find a formula for the volume and use part (a).]

76. A rectangular region of 6000 square feet is to be fenced in on three sides with fencing costing $3.75 per foot and on the fourth side with fencing costing $2.00 per foot. Express the cost of the fence as a function of the length x of the fourth side.

77. A box with a square base measuring $t \times t$ ft is to be made of three kinds of wood. The cost of the wood for the base is 85¢ per square foot; the wood for the sides costs 50¢ per square foot, and the wood for the top $1.15 per square foot. The volume of the box is to be 10 cubic feet. Express the total cost of the box as a function of the length t.

78. Average tuition and fees in private four-year colleges in recent years were as follows.*

Year	Tuition and Fees	Year	Tuition and Fees
1995	$12,432	1999	$15,380
1996	$12,823	2000	$16,332
1997	$13,664	2001	$17,727
1998	$14,709	2002	$18,723

(a) Use linear regression to find the rule of a function f that gives the approximate average tuition in year x, where $x = 0$ corresponds to 1990.

*The College Board.

(b) Find $f(6)$, $f(8)$, and $f(10)$. How do they compare with the actual figures?
(c) Use f to estimate tuition and fees in 2011.

79. The number of U.S. commercial radio stations whose primary format is top 40 hits has been increasing in recent years, as shown in the table.*

Year	Number of Stations
1998	379
1999	401
2001	468
2002	474
2003	491
2004	497
2005	502

(a) Use linear regression to find the rule of a function g that gives the number of top-40 stations in year x, where $x = 0$ corresponds to 1990.
(b) Find $g(8)$ and $g(11)$. How do they compare with the actual figures?
(c) Data for the year 2000 is missing. Estimate the number of stations in 2000.
(d) Assuming that this function remains accurate, estimate the number of stations in 2011.

*World Almanac and Book of Facts 2006.

3.3 Graphs of Functions

Section Objectives

- Recognize the general shape and behavior of graphs of basic functions.
- Graph step functions and piecewise-defined functions.
- Find local maxima and minima.
- Determine intervals on which a function is increasing or decreasing.
- Use the vertical line test to identify the graph of a function.
- Interpret information presented in a graph.

The graph of a function f is the graph of the equation $y = f(x)$. Hence

The graph of the function f consists of the points $(x, f(x))$ for every number x in the domain of f.

When the rule of a function is given by an algebraic formula, the graph is easily obtained with technology. However, machine-generated graphs can sometimes be

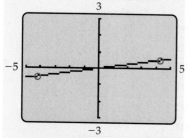
incomplete or misleading. So the emphasis here is on using your algebraic knowledge *before* reaching for a calculator. Doing so will often tell you that a calculator is inappropriate or help you to interpret screen images when a calculator is used.

Some functions appear so frequently that you should memorize the shapes of their graphs, which are easily obtained by hand-graphing or by using technology. Regardless of how you first find these graphs, you should be able to reproduce them without looking them up or resorting to technology. These basic graphs are summarized in the **catalog of functions** at the end of this section. A title on an example (such as "linear function" in Example 1) indicates a function that is in the catalog.

EXAMPLE 1

Linear Functions   The graph of a function of the form

$$f(x) = mx + b \qquad \text{(with m and b constants)}$$

is the graph of the equation $y = mx + b$. As we saw in Section 1.4, the graph is a straight line with slope m and y-intercept b that can easily be obtained by hand. Some typical linear functions are graphed in Figure 3–8 (several of them have special names). ∎

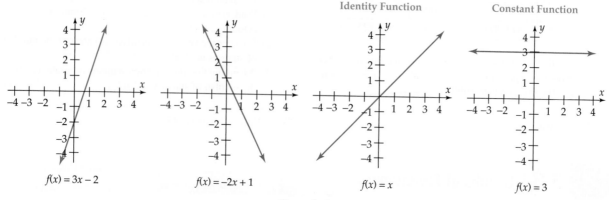

$$f(x) = 3x - 2 \qquad\qquad f(x) = -2x + 1 \qquad\qquad f(x) = x \qquad\qquad f(x) = 3$$

Figure 3–8

EXAMPLE 2

Square and Cube Functions   Figure 3–9 shows the graphs of $f(x) = x^2$ and $g(x) = x^3$. They can be obtained by plotting points (as was done for f in Example 1 of Section 2.1) or by using technology. ∎

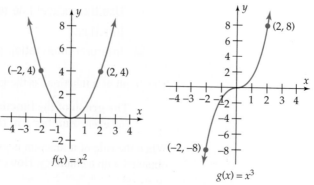

$$f(x) = x^2 \qquad\qquad\qquad g(x) = x^3$$

Figure 3–9

EXERCISES 3.3

In Exercises 1–4, state whether or not the graph is the graph of a function. If it is, find $f(3)$.

1.

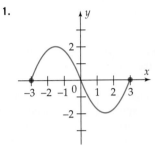

2.

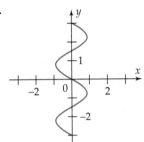

3.

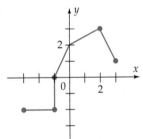

4.

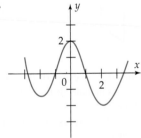

In Exercises 5–11, sketch the graph of the function, being sure to indicate which endpoints are included and which ones are excluded.

5. $f(x) = 2[\![x]\!]$ **6.** $f(x) = -[\![x]\!]$

7. $g(x) = [\![-x]\!]$ [This is *not* the same function as in Exercise 6.]

8. $f(x) = \begin{cases} x^2 & \text{if } x \geq -1 \\ 2x + 3 & \text{if } x < -1 \end{cases}$

9. $k(u) = \begin{cases} -2u - 2 & \text{if } u < -3 \\ u - [\![u]\!] & \text{if } -3 \leq u \leq 1 \\ 2u^2 & \text{if } u > 1 \end{cases}$

10. As of this writing, U.S. postage rates for large envelopes are 80 cents for the first ounce (or fraction thereof) plus 17 cents for each additional ounce or fraction thereof (see Example 5). Assume that each large envelope carries one 80 cent stamp and as many 17 cent stamps as necessary. Then the *number* of stamps required for a large envelope is a function of the weight of the envelope in ounces. Call this function the *postage stamp function.*

    (a) Describe the rule of the postage stamp function algebraically.

    (b) Sketch the graph of the postage stamp function.

11. A common mistake is to graph the function f in Example 6 by graphing both $y = x^2$ and $y = x + 2$ on the same screen (with no restrictions on x). Explain why this graph could not possibly be the graph of a function.

Exercises 12–21 deal with the graph of g shown in the figure.

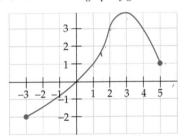

12. Is g a function? Why or why not?

13. What is the domain of g?

14. What is the range of g?

15. Find the approximate intervals where g is increasing.

16. Find the approximate intervals where g is decreasing.

17. If $t = 2$, then $g(t + 1.5) = ?$

18. If $t = 2$, then $g(t) + g(1.5) = ?$

19. If $t = 2$, then $g(t) + 1.5 = ?$

20. For what values of x is $g(x) < 0$?

21. For what values of a is $g(a) = 1?$.

In Exercises 22–24, (a) Use the fact that the absolute value function is piecewise-defined (see Example 7) to write the rule of the given function as a piecewise-defined function whose rule does not include any absolute value bars. (b) Graph the function.

22. $g(x) = |x| - 4$ **23.** $h(x) = |x|/2 - 2$

24. $g(x) = |x + 3|$

25. Show that the function $f(x) = |x| + |x - 2|$ is constant on the interval $[0, 2]$. [*Hint:* Use the definition of absolute value (see Example 7) to compute $f(x)$ when $0 \leq x \leq 2$.]

In Exercises 26–31, find the approximate location of all local maxima and minima of the function.

26. $f(x) = x^3 - x$

27. $g(t) = -\sqrt{16 - t^2}$

28. $h(x) = \dfrac{x}{x^2 + 1}$

29. $k(x) = x^3 - 3x + 1$

30. $l(x) = \dfrac{1}{1 + x^2}$

31. $m(x) = x^3$

In Exercises 32–35, find the approximate intervals on which the function is increasing, those on which it is decreasing, and those on which it is constant.

32. $f(x) = |x - 1| - |x + 1|$

33. $f(x) = -x^3 - 8x^2 + 8x + 5$

34. $f(x) = \sqrt{x}$

35. $f(x) = \dfrac{1}{x}$

36. Let $F(x)$ = the U.S. federal debt in year x, and let $p(x)$ = the federal debt as a percent of the gross domestic product in year x. The graphs of these functions appear below.* Explain why the graph of F is increasing from 1996–2001, while the graph of p is decreasing during that period.

Gross Federal Debt

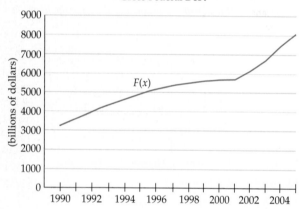

Federal Debt as a Percent of Gross Domestic Product

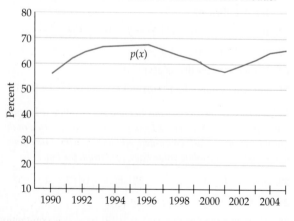

*Graphs prepared by U.S. Census Bureau, based on data from the U.S. Office of Management and Budget.

37. Find the dimensions of the rectangle with perimeter 100 inches and largest possible area, as follows.

(a) Use the figure to write an equation in x and z that expresses the fact that the perimeter of the rectangle is 100.

(b) The area A of the rectangle is given by $A = xz$ (why?). Write an equation that expresses A as a function of x. [*Hint:* Solve the equation in part (a) for z, and substitute the result in the area equation.]

(c) Graph the function in part (b), and find the value of x that produces the largest possible value of A. What is z in this case?

38. Find the dimensions of the rectangle with area 240 square inches and smallest possible perimeter, as follows.

(a) Using the figure for Exercise 37, write an equation for the perimeter P of the rectangle in terms of x and z.

(b) Write an equation in x and z that expresses the fact that the area of the rectangle is 240.

(c) Write an equation that expresses P as a function of x. [*Hint:* Solve the equation in part (b) for z, and substitute the result in the equation of part (a).]

(d) Graph the function in part (c), and find the value of x that produces the smallest possible value of P. What is z in this case?

39. Find the dimensions of a box with a square base that has a volume of 867 cubic inches and the smallest possible surface area, as follows.

(a) Write an equation for the surface area S of the box in terms of x and h. [Be sure to include all four sides, the top, and the bottom of the box.]

(b) Write an equation in x and h that expresses the fact that the volume of the box is 867.

(c) Write an equation that expresses S as a function of x. [*Hint:* Solve the equation in part (b) for h, and substitute the result in the equation of part (a).]

(d) Graph the function in part (c), and find the value of x that produces the smallest possible value of S. What is h in this case?

40. Find the radius r and height h of a cylindrical can with a surface area of 60 square inches and the largest possible volume, as follows.

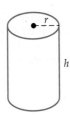

(a) Write an equation for the volume V of the can in terms of r and h.

(b) Write an equation in r and h that expresses the fact that the surface area of the can is 60. [*Hint:* Think of cutting the top and bottom off the can; then cut the side of the can lengthwise and roll it out flat; it's now a rectangle. The surface area is the area of the top and bottom plus the area of this rectangle. The length of the rectangle is the same as the circumference of the original can (why?).]

(c) Write an equation that expresses V as a function of r. [*Hint:* Solve the equation in part (b) for h, and substitute the result in the equation of part (a).]

(d) Graph the function in part (c), and find the value of r that produces the largest possible value of V. What is h in this case?

41. Match each of the functions (a)–(e) with the graph that best fits the situation.

(a) The phases of the moon as a function of time;

(b) The demand for a product as a function of its price;

(c) The height of a ball thrown from the top of a building as a function of time;

(d) The distance a woman runs at constant speed as a function of time;

(e) The temperature of an oven turned on and set to 350° as a function of time.

In Exercises 42 and 43, sketch a plausible graph of the given function. Label the axes and specify a reasonable domain and range.

42. The distance from the top of your head to the ground as you jump on a trampoline as a function of time.

43. The temperature of an oven that is turned on, set to 350°, and 45 minutes later turned off as a function of time.

44. A plane flies from Austin, Texas, to Cleveland, Ohio, a distance of 1200 miles. Let f be the function whose rule is $f(t) =$ distance (in miles) from Austin at time t hours. Draw a plausible graph of f under the given circumstances. [There are many possible correct answers for each part.]

(a) The flight is nonstop and takes less than 4 hours.

(b) Bad weather forces the plane to land in Dallas (about 200 miles from Austin), remain overnight (for 8 hours), and continue the next day.

(c) The flight is nonstop, but owing to heavy traffic, the plane must fly in a holding pattern over Cincinnati (about 200 miles from Cleveland) for an hour before going on to Cleveland.

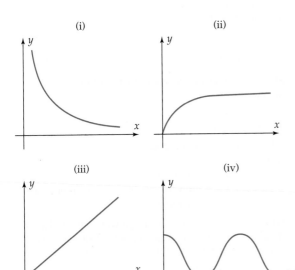

In Exercises 45–46, the graph of a function f is shown. Find and label the given points on the graph.

45. (a) $(k, f(k))$
   (b) $(-k, f(-k))$
   (c) $(k, -f(k))$

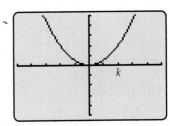

46. (a) $(k, f(k))$
   (b) $(k, .5f(k))$
   (c) $(.5k, f(.5k))$
   (d) $(2k, f(2k))$

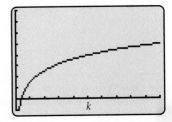

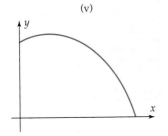

47. The graph of the function f, whose rule is $f(x) =$ average interest rate on a 30-year fixed-rate mortgage in year x, is shown in the figure.* Use it to answer these questions (reasonable approximations are OK).

(a) Compute $f(1977)$, $f(1982)$ and $f(2000)$
(b) In what year between 1990 and 2006 were rates the lowest? The highest?
(c) During what three-year period were rates changing the fastest? How do you determine this from the graph?

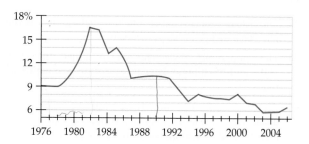

48. The annual percentage changes in various consumer price indexes (CPIs) are shown in the figure.† Use it to answer the following questions. In each case, explain how you got your answer from the graph.

(a) Did the CPI for medical care increase or decrease from 1990 to 1996?
(b) During what time intervals was the CPI for fuel oil increasing?
(c) If the CPI for fuel oil stood at 91 at the beginning of 1999, approximately what was it at the beginning of 2000?

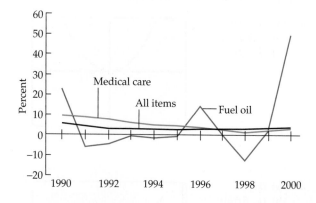

49. The Cleveland temperature graph from Example 2 of Section 3.1, is reproduced below. Let $T(x)$ denote the temperature at time x hours after midnight.
   Determine whether the following statements are true or false.

(a) $T(4 \cdot 3) = T(4) \cdot T(3)$
(b) $T(4 \cdot 3) = 4 \cdot T(3)$
(c) $T(4 + 14) = T(4) + T(14)$

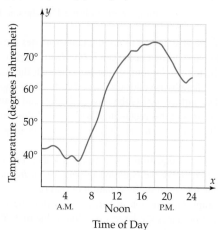

Time of Day

50. Draw the graph of a function f that satisfies the following four conditions:

(i) domain $f = [-2, 4]$
(ii) range $f = [-5, 6]$
(iii) $f(-1) = f(3)$
(iv) $f\left(\dfrac{1}{2}\right) = 0$

51. Sketch the graph of a function f that satisfies these five conditions:

(i) $f(-1) = 2$
(ii) $f(x) \geq 2$ when x is in the interval $\left(-1, \frac{1}{2}\right)$
(iii) $f(x)$ starts decreasing when $x = 1$
(iv) $f(3) = 3 = f(0)$
(v) $f(x)$ starts increasing when $x = 5$
[*Note:* The function whose graph you sketch need not be given by an algebraic formula.]

52. Wireless telephone services are growing rapidly. The table shows the industry's revenue (in billions of dollars) over a five-year period.*

Year	Revenue
1999	40.018
2000	52.966
2002	76.508
2003	87.624
2004	100.600

---

*Federal Home Mortgage Corporation.
†Graph prepared by U.S. Census Bureau, based on data from the Bureau of Labor Statistics.

*New York Times 2006 Almanac.

(a) Make a scatter plot of the data, with $x = 0$ correspon-ding to 1999.
(b) Use linear regression to find a function that models this data. Assume that the model remains accurate.
(c) Use the model to estimate the revenue in 2001.
(d) When will revenue reach $170 billion?

53. The percentage of adults in the United States who smoke has been decreasing, as shown in the table.*

Year	Percent Who Smoke
1965	42.5
1974	37.0
1980	33.3
1987	29.2
1994	25.4
2000	22.9
2005	20.9

(a) Make a scatter plot of the data, with $x = 0$ correspon-ding to 1965.
(b) Use linear regression to find a function that models this data.
(c) Use the model to estimate the percentage of smokers in 1991 and 2013. [For comparison purposes, the actual figure for 1991 is 25.8%.]
(d) If this model remains accurate, when will less than 15% of adults smoke?
(e) According to this model, will smoking even disappear entirely? If so, when?

In Exercises 54 and 55, sketch the graph of the equation.

54. $|x| + |y| = 1$ **55.** $|y| = x^2$

THINKERS

56. Assume that on Sunday you read a long book containing a lot of factual material. Assume that by Monday you only remember 2/3 of the material. On Tuesday you remember 2/3 of what you remembered on Monday. On Wednesday

you remember 2/3 of what you remembered on Tuesday, and so on. Let $f(t)$ be the percent of the material you remember t days after Sunday. (So $f(0) = 100$, and $f(1) = 66\frac{2}{3}$). Sketch $f(t)$ from $t = 0$ to $t = 12$.

57. For each m, let $f(m)$ be the largest real solution to this equation: $x^2 - 4x + m = 0$.
(a) Find the domain of f
(b) Find the range of f
(c) Sketch a graph of f

58. Let $f(x) = \begin{cases} x^2 & \text{if } x \text{ is an integer} \\ x & \text{if } x \text{ is not an integer} \end{cases}$

Sketch f.

59. A jogger begins her daily run from her home. The graph shows her distance from home at time t minutes. The graph shows, for example, that she ran at a slow but steady pace for 10 minutes, then increased her pace for 5 minutes, all the time moving farther from home. Describe the rest of her run.

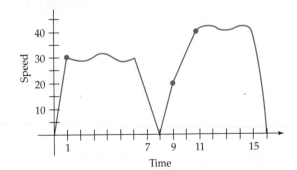

60. The graph shows the speed (in mph) at which a driver is going at time t minutes. Describe his journey.

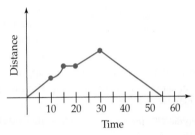

3.3.A SPECIAL TOPICS   Parametric Graphing

Section Objectives
■ Obtain graphs of parametric equations.
■ Graph equations that define x as a function of y.

As we have seen, functional notation is an excellent way to describe certain kinds of relationships and curves. It is less helpful, however, when describing curves that fail the vertical line test. For example, Figure 3–30 shows a curve (called a *Lissajous Figure*) that is important in electrical engineering, yet would be troublesome to describe in functional notation.

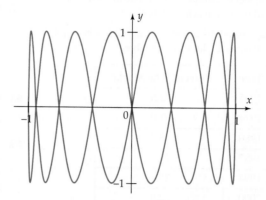

Figure 3–30

When we describe a curve by $y = f(x)$, we are thinking of the curve as the set of points satisfying that equation—the set of all the points where the y coordinate is equal to the value of the function f at the x coordinate. Now we take a different approach. We picture a point moving on the plane, and let the curve be the path the point has taken. The equations that describe the coordinates of this point at a given time t are called **parametric equations,** and the variable t is called the **parameter.**

EXAMPLE 1

Let $x = \dfrac{1}{2}t + 1$ and $y = t^2$.

(a) Make a table of values of x and y for $t = -3, -2, -1, 0, 1, 2,$ and 3.

(b) Plot the points in the table.

(c) Connect the points to find the curve traced out by this set of parametric equations.

SOLUTION

(a)

t	-3	-2	-1	0	1	2	3
$x = \dfrac{1}{2}t + 1$	$-.5$	0	$.5$	1	1.5	2	2.5
$y = t^2$	9	4	1	0	1	4	9
The point (x, y)	$(-.5, 9)$	$(0, 4)$	$(.5, 1)$	$(1, 0)$	$(1.5, 1)$	$(2, 4)$	$(2.5, 9)$

(b) We plot the points from the final row of the table to obtain Figure 3–31.

(c) We may plot a few more points to see the shape of the curve before we connect them all. The resultant curve is shown in Figure 3–32.

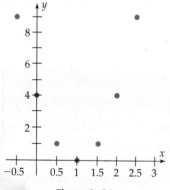

Figure 3–31

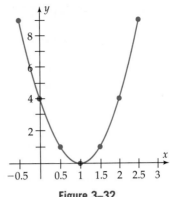

Figure 3–32

EXAMPLE 2

Use a calculator to graph the parametric equations

$$x = t^3 - t \qquad y = 4 - t^2$$

in the window $-7 \leq x \leq 7$, $-2 \leq y \leq 5$.

SOLUTION We change to parametric graphing mode, as suggested in the Technology Tip, and enter the equations (Figure 3–33). Setting up the viewing window requires some additional steps (first three lines of Figure 3–34). We don't know a suitable t-range, so we choose $-10 \leq t \leq 10$. The t-step (called t-pitch on Casio) determines how much t changes after a point is plotted; we set it at .15*. Both the t-range and t-step can be adjusted later if necessary.

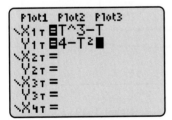

Figure 3–33

WINDOW
Tmin=-10
Tmax=10
Tstep=.15
Xmin=-7
Xmax=7
Xscl=1
↓Ymin=-2

Figure 3–34

WINDOW
↑Tstep=.15
Xmin=-7
Xmax=7
Xscl=1
Ymin=-2
Ymax=5
Yscl=1

Figure 3–35

Finally, we obtain the graph in Figure 3–36. ∎

EXAMPLE 3

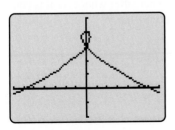

Figure 3–36

Graph the curve given by

$$x = t^2 - t - 1 \qquad \text{and} \qquad y = t^3 - 4t - 6 \quad (-2 \leq t \leq 3).$$

SOLUTION Using the standard viewing window, we obtain the graph in Figure 3–37 on the next page. Note that the graph crosses over itself at one point and that it does not extend forever to the left and right but has endpoints.

---

*If the t-step is much smaller than .15, the graph may take a long time to plot. If it is too large, the graph may look like a series of connected line segments rather than a smooth curve.

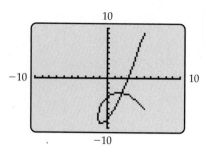

Figure 3–37

GRAPHING EXPLORATION

Graph these same parametric equations, but set the range of t values so that $-4 \le t \le 4$. What happens to the graph? Now change the range of t values so that $-10 \le t \le 10$. Find a viewing window large enough to show the entire graph, including endpoints. ∎

As we have seen, when given y as a function of x, we can graph it using the standard mode on our calculator. When we have x as a function of y, we can graph the curve using parametric equations.

EXAMPLE 4

Graph $x = y^3 - 3y^2 - 4y + 7$

SOLUTION Let t be any real number. If $y = t$, then $x = t^3 - 3t^2 - 4t + 7$. So the graph of $x = y^3 - 3y^2 - 4y + 7$ is the same as the graph of the parametric equations

$$x = t^3 - 3t^2 - 4t + 7 \quad y = t$$

As before, we change to parametric mode, enter the equations, set up the viewing window, and graph (Figures 3–38, 3–39). ∎

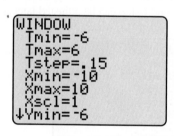

Figure 3–38

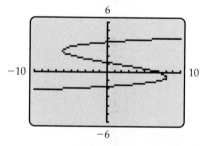

Figure 3–39

TECHNOLOGY TIP

If you have trouble finding appropriate ranges for t, x, and y, it might help to use the TABLE feature to display a table of t-x-y values produced by the parametric equations.

Any function of the form $y = f(x)$ can be expressed in terms of parametric equations and graphed that way. For instance, to graph $f(x) = x^2 + 1$, let $x = t$ and $y = f(x) = t^2 + 1$. Parametric graphing will be used hereafter whenever it is convenient and will be studied more thoroughly in Section 10.5.

EXERCISES 3.3.A

In Exercises 1–6, find a viewing window that shows a complete graph of the curve determined by the parametric equations.

1. $x = 3t^2 - 5$ and $y = t^2$ $(-4 \le t \le 4)$

2. The Zorro curve: $x = .1t^3 - .2t^2 - 2t + 4$ and $y = 1 - t$ $(-5 \le t \le 6)$

3. $x = t^2 - 3t + 2$ and $y = 8 - t^3$ $(-4 \le t \le 4)$

4. $x = t^2 - 6t$ and $y = \sqrt{t + 7}$ $(-5 \le t \le 9)$

5. $x = 1 - t^2$ and $y = t^3 - t - 1$ $(-4 \le t \le 4)$

6. $x = t^2 - t - 1$ and $y = 1 - t - t^2$

In Exercises 7–12, use parametric graphing. Find a viewing window that shows a complete graph of the equation.

7. $x = y^3 + 5y^2 - 4y - 5$

8. $\sqrt[3]{y^2 - y - 1} - x + 2 = 0$

9. $xy^2 + xy + x = y^3 - 2y^2 + 4$
  [*Hint:* First solve for x.]

10. $2y = xy^2 + 180x$ **11.** $x - \sqrt{y} + y^2 + 8 = 0$

12. $y^2 - x - \sqrt{y + 5} + 4 = 0$

14. Use parametric equations to describe a curve that crosses itself more times than the curve in Exercise 13. [Many correct answers are possible.]

THINKERS

13. Graph the curve given by

$$x = (t^2 - 1)(t^2 - 4)(t + 5) + t + 3$$

$$y = (t^2 - 1)(t^2 - 4)(t^3 + 4) + t - 1$$

$$(-2.5 \le t \le 2.5)$$

How many times does this curve cross itself?

3.4   Graphs and Transformations

Section Objectives
- Recognize the basic geometric transformations of a graph.
- Explore the relationship between algebraic changes in the rule of a function and geometric transformations of its graph.

If we know the rule of a function, then we can obtain new, related functions by carefully modifying the rule. In this section, we will explore how certain algebraic changes to a function's rule affect its graph. The same format will be used for each kind of change.

First: You will assemble some evidence by doing a graphing exploration on your calculator.

Second: We will draw some general conclusions from the evidence you obtained.

Third: We may discuss how these conclusions can be proved.

GRAPHING EXPLORATION

Consider the functions

$$f(x) = x^2 \qquad g(x) = x^2 + 5 \qquad h(x) = x^2 - 7$$

Graph f in the standard window and look at the graph, then graph g and see how the 5 changed the basic graph. Then graph h and notice the change the -7 made. Now answer these questions:

Do the graphs of g and h look very similar to the graph of f in *shape*?

How do their vertical positions differ?

Where would you predict that the graph of $k(x) = x^2 - 9$ is located relative to the graph $f(x) = x^2$, and what is its shape?

Confirm your prediction by graphing k on the same screen as f, g, and h.

The results of this Exploration should make the following statements plausible.

Vertical Shifts

Let f be a function and c a positive constant.

The graph of $g(x) = f(x) + c$ is the graph of f shifted c units upward.

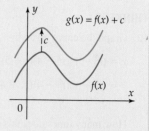

The graph of $h(x) = f(x) - c$ is the graph of f shifted c units downward.

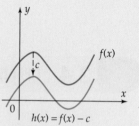

To see why these statements are true, suppose $f(x) = x^2$ and $g(x) = x^2 + 5$. For any given value of x, consider the points

$$P = (x, x^2) \text{ on the graph of } f \qquad \text{and} \qquad Q = (x, x^2 + 5) \text{ on the graph of } g.$$

The x-coordinates show that P and Q lie in the same vertical line. The y-coordinates show that Q lies 5 units directly above P. Thus, the graph of $g(x) = f(x) + 5$ is just the graph of f shifted 5 units upward.

EXAMPLE 1

A calculator was used to obtain a graph of $f(x) = .04x^3 - x - 3$ in Figure 3–40. The graph of

$$h(x) = f(x) - 4 = (.04x^3 - x - 3) - 4$$

is the graph of f shifted 4 units downward, as shown in Figure 3–41.

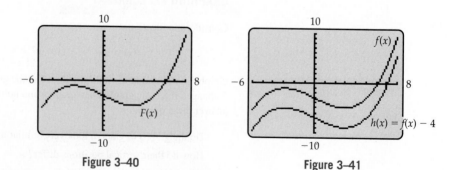

Figure 3–40   Figure 3–41

Although it may appear that the graph of h is closer to the graph of f at the right edge of Figure 3–41 than in the center, this is an optical illusion. The *vertical* distance between the graphs is always 4 units.

Use the trace feature of your calculator as follows to confirm that the vertical distance is always 4:*

> Move the cursor to any point on the graph of f, and note its coordinates.
>
> Use the down arrow to drop the cursor to the graph of h, and note the coordinates of the cursor in its new position.

The x-coordinates will be the same in both cases, and the new y-coordinate will be 4 less than the original y-coordinate.

∎

◗ HORIZONTAL SHIFTS

Consider the functions

$$f(x) = 2x^3 \qquad g(x) = 2(x + 6)^3 \qquad h(x) = 2(x - 8)^3$$

Graph f in the standard window and look at the graph, then graph g and see how the 6 changed the basic graph. Then graph h and notice the change the -8 made. Now answer these questions:

> Do the graphs of g and h look very similar to the graph of f in *shape*?
>
> How do their horizontal positions differ?
>
> Where would you predict that the graph of $k(x) = 2(x + 2)^3$ is located relative to the graph of $f(x) = 2x^3$, and what is its shape?

Confirm your prediction by graphing k on the same screen as f, g, and h.

The results of this Exploration should make the following statements plausible.

Horizontal Shifts

Let f be a function and c a positive constant.

The graph of $g(x) = f(x + c)$ is the graph of f shifted horizontally c units to the left.

The graph of $h(x) = f(x - c)$ is the graph of f shifted horizontally c units to the right.

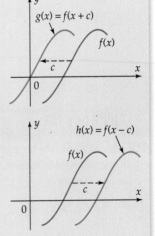

---

*The trace cursor can be moved vertically from graph to graph by using the up and down arrows.

To see why the first statement is true, suppose $g(x) = f(x + 4)$. Then the value of g at x is the same as the value of f at $x + 4$, which is 4 units to the right of x on the horizontal axis. So the graph of f is the graph of g shifted 4 units to the *right*, which means that the graph of g is the graph of f shifted 4 units to the *left*. An analogous argument works for the second statement in the box.

EXAMPLE 2

In some cases, shifting the graph of a function f horizontally may produce a graph that overlaps the graph of f. For instance, a graph of $f(x) = x^2 - 7$ is shown in red in Figure 3–42. The graph of

$$g(x) = f(x + 5) = (x + 5)^2 - 7$$

is the graph of f shifted 5 units to the left, and the graph of

$$h(x) = f(x - 4) = (x - 4)^2 - 7$$

is the graph of f shifted 4 units to the right, as shown in Figure 3–42. ∎

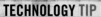
TECHNOLOGY TIP

If the function f of Example 2 is entered as $y_1 = x^2 - 7$, then the functions g and h can be entered as $y_2 = y_1(x + 5)$ and $y_3 = y_1(x - 4)$ on calculators other than TI-86 and Casio.

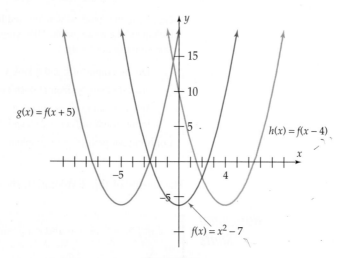

Figure 3–42

EXPANSIONS AND CONTRACTIONS

TECHNOLOGY TIP

On most calculators, you can graph both functions in the Exploration at the same time by keying in

$$y = \{1, 3\}(x^2 - 4).$$

GRAPHING EXPLORATION

In the viewing window with $-5 \leq x \leq 5$ and $-15 \leq y \leq 15$, graph these functions on the same screen:

$$f(x) = x^2 - 4 \qquad g(x) = 3f(x) = 3(x^2 - 4).$$

The table of values in Figure 3–43 shows that the y-coordinates on the graph of $Y_2 = g(x)$ are always 3 times the y-coordinates of the corresponding points on the graph of $Y_1 = f(x)$. To translate this into visual terms, imagine that the graph of f is nailed to the x-axis at its intercepts (± 2). The graph of g is then obtained by

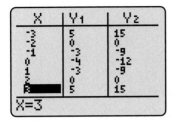

Figure 3–43

"stretching" the graph of f away from the x-axis (with the nails holding the x-intercepts in place) by a factor of 3, as shown in Figure 3–44.

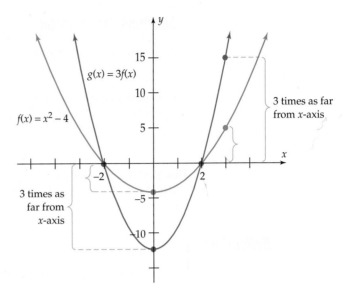

Figure 3–44

GRAPHING EXPLORATION

In the viewing window with $-4 \leq x \leq 4$ and $-5 \leq y \leq 12$, graph these functions on the same screen:

$$f(x) = x^2 - 4 \qquad h(x) = \frac{1}{4}(x^2 - 4).$$

Your screen should suggest that the graph of h is the graph of f "shrunk" vertically toward the x-axis by a factor of $1/4$.

Analogous facts are true in the general case.

Expansions and Contractions

If $c > 1$, then the graph of $g(x) = cf(x)$ is the graph of f stretched vertically away from the x-axis by a factor of c.

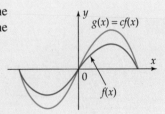

If $0 < c < 1$, then the graph of $h(x) = cf(x)$ is the graph of f shrunk vertically toward the x-axis by a factor of c.

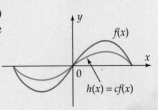

REFLECTIONS

In the standard viewing window, graph these functions on the same screen.

$$f(x) = .04x^3 - x \qquad g(x) = -f(x) = -(.04x^3 - x).$$

By moving your trace cursor from graph to graph, verify that for every point on the graph of f, there is a point on the graph of g with the same first coordinate that is on the opposite side of the x-axis, the same distance from the x-axis.

This Exploration shows that the graph of g is the mirror image (reflection) of the graph of f, with the x-axis being the mirror. The same thing is true in the general case.

Reflections

Let f be a function. The graph of $g(x) = -f(x)$ is the graph of f reflected in the x-axis.

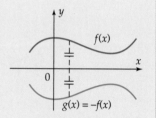

EXAMPLE 3

If $f(x) = x^2 - 3$, then the graph of

$$g(x) = -f(x) = -(x^2 - 3)$$

is the reflection of the graph of f in the x-axis, as shown in Figure 3–45.   ■

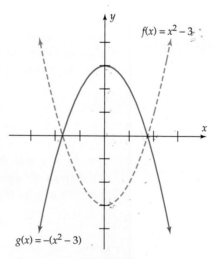

Figure 3–45

We now examine a different kind of reflection.

GRAPHING EXPLORATION

In the standard viewing window, graph these functions on the same screen:

$$f(x) = \sqrt{5x + 10} \quad \text{and} \quad h(x) = f(-x) = \sqrt{5(-x) + 10}.$$

Think carefully: How are the two graphs related to the y-axis? Now graph these two functions on the same screen:

$$f(x) = x^2 + 3x - 3$$

$$h(x) = f(-x) = (-x)^2 + 3(-x) - 3 = x^2 - 3x - 3.$$

Are the graphs of f and h related in the same way as the first pair?

This Exploration shows that the graph of h in each case is the mirror image (reflection) of the graph of f, with the y-axis as the mirror. The same thing is true in the general case.

Reflections

Let f be a function. The graph of $h(x) = f(-x)$ is the graph of f reflected in the y-axis.

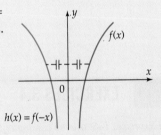

To see why this is true, let a be any number. Then

$(a, f(a))$ is on the graph of f and $(-a, h(-a))$ is on the graph of h.

However, $h(-a) = f(-(-a)) = f(a)$, so the two points are

$(a, f(a))$ on the graph of f and $(-a, f(a))$ on the graph of h.

These points lie on opposite sides of the y-axis, the same distance from the axis, because their first coordinates are negatives of each other. The points are on the same horizontal line because their second coordinates are the same. Thus, every point on the graph of f has a mirror-image point on the graph of h, the y-axis being the mirror.

Other algebraic operations and their graphical effects are considered in Exercises 48–61.

COMBINING TRANSFORMATIONS

The transformations described above may be used in sequence to analyze the graphs of functions whose rules are algebraically complicated.

EXAMPLE 4

To understand the graph of $g(x) = 2(x - 3)^2 - 1$, note that the rule of g may be obtained from the rule of $f(x) = x^2$ in three steps:

$$f(x) = x^2 \xrightarrow{\text{Step 1}} (x - 3)^2 \xrightarrow{\text{Step 2}} 2(x - 3)^2 \xrightarrow{\text{Step 3}} 2(x - 3)^2 - 1 = g(x).$$

Step 1 shifts the graph of f horizontally 3 units to the right; step 2 stretches the resulting graph away from the x-axis by a factor of 2; step 3 shifts this graph 1 unit downward, thus producing the graph of g in Figure 3–46. ∎

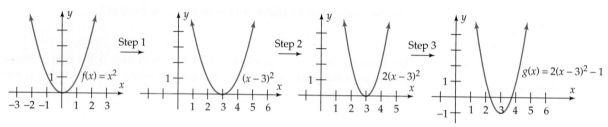

Figure 3–46

EXERCISES 3.4

In Exercises 1–8, use the catalog of functions at the end of Section 3.3 and information from this section to match each function with its graph, which is one of A–L.

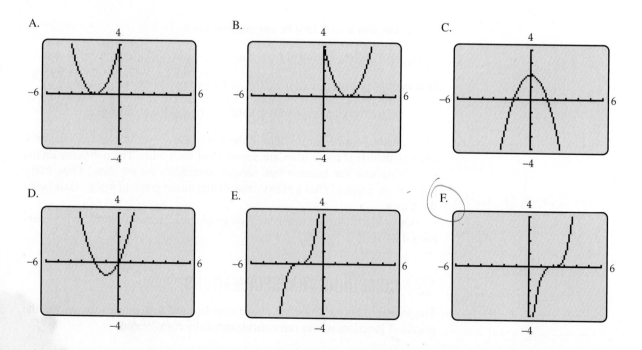

G.
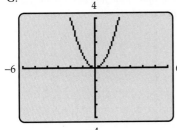

H.
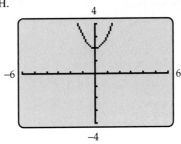

I.
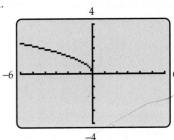

J.
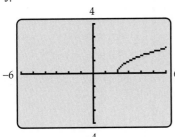

K.
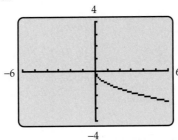

L.
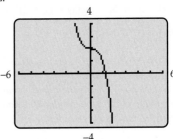

1. $f(x) = x^2 + 2$

2. $f(x) = \sqrt{x - 2}$

3. $g(x) = (x - 2)^3$

4. $g(x) = (x - 2)^2$

5. $f(x) = -\sqrt{x}$

6. $f(x) = (x + 1)^2 - 1$

7. $g(x) = -x^2 + 2$

8. $g(x) = -x^3 + 2$

9. The figure shows the graphs of $f(x) - 2$, $2f(x)$, and $f(x + 1)$. Sketch a graph of the function f.

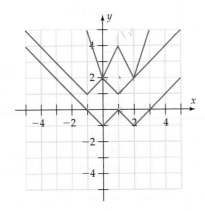

10. Fill in the entries in the following table

x	$f(x)$	$g(x) = f(x) + 2$	$h(x) = \frac{1}{2}f(x)$	$i(x) = 3f(x) - 2$
-1	$-1/2$			
0	1			
1	2			
2	6			
3	8			

11. Fill in the entries in the following table. If it is impossible to fill in an entry, put an X in it.

t	$f(t)$	$g(t) = f(t) - 3$	$h(t) = 4f(-t)$	$i(t) = f(t - 1) - 2$
-2	3			
-1	6			
0	8			
1	0			
2	5			

In Exercises 12–15, use the graph of $y = |x|$ and information from this section (but not a calculator) to sketch the graph of the function.

12. $f(x) = |x - 2|$

13. $g(x) = |x| - 2$

14. $g(x) = -|x|$

15. $f(x) = |x + 2| - 2$

In Exercises 16–19, find a single viewing window that shows complete graphs of the functions f, g, and h.

16. $f(x) = .25x^3 - 9x + 5$; $g(x) = f(x) + 15$; $h(x) = f(x) - 20$

17. $f(x) = \sqrt{x^2 - 9} - 5$; $g(x) = 3f(x)$; $h(x) = .5f(x)$

18. $f(x) = |x^2 - 5|$; $g(x) = f(x + 8)$; $h(x) = f(x - 6)$

19. $f(x) = .125x^3 - .25x^2 - 1.5x + 5$; $g(x) = f(x) - 5$; $h(x) = 5 - f(x)$

In Exercises 20 and 21, find complete graphs of the functions f and g in the same viewing window.

20. $f(x) = \dfrac{4 - 5x^2}{x^2 + 1}$; $g(x) = -f(x)$

21. $f(x) = x^4 - 4x^3 + 2x^2 + 3$; $g(x) = f(-x)$

In Exercises 22–25, describe a sequence of transformations that will transform the graph of the function f into the graph of the function g.

22. $f(x) = x^2 + x$; $g(x) = (x - 3)^2 + (x - 3) + 2$

23. $f(x) = x^2 + 5$; $g(x) = (x + 2)^2 + 10$

24. $f(x) = \sqrt{x^3 + 5}$; $g(x) = -\dfrac{1}{2}\sqrt{x^3 + 5} - 6$

25. $f(x) = \sqrt{x^4 + x^2 + 1}$; $g(x) = 10 - \sqrt{4x^4 + 4x^2 + 4}$

In Exercises 26–29, write the rule of a function g whose graph can be obtained from the graph of the function f by performing the transformations in the order given.

26. $f(x) = x^2 + 2$; shift the graph horizontally 5 units to the left and then vertically upward 4 units.

27. $f(x) = x^2 - x + 1$; reflect the graph in the x-axis, then shift it vertically upward 3 units.

28. $f(x) = \sqrt{x}$; shift the graph horizontally 6 units to the right, stretch it away from the x-axis by a factor of 2, and shift it vertically downward 3 units.

29. $f(x) = \sqrt{-x}$; shift the graph horizontally 3 units to the left, then reflect it in the x-axis, and shrink it toward the x-axis by a factor of $1/2$.

30. Let $f(x) = x^2 + 3x$, and let $g(x) = f(x) + 2$.

   (a) Write the rule of $g(x)$.

   (b) Find the difference quotients of $f(x)$ and $g(x)$. How are they related?

31. Let $f(x) = x^2 + 5$, and let $g(x) = f(x - 1)$.

   (a) Write the rule of $g(x)$ and simplify.

   (b) Find the difference quotients of $f(x)$ and $g(x)$.

   (c) Let $d(x)$ denote the difference quotient of $f(x)$. Show that the difference quotient of $g(x)$ is $d(x - 1)$.

In Exercises 32–35, use the graph of the function f in the figure to sketch the graph of the function g.

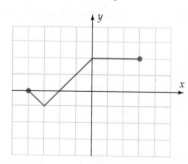

32. $g(x) = f(x) + 3$

33. $g(x) = f(x) - 1$

34. $g(x) = 3f(x)$

35. $g(x) = .25f(x)$

In Exercises 36–39, use the graph of the function f in the figure to sketch the graph of the function h.

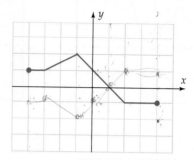

36. $h(x) = -f(x)$

37. $h(x) = -4f(x)$

38. $h(x) = f(-x)$

39. $h(x) = f(-x) + 2$

In Exercises 40–45, use the graph of the function f in the figure to sketch the graph of the function g.

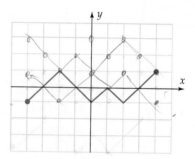

40. $g(x) = f(x + 3)$

41. $g(x) = f(x - 2)$

42. $g(x) = f(x - 2) + 3$

43. $g(x) = f(x + 1) - 3$

44. $g(x) = 2 - f(x)$

45. $g(x) = f(-x) + 2$

46. Graph $f(x) = -|x - 3| - |x - 17| + 20$ in the window with $0 \le x \le 20$ and $-2 \le y \le 12$. Think of the x-axis as a table and the graph as a side view of a fast-food carton placed upside down on the table (the flat part of the graph is the bottom of the carton). Find the rule of a function g whose graph (in this viewing window) looks like another fast-food carton, which has been placed right side up on top of the first one.

47. A factory has a linear cost function $c(x) = ax + b$, where b represents fixed costs and a represents the variable costs (labor and materials) of making one item, both in thousands of dollars.

   (a) If property taxes (part of the fixed costs) are increased by \$35,000 per year, what effect does this have on the graph of the cost function?

   (b) If variable costs increase by 12 cents per item, what effect does this have on the graph of the cost function?

In Exercises 48–50, assume $f(x) = (.2x)^6 - 4$. Use the standard viewing window to graph the functions f and g on the same screen.

48. $g(x) = f(2x)$ **49.** $g(x) = f(3x)$ **50.** $g(x) = f(4x)$

51. On the basis of the results of Exercises 48–50, describe the transformation that transforms the graph of a function $f(x)$

into the graph of the function $f(cx)$, where c is a constant with $c > 1$. [*Hint:* How are the two graphs related to the y-axis? Stretch your mind.]

In Exercises 52–55, assume $f(x) = x^2 - 3$. Use the standard viewing window to graph the functions f and g on the same screen.

52. $g(x) = f\left(\dfrac{1}{2}x\right)$

53. $g(x) = f\left(\dfrac{1}{3}x\right)$

54. $g(x) = f\left(\dfrac{1}{4}x\right)$

55. $g(x) = f\left(\dfrac{1}{10}x\right)$

56. On the basis of the results of Exercises 52–55, describe the transformation that transforms the graph of a function $f(x)$ into the graph of the function $f(cx)$, where c is a constant with $0 < c < 1$. [*Hint:* How are the two graphs related to the y-axis?]

In Exercises 57–60, use the standard viewing window to graph the function f and the function $g(x) = f(|x|)$ on the same screen.

57. $f(x) = x - 4$

58. $f(x) = x^3 - 3$

59. $f(x) = .5(x - 4)^2 - 9$

60. $f(x) = x^3 - 2x$

61. On the basis of the results of Exercises 57–60, describe the relationship between the graph of a function $f(x)$ and the graph of the function $f(|x|)$.

In Exercises 62–65, use the standard viewing window to graph the function f and the function $g(x) = |f(x)|$ on the same screen. Exercise 66 may be helpful for interpreting the results.

62. $f(x) = .5x^2 - 5$

63. $f(x) = x^3 - 4x^2 + x + 3$

64. $f(x) = x + 3$

65. $f(x) = x^3 - 2x$

66. (a) Let f be a function, and let g be the function defined by $g(x) = |f(x)|$. Use the definition of absolute value (page 9) to explain why the following statement is true:

$$g(x) = \begin{cases} f(x) & \text{if } f(x) \geq 0 \\ -f(x) & \text{if } f(x) < 0 \end{cases}$$

(b) Use part (a) and your knowledge of transformations to explain why the graph of g consists of those parts of the graph of f that lie above the x-axis together with the reflection in the x-axis of those parts of the graph of f that lie below the x-axis.

67. Because of a calculator's small screen size, it is not always easy (or even possible!) to find a viewing window that displays what its user desires.

(a) Graph $f(x) = x^2 - x - 6$ in the standard viewing window. Let $h(x) = f(x - 1000)$. What should $h(x)$ look like?

(b) Find an appropriate viewing window for the graph of $h(x)$.

(c) Try to find a viewing window that clearly displays both the graph of f and the graph of h. What makes this problem difficult?

(d) Let $g(x) = 1000\,f(x)$. What should $g(x)$ look like?

(e) Find an appropriate viewing window for the graph of $g(x)$. Can you find a viewing window that clearly displays both the graph of f and the graph of g?

3.4.A *SPECIAL TOPICS* Symmetry

Section Objectives

■ Recognize symmetries of a graph with respect to the x- and y-axes and the origin.

■ Identify even and odd functions, given a formula or a graph.

A graph is **symmetric with respect to the y-axis** if the part of the graph on the right side of the y-axis is the mirror image of the part on the left side of the y-axis (with the y-axis being the mirror), as shown in Figure 3–47.

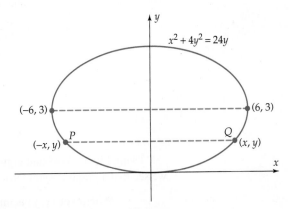

Figure 3–47

Each point P on the left side of the graph has a mirror image point Q on the right side of the graph, as indicated by the dashed lines. Note that:

Their second coordinates are the same (P and Q are on the same side of the x-axis and the same distance from it);

Their first coordinates are negatives of each other (P and Q lie on opposite sides of the y-axis and the same distance from it).

Thus, a graph is symmetric with respect to the y-axis provided that

Whenever (x, y) is on the graph, then $(-x, y)$ is also on it.

In algebraic terms, this means that replacing x by $-x$ in the equation leads to the same number y. In other words, replacing x by $-x$ produces an equivalent equation.

EXAMPLE 1

Replacing x by $-x$ in the equation $y = x^4 - 5x^2 + 3$ produces

$$y = (-x)^4 - 5(-x)^2 + 3,$$

which is the same equation because $(-x)^2 = x^2$ and $(-x)^4 = x^4$. Therefore, the graph is symmetric with respect to the y-axis.

GRAPHING EXPLORATION

Confirm this fact by graphing the equation $y = x^4 - 5x^2 + 3$.

■

 ## x-AXIS SYMMETRY

A graph is **symmetric with respect to the x-axis** if the part of the graph above the x-axis is the mirror image of the part below the x-axis (the x-axis being the mirror), as shown in Figure 3–48.

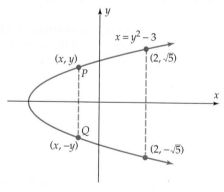

Figure 3–48

Using Figure 3–48 and argument analogous to the one preceding Example 1, we see that a graph is symmetric with respect to the x-axis provided that

Whenever (x, y) is on the graph, then $(x, -y)$ is also on it.

In algebraic terms, this means that replacing y by $-y$ in the equation leads to the same number x. In other words, replacing y by $-y$ produces an equivalent equation.

EXAMPLE 2

Replacing y by $-y$ in the equation $y^2 = 4x - 12$ produces $(-y)^2 = 4x - 12$, which is the same equation, so the graph is symmetric with respect to the x-axis.

GRAPHING EXPLORATION

Confirm this fact by graphing the equation. To do this, note that every point on the graph of $y^2 = 4x - 12$ is also on the graph of either $y = \sqrt{4x - 12}$ or $y = -\sqrt{4x - 12}$. Each of these latter equations defines a function; graph them both on the same screen.

■

ORIGIN SYMMETRY

A graph is **symmetric with respect to the origin** if a straight line through the origin and any point P on the graph also intersects the graph at a point Q such that the origin is the midpoint of segment PQ, as shown in Figure 3–49.

Here is a way to visualize origin symmetry: Picture hammering a nail into the origin and rotating the graph $180°$. If the graph winds up looking the same, then it is symmetric about the origin.

Using Figure 3–49, we can also describe symmetry with respect to the origin in terms of coordinates and equations (as proved in Exercise 38):

Whenever (x, y) is on the graph, then $(-x, -y)$ is also on it.

In algebraic terms, this means that replacing x by $-x$ and y by $-y$ in the equation produces an equivalent equation.

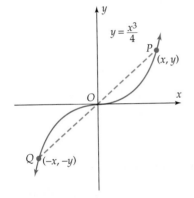

Figure 3–49

EXAMPLE 3

Replacing x with $-x$ and y with $-y$ in the equation $y = \dfrac{x^3}{10} - x$ yields

$$-y = \frac{(-x)^3}{10} - (-x)$$

$$-y = \frac{-x^3}{10} + x$$

$$y = \frac{x^3}{10} - x$$

Therefore the graph of $y = \dfrac{x^3}{10} - x$ is symmetric with respect to the origin.

GRAPHING EXPLORATION

Confirm this fact by graphing the equation $y = \dfrac{x^3}{10} - x$.

Here is a summary of the various tests for symmetry:

Symmetry Tests

Symmetry with Respect to	Coordinate Test for Symmetry	Algebraic Test for Symmetry
y-axis	(x, y) on graph implies $(-x, y)$ on graph.	Replacing x by $-x$ produces an equivalent equation.
x-axis	(x, y) on graph implies $(x, -y)$ on graph.	Replacing y by $-y$ produces an equivalent equation.
origin	(x, y) on graph implies $(-x, -y)$ on graph.	Replacing x by $-x$ and y by $-y$ produces an equivalent equation.

EVEN AND ODD FUNCTIONS

For *functions,* the algebraic description of symmetry takes a different form. A function f whose graph is symmetric with respect to the y-axis is called an **even function.** To say that the graph of $y = f(x)$ is symmetric with respect to the y-axis means that replacing x by $-x$ produces the same y value. In other words, the function takes the same value at both x and $-x$. Therefore,

Even Functions

A function f is even provided that

$$f(x) = f(-x) \text{ for every number } x \text{ in the domain of } f.$$

The graph of an even function is symmetric with respect to the y-axis.

For example, $f(x) = x^4 + x^2$ is even because

$$f(-x) = (-x)^4 + (-x)^2 = x^4 + x^2 = f(x).$$

Thus, the graph of f is symmetric with respect to the y-axis, as you can easily verify with your calculator (do it!).

Except for zero functions ($f(x) = 0$ for every x in the domain), *the graph of a function is never symmetric with respect to the x-axis.* The reason is the vertical line test: The graph of a function never contains two points with the same first coordinate. If both $(5, 3)$ and $(5, -3)$, for instance, were on the graph, this would say that $f(5) = 3$ and $f(5) = -3$, which is impossible when f is a function.

A function whose graph is symmetric with respect to the origin is called an **odd function.** If both (x, y) and $(-x, -y)$ are on the graph of such a function f, then we must have both

$$y = f(x) \qquad \text{and} \qquad -y = f(-x),$$

so $f(-x) = -y = -f(x)$. Therefore,

Odd Functions

A function f is odd provided that

$$f(-x) = -f(x) \text{ for every number } x \text{ in the domain of } f.$$

The graph of an odd function is symmetric with respect to the origin.

For example, $f(x) = x^3$ is an odd function because

$$f(-x) = (-x)^3 = -x^3 = -f(x).$$

Hence, the graph of f is symmetric with respect to the origin (verify this with your calculator).

EXERCISES 3.4.A

In Exercises 1–4, sketch the graph of the equation. If the graph is symmetric with respect to the x-axis, the y-axis, or the origin, say so.

1. $y = x^2 + 2$　　**2.** $x = (y - 3)^2$

3. $y = x^3 + 2$　　**4.** $y = (x + 2)^3$

In Exercises 5–16, determine whether the given function is even, odd, or neither.

5. $f(x) = 4x$　　**6.** $k(t) = -5t$

7. $f(x) = x^2 - |x|$　　**8.** $h(u) = |3u|$

9. $k(t) = t^4 - 6t^2 + 5$

10. $f(x) = x(x^4 - x^2) + 4$

11. $f(x) = x(x^4 - x^2) + 4x$

12. $f(t) = \sqrt{t^2 - 5}$　　**13.** $h(x) = \sqrt{7 - 2x^2}$

14. $f(x) = \dfrac{x^2 + 2}{x - 7}$　　**15.** $g(x) = \dfrac{x^2 + 1}{x^2 - 1}$

16. $L(a) = \dfrac{|a|}{a}$

In Exercises 17–20, determine algebraically whether or not the graph of the given equation is symmetric with respect to the x-axis.

17. $x^2 - 6x + y^2 + 8 = 0$

18. $x^2 + 8x + y^2 = -15$

19. $x^2 - 2x + y^2 + 2y = 2$

20. $x^2 - x + y^2 - y = 0$

In Exercises 21–28, determine whether the given graph is symmetric with respect to the y-axis, the x-axis, or the origin.

21.

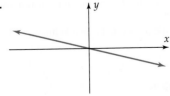

22.

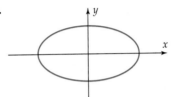

23.

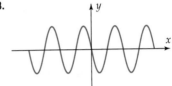

24.

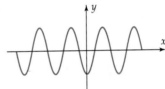

25.

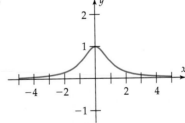

26.

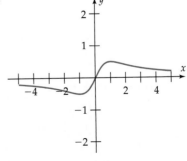

27.

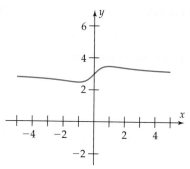

28.

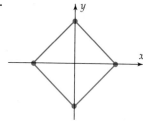

In Exercises 29–32, complete the graph of the given function, assuming that it satisfies the given symmetry condition.

29. Even

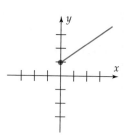

30. Even

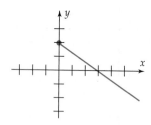

31. Odd

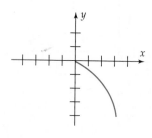

32. Odd

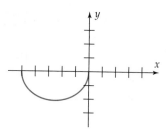

33. (a) Draw some coordinate axes, and plot the points $(0, 1)$, $(1, -3)$, $(-5, 2)$, $(-3, 5)$, $(2, 3)$, and $(4, 1)$.
 (b) Suppose the points in part (a) lie on the graph of an *even* function f. Plot the points $(0, f(0))$, $(-1, f(-1))$, $(5, f(5))$, $(3, f(3))$, $(-2, f(-2))$, and $(-4, f(-4))$.

34. Draw the graph of an *even* function that includes the points $(0, -3)$, $(-3, 0)$, $(2, 0)$, $(1, -4)$, $(2.5, -1)$, $(-4, 3)$, and $(-5, 3)$.*

35. (a) Plot the points $(0, 0)$; $(2, 3)$; $(3, 4)$; $(5, 0)$; $(7, -3)$; $(-1, -1)$; $(-4, -1)$; $(-6, 1)$.
 (b) Suppose the points in part (a) lie on the graph of an *odd* function f. Plot the points $(-2, f(-2))$; $(-3, f(-3))$; $(-5, f(-5))$; $(-7, f(-7))$; $(1, f(1))$; $(4, f(4))$; $(6, f(6))$.
 (c) Draw the graph of an odd function f that includes all the points plotted in parts (a) and (b).*

36. Draw the graph of an odd function that includes the points $(-3, 5)$, $(-1, 1)$, $(2, -6)$, $(4, -9)$, and $(5, -5)$.*

37. Show that any graph that has two of the three types of symmetry (x-axis, y-axis, origin) necessarily has the third type also.

38. Use the midpoint formula to show that $(0, 0)$ is the midpoint of the segment joining (x, y) and $(-x, -y)$. Conclude that the coordinate test for symmetry with respect to the origin (page 192) is correct.

THINKER

39. In the first half of 2005, Ford Motor company spent 768 million dollars on advertising. Presumably, if they had spent more money, they would have sold more cars, and if they had spent less money, they would have sold fewer cars. Let x be the change in their advertising budget for the first half of 2006, in millions of dollars. If x is 10, that would correspond to Ford spending 778 million dollars, for example, and if x is -10 then that corresponds to Ford spending 758 million.
  Let $f(x)$ be the number of cars that Ford sells in the first half of 2006, with $-100 \leq x \leq 100$.

 (a) Would it be realistic for $f(x)$ to be an even function? Why or why not?
 (b) Would it be realistic for $f(x)$ to be an odd function? Why or why not?

*There are many correct answers.

3.5   Operations on Functions

Section Objectives
- Find the sum, difference, product, and quotient of two functions.
- Compose functions to create a new function.
- Write a function as the composite of two or more functions.

We now examine ways in which two or more given functions can be used to create new functions. If f and g are functions, then their **sum** is the function h defined by the rule

$$h(x) = f(x) + g(x).$$

For example, if $f(x) = 3x^2 + x$ and $g(x) = 4x - 2$, then

$$\begin{aligned} h(x) &= f(x) + g(x) \\ &= (3x^2 + x) + (4x - 2) \\ &= 3x^2 + 5x - 2. \end{aligned}$$

Instead of using a different letter h for the sum function, we shall usually denote it by $f + g$. Thus, the sum $f + g$ is defined by the rule

$$(f + g)(x) = f(x) + g(x).$$

This rule is *not* just a formal manipulation of symbols. If x is a number, then so are $f(x)$ and $g(x)$. The plus sign in $f(x) + g(x)$ is addition of *numbers*, and the result is a number. But the plus sign in $f + g$ is addition of *functions*, and the result is a new function.

The **difference** $f - g$ is the function defined by the rule

$$(f - g)(x) = f(x) - g(x).$$

The domain of the sum and difference functions is the set of all real numbers that are in both the domain of f and the domain of g.

EXAMPLE 1

If $f(x) = \sqrt{9 - x^2}$ and $g(x) = \sqrt{x - 2}$, find the rules of the functions $f + g$ and $f - g$ and their domains.

TECHNOLOGY TIP

If you have two functions entered in the equation memory as y_1 and y_2, you can graph their sum by entering $y_1 + y_2$ as y_3 in the equation memory and graphing y_3. Differences, products, and quotients are graphed similarly. To find the correct keys for y_1 and y_2, see the Tip on page 153.

SOLUTION   We have

$$(f + g)(x) = f(x) + g(x) = \sqrt{9 - x^2} + \sqrt{x - 2};$$
$$(f - g)(x) = f(x) - g(x) = \sqrt{9 - x^2} - \sqrt{x - 2}.$$

The domain of f consists of all x such that $9 - x^2 \geq 0$ (so that the square root will be defined), that is, all x with $-3 \leq x \leq 3$. Similarly, the domain of g consists of all x such that $x \geq 2$. The domain of $f + g$ and $f - g$ consists of all real numbers in both the domain of f and the domain of g, namely, all x such that $2 \leq x \leq 3$. ∎

The **product** and **quotient** of functions f and g are the functions defined by the rules

$$(fg)(x) = f(x)g(x) \qquad \text{and} \qquad \left(\frac{f}{g}\right)(x) = \frac{f(x)}{g(x)}.$$

The domain of fg consists of all real numbers in both the domain of f and the domain of g. The domain of f/g consists of all real numbers x in both the domain of f and the domain of g such that $g(x) \neq 0$.

EXAMPLE 2

If $f(x) = \sqrt{3x}$ and $g(x) = x^2 - 1$, find the rules of the functions fg and f/g and their domains.

SOLUTION   The rules are

$$(fg)(x) = f(x)g(x) \qquad\qquad \left(\frac{f}{g}\right)(x) = \frac{f(x)}{g(x)}$$

$$= \sqrt{3x}\,(x^2 - 1) \qquad\qquad\qquad = \frac{\sqrt{3x}}{x^2 - 1}$$

$$= (\sqrt{3x})\,x^2 - \sqrt{3x}$$

The domain of fg consists of all numbers x in both the domain of f (all nonnegative real numbers) and the domain of g (all real numbers), that is, all $x \geq 0$. The domain of f/g consists of all these x for which $g(x) \neq 0$, that is, all nonnegative real numbers *except* $x = 1$. ∎

If c is a real number and f is a function, then the product of f and the constant function $g(x) = c$ is usually denoted cf. For example, if the function

$$f(x) = x^3 - x + 2,$$

and $c = 5$, then $5f$ is the function given by

$$(5f)(x) = 5 \cdot f(x)$$

$$= 5(x^3 - x + 2)$$

$$= 5x^3 - 5x + 10$$

■ COMPOSITION OF FUNCTIONS

Another way of combining functions is illustrated by the function $h(x) = \sqrt{x^3}$. To compute $h(4)$, for example, you first find $4^3 = 64$ and then take the square root $\sqrt{64} = 8$. So the rule of h may be rephrased as follows:

First apply the function $f(x) = x^3$,

Then apply the function $g(t) = \sqrt{t}$ to the result.

The same idea can be expressed in functional notation like this:

$$x \xrightarrow{\text{first apply } f} f(x) \xrightarrow{\text{then apply } g \text{ to the result}} g(f(x)).$$

$$x \qquad\qquad\qquad x^3 \qquad\qquad\qquad\qquad \sqrt{x^3}$$

apply h

So the rule of h may be written as $h(x) = g(f(x))$, where $f(x) = x^3$ and $g(t) = \sqrt{t}$. We can think of h as being made up of two simpler functions f and g, or we can think of f and g being "composed" to create the function h. Both viewpoints are useful.

EXAMPLE 3

Suppose $f(x) = 4x^2 + 1$ and $g(t) = \dfrac{1}{t + 2}$. Define a new function h whose rule is "first apply f; then apply g to the result." In functional notation,

$$x \xrightarrow{\text{first apply } f} f(x) \xrightarrow{\text{then apply } g \text{ to the result}} g(f(x)).$$

So the rule of the function h is $h(x) = g(f(x))$. Evaluating $g(f(x))$ means that whenever t appears in the formula for $g(t)$, we must replace it by $f(x) = 4x^2 + 1$:

$$h(x) = g(f(x)) = \frac{1}{f(x) + 2}$$

$$= \frac{1}{(4x^2 + 1) + 2}$$

$$= \frac{1}{4x^2 + 3} \qquad \blacksquare$$

The function h in Example 3 is an illustration of the following definition.

Composite Functions

Let f and g be functions. The **composite function** of f and g defined as follows.

For input x, the output is $g(f(x))$.

This composite function is denoted $g \circ f$.

The symbol "$g \circ f$" is read "g circle f" or "f followed by g." (Note the order carefully; the functions are applied *right* to *left*.) So the rule of the composite function is

$$(g \circ f)(x) = g(f(x)).$$

EXAMPLE 4

If $f(x) = 2x + 5$ and $g(t) = 3t^2 + 2t + 4$, then find

$$(f \circ g)(2), \qquad (g \circ f)(-1), \qquad (g \circ f)(5), \qquad (g \circ f)(x).$$

SOLUTION

$(f \circ g)(2) = f(g(2))$

$\qquad = f(3 \cdot 2^2 + 2 \cdot 2 + 4)$

$\qquad = f(20)$

$\qquad = 2 \cdot 20 + 5$

$\qquad = 45.$

Similarly, $\quad (g \circ f)(-1) = g(f(-1))$

$\qquad\qquad = g(2(-1) + 5)$

$\qquad\qquad = g(3)$

$\qquad\qquad = 3 \cdot 3^2 + 2 \cdot 3 + 4$

$\qquad\qquad = 37.$

The value of a composite function can also be computed like this:

$$(g \circ f)(5) = g(f(5)) = 3(f(5)^2) + 2(f(5)) + 4 = 3(15^2) + 2(15) + 4 = 709.$$

and

$$(g \circ f)(x) = g(f(x)) = 3(2x + 5)^2 + 2(2x + 5) + 4 = 12x^2 + 64x + 89. \quad \blacksquare$$

The domain of $g \circ f$ is determined by this convention.

Domain of
$g \circ f$

> The domain of the composite function $g \circ f$ is the set of all real numbers x such that x is in the domain of f and $f(x)$ is in the domain of g.

EXAMPLE 5

Find the rule and domain of $g \circ f$, when $f(x) = \sqrt{x}$ and $g(t) = t^2 - 5$.

SOLUTION

$$(g \circ f)(x) = g(f(x)) = (f(x))^2 - 5 = (\sqrt{x})^2 - 5 = x - 5.$$

Although $x - 5$ is defined for every real number x, the domain of $g \circ f$ is *not* the set of all real numbers. The domain of g is the set of all real numbers, but the function $f(x) = \sqrt{x}$ is defined only when $x \geq 0$. So the domain of $g \circ f$ is the set of nonnegative real numbers, that is, the interval $[0, \infty)$. $\quad \blacksquare$

EXAMPLE 6

Write the function $h(x) = \sqrt{3x^2 + 1}$ in two different ways as the composite of two functions.

SOLUTION   Let $f(x) = 3x^2 + 1$ and $g(x) = \sqrt{x}$.* Then

$$(g \circ f)(x) = g(f(x)) = g(3x^2 + 1) = \sqrt{3x^2 + 1} = h(x).$$

Similarly, h is also the composite $j \circ k$, where $j(x) = \sqrt{x + 1}$ and $k(x) = 3x^2$:

$$(j \circ k)(x) = j(k(x)) = j(3x^2) = \sqrt{3x^2 + 1} = h(x). \quad \blacksquare$$

TECHNOLOGY TIP

Evaluating composite functions is easy on calculators other than TI-86 and most Casio calculators. If the functions are entered in the equation memory as $y_1 = g(x)$ and $y_2 = h(x)$ (with f in place of y on HP-39gs), then keying in $y_2(y_1(5))$ ENTER produces the number $h(g(5))$.

On TI-86 and most Casio calculators, this syntax does *not* produce $h(g(5))$; it produces

$$h(x) \cdot g(x) \cdot 5$$

for whatever number is stored in the x-memory.

EXAMPLE 7

If $k(x) = (x^2 - 2x + \sqrt{x})^3$, then k is $g \circ f$, where $f(x) = x^2 - 2x + \sqrt{x}$ and $g(x) = x^3$ because

$$(g \circ f)(x) = g(f(x)) = g(x^2 - 2x + \sqrt{x}) = (x^2 - 2x + \sqrt{x})^3 = k(x). \quad \blacksquare$$

---

*Now that you have the idea of composite functions, we'll use the same letter for the variable in both functions.

By using the function operations above, a complicated function may be considered as being built up from simple parts.

EXAMPLE 8

The function

$$f(x) = \sqrt{\frac{3x^2 - 4x + 5}{x^3 + 1}}$$

may be considered as the composite $f = g \circ h$, where

$$h(x) = \frac{3x^2 - 4x + 5}{x^3 + 1} \quad \text{and} \quad g(x) = \sqrt{x},$$

since

$$(g \circ h)(x) = g(h(x)) = g\left(\frac{3x^2 - 4x + 5}{x^3 + 1}\right) = \sqrt{\frac{3x^2 - 4x + 5}{x^3 + 1}} = f(x).$$

The function

$$h(x) = \frac{3x^2 - 4x + 5}{x^3 + 1}$$

is the quotient $\dfrac{p}{q}$, where

$$p(x) = 3x^2 - 4x + 5 \quad \text{and} \quad q(x) = x^3 + 1.$$

The function $p(x) = 3x^2 - 4x + 5$ may be written $p = k - s + r$, where

$$k(x) = 3x^2, \quad s(x) = 4x, \quad r(x) = 5.$$

The function k, in turn, can be considered as the product $3I^2$, where I is the *identity function* [whose rule is $I(x) = x$]:

$$(3I^2)(x) = 3(I^2(x)) = 3(I(x)I(x)) = 3 \cdot x \cdot x = 3x^2 = k(x).$$

Similarly, $s(x) = (4I)(x) = 4I(x) = 4x$. The function $q(x) = x^3 + 1$ may be "decomposed" in the same way.

Thus, the complicated function f is just the result of performing suitable operations on the identity function I and various constant functions. ∎

As you may have noticed, there are two possible ways to form a composite function from two given functions. If f and g are functions, we can consider either

$$(g \circ f)(x) = g(f(x)), \quad \text{[the composite of } f \text{ and } g]$$
$$(f \circ g)(x) = f(g(x)). \quad \text{[the composite of } g \text{ and } f]$$

The *order is important*, as we shall now see:

$g \circ f$ **and** $f \circ g$ **usually are** *not* **the same function.**

EXAMPLE 9

If $f(x) = x^2$ and $g(x) = x + 3$, then

$$(f \circ g)(x) = f(g(x)) \qquad\qquad (g \circ f)(x) = g(f(x))$$
$$= f(x + 3) \qquad\qquad\qquad = g(x^2)$$
$$= (x + 3)^2 \qquad\qquad\qquad = x^2 + 3,$$
$$= x^2 + 6x + 9.$$

Obviously, $g \circ f \neq f \circ g$, since, for example, they have different values at $x = 0$. ∎

CAUTION

Don't confuse the product function fg with the composite function $f \circ g$ (g followed by f). For instance, if $f(x) = 2x^2$ and $g(x) = x - 3$, then the product fg is given by

$$(fg)(x) = f(x)g(x) = 2x^2(x - 3) = 2x^3 - 6x^2.$$

It is *not* the same as the composite $f \circ g$ because

$$(f \circ g)(x) = f(g(x)) = f(x - 3) = 2(x - 3)^2 = 2x^2 - 12x + 18.$$

APPLICATIONS

Compositions of functions arise in applications involving several functional relationships simultaneously. In such cases, one quantity may have to be expressed as a function of another.

EXAMPLE 10

A circular puddle of liquid is evaporating and slowly shrinking in size. After t minutes, the radius r of the puddle measures $\dfrac{18}{2t + 3}$ inches; in other words, the radius is a function of time. The area A of the puddle is given by $A = \pi r^2$, that is, area is a function of the radius r. We can express the area as a function of time by substituting $r = \dfrac{18}{2t + 3}$ in the area equation:

$$A = \pi r^2 = \pi \left(\frac{18}{2t + 3}\right)^2.$$

This amounts to forming the composite function $f \circ g$, where $f(r) = \pi r^2$ and $g(t) = \dfrac{18}{2t + 3}$:

$$(f \circ g)(t) = f(g(t)) = f\left(\frac{18}{2t + 3}\right) = \pi\left(\frac{18}{2t + 3}\right)^2.$$

When area is expressed as a function of time, it is easy to compute the area of the puddle at any time. For instance, after 12 minutes, the area of the puddle is

$$A = \pi\left(\frac{18}{2t + 3}\right)^2 = \pi\left(\frac{18}{2 \cdot 12 + 3}\right)^2 = \frac{4\pi}{9} \approx 1.396 \text{ square inches.} \quad ∎$$

EXAMPLE 11

At noon, a car leaves Podunk on a straight road, heading south at 45 mph, and a plane 3 miles above the ground passes over Podunk heading east at 350 mph.

(a) Express the distance r traveled by the car and the distance s traveled by the plane as functions of time.

(b) Express the distance d between the plane and the car in terms of r and s.

(c) Express d as a function of time.

(d) How far apart were the plane and the car at 1:30 P.M.?

SOLUTION

(a) Traveling at 45 mph for t hours, the car will go a distance of $45t$ miles. Hence, the equation $r = 45t$ expresses the distance r as a function of the time t. Similarly, the equation $s = 350t$ expresses the distance s as a function of the time t.

(b) To express the distance d as a function of r and s, consider Figure 3–50.

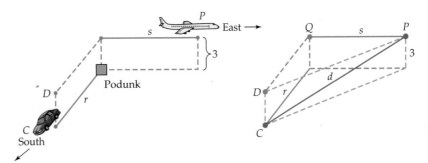

Figure 3–50

Right triangle PQD and the Pythagorean Theorem show that $(PD)^2 = r^2 + s^2$; hence, $PD = \sqrt{r^2 + s^2}$. Applying the Pythagorean Theorem to right triangle PDC, we have

$$d^2 = 3^2 + (PD)^2$$
$$d^2 = 3^2 + (\sqrt{r^2 + s^2})^2$$
$$d^2 = 9 + r^2 + s^2$$
$$d = \sqrt{9 + r^2 + s^2}.$$

(c) The preceding equation expresses d in terms of r and s. By substituting $r = 45t$ and $s = 350t$ in this equation, we can express d as a function of the time t:

$$d = \sqrt{9 + r^2 + s^2}$$
$$d = \sqrt{9 + (45t)^2 + (350t)^2}$$
$$d = \sqrt{9 + 2025t^2 + 122{,}500t^2} = \sqrt{9 + 124{,}525t^2}.$$

(d) At 1:30 P.M., we have $t = 1.5$ (since noon is $t = 0$). At this time,

$$d = \sqrt{9 + 124{,}525t^2} = \sqrt{9 + 124{,}525(1.5)^2} = \sqrt{280{,}190.25}$$
$$\approx 529.33 \text{ miles.}$$

EXERCISES 3.5

In Exercises 1–4, find $(f + g)(x)$, $(f - g)(x)$, *and* $(g - f)(x)$.

1. $f(x) = -3x + 2$, $\quad g(x) = x^3$

2. $f(x) = x^2 + 2$, $\quad g(x) = x^2 - 4x - 2$

3. $f(x) = 1/x$, $\quad g(x) = x^2 + 2x - 5$

4. $f(x) = \sqrt{x}$, $\quad g(x) = x^2 + 1 + \sqrt{x}$

In Exercises 5–8, find $(fg)(x)$, $(f/g)(x)$, *and* $(g/f)(x)$.

5. $f(x) = -3x + 2$, $\quad g(x) = x^3$

6. $f(x) = 4x^2 + x^4$, $\quad g(x) = \sqrt{x^2 + 4}$

7. $f(x) = x + 5$, $\quad g(x) = x - 5$

8. $f(x) = \sqrt{x^2 - 1}$, $\quad g(x) = \sqrt{x - 1}$

In Exercises 9–12, find the domains of fg *and* f/g.

9. $f(x) = x^2 + 1$, $\quad g(x) = 1/x$

10. $f(x) = x^2 + 2$, $\quad g(x) = \dfrac{1}{x^2 + 2}$

11. $f(x) = \sqrt{4 - x^2}$, $\quad g(x) = \sqrt{3x + 4}$

12. $f(x) = 3x^2 + x^4 + 2$, $\quad g(x) = 4x - 3$

In Exercises 13–16, find the indicated values, where

$$g(t) = t^2 - t \text{ and } f(x) = 1 + x.$$

13. $g(f(0))$

14. $(f \circ g)(3)$

15. $g(f(2) + 3)$

16. $f(2g(1))$

In Exercises 17–20, find $(g \circ f)(3)$, $(f \circ g)(1)$, *and* $(f \circ f)(0)$.

17. $f(x) = 3x - 2$, $\quad g(x) = x^2$

18. $f(x) = |x + 2|$, $\quad g(x) = -x^2$

19. $f(x) = x$, $\quad g(x) = -3$

20. $f(x) = x^2 - 1$, $\quad g(x) = \sqrt{x}$

In Exercises 21–24, find the rule of the function $f \circ g$, *the domain of* $f \circ g$, *the rule of* $g \circ f$, *and the domain of* $g \circ f$.

21. $f(x) = -3x + 2$, $\quad g(x) = x^3$

22. $f(x) = 1/x$, $\quad g(x) = \sqrt{x}$

23. $f(x) = \dfrac{1}{2x + 1}$, $\quad g(x) = x^2 - 1$

24. $f(x) = (x - 3)^2$, $\quad g(x) = \sqrt{x} + 3$

In Exercises 25–28, find the rules of the functions ff *and* $f \circ f$.

25. $f(x) = x^3$

26. $f(x) = (x - 1)^2$

27. $f(x) = 1/x$

28. $f(x) = \dfrac{1}{x - 1}$

In Exercises 29–32, verify that $(f \circ g)(x) = x$ *and* $(g \circ f)(x) = x$ *for every x.*

29. $f(x) = 9x + 8$, $\quad g(x) = \dfrac{x - 8}{9}$

30. $f(x) = \sqrt[3]{x - 1}$, $\quad g(x) = x^3 + 1$

31. $f(x) = \sqrt[3]{x} + 2$, $\quad g(x) = (x - 2)^3$

32. $f(x) = 2x^3 - 5$, $\quad g(x) = \sqrt[3]{\dfrac{x + 5}{2}}$

Exercises 33 and 34 refer to the function f whose graph is shown in the figure.

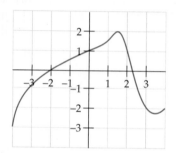

33. Let g be the composite function $f \circ f$. Use the graph of f to fill in the following table (approximate where necessary).

x	$f(x)$	$g(x) = f(f(x))$
-4	-3	-1
-3	-1	$1/2$
-2	0	1
-1	$1/2$	$5/4$
0	1	$3/2$
1	$3/2$	2
2	1	$3/2$
3	-2	0
4	-2	0

34. Use the information obtained in Exercise 33 to sketch the graph of the function g.

In Exercises 35–38, fill the blanks in the given table. In each case the values of the functions f and g are given by these tables:

x	$f(x)$
1	3
2	5
3	1
4	2
5	3

t	$g(t)$
1	5
2	4
3	4
4	3
5	2

35.

x	$(g \circ f)(x)$
1	4
2	
3	5
4	
5	

36.

t	$(f \circ g)(t)$
1	3
2	2
3	2
4	1
5	6

37.

x	$(f \circ f)(x)$
1	
2	
3	3
4	
5	

38.

t	$(g \circ g)(t)$
1	
2	
3	
4	4
5	

In Exercises 39–42, write the given function as the composite of two functions, neither of which is the identity function, as in Examples 6 and 7. (There may be more than one way to do this.)

39. $f(x) = \sqrt[3]{x^2 + 2}$

40. $g(x) = \sqrt{x + 3} - \sqrt[3]{x + 3}$

41. $h(x) = (7x^3 - 10x + 17)^7$

42. $f(x) = \dfrac{1}{3x^2 + 5x - 7}$

43. If $f(x) = x + 1$ and $g(t) = t^2$, then

$$(g \circ f)(x) = g(f(x)) = g(x + 1) = (x + 1)^2$$
$$= x^2 + 2x + 1$$

Find two other functions $h(x)$ and $k(t)$ such that $(k \circ h)(x) = x^2 + 2x + 1$.

44. If f is any function and I is the identity function, what are $f \circ I$ and $I \circ f$?

In Exercises 45–48, determine whether the functions $f \circ g$ and $g \circ f$ are defined. If a composite is defined, find its domain.

45. $f(x) = x^3$, $g(x) = \sqrt{x}$

46. $f(x) = x^2 + 1$, $g(x) = \sqrt{x}$

47. $f(x) = \sqrt{x + 10}$, $g(x) = 5x$

48. $f(x) = -x^2$, $g(x) = \sqrt{x}$

49. (a) If $f(x) = 2x^3 + 5x - 1$, find $f(x^2)$.
(b) If $f(x) = 2x^3 + 5x - 1$, find $(f(x))^2$.
(c) Are the answers in parts (a) and (b) the same? What can you conclude about $f(x^2)$ and $(f(x))^2$?

50. Give two examples of functions f such that

$$f\left(\frac{1}{x}\right) \neq \frac{1}{f(x)}.$$

In Exercises 51 and 52, graph both $f \circ g$ and $g \circ f$ on the same screen. Use the graphs to determine whether $f \circ g$ is the same function as $g \circ f$.

51. $f(x) = x^5 - x^3 - x$; $g(x) = x - 2$

52. $f(x) = x^3 + x$; $g(x) = \sqrt[3]{x - 1}$

In Exercises 53–56, find $g \circ f$, and find the difference quotient of the function $g \circ f$.

53. $f(x) = x + 3$; $g(x) = x^2 + 1$

54. $f(x) = 2x$; $g(x) = 8x$

55. $f(x) = x + 1$; $g(x) = \dfrac{2}{x - 1}$

56. $f(x) = x^3$; $g(x) = x + 2$

57. (a) What is the area of the puddle in Example 10 after one day? After a week? After a month?
(b) Does the puddle ever totally evaporate? Is this realistic? Under what circumstances might this area function be an accurate model of reality?

58. In a laboratory culture, the number $N(d)$ of bacteria (in thousands) at temperature d degrees Celsius is given by the function

$$N(d) = \frac{-90}{d + 1} + 20 \quad (4 \le d \le 32).$$

The temperature $D(t)$ at time t hours is given by the function $D(t) = 2t + 4 \quad (0 \le t \le 14)$.

(a) What does the composite function $N \circ D$ represent?
(b) How many bacteria are in the culture after 4 hours? After 10 hours?

59. A certain fungus grows in a circular shape. Its diameter after t weeks is $6 - \dfrac{50}{t^2 + 10}$ inches.

(a) Express the area covered by the fungus as a function of time.
(b) What is the area covered by the fungus when $t = 0$? What area does it cover at the end of 8 weeks?
(c) When is its area 25 square inches?

60. Tom left point P at 6 A.M. walking south at 4 mph. Anne left point P at 8 A.M. walking west at 3.2 mph.

(a) Express the distance between Tom and Anne as a function of the time t elapsed since 6 A.M.
(b) How far apart are Tom and Anne at noon?
(c) At what time are they 35 miles apart?

61. As a weather balloon is inflated, its radius increases at the rate of 4 centimeters per second. Express the volume of the balloon as a function of time and determine the volume of the balloon after 4 seconds. [*Hint:* The volume of a sphere of radius r is $4\pi r^3/3$.]

62. Express the surface area of the weather balloon in Exercise 61 as a function of time. [*Hint:* The surface area of a sphere of radius r is $4\pi r^2$.]

63. Charlie, who is 6 feet tall, walks away from a streetlight that is 15 feet high at a rate of 5 feet per second, as shown in the figure on the next page. Express the length s of Charlie's shadow as a function of time. [*Hint:* First use similar triangles to express *s* as a function of the distance *d* from the streetlight to Charlie.]

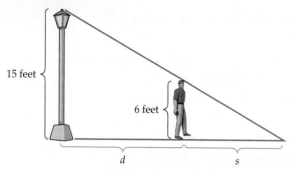

15 feet

6 feet

d

s

64. A water-filled balloon is dropped from a window 120 feet above the ground. Its height above the ground after *t* seconds is $120 - 16t^2$ feet. Laura is standing on the ground 40 feet from the point where the balloon will hit the ground, as shown in the figure.

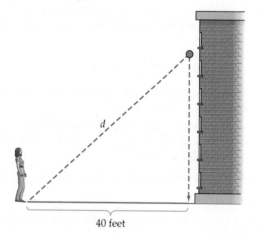

d

40 feet

(a) Express the distance *d* between Laura and the balloon as a function of time.

(b) When is the balloon exactly 90 feet from Laura?

THINKER

65. Find a function *f* (other than the identity function) such that $(f \circ f \circ f)(x) = x$ for every *x* in the domain of *f*. [Several correct answers are possible.]

66. If *f* is an increasing function, does $f \circ f$ have to be increasing? Why or why not?

67. Let $f(x) = x^2 - .2$

(a) Using a calculator, compute $f(0)$, $(f \circ f)(0)$, $(f \circ f \circ f)(0)$, ... etc. What happens as you keep going?

(b) Does the same thing happen if you look at $f(1)$, $(f \circ f)(1)$, $(f \circ f \circ f)(1)$, ... ?

(c) Repeat parts (a) and (b) using $f(x) = x^2 - .9$.

(d) Repeat parts (a) and (b) using $f(x) = x^2 - 1.3$.

[*Hint:* You may be able to save yourself some keystrokes using the ANS and ENTRY keys on your calculator.]

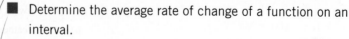

3.6   Rates of Change

Section Objectives

- ✓ ■ Determine the average rate of change of a function on an interval.
- ✓ ■ Understand average rate of change as applied to real-life situations.
- ✓ ■ Interpret average rate of change graphically.
- ■ Use the difference quotient to find the average rate of change over very small intervals.
- ■ Approximate the instantaneous rate of change of a function at a point.

Rates of change play a central role in the analysis of many real-world situations. To understand the basic ideas involved in rates of change, we take another look at the falling rock from Sections 3.1 and 3.2. We saw that when the rock is dropped from a high place, then the distance it travels (ignoring wind resistance) is given by the function

$$d(t) = 16t^2$$

with distance $d(t)$ measured in feet and time t in seconds. The following table shows the distance the rock has fallen at various times:

Time t	0	1	2	3	3.5	4	4.5	5
Distance $d(t)$	0	16	64	144	196	256	324	400

To find the distance the rock falls from time $t = 1$ to $t = 3$, we note that at the end of three seconds, the rock has fallen $d(3) = 144$ feet, whereas it had fallen only $d(1) = 16$ feet at the end of one second. So during this time interval, the rock traveled

$$d(3) - d(1) = 144 - 16 = 128 \text{ feet.}$$

The distance traveled by the rock during other time intervals can be found similarly:

Time Interval	Distance Traveled
$t = 1$ to $t = 4$	$d(4) - d(1) = 256 - 16 = 240$
$t = 2$ to $t = 3.5$	$d(3.5) - d(2) = 196 - 64 = 132$
$t = 2$ to $t = 4.5$	$d(4.5) - d(2) = 324 - 64 = 260$

The same procedure works in general:

The distance traveled from time $t = a$ to time $t = b$ is $d(b) - d(a)$ feet.

In the preceding chart, the length of each time interval can be computed by taking the difference between the two times. For example, from $t = 1$ to $t = 4$ is a time interval of length $4 - 1 = 3$ seconds. Similarly, the interval from $t = 2$ to $t = 3.5$ is of length $3.5 - 2 = 1.5$ seconds, and in general,

The time interval from $t = a$ to $t = b$ is an interval of $b - a$ seconds.

Since Distance = Average speed × Time,

$$\text{Average speed} = \frac{\text{Distance traveled}}{\text{Time interval}}.$$

Hence, the average speed over the time interval from $t = a$ to $t = b$ is

$$\text{Average speed} = \frac{\text{Distance traveled}}{\text{Time interval}} = \frac{d(b) - d(a)}{b - a}.$$

For example, to find the average speed from $t = 1$ to $t = 4$, apply the preceding formula with $a = 1$ and $b = 4$:

$$\text{Average speed} = \frac{d(4) - d(1)}{4 - 1} = \frac{256 - 16}{4 - 1} = \frac{240}{3} = 80 \text{ ft per second.}$$

Similarly, the average speed from $t = 2$ to $t = 4.5$ is

$$\frac{d(4.5) - d(2)}{4.5 - 2} = \frac{324 - 64}{4.5 - 2} = \frac{260}{2.5} = 104 \text{ ft per second.}$$

The units in which average speed is measured here (feet per second) indicate the number of units of distance traveled during each unit of time, that is, the *rate of change* of distance (feet) with respect to time (seconds). The preceding discussion can be summarized by saying that the average speed (rate of change of distance with respect to time) as time changes from $t = a$ to $t = b$ is given by

$$\text{Average speed} = \text{Average rate of change}$$

$$= \frac{\text{Change in distance}}{\text{Change in time}} = \frac{d(b) - d(a)}{b - a}.$$

Although speed is the most familiar example, rates of change play a role in many other situations as well, as illustrated in Examples 1–3 below. Consequently, we define the average rate of change of any function as follows.

Average Rate of Change

> Let f be a function. The **average rate of change of $f(x)$** with respect to x as x changes from a to b is the number
>
> $$\frac{\text{Change in } f(x)}{\text{Change in } x} = \frac{f(b) - f(a)}{b - a}.$$

EXAMPLE 1

Heidi started a big driving trip at 3 P.M. The odometer reading on her car said 103,846. She finished her trip at 9 P.M., and now the odometer read 104,176. What was her average speed during the trip?

SOLUTION

Her average speed was $\dfrac{\text{Distance traveled}}{\text{Time interval}}$. Her distance traveled was $104,176 - 103,846$ miles. Her time interval was $9 - 3$ hours. So her average speed was

$$\frac{104,176 - 103,846}{9 - 3} = \frac{330 \text{ miles}}{6 \text{ hours}} = 55 \text{ miles per hour.} \qquad \blacksquare$$

EXAMPLE 2

A large heavy-duty balloon is being filled with water. Its approximate volume (in gallons) is given by

$$V(x) = \frac{x^3}{55},$$

where x is the radius of the balloon (in inches). Find the average rate of change of the volume of the balloon as the radius increases from 5 to 10 inches.

SOLUTION

$$\frac{\text{Change in volume}}{\text{Change in radius}} = \frac{V(10) - V(5)}{10 - 5} \approx \frac{18.18 - 2.27}{10 - 5} = \frac{15.91}{5}$$

$$= 3.182 \text{ gallons per inch.} \qquad \blacksquare$$

EXAMPLE 3

According to the *Encyclopedia Britannica* almanac, these are the estimated number of cell phone users in the United States, from 1993 to 2004.

Year	1993	1994	1995	1996	1997	1998
Millions of subscribers	16.009	24.134	33.786	44.043	55.312	69.209

Year	1999	2000	2001	2002	2003	2004
Millions of subscribers	86.047	109.478	128.375	140.767	158.722	182.140

Let $f(t)$ be the number of cell phone users in year t. Find the average rate of change in cell-phone use during the following time periods:

(a) 1993–1998   (b) 2002–2004

SOLUTION

(a) Average rate of change $= \dfrac{f(1998) - f(1993)}{1998 - 1993}$

$$= \frac{69{,}209{,}000 - 16{,}009{,}000}{1998 - 1993}$$

$$= \frac{53{,}200{,}000}{5} = 10{,}640{,}000 \text{ users/year.}$$

(b) Average rate of change $= \dfrac{f(2004) - f(2002)}{2004 - 2002}$

$$= \frac{182{,}140{,}000 - 140{,}767{,}000}{2004 - 2002}$$

$$= \frac{41{,}373{,}000}{2}$$

$$= 20{,}686{,}500 \text{ users/year.} \qquad \blacksquare$$

EXAMPLE 4

Figure 3–51 is the graph of the temperature function f during a particular day; $f(x)$ is the temperature at x hours after midnight. What is the average rate of change of the temperature (a) from 4 A.M. to noon? (b) from 3 P.M. to 8 P.M.?

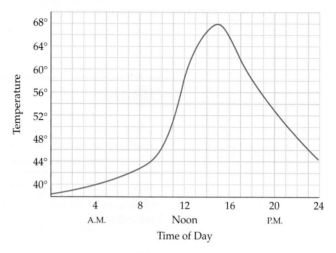

Figure 3–51

SOLUTION

(a) The graph shows that the temperature at 4 A.M. is $f(4) = 40°$ and the temperature at noon is $f(12) = 58°$. The average rate of change of temperature is

$$\frac{\text{Change in temperature}}{\text{Change in time}} = \frac{f(12) - f(4)}{12 - 4} = \frac{58 - 40}{12 - 4} = \frac{18}{8}$$

$$= 2.25° \text{ per hour.}$$

The rate of change is positive because the temperature is increasing at an average rate of 2.25° per hour.

(b) Now 3 P.M. corresponds to $x = 15$ and 8 P.M. to $x = 20$. The graph shows that $f(15) = 68°$ and $f(20) = 53°$. Hence, the average rate of change of temperature is

$$\frac{\text{Change in temperature}}{\text{Change in time}} = \frac{f(20) - f(15)}{20 - 15} = \frac{53 - 68}{20 - 15} = \frac{-15}{5}$$

$$= -3° \text{ per hour.}$$

The rate of change is negative because the temperature is decreasing at an average rate of 3° per hour. ■

 ## GEOMETRIC INTERPRETATION OF AVERAGE RATE OF CHANGE

If P and Q are points on the graph of a function f, then the straight line determined by P and Q is called a **secant line.** Figure 3–52 shows the secant line joining the points $(4, 40)$ and $(12, 58)$ on the graph of the temperature function f of Example 4.

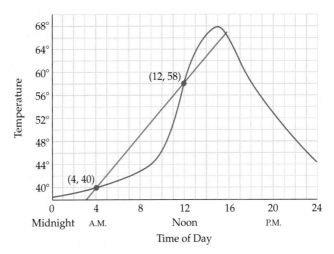

Figure 3–52

Using the points (4, 40) and (12, 58), we see that the slope of this secant line is

$$\frac{58 - 40}{12 - 4} = \frac{18}{8} = 2.25.$$

To say that (4, 40) and (12, 58) are on the graph of f means that $f(4) = 40$ and $f(12) = 58$. Thus,

$$\text{Slope of secant line} = 2.25 = \frac{58 - 40}{12 - 4} = \frac{f(12) - f(4)}{12 - 4}$$

$$= \text{Average rate of change as } x \text{ goes from 4 to 12.}$$

The same thing happens in the general case.

Secant Lines and Average Rates of Change

> If f is a function, then the average rate of change of $f(x)$ with respect to x as x changes from $x = a$ to $x = b$ is the slope of the secant line joining the points $(a, f(a))$ and $(b, f(b))$ on the graph of f.

EXAMPLE 5

Assume we have 20 liters of oxygen gas in a thick, unmoving container. The pressure exerted on the walls of the container is a function of the temperature in the room:

$$p(T) = .00411T + 1.1213$$

where T is in degrees Celsius and $p(T)$ is in atmospheres. So at zero degrees Celsius (the temperature at which water freezes) the pressure is 1.1213 atmospheres, and at 100 degrees Celsius (the temperature at which water boils) the pressure is 1.5323 atmospheres.

At what average rate does the pressure change as the temperature increases?

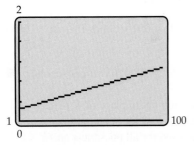

Figure 3–53

SOLUTION   The graph of $p(T) = .00411T + 1.1213$ is a straight line (Figure 3–53). So the secant line joining any two points on the graph is just the graph itself, the line $y = .00411T + 1.1213$. As we saw in Section 1.4, the slope of this line is .00411. Therefore, the average rate of change of the pressure function between any two values of T is .00411. In other words, at any temperature, the pressure will increase at a rate of .00411 atmospheres per degree. ■

The argument used in Example 5 works for any function whose graph is a straight line and leads to this conclusion.

Rates of Change of Linear Functions

> The average rate of change of a linear function $f(x) = mx + b$. as x changes from one value to another, is the slope m of the line.

 ## THE DIFFERENCE QUOTIENT

Average rates of change are often computed for very small intervals. For instance, we might compute the rate from 4 to 4.01 or from 4 to 4.001. Since $4.01 = 4 + .01$ and $4.001 = 4 + .001$, we are doing essentially the same thing in both cases: computing the rate of change over the interval from 4 to $4 + h$ for some small nonzero quantity h. Furthermore, it's often possible to use a single calculation to determine the average rate for all possible values of h.

EXAMPLE 6

Consider the falling rock with which this section began. The distance the rock has traveled at time t is given by $d(t) = 16t^2$, and its average speed (rate of change) from $t = 4$ to $t = 4 + h$ is

$$\text{Average speed} = \frac{d(4 + h) - d(4)}{(4 + h) - 4} = \frac{16(4 + h)^2 - 16 \cdot 4^2}{h}$$

$$= \frac{16(16 + 8h + h^2) - 256}{h} = \frac{256 + 128h + 16h^2 - 256}{h}$$

$$= \frac{128h + 16h^2}{h} = \frac{h(128 + 16h)}{h} = 128 + 16h.$$

Thus, we can quickly compute the average speed over the interval from 4 to $4 + h$ seconds for any value of h by using the formula

$$\text{Average speed} = 128 + 16h.$$

For example, the average speed from 4 seconds to 4.001 seconds (here $h = .001$) is

$$128 + 16h = 128 + 16(.001) = 128 + .016 = 128.016 \text{ feet per second.} \quad ■$$

Similar calculations can be done with any number in place of 4. In each such case, we are dealing with an interval from x to $x + h$ for some number x. As in Example 6, a single computation can often be used for all possible x and h.

EXAMPLE 7

The average speed of the falling rock of Example 6 from time x to time $x + h$ is:*

$$\text{Average speed} = \frac{d(x + h) - d(x)}{(x + h) - x} = \frac{16(x + h)^2 - 16x^2}{h}$$

$$= \frac{16(x^2 + 2xh + h^2) - 16x^2}{h} = \frac{16x^2 + 32xh + 16h^2 - 16x^2}{h}$$

$$= \frac{32xh + 16h^2}{h} = \frac{h(32x + 16h)}{h} = 32x + 16h.$$

When $x = 4$, then this result states that the average speed from 4 to $4 + h$ is $32(4) + 16h = 128 + 16h$, which is exactly what we found in Example 4. To find the average speed from 3 to 3.1 seconds, apply the formula

$$\text{Average speed} = 32x + 16h$$

with $x = 3$ and $h = .1$:

$$\text{Average speed} = 32 \cdot 3 + 16(.1) = 96 + 1.6 = 97.6 \text{ feet per second.} \qquad \blacksquare$$

More generally, we can compute the average rate of change of any function f over the interval from x to $x + h$ just as we did in Example 7: Apply the definition of average rate of change in the box on page 206 with x in place of a and $x + h$ in place of b:

$$\textbf{Average rate of change} = \frac{f(b) - f(a)}{b - a} = \frac{f(x + h) - f(x)}{(x + h) - x}$$

$$= \frac{f(x + h) - f(x)}{h}.$$

This last quantity is just the difference quotient of f (see page 154). Therefore,

Difference Quotients and Rates of Change

> If f is a function, then the average rate of change of f over the interval from x to $x + h$ is given by the difference quotient
> $$\frac{f(x + h) - f(x)}{h}.$$

EXAMPLE 8

Find the difference quotient of $V(x) = x^3/55$, and use it to find the average rate of change of V as x changes from 8 to 8.01.

---

*Note that this calculation is the same as in Example 6 except that 4 has been replaced by x.

SOLUTION Use the definition of the difference quotient and algebra:

$$\frac{V(x + h) - V(x)}{h} = \frac{\overbrace{\frac{(x + h)^3}{55}}^{V(x+h)} - \overbrace{\frac{x^3}{55}}^{V(x)}}{h} = \frac{\frac{1}{55}[(x + h)^3 - x^3]}{h}$$

$$= \frac{1}{55} \cdot \frac{(x + h)^3 - x^3}{h} = \frac{1}{55} \cdot \frac{x^3 + 3x^2h + 3xh^2 + h^3 - x^3}{h}$$

$$= \frac{1}{55} \cdot \frac{3x^2h + 3xh^2 + h^3}{h} = \frac{1}{55} \cdot \frac{h(3x^2 + 3xh + h^2)}{h}$$

$$= \frac{3x^2 + 3xh + h^2}{55}.$$

When x changes from 8 to 8.01 = 8 + .01, we have $x = 8$ and $h = .01$. So the average rate of change is

$$\frac{3x^2 + 3xh + h^2}{55} = \frac{3 \cdot 8^2 + 3 \cdot 8(.01) + (.01)^2}{55} \approx 3.495. \qquad \blacksquare$$

If you've been reading carefully, you might be thinking that this process makes something simple (computing the average rate of change of a function from $x = 8$ to $x = 8.01$) into something difficult (FIRST computing the average rate of change from x to $x + h$, using some messy algebra, and then letting $x = 8$ and $h = .01$). If all we wanted to do was compute one such rate, you would be right. We usually use this technique when we want to look at the same function for many values of h, as we will in the next example. We basically are doing an ugly calculation one time to make a series of future calculations simpler.

 INSTANTANEOUS RATE OF CHANGE

Rates of change are a major theme in calculus—not just the average rate of change discussed above, but also the *instantaneous rate of change* of a function (that is, its rate of change at a particular instant). Even without calculus, however, we can obtain quite accurate approximations of instantaneous rates of change by using average rates appropriately.

EXAMPLE 9

A rock is dropped from a high place. What is its speed exactly 3 seconds after it is dropped?

SOLUTION The distance the rock has fallen at time t is given by the function $d(t) = 16t^2$. The exact speed at $t = 3$ can be approximated by finding the average speed over very small time intervals, say, 3 to 3.01 or even shorter intervals. Over a very short time span, such as a hundredth of a second, the rock cannot change speed very much, so these average speeds should be a reasonable approximation of its speed at the instant $t = 3$. Example 7 shows that the average speed is given by the difference quotient $32x + 16h$. When $x = 3$, the difference quotient is $32 \cdot 3 + 16h = 96 + 16h$, and we have the following:

Change in Time 3 to 3 + h	h	Average Speed [Difference Quotient at x = 3] 96 + 16h
3 to 3.1	.1	$96 + 16(.1) = 97.6$ ft per second
3 to 3.01	.01	$96 + 16(.01) = 96.16$ ft per second
3 to 3.005	.005	$96 + 16(.005) = 96.08$ ft per second
3 to 3.00001	.00001	$96 + 16(.00001) = 96.00016$ ft per second

The table suggests that the exact speed of the rock at the instant $t = 3$ seconds is very close to 96 feet per second. ∎

EXAMPLE 10

A balloon is being filled with water in such a way that when its radius is x inches, then its volume is $V(x) = x^3/55$ gallons. In Example 2, we saw that the average rate of change of the volume as the radius increases from 5 inches to 10 inches is 3.182 gallons per inch. What is the rate of change at the instant when the radius is 7 inches?

SOLUTION   The average rate of change when the radius goes from x to $x + h$ inches is given by the difference quotient of $V(x)$, which was found in Example 8:

$$\frac{V(x + h) - V(x)}{h} = \frac{3x^2 + 3xh + h^2}{55}.$$

Therefore, when $x = 7$, the difference quotient is

$$\frac{3 \cdot 7^2 + 3 \cdot 7 \cdot h + h^2}{55} = \frac{147 + 21h + h^2}{55},$$

and we have these average rates of change over small intervals near 7:

Change in Radius 7 to 7 + h	h	Average Rate of Change of Volume [Difference Quotient at x = 7] $\frac{147 + 21h + h^2}{55}$
7 to 7.01	.01	2.6765 gallons per inch
7 to 7.001	.001	2.6731 gallons per inch
7 to 7.0001	.0001	2.6728 gallons per inch
7 to 7.00001	.00001	2.6727 gallons per inch

The chart suggests that at the instant the radius is 7 inches, the volume is changing at a rate of approximately 2.673 gallons per inch. ∎

EXERCISES 3.6

1. A car moves along a straight test track. The distance traveled by the car at various times is shown in this table:

Time (seconds)	0	5	10	15	20	25	30
Distance (feet)	0	20	140	400	680	1400	1800

Find the average speed of the car over the interval from

(a) 0 to 10 seconds   (b) 10 to 20 seconds
(c) 20 to 30 seconds   (d) 15 to 30 seconds

2. Find the average rate of change of the volume of the balloon in Example 2 as the radius increases from

(a) 2 to 5 inches   (b) 4 to 8 inches

3. Use the function of Example 3 to find the average rate of change in cell-phone use from

(a) 1994–1996   (b) 1999–2004

4. The graph shows the total amount spent on advertising (in millions of dollars) in the United States, as estimated by a leading advertising publication.* Find the average rate of change in advertising expenditures over the following time periods:

(a) 1950–1970   (b) 1970–1980
(c) 1980–2000   (d) 1950–2000
(e) During which of these periods were expenditures increasing at the fastest rate?

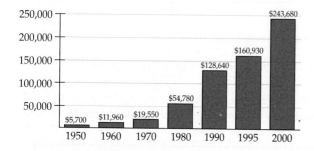

5. The following table shows the total projected elementary and secondary school enrollment (in thousands) for selected years.[†] Find the average rate of change of enrollment from

(a) 1980 to 1985   (b) 1985 to 1995
(c) 1995 to 2005   (d) 2005 to 2014
(e) During which of these periods was enrollment increasing at the fastest rate? At the slowest rate?

Year	Enrollment
1980	40,877
1985	39,422
1990	41,217
1995	44,840
2000	47,204
2005	48,375
2010	48,842
2014	49,993

6. The graph shows the minimum wage (in constant 2000 dollars), with $x = 0$ corresponding to 1950.* Find the average rate of change in the minimum wage from

(a) 1950 to 1976   (b) 1976 to 1995
(c) 1995 to 2000   (d) 1976 to 2000

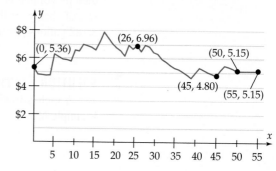

7. The table shows the total number of shares traded (in billions) on the New York Stock Exchange in selected years.[†]

Year	1995	1997	1999	2001	2003	2005
Volume	87.2	133.3	203.9	307.5	352.4	403.8

Find the average rate of change in share volume from
(a) 1995 to 1999   (b) 1999 to 2001
(c) 2001 to 2005   (d) 1995 to 2005
(e) During which of these periods did share volume increase at the fastest rate?

8. The graph in the figure[‡] shows the popularity of the name Frances since the 1880s, when records started to be kept. The y-axis shows the usage of "Frances" per million babies. The name increased in popularity until the 1910s, and then its popularity decreased. Estimate the average rate of change of popularity (in usage per million per year) over the interval:

(a) 1880 to 1890   (b) 1880 to 1930
(c) 1930 to 1950   (d) 1950 to 1990

---

*Advertising Age.
[†]U.S. National Center for Education Statistics.

*U.S. Employment Standards Administration.
[†]NYSE Factbook.
[‡]Baby Name Wizard.

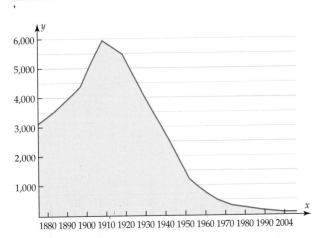

1880 1890 1900 1910 1920 1930 1940 1950 1960 1970 1980 1990 2004

9. The Pennyfarthing Bicycle Company has found that its sales are related to the amount of advertising it does in trade magazines. The graph in the figure shows the sales (in thousands of dollars) as a function of the amount of advertising (in number of magazine ad pages). Find the average rate of change of sales when the number of ad pages increases from

(a) 10 to 20   (b) 20 to 60
(c) 60 to 100   (d) 0 to 100
(e) Is it worthwhile to buy more than 70 pages of ads, if the cost of a one-page ad is $2000? If the cost is $5000? If the cost is $8000?

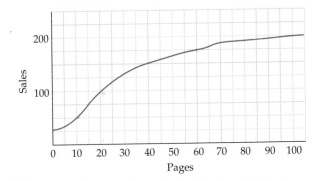

10. When blood flows through an artery (which can be thought of as a cylindrical tube) its velocity is greatest at the center of the artery. Because of friction along the walls of the tube, the blood's velocity decreases as the distance r from the center of the artery increases, finally becoming 0 at the wall of the artery. The velocity (in centimeters per second) is given by the function $v = 18{,}500(.000065 - r^2)$, where r is measured in centimeters. Find the average rate of change of the velocity as the distance from the center changes from

(a) $r = .001$ to $r = .002$ (b) $r = .002$ to $r = .003$
(c) $r = 0$ to $r = .005$

In Exercises 11–17, find the average rate of change of the function f over the given interval.

11. $f(x) = 3 + x^3$ from $x = 0$ to $x = 2$

12. $f(x) = .25x^4 - x^2 - 2x + 4$ from $x = -1$ to $x = 4$

13. $f(x) = x^3 - 3x^2 - 8x + 6$ from $x = -1$ to $x = 3$

14. $f(x) = -\sqrt{x^4 - x^3 + 2x^2 - x + 4}$ from $x = 0$ to $x = 3$

15. $f(x) = \sqrt{x^3 + 2x^2 - 6x + 5}$ from $x = 1$ to $x = 1.01$

16. $f(x) = \sqrt{x^3 + 2x^2 - 6x + 5}$ from $x = 1$ to $x = 1.00001$

17. $f(x) = \dfrac{x^2 - 3}{2x - 4}$ from $x = 3$ to $x = 8$

In Exercises 18–25, compute and simplify the difference quotient of the function.

18. $f(x) = x + 5$ 19. $f(x) = 7x + 2$

20. $f(x) = x^2 + 3$ 21. $f(x) = x^2 + 3x - 1$

22. $f(t) = 160{,}000 - 8000t + t^2$

23. $V(x) = x^3$

24. $A(r) = \pi r^2$ 25. $V(p) = 5/p$

26. Water is draining from a large tank. After t minutes, there are $160{,}000 - 8000t + t^2$ gallons of water in the tank.

(a) Use the results of Exercise 22 to find the average rate at which the water runs out in the interval from 10 to 10.1 minutes.
(b) Do the same for the interval from 10 to 10.01 minutes.
(c) Estimate the rate at which the water runs out after exactly 10 minutes.

27. Use the results of Exercise 23 to find the average rate of change of the volume of a cube whose side has length x as x changes from

(a) 4 to 4.1   (b) 4 to 4.01   (c) 4 to 4.001
(d) Estimate the rate of change of the volume at the instant when $x = 4$.

28. Use the results of Exercise 24 to find the average rate of change of the area of a circle of radius r as r changes from

(a) 4 to 4.5   (b) 4 to 4.2   (c) 4 to 4.1
(d) Estimate the rate of change at the instant when $r = 4$.
(e) How is your answer in part (d) related to the circumference of a circle of radius 4?

29. Under certain conditions, the volume V of a quantity of air is related to the pressure p (which is measured in kilopascals) by the equation $V = 5/p$. Use the results of Exercise 25 to estimate the rate at which the volume is changing at the instant when the pressure is 50 kilopascals.

30. Two cars race on a straight track, beginning from a dead stop. The distance (in feet) each car has covered at each time during the first 16 seconds is shown in the figure.

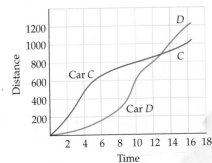

(a) What is the average speed of each car during this 16-second interval?

(b) Find an interval beginning at $t = 4$ during which the average speed of car D was approximately the same as the average speed of car C from $t = 2$ to $t = 10$.

(c) Use secant lines and slopes to justify the statement "car D traveled at a higher average speed than car C from $t = 4$ to $t = 10$."

31. The figure shows the profits earned by Soupy Soy Sauce during the last quarters of three consecutive years.

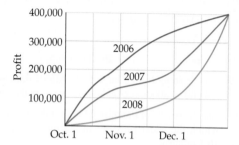

(a) Explain why the average rate of change of profits from October 1 to December 31 was the same in all three years.

(b) During what month in what year was the average rate of change of profits the greatest?

32. The graph in the figure shows the chipmunk population in a certain wilderness area. The population increases as the chipmunks reproduce but then decreases sharply as predators move into the area.

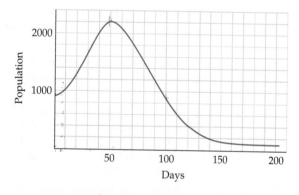

(a) During what approximate time period, beginning on day 0, is the average growth rate of the chipmunk population positive?

(b) During what approximate time period, beginning on day 0, is the average growth rate of the chipmunk population 0?

(c) What is the average growth rate of the chipmunk population from day 50 to day 100? What does this number mean?

(d) What is the average growth rate from day 45 to day 50? From day 50 to day 55? What is the approximate average growth rate from day 49 to day 51?

33. The following is a graph of the price of a share of stock in the International House of Pancakes from June of 2001 through June of 2006

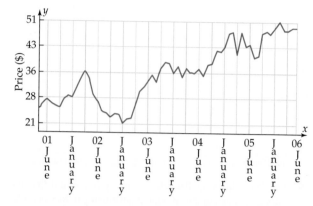

(a) What was the average rate of change of IHOP stock (in dollars per year) between September of 2001 and March of 2002?

(b) What was the average rate of change of IHOP stock between March of 2002 and January of 2003?

(c) What was the average rate of change of IHOP stock between June of 2001 and June of 2006?

34. Lucy has a viral flu. How bad she feels depends primarily on how fast her temperature is rising at that time. The figure shows her temperature during the first day of the flu.

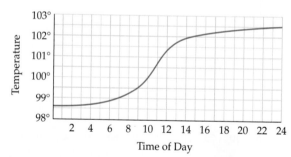

(a) At what average rate does her temperature rise during the entire day?

(b) During what two-hour period during the day does she feel worst?

(c) Find two time intervals, one in the morning and one in the afternoon, during which she feels about the same (that is, during which her temperature is rising at the same average rate).

35. The table shows the average weekly earnings (including overtime) of production workers and nonsupervisory employees in industry (excluding agriculture) in selected years.*

Year	1980	1985	1990	1995	2000	2005
Weekly Earnings	$191	$257	$319	$369	$454	$538

*U.S. Bureau of Labor Statistics.

(a) Use linear regression to find a function that models this data, with $x = 0$ corresponding to 1980.
(b) According to your function, what is the average rate of change in earnings over any time period between 1980 and 2005?
(c) Use the data in the table to find the average rate of change in earnings from 1980 to 1990 and from 2000–2005. How do these rates compare with the ones given by the model?
(d) If the model remains accurate, when will average weekly earnings reach $600?

36. The estimated number of 15- to 24-year-old people world-wide (in millions) who are living with HIV/AIDS in selected years is given in the table.*

Year	2001	2003	2005	2007	2009
15- to 24-year-olds with HIV/AIDS	12	14.5	17	19	20.5

(a) Use linear regression to find a function that models this data, with $x = 0$ corresponding to 2000.
(b) According to your function, what is the average rate of change in this HIV/AIDS population over any time period between 2001 and 2009?
(c) Use the data in the table to find the average rate of change in this HIV/AIDS population from 2001 to 2009. How does this rate compare with the one given by the model?
(d) If the model remains accurate, when will the number of people in this age group with HIV/AIDS reach 25 million?

*Kaiser Family Foundation, UNICEF, U.S. Census Bureau.

3.7 Inverse Functions*

Section Objectives
- Determine graphically if a function is one-to-one.
- Find inverse functions algebraically.
- Explore the properties of inverse functions.
- Graph inverse functions.

Consider the functions f and h given by these tables:

f-input	-2	-1	0	1	2
f-output	-3	-2	1	4	5

h-input	1	2	3	4	5
h-output	-1	3	0	3	2

With the function h, two different inputs (2 and 4) produce the same output 3. With the function f, however, different inputs always produce different outputs. Functions with this property have a special name. A function f is said to be **one-to-one** if distinct inputs always produce distinct outputs, that is,

$$\text{if } a \neq b, \text{ then } f(a) \neq f(b)$$

In graphical terms, this means that two points on the graph, $(a, f(a))$ and $(b, f(b))$, that have different x-coordinates $[a \neq b]$ must also have different y-coordinates $[f(a) \neq f(b)]$. Consequently, these points cannot lie on the same horizontal line because all points on a horizontal line have the same y-coordinate. Therefore, we have this geometric test to determine whether a function is one-to-one.

*This section is used only in Section 5.3, Special Topics 5.3.A, and Section 7.4. It may be postponed until then.

*The Horizontal
Line Test*

If a function f is one-to-one, then it has this property:

> No horizontal line intersects the graph of f more than once.

Conversely, if the graph of a function has this property, then the function is one-to-one.

EXAMPLE 1

Which of the following functions are one-to-one?

(a) $f(x) = 7x^5 + 3x^4 - 2x^3 + 2x + 1$

(b) $g(x) = x^3 - 3x - 1$

(c) $h(x) = 1 - .2x^3$

SOLUTION   Complete graphs of each function are shown in Figure 3–54.

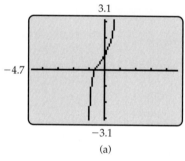
(a)

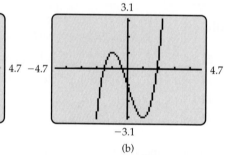
(b)

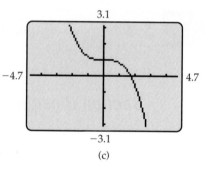
(c)

Figure 3–54

(a) The graph of f in Figure 3–54(a) passes the horizontal line test, since no horizontal line intersects the graph more than once. Hence, f is one-to-one.

(b) The graph of g in Figure 3–54(b) obviously fails the horizontal line test because many horizontal lines (including the x-axis) intersect the graph more than once. Therefore, g is not one-to-one.

(c) The graph of h in Figure 3–54(c) appears to contain a horizontal line segment. So h appears to fail the Horizontal Line Test because the horizontal line through $(0, 1)$ seems to intersect the graph infinitely many times. But appearances are deceiving.

TECHNOLOGY TIP

Although a horizontal segment may appear on a calculator screen when the graph is actually rising or falling, there is another possibility. The graph may have a tiny wiggle (less than the height of a pixel) and thus fail the horizontal line test:

You can usually detect such a wiggle by zooming in to magnify that portion of the graph or by using the trace feature to see whether the y-coordinates increase and then decrease (or vice versa) along the "horizontal" segment.

GRAPHING EXPLORATION

Graph $h(x) = 1 - .2x^3$ and use the trace feature to move from left to right along the "horizontal" segment. Do the y-coordinates stay the same, or do they decrease?

The Exploration shows that the graph is actually falling from left to right, so that each horizontal line intersects it only once. (It appears to have a horizontal segment because the amount the graph falls there is less than the height of a pixel on the screen.) Therefore, h is a one-to-one function.   ■

The function f in Example 1 is an **increasing function** (its graph is always rising from left to right), and the function h is a **decreasing function** (its graph is always falling from left to right). Every increasing or decreasing function is necessarily one-to-one because its graph can never touch the same horizontal line twice (it would have to change from rising to falling, or vice versa, to do so).

INVERSE FUNCTIONS

We begin with a simple example that illustrates the basic idea of an inverse function. Consider the one-to-one function f introduced at the beginning of this section.

f-input	-2	-1	0	1	2
f-output	-3	-2	1	4	5

Now define a new function g by the following table (which simply *switches* the rows in the f table).

g-input	-3	-2	1	4	5
g-output	-2	-1	0	1	2

Note that the inputs of f are the outputs of g and the outputs of f are the inputs of g. In other words,

$$\text{Domain of } f = \text{Range of } g \quad \text{and} \quad \text{Range of } f = \text{Domain of } g.$$

The rule of g *reverses* the action of f by taking each output of f back to the input it came from. For instance,

$$g(4) = 1 \quad \text{and} \quad f(1) = 4$$
$$g(-3) = -2 \quad \text{and} \quad f(-2) = -3$$

and in general,

$$g(y) = x \quad \text{exactly when} \quad f(x) = y.$$

We say that g is the *inverse function* of f.

The preceding construction works for any one-to-one function f. Each output of f comes from exactly one input (because different inputs produce different outputs). Consequently, we can define a new function g that reverses the action of f by sending each output back to the unique input it came from. For instance, if $f(7) = 11$, then $g(11) = 7$. Thus, the outputs of f become the inputs of g, and we have this definition.

Inverse Functions

> Let f be a one-to-one function. Then the **inverse function** of f is the function g whose rule is
>
> $$g(y) = x \quad \text{exactly when} \quad f(x) = y.$$
>
> The domain of g is the range of f and the range of g is the domain of f.

EXAMPLE 2

The graph of $f(x) = 3x - 2$ is a straight line that certainly passes the Horizontal Line Test, so f is one-to-one and has an inverse function g. From the definition of g we know that

$$g(y) = x \qquad \text{exactly when} \qquad f(x) = y$$

that is,

$$g(y) = x \qquad \text{exactly when} \qquad 3x - 2 = y.$$

To find the rule of g, we need only solve this last equation for x:

$$3x - 2 = y$$

Add 2 to both sides: $\qquad\qquad 3x = y + 2$

Divide both sides by 3: $\qquad\qquad x = \dfrac{y + 2}{3}$

Since $g(y) = x$, we see that the rule of g is $g(y) = \dfrac{y + 2}{3}$. ∎

Recall that the letter used for the variable of a function doesn't matter. For instance, $h(x) = x^2$ and $h(t) = t^2$ and $h(u) = u^2$ all describe the same function, whose rule is "square the input." When dealing with inverse functions, it is customary to use the same variable for both f and its inverse g. Consequently, the inverse function in Example 2 would normally be written as

$$g(x) = \frac{x + 2}{3}.$$

We can summarize this procedure as follows.

Finding Inverse Functions Algebraically

To find the inverse function of a one-to-one function f:

1. Solve the equation $f(x) = y$ for x.

2. The solution is an expression in y, which is the rule of the inverse function g, that is, $x = g(y)$.

3. Rewrite the rule of $x = g(y)$ by interchanging x and y.

EXAMPLE 3

Use your calculator to verify that the function $f(x) = x^3 + 5$ passes the Horizontal Line Test and hence is one-to-one. Its inverse can be found by solving for x in the equation $x^3 + 5 = y$:

Subtract 5 from both sides: $\qquad x^3 = y - 5$

Take cube roots on both sides: $\qquad x = \sqrt[3]{y - 5}.$

Therefore, $g(y) = \sqrt[3]{y - 5}$ is the inverse function of f. Interchanging x and y, we write this rule as $g(x) = \sqrt[3]{x - 5}$. ∎

The second method of graphing inverse functions by reversing coordinates depends on this geometric fact, which is proved in Exercise 49:

The line $y = x$ is the perpendicular bisector of
the line segment from (a, b) to (b, a),

as shown in Figure 3–57 when $a = 7, b = 2$.

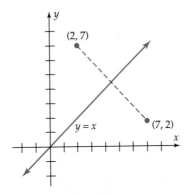

Figure 3–57

Thus (a, b) and (b, a) lie on opposite sides of $y = x$, the same distance from it: They are mirror images of each other, with the line $y = x$ being the mirror.* Consequently, the graph of the inverse function g is the mirror image of the graph of f. In formal terms,

**Inverse Function
Graphs**

If g is the inverse function of f, then the graph of g is the reflection of the graph of f in the line $y = x$.

GRAPHING EXPLORATION

Illustrate this fact by graphing the line $y = x$, the function

$$f(x) = x^3 + 5$$

of Example 3, and its inverse $g(x) = \sqrt[3]{x - 5}$ on the same screen (use a square viewing window so that the mirror effect won't be distorted).

NOTE

In many texts, the inverse function of a function f is denoted f^{-1}. In this notation, for instance, the inverse of the function $f(x) = x^3 + 5$ in Example 3 would be written as $f^{-1}(x) = \sqrt[3]{x - 5}$. Similarly, the reversal properties of inverse functions become

$$f^{-1}(f(x)) = x \text{ for every } x \text{ in the domain of } f; \text{ and}$$
$$f(f^{-1}(x)) = x \text{ for every } x \text{ in the domain of } f^{-1}.$$

In this context, f^{-1} does *not* mean $1/f$ (see Exercise 45).

---

*In technical terms, (a, b) and (b, a) are **symmetric with respect to the line** $y = x$.

EXERCISES 3.7

In Exercises 1–8, use a calculator and the Horizontal Line Test to determine whether or not the function f is one-to-one.

1. $f(x) = x^4 - 4x^2 + 3$

2. $f(x) = x^4 - 4x + 3$

3. $f(x) = x^3 + x - 5$

4. $f(x) = \begin{cases} x - 3 & \text{if } x \le 3 \\ 2x - 6 & \text{if } x > 3 \end{cases}$

5. $f(x) = x^5 + 2x^4 - x^2 + 4x - 5$

6. $f(x) = x^3 - 4x^2 + x - 10$

7. $f(x) = .1x^3 - .1x^2 - .005x + 1$

8. $f(x) = .1x^3 + .005x + 1$

In Exercises 9–22, use algebra to find the inverse of the given one-to-one function.

9. $f(x) = -x$

10. $f(x) = -x + 1$

11. $f(x) = 5x - 4$

12. $f(x) = -3x + 5$

13. $f(x) = 5 - 2x^3$

14. $f(x) = (x^5 + 1)^3$

15. $f(x) = \sqrt{4x - 7}$

16. $f(x) = 5 + \sqrt{3x - 2}$

17. $f(x) = 1/x$

18. $f(x) = 1/\sqrt{x}$

19. $f(x) = \dfrac{1}{2x + 1}$

20. $f(x) = \dfrac{x}{x + 1}$

21. $f(x) = \dfrac{x^3 - 1}{x^3 + 5}$

22. $f(x) = \sqrt[5]{\dfrac{3x - 1}{x - 2}}$

In Exercises 23–28, use the Round-Trip Theorem on page 223 to show that g is the inverse of f.

23. $f(x) = x + 1, \qquad g(x) = x - 1$

24. $f(x) = 2x - 6, \qquad g(x) = \dfrac{x}{2} + 3$

25. $f(x) = \dfrac{1}{x + 1}, \qquad g(x) = \dfrac{1 - x}{x}$

26. $f(x) = \dfrac{-3}{2x + 5}, \qquad g(x) = \dfrac{-3 - 5x}{2x}$

27. $f(x) = x^5, \qquad g(x) = \sqrt[5]{x}$

28. $f(x) = x^3 - 1, \qquad g(x) = \sqrt[3]{x + 1}$

29. Show that the inverse function of the function f whose rule is $f(x) = \dfrac{2x + 1}{3x - 2}$ is f itself.

30. List three different functions (other than the ones in Example 6 and Exercise 29), each of which is its own inverse. [Many correct answers are possible.]

31. Let $f(t)$ be the population of rabbits on Christy's property t years after she received 10 of them as a gift.

t	$f(t)$
0	10
1	23
2	48
3	64
4	70
5	71

Compute the following, including units, or write "not enough information to tell." f^{-1} denotes the inverse function of f.

(a) $f(2)$ (b) $f^{-1}(48)$

(c) $f^{-1}(71)$ (d) $3 \cdot f^{-1}(70)$

(e) $f^{-1}(2 \cdot 48)$ (f) $f(70)$

(g) $f^{-1}(4)$

In Exercises 32 and 33, the graph of a function f is given. Sketch the graph of the inverse function of f. [Reflect carefully.]

32.

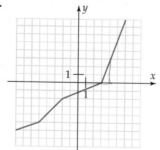

33.

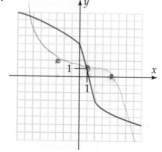

In Exercises 34–39, each given function has an inverse function. Sketch the graph of the inverse function.

34. $f(x) = \sqrt{x + 3}$

35. $f(x) = \sqrt{3x - 2}$

36. $f(x) = .3x^5 + 2$

37. $f(x) = \sqrt[3]{x + 3}$

38. $f(x) = \sqrt[5]{x^3 + x - 2}$

39. $f(x) = \begin{cases} x^2 - 1 & \text{if } x \le 0 \\ -.5x - 1 & \text{if } x > 0 \end{cases}$

In Exercises 40–42, none of the functions has an inverse. State at least one way of restricting the domain of the function (that is, find a function with the same rule and a smaller domain) so that the restricted function has an inverse. Then find the rule of the inverse function.

Example: $f(x) = x^2$ has no inverse. But the function h with domain all $x \ge 0$ and rule $h(x) = x^2$ is increasing (its graph is the right half of the graph of f—see Figure 2–2 on page 78)—and therefore has an inverse.

40. $f(x) = |x|$

41. $f(x) = -x^2$

42. $f(x) = \sqrt{4 - x^2}$

43. $f(x) = \dfrac{1}{x^2 + 1}$

44. $f(x) = 3(x + 5)^2 + 2$

45. (a) Using the f^{-1} notation for inverse functions, find $f^{-1}(x)$ when $f(x) = 3x + 2$.
  (b) Find $f^{-1}(1)$ and $1/f(1)$. Conclude that f^{-1} is not the same function as $1/f$.

46. Let C be the temperature in degrees Celsius. Then the temperature in degrees Fahrenheit is given by $f(C) = \frac{9}{5}C + 32$. Let g be the function that converts degrees Fahrenheit to degrees Celsius. Show that g is the inverse function of f and find the rule of g.

THINKERS

47. Let m and b be constants with $m \ne 0$. Show that the function $f(x) = mx + b$ has an inverse function g and find the rule of g.

48. Prove that the function $h(x) = 1 - .2x^3$ of Example 1(c) is one-to-one by showing that it satisfies the definition:

$$\text{If } a \ne b, \text{ then } h(a) \ne h(b).$$

[*Hint:* Use the rule of h to show that when $h(a) = h(b)$, then $a = b$. If this is the case, then it is impossible to have $h(a) = h(b)$ when $a \ne b$.]

49. Show that the points $P = (a, b)$ and $Q = (b, a)$ are symmetric with respect to the line $y = x$ as follows.
  (a) Find the slope of the line through P and Q.
  (b) Use slopes to show that the line through P and Q is perpendicular to $y = x$.
  (c) Let R be the point where the line $y = x$ intersects line segment PQ. Since R is on $y = x$, it has coordinates (c, c) for some number c, as shown in the figure. Use the distance formula to show that segment PR has the same length as segment RQ. Conclude that the line $y = x$ is the perpendicular bisector of segment PQ. Therefore, P and Q are symmetric with respect to the line $y = x$.

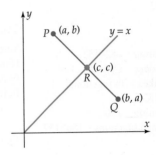

50. Suppose that functions f and g have these round-trip properties:

  (1) $g(f(x)) = x$ for every x in the domain of f.
  (2) $f(g(y)) = y$ for every y in the domain of g.

To complete the proof of the Round-Trip Theorem, we must show that g is the inverse function of f. Do this as follows.

  (a) Prove that f is one-to-one by showing that

$$\text{if } a \ne b, \quad \text{then} \quad f(a) \ne f(b).$$

  [*Hint:* If $f(a) = f(b)$, apply g to both sides and use (1) to show that $a = b$. Consequently, if $a \ne b$, it is impossible to have $f(a) = f(b)$.]
  (b) If $g(y) = x$, show that $f(x) = y$. [*Hint:* Use (2).]
  (c) If $f(x) = y$, show that $g(y) = x$. [*Hint:* Use (1).]

Parts (b) and (c) prove that

$$g(y) = x \quad \text{exactly when} \quad f(x) = y.$$

Hence, g is the inverse function of f (see page 219).

51. Prove that every function f that has an inverse function g is one-to-one. [*Hint:* The proof of the Round-Trip Theorem on page 223 shows that f and g have the round-trip properties; use Exercise 50(a).]

52. True or false: If a function has an inverse, then its inverse has an inverse. Justify your answer.

53. True or false: If a one-to-one function is increasing, then its inverse is increasing. Justify your answer.

Chapter 3 Review

IMPORTANT CONCEPTS

IMPORTANT FACTS & FORMULAS

■ The average rate of change of a function f as x changes from a to b is the number

$$\frac{f(b) - f(a)}{b - a}.$$

■ The average rate of change of a function f as x changes from a to b is the slope of the secant line joining
   the points $(a, f(a))$ and $(b, f(b))$.

■ The difference quotient of the function f is the quantity

$$\frac{f(x + h) - f(x)}{h}.$$

CATALOG OF BASIC FUNCTIONS—PART 1

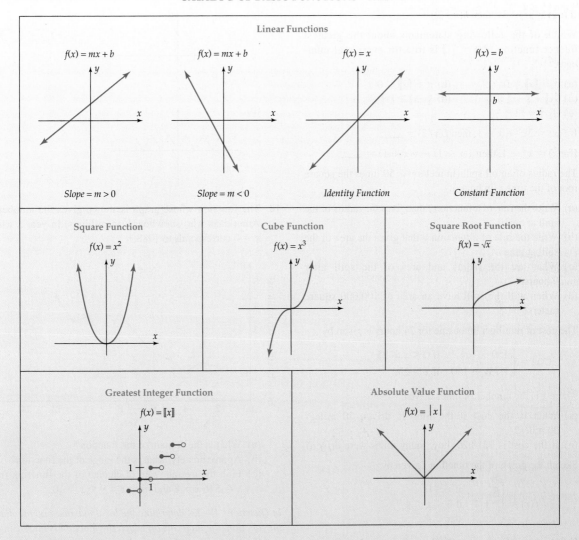

REVIEW QUESTIONS

1. Let $[\![x]\!]$ denote the greatest integer function and evaluate
   (a) $[\![-5/2]\!] = $ ____.   (b) $[\![1755]\!] = $ ____.
   (c) $[\![18.7]\!] + [\![-15.7]\!] = $ ____.
   (d) $[\![-7]\!] - [\![7]\!] = $ ____.

2. If $f(x) = x + |x| + [\![x]\!]$, then find $f(0)$, $f(-1)$, $f(1/2)$, and $f(-3/2)$.

3. Let f be the function given by the rule $f(x) = 7 - 2x$. Complete this table:

x	0	1	2	-4	t	k	$b-1$	$1-b$	$6-2u$
$f(x)$	7								

4. What is the domain of the function g given by
$$g(t) = \frac{\sqrt{t-2}}{t-3}?$$

5. In each case, give a *specific* example of a function and numbers a, b to show that the given statement may be *false*.
   (a) $f(a+b) = f(a) + f(b)$ (b) $f(ab) = f(a)f(b)$

6. If $f(x) = |3 - x|\sqrt{x-3} + 7$, then
$$f(7) - f(4) = \underline{\qquad}.$$

7. What is the domain of the function given by
$$g(r) = \sqrt{r-4} + \sqrt{r-2}?$$

8. What is the domain of the function $f(x) = \sqrt{-x + 2}$?

9. If $h(x) = x^2 - 3x$, then $h(t + 2) = $ _____.

10. Which of the following statements about the greatest integer function $f(x) = [\![x]\!]$ is true for *every* real number x?

(a) $x - [\![x]\!] = 0$ (b) $x - [\![x]\!] \leq 0$
(c) $[\![x]\!] + [\![-x]\!] \leq 0$ (d) $[\![-x]\!] \geq [\![x]\!]$
(e) $3[\![x]\!] = [\![3x]\!]$

11. If $f(x) = 2x^3 + x + 1$, then $f(x/2) = $ _____.

12. If $g(x) = x^2 - 1$, then $g(x - 1) - g(x + 1) = $ _____.

13. The radius of an oil spill (in meters) is 50 times the square root of the time t (in hours).

(a) Write the rule of a function f that gives the radius of the spill at time t.
(b) Write the rule of a function g that gives the area of the spill at time t.
(c) What are the radius and area of the spill after 9 hours?
(d) When will the spill have an area of 100,000 square meters?

14. The cost of renting a limousine for 24 hours is given by

$$C(x) = \begin{cases} 150 & \text{if } 0 < x \leq 25 \\ 1.75x + 150 & \text{if } x > 25, \end{cases}$$

where x is the number of miles driven.

(a) What is the cost if the limo is driven 20 miles? 30 miles?
(b) If the cost is $218.25, how many miles were driven?

15. Sketch the graph of the function f given by

$$f(x) = \begin{cases} x^2 & \text{if } x \leq 0 \\ x + 1 & \text{if } 0 < x < 4 \\ \sqrt{x} & \text{if } x \geq 4. \end{cases}$$

16. U.S. Express Mail rates in 2003 are shown in the following table. Sketch the graph of the function e, whose rule is $e(x) = $ cost of sending a package weighing x pounds by Express Mail.

EXPRESS MAIL Letter Rate—Post Office to Addressee Service	
Up to 8 ounces	$13.65
Over 8 ounces to 2 pounds	$17.85
Up to 3 pounds	$21.05
Up to 4 pounds	$24.20
Up to 5 pounds	$27.30
Up to 6 pounds	$30.40
Up to 7 pounds	$33.45

17. Which of the following are graphs of functions of x?

(a)

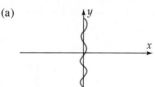

(b)

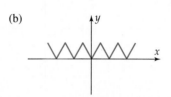

18. The function whose graph is shown gives the number of Americans who snowboard (in millions) in year x where $x = 0$ corresponds to 1988.*

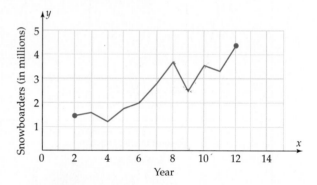

(a) What is the domain of the function?
(b) Approximately, what is the range of the function?
(c) Find the average rate of change of the function from $x = 6$ to $x = 8$ and from $x = 9$ to $x = 10$.

In Questions 19–22, determine the local maxima and minima of the function, the intervals on which the function is increasing, and the intervals on which it is decreasing.

19. $g(x) = \sqrt{x^2 + x + 1}$

20. $f(x) = 2x^3 - 5x^2 + 4x - 3$

21. $g(x) = x^3 + 8x^2 + 4x - 3$

22. $f(x) = .5x^4 + 2x^3 - 6x^2 - 16x + 2$

In Questions 23 and 24, sketch the graph of the curve given by the parametric equations.

23. $x = t^2 - 4$ and $y = 2t + 1$ $(-3 \leq t \leq 3)$

24. $x = t^3 + 3t^2 - 1$ and $y = t^2 + 1$ $(-3 \leq t \leq 2)$

25. Sketch a graph that is symmetric with respect to both the x-axis and the y-axis. (There are many correct answers and your graph need not be the graph of an equation.)

---

*National Sporting Goods Association, TransWorld Snowboarding Business.

26. Sketch the graph of a function that is symmetric with respect to the origin. (There are many correct answers and you don't have to state the rule of your function.)

In Questions 27 and 28, determine algebraically whether the graph of the given equation is symmetric with respect to the x-axis, the y-axis, or the origin.

27. $x^2 = y^2 + 2$

28. $5y = 7x^2 - 2x$

In Questions 29–31, determine whether the given function is even, odd, or neither.

29. $g(x) = 9 - x^2$ 30. $f(x) = |x|x + 1$

31. $h(x) = 3x^5 - x(x^4 - x^2)$

32. (a) Draw some coordinate axes and plot the points $(-2, 1)$, $(-1, 3)$, $(0, 1)$, $(3, 2)$, $(4, 1)$.
    (b) Suppose the points plotted in part (a) lie on the graph of an *even* function f. Plot these points: $(2, f(2))$, $(1, f(1))$, $(0, f(0))$, $(-3, f(-3))$, $(-4, f(-4))$.

33. Determine whether the circle with equation

$$x^2 + y^2 + 6y = -5$$

is symmetric with respect to the x-axis, the y-axis, or the origin.

34. Sketch the graph of a function f that satisfies all of these conditions:

   (i) domain of $f = [-3, 4]$
   (ii) range of $f = [-2, 5]$
   (iii) $f(-2) = 0$
   (iv) $f(1) > 2$

   [*Note:* There are many possible correct answers and the function whose graph you sketch need *not* have an algebraic rule.]

35. Sketch the graph of $g(x) = 5 + \dfrac{4}{x - 5}$.

Use the graph of the function f in the figure to answer Questions 36–39.

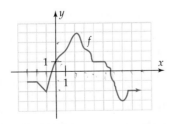

36. What is the domain of f?

37. What is the range of f?

38. Find all numbers x such that $f(x) = 1$.

39. Find a number x such that $f(x + 1) < f(x)$. (Many correct answers are possible.)

Use the graph of the function f in the figure to answer Questions 40–46.

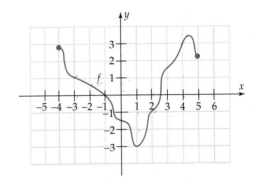

40. What is the domain of f?   41. $f(-3) =$ _____.

42. $f(2 + 2) =$ _____.

43. $f(-1) + f(1) =$ _____.

44. True or false: $2f(2) = f(4)$.

45. True or false: $3f(2) = -f(4)$.

46. True or false: $f(x) = 3$ for exactly one number x.

Use the graphs of the functions f and g in the figure to answer Questions 47–52.

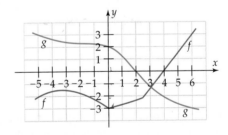

47. For which values of x is $f(x) = 0$?

48. True or false: If a and b are numbers such that $-5 \le a < b \le 6$, then $g(a) < g(b)$.

49. For which values of x is $g(x) \ge f(x)$?

50. Find $f(0) - g(0)$.

51. For which values of x is $f(x + 1) < 0$?

52. What is the distance from the point $(-5, g(-5))$ to the point $(6, g(6))$?

53. Fireball Bob and King Richard are two NASCAR racers. The graph on the next page shows their distance traveled in a recent race as a function of time.

   (a) Which car made the most pit stops?
   (b) Which car started out the fastest?

(c) Which car won the race?

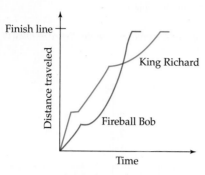

In Questions 54–57, list the transformations, in the order in which they should be performed on the graph of $g(x) = x^2$, so as to produce a complete graph of the function f.

54. $f(x) = (x - 2)^2$ **55.** $f(x) = .25x^2 + 2$

56. $f(x) = -(x + 4)^2 - 5$ **57.** $f(x) = -3(x - 7)^2 + 2$

58. The graph of a function f is shown in the figure. On the same coordinate plane, carefully draw the graphs of the functions g and h whose rules are

$$g(x) = -f(x) \qquad \text{and} \qquad h(x) = 1 - f(x)$$

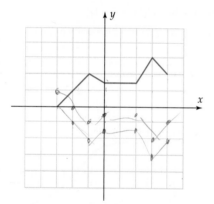

59. The figure shows the graph of a function f. If g is the function given by $g(x) = f(x + 2)$, then which of these statements about the graph of g is true?

(a) It does not cross the x-axis.
(b) It does not cross the y-axis.
(c) It crosses the y-axis at $y = 4$.
(d) It crosses the y-axis at the origin.
(e) It crosses the x-axis at $x = -3$.

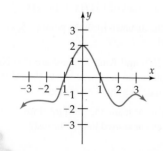

60. If $f(x) = 3x + 2$ and $g(x) = x^3 + 1$, find:

(a) $(f + g)(-1)$ (b) $(f - g)(2)$ (c) $(fg)(0)$

61. If $f(x) = \dfrac{1}{x - 1}$ and $g(x) = \sqrt{x^2 + 5}$, find:

(a) $(f/g)(2)$ (b) $(g/f)(x)$
(c) $(fg)(c + 1)$ $(c \neq 1)$

62. Find two functions f and g such that neither is the identity function, and

$$(f \circ g)(x) = (2x + 1)^2.$$

63. Use the graph of the function g in the figure to fill in the following table, in which h is the composite function $g \circ g$.

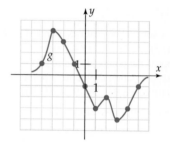

x	-4	-3	-2	-1	0	1	2	3	4
$g(x)$					-1				
$h(x) = g(g(x))$									

Questions 64–69 refer to the functions $f(x) = \dfrac{1}{x + 1}$ and $g(t) = t^3 + 3$.

64. $(f \circ g)(1) =$ _____ .   **65.** $(g \circ f)(2) =$ _____ .

66. $g(f(-2)) =$ _____ .   **67.** $(g \circ f)(x - 1) =$ _____ .

68. $g(2 + f(0)) =$ _____ .   **69.** $f(g(1) - 1) =$ _____ .

70. Let f and g be the functions given by

$$f(x) = 4x + x^4 \qquad \text{and} \qquad g(x) = \sqrt{x^2 + 1}$$

(a) $(f \circ g)(x) =$ _____ .   (b) $(g - f)(x) =$ _____ .

71. If $f(x) = \dfrac{1}{x}$ and $g(x) = x^2 - 1$, then

$$(f \circ g)(x) = \_\_\_\_ \qquad \text{and} \qquad (g \circ f)(x) = \_\_\_\_.$$

72. Let $f(x) = x^2$. Give an example of a function g with domain all real numbers such that $g \circ f \neq f \circ g$.

73. If $f(x) = \dfrac{1}{1 - x}$ and $g(x) = \sqrt{x}$, then find the domain of the composite function $f \circ g$.

74. These tables show the values of the functions f and g at certain numbers:

x	-1	0	1	2	3
$f(x)$	1	0	1	3	5

and

t	0	1	2	3	4
$g(t)$	−1	0	1	2	5

Which of the following statements are *true*?

(a) $(g − f)(1) = 1$ (b) $(f \circ g)(2) = (f − g)(0)$
(c) $f(1) + f(2) = f(3)$ (d) $(g \circ f)(2) = 1$
(e) None of the above is true.

75. Find the average rate of change of the function

$$g(x) = \frac{x^3 − x + 1}{x + 2}$$

as x changes from

(a) −1 to 1   (b) 0 to 2

76. Find the average rate of change of the function
$f(x) = \sqrt{x^2 − x + 1}$ as x changes from

(a) −3 to 0   (b) −3 to 3.5   (c) −3 to 5

77. If $f(x) = 2x + 1$ and $g(x) = 3x − 2$, find the average rate of change of the composite function $f \circ g$ as x changes from 3 to 5.

78. If $f(x) = x^2 + 1$ and $g(x) = x − 2$, find the average rate of change of the composite function $f \circ g$ as x changes from −1 to 1.

In Questions 79–82, find the difference quotient of the function.

79. $f(x) = 3x + 4$ **80.** $g(x) = \sqrt{x}$

81. $g(x) = x^2 − 1$ **82.** $f(x) = x^2 + x$

83. The profit (in hundreds of dollars) from selling x tons of Wonderchem is given by

$$P(x) = .2x^2 + .5x − 1.$$

What is the average rate of change of profit when the number of tons of Wonderchem sold increases from

(a) 4 to 8 tons? (b) 4 to 5 tons? (c) 4 to 4.1 tons?

84. On the planet Mars, the distance traveled by a falling rock (ignoring atmospheric resistance) in t seconds is $6.1t^2$ feet. How far must a rock fall in order to have an average speed of 25 feet per second over that time interval?

85. The graph in the figure shows the population of fruit flies during a 50-day experiment in a controlled atmosphere.

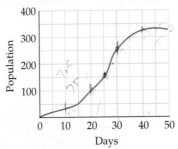

(a) During what 5-day period is the average rate of population growth the slowest?

(b) During what 10-day period is the average rate of population growth the fastest?
(c) Find an interval beginning at the 30th day during which the average rate of population growth is the same as the average rate from day 10 to day 20.

86. The graph of the function g in the figure consists of straight line segments. Find an interval over which the average rate of change of g is

(a) 0   (b) −3   (c) .5
(d) Explain why the average rate of change of g is the same from −3 to −1 as it is from −2.5 to 0.

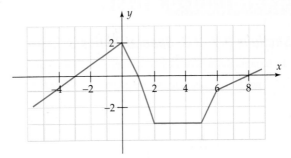

87. The table shows the median sale price for single-family homes in the western United States in selected years.

Year	1985	1990	1995	1998	2000	2001
Median Price	$95,400	$129,600	$141,000	$164,800	$183,000	$194,500

(a) Find the average rate of change of the median price from 1985 to 1998, from 1998 to 2001, and from 1985 to 2001.
(b) If the average rate of change from 1998 to 2001 continues in years after 2001, what is the median price in 2005?

88. Find the inverse of the function $f(x) = 2x + 1$.

89. Find the inverse of the function $f(x) = \sqrt{5 − x} + 7$,

90. Find the inverse of the function $f(x) = \sqrt[5]{x^3 + 1}$.

91. The graph of a function f is shown in the figure. Sketch the graph of the inverse function of f.

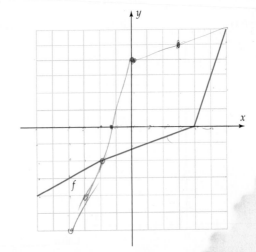

92. Which of the following functions have inverse functions (give reasons for your answers):

   (a) $f(x) = x^3$ (b) $f(x) = 1 - x^2$, $x \le 0$

   (c) $f(x) = |x|$

In Exercises 93–95, determine whether or not the given function has an inverse function (give reasons for your answer). If it does, find the graph of the inverse function.

93. $f(x) = 1/x$

94. $f(x) = .02x^3 - .04x^2 + .6x - 4$

95. $f(x) = .2x^3 - 4x^2 + 6x - 15$

Chapter 3 Test

Sections 3.1–3.3; Special Topics 3.3.A

1. Express the area of a circle as a function of its radius.

2. Consider the function f whose graph is shown below:

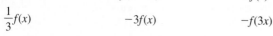

Label each of the following graphs with one of the labels below, and explain your reasoning. (There are more labels than there are graphs. Each graph gets one label, and there will be some labels left over.)

$f(-x)$	$-f(x)$	$f(3x)$
$f\left(\frac{1}{3}x\right)$	$f(x) + 3$	$f(x) - 3$
$f(x + 3)$	$f(x - 3)$	$3f(x)$
$\frac{1}{3}f(x)$	$-3f(x)$	$-f(3x)$

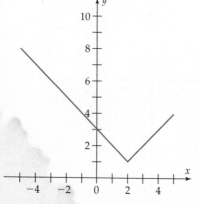

Label:_____

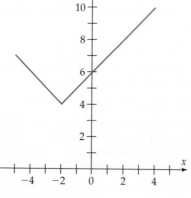

Label:_____

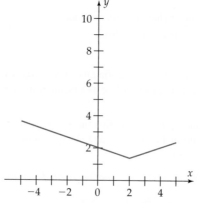

Label:_____

3. The function g is given by the following graph

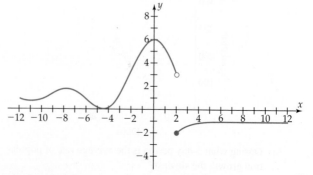

(a) Find the domain of g (b) Find the range of g
(c) Find $g(0)$

4. Does the equation $y = [\![x + 5]\!]$ define y as a function of x? Why or why not?

5. Compute and simplify the difference quotient
$$\dfrac{f(x + h) - f(x)}{h} \text{ for the function } f(x) = \dfrac{2}{x} + 3. \text{ Assume}$$
that $h \neq 0$

6. Graph the curve given by
$$x = t^3 - t$$
$$y = \sqrt[3]{t}$$
with $-1.5 \leq t \leq 1.5$. Choose your viewing window carefully.

Sections 3.4–3.7; Special Topics 3.4.A

7. A child's score, s, on a standardized test is a function of the number of books, b, he or she has read. The formula for s is given below:
$$s(b) = 15 + \sqrt{10b + 100}$$

(a) Explain in terms of books and test scores the meaning of the following statements: $s(0) = 25$, $s^{-1}(35) = 30$.
(b) How many books would a child have to read to get a score of 45 on the test?

8. Determine whether the given graph is symmetric with respect to the y-axis, the x-axis, the origin, or none of these:

(a)

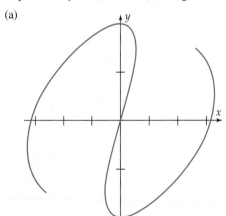

(b)

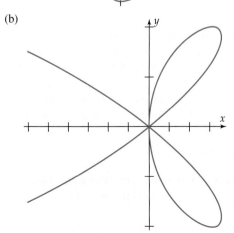

(c)

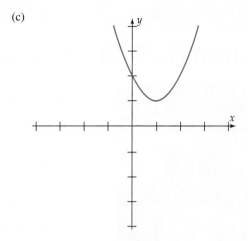

9. Describe a sequence of transformations that will transform the graph of the function f into the graph of the function g
$$f(x) = x^2 + 2 \qquad g(x) = (x + 5)^2 + 9$$

10. The height in feet of a dropped ball after t seconds is given by $h(t) = -16t^2 + 500$.

(a) Find the average rate at which the ball falls in the interval from 3 to 3.1 seconds.
(b) Find the average rate at which the ball falls in the interval from 3 to 3.01 seconds.
(c) Estimate the rate at which the ball falls after exactly 3 seconds.

11. The table below shows the total number of shares traded (in millions) on the New York Stock Exchange in the year 2007. Find the average rate of change in monthly share volume from January through April. Include units in your answer.

Month	January	February	March	April
Volume	110.15	106.07	139.35	108.24

12. The given function has an inverse function. Sketch the graph of the inverse function:
$$f(x) = \begin{cases} x^2 & \text{if } x \leq 0 \\ -.6x & \text{if } x > 0 \end{cases}$$

13. Use the round-trip theorem to show that g is the inverse of f, where
$$f(x) = \sqrt[5]{x - 1} \qquad \text{and} \qquad g(x) = x^5 + 1$$

DISCOVERY PROJECT 3 Feedback—Good and Bad

Whether it's for a concert or a worship service, a sporting event or a graduation ceremony, a shopping mall or a lecture hall, audio engineers are concerned to avoid audio feedback. If the system is not set up correctly, then sound from the speakers reaches the microphone with enough clarity that it is fed back into the audio system and amplified. This feedback cycle repeats again and again. Each time the amplified sound follows very quickly behind the previous one so that the audience hears an unpleasant hum that rapidly becomes a loud screech. This is feedback we'd prefer to avoid.

Other types of feedback are considerably more pleasing. Consider, for example, an investment of x dollars that earns 4% interest compounded annually. Then the amount in the account (principal plus interest) after one year is $1.04x$. Assume the initial deposit is left in the account without adding deposits or withdrawing funds. If $1000 is invested, then the balance (rounded to the nearest penny) is

Beginning	1000
End of year 1	$1.04 \cdot 1000 = 1040.00$
End of year 2	$1.04 \cdot 1040.00 = 1081.60$
End of year 3	$1.04 \cdot 1081.60 = 1124.86$

and so on. You can easily carry out this process on a calculator as follows: key in 1000 and press ENTER; then key in "$\times 1.04$" and press ENTER repeatedly. Each time you press ENTER, the calculator computes as shown in Figure 1. The result is an ever-increasing bank balance.

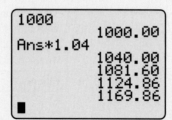
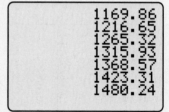

Figure 1

In each of the preceding examples, the output of the process was fed back as input and the process repeated to generate the result—unpleasant in the first case and quite pleasant in the second. Now we strip away the specifics and look at the process in more general terms. Suppose we have a function f and x is in its domain. We find $f(x)$ and feed that value back to the function to find $f(f(x))$. Then feed that value back to find $f(f(f(x)))$. The sequence of values obtained by continuing this process

$$x, f(x), f(f(x)), f(f(f(x))), f(f(f(f(x)))), \ldots$$

is called the **orbit** of x and the feedback process is called **iteration.** The functions themselves $(f, f^2 = f \circ f, f^3 = f \circ f \circ f, \ldots)$ are called **iterates** of f.

Tom Sobolik/Black Star Publishing/PictureQuest

(b) If the car takes 60 meters to come to a complete stop, what was its speed?

48. Jack throws a baseball. Its height above the ground (in feet) is given by

$$h(x) = -.0013x^2 + .26x + 5.5$$

where x is the distance (in feet) from Jack to a point on the ground directly below the ball.

(a) How far from Jack is the ball when it reaches the highest point on its flight? How high is the ball at that point?

(b) How far from Jack does the ball hit the ground?

In Exercises 49–52, use the formula for the height h of an object that is traveling vertically (subject only to gravity) at time t:

$$h = -16t^2 + v_0 t + h_0,$$

where h_0 is the initial height and v_0 is the initial velocity; t is measured in seconds and h in feet.

49. A ball is thrown upward from the top of a 96-foot-high tower with an initial velocity of 80 feet per second. When does the ball reach its maximum height and how high is it at that time?

50. A penny is dropped from the top of Bank of America building in Atlanta, Georgia. How long does it take to reach the ground? (Assume the Bank of America building is 1024 feet high.)

51. A ball is thrown upward from a height of 5 feet with an initial velocity of 11 feet per second. Find its maximum height.

52. A bullet is fired upward from ground level with an initial velocity of 1800 feet per second. How high does it go?

53. The sum of the height h and the base b of a triangle is 30. What height and base will produce a triangle of maximum area?

54. A gutter is to be made by bending up the edge of a 20-inch-wide piece of aluminum. What depth should the gutter be to have the maximum possible cross-sectional area?

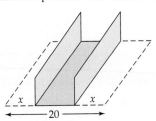

55. A field bounded on one side by a river is to be fenced on three sides so as to form a rectangular enclosure. If 200 feet of fencing is to be used, what dimensions will yield an enclosure of the largest possible area?

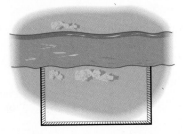

56. A rectangular box (with top) has a square base. The sum of the lengths of its 12 edges is 8 feet. What dimensions should the box have so that its surface area is as large as possible?

57. A gardener wants to use 130 feet of fencing to enclose a rectangular garden and divide it into two plots, as shown in the figure. What is the largest possible area for such a garden?

58. A rectangular garden next to a building is to be fenced on three sides. Fencing for the side parallel to the building costs $80 per foot, and material for the other two sides costs $20 per foot. If $1800 is to be spent on fencing, what are the dimensions of the garden with the largest possible area?

59. At Middleton Place, a plantation near Charleston, South Carolina, there is a "joggling board" that was once used for courting. A young girl would sit at one end, her suitor at the other end, and her mother in the center. The mother would bounce on the board, thus causing the girl and her suitor to move closer together. A joggling board is 8 feet long and an average mother sitting at its center causes the board to deflect 2 inches, as shown in the figure. The shape of the deflected board is parabolic.

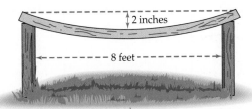

(a) Find the equation of the parabola, assuming that the joggling board lies on the x-axis with its center at the origin.

(b) How far from the center of the board is the deflection 1 inch?

60. A salesperson finds that her sales average 40 cases per store when she visits 20 stores a week. Each time she visits an additional store per week, the average sales per store decrease by 1 case. How many stores should she visit each week if she wants to maximize her sales?

61. A potter can sell 120 bowls per week at $5 per bowl. For each 50¢ decrease in price, 20 more bowls are sold. What price should be charged to maximize sales income?

62. A vendor can sell 200 souvenirs per day at a price of $2 each. Each 10¢ price increase decreases the number of sales by 25 per day. Souvenirs cost the vendor $1.50 each. What price should be charged to maximize the profit?

63. When a basketball team charges $10 per ticket, average attendance is 500 people. For each 25¢ decrease in ticket price, average attendance increases by 30 people. What should the ticket price be to ensure maximum income?

THINKERS

64. The *discriminant* of a quadratic function

$$f(x) = ax^2 + bx + c$$

is the number $b^2 - 4ac$. For each of the discriminants listed, state which graphs could possibly be the graph of f.

(a) $b^2 - 4ac = 25$
(b) $b^2 - 4ac = 0$
(c) $b^2 - 4ac = -49$
(d) $b^2 - 4ac = 72$

(i)
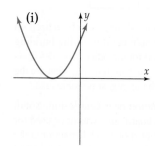

(ii)
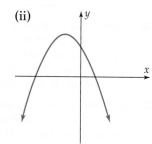

(iii)

(iv)

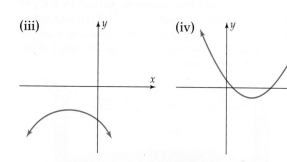

65. We are designing a track for a 200 meter race. The shape of the track will be two x-meter, long straight stretches along with two semicircular caps of radius r, as shown below:

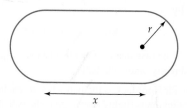

Inside the track, we are going to have to plant grass, pull weeds, etc. So we would like to minimize the total area of the track. What is the minimum possible area? [*Hint:* Find an expression for the area, and get it in terms of one variable.]

66. According to the "logistic growth" model, the rate at which a population of bunnies grows is a function of x, the number of bunnies there already are

$$f(x) = kx(C - x) \text{ bunnies/year}$$

where $C > 0$ is the "carrying capacity" of the bunnies' environment and k is a positive constant that can be determined experimentally. If $f(x)$ is big, that means the bunny population is growing quickly. If $f(x)$ is negative, it means the bunny population is declining.

(a) What bunny populations will yield a growth rate of zero? (These are called "stable populations.")
(b) For what bunny population is the growth rate largest?
(c) What bunny populations will yield a positive growth rate?

4.2 Polynomial Functions

Section Objectives
- Recognize the algebraic forms of a polynomial.
- Use the Division Algorithm.
- Apply the Remainder Theorem.
- Apply the Factor Theorem.
- Find the rule of a polynomial with given degree and roots.
- Determine the maximum possible number of roots a polynomial may have.

Informally, a **polynomial** is an algebraic expression such as

$$x^3 - 6x^2 + \tfrac{1}{2} \quad \text{or} \quad x^{15} + x^{10} + 7 \quad \text{or} \quad x - 6.7 \quad \text{or} \quad 12.$$

Formally, a **polynomial in x** is an algebraic expression that can be written in the form

$$a_n x^n + a_{n-1} x^{n-1} + \cdots + a_3 x^3 + a_2 x^2 + a_1 x + a_0,$$

where n is a nonnegative integer, x is a variable,* and each of $a_0, a_1, \ldots, a_n$ is a constant, called a **coefficient.** The coefficient a_0 is called the **constant term.** A polynomial that consists only of a constant term, such as 12, is called a **constant polynomial.** The **zero polynomial** is the constant polynomial 0.

The *exponent* of the highest power of x that appears with *nonzero* coefficient is the **degree** of the polynomial, and the nonzero coefficient of this highest power of x is the **leading coefficient.** For example,

Polynomial	Degree	Leading Coefficient	Constant Term
$6x^7 + 4x^3 + 5x^2 - 7x + 10$	7	6	10
x^3	3	1	0
12 (think of this as $12x^0$)	0	12	12
$0x^9 + 2x^6 + 3x^7 + x^8 - 2x - 4$	8	1	-4

The degree of the zero polynomial is *not defined* since no exponent of x occurs with nonzero coefficient.

EXAMPLE 1

Which of the following are polynomials?

(a) $x^2 - x^3 + 2x - x^4$

(b) $3x^4 - 2x^2 - \dfrac{1}{x} + 3$

(c) $(x^2 + 5)(3x^2 - 2)$

(d) $x^2 + 3x + 5^x$

(e) $x^2 + 3x + \pi^3$

(f) $x + x^{3/2} + 1$

SOLUTION

(a) $x^2 - x^3 + 2x - x^4$ is a polynomial. The order in which we write the terms doesn't change whether or not an expression is a polynomial.

(b) $3x^4 - 2x^2 - \dfrac{1}{x} + 3$ is not a polynomial. The term $-\dfrac{1}{x}$ cannot be written in the form ax^n for any positive integer n.

(c) $(x^2 + 5)(3x^2 - 2)$ is a polynomial. Its expanded form is $3x^4 + 13x^2 - 10$.

(d) $x^2 + 3x + 5^x$ is not a polynomial. The exponents in a polynomial cannot be variables.

(e) $x^2 + 3x + \pi^3$ is a polynomial. π^3 is just a constant.

(f) $x + x^{3/2} + 1$ is not a polynomial. The exponents of x must be whole numbers, and 3/2 is not a whole number. ∎

---

*Any letter may be used as the variable in a polynomial.

A **polynomial function** is a function whose rule is given by a polynomial, such as $f(x) = x^5 + 3x^2 - 2$. First-degree polynomial functions, such as $g(x) = 3x - 4$, are called **linear functions,** and, as we saw in Section 4.1, second-degree polynomial functions are called **quadratic functions.**

POLYNOMIAL ARITHMETIC

You should be familiar with addition, subtraction, and multiplication of polynomials, which are presented in the Algebra Review Appendix. Long division of polynomials is quite similar to long division of numbers, as we now see.

EXAMPLE 2

Divide $8x^3 + 2x^2 + 1$ by $2x^2 - x$.

SOLUTION   We set up the division in the same way that is used for numbers.

$$\text{Divisor} \rightarrow \;\; 2x^2 - x \,\overline{\big)\, 8x^3 + 2x^2 + 1} \;\; \leftarrow \text{Dividend}$$

Begin by dividing the first term of the divisor $(2x^2)$ into the first term of the dividend $(8x^3)$ and putting the result $\left(\text{namely, } \dfrac{8x^3}{2x^2} = 4x\right)$ on the top line, as shown below. Then multiply $4x$ times the entire divisor, put the result on the third line, and subtract

$$
\begin{array}{r}
4x \quad\quad\quad \leftarrow \text{Partial Quotient} \\
2x^2 - x \,\overline{\big)\, 8x^3 + 2x^2 + 1} \\
\underline{8x^3 - 4x^2} \quad\quad \leftarrow 4x(2x^2 - x) \\
6x^2 + 1 \quad \leftarrow \text{Subtraction*}
\end{array}
$$

Now divide the first term of the divisor $(2x^2)$ into $6x^2$ and put the result $\left(\dfrac{6x^2}{2x^2} = 3\right)$ on the top line, as shown below. Then multiply 3 times the entire divisor, put the result on the fifth line, and subtract

$$
\begin{array}{r}
4x \;\; + 3 \quad\quad\quad \leftarrow \text{Quotient} \\
2x^2 - x \,\overline{\big)\, 8x^3 + 2x^2 \quad\quad + 1} \\
\underline{8x^3 - 4x^2} \quad\quad\quad\; \leftarrow 4x(2x^2 - x) \\
6x^2 \quad\quad + 1 \;\; \leftarrow \text{Subtraction} \\
\underline{6x^2 - 3x} \quad\quad\; \leftarrow 3(2x^2 - x) \\
\text{Remainder} \rightarrow 3x \;\; + 1 \quad \leftarrow \text{Subtraction}
\end{array}
$$

The division process stops when the remainder is 0 or has smaller degree than the divisor, which is the case here.   ∎

---

*If this subtraction is confusing, write it out horizontally and watch the signs:

$$(8x^3 + 2x^2 + 1) - (8x^3 - 4x^2) = 8x^3 + 2x^2 + 1 - 8x^3 + 4x^2 = 6x^2 + 1.$$

We review the process of checking a long division problem by computing 4509/31:

$$
\begin{array}{r}
145 \\
31\overline{)4509} \\
31 \\
\hline
140 \\
124 \\
\hline
169 \\
155 \\
\hline
14
\end{array}
$$

Check:

$$
\begin{array}{rl}
145 & \leftarrow \text{Quotient} \\
\times\ 31 & \leftarrow \text{Divisor} \\
\hline
4495 & \\
+\ 14 & \leftarrow \text{Remainder} \\
\hline
4509 & \leftarrow \text{Dividend}
\end{array}
$$

We can summarize this process in one line:

$$\text{Divisor} \cdot \text{Quotient} + \text{Remainder} = \text{Dividend}.$$

The same thing works for division of polynomials, as you can see by examining the division problem from Example 2.

$$
\underset{\text{Divisor}}{(2x^2 - x)} \cdot \underset{\text{Quotient}}{(4x + 3)} + \underset{\text{Remainder}}{(3x + 1)} = (8x^3 + 2x^2 - 3x) + (3x + 1)
$$

$$
= \underset{\text{Dividend}}{8x^3 + 2x^2 + 1}
$$

This fact is so important that it is given a special name and a formal statement.

The Division Algorithm

If a polynomial $f(x)$ is divided by a nonzero polynomial $h(x)$, then there is a quotient polynomial $q(x)$ and a remainder polynomial $r(x)$ such that

$$\text{Dividend} = \text{Divisor} \cdot \text{Quotient} + \text{Remainder}$$

$$f(x) = h(x)\, q(x) + r(x),$$

where either $r(x) = 0$ or $r(x)$ has degree less than the degree of the divisor $h(x)$.

EXAMPLE 3

Show that $2x^2 + 1$ is a factor of $6x^3 - 4x^2 + 3x - 2$.

SOLUTION   We divide $6x^3 - 4x^2 + 3x - 2$ by $2x^2 + 1$ and find that the remainder is 0.

$$
\begin{array}{r}
3x\ -\ 2 \\
2x^2 + 1\overline{)6x^3 - 4x^2 + 3x - 2} \\
6x^3\quad\ \ + 3x \\
\hline
-4x^2\qquad -2 \\
-4x^2\qquad -2 \\
\hline
0.
\end{array}
$$

TECHNOLOGY TIP

The TI-89 does polynomial division (use PROPFRAC in the ALGEBRA menu). It dispays the answer as the sum of a fraction and a polynomial:

$$\frac{\text{Remainder}}{\text{Divisor}} + \text{Quotient}.$$

Since the remainder is 0, the Division Algorithm tells us that

$$\text{Dividend} = \text{Divisor} \cdot \text{Quotient} + \text{Remainder}$$

$$6x^3 - 4x^2 + 3x - 2 = (2x^2 + 1)(3x - 2) + 0$$

$$= (2x^2 + 1)(3x - 2).$$

Therefore, $2x^2 + 1$ is a factor of $6x^3 - 4x^2 + 3x - 2$, and the other factor is the quotient $3x - 2$. ∎

Example 3 illustrates this fact.

Remainders and Factors

> The remainder in polynomial division is 0 exactly when the divisor is a factor of the dividend. In this case, the quotient is the other factor.

REMAINDERS AND ROOTS

When a polynomial $f(x)$ is divided by a first-degree polynomial, such as $x - 3$ or $x + 5$, the remainder is a constant (because constants are the only polynomials of degree less than 1). This remainder has an interesting connection with the values of the polynomial function $f(x)$.

EXAMPLE 4

Let $f(x) = x^3 - 2x^2 - 4x + 5$.

(a) Find the quotient and remainder when $f(x)$ is divided by $x - 3$.
(b) Find $f(3)$.

NOTE

When the divisor is a first-degree polynomial such as $x - 3$, there is a convenient shorthand method of division, called **synthetic division**. See Special Topics 4.2.A for details.

SOLUTION

(a) Using long division, we have

$$
\begin{array}{r}
x^2 + x - 1 \\
x - 3 \overline{\smash{\big)}\ x^3 - 2x^2 - 4x + 5} \\
\underline{x^3 - 3x^2\phantom{{}- 4x + 5}} \\
x^2 - 4x + 5 \\
\underline{x^2 - 3x\phantom{{}+ 5}} \\
-x + 5 \\
\underline{-x + 3} \\
2.
\end{array}
$$

Therefore, the quotient is $x^2 + x - 1$, and the remainder is 2.

(b) Using the Division Algorithm, we can write the dividend
$f(x) = x^3 - 2x^2 - 4x + 5$ as

$$\text{Dividend} = \text{Divisor} \cdot \text{Quotient} + \text{Remainder}$$

$$f(x) = (x - 3)(x^2 + x - 1) + 2.$$

Hence,

$$f(3) = (3 - 3)(3^2 + 3 - 1) + 2 = 0 + 2 = 2.$$

Note that the number $f(3)$ is the same as the remainder when $f(x)$ is divided by $x - 3$. ∎

The argument used in Example 4 to show that $f(3)$ is the remainder when $f(x)$ is divided by $x - 3$ also works in the general case and proves this fact.

Remainder Theorem

> If a polynomial $f(x)$ is divided by $x - c$, then the remainder is the number $f(c)$.

EXAMPLE 5

To find the remainder when $f(x) = x^{79} + 3x^{24} + 5$ is divided by $x - 1$, we apply the Remainder Theorem with $c = 1$. The remainder is

$$f(1) = 1^{79} + 3 \cdot 1^{24} + 5 = 1 + 3 + 5 = 9.$$ ∎

EXAMPLE 6

To find the remainder when $f(x) = 3x^4 - 8x^2 + 11x + 1$ is divided by $x + 2$, we must apply the Remainder Theorem *carefully*. The divisor in the theorem is $x - c$, not $x + c$. So we rewrite $x + 2$ as $x - (-2)$ and apply the theorem with $c = -2$. The remainder is

$$f(-2) = 3(-2)^4 - 8(-2)^2 + 11(-2) + 1 = 48 - 32 - 22 + 1 = -5.$$ ∎

If $f(x)$ is a polynomial, then a solution of the equation $f(x) = 0$ is called a **root** or **zero** of $f(x)$. Thus, a number c is a root of $f(x)$ if $f(c) = 0$. A root that is a real number is called a **real root.** For example, 4 is a real root of the polynomial $f(x) = 3x - 12$ because $f(4) = 3 \cdot 4 - 12 = 0$. There is an interesting connection between the roots of a polynomial and its factors.

EXAMPLE 7

Let $f(x) = x^3 - 4x^2 + 2x + 3$.

(a) Show that 3 is a root of $f(x)$.

(b) Show that $x - 3$ is a factor of $f(x)$.

SOLUTION

(a) Evaluating $f(x)$ at 3 shows that

$$f(3) = 3^3 - 4(3^2) + 2(3) + 3 = 0.$$

Therefore, 3 is a root of $f(x)$.

(b) If $f(x)$ is divided by $x - 3$, then by the Division Algorithm, there is a quotient polynomial $q(x)$ such that

$$f(x) = (x - 3)q(x) + \text{remainder}.$$

The remainder Theorem shows that the remainder when $f(x)$ is divided by $x - 3$ is the number $f(3)$, which is 0, as we saw in part (a). Therefore,

$$f(x) = (x - 3)q(x) + 0 = (x - 3)q(x).$$

Thus, $x - 3$ is a factor of $f(x)$. [To determine the other factor, the quotient $q(x)$, you have to perform the division.]

Example 7 illustrates this fact, which can be proved by the same argument used in the example.

Factor Theorem

The number c is a root of the polynomial $f(x)$ exactly when $x - c$ is a factor of $f(x)$.

EXAMPLE 8

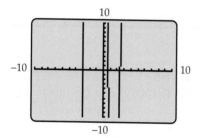

Figure 4–8

The graph of $f(x) = 15x^3 - x^2 - 114x + 72$ in the standard viewing window (Figure 4–8) is obviously not complete but suggests that -3 is an x-intercept, and hence a root of $f(x)$. It is easy to verify that this is indeed the case.

$$f(-3) = 15(-3)^3 - (-3)^2 - 114(-3) + 72 = -405 - 9 + 342 + 72 = 0.$$

Since -3 is a root, $x - (-3) = x + 3$ is a factor of $f(x)$. Use synthetic or long division to verify that the other factor is $15x^2 - 46x + 24$. By factoring this quadratic, we obtain a complete factorization of $f(x)$.

$$f(x) = (x + 3)(15x^2 - 46x + 24) = (x + 3)(3x - 2)(5x - 12).$$

EXAMPLE 9

Find three polynomials of different degrees that have 1, 2, 3, and -5 as roots.

SOLUTION   A polynomial that has 1, 2, 3, and -5 as roots must have $x - 1$, $x - 2$, $x - 3$, and $x - (-5) = x + 5$ as factors. Many polynomials satisfy these conditions, such as

$$g(x) = (x - 1)(x - 2)(x - 3)(x + 5) = x^4 - x^3 - 19x^2 + 49x - 30$$

$$h(x) = 8(x - 1)(x - 2)(x - 3)^2(x + 5)$$

$$k(x) = 2(x + 4)^2(x - 1)(x - 2)(x - 3)(x + 5)(x^2 + x + 1).$$

Note that g has degree 4. When h is multiplied out, its leading term is $8x^5$, so h has degree 5. Similarly, k has degree 8 since its leading term is $2x^8$.

If a polynomial $f(x)$ has four roots, say a, b, c, d, then by the same argument used in Example 8, it must have

$$(x - a)(x - b)(x - c)(x - d)$$

as a factor. Since $(x - a)(x - b)(x - c)(x - d)$ has degree 4 (multiply it out—its leading term is x^4), $f(x)$ must have degree at least 4. In particular, this means that no polynomial of degree 3 can have four or more roots. A similar argument works in the general case.

Number of Roots

A polynomial of degree n has at most n distinct roots.

EXERCISES 4.2

In Exercises 1–10, determine whether the given algebraic expression is a polynomial. If it is, list its leading coefficient, constant term, and degree.

1. $1 + x^3$

2. -7

3. $(x - 1)(x^2 + 1)$

4. $7^x + 2x + 1$

5. $(x + \sqrt{3})(x - \sqrt{3})$

6. $x^3 + 3x^2 + \pi^3$

7. $x^3 + 3x^2 + \pi^x$

8. $4x^2 + 3\sqrt{x} + 5$

9. $\dfrac{7}{x^2} + \dfrac{5}{x} - 15$

10. $(x - 1)^k$ (where k is a fixed positive integer)

In Exercises 11–18, state the quotient and remainder when the first polynomial is divided by the second. Check your division by calculating (Divisor)(Quotient) + Remainder.

11. $3x^4 + 8x^2 - 6x + 1$; $x + 1$

12. $x^5 - x^3 + x - 5$; $x - 2$

13. $x^5 + 2x^4 - 6x^3 + x^2 - 5x + 1$; $x^3 + 1$

14. $2x^5 + 5x^4 + x^3 - 7x^2 - 13x + 12$; $x^2 + 2x + 3$

15. $2x^5 + 5x^4 + x^3 - 7x^2 - 13x + 12$; $2x^3 + x^2 - 7x + 4$

16. $3x^4 - 3x^3 - 11x^2 + 6x - 1$; $x^3 + x^2 - 2$

17. $5x^4 + 5x^2 + 5$; $x^2 - x + 1$

18. $x^5 - 1$; $x - 1$

In Exercises 19–22, determine whether the first polynomial is a factor of the second.

19. $x^2 + 5x - 1$; $x^3 + 2x^2 - 5x - 6$

20. $x^2 + 9$; $4x^5 + 13x^4 + 36x^3 + 108x^2 - 81$

21. $x^2 + 3x - 1$; $x^4 + 3x^3 - 2x^2 - 3x + 1$

22. $x^2 - 4x + 7$; $x^3 - 3x^2 - 3x + 9$

In Exercises 23–27, determine which of the given numbers are roots of the given polynomial.

23. $2, 3, -5, 1$; $g(x) = x^4 + 6x^3 - x^2 - 30x$

24. $1, 1/2, 2, -1/2, 1/3$; $f(x) = 6x^2 + x - 1$

25. $2\sqrt{2}, \sqrt{2}, -\sqrt{2}, 1, -1$; $h(x) = x^3 + x^2 - 8x - 8$

26. $\sqrt{3}, -\sqrt{3}, 1, -1$; $k(x) = 8x^3 - 12x^2 - 6x + 9$

27. $3, -3, 0$; $l(x) = x(x + 3)^{27}$

In Exercises 28–38, find the remainder when $f(x)$ is divided by $g(x)$, without using division.

28. $f(x) = x^2 - 1$; $g(x) = x - 1$

29. $f(x) = x^{10} + x^8$; $g(x) = x - 1$

30. $f(x) = x^6 - 10$; $g(x) = x + 2$

31. $f(x) = 3x^4 - 6x^3 + 2x - 1$; $g(x) = x + 3/2$

32. $f(x) = x^5 - 3x^2 + 2x - 1$; $g(x) = x - 3$

33. $f(x) = x^3 - 2x^2 + 8x - 4$; $g(x) = x + 2$

34. $f(x) = 10x^{70} - 8x^{60} + 6x^{40} + 4x^{32} - 2x^{15} + 5$; $g(x) = x - 1$

35. $f(x) = 2x^5 - \sqrt{3}x^4 + x^3 - \sqrt{3}x^2 + \sqrt{3}x - 100$; $g(x) = x - 10$

36. $f(x) = x^3 + 8x^2 - 29x + 44$; $g(x) = x + 11$

37. $f(x) = 2\pi x^5 - 3\pi x^4 + 2\pi x^3 - 8\pi x - 8\pi$; $g(x) = x - 20$

38. $f(x) = x^5 - 10x^4 + 20x^3 - 5x - 95$; $g(x) = x + 10$

In Exercises 39–46, use the Factor Theorem to determine whether or not $h(x)$ is a factor of $f(x)$.

39. $h(x) = x - 1$; $f(x) = x^5 + 1$

40. $h(x) = x - 1/2$; $f(x) = 2x^4 + x^3 + x - 3/4$

41. $h(x) = x + 3$; $f(x) = x^3 - 3x^2 - 4x - 12$

42. $h(x) = x + 1$; $f(x) = x^3 - 4x^2 + 3x + 8$

43. $h(x) = x - 1$; $f(x) = 14x^{99} - 65x^{56} + 51$

44. $h(x) = x - 2$; $f(x) = x^3 + x^2 - 4x + 4$

45. $h(x) = x - \sqrt{2}$; $f(x) = 3x^3 - 4x^2 - 6x + 8$

46. $h(x) = x - 2$; $f(x) = x^3 - \sqrt{2}x^2 - (6 + \sqrt{2})x + 6\sqrt{2}$

In Exercises 47–50, use the Factor Theorem and a calculator to factor the polynomial, as in Example 7.

47. $f(x) = 6x^3 - 7x^2 - 89x + 140$

48. $g(x) = x^3 - 5x^2 - 5x - 6$

49. $h(x) = 4x^4 + 4x^3 - 35x^2 - 36x - 9$

50. $f(x) = x^5 - 5x^4 - 5x^3 + 25x^2 + 6x - 30$

In Exercises 51–54, each graph is of a polynomial function $f(x)$ of degree 5 whose leading coefficient is 1. The graph is not drawn to scale. Use the Factor Theorem to find the polynomial. [Hint: What are the roots of $f(x)$? What does the Factor Theorem tell you?]

51.

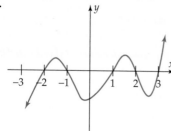

52.

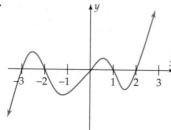

53.

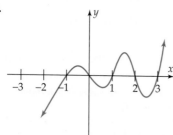

54.

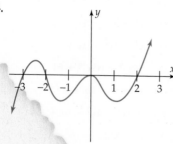

In Exercises 55–60, find a polynomial with the given degree n, the given roots, and no other roots.

55. $n = 3$;   roots 1, 7, -4

56. $n = 3$;   roots 1, -1

57. $n = 2$;   roots 1, -1

58. $n = 1$;   root 5

59. $n = 6$;   roots 1, 2, π

60. $n = 5$;   root 3

61. Find a polynomial function f of degree 3 such that

$$f(10) = 25$$

and the roots of $f(x)$ are 0, 5, and 8.

62. Find a polynomial function g of degree 4 such that the roots of g are 0, -1, 2, -3, and $g(3) = 288$.

In Exercises 63–66, find a number k satisfying the given condition.

63. $x - 2$ is a factor of $x^3 + 3x^2 + kx - 2$.

64. $x - 3$ is a factor of $x^4 - 5x^3 - kx^2 + 18x + 18$.

65. $x - 1$ is a factor of $k^2x^4 - 2kx^2 + 1$.

66. $x + 2$ is a factor of $x^3 - kx^2 + 3x + 7k$.

67. Use the Factor Theorem to show that for every real number c, $x - c$ is *not* a factor of $x^4 + x^2 + 1$.

68. Let c be a real number and n a positive integer.
   (a) Show that $x - c$ is a factor of $x^n - c^n$.
   (b) If n is even, show that $x + c$ is a factor of $x^n - c^n$. [*Remember:* $x + c = x - (-c)$.]

69. (a) If c is a real number and n an odd positive integer, give an example to show that $x + c$ may not be a factor of $x^n - c^n$.
   (b) If c and n are as in part (a), show that $x + c$ is a factor of $x^n + c^n$.

THINKERS

70. For what value of k is the difference quotient of

$$g(x) = kx^2 + 2x + 1$$

equal to $7x + 2 + 3.5h$?

71. For what value of k is the difference quotient of

$$f(x) = x^2 + kx$$

equal to $2x + 5 + h$?

72. Use the fact that $x - 2$ is a factor of $x^3 - 6x^2 + 9x - 2$ to find all the roots of

$$f(x) = x^3 - 6x^2 + 9x - 2.$$

73. Use the fact that $(x + 3)^2$ is a factor of $x^4 + 2x^3 - 91x^2 - 492x - 684$ to find all the roots of

$$f(x) = x^4 + 2x^3 - 91x^2 - 492x - 684.$$

4.2.A *SPECIAL TOPICS*   Synthetic Division

Section Objectives
- Use synthetic division to divide a polynomial by a polynomial of the form $x - c$.
- Find factors of a polynomial using synthetic division.

Synthetic division is a fast method of doing polynomial division when the divisor is a first-degree polynomial of the form $x - c$ for some real number c. To see how it works, we first consider an example of ordinary long division.

$$
\begin{array}{r}
3x^3 + 6x^2 + 4x - 3 \quad \leftarrow \text{Quotient}\\
x - 2\,\overline{\big)\,3x^4 \phantom{{}+ 6x^3} - 8x^2 - 11x + 1} \quad \leftarrow \text{Dividend}\\
3x^4 - 6x^3 \phantom{{}- 8x^2 - 11x + 1}\\
\hline
6x^3 - 8x^2 \phantom{{}- 11x + 1}\\
6x^3 - 12x^2 \phantom{{}- 11x + 1}\\
\hline
4x^2 - 11x \phantom{{}+ 1}\\
4x^2 - 8x \phantom{{}+ 1}\\
\hline
-3x + 1\\
-3x + 6\\
\hline
-5 \quad \leftarrow \text{Remainder}
\end{array}
$$

This calculation involves a lot of redundancy. If we insert 0 coefficients for terms that don't appear above and keep the various coefficients in the proper columns, we can eliminate the repetition and all the x's.

Divisor →

$$
\begin{array}{r}
3 \quad 6 \quad 4 \quad -3 \quad \leftarrow \text{Quotient}\\
1 - 2\,\overline{\big)\,3 \quad 0 \quad -8 \quad -11 \quad 1} \quad \leftarrow \text{Dividend}\\
-6\\
\hline
6\\
-12\\
\hline
4\\
-8\\
\hline
-3\\
+6\\
\hline
-5 \quad \leftarrow \text{Remainder}
\end{array}
$$

We can make our work cleaner by moving the lower lines upward and writing 2 in the divisor position (since that's enough to remind us that the divisor is $x - 2$).

$$
\begin{array}{r}
3 \quad 6 \quad 4 \quad -3 \quad\quad \leftarrow \text{Quotient}\\
2\,\overline{\big)\,3 \quad 0 \quad -8 \quad -11 \quad 1} \quad \leftarrow \text{Dividend}\\
-6 \quad -12 \quad -8 \quad 6\\
\hline
6 \quad 4 \quad -3 \;\boxed{-5} \quad \leftarrow \text{Remainder}
\end{array}
$$

Since the last line contains most of the quotient line, we can save more space and still preserve the essential information by inserting a 3 in the last line and omitting the top line.

$$
\begin{array}{r}
\text{Divisor} \rightarrow 2\, \underline{|\,3 \quad\ \ 0 \quad -8 \quad -11 \quad\ \ 1\,} \leftarrow \text{Dividend} \\
-6 \quad -12 \quad -8 \quad\ \ 6 \\
\hline
\underbrace{3 \quad\ \ 6 \quad\ \ 4 \quad -3}_{\text{Quotient}} \;\boxed{-5} \leftarrow \text{Remainder}
\end{array}
$$

Synthetic division is a quick method for obtaining the last row of this array. Here is a step-by-step explanation of the division of $3x^4 - 8x^2 - 11x + 1$ by $x - 2$.

Step 1 In the first row, list the 2 from the divisor and the coefficients of the dividend in order of decreasing powers of x (insert 0 coefficients for missing powers of x).

$$2\,\underline{|\ \ 3 \quad 0 \quad -8 \quad -11 \quad 1\ }$$

Step 2 Bring down the first dividend coefficient (namely, 3) to the third row

$$
\begin{array}{l}
2\,\underline{|\ \ 3 \quad 0 \quad -8 \quad -11 \quad 1\ } \\
\quad\ \ 3 \qquad\qquad\qquad\qquad \boxed{}
\end{array}
$$

Step 3 Multiply $2 \cdot 3$ and insert the answer 6 in the second row, in the position shown here.

$$
\begin{array}{l}
2\,\underline{|\ \ 3 \quad 0 \quad -8 \quad -11 \quad 1\ } \\
\qquad\quad\ \ 6 \\
\quad\ \ 3 \qquad\qquad\qquad\qquad \boxed{}
\end{array}
$$

Step 4 Add $0 + 6$ and write the answer 6 in the third row.

$$
\begin{array}{l}
2\,\underline{|\ \ 3 \quad 0 \quad -8 \quad -11 \quad 1\ } \\
\qquad\quad\ \ 6 \\
\quad\ \ 3 \quad 6 \qquad\qquad\qquad \boxed{}
\end{array}
$$

Step 5 Multiply $2 \cdot 6$ and insert the answer 12 in the second row.

$$
\begin{array}{l}
2\,\underline{|\ \ 3 \quad 0 \quad -8 \quad -11 \quad 1\ } \\
\qquad\quad\ \ 6 \quad 12 \\
\quad\ \ 3 \quad 6 \qquad\qquad\qquad \boxed{}
\end{array}
$$

Step 6 Add $-8 + 12$ and write the answer 4 in the third row.

$$
\begin{array}{l}
2\,\underline{|\ \ 3 \quad 0 \quad -8 \quad -11 \quad 1\ } \\
\qquad\quad\ \ 6 \quad 12 \\
\quad\ \ 3 \quad 6 \quad\ \ 4 \qquad\qquad \boxed{}
\end{array}
$$

Step 7 Multiply $2 \cdot 4$ and insert the answer 8 in the second row.

$$
\begin{array}{l}
2\,\underline{|\ \ 3 \quad 0 \quad -8 \quad -11 \quad 1\ } \\
\qquad\quad\ \ 6 \quad 12 \quad\ \ 8 \\
\quad\ \ 3 \quad 6 \quad\ \ 4 \qquad\qquad \boxed{}
\end{array}
$$

Step 8 Add $-11 + 8$ and write the answer -3 in the third row.

$$
\begin{array}{l}
2\,\underline{|\ \ 3 \quad 0 \quad -8 \quad -11 \quad 1\ } \\
\qquad\quad\ \ 6 \quad 12 \quad\ \ 8 \\
\quad\ \ 3 \quad 6 \quad\ \ 4 \quad -3 \quad\ \boxed{}
\end{array}
$$

Step 9 Multiply $2 \cdot (-3)$ and insert the answer -6 in the second row.

$$
\begin{array}{l}
2\,\underline{|\ \ 3 \quad 0 \quad -8 \quad -11 \quad 1\ } \\
\qquad\quad\ \ 6 \quad 12 \quad\ \ 8 \quad -6 \\
\quad\ \ 3 \quad 6 \quad\ \ 4 \quad -3 \quad\ \boxed{}
\end{array}
$$

Step 10 Add $1 + (-6)$ and write the answer -5 in the third row.

$$
\begin{array}{l}
2\,\underline{|\ \ 3 \quad 0 \quad -8 \quad -11 \quad 1\ } \\
\qquad\quad\ \ 6 \quad 12 \quad\ \ 8 \quad -6 \\
\quad\ \ 3 \quad 6 \quad\ \ 4 \quad -3 \quad \boxed{-5}
\end{array}
$$

Except for the signs in the second row, this last array is the same as the array obtained from the long division process, and we can read off the quotient and remainder:

The last number in the third row is the remainder.

The other numbers in the third row are the coefficients of the quotient (arranged in order of decreasing powers of x).

Since we are dividing the *fourth*-degree polynomial $3x^4 - 8x^2 - 11x + 1$ by the *first*-degree polynomial $x - 2$, the quotient must be a polynomial of degree *three* with coefficients 3, 6, 4, -3, namely, $3x^3 + 6x^2 + 4x - 3$. The remainder is -5.

CAUTION

Synthetic division can be used *only* when the divisor is a first-degree polynomial of the form $x - c$. In the example above, $c = 2$. If you want to use synthetic, division with a divisor such as $x + 3$, you must write it as $x - (-3)$, which is of the form $x - c$ with $c = -3$.

EXAMPLE 1

To divide $x^5 + 5x^4 + 6x^3 - x^2 + 4x + 29$ by $x + 3$, we write the divisor as $x - (-3)$ and proceed as above.

TECHNOLOGY TIP

Synthetic division programs are in the Program Appendix.

$$
\begin{array}{r|rrrrrr}
-3 & 1 & 5 & 6 & -1 & 4 & 29 \\
   &   & -3 & -6 & 0 & 3 & -21 \\
\hline
   & 1 & 2 & 0 & -1 & 7 & \underline{8} \\
\end{array}
$$

The last row shows that the quotient is $x^4 + 2x^3 - x + 7$ and the remainder is 8. ∎

EXAMPLE 2

Show that $x - 7$ is a factor of $8x^5 - 52x^4 + 2x^3 - 198x^2 - 86x + 14$ and find the other factor.

SOLUTION $x - 7$ is a factor exactly when division by $x - 7$ leaves remainder 0, in which case the quotient is the other factor. Using synthetic division, we have

$$
\begin{array}{r|rrrrrr}
7 & 8 & -52 & 2 & -198 & -86 & 14 \\
  &   & 56 & 28 & 210 & 84 & -14 \\
\hline
  & 8 & 4 & 30 & 12 & -2 & \underline{0} \\
\end{array}
$$

Since the remainder is 0, the divisor $x - 7$ and the quotient

$$8x^4 + 4x^3 + 30x^2 + 12x - 2$$

are factors.

$$8x^5 - 52x^4 + 2x^3 - 198x^2 - 86x + 14$$
$$= (x - 7)(8x^4 + 4x^3 + 30x^2 + 12x - 2). \quad ∎$$

EXERCISES 4.2.A

In Exercises 1–8, use synthetic division to find the quotient and remainder.

1. $(3x^4 - 8x^3 + 9x + 8) \div (x - 2)$

2. $(4x^3 - 3x^2 + 6x + 7) \div (x - 2)$

3. $(2x^4 + 7x^3 - 2x - 8) \div (x + 3)$

4. $(3x^3 - 2x^2 - 8) \div (x + 5)$

5. $(5x^4 - 3x^2 - 4x + 6) \div (x - 7)$

6. $(3x^4 - 2x^3 + 7x - 4) \div (x - 3)$

7. $(x^6 - 1) \div (x - 1)$

8. $(x^6 - x^5 + x^4 - x^3 + x^2 - x + 1) \div (x + 3)$

In Exercises 9–12, use synthetic division to find the quotient and the remainder. In each divisor $x - c$, the number c is not an integer, but the same technique will work.

9. $(3x^4 - 2x^2 + 2) \div \left(x - \dfrac{1}{4}\right)$

10. $(2x^4 - 3x^2 + 1) \div \left(x - \dfrac{1}{2}\right)$

11. $(x^3 - x^2 - 2x + 2) \div (x - \sqrt{2})$

12. $\left(10x^5 - 3x^4 + 14x^3 + 13x^2 - \dfrac{4}{3}x + \dfrac{7}{3}\right) \div \left(x + \dfrac{1}{5}\right)$

In Exercises 13–16, use synthetic division to show that the first polynomial is a factor of the second and find the other factor.

13. $x + 4;$ $x^3 + 3x^2 - 34x - 120$

14. $x - 5;$ $x^5 - 8x^4 + 17x^2 + 293x - 15$

15. $x - 1/2;$ $2x^5 - 7x^4 + 15x^3 - 6x^2 - 10x + 5$

16. $x + 1/3;$ $3x^6 + x^5 - 6x^4 + 7x^3 + 3x^2 - 15x - 5$

In Exercises 17 and 18, use a calculator and synthetic division to find the quotient and remainder.

17. $(x^3 - 5.27x^2 + 10.708x - 10.23) \div (x - 3.12)$

18. $(2.79x^4 + 4.8325x^3 - 6.73865x^2 + .9255x - 8.125)$
    $\div (x - 1.35)$

19. When $x^3 + cx + 4$ is divided by $x + 2$, the remainder is 4. Find c.

20. If $x - d$ is a factor of $2x^3 - dx^2 + (1 - d^2)x + 5$, what is d?

THINKERS

21. Let $g(x) = x^5 - 2x^4 - x^3 + 3x + 1$.

    (a) Show that when we use synthetic division to divide by $(x - 3)$, the quotient's coefficients are all positive.

    (b) Show that when we divide by $(x - a)$, where $a > 3$, the quotient's coefficients are all positive.

    (c) Use part (b) to show that g has no root greater than three.

22. Let $g(x)$ be a polynomial function. Assume that dividing g by $x - a$ gives a quotient with all positive terms and a positive remainder. What does this tell us about the possible roots of $g(x)$?

4.3   Real Roots of Polynomials

Section Objectives

■ Use the Rational Root Test to find the rational roots of a polynomial.

■ Use the Bounds Test and a graphing calculator to find an interval that contains all the real roots of a polynomial.

Finding the real roots of polynomials is the same as solving polynomial equations. The root of a first-degree polynomial, such as $5x - 3$, can be found by solving the equation $5x - 3 = 0$. Similarly, the roots of any second-degree polynomial can be found by using the quadratic formula (Section 1.2). Although the roots of higher-degree polynomials can always be approximated graphically as in Section 2.2, it is better to find exact solutions, if possible.

RATIONAL ROOTS

When a polynomial has integer coefficients, all of its **rational roots** (roots that are rational numbers) can be found exactly by using the following result.

The Rational Root Test

> If a rational number r/s (in lowest terms) is a root of the polynomial
>
> $$a_n x^n + \cdots + a_1 x + a_0,$$
>
> where the coefficients $a_n, \ldots, a_1, a_0$ are integers with $a_n \neq 0$, $a_0 \neq 0$, then
>
> r is a factor of the constant term a_0 and
>
> s is a factor of the leading coefficient a_n.

The test states that every rational root must satisfy certain conditions.* By finding all the numbers that satisfy these conditions, we produce a list of *possible* rational roots. Then we must evaluate the polynomial at each number on the list to see whether the number actually is a root. This testing process can be considerably shortened by using a calculator, as in the next example.

EXAMPLE 1

Find the rational roots of

$$f(x) = 2x^4 + x^3 - 17x^2 - 4x + 6.$$

SOLUTION   If $f(x)$ has a rational root r/s, then by the Rational Root Test r must be a factor of the constant term 6. Therefore, r must be one of $\pm 1, \pm 2, \pm 3$, or ± 6 (the only factors of 6). Similarly, s must be a factor of the leading coefficient 2, so s must be one of ± 1 or ± 2 (the only factors of 2). Consequently, the only *possibilities* for r/s are

$$\frac{\pm 1}{\pm 1}, \frac{\pm 2}{\pm 1}, \frac{\pm 3}{\pm 1}, \frac{\pm 6}{\pm 1}, \frac{\pm 1}{\pm 2}, \frac{\pm 2}{\pm 2}, \frac{\pm 3}{\pm 2}, \frac{\pm 6}{\pm 2}.$$

Eliminating duplications from this list, we see that the only *possible* rational roots are

$$1, -1, 2, -2, 3, -3, 6, -6, \frac{1}{2}, -\frac{1}{2}, \frac{3}{2}, -\frac{3}{2}.$$

Now graph $f(x)$ in a viewing window that includes all of these numbers on the x-axis, say $-7 \leq x \leq 7$ and $-5 \leq y \leq 5$ (Figure 4–9 on the next page). A complete graph isn't necessary, since we are interested only in the x-intercepts.

---

*Since the proof of the Rational Root Test sheds no light on how the test is actually used to solve equations, it will be omitted.

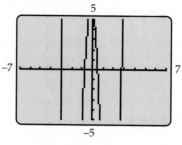

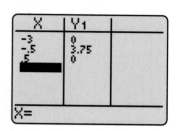

Figure 4–9   Figure 4–10

Figure 4–9 shows that the only numbers on our list that could possibly be roots (x-intercepts) are −3, −1/2, and 1/2, so these are the only ones that need be tested. We use the table feature to evaluate $f(x)$ at these three numbers (Figure 4–10). The table shows that −3 and 1/2 are the only rational roots of $f(x)$. Its other roots (x-intercepts) in Figure 4–9 must be irrational numbers.   ∎

ROOTS AND THE FACTOR THEOREM

Once some roots of a polynomial have been found, the Factor Theorem can be used to factor the polynomial, which may lead to additional roots.

EXAMPLE 2

Find all the roots of $f(x) = 2x^4 + x^3 - 17x^2 - 4x + 6$.

SOLUTION   In Example 1, we saw that −3 and 1/2 are the rational roots of $f(x)$. By the Factor Theorem, $x - (-3) = x + 3$ and $x - 1/2$ are factors of $f(x)$. Using synthetic or long division twice, we have

$$2x^4 + x^3 - 17x^2 - 4x + 6 = (x + 3)(2x^3 - 5x^2 - 2x + 2)$$

$$= (x + 3)(x - .5)(2x^2 - 4x - 4).$$

The remaining roots of $f(x)$ are the roots of $2x^2 - 4x - 4$, that is, the solutions of

$$2x^2 - 4x - 4 = 0$$

$$x^2 - 2x - 2 = 0.$$

They are easily found by the quadratic formula.

$$x = \frac{-(-2) \pm \sqrt{(-2)^2 - 4 \cdot 1 \cdot (-2)}}{2 \cdot 1}$$

$$= \frac{2 \pm \sqrt{12}}{2} = \frac{2 \pm 2\sqrt{3}}{2} = 1 \pm \sqrt{3}.$$

Therefore, $f(x)$ has rational roots −3 and 1/2, and irrational roots $1 + \sqrt{3}$ and $1 - \sqrt{3}$.   ∎

EXAMPLE 3

Factor $f(x) = 2x^5 - 10x^4 + 7x^3 + 13x^2 + 3x + 9$ completely.

SOLUTION   We begin by finding as many roots of $f(x)$ as we can. By the Rational Root Test, every rational root is of the form r/s where $r = \pm 1, \pm 3,$ or ± 9 and $s = \pm 1$ or ± 2. Thus, the possible rational roots are

$$\pm 1, \quad \pm 3, \quad \pm 9, \quad \pm \frac{1}{2}, \quad \pm \frac{3}{2}, \quad \pm \frac{9}{2}.$$

The partial graph of $f(x)$ in Figure 4–11 shows that the only possible roots (x-intercepts) are -1 and 3. You can easily verify that both -1 and 3 are roots of $f(x)$.

Since -1 and 3 are roots, $x - (-1) = x + 1$ and $x - 3$ are factors of $f(x)$ by the Factor Theorem. Division shows that

$$\begin{aligned} f(x) &= 2x^5 - 10x^4 + 7x^3 + 13x^2 + 3x + 9 \\ &= (x + 1)(2x^4 - 12x^3 + 19x^2 + 3x + 9) \\ &= (x + 1)(x - 3)(2x^3 - 6x^2 + x - 3). \end{aligned}$$

The other roots of $f(x)$ are the roots of $g(x) = 2x^3 - 6x^2 + x - 3$. We first check for rational roots of $g(x)$. Since every root of $g(x)$ is also a root of $f(x)$ (why?), the only possible rational roots of $g(x)$ are -1 and 3 [the rational roots of $f(x)$]. We have

$$g(-1) = 2(-1)^3 - 6(-1)^2 + (-1) - 3 = 12;$$
$$g(3) = 2(3^3) - 6(3^2) + 3 - 3 = 0.$$

So -1 is not a root, but 3 is a root of $g(x)$. By the Factor Theorem, $x - 3$ is a factor of $g(x)$. Division shows that

$$\begin{aligned} f(x) &= (x + 1)(x - 3)(2x^3 - 6x^2 + x - 3) \\ &= (x + 1)(x - 3)(x - 3)(2x^2 + 1). \end{aligned}$$

Since $2x^2 + 1$ has no real roots, it cannot be factored. So the factorization of $f(x)$ is complete. ■

◗ BOUNDS

The polynomial $f(x)$ in Example 2 had degree 4 and had four real roots. Since a polynomial of degree n has at most n roots, we know that we found all the roots of $f(x)$. In other cases, however, special techniques may be needed to guarantee that we have found all the roots.

EXAMPLE 4

Prove that all the real roots of

$$g(x) = x^5 - 2x^4 - x^3 + 3x + 1$$

lie between -1 and 3. Then find all the real roots of $g(x)$.

SOLUTION   We first prove that $g(x)$ has no root larger than 3, as follows. Use synthetic division to divide $g(x)$ by $x - 3$.*

$$\begin{array}{r|rrrrrr} 3 & 1 & -2 & -1 & 0 & 3 & 1 \\ & & 3 & 3 & 6 & 18 & 63 \\ \hline & 1 & 1 & 2 & 6 & 21 & \boxed{64} \end{array}$$

Thus, the quotient is $x^4 + x^3 + 2x^2 + 6x + 21$, and the remainder is 64. Applying the Division Algorithm, we have

$$f(x) = (x - 3)(x^4 + x^3 + 2x^2 + 6x + 21) + 64.$$

---

*If you haven't read Special Topics 4.2.A, use long division to find the quotient and remainder.

When $x > 3$, then the factor $x - 3$ is positive, and the quotient

$$x^4 + x^3 + 2x^2 + 6x + 21$$

is also positive (because all its coefficients are). The remainder 64 is also positive. Therefore, $f(x)$ is positive whenever $x > 3$. In particular, $f(x)$ is never zero when $x > 3$, and so there are no roots of $f(x)$ greater than 3.

Now we show that $g(x)$ has no root less than -1. Divide $g(x)$ by

$$x - (-1) = x + 1:$$

$$
\begin{array}{r|rrrrrr}
-1 & 1 & -2 & -1 & 0 & 3 & 1 \\
   &   & -1 & 3 & -2 & 2 & -5 \\
\hline
   & 1 & -3 & 2 & -2 & 5 & \boxed{-4}
\end{array}
$$

Read off the quotient and remainder and apply the Division Algorithm:

$$f(x) = (x + 1)(x^4 - 3x^3 + 2x^2 - 2x + 5) - 4.$$

When $x < -1$, then the factor $x + 1$ is negative. When x is negative, its odd powers are negative and its even powers are positive. Consequently, the quotient $x^4 - 3x^3 + 2x^2 - 2x + 5$ is positive (because the odd powers of x are multiplied by negative coefficients). The product of the positive quotient with the negative factor $x + 1$ is negative. The remainder -4 is also negative. Hence, $f(x)$ is negative whenever $x < -1$. So there are no real roots less than -1. Therefore, all the real roots of $g(x)$ lie between -1 and 3.

Finally, we find the roots of $g(x) = x^5 - 2x^4 - x^3 + 3x + 1$. The only possible rational roots are ± 1 (why?) and it is easy to verify that neither is actually a root. The graph of $g(x)$ in Figure 4–12 shows that there are exactly three real roots (x-intercepts) between -1 and 3. Since all the real roots of $g(x)$ lie between -1 and 3, $g(x)$ has only these three real roots. They are readily approximated by a root finder:

$$x \approx -.3361, \qquad x \approx 1.4268, \qquad \text{and} \qquad x \approx 2.2012. \qquad \blacksquare$$

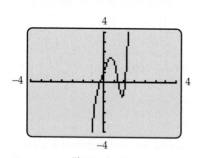

Figure 4–12

Suppose $f(x)$ is a polynomial and r and s are real numbers with $r < s$. If all the real roots of $f(x)$ are between r and s, we say that r is a **lower bound** and s is an **upper bound** for the real roots of $f(x)$.* Example 4 shows that -1 is a lower bound and 3 is an upper bound for the real roots of $g(x) = x^5 - 2x^4 - x^3 + 3x + 1$.

If you know lower and upper bounds for the real roots of a polynomial, you can usually determine the number of real roots the polynomial has, as we did in Example 4. The technique used in Example 4 to test possible lower and upper bounds works in the general case:

Bounds Test

Let $f(x)$ be a polynomial with positive leading coefficient.

If $d > 0$ and every number in the last row in the synthetic division of $f(x)$ by $x - d$ is nonnegative,[†] then d is an upper bound for the real roots of $f(x)$.

If $c < 0$ and the numbers in the last row of the synthetic division of $f(x)$ by $x - c$ are alternately positive and negative [with 0 considered as either][‡] then c is a lower bound for the real roots of $f(x)$.

---

*The bounds are not unique. Any number smaller than r is also a lower bound, and any number larger than s is also an upper bound.

[†]Equivalently, all the coefficients of the quotient and the remainder are nonnegative.

[‡]Equivalently, the coefficients of the quotient are alternatively positive and negative, with the last one and the remainder having opposite signs.

EXAMPLE 5

Find all real roots of

$$f(x) = x^7 - 6x^6 + 9x^5 + 7x^4 - 28x^3 + 33x^2 - 36x + 20.$$

SOLUTION By the Rational Root Test, the only possible roots are

$$\pm 1, \quad \pm 2, \quad \pm 4, \quad \pm 5, \quad \pm 10, \quad \text{and} \quad \pm 20.$$

The graph of $f(x)$ in Figure 4–13 is hard to read but shows that the possible roots are quite close to the origin. Changing the window (Figure 4–14), we see that the only numbers on the list that could possibly be roots (x-intercepts) are 1 and 2. You can easily verify that both 1 and 2 are roots of $f(x)$.

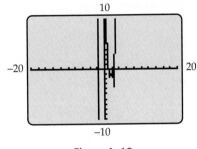

Figure 4–13

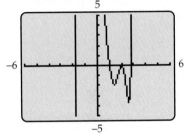

Figure 4–14

In Figures 4–13 and 4–14, all the real roots of $f(x)$ lie between -2 and 6, which suggests that these numbers might be lower and upper bounds for the real roots of $f(x)$.* The Bounds Test shows that this is indeed the case:

-2	1	-6	9	7	-28	33	-36	20
		-2	16	-50	86	-116	172	-272
	1	-8	25	-43	58	-86	136	-252

Alternating Signs
-2 is a lower bound

6	1	-6	9	7	-28	33	-36	20
		6	0	54	366	2028	12,366	73,980
	1	0	9	61	338	2061	12,330	74,000

All Nonnegative
6 is an upper bound

Therefore, the four x-intercepts in Figure 4–14 are the only real roots of $f(x)$. We have seen that two of these are the rational roots, 1 and 2. A root finder shows that the other roots are $x \approx -1.7913$ and $x \approx 2.7913$. ∎

SUMMARY

The examples above illustrate the following guidelines for finding all the real roots of a polynomial $f(x)$.

1. Use the Rational Root Test to find all the rational roots of $f(x)$. [*Examples 1, 3, 5*]

2. Write $f(x)$ as the product of linear factors (one for each rational root) and another factor $g(x)$. [*Examples 2, 3*]

3. If $g(x)$ has degree 2, find its roots by factoring or the quadratic formula. [*Example 2*]

---

*If you are wondering why we don't test 3, 4, or 5 as upper bounds, we did—but the Bounds Test is inconclusive for these numbers, as you can easily verify. See the Note at the top of page.

4. If $g(x)$ has degree 3 or more, use the Bounds Test, if possible, to find lower and upper bounds for the roots of $g(x)$ and approximate the remaining roots graphically. [*Examples 4, 5*]

Shortcuts and variations are always possible. For instance, if the graph of a cubic shows three x-intercepts, then it has three real roots (the maximum possible) and there is no point in finding bounds on the roots. In order to find as many roots as possible exactly in guideline 4, check to see if the rational roots of $f(x)$ are also roots of $g(x)$ and factor $g(x)$ accordingly, as in Example 3.

EXERCISES 4.3

Directions: *When asked to find the roots of a polynomial, find exact roots whenever possible and approximate the other roots. In Exercises 1–15, find all the rational roots of the polynomial.*

1. $x^3 - 3x^2 - x + 3$

2. $x^3 - x^2 - 3x + 3$

3. $x^3 - 3x^2 - 6x + 8$

4. $3x^3 + 17x^2 + 35x + 25$

5. $6x^3 - 11x^2 - 19x - 6$

6. $x^4 - x^2 - 2$

7. $f(x) = 2x^5 - 3x^4 - 11x^3 + 6x^2$ [*Hint:* The Rational Root Test can only be used on polynomials with nonzero constant terms. Factor $f(x)$ as a product of a power of x and a polynomial $g(x)$ with nonzero constant term. Then use the Rational Root Test on $g(x)$.]

8. $2x^6 - 3x^5 - 7x^4 - 6x^3$

9. $f(x) = \frac{1}{12}x^3 - \frac{1}{12}x^2 - \frac{2}{3}x + 1$ [*Hint:* The Rational Root

Test can only be used on polynomials with integer coefficients. Note that $f(x)$ and $12f(x)$ have the same roots (why?).]

10. $\frac{1}{2}x^4 + \frac{5}{2}x^3 + 3x^2 - 2x - 4$

11. $\frac{1}{3}x^3 - \frac{5}{6}x^2 - \frac{1}{6}x + 1$

12. $\frac{1}{3}x^7 - \frac{1}{2}x^6 - \frac{1}{6}x^5 + \frac{1}{6}x^4$

13. $.1x^3 - 1.9x + 3$

14. $.05x^4 + .45x^2 - .4x + 1$

15. $x^{10} - 10x^9 + 45x^8 - 120x^7 + 210x^6 - 252x^5 + 210x^4 - 120x^3 + 45x^2 - 10x + 1$

In Exercises 16–22, factor the polynomial as a product of linear factors and a factor $g(x)$ such that $g(x)$ is either a constant or a polynomial that has no rational roots.

16. $x^{15} - x - 1$

17. $2x^3 - 2x^2 + 3x - 3$

18. $12x^3 - 10x^2 + 6x - 2$

19. $x^6 - 4x^5 + 3x^4 - 12x^3$

20. $x^5 - 2x^4 + 2x^3 - 3x + 2$

21. $x^5 - 6x^4 + 5x^3 + 34x^2 - 84x + 56$

22. $x^5 + 4x^3 + x^2 + 6x$

In Exercises 23–28, use the Bounds Test to find lower and upper bounds for the real roots of the polynomial.

23. $x^3 + 2x^2 - 7x + 20$

24. $x^3 - 15x^2 - 16x + 12$

25. $x^3 - 5x^2 + 5x + 3$

26. $x^4 - 2x^3 - 3x^2 + 4x + 4$

27. $-x^5 - 5x^4 + 9x^3 + 18x^2 - 68x + 176$ [*Hint:* The Bounds Test applies only to polynomials with positive leading coefficient. The polynomial $f(x)$ has the same roots as $-f(x)$ (why?).]

28. $-.002x^3 - 5x^2 + 8x - 3$

In Exercises 29–40, find all real roots of the polynomial.

29. $2x^3 - x^2 - 13x - 6$

30. $t^4 - 3t^3 + 5t^2 - 9t + 6$

31. $6x^3 - 13x^2 + x + 2$

32. $z^3 + z^2 + 2z + 2$

33. $x^4 + x^3 - 19x^2 + 32x - 12$

34. $3x^6 - 7x^5 - 22x^4 + 8x^3$

35. $2x^5 - x^4 - 10x^3 + 5x^2 + 12x - 6$

36. $x^{10} - 10x^9 + 45x^8 - 120x^7 + 210x^6 - 252x^5 + 210x^4 - 120x^3 + 45x^2 - 10x + 1$

37. $x^6 - 3x^5 - 4x^4 - 9x^2 + 27x + 36$

38. $x^5 + 8x^4 + 20x^3 + 9x^2 - 27x - 27$

39. $x^4 - 48x^3 - 101x^2 + 49x + 50$

40. $3x^7 + 8x^6 - 13x^5 - 36x^4 - 10x^3 + 21x^2 + 41x + 10$

41. (a) Show that $\sqrt{2}$ is an irrational number. [*Hint:* $\sqrt{2}$ is a root of $x^2 - 2$. Does this polynomial have any rational roots?]

(b) Show that $\sqrt{3}$ is irrational.

(c) What would happen if you tried to use the techniques from the previous parts of this question to show $\sqrt{4}$ is irrational?

42. Graph $f(x) = .001x^3 - .199x^2 - .23x + 6$ in the standard viewing window.

(a) How many roots does $f(x)$ appear to have? Without changing the viewing window, explain why $f(x)$ must have an additional root. [*Hint:* Each root corresponds to a factor of $f(x)$. What does the rest of the factorization consist of?]

(b) Find all the roots of $f(x)$.

43. According to data from the FBI, the number of people murdered each year per 100,000 can be approximated by the polynomial function

$$f(x) = -.0002724x^5 + .005237x^4 - .03027x^3$$
$$+ .1069x^2 - .9062x + 9.003$$
$$(0 \le x \le 10)$$

where $x = 0$ corresponds to 1995.

(a) What was the murder rate in 2000 and in 2003?

(b) According to this model, in what year was the murder rate 7 people per 100,000?

(c) According to this model, in what year between 1995 and 2005 was the murder rate the highest?

(d) According to this model, in what year between 1995 and 2005 was the murder rate the lowest? (Be careful! This one isn't immediate.)

(e) According to this model, during what time interval between 1995 and 2005 was the murder rate increasing?

44. During the first 150 hours of an experiment, the growth rate of a bacteria population at time t hours is

$$g(t) = -.0003t^3 + .04t^2 + .3t + .2 \text{ bacteria per hour.}$$

(a) What is the growth rate at 50 hours? At 100 hours?

(b) What is the growth rate at 145 hours? What does this mean?

(c) At what time is the growth rate 0?

(d) At what time is the growth rate -50 bacteria per hour?

(e) At what time does the highest growth rate occur?

45. An open-top reinforced box is to be made from a 12- by 36-inch piece of cardboard by cutting along the marked lines, discarding the shaded pieces, and folding as shown in the figure. If the box must be less than 2.5 inches high, what size squares should be cut from the corners in order for the box to have a volume of 448 cubic inches?

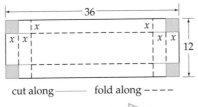

cut along ——— fold along - - - -

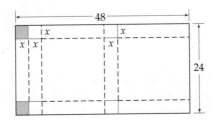

46. A box with a lid is to be made from a 48- by 24-inch piece of cardboard by cutting and folding, as shown in the figure. If the box must be at least 6 inches high, what size squares should be cut from the two corners in order for the box to have a volume of 1000 cubic inches?

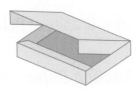

47. In a sealed chamber where the temperature varies, the instantaneous rate of change of temperature with respect to time over an 11-day period is given by

$$F(t) = .0035t^4 - .4t^2 - .2t + 6,$$

where time is measured in days and temperature in degrees Fahrenheit (so that rate of change is in degrees per day).

(a) At what rate is the temperature changing at the beginning of the period $(t = 0)$? At the end of the period $(t = 11)$?

(b) When is the temperature increasing at a rate of $4°F$ per day?

(c) When is the temperature decreasing at a rate of $3°F$ per day?

(d) When is the temperature decreasing at the fastest rate?

48. (a) If c is a root of

$$f(x) = 5x^4 - 4x^3 + 3x^2 - 4x + 5,$$

show that $1/c$ is also a root.

(b) Do part (a) with $f(x)$ replaced by

$$g(x) = 2x^6 + 3x^5 + 4x^4 - 5x^3 + 4x^2 + 3x + 2.$$

(c) Let $f(x) = a_{12}x^{12} + a_{11}x^{11} + \cdots + a_2x^2 + a_1x + a_0$. What conditions must the coefficients a_i satisfy in order that this statement be true: If c is a root of $f(x)$, then $1/c$ is also a root?

49. According to the "modified logistic growth" model, the rate at which a population of bunnies grows is a function of x, the number of bunnies there already are:

$$f(x) = k(-x^3 + x^2(T + C) - CTx) \text{ bunnies/year}$$

where C is the "carrying capacity" of the bunnies' environment, T is the "threshold population" of bunnies necessary for them to thrive and survive, and k is a positive constant that can be determined experimentally. If $f(x)$ is big, that means the bunny population is growing quickly. If $f(x)$ is negative, it means the bunny population is declining.

(a) Why can we assume $T < C$?

(b) What is happening to the bunny population if x is between T and C?

(c) What is happening to the bunny population if $x < T$?

(d) What is happening to the bunny population if $x > C$?

(e) Factor $k(-x^3 + x^2(T + C) - CTx)$

(f) What bunny populations will remain stable (unchanging)?

4.4 Graphs of Polynomial Functions

Section Objectives
- Understand the properties of the graph of a polynomial.
- Find a complete graph of a polynomial.
- Use polynomial graphs in applications.

The graphs of first- and second-degree polynomial functions are straight lines and parabolas respectively (Sections 1.4 and 4.1). What happens when the degree is higher?

The simplest polynomial functions are those of the form $f(x) = ax^n$ (where a is a constant). Their graphs are of four types, as shown in the following chart.

GRAPH OF $f(x) = ax^n$

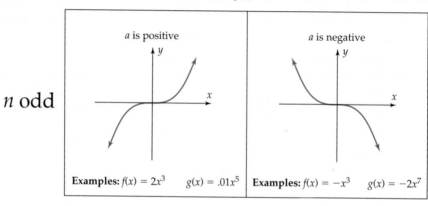

n odd

a is positive

a is negative

Examples: $f(x) = 2x^3$ $g(x) = .01x^5$ **Examples:** $f(x) = -x^3$ $g(x) = -2x^7$

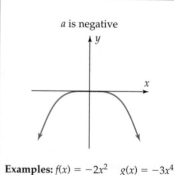

n even

a is positive

a is negative

Examples: $f(x) = 2x^4$ $g(x) = 2x^6$ **Examples:** $f(x) = -2x^2$ $g(x) = -3x^4$

GRAPHING EXPLORATION

Verify the accuracy of the preceding summary by graphing each of the examples in the window with $-5 \le x \le 5$ and $-30 \le y \le 30$. What effect does increasing the value of n have on these graphs?

The graphs of more complicated polynomial functions can vary considerably in shape. Understanding the properties discussed below should assist you to interpret screen images correctly and to determine when a polynomial graph is complete.

CONTINUITY

Every polynomial graph is **continuous,** meaning that it is an unbroken curve, with no jumps, gaps, or holes. Furthermore, polynomial graphs have no sharp corners. Thus, neither of the graphs in Figure 4–15 is the graph of a polynomial function.

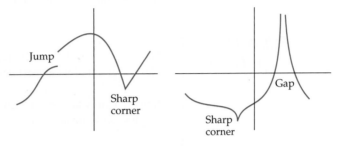

Figure 4–15

 SHAPE OF THE GRAPH WHEN $|x|$ IS LARGE

The shape of a polynomial graph at the far left and far right of the coordinate plane is easily determined by using our knowledge of graphs of functions of the form $f(x) = ax^n$.

EXAMPLE 1

Consider the function $f(x) = 2x^3 + x^2 - 6x$ and the function determined by its leading term $g(x) = 2x^3$.

GRAPHING EXPLORATION

Using the standard viewing window, graph f and g on the same screen.

Do the two graphs look different? Now graph f and g in the viewing window with $-20 \le x \le 20$ and $-10{,}000 \le y \le 10{,}000$. Do the graphs look almost the same? Finally, graph f and g in the viewing window with

$$-100 \le x \le 100 \text{ and } -1{,}000{,}000 \le y \le 1{,}000{,}000.$$

Do the graphs look virtually identical?

The reason the answer to the last question is "yes" can be understood from this table.

x	-100	-50	70	100
$-6x$	600	300	-420	-600
x^2	10,000	2,500	4,900	10,000
$g(x) = 2x^3$	$-2{,}000{,}000$	$-250{,}000$	686,000	2,000,000
$f(x) = 2x^3 + x^2 - 6x$	$-1{,}989{,}400$	$-247{,}200$	690,480	2,009,400

It shows that when $|x|$ is large, the terms x^2 and $-6x$ are insignificant compared with $2x^3$ and barely affect the value of $f(x)$. Hence, the values of $f(x)$ and $g(x)$ are relatively close. ∎

Example 1 is typical of what happens in every case: When $|x|$ is very large, the highest power of x totally overwhelms all lower powers and plays the greatest role in determining the value of the function.

Behavior When $|x|$ Is Large

When $|x|$ is very large, the graph of a polynomial function closely resembles the graph of its highest degree term.

In particular, when the polynomial function has odd degree, one end of its graph shoots upward and the other end downward.

When the polynomial function has even degree, both ends of its graph shoot upward or both ends shoot downward.

EXAMPLE 2

Graph $f(x) = \dfrac{x^5}{120} - \dfrac{x^3}{6} + x$ and $g(x) = \dfrac{x^5}{120}$ on the same axes, first using the window $-4 \le x \le 4$ and then using the window $-12 \le x \le 12$ (Figure 4–16).

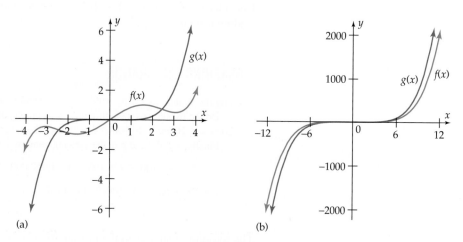

(a)   (b)

Figure 4–16

Notice how the first window shows that these are very different functions, but the second window shows that as $|x|$ gets larger, the graph of $\dfrac{x^5}{120} - \dfrac{x^3}{6} + x$ closely resembles the graph of $\dfrac{x^5}{120}$. If you were to graph these functions on the interval $-1000 \le x \le 1000$, you would be hard pressed to tell the difference between these curves. ∎

⬤ x-INTERCEPTS

As we saw in Section 4.3, the x-intercepts of the graph of a polynomial function are the real roots of the polynomial. Since a polynomial of degree n has at most n distinct roots (page 257), we have the following fact.

x-Intercepts

> The graph of a polynomial function of degree n meets the x-axis at most n times.

There is another connection between roots and graphs. For example, it is easy to see that the roots of

$$f(x) = (x + 3)^2(x + 1)(x - 1)^3$$

are -3, -1, and 1. We say that

$$-3 \text{ is a root of multiplicity 2;}$$

$$-1 \text{ is a root of multiplicity 1;}$$

$$1 \text{ is a root of multiplicity 3.}$$

Observe that the graph of $f(x)$ in Figure 4–17 does not cross the x-axis at -3 (a root whose multiplicity is an *even* number) but does cross the x-axis at -1 and 1 (roots of *odd* multiplicity).

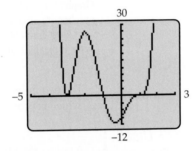

Figure 4–17

More generally, a number c is a **root of multiplicity k** of a polynomial $f(x)$ if $(x - c)^k$ is a factor of $f(x)$ and no higher power of $(x - c)$ is a factor, and we have this fact.

Multiplicity and Graphs

> Let c be a root of multiplicity k of a polynomial function f.
>
> If k is odd, the graph of f crosses the x-axis at c.
>
> If k is even, the graph of f touches, but does not cross, the x-axis at c.

EXAMPLE 3

It is a fact that $-x^4 + 2x^3 + 3x^2 - 4x - 4 = -(x + 1)^2 (x - 2)^2$. Use this fact to sketch the graph of

$$f(x) = -x^4 + 2x^3 + 3x^2 - 4x - 4$$

SOLUTION f has double roots at $x = -1$ and $x = 2$. The leading coefficient is negative and even, which tells us that both ends of the graph shoot downward. We find the y-intercept by observing that $f(0) = -4$. We obtain the graph shown in Figure 4–18 on the next page. ■

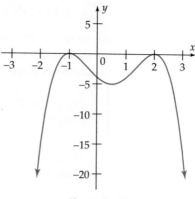

Figure 4–18

LOCAL EXTREMA

The term **local extremum** (plural, extrema) refers to either a local maximum or a local minimum, that is, a point where the graph is a peak or a valley.

GRAPHING EXPLORATION

Graph $f(x) = x^3 + 2x^2 - 4x - 3$ in the standard viewing window. What is the total number of peaks and valleys on the graph? What is the degree of $f(x)$?

Now graph $g(x) = x^4 - 3x^3 - 2x^2 + 4x + 5$ in the standard viewing window. What is the total number of peaks and valleys on the graph? What is the degree of $g(x)$?

The two polynomials you have just graphed are illustrations of the following fact, which is proved in calculus.

Local Extrema

A polynomial function of degree n has at most $n - 1$ local extrema. In other words, the total number of peaks and valleys on the graph is at most $n - 1$.

BENDING

A polynomial graph may bend upward or downward as indicated here by the vertical arrows (Figure 4–19):

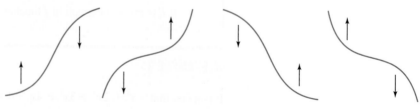

Figure 4–19

A point at which the graph changes from bending downward to bending upward (or vice versa) is called a **point of inflection.** The direction in which a graph bends may not always be clear on a calculator screen, and calculus is usually required to determine the exact location of points of inflection. The number of inflection points and hence the amount of bending in the graph are governed by these facts, which are proved in calculus.

Points
of Inflection

The graph of a polynomial function of degree n (with $n \geq 2$) has at most $n - 2$ points of inflection.

The graph of a polynomial function of odd degree n (with $n \geq 3$) has at least one point of inflection.

Thus, the graph of a quadratic function (degree 2) has no points of inflection ($n - 2 = 2 - 2 = 0$), and the graph of a cubic has exactly one (since it has at least one and at most $3 - 2 = 1$). Figure 4–20 shows several cubic graphs, with their inflection point marked.

TECHNOLOGY TIP

Points of inflection may be found by using INFLC in the TI-86/89 GRAPH MATH menu.

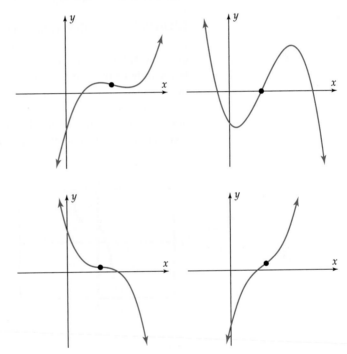

Figure 4–20

COMPLETE GRAPHS OF POLYNOMIAL FUNCTIONS

By using the facts discussed earlier, you can often determine whether or not the graph of a polynomial function is complete (that is, shows all the important features).

EXAMPLE 4

Find a complete graph of

$$f(x) = x^4 + 10x^3 + 21x^2 - 40x - 80.$$

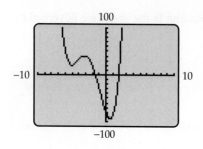

Figure 4–21

SOLUTION Since $f(0) = -80$, the standard viewing window probably won't show a complete graph, so we try the window with

$$-10 \le x \le 10 \qquad \text{and} \qquad -100 \le y \le 100$$

and obtain Figure 4–21. The three peaks and valleys shown here are the only ones because a fourth-degree polynomial graph has at most three local extrema. There cannot be more x-intercepts than the two shown here because if the graph turned toward the x-axis farther out, there would be an additional peak, which is impossible. Finally, the outer ends of the graph resemble the graph of x^4, the highest-degree term (see the chart on page 270). Hence, Figure 4–21 includes all the important features of the graph and is therefore complete. ∎

EXAMPLE 5

Find a complete graph of $f(x) = x^3 - 1.8x^2 + x + 2$.

SOLUTION We first try the standard window (Figure 4–22). The graph is similar to the graph of the leading term $y = x^3$ but does not appear to have any local extrema. However, if you use the trace feature on the flat portion of the graph to the right of the x-axis, you see that the y-coordinates increase, then decrease, then increase (try it!). Zooming in on the portion of the graph between 0 and 1 (Figure 4–22), we see that the graph actually has a tiny peak and valley (the maximum possible number of local extrema for a cubic). So Figures 4–22 and 4–23 together provide a complete graph of f. ∎

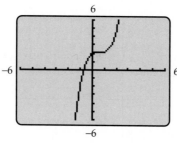

Figure 4–22

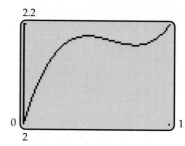

Figure 4–23

The following fact is proved in Exercise 59.

No nonconstant polynomial graph contains any horizontal line segments.

However, a calculator may erroneously show horizontal segments, as in Figure 4–22. So always investigate such segments, by using trace or zoom-in, to determine any hidden behavior, such as that in Example 5.

EXAMPLE 6

The graph of $f(x) = .01x^5 + x^4 - x^3 - 6x^2 + 5x + 4$ in the standard window is shown in Figure 4–24. Explain why this is *not* a complete graph and find a complete graph of f.

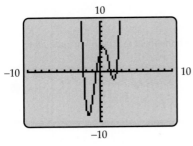

Figure 4–24

SOLUTION   When $|x|$ is large, the graph of f must resemble the graph of $y = .01x^5$, whose left end goes downward (see the chart on page 270). Since Figure 4–24 does not show this, it is not a complete graph. To have the same shape as the graph of $y = .01x^5$, the graph of f must turn downward and cross the x-axis somewhere to the left of the origin. Figure 4–24 shows three local extrema. Even without graphing, we can see that there must be one more peak (where the graph turns downward on the left), making a total of four local extrema (the most a fifth-degree polynomial can have), and another x-intercept, for a total of five. When these additional features are shown, we will have a complete graph.

GRAPHING EXPLORATION

Find a viewing window that includes the local maximum and x-intercept not shown in Figure 4–24. When you do, the scale will be such that the local extrema and x-intercepts shown in Figure 4–24 will no longer be visible.

Consequently, a complete graph of $f(x)$ requires several viewing windows in order to see all the important features.   ■

The graphs obtained in Examples 4–6 were known to be complete because in each case, they included the maximum possible number of local extrema. In many cases, however, a graph may not have the largest possible number of peaks and valleys. In such cases, use any available information and try several viewing windows to obtain the most likely complete graph.

 APPLICATIONS

The solution of many applied problems reduces to finding a local extremum of a polynomial function.

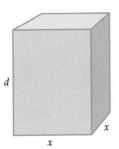

Figure 4–25

EXAMPLE 7

A rectangular box with a square base (Figure 4–25) is to be mailed. The sum of the height of the box and the perimeter of the base is to be 84 inches, the maximum allowable under postal regulations. What are the dimensions of the box with largest possible volume that meets these conditions?

SOLUTION   If the length of one side of the base is x, then the perimeter of the base (the sum of the length of its four sides) is $4x$. If the height of the box is d, then $4x + d = 84$, so $d = 84 - 4x$, and hence, the volume is

$$V = x \cdot x \cdot d = x \cdot x \cdot (84 - 4x) = 84x^2 - 4x^3.$$

The graph of the polynomial function $V(x) = 84x^2 - 4x^3$ in Figure 4–26 is complete (why?). However, the only relevant part of the graph in this situation is the portion with x and $V(x)$ positive (because x is a length and $V(x)$ is a volume). The graph of $V(x)$ has a local maximum between 10 and 20, and this local maximum value is the largest possible volume for the box.

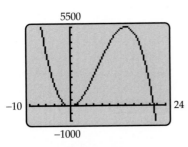

Figure 4–26

Use a maximum finder to find the x-value at which the local maximum occurs. State the dimensions of the box in this case.

EXAMPLE 8

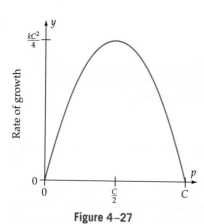

Figure 4–27

Assume we let some box-elder bugs loose in a neighborhood, and want to model their population growth. The "carrying capacity" of an environment is the number of box-elder bugs that it can support. According to one model, the rate at which a population grows is proportional to its size, and to the difference between that size and the environment's carrying capacity. We can write this model as an equation:

$$R = kP(C - P)$$

where R is the rate of growth, P is the population, C is the carrying capacity, and k is a constant of proportionality. If we multiply the equation out we get a polynomial, where P is our variable:

$$R = (-k)P^2 + (kC)P$$

The graph of this polynomial is given in Figure 4–27. If we wish to graph this equation on a calculator, we would have to choose sample values for C and k.

EXERCISES 4.4

In Exercises 1–6, decide whether the given graph could possibly be the graph of a polynomial function.

1.

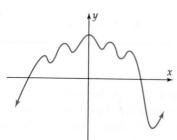

2.

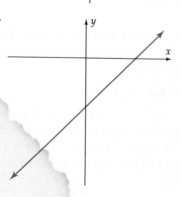

3.

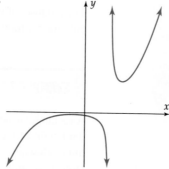

4.

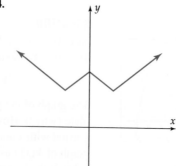

5.

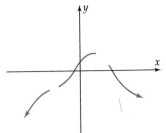

6.

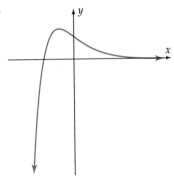

10.

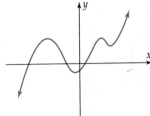

11.

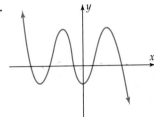

12.

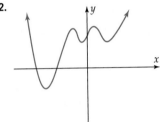

In Exercises 7–12, determine whether the given graph could possibly be the graph of a polynomial function of degree 3, of degree 4, or of degree 5.

7.

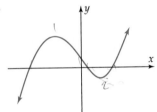

8.

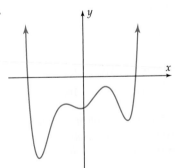

9.

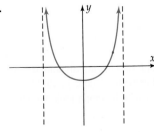

In Exercises 13 and 14, find a viewing window in which the graph of the given polynomial function f appears to have the same general shape as the graph of its leading term.

13. $f(x) = x^4 - 6x^3 + 9x^2 - 3$

14. $f(x) = x^3 - 5x^2 + 4x - 2$

In Exercises 15–18, a complete graph of a polynomial function is shown. List each root of the polynomial and state whether its multiplicity is even or odd.

15.

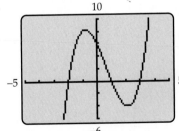

16.

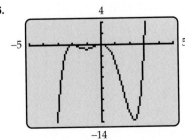

17.

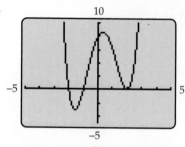

18.

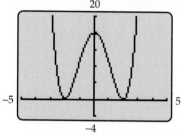

In Exercises 19–24, use your knowledge of polynomial graphs, not a calculator, to match the given function with its graph, which is one of (a)–(f).

(a)

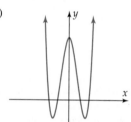

(b)

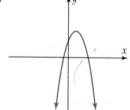

(c)

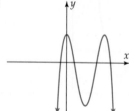

(d)

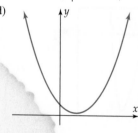

(e)

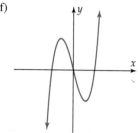

(f)

19. $f(x) = 2x - 3$

20. $g(x) = x^2 - 4x + 7$

21. $g(x) = x^3 - 4x$

22. $f(x) = x^4 - 5x^2 + 4$

23. $f(x) = -x^4 + 6x^3 - 9x^2 + 2$

24. $g(x) = -2x^2 + 3x + 1$

In Exercises 25–28, graph the function in the standard viewing window and explain why that graph cannot possibly be complete.

25. $f(x) = .01x^3 - .2x^2 - .4x + 7$

26. $g(x) = .01x^4 + .1x^3 - .8x^2 - .7x + 9$

27. $h(x) = .005x^4 - x^2 + 5$

28. $f(x) = .001x^5 - .01x^4 - .2x^3 + x^2 + x - 5$

In Exercises 29–32 find a single viewing window that shows a complete graph of the function.

29. $f(x) = x^3 + 8x^2 + 20x - 15$

30. $f(x) = 10x^3 - 12x^2 + 2x$

31. $f(x) = 10x^4 - 80x^3 + 239x^2 - 316x + 155$

32. $f(x) = 10x^4 - 80x^3 + 241x^2 - 324x + 163$

In Exercises 33–36, find a complete graph of the function and list the viewing window(s) that show this graph. (It may not be possible to obtain a complete graph in a single window.)

33. $f(x) = .1x^5 + 3x^4 - 4x^3 - 11x^2 + 3x + 5$

34. $f(x) = x^4 - 48x^3 - 101x^2 + 49x + 50$

35. $f(x) = .03x^3 - 1.5\,x^2 - 200x$

36. $f(x) = .3x^5 + 2x^4 - 7x^3 + 2x^2$

37. (a) Explain why the graph of a cubic polynomial function has either two local extrema or none at all. [*Hint:* If it had only one, what would the graph look like when $|x|$ is very large?]

(b) Explain why the general shape of the graph of a cubic polynomial function must be one of the following:

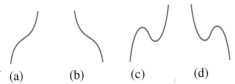

(a) (b) (c) (d)

38. The figure shows an incomplete graph of a fourth-degree even polynomial function f. (Even functions were defined in Special Topics 3.4.A.)

(a) Find the roots of f.
(b) Draw a complete graph of f.
(c) Explain why
$$f(x) = k(x - a)(x - b)(x - c)(x - d),$$
where a, b, c, d are the roots of f.
(d) Experiment with your calculator to find the value of k that produces the graph in the figure.
(e) List the approximate intervals on which f is increasing and those on which it is decreasing.

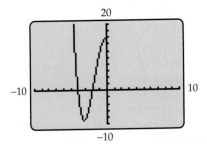

39. A complete graph of a polynomial function g is shown below.

(a) Is the degree of $g(x)$ even or odd?
(b) Is the leading coefficient of $g(x)$ positive or negative?
(c) What are the real roots of $g(x)$?
(d) What is the smallest possible degree of $g(x)$?

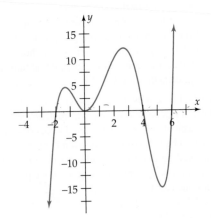

40. Do Exercise 39 for the polynomial function g whose complete graph is shown here.

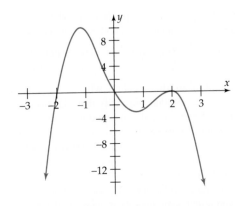

41. f is a third degree polynomial function whose leading coefficient is negative. Gordon graphs the function on his calculator, without being careful about choosing a window, and gets the plot shown below. Which of the patterns shown in Exercise 37 does this graph have?

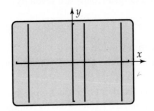

42. f is a fourth degree polynomial function. Madison graphs the function on her calculator, without being careful about choosing a window, and gets the plot shown below. Sketch the general shape of the graph and state whether the leading coefficient is positive or negative.

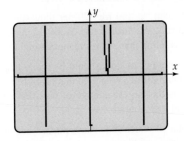

In Exercises 43–48, sketch a complete graph of the function. Label each x-intercept and the coordinates of each local extremum; find intercepts and coordinates exactly when possible and otherwise approximate them.

43. $f(x) = -x^3 + 3x^2 - 2$
44. $f(x) = .25x^4 + 2x^3 + 4x^2$
45. $f(x) = x^4 - 9x^3 + 30x^2 - 44x + 24$
46. $f(x) = 3x^3 - 18.5x^2 - 4.5x - 45$
47. $f(x) = x^5 - 3x^3 + x$
48. $f(x) = x^6 - 3x^3 + \dfrac{9}{4}$

49. The sales $f(x)$ of a certain product (in dollars) are related to the amount x (in thousands of dollars) spent on advertising by

$$f(x) = -3x^3 + 135x^2 + 3600x + 12{,}000$$

$$(0 \le x \le 40).$$

(a) Graph f in the window with $0 \le x \le 40$ and $0 \le y \le 180{,}000$ and verify that f is concave upward near the origin and concave downward near $x = 40$.

(b) Compute the average rate of change of $f(x)$ from $x = 0$ to $x = 15$ and from $x = 15$ to $x = 40$. What do these numbers tell you about the rate at which sales are increasing in each interval?

(c) This function has an inflection point at $x = 15$ (a fact you might want to verify if your calculator can find inflection points). Use the results of part (b) to explain why the inflection point is sometimes called the **point of diminishing returns.**

50. The profits (in thousands of dollars) from producing x hundred thousand tungsten darts are given by

$$g(x) = -x^3 + 27x^2 + 20x - 60 \qquad (0 \le x \le 20).$$

(a) Graph g in a window with $0 \le x \le 20$. If you have an appropriate calculator, verify that there is a point of inflection when $x = 9$.

(b) Verify that the point of inflection is the point of diminishing returns (see Exercise 49) by computing the average rate of change of profit from $x = 0$ to $x = 9$ and from $x = 9$ to $x = 20$.

51. When there are 22 apple trees per acre, the average yield has been found to be 500 apples per tree. For each additional tree planted per acre, the yield per tree decreases by 15 apples per tree. How many additional trees per acre should be planted to maximize the yield?

52. Name tags can be sold for $29 per thousand. The cost of manufacturing x thousand tags is $.001x^3 + .06x^2 - 1.5x$ dollars. Assuming that all tags manufactured are sold,

(a) What number of tags should be made to guarantee a maximum profit? What will that profit be?

(b) What is the largest number of tags that can be made without losing money?

53. The top of a 12-ounce can of soda pop is three times thicker than the sides and bottom (so that the flip-top opener will work properly), and the can has a volume of 355 cubic centimeters. What should the radius and height of the can be in order to use the least possible amount of metal? [Assume that the entire can is made from a single sheet of metal, with three layers being used for the top. Example 4 in Section 2.4 may be helpful.]

54. An open-top reinforced box is to be made from a 12-by-36-inch piece of cardboard as in Exercise 45 of Section 4.3. What size squares should be cut from the corners in order to have a box with maximum volume?

In calculus, you will learn that many complicated functions can be approximated by polynomials. For Exercises 55–58, use a

*calculator to graph the function and the polynomial on the same axes in the given window, and determine where the polynomial is a good approximation.**

55. $f(x) = \sin(x)$, $p(x) = -\dfrac{1}{5040}x^7 + \dfrac{1}{120}x^5 - \dfrac{1}{6}x^3 + x$,

$-6 \le x \le 6$, $-4 \le y \le -4$

56. $f(x) = \dfrac{1}{1-x}$, $p(x) = x^7 + x^6 + x^5 + x^4 + x^3 + x^2 + x + 1$,

$-3 \le x \le 3$, $-6 \le y \le 6$

57. (a) $f(x) = e^x$, $p(x) = \dfrac{1}{5040}x^7 + \dfrac{1}{720}x^6 + \dfrac{1}{120}x^5 +$

$\dfrac{1}{24}x^4 + \dfrac{1}{6}x^3 + \dfrac{1}{2}x^2 + x + 1$, $-6 \le x \le 6$, $-3 \le y \le 10$

(b) $f(x) = e^x$, $p(x) = \dfrac{1}{5040}x^7 + \dfrac{1}{720}x^6 + \dfrac{1}{120}x^5 + \dfrac{1}{24}x^4 +$

$\dfrac{1}{6}x^3 + \dfrac{1}{2}x^2 + x + 1$, $-6 \le x \le 6$, $-50 \le y \le 400$

[*Hint:* On some calculators, the "e^x" key is labelled "exp(x)."]

58. $f(x) = \tan(x)$, $p(x) = \dfrac{17}{315}x^7 + \dfrac{2}{15}x^5 + \dfrac{1}{3}x^3 + x$, $-3 \le x \le 3$,

$-10 \le y \le 10$

THINKERS

59. (a) Graph $g(x) = .01x^3 - .06x^2 + .12x + 3.92$ in the viewing window with $-3 \le x \le 3$ and $0 \le y \le 6$ and verify that the graph appears to coincide with the horizontal line $y = 4$ between $x = 1$ and $x = 3$. In other words, it appears that every x with $1 \le x \le 3$ is a solution of the equation

$$.01x^3 - .06x^2 + .12x + 3.92 = 4.$$

Explain why this is impossible. Conclude that the actual graph is not horizontal between $x = 1$ and $x = 3$.

(b) Use the trace feature to verify that the graph is actually rising from left to right between $x = 1$ and $x = 3$. Find a viewing window that shows this.

(c) Show that it is not possible for the graph of a polynomial $f(x)$ to contain a horizontal segment. [*Hint:* A horizontal line segment is part of the horizontal line $y = k$ for some constant k. Adapt the argument in part (a), which is the case $k = 4$.]

60. (a) Let $f(x)$ be a polynomial of odd degree. Explain why $f(x)$ must have at least one real root. [*Hint:* Why must the graph of f cross the x-axis, and what does this mean?]

(b) Let $g(x)$ be a polynomial of even degree, with a negative leading coefficient and a positive constant term. Explain why $g(x)$ must have at least one positive and at least one negative root.

_____

*You don't need to know what the SIN and e^x keys mean to do this exercise.

61. The graph of

$$f(x) = (x + 18)(x^2 - 20)(x - 2)^2(x - 10)$$

has x-intercepts at each of its roots, that is, at $x = -18$, $\pm\sqrt{20} \approx \pm 4.472$, 2, and 10. It is also true that $f(x)$ has a relative minimum at $x = 2$.

(a) Draw the x-axis and mark the roots of $f(x)$. Then use the fact that $f(x)$ has degree 6 (why?) to sketch the general shape of the graph (as was done for cubics in Exercise 37).

(b) Now graph $f(x)$ in the standard viewing window. Does the graph resemble your sketch? Does it even show all the x-intercepts between -10 and 10?

(c) Graph $f(x)$ in the viewing window with $-19 \le x \le 11$ and $-10 \le y \le 10$. Does this window include all the x-intercepts as it should?

(d) List viewing windows that give a complete graph of $f(x)$.

62. (a) Graph $f(x) = x^3 - 4x$ in the viewing window with $-3 \le x \le 3$ and $-5 \le y \le 5$.

(b) Graph the difference quotient of $f(x)$ (with $h = .01$) on the same screen.

(c) Find the x-coordinates of the relative extrema of $f(x)$. How do these numbers compare with the x-intercepts of the difference quotient?

(d) Repeat this problem with the function $f(x) = x^4 - x^2$.

4.4.A SPECIAL TOPICS   Polynomial Models*

Section Objective ■ Use regression to find polynomial models to fit real-life data.

In section 2.5 we started with a set of data points, and we found the best linear model for those points. Our method was to use the calculator to compute the "least squares regression line," the line that minimized the sum of the squares of the error terms. When the scatter plot of the data points looks more like a higher-degree polynomial graph, we use a similar procedure to find a polynomial model for the points. Most calculators have regression procedures that allow us to construct quadratic, cubic and quartic (fourth-degree) models.

EXAMPLE 1

The table shows the population of San Francisco in selected years.[†]

Year	1950	1960	1970	1980	1990	2000
Population	775,357	740,316	715,674	678,974	723,959	776,733

(a) Find a polynomial model for this data.

(b) Use the model to estimate the population of San Francisco in 1995 and 2005.

SOLUTION

(a) Let $x = 0$ correspond to 1950 and plot the data points, as in Figure 4–28 on the next page. We use quartic regression to obtain the following function:

$$f(x) = -.1072x^4 + 13.2x^3 - 387.24x^2 - 168.65x + 774,231.^{‡}$$

TECHNOLOGY TIP

The quadratic, cubic, and quartic regression commands are in the same menu as the linear regression command and are labeled as follows:

TI-84+/89: QuadReg, CubicReg, QuartReg

TI-86: P2Reg, P3Reg, P4Reg.

Casio 9850: $x\wedge 2$, $x\wedge 3$, $x\wedge 4$,

HP-39gs: quadratic, cubic

Quartic regression is not available on HP-39gs.

---

*Section 2.5 is a prerequisite for this optional section. This material will be used in the optional Section 5.5 and in clearly identified exercises but not elsewhere.

[†]U.S. Census Bureau.

[‡]Here and later, coefficients are rounded for convenient reading, but the full coefficients are used to produce the graphs and estimates.

(The procedure is the same as for linear regression; see Example 3 on page 123 and the Technology Tip in the margin of preceding page. The graph of *f* in Figure 4–29 appears to fit the data well.

(b) To estimate the population in 1995 and 2005, we evaluate *f* at *x* = 45 and *x* = 55, as shown in Figure 4–30. According to this model, the 1995 population was about 745,844 and the 2005 population is about 809,002. ∎

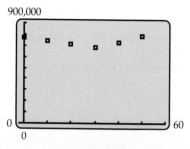

Figure 4–28

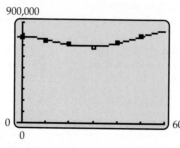

Figure 4–29

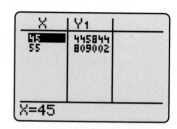

Figure 4–30

The actual population of San Francisco in 1995 was 730,628 and in 2005 was 739,426. Our estimate for 1995 in Example 1 was more accurate than the estimate for 2005. This is because, if we don't know a lot about a data set, it tends to be safer to *interpolate* (use a model to fill in data between points) than to *extrapolate* (use a model to predict data outside the data points). As we shall see later, population data tends to lend itself best to exponential models, so our polynomial model is not going to give very accurate extrapolations.

GRAPHING EXPLORATION

Use your minimum finder and the population function in Example 1 to estimate the year since 1950 when the population of San Francisco was smallest.

EXAMPLE 2

The table below gives the average price of gasoline during various years*:

Year	Gas Price ($/gal)	Year	Gas Price ($/gal)
1970	0.357	1990	1.127
1974	0.524	1994	1.075
1978	0.630	1998	1.030
1982	1.259	2002	1.341
1986	0.890	2006	2.594

Find a suitable polynomial model for this data.

SOLUTION   Letting *x* = 0 correspond to 1970 and plotting the data points we obtain the scatter plot in Figure 4–31. The points are not in a straight line but could be part of a polynomial graph of higher degree.

*From Bureau of Labor and Statistics. 2006 data averaged through part of the year.

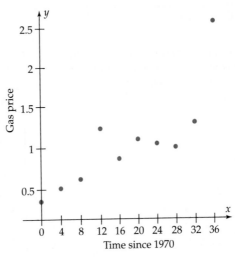

Figure 4–31

Since the data points suggest a curve that is concave upward on the left, concave downward in the middle, and concave upward again on the right, a fourth-degree polynomial might provide a reasonable model. We use the regression feature of a calculator to find this model:

$$f(x) = .00001547x^4 - .0008969x^3 + .01441x^2 - .01676x + .3672$$

The graph of f is shown in Figure 4–32.

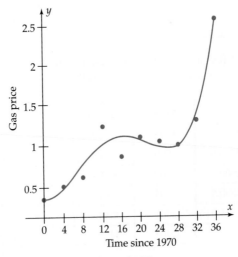

Figure 4–32

> **NOTE**
> You must have at least three data points for quadratic regression, at least four for cubic regression, and at least five for quartic regression. If you have exactly the required minimum number data points, no two of them can have the same first coordinate. In this case, the polynomial regression function will pass through all of the data points (an exact fit). When you have more than the minimum number of data points required, the fit will generally be approximate rather than exact.

Although this function provides a reasonable model from 1970–2006, our knowledge of polynomial graphs suggests that they may not be accurate in the future. You can find the today's average price of gasoline on the world-wide-web. Compare the predicted value to the actual value. ∎

EXERCISES 4.4.A

In Exercises 1–4, a scatter plot of data is shown. State the type of polynomial model that seems most appropriate for the data (linear, quadratic, cubic, or quartic). If none of them is likely to provide a reasonable model, say so.

1.

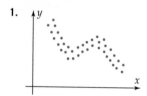

2.

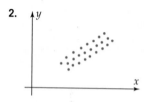

3.

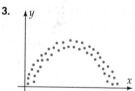

4.

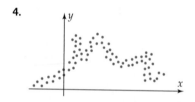

5. The table, which is based on the FBI Uniform Crime Reports, shows the rate of property crime per 100,000 population.

Year	Crimes	Year	Crimes
1992	4903.7	1999	3743.6
1993	4740.0	2000	3618.3
1994	4660.2	2002	3630.6
1995	4590.5	2003	3591.2
1996	4451.0	2004	3514.2
1997	4316.3	2005	3429.8

(a) Use cubic regression to find a polynomial function that models this data, with $x = 0$ corresponding to 1990.
(b) According to this model, what was the property crime rate in 1998 and 2001?
(c) The actual crime rates in 1998 and 2001 were 4052.5 and 3658.1 respectively. Was the model accurate?
(d) For how many years in the future is this model likely to be a reasonable one?

6. The table shows actual and projected enrollment (in millions) in public high schools in selected years.*

Year	Enrollment	Year	Enrollment
1975	14.3	1995	12.5
1980	13.2	2000	13.5
1985	12.4	2002	12.8
1990	11.3	2014	14.9

(a) Use quartic regression to find a polynomial function that models this data, with $x = 0$ corresponding to 1975.
(b) According to the model, what was enrollment in 1998 and in 1999?
(c) Estimate the year between 1975 and 2000 in which enrollment was the lowest. Does this estimate appear to be accurate?

7. The table shows the air temperature at various times during a spring day in Gainesville, Florida.

Time	Temp (F°)	Time	Temp (F°)
6 A.M.	52	1 P.M.	82
7 A.M.	56	2 P.M.	86
8 A.M.	61	3 P.M.	85
9 A.M.	67	4 P.M.	83
10 A.M.	72	5 P.M.	78
11 A.M.	77	6 P.M.	72
noon	80		

(a) Sketch a scatter plot of the data, with $x = 0$ corresponding to midnight.
(b) Find a quadratic polynomial model for the data.
(c) What is the predicted temperature for noon? For 9 A.M.? For 2 P.M.?

8. Right-fielder Bobby Abreu is an excellent batter. The following table shows his batting average for various years:

Year	Batting Average	Year	Batting Average
1996	.227	2001	.289
1997	.250	2002	.308
1998	.312	2003	
1999	.335	2004	.301
2000	.316	2005	.286

*U.S. National Center for Educational Statistics.

(a) Sketch a scatter plot of the data from 1996 to 2005, with $x = 0$ corresponding to 1996.

(b) Find a good polynomial model for this data.

(c) Use your model to estimate Bobby Abreu's batting average in 2003.

(d) Keeping in mind that baseball players' batting averages tend to decline as they get older, and that they tend to stay above .100, how accurate do you think your model would be if he were still playing baseball in 2016? Why?

*Use the following table in Exercises 9 and 10. It shows the median income of U.S. households in constant 2004 dollars.**

Year	Median Income	Year	Median Income
1990	$42,086	1998	$43,659
1992	$41,043	2000	$45,730
1994	$40,439	2002	$45,222
1996	$41,722	2004	$44,473

9. (a) Sketch a scatter plot of the data from 1990 to 2004, with $x = 0$ corresponding to 1990 and y measured in thousands.

(b) Decide whether a quadratic or a cubic model seems more appropriate and state its rule.

(c) Use the model to predict the median income in 2012.

(d) Does this model seem reasonable after 2012?

10. (a) Sketch a scatter plot of the data from 1994 to 2004, with $x = 0$ corresponding to 1994 and y measured in thousands.

(b) Find both cubic and quartic models for this data.

(c) Is there any significant difference between the models from 1994 to 2004? What about from 2004 to 2012?

(d) According to these models, when will the median income reach $60,000?

11. The table shows the percentage of male high school students who currently smoke cigarettes in various years.[†]

Year	Percent	Year	Percent
1991	27.6	1999	34.7
1993	29.8	2001	29.2
1995	35.4	2003	21.8
1997	37.7		

(a) Sketch a scatter plot of the data, with $x = 0$ corresponding to 1990.

(b) Find quadratic, cubic, and quartic polynomial models for the data.

(c) Which model seems to fit this data best? Which one seems most reasonable for future years.

12. The table shows the U.S. public debt per person (in dollars) in selected years.*

Year	Debit	Year	Debit
1981	$4,338	1993	$17,105
1983	$5,887	1995	$18,930
1985	$7,598	1997	$20,026
1987	$9,615	1999	$20,746
1989	$11,545	2001	$20.353
1991	$14,436	2003	$21,459

(a) Sketch a scatter plot of the data.

(b) Find a quartic polynomial model for the data.

(c) Use the model to estimate the public debt per person in 1996. How does your estimate compare with the actual figure of $19,805?

(d) Is this model likely to be accurate after 2003? Why?

13. (a) Find both a cubic and a quartic model for the data on the number of unemployed people in the labor force in Example 6 of Section 2.5.

(b) Does either model seem likely to be accurate in the future?

14. The table shows the total advertising expenditures in the United States (in billions of dollars) in selected years.[†]

Year	Expenditures
1990	129.59
1992	132.65
1994	151.68
1996	175.23
1998	201.59
2000	236.33

(a) Sketch a scatter plot of the dat͏ ding to 1990.

(b) Find a quadr͏

(c) Use t͏ 2002.

(d) If this ͏ reach $3͏

*U.S. Census Bureau.
[†]Youth at Risk Behavior Survey.

*U.S. Bureau of Publi͏
[†]*Statistical Abstract of* ͏

4.5   Rational Functions

Section Objectives

- ■ Find the domain of a rational function.
- ■ Find the asymptotes of a linear rational function.
- ■ Analyze the graph of a rational function algebraically and graphically.
- ■ Use rational functions to solve applied problems.

A **rational function** is a function whose rule is the quotient of two polynomials, such as

$$f(x) = \frac{1}{x}, \qquad t(x) = \frac{4x - 3}{2x + 1}, \qquad k(x) = \frac{2x^3 + 5x + 2}{x^2 - 7x + 6}.$$

A polynomial function is defined for every real number, but the rational function $f(x) = g(x)/h(x)$ is defined only when its denominator is nonzero. Hence,

Domain

> The domain of the rational function $f(x) = \dfrac{g(x)}{h(x)}$ is the set of all real numbers that are *not* roots of the denominator $h(x)$.

For instance, the domain of

$$f(x) = \frac{x^2 + 3x + 1}{x^2 - x - 6}$$

can be found by determining the roots of the denominator. It factors as

$$x^2 - x - 6 = (x + 2)(x - 3),$$

so its roots are -2 and 3. Hence, the domain of f is the set of all real numbers except -2 and 3.

The key to understanding the behavior of rational numbers is the following fact from arithmetic.

The Big-Little Principle

> If c is a number far from 0, then $1/c$ is a number close to 0. Conversely, if c is close to 0, then $1/c$ is far from 0. In less precise but more suggestive terms:
>
> $$\frac{1}{\text{big}} = \text{little} \qquad \text{and} \qquad \frac{1}{\text{little}} = \text{big.}$$

For example, 5000 is big (far from 0), and 1/5000 is little (close to 0). Similarly, $-1/1000$ is very close to 0, but

$$\frac{1}{-1/1000} = -1000$$

is far from 0. To see the role played by the Big-Little Principle, we examine two rational functions that are part of the catalog of basic functions.

To see this hidden behavior, graph both f and the line $y = 2$ in the viewing window with $1 \le x \le 50$ and $1.7 \le y \le 2.1$.

This Exploration shows that the graph has a local minimum near $x = 4$ and then stays below the asymptote, moving closer and closer to it as x takes larger values. ∎

APPLICATIONS

Several applications of rational functions were considered in Section 2.4. Here is another one.

EXAMPLE 7

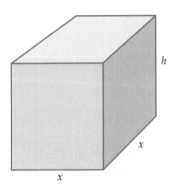

A cardboard box with a square base and a volume of 1000 cubic inches is to be constructed (Figure 4–45). The box must be at least 2 inches in height.

(a) What are the possible lengths for a side of the base if no more than 1100 square inches of cardboard can be used to construct the box?

(b) What is the least possible amount of cardboard that can be used?

(c) What are the dimensions of the box that uses the least possible amount of cardboard?

Figure 4–45

SOLUTION   The amount of cardboard needed is given by the surface area S of the box. From Figure 4–45, we have

$$S = \underset{top}{area\ of} + \underset{bottom}{area\ of} + \underset{side}{area\ of\ each}$$

$$S = x^2 + x^2 + xh + xh + xh + xh = 2x^2 + 4xh.$$

Since the volume of the box is given by

$$\text{Length} \times \text{Width} \times \text{Height} = \text{Volume},$$

we have

$$x \cdot x \cdot h = 1000 \qquad \text{or, equivalently,} \qquad h = \frac{1000}{x^2}.$$

Substituting the above into the surface area formula allows us to express the surface area as a function of one variable, x:

$$S(x) = 2x^2 + 4xh = 2x^2 + 4x\left(\frac{1000}{x^2}\right) = 2x^2 + \frac{4000}{x} = \frac{2x^3 + 4000}{x}.$$

Although the rational function $S(x)$ is defined for all nonzero real numbers, x is a length here and must be positive. Furthermore, $x^2 \le 500$ because if $x^2 > 500$, then $h = \dfrac{1000}{x^2}$ would be less than 2, contrary to specifications. Hence, the only values of x that make sense in this context are those with $0 < x \le \sqrt{500}$. Since $\sqrt{500} \approx 22.4$, we choose the viewing window in Figure 4–46. For each point

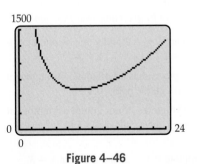

Figure 4–46

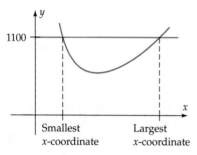

Figure 4–47

(x, y) on the graph, x is a possible side length for the base of the box, and y is the corresponding surface area.

(a) The points on the graph corresponding to the requirement that no more than 1100 square inches of cardboard be used are those whose y-coordinates are less than or equal to 1100. The x-coordinates of these points are the possible side lengths. The x-coordinates of the points where the graph of S meets the horizontal line y = 1100 are the smallest and largest possible values for x, as indicated in Figure 4–47.

GRAPHING EXPLORATION

Graph S(x) and y = 1100 on the same screen. Use an intersection finder to show that the possible side lengths that use no more than 1100 square inches of cardboard are those with 3.73 ≤ x ≤ 21.36.

(b) The least possible amount of cardboard corresponds to the point on the graph of S(x) with the smallest y-coordinate.

GRAPHING EXPLORATION

Show that the graph of S has a local minimum at the point (10.00, 600.00). Consequently, the least possible amount of cardboard is 600 square inches and this occurs when x = 10.

(c) When x = 10, h = $1000/10^2$ = 10. So the dimensions of the box using the least amount of cardboard are 10 × 10 × 10. ∎

EXERCISES 4.5

In Exercises 1–6, find the domain of the function. You may need to use some of the techniques of Section 4.3

1. $f(x) = \dfrac{2x}{3x - 4}$

2. $g(x) = \dfrac{x + 1}{2x^2 - x - 3}$

3. $h(x) = \dfrac{x^2 + 4}{x^2 + 9}$

4. $i(x) = \dfrac{x^4 + 2x^3}{x^5 - 81x}$

5. $j(x) = \dfrac{x}{\pi x^3 + \pi x^2 - 9\pi x - 9\pi}$

6. $k(x) = \dfrac{2}{x^5 + 4x^4 - 4}$

In Exercises 7–10, find equations of graphs with the given properties. Check your answer by graphing your function.

7. f has vertical asymptotes at x = 3 and x = −3, and a horizontal asymptote at y = 2

8. f has no vertical asymptotes, has a horizontal asymptote at the x axis, and goes through the point (0,2)

9. f has four vertical asymptotes, a horizontal asymptote at y = −1, goes through the point (0,4) and is an even function.

10. f has a vertical asymptote at x = 2, and a hole at x = 3.

In Exercises 11–14, use the graphs in Example 1 and the information in Section 3.4 to match the function with its graph, which is one of those shown here.

A.

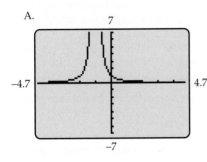

B.

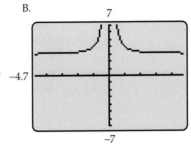

C.

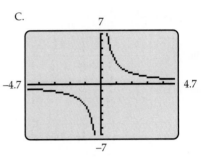

D.

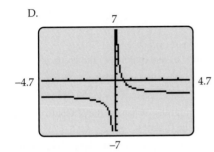

E.

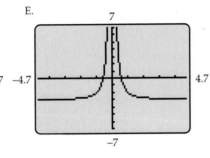

F.

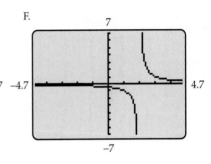

11. $f(x) = \dfrac{1}{x - 2}$

12. $g(x) = \dfrac{3}{x}$

13. $h(x) = \dfrac{1}{(x + 1)^2}$

14. $f(x) = \dfrac{1}{x^2} - 3$

In Exercises 15–18, use algebra to determine the location of the vertical asymptotes and holes in the graph of the function.

15. $f(x) = \dfrac{x^3 + 6x^2 + 11x + 6}{x^3 - x}$

16. $g(x) = \dfrac{x^2}{x^4 - x^2}$

17. $f(x) = \dfrac{x^2 + 8x + 2}{x^2 + 7x + 2}$

18. $g(x) = \dfrac{x^3 + 5x^2 + 8x + 4}{x^3 + 4x^2 + 5x + 2}$

In Exercises 19–24, find the horizontal asymptote, if any, of the graph of the given function. If there is a horizontal asymptote, find a viewing window in which the ends of the graph are within .1 of this asymptote.

19. $f(x) = \dfrac{x + 1}{x^6 + 20}$

20. $g(x) = \dfrac{3x^5 + 2x^4 + 1}{6x^5 + 8x^4 - 3x^2 + 2x + 1}$

21. $a(x) = \dfrac{x^5 - x^2 + x}{x^4 - 2x + 3}$

22. $m(x) = \dfrac{x^2 + 2x + 1}{x - 6}$

23. $f(x) = \dfrac{2x^3 + 4x^2 + 2x + 1}{3x^3 - 4x^2 - 2x}$

24. $r(x) = \dfrac{x^4 - 2x + 3}{x^5 - x^2 + x}$

In Exercises 25–36, analyze the function algebraically. List its vertical asymptotes, holes, y-intercept, and horizontal asymptote, if any. Then sketch a complete graph of the function.

25. $f(x) = \dfrac{1}{x - 2}$

26. $f(x) = \dfrac{-7}{x - 6}$

27. $f(x) = \dfrac{2x}{x + 1}$

28. $f(x) = \dfrac{2x - 3}{2x}$

29. $f(x) = \dfrac{x}{x(x - 2)(x - 3)}$

30. $f(x) = \dfrac{5x^2}{(x + 2)(x - 3)}$

31. $f(x) = \dfrac{2}{x^2 + 1}$

32. $f(x) = \dfrac{5}{(x + 1)^2(x - 4)}$

33. $f(x) = \dfrac{(x^2 + 6x + 5)(x + 5)}{(x + 5)^3(x - 1)}$

34. $f(x) = \dfrac{x^3 + 2x^2 - x - 2}{x^2 - x - 12}$

35. $f(x) = \dfrac{2x^3 + 3x^2 - 3x - 2}{x^3 + x^2 - 4x - 4}$

36. $f(x) = \dfrac{(x - 1)(x - 2)(x - 3)}{(4x + 1)}$

In Exercises 37–42, find a viewing window, or windows, that shows a complete graph of the function. Be alert for hidden behavior, such as that in Example 6.

37. $f(x) = \dfrac{x^3 + 4x^2 - 5x}{(x^2 - 4)(x^2 - 9)}$

38. $g(x) = \dfrac{x^4 + 2x^3 - 13x^2 + 10x}{x + 7}$

39. $h(x) = \dfrac{2x^2 - x - 6}{x^3 + x^2 - 6x}$

40. $f(x) = \dfrac{x^3 - x + 1}{x^4 - 2x^3 - 2x^2 + x - 1}$

41. $g(x) = \dfrac{x - 2}{x^3 - 11x^2 - x + 11}$

42. $h(x) = \dfrac{x^2 - 9}{x^3 + 2x^2 - 23x - 60}$

In Exercises 43–48, find and simplify the difference quotient of the function. [See Sections 3.2 and 3.6]

43. $f(x) = \dfrac{1}{x}$

44. $g(x) = \dfrac{2}{3x}$

45. $f(x) = \dfrac{3}{x - 2}$

46. $h(x) = \dfrac{1}{x^2}$

47. $g(x) = \dfrac{3}{x^2}$

48. $f(x) = \dfrac{m}{nx^2}$

49. (a) Use the difference quotient in Exercise 43 to determine the average rate of change of $f(x) = 1/x$ as x changes from 2 to 2.1, from 2 to 2.01, and from 2 to 2.001. Estimate the instantaneous rate of change of f at $x = 2$.

    (b) Determine the average rate of change of $f(x) = 1/x$ as x changes from 3 to 3.1, from 3 to 3.01, and from 3 to 3.001. Estimate the instantaneous rate of change of f at $x = 3$.

    (c) How are the instantaneous rates of change of f at $x = 2$ and $x = 3$ related to the values of the function $g(x) = -1/x^2$ at $x = 2$ and $x = 3$?

50. One way to limit current in a circuit is to add a "resistor." Resistance is measured in Ohms, and can never be negative. If two resistors are wired "in series" the total resistance is simply the sum $R_1 + R_2$.

It is more interesting if we wire them "in parallel."

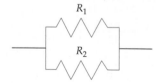

In that case we get: Total Resistance $= \dfrac{1}{\dfrac{1}{R_1} + \dfrac{1}{R_2}}$.

Assume that two resistors are wired in parallel and that R_1 is 5 Ohms.

(a) Write the total resistance as a rational function of R_2.

(b) If we allow R_2 to get larger and larger, the total resistance approaches a value. Compute that value.

(c) Is the total resistance defined if $R_2 = 0$?

(d) What happens to the total resistance if R_2 gets closer and closer to zero?

(e) Sketch a graph of total resistance vs. R_2.

51. (a) When $x \geq 0$, what rational function has the same graph as

$$f(x) = \frac{x - 1}{|x| - 2}?$$

[*Hint:* Use the definition of absolute value on page 9.]

(b) When $x < 0$, what rational function has the same graph as

$$f(x) = \frac{x - 1}{|x| - 2}?$$

[See the hint for part (a).]

(c) Use parts (a) and (b) to explain why the graph of

$$f(x) = \frac{x - 1}{|x| - 2}$$

has two vertical asymptotes. What are they? Confirm your answer by graphing the function.

52. Newton's law of gravitation states that every object in the universe attracts every other object, to some extent. If one object has mass a and the other has mass b then the force that they exert on each other is given by the equation

$F = G\dfrac{ab}{d^2}$, where d is the distance between the objects, and G is a constant called, appropriately, the Gravitational Constant. We can approximate G rather well by $G \approx 6.673 \times 10^{-11}$. So if we put a 90 kilogram person about 2 meters from a 80 kilogram person, there would be a force of

$(6.673 \times 10^{-11})\dfrac{(90)(80)}{2^2} \approx 1.201 \times 10^{-7}$ Newtons between them, or about .000000027 pounds.

(a) Use the force equation to determine what happens to the force between two objects as they get farther and farther apart.

(b) Use the force equation to determine what happens to the force between two objects as they get closer and closer together.

(c) The mass of the moon is approximately 7.36×10^{22} kilograms. The mass of the Earth is approximately 5.97×10^{24} kilograms. The distance from the moon to the Earth ranges from 357,643 km to 406,395 km. Draw a graph of the force that the Earth exerts on the moon versus their distance apart.

53. It costs 2.5 cents per square inch to make the top and bottom of the box in Example 7. The sides cost 1.5 cents per square inch. What are the dimensions of the cheapest possible box?

54. A box with a square base and a volume of 1000 cubic inches is to be constructed. The material for the top and bottom of the box costs $3 per 100 square inches, and the material for the sides costs $1.25 per 100 square inches.

(a) If x is the length of a side of the base, express the cost of constructing the box as a function of x.

(b) If the side of the base must be at least 6 inches long, for what value of x will the cost of the box be $7.50?

55. Our friend Joseph collects "action figures." In 1980, his annual action figure budget was $20, but it has gone up by $5 every year after that. The cost of action figures has risen as well. In 1980, they cost an average of $2 per figure, but that number has gone up by .25 every year after that.

(a) How many figures could Joseph buy in 1980?

(b) How many figures will he be able to buy in 2010?

(c) Write the number of figures he can buy in a given year as a function of time. Let $t = 0$ correspond to 1980.

(d) Graph the function you found in part (c). Use your graph to check your answer to part (b).

(e) Will he ever be able to buy 18? If so, when? Will he ever be able to buy 21? If so, when?

56. Pure alcohol is being added to 100 gallons of a coolant mixture that is 40% alcohol.

(a) Find the rule of the concentration function $c(x)$ that expresses the percentage of alcohol in the resulting mixture as a function of the number x of gallons of pure alcohol that are added. [*Hint:* The final mixture contains $100 + x$ gallons (why?). So $c(x)$ is the amount of alcohol in the final mixture divided by the total amount $100 + x$. How much alcohol is in the original 100-gallon mixture? How much is in the final mixture?]

(b) How many gallons of pure alcohol should be added to produce a mixture that is at least 60% alcohol and no more than 80% alcohol? Your answer will be a range of values.

(c) Determine algebraically the exact amount of pure alcohol that must be added to produce a mixture that is 70% alcohol.

57. A rectangular garden with an area of 200 square meters is to be located next to a building and fenced on three sides, with the building acting as a fence on the fourth side.

(a) If the side of the garden parallel to the building has length x meters, express the amount of fencing needed as a function of x.

(b) For what values of x will less than 60 meters of fencing be needed?

(c) What value of x will result in the least possible amount of fencing being used? What are the dimensions of the garden in this case?

58. A certain company has fixed costs of $40,000 and variable costs of $2.60 per unit.

(a) Let x be the number of units produced. Find the rule of the average cost function. [The average cost is the cost of the units divided by the number of units.]

(b) Graph the average cost function in a window with $0 \leq x \leq 100,000$ and $0 \leq y \leq 20$.

(c) Find the horizontal asymptote of the average cost function. Explain what the asymptote means in this situation. [How low can the average cost possibly be?]

59. Radioactive waste is stored in a cylindrical tank; the exterior has radius r and height h as shown in the figure. The sides, top, and bottom of the tank are 1 foot thick, and the tank has a volume of 150 cubic feet (including top, bottom, and walls).

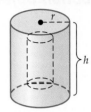

(a) Express the interior height h_1 (that is, the height of the storage area) as a function of h.

(b) Express the interior height as a function of r.

(c) Express the volume of the interior as a function of r.

(d) Explain why r must be greater than 2.

(e) What should the dimensions of the tank be in order for it to hold as much as possible?

60. The relationship between the fixed focal length F of a camera, the distance u from the object being photographed to the lens, and the distance v from the lens to the film is given by

$$\frac{1}{F} = \frac{1}{u} + \frac{1}{v}.$$

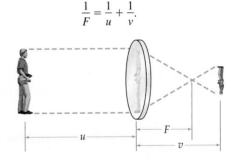

(a) If the focal length is 50 millimeters, express v as a function of u.

(b) What is the horizontal asymptote of the graph of the function in part (a)?

(c) Graph the function in part (a) when 50 millimeters $< u <$ 35,000 millimeters.

(d) When you focus the camera on an object, the distance between the lens and the film is changed. If the distance from the lens to the film changes by less than .1 millimeter, the object will remain in focus. Explain why you have more latitude in focusing on distant objects than on very close ones.

61. The formula for the gravitational acceleration (in units of meters per second squared) of an object relative to the earth is

$$g(r) = \frac{3.987 \times 10^{14}}{(6.378 \times 10^6 + r)^2},$$

where r is the distance in meters above the earth's surface.

(a) What is the gravitational acceleration at the earth's surface?

(b) Graph the function $g(r)$ for $r \geq 0$.

(c) Can you ever escape the pull of gravity? [Does the graph have any r-intercepts?]

4.5.A *SPECIAL TOPICS* Other Rational Functions

Section Objective ■ Find the graph of a rational function whose numerator has larger degree than its denominator.

We now take a closer look at the graphs of rational functions in which the degree of the numerator is larger than the degree of the denominator. We have seen that such graphs do not have horizontal asymptotes. It turns out we can use polynomial division to find out what these graphs look like when $|x|$ is very large.

EXAMPLE 1

Describe the behavior of $f(x) = \dfrac{x^3 + 3x^2 + x + 1}{x^2 + 2x - 1}$ when $|x|$ is very large.

SOLUTION   We divide the numerator of $f(x)$ by its denominator:

$$
\begin{array}{r}
x + 1 \\
x^2 + 2x - 1 \overline{\smash{\big)}\ x^3 + 3x^2 + x + 1} \\
\underline{x^3 + 2x^2 - x} \\
x^2 + 2x + 1 \\
\underline{x^2 + 2x - 1} \\
2.
\end{array}
$$

By the Division Algorithm,

$$x^3 + 3x^2 + x + 1 = (x^2 + 2x - 1)(x + 1) + 2.$$

Dividing both sides by $x^2 + 2x - 1$, we have

$$\frac{x^3 + 3x^2 + x + 1}{x^2 + 2x - 1} = \frac{(x^2 + 2x - 1)(x + 1) + 2}{x^2 + 2x - 1}$$

$$f(x) = (x + 1) + \frac{2}{x^2 + 2x - 1}.$$

Now when x is very large in absolute value, so is $x^2 + 2x - 1$. Hence, $2/(x^2 + 2x - 1)$ is very close to 0 by the Big-Little Principle, and $f(x)$ is very close to $(x + 1) + 0$. Therefore, as x gets larger in absolute value, the graph of $f(x)$ gets closer and closer to the line $y = x + 1$. An asymptote such as this (that is, a nonvertical and non horizontal straight line) is called an **oblique asymptote.**

The graph of $f(x)$ is shown in Figure 4–48. Notice that there two vertical asymptotes, corresponding to the roots of its denominator, and that as x gets very large, it gets closer and closer to the graph of $y = x + 1$, as predicted. Except near the vertical asymptotes of $f(x)$, the two graphs are virtually identical.

GRAPHING EXPLORATION

Verify the last sentence above as follows. Using the viewing window with $-20 \le x \le 20$ and $-20 \le y \le 20$, graph both $f(x)$ and $y = x + 1$ on the same screen.

■

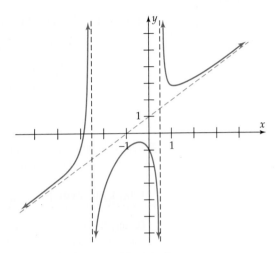

Figure 4–48

The long division process shown in Example 1 works for a general rational function. As x gets larger in absolute value, the graph of a rational function gets closer and closer to the quotient obtained when its numerator is divided by its denominator.

EXAMPLE 2

Graph

$$g(x) = \frac{x^3 + 2x^2 - 7x + 5}{x - 1}.$$

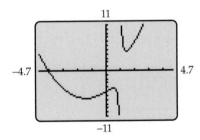

Figure 4–49

SOLUTION   We first note that there is a vertical asymptote at $x = 1$ (root of the denominator, but not the numerator). The y-intercept is at $g(0) = -5$. By carefully choosing a viewing window that accurately portrays the behavior of $g(x)$ near its vertical asymptote, we obtain Figure 4–49.

GRAPHING EXPLORATION

Verify that the x-intercept near $x = -4$ is the only one by showing graphically that the numerator of $g(x)$ has exactly one real root.

To confirm that Figure 4–49 is a complete graph, we find its asymptote when $|x|$ is large. Divide the numerator by the denominator.

$$
\require{enclose}
\begin{array}{r}
x^2 + 3x - 4 \\
x - 1 \enclose{longdiv}{x^3 + 2x^2 - 7x + 5} \\
\underline{x^3 - x^2} \\
3x^2 - 7x + 5 \\
\underline{3x^2 - 3x} \\
-4x + 5 \\
\underline{-4x + 4} \\
1
\end{array}
$$

Hence, by the Division Algorithm,

$$x^3 + 2x^2 - 7x + 5 = (x - 1)(x^2 + 3x - 4) + 1$$

$$\frac{x^3 + 2x^2 - 7x + 5}{x - 1} = \frac{(x - 1)(x^2 + 3x - 4) + 1}{x - 1}$$

$$g(x) = (x^2 + 3x - 4) + \frac{1}{x - 1}.$$

When $|x|$ is large, $1/(x - 1)$ is very close to 0 (why?), so that $y = x^2 + 3x - 4$ is the asymptote. Once again, the asymptote is given by the quotient of the division.

GRAPHING EXPLORATION

Graph $g(x)$ and $y = x^2 + 3x - 4$ on the same screen to show that the graph of $g(x)$ does get very close to the asymptote when $|x|$ is large. Then find a large enough viewing window that the two graphs appear to be identical.

EXAMPLE 3

Graph $h(x) = \dfrac{x^5 - 5x^3 + 4x + 1}{x^2 - 1}$.

SOLUTION   We first note there are vertical asymptotes at $x = 1$ and $x = -1$, and the y-intercept is at -1. (Why?) Because the numerator is larger than the denominator, we divide the numerator by the denominator to find the (non-horizontal) asymptote.

$$
\begin{array}{r}
x^3 \qquad\quad - 4x \\
x^2 - 1 \overline{\smash{\big)}\, x^5 + 0x^4 - 5x^3 + 0x^2 + 4x + 1} \\
\underline{x^5 \qquad\quad - x^3} \\
- 4x^3 \qquad\quad + 4x \\
\underline{- 4x^3 \qquad\quad + 4x} \\
1
\end{array}
$$

When $|x|$ is large, $h(x)$ will approach the quotient, $y = x^3 - 4x$, as we see in Figures 4–50 and 4–51. (Notice the calculator graph's inaccuracy near the vertical asymptotes)

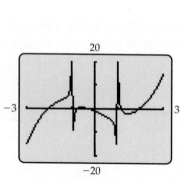

Figure 4–50

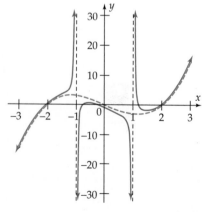

Figure 4–51

The procedures used in the preceding examples may be summarized as follows.

Graphing $f(x) = \dfrac{g(x)}{h(x)}$ *When*

Degree $g(x) >$ *Degree* $h(x)$

1. Analyze the function algebraically to determine its vertical asymptotes, holes, and intercepts.

2. Divide the numerator $g(x)$ by the denominator $h(x)$. The quotient $q(x)$ is the nonvertical asymptote of the graph, which describes the behavior of the graph when $|x|$ is large.

3. Use the preceding information to select an appropriate viewing window (or windows), to interpret the calculator's version of the graph (if necessary), and to sketch an accurate graph.

EXERCISES 4.5.A

In Exercises 1–4, find the nonvertical asymptote of the graph of the function and find a viewing window in which the ends of the graph are within .1 of this asymptote.

1. $f(x) = \dfrac{x^3 - 1}{x^2 - 4}$

2. $g(x) = \dfrac{x^3 - 4x^2 + 6x + 5}{x - 2}$

3. $f(x) = \dfrac{2x^3 - 3x^2 + 7x + 2}{x^2 - 3x + 8}$

4. $m(x) = \dfrac{x^5 + 2x^4 + 2x^3 + x^2 + 2x + 1}{x^3 + x^2}$

In Exercises 5–12, analyze the function algebraically. List its vertical asymptotes and holes, and determine its nonvertical asymptote. Then sketch a complete graph of the function.

5. $f(x) = \dfrac{x^2 - x - 6}{x - 2}$

6. $f(x) = \dfrac{x^2 + 1}{x - 1}$

7. $Q(x) = \dfrac{4x^2 - 8x - 21}{2x - 5}$

8. $K(x) = \dfrac{3x^2 - 12x + 15}{3x + 6}$

9. $f(x) = \dfrac{x^3 - 2}{x - 1}$

10. $p(x) = \dfrac{x^3 + 8}{x + 1}$

11. $q(x) = \dfrac{x^5 + 3x^4 - 11x^3 - 3x^2 + 10x + 1}{x^3 - 2x^2 - x + 2}$

12. $r(x) = \dfrac{x^3 - 4x^2 + x + 6}{x^2 - 5x + 6}$

In Exercises 13–18, find a viewing window (or windows) that shows a complete graph of the function (if possible, with no erroneous vertical line segments). Be alert for hidden behavior.

13. $f(x) = \dfrac{2x^2 + 5x + 2}{2x + 7}$

14. $g(x) = \dfrac{2x^3 + 1}{x^2 - 1}$

15. $h(x) = \dfrac{x^3 - 32x^2 + 341x - 1212}{x^2 - 20x + 99}$

16. $f(x) = \dfrac{3x^3 - 11x - 1}{x^2 - 4}$

17. $g(x) = \dfrac{2x^4 + 7x^3 + 7x^2 + 2x}{x^3 - x + 50}$

18. $h(x) = \dfrac{10x^3 + 7x^2 - 4}{x^2 + 2x - 3}$

19. (a) Show that when $0 < x < 4$, the rational function

$$r(x) = \dfrac{4096x^3 + 34{,}560x^2 + 19{,}440x + 729}{18{,}432x^2 + 34{,}560x + 5832}$$

is a good approximation of the function $s(x) = \sqrt{x}$ by graphing both functions in the viewing window with $0 \le x \le 4$ and $0 \le y \le 2$.

(b) For what values of x is $r(x)$ within .01 of $s(x)$?

20. Find a rational function f that has these properties:

(i) The curve $y = x^3 - 8$ is an asymptote of the graph of f.

(ii) $f(2) = 1$.

(iii) The line $x = 1$ is a vertical asymptote of the graph of f.

THINKERS

21. Determine the nonvertical asymptote of the following function:

$$f(x) = \dfrac{rx^3 - 2rx^2 - 10rx + bx^2 - 2bx + 1}{x^2 - 2x - 10},$$

where r and b are constants.

4.6   Polynomial and Rational Inequalities

Section Objectives
- Solve linear inequalities algebraically.
- Solve polynomial inequalities algebraically and graphically.
- Solve quadratic and factorable inequalities.
- Solve rational inequalities algebraically and graphically.

Inequalities may be solved by using algebraic or geometric methods, both of which are discussed here. Whenever possible, we shall use algebra to obtain exact solutions. When algebraic methods are too difficult, approximate graphical solutions will be found. The basic tools for working with inequalities are the following principles.

Basic Principles for Solving Inequalities

Performing any of the following operations on an inequality produces an equivalent inequality:*

1. Add or subtract the same quantity on both sides of the inequality.

2. Multiply or divide both sides of the inequality by the same *positive* quantity.

3. Multiply or divide both sides of the inequality by the same *negative* quantity and *reverse the direction of the inequality.*

Note principle 3 carefully. It says, for example, that if you multiply both sides of $-3 < 5$ by -2, the equivalent inequality is $6 > -10$ (direction of inequality is reversed).

 LINEAR INEQUALITIES

EXAMPLE 1

Solve $3x + 2 > 8$.

SOLUTION   We use the basic principles to transform the inequality into one whose solutions are obvious.

$$3x + 2 > 8$$

Subtract 2 from both sides: $\qquad 3x > 6$

Divide both sides by 3: $\qquad x > 2$

Therefore, the solutions are all real numbers greater than 2. In interval notation, we say the solutions are the numbers in the interval $(2, \infty)$. ∎

---

*Two inequalities are **equivalent** if they have the same solutions.

EXAMPLE 2

Solve $5x + 3 \le 6 + 7x$.

SOLUTION   We again use the basic principles to transform the inequality into one whose solutions are obvious.

$$5x + 3 \le 6 + 7x$$

Subtract $7x$ from both sides: $-2x + 3 \le 6$

Subtract 3 from both sides: $-2x \le 3$

Divide both sides by -2 and reverse
the direction of the inequality: $x \ge -3/2$

Therefore, the solutions are all real numbers greater than or equal to $-3/2$, that is, the interval $[-3/2, \infty)$.   ■

EXAMPLE 3

A solution of the inequality $2 \le 3x + 5 < 2x + 11$ is any number that is a solution of *both* of these inequalities:

$$2 \le 3x + 5 \qquad \text{and} \qquad 3x + 5 < 2x + 11.$$

Each of these inequalities can be solved by the methods used earlier. For the first one, we have

$$2 \le 3x + 5$$

Subtract 5 from both sides: $-3 \le 3x$

Divide both sides by 3: $-1 \le x$.

The second inequality is solved similarly:

$$3x + 5 < 2x + 11$$

Subtract 5 from both sides. $3x < 2x + 6$

Subtract $2x$ from both sides: $x < 6$.

The solutions of the original inequality are the numbers x that satisfy *both* $-1 \le x$ *and* $x < 6$, that is, all x with $-1 \le x < 6$. Thus, the solutions are the numbers in the interval $[-1, 6)$, as shown in Figure 4–52.   ■

Figure 4–52

EXAMPLE 4

When solving the inequality $4 < 3 - 5x < 18$, in which the variable appears only in the middle part, you can proceed as follows.

$$4 < 3 - 5x < 18$$

Subtract 3 from each part: $1 < -5x < 15$

Divide each part by -5 and reverse
the directions of the inequalities: $-\dfrac{1}{5} > x > -3$.

Reading this last inequality from right to left we see that

$$-3 < x < -1/5,$$

so the solutions are the numbers in the interval $(-3, -1/5)$.   ■

POLYNOMIAL INEQUALITIES

Although the basic principles play a role in the solution of nonlinear inequalities, the key to solving such inequalities is this geometric fact.

> **The graph of $y = f(x)$ lies above the x-axis exactly when**
> **$f(x) > 0$ and below the x-axis exactly when $f(x) < 0$.**

Consequently, the solutions of $f(x) > 0$ are the numbers x for which the graph of f lies above the x-axis and the solutions $f(x) < 0$ are the numbers x for which the graph of f lies below the x-axis.

EXAMPLE 5

Solve $2x^3 - 15x < x^2$.

SOLUTION   We replace the inequality by an equivalent one,

$$2x^3 - x^2 - 15x < 0,$$

and consider the graph of the function $f(x) = 2x^3 - x^2 - 15x$ (Figure 4–53). Since $f(x)$ factors as

$$f(x) = 2x^3 - x^2 - 15x = x(2x^2 - x - 15) = x(2x + 5)(x - 3),$$

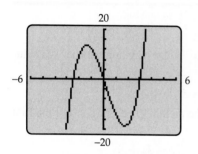

Figure 4–53

its roots (the x-intercepts of its graph) are $x = 0$, $x = -5/2$, and $x = 3$. The graph of $f(x) = 2x^3 - x^2 - 15x$ in Figure 4–53 is complete (why?) and lies below the x-axis when $x < -5/2$ or $0 < x < 3$. Therefore, the solutions of

$$2x^3 - x^2 - 15x < 0,$$

and hence of the original inequality, are all numbers x such that $x < -5/2$ or $0 < x < 3$. ∎

EXAMPLE 6

Solve $2x^3 - x^2 - 15x \geq 0$.

SOLUTION   Figure 4–53 shows that the solutions of $2x^3 - x^2 - 15x > 0$ (that is, the numbers x for which the graph of $f(x) = 2x^3 - x^2 - 15x$ lies above the x-axis) are all x such that $-5/2 < x < 0$ or $x > 3$. The solutions of the equation $2x^3 - x^2 - 15x = 0$ are the roots of $f(x) = 2x^3 - x^2 - 15x$, namely, 0, $-5/2$, and 3 as we saw in Example 5. Therefore, the solutions of the given inequality are all numbers x such that $-5/2 \leq x \leq 0$ or $x \geq 3$. ∎

When the roots of a polynomial $f(x)$ cannot be determined exactly, a root finder can be used to approximate them and to find approximate solutions of the inequalities $f(x) > 0$ and $f(x) < 0$.

EXAMPLE 7

Solve $x^4 + 10x^3 + 21x^2 + 8 > 40x + 88$.

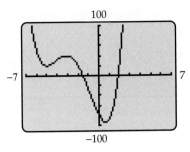

Figure 4–54

SOLUTION   This inequality is equivalent to

$$x^4 + 10x^3 + 21x^2 - 40x - 80 > 0.$$

The graph $f(x) = x^4 + 10x^3 + 21x^2 - 40x - 80$ in Figure 4–54 is complete (why?) and shows that $f(x)$ has two roots, one between -2 and -1 and the other near 2.

GRAPHING EXPLORATION

Use a root finder to show that the approximate roots of $f(x)$ are -1.53 and 1.89.

Therefore, the approximate solutions of the inequality (the numbers x for which the graph is above the x-axis) are all numbers x such that $x < -1.53$ or $x > 1.89$. ■

CAUTION

Do not attempt to write the solution in Example 7, namely, "$x < -1.53$ or $x > 1.89$" as a single inequality. If you do, the result will be a *nonsense statement* such as $-1.53 > x > 1.89$ (which says, among other things, that $-1.53 > 1.89$).

QUADRATIC AND FACTORABLE INEQUALITIES

The preceding examples show that solving a polynomial inequality depends only on knowing the roots of a polynomial and the places where its graph is above or below the x-axis. In the case of quadratic inequalities or completely factored polynomial inequalities, a calculator is not needed to determine this information.

EXAMPLE 8

The solutions of $2x^2 + 3x - 4 \le 0$ are the numbers x at which the graph of $f(x) = 2x^2 + 3x - 4$ lies on or below the x-axis. The points where the graph meets the x-axis are the roots of $f(x) = 2x^2 + 3x - 4$, which can be found by means of the quadratic formula:

$$x = \frac{-3 \pm \sqrt{3^2 - 4 \cdot 2(-4)}}{2 \cdot 2} = \frac{-3 \pm \sqrt{41}}{4}.$$

From Section 4.1, we know that the graph of $f(x)$ is an upward-opening parabola, so the graph must have the general shape shown in Figure 4–55.

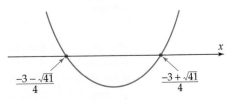

Figure 4–55

The graph lies below the x-axis between the two roots. Therefore, the solutions of the original inequality are all numbers x such that

$$\frac{-3 - \sqrt{41}}{4} \le x \le \frac{-3 + \sqrt{41}}{4}. \qquad \blacksquare$$

EXAMPLE 9

Solve $(x + 15)(x - 2)^6(x - 10) \le 0$.

SOLUTION   The roots of $f(x) = (x + 15)(x - 2)^6(x - 10)$ are easily read from the factored form: -15, 2, and 10. So we need only determine where the graph of $f(x)$ is on or below the x-axis. To do this without a calculator, note that the three roots of $f(x)$ divide the x-axis into four intervals:

$$x < -15, \qquad -15 < x < 2, \qquad 2 < x < 10, \qquad x > 10.$$

For each of these intervals, we shall determine whether the graph is above or below the x-axis.

Consider, for example, the interval between the roots 2 and 10. The graph of $f(x)$ touches the x-axis at $x = 2$ and $x = 10$ but does not touch the axis at any point in between, since the only other root (x-intercept) is -15. Since a polynomial graph is continuous—it has no gaps or holes—the graph of $f(x)$ cannot "jump over" the x-axis between $x = 2$ and $x = 10$. It must be either entirely above the x-axis there or entirely below it.

To determine which is the case, choose any number between 2 and 10, say, $x = 4$, and test $f(4)$.

$$f(4) = (4 + 15)(4 - 2)^6(4 - 10) = 19(2^6)(-6).$$

You don't even have to finish the computation to see that $f(4)$ is a negative number. Therefore, the point $(4, f(4))$ on the graph of $f(x)$ lies below the x-axis. Since one point of the graph between 2 and 10 lies below the x-axis, the entire graph must be below the x-axis between 2 and 10.

The location of the graph on the other intervals can be determined similarly, by choosing a test number in each interval, as summarized in this chart.

Interval	$x < -15$	$-15 < x < 2$	$2 < x < 10$	$x > 10$
Test number in this interval	-20	0	4	11
Value of $f(x)$ at test number	$(-5)(-22)^6(-30)$	$15(-2)^6(-10)$	$19(2^6)(-6)$	$26(9^6)(1)$
Sign of $f(x)$ at test number	$+$	$-$	$-$	$+$
Graph	Above x-axis	Below x-axis	Below x-axis	Above x-axis

The last line of the chart shows that the intervals where the graph is below the x-axis are $-15 < x < 2$ and $2 < x < 10$. Since the graph touches the x-axis at the roots -15, 2, and 10, the solutions of the original inequality (the numbers x for which the graph is on or below the x-axis) are all numbers x such that $-15 \le x \le 10$. $\qquad \blacksquare$

The procedures used in Examples 5–9 may be summarized as follows.

Solving Polynomial Inequalities

1. Write the inequality in one of these forms:

$$f(x) > 0, \qquad f(x) \geq 0, \qquad f(x) < 0, \qquad f(x) \leq 0.$$

2. Determine the roots of $f(x)$, exactly if possible, approximately otherwise.

3. Use a calculator (as in Examples 5–7), your knowledge of quadratic functions (as in Example 8), or a sign chart (as in Example 9) to determine whether the graph of $f(x)$ is above or below the x-axis on each of the intervals determined by the roots.

4. Use the information in step 3 to find the solutions of the inequality.

RATIONAL INEQUALITIES

Rational inequalities are solved in essentially the same way that polynomial inequalities are solved, with one difference. The graph of a rational function may cross the x-axis at an x-intercept, but there is another possibility: The graph may be above the x-axis on one side of a vertical asymptote and below it on the other side (see, for instance, Examples 5–6 in Section 4.5). Since the x-intercepts of the graph of the rational function $g(x)/h(x)$ are determined by the roots of its numerator $g(x)$ and the vertical asymptotes by the roots of its denominator $h(x)$, all of these roots must be considered in determining the solution of an inequality involving $g(x)/h(x)$.

EXAMPLE 10

Solve $\dfrac{x}{x-1} > -6$.

SOLUTION   There are three ways to solve this inequality.

Geometric: The fastest way to get an approximate solution is to replace the given inequality by an equivalent one,

$$\frac{x}{x-1} + 6 > 0.$$

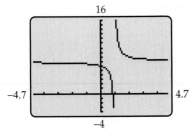

Figure 4–56

and graph the function $f(x) = \dfrac{x}{x-1} + 6$ as in Figure 4–56.

The graph is above the x-axis everywhere except between the x-intercept and the vertical asymptote $x = 1$. Using a root finder, we see that the x-intercept is approximately .857. Therefore, the approximate solutions of the original inequality are all numbers x such that $x < .857$ or $x > 1$.

Algebraic/Geometric: Proceed as above, but rewrite the rule of the function f as a single rational expression before graphing.

$$f(x) = \frac{x}{x-1} + 6 = \frac{x}{x-1} + \frac{6(x-1)}{x-1} = \frac{x + 6x - 6}{x-1} = \frac{7x - 6}{x-1}.$$

When the rule of f is written in this form, it is easy to see that the x-intercept of the graph (the root of the numerator) is $x = 6/7$ (whose decimal approximation begins .857). Therefore, the exact solutions of the original inequality (the numbers x for which the graph in Figure 4–56 is above the x-axis) are all numbers x such that $x < 6/7$ or $x > 1$.

Algebraic: Write the rule of the function f as a single rational expression $f(x) = \dfrac{7x - 6}{x - 1}$. The roots of the numerator and denominator (6/7 and 1) divide the x-axis into three intervals. Use test numbers and a sign chart instead of graphing to determine the location of the graph on each interval:*

Interval	$x < 6/7$	$6/7 < x < 1$	$x > 1$
Test number in this interval	0	.9	2
Value of $f(x)$ at test number	$\dfrac{7 \cdot 0 - 6}{0 - 1}$	$\dfrac{7(.9) - 6}{.9 - 1}$	$\dfrac{7 \cdot 2 - 6}{2 - 1}$
Sign of $f(x)$ at test number	+	−	+
Graph	Above x-axis	Below x-axis	Above x-axis

The last line of the chart shows that the solutions of the original inequality (the numbers x for which the graph is above the x-axis) are all such that $x < 6/7$ or $x > 1$. ∎

The algebraic technique of writing the left side of the inequality as a single rational expression is useful whenever the resulting numerator has low degree (so that its roots can be found exactly), but can usually be omitted when the roots of the numerator must be approximated.

CAUTION

Don't treat rational inequalities as if they are equations, as in this *incorrect* "solution" of the preceding example:

$$\frac{x}{x - 1} > -6$$

$$x > -6(x - 1) \qquad \text{[Both sides multiplied by } x - 1]$$

$$x > -6x + 6$$

$$7x > 6$$

$$x > \frac{6}{7}$$

According to this, the inequality has no negative solution and $x = 1$ is a solution, but as we saw in Example 10, *every* negative number is a solution and $x = 1$ is not.[†]

---

*The justification for this approach is essentially the same as that in Example 9: Because f is continuous everywhere that it is defined, the graph can change from one side of the x-axis to the other only at x-intercepts or vertical asymptotes, so testing one number in each interval is sufficient to determine the side on which the graph lies.

[†]The source of the error is multiplying by $x - 1$. This quantity is negative for some values of x and positive for others. To do this calculation correctly, you must consider two separate cases and reverse the direction of the inequality when $x - 1$ is negative.

APPLICATIONS

> ### EXAMPLE 11

A computer store has determined that the cost C of ordering and storing x laser printers is given by

$$C = 2x + \frac{300{,}000}{x}.$$

If the delivery truck can bring at most 450 printers per order, how many printers should be ordered at a time to keep the cost below $1600?

SOLUTION To find the values of x that make C less than 1600, we must solve the inequality

$$2x + \frac{300{,}000}{x} < 1600 \qquad \text{or, equivalently,} \qquad 2x + \frac{300{,}000}{x} - 1600 < 0.$$

We shall solve this inequality graphically, although it can also be solved algebraically. In this context, the only solutions that make sense are those between 0 and 450. So we choose the viewing window in Figure 4–57 and graph

$$f(x) = 2x + \frac{300{,}000}{x} - 1600.$$

Figure 4–57 is consistent with the fact that $f(x)$ has a vertical asymptote at $x = 0$ and shows that the desired solutions (numbers where the graph is below the x-axis) are all numbers x between the root and 450. A root finder shows that the root is $x \approx 300$. In fact, this is the exact root, since a simple computation shows that $f(300) = 0$. (Do it!) Therefore, to keep costs under $1600, x printers should be ordered each time, with $300 < x \leq 450$.

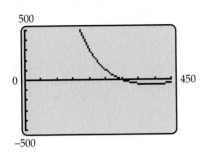

Figure 4–57

EXERCISES 4.6

In Exercises 1–20, solve the inequality and express your answer in interval notation.

1. $2x + 4 \leq 7$

2. $4x - 3 > -12$

3. $3 - 5x < 13$

4. $2 - 3x < 11$

5. $6x + 3 \leq x - 5$

6. $5x + 3 \leq 2x + 7$

7. $5 - 7x < 2x - 4$

8. $8 - 4x > 7x + 2$

9. $2 < 3x - 4 < 8$

10. $4 < 9x + 2 < 10$

11. $0 < 5 - 2x \leq 11$

12. $-4 \leq 7 - 3x < 0$

13. $5x + 6(-8x - 1) < 2(x - 1)$

14. $x + 3(x - 5) \geq 3x + 2(x + 1)$

15. $\dfrac{x + 1}{2} - 3x \leq \dfrac{x + 5}{3}$

16. $\dfrac{x - 1}{4} + 2x \geq \dfrac{2x - 1}{3} + 2$

17. $2x + 3 \leq 5x + 6 < -3x + 7$

18. $2x - 1 < x + 4 < 9x + 2$

19. $3 - x < 2x + 1 \leq 3x - 4$

20. $2x + 5 \leq 4 - 3x < 1 - 4x$

In Exercises 21–24, a, b, c, and d are positive constants. Solve the inequality for x.

21. $ax - b < c$

22. $d - cx > a$

23. $0 < x - c < a$

24. $-d < x - c < d$

In Exercises 25–46, solve the inequality. Find exact solutions when possible and approximate ones otherwise.

25. $x^2 - 4x + 3 \leq 0$

26. $x^2 - 7x + 10 \leq 0$

27. $8 + x - x^2 \leq 0$

28. $x^2 + 8x + 20 \geq 0$

29. $x^3 - x \geq 0$

30. $x^3 + 2x^2 + x > 0$

31. $x^3 + 3x^2 - x - 3 < 0$

32. $x^4 - 14x^3 + 48x^2 \geq 0$

33. $x^4 - 5x^2 + 4 < 0$

34. $x^4 - 10x^2 + 9 \leq 0$

35. $2x^4 + 3x^3 < 2x^2 + 4x - 2$

36. $x^5 + 5x^4 > 4x^3 - 3x^2 + 2$

37. $\dfrac{3x + 1}{2x - 4} > 0$

38. $\dfrac{2x^2 + x - 1}{x^2 - 4x + 4} \geq 0$

39. $\dfrac{x - 2}{x - 1} < 1$

40. $\dfrac{-x + 5}{2x + 3} \geq 2$

41. $\dfrac{2}{x + 3} \geq \dfrac{1}{x - 1}$

42. $\dfrac{1}{x - 1} < \dfrac{-1}{x + 2}$

43. $\dfrac{x^3 - 3x^2 + 5x - 29}{x^2 - 7} > 3$

44. $\dfrac{x^4 - 3x^3 + 2x^2 + 2}{x - 2} > 15$

45. $\dfrac{2x^2 + 6x - 8}{2x^2 + 5x - 3} < 1$ [Be alert for hidden behavior.]

46. $\dfrac{1}{x^2 + x - 6} + \dfrac{x - 2}{x + 3} > \dfrac{x + 3}{x - 2}$

In Exercises 47–48, solve the inequality using the method of Example 9.

47. $x(x - 1)^3 (x - 2)^4 \geq 0$

48. $x^4 - 5x^3 + 9x^2 - 7x + 2 < 0$

   [*Hint:* First find the rational roots; then factor.]

In Exercises 49–51, read the solution of the inequality from the given graph.

49. $3 - 2x < .8x + 7$

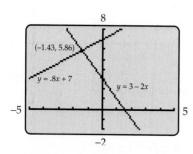

50. $8 - |7 - 5x| > 3$

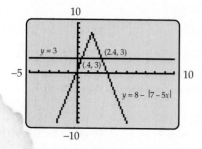

51. $x^2 + 3x + 1 \geq 4$

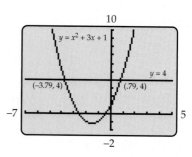

52. The graphs of the revenue and cost functions for a manufacturing firm are shown in the figure.

   (a) What is the break-even point?

   (b) Shade in the region representing profit.

   (c) What does the y-intercept of the cost graph represent? Why is the y-intercept of the revenue graph 0?

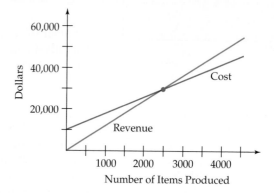

53. One freezer costs \$723.95 and uses 90 kilowatt hours (kwh) of electricity each month. A second freezer costs \$600 and uses 100 kwh of electricity each month. The expected life of each freezer is 12 years. What is the minimum electric rate (in *cents* per kwh) for which the 12-year total cost (purchase price + electricity costs) will be less for the first freezer?

54. A business executive leases a car for \$300 per month. She decides to lease another brand for \$250 per month but has to pay a penalty of \$1000 for breaking the first lease. How long must she keep the second car to come out ahead?

55. A sales agent is given a choice of two different compensation plans. The first plan has no salary, but a 10% commission on total sales. The second plan has a salary of \$3000 per month, plus a 2% commission on total sales. What range of monthly sales will make the first plan a better choice for the sales agent?

56. A developer subdivided 60 acres of a 100-acre tract, leaving 20% of the 60 acres as a park. Zoning laws require that at least 25% of the total tract be set aside for parks. For financial reasons, the developer wants to have no more than 30% of the tract as parks. How many one-quarter-acre lots can the developer sell in the remaining 40 acres and still meet the requirements for the whole tract?

57. Emma and Aidan currently pay $60 per month for phone service from AT&T. This fee gets them 900 minutes per month. They look at their phone bills and realize that, at most, they talk for 100 minutes per month. They find out that they can go with Virgin Mobile and pay 18 cents per minute. If they choose to switch services, they will have to buy two new phones at $40 each, and pay a $175 "cancellation fee" to AT&T.

(a) Assuming that they talk for 100 minutes per month, how many months would they have to talk before they would be saving money?

(b) Assume they make the switch, and talk between zero and 100 minutes per month. What is the range of possible savings?

58. How many gallons of a 12% salt solution should be added to 10 gallons of an 18% salt solution to produce a solution whose salt content is between 14% and 16%?

59. Find all pairs of numbers that satisfy these two conditions: Their sum is 20, and the sum of their squares is less than 362.

60. The length of a rectangle is 6 inches longer than its width. What are the possible widths if the area of the rectangle is at least 667 square inches?

61. It costs a craftsman $5 in materials to make a medallion. He has found that if he sells the medallions for $50 - x$ dollars each, where x is the number of medallions produced each week, then he can sell all that he makes. His fixed costs are $350 per week. If he wants to sell all he makes and show a profit each week, what are the possible numbers of medallions he should make?

62. A retailer sells file cabinets for $80 - x$ dollars each, where x is the number of cabinets she receives from the supplier each week. She pays $10 for each file cabinet and has fixed costs of $600 per week. How many file cabinets should she order from the supplier each week to guarantee that she makes a profit?

In Exercises 63–66, you will need the formula for the height h of an object above the ground at time t seconds:

$$h = -16t^2 + v_0t + h_0;$$

this formula was explained on page 249.

63. A toy rocket is fired straight up from ground level with an initial velocity of 80 feet per second. During what time interval will it be at least 64 feet above the ground?

64. A projectile is fired straight up from ground level with an initial velocity of 72 feet per second. During what time interval is it at least 37 feet above the ground?

65. A ball is dropped from the roof of a 120-foot-high building. During what time period will it be strictly between 56 feet and 39 feet above the ground?

66. A ball is thrown straight up from a 40-foot-high tower with an initial velocity of 56 feet per second.

(a) During what time interval is the ball at least 8 feet above the ground?

(b) During what time interval is the ball between 53 feet and 80 feet above the ground?

67. (a) Solve the inequalities $x^2 < x$ and $x^2 > x$.

(b) Use the results of part (a) to show that for any nonzero real number c with $|c| < 1$, it is always true that $c^2 < |c|$.

(c) Use the results of part (a) to show that for any nonzero real number c with $|c| > 1$, it is always true that $c^2 > c$.

68. (a) If $0 < a \le b$, prove that $1/a \ge 1/b$.

(b) If $a \le b < 0$, prove that $1/a \ge 1/b$.

(c) If $a < 0 < b$, how are $1/a$ and $1/b$ related?

THINKERS

In Exercises 69–77, solve the inequality.

69. $4x - 5 \ge 4x + 2$

70. $3x - 4 < 3x - 4$

71. $3x - 4 \ge 3x - 4$

72. $(x - \pi)^2 \ge 0$

73. $(x + 2)^2(x - 3)^2 < 0$

74. $(2x - 5)^2 > 0$

75. $(x + 1)^2 < 0$

76. $3 < 6x + 6 < 2$

77. $8 \le 4x - 2 \le 8$

78. We know that for large values of x, we can approximate $x^2 - 2x^2 + x - 1$ by using x^3.

(a) Compute the percent error in this approximation when $x = 50$ and when $x = 100$.

(b) For what positive values of x is the error less than 10%?

4.6.A *SPECIAL TOPICS* Absolute Value Inequalities

Section Objective ■ Solve absolute value inequalities algebraically and graphically.

Polynomial and rational inequalities involving absolute value can be solved graphically, just as was done earlier: Rewrite the inequality in an equivalent form that has 0 on the right side of the inequality sign; then graph the function whose rule is given by the left side and determine where the graph is above or below the x-axis.

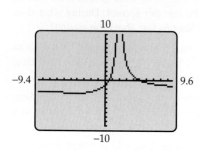

Figure 4–58

EXAMPLE 1

Solve $\left|\dfrac{x+4}{x-2}\right| > 3$

SOLUTION We use the equivalent inequality

$$\left|\dfrac{x+4}{x-2}\right| - 3 > 0$$

and graph the function $f(x) = \left|\dfrac{x+4}{x-2}\right| - 3$ (Figure 4–58). The graph is above the x-axis between the two x-intercepts, which can be found algebraically or graphically.

GRAPHING EXPLORATION

Verify that the x-intercepts are $x = 1/2$ and $x = 5$.

Since $f(x)$ is not defined at $x = 2$ (where the graph has a vertical asymptote), the solutions of the original inequality are all x such that $1/2 < x < 2$ or $2 < x < 5$. ∎

EXAMPLE 2

Solve $|x^4 + 2x^2 - x + 2| < 11x$.

SOLUTION We determine the numbers for which the graph of

$$f(x) = |x^4 + 2x^2 - x + 2| - 11x$$

lies below the x-axis. (Why?) Convince yourself that the graph of $f(x)$ in Figure 4–59 is complete.

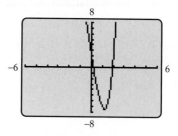

Figure 4–59

A root finder shows that the approximate x-intercepts are $x = .17$ and $x = 1.92$. Therefore, the approximate solutions of the original inequality (the numbers where the graph is below the x-axis) are all x such that $.17 < x < 1.92$. ∎

ALGEBRAIC METHODS

Most linear and quadratic inequalities involving absolute values can be solved exactly by algebraic means. In fact, this is often the easiest way to solve such inequalities. The key to the algebraic method is the fact that the absolute value of a number can be interpreted as distance on the number line. For example, the inequality $|r| \leq 5$ states that the distance from r to 0 (namely, $|r|$) is 5 units or less. A glance at the number line in Figure 4–60 shows that these are the numbers r with $-5 \leq r \leq 5$.

Figure 4–60

Similarly, the numbers r such that $|r| \geq 5$ are those whose distance to 0 is 5 or more units, that is, the numbers r with $r \leq -5$ or $r \geq 5$. This argument works with any positive number k in place of 5 and proves the following facts (which are also true with $<$ and $>$ in place of $\leq$ and $\geq$).

**Absolute Value
Inequalities**

> Let k be a positive number and r any real number.
>
> $|r| \leq k$ is equivalent to $-k \leq r \leq k.$
>
> $|r| \geq k$ is equivalent to $r \leq -k$ or $r \geq k.$

EXAMPLE 3

To solve $|3x - 7| \leq 11$, apply the first fact in the box, with $3x - 7$ in place of r and 11 in place of k, and obtain this equivalent inequality: $-11 \leq 3x - 7 \leq 11$. Then

Add 7 to each part: $-4 \leq 3x \leq 18$

Divide each part by 3: $-4/3 \leq x \leq 6.$

Therefore, the solutions of the original inequality are all numbers in the interval $[-4/3, 6]$. ∎

EXAMPLE 4

To solve $|5x + 2| > 3$, apply the second fact in the box, with $5x + 2$ in place of r, and 3 in place of k, and $>$ in place of $\geq$. This produces the equivalent statement:

$$5x + 2 < -3 \qquad \text{or} \qquad 5x + 2 > 3$$
$$5x < -5 \qquad\qquad\qquad 5x > 1$$
$$x < -1 \qquad \text{or} \qquad x > 1/5.$$

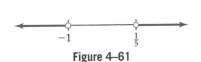

Figure 4–61

Therefore, the solutions of the original inequality are the numbers in *either* of the intervals $(-\infty, -1)$ or $(1/5, \infty)$, as shown in Figure 4–61. ∎

EXAMPLE 5

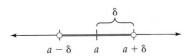

Figure 4–62

If a and δ are real numbers with δ positive, then the inequality $|x - a| < \delta$ is equivalent to $-\delta < x - a < \delta$. Adding a to each part shows that

$$a - \delta < x < a + \delta \text{ as shown in Figure 4–62.}$$ ∎

EXAMPLE 6

To solve $|x^2 - x - 4| \geq 2$, we use the fact in the box on the preceding page to replace it by an equivalent inequality:

$$x^2 - x - 4 \leq -2 \qquad \text{or} \qquad x^2 - x - 4 \geq 2,$$

which is the same as

$$x^2 - x - 2 \leq 0 \qquad \text{or} \qquad x^2 - x - 6 \geq 0.$$

The solutions are all numbers that are solutions of *either one* of the two inequalities. To solve the first of these inequalities, note that the graph of

$$f(x) = x^2 - x - 2 = (x + 1)(x - 2)$$

is an upward-opening parabola that crosses the x-axis at -1 and 2 (the roots of $f(x)$). Therefore, the solutions of

$$x^2 - x - 2 \leq 0$$

(the numbers for which the graph of $f(x)$ is on or below the x-axis) are all x with $-1 \leq x \leq 2$. The second inequality above, $x^2 - x - 6 \geq 0$, is solved similarly.

GRAPHING EXPLORATION

What is the shape of the graph of $g(x) = x^2 - x - 6$ and what are its x-intercepts?

This Graphing Exploration shows that the solutions of the second inequality (the numbers for which the graph of $g(x)$ is on or above the x-axis) are all x with $x \leq -2$ or $x \geq 3$.

Consequently, the solutions of the original inequality are all numbers x such that $x \leq -2$ or $-1 \leq x \leq 2$ or $x \geq 3$, as shown in Figure 4–63. ∎

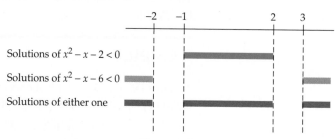

Figure 4–63

EXERCISES 4.6.A

In Exercises 1–30, solve the inequality. Find exact solutions when possible and approximate ones otherwise.

1. $|3x + 2| \leq 2$

2. $|4x + 2| < 8$

3. $|3 - 2x| < 2/3$

4. $|4 - 5x| \leq 4$

5. $|5x + 2| \geq \dfrac{3}{4}$

6. $|5 - 2x| > 7$

7. $\left|\dfrac{12}{5} + 2x\right| > \dfrac{1}{4}$

8. $\left|\dfrac{5}{6} + 3x\right| < \dfrac{7}{6}$

9. $\left|\dfrac{x - 1}{x + 2}\right| \leq 3$

10. $\left|\dfrac{x + 1}{3x + 5}\right| < 2$

11. $\left|\dfrac{1 - 4x}{2 + 3x}\right| < 1$

12. $\left|\dfrac{3x + 1}{1 - 2x}\right| \geq 2$

The familiar multiplication patterns and exponent laws for integer exponents hold in the complex number system.

EXAMPLE 2

(a) $(3 + 2i)(3 - 2i) = 3^2 - (2i)^2$

$$= 9 - 4i^2$$
$$= 9 - 4(-1)$$
$$= 9 + 4 = 13.$$

(b) $(4 + i)^2 = 4^2 + 2 \cdot 4 \cdot i + i^2$

$$= 16 + 8i + (-1)$$
$$= 15 + 8i.$$

(c) To find i^{54}, we first note that $i^4 = i^2 i^2 = (-1)(-1) = 1$ and that

$$54 = 52 + 2 = 4 \cdot 13 + 2.$$

Consequently,

$$i^{54} = i^{52+2} = i^{52} i^2 = i^{4 \cdot 13} i^2 = (i^4)^{13} i^2 = 1^{13}(-1) = -1. \qquad \blacksquare$$

The **conjugate** of the complex number $a + bi$ is the number $a - bi$, and the conjugate of $a - bi$ is $a + bi$. For example, the conjugate of $3 + 4i$ is $3 - 4i$ and the conjugate of $-3i = 0 - 3i$ is $0 + 3i = 3i$. *Every real number is its own conjugate;* for instance, the conjugate of $17 = 17 + 0i$ is $17 - 0i = 17$.

For any complex number $a + bi$, we have

$$(a + bi)(a - bi) = a^2 - (bi)^2 = a^2 - b^2 i^2 = a^2 - b^2(-1) = a^2 + b^2.$$

Since a^2 and b^2 are nonnegative real numbers, so is $a^2 + b^2$. Therefore, *the product of a complex number and its conjugate is a nonnegative real number.* This fact enables us to express quotients of complex numbers in standard form.

TECHNOLOGY TIP

To do complex arithmetic on TI-86 and HP-39gs, enter $a + bi$ as (a, b). On other calculators, use the special i key whose location is

TI-84+/89: keyboard

Casio: 9850: OPTN/CPLX

EXAMPLE 3

To express $\dfrac{3 + 4i}{1 + 2i}$ in the form $a + bi$, *multiply both numerator and denominator by the conjugate of the denominator,* namely, $1 - 2i$:

$$\frac{3 + 4i}{1 + 2i} = \frac{3 + 4i}{1 + 2i} \cdot \frac{1 - 2i}{1 - 2i}$$

$$= \frac{(3 + 4i)(1 - 2i)}{(1 + 2i)(1 - 2i)}$$

$$= \frac{3 + 4i - 6i - 8i^2}{1^2 - (2i)^2}$$

$$= \frac{3 + 4i - 6i - 8(-1)}{1 - 4i^2}$$

$$= \frac{11 - 2i}{1 - 4(-1)}$$

$$= \frac{11 - 2i}{5} = \frac{11}{5} - \frac{2}{5}i.$$

This is the form $a + bi$ with $a = 11/5$ and $b = -2/5$. $\qquad \blacksquare$

EXAMPLE 4

Express $\dfrac{1}{1-i}$ in standard form.

SOLUTION   We note that the conjugate of the denominator is $1 + i$, and therefore

$$\frac{1}{1-i} = \frac{1 \cdot (1+i)}{(1-i)(1+i)}$$

$$= \frac{1+i}{1^2 - i^2}$$

$$= \frac{1+i}{1-(-1)}$$

$$= \frac{1+i}{2} = \frac{1}{2} + \frac{1}{2}i.$$

We can check this result by multiplying $\dfrac{1}{2} + \dfrac{1}{2}i$ by $1 - i$ to see whether the product is 1 $\left(\text{which it should be if } \dfrac{1}{2} + \dfrac{1}{2}i = \dfrac{1}{1-i}\right)$:

$$\left(\frac{1}{2} + \frac{1}{2}i\right)(1-i) = \frac{1}{2} \cdot 1 - \frac{1}{2}i + \frac{1}{2}i \cdot 1 - \frac{1}{2}i^2 = \frac{1}{2} - \frac{1}{2}(-1) = 1. \quad \blacksquare$$

Since $i^2 = -1$, we define $\sqrt{-1}$ to be the complex number i. Similarly, since $(5i)^2 = 5^2 i^2 = 25(-1) = -25$, we define $\sqrt{-25}$ to be $5i$. In general,

***Square Roots
of Negative Numbers***

Let b be a positive real number.
$$\sqrt{-b} \text{ is defined to be } \sqrt{b}\,i$$
because $(\sqrt{b}\,i)^2 = (\sqrt{b})^2 i^2 = b(-1) = -b.$

CAUTION

$\sqrt{b}\,i$ is *not* the same as $\sqrt{bi}$. To avoid confusion it may help to write $\sqrt{b}\,i$ as $i\sqrt{b}$.

EXAMPLE 5

Express the following in the form $a + bi$:   (a) $\sqrt{-3}$ (b) $\dfrac{1 - \sqrt{-7}}{3}$

SOLUTION

(a) $\sqrt{-3} = \sqrt{3}\,i = 0 + \sqrt{3}\,i$ or $0 + i\sqrt{3}$.

(b) $\dfrac{1 - \sqrt{-7}}{3} = \dfrac{1 - \sqrt{7}\,i}{3} = \dfrac{1}{3} - \dfrac{\sqrt{7}}{3}i.$ $\blacksquare$

CAUTION

The property $\sqrt{cd} = \sqrt{c}\sqrt{d}$ (or equivalently in exponential notation, $(cd)^{1/2} = c^{1/2}d^{1/2}$), which is valid for positive real numbers, does *not hold* when both c and d are negative.

$$\sqrt{-20}\,\sqrt{-5} = \sqrt{20}i \cdot \sqrt{5}i = \sqrt{20}\,\sqrt{5} \cdot i^2 = \sqrt{20 \cdot 5}(-1)$$

$$= \sqrt{100}(-1) = -10.$$

But $\sqrt{(-20)(-5)} = \sqrt{100} = 10$, so that

$$\sqrt{(-20)(-5)} \neq \sqrt{-20}\,\sqrt{-5}.$$

To avoid difficulty, *always write square roots of negative numbers in terms of i before doing any simplification.*

TECHNOLOGY TIP

Most calculators that do complex number arithmetic automatically return a complex number when asked for the square root of a negative number. On TI-84+/89, however, the MODE must be set to "rectangular" or "$a + bi$."

EXAMPLE 6

$$(7 - \sqrt{-4})(5 + \sqrt{-9}) = (7 - \sqrt{4}i)(5 + \sqrt{9}i)$$

$$= (7 - 2i)(5 + 3i)$$

$$= 35 + 21i - 10i - 6i^2$$

$$= 35 + 11i - 6(-1) = 41 + 11i. \qquad \blacksquare$$

Since every negative real number has a square root in the complex number system, we can now find complex solutions for equations that have no real solutions. For example, the solutions of $x^2 = -25$ are $x = \pm\sqrt{-25} = \pm 5i$. In fact,

Every quadratic equation with real coefficients has solutions in the complex number system.

EXAMPLE 7

To solve the equation $2x^2 + x + 3 = 0$, we apply the quadratic formula.

$$x = \frac{-1 \pm \sqrt{1^2 - 4 \cdot 2 \cdot 3}}{2 \cdot 2} = \frac{-1 \pm \sqrt{-23}}{4}.$$

Since $\sqrt{-23}$ is not a real number, this equation has no real number solutions. But $\sqrt{-23}$ *is* a complex number, namely, $\sqrt{-23} = \sqrt{23}i$. Thus, the equation does have solutions in the complex number system.

$$x = \frac{-1 \pm \sqrt{-23}}{4} = \frac{-1 \pm \sqrt{23}i}{4} = -\frac{1}{4} \pm \frac{\sqrt{23}}{4}i.$$

Note that the two solutions, $-\frac{1}{4} + \frac{\sqrt{23}}{4}i$ and $-\frac{1}{4} - \frac{\sqrt{23}}{4}i$, are conjugates of each other. $\qquad \blacksquare$

EXAMPLE 8

To find *all* solutions of $x^3 = 1$, we rewrite the equation and use the Difference of Cubes pattern (see the Algebra Review Appendix) to factor:

$$x^3 = 1$$

$$x^3 - 1 = 0$$

$$(x - 1)(x^2 + x + 1) = 0$$

$$x - 1 = 0 \quad \text{or} \quad x^2 + x + 1 = 0.$$

The solution of the first equation is $x = 1$. The solutions of the second can be obtained from the quadratic formula.

$$x = \frac{-1 \pm \sqrt{1^2 - 4 \cdot 1 \cdot 1}}{2 \cdot 1} = \frac{-1 \pm \sqrt{-3}}{2} = \frac{-1 \pm \sqrt{3}i}{2} = -\frac{1}{2} \pm \frac{\sqrt{3}}{2}i.$$

Therefore, the equation $x^3 = 1$ has one real solution ($x = 1$) and two nonreal complex solutions [$x = -1/2 + (\sqrt{3}/2)i$ and $x = -1/2 - (\sqrt{3}/2)i$]. Each of these solutions is said to be a **cube root of 1** or a **cube root of unity.** Observe that the two nonreal complex cube roots of unity are conjugates of each other. ∎

The preceding examples illustrate this useful fact (whose proof is discussed in Section 4.8).

Conjugate Solutions

If $a + bi$ is a solution of a polynomial equation with *real* coefficients, then its conjugate $a - bi$ is also a solution of this equation.

EXERCISES 4.7

In Exercises 1–54, perform the indicated operation and write the result in the form $a + bi$.

1. $(2 + 3i) + (6 - i)$

2. $(-3 + 2i) + (8 + 6i)$

3. $(2 - 8i) - (4 + 2i)$

4. $(3 + 5i) + (2 - 5i)$

5. $\frac{5}{4} - \left(\frac{7}{4} + 2i\right)$

6. $(\sqrt{3} + i) + (\sqrt{5} - 2i)$

7. $\left(\frac{\sqrt{2}}{2} + i\right) - \left(\frac{\sqrt{3}}{2} - i\right)$

8. $\left(\frac{1}{2} + \frac{\sqrt{3}i}{2}\right) + \left(\frac{3}{4} - \frac{5\sqrt{3}i}{2}\right)$

9. $(2 + i)(3 + 5i)$

10. $(2 - i)(5 + 2i)$

11. $(0 - 6i)(5 + 0i)$

12. $(4 + 3i)(4 - 3i)$

13. $(2 - 5i)^2$

14. $(3 + i)(5 - i)i$

15. $(\sqrt{3} + i)(\sqrt{3} - i)$

16. $\left(\frac{1}{2} - i\right)\left(\frac{1}{4} + 2i\right)$

17. i^{19}

18. i^{26}

19. i^{33}

20. $(-i)^{53}$

21. $(-i)^{107}$

22. $(-i)^{213}$

23. $\frac{1}{3 + 2i}$

24. $\frac{1}{i}$

25. $\frac{4}{3i}$

26. $\frac{i}{2 + i}$

27. $\frac{3}{4 + 5i}$

28. $\frac{2 + 3i}{i}$

29. $\frac{1}{i(4 + 5i)}$

30. $\frac{1}{(2 - i)(2 + i)}$

31. $\frac{2 + 3i}{i(4 + i)}$

32. $\frac{2}{(2 + 3i)(4 + i)}$

33. $\frac{2 + i}{1 - i} + \frac{1}{1 + 2i}$

34. $\frac{1}{2 - i} + \frac{3 + i}{2 + 3i}$

35. $\frac{i}{3 + i} - \frac{3 + i}{4 + i}$

36. $6 + \frac{2i}{3 + i}$

37. $\sqrt{-36}$

38. $\sqrt{-121}$

39. $\sqrt{-14}$

40. $\sqrt{-800}$

41. $-\sqrt{-16}$

42. $-\sqrt{-12}$

43. $\sqrt{-16} + \sqrt{-49}$

44. $\sqrt{-25} - \sqrt{9}$

45. $\sqrt{-15} - \sqrt{-18}$

46. $\sqrt{-12}\sqrt{-3}$

47. $\sqrt{-16}/\sqrt{-36}$

48. $-\sqrt{-64}/\sqrt{-4}$

49. $(\sqrt{-25} + 2)(\sqrt{-49} - 3)$

50. $(5 - \sqrt{-3})(-1 + \sqrt{-9})$

51. $(2 + \sqrt{-5})(1 - \sqrt{-10})$

52. $\sqrt{-3}(3 - \sqrt{-27})$

53. $1/(1 + \sqrt{-5})$

54. $(1 + \sqrt{-4})(3 - \sqrt{-9})$

In Exercises 55–58, find x and y. Remember that

$$a + bi = c + di$$

exactly when a = c and b = d.

55. $3x - 4i = 6 + 2yi$ **56.** $5 + 3yi = 10x + 36i$

57. $3 + 4xi = 2y - 3i$ **58.** $10 = (6 + 8i)(x + yi)$

In Exercises 59–70, solve the equation and express each solution in the form a + bi.

59. $3x^2 - 2x + 5 = 0$ **60.** $5x^2 + 2x + 1 = 0$

61. $x^2 + 5x + 6 = 0$ **62.** $x^2 + 6x + 25 = 0$

63. $2x^2 - x = -4$ **64.** $x^2 + 1 = 4x$

65. $x^2 + 1770.25 = -84x$ **66.** $3x^2 + 4 = -5x$

67. $x^3 - 8 = 0$ **68.** $x^3 + 125 = 0$

69. $x^4 - 1 = 0$ **70.** $x^4 - 81 = 0$

71. Simplify: $i + i^2 + i^3 + \cdots + i^{15}$

72. Simplify: $i - i^2 + i^3 - i^4 + i^5 - \cdots + i^{15}$

THINKERS

73. It is easy to compare two real numbers. For instance, $5 < 8$, $4 = \dfrac{28}{7}$, and $-3 > -10$. It is harder to compare two complex numbers. Is $5 + 12i$ less than, greater than, or equal to $11 + 6i$? On the face of it, this question is not possible to answer. When comparing complex numbers, mathematicians look at their *moduli*, a measure of how "far away" they are from $0 + 0i$, or zero. The modulus of a complex number is defined this way:

$$\text{mod}(a + bi) = \sqrt{a^2 + b^2}$$

(a) Compute the modulus of the following complex numbers:
- (i) $3 - 4i$
- (ii) $24 + 7i$
- (iii) $8 + 0i$
- (iv) $-8 + 0i$
- (v) $0 + 8i$

(b) Which is larger, $\text{mod}(5 + 12i)$ or $\text{mod}(11 + 6i)$?

If $z = a + bi$ is a complex number, then its conjugate is usually denoted $\bar{z}$, that is, $\bar{z} = a - bi$. In Exercises 74–78, prove that for any complex numbers $z = a + bi$ and $w = c + di$:

74. $\overline{z + w} = \bar{z} + \bar{w}$ **75.** $\overline{zw} = \bar{z} \cdot \bar{w}$

76. $\overline{\left(\dfrac{z}{w}\right)} = \dfrac{\bar{z}}{\bar{w}}$ **77.** $\bar{\bar{z}} = z$

78. z is a real number exactly when $\bar{z} = z$.

79. The **real part** of the complex number $a + bi$ is defined to be the real number a. The **imaginary part** of $a + bi$ is defined to the real number b (*not bi*).

(a) Show that the real part of $z = a + bi$ is $\dfrac{z + \bar{z}}{2}$.

(b) Show that the imaginary part of $z = a + bi$ is $\dfrac{z - \bar{z}}{2i}$.

80. If $z = a + bi$ (with a, b real numbers, not both 0), express $1/z$ in standard form.

81. Construction of the Complex Numbers. We assume that the real number system is known. To construct a new number system with the desired properties, we must do the following:
- (i) Define a set C (whose elements will be called complex numbers).
- (ii) Ensure that the set C contains the real numbers or at least a copy of them.
- (iii) Define addition and multiplication in the set C in such a way that the usual laws of arithmetic are valid.
- (iv) Show that C has the other properties listed in the box on page 322.

We begin by defining C to be the set of all ordered pairs of real numbers. Thus, $(1, 5)$, $(-6, 0)$, $(4/3, -17)$, and $(\sqrt{2}, 12/5)$ are some of the elements of the set C. More generally, a complex number (= element of C) is any pair (a, b), where a and b are real numbers. By definition, two complex numbers are *equal* exactly when they have the same first and the same second coordinate.

(a) *Addition in C* is defined by this rule:

$$(a, b) + (c, d) = (a + c, b + d)$$

For example,

$$(3, 2) + (5, 4) = (3 + 5, 2 + 4) = (8, 6).$$

Verify that this addition has the following properties. For any complex numbers $(a, b), (c, d), (e, f)$ in C:
- (i) $(a, b) + (c, d) = (c, d) + (a, b)$
- (ii) $[(a, b) + (c, d)] + (e, f) = (a, b) + [(c, d) + (e, f)]$
- (iii) $(a, b) + (0, 0) = (a, b)$
- (iv) $(a, b) + (-a, -b) = (0, 0)$

(b) *Multiplication in C* is defined by this rule:

$$(a, b)(c, d) = (ac - bd, bc + ad)$$

For example,

$$(3, 2)(4, 5) = (3 \cdot 4 - 2 \cdot 5, 2 \cdot 4 + 3 \cdot 5)$$
$$= (12 - 10, 8 + 15) = (2, 23).$$

Verify that this multiplication has the following properties. For any complex numbers $(a, b), (c, d), (e, f)$ in C:
- (i) $(a, b)(c, d) = (c, d)(a, b)$
- (ii) $[(a, b)(c, d)](e, f) = (a, b)[(c, d)(e, f)]$
- (iii) $(a, b)(1, 0) = (a, b)$
- (iv) $(a, b)(0, 0) = (0, 0)$

(c) Verify that for any two elements of C with second coordinate zero:
- (i) $(a, 0) + (c, 0) = (a + c, 0)$
- (ii) $(a, 0)(c, 0) = (ac, 0)$

Identify $(t, 0)$ with the real number t. Statements (i) and (ii) show that when addition or multiplication in C is performed on two real numbers (that is, elements of C with second coordinate 0), the result is the usual sum or product of real numbers. Thus, C contains (a copy of) the real number system.

(d) *New Notation.* Since we are identifying the complex number $(a, 0)$ with the real number a, we shall hereafter denote $(a, 0)$ simply by the symbol a. Also, let i denote the complex number $(0, 1)$.

(i) Show that $i^2 = -1$, that is,

$$(0, 1)(0, 1) = (-1, 0).$$

(ii) Show that for any complex number $(0, b)$, $(0, b) = bi$ [that is, $(0, b) = (b, 0)(0, 1)$].

(iii) Show that any complex number (a, b) can be written: $(a, b) = a + bi$, that is,

$$(a, b) = (a, 0) + (b, 0)(0, 1).$$

In this new notation, every complex number is of the form $a + bi$ with a, b real and $i^2 = -1$, and our construction is finished.

4.8   Theory of Equations*

Section Objectives

- Understand the Fundamental Theorem of Algebra.
- Analyze the roots of polynomials with real coefficients.
- Factor a polynomial completely over the real numbers.

The complex numbers were constructed to obtain a solution for equations like $x^2 = -1$, that is, a root of the polynomial $x^2 + 1$. In Section 4.7, we saw that *every* quadratic polynomial with real coefficients has roots in the complex number system. A natural question now arises:

> Do we have to enlarge the complex number system (perhaps many times) to find roots for higher degree polynomials?

In this section, we shall see that the somewhat surprising answer is no.

To give the full answer, we shall consider not just polynomials with real coefficients, but also those with complex coefficients, such as

$$x^3 - ix^2 + (4 - 3i)x + 1 \qquad \text{or} \qquad (-3 + 2i)x^6 - 3x + (5 - 4i).$$

The discussion of polynomial division in Section 4.2 can easily be extended to include polynomials with complex coefficients. In fact, *all of the results in Section 4.2 are valid for polynomials with complex coefficients.* For example, you can check that i is a root of $f(x) = x^2 + (i - 1)x + (2 + i)$ and that $x - i$ is a factor of $f(x)$:

$$f(x) = x^2 + (i - 1)x + (2 + i) = (x - i)[x - (1 - 2i)].$$

Since every real number is also a complex number, polynomials with real coefficients are just special cases of polynomials with complex coefficients. So in the rest of this section, "polynomial" means "polynomial with complex (possibly real) coefficients" unless specified otherwise. We can now answer the question posed in the first paragraph.

Fundamental Theorem of Algebra

Every nonconstant polynomial has a root in the complex number system.

---

*Section 4.7 is a prerequisite for this section.

Although this is obviously a powerful result, neither the Fundamental Theorem nor its proof provides a practical method for *finding* a root of a given polynomial.* The proof of the Fundamental Theorem is beyond the scope of this book, but we shall explore some of the useful implications of the theorem, such as this one.

Factorization over the Complex Numbers

> Let $f(x)$ be a polynomial of degree $n > 0$ with leading coefficient d. Then there are (not necessarily distinct) complex numbers $c_1, c_2, \ldots, c_n$ such that
> $$f(x) = d(x - c_1)(x - c_2)(x - c_3) \cdots (x - c_n).$$
> Furthermore, $c_1, c_2, \ldots, c_n$ are the only roots of $f(x)$.

Proof   By the Fundamental Theorem, $f(x)$ has a complex root c_1. The Factor Theorem shows that $x - c_1$ must be a factor of $f(x)$, say,

$$f(x) = (x - c_1)g(x),$$

where $g(x)$ has degree $n - 1$.[†] If $g(x)$ is nonconstant, then it has a complex root c_2 by the Fundamental Theorem. Hence, $x - c_2$ is a factor of $g(x)$, so

$$f(x) = (x - c_1)(x - c_2)h(x)$$

for some $h(x)$ of degree $n - 2$ [1 less than the degree of $g(x)$]. If $h(x)$ is nonconstant, then it has a complex root c_3, and the argument can be repeated. Continuing in this way, with the degree of the last factor going down by 1 at each step, we reach a factorization in which the last factor is a constant (degree 0 polynomial):

$$(*) \qquad f(x) = (x - c_1)(x - c_2)(x - c_3) \cdots (x - c_n)d.$$

If the right side were multiplied out, it would look like

$$dx^n + \text{lower degree terms.}$$

So the constant factor d is the leading coefficient of $f(x)$.

It is easy to see from the factored form $(*)$ that the numbers $c_1, c_2, \ldots, c_n$ are roots of $f(x)$. If k is *any* root of $f(x)$, then

$$0 = f(k) = d(k - c_1)(k - c_2)(k - c_3) \cdots (k - c_n).$$

The product on the right is 0 only when one of the factors is 0. Since the leading coefficient d is nonzero, we must have

$$k - c_1 = 0 \quad \text{or} \quad k - c_2 = 0 \quad \text{or} \quad \cdots \quad k - c_n = 0$$
$$k = c_1 \quad \text{or} \quad k = c_2 \quad \text{or} \quad \cdots \quad k = c_n.$$

Therefore, k is one of the c's, and $c_1, \ldots, c_n$ are the only roots of $f(x)$. This completes the proof. ∎

Since the n roots $c_1, \ldots, c_n$ of $f(x)$ might not all be distinct, we see that

TECHNOLOGY TIP

To find all roots of a polynomial, see the Technology Tip on page 326.

To factor a polynomial as a product of linear factors on TI-89, try

ALGEBRA/COMPLEX/cFACTOR.

---

*It might seem strange that you can prove that a root exists without actually exhibiting one. But such "existence theorems" are quite common. A rough analogy is the situation that occurs when someone is killed by a sniper's bullet. The police know that there is a killer, but *finding* the killer may be impossible.
[†]The degree of $g(x)$ is 1 less than the degree n of $f(x)$ because $f(x)$ is the product of $g(x)$ and $x - c_1$ (which has degree 1).

Number of Roots

Every polynomial of degree $n > 0$ has at most n different roots in the complex number system.

Suppose $f(x)$ has repeated roots, meaning that some of the $c_1, c_2, \ldots, c_n$ are the same in factorization (∗). Recall that a root c is said to have multiplicity k if $(x - c)^k$ is a factor of $f(x)$ but no higher power of $(x - c)$ is a factor. Consequently, if every root is counted as many times as its multiplicity, then the statement in the preceding box implies that

A polynomial of degree n has exactly n roots.

EXAMPLE 1

Find a polynomial $f(x)$ of degree 5 such that 1, -2, and 5 are roots, 1 is a root of multiplicity 3 and $f(2) = -24$.

SOLUTION   Since 1 is a root of multiplicity 3, $(x - 1)^3$ must be a factor of $f(x)$. There are at least two other factors corresponding to the roots -2 and 5: $x - (-2) = x + 2$ and $x - 5$. The product of these factors $(x - 1)^3(x + 2)(x - 5)$ has degree 5, as does $f(x)$, so $f(x)$ must look like this:

$$f(x) = d(x - 1)^3(x + 2)(x - 5),$$

where d is the leading coefficient. Since $f(2) = -24$, we have

$$d(2 - 1)^3(2 + 2)(2 - 5) = f(2) = -24,$$

which reduces to $-12d = -24$. Therefore, $d = (-24)/(-12) = 2$, and

$$f(x) = 2(x - 1)^3(x + 2)(x - 5)$$
$$= 2x^5 - 12x^4 + 4x^3 + 40x^2 - 54x + 20. \qquad ■$$

POLYNOMIALS WITH REAL COEFFICIENTS

Recall that the **conjugate** of the complex number $a + bi$ is the number $a - bi$ (see page 323). We usually write a complex number as a single letter, say z, and indicate its conjugate by $\bar{z}$ (sometimes read "z bar"). For instance, if $z = 3 + 7i$, then $\bar{z} = 3 - 7i$. Conjugates play a role whenever a quadratic polynomial with real coefficients has complex roots.

EXAMPLE 2

The quadratic formula shows that $x^2 - 6x + 13$ has two complex roots.

$$\frac{-(-6) \pm \sqrt{(-6)^2 - 4 \cdot 1 \cdot 13}}{2 \cdot 1} = \frac{6 \pm \sqrt{-16}}{2} = \frac{6 \pm 4i}{2} = 3 \pm 2i.$$

The complex roots are $z = 3 + 2i$ and its conjugate $\bar{z} = 3 - 2i$. ■

The preceding example is a special case of a more general theorem, whose proof is outlined in Exercises 55 and 56.

Conjugate Roots Theorem

> Let $f(x)$ be a polynomial with *real* coefficients. If the complex number z is a root of $f(x)$, then its conjugate $\bar{z}$ is also a root of $f(x)$.

EXAMPLE 3

Find a polynomial with real coefficients whose roots include the numbers 2 and $3 + i$.

SOLUTION   Since $3 + i$ is a root, its conjugate $3 - i$ must also be a root. Consider the polynomial

$$f(x) = (x - 2)[(x - (3 + i)][x - (3 - i)].$$

Obviously, 2, $3 + i$, and $3 - i$ are roots of $f(x)$. Multiplying out this factored form shows that $f(x)$ *does* have real coefficients.

$$
\begin{aligned}
f(x) &= (x - 2)[x^2 - (3 - i)x - (3 + i)x + (3 + i)(3 - i)] \\
&= (x - 2)(x^2 - 3x + ix - 3x - ix + 9 - i^2) \\
&= (x - 2)(x^2 - 6x + 10) \\
&= x^3 - 8x^2 + 22x - 20.
\end{aligned}
$$

The next-to-last line of this calculation also shows that $f(x)$ can be factored as a product of a linear and a quadratic polynomial, each with *real* coefficients. ∎

The technique in Example 3 works because the polynomial

$$[x - (3 + i)][x - (3 - i)]$$

turns out to have real coefficients. The proof of the following result shows why this must always be the case.

Factorization over the Real Numbers

> Every nonconstant polynomial with real coefficients can be factored as a product of linear and quadratic polynomials with real coefficients in such a way that the quadratic factors, if any, have no real roots.

Proof   The box on page 329 shows that

$$f(x) = d(x - c_1)(x - c_2) \cdots (x - c_n),$$

where $c_1. \ldots, c_n$ are the roots of $f(x)$. If some c_i is a real number, then the factor $x - c_i$ is a linear polynomial with real coefficients.* If some c_j is a nonreal

---

*In Example 3, for instance, 2 is a real root and $x - 2$ a linear factor.

complex root, then its conjugate must also be a root. Thus some c_k is the conjugate of c_j, say, $c_j = a + bi$ (with a, b real) and $c_k = a - bi$.* In this case,

$$(x - c_j)(x - c_k) = [x - (a + bi)][x - (a - bi)]$$
$$= x^2 - (a - bi)x - (a + bi)x + (a + bi)(a - bi)$$
$$= x^2 - ax + bix - ax - bix + a^2 - (bi)^2$$
$$= x^2 - 2ax + (a^2 + b^2).$$

Therefore, the factor $(x - c_j)(x - c_k)$ of $f(x)$ is a quadratic with real coefficients (because a and b are real numbers). Its roots (c_j and c_k) are nonreal. By taking the real roots of $f(x)$ one at a time and the nonreal ones in conjugate pairs in this fashion, we obtain the desired factorization of $f(x)$.

EXAMPLE 4

Given that $1 + i$ is a root of $f(x) = x^4 - 2x^3 - x^2 + 6x - 6$, factor $f(x)$ completely over the real numbers.

SOLUTION   Since $1 + i$ is a root of $f(x)$, so is its conjugate $1 - i$, and hence $f(x)$ has this quadratic factor:

$$[x - (1 + i)][x - (1 - i)] = x^2 - 2x + 2.$$

Dividing $f(x)$ by $x^2 - 2x + 2$ shows that the other factor is $x^2 - 3$, which factors as $(x + \sqrt{3})(x - \sqrt{3})$. Therefore,

$$f(x) = (x + \sqrt{3})(x - \sqrt{3})(x^2 - 2x + 2).$$

*In Example 3, for instance, $c_j = 3 + i$ and $c_k = 3 - i$ are conjugate roots.

EXERCISES 4.8

In Exercises 1–4, find the remainder when $f(x)$ is divided by $g(x)$ without using synthetic or long division.

1. $f(x) = x^{15} + x^{10} + x^2$; $g(x) = x - 1$

2. $f(x) = x^4 + 3x^2 - 10$; $g(x) = x - 2$

3. $f(x) = x^5 + x^4 + x^3 + x^2$; $g(x) = x + 1$

4. $f(x) = 2x^{27} + x^2 + x + 8$; $g(x) = x$

In Exercises 5–8, determine if $g(x)$ is a factor of $f(x)$ without using synthetic division or long division.

5. $f(x) = x^5 - 3x^3 - 2x^2$; $g(x) = x - 2$

6. $f(x) = 3x^3 + 5x^2 - 2x + 3$; $g(x) = x + 1$

7. $f(x) = (3 + i)x^3 + (1 - 2i)x^2 + (2 + i)x + (1 - i)$; $g(x) = x - i$

8. $f(x) = x^2 + 2x + 2$; $g(x) = x + (1 + i)$

In Exercises 9–12, list the roots of the polynomial and state the multiplicity of each root.

9. $f(x) = x^{54}\left(x + \dfrac{4}{5}\right)$

10. $g(x) = 3\left(x + \dfrac{1}{6}\right)\left(x - \dfrac{1}{5}\right)\left(x + \dfrac{1}{4}\right)$

11. $h(x) = 2x^{15}(x - \pi)^{14}[x - (\pi + 1)]^{13}$

12. $k(x) = (x - \sqrt{7})^7(x - \sqrt{5})^5(2x - 1)$

In Exercises 13–24, find all the roots of $f(x)$ in the complex number system; then write $f(x)$ as a product of linear factors.

13. $f(x) = x^2 - 2x + 5$

14. $f(x) = x^2 - 4x + 13$

15. $f(x) = 3x^2 + 18x + 27$

16. $f(x) = 6x^2 - 9x - 60$

17. $f(x) = x^3 - 27$ [*Hint:* Factor first.]

18. $f(x) = x^3 + 125$

19. $f(x) = x^3 + 8$

20. $f(x) = x^6 - 64$ [*Hint:* Let $u = x^3$ and factor $u^2 - 64$ first.]

21. $f(x) = x^4 - 1$

22. $f(x) = x^4 - x^2 - 6$

23. $f(x) = x^4 + 2x^3 + x^2 - 2x - 2$

24. $f(x) = x^4 + 2x^3 + 9x^2 + 18x$

In Exercises 25–42, find a polynomial $f(x)$ with real coefficients that satisfies the given conditions. Some of these problems have many correct answers.

25. Degree 3; only roots are $1, 7, -4$.

26. Degree 3; only roots are 1 and -1.

27. Degree 6; only roots are $1, 2, \pi$.

28. Degree 5; only root is 2.

29. Degree 3; roots $-3, 0, 4; f(5) = 80$.

30. Degree 3; roots $-1, 1/2, 2; f(0) = 2$.

31. Roots include $2 + i$ and $2 - i$.

32. Roots include 3 and $4i - 1$.

33. Roots include $-3, 1 - i, 1 + 2i$.

34. Roots include $1, 2 + i, 3i, -1$.

35. Degree 2; roots $1 + 2i$ and $1 - 2i$.

36. Degree 4; roots $3i$ and $-3i$, each of multiplicity 2.

37. Degree 4; only roots are $4, 3 + i$, and $3 - i$.

38. Degree 5; roots 2 (of multiplicity 3), i, and $-i$.

39. Degree 6; roots 0 (of multiplicity 3) and $3, 1 + i, 1 - i$, each of multiplicity 1.

40. Degree 6; roots include i (of multiplicity 2) and 3.

41. Degree 2; roots include $1 + i; f(0) = 6$.

42. Degree 3; roots include $2 + 3i$ and $-2; f(2) = -3$.

In Exercises 43–46, find a polynomial with complex coefficients that satisfies the given conditions.

43. Degree 2; roots i and $1 - 2i$.

44. Degree 2; roots $2i$ and $1 + i$.

45. Degree 3; roots $3, i$, and $2 - i$.

46. Degree 4; roots $\sqrt{2}, -\sqrt{2}, 1 + i$, and $1 - i$.

In Exercises 47–52, one root of the polynomial is given; find all the roots.

47. $x^3 - 2x^2 - 2x - 3$; root 3.

48. $x^3 + x^2 + x + 1$; root i.

49. $x^4 + 3x^3 + 3x^2 + 3x + 2$; root i.

50. $x^4 - 6x^3 + 29x^2 - 76x + 68$; root 2 of multiplicity 2.

51. $x^4 - 4x^3 + 6x^2 - 4x + 5$; root $2 - i$.

52. $x^4 - 5x^3 + 10x^2 - 20x + 24$; root $2i$.

53. Let $z = a + bi$ and $w = c + di$ be complex numbers (a, b, c, d are real numbers). Prove the given equality by computing each side and comparing the results.

(a) $\overline{z + w} = \overline{z} + \overline{w}$ (The left side says: First find $z + w$ and then take the conjugate. The right side says: First take the conjugates of z and w and then add.)

(b) $\overline{z \cdot w} = \overline{z} \cdot \overline{w}$

54. Let $g(x)$ and $h(x)$ be polynomials of degree n and assume that there are $n + 1$ numbers $c_1, c_2, \ldots, c_n, c_{n+1}$ such that

$$g(c_i) = h(c_i) \quad \text{for every } i.$$

Prove that $g(x) = h(x)$. [*Hint:* Show that each c_i is a root of $f(x) = g(x) - h(x)$. If $f(x)$ is nonzero, what is its largest possible degree? To avoid a contradiction, conclude that $f(x) = 0$.]

55. Suppose $f(x) = ax^3 + bx^2 + cx + d$ has real coefficients and z is a complex root of $f(x)$.

(a) Use Exercise 53 and the fact that $\overline{r} = r$, when r is a real number, to show that

$$\overline{f(z)} = \overline{az^3 + bz^2 + cz + d}$$
$$= a\overline{z}^3 + b\overline{z}^2 + c\overline{z} + d = f(\overline{z}).$$

(b) Conclude that $\overline{z}$ is also a root of $f(x)$. [*Note:* $f(\overline{z}) = \overline{f(z)} = \overline{0} = 0$.]

56. Let $f(x)$ be a polynomial with real coefficients and z a complex root of $f(x)$. Prove that the conjugate $\overline{z}$ is also a root of $f(x)$. [*Hint:* Exercise 55 is the case when $f(x)$ has degree 3; the proof in the general case is similar.]

57. Use the statement in the second box on page 331 to show that every polynomial with real coefficients and *odd* degree must have at least one real root.

58. Give an example of a polynomial $f(x)$ with complex, non-real coefficients and a complex number z such that z is a root of $f(x)$, but its conjugate is not. Hence, the conclusion of the Conjugate Roots Theorem (page 331) may be *false* if $f(x)$ doesn't have real coefficients.

Chapter 4 Review

IMPORTANT CONCEPTS

IMPORTANT FACTS & FORMULAS

■ The graph of $f(x) = ax^2 + bx + c$ is a parabola whose vertex has x-coordinate $-b/2a$.

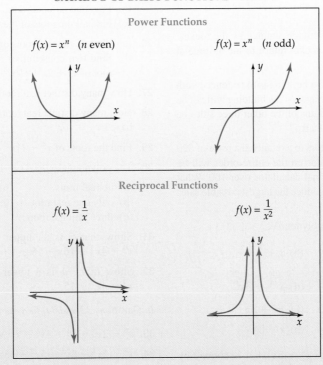

CATALOG OF BASIC FUNCTIONS—PART 2

REVIEW QUESTIONS

In Questions 1–5, find the vertex of the graph of the quadratic function.

1. $f(x) = (x - 2)^2 + 3$
2. $f(x) = 2(x + 1)^2 - 1$

3. $f(x) = x^2 - 8x + 12$
4. $f(x) = x^2 - 6x + 9$

5. $f(x) = 3x^2 - 9x + 1$

6. Which of the following statements about the functions

$$f(x) = 3x^2 + 2 \quad \text{and} \quad g(x) = -3x^2 + 2$$

is *false*?

(a) The graphs of f and g are parabolas.
(b) The graphs of f and g have the same vertex.
(c) The graphs of f and g open in opposite directions.
(d) The graph of f is the graph of $y = 3x^2$ shifted 2 units to the right.

7. A preschool wants to construct a fenced playground. The fence will be attached to the building at two corners, as shown in the figure. There is 400 feet of fencing available, all of it to be used.

(a) Write an equation in x and y that gives the amount of fencing to be used. Solve the equation for y.
(b) Write the area of the playground as a function of x. [Part (a) may be helpful.]
(c) What are the dimensions of the playground with the largest possible area?

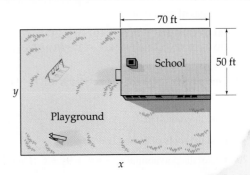

8. A model rocket is launched straight up from a platform at time $t = 0$ (where t is time measured in seconds). The

altitude $h(t)$ of the rocket above the ground at given time t is given by $h(t) = 10 + 112t - 16t^2$ (where $h(t)$ is measured in feet).
(a) What is the altitude of the rocket the instant it is launched?
(b) What is the altitude of the rocket 2 seconds after launching?
(c) What is the maximum altitude attained by the rocket?
(d) At what time does the rocket return to the altitude at which it was launched?

9. A rectangular garden next to a building is to be fenced with 120 feet of fencing. The side against the building will not be fenced. What should the lengths of the other three sides be to ensure the largest possible area?

10. A factory offers 100 calculators to a retailer at a price of $20 each. The price per calculator on the entire order will be reduced 5¢ for each additional calculator over 100. What number of calculators will produce the largest possible sales revenue for the factory?

11. Which of the following are polynomials?

(a) $2^3 + x^2$

(b) $x + \dfrac{1}{x}$

(c) $(\sqrt{2})x^3 + 2x^2 + 1$

(d) $x^{\frac{120}{10}} - 3x^{\frac{60}{10}}$

(e) $\sqrt[5]{2x^8}$

(f) $\pi x + 2\pi^3$

(g) $3x^3 + 2x^2 + \dfrac{1}{2}x^{1/2} + 1$

(h) $x - |x|$

12. What is the remainder when $x^4 + 3x^3 + 1$ is divided by $x^2 + 1$?

13. What is the remainder when

$$x^{112} - 2x^8 + 9x^5 - 4x^4 + x - 5$$

is divided by $x - 1$?

14. Is $x - 1$ a factor of $f(x) = 14x^{87} - 65x^{56} + 51$? Justify your answer.

15. Use synthetic division to show that $x - 2$ is a factor of $x^6 - 5x^5 + 8x^4 + x^3 - 17x^2 + 16x - 4$ and find the other factor.

16. List the roots of the following polynomials and the multiplicities of each root:

(a) $f(x) = \pi(x - 1)^3(x^2 - 9)^2(x + 2)$
(b) $g(x) = x^4 - 8x^3 + 18x^2 - 27$ [*Hint:* Try computing $g(-1)$.]

17. Find a polynomial f of degree 3 such that $f(-1) = 0, f(1) = 0$, and $f(0) = 5$.

18. Find the root(s) of $2\left(\dfrac{x}{5} + 7\right) - 3x - \dfrac{x + 2}{5} + 4$.

19. Find the roots of $3x^2 - 2x - 5$.

20. Factor the polynomial $x^3 - 8x^2 + 9x + 6$. [*Hint:* 2 is a root.]

21. Find all real roots of $x^6 - 4x^3 + 4$.

22. Find all real roots of $9x^3 - 6x^2 - 35x + 26$. [*Hint:* Try $x = -2$.]

23. Find all real roots of $3y^3(y^4 - y^2 - 5)$.

24. Find the rational roots of $x^4 - 2x^3 - 4x^2 + 1$.

25. Consider the polynomial $2x^3 - 8x^2 + 5x + 3$.
(a) List the only *possible* rational roots.
(b) Find one rational root.
(c) Find all the roots of the polynomial.

26. (a) Find all rational roots of $x^3 + 2x^2 - 2x - 2$.
(b) Find two consecutive integers such that an irrational root of $x^3 + 2x^2 - 2x - 2$ lies between them.

27. How many distinct real roots does $x^3 + 4x$ have?

28. How many distinct real roots does $f(x) = x^3 - x^2 - 3x + 3$ have?

29. Find the roots of $x^4 - 11x^2 + 18$.

30. The polynomial $x^3 - 2x + 1$ has
(a) no real roots.   (d) only one rational root.
(b) only one real root.   (e) none of the above.
(c) three rational roots.

31. Show that 5 is an upper bound for the real roots of $x^4 - 4x^3 + 16x - 16$.

32. Show that -1 is a lower bound for the real roots of $x^4 - 4x^3 + 15$.

In Questions 33 and 34, find the real roots of the polynomial.

33. $x^6 - 2x^5 - x^4 + 3x^3 - x^2 - x + 1$

34. $x^5 - 5x^4 + 8x^3 - 5x^2 - 13x + 30$

In Questions 35 and 36, compute and simplify the difference quotient of the function.

35. $f(x) = x^2 + x$ 36. $g(x) = 3x^3 + 2$

37. Draw the graph of a function that could not possibly be the graph of a polynomial function and explain why.

38. Draw a graph that could be the graph of a polynomial function of degree 5. You need not list a specific polynomial, nor do any computation.

39. Which of the statements (a)–(c) is *not* true about the polynomial function f whose graph is shown in the figure?

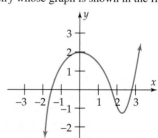

(a) f has three roots between -2 and 3.
(b) $f(x)$ could possibly be a fifth-degree polynomial.
(c) $(f \circ f)(0) > 0$.
(d) $f(2) - f(-1) < 3$.
(e) $f(x)$ is positive for all x in the interval $[-1, 0]$.

40. Which of the statements (i)–(v) about the polynomial function f whose graph is shown in the figure are *false*?

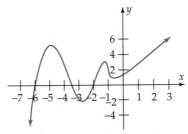

(i) f has 2 roots in the interval $[-6, -3)$.
(ii) $f(-3) - f(-6) < 0$.
(iii) $f(0) < f(1)$.
(iv) $f(2) - 2 = 0$.
(v) f has degree ≤ 4.

In Questions 41–44, find a viewing window (or windows) that shows a complete graph of the function. Be alert for hidden behavior.

41. $f(x) = 32x^3 - 99x^2 + 100x + 2$

42. $g(x) = .3x^5 - 4x^4 + x^3 - 4x^2 + 5x + 1$

43. $h(x) = 4x^3 - 100x^2 + 600x$

44. $f(x) = x^7 - 11x^6 + 50x^5 - 120x^4 + 160x^3 - 112x^2 + 32x$

45. HomeArt makes plastic replicas of famous statues. Their total cost to produce copies of a particular statue are shown in the table below.

(a) Make a scatter plot of the data.
(b) Use cubic regression to find a function $C(x)$ that models the data [that is, $C(x)$ is the cost of making x statues]. Assume that C is reasonably accurate when $x \leq 100$.
(c) Use C to estimate the cost of making the seventy-first statue.
(d) Use C to approximate the average cost per statue when 35 are made and when 75 are made. [Recall that the average cost of x statues is $C(x)/x$.]

Number of Statues	Total Cost
0	$2,000
10	2,519
20	2,745
30	2,938
40	3,021
50	3,117
60	3,269
70	3,425

46. The table gives the estimated cost of a college education at a public institution. Costs include tuition, fees, books, and room and board for four years. (*Source:* Teachers Insurance and Annuity Association College Retirement Equities Fund)

Enrollment Year	Costs
1998	$46,691
2000	52,462
2002	58,946
2004	66,232
2006	74,418

Enrollment Year	Costs
2008	$83,616
2010	93,951
2012	105,564
2014	118,611

(a) Make a scatter plot of the data (with $x = 0$ corresponding to 1990).
(b) Use quartic regression to find a function C that models the data.
(c) Estimate the cost of a college education in 2007 and in 2015.

In Questions 47–54, sketch a complete graph of the function. Label the x-intercepts, all local extrema, holes, and asymptotes.

47. $f(x) = x^3 - 9x$

48. $g(x) = x^3 - 2x^2 + 3$

49. $h(x) = x^4 + 3x^3 - 12x^2 - 20x + 48$

50. $f(x) = x^4 - 3x - 2$

51. $g(x) = \dfrac{-2}{x + 4}$

52. $\dfrac{3x^3 + 2x^2 - 8x}{x^3 - 3x^2 + 2x}$

53. $k(x) = \dfrac{4x + 10}{3x - 9}$

54. $f(x) = \dfrac{x + 1}{x^2 - 1}$

In Questions 55 and 56, list all asymptotes of the graph of the function.

55. $f(x) = \dfrac{x^2 - 1}{x^3 - 2x^2 - 5x + 6}$

56. $g(x) = \dfrac{x^4 - 6x^3 + 2x^2 - 6x + 2}{x^2 - 3}$

In Questions 57–60, find a viewing window (or windows) that shows a complete graph of the function. Be alert for hidden behavior.

57. $f(x) = \dfrac{x - 3}{x^2 + x - 2}$

58. $g(x) = \dfrac{x^2 - x - 6}{x^3 - 3x^2 + 3x - 1}$

59. $h(x) = \dfrac{x^4 + 4}{x^4 - 99x^2 - 100}$

60. $k(x) = \dfrac{x^3 - 2x^2 - 4x + 8}{x - 10}$

61. It costs the Junkfood Company 50¢ to produce a bag of Munchies. There are fixed costs of $500 per day for building, equipment, etc. The company has found that if the price of a bag of Munchies is set at $1.95 - \dfrac{x}{2000}$ dollars, where x is the number of bags produced per day, then all the bags that are produced will be sold. What number of bags can be produced each day if all are to be sold and the company is to make a profit? What are the possible prices?

62. Sunnyvale village is proud of its Main Street, which is a 33 mile long street running from the northernmost point of the village to its southernmost point. There is a 3 mile long stretch that goes through "Downtown Sunnyvale." They have a law that requires the speed limit through downtown to be 15 miles/hour less than the speed limit everywhere else.

(a) If the speed limit through downtown is 20 miles/hour, how long does it take to get from the northernmost point of Sunnyvale to its southernmost point?

(b) Express the total time for the north-to-south journey as a function of the speed limit through downtown.

(c) The Sunnyvale Chamber of Commerce would like the time for a trip through Sunnyvale to take 1.5 hours. What would the downtown speed limit have to be for this to be the case? Do you think the citizens of Sunnyvale would agree to their proposal?

63. The survival rate s of seedlings in the vicinity of a parent tree is given by

$$s = \frac{.5x}{1 + .4x^2},$$

where x is the distance from the seedling to the tree (in meters) and $0 < x \le 10$.

(a) For what distances is the survival rate at least .21?

(b) What distance produces the maximum survival rate?

In Questions 64 and 65, find the average rate of change of the function between x and $x + h$.

64. $f(x) = \dfrac{x}{x + 1}$

65. $g(x) = \dfrac{1}{x^2 + 1}$

66. Which of these statements about the graph of

$$f(x) = \frac{(x - 1)(x + 3)}{(x^2 + 1)(x^2 - 1)}$$

is *true*?

(a) The graph has two vertical asymptotes.

(b) The graph touches the x-axis at $x = 3$.

(c) The graph lies above the x-axis when $x < -1$.

(d) The graph has a hole at $x = 1$.

(e) The graph has no horizontal asymptotes.

67. Solve for x: $-3(x - 4) \le 5 + x$.

68. Solve for y: $\left|\dfrac{y + 2}{3}\right| \ge 5$.

69. Solve for x: $-4 < 2x + 5 < 9$.

70. On which intervals is $\dfrac{2x - 1}{3x + 1} < 1$?

71. On which intervals is $\dfrac{2}{x + 1} < x$?

72. Solve for x: $\left|\dfrac{1}{1 - x^2}\right| \ge \dfrac{1}{2}$.

73. Solve for x: $x^2 + x > 12$.

74. Solve for x: $(x - 1)^2(x^2 - 1)x \le 0$.

75. If $0 < r \le s - t$, then which of these statements is *false*?

(a) $s \ge r + t$ (b) $t - s \le -r$

(c) $-r \ge s - t$ (d) $\dfrac{s - t}{r} > 0$

(e) $s - r \ge t$

76. Solve and express your answer in interval notation:

$$2x - 3 \le 5x + 9 < -3x + 4.$$

In Questions 77–84, solve the inequality.

77. $|3x + 2| \ge 2$

78. $x^2 + x - 20 > 0$

79. $\dfrac{x - 2}{x + 4} \le 3$

80. $(x + 1)^2(x - 3)^4(x + 2)^3(x - 7)^5 > 0$

81. $\dfrac{x^2 + x - 9}{x + 3} < 1$

82. $\dfrac{x^2 - x - 6}{x - 3} > 1$

83. $\dfrac{x^2 - x - 5}{x^2 + 2} > -2$

84. $\dfrac{x^4 - 3x^2 + 2x - 3}{x^2 - 4} < -1$

In Questions 85–92, solve the equation in the complex number system.

85. $x^2 + 3x + 10 = 0$

86. $x^2 + 2x + 5 = 0$

87. $5x^2 + 2 = 3x$

88. $-3x^2 + 4x - 5 = 0$

89. $3x^4 + x^2 - 2 = 0$

90. $8x^4 + 10x^2 + 3 = 0$

91. $x^3 + 8 = 0$

92. $x^3 - 27 = 0$

93. One root of $x^4 - x^3 - x^2 - x - 2$ is i. Find all the roots.

94. One root of $x^4 + x^3 - 5x^2 + x - 6$ is i. Find all the roots.

95. Give an example of a fourth-degree polynomial with real coefficients whose roots include 0 and $1 + i$.

96. Find a fourth-degree polynomial f whose only roots are $2 + i$ and $2 - i$, such that $f(-1) = 50$.

Chapter 4 Test

Sections 4.1–4.4 including Special Topics 4.2.A and 4.4.A.

1. An object moves back and forth along a straight track. Its distance in meters from its starting point is given by the function $f(t) = 15t^4 - 5t^2 + 6t^3 - 2t$, where t is in minutes. At what time does it return to its starting point?

2. State the quotient and remainder when the first polynomial is divided by the second.

   (a) $x^3 + 5x^2 + 5x + 7, x + 4$
   (b) $2x^4 - 5x^3 + 2x^2 - x + 2, x - 2$
   (c) $x^5 - 2x^3 - 2x^4 + 5x^2 - 7x + 3, x^2 - 3$

3. Without using a calculator, match the following functions with their graphs:

   (a) $f(x) = (x - 1)^2 + 2$ (b) $f(x) = (x + 1)^2 + 2$
   (c) $f(x) = (x - 1)^2 - 2$ (d) $f(x) = (x + 1)^2 - 2$
   (e) $f(x) = -(x - 1)^2 + 2$ (f) $f(x) = -(x + 1)^2 + 2$
   (g) $f(x) = -(x - 1)^2 - 2$ (h) $f(x) = -(x + 1)^2 - 2$

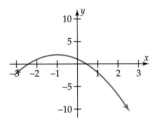
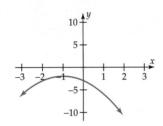
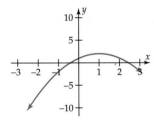
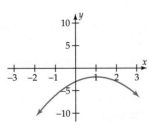
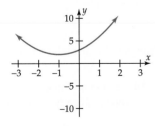
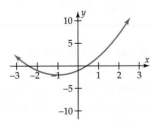
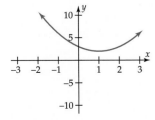
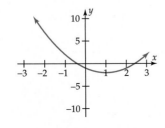

4. A mime troupe decides to start charging admission to their performances. They figure that if they charge $10 per ticket, they can sell 75 tickets. For every quarter decrease in price, they can sell 5 more tickets. Conversely, for every quarter increase in price, they can sell 5 fewer tickets.

   (a) Write a function that gives their total income if they charge $x for a ticket.
   (b) What should they charge to make as much money as possible?

5. Find all real roots of the following polynomials

   (a) $3x^3 - x^2 - 3x + 1$
   (b) $x^4 - 11x^2 + 18$
   (c) $x^4 + x^3 + x^2 + 3x - 6$

Sections 4.5–4.8 including Special Topics 4.5.A and 4.6.A

6. Two positive numbers, x and y sum to 30. If $xy > 200$, what are the possible values of x? (x need not be an integer)

7. Solve the following equations and write the answers in the form $a + bi$

   (a) $2x + 3 - 4i = 3x + 6 + 2i$
   (b) $x^2 - 4x + 13 = 0$
   (c) $4 + x^2 = 5x^2 + 5$

8. Consider the following function: $f(x) = \dfrac{x + 1}{x^2 + 2x - 8}$

   (a) Find the y intercept of this function
   (b) List the vertical and horizontal asymptotes of this function.
   (c) Find the zeros of this function.
   (d) Sketch a complete graph of this function.

9. Solve the following inequalities

   (a) $3x - 6 \geq 4 - 2x$
   (b) $x^2 - 3x + 2 < 0$
   (c) $\dfrac{3x - 1}{x + 2} > \dfrac{1}{x + 2}$
   (d) $|2x - 1| < 5$
   (e) $|4x - 6| \geq 12$

10. Find the nonvertical asymptote of the following function, and sketch its graph:

$$\frac{3x^2 - 3x + 1}{x - 1}$$

11. Find a 5th degree polynomial with real coefficients whose roots include $1 + 2i$, $5 + i$ and 0. You may leave it in factored form if you choose.

DISCOVERY PROJECT 4   Architectural Arches

Mike Mazzaschi / Stock, Boston Inc. / PictureQuest

You can see arches almost everywhere you look—in windows, entryways, tunnels, and bridges. Common arch shapes are semicircles, semiellipses, and parabolas. When constructed on a level base, arches are symmetric from left to right. This means that a mathematical function that describes an arch must be an even function.

A *semicircular arch* always has the property that its base is twice as wide as its height. This ratio can be modified by placing a rectangular area under the semicircle, giving a shape known as a *Norman arch*. This approach gives a tunnel or room a vaulted ceiling. *Parabolic arches* can also be created to give a more vaulted appearance.

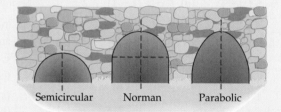

Semicircular   Norman   Parabolic

For the following exercises in arch modeling, you should always set the origin of your coordinate system to be the center of the base of the arch.

1. Show that the function that models a semicircular arch of radius r is
    $$h(x) = \sqrt{r^2 - x^2}.$$

2. Write a function $h(x)$ that models a semicircular arch that is 15 feet tall. How wide is the arch?

3. Write a function $n(x)$ that models a Norman arch that is 15 feet tall and 16 feet wide at the base.

4. Parabolic arches are typically modeled by using the function
    $$p(x) = H - ax^2,$$
    where H is the height of the arch. Write a function $p(x)$ for an arch that is 15 feet tall and 16 feet wide at the base.

5. Would a truck that is 12 feet tall and 9 feet wide fit through all three arches? How could you fix any of the arches that is too small so that the truck would fit through?

EXPONENTIAL AND LOGARITHMIC FUNCTIONS

How old is that dinosaur?

Population growth (of humans, fish, bacteria, etc.), compound interest, radioactive decay, and a host of other phenomena can be mathematically described by exponential functions (see Exercises 52, 65 and 67 on pages 366 and 368). Archeologists sometimes use carbon-14 dating to determine the approximate age of an artifact (such as a dinosaur skeleton, a mummy, or a wooden statue). This involves using logarithms to solve an appropriate exponential equation. See Exercise 75 on page 369.

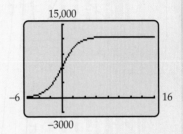

© Rudi Von Briel/PhotoEdit

Chapter Outline

Exponential and logarithmic functions provide mathematical models of many natural and scientific phenomena. Technology is usually needed to evaluate these functions, but the calculations can sometimes be done by hand. Knowing the domains and ranges of these functions and the unique shapes of their graphs is essential for understanding why these functions are so common in nature.

5.1   Radicals and Rational Exponents

Section Objectives
- ■ Find exact and approximate roots of real numbers.
- ■ Simplify expressions with rational exponents.
- ■ Interpret radical notation.
- ■ Rationalize numerators and denominators.

A **square root** of a nonnegative number c is a number whose square is c, that is, a solution of $x^2 = c$. For a real number c and a positive integer n, we define the **nth roots** of c as the solutions of the equation $x^n = c$. The graphs on the next page show that this equation may have two, one, or no solutions, depending on whether n is even or odd and whether c is positive or negative.*

---

*The solutions of $x^n = c$ are the x-coordinates of the intersection points of the graph of $y = x^n$ and the horizontal line $y = c$. The graph of $y = x^n$ was discussed on page 270.

$x^n = c$	$x^n = c$		
n odd	n even		
Exactly one solution for any c	$c > 0$ One positive and one negative solution	$c = 0$ One solution $x = 0$	$c < 0$ No solution

Consequently, we have the following definition.

nth Roots

Let c be a real number and n a positive integer. The principal **nth root of c** is denoted by either of the symbols

$$\sqrt[n]{c} \quad \text{or} \quad c^{1/n}$$

and is defined to be

The solution of $x^n = c$, when n is odd;

The nonnegative solution of $x^n = c$, when n is even and $c \geq 0$.

As before, principal square roots are denoted $\sqrt{}$ rather than $\sqrt[2]{}$.

EXAMPLE 1

Find the following roots.

(a) $\sqrt[3]{-8}$ (b) $81^{1/4}$ (c) $\sqrt[5]{32}$ (d) $\sqrt{\dfrac{1}{4}}$

SOLUTION

(a) $\sqrt[3]{-8} = (-8)^{1/3} = -2$ because -2 is the solution of $x^3 = -8$.

(b) $81^{1/4} = \sqrt[4]{81} = 3$ because 3 is the positive solution of $x^4 = 81$.

(c) $\sqrt[5]{32} = 32^{1/5} = 2$ because 2 is the solution of $x^5 = 32$.

(d) $\sqrt{\dfrac{1}{4}} = \dfrac{1}{2}$ because $\dfrac{1}{2}$ is the positive solution of $x^2 = \dfrac{1}{4}$.

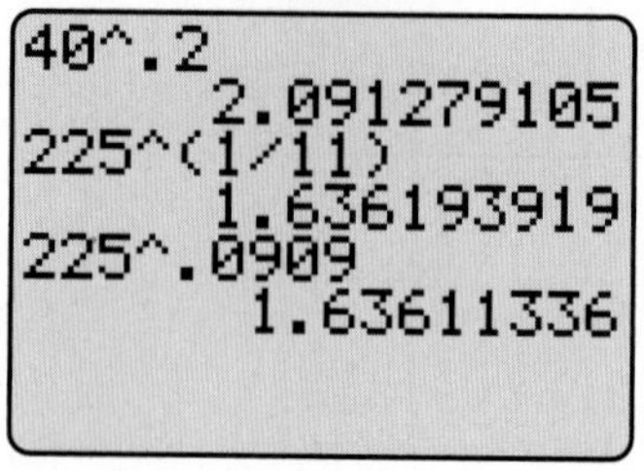

```
40^.2
       2.091279105
225^(1/11)
       1.636193919
225^.0909
       1.63611336
```

Figure 5-1

EXAMPLE 2

Use a calculator to approximate

(a) $40^{1/5}$ (b) $225^{1/11}$

SOLUTION

(a) Since $1/5 = .2$, you can compute $40^{.2}$, as in Figure 5–1.

(b) $1/11$ is the infinite decimal $.090909 \ldots$. In such cases, it is best to leave the exponent in fractional form and use parentheses, as shown in Figure 5–1. If you round off the decimal equivalent of $1/11$, say as $.0909$, you will not get the same answer, as Figure 5–1 shows. ∎

RATIONAL EXPONENTS

The next step is to give a meaning to fractional exponents for any fraction, not just those of the form $1/n$. If possible, they should be defined in such a way that the various exponent rules continue to hold. Consider, for example, how $4^{3/2}$ might be defined. The exponent $3/2$ can be written as either

$$3 \cdot \left(\frac{1}{2}\right) \qquad \text{or} \qquad \left(\frac{1}{2}\right) \cdot 3.$$

If the power of a power property $(c^m)^n = c^{mn}$ is to hold, we might define $4^{3/2}$ as either $(4^3)^{1/2}$ or $(4^{1/2})^3$. The result is the same in both cases:

$$(4^3)^{1/2} = 64^{1/2} = \sqrt{64} = 8 \qquad \text{and} \qquad (4^{1/2})^3 = (\sqrt{4})^3 = 2^3 = 8.$$

It can be proved that the same thing is true in the general case, which leads to this definition.

*Rational
Exponents*

> Let c be a positive real number and let t/k be a rational number in lowest terms with positive denominator. Then,
>
> $c^{t/k}$ is defined to be the number $(c^t)^{1/k} = (c^{1/k})^t$.

Since every terminating decimal is a rational number, expressions such as $13^{3.77}$ now have a meaning, namely, $13^{377/100}$. (Actually, we used this fact earlier when we computed $40^{.2}$ and $225^{.0909}$ in Figure 5–1.)

TECHNOLOGY TIP

TI-89 will produce real number answers to computations like $(-8)^{2/3}$ if you select "Real" in the COMPLEX FORMAT submenu of the MODE menu.

NOTE

The preceding definition is also valid when c is negative, *provided that* the exponent has an odd denominator, such as $(-8)^{2/3}$ (see Exercise 69). Nevertheless, if you try to compute $(-8)^{2/3}$ on your calculator, you may get an error message or a complex number (indicated by an ordered pair or an expression involving i) instead of the correct answer,

$$(-8)^{2/3} = [(-8)^2]^{1/3} = \sqrt[3]{(-8)^2} = \sqrt[3]{64} = 4.$$

If this happens, you can probably get the correct answer by keying in either $[(-8)^2]^{1/3}$ or $[(-8)^{1/3}]^2$, each of which is equal to $(-8)^{2/3}$.

Rational exponents were defined in a way that guaranteed that one of the familiar exponent laws would remain valid. In fact, all exponent laws developed for integer exponents are valid for rational exponents, as summarized here and illustrated in Examples 3 through 5.

Exponent Laws

Let c and d be nonnegative real numbers, and let r and s be any rational numbers. Then

1. $c^r c^s = c^{r+s}$

2. $\dfrac{c^r}{c^s} = c^{r-s}$ $(c \neq 0)$

3. $(c^r)^s = c^{rs}$

4. $(cd)^r = c^r d^r$

5. $\left(\dfrac{c}{d}\right)^r = \dfrac{c^r}{d^r}$ $(d \neq 0)$

6. $c^{-r} = \dfrac{1}{c^r}$ $(c \neq 0)$

CAUTION

The exponent laws deal only with products and quotients. There are no analogous properties for sums. In particular, if both c and d are nonzero, then

$(c + d)^r$ is **not** equal to $c^r + d^r$.

EXAMPLE 3

Compute the product $x^{1/2}(x^{3/4} - x^{3/2})$.

SOLUTION

Distributive law:

$$x^{1/2}(x^{3/4} - x^{3/2}) = x^{1/2}x^{3/4} - x^{1/2}x^{3/2}$$

Multiplication with exponents (law 1):

$$= x^{1/2+3/4} - x^{1/2+3/2}$$

$\dfrac{1}{2} + \dfrac{3}{4} = \dfrac{5}{4}$ and $\dfrac{1}{2} + \dfrac{3}{2} = 2$:

$$= x^{5/4} - x^2. \qquad \blacksquare$$

EXAMPLE 4

Simplify $(8r^{3/4}s^{-3})^{2/3}$, and express it without negative exponents.

SOLUTION

Product to a power (law 4):

$$(8r^{3/4}s^{-3})^{2/3} = 8^{2/3}(r^{3/4})^{2/3}(s^{-3})^{2/3}$$

Power of a power (law 3):

$$= 8^{2/3}r^{(3/4)(2/3)}s^{(-3)(2/3)}$$

$\dfrac{3}{4} \cdot \dfrac{2}{3} = \dfrac{1}{2}$ and $(-3)\dfrac{2}{3} = -2$:

$$= 8^{2/3}r^{1/2}s^{-2}$$

Negative exponents (law 6):

$$= \dfrac{8^{2/3}r^{1/2}}{s^2}$$

$8^{2/3} = \sqrt[3]{8^2} = \sqrt[3]{64} = 4$:

$$= \dfrac{4r^{1/2}}{s^2} \quad \text{or} \quad \dfrac{4\sqrt{r}}{s^2}. \qquad \blacksquare$$

EXAMPLE 5

Simplify $\dfrac{x^{7/2}y^3}{(xy^{7/4})^2}$ and express it without negative exponents.

SOLUTION

Product to a power (law 4):

$$\frac{x^{7/2}y^3}{(xy^{7/4})^2} = \frac{x^{7/2}y^3}{x^2(y^{7/4})^2}$$

$$\frac{7}{4} \cdot 2 = \frac{7}{2}:$$

$$= \frac{x^{7/2}y^3}{x^2y^{7/2}}$$

Division with exponents (law 2):

$$= x^{7/2-2}y^{3-7/2}$$

$$\frac{7}{2} - 2 = \frac{3}{2} \quad \text{and} \quad 3 - \frac{7}{2} = -\frac{1}{2}:$$

$$= x^{3/2}y^{-1/2}$$

Negative exponents (law 6):

$$= \frac{x^{3/2}}{y^{1/2}} \qquad\blacksquare$$

RADICAL NOTATION

As we saw above, there are two notations for nth roots: $\sqrt[n]{c}$ and $c^{1/n}$. Similarly, $c^{t/k}$ can be expressed in terms of radicals.

$$c^{t/k} = (c^t)^{1/k} = \sqrt[k]{c^t} \quad \text{and} \quad c^{t/k} = (c^{1/k})^t = (\sqrt[k]{c})^t.$$

When radical notation is used, the exponent laws have a different form. For instance,

$$(cd)^{1/n} = c^{1/n}d^{1/n} \quad \text{means the same thing as} \quad \sqrt[n]{cd} = \sqrt[n]{c}\,\sqrt[n]{d};$$

$$\left(\frac{c}{d}\right)^{1/n} = \frac{c^{1/n}}{d^{1/n}} \quad \text{means the same thing as} \quad \sqrt[n]{\frac{c}{d}} = \frac{\sqrt[n]{c}}{\sqrt[n]{d}}.$$

When simplifying an expression involving radicals, either you can change to rational exponents and proceed as in Examples 3–5, or you can use the exponent laws in their radical notation, as in the next example.

EXAMPLE 6

Simplify each of the following.

(a) $\sqrt{63}$ (b) $\sqrt{12} - \sqrt{75}$ (c) $\sqrt[3]{40k^4}$

SOLUTION

(a) Look for perfect squares.

Factor a perfect square out of 63: $\qquad \sqrt{63} = \sqrt{9 \cdot 7}$

Write as a product of roots: $\qquad = \sqrt{9}\,\sqrt{7}$

$$= 3\sqrt{7}$$

(b) Look for perfect squares.

Factor perfect squares out of 12 and 75: $\quad \sqrt{12} - \sqrt{75} = \sqrt{4 \cdot 3} - \sqrt{25 \cdot 3}$

Product of roots: $\qquad = \sqrt{4}\,\sqrt{3} - \sqrt{25}\,\sqrt{3}$

$$= 2\sqrt{3} - 5\sqrt{3} = -3\sqrt{3}.$$

(c) Look for perfect cubes. We note that $8k^3 = (2k)^3$ is a perfect cube.

Factor out $8k^3$: $\qquad \sqrt[3]{40k^4} = \sqrt[3]{8k^3 \cdot 5k}$

Product of roots: $\qquad = \sqrt[3]{8k^3}\,\sqrt[3]{5k}$

$$= \sqrt[3]{(2k)^3}\,\sqrt[3]{5k} = 2k\sqrt[3]{5k}. \qquad\blacksquare$$

EXAMPLE 7

EXAMPLE 7

Compute each of the following

(a) $27^{-5/3}$ (b) $2^{3/2}$ (c) $256^{3/4}$ (d) $(-1)^{2/6}$

SOLUTION

(a) $27^{-5/3} = \dfrac{1}{27^{5/3}} = \dfrac{1}{(\sqrt[3]{27})^5} = \dfrac{1}{3^5} = \dfrac{1}{243}$

(b) $2^{3/2} = (\sqrt{2})^3 = 2\sqrt{2}$

(c) $256^{3/4} = (\sqrt[4]{256})^3 = 4^3 = 64$

(d) Recall that, in order to use our definition of rational exponents, the fraction must be in lowest terms. So $(-1)^{2/6} = (-1)^{1/3} = \sqrt[3]{-1} = -1$. Notice that we get an *incorrect* answer if we do not first reduce the fraction, because $\sqrt[6]{(-1)^2} = \sqrt[6]{1} = 1$. ∎

RATIONALIZING DENOMINATORS AND NUMERATORS

When dealing with fractions in the days before calculators, it was customary to *rationalize the denominators*, that is, write equivalent fractions with no radicals in the denominator, because this made many computations easier. With calculators, of course, there is no computational advantage to rationalizing denominators. Nevertheless, rationalizing denominators or numerators is sometimes needed to simplify expressions and to derive useful formulas. For example, several key calculus formulas can be derived by rationalizing the numerator of a rational expression.

EXAMPLE 8

(a) To rationalize the denominator of $\dfrac{7}{\sqrt{5}}$, multiply it by 1, with 1 written as a suitable radical fraction.

$$\frac{7}{\sqrt{5}} = \frac{7}{\sqrt{5}} \cdot 1 = \frac{7}{\sqrt{5}} \cdot \frac{\sqrt{5}}{\sqrt{5}} = \frac{7\sqrt{5}}{5}.$$

(b) To rationalize the denominator of $\dfrac{2}{3 + \sqrt{6}}$, multiply by 1 written as a radical fraction and use the multiplication pattern $(a + b)(a - b) = a^2 - b^2$.

$$\frac{2}{3 + \sqrt{6}} = \frac{2}{3 + \sqrt{6}} \cdot 1$$

$$= \frac{2}{3 + \sqrt{6}} \cdot \frac{3 - \sqrt{6}}{3 - \sqrt{6}} = \frac{2(3 - \sqrt{6})}{(3 + \sqrt{6})(3 - \sqrt{6})}$$

$$= \frac{6 - 2\sqrt{6}}{3^2 - (\sqrt{6})^2} = \frac{6 - 2\sqrt{6}}{9 - 6} = \frac{6 - 2\sqrt{6}}{3}.$$

∎

EXAMPLE 9

Assume that $h \neq 0$, and rationalize the numerator of $\dfrac{\sqrt{x+h} - \sqrt{x}}{h}$; that is, write an equivalent fraction with no radicals in the numerator.

SOLUTION   Again, the technique is to multiply the fraction by 1, with 1 written as a suitable radical fraction:

$$\frac{\sqrt{x+h} - \sqrt{x}}{h} = \frac{\sqrt{x+h} - \sqrt{x}}{h} \cdot 1$$

$$= \frac{\sqrt{x+h} - \sqrt{x}}{h} \cdot \frac{\sqrt{x+h} + \sqrt{x}}{\sqrt{x+h} + \sqrt{x}} = \frac{(\sqrt{x+h})^2 - (\sqrt{x})^2}{h(\sqrt{x+h} + \sqrt{x})}$$

$$= \frac{x+h-x}{h(\sqrt{x+h} + \sqrt{x})} = \frac{h}{h(\sqrt{x+h} + \sqrt{x})} = \frac{1}{\sqrt{x+h} + \sqrt{x}}.$$

Notice that in its original form, it is hard to determine what happens to this fraction when h is close to zero. In the final form, we can see that when h gets close to zero, the expression gets close to $1/2\sqrt{x}$. ∎

IRRATIONAL EXPONENTS

An example will illustrate how a^t is defined when t is an irrational number.*
To compute $10^{\sqrt{2}}$ we use the infinite decimal expansion $\sqrt{2} \approx 1.414213562 \cdots$
(see Special Topics 1.1.A). Each of

$$1.4, \ 1.41, \ 1.414, \ 1.4142, \ 1.41421, \ldots$$

is a rational number approximation of $\sqrt{2}$, and each is a more accurate approximation than the preceding one. We know how to raise 10 to each of these rational numbers:

$$10^{1.4} = 10^{14/100} \approx 25.1189$$

$$10^{1.41} = 10^{141/1{,}000} \approx 25.7040$$

$$10^{1.414} = 10^{1{,}414/10{,}000} \approx 25.9418$$

$$10^{1.4142} = 10^{14{,}142/100{,}000} \approx 25.9537$$

$$10^{1.41421} = 10^{141{,}421/1{,}000{,}000} \approx 25.9543$$

$$10^{1.414213} = 10^{1{,}414{,}213/10{,}000{,}000} \approx 25.9545$$

It appears that as the exponent r gets closer and closer to $\sqrt{2}$, 10^r gets closer and closer to a real number whose decimal expansion begins 25.954 . . . . We define $10^{\sqrt{2}}$ to be this number.

Similarly, for any $a > 0$,

a^t is a well-defined *positive* number for each real exponent t.

We shall also assume this fact.

The exponent laws (page 345) are valid for *all* real exponents.

---

*This example is not a proof but should make the idea plausible. Calculus is required for a rigorous proof.

EXERCISES 5.1

Note: *Unless directed otherwise, assume that all letters represent positive real numbers.*

In Exercises 1–10, simplify the expression. Assume a, b, c, d > 0.

1. $(25k^2)^{3/2}(16k^{1/3})^{3/4}$

2. $(4x^{5/6})(2y^{3/4})(x^{7/6})(3y^{-1/4})$

3. $(c^{2/5}d^{-2/3})(c^6d^3)^{4/3}$

4. $(\sqrt[3]{3}x^2y)(\sqrt[3]{9}x^{-1/3}y^{3/5})^{-2}$

5. $\dfrac{(x^2)^{1/3}(y^2)^{2/3}}{3x^{2/3}y^2}$

6. $\dfrac{(a^{1/2}b^2)^3(a^{1/2}b^0c)}{(ab^2)^2(bc^5)^0}$

7. $\dfrac{(7a)^2(5b)^{3/2}}{(5a)^{3/2}(7b)^4}$

8. $\dfrac{\sqrt{ab}\,\sqrt[3]{ab^4}}{\sqrt{a}\,(\sqrt[3]{b})^4}$

9. $(a^{x^2})^{1/x}$

10. $\dfrac{(b^x)^{x-1}}{b^{-x}}$

In Exercises 11–16, compute and simplify.

11. $x^{1/2}(x^{2/3} - x^{4/3})$

12. $x^{1/2}(3x^{3/2} + 2x^{-1/2})$

13. $(x^{1/2} + y^{1/2})(x^{1/2} - y^{1/2})$

14. $(x^{1/3} + y^{1/2})(2x^{1/3} - y^{3/2})$

15. $(x + y)^{1/2}[(x + y)^{1/2} - (x + y)]$

16. $(x^{1/3} + y^{1/3})(x^{2/3} - x^{1/3}y^{1/3} + y^{2/3})$

In Exercises 17–22, factor the given expression. For example,

$$x - x^{1/2} - 2 = (x^{1/2} - 2)(x^{1/2} + 1).$$

17. $x^{2/3} + x^{1/3} - 6$

18. $x^{2/7} - 2x^{1/7} - 15$

19. $x + 4x^{1/2} + 3$

20. $x^{1/3} + 11x^{1/6} + 24$

21. $x^{4/5} - 81$

22. $x + 3x^{2/3} + 3x^{1/3} + 1$

In Exercises 23–28, write the given expression without using radicals.

23. $\dfrac{1}{\sqrt{x}}$

24. $\sqrt[5]{x^2}$

25. $a\sqrt{a+b}$

26. $\sqrt{\sqrt{\sqrt[3]{a^3b^4}}}$

27. $\sqrt[5]{t}\sqrt{16t^5}$

28. $\sqrt{x}(\sqrt[3]{x^2})(\sqrt[4]{x^3})$

In Exercises 29–44, simplify the expression without using a calculator.

29. $\sqrt{80}$

30. $\sqrt{120}$

31. $\sqrt{6}\sqrt{12}$

32. $\sqrt[3]{12}\sqrt[3]{10}$

33. $\dfrac{-6 + \sqrt{99}}{15}$

34. $\dfrac{18 - \sqrt{126}}{3}$

35. $\sqrt{50} - \sqrt{72}$

36. $\sqrt{150} + \sqrt{24}$

37. $5\sqrt{20} - \sqrt{45} + 2\sqrt{80}$

38. $\sqrt[3]{40} + 2\sqrt[3]{135} - 5\sqrt[3]{320}$

39. $\sqrt{16a^8b^{-2}}$

40. $\sqrt{54m^{-6}n^3}$

41. $\dfrac{\sqrt{c^2d^6}}{\sqrt{4c^3d^{-4}}}$

42. $\dfrac{\sqrt{a^{-10}b^{-12}}}{\sqrt{a^{14}d^{-4}}}$

43. $\dfrac{\sqrt[3]{a^5b^4c^3}}{\sqrt[3]{a^{-1}b^2c^6}}$

44. $\dfrac{\sqrt[5]{16a^4b^2}}{\sqrt[5]{2^{-1}a^{14}b^{-3}}}$

In Exercises 45–52, rationalize the denominator and simplify your answer.

45. $\dfrac{3}{\sqrt{8}}$

46. $\dfrac{2}{\sqrt{6}}$

47. $\dfrac{3}{2 + \sqrt{12}}$

48. $\dfrac{1 + \sqrt{3}}{5 + \sqrt{10}}$

49. $\dfrac{2}{\sqrt{x} + 2}$

50. $\dfrac{\sqrt{x}}{\sqrt{x} - \sqrt{c}}$

51. $\dfrac{10}{\sqrt[3]{2}}$

52. $\dfrac{-6}{\sqrt[3]{4}}$

In Exercises 53–56, use the fact that $x^3 + y^3 = (x + y)(x^2 - xy + y^2)$ to rationalize the denominator.

53. $\dfrac{1}{\sqrt[3]{3} + 1}$

54. $\dfrac{5}{6 - \sqrt[3]{5}}$

55. $\dfrac{1}{\sqrt[3]{4} - \sqrt[3]{2} + 1}$

56. $\dfrac{3}{\sqrt[3]{2} + \sqrt[3]{3}}$

In Exercises 57–60, find the difference quotient of the given function. Then rationalize its numerator and simplify.

57. $f(x) = \sqrt{x + 1}$

58. $g(x) = 2\sqrt{x + 3}$

59. $f(x) = \sqrt{x^2 + 1}$

60. $g(x) = \sqrt{x^2 - x}$

In Exercises 61–64, use the equation $y = 92.8935 \cdot x^{.6669}$ which gives the approximate distance y (in millions of miles) from the sun to a planet that takes x earth years to complete one orbit of the sun. Find the distance from the sun to the planet whose orbit time is given.

61. Mercury (.24 years)

62. Mars (1.88 years)

63. Saturn (29.46 years)

64. Pluto (247.69 years)

Between 1790 and 1860, the population y of the United States (in millions) in year x was given by $y = 3.9572(1.0299^x)$, where $x = 0$ corresponds to 1790. In Exercises 65–68, find the U.S. population in the given year.

65. 1800

66. 1817

67. 1845

68. 1859

69. Here are some of the reasons why restrictions are necessary when defining fractional powers of a negative number.

(a) Explain why the equations $x^2 = -4$, $x^4 = -4$, $x^6 = -4$, etc., have no real solutions. Hence, we cannot define $c^{1/2}$, $c^{1/4}$, $c^{1/6}$ when $c = -4$.

(b) Since $1/3$ is the same as $2/6$, it should be true that $c^{1/3} = c^{2/6}$, that is, that $\sqrt[3]{c} = \sqrt[6]{c^2}$. Show that this is false when $c = -8$.

70. Use a calculator to find $(3141)^{-3141}$. Explain why your answer cannot possibly be the number $(3141)^{-3141}$. Why does your calculator behave the way that it does?

71. (a) Graph $f(x) = x^5$ and explain why this function has an inverse function.

(b) Show algebraically that the inverse function is $g(x) = x^{1/5}$.

(c) Does $f(x) = x^6$ have an inverse function? Why or why not?

72. If n is an odd positive integer, show that $f(x) = x^n$ has an inverse function and find the rule of the inverse function. [*Hint:* Exercise 71 is the case when $n = 5$.]

In Exercises 73–75, use the catalog of basic functions (page 170) and Section 3.4 to describe the graph of the given function.

73. $g(x) = \sqrt{x} + 3$

74. $h(x) = \sqrt{x} - 2$

75. $k(x) = \sqrt{x + 4} - 4$

76. (a) Suppose r is a solution of the equation $x^n = c$ and s is a solution of $x^n = d$. Verify that rs is a solution of $x^n = cd$.

(b) Explain why part (a) shows that $\sqrt[n]{cd} = \sqrt[n]{c}\,\sqrt[n]{d}$.

77. The output Q of an industry depends on labor L and capital C according to the equation

$$Q = L^{1/4}C^{3/4}.$$

(a) Use a calculator to determine the output for the following resource combinations.

L	C	$Q = L^{1/4}C^{3/4}$
10	7	
20	14	
30	21	
40	28	
60	42	

(b) When you double both labor and capital, what happens to the output? When you triple both labor and capital, what happens to the output?

78. Do Exercise 77 when the equation relating output to resources is $Q = L^{1/4}C^{1/2}$.

79. Do Exercise 77 when the equation relating output to resources is $Q = L^{1/2}C^{3/4}$.

80. In Exercises 77–79, how does the sum of the exponents on L and C affect the increase in output?

5.1.A *SPECIAL TOPICS* Radical Equations

Section Objectives
- Use algebraic and graphical methods to solve radical equations.
- Solve applied problems that involve radicals.

The algebraic solution of equations involving radicals depends on this fact: If two quantities are equal, say,

$$x - 2 = 3,$$

then their squares are also equal:

$$(x - 2)^2 = 9.$$

Thus,

every solution of $x - 2 = 3$ is also a solution of $(x - 2)^2 = 9$.

But *be careful:* This works only in one direction. For instance, -1 is a solution of $(x - 2)^2 = 9$, but not of $x - 2 = 3$. This is an example of the Power Principle.

Power Principle

If both sides of an equation are raised to the same positive integer power, then every solution of the original equation is also a solution of the new equation. But the new equation may have solutions that are *not* solutions of the original one.

Consequently, if you raise both sides of an equation to a power, you must *check your solutions* in the *original* equation. Graphing provides a quick way to eliminate most extraneous solutions. But only an algebraic computation can confirm an exact solution.

EXAMPLE 1

Solve $5 + \sqrt{3x - 11} = x$.

SOLUTION   We first rearrange terms to get the radical expression alone on one side.

$$\sqrt{3x - 11} = x - 5.$$

Then we square both sides and solve the resulting equation.

$$(\sqrt{3x - 11})^2 = (x - 5)^2$$
$$3x - 11 = x^2 - 10x + 25$$
$$0 = x^2 - 13x + 36$$
$$0 = (x - 4)(x - 9)$$
$$x - 4 = 0 \quad \text{or} \quad x - 9 = 0$$
$$x = 4 \qquad\qquad x = 9.$$

These are *possible* solutions. We must check each one in the original equation. If $x = 9$ we have

Left side:	$5 + \sqrt{3x - 11}$	Right side:	x
	$5 + \sqrt{3 \cdot 9 - 11}$		9
	$5 + \sqrt{16}$		
	9		

Hence, $x = 9$ is a solution of the original equation. When we try the same calculations with $x = 4$, we obtain

Left side:	$5 + \sqrt{3x - 11}$	Right side:	x
	$5 + \sqrt{3 \cdot 4 - 11}$		4
	$5 + \sqrt{1}$		
	6		

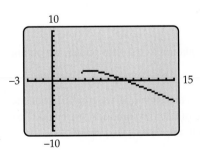

Figure 5–2

The two sides are not the same, so $x = 4$ is not a solution of the original equation.

These results can be confirmed graphically by graphing $y = 5 + \sqrt{3x - 11} - x$, as in Figure 5–2. The x-intercepts of this graph are the solutions of the equation (why?). There is an x-intercept at $x = 9$, but none at $x = 4$, indicating that $x = 9$ is a solution and $x = 4$ is not. ∎

EXAMPLE 2

Solve $\sqrt{2x - 3} - \sqrt{x + 7} = 2$.

SOLUTION   We first rearrange terms so that one side contains only a single radical term.

$$\sqrt{2x - 3} = \sqrt{x + 7} + 2.$$

Then we square both sides and simplify.

$$(\sqrt{2x-3})^2 = (\sqrt{x+7}+2)^2$$

$$2x-3 = (\sqrt{x+7})^2 + 2 \cdot 2 \cdot \sqrt{x+7} + 2^2$$

$$2x-3 = x+7+4\sqrt{x+7}+4$$

$$x-14 = 4\sqrt{x+7}.$$

Now we square both sides and solve the resulting equation.

$$(x-14)^2 = (4\sqrt{x+7})^2$$

$$x^2 - 28x + 196 = 4^2 \cdot (\sqrt{x+7})^2$$

$$x^2 - 28x + 196 = 16(x+7)$$

$$x^2 - 28x + 196 = 16x + 112$$

$$x^2 - 44x + 84 = 0$$

$$(x-2)(x-42) = 0$$

$$x-2 = 0 \quad \text{or} \quad x-42 = 0$$

$$x = 2 \qquad\qquad x = 42.$$

Substituting 2 and 42 in the left side of the original equation shows that

$$\sqrt{2 \cdot 2 - 3} - \sqrt{2+7} = \sqrt{1} - \sqrt{9} = 1 - 3 = -2;$$

$$\sqrt{2 \cdot 42 - 3} - \sqrt{42+7} = \sqrt{81} - \sqrt{49} = 9 - 7 = 2.$$

Therefore, 42 is the only solution of the equation. ∎

Many radical equations are not amenable to algebraic techniques. In such cases, graphical or numerical means must be used to approximate the solutions.

EXAMPLE 3

Solve $\sqrt[5]{x^2 - 6x + 2} = x - 4$.

SOLUTION   If we raise both sides of the equation to the fifth power, we obtain

$$x^2 - 6x + 2 = (x-4)^5.$$

Unfortunately, this equation is not readily solvable, even if we multiply out the right hand side. The best we can do is to approximate the solutions. We can do this graphically as follows: We rewrite the equation as

$$\sqrt[5]{x^2 - 6x + 2} - x + 4 = 0,$$

then the solutions are the x-intercepts of the graph of

$$h(x) = \sqrt[5]{x^2 - 6x + 2} - x + 4$$

(see Figure 5–3). Alternatively, we can graph $f(x) = \sqrt[5]{x^2 - 6x + 2}$ and

$$g(x) = x - 4$$

on the same screen and find the x-coordinate of their intersection (Figure 5–4).

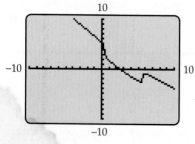

Figure 5–3

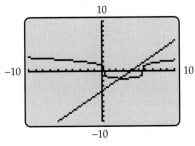

Figure 5–4

Equations involving rational exponents can be solved graphically, provided that the function to be graphed is entered properly. Many such equations can also be solved algebraically by making an appropriate substitution.

GRAPHING EXPLORATION

Use a root finder in Figure 5–3 or an intersection finder in Figure 5–4 to verify that $x \approx 2.534$ is the solution of the equation.

■

EXAMPLE 4

Solve $x^{2/3} - 2x^{1/3} - 15 = 0$ both algebraically and graphically.

SOLUTION

Algebraic: Let $u = x^{1/3}$, rewrite the equation, and solve:

$$x^{2/3} - 2x^{1/3} - 15 = 0$$

$$(x^{1/3})^2 - 2x^{1/3} - 15 = 0$$

$$u^2 - 2u - 15 = 0$$

$$(u + 3)(u - 5) = 0$$

$$u + 3 = 0 \quad \text{or} \quad u - 5 = 0$$

$$u = -3 \qquad\qquad u = 5$$

$$x^{1/3} = -3 \qquad\qquad x^{1/3} = 5.$$

Cubing both sides of these last equations shows that

$$(x^{1/3})^3 = (-3)^3 \quad \text{or} \quad (x^{1/3})^3 = 5^3$$

$$x = -27 \qquad\qquad x = 125.$$

Since we cubed both sides, we must check these numbers in the original equation. Verify that both *are* solutions.

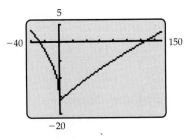

Figure 5–5

Graphical: Graph the function $f(x) = x^{2/3} - 2x^{1/3} - 15$ and find the x-intercepts, namely, $x = -27$ and $x = 125$. The only difficulty is the one mentioned in the Note on page 344. Depending on your calculator, you might have to enter the function f in one of these forms:

$$f(x) = (x^2)^{1/3} - 2x^{1/3} - 15 \quad \text{or} \quad f(x) = (x^{1/3})^2 - 2x^{1/3} - 15.$$

Otherwise, the calculator might not produce a graph when x is negative. Your result should resemble Figure 5–5.

■

EXAMPLE 5

Hoa, who is standing at point A on the bank of a 2.5-kilometer-wide river wants to reach point B, 15 kilometers downstream on the opposite bank. She plans to row to a point C on the opposite shore and then run to B, as shown in Figure 5–6. She can row at a rate of 4 kilometers per hour and can run at 8 kilometers per hour.

(a) If her trip took 3 hours, how far from B did she land?

(b) How far from B should she land to make the time for the trip as short as possible?

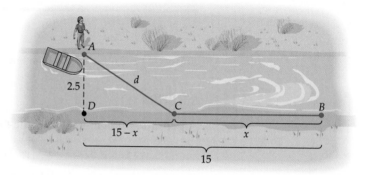

Figure 5–6

SOLUTION   Let x be the distance that Hoa ran from C to B. Using the basic formula for distance, we have

$$\text{Rate} \times \text{Time} = \text{Distance}$$

$$\text{Time} = \frac{\text{Distance}}{\text{Rate}} = \frac{x}{8}.$$

Similarly, the time required to row distance d is

$$\text{Time} = \frac{\text{Distance}}{\text{Rate}} = \frac{d}{4}.$$

Since $15 - x$ is the distance from D to C, the Pythagorean Theorem applied to right triangle ADC shows that

$$d^2 = (15 - x)^2 + 2.5^2 \qquad \text{or, equivalently,} \qquad d = \sqrt{(15 - x)^2 + 6.25}.$$

Therefore, the total time for the trip is given by

$$T(x) = \text{Rowing time} + \text{Running time} = \frac{d}{4} + \frac{x}{8} = \frac{\sqrt{(x - 15)^2 + 6.25}}{4} + \frac{x}{8}.$$

(a) If the trip took 3 hours, then $T(x) = 3$, and we must solve the equation

$$\frac{\sqrt{(x - 15)^2 + 6.25}}{4} + \frac{x}{8} = 3.$$

Using the viewing window with $0 \leq x \leq 15$ and $-2 \leq y \leq 2$, graph the function

$$f(x) = \frac{\sqrt{(x-15)^2 + 6.25}}{4} + \frac{x}{8} - 3$$

and use a root finder to find its x-intercept (the solution of the equation).

This Graphing Exploration shows that Hoa should land approximately 6.74 kilometers from B to make the trip in 3 hours.

(b) To find the shortest possible time, we must find the value of x that makes

$$T(x) = \frac{\sqrt{(x-15)^2 + 6.25}}{4} + \frac{x}{8}$$

as small as possible.

Using the viewing window with $0 \leq x \leq 15$ and $0 \leq y \leq 4$, graph $T(x)$ and use a minimum finder to verify that the lowest point on the graph (that is, the point with the y-coordinate $T(x)$ as small as possible) is approximately $(13.56, 2.42)$.

Therefore, the shortest time for the trip will be 2.42 hours and will occur if Hoa rows to a point 13.56 kilometers from B.   ■

EXERCISES 5.1.A

In Exercises 1–26, find all real solutions of each equation. Find exact solutions when possible and approximate ones otherwise.

1. $\sqrt{x+2} = 3$

2. $\sqrt{x-7} = 4$

3. $\sqrt{4x+9} = 0$

4. $\sqrt{4x+9} = -1$

5. $\sqrt[3]{5-11x} = 3$

6. $\sqrt[3]{6x-10} = 2$

7. $\sqrt[3]{x^2-1} = 2$

8. $(x-2)^{2/3} = 9$

9. $\sqrt{x^2-x-1} = 1$

10. $\sqrt{x^2-5x+4} = 2$

11. $\sqrt{x+7} = x-5$

12. $\sqrt{x+5} = x-1$

13. $(3x^2 + 7x - 2)^{1/2} = x + 1$

14. $\sqrt{4x^2 - 10x + 5} = x - 3$

15. $(x^3 + x^2 - 4x + 5)^{1/3} = x + 1$

16. $\sqrt[3]{x^3 - 6x^2 + 2x + 3} = x - 1$

17. $\sqrt[5]{9 - x^2} = x^2 + 1$

18. $(x^3 - x + 1)^{1/4} = x^2 - 1$

19. $\sqrt[3]{x^5 - x^3 - x} = x + 2$

20. $\sqrt{x^3 + 2x^2 - 1} = x^3 + 2x - 1$

21. $\sqrt{x^2 + x - 1} = \sqrt{14 - x}$

22. $\sqrt[3]{x^3 + 3x} = \sqrt[3]{3x^2 + 1}$

23. $\sqrt{5x + 6} = 3 + \sqrt{x + 3}$

24. $\sqrt{3y + 1} - 1 = \sqrt{y + 4}$

25. $\sqrt{2x - 5} = 1 + \sqrt{x - 3}$

26. $\sqrt{x - 3} + \sqrt{x + 5} = 4$

27. The surface area S of the right circular cone in the figure is given by $S = \pi r \sqrt{r^2 + h^2}$. What radius should be used to produce a cone of height 5 inches and surface area 100 square inches?

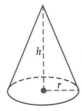

28. What is the radius of the base of a cone whose surface area is 18π square centimeters and whose height is 4 cm?

29. Find the radius of the base of a conical container whose height is $1/3$ of the radius and whose volume is 180 cubic inches. [*Note:* The volume of a cone of radius r and height h is $\pi r^2 h/3$.]

30. The surface area of the right square pyramid in the figure is given by $S = b\sqrt{b^2 + 4h^2}$. If the pyramid has height 10 feet and surface area 100 square feet, what is the length of a side b of its base?

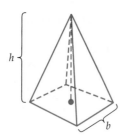

In Exercises 31–34, assume that all letters represent positive numbers and solve each equation for the required letter.

31. $A = \sqrt{1 + \dfrac{a^2}{b^2}}$ for b

32. $T = 2\pi\sqrt{\dfrac{m}{g}}$ for g

33. $y = \dfrac{1}{\sqrt{1 - x^2}}$

34. $R = \sqrt{d^2 + k^2}$ for d

In Exercises 35–42, solve each equation algebraically.

35. $x - 4x^{1/2} + 4 = 0$ [*Hint:* Let $u = x^{1/2}$.]

36. $x - x^{1/2} - 12 = 0$

37. $2x - 8\sqrt{x} - 24 = 0$

38. $3x - 11\sqrt{x} - 4 = 0$

39. $x^{2/3} + 3x^{1/3} + 2 = 0$ [*Hint:* Let $u = x^{1/3}$.]

40. $x^{4/3} - 4x^{2/3} + 3 = 0$

41. $x^{1/2} - x^{1/4} - 2 = 0$ [*Hint:* Let $u = x^{1/4}$.]

42. $x^{1/3} + x^{1/6} - 2 = 0$

In Exercises 43–46, solve each equation graphically.

43. $x^{3/5} - 2x^{2/5} + x^{1/5} - 6 = 0$

44. $x^{5/3} + x^{4/3} - 3x^{2/3} + x = 5$

45. $x^{-3} + 2x^{-2} - 4x^{-1} + 5 = 0$

46. $x^{-2/3} - 3x^{-1/3} = 4$

47. A rope is to be stretched at uniform height from a tree to a 35-foot-long fence, which is 20 feet from the tree, and then to the side of a building at a point 30 feet from the fence, as shown in the figure.

(a) If 63 feet of rope is to be used, how far from the building wall should the rope meet the fence?

(b) How far from the building wall should the rope meet the fence if as little rope as possible is to be used?

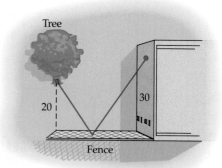

48. Anne is standing on a straight road and wants to reach her helicopter, which is located 2 miles down the road from her, a mile from the road in a field (see the figure). She can run 5 miles per hour on the road and 3 miles per hour in the field. She plans to run down the road, then cut diagonally across the field to reach the helicopter.

(a) If she reaches the helicopter in exactly 42 minutes (.7 hours) where did she leave the road?

(b) Where should she leave the road to reach the helicopter as soon as possible?

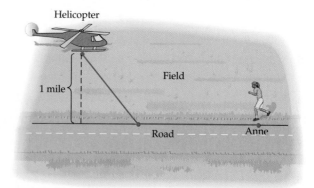

49. A power plant is located on the bank of a river that is $\frac{1}{2}$ mile wide. Wiring is to be laid across the river and then along the shore to a substation 8 miles downstream, as shown in the figure. It costs $12,000 per mile for underwater wiring and $8000 per mile for wiring on land. If $72,000 is to be spent on the project, how far from the substation should the wiring come to shore?

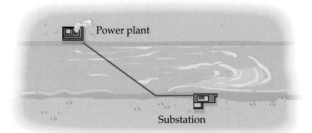
Power plant

Substation

50. A spotlight is to be placed on a building wall to illuminate a bench that is 32 feet from the base of the wall. The intensity I of the light at the bench is known to be x/d^3, where x is the height of the spotlight above the ground and d is the distance from the bench to the spotlight.

(a) Express I as a function of x. [It may help to draw a picture.]

(b) How high should the spotlight be in order to provide maximum illumination at the bench?

51. If an object is dropped from a height h_0 feet, it will take $\frac{1}{4}\sqrt{h_0}$ seconds to hit the ground, assuming that h_0 is small enough that air resistance is negligible.

(a) We wish to make a movie by dropping a running camcorder off of a building. From how high would we have to drop it to make a 10-second film?

(b) How long would the camera take to hit the ground if dropped off of the Sears Tower in Chicago? (See example 4 of Section 1.1.)

52. In an attempt to steady a tottering, old, statue, two ropes are tied to the top, and secured firmly to the ground. The first rope winds up three feet from the base of the statue. The second rope winds up five feet from the base of the statue. If 10 total feet of rope are used, how tall is the statue?

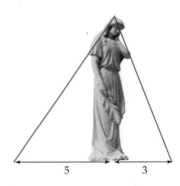

5   3

5.2   Exponential Functions

Section Objectives
- Explore graphs of exponential functions.
- Use exponential functions to model growth and decay.
- Use the natural exponential function.

For each positive real number a there is a function (called the **exponential function with base a**) whose domain is all real numbers and whose rule is $f(x) = a^x$. For example,

$$f(x) = 10^x, \qquad g(x) = 2^x, \qquad h(x) = \left(\frac{1}{2}\right)^x, \qquad k(x) = \left(\frac{3}{2}\right)^x.$$

The graph of $f(x) = a^x$ is the next entry in the catalog of basic functions. To see how its graph depends on the size of the base a, do the following Graphing Exploration.

GRAPHING EXPLORATION

Using viewing window with $-3 \le x \le 7$ and $-2 \le y \le 18$, graph

$$f(x) = 1.3^x, \qquad g(x) = 2^x, \qquad \text{and} \qquad h(x) = 10^x$$

on the same screen. How is the steepness of the graph of $f(x) = a^x$ related to the size of a?

The Graphing Exploration illustrates these facts:

The Exponential Function

$$f(x) = a^x \, (a > 1)$$

When $a > 1$, the graph of $f(x) = a^x$ has the shape shown here and the properties listed below.

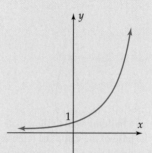

The graph is above the x-axis.   The negative x-axis is a horizontal

The y-intercept is 1.   asymptote.

$f(x)$ is an increasing function.   The larger the base a, the more steeply the graph rises to the right.

EXAMPLE 1

Graph $f(x) = 2^x$ and estimate the height of the graph when $x = 50$.

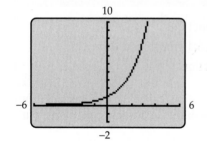

Figure 5–7

SOLUTION   A small portion of the graph is shown in Figure 5–7. If the x-axis were to be extended with the same scale, $x = 50$ would be at the right edge of the page. At that point, the height of the graph is $f(50) = 2^{50}$. Now the y-axis scale in Figure 5–7 is approximately 12 units to the inch, which is equivalent to 760,320 units per mile, as you can readily verify. Therefore, the height of the graph at $x = 50$ is

$$\frac{2^{50}}{760,320} = 1,480,823,741 \text{ MILES},$$

which would put that part of the graph well beyond the planet Saturn!   ■

When the base a is between 0 and 1, then the graph of $f(x) = a^x$ has a different shape.

GRAPHING EXPLORATION

Using viewing window with $-4 \le x \le 4$ and $-1 \le y \le 4$, graph

$$f(x) = .2^x, \qquad g(x) = .4^x, \qquad h(x) = .6^x, \qquad \text{and} \qquad k(x) = .8^x$$

on the same screen. How is the steepness of the graph of $f(x) = a^x$ related to the size of a?

The exploration supports this conclusion.

The Exponential Function

$$f(x) = a^x \, (0 < a < 1)$$

When $0 < a < 1$, the graph of $f(x) = a^x$ has the shape shown here and the properties listed below.

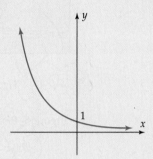

The graph is above the x-axis.   The positive x-axis is a horizontal
The y-intercept is 1.   asymptote.
$f(x)$ is a decreasing function.   The closer the base a is to 0, the more
                                                      steeply the graph falls to the right.

EXAMPLE 2

Without graphing, describe the graph of $g(x) = 3^{-x}$.

SOLUTION   Note that

$$g(x) = 3^{-x} = (3^{-1})^x = \left(\frac{1}{3}\right)^x.$$

GRAPHING EXPLORATION

Verify the analysis in Example 2
by graphing $g(x) = 3^{-x}$ in the
viewing window with $-4 \le x \le 4$
and $-2 \le y \le 18$.

So $g(x)$ is an exponential function with a positive base less than 1. Its graph has the shape shown in the preceding box: It falls quickly to the right and rises very steeply to the left of the y-axis.   ■

Exponential functions that model real-life situations generally have the form $f(x) = Pa^{kx}$, such as

$$f(x) = 5 \cdot 2^{.45x}, \qquad g(x) = 3.5(10^{-.03x}), \qquad h(x) = (-6)(1.076^{2x}).$$

Their graphs have the same basic shape as the graph of $f(x) = a^x$, but rise or fall at different rates, depending on the constants P, a, and k.

EXAMPLE 3

Figure 5–8 on the next page shows the graphs of

$$f(x) = 3^x, \qquad g(x) = 3^{.15x}, \qquad h(x) = 3^{.35x}, \qquad k(x) = 3^{-x}, \qquad p(x) = 3^{-.4x}.$$

Note how the coefficient of x determines the steepness of the graph. When this coefficient is positive, the graph rises, and when it is negative, the graph falls from left to right.

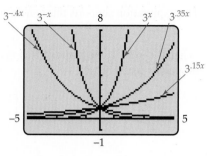

Figure 5–8

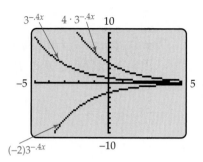

Figure 5–9

Figure 5–9 shows the graphs of

$$p(x) = 3^{-.4x}, \qquad q(x) = 4 \cdot 3^{-.4x}, \qquad r(x) = (-2)3^{-.4x}.$$

As we saw in Section 3.4, the graph of $q(x) = 4 \cdot 3^{-.4x}$ is the graph of $p(x) = 3^{-.4x}$ stretched away from the x-axis by a factor of 4. The graph of $r(x) = (-2)3^{-.4x}$ is the graph of $p(x) = 3^{-.4x}$ stretched away from the x-axis by a factor of 2 *and* reflected in the x-axis. ∎

EXPONENTIAL GROWTH

Exponential functions are useful for modeling situations in which a quantity increases by a fixed multiplier.

EXAMPLE 4

If you deposit \$5000 in a savings account that pays 3% interest, compounded annually, how much money is in the account after nine years?

SOLUTION   After one year, the account balance is

$$5000 + 3\% \text{ of } 5000 = 5000 + (.03)5000 = 5000(1 + .03)$$

$$= 5000(1.03) = \$5150.$$

The initial balance has grown by a factor of 1.03. If you leave the \$5150 in the account, then at the end of the second year, the balance is

$$5150 + 3\% \text{ of } 5150 = 5150 + (.03)5150 = 5150(1 + .03) = 5150(1.03).$$

Once again, the amount at the beginning of the year has grown by a factor of 1.03. The same thing happens every year. A balance of P dollars at the beginning of the year grows to $P(1.03)$. So the balance grows like this:

Year 1   Year 2   Year 3

$$5000 \rightarrow 5000(1.03) \rightarrow \underbrace{[5000(1.03)](1.03)}_{5000(1.03)^2} \rightarrow \underbrace{[5000(1.03)(1.03)](1.03)}_{5000(1.03)^3} \rightarrow \cdots$$

Consequently, the balance at the end of year x is given by

$$f(x) = 5000 \cdot 1.03^x.$$

The balance at the end of nine years is $f(9) = 5000(1.03^9) = \$6523.87$ (rounded to the nearest penny). ∎

EXAMPLE 5

The world population in 1980 was about 4.5 billion people and has been increasing at approximately 1.5% per year.

(a) Estimate the world population in 2010.

(b) In what year will the population be double what it is in 2010?

SOLUTION

(a) The world population in 1981 was

$$4.5 + 1.5\% \text{ of } 4.5 = 4.5 + .015(4.5) = 4.5(1 + .015) = 4.5(1.015).$$

Similarly, in each successive year, the population increased by a factor of 1.015, so the population (in billions) in year x is given by

$$g(x) = 4.5(1.015^x),$$

where $x = 0$ corresponds to 1980. The year 2010 corresponds to $x = 30$, so the population then is $g(30) = 4.5(1.015^{30}) \approx 7.03$ billion people.

(b) Twice the population in 2010 is $2(7.03) = 14.06$ billion. We must find the number x such that $g(x) = 14.06$; that is, we must solve the equation

$$4.5(1.015^x) = 14.06.$$

This can be done with an equation solver or by graphical means, as in Figure 5–10, which shows the intersection point of $y = 4.5(1.015^x)$ and $y = 14.06$. The solution is $x \approx 76.5$, which corresponds to the year 2056. Thus, according to this model, the world population will double in your lifetime. This is what is meant by the population explosion. ■

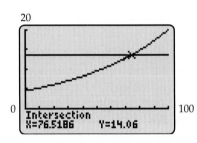

Figure 5–10

Examples 4 and 5 illustrate **exponential growth.** The functions developed there, $f(x) = 5000(1.03^x)$ and $g(x) = 4.5(1.015^x)$, are typical of the general case.

Exponential Growth

> Exponential growth can be described by a function of the form
>
> $$f(x) = Pa^x,$$
>
> where $f(x)$ is the quantity at time x, P is the initial quantity (when $x = 0$) and $a > 1$ is the factor by which the quantity changes when x increases by 1.
>
> If the quantity is growing at the rate r per time period, then $a = 1 + r$, and
>
> $$f(x) = Pa^x = P(1 + r)^x.$$

EXAMPLE 6

At the beginning of an experiment, a culture contains 1000 bacteria. Five hours later, there are 7600 bacteria. Assuming that the bacteria grow exponentially, how many will there be after 24 hours?

SOLUTION   The bacterial population is given by $f(x) = Pa^x$, where P is the initial population, a is the change factor, and x is the time in hours. We are given that $P = 1000$, so $f(x) = 1000a^x$. The next step is to determine a. Since there are 7600 bacteria when $x = 5$, we have

$$7600 = f(5) = 1000a^5,$$

so

$$1000\,a^5 = 7600$$

$$a^5 = 7.6$$

$$a = \sqrt[5]{7.6} = (7.6)^{.2}.$$

Therefore, the population function is $f(x) = 1000(7.6^{.2})^x = 1000 \cdot (7.6)^{.2x}$. After 24 hours, the bacteria population will be

$$f(24) = 1000(7.6)^{.2(24)} \approx 16,900,721. \qquad \blacksquare$$

 EXPONENTIAL DECAY

In some situations, a quantity *decreases* by a fixed multiplier as time goes on.

EXAMPLE 7

When tap water is filtered through a layer of charcoal and other purifying agents, 30% of the chemical impurities in the water are removed, and 70% remain. If the water is filtered through a second purifying layer, then the amount of impurities remaining is 70% of 70%, that is, $(.7)(.7) = .7^2 = .49$ or 49%. A third layer results in $.7^3$ of the impurities remaining. Thus, the function

$$f(x) = .7^x$$

gives the percentage of impurities remaining in the water after it passes through x layers of purifying material. How many layers are needed to ensure that 95% of the impurities are removed from the water?

SOLUTION   If 95% of the impurities are removed, then 5% will remain. Hence, we must find x such that $f(x) = .05$, that is, we must solve the equation $.7^x = .05$. This can be done numerically or graphically. Figure 5–11 shows that the solution is $x \approx 8.4$, so 8.4 layers of material are needed. $\qquad \blacksquare$

Example 7 illustrates **exponential decay.** Note that the impurities were removed at a rate of $30\% = .3$ and that the amount of impurities remaining in the water was changing by a factor of $1 - .30 = .7$. The same thing is true in the general case.

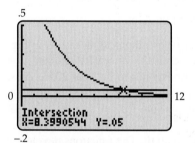

Figure 5–11

GRAPHING EXPLORATION

Determine how many layers are needed to ensure that 99% of the impurities are removed.

Exponential Decay

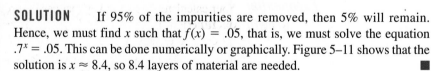

Exponential decay can be described by a function of the form

$$f(x) = Pa^x,$$

where $f(x)$ is the quantity at time x, P is the initial quantity (when $x = 0$) and $0 < a < 1$. Here, a is the factor by which the quantity changes when x increases by 1.

If the quantity is decaying at the rate r per time period, then $a = 1 - r$, and

$$f(x) = Pa^x = P(1 - r)^x.$$

One of the important uses of exponential functions is to describe radioactive decay. The **half-life** of a radioactive element is the time it takes a given quantity to decay to one-half of its original mass. The half-life depends only on the substance and not on the size of the sample. Exercise 82 proves the following result.

Radioactive
Decay

The mass $M(x)$ of a radioactive element at time x is given by

$$M(x) = c(.5^{x/h}),$$

where c is the original mass and h is the half-life of the element.

EXAMPLE 8

Plutonium (^{239}Pu) has a half-life of 24,360 years. So the amount remaining from 1 kilogram after x years is given by

$$M(x) = 1(.5^{x/24360}) = (.5^{1/24360})^x \approx .99997^x.$$

Thus, M is an exponential function whose base is very close to 1. Its graph falls *very slowly* from left to right, as you can easily verify by graphing M in a window with $0 \le x \le 2000$. This means that even after an extremely long time, a substantial amount of plutonium will remain. In fact, most of the original kilogram is still there after *ten thousand years* because $M(10,000) \approx .75$ kg. This is the reason that nuclear waste disposal is such a serious problem. ■

THE NUMBER *e* AND THE NATURAL EXPONENTIAL FUNCTION

There is an irrational number, denoted e, that arises naturally in a variety of phenomena and plays a central role in the mathematical description of the physical universe. Its decimal expansion begins

$$e = \mathbf{2.718281828459045} \ldots.$$

Your calculator has an e^x key that can be used to evaluate the **natural exponential function** $f(x) = e^x$. If you key in e^1, the calculator will display the first part of the decimal expansion of e.

The graph of $f(x) = e^x$ has the same shape as the graph of $g(x) = 2^x$ in Figure 5–7 but climbs more steeply.

GRAPHING EXPLORATION

Graph $f(x) = e^x$, $g(x) = 2^x$, and $h(x) = 3^x$ on the same screen in a window with $-5 \le x \le 5$. The Technology Tip in the margin may be helpful.

EXAMPLE 9

Population Growth   If the population of the United States continues to grow as it has recently, then the approximate population of the United States (in millions) in year t will be given by the function

$$P(t) = 227e^{.0093t},$$

where 1980 corresponds to $t = 0$.

(a) Estimate the population in 2015.

(b) When will the population reach half a billion?

SOLUTION

(a) The population in 2015 (that is, $t = 35$) will be approximately

$$P(35) = 227e^{.0093(35)} \approx 314.3 \text{ million people.}$$

(b) Half a billion is 500 million people. So we must find the value of t for which $P(t) = 500$, that is, we must solve the equation

$$227e^{.0093t} = 500.$$

This can be done graphically by finding the intersection of the graph of $P(t)$ and the horizontal line $y = 500$, which occurs when $t \approx 84.9$ (Figure 5–12). Therefore, the population will reach half a billion late in the year 2064. ∎

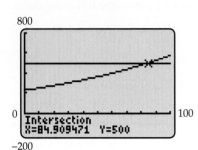

800

0

100

−200

Intersection
X=84.909471   Y=500

Figure 5–12

🔲 OTHER EXPONENTIAL FUNCTIONS

The population growth models in earlier examples do not take into account factors that may limit population growth in the future (wars, new diseases, etc.). Example 10 illustrates a function, called a **logistic model,** that is designed to model such situations more accurately.

EXAMPLE 10

Inhibited Population Growth   There is an upper limit on the fish population in a certain lake due to the oxygen supply, available food, etc. The population of fish in this lake at time t months is given by the function

$$p(t) = \frac{20,000}{1 + 24e^{-t/4}} \qquad (t \geq 0).$$

What is the upper limit on the fish population?

SOLUTION   The graph of $p(t)$ in Figure 5–13 suggests that the horizontal line $y = 20,000$ is a horizontal asymptote of the graph.

In other words, the fish population never goes above 20,000. You can confirm this algebraically by rewriting the rule of p in this form.

$$p(t) = \frac{20,000}{1 + 24e^{-t/4}} = \frac{20,000}{1 + \dfrac{24}{e^{t/4}}}.$$

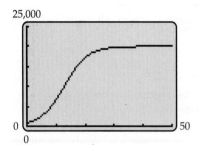

25,000

0

50

0

Figure 5–13

When t is very large, so is $t/4$, which means that $e^{t/4}$ is huge. Hence, by the Big-Little Principle (page 288), $\dfrac{24}{e^{t/4}}$ is very close to 0, and $p(t)$ is very close to $\dfrac{20,000}{1 + 0} = 20,000$. Since $e^{t/4}$ is positive, the denominator of $p(t)$ is slightly bigger than 1, so $p(t)$ is always less than 20,000. ∎

When a cable, such as a power line, is suspended between towers of equal height as in Figure 5–14, it forms a curve called a **catenary,** which is the graph of a function of the form

$$f(x) = A(e^{kx} + e^{-kx})$$

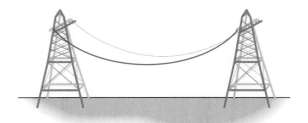

Figure 5–14

for suitable constants A and k. The Gateway Arch in St. Louis (Figure 5–15) has the shape of an inverted catenary, which was chosen because it evenly distributes the internal structural forces.

Figure 5–15

GRAPHING EXPLORATION

Graph each of the following functions in the window with $-5 \le x \le 5$ and $-10 \le y \le 80$.

$$y_1 = 10(e^{.4x} + e^{-.4x}), \qquad y_2 = 10(e^{2x} + e^{-2x}),$$
$$y_3 = 10(e^{3x} + e^{-3x}).$$

How does the coefficient of x affect the shape of the graph?

Predict the shape of the graph of $y = -y_1 + 80$. Confirm your answer by graphing.

EXERCISES 5.2

In Exercises 1–10, sketch a complete graph of the function.

1. $f(x) = 3^{-x}$

2. $f(x) = (1.001)^{-x}$

3. $g(x) = (5/2)^x$

4. $g(x) = (1.001)^x$

5. $h(x) = (1/\pi)^x$

6. $h(x) = (1/e)^{-x}$

7. $f(x) = 1 - 2^{-x}$

8. $g(x) = (1.2)^x + (.8)^{-x}$

9. $h(x) = 2^{x^2}$

10. $h(x) = 2^{-x^2}$

In Exercises 11–16, list the transformations needed to transform the graph of $h(x) = 2^x$ into the graph of the given function. [Section 3.4 may be helpful.]

11. $f(x) = 2^x - 5$

12. $g(x) = -(2^x)$

13. $k(x) = 3(2^x)$

14. $g(x) = 2^{x-1}$

15. $f(x) = 2^{x+2} - 5$

16. $g(x) = -5(2^{x-1}) + 7$

In Exercises 17 and 18, match the functions to the graphs. Assume $a > 1$ and $c > 1$.

17. $f(x) = a^x$

   $g(x) = a^{x+1}$

   $h(x) = a^x + 1$

   $j(x) = (a + 1)^x$

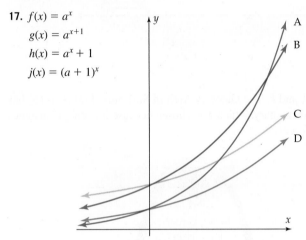

18. $f(x) = c^x$

   $g(x) = \left(\dfrac{1}{c}\right)^x$

   $h(x) = c^{1/x}$

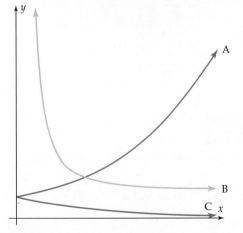

In Exercises 19–23, determine whether the function is even, odd, or neither (see Special Topics 3.4.A).

19. $f(x) = 10^x$

20. $g(x) = 2^x - x$

21. $f(x) = \dfrac{e^x + e^{-x}}{2}$

22. $f(x) = \dfrac{e^x - e^{-x}}{2}$

23. $f(x) = e^{-x^2}$

24. Use the Big-Little Principle to explain why $e^x + e^{-x}$ is approximately equal to e^x when x is large.

In Exercises 25–32, find the average rate of change of the function.

25. $f(x) = 3(4^x)$ as x goes from 1 to 3

26. $f(x) = 3(4^x)$ as x goes from 10 to 12

27. $g(x) = 3^{x^2-x-3}$ as x goes from -1 to 1

28. $h(x) = 2^x$ as x goes from 1 to 2

29. $h(x) = 2^x$ as x goes from 1 to 1.001

30. $h(x) = e^x$ as x goes from 1 to 2

31. $h(x) = e^x$ as x goes from 1 to 1.001

32. $f(x) = a^x$, $a > 0$, as x goes from 0 to 0.001

In Exercises 33–36, find the difference quotient of the function.

33. $f(x) = 10^x$

34. $g(x) = 5^{x^2}$

35. $f(x) = 2^x + 2^{-x}$

36. $f(x) = e^x - e^{-x}$

In Exercises 37–44, find a viewing window (or windows) that shows a complete graph of the function.

37. $k(x) = e^{-x}$

38. $f(x) = e^{-x^2}$

39. $f(x) = \dfrac{e^x + e^{-x}}{2}$

40. $h(x) = \dfrac{e^x - e^{-x}}{2}$

41. $g(x) = 2^x - x$

42. $k(x) = \dfrac{2}{e^x + e^{-x}}$

43. $f(x) = \dfrac{5}{1 + e^{-x}}$

44. $g(x) = \dfrac{10}{1 + 9e^{-x/2}}$

In Exercises 45–50, list all asymptotes of the graph of the function and the approximate coordinates of each local extremum.

45. $f(x) = x2^x$

46. $g(x) = x2^{-x}$

47. $h(x) = e^{x^2/2}$

48. $k(x) = 2^{x^2-6x+2}$

49. $f(x) = e^{-x^2}$

50. $g(x) = -xe^{x^2/20}$

51. There is a colony of fruit flies in Andy's kitchen. Assume we can model the population t days from now by the function $p(t) = 100 \cdot (12)^{t/10}$. An average fruit fly is about .1 inches long.

   (a) How many fruit flies are currently in Andy's kitchen?

   (b) How many will there be at this time next week? In two weeks?

   (c) In how many days will the population reach 2500?

   (d) Is it realistic to assume that this model will remain valid for a year? Justify your answer. [*Hint:* According to the model, what will the population be in a year?]

52. If current rates of deforestation and fossil fuel consumption continue, then the amount of atmospheric carbon dioxide in parts per million (ppm) will be given by $f(x) = 375e^{.00609x}$, where $x = 0$ corresponds to 2000.*

   (a) What is the amount of carbon dioxide in 2003? In 2022?

   (b) In what year will the amount of carbon dioxide reach 500 ppm?

53. The pressure of the atmosphere $p(x)$ (in pounds per square inch) is given by

$$p(x) = ke^{-.0000425x},$$

---

*Based on projections from the International Panel on Climate Change.

where x is the height above sea level (in feet) and k is a constant.

(a) Use the fact that the pressure at sea level is 15 pounds per square inch to find k.
(b) What is the pressure at 5000 feet?
(c) If you were in a spaceship at an altitude of 160,000 feet, what would the pressure be?

54. (a) The function $g(t) = .6 - e^{-0.479t}$ gives the percentage of the United States population (expressed as a decimal) that has seen a new television show t weeks after it goes on the air. According to this model, what percentage of people have seen the show after 24 weeks?
(b) The show will be renewed if over half the population has seen it at least once. Approximately when will 50% of the people have seen the show?
(c) According to this model, when will 59.9% of the people have seen it? When will 60% have seen it?

55. According to data from the National Center for Health Statistics, the life expectancy at birth for a person born in a year x is approximated by the function

$$D(x) = \frac{79.257}{1 + 9.7135 \times 10^{24} \cdot e^{-.0304x}}$$

$$(1900 \le x \le 2050).$$

(a) What is the life expectancy of someone born in 1980? in 2000?
(b) In what year was life expectancy at birth 60 years?

56. The number of subscribers to basic cable TV (in millions) can be approximated by

$$g(x) = \frac{76.7}{1 + 16(.8444^x)},$$

where $x = 0$ corresponds to 1970.*

(a) Estimate the number of subscribers in 2005 and 2010.
(b) When does the number of subscribers reach 70 million?
(c) According to this model, will the number of subscribers ever each 90 million?

57. (a) The beaver population near a certain lake in year t is approximately

$$p(t) = \frac{2000}{1 + 199e^{-.5544t}}.$$

What is the population now ($t = 0$) and what will it be in 5 years?
(b) Approximately when will there be 1000 beavers?

58. The Gateway Arch (Figure 5–15) is 630 feet high and 630 feet wide at ground level. Suppose it were placed on a coordinate plane with the x-axis at ground level and the y-axis going through the center of the arch. Find a catenary

*Based on data from *The Cable TV Financial Datebook* and *The Pay TV Newsletter.*

function $g(x) = A(e^{kx} + e^{-kx})$ and a constant C such that the graph of the function $f(x) = g(x) + C$ provides a model of the arch. [*Hint:* Experiment with various values of A, k, C as in the Graphing Exploration on page 365. Many correct answers are possible.]

59. (a) A genetic engineer is growing cells in a fermenter. The cells multiply by splitting in half every 15 minutes. The new cells have the same DNA as the original ones. Complete the following table.

Time (hours)	Number of Cells
0	1
.25	2
.5	4
.75	
1	

(b) Write the rule of the function that gives the number of C cells at time t hours.

60. Do Exercise 59, using the following table, instead of the given one.

Time (hours)	Number of Cells
0	300
.25	600
.5	1200
.75	
1	

61. A weekly census of the tree-frog population in Frog Hollow State Park produces the following results.

Week	1	2	3	4	5	6
Population	18	54	162	486	1458	4374

(a) Find a function of the form $f(x) = Pa^x$ that describes the frog population at time x weeks.
(b) What is the growth factor in this situation (that is, by what number must this week's population be multiplied to obtain next week's population)?
(c) Each tree frog requires 10 square feet of space and the park has an area of 6.2 square miles. Will the space required by the frog population exceed the size of the park in 12 weeks? In 14 weeks? [*Remember:* 1 square mile $= 5280^2$ square feet.]

62. An eccentric billionaire offers you a job for the month of September. She says that she will pay you 2¢ on the first day, 4¢ on the second day, 8¢ on the third day, and so on, doubling your pay on each successive day.

(a) Let $P(x)$ denote your salary in *dollars* on day x. Find the rule of the function P.
(b) Would you be better off financially if instead you were paid $10,000 per day? [*Hint:* Consider $P(30)$.]

63. Take an ordinary piece of typing paper and fold it in half; then the folded sheet is twice as thick as the single sheet was. Fold it in half again so that it is twice as thick as before. Keep folding it in half as long as you can. Soon the folded paper will be so thick and small that you will be unable to continue, but suppose you could keep folding the paper as many times as you wanted. Assume that the paper is .002 inches thick.

   (a) Make a table showing the thickness of the folded paper for the first four folds (with fold 0 being the thickness of the original unfolded paper).

   (b) Find a function of the form $f(x) = Pa^x$ that describes the thickness of the folded paper after x folds.

   (c) How thick would the paper be after 20 folds?

   (d) How many folds would it take to reach the moon (which is 243,000 miles from the earth)? [*Hint:* One mile is 5280 feet.]

64. The figure is the graph of an exponential growth function $f(x) = Pa^x$.

   (a) In this case, what is P? [*Hint:* What is $f(0)$?]

   (b) Find the rule of the function f by finding a. [*Hint:* What is $f(2)$?]

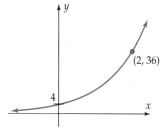

65. Suppose you invest $1200 in an account that pays 4% interest, compounded annually and paid from date of deposit to date of withdrawal.

   (a) Find the rule of the function f that gives the amount you would receive if you closed the account after x years.

   (b) How much would you receive after 3 years? After 5 years and 9 months?

   (c) When should you close the account to receive $1850?

66. Anne now has a balance of $800 on her credit card, on which 1.5% interest per month is charged. Assume that she makes no further purchases or payments (and that the credit card company doesn't turn her account over to a bill collector).

   (a) Find the rule of the function g that gives Anne's total credit card debt after x months.

   (b) How much will Anne owe after one year? After two years?

   (c) When will she owe twice the amount she owes now?

67. The population of Mexico was 100.4 million in 2000 and is expected to grow at the rate of 1.4% per year.

   (a) Find the rule of the function f that gives Mexico's population (in millions) in year x, with $x = 0$ corresponding to 2000.

   (b) Estimate Mexico's population in 2010.

   (c) When will the population reach 125 million people?

68. The number of digital devices (such as MP3 players, hand-held computers, cell phones, and PCs) in the world was approximately .94 billion in 1999 and is growing at a rate of 28.3% a year.*

   (a) Find the rule of a function that gives the number of digital devices (in billions) in year x, with $x = 0$ corresponding to 1999.

   (b) Approximately how many digital devices will be in use in 2010?

   (c) If this model remains accurate, when will the number of digital devices reach 6 billion?

69. The U.S. Census Bureau estimates that the Hispanic population in the United States will increase from 32.44 million in 2000 to 98.23 million in 2050.†

   (a) Find an exponential function that gives the Hispanic population in year x, with $x = 0$ corresponding to 2000.

   (b) What is the projected Hispanic population in 2010 and 2025?

   (c) In what year will the Hispanic population reach 55 million?

70. The U.S. Department of Commerce estimated that there were 54 million Internet users in the United States in 1999 and 85 million in 2002.

   (a) Find an exponential function that models the number of Internet users in year x, with $x = 0$ corresponding to 1999.

   (b) For how long is this model likely to remain accurate? [*Hint:* The current U.S. population is about 305 million.]

71. At the beginning of an experiment, a culture contains 200 *H. pylori* bacteria. An hour later there are 205 bacteria. Assuming that the *H. pylori* bacteria grow exponentially, how many will there be after 10 hours? After 2 days?

72. The population of India was approximately 1030 million in 2001 and was 967 million in 1997. If the population continues to grow exponentially at the same rate, what will it be in 2010?

73. Kerosene is passed through a pipe filled with clay to remove various pollutants. Each foot of pipe removes 25% of the pollutants.

   (a) Write the rule of a function that gives the percentage of pollutants remaining in the kerosene after it has passed through x feet of pipe. [See Example 7.]

   (b) How many feet of pipe are needed to ensure that 90% of the pollutants have been removed from the kerosene?

74. If inflation runs at a steady 3% per year, then the amount a dollar is worth decreases by 3% each year.

*Based on data and projections from IDC.
†*Statistical Abstract of the United States: 2007.*

(a) Write the rule of a function that gives the value of a dollar in year x.

(b) How much will the dollar be worth in 5 years? In 10 years?

(c) How many years will it take before today's dollar is worth only a dime?

75. You have 5 grams of carbon-14, whose half-life is 5730 years.

(a) Write the rule of the function that gives the amount of carbon-14 remaining after x years. [See the box preceding Example 8.]

(b) How much carbon-14 will be left after 4000 years? After 8000 years?

(c) When will there be just 1 gram left?

76. (a) The half-life of radium is 1620 years. If you start with 100 milligrams of radium, what is the rule of the function that gives the amount remaining after t years?

(b) How much radium is left after 800 years? After 1600 years? After 3200 years?

THINKERS

77. Find a function $f(x)$ with the property $f(r + s) = f(r)f(s)$ for all real numbers r and s.

78. Find a function $g(x)$ with the property $g(2x) = (g(x))^2$ for every real number x.

79. (a) Using the viewing window with $-4 \leq x \leq 4$ and $-1 \leq y \leq 8$, graph $f(x) = \left(\frac{1}{2}\right)^x$ and $g(x) = 2^x$ on the same screen. If you think of the y-axis as a mirror, how would you describe the relationship between the two graphs?

(b) Without graphing, explain how the graphs of $g(x) = 2^x$ and $k(x) = 2^{-x}$ are related.

80. Look back at Section 4.4, where the basic properties of graphs of polynomial functions were discussed. Then review the basic properties of the graph of $f(x) = a^x$ discussed in this section. Using these various properties, give an argument to show that for any fixed positive number $a(\neq 1)$, it is *not* possible to find a polynomial function

$g(x) = c_n x^n + \cdots + c_1 x + c_0$ such that $a^x = g(x)$ for *all* numbers x. In other words, *no exponential function is a polynomial function*. However, see Exercise 81.

81. Approximating exponential functions by polynomials. For each positive integer n, let f_n be the polynomial function whose rule is

$$f_n(x) = 1 + x + \frac{x^2}{2!} + \frac{x^3}{3!} + \frac{x^4}{4!} + \cdots + \frac{x^n}{n!},$$

where $k!$ is the product $1 \cdot 2 \cdot 3 \cdots k$.

(a) Using the viewing window with $-4 \leq x \leq 4$ and $-5 \leq y \leq 55$, graph $g(x) = e^x$ and $f_4(x)$ on the same screen. Do the graphs appear to coincide?

(b) Replace the graph of $f_4(x)$ by that of $f_5(x)$, then by $f_6(x)$, $f_7(x)$, and so on until you find a polynomial $f_n(x)$ whose graph appears to coincide with the graph of $g(x) = e^x$ in this viewing window. Use the trace feature to move from graph to graph at the same value of x to see how accurate this approximation is.

(c) Change the viewing window so that $-6 \leq x \leq 6$ and $-10 \leq y \leq 400$. Is the polynomial you found in part (b) a good approximation for $g(x)$ in this viewing window? What polynomial is?

82. This exercise provides a justification for the claim that the function $M(x) = c(.5)^{x/h}$ gives the mass after x years of a radioactive element with half-life h years. Suppose we have c grams of an element that has a half-life of 50 years. Then after 50 years, we would have $c\left(\frac{1}{2}\right)$ grams. After another 50 years, we would have half of that, namely, $c\left(\frac{1}{2}\right)\left(\frac{1}{2}\right) = c\left(\frac{1}{2}\right)^2$.

(a) How much remains after a third 50-year period? After a fourth 50-year period?

(b) How much remains after t 50-year periods?

(c) If x is the number of years, then $x/50$ is the number of 50-year periods. By replacing the number of periods t in part (b) by $x/50$, you obtain the amount remaining after x years. This gives the function $M(x)$ when $h = 50$. The same argument works in the general case (just replace 50 by h). Find $M(x)$.

5.2.A *SPECIAL TOPICS* Compound Interest and the Number *e*

Section Objective ■ Apply compound interest formulas to financial situations.

When money earns compound interest, as in Example 4 on page 360, the exponential growth function can be described as follows.

Compound Interest Formula

> If P dollars is invested at interest rate r per time period (expressed as a decimal), then the amount A after t periods is
>
> $$A = P(1 + r)^t$$

EXAMPLE 1

Suppose you borrow $50 from your friendly neighborhood loan shark, who charges 18% interest per week. How much do you owe after one year (assuming that he lets you wait that long to pay)?

SOLUTION   You use the compound interest formula with $P = 50$, $r = .18$, and $t = 52$ (because interest is compounded weekly and there are 52 weeks in a year). So you figure that you owe

$$A = P(1 + r)^t = 50(1 + .18)^{52} = 50 \cdot 1.18^{52} = \$273,422.58.*$$

When you try to pay the loan shark this amount, however, he points out that a 365-day year has more than 52 weeks, namely, $\dfrac{365}{7} = 52\dfrac{1}{7}$ weeks. So you recalculate with $t = 365/7$ (and careful use of parentheses, as shown in Figure 5–16) and find that you actually owe

$$A = P(1 + r)^t = 50(1 + .18)^{365/7} = 50 \cdot 1.18^{365/7} = \$279,964.68.$$

Ouch! ∎

```
50*1.18^(365/7)
         279964.6785
■
```

Figure 5–16

As Example 1 illustrates, the compound interest formula can be used even when the number of periods t is not an integer. You must also learn how to read "financial language" to apply the formula correctly, as shown in the following example.

EXAMPLE 2

Determine the amount a $3500 investment is worth after three and a half years at the following interest rates:

(a) 6.4% compounded annually;

(b) 6.4% compounded quarterly;

(c) 6.4% compounded monthly.

SOLUTION

(a) Using the compound interest formula with $P = 3500$, $r = .064$, and $t = 3.5$, we have

$$A = 3500(1 + .064)^{3.5} = \$4348.74.$$

(b) "6.4% interest, compounded quarterly" means that the interest period is one-fourth of a year and the interest rate per period is $.064/4 = .016$. Since there are four interest periods per year, the number of periods in 3.5 years is $4(3.5) = 14$, so

$$A = 3500\left(1 + \frac{.064}{4}\right)^{14} = 3500(1 + .016)^{14} = \$4370.99.$$

---

*Here and below, all financial answers are rounded to the nearest penny.

(c) Similarly, "6.4% compounded monthly" means that the interest period is one month (1/12 of a year) and the interest rate per period is .064/12. The number of periods (months) in 3.5 years is 42, so

$$A = 3500\left(1 + \frac{.064}{12}\right)^{42} = \$4376.14.$$

Note that the more often interest is compounded, the larger the final amount ▪

EXAMPLE 3

If $5000 is invested at 6.5% annual interest, compounded monthly, how long will it take for the investment to double?

SOLUTION

The compound interest formula (with $P = 5000$ and $r = .065/12$) shows that the amount in the account after t months is $5000\left(1 + \frac{.065}{12}\right)^t$. We must find the value of t such that

$$5000\left(1 + \frac{.065}{12}\right)^t = 10{,}000.$$

Algebraic methods for solving this equation will be considered in Section 5.5. For now, we use technology.

GRAPHING EXPLORATION

Solve the equation, either by using an equation solver, or by graphical means as follows. Graph

$$y = 5000\left(1 + \frac{.065}{12}\right)^t - 10{,}000$$

in a viewing window with $0 \le t \le 240$ (that's 20 years) and find the t-intercept.

The exploration shows that it will take 128.3 months (approximately 10.7 years) for the investment to double. ▪

EXAMPLE 4

What interest rate, compounded annually, is needed for a $16,000 investment to grow to $50,000 in 18 years?

SOLUTION   In the compound interest formula, we have $A = 50{,}000$, $P = 16{,}000$ and $t = 18$. We must find r in the equation

$$16{,}000(1 + r)^{18} = 50{,}000.$$

The equation can be solved numerically with an equation solver or by one of the following methods.

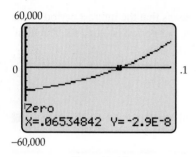

60,000

0 ─────────────────────── .1

Zero
X=.06534842 Y=-2.9E-8

−60,000

Figure 5–17

Graphical: Rewrite the equation as $16{,}000(1 + r)^{18} - 50{,}000 = 0$. Then the solution is the *r*-intercept of the graph of $y = 16{,}000(1 + r)^{18} - 50{,}000$, as shown in Figure 5–17.

Algebraic:

$$16{,}000(1 + r)^{18} = 50{,}000$$

Divide both sides by 16,000: $(1 + r)^{18} = \dfrac{50{,}000}{16{,}000} = 3.125$

Take 18th roots on both sides: $\sqrt[18]{(1 + r)^{18}} = \sqrt[18]{3.125}$

$$1 + r = \sqrt[18]{3.125}$$

$$r = \sqrt[18]{3.125} - 1 \approx .06535.$$

So the necessary interest rate is about 6.535%.   ■

CONTINUOUS COMPOUNDING AND THE NUMBER *e*

As a general rule, the more often interest is compounded, the better off you are, as we saw in Example 2. But there is, alas, a limit.

EXAMPLE 5

The Number *e*   You have \$1 to invest for 1 year. The Exponential Bank offers to pay 100% annual interest, compounded *n* times per year and rounded to the nearest penny. You may pick any value you want for *n*. We have already seen that the larger *n* is, the more money you wind up earning. How large should you choose *n* in order to make your \$1 grow to \$5?

SOLUTION   Since interest rate is compounded *n* times per year and the annual rate is 100% (= 1.00), the interest rate per period is $r = 1/n$ and the number of periods in 1 year is *n*. According to the formula, the amount at the end of the year will be $A = \left(1 + \dfrac{1}{n}\right)^n$. Here's what happens for various values of *n*:

Interest Is Compounded	$n =$	$\left(1 + \dfrac{1}{n}\right)^n =$
Annually	1	$\left(1 + \frac{1}{1}\right)^1 = 2$
Semiannually	2	$\left(1 + \frac{1}{2}\right)^2 = 2.25$
Quarterly	4	$\left(1 + \frac{1}{4}\right)^4 \approx 2.4414$
Monthly	12	$\left(1 + \frac{1}{12}\right)^{12} \approx 2.6130$
Daily	365	$\left(1 + \frac{1}{365}\right)^{365} \approx 2.71457$
Hourly	8760	$\left(1 + \frac{1}{8760}\right)^{8760} \approx 2.718127$
Every minute	525,600	$\left(1 + \frac{1}{525{,}600}\right)^{525{,}600} \approx 2.7182792$
Every second	31,536,000	$\left(1 + \frac{1}{31{,}536{,}000}\right)^{31{,}536{,}000} \approx 2.7182818$

Since interest is rounded to the nearest penny, your dollar will grow no larger than \$2.72, no matter how big *n* is. You will not be able to make your dollar grow to \$5 at that interest rate.   ■

The last entry in the preceding table, 2.7182818, is the number *e* to seven decimal places. This is just one example of how *e* arises naturally in real-world situations. In calculus, it is provided that *e* is the *limit* of $\left(1 + \dfrac{1}{n}\right)^n$, meaning that as *n* gets larger and larger, $\left(1 + \dfrac{1}{n}\right)^n$ gets closer and closer to *e*.

GRAPHING EXPLORATION

Confirm this fact graphically by graphing the function

$$f(x) = \left(1 + \frac{1}{x}\right)^x$$

and the horizontal line $y = e$ in the viewing window with $0 \le x \le 5000$ and $-1 \le y \le 4$ and noting that the two graphs appear to be identical.

When interest is compounded *n* times per year for larger and larger values of *n*, as in Example 5, we say that the interest is **continuously compounded.** In this terminology, Example 5 says that $1 will grow to $2.72 in 1 year at an interest rate of 100% compounded continuously. A similar argument with more realistic interest rates (see Exercise 30) produces the following result (Example 5 is the case when $P = 1$, $r = 1$, and $t = 1$).

Continuous Compounding

> If *P* dollars is invested at interest rate *r*, compounded continuously, then the amount *A* after *t* years is
>
> $$A = Pe^{rt}.$$

EXAMPLE 6

If $3800 is invested in a CD with a 3.8% interest rate, compounded continuously, find:

(a) The amount in the account after seven and a half years.

(b) The number of years for the account balance to reach $5000.

SOLUTION

(a) Apply the continuous compounding formula with $P = 3800$, $r = .038$, and $t = 7.5$.

$$A = 3800e^{(.038)7.5} = 3800e^{.285} \approx \$5053.10.$$

(b) We must solve the equation

$$3800e^{.038t} = 5000, \quad \text{or, equivalently,} \quad 3800e^{.038t} - 5000 = 0.$$

GRAPHING EXPLORATION

Solve the equation graphically and verify that it will take a bit more than seven years for the investment to be worth $5000.

EXERCISES 5.2.A

1. If $1,000 is invested at 8%, find the value of the investment after 5 years if interest is compounded

   (a) annually.   (b) quarterly.   (c) monthly.
   (d) weekly.

2. If $2500 is invested at 11.5%, what is the value of the investment after 10 years if interest is compounded

   (a) annually?   (b) monthly?   (c) daily?

In Exercises 3–10, determine how much money will be in a savings account if the initial deposit was $500 and the interest rate is:

3. 3% compounded annually for 8 years.

4. 3% compounded annually for 10 years.

5. 3% compounded quarterly for 10 years.

6. 2.5% compounded annually for 20 years.

7. 2.477% compounded quarterly for 20 years.

8. 2.469% compounded continuously for 20 years.

9. 3% compounded continuously for 10 years, 7 months.

10. 3% compounded continuously for 30 years.

*A sum of money P that can be deposited today to yield some larger amount A in the future is called the **present value** of A. In Exercises 11–14, find the present value of the given amount A. [Hint: Substitute A, the interest rate per period r, and the number t of periods in the compound interest formula and solve for P.]*

11. $5000 at 6% compounded annually for 7 years.

12. $3500 at 5.5% compounded annually for 4 years.

13. $4800 at 7.2% compounded quarterly for 5 years.

14. $7400 at 5.9% compounded quarterly for 8 years.

15. You are to receive an insurance settlement in the amount of $8000. Because of various bureaucratic delays, it will take you about three years to collect your money.

    (a) Assuming that your bank offers you an interest rate of 4 percent, compounded continuously, what is the present value of your settlement?

    (b) If your insurance agent offers you $7050, payable immediately, to give up the settlement, is it best to take the deal?

16. You win a lawsuit, and the defendant is ordered to pay you $5000, and has up to eight years to pay you. We can assume that the defendant will probably wait until the last possible minute to give you your check.

    (a) If you can get an interest rate of 3.75 percent on your money, compounded continuously, what is the present value of the money in question?

    (b) If the defendant offers you $4000 (paid immediately) to forgive the debt, is it best to take the deal?

17. You have $10,000 to invest for two years. Fund *A* pays 13.2% interest, compounded annually. Fund *B* pays 12.7% interest, compounded quarterly. Fund *C* pays 12.6% interest, compounded monthly. Which fund will return the most money?

18. If you invest $7400 for five years, are you better off with an interest rate of 5% compounded quarterly or 4.8% compounded continuously?

19. If you borrow $1200 at 14% interest, compounded monthly, and pay off the loan (principal and interest) at the end of two years, how much interest will you have paid?

20. A developer borrows $150,000 at 6.5% interest, compounded quarterly, and agrees to pay off the loan in four years. How much interest will she owe?

21. A manufacturer has settled a lawsuit out of court by agreeing to pay $1.5 million four years from now. At this time, how much should the company put in an account paying 6.4% annual interest, compounded monthly, to have $1.5 million in four years? [*Hint:* See Exercises 11–14.]

22. Lisa Chow wants to have $30,000 available in five years for a down payment on a house. She has inherited $25,000. How much of the inheritance should be invested at 5.7% annual interest, compounded quarterly, to accumulate the $30,000?

23. If an investment of $1000 grows to $1407.10 in seven years with interest compounded annually, what is the interest rate?

24. If an investment of $2000 grows to $2700 in three and a half years, with an annual interest rate that is compounded quarterly, what is the annual interest rate?

25. If you put $3000 in a savings account today, what interest rate (compounded annually) must you receive in order to have $4000 after five years?

26. If interest is compounded continuously, what annual rate must you receive if your investment of $1500 is to grow to $2100 in six years?

27. At an interest rate of 8% compounded annually, how long will it take to double an investment of

    (a) $100   (b) $500   (c) $1200?
    (d) What conclusion about doubling time do parts (a)–(c) suggest?

28. At an interest rate of 6% compounded annually, how long will it take to double an investment of *P* dollars?

29. How long will it take to double an investment of $500 at 7% annual interest, compounded continuously?

THINKERS

30. This exercise provides an illustration of why the continuous compounding formula (page 373) is valid, using a realistic interest rate. We shall determine the value of $4000 deposited for three years at 5% interest compounded *n* times per year for larger and larger values of *n*. In this case, the interest rate

per period is $.05/n$ and the number of periods in three years is $3n$. So the amount in the account at the end of three years is:

$$A = 4000\left(1 + \frac{.05}{n}\right)^{3n} = 4000\left[\left(1 + \frac{.05}{n}\right)^{n}\right]^{3}.$$

(a) Fill in the missing entries in the following table.

n	$\left(1 + \dfrac{.05}{n}\right)^{n}$
1,000	
10,000	
500,000	
1,000,000	
5,000,000	
10,000,000	

(b) Compare the entries in the second column of the table with the number $e^{.05}$ and fill the blank in the following sentence:

As n gets larger and larger, the value of $\left(1 + \dfrac{.05}{n}\right)^{n}$

gets closer and closer to the number _____.

(c) Use you answer to part (b) to fill the blank in the following sentence:

As n gets larger and larger, the value of

$$A = 4000\left[\left(1 + \frac{.05}{n}\right)^{n}\right]^{3}$$

gets closer and closer to _____.

(d) Compare your answer in part (c) to the value of the investment given by the continuous compounding formula.

31. Municipal bonds are investments issued by cities, states, or counties that wish to raise money to build things like schools, highways and hospitals. You buy a bond for a certain amount, and you get an interest payment every six months. Then, at a certain time (the "maturity date") you get your principal (the amount you paid for the bond) back. For example, if you bought a $10000 bond with a 10% interest rate, you would get payments of $500 every six months for a while, and then you would get a payment for $10500 back at the maturity date.

(a) In 2007 Sioux City, Iowa issued $5000 bonds for their community school district at an interest rate of 3.63%. The maturity date is October 1, 2012. The interest was to be paid every April 1 and October 1. If you bought one of these bonds on April 1, 2007, and held it until the maturity date, how much total interest will you have earned?

(b) What if, instead of buying the Sioux City bond, you could buy a CD (Certificate of Deposit) from a local bank that paid 3.4%, compounded semi-annually. Again, assuming you were going to save $5000 from April 1, 2007 though October 1, 2012, which would be the better choice and why?

(c) As discussed above, when you buy the Sioux City municipal bond, you are getting a payment every six months. What if you took those interest payments, and put them in a bank account that pays 3% interest, compounded semi-annually? Now how much total interest will you have earned on October 1, 2012? Would you make more money doing this, or buying the CD?

5.3 Common and Natural Logarithmic Functions*

Section Objectives
- ■ Evaluate common and natural logarithms.
- ■ Translate logarithmic statements in exponential statements, and vice-versa.
- ■ Use the properties of logarithms.
- ■ Find the graphs of logarithmic functions.

Roadmap

We begin with the only logarithms that are in widespread use, common and natural logarithms. Natural logarithms are emphasized because of their central role in calculus. Those who prefer to begin with logarithms to an arbitrary base b should cover Special Topics 5.4.A before reading this section.

The discovery of logarithms in the seventeenth century allowed scientists to perform many crucial computations that previously had been too difficult to be practical. Although computers now handle these computations, logarithms are still extremely useful in the sciences and engineering. Logarithmic functions provide excellent models of different phenomena, including sound volume, earthquake intensity, the perceived brightness of stars, computational complexity, the spread of certain kinds of diseases, the growth of rumors, and much more. Logarithms also have properties that make it possible to solve certain types of equations more easily.

---

*Section 3.7 (Inverse Functions) is a prerequisite for this section.

COMMON LOGARITHMS

The exponential function $f(x) = 10^x$, whose graph is shown in Figure 5–18, is an increasing function and hence is one-to-one (as explained on page 219). Therefore, f has an inverse function g whose graph is the reflection of the graph of f in the line $y = x$ (see page 225), as shown in Figure 5–19.*

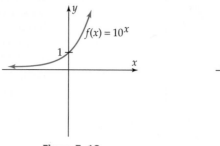

Figure 5–18   **Figure 5–19**

This inverse function g is called the **common logarithmic function.** The value of this function at the number x is denoted **log x** and called the **common logarithm** of the number x. Every calculator has a LOG key for evaluating the function $g(x) = \log x$. For instance,

$$\log .01 = -2, \qquad \log .6 = -.2218, \qquad \text{and} \qquad \log 10000 = 5^\dagger$$

As we saw in Section 3.7, the relationship between a function f and its inverse function g is given by

$$g(v) = u \qquad \text{exactly when} \qquad f(u) = v.$$

When $f(x) = 10^x$ and $g(x) = \log x$, this statement takes the following form.

Definition of Common Logarithms

> Let u and v be real numbers, with $v > 0$. Then
>
> $$\log v = u \qquad \text{exactly when} \qquad 10^u = v.$$
>
> In other words,
>
> $$\log v \text{ is the exponent to which 10 must be raised to produce } v.$$

EXAMPLE 1

Without using a calculator, find

(a) $\log 1000$ (b) $\log 1$ (c) $\log \sqrt{10}$ (d) $\log \left(\dfrac{1}{\sqrt{10}} \right)$

---

*Parametric equations for the graph of $f(x) = 10^x$ can be obtained by letting

$$x = t \qquad \text{and} \qquad y = 10^t \quad (t \text{ any real number}).$$

As explained on page 224, parametric equations for the graph of the inverse function g can then be obtained by letting

$$x = 10^t \qquad \text{and} \qquad y = t \quad (t \text{ any real number}).$$

This trick will allow you to display the graphs of Figure 5–19 on your calculator in parametric mode.
†Here and below, all logarithms are rounded to four decimal places, and an equal sign is used rather than the more correct "approximately equal." The word "common" will be omitted except when it is necessary to distinguish these logarithms from other types that are introduced below.

SOLUTION

(a) To find log 1000, ask yourself, "What power of 10 equals 1000?" The answer is 3 because $10^3 = 1000$. Therefore, log 1000 = 3.

(b) To what power must 10 be raised to produce 1? Since $10^0 = 1$, we conclude that log 1 = 0.

(c) Log $\sqrt{10} = 1/2$ because 1/2 is the exponent to which 10 must be raised to produce $\sqrt{10}$, that is $10^{1/2} = \sqrt{10}$.

(d) Log $\dfrac{1}{\sqrt{10}} = -1/2$ because $10^{-1/2} = \dfrac{1}{\sqrt{10}}$. ∎

EXAMPLE 2

Translate each of the following logarithmic statements into an equivalent exponential statement.

$$\log 29 = 1.4624 \qquad \log .47 = -.3279 \qquad \log (k + t) = d.$$

SOLUTION Using the definition above, we have these translations.

Logarithmic Statement	Equivalent Exponential Statement
$\log 29 = 1.4624$	$10^{1.4624} = 29$
$\log .47 = -.3279$	$10^{-.3279} = .47$
$\log (k + t) = d$	$10^d = k + t$

∎

EXAMPLE 3

Translate each of the following exponential statements into an equivalent logarithmic statement.

$$10^{5.5} = 316{,}227.766 \qquad 10^{.66} = 4.5708819 \qquad 10^{rs} = t$$

SOLUTION Translate as follows.

Exponential Statement	Equivalent Logarithmic Statement
$10^{5.5} = 316{,}277.766$	$\log 316{,}277.766 = 5.5$
$10^{.66} = 4.5708819$	$\log 4.5708819 = .66$
$10^{rs} = t$	$\log t = rs$

∎

EXAMPLE 4

Solve the equation log $x = 4$.

SOLUTION log $x = 4$ is equivalent to $10^4 = x$. So the solution is $x = 10{,}000$. ∎

NATURAL LOGARITHMS

Common logarithms are closely related to the exponential function $f(x) = 10^x$. With the advent of calculus, however, it became clear that the most useful exponential function in science and engineering is $g(x) = e^x$. Consequently, a new type of logarithm, based on the number e instead of 10, was developed. This development is essentially a copy of what was done above, with some minor changes in notation.

The exponential function $f(x) = e^x$ whose graph is shown in Figure 5–20 is increasing and hence one-to-one, so f has an inverse function g whose graph is the reflection of the graph of f in the line $y = x$, as shown in Figure 5–21.

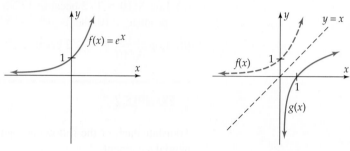

Figure 5–20   Figure 5–21

This inverse function g is called the **natural logarithmic function.** The value $g(x)$ of this function at a number x is denoted **ln x** and called the **natural logarithm** of the number x. Every calculator has an LN key for evaluating natural logarithms. For instance,

$$\ln .15 = -1.8971, \quad \ln 186 = 5.2257, \quad \text{and} \quad \ln 2.718 = .9999.$$

When the relationship of inverse functions (Section 3.7)

$$g(v) = u \quad \text{exactly when} \quad f(u) = v$$

is applied to the function $f(x) = e^x$ and its inverse $g(x) = \ln x$, it says the following.

Definition of Natural Logarithms

Let u and v be real numbers, with $v > 0$. Then

$$\ln v = u \quad \text{exactly when} \quad e^u = v.$$

In other words,

$\ln v$ is the exponent to which e must be raised to produce v.

EXAMPLE 5

Translate:

(a) $\ln 14 = 2.6391$ into an equivalent exponential statement.

(b) $e^{5.0626} = 158$ into an equivalent logarithmic statement.

SOLUTION

(a) Using the preceding definition, we see that $\ln 14 = 2.6391$ is equivalent to $e^{2.6391} = 14$.

(b) Similarly, $e^{5.0626} = 158$ is equivalent to $\ln 158 = 5.0626$. ■

PROPERTIES OF LOGARITHMS

Since common and natural logarithms have almost identical definitions (just replace 10 by e), it is not surprising that they share the same essential properties. You don't need a calculator to understand these properties. You need only use the definition of logarithms or translate logarithmic statements into equivalent exponential ones (or vice versa).

EXAMPLE 6

What is $\ln(-10)$?

Translation: To what power must e be raised to produce -10?

Answer: The graph of $f(x) = e^x$ in Figure 5–20 shows that every power of e is *positive*. So e^x can *never* be -10 or any negative number or zero, and hence, $\ln(-10)$ is not defined. Similarly, $\log(-10)$ is not defined because every power of 10 is positive. Therefore,

$$\textbf{ln } v \textbf{ and log } v \textbf{ are defined only when } v > 0.$$ ■

EXAMPLE 7

What is $\ln 1$?

Translation: To what power must e be raised to produce 1?

Answer: We know that $e^0 = 1$, which means that $\ln 1 = 0$. Combining this fact with Example 1(b), we have

$$\textbf{ln 1} = \textbf{0} \qquad \textbf{and} \qquad \textbf{log 1} = \textbf{0}.$$ ■

EXAMPLE 8

What is $\ln e^9$?

Translation: To what power must e be raised to produce e^9?

Answer: Obviously, the answer is 9. So $\ln e^9 = 9$ and in general

$$\textbf{ln } e^k = k \qquad \textbf{for every real number } k.$$

Similarly,

$$\textbf{log } 10^k = k \qquad \textbf{for every real number } k$$

because k is the exponent to which 10 must be raised to produce 10^k. In particular, when $k = 1$, we have

$$\textbf{ln } e = \textbf{1} \qquad \textbf{and} \qquad \textbf{log 10} = \textbf{1}.$$ ■

EXAMPLE 9

Find $10^{\log 678}$ and $e^{\ln 678}$.

SOLUTION By definition, log 678 is the exponent to which 10 must be raised to produce 678. So if you raise 10 to this exponent, the answer will be 678, that is, $10^{\log 678} = 678.$* Similarly, ln 678 is the exponent to which e must be raised to produce 678, so that $e^{\ln 678} = 678$. The same argument works with any positive number v in place of 678:

$$e^{\ln v} = v \qquad \text{and} \qquad 10^{\log v} = v \quad \text{for every } v > 0. \qquad \blacksquare$$

The facts presented in the preceding examples may be summarized as follows.

Properties of Logarithms

Natural Logarithms	**Common Logarithms**
1. ln v is defined only when $v > 0$;	log v is defined only when $v > 0$.
2. ln 1 = 0 and ln e = 1;	log 1 = 0 and log 10 = 1.
3. ln $e^k = k$ for every real number k;	log $10^k = k$ for every real number k.
4. $e^{\ln v} = v$ for every $v > 0$;	$10^{\log v} = v$ for every $v > 0$.

EXAMPLE 10

Applying Property 3 with $k = 2x^2 + 7x + 9$ shows that

$$\ln e^{2x^2+7x+9} = 2x^2 + 7x + 9. \qquad \blacksquare$$

EXAMPLE 11

Solve the equation $\ln(x + 1) = 2$.

SOLUTION Since $\ln(x + 1) = 2$, we have

$$e^{\ln(x+1)} = e^2.$$

Applying Property 4 with $v = x + 1$ shows that

$$x + 1 = e^{\ln(x+1)} = e^2$$

$$x = e^2 - 1 \approx 6.3891. \qquad \blacksquare$$

Property 4 has another interesting consequence. If a is any positive number, then $e^{\ln a} = a$. Hence, the rule of the exponential function $f(x) = a^x$ can be written as

$$f(x) = a^x = (e^{\ln a})^x = e^{(\ln a)x}.$$

---

*This is equivalent, in a sense, to answering the question "Who is the author whose name is Stephen King?" The answer is described in the question!

For example, $f(x) = 2^x = e^{(\ln 2)x} \approx e^{.6931x}$. Thus, we have this useful result.

Exponential
Functions

Every exponential growth or decay function can be written in the form

$$f(x) = Pe^{kx},$$

where $f(x)$ is the amount at time x, P is the initial quantity, and k is positive for growth and negative for decay.

EXAMPLE 12

Write $f(x) = 3 \cdot 5^x$ in the form $f(x) = Pe^{kx}$.

SOLUTION $3 \cdot 5^x = 3 \cdot (e^{\ln 5})^x$

$$= 3e^{\ln 5 \cdot x}$$

$$\approx 3e^{1.6094x}$$ ∎

GRAPHS OF LOGARITHMIC FUNCTIONS

Figure 5–22 shows the graphs of two more entries in the catalog of basic functions, $f(x) = \log x$ and $g(x) = \ln x$. Both are increasing functions with these four properties:

Domain: all positive real numbers   ***x*-intercept:** 1
Range: all real numbers   **Vertical Asymptote:** *y*-axis

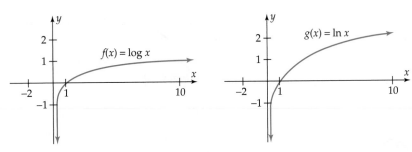

Figure 5–22

Calculators and computers do not accurately show that the *y*-axis is a vertical asymptote of these graphs. By evaluating the functions at very small numbers (such as $x = 1/10^{500}$), you can see that the graphs go lower and lower as x gets closer to 0. On a calculator, however, the graph will appear to end abruptly near the *y*-axis (try it!).

Some viewing windows may give the impression that logarithmic graphs (such as those in Figures 5–20 and 5–21) have horizontal asymptotes. Don't be fooled! These graphs have no horizontal asymptotes—the *y*-values get arbitrarily large.

EXAMPLE 13

Sketch the graph of $f(x) = \ln (x - 2)$.

SOLUTION   Using a calculator to graph $f(x) = \ln (x - 2)$, we obtain Figure 5–23, in which the graph appears to end abruptly near $x = 2$. Fortunately, however, we have read Section 3.4, so we know that this is *not* how the graph looks. From Section 3.4, we know that the graph of $f(x) = \ln (x - 2)$ is the graph of $g(x) = \ln x$ shifted horizontally 2 units to the right, as shown in Figure 5–24. In particular, the graph of f has a vertical asymptote at $x = 2$ and drops sharply downward there.   ■

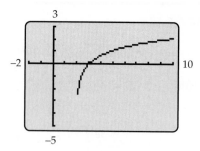

Figure 5–23   **Figure 5–24**

GRAPHING EXPLORATION

Graph $y_1 = \ln (x - 2)$ and $y_2 = -5$ in the same viewing window and verify that the graphs do not appear to intersect, as they should. Nevertheless, try to solve the equation $\ln (x - 2) = -5$ by finding the intersection point of y_1 and y_2. Some calculators will find the intersection point even through it does not show on the screen. Others produce an error message, in which case the SOLVER feature should be used instead of a graphical solution.

Although logarithms are only defined for positive numbers, many logarithmic functions include negative numbers in their domains.

EXAMPLE 14

Find the domain of each of the following functions.

(a) $f(x) = \ln (x + 4)$ (b) $g(x) = \log x^2$

SOLUTION

(a) $f(x) = \ln (x + 4)$ is defined only when $x + 4 > 0$, that is, when $x > -4$. So the domain of f consists of all real numbers greater than -4.

(b) Since $x^2 > 0$ for all nonzero x, the domain of $g(x) = \log x^2$ consists of all real numbers except 0.   ■

GRAPHING EXPLORATION

Verify the conclusions of Example 14 by graphing each of the functions. What is the vertical asymptote of each graph?

EXERCISES 5.3

Unless stated otherwise, all letters represent positive numbers.

In Exercises 1–4, find the logarithm, without using a calculator.

1. log 10,000

2. log .001

3. $\log \dfrac{\sqrt{10}}{1000}$

4. $\log \sqrt[3]{.01}$

In Exercises 5–14, translate the given logarithmic statement into an equivalent exponential statement.

5. log 1000 = 3

6. log .001 = −3

7. log 750 = 2.88

8. log (.8) = −.097

9. ln 3 = 1.0986

10. log (log(x)) = 1

11. ln .01 = −4.6052

12. ln s = r

13. ln (x^2 + 2y) = z + w

14. log (a + c) = d

In Exercises 15–24, translate the given exponential statement into an equivalent logarithmic statement.

15. 10^{-2} = .01

16. 10^3 = 1000

17. $10^{.4771}$ = 3

18. 10^{3k} = 6r

19. $e^{3.25}$ = 25.79

20. $e^{3.14}$ = 23.1039

21. $e^{12/7}$ = 5.5527

22. e^k = t

23. $e^{2/r}$ = w

24. e^e = 15.1543

In Exercises 25–36, evaluate the given expression without using a calculator.

25. $\log 10^{\sqrt{43}}$

26. $\log 10^{\sqrt[3]{r^2-s^2}}$

27. $\ln e^{15}$

28. $e^{\ln \pi}$

29. $\ln \sqrt{e}$

30. $\ln \sqrt[5]{e}$

31. $e^{\ln 931}$

32. log (log(10,000,000,000))

33. $\ln e^{x+y}$

34. $\ln e^{x^2+2y}$

35. $e^{\ln x^2}$

36. $e^{\ln(\ln 2)}$

In Exercises 37–40, write the rule of the function in the form $f(x) = Pe^{kx}$. (See the discussion and box after Example 11.)

37. $f(x) = 4(25^x)$

38. $g(x) = 3.9(1.03^x)$

39. $g(x) = -16(30.5^x)$

40. $f(x) = -2.2(.75^x)$

In Exercises 41–42, write the rule of the function in the form $f(x) = a^x$. (See the discussion and box after Example 11.)

41. $g(x) = e^{-3x}$

42. $f(x) = e^{1.6094x}$

In Exercises 43–46, find the domain of the given function (that is, the largest set of real numbers for which the rule produces well-defined real numbers).

43. $f(x) = \ln (x + 1)$

44. $g(x) = \ln (x + 2)$

45. $h(x) = \log (-x)$

46. $k(x) = \log (\ln (2) - x)$

47. (a) Graph $y = x$ and $y = e^{\ln x}$ in separate viewing windows [or use a split-screen if your calculator has that feature]. For what values of x are the graphs identical?
    (b) Use the properties of logarithms to explain your answer in part (a).

48. (a) Graph $y = x$ and $y = \ln (e^x)$ in separate viewing windows [or a split-screen if your calculator has that feature]. For what values of x are the graphs identical?
    (b) Use the properties of logarithms to explain your answer in part (a).

49. Do the graphs of $f(x) = \log x^2$ and $g(x) = 2 \log x$ appear to be the same? How do they differ?

50. Do the graphs of $h(x) = \log x^3$ and $k(x) = 3 \log x$ appear to be the same?

In Exercises 51–56, list the transformations that will change the graph of $g(x) = \ln x$ into the graph of the given function. [Section 3.4 may be helpful.]

51. $f(x) = 2 \cdot \ln x$

52. $f(x) = \ln x - 7$

53. $h(x) = \ln (x - 4)$

54. $k(x) = \ln (x + 2)$

55. $h(x) = \ln (x + 3) - 4$

56. $k(x) = \ln (x - 2) + 2$

In Exercises 57–60, sketch the graph of the function.

57. $f(x) = \log (x - 3)$

58. $g(x) = 2 \ln x + 3$

59. $h(x) = -2 \log x$

60. $f(x) = \ln (-x) - 3$

In Exercises 61–68, find a viewing window (or windows) that shows a complete graph of the function.

61. $f(x) = \dfrac{x}{\ln x}$

62. $g(x) = \dfrac{\ln x}{x}$

63. $h(x) = \dfrac{\ln x^2}{x}$

64. $k(x) = e^{2/\ln x}$

65. $f(x) = 10 \log x - x$

66. $f(x) = \dfrac{\log x}{x}$

67. $l(x) = e^{e^x}$

68. $r(x) = \ln(e^x)$

In Exercises 69–72, find the average rate of change of the function.

69. $f(x) = \ln (x - 2)$, as x goes from 3 to 5.

70. $g(x) = x - \ln x$, as x goes from .5 to 1.

71. $g(x) = \log (x^2 + x + 1)$, as x goes from -5 to -3.

72. $f(x) = x \log |x|$, as x goes from 1 to 4.

73. (a) What is the average change of $f(t) = \ln t$, as t goes from 2 to $2 + h$?

(b) What is the average change of $f(t) = \ln t$, as t goes from 2 to $2 + h$ when h is .01? When h is .001? .0001? .00001?

(c) What is the average change of $f(t) = \ln t$, as t goes from 4 to $4 + h$ when h is .01? When h is .001? .0001? .00001?

(d) Approximate the average change of $f(t) = \ln t$, as t goes from 5 to $5 + h$ for very small values of h.

(e) Work some more examples like those above. What is the average rate of change of $f(t) = \ln t$, as t goes from x to $x + h$ for very small values of h?

74. (a) Find the average rate of change of $f(x) = \ln x^2$, as x goes from .5 to 2.

(b) Find the average rate of change of $g(x) = \ln (x - 3)^2$, as x goes from 3.5 to 5.

(c) What is the relationship between your answers in parts (a) and (b) and why is this so?

75. Show that $g(x) = \ln \left(\dfrac{x}{1 - x} \right)$ is the inverse function of

$f(x) = \dfrac{1}{1 + e^{-x}}$. (See Section 3.7.)

76. The doubling function $D(x) = \dfrac{\ln 2}{\ln (1 + x)}$ gives the years required to double your money when it is invested at interest rate x (expressed as a decimal), compounded annually.

(a) Find the time it takes to double your money at each of these interest rates: 4%, 6%, 8%, 12%, 18%, 24%, 36%.

(b) Round the answers in part (a) to the nearest year and compare them with these numbers: 72/4, 72/6, 72/8, 72/12, 72/18, 72/24, 72/36. Use this evidence to state a rule of thumb for determining approximate doubling time, without using the function D. This rule of thumb, which has long been used by bankers, is called the **rule of 72.**

77. Suppose $f(x) = A \ln x + B$, where A and B are constants. If $f(1) = 10$ and $f(e) = 1$, what are A and B?

78. If $f(x) = A \ln x + B$ and $f(e) = 5$ and $f(e^2) = 8$, what are A and B?

79. The height h above sea level (in meters) is related to air temperature t (in degrees Celsius), the atmospheric pressure p (in centimeters of mercury at height h), and the atmospheric pressure c at sea level by

$$h = (30t + 8000) \ln (c/p).$$

If the pressure at the top of Mount Rainier is 44 centimeters on a day when sea level pressure is 75.126 centimeters and the temperature is 7°C, what is the height of Mount Rainier?

80. Mount Everest is 8850 meters high. What is the atmospheric pressure at the top of the mountain on a day when the temperature is -25°C and the atmospheric pressure at sea level is 75 centimeters? [See Exercise 79.]

81. Beef consumption in the United States (in billions of pounds) in year x can be approximated by the function

$$f(x) = -154.41 + 39.38 \ln x \qquad (x \geq 90).$$

where $x = 90$ corresponds to 1990.*

(a) How much beef was consumed in 1999 and in 2002?

(b) According to this model when will beef consumption reach 35 billion pounds per year?

82. Students in a precalculus class were given a final exam. Each month thereafter, they took an equivalent exam. The class average on the exam taken after t months is given by

$$F(t) = 82 - 8 \cdot \ln (t + 1).$$

(a) What was the class average after six months?

(b) After a year?

(c) When did the class average drop below 55?

83. One person with a flu virus visited the campus. The number T of days it took for the virus to infect x people was given by:

$$T = -.93 \ln \left[\frac{7000 - x}{6999x} \right].$$

(a) How many days did it take for 6000 people to become infected?

(b) After two weeks, how many people were infected?

84. The population of St. Petersburg, Florida (in thousands) can be approximated by the function

$$g(x) = -127.9 + 81.91 \ln x \qquad (x \geq 70),$$

where $x = 70$ corresponds to 1970.

(a) Estimate the population in 1995 and 2003.

(b) If this model remains accurate, when will the population be 260,000?

85. A bicycle store finds that the number N of bikes sold is related to the number d of dollars spent on advertising by $N = 51 + 100 \cdot \ln (d/100 + 2)$.

(a) How many bikes will be sold if nothing is spent on advertising? If $1000 is spent? If $10,000 is spent?

(b) If the average profit is $25 per bike, is it worthwhile to spend $1000 on advertising? What about $10,000?

(c) What are the answers in part (b) if the average profit per bike is $35?

86. **Approximating Logarithmic Functions by Polynomials.** For each positive integer n, let f_n be the polynomial function whose rule is

$$f_n(x) = x - \frac{x^2}{2} + \frac{x^3}{3} - \frac{x^4}{4} + \frac{x^5}{5} - \cdots \pm \frac{x^n}{n}$$

---

*Based on data from the U.S. Department of Agriculture.

where the sign of the last term is + if n is odd and − if n is even. In the viewing window with $-1 \le x \le 1$ and $-4 \le y \le 1$, graph $g(x) = \ln(1 + x)$ and $f_4(x)$ on the same screen. For what values of x does f_4 appear to be a good approximation of g?

87. Using the viewing window in Exercise 86, find a value of n for which the graph of the function f_n (as defined in Exercise 86) appears to coincide with the graph of $g(x) = \ln(1 + x)$. Use the trace feature to move from graph to graph to see how good this approximation actually is.

88. A *harmonic sum* is a sum of this form:
$$1 + \frac{1}{2} + \frac{1}{3} + \frac{1}{4} + \cdots + \frac{1}{k}.$$
    (a) Compute $1 + \frac{1}{2} + \frac{1}{3} + \frac{1}{4}$, $1 + \frac{1}{2} + \frac{1}{3} + \frac{1}{4} + \frac{1}{5}$, and $1 + \frac{1}{2} + \frac{1}{3} + \frac{1}{4} + \frac{1}{5} + \frac{1}{6}$
    (b) How many terms do you need in a harmonic sum for it to exceed three?
    (c) It turns out to be hard to determine how many terms you would need for the sum to exceed 10. It will take

thousands of terms, more than you would want to plug into a calculator. Using calculus, we can derive this lower-bound formula: $\displaystyle\sum_{i=1}^{n} \frac{1}{i} > \ln n$. It means that the harmonic sum with n terms is always greater than $\ln n$. Use this formula to find a value of n such that the harmonic sum with n terms is greater than ten.
    (d) Calculus also gives us an upper-bound formula:
$$\sum_{i=1}^{n} \frac{1}{i} < \ln n + 1.$$
Estimate the harmonic sum with 100,000 terms. How close is your estimate to the real number?

89. The ancient Sumerians started using a place-value system around 3000 BC. Assume that in 3000 BC you started adding $1 + \frac{1}{2} + \frac{1}{3} + \frac{1}{4} + \cdots$ at the rate of ten additions per second.
    (a) What would the value be today? Make your best guess.
    (b) Use the upper-bound and lower-bound formulas given in Exercise 88 to estimate what the value would be today. Was your guess close?
    (c) In what year are you guaranteed to be above 28.187?

5.4 Properties of Logarithms

Section Objectives
- ■ Use the Product, Quotient, and Power Laws for logarithms to simplify logarithmic expressions.
- ■ Use logarithms to solve applied problems.

Logarithms have several important properties beyond those presented in Section 5.3. These properties, which we shall call *logarithm laws,* arise from the fact that logarithms are exponents. Essentially, they are properties of exponents translated into logarithmic language.

The first law of exponents says that $b^m b^n = b^{m+n}$, or in words,

The exponent of a product is the sum of the exponents of the factors.

Since logarithms are just particular kinds of exponents, this statement translates as follows.

The logarithm of a product is the sum of the logarithms of the factors.

Here is the same statement in symbolic language.

Product Law for Logarithms

For all $v, w > 0$,
$$\ln(vw) = \ln v + \ln w$$
and
$$\log(vw) = \log v + \log w.$$

Before proving the Product Law, we illustrate it in the case when $v = 10^2$ and $w = 10^3$.

We have

$$\log v = \log 10^2 = 2 \quad \text{and} \quad \log w = \log 10^3 = 3,$$

so that

$$\log v + \log w = 2 + 3 = 5.$$

We also have

$$\log vw = \log(10^2 \, 10^3)$$
$$= \log(10^5) = 5.$$

Hence, $\log vw = \log v + \log w$ in this case.

Here is the formal proof of the Product Law for natural logarithms.

Proof According to Property 4 of logarithms (in the box on page 380),

$$e^{\ln v} = v \quad \text{and} \quad e^{\ln w} = w.$$

Therefore, by the first law of exponents (with $m = \ln v$ and $n = \ln w$),

$$vw = e^{\ln v} e^{\ln w} = e^{\ln v + \ln w}.$$

So raising e to the exponent $(\ln v + \ln w)$ produces vw. But the definition of logarithm says that $\ln vw$ is the exponent to which e must be raised to produce vw. Therefore, we must have $\ln vw = \ln v + \ln w$. A similar argument works for common logarithms. ∎

EXAMPLE 1

A calculator shows that $\ln 7 = 1.9459$ and $\ln 9 = 2.1972$. Therefore,

$$\ln 63 = \ln (7 \cdot 9) = \ln 7 + \ln 9 = 1.9459 + 2.1972 = 4.1341. \quad \blacksquare$$

CALCULATOR EXPLORATION

We know that $5 \cdot 7 = 35$. Key in LOG(35) ENTER. Then key in LOG(5) + LOG(7) ENTER. The answers are the same by the Product Law. Do you get the same answer if you key in LOG(5) × LOG(7) ENTER?

EXAMPLE 2

Use the Product Law to write

(a) $\log (7xy)$ as a sum of three logarithms.

(b) $\log x^2 + \log y + 1$ as a single logarithm.

SOLUTION

(a) $\log (7xy) = \log 7x + \log y = \log 7 + \log x + \log y$

(b) Note that $\log 10 = 1$ (why?). Hence,

$$\log x^2 + \log y + 1 = \log x^2 + \log y + \log 10$$
$$= \log (x^2 y) + \log 10$$
$$= \log (10x^2 y). \quad \blacksquare$$

EXAMPLE 3

If a population of cells grows by a factor of ten every year, what do we know about the common logarithm of the population?

SOLUTION Assume the population is P. Then, next year, the population will be $10P$. The logarithm of the population will be

$$\log(10P) = \log(10) + \log(P)$$
$$= 1 + \log(P).$$

So the logarithm of the population will increase by one every year. ■

CAUTION

A common error in applying the Product Law for Logarithms is to write the *false* statement

$$\ln 7 + \ln 9 = \ln (7 + 9)$$
$$= \ln 16$$

instead of the correct statement

$$\ln 7 + \ln 9 = \ln (7 \cdot 9)$$
$$= \ln 63.$$

GRAPHING EXPLORATION

Illustrate the Caution in the margin graphically by graphing both

$$f(x) = \ln x + \ln 9 \qquad \text{and} \qquad g(x) = \ln(x + 9)$$

in the standard viewing window and verifying that the graphs are not the same. In particular, the functions have different values at $x = 7$.

The second law of exponents, namely, $b^m / b^n = b^{m-n}$, may be roughly stated in words as follows.

The exponent of the quotient is the difference of exponents.

When the exponents are logarithms, this says

The logarithm of a quotient is the difference of the logarithms.

In other words,

Quotient Law for Logarithms

For all $v, w > 0$,

$$\ln\left(\frac{v}{w}\right) = \ln v - \ln w$$

and

$$\log\left(\frac{v}{w}\right) = \log v - \log w.$$

The proof of the Quotient Law is very similar to the proof of the Product Law (see Exercise 27).

```
log(297/39)
         .8816918423
log(297)-log(39)
         .8816918423
```

Figure 5–25

EXAMPLE 4

Figure 5–25 illustrates the Quotient Law by showing that

$$\log\left(\frac{297}{39}\right) = \log 297 - \log 39.$$

■

EXAMPLE 5

For any $w > 0$.

$$\ln\left(\frac{1}{w}\right) = \ln 1 - \ln w = 0 - \ln w = -\ln w$$

and

$$\log\left(\frac{1}{w}\right) = \log 1 - \log w = 0 - \log w = -\log w.\ \blacksquare$$

GRAPHING EXPLORATION

Illustrate the Caution graphically by graphing both $f(x) = \ln (x/3)$ and $g(x) = (\ln x)/(\ln 3)$ and verifying that the graphs are not the same at $x = 36$.

CAUTION

Do not confuse $\ln\left(\dfrac{v}{w}\right)$ with the quotient $\dfrac{\ln v}{\ln w}$. They are *different* numbers. For example,

$$\ln\left(\frac{36}{3}\right) = \ln (12) = 2.4849, \quad \text{but} \quad \frac{\ln 36}{\ln 3} = \frac{3.5835}{1.0986} = 3.2619.$$

The third law of exponents, namely, $(b^m)^k = b^{mk}$, can also be translated into logarithmic language.

Power Law for Logarithms

For all k and all $v > 0$,

$$\ln (v^k) = k(\ln v)$$

and

$$\log (v^k) = k(\log v).$$

Proof   Since $v = 10^{\log v}$ (why?), the third law of exponents (with $b = 10$ and $m = \log v$) shows that

$$v^k = (10^{\log v})^k = 10^{(\log v)k} = 10^{k(\log v)}.$$

So raising 10 to the exponent $k(\log v)$ produces v^k. But the exponent to which 10 must be raised to produce v^k is, by definition, $\log (v^k)$. Therefore, $\log (v^k) = k(\log v)$, and the proof is complete. A similar argument with e in place of 10 and "ln" in place of "log" works for natural logarithms. $\blacksquare$

EXAMPLE 6

Express $\ln \sqrt{19}$ without radicals or exponents.

SOLUTION   First write $\sqrt{19}$ in exponent notation, then use the Power Law:

$$\ln \sqrt{19} = \ln 19^{1/2}$$

$$= \frac{1}{2}\ln 19 \quad \text{or} \quad \frac{\ln 19}{2}.\ \blacksquare$$

EXAMPLE 7

Express as a single logarithm: $\dfrac{\log(x^2 + 1)}{3} - \log x$.

SOLUTION

$$\frac{\log(x^2 + 1)}{3} - \log x = \frac{1}{3} \log (x^2 + 1) - \log x$$

$$= \log (x^2 + 1)^{1/3} - \log x \qquad \text{[Power Law]}$$

$$= \log \sqrt[3]{x^2 + 1} - \log x$$

$$= \log \left(\frac{\sqrt[3]{x^2 + 1}}{x} \right) \qquad \text{[Quotient Law]} \qquad ■$$

EXAMPLE 8

Express as a single logarithm: $\ln 3x + 4 \ln x - \ln 3xy$.

SOLUTION

$$\ln 3x + 4 \cdot \ln x - \ln 3xy = \ln 3x + \ln x^4 - \ln 3xy \qquad \text{[Power Law]}$$

$$= \ln (3x \cdot x^4) - \ln 3xy \qquad \text{[Product Law]}$$

$$= \ln \frac{3x^5}{3xy} \qquad \text{[Quotient Law]}$$

$$= \ln \frac{x^4}{y} \qquad \text{[Cancel } 3x] \qquad ■$$

EXAMPLE 9

Simplify: $\ln \left(\dfrac{\sqrt{x}}{x} \right) + \ln \sqrt[4]{ex^2}$.

SOLUTION   Begin by changing to exponential notation.

$$\ln \left(\frac{x^{1/2}}{x} \right) + \ln (ex^2)^{1/4} = \ln (x^{-1/2}) + \ln (ex^2)^{1/4}$$

$$= -\frac{1}{2} \cdot \ln x + \frac{1}{4} \cdot \ln ex^2 \qquad \text{[Power Law]}$$

$$= -\frac{1}{2} \cdot \ln x + \frac{1}{4}(\ln e + \ln x^2) \qquad \text{[Product Law]}$$

$$= -\frac{1}{2} \cdot \ln x + \frac{1}{4}(\ln e + 2 \cdot \ln x) \qquad \text{[Power Law]}$$

$$= -\frac{1}{2} \cdot \ln x + \frac{1}{4} \cdot \ln e + \frac{1}{2} \cdot \ln x$$

$$= \frac{1}{4} \cdot \ln e = \frac{1}{4} \qquad \text{[ln } e = 1] \qquad ■$$

APPLICATIONS

Because logarithmic growth is slow, measurements on a logarithmic scale (that is, on a scale determined by a logarithmic function) can sometimes be deceptive.

EXAMPLE 10

Earthquakes   The magnitude $R(i)$ of an earthquake on the Richter scale is given by $R(i) = \log(i/i_0)$, where i is the amplitude of the ground motion of the earthquake and i_0 is the amplitude of the ground motion of the so-called zero earthquake.* A moderate earthquake might have 1000 times the ground motion of the zero earthquake (that is, $i = 1000i_0$). So its magnitude would be

$$\log(1000i_0/i_0) = \log 1000 = \log 10^3 = 3.$$

An earthquake with 10 times this ground motion (that is, $i = 10 \cdot 1000i_0 = 10{,}000i_0$) would have a magnitude of

$$\log(10{,}000i_0/i_0) = \log 10{,}000 = \log 10^4 = 4.$$

So a *tenfold* increase in ground motion produces only a one-point change on the Richter scale. In general,

> **Increasing the ground motion by a factor of 10^k increases the Richter magnitude by k units.**[†]

For instance, the 1989 World Series earthquake in San Francisco measured 7.0 on the Richter scale, and the great earthquake of 1906 measured 8.3. The difference of 1.3 points means that the 1906 quake was $10^{1.3} \approx 20$ times more intense than the 1989 one in terms of ground motion.   ∎

---

*The zero earthquake has ground motion amplitude of less than 1 micron on a standard seismograph 100 kilometers from the epicenter.

[†]*Proof:* If one quake has ground motion amplitude i and the other $10^k i$, then

$$R(10^k i) = \log(10^k i/i_0) = \log 10^k + \log(i/i_0)$$

$$= k + \log(i/i_0) = k + R(i).$$

EXERCISES 5.4

In Exercises 1–10, write the given expression as a single logarithm.

1. $\ln x^2 + 3 \ln y$

2. $-5(\ln x) + \ln 4y - \ln 3z$

3. $\log(x^2 - 9) - \log(x + 3)$

4. $3(\log 2x) - 4[\log x - \log(y - 5)]$

5. $2(\ln x) - 3(\ln x^2 + \ln x)$

6. $-\log\left(\dfrac{3\sqrt{x}}{2}\right) - \log(\sqrt{5x})$

7. $3 \ln(e^2 - e) - 3$

8. $3 \log(7) - 4$

9. $\log(10x) + \log(20y) - 1$

10. $\ln(e^3 x^2) - \ln(ey^3) + 2$

In Exercises 11–16, let $u = \ln x$ and $v = \ln y$. Write the given expression in terms of u and v. For example,

$$\ln x^3 y = \ln x^3 + \ln y = 3 \ln x + \ln y = 3u + v.$$

11. $\ln(x^2 y^5)$

12. $\ln(x^4 y^3)$

13. $\ln(\sqrt{x} \cdot y^2)$

14. $\ln\left(\dfrac{\sqrt{xy}}{y^2}\right)$

15. $\ln\left(\sqrt[3]{x^2}\sqrt{y}\right)$

16. $\ln\left(\dfrac{\sqrt[3]{x^2 y^2}}{x^5}\right)$

In Exercises 17–23, use graphical or algebraic means to determine whether the statement is true or false.

17. $\ln |x| = |\ln x|$?

18. $\ln \left(\dfrac{1}{x}\right) = \dfrac{1}{\ln x}$?

19. $\log x^5 = 5(\log x)$?

20. $e^{x \ln x} = x^x \quad (x > 0)$?

21. $\ln x^3 = (\ln x)^3$?

22. $\log \sqrt{x} = \sqrt{\log x}$?

23. $\ln (x + 5) = \ln(x) + \ln 5$?

In Exercises 24 and 25, find values of a and b for which the statement is false.

24. $\dfrac{\log a}{\log b} = \log \left(\dfrac{a}{b}\right)$

25. $\log (a + b) = \log a + \log b$

26. If $\ln b^{10} = 10$, what is b?

27. Prove the Quotient Law for Logarithms: For $v, w > 0$,
$\ln \left(\dfrac{v}{w}\right) = \ln v - \ln w$. (Use properties of exponents and the fact that $v = e^{\ln v}$ and $w = e^{\ln w}$.)

In Exercises 28–31, state the magnitude on the Richter scale of an earthquake that satisfies the given condition.

28. 100 times stronger than the zero quake.

29. $10^{4.7}$ times stronger than the zero quake.

30. 250 times stronger than the zero quake.

31. 1500 times stronger than the zero quake.

Exercises 32–35 deal with the energy intensity i of a sound, which is related to the loudness of the sound by the function $L(i) = 10 \cdot \log (i/i_0)$, where i_0 is the minimum intensity detectable by the human ear and $L(i)$ is measured in decibels. Find the decibel measure of the sound.

32. Ticking watch (intensity is 100 times i_0).

33. Soft music (intensity is 10,000 times i_0).

34. Loud conversation (intensity is 4 million times i_0).

35. Victoria Falls in Africa (intensity is 10 billion times i_0).

36. How much louder is the sound in Exercise 33 than the sound in Exercise 32?

37. The perceived loudness L of a sound of intensity I is given by $L = k \cdot \ln I$, where k is a certain constant. By how much must the intensity be increased to double the loudness? (That is, what must be done to I to produce $2L$?)

THINKERS

38. Compute each of the following pairs of numbers.

   (a) $\log 18$ and $\dfrac{\ln 18}{\ln 10}$

   (b) $\log 456$ and $\dfrac{\ln 456}{\ln 10}$

   (c) $\log 8950$ and $\dfrac{\ln 8950}{\ln 10}$

   (d) What do these results suggest?

39. Prove that for any positive number c, $\log c = \dfrac{\ln c}{\ln 10}$. [*Hint:* We know that $10^{\log c} = c$ (why?). Take natural logarithms on both sides and use a logarithm law to simplify and solve for $\log c$.]

40. Find each of the following logarithms.

   (a) $\log 8.753$ (b) $\log 87.53$ (c) $\log 875.3$
   (d) $\log 8753$ (e) $\log 87{,}530$
   (f) How are the numbers 8.753, 87.53, . . . , 87,530 related to one another? How are their logarithms related? State a general conclusion that this evidence suggests.

41. Prove that for every positive number c, $\log c$ can be written in the form $k + \log b$, where k is an integer and $1 \le b < 10$. [*Hint:* Write c in scientific notation and use logarithm laws to express $\log c$ in the required form.]

42. A scientist is measuring the spread of a rumor over time. She notices a nice pattern when she graphs the natural logarithm of the number of people who know the rumor after t days:

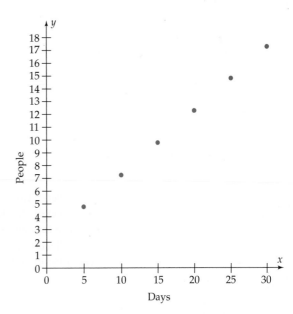

   (a) Find a good model for the number of people who know the rumor at a given time t, where $0 \le t \le 30$.

   (b) A friend of the scientist wonders why she didn't just graph the number of people instead of the logarithm of the number of people. What was the advantage of using the logarithm in the graph?

43. Wayland and Christy have been tracking the number of cases of flu in their city:

Weeks since January 1	0	1	2	3	4	5	6
Number of cases	10	13	16	20	24	31	38

Wayland thinks this is exponential growth. Christy doesn't think so. After playing around with the data, they plot the points and still disagree.

(a) Plot the points. Do you agree with Wayland or with Christy?

(b) They create a new plot, this time using the natural logarithms of the number of cases. So they plot the points $(0, \ln(10))$, $(2, \ln(13))$, etc. As soon as they see this new plot, they agree! Construct this new plot.

(c) Who was right, Wayland or Christy? Why?

5.4.A *SPECIAL TOPICS*   Logarithmic Functions to Other Bases*

Section Objectives
- ◼ Learn the definition and properties of logarithms to any base b.
- ◼ Use the Change of Base formula to evaluate logarithms to base b.

The same procedure used in Sections 5.3–5.4 can be carried out with any positive number b in place of 10 and e.

> **Throughout this section, b is a fixed positive number with $b > 1$.**[†]

The exponential function $f(x) = b^x$, whose graph is shown in Figure 5–26, is an increasing function and hence is one-to-one (as explained on page 219). Therefore, f has an inverse function g whose graph is the reflection of the graph of f in the line $y = x$ (see page 225), as shown in Figure 5–27.

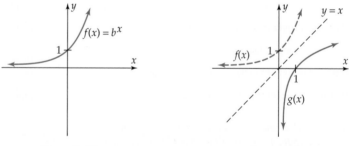

Figure 5–26   Figure 5–27

This inverse function g is called the **logarithmic function to the base b.** The value $g(x)$ of this function at a number x is denoted **$\log_b x$** and called the **logarithm to the base b of the number x.**

In Section 3.7, we saw that the relationship between a function f and its inverse function g is given by

$$g(v) = u \qquad \text{exactly when} \qquad f(u) = v.$$

When $f(x) = b^x$ and $g(x) = \log_b x$, this statement takes the following form.

---

*This material is not needed in the sequel and may be read before Section 5.3 if desired. Section 3.7 (Inverse Functions) is a prerequisite for this section, which replicates the discussion of Sections 5.3 and 5.4 in a more general context.

[†]The discussion is also valid when $0 < b < 1$, but in that case, the graphs have a different shape.

Definition of Logarithms to Base b

Let u and v be real numbers, with $v > 0$. Then

$$\log_b v = u \qquad \text{exactly when} \qquad b^u = v.$$

In other words,

$\log_b v$ is the exponent to which b must be raised to produce v.

EXAMPLE 1

Translate:

(a) $\log_3(14) = 2.4022$* into an equivalent exponential statement
(b) $4^{2.67} = 40.504$ into an equivalent logarithmic statement

SOLUTION

(a) Using the preceding definition, we see that $\log_3(14) = 2.4022$ is equivalent to $3^{2.4022} = 14$.

(b) Similarly, $4^{2.67} = 40.504$ is equivalent to $\log_4(40.504)) = 2.67$. ∎

EXAMPLE 2

Compute the following logarithms

(a) $\log_2 16$ (b) $\log_5 1/5$ (c) $\log_7 1$
(d) $\log_b(-10)$ where b is greater than 1

SOLUTION

(a) To find $\log_2 16$, ask yourself, "What power of 2 equals 16?" Since $2^4 = 16$, we see that $\log_2 16 = 4$.

(b) Similarly, $\log_5(1/5) = -1$ because $5^{-1} = 1/5$.

(c) We know that $7^0 = 1$, so $\log_7 1 = 0$.

(d) What power of b equals -10? The graph of $f(x) = b^x$ in figure 5–26 shows that every power of b is *positive*. So b^x can *never* be -10 or any negative number or zero, and hence $\log_b(-10)$ is not defined. ∎

EXAMPLE 3

Logarithms to the base 10 are called **common logarithms.** Is it customary to write $\log v$ instead of $\log_{10} v$. Then

$$\log 100 = 2 \qquad \text{because} \qquad 10^2 = 100;$$

$$\log .001 = -3 \qquad \text{because} \qquad 10^{-3} = \frac{1}{10^3} = \frac{1}{1000} = .001.$$

*Here and below, all logarithms are rounded to four decimal places. So, strictly speaking, the equal sign should be replaced by an "approximately equal" sign ($\approx$).

Calculators have a LOG key for evaluating common logarithms. For instance,

$$\log .4 = -.3979, \qquad \log 45.3 = 1.6561, \qquad \log 685 = 2.8357. \qquad \blacksquare$$

EXAMPLE 4

The most frequently used base for logarithms in modern applications is the number e ($\approx 2.71828 \ldots$). Logarithms to the base e are called **natural logarithms** and use a different notation: We write $\ln v$ instead of $\log_e v$. Calculators also have an LN key for evaluating natural logarithms. For example,

$$\ln .5 = -0.6931, \qquad \ln 65 = 4.1744, \qquad \ln 158 = 5.0626. \qquad \blacksquare$$

You *don't* need a calculator to understand the essential properties of logarithms. You need only translate logarithmic statements into exponential ones (or vice versa).

EXAMPLE 5

Assume $b > 1$.

(a) Compute $\log_b 1$ (b) Compute $\log_b b$ (c) Compute $\log_b b^k$

SOLUTION

(a) To what power must b be raised to produce 1? The answer is that $b^0 = 1$, therefore $\log_b 1 = 0$.

(b) To what power must b be raised to produce b? We know that $b^1 = b$; therefore $\log_b b = 1$.

(c) To what power must b be raised to produce b^k? Obviously, $b^k = b^k$, therefore $\log_b b^k = k$. This property holds even when k is a complicated expression. For instance, if x and y are positive, then $\log_6 6^{\sqrt{3x+y}} = \sqrt{3x + y}$ (here $k = \sqrt{3x + y}$). $\qquad \blacksquare$

EXAMPLE 6

Compute $10^{\log 439}$

SOLUTION   By definition, $\log 439$ is the power to which 10 must be raised to produce 439. So $10^{\log 439} = 439$. Similarly, if $b > 1$, $b^{\log_b v} = v$, **for every $v > 0$.** $\qquad \blacksquare$

Here is a summary of the facts illustrated in Examples 2(d), 5 and 6.

Properties of Logarithms

1. $\log_b v$ is defined only when $v > 0$.

2. $\log_b 1 = 0$ and $\log_b b = 1$.

3. $\log_b (b^k) = k$ for every real number k.

4. $b^{\log_b v} = v$ for every $v > 0$.

LOGARITHM LAWS

The first law of exponents states that $b^m b^n = b^{m+n}$, or in words,

> The exponent of a product is the sum of the exponents of the factors.

Since logarithms are just particular kinds of exponents, this statement translates as follows.

The logarithm of a product is the sum of the logarithms of the factors.

The second and third laws of exponents, namely, $b^m/b^n = b^{m-n}$ and $(b^m)^k = b^{mk}$, can also be translated into logarithmic language.

Logarithm Laws

Let b, v, w, k be real numbers, with b, v, w positive and $b \neq 1$.

Product Law: $\log_b (vw) = \log_b v + \log_b w.$

Quotient Law: $\log_b\left(\dfrac{v}{w}\right) = \log_b v - \log_b w.$

Power Law: $\log_b (v^k) = k(\log_b v).$

Proof of the Quotient Law   According to Property 4 in the box on page 394,

$$b^{\log_b v} = v \qquad \text{and} \qquad b^{\log_b w} = w.$$

Therefore, by the second law of exponents (with $m = \log_b v$ and $n = \log_b w$), we have

$$\frac{v}{w} = \frac{b^{\log_b v}}{b^{\log_b w}} = b^{\log_b v - \log_b w}.$$

Since $\log_b (v/w)$ is the exponent to which b must be raised to produce v/w, we must have $\log_b (v/w) = \log_b v - \log_b w$. This proves the Quotient Law. The Product and Power Laws are proved in a similar fashion.

EXAMPLE 7

Simplify and write as a single logarithm.

(a) $\log_3 (x + 2) + \log_3 y - \log_3 (x^2 - 4)$

(b) $3 - \log_5 (125x)$

SOLUTION

(a) $\log_3 (x + 2) + \log_3 y - \log_3 (x^2 - 4)$

$\quad = \log_3[(x + 2)y] - \log_3 (x^2 - 4)$ [Product Law]

$\quad = \log_3\left(\dfrac{(x + 2)y}{x^2 - 4}\right)$ [Quotient Law]

$\quad = \log_3\left(\dfrac{(x + 2)y}{(x + 2)(x - 2)}\right)$ [Factor denominator]

$\quad = \log_3\left(\dfrac{y}{x - 2}\right)$ [Cancel common factor]

(b) $3 - \log_5 (125x)$

$$= 3 - (\log_5 125 + \log_5 x) \qquad \text{[Product Law]}$$
$$= 3 - \log_5 125 - \log_5 x$$
$$= 3 - 3 - \log_5 x \qquad [\log_5 125 = 3 \text{ because } 5^3 = 125]$$
$$= -\log_5 x$$
$$= \log_5 x^{-1} = \log_5 \left(\frac{1}{x}\right) \qquad \text{[Power Law]} \qquad \blacksquare$$

CAUTION

1. A common error in using the Product Law is to write something like $\log 6 + \log 7 = \log (6 + 7) = \log 13$ instead of the correct statement $\log 6 + \log 7 = \log (6 \cdot 7) = \log 42$.

2. Do not confuse $\log_b\left(\dfrac{v}{w}\right)$ with the quotient $\dfrac{\log_b v}{\log_b w}$. They are *different* numbers. For example, when $b = 10$

$$\log\left(\frac{48}{4}\right) = \log 12 = 1.0792 \qquad \text{but} \qquad \frac{\log 48}{\log 4} = \frac{1.6812}{0.6021} = 2.7922.$$

For graphic illustrations of the errors mentioned in the Caution, see Exercises 82 and 83.

EXAMPLE 8

Given that

$$\log_7 2 = .3562, \qquad \log_7 3 = .5646, \qquad \text{and} \qquad \log_7 5 = .8271,$$

find:

(a) $\log_7 10$;   (b) $\log_7 2.5$;   (c) $\log_7 48$.

SOLUTION

(a) By the Product Law,

$$\log_7 10 = \log_7 (2 \cdot 5) = \log_7 2 + \log_7 5 = .3562 + .8271 = 1.1833.$$

(b) By the Quotient Law,

$$\log_7 2.5 = \log_7 \left(\frac{5}{2}\right) = \log_7 5 - \log_7 2 = .8271 - .3562 = .4709.$$

(c) By the Product and Power Laws,

$$\log_7 48 = \log_7 (3 \cdot 16) = \log_7 3 + \log_7 16 = \log_7 3 + \log_7 2^4$$
$$= \log_7 3 + 4 \cdot \log_7 2 = .5646 + 4(.3562)$$
$$= 1.9894. \qquad \blacksquare$$

Example 8 worked because we were *given* several logarithms to base 7. But there's no $\log_7$ key on the calculator, so how do you find logarithms to base 7 or to any base other than e or 10? *Answer:* Use the LN key on the calculator and the following formula.

Change of Base Formula

For any positive numbers b and v,

$$\log_b v = \frac{\ln v}{\ln b}.$$

Proof   By Property 4 in the box on page 394, $b^{\log_b v} = v$. Take the natural logarithm of each side of this equation:

$$\ln (b^{\log_b v}) = \ln v.$$

Apply the Power Law for natural logarithms on the left side.

$$(\log_b v)(\ln b) = \ln v.$$

Dividing both sides by $\ln b$ finishes the proof.

$$\log_b v = \frac{\ln v}{\ln b}.$$ ■

EXAMPLE 9

To find $\log_7 3$, apply the change of base formula with $b = 7$.

$$\log_7 3 = \frac{\ln 3}{\ln 7} = \frac{1.0986}{1.9459} = .5646.$$ ■

EXERCISES 5.4.A

Note: *Unless stated otherwise, all letters represent positive numbers and $b \ne 1$.*

In Exercises 1–8, fill in the missing entries in each table.

1.

x	0	1	2	4
$f(x) = \log_4 x$				

2.

x	1/36	6	36	$\sqrt{6}$
$g(x) = \log_6 x$	-2	1		

3.

x		1/6	1	216
$h(x) = \log_6 x$	-2			

4.

x	-4	$-3\frac{7}{8}$	4	12
$k(x) = \log_2 (x + 4)$				

5.

x	0	1/7	$\sqrt{7}$	49
$f(x) = 2 \log_7 x$				

6.

x	.001		10	
$g(x) = -2 \log x$		2		-4

7.

x	-2.75	-1	1	29
$h(x) = 3 \log_2 (x + 3)$				

8.

x	$1/e^2$	1	e	e^3
$k(x) = 3 \ln x$				

In Exercises 9–18, translate the given exponential statement into an equivalent logarithmic one.

9. $10^{-2} = .01$

10. $10^4 = 10,000$

11. $\sqrt[3]{10} = 10^{1/3}$

12. $10^{0.6990} \approx 5$

13. $10^{7k} = r$

14. $10^{(2a-b)} = c$

15. $7^8 = 5,764,801$

16. $5^{-3} = 1/125$

17. $3^{-3} = 1/9$

18. $k^{27} = 10,134$

In Exercises 19–28, translate the given logarithmic statement into an equivalent exponential one.

19. $\log 10{,}000 = 4$

20. $\log .0001 = -4$

21. $\log 750 \approx 2.88$

22. $\log (.69) \approx -.1612$

23. $\log_5 125 = 3$

24. $\log_{25} (1/125) = -3/2$

25. $\log_2 (1/4) = -2$

26. $\log_5 \sqrt[4]{5} = 1/4$

27. $\log (x^2 + 2y) = z + w$

28. $\log (p - q) = r$

In Exercises 29–36, evaluate the given expression without using a calculator.

29. $\log 10^{\sqrt{97}}$

30. $\log_{13} (13^{\sqrt{2}})$

31. $\log 10^{x^2 + y^2}$

32. $\log_{2.7} [2.7^{(3x^2 + 1)}]$

33. $\log_{16} 4$

34. $\log_3 81$

35. $\log_{\sqrt{3}} (27)$

36. $\log_{\sqrt[3]{2}} (1/16)$

In Exercises 37–40, a graph or a table of values for the function $f(x) = \log_b x$ is given. Find b.

37.

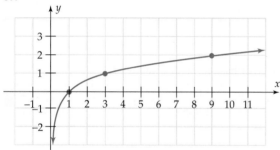

38.

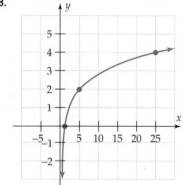

39.

x	.05	1	400	$2\sqrt{5}$
$f(x)$	-1	0	2	$1/2$

40.

x	1/25	1	5	125
$f(x)$	-4	0	2	6

In Exercises 41–46, find x.

41. $\log_3 243 = x$

42. $\log_{243} 81$

43. $\log_{27} x = 1/3$

44. $\log_2 x = -6$

45. $\log_x 64 = 3$

46. $\log_x (1/64) = -3/4$

In Exercises 47–60, write the given expression as the logarithm of a single quantity, as in Example 7.

47. $2 \log x + 3 \log y - 6 \log z$

48. $2 \log_5 x - 7 \log_5 y + 3 \log_5 (2z)$

49. $\log x - \log (x + 3) + \log (x^2 - 9)$

50. $\log_4 (z^2 + 7z) + \log_4 (z - 4) - \log_4 (2z)$

51. $\dfrac{1}{2} \log_2 (25c^2)$

52. $-\dfrac{1}{4} \log_6 (16p^{12})$

53. $-2 \log_4 (7c)$

54. $\dfrac{2}{3} \log_4 (3x + 2)$

55. $2 \ln (x + 1) - \ln (x + 2)$

56. $5 \ln (2x - 1) - 2 \ln (3x + 5)$

57. $\log_2 (2x) - 1$

58. $3 - \log_6 (36y)$

59. $2 \ln (e^2 - e) - 2$

60. $3 - 2 \log_4 (10)$

In Exercises 61–68, use a calculator and the change of base formula to find the logarithm.

61. $\log_2 10$

62. $\log_3 18$

63. $\log_7 5$

64. $\log_{16} 27$

65. $\log_{500} 1000$

66. $\log_8 3$

67. $\log_{12} 56$

68. $\log_{12} 5$

In Exercises 69–74, answer true or false and give reasons for your answer.

69. $\log_b (r/5) = \log_b r - \log_b 5$

70. $\log_2 (3p + q) = \log_2 (3p) + \log_2 q$

71. $(\log_b r)/t = \log_b (r^{1/t})$

72. $\log_a (3bc) = \log_a (3b) + \log_a c$

73. $\log_5 (5x) = 5(\log_5 x)$

74. $1 + 2 \log t + \log 3 = \log (30t^2)$

THINKERS

75. Which is larger: 397^{398} or 398^{397}? [*Hint:* $\log 397 \approx 2.5988$ and $\log 398 \approx 2.5999$ and $f(x) = 10^x$ is an increasing function.]

76. If $\log_b 9.21 = 7.4$ and $\log_b 359.62 = 19.61$, then what is $\log_b 359.62 / \log_b 9.21$?

In Exercises 77–80, assume that a and b are positive, with $a \neq 1$ and $b \neq 1$.

77. Express $\log_b u$ in terms of logarithms to the base a.

78. Show that $\log_b a = 1/\log_a b$.

79. How are $\log_{10} u$ and $\log_{100} u$ related?

80. Show that $a^{\log b} = b^{\log a}$.

81. If $\log_b x = \dfrac{1}{2} \log_b v + 3$, show that $x = (b^3)\sqrt{v}$.

82. Graph the functions $f(x) = \log x + \log 7$ and $g(x) = \log (x + 7)$ on the same screen. For what values of x is it true that $f(x) = g(x)$? What do you conclude about the statement

$$\log 6 + \log 7 = \log (6 + 7)?$$

83. Graph the functions $f(x) = \log (x/4)$ and $g(x) = (\log x)/(\log 4)$. Are they the same? What does this say about a statement such as

$$\log \left(\frac{48}{4}\right) = \frac{\log 48}{\log 4}?$$

In Exercises 84–86, sketch a complete graph of the function, labeling any holes, asymptotes, or local extrema.

84. $f(x) = \log_5 x + 2$ **85.** $h(x) = x \log x^2$

86. $g(x) = \log_{20} x^2$

87. The number 10^{100} (The number 1 followed by one-hundred zeros) is called a googol.

  (a) It is a fact that $\log_5(10^{100}) = 143.0677$. Is 5^{150} greater or smaller than one googol? How do you know?

  (b) Which is bigger, 2^{3322} or (googol)^{10} (one googol to the tenth power)?

88. Assume a and b are constants with $a > 1$ and $b > 1$. For a particular positive number x we know that $\log_a x > \log_b x$. Is it possible to tell if $a > b$ or if $b > a$? Why or why not?

5.5   Algebraic Solutions of Exponential and Logarithmic Equations

Section Objectives

■ Solve exponential and logarithmic equations algebraically.

■ Use exponential and logarithmic equations to solve applied problems.

Most of the exponential and logarithmic equations solved by graphical means earlier in this chapter could also have been solved algebraically. The algebraic techniques for solving such equations are based on the properties of logarithms.

 EXPONENTIAL EQUATIONS

The easiest exponential equations to solve are those in which both sides are powers of the same base.

EXAMPLE 1

Solve $8^x = 2^{x+1}$.

SOLUTION   Using the fact that $8 = 2^3$, we rewrite the equation as follows.

$$8^x = 2^{x+1}$$
$$(2^3)^x = 2^{x+1}$$
$$2^{3x} = 2^{x+1}$$

Since the powers of 2 are equal, the exponents must be the same, that is,

$$3x = x + 1$$
$$2x = 1$$
$$x = \frac{1}{2}.$$
■

When different bases are involved in an exponential equation, a different solution technique is needed.

EXAMPLE 2

Solve $5^x = 2$.

SOLUTION

Take logarithms on each side:* $\ln 5^x = \ln 2$

Use the Power Law: $x(\ln 5) = \ln 2$

Divide both sides by $\ln 5$: $x = \dfrac{\ln 2}{\ln 5} \approx \dfrac{.6931}{1.6094} \approx .4307.$

Remember: $\dfrac{\ln 2}{\ln 5}$ is neither $\ln \dfrac{2}{5}$ nor $\ln 2 - \ln 5$. ∎

EXAMPLE 3

Solve $2^{4x-1} = 3^{1-x}$,

SOLUTION

Take logarithms of each side: $\ln 2^{4x-1} = \ln 3^{1-x}$

Use the Power Law: $(4x - 1)(\ln 2) = (1 - x)(\ln 3)$

Multiply out both sides: $4x(\ln 2) - \ln 2 = \ln 3 - x(\ln 3)$

Rearrange terms: $4x(\ln 2) + x(\ln 3) = \ln 2 + \ln 3$

Factor left side: $(4 \cdot \ln 2 + \ln 3)x = \ln 2 + \ln 3$

Divide both sides by $(4 \cdot \ln 2 + \ln 3)$: $x = \dfrac{\ln 2 + \ln 3}{4 \cdot \ln 2 + \ln 3} \approx .4628.$ ∎

 ## APPLICATIONS OF EXPONENTIAL EQUATIONS

As we saw in Section 5.2, the mass of a radioactive element at time x is given by

$$M(x) = c(.5^{x/h}),$$

where c is the initial mass and h is the half-life of the element.

EXAMPLE 4

After 43 years, a 20-milligram sample of strontium-90 (^{90}Sr) decays to 6.071 mg. What is the half-life of strontium-90?

SOLUTION   The mass of the sample at time x is given by

$$f(x) = 20(.5^{x/h}),$$

---

*We shall use natural logarithms, but the same techniques are valid for logarithms to other bases (Exercise 34).

where h is the half-life of strontium-90. We know that $f(x) = 6.071$ when $x = 43$, that is, $6.071 = 20(.5^{43/h})$. We must solve this equation for h.

Divide both sides by 20:
$$\frac{6.071}{20} = .5^{43/h}$$

Take logarithms on both sides:
$$\ln \frac{6.071}{20} = \ln .5^{43/h}$$

Use the Power Law:
$$\ln \frac{6.071}{20} = \frac{43}{h} \ln .5$$

Multiply both sides by h:
$$h \ln \frac{6.071}{20} = 43 \ln .5$$

Divide both sides by $\ln \frac{6.071}{20}$:
$$h = \frac{43 \ln .5}{\ln(6.071/20)} \approx 25.$$

Therefore, strontium-90 has a half-life of 25 years. ■

EXAMPLE 5

When a living organism dies, its carbon-14 decays. The half-life of carbon-14 is 5730 years. If the skeleton of a mastodon has lost 58% of its original carbon-14, when did the mastodon die?*

SOLUTION   Time is measured from the death of the mastodon. The amount of carbon-14 left in the skeleton at time x is given by

$$M(x) = c(.5^{x/5730}),$$

where c is the original mass of carbon-14. The skeleton has lost 58% of c, that is, $.58c$. So the present value of $M(x)$ is $c - .58c = .42c$, and we have

$$M(x) = c(.5^{x/5730})$$

$$.42c = c(.5^{x/5730})$$

$$.42 = .5^{x/5730}.$$

The solution of this equation is the time elapsed from the mastodon's death to the present. It can be solved as above.

$$\ln .42 = \ln (.5)^{x/5730}$$

$$\ln .42 = \frac{x}{5730} (\ln .5)$$

$$5730(\ln .42) = x(\ln .5)$$

$$x = \frac{5730(\ln .42)}{\ln .5} \approx 7171.32.$$

Therefore, the mastodon died approximately 7200 years ago. ■

---

*Archeologists can determine how much carbon-14 has been lost by a technique that involves measuring the ratio of carbon-14 to carbon-12 in the skeleton.

EXAMPLE 6

A certain bacteria is known to grow exponentially, with the population at time t given by a function of the form $g(t) = Pe^{kt}$, where P is the original population and k is the continuous growth rate. A culture shows 1000 bacteria present. Seven hours later, there are 5000.

(a) Find the continuous growth rate k.

(b) Determine when the population will reach one billion.

SOLUTION

(a) The original population is $P = 1000$, so the growth function is $g(t) = 1000e^{kt}$. We know that $g(7) = 5000$, that is,

$$1000e^{k \cdot 7} = 5000.$$

To determine the growth rate, we solve this equation for k.

Divide both sides by 1000:	$e^{7k} = 5$
Take logarithms of both sides:	$\ln e^{7k} = \ln 5$
Use the Power Law:	$7k \ln e = \ln 5.$

Since $\ln e = 1$ (why?), this equation becomes

$$7k = \ln 5$$

Divide both sides by 7:	$k = \dfrac{\ln 5}{7} \approx .22992.$

Therefore, the growth function is $g(t) \approx 1000e^{.22992t}$.

(b) The population will reach one billion when $g(t) = 1,000,000,000$, that is, when

$$1000e^{.22992t} = 1,000,000,000.$$

So we solve this equation for t:

Divide both sides by 1000:	$e^{.22992t} = 1,000,000$
Take logarithms on both sides:	$\ln e^{.22992t} = \ln 1,000,000$
Use the Power Law:	$.22992t \ln e = \ln 1,000,000$
Remember $\ln e = 1$:	$.22992t = \ln 1,000,000$
Divide both sides by .22992:	$t = \dfrac{\ln 1,000,000}{.22992} \approx 60.09.$

Therefore, it will take a bit more than 60 hours for the culture to grow to one billion.   ■

EXAMPLE 7

Inhibited Population Growth   The population of fish in a lake at time t months is given by the function

$$p(t) = \frac{20,000}{1 + 24e^{-t/4}}.$$

How long will it take for the population to reach 15,000?

SOLUTION   We must solve this equation for t.

$$15,000 = \frac{20,000}{1 + 24e^{-t/4}}$$

$$15,000(1 + 24e^{-t/4}) = 20,000$$

$$1 + 24e^{-t/4} = \frac{20,000}{15,000} = \frac{4}{3}$$

$$24e^{-t/4} = \frac{1}{3}$$

$$e^{-t/4} = \frac{1}{3} \cdot \frac{1}{24} = \frac{1}{72}$$

$$\ln e^{-t/4} = \ln\left(\frac{1}{72}\right)$$

$$\left(-\frac{t}{4}\right)(\ln e) = \ln 1 - \ln 72$$

$$-\frac{t}{4} = -\ln 72 \qquad\qquad \text{[}\ln e = 1 \text{ and } \ln 1 = 0\text{]}$$

$$t = 4(\ln 72) \approx 17.1067.$$

So the population reaches 15,000 in a little over 17 months.   ∎

LOGARITHMIC EQUATIONS

Equations that involve only logarithmic terms may be solved by using the following fact, which is proved in Exercise 33 (and is valid with log replaced by ln).

If $\log u = \log v$, then $u = v$.

EXAMPLE 8

Solve $\log (3x + 2) + \log (x + 2) = \log (7x + 6)$.

SOLUTION   First we write the left side as a single logarithm.

$$\log (3x + 2) + \log (x + 2) = \log (7x + 6)$$

Use the Product Law: $$\log[(3x + 2)(x + 2)] = \log (7x + 6)$$

Multiply out left side: $$\log (3x^2 + 8x + 4) = \log (7x + 6).$$

Since the logarithms are equal, we must have

$$3x^2 + 8x + 4 = 7x + 6$$

Subtract $7x + 6$ from both sides: $$3x^2 + x - 2 = 0$$

Factor: $$(3x - 2)(x + 1) = 0$$

$$3x - 2 = 0 \quad \text{or} \quad x + 1 = 0$$

$$3x = 2 \qquad\qquad\qquad x = -1$$

$$x = \frac{2}{3}$$

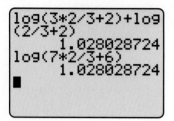

Figure 5–28

Thus, $x = 2/3$ and $x = -1$ are the *possible* solutions and must be checked in the *original* equation. When $x = 2/3$, both sides of the original equation have the same value, as shown in Figure 5–28. So $2/3$ is a solution. When $x = -1$, however, the right side of the equation is

$$\log (7x + 6) = \log [7(-1) + 6] = \log (-1),$$

which is not defined. So -1 is not a solution. ∎

Equations that involve both logarithmic and constant terms may be solved by using the basic property of logarithms (see page 380).

$$(*) \qquad 10^{\log v} = v \qquad \text{and} \qquad e^{\ln v} = v.$$

EXAMPLE 9

Solve $7 + 2 \log 5x = 11$.

SOLUTION We start by getting all the logarithmic terms on one side and the constant on the other.

Subtract 7 from both sides:	$2 \log 5x = 4$
Divide both sides by 2:	$\log 5x = 2.$

We know that if two quantities are equal, say $a = b$, then $10^a = 10^b$. We use this fact here, with the two sides of the preceding equation as a and b.

Exponentiate both sides:	$10^{\log 5x} = 10^2$
Use the basic logarithm property ($*$):	$5x = 100$
Divide both sides by 5:	$x = 20.$

Verify that 20 is actually a solution of the original equation. ∎

EXAMPLE 10

Solve $\ln (x - 3) = 5 - \ln (x - 3)$.

SOLUTION We proceed as in Example 9, but since the base for these logarithms is e, we use e rather than 10 when we exponentiate.

	$\ln (x - 3) = 5 - \ln (x - 3)$
Add $\ln (x - 3)$ to both sides:	$2 \ln (x - 3) = 5$
Divide both sides by 2:	$\ln (x - 3) = \dfrac{5}{2}$
Exponentiate both sides:	$e^{\ln(x-3)} = e^{5/2}$
Use the basic property of logarithms ($*$):	$x - 3 = e^{5/2}$
Add 3 to both sides:	$x = e^{5/2} + 3 \approx 15.1825.$

This is the only possibility for a solution. A calculator shows that it actually is a solution of the original equation. ∎

EXAMPLE 11

Solve $\log (x - 16) = 2 - \log (x - 1)$.

SOLUTION

$$\log (x - 16) = 2 - \log (x - 1)$$

Add $\log (x - 1)$ to both sides: $\log (x - 16) + \log (x - 1) = 2$

Use the Product Law: $\log [(x - 16)(x - 1)] = 2$

Multiply out left side: $\log (x^2 - 17x + 16) = 2$

Exponentiate both sides: $10^{\log (x^2 - 17x + 16)} = 10^2$

Use the basic logarithm property ($*$): $x^2 - 17x + 16 = 100$

Subtract 100 from both sides: $x^2 - 17x - 84 = 0$

Factor: $(x + 4)(x - 21) = 0$

$$x + 4 = 0 \qquad \text{or} \qquad x - 21 = 0$$

$$x = -4 \quad \text{or} \qquad\qquad x = 21.$$

You can easily verify that 21 is a solution of the original equation, but -4 is not [when $x = -4$, then $\log (x - 16) = \log (-20)$, which is not defined]. ∎

EXAMPLE 12

To solve $\log (x + 5) = 1 - \log (x - 2)$,

Rearrange terms: $\log (x + 5) + \log (x - 2) = 1$

Use the Product Law: $\log [(x + 5)(x - 2)] = 1$

$$\log (x^2 + 3x - 10) = 1$$

Exponentiate both sides: $10^{\log (x^2 + 3x - 10)} = 10^1$

Use the basic logarithm property ($*$): $x^2 + 3x - 10 = 10$

$$x^2 + 3x - 20 = 0.$$

This equation can be solved with the quadratic formula.

$$x = \frac{-3 \pm \sqrt{3^2 - 4 \cdot 1 \cdot (-20)}}{2 \cdot 1} = \frac{-3 \pm \sqrt{89}}{2}.$$

An easy way to verify that

$$x = \frac{-3 + \sqrt{89}}{2}$$

```
(-3+√(89))/2→A
           3.216990566
log(A+5)
           .9147127884
1-log(A-2)
           .9147127884
■
```

Figure 5-29

is a solution is to store this number in your calculator as A and then evaluate both sides of the original equation at $x = A$, as shown in Figure 5–29. The other possibility, however, is not a solution because

$$x = \frac{-3 - \sqrt{89}}{2}$$

is negative, so $\log (x - 2)$ is not defined. ∎

EXAMPLE 13

The number of pounds of fish (in billions) used for human consumption in the United States in year x is approximated by the function

$$f(x) = 10.57 + 1.75 \ln x,$$

where $x = 5$ corresponds to 1995.*

(a) How many pounds of fish were used in 2004?

(b) When will fish consumption reach 16 billion pounds?

SOLUTION

(a) Since 2004 corresponds to $x = 14$, we evaluate $f(x)$ at 14.

$$f(14) = 10.57 + 1.75 \ln 14 \approx 15.19 \text{ billion pounds.}$$

(b) Fish consumption is 16 billion pounds when $f(x) = 16$, so we must solve the equation

$$10.57 + 1.75 \ln x = 16$$

Subtract 10.57 from both sides: $1.75 \ln x = 5.43$

Divide both sides by 1.75: $\ln x = \dfrac{5.43}{1.75}$

Exponentiate both sides: $e^{\ln x} = e^{5.43/1.75}$

Use the basic property of logarithms (∗): $x \approx 22.26.$

Since $x = 22$ corresponds to 2012, fish consumption will reach 16 billion pounds in 2012.   ∎

---

*Based on data from the U.S. National Oceanic and Atmospheric Administration and the National Marine Fisheries Service.

EXERCISES 5.5

In Exercises 1–8, solve the equation without using logarithms.

1. $3^x = 81$ **2.** $5^x - 2 = 23$ **3.** $3^{x+1} = 9^{5x}$

4. $3^{7x} = 9^{2x-5}$ **5.** $3^{5x}9^{x^2} = 27$

6. $7^{x^2+3x} = 1/49$ **7.** $9^{x^2} = 3^{-5x-2}$

8. $5^{2x^2+3x} = 25^{6-x}$

In Exercises 9–22, solve the equation. First express your answer in terms of natural logarithms (for instance, $x = (2 + \ln 5)/(\ln 3)$). Then use a calculator to find an approximation for the answer.

9. $3^x = 5$ **10.** $2^x = 9$ **11.** $2^x = 3^{x-1}$

12. $9^{x-1} = 8^{x-3}$ **13.** $3^{1-2x} = 5^{x+5}$

14. $2^{1-3x} = 7^{x+3}$ **15.** $2^{1-3x} = 3^{x+1}$

16. $5^{x+3} = 2^x$ **17.** $e^{2x} = 5$

18. $e^{-9x} = 3$ **19.** $6e^{-1.4x} = 21$

20. $27e^{-x/4} = 67.5$ **21.** $2.1e^{(x/2)\ln 3} = 5$

22. $2.7e^{(-x/3)\ln 7} = 21$

In Exercises 23–29, solve the equation for x by first making an appropriate substitution, as in the Hint for Exercise 23.

23. $9^x - 4 \cdot 3^x + 3 = 0$ [*Hint:* Let $u = 3^x$ and note that $9^x = (3^2)^x = 3^{2x} = (3^x)^2$. Hence, the equation becomes $u^2 - 4u + 3 = 0$. Solve this equation for u. In each solution, replace u by 3^x and solve for x.]

24. $25^x - 8 \cdot 5^x = -12$

25. $e^{2x} - 5e^x + 6 = 0$ [*Hint:* Let $u = e^x$.]

26. $3e^{2x} - 16e^x + 5 = 0$ **27.** $6e^{2x} - 16e^x = 6$

28. $6e^{2x} + 7e^x = 10$ **29.** $4^x + 6 \cdot 4^{-x} = 5$

In Exercises 30–32, solve the equation for x.

30. $\dfrac{e^x + e^{-x}}{e^x - e^{-x}} = t$ **31.** $\dfrac{e^x - e^{-x}}{2} = t$

32. $\dfrac{e^x - e^{-x}}{e^x + e^{-x}} = t$

33. (a) Prove that if $\ln u = \ln v$, then $u = v$. [*Hint:* Property ($*$) on page 404.]
(b) Is it always the case that if $u = v$ then $\ln u = \ln v$? Why or why not?

34. (a) Solve $7^x = 3$, using natural logarithms. Leave your answer in logarithmic form; don't approximate with a calculator.
(b) Solve $7^x = 3$, using common (base 10) logarithms. Leave your answer in logarithmic form.
(c) Use the change of base formula in Special Topics 5.4.A to show that your answers in parts (a) and (b) are the same.

In Exercises 35–44, solve the equation as in Example 8.

35. $\ln (3x - 5) = \ln 11 + \ln 2$

36. $\log (3x + 8) = \log (2x + 1) + \log 3$

37. $\log (3x - 1) + \log 2 = \log 4 + \log (x + 2)$

38. $\ln (2x - 1) - \ln 2 = \ln (3x + 6) - \ln 6$

39. $2 \ln x = \ln 36$

40. $2 \log x = 3 \log 9$

41. $\ln x + \ln (x + 1) = \ln 3 + \ln 4$

42. $\ln (5x - 2) + \ln x = \ln 3$

43. $\ln x = \ln 3 - \ln (x + 5)$

44. $\ln (3x - 4) + \ln x = \ln e$

In Exercises 45–52, solve the equation.

45. $\ln (x + 9) - \ln x = 1$

46. $\ln (3x + 5) - 1 = \ln (2x - 3)$

47. $\log x + \log (x - 3) = 1$

48. $\log (x - 4) + \log (x - 1) = 1$

49. $\log \sqrt{x^2 - 1} = 2$

50. $\log \sqrt[5]{x^2 + 15x} = 2/5$

51. $\ln (x^2 + 1) - \ln (x - 1) = 1 + \ln (x + 1)$

52. $\dfrac{\ln (2x + 1)}{\ln (3x - 1)} = 2$

Exercises 53–62 deal with radioactive decay and the function $M(x) = c(.5^{x/h})$; see Examples 4 and 5.

53. A sample of 300 grams of uranium decays to 200 grams in .26 billion years. Find the half-life of uranium.

54. It takes 1000 years for a sample of 300 mg of radium-226 to decay to 195 mg. Find the half-life of radium-226.

55. A 3-gram sample of an isotope of sodium decays to 1 gram in 23.7 days. Find the half-life of the isotope of sodium.

56. The half-life of cobalt-60 is 5.3 years. How long will it take for 100 grams to decay to 33 grams?

57. After six days a sample of radon-222 decayed to 33.6% of its original mass. Find the half-life of radon-222. [*Hint:* When $x = 6$, then $M(x) = .336P$.]

58. Krypton-85 loses 6.44% of its mass each year. What is its half-life?

59. How old is a piece of ivory that has lost 36% of its carbon-14?

60. How old is a mummy that has lost 62% of its carbon-14?

61. A Native American mummy was found recently. If it has lost 26.4% of its carbon-14, approximately how long ago did the Native American die?

62. How old is a wooden statue that has only one-third of its original carbon-14?

Exercises 63–68 deal with the compound interest formula $A = P(1 + r)^t$, which was discussed in Special Topics 5.2.A.

63. At what annual rate of interest should $1000 be invested so that it will double in 10 years if interest is compounded quarterly?

64. How long does it take $500 to triple if it is invested at 6% compounded: (a) annually, (b) quarterly, (c) daily?

65. (a) How long will it take to triple your money if you invest $500 at a rate of 5% per year compounded annually?
(b) How long will it take at 5% compounded quarterly?

66. At what rate of interest (compounded annually) should you invest $500 if you want to have $1500 in 12 years?

67. How much money should be invested at 5% interest, compounded quarterly, so that 9 years later the investment will be worth $5000? This amount is called the **present value** of $5000 at 5% interest.

68. Find a formula that gives the time needed for an investment of P dollars to double, if the interest rate is r% compounded annually. [*Hint:* Solve the compound interest formula for t, when $A = 2P$.]

Exercises 69–76 deal with functions of the form $f(x) = Pe^{kx}$, where k is the continuous exponential growth rate (see Example 6).

69. The present concentration of carbon dioxide in the atmosphere is 364 parts per million (ppm) and is increasing exponentially at a continuous yearly rate of .4% (that is, $k = .004$). How many years will it take for the concentration to reach 500 ppm?

70. The amount P of ozone in the atmosphere is currently decaying exponentially each year at a continuous rate of $\frac{1}{4}\%$ (that is, $k = -.0025$). How long will it take for half the ozone to disappear (that is, when will the amount be $P/2$)? [Your answer is the half-life of ozone.]

71. The population of Brazil increased from 151 million in 1990 to 180 million in 2002.*

    (a) At what continuous rate was the population growing during this period?

    (b) Assuming that Brazil's population continues to increase at this rate, when will it reach 250 million?

72. Between 1996 and 2004, the number of United States subscribers to cell-phone plans has grown nearly exponentially. In 1996 there were 44,043,000 subscribers and in 2004 there were 182,140,000†.

    (a) What is the continuous growth rate of the number of cell-phone subscribers?

    (b) In what year were there 60,000,000 cell-phone subscribers?

    (c) Assuming that this rate continuous, in what year will there be 350,000,000 subscribers?

    (d) In 2007 the United States population was approximately 300 million. Is your answer to part (c) realistic? If not, what could have gone wrong?

73. The probability P percent of having an accident while driving a car is related to the alcohol level of the driver's blood by the formula $P = e^{kt}$, where k is a constant. Accident statistics show that the probability of an accident is 25% when the blood alcohol level is $t = .15$.

    (a) Find k. [Use $P = 25$, not .25.]

    (b) At what blood alcohol level is the probability of having an accident 50%?

74. Under normal conditions, the atmospheric pressure (in millibars) at height h feet above sea level is given by $P(h) = 1015e^{-kh}$, where k is a positive constant.

    (a) If the pressure at 18,000 feet is half the pressure at sea level, find k.

    (b) Using the information from part (a), find the atmospheric pressure at 1000 feet, 5000 feet, and 15,000 feet.

75. One hour after an experiment begins, the number of bacteria in a culture is 100. An hour later, there are 500.

    (a) Find the number of bacteria at the beginning of the experiment and the number three hours later.

    (b) How long does it take the number of bacteria at any given time to double?

76. If the population at time t is given by $S(t) = ce^{kt}$, find a formula that gives the time it takes for the population to double.

77. The spread of a flu virus in a community of 45,000 people is given by the function

$$f(t) = \frac{45{,}000}{1 + 224e^{-.899t}},$$

where $f(t)$ is the number of people infected in week t.

    (a) How many people had the flu at the outbreak of the epidemic? After three weeks?

    (b) When will half the town be infected?

78. The beaver population near a certain lake in year t is approximately

$$p(t) = \frac{2000}{1 + 199e^{-.5544t}}.$$

    (a) When will the beaver population reach 1000?

    (b) Will the population ever reach 2000? Why?

79. Assume that you watched 1000 hours of television this year, and will watch 750 hours next year, and will continue to watch 75% as much every year thereafter.

    (a) In what year will you be down to ten hours per year?

    (b) In what year would you be down to one hour per year?

80. In the year 2009, Olivia's bank balance is $1000. In the year 2010, her balance is $1100.

    (a) If her balance is growing exponentially, in what year will it reach $2500?

    (b) If her balance is instead growing linearly, in what year will it reach $2500?

THINKERS

81. According to one theory of learning, the number of words per minute N that a person can type after t weeks of practice is given by $N = c(1 - e^{-kt})$, where c is an upper limit that N cannot exceed and k is a constant that must be determined experimentally for each person.

    (a) If a person can type 50 wpm (words per minute) after four weeks of practice and 70 wpm after eight weeks, find the values of k and c for this person. According to the theory, this person will never type faster than c wpm.

    (b) Another person can type 50 wpm after four weeks of practice and 90 wpm after eight weeks. How many weeks must this person practice to be able to type 125 wpm?

82. Kate has been offered two jobs, each with the same starting salary of $32,000 and identical benefits. Assuming satisfactory performance, she will receive a $1600 raise each year at the Great Gizmo Company, whereas the Wonder Widget Company will give her a 4% raise each year.

    (a) In what year (after the first year) would her salary be the same at either company? Until then, which company pays better? After that, which company pays better?

    (b) Answer the questions in part (a) assuming that the annual raise at Great Gizmo is $2000.

---

*U.S. Census Bureau, International Data Base.
†www.census.gov

5.6   Exponential, Logarithmic, and Other Models*

Section Objectives

■ Determine the best type of function to model a set of data.
■ Find a reasonable model for a set of data.

Many data sets can be modeled by suitable exponential, logarithmic, and related functions. Most calculators have regression procedures for constructing the following models.

Model	Equation	Examples	
Power	$y = ax^r$	$y = 5x^{2.7}$	$y = 3.5x^{-.045}$
Exponential	$y = ab^x$ or $y = ae^{kx}$	$y = 2(1.64)^x$	$y = 2 \cdot e^{.4947x}$
Logistic	$y = \dfrac{a}{1 + be^{-kx}}$	$y = \dfrac{20{,}000}{1 + 24e^{-.25x}}$	$y = \dfrac{650}{1 + 6e^{.3x}}$
Logarithmic	$y = a + b \ln x$	$y = 5 + 4.2 \ln x$	$y = 2 - 3 \ln x$

We begin by examining exponential models, such as $y = 3 \cdot 2^x$. A table of values for this model is shown below. Look carefully at the ratio of successive entries (that is, each entry divided by its predecessor).

x	$y = 3 \cdot 2^x$
0	3
	$\dfrac{48}{3} = 16$
4	48
	$\dfrac{768}{48} = 16$
8	768
	$\dfrac{12{,}288}{768} = 16$
12	12,288
	$\dfrac{196{,}608}{12{,}228} = 16$
16	196,608

It should not be a surprise that the ratio of successive entries is constant. For at each step, x changes from x to $x + 4$ (from 0 to 4, from 4 to 8, and so on) and y changes from $3 \cdot 2^x$ to $3 \cdot 2^{x+4}$. Hence, the ratio of successive terms is always

$$\frac{3 \cdot 2^{x+4}}{3 \cdot 2^x} = \frac{3 \cdot 2^x \cdot 2^4}{3 \cdot 2^x} = 2^4 = 16.$$

A similar argument applies to any exponential model $y = ab^x$ and shows that if x changes by a fixed amount k, then the ratio of the corresponding y values is the constant b^k (in our example b was 2 and k was 4). This suggests that

when the ratio of successive, equally spaced, entries in a table of data is approximately constant, an exponential model is appropriate.

*This section is optional; its prerequisites are Section 2.5 and Special Topics 4.4.A. It will be used in clearly identifiable exercises but not elsewhere in the text.

EXAMPLE 1

In the years before the Civil War, the population of the United States grew rapidly, as shown in the following table from the U.S. Bureau of the Census. Find a model for this growth.

Year	Population in Millions	Year	Population in Millions
1790	3.93	1830	12.86
1800	5.31	1840	17.07
1810	7.24	1850	23.19
1820	9.64	1860	31.44

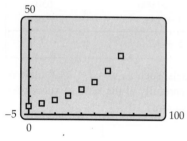

Figure 5–30

SOLUTION   The data points (with $x = 0$ corresponding to 1790) are shown in Figure 5–30. Their shape suggests either a polynomial graph of even degree or an exponential graph. Since populations generally grow exponentially, an exponential model is likely to be a good choice. We can confirm this by looking at the ratios of successive entries in the table.

Year	Population		Year	Population	
1790	3.93		1830	12.86	
		$\dfrac{5.31}{3.93} \approx 1.351$			$\dfrac{17.07}{12.86} \approx 1.327$
1800	5.31		1840	17.07	
		$\dfrac{7.24}{5.31} \approx 1.363$			$\dfrac{23.19}{17.07} \approx 1.359$
1810	7.24		1850	23.19	
		$\dfrac{9.64}{7.24} \approx 1.331$			$\dfrac{31.44}{23.19} \approx 1.356$
1820	9.64		1860	31.44	
		$\dfrac{12.86}{9.64} \approx 1.334$			
1830	12.86				

The ratios are almost constant, as they would be in an exponential model. So we use regression to find such an exponential model. The procedure is the same as for linear and polynomial regression (see the Tips on pages 128 and 283). It produces this model:*

$$y = 3.9572 \, (1.0299^x).$$

The graph in Figure 5–31 appears to fit the data quite well. In fact, you can readily verify that the model has an error of less than 1% for each of the data points. Furthermore, as discussed before the example, when x changes by 10, the value of y changes by approximately $1.0299^{10} \approx 1.343$, which is very close to the successive ratios of the data that were computed above. ∎

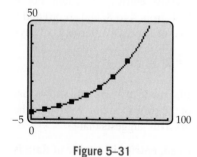

Figure 5–31

---

*Throughout this section, coefficients are rounded for convenient reading, but the full expansion is used for calculations and graphs.

EXAMPLE 2

After the Civil War, the U.S. population continued to increase, as shown below.

Year	Population in Millions	Year	Population in Millions	Year	Population in Millions
1870	38.56	1920	106.02	1960	179.32
1880	50.19	1930	123.20	1970	202.30
1890	62.98	1940	132.16	1980	226.54
1900	76.21	1950	151.33	1990	248.72
1910	92.23			2000	281.42

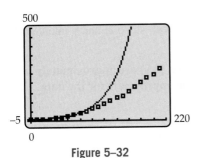

Figure 5–32

However, the model from Example 1 does not remain valid, as can be seen in Figure 5–32, which shows its graph together with all the data points from 1790 through 2000 ($x = 0$ corresponds to 1790).

The problem is that the *rate* of growth has steadily decreased since the Civil War. For instance, the ratio of the first two entries in the preceding table is

$$\frac{50.19}{38.56} \approx 1.302,$$

and the ratio of the last two is

$$\frac{281.42}{248.72} \approx 1.13.$$

So an exponential model may not be the best choice now. Other possibilities are polynomial models (which grow at a slower rate) or logistic models (in which the growth rate decreases with time). Figure 5–33 shows three possible models, each obtained by using the appropriate regression program on a calculator, with all the data points from 1790 through 2000.

Exponential Model

$$y = 6.06616 \cdot 1.02039^x$$

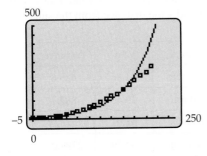

Polynominal Model

$$y = (7.94 \times 10^{-8})x^4 - (2.76 \times 10^{-5})x^3 + .0093x^2 - .1621x + 5.462$$

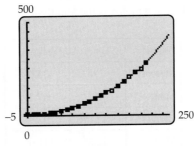

Logistic Model*

$$y = \frac{442.1}{1 + 56.33e^{-.0216x}}$$

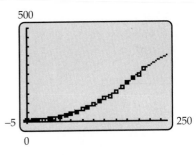

Figure 5–33

*This model was obtained on a TI-83. Other calculators may produce a slightly different model or an error message.

As expected, the polynomial and logistic models fit the data better than does the exponential model. The main difference between them is that the polynomial model indicates unlimited future growth, whereas the logistic model has the population growing more slowly in the future (and eventually leveling off—see Exercise 13). ∎

In Example 1, we used the ratios of successive entries of the data table to determine that an exponential model was appropriate. Here is another way to make that determination. Consider the exponential function $y = ab^x$. Taking natural logarithms of both sides and using the logarithm laws on the right side shows that

$$\ln y = \ln (ab^x) = \ln a + \ln b^x = \ln a + x \ln b.$$

Now $\ln a$ and $\ln b$ are constants, say, $k = \ln a$ and $m = \ln b$, so

$$\ln y = mx + k.$$

Thus, the points $(x, \ln y)$ lie on the straight line with slope m and y-intercept k. Consequently, we have this guideline.

> **If (x, y) are data points and if the points $(x, \ln y)$ are approximately linear, then an exponential model may be appropriate for the data.**

Similarly, if $y = ax^r$ is a power function, then

$$\ln y = \ln (ax^r) = \ln a + r \ln x.$$

Since $\ln a$ is a constant, say $k = \ln a$, we have

$$\ln y = r \ln x + k,$$

which means that the points $(\ln x, \ln y)$ lie on a straight line with slope r and y-intercept k. Consequently, we have this guideline.

> **If (x, y) are data points and if the points $(\ln x, \ln y)$ are approximately linear, then a power model may be appropriate for the data.**

EXAMPLE 3

The length of time that a planet takes to make one complete rotation around the sun is its year. The table shows the length (in earth years) of each planet's year and the distance of that planet from the sun (in millions of miles).* Find a model for this data in which x is the length of the year and y the distance from the sun.

Planet	Year	Distance	Planet	Year	Distance
Mercury	.24	36.0	Jupiter	11.86	483.6
Venus	.62	67.2	Saturn	29.46	886.7
Earth	1	92.9	Uranus	84.01	1783.0
Mars	1.88	141.6	Neptune	164.79	2794.0

---

*Since the orbit of a planet around the sun is not circular, its distance from the sun varies through the year. Each given number is the average of its maximum and minimum distances from the sun.

SOLUTION   Figure 5–34 shows the data points for the five planets with the shortest years. Figure 5–35 shows all the data points, but on this scale, the first four points look like a single large one near the origin.

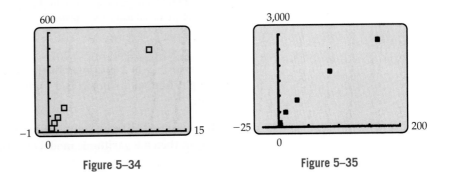

Figure 5–34   Figure 5–35

TECHNOLOGY TIP

Suppose the x- and y-coordinates of the data points are stored in lists L_1 and L_2, respectively. On calculators other than TI-89, keying in

$$\ln L_2 \text{ STO} \rightarrow L_4$$

produces the list L_4, whose entries are the natural logarithms of the numbers in list L_2, and stores it in the statistics editor. You can then use lists L_1 and L_4 to plot the points $(x, \ln y)$. For TI-89, check your instruction manual.

Plotting the point $(x, \ln y)$ for each data point (x, y) (see the Technology Tip in the margin) produces Figure 5–36. Its points do not form a linear pattern (four of them are almost vertical near the y-axis and the other five almost horizontal), so an exponential function is not an appropriate model. On the other hand, the points $(\ln x, \ln y)$ in Figure 5–37 do form a linear pattern, which suggests that a power model will work.

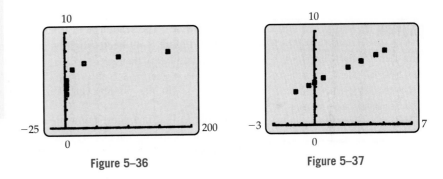

Figure 5–36   Figure 5–37

A calculator's power regression feature produces this model.

$$y = 92.8982 \, x^{.6668}.$$

Its graph in Figure 5–38 shows that it fits the original data points quite well.   ■

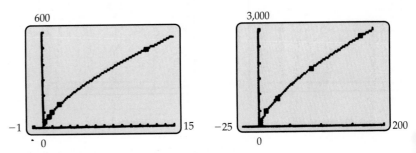

Figure 5–38

GRAPHING EXPLORATION

There are three dwarf-planets in our solar system—Ceres, Pluto, and Eris. From its discovery in 1930 until 2006, Pluto was considered a planet. The length of Pluto's year is 247.69 Earth years, and it is 3674.5 million miles from the sun. Does Pluto fit the model we found using the 8 planets?

If $y = a + b \ln x$ is a logarithmic model, then the points $(\ln x, y)$ lie on the straight line with slope b and y-intercept a (why?). Thus, we have this guideline.

If (x, y) are data points and if the points $(\ln x, y)$ are approximately linear, then a logarithmic model may be appropriate for the data.

EXAMPLE 4

Find a model for population growth in El Paso, Texas, given the following data.*

Year	1950	1970	1980	1990	2000	2005
Population	130,485	322,261	425,259	515,342	563,662	598,590

> **CAUTION**
>
> When using logarithmic models, you must have data points with positive first coordinates (since logarithms of negative numbers and 0 are not defined).

SOLUTION   The scatter plot of the data points (with $x = 50$ corresponding to 1950) in Figure 5–39 bends slightly, suggesting a logarithmic curve. So we plot the points $(\ln x, y)$, that is,

$$(\ln 50, 130485), (\ln 70, 322261), \ldots, (\ln 105, 598590),$$

in Figure 5–40. Since these points lie approximately on a straight line, a logarithmic model is appropriate. Using logarithmic regression on a calculator, we obtain this model.

$$-2{,}380{,}341.254 + 640{,}178.5447 \ln x.$$

Its graph in Figure 5–41 is a good fit for the data. ∎

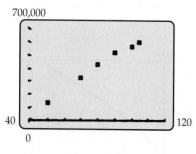

Figure 5–39

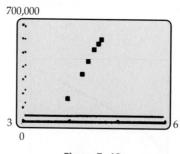

Figure 5–40

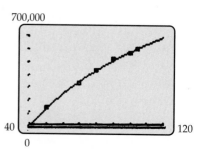

Figure 5–41

*U.S. Bureau of the Census.

EXERCISES 5.6

In Exercises 1–10, state which of the following models might be appropriate for the given scatter plot of data (more than one model may be appropriate).

Model	Corresponding Function
A. Linear	$y = ax + b$
B. Quadratic	$y = ax^2 + bx + c$
C. Power	$y = ax^r$
D. Cubic	$y = ax^3 + bx^2 + cx + d$
E. Exponential	$y = ab^x$
F. Logarithmic	$y = a + b \ln x$
G. Logistic	$y = \dfrac{a}{1 + be^{-kx}}$

1.

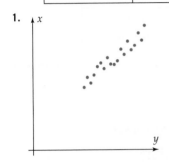

2.

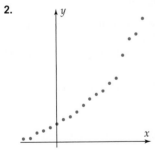

3.

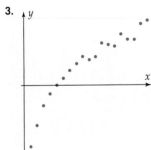

4.

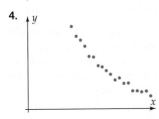

5.

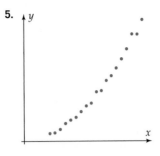

6.

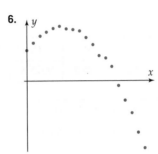

7.

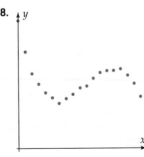

8.

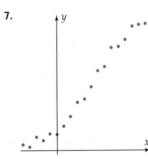

9.

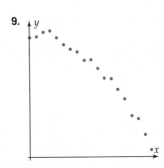

10.

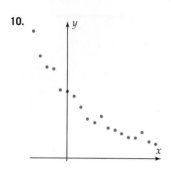

In Exercises 11 and 12, compute the ratios of successive entries in the table to determine whether or not an exponential model is appropriate for the data.

11.

x	0	2	4	6	8	10
y	3	15.2	76.9	389.2	1975.5	9975.8

12.

x	1	3	5	7	9	11
y	3	21	55	105	171	253

13. (a) Show algebraically that in the logistic model for the U.S. population in Example 2, the population can never exceed 442.1 million people.

  (b) Confirm your answer in part (a) by graphing the logistic model in a window that includes the next three centuries.

14. According to estimates by the U.S. Bureau of the Census, the U.S. population was 287.7 million in 2002. On the basis of this information, which of the models in Example 2 appears to be the most accurate predictor?

15. Graph each of the following power functions in a window with $0 \le x \le 20$.

  (a) $f(x) = x^{-1.5}$ (b) $g(x) = x^{.75}$ (c) $h(x) = x^{2.4}$

16. On the basis of your graphs in Exercise 15, describe the general shape of the graph of $y = ax^r$ when $a > 0$ and

  (a) $r < 0$ (b) $0 < r < 1$ (c) $r > 1$

In Exercises 17–20, determine whether an exponential, power, or logarithmic model (or none or several of these) is appropriate for the data by determining which (if any) of the following sets of points are approximately linear:

$$\{(x, \ln y)\}, \qquad \{(\ln x, \ln y)\}, \qquad \{(\ln x, y)\}.$$

where the given data set consists of the points $\{(x, y)\}$.

17.

x	1	3	5	7	9	11
y	2	25	81	175	310	497

18.

x	3	6	9	12	15	18
y	385	74	14	2.75	.5	.1

19.

x	5	10	15	20	25	30
y	17	27	35	40	43	48

20.

x	5	10	15	20	25	30
y	2	110	460	1200	2500	4525

21. The table shows the number of babies born as twins, triplets, quadruplets, etc., over a 7-year period.

Year	Multiple Births
1989	92,916
1990	96,893
1991	98,125
1992	99,255
1993	100,613
1994	101,658
1995	101,709

  (a) Sketch a scatter plot of the data, with $x = 1$ corresponding to 1989.

  (b) Plot each of the following models on the same screen as the scatter plot.

$$f(x) = 93,201.973 + 4,545.977 \ln x$$

$$g(x) = \frac{102,519.98}{1 + .1536e^{-.4263x}}.$$

  (c) Use the table feature to estimate the number of multiple births in 2000 and 2010.

  (d) Over the long run, which model do you think is the better predictor?

22. The graph shows the U.S. Census Bureau estimates of future U.S. population.

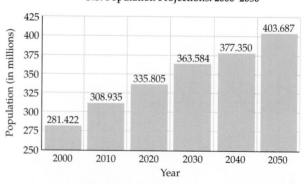

U.S. Population Projections: 2000–2050

  (a) How well do the projections in the graph compare with those given by the logistic model in Example 2?

  (b) Find a logistic model of the U.S. population, using the data given in Example 2 for the years from 1900 to 2000.

  (c) How well do the projections in the graph compare with those given by the model in part (b)?

23. Infant mortality rates in the United States are shown in the table.

Year	Infant Mortality Rate*	Year	Infant Mortality Rate*
1920	76.7	1985	10.6
1930	60.4	1990	9.2
1940	47.0	1995	7.6
1950	29.2	2000	6.9
1960	26.0	2001	6.8
1970	20.0	2002	7.0
1980	12.6	2003	6.9

(a) Sketch a scatter plot of the data, with $x = 0$ corresponding to 1900.
(b) Verify that the set of points $(x, \ln y)$, where (x, y) are the original data points, is approximately linear.
(c) On the basis of part (b), what type of model would be appropriate for this data? Find such a model.

24. The number of children who were home schooled in the United States in selected years is shown in the table.[†]

(a) Sketch a scatter plot of the data, with $x = 0$ corresponding to 1980.
(b) Find a quadratic model for the data.
(c) Find a logistic model for the data.
(d) What is the number of home-schooled children predicted by each model for the year 2015?
(e) What are the limitations of each model?

Fall of School Year	Number of Children (in thousands)
1985	183
1988	225
1990	301
1992	470
1993	588
1994	735
1995	800
1996	920
1997	1100
1999	1400
2000	1700
2005	1900

25. The average number of students per computer in the U.S. public schools (elementary through high school) is shown in the table at the top of the next column.

Fall of School Year	Students/Computer
1987	32
1988	25
1989	22
1990	20
1991	18
1992	16
1993	14
1994	10.5
1995	10
1996	7.8
1997	6.1
1998	5.7
1999	5.4
2000	5
2002	4.9
2003	4.8

(a) Sketch a scatter plot of the data, with $x = 7$ corresponding to 1987.
(b) Find an exponential model for the data.
(c) Use the model to estimate the number of students per computer in 2015.
(d) In what year, according to this model, will each student have his or her own computer in school?
(e) What are the limitations of this model?

26. (a) Find an exponential model for the federal debt, based on the data in the table.* Let $x = 0$ correspond to 1960.
(b) Use the model to estimate the federal debt in 2010.

GROSS FEDERAL DEBT	
End of Fiscal Year	Amount (in billions of dollars)
1960	290.5
1965	322.3
1970	380.9
1975	541.9
1980	909.0
1985	1817.4
1990	3206.3
1995	4920.6
2000	5628.7
2005	8031.4[†]

*Rates are infant (under 1 year) deaths per 1000 live births.
[†]National Home Education Research Institute

*http//:www.whitehouse.gov/omb/budget/fy2006/pdf/hist.pdf (pages 118–119)
[†]Estimated

27. The number of U.S. adults on probation from 2000 to 2004 is shown in the first two columns of the following table.*

Year	Adults on Probation	Predicted Number of Adults on Probation	Ratio
2000	1,316,333		
2001	1,330,007		
2002	1,367,547		
2003	1,392,796		
2004	1,421,911		

(a) Sketch a scatter plot of the data with $x = 0$ corresponding to 2000.

(b) Fill in column 4 by dividing each entry in column 2 by the preceding one; round your answers to two decimal places. In view of column 4, what type of model is appropriate here?

(c) Find an appropriate model for the data.

(d) Use the model to complete column 3 in the table. How well does the model fit the data?

(e) If the model remains accurate, how many adults will be on probation in 2010?

28. In the past two decades, more women than men have been entering college. The table shows the percentage of male first-year college students in selected years.[†]

Year	1985	1990	1995	1997	1998	1999	2003	2004	2005
Percent	48.9	46.9	45.6	45.5	45.5	45.3	45.1	44.9	45.0

(a) Find three models for this data: exponential, logarithmic, and power, with $x = 5$ corresponding to 1985.

(b) For the years 1985–2005, is there any significant difference among the models?

(c) Assume that the models remain accurate. What year does each predict as the first year in which fewer than 43% of first-year college students will be male?

(d) We actually have some additional data:

Year	2000	2001	2002	2006
Percent	45.2	44.9	45.0	45.1

Which model did the best job of predicting the new data?

29. The table gives the life expectancy (at birth) of a woman born in the given year.*

Year	Life Expectancy (in years)
1910	51.8
1930	61.6
1950	71.1
1970	74.7
1990	78.8
1995	78.9
1997	79.4
1999	79.4
2000	79.7
2001	79.8
2002	79.9
2003	80.1

(a) Find a logarithmic model for the data, with $x = 10$ corresponding to 1910.

(b) Use the model to find the life expectancy of a woman born in 1996. [For comparison, the actual expectancy is 79.1 years.]

(c) Assume that the model remains accurate. In what year will the life expectancy of a woman born in that year be at least 83 years?

30. The table gives the death rate in motor vehicle accidents (per 100,000 population) in selected years.

Year	1970	1980	1985	1990	1995	2000	2003
Death Rate	26.8	23.4	19.3	18.8	16.5	15.6	15.4

(a) Find an exponential model for the data, with $x = 0$ corresponding to 1970.

(b) What was the death rate in 1998 and in 2002?

(c) Assume that the model remains accurate, when will the death rate drop to 13 per 100,000?

*Statistical Abstract of the United States.
†Higher Education Research Institute at UCLA.

*National Center for Health Statistics.

Chapter 5 Review

IMPORTANT CONCEPTS

IMPORTANT FACTS & FORMULAS

■ *Laws of Exponents:*

$$c^r c^s = c^{r+s} \qquad (cd)^r = c^r d^r$$
$$\frac{c^r}{c^s} = c^{r-s} \qquad \left(\frac{c}{d}\right)^r = \frac{c^r}{d^r}$$
$$(c^r)^s = c^{rs} \qquad c^{-r} = \frac{1}{c^r}$$

■ $g(x) = \log x$ is the inverse function of $f(x) = 10^x$:

$$10^{\log v} = v \text{ for all } v > 0 \qquad \text{and} \qquad \log 10^u = u \text{ for all } u.$$

■ $g(x) = \ln x$ is the inverse function of $f(x) = e^x$:

$$e^{\ln v} = v \text{ for all } v > 0 \qquad \text{and} \qquad \ln(e^u) = u \text{ for all } u.$$

■ $h(x) = \log_b x$ is the inverse function of $k(x) = b^x$:

$$b^{\log_b v} = v \text{ for all } v > 0 \qquad \text{and} \qquad \log_b(b^u) = u \text{ for all } u.$$

■ *Logarithm Laws:* For all $v, w > 0$ and any k:

$$\ln(vw) = \ln v + \ln w \qquad \log_b(vw) = \log_b v + \log_b w$$

$$\ln\left(\frac{v}{w}\right) = \ln v - \ln w \qquad \log_b\left(\frac{v}{w}\right) = \log_b v - \log_b w$$

$$\ln(v^k) = k(\ln v) \qquad \log_b(v^k) = k(\log_b v)$$

■ *Exponential Growth Functions:*

$$f(x) = P(1 + r)^x \quad (0 < r < 1),$$

$$f(x) = Pa^x \quad (a > 1),$$

$$f(x) = Pe^{kx} \quad (k > 0)$$

■ *Exponential Decay Functions:*

$$f(x) = P(1 - r)^x \quad (0 < r < 1),$$

$$f(x) = Pa^x \quad (0 < a < 1),$$

$$f(x) = Pe^{kx} \quad (k < 0)$$

■ *Compound Interest Formula:* $A = P(1 + r)^t$

■ *Change of Base Formula:* $\log_b v = \dfrac{\ln v}{\ln b}$

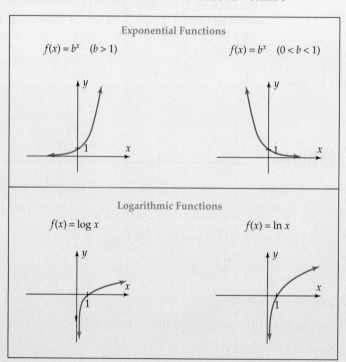

CATALOG OF BASIC FUNCTIONS—PART 3

Exponential Functions

$f(x) = b^x \quad (b > 1)$ $f(x) = b^x \quad (0 < b < 1)$

Logarithmic Functions

$f(x) = \log x$ $f(x) = \ln x$

REVIEW QUESTIONS

In Questions 1–6, simplify the expression.

1. $\sqrt{\sqrt[5]{a^{20}}}$

2. $(\sqrt[4]{5}a^2b)^4 (a^3\sqrt{b})^2$

3. $(x^{2/5}y^{-2/3})(x^{2/3}y^4)^{1/2}$

4. $\dfrac{(2x)^{1/3}(2y)^{-3}(9x)^{1/2}}{(3x)^{2/3}(3y)^2x^{-5/2}}$

5. $(x^{1/3} + y^{1/3})(x^{1/3} - y^{1/3})$ **6.** $2a^{3/5}(5a^{7/5} - 4a^{-3/5})$

In Exercises 7 and 8, simplify and write the expression without radicals or negative exponents:

7. $\dfrac{\sqrt[3]{7x^{-4}y^{11}}}{\sqrt[3]{56x^2y^2}}$

8. $\dfrac{(9x^4)^{1/3}\,3^{-2}\,x^{-5}}{3x^9}$

9. Rationalize the numerator and simplify:

$$\dfrac{\sqrt{3x + 3h - 5} - \sqrt{3x - 5}}{h}$$

10. Rationalize the denominator: $\dfrac{6}{\sqrt{2x} + 8}$

In Questions 11–16, find all real solutions of the equation.

11. $\sqrt{2x - 1} = 3x - 4$ 12. $\sqrt[3]{9 - y^2} = -3$

13. $3\sqrt{x + 3} + \sqrt{5x - 1} = 2$

14. $x^{2/3} - 3x^{1/3} - 18 = 0$

15. $\sqrt[3]{x^4 - 2x^3 + 6x - 7} = x + 3$

16. $x^{4/3} + x - 2x^{2/3} + x^{1/3} = 4$

In Questions 17 and 18, find a viewing window (or windows) that shows a complete graph of the function.

17. $f(x) = 2^{x^3 - x - 2}$ 18. $g(x) = \ln\left(\dfrac{x}{x - 2}\right)$

In Questions 19–22, sketch a complete graph of the function. Indicate all asymptotes clearly.

19. $g(x) = 2^x - 1$ 20. $f(x) = 2^{x-1}$

21. $h(x) = \ln(x + 4) - 2$ 22. $k(x) = \ln\left(\dfrac{x + 4}{x}\right)$

23. A computer software company claims the following function models the "learning curve" for their mathematical software.

$$P(t) = \dfrac{100}{1 + 48.2e^{-.52t}},$$

where t is measured in months and $P(t)$ is the average percent of the software program's capabilities mastered after t months.

(a) Initially, what percent of the program is mastered?
(b) After six months, what percent of the program is mastered?
(c) Roughly, when can a person expect to "learn the most in the least amount of time"?
(d) If the company's claim is true, how many months will it take to have completely mastered the program?

24. Compunote has offered you a starting salary of $60,000 with $1000 yearly raises. Calcuplay offers you an initial salary of $30,000 and a guaranteed 6% raise each year.

(a) Complete the following table for each company.

Year	Compunote
1	$60,000
2	$61,000
3	62000
4	63000
5	64000

Year	Calcuplay
1	$30,000
2	$31,800
3	33707
4	35736
5	37874

(b) For each company, write a function that gives your salary in terms of years employed.

(c) If you plan on staying with the company for only five years, which job should you take to earn the most money?
(d) In what year does the salary at Calcuplay exceed the salary at Compunote?

In Questions 25–30, translate the given exponential statement into an equivalent logarithmic one.

25. $e^{3.14} = 23$ 26. $\dfrac{e^{1.80001}}{2} = 3$

27. $e^{x-7y} = 4a - b$ 28. $e^{2a^2-9} = 3.16$

29. $10^{2.248} = 177$ 30. $10^{4x-1} = y$

In Questions 31–36, translate the given logarithmic statement into an equivalent exponential one.

31. $\ln 404 = 6.0014$ 32. $\ln(3x - 2y) = b$

33. $\ln(4xy) = 15t$ 34. $\log 3675 = 3.565258$

35. $\log_6(3x - 4) = y$ 36. $\log_a(uv) = w^2$

In Questions 37–40, evaluate the given expression without using a calculator.

37. $\ln e^{2/3}$ 38. $\ln \sqrt[4]{e^5}$

39. $e^{\ln(x/2)}$ 40. $e^{3\ln(3x^2-7y)}$

41. Simplify: $4\ln \sqrt[5]{x} + (1/5)\ln x$

42. Simplify: $\ln(e^{5e^2})^{-2} + 4e^2$

In Questions 43–45, write the given expression as a single logarithm.

43. $\ln 6a - 4\ln b - \ln 2a$

44. $\log_4 16x^2 + 2\log_4 y - 2$

45. $3\ln x - 4(\ln x^3 - 5\ln x)$

46. $\log(-.001) = ?$

47. $\log_{30} 900 = ?$

48. You are conducting an experiment about memory. The people who participate agree to take a test at the end of your course and every month thereafter for a period of two years. The average score for the group is given by the model

$$M(t) = 91 - 14\ln(t + 1) \qquad (0 \le t \le 24)$$

where t is time in months after the first test.

(a) What is the average score on the initial exam?
(b) What is the average score after three months?
(c) When will the average drop below 50%?
(d) Is the magnitude of the rate of memory loss greater in the first month after the course (from $t = 0$ to $t = 1$) or after the first year (from $t = 12$ to $t = 13$)?
(e) Hypothetically, if the model could be extended past $t = 24$ months, would it be possible for the average score to be 0%?

49. Which of the following statements are *true*?

(a) $\ln 10 = (\ln 2)(\ln 5)$ (b) $\ln (e/6) = \ln e + \ln 6$
(c) $\ln (1/7) + \ln 7 = 0$ (d) $\ln (-e) = -1$
(e) None of the above is true.

50. Which of the following statements are *false*?

(a) $10 (\log 5) = \log 50$ (b) $\log 100 + 3 = \log 10^5$
(c) $\log 1 = \ln 1$ (d) $\log 6 / \log 3 = \log 2$
(e) All of the above are false.

Use the following six graphs for Questions 51 and 52.

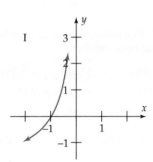

I

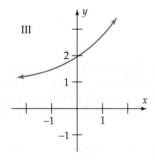

II

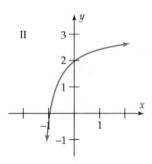

III

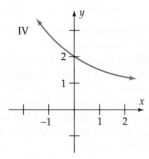

IV

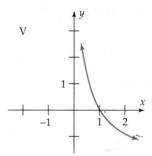

V

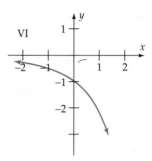

VI

51. If $b > 1$, then the graph of $f(x) = -\log_b x$ could possibly be:

(a) I   (b) IV
(c) V   (d) VI
(e) none of these

52. If $0 < b < 1$, then the graph of $g(x) = b^x + 1$ could possibly be:

(a) II   (b) III
(c) IV   (d) VI
(e) none of these

53. If $\log_4 16^{5x^2} = 160$, then what is x?

54. What is the domain of the function $f(x) = \ln \left(\dfrac{x}{x^2 - 1} \right)$?

In Questions 55–63, solve the equation for x.

55. $9^{9-x} = 3^{x^2-5x}$

56. $e^{5x} = 14$

57. $2 \cdot 27^x - 7 = -19/3$

58. $248e^{-3x} = 620$

59. $2a = 3b - c \ln x$

60. $4^x = 5^{2x+1}$

61. $\ln 5x + \ln (2x - 4) = \ln 30$

62. $\ln (2x - 5) - \ln 4x = 2$

63. $\log (4x^2 - 9) = 2 + \log (2x - 3)$

64. At a small community college the spread of a rumor through the population of 500 faculty and students can be modeled by.

$$\ln (n) - \ln (1000 - 2n) = .65t - \ln 998,$$

where n is the number of people who have heard the rumor after t days.

(a) How many people know the rumor initially? (at $t = 0$)
(b) How many people have heard the rumor after four days?

(c) Roughly, in how many weeks will the entire population have heard the rumor?

(d) Use the properties of logarithms to write n as a function of t; in other words solve the model above for n in terms of t.

(e) Enter the function you found in part (d) into your calculator and use the table feature to check your answers to parts (a), (b), and (c). Do they agree?

(f) Now graph the function. Roughly over what time interval does the rumor seem to "spread" the fastest?

65. The half-life of polonium (^{210}Po) is 140 days. If you start with 10 milligrams, how much will be left at the end of a year?

66. An insect colony grows exponentially from 100 to 1500 in 2 months time. If this growth pattern continues, how long will it take the insect population to reach 100,000?

67. Hydrogen-3 decays at a rate of 5.59% per year. Find its half-life.

68. The half-life of radium-88 is 1590 years. How long will it take for 10 grams to decay to 1 gram?

69. How much money should be invested at 4.5% per year, compounded quarterly, in order to have $5000 in 6 years?

70. At what annual rate should you invest your money if you want it to triple in 20 years (assume continuous compounding)?

71. One earthquake measures 4.6 on the Richter scale. A second earthquake is 1000 times more intense than the first. What does it measure on the Richter scale?

72. The table gives the population of Austin, Texas.*

Year	Population
1950	132,459
1970	253,539
1980	345,890
1990	465,622
2000	656,562

(a) Sketch a scatter plot of the data, with $x = 0$ corresponding to 1950.

(b) Find an exponential model for the data.

(c) Use the model to estimate the population of Austin in 1960 and 2005.

(d) It turns out that the 2004 population is 681,804. If we wanted to estimate the 2009 population, should we refine our exponential model, or should we decide that an exponential model isn't a good one? Why or why not?

73. The wind-chill factor is the temperature that would produce the same cooling effect on a person's skin if there were no wind. The table shows the wind-chill factors for various wind speeds when the temperature is 25°F.*

Wind Speed (mph)	Wind Chill Temperature (in °F)
0	25
5	19
10	15
15	13
20	11
25	9
30	8
35	7
40	6
45	5

(a) What does a 20-mph wind make 25°F feel like?

(b) Sketch a scatter plot of the data.

(c) Explain why an exponential model would be appropriate.

(d) Find an exponential model for the data.

(e) According to the model, what is the wind-chill factor for a 23-mph wind?

74. Cigarette consumption in the United States has been decreasing for some time, as shown in the table (in which the number of cigarettes each year is in billions).†

Year	Cigarettes
2000	565
2001	562
2002	532
2003	499
2004	494
2005	489

(a) Let $x = 10$ correspond to 1995 and find exponential, logarithmic, and power models for the data.

(b) Which of the three models do you think is most appropriate? Justify your answer.

(c) If you were a doctor, which model would you prefer for the future? Which one would you prefer if you manufactured cigarettes?

*National Weather Service.
†U.S. Dept. of Agriculture.

*U.S. Census Bureau.

Chapter 5 Test

Sections 5.1 and 5.2; Special Topics 5.1.A and 5.2.A

1. Rationalize the denominator. Assume all constants are positive, real numbers.

   (a) $\dfrac{1}{\sqrt{2+h}}$

   (b) $\dfrac{a-b}{\sqrt{a}+\sqrt{b}} \ (a \neq b)$

   (c) $\dfrac{3}{\sqrt{x}-4}$

2. Find all real solutions of each equation.

   (a) $\sqrt{x^3 + x^2 - 4x + 5} = x + 1$

   (b) $(x+1)^{2/3} = 4$

3. A population grows at the rate $P = P_0 e^{.01t}$ where t is in years, and P_0 is a positive constant.

   (a) When $t = 10$, the population is 1,000. Find P_0.

   (b) Use your answer to part (a) to find the initial population.

   (c) Use your answer to part (a) to find the population when $t = 20$.

4. The following is a graph of $y = Ae^{kt}$

   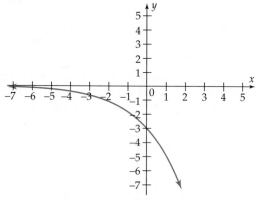

   (a) Is k positive or negative? How do you know?

   (b) Find the value of A

5. Simplify the following expressions. Assume all constants are positive, real numbers.

   (a) $\dfrac{10 + \sqrt{75}}{5}$

   (b) $\dfrac{3ab^2 - \sqrt[3]{a^4 b^7}}{a}$

   (c) $\dfrac{a^{1/2}(a^{3/2} b^{5/2})}{b^{-1/2}}$

6. In 1965, Gordon Moore (a cofounder of Intel) observed that the amount of computing power possible to put on a chip doubles every two years.* In 1990 there were about 1,000,000 transistors per chip. Assuming that Moore's law is true:

   (a) How many transistors per chip were there in 2000?

   (b) How many were there in 1985?

   (c) How many will there be in 2020?

   ---
   *Intel Corporation.

7. How much money will be in a savings account after one year if the initial deposit is $3000 and the interest rate is 8% compounded quarterly?

Sections 5.3–5.6; Special Topics 5.4.A

8. (a) $\log_5 625$ (b) $11^{\log_{11}(x+h)}$ (c) $\log_3(9^{1138})$

9. The half-life of Carbon 14 is 5730 years. If a sample of bone has lost 10% of its original C-14, roughly how old is the bone?

10. Simplify the following expressions

    (a) $\ln(e^{x^2 + y^2})$ (b) $e^{\ln(a) - 2\ln(b)}$

    (c) $\ln\left(\dfrac{1}{\sqrt[3]{e}}\right)$

11. The following is a graph of $y = A\log(x) + B$. Find A and B.

    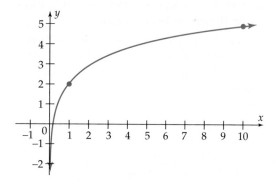

12. Simplify the following expressions, if possible. Assume all constants are positive real numbers

    (a) $\dfrac{\ln(x) + \ln(a)}{\ln(ax)}$ (b) $\ln\left(\dfrac{2}{b^4}\right) + 4\ln(b)$

    (c) $\ln(3 + x^2)$

13. A population of mayflies grows according to the equation $P = 230e^{.6t}$ where t is in days.

    (a) How many mayflies were there initially?

    (b) In how many days will there be 10,000 mayflies?

14. The following data show Madison's average score at the game of Fizzbin:

Month	Score
January	120.00
March	136.48
April	140.79
May	144.14
June	146.88

    (a) Find a logarithmic model ($y = a + b \ln x$) for the data

    (b) What does the model predict her score was for February? What will it be in December?

Exponential and Logistic Modeling of Diseases

Greg Pease / Getty Images

Diseases that are contagious and are transmitted homogeneously through a population often appear to be spreading exponentially. That is, the rate of spread is proportional to the number of people in the population who are already infected. This is a reasonable model as long as the number of infected people is relatively small in comparison with the number of people in the population who can be infected. The standard exponential model looks like this:

$$f(t) = Y_0 e^{rt}.$$

Y_0 is the initial number of infected people (the number on the arbitrarily decided day 0), and r is the rate by which the disease spreads through the population. If the time t is measured in days, then r is the ratio of new infections to current infections each day.

Suppose that in Big City, population 3855, there is an outbreak of dingbat disease. On the first Monday after the outbreak was discovered (day 0), 72 people have dingbat disease. On the following Monday (day 7), 193 people have dingbat disease.

1. Using the exponential model, Y_0 is clearly 72. Calculate the value of r.

2. Using your values of Y_0 and r, predict the number of cases of dingbat disease that will be reported on day 14.

It turns out that eventually, the spread of disease must slow as the number of infected people approaches the number of susceptible people. What happens is that some of the people to whom the disease would spread are already infected. As time goes on, the spread of the disease becomes proportional to the number of susceptible and uninfected people. The disease then follows the logistic model:

$$g(t) = \frac{rY_0}{aY_0 + (r - aY_0)e^{-rt}}.$$

Y_0 is still the initial value, and r serves the same function as before, at least at the initial time. The extra parameter a is not so obvious, but it is inversely related to the number of people susceptible to the disease. Unfortunately, the algebra to solve for a is quite complicated. It is much easier to approximate a using the same r from the exponential model.

3. On day 14 in Big City, 481 people have dingbat disease. Using the values of Y_0 and r from Exercise 1 in the rule of the function g, determine the value of a. [*Hint:* $g(14) = 481$.] Does g overestimate or underestimate the number of people with dingbat disease on day 7?

4. Use the function g from Exercise 3 to approximate the number of people in Big City who are susceptible to the disease. Does this model make sense? [Remember, as time goes on, the number of people infected approaches the number of people susceptible.]

DISCOVERY PROJECT 5

5. In the logistic model, the rate at which the disease spreads tends to fall over time. This means that the value of r you calculated in Exercise 1 is a little low. Raise the value of r and find the new value of a as in Exercise 3. Experiment until you find a value of r for which $g(7) = 193$ (meaning that the model g matches the data on day 7).

6. Using the function g from Exercise 5, repeat Exercise 4.

Chapter 6

TRIGONOMETRIC FUNCTIONS

Don't touch that dial!

Radio stations transmit by sending out a signal in the form of an electromagnetic wave that can be described by a trigonometric function. The shape of this signal is modified by the sounds being transmitted. AM radio signals are modified by varying the "height," or amplitude, of the waves, whereas FM signals are modified by varying the frequency of the waves. See Exercise 47 on page 487.

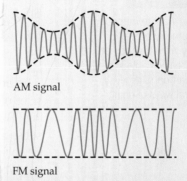

AM signal

FM signal

© Eduardo Garcia/Getty Images

Chapter Outline

Roadmap

Instructors who want to introduce triangle trigonometry before (or simultaneously with) trigonometric functions of a real variable should consult the chart on page xiv. It provides several ways of doing this.

The ancient Greeks developed trigonometry for measuring angles and sides of triangles to solve problems in astronomy, navigation, and surveying.[†] But with the invention of calculus in the seventeenth century and the subsequent explosion of knowledge in the physical sciences, a different viewpoint toward trigonometry arose.

Whereas the ancients dealt only with *angles*, the classical trigonometric concepts of sine and cosine are now considered as *functions of real numbers*. The advantage of this switch in viewpoint is that almost any phenomenon involving rotation or repetition can be described in terms of trigonometric functions, including light rays, sound waves, planetary orbits, weather, animal populations, radio transmission, guitar strings, pendulums, and many more.

The presentation of trigonometry here reflects this modern viewpoint. Nevertheless, angles still play an important role in defining the trigonometric functions, so the chapter begins with them.

6.1 Angles and Their Measurement

Section Objectives
- Use basic terminology to describe angles.
- Learn radian measure for angles
- Convert the measure of an angle from radians to degree and vice versa.

In trigonometry an **angle** is formed by rotating a half-line around its endpoint (the **vertex**), as shown in Figure 6–1, where the arrow indicates the direction of rotation. The position of the half-line at the beginning is the **initial side,** and its final position is the **terminal side** of the angle.

---

*Parts of Section 6.6 may be covered much earlier; see the Roadmap at the beginning of Section 6.2.
[†]In fact, "trigonometry" means "triangle measurement."

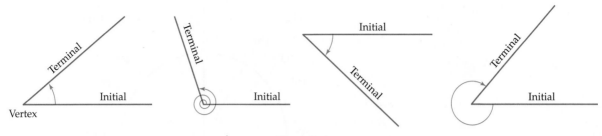

Figure 6-1

Figure 6–2 shows that different angles (that is, angles obtained by different rotations) may have the same initial and terminal side.* Such angles are said to be **coterminal.**

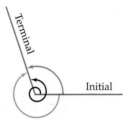

Figure 6-2

An angle in the coordinate plane is said to be in **standard position** if its vertex is at the origin and its initial side on the positive *x*-axis, as in Figure 6–3. When measuring angles in standard position, we use positive numbers for angles obtained by counterclockwise rotation (**positive angles**) and negative numbers for ones obtained by clockwise rotation (**negative angles**).

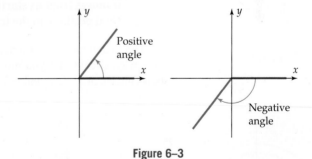

Figure 6-3

The classical unit for angle measurement is the **degree** (in symbols, °), as explained in the Geometry Review Appendix. You should be familiar with the positive angles in standard position shown in Figure 6–4 on the next page. Note that a 360° angle corresponds to one full revolution and thus is coterminal with an angle of 0°.

---

*They are *not* the same angle, however. For instance, both $\frac{1}{2}$ turn and $1\frac{1}{2}$ turns put a circular faucet handle in the same position, but the water flow is quite different.

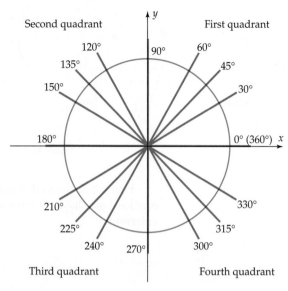

Figure 6–4

RADIAN MEASURE

Because it simplifies many formulas in calculus and physics, a different unit of angle measurement is used in mathematical and scientific applications. Recall that the unit circle is the circle of radius 1 with center at the origin; its equation is $x^2 + y^2 = 1$. When a positive angle in standard position is formed by rotating the initial side (the positive x-axis) counterclockwise, then the point $P = (1, 0)$ moves along the unit circle, as in Figure 6–5. The **radian measure** of the angle is defined to be

> **the distance traveled along the unit circle by the point P as it moves from its starting position on the initial side to its final position on the terminal side of the angle.**

The radian measure of a negative angle in standard position is found in the same way, except that you move clockwise along the unit circle. Figure 6–6 shows angles of 3.75, 7, and -2 radians, respectively.

Figure 6–5

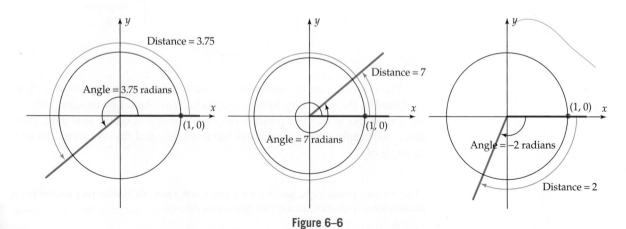

Figure 6–6

To become comfortable with radian measure, think of the terminal side of the angle revolving around the origin: When it makes one full revolution, it produces an angle of 2π radians (because the circumference of the unit circle is 2π). When it makes half a revolution, it forms an angle whose radian measure is $1/2$ of 2π, that is, π radians, and so on, as illustrated in Figure 6–7 and the table below.

1 revolution	3/4 revolution	1/2 revolution	$1\frac{1}{4}$ revolutions
2π radians	$\frac{3}{4} \cdot 2\pi = \frac{3\pi}{2}$ radians	$\frac{1}{2} \cdot 2\pi = \pi$ radians	$\frac{5}{4} \cdot 2\pi = \frac{5\pi}{2}$ radians

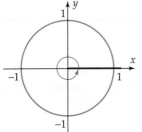 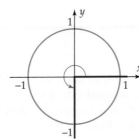 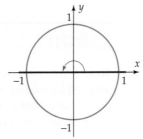 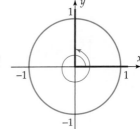

Figure 6–7

Terminal Side	Radian Measure of Angle	Equivalent Degree Measure
1 revolution	2π	$360°$
$\frac{7}{8}$ revolution	$\frac{7}{8} \cdot 2\pi = \frac{7\pi}{4}$	$\frac{7}{8} \cdot 360 = 315°$
$\frac{3}{4}$ revolution	$\frac{3}{4} \cdot 2\pi = \frac{3\pi}{2}$	$\frac{3}{4} \cdot 360 = 270°$
$\frac{2}{3}$ revolution	$\frac{2}{3} \cdot 2\pi = \frac{4\pi}{3}$	$\frac{2}{3} \cdot 360 = 240°$
$\frac{1}{2}$ revolution	$\frac{1}{2} \cdot 2\pi = \pi$	$\frac{1}{2} \cdot 360 = 180°$
$\frac{1}{3}$ revolution	$\frac{1}{3} \cdot 2\pi = \frac{2\pi}{3}$	$\frac{1}{3} \cdot 360 = 120°$
$\frac{1}{4}$ revolution	$\frac{1}{4} \cdot 2\pi = \frac{\pi}{2}$	$\frac{1}{4} \cdot 360 = 90°$
$\frac{1}{6}$ revolution	$\frac{1}{6} \cdot 2\pi = \frac{\pi}{3}$	$\frac{1}{6} \cdot 360 = 60°$
$\frac{1}{8}$ revolution	$\frac{1}{8} \cdot 2\pi = \frac{\pi}{4}$	$\frac{1}{8} \cdot 360 = 45°$
$\frac{1}{12}$ revolution	$\frac{1}{12} \cdot 2\pi = \frac{\pi}{6}$	$\frac{1}{12} \cdot 360 = 30°$

Although equivalent degree measures are given in the table, you should learn to "think in radians" as much as possible rather than mentally translating from radians to degrees.

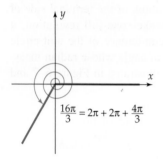

$$\frac{16\pi}{3} = 2\pi + 2\pi + \frac{4\pi}{3}$$

Figure 6–8

EXAMPLE 1

To construct an angle of $16\pi/3$ radians in standard position, note that

$$\frac{16\pi}{3} = \frac{6\pi}{3} + \frac{6\pi}{3} + \frac{4\pi}{3} = 2\pi + 2\pi + \frac{4\pi}{3}.$$

So the terminal side must be rotated counterclockwise through two complete revolutions (each full-circle revolution is 2π radians) and then rotated an additional $2/3$ of a revolution (since $4\pi/3$ is $2/3$ of a complete revolution of 2π radians), as shown in Figure 6–8. ■

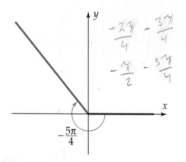

Figure 6–9

EXAMPLE 2

Since $-5\pi/4 = -\pi - \pi/4$, an angle of $-5\pi/4$ radians in standard position is obtained by rotating the terminal side *clockwise* for half a revolution (π radians) plus an additional $1/8$ of a revolution (since $\pi/4$ is $1/8$ of a full-circle revolution of 2π radians), as shown in Figure 6–9. ■

Consider an angle of t radians in standard position (Figure 6–10). Since 2π radians corresponds to a full revolution of the terminal side, this angle has the same terminal side as an angle of $t + 2\pi$ radians or $t - 2\pi$ radians or $t + 4\pi$ radians.

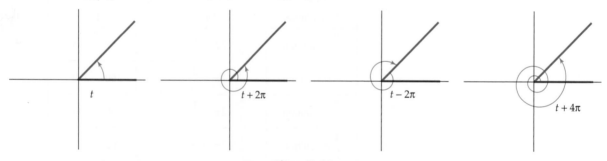

Figure 6–10

The same thing is true in general.

Coterminal Angles

Increasing or decreasing the radian measure of an angle by an integer multiple of 2π results in a coterminal angle.

EXAMPLE 3

Find angles in standard position that are coterminal with an angle of

(a) $23\pi/5$ radians

(b) $-\pi/12$ radians.

SOLUTION

(a) We can subtract 2π to obtain a coterminal angle whose measure is

$$\frac{23\pi}{5} - 2\pi = \frac{23\pi}{5} - \frac{10\pi}{5} = \frac{13\pi}{5} \text{ radians},$$

or we can subtract 4π to obtain a coterminal angle of measure

$$\frac{23\pi}{5} - 4\pi = \frac{3\pi}{5} \text{ radians}.$$

Subtracting 6π produces a coterminal angle of

$$\frac{23\pi}{5} - 6\pi = -\frac{7\pi}{5} \text{ radians}.$$

(b) An angle of $\frac{-\pi}{12}$ radians is coterminal with an angle of

$$-\frac{\pi}{12} + 2\pi = \frac{23\pi}{12} \text{ radians}$$

and with an angle of

$$-\frac{\pi}{12} - 2\pi = -\frac{25\pi}{12} \text{ radians}. \qquad ■$$

 ## RADIAN/DEGREE CONVERSION

Although we shall generally work with radians, it may occasionally be necessary to convert from radian to degree measure or vice versa. The key to doing this is the fact that

(∗) $$\pi \text{ radians} = 180°.$$

Dividing both sides of (∗) by π shows that

$$1 \text{ radian} = \frac{180}{\pi} \text{ degrees} \approx 57.3°,$$

and dividing both sides of (∗) by 180 shows that

$$1° = \frac{\pi}{180} \text{ radians} \approx .0175 \text{ radians}.$$

Consequently, we have these rules.

Radian/Degree Conversion

To convert radians to degrees, multiply by $\dfrac{180}{\pi}$.

To convert degrees to radians, multiply by $\dfrac{\pi}{180}$.

EXAMPLE 4

Find the degree measure of the angles with radian measure:

(a) 2.4 radians   (b) $\pi/60$ radians   (c) $-.3$ radians.

If your TI or Casio calculator is in radian mode, you can convert an angle from degrees to radians by using ° in the menu/submenu listed below.

 TI-84+: ANGLE

 TI-86/89: MATH/ANGLE

 Casio: OPTN/ANGLE

For example, keying in 180° and pressing ENTER, produces

$$3.141592654,$$

the decimal approximation of π.

  Similarly, if the calculator is in degree mode, you can convert an angle from radians to degrees by using r in the same menu.

  To make conversions on HP-39gs, use *DEG→RAD* or *RAD→DEG* in the MATH/REAL menu.

SOLUTION In each case, multiply the given radian measure by $\dfrac{180}{\pi}$.

(a) $2.4\left(\dfrac{180}{\pi}\right) = \dfrac{432}{\pi} \approx 137.51°.$

(b) $\dfrac{\pi}{60}\left(\dfrac{180}{\pi}\right) = \dfrac{180}{60} = 3°.$

(c) $(-.3)\left(\dfrac{180}{\pi}\right) = \dfrac{-54}{\pi} \approx -17.19°.$ ∎

EXAMPLE 5

Find the radian measure of the angles with degree measure:

(a) $12°$ (b) $-150°$ (c) $236°$

SOLUTION In each case, multiply by $\dfrac{\pi}{180}$.

(a) $12\left(\dfrac{\pi}{180}\right) = \dfrac{\pi}{15}$ radians.

(b) $-150\left(\dfrac{\pi}{180}\right) = \dfrac{-5\pi}{6}$ radians.

(c) $236\left(\dfrac{\pi}{180}\right) = \dfrac{59\pi}{45} \approx 4.12$ radians. ∎

EXERCISES 6.1

In Exercises 1–5, find the radian measure of the angle in standard position formed by rotating the terminal side by the given amount.

1. 1/9 of a circle

2. 1/24 of a circle

3. 1/18 of a circle

4. 1/72 of a circle

5. 1/36 of a circle

6. State the radian measure of *every* standard position angle in the figure.*

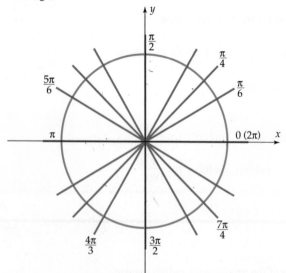

*This is the same diagram that appears in Figure 6–4 on page 430, showing positive angles in standard position.

In Exercises 7–10, estimate the radian measure of the angle.

7.

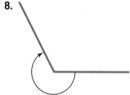

8.

9.

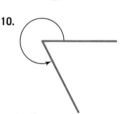

10.

In Exercises 11–14, find the radian measure of four angles in standard position that are coterminal with the angle in standard position whose measure is given.

11. $\pi/4$

12. $7\pi/5$

13. $-\pi/6$

14. $-9\pi/7$

In Exercises 15–18, determine whether or not the given angles in standard position are coterminal.

15. $\dfrac{5\pi}{12}, \dfrac{17\pi}{12}$

16. $\dfrac{7\pi}{6}, -\dfrac{5\pi}{6}$

17. $117°, 837°$

18. $170°, -550°$

In Exercises 19–26, find the radian measure of an angle in standard position that has measure between 0 and 2π and is coterminal with the angle in standard position whose measure is given.

19. $-\pi/3$ 20. $-3\pi/4$ 21. $19\pi/4$
22. $16\pi/3$ 23. $-7\pi/5$ 24. $45\pi/8$
25. 7 26. 18.5

In Exercises 27–38, convert the given degree measure to radians.

27. $6°$ 28. $-10°$ 29. $-12°$
30. $36°$ 31. $75°$ 32. $-105°$
33. $135°$ 34. $-165°$ 35. $-225°$
36. $252°$ 37. $930°$ 38. $-585°$

In Exercises 39–50, convert the given radian measure to degrees.

39. $\pi/5$ 40. $-\pi/6$ 41. $-\pi/10$
42. $2\pi/5$ 43. $3\pi/4$ 44. $-5\pi/3$
45. $\pi/45$ 46. $-\pi/60$ 47. $-5\pi/12$
48. $7\pi/15$ 49. $27\pi/5$ 50. $-41\pi/6$

In Exercises 51–56, determine the positive radian measure of the angle that the second hand of a clock traces out in the given time.

51. 40 seconds   52. 50 seconds
53. 35 seconds   54. 2 minutes and 15 seconds.
55. 3 minutes and 25 seconds   56. 1 minute and 55 seconds

In Exercises 57–64, a wheel is rotating around its axle. Find the angle (in radians) through which the wheel turns in the given time when it rotates at the given number of revolutions per minute (rpm). Assume that $t > 0$ and $k > 0$.

57. 3.5 minutes, 1 rpm   58. t minutes, 1 rpm
59. 1 minute, 2 rpm   60. 3.5 minutes, 2 rpm
61. 4.25 minutes, 5 rpm   62. t minutes, 5 rpm
63. 1 minute, k rpm   64. t minutes, k rpm

6.1.A SPECIAL TOPICS   Arc Length and Angular Speed

Section Objectives
- Find the length of an arc intercepted by a given angle.
- Find the area of a circular sector.
- Find the linear and angular speeds of an object moving around a circle.

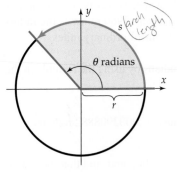

Figure 6–11

Consider a circle of radius r and an angle of θ radians, as shown in Figure 6–11. The sides of the angle of θ radians determine an arc of length s on the circle. We say that the **central angle** of θ radians **intercepts an arc** of length s on the circle (or that an arc of length s is intercepted by a central angle of θ radians).

It can be shown that the ratio of the arc length s to the circumference $2\pi r$ of the entire circle in Figure 6–11 is the same as the ratio of the central angle of θ radians to the full-circle angle of 2π radians; that is,

$$\frac{s}{2\pi r} = \frac{\theta}{2\pi}.$$

Solving this equation for s, we obtain the following fact.

Arc Length

A central angle of θ radians in a circle of radius r intercepts an arc of length

$$s = \theta r,$$

that is, the length s of the arc is the product of the radian measure of the angle and the radius of the circle.

EXAMPLE 1

(a) Find the length of an arc of a circle of radius 8 ft that is intercepted by a central angle of 2.3 radians.

(b) Find the length of an arc of a circle of radius 12 inches that is intercepted by a central angle of 60°.

SOLUTION

(a) We apply the arc length formula with $\theta = 2.3$ and $r = 8$:

$$s = \theta r = (2.3)8 = 18.4 \text{ ft}.$$

(b) The arc length formula requires that the angle be in radians. Since 60° is $\pi/3$ radians, we have

$$s = \theta r = \frac{\pi}{3} \cdot 12 = 4\pi \approx 12.57 \text{ inches}. \qquad \blacksquare$$

EXAMPLE 2

A central angle intercepts an arc of length 14.64 m on a circle of radius 6 m. Find the radian measure of the angle.

SOLUTION
We solve the arc length formula $s = \theta r$ for θ:

$$\theta = \frac{s}{r} = \frac{14.64}{6} = 2.44 \text{ radians}. \qquad \blacksquare$$

EXAMPLE 3

Figure 6–12

A **nautical mile** is defined to be the length of an arc on the surface of the earth that is intercepted by an angle θ of measure 1/60 of 1° whose vertex is at the center of the earth (see Figure 6–12, which is not to scale). Using 3960 miles as the radius of the earth, find the length of a nautical mile in terms of ordinary miles.

SOLUTION
First we express the angle θ in radians by using the formula in the box on page 433:

$$\left(\frac{1}{60}\right)^{\circ} = \frac{1}{60} \cdot \frac{\pi}{180} \text{ radians} = \frac{\pi}{10{,}800} \text{ radians} \approx .0002908882 \text{ radians}.$$

Now we use the arc length formula in the preceding box and find that the length of a nautical mile is

$$s = \theta r = .0002908882(3960) \approx 1.15 \text{ miles}. \qquad \blacksquare$$

EXAMPLE 4

The second hand on a large clock is 6 inches long. How far does the tip of the second hand move in 15 seconds?

SOLUTION   The second hand makes a full revolution every 60 seconds, that is, it moves through an angle of 2π radians. During a 15-second interval it will make $\dfrac{15}{60} = \dfrac{1}{4}$ of a revolution, moving through an angle of $\pi/2$ radians (Figure 6–13). If we think of the second hand as the radius of a circle, then during a 15-second interval, its tip travels along the arc intercepted by an angle of $\pi/2$ radians. Therefore, the distance (arc length) traveled by the tip of the second hand is

$$s = \theta r = \frac{\pi}{2} \cdot 6 = 3\pi \approx 9.425 \text{ inches.} \qquad \blacksquare$$

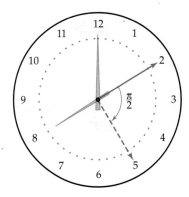

Figure 6–13

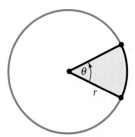

Figure 6–14

AREA OF A CIRCULAR SECTOR

Look at the shaded sector with a central angle of θ radians in Figure 6–14. Since a full circle angle is 2π, the area A of the shaded sector is $\dfrac{\theta}{2\pi}$ of the area of the entire circular disc (which is, as usual, πr^2). Hence,

$$A = \frac{\theta}{2\pi} (\pi r^2) = \frac{\theta r^2}{2} = \frac{1}{2} r^2 \theta.$$

Area of
a Sector

> The area A of a sector with central angle θ radians in a circle of radius r is
>
> $$A = \frac{1}{2} r^2 \theta.$$

EXAMPLE 5

In a circle of radius 5 ft, find the area of a sector with central angle

(a) 2.4 radians   (b) 45°.

SOLUTION

(a) Using the area formula in the preceding box, we have

$$A = \frac{1}{2} r^2 \theta = \frac{1}{2} \cdot 5^2 \cdot 2.4 = \frac{60}{2} = 30 \text{ square feet.}$$

(b) The formula is valid only when the angle is expressed in radians. Since 45° is $\pi/4$, the area is

$$A = \frac{1}{2}r^2\theta = \frac{1}{2} \cdot 5^2 \cdot \left(\frac{\pi}{4}\right) = \frac{25\pi}{8} \approx 9.8175 \text{ square feet.}\quad\blacksquare$$

ANGULAR SPEED

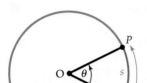

Figure 6–15

Suppose a point P moves along a circle of radius r at a constant rate, tracing out an arc of length s, as in Figure 6–15. The **linear speed** of P is the rate at which the distance traveled by P along the circle changes with respect to time, namely,

$$\frac{\text{Distance traveled by } P}{\text{Time elapsed}} = \frac{s}{t}.$$

The **angular speed** of P is the rate at which the angle θ changes with respect to time:

$$\frac{\text{Angle traced out by } OP}{\text{Time elapsed}} = \frac{\theta}{t}.$$

The linear speed of P is denoted by v and the angular speed by ω (the lower case Greek letter omega). Using the arc length formula $s = \theta r$, we have this summary:

Linear and Angular Speed

Suppose a point P moves along a circle of radius r and center O for a period of time t. If the arc traveled by P has length s and the angle traced out by OP measures θ radians, then

$$v = \frac{s}{t} = \frac{\theta r}{t} \qquad \text{and} \qquad \omega = \frac{\theta}{t}.$$

Linear Speed $\qquad\qquad$ Angular Speed

EXAMPLE 6

A merry-go-round makes eight revolutions per minute.

(a) What is the angular speed of the merry-go-round in radians per minute?
(b) How fast is a horse 12 feet from the center traveling?
(c) How fast is a horse 4 feet from the center traveling?

SOLUTION

(a) Each revolution of the merry-go-round corresponds to a central angle of 2π radians, so it travels through an angle of $8 \cdot 2\pi = 16\pi$ radians in one minute. Therefore, its angular speed is

$$\omega = \frac{\theta}{t} = \frac{16\pi}{1} = 16\pi \text{ radians per minute.}$$

(b) The horse that is 12 feet from the center travels along a circle of radius 12 and travels through an angle of 16π radians in 1 minute. Therefore, its linear speed is

$$v = \frac{\theta r}{t} = \frac{16\pi \cdot 12}{1} = 192\pi \text{ feet per minute,}$$

which is approximataely 6.9 miles per hour.

(c) For the horse that is 4 feet from the center, $r = 4$, $\theta = 16\pi$, and $t = 1$. Hence, the linear speed is

$$v = \frac{\theta r}{t} = \frac{16\pi \cdot 4}{1} = 64\pi \text{ feet per minute (about 2.28 mph).} \qquad \blacksquare$$

By rewriting the linear speed equation, we obtain the relationship between linear and angular speed.

$$v = \frac{s}{t} = \frac{\theta r}{t} = r \cdot \frac{\theta}{t} = r\omega.$$

Linear and Angular Speed

If a point P moves along a circle of radius r with linear speed v and angular speed ω, then

$$v = r\omega.$$

EXAMPLE 7

A bicycle, each of whose wheels is 26 inches in diameter, travels at a constant 14 mph.

(a) What is the angular speed of one of its wheels (in radians per hour)?

(b) How many revolutions per minute does each wheel make?

SOLUTION   Within each computation, we must be careful to use comparable units.

(a) Since the wheel size is given in inches, we also express the speed in terms of inches. Recall that there are 5280 feet in a mile and 12 inches in a foot, which means that

$$1 \text{ mile} = 5280 \cdot 12 = 63{,}360 \text{ inches.}$$

Thus, the bike's linear speed is

$$v = 14 \text{ miles per hour} = 14(63{,}360) \text{ inches per hour.}$$

The radius of the wheel is $\frac{26}{2} = 13$ inches. As we saw above, $v = r\omega$, so the angular speed is

$$\omega = \frac{v}{r} = \frac{14(63{,}360)}{13} \approx 68{,}233.85 \text{ radians per hour.}$$

(b) Each revolution corresponds to an angle of 2π radians, so the number of revolutions per hour is $\dfrac{68{,}233.85}{2\pi}$. Dividing this number by 60 gives the revolutions per minute.

$$\frac{1}{60}\left(\frac{68{,}233.85}{2\pi}\right) \approx 181 \text{ revolutions per minute.} \qquad \blacksquare$$

EXERCISES 6.1.A

In Exercises 1–4, find the length of the circular arc subtended by the central angle whose radian measure is given. Assume the circle has diameter 10.

1. 1 radian
2. 2 radians
3. 1.75 radians
4. 2.2 radians

5. The second hand on a clock is 6 centimeters long. How far does its tip travel in 40 seconds? (See Exercise 51 in section 6.1.)

6. The second hand on a clock is 5 centimeters long. How far does its tip travel in 2 minutes and 15 seconds? (See Exercise 54 in section 6.1.)

In Exercises 7–10, find the radian measure of the angle θ.

7.
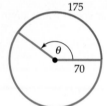

8.
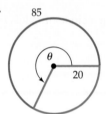

9.
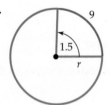

10.
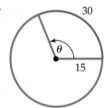

In Exercises 11–14, an arc length and angle (in radian measure) are given. Find the radius r of the circle.

11.
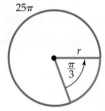

12.
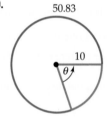

13.
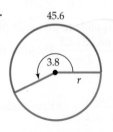

14.
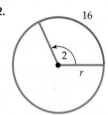

In Exercises 15–18, assume that a wheel on a car has radius 36 centimeters. Find the angle (in radians) that the wheel turns while the car travels the given distance.

15. 2 meters (= 200 centimeters)
16. 5 meters
17. 720 meters
18. 1 kilometer (= 1000 meters)

*In Exercises 19–22, the latitudes of a pair of cities are given. Assume that one city is directly south of the other and that the earth is a perfect sphere of radius 3960 miles. Find the distance between the two cities. [The **latitude** of a point P on the earth is the degree measure of the angle θ between the point and the plane of the equator (with the center of the earth being the vertex), as shown in the figure. Remember that angles are measured in radians in the arc length formula.]*

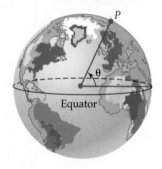

19. The North Pole (latitude 90° north) and Springfield, Illinois (latitude 40° north).

20. San Antonio, Texas (latitude 29.5° north), and Mexico City, Mexico (latitude 20° north).

21. Cleveland, Ohio (latitude 41.5° north), and Tampa, Florida (latitude 28° north).

22. Copenhagen, Denmark (latitude 54.3° north), and Rome, Italy (latitude 42° north).

In Exercises 23–26, find the area of the sector with the given central angle and the given radius.

23. Angle 1 radian; radius 5 ft
24. Angle 1.8 radians; radius 10 in
25. Angle 2.4 radians; radius 8 m
26. Angle 4 radians; radius 9 cm

27. If a sector of a circle of radius 2 ft has an area of 4 sq ft, find the radian measure of the central angle.

28. If a sector of a circle of radius 5 m has an area of 37.5 sq m, find the radian measure of the central angle.

29. If the area of a sector of a circle with a central angle of 1.5 radians is 4.32 sq ft, find the radius of the circle.

30. If the area of a sector of a circle with a central angle of 2.2 radians is 23.276 sq in, find the radius of the circle.

31. One day in Syene (modern day Aswan), the Greek mathematician Eratosthenes (c. 276–194 B.C.E.) noted that the sun's rays shown directly down a deep well. One year later at the same time in Alexandria (which is 500 miles due north of Syene), he determined that the sun's rays were at an angle of approximately 7.2°, as shown in the figure. From this information Eratosthenes was able to find the approximate radius and circumference of the earth. You should do the same.

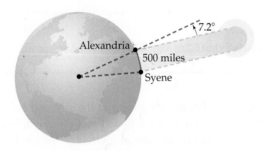

32. A person is standing on the equator. The earth is making one full revolution each 24 hours. Assuming that the radius of the earth is 3960 miles, find the person's approximate

(a) linear velocity (in mph)
(b) angular velocity (in radians per hour)

33. One end of a rope is attached to a winch (circular drum) of radius 2 feet and the other to a steel beam on the ground. When the winch is rotated, the rope wraps around the drum and pulls the object upward (see the figure). Through what angle must the winch be rotated to raise the beam 6 feet above the ground?

34. A circular saw blade has an angular speed of 15,000 radians per minute.

(a) How many revolutions per minute does the saw make?
(b) How long will it take the saw to make 6000 revolutions?

35. A circular gear rotates at the rate of 200 revolutions per minute (rpm).

(a) What is the angular speed of the gear in radians per minute?
(b) What is the linear speed of a point on the gear 2 inches from the center in inches per minute and in feet per minute?

36. A wheel in a large machine is 2.8 feet in diameter and rotates at 1200 rpm.

(a) What is the angular speed of the wheel?
(b) How fast is a point on the circumference of the wheel traveling in feet per minute? In miles per hour?

37. A riding lawn mower has wheels that are 15 inches in diameter. If the wheels are making 2.5 revolutions per second.

(a) What is the angular speed of a wheel?
(b) How fast is the lawn mower traveling in miles per hour?

38. A weight is attached to the end of a 4-foot-long leather cord. A boy swings the cord in a circular motion over his head. If the weight makes 12 revolutions every 10 seconds, what is its linear speed?

39. A merry-go-round horse is traveling at 10 feet per second, and the merry-go-round is making 6 revolutions per minute. How far is the horse from the center of the merry-go-round?

40. The pedal sprocket of a bicycle has radius 4.5 inches, and the rear wheel sprocket has radius 1.5 inches (see figure). If the rear wheel has a radius of 13.5 inches and the cyclist is pedaling at the rate of 80 rpm, how fast is the bicycle traveling in feet per minute? In miles per hour?

6.2   The Sine, Cosine, and Tangent Functions

Section Objectives

- Use the unit circle to define the sine, cosine and tangent functions.
- Find the exact values of sine, cosine, and tangent at $\pi/3$, $\pi/4$, $\pi/6$, and integer multiples of these numbers.
- Use the point-in-the-plane description to evaluate the trigonometric functions.

NOTE

If you have read Chapter 8, use Alternate Section 6.2 on page 452 in place of this section.

Roadmap

Instructors who wish to cover all six trigonometric functions simultaneously should incorporate Section 6.6 into Sections 6.2–6.4, as follows.

Subsection of Section 6.6	Cover at the end of
Part I	Section 6.2
Part II	Section 6.3
Part III	Section 6.4

Instructors who prefer to introduce triangle trigonometry early should consult the chart on page xiv.

Unlike most of the functions seen thus far, the definitions of the sine and cosine functions do not involve algebraic formulas. Instead, these functions are defined geometrically, using the unit circle.* Recall that the unit circle is the circle of radius 1 with center at the origin, whose equation is $x^2 + y^2 = 1$.

Both the sine and cosine functions have the set of all real numbers as domain. Their rules are given by the following three-step geometric process:

1. Given a real number t, construct an angle of t radians in standard position.

2. Find the coordinates of the point P where the terminal side of this angle meets the unit circle $x^2 + y^2 = 1$, say $P = (a, b)$.

3. The value of the **cosine function** at t (denoted cos t) is the x-coordinate of P:

$$\cos t = a.$$

The value of the **sine function** at t (denoted sin t) is the y-coordinate of P:

$$\sin t = b.$$

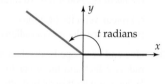

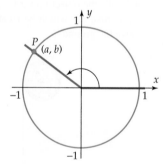

Sine and Cosine

If P is the point where the terminal side of an angle of t radians in standard position meets the unit circle, then

$$P \text{ has coordinates } (\cos t, \sin t).$$

---

*If you have previously studied the trigonometry of triangles, the definition given here may not look familiar. If this is the case, just concentrate on this definition and don't worry about relating it to any definition you remember from the past. The connection between this definition and the trigonometry of triangles will be explained in Chapter 8.

With your calculator in radian mode and parametric graphing mode, set the range values as follows:

$$0 \le t \le 2\pi, \qquad -1.8 \le x \le 1.8, \qquad -1.2 \le y \le 1.2*$$

Then graph the curve given by the parametric equations

$$x = \cos t \quad \text{and} \quad y = \sin t.$$

The graph is the unit circle. Use the trace to move around the circle. At each point, the screen will display three numbers: the values of t, x, and y. For each t, the cursor is on the point where the terminal side of an angle of t radians meets the unit circle, so the corresponding x is the number $\cos t$ and the corresponding y is the number $\sin t$.

The **tangent function** is defined as the quotient of the sine and cosine functions. Its value at the number t, denoted $\tan t$, is given by

$$\tan t = \frac{\sin t}{\cos t}.$$

EXAMPLE 1

Evaluate the three trigonometric functions at

(a) $t = \pi$ (b) $t = \pi/2$.

SOLUTION

(a) Construct an angle of π radians, as in Figure 6–16. Its terminal side lies on the negative x-axis and intersects the unit circle at $P = (-1, 0)$. Hence,

$\sin \pi = y$-coordinate of $P = 0$

$\cos \pi = x$-coordinate of $P = -1$

$\tan \pi = \dfrac{\sin \pi}{\cos \pi} = \dfrac{0}{-1} = 0$

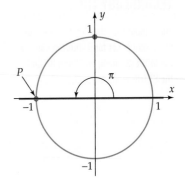

Figure 6–16

---

*Parametric graphing is explained in *Special Topics 3.3.A*. These settings give a square viewing window on calculators with a screen measuring approximately 95 by 63 pixels (such as TI-84+), and hence the unit circle will look like a circle. For wider screens, adjust the x range settings to obtain a square window.

(b) An angle of $\pi/2$ radians (Figure 6–17) has its terminal side on the positive y-axis and intersects the unit circle at $P = (0, 1)$.

$$\cos \frac{\pi}{2} = x\text{-coordinate of } P = 0$$

$$\sin \frac{\pi}{2} = y\text{-coordinate of } P = 1$$

$$\tan \frac{\pi}{2} = \frac{\sin(\pi/2)}{\cos(\pi/2)} = \frac{1}{0} \text{ undefined}$$

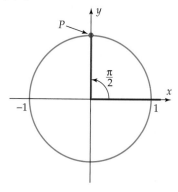

Figure 6–17   ∎

The definitions of sine and cosine show that $\sin t$ and $\cos t$ are defined for every real number t. Example 1(b), however, shows that $\tan t$ is *not* defined when the x-coordinate of the point P is 0. This occurs when P has coordinates $(0, 1)$ or $(0, -1)$, that is, when $t = \pm\pi/2, \pm3\pi/2, \pm5\pi/2$, etc. Consequently, the domain (set of inputs) of each trigonometric function is as follows.

Function	Domain
$f(t) = \sin t$	All real numbers
$g(t) = \cos t$	All real numbers
$h(t) = \tan t$	All real numbers except $\pm\pi/2, \pm3\pi/2, \pm5\pi/2, \ldots$

In most cases, evaluating trigonometric functions is not as simple as in Example 1. Usually, you must use the SIN, COS, and TAN keys on a calculator (in radian mode) to approximate $\sin t$, $\cos t$, and $\tan t$, as illustrated in Figure 6–18.

```
sin(2.5)
          .5984721441
cos(-6)
          .9601702867
tan(15)
          -.8559934009
■
```

Figure 6–18

EXAMPLE 2

When a baseball is hit by a bat, the horizontal distance d traveled by the ball is approximated by

$$d = \frac{v^2 \sin t \cos t}{16},$$

where the ball leaves the bat at an angle of t radians and has initial velocity v feet per second, as shown in Figure 6–19.

(a) How far does the ball travel when the initial velocity is 90 feet per second and $t = .7$?

(b) If the initial velocity is 105 feet per second and $t = 1$, how far does the ball travel?

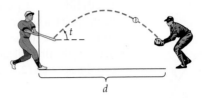

Figure 6–19

SOLUTION

(a) Let $v = 90$ and $t = .7$ in the formula for d. Then a calculator (in radian mode) shows that

$$d = \frac{v^2 \sin t \cos t}{16} = \frac{90^2 \sin .7 \cos .7}{16} \approx 249.44 \text{ feet.}$$

(b) Now let $v = 105$ and $t = 1$. Then

$$d = \frac{v^2 \sin t \cos t}{16} = \frac{105^2 \sin 1 \cos 1}{16} \approx 313.28. \qquad \blacksquare$$

SPECIAL VALUES

The trigonometric functions can be evaluated exactly at $t = \pi/3$, $t = \pi/4$, $t = \pi/6$, and any integer multiples of these numbers by using the following facts (which are explained in Examples 2–4 of the Geometry Review Appendix):*

A right triangle with hypotenuse 1 and angles of $\pi/6$ and $\pi/3$ radians has sides of lengths $1/2$ (opposite the angle of $\pi/6$) and $\sqrt{3}/2$ (opposite the angle of $\pi/3$).

A right triangle with hypotenuse 1 and two angles of $\pi/4$ radians has two sides of length $\sqrt{2}/2$.

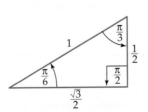

Figure 6–20

*Angles in the Geometry Review are given in degree measure: $60°, 45°, 30°$ instead of radian measure $\pi/3$, $\pi/4$, $\pi/6$, as is done here.

EXAMPLE 3

Evaluate the three trigonometric functions at $t = \pi/6$.

SOLUTION   Construct an angle of $\pi/6$ radians in standard position and let P be the point where its terminal side intersects the unit circle. Draw a vertical line from P to the x-axis, as shown in Figure 6–21, forming a right triangle that matches the first triangle in Figure 6–20. The sides of this triangle show that P has coordinates $(\sqrt{3}/2, 1/2)$. By the definition,

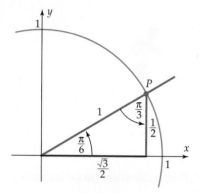

Figure 6–21

$$\sin \frac{\pi}{6} = y\text{-coordinate of } P = \frac{1}{2}$$

$$\cos \frac{\pi}{6} = x\text{-coordinate of } P = \frac{\sqrt{3}}{2}$$

$$\tan \frac{\pi}{6} = \frac{\sin(\pi/6)}{\cos(\pi/6)} = \frac{1/2}{\sqrt{3}/2} = \frac{1}{\sqrt{3}} = \frac{\sqrt{3}}{3}. \qquad \blacksquare$$

EXAMPLE 4

Evaluate the trigonometric functions at $t = \pi/4$.

SOLUTION   Construct an angle of $\pi/4$ radians in standard position whose terminal side intersects the unit circle at P. Draw a vertical line from P to the x-axis to form a right triangle that matches the second triangle in Figure 6–20. As Figure 6–22 shows, P has coordinates $(\sqrt{2}/2, \sqrt{2}/2)$ so that

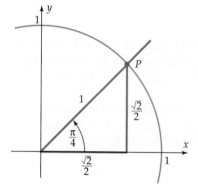

Figure 6–22

$$\sin \frac{\pi}{4} = y\text{-coordinate of } P = \frac{\sqrt{2}}{2}$$

$$\cos \frac{\pi}{4} = x\text{-coordinate of } P = \frac{\sqrt{2}}{2}$$

$$\tan \frac{\pi}{4} = \frac{\sin(\pi/6)}{\cos(\pi/6)} = \frac{\sqrt{2}/2}{\sqrt{2}/2} = 1. \qquad \blacksquare$$

EXAMPLE 5

Evaluate the trigonometric functions at $-5\pi/4$.

SOLUTION   Construct an angle of $-5\pi/4$ radians in standard position and let P be the point where the terminal side intersects the unit circle. Draw a vertical line from P to the x-axis, as shown in Figure 6–23, forming a right triangle that matches

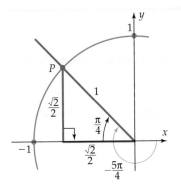

Figure 6–23

the second triangle in Figure 6–20. The sides of the triangle in Figure 6–23 show that P has coordinates $(-\sqrt{2}/2, \sqrt{2}/2)$. Hence,

$$\sin \frac{-5\pi}{4} = y\text{-coordinate of } P = \frac{\sqrt{2}}{2}$$

$$\cos \frac{-5\pi}{4} = x\text{-coordinate of } P = -\frac{\sqrt{2}}{2}$$

$$\tan \frac{-5\pi}{4} = \frac{\sin t}{\cos t} = \frac{\sqrt{2}/2}{-\sqrt{2}/2} = -1. \qquad \blacksquare$$

EXAMPLE 6

Evaluate the trigonometric functions at $11\pi/3$.

SOLUTION   Construct an angle of $11\pi/3$ radians in standard position and draw a vertical line from the x-axis to the point P where the terminal side of the angle meets the unit circle, as shown in Figure 6–24. The right triangle formed in this way matches the first triangle in Figure 6–20. The sides of the triangle in Figure 6–24 show that the coordinates of P are $(1/2, -\sqrt{3}/2)$. Therefore,

$$\sin \frac{11\pi}{3} = y\text{-coordinate of } P = -\frac{\sqrt{3}}{2}$$

$$\cos \frac{11\pi}{3} = x\text{-coordinate of } P = \frac{1}{2}$$

$$\tan \frac{11\pi}{3} = \frac{(\sin 11\pi/3)}{(\cos 11\pi/3)} = \frac{-\sqrt{3}/2}{1/2} = -\sqrt{3}. \qquad \blacksquare$$

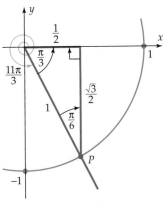

Figure 6–24

POINT-IN-THE-PLANE DESCRIPTION OF TRIGONOMETRIC FUNCTIONS

In evaluating $\sin t$, $\cos t$ and $\tan t$, from the definition, we use the point where the unit circle intersects the terminal side of an angle of t radians in standard position. Here is an alternative method of evaluating the trigonometric functions that uses *any* point on the terminal side of the angle (except the origin); it is proved at the end of this section.

Point-in-the-Plane Description

Let t be a real number. Let (x, y) be any point (except the origin) on the terminal side of an angle of t radians in standard position. Then,

$$\sin t = \frac{y}{r} \qquad \cos t = \frac{x}{r} \qquad \tan t = \frac{y}{x}$$

where $r = \sqrt{x^2 + y^2}$ is the distance from (x, y) to the origin.

Figure 6-25

EXAMPLE 7

Figure 6–25 shows an angle of t radians in standard position. Evaluate the three trigonometric functions at t.

SOLUTION   Apply the facts in the box above with $(x, y) = (3, -4)$ and

$$r = \sqrt{x^2 + y^2} = \sqrt{3^2 + (-4)^2} = \sqrt{25} = 5.$$

Then we have

$$\sin t = \frac{y}{r} = \frac{-4}{5} = -\frac{4}{5}, \qquad \cos t = \frac{x}{r} = \frac{3}{5}, \qquad \tan t = \frac{y}{x} = \frac{-4}{3} = -\frac{4}{3}. \qquad ■$$

EXAMPLE 8

The terminal side of a first-quadrant angle of t radians in standard position lies on the line with equation $2x - 3y = 0$. Evaluate the three trigonometric functions at t.

SOLUTION   Verify that the point $(3, 2)$ satisfies the equation and hence lies on the terminal side of the angle (Figure 6–26). Now we have

$$(x, y) = (3, 2) \qquad \text{and} \qquad r = \sqrt{x^2 + y^2} = \sqrt{3^2 + 2^2} = \sqrt{13}.$$

Therefore,

$$\sin t = \frac{y}{r} = \frac{2}{\sqrt{13}}, \qquad \cos t = \frac{x}{r} = \frac{3}{\sqrt{13}}, \qquad \tan t = \frac{y}{x} = \frac{2/\sqrt{13}}{3/\sqrt{13}} = \frac{2}{3}. \qquad ■$$

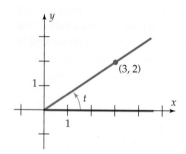

Figure 6-26

Proof of the Point-in-the-Plane Description   Let Q be the point on the terminal side of the standard position angle of t radians and let P be the point where the terminal side meets the unit circle, as in Figure 6–27. The definition of sine and cosine shows that P has coordinates $(\cos t, \sin t)$. The distance formula shows that the segment OQ has length $\sqrt{x^2 + y^2}$, which we denote by r.

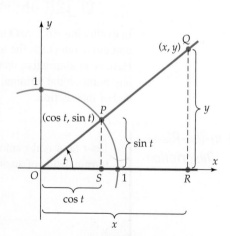

Figure 6-27

Both triangles QOR and POS are right triangles containing an angle of t radians. Therefore, these triangles are *similar*.* Consequently,

$$\frac{\text{length } OP}{\text{length } OQ} = \frac{\text{length } PS}{\text{length } QR} \quad \text{and} \quad \frac{\text{length } OP}{\text{length } OQ} = \frac{\text{length } OS}{\text{length } OR}.$$

Figure 6–23 shows what each of these lengths is. Hence,

$$\frac{1}{r} = \frac{\sin t}{y} \qquad \text{and} \qquad \frac{1}{r} = \frac{\cos t}{x}$$

$$r \sin t = y \qquad\qquad r \cos t = x$$

$$\sin t = \frac{y}{r} \qquad\qquad \cos t = \frac{x}{r}.$$

Similar arguments work when the terminal side is not in the first quadrant. In every case, $\tan t = \dfrac{\sin t}{\cos t} = \dfrac{y/r}{x/r} = \dfrac{y}{x}$. This completes the proof of the statements in the box on page 447. ■

---

*See the Geometry Review Appendix for the basic facts about similar triangles.

EXERCISES 6.2

Note: *Unless stated otherwise, all angles are in standard position.*

In Exercises 1–10, use the definition (not a calculator) to find the function value.

1. $\sin(3\pi/2)$ **2.** $\sin(-\pi)$ **3.** $\cos(3\pi/2)$

4. $\cos(-\pi/2)$ **5.** $\tan(4\pi)$ **6.** $\tan(-\pi)$

7. $\cos(-3\pi/2)$ **8.** $\sin(9\pi/2)$ **9.** $\cos(-11\pi/2)$

10. $\tan(-13\pi)$

In Exercises 11–14, assume that the terminal side of an angle of t radians passes through the given point on the unit circle. Find $\sin t$, $\cos t$, $\tan t$.

11. $(-2/\sqrt{5}, 1/\sqrt{5})$ **12.** $(1/\sqrt{10}, -3/\sqrt{10})$

13. $(-3/5, -4/5)$ **14.** $(.6, -.8)$

In Exercises 15–29, find the exact value of the sine, cosine, and tangent of the number, without using a calculator.

15. $\pi/3$ **16.** $2\pi/3$ **17.** $7\pi/4$

18. $5\pi/4$ **19.** $3\pi/4$ **20.** $-7\pi/3$

21. $5\pi/6$ **22.** 3π **23.** $-23\pi/6$

24. $11\pi/6$ **25.** $-19\pi/3$ **26.** $-10\pi/3$

27. $-15\pi/4$ **28.** $-25\pi/4$ **29.** $-17\pi/2$

30. Fill the blanks in the following table. Write each entry as a fraction with denominator 2 and with a radical in the numerator. For example,

$$\sin\frac{\pi}{2} = 1 = \frac{\sqrt{4}}{2}.$$

Some students find the resulting pattern an easy way to remember these functional values.

t	0	$\pi/6$	$\pi/4$	$\pi/3$	$\pi/2$
$\sin t$					
$\cos t$					

In Exercises 31–36, write the expression as a single real number. Do not use decimal approximations. [Hint: Exercises 15–21 may be helpful.]

31. $\sin(\pi/3)\cos(\pi) + \sin(\pi)\cos(\pi/3)$

32. $\sin(\pi/6)\cos(\pi/2) - \cos(\pi/6)\sin(\pi/2)$

33. $\cos(\pi/2)\cos(\pi/4) - \sin(\pi/2)\sin(\pi/4)$

34. $\cos(2\pi/3)\cos(\pi) + \sin(2\pi/3)\sin(\pi)$

35. $\sin(3\pi/4)\cos(5\pi/6) - \cos(3\pi/4)\sin(5\pi/6)$

36. $\sin(-7\pi/3)\cos(5\pi/4) + \cos(-7\pi/3)\sin(5\pi/4)$

In Exercises 37–42, find sin t, cos t, tan t when the terminal side of an angle of t radians in standard position passes through the given point.

37. $(3, 5)$ **38.** $(-2, 1)$

39. $(-4, -5)$ **40.** $(3, -4)$

41. $(\sqrt{3}, -8)$ **42.** $(-2, \pi)$

In Exercises 43–46, use a calculator in radian mode.

43. When a plane flies faster than the speed of sound, the sound waves it generates trail the plane in a cone shape, as shown in the figure. When the bottom part of the cone hits the ground, you hear a sonic boom. The equation that describes this situation is

$$\sin\left(\frac{t}{2}\right) = \frac{w}{p},$$

where t is the radian measure of the angle of the cone, w is the speed of the sound wave, p is the speed of the plane, and $p > w$.

(a) Find the speed of the sound wave when the plane flies at 1200 mph and $t = .8$.

(b) Find the speed of the plane if the sound wave travels at 500 mph and $t = .7$.

44. Suppose the batter in Example 2 hits the ball with an initial velocity of 100 feet per second.

(a) Complete this table.

t	.5	.6	.7	.8	.9
d					

(b) By experimentation, find the value of t (to two decimal places) that produces the longest distance.

(c) If $t = 1.6$, what is d? Explain your answer.

45. The average daily temperature in St. Louis, Missouri (in degrees Fahrenheit), is approximated by the function

$$T(x) = 24.6 \sin(.522x - 2.1) + 56.3 \qquad (1 \le x < 13),$$

where $x = 1$ corresponds to January 1, $x = 2$ to February 1, etc.*

---

*Based on data from the National Climatic Data Center.

(a) Complete this table.

Date	Average Temperature
Jan. 1	
Mar. 1	
May 1	
July 1	
Sept. 1	
Nov. 1	

(b) Make a table that shows the average temperature every third day in June, beginning on June 1. [Assume that three days = 1/10 of a month.]

46. A regular polygon has n equal sides and n equal angles formed by the sides. For example, a regular polygon of three sides is an equilateral triangle, and a regular polygon of four sides is a square. If a regular polygon of n sides is circumscribed around a circle of radius r, as shown in the figure for $n = 4$ and $n = 5$, then the area of the polygon is given by

$$A = nr^2 \tan\left(\frac{\pi}{n}\right).$$

(a) Find the area of a regular polygon of 12 sides circumscribed around a circle of radius 5.

(b) Complete the following table for a regular polygon of n sides circumscribed around the unit circle (which, as you recall, has radius 1).

n	5	50	500	5000	10,000
Area					

(c) As n gets larger and larger, what number does the area get very close to? [*Hint:* What is the area of the unit circle?]

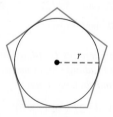

$n = 4$ $n = 5$

In Exercises 47–54, find the average rate of change of the function over the given interval. Exact answers are required.

47. $f(t) = \cos t$ from $t = \pi/2$ to $t = \pi$

48. $g(t) = \sin t$ from $t = \pi/2$ to $t = \pi$

49. $g(t) = \sin t$ from $t = \pi/6$ to $t = 11\pi/3$

50. $h(t) = \tan t$ from $t = \pi/6$ to $t = 11\pi/3$

51. $f(t) = \cos t$ from $t = -5\pi/4$ to $t = \pi/4$

52. $g(t) = \sin t$ from $t = -5\pi/4$ to $t = \pi/4$

53. $h(t) = \tan t$ from $t = \pi/6$ to $t = \pi/3$

54. $f(t) = \cos t$ from $t = \pi/4$ to $t = \pi/3$

55. (a) Use a calculator to find the average rate of change of $g(t) = \sin t$ from 2 to $2 + h$, for each of these values of h: .01, .001, .0001, and .00001.

    (b) Compare your answers in part (a) with the number $\cos 2$. What would you guess that the instantaneous rate of change of $g(t) = \sin t$ is at $t = 2$?

56. (a) Use a calculator to find the average rate of change of $f(t) = \cos t$ from 5 to $5 + h$, for each of these values of h: .01, .001, .0001, and .00001.

    (b) Compare your answers in part (a) with the number $-\sin 5$. What would you guess that the instantaneous rate of change of $f(t) = \cos t$ is at $t = 5$?

In Exercises 57–62, assume that the terminal side of an angle of t radians in standard position lies in the given quadrant on the given straight line. Find sin t, cos t, tan t. [Hint: Find a point on the terminal side of the angle.]

57. Quadrant IV; line with equation $y = -2x$.

58. Quadrant III; line with equation $2y - 5x = 0$.

59. Quadrant IV; line through $(-3, 5)$ and $(-9, 15)$.

60. Quadrant III; line through the origin parallel to
$$7x - 2y = -6.$$

61. Quadrant II; line through the origin parallel to
$$2y + x = 6.$$

62. Quadrant I; line through the origin perpendicular to
$$3y + x = 6.$$

63. The values of $\sin t$, $\cos t$, and $\tan t$ are determined by the point (x, y) where the terminal side of an angle of t radians in standard position intersects the unit circle. The coordinates x and y are positive or negative, depending on what quadrant (x, y) lies in. For instance, in the second quadrant x is negative and y is positive, so that $\cos t$ (which is x by definition) is negative. Fill the blanks in this chart with the appropriate sign ($+$ or $-$).

Quadrant II $\pi/2 < t < \pi$		Quadrant I $0 < t < \pi/2$	
$\sin t$		$\sin t$	$+$
$\cos t$	$-$	$\cos t$	
$\tan t$		$\tan t$	

Quadrant III $\pi < t < 3\pi/2$		Quadrant IV $3\pi/2 < t < 2\pi$	
$\sin t$		$\sin t$	
$\cos t$		$\cos t$	
$\tan t$		$\tan t$	

64. (a) Find two numbers c and d such that
$$\sin(c + d) \neq \sin c + \sin d.$$

    (b) Find two numbers c and d such that
$$\cos(c + d) \neq \cos c + \cos d.$$

In Exercises 65–70, draw a rough sketch to determine if the given number is positive.

65. $\sin 1$ [*Hint:* The terminal side of an angle of 1 radian lies in the first quadrant (why?), so any point on it will have a positive y-coordinate.]

66. $\cos 2$ **67.** $\tan 3$ **68.** $(\cos 2)(\sin 2)$

69. $\tan 1.5$ **70.** $\cos 3 + \sin 3$

In Exercises 71–76, find all the solutions of the equation.

71. $\sin t = 1$ **72.** $\cos t = -1$ **73.** $\tan t = 0$

74. $\sin t = -1$ **75.** $|\sin t| = 1$ **76.** $|\cos t| = 1$

THINKERS

77. Using only the definition and no calculator, determine which number is larger: $\sin(\cos 0)$ or $\cos(\sin 0)$.

78. With your calculator in radian mode and function graphing mode, graph the following functions on the same screen, using the viewing window with $0 \leq x \leq 2\pi$ and $-3 \leq y \leq 3$: $f(x) = \cos x^3$ and $g(x) = (\cos x)^3$. Are the graphs the same? What do you conclude about the statement $\cos x^3 = (\cos x)^3$?

79. Figure R is a diagram of a merry-go-round that includes horses A through F. The distance from the center P to A is 1 unit and the distance from P to D is 5 units. Define six functions as follows:

$$A(t) = \text{vertical distance from horse } A \text{ to the } x\text{-axis at time } t \text{ minutes};$$

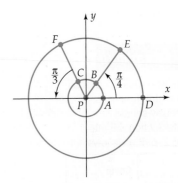

Figure R

and similarly for $B(t)$, $C(t)$, $D(t)$, $E(t)$, $F(t)$. The merry-go-round rotates counterclockwise at a rate of 1 revolution per minute, and the horses are in the positions shown in Figure R at the starting time $t = 0$. As the merry-go-round rotates, the horses move around the circles shown in Figure R.

(a) Show that $B(t) = A(t + 1/8)$ for every t.

(b) In a similar manner, express $C(t)$ in terms of the function $A(t)$.

(c) Express $E(t)$ and $F(t)$ in terms of the function $D(t)$.

(d) Explain why Figure S is valid and use it and similar triangles to express $D(t)$ in terms of $A(t)$.

(e) In a similar manner, express $E(t)$ and $F(t)$ in terms of $A(t)$.

(f) Show that $A(t) = \sin(2\pi t)$ for every t. [*Hint:* Exercises 57–64 in Section 6.1 may be helpful.]

(g) Use parts (a), (b), and (f) to express $B(t)$ and $C(t)$ in terms of the sine function.

(h) Use parts (d), (e), and (f) to express $D(t)$, $E(t)$, and $F(t)$ in terms of the sine function.

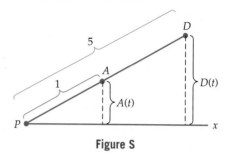

Figure S

6.2 *ALTERNATE* The Sine, Cosine, and Tangent Functions

Section Objectives

■ Use the point-in-the-plane description to evaluate trigonometric functions of real numbers.

■ Use the unit circle to define the sine, cosine and tangent functions.

■ Find the exact values of sine, cosine, and tangent a $\pi/3$, $\pi/4$, $\pi/6$, and integer multiples of these numbers.

> **NOTE**
>
> If you have not read Chapter 8, omit this section. If you have read Chapter 8, use this section in place of Section 6.2.

Trigonometric functions of any angle are defined in Alternate Section 8.1, using a point on the terminal side of the angle. According to that definition, the domains of the sine, cosine, and tangent functions consist of *angles*. We now define trigonometric functions whose domains consist of *real numbers*. The basic idea is quite simple: If t is a real number, then

sin t is defined to be the sine of an angle of t radians.

cos t is defined to be the cosine of an angle of t radians.

tan t is defined to be the tangent of an angle of t radians.

In other words, instead of starting with an angle as in Chapter 8, we now start with a number t and then move to an angle of t radians, as summarized here:

> *Roadmap*
>
> Instructors who wish to cover all six trigonometric functions simultaneously should incorporate Section 6.6 into Sections 6.2–6.4, as follows.
>
> | Subsection of Section 6.6 | Cover at the end of |
> |---|---|
> | Part I | Section 6.2 |
> | Part II | Section 6.3 |
> | Part III | Section 6.4 |

Trigonometric Functions of Real Numbers

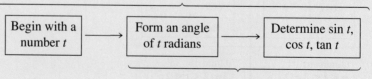

Trigonometric Functions of Angles

Adapting the definition of Alternate Section 8.1 to this new viewpoint, we have the following.

Point-in-the-Plane Description

Let t be a real number. Let (x, y) be any point (except the origin) on the terminal side of an angle of t radians in standard position. Then

$$\sin t = \frac{y}{r}, \qquad \cos t = \frac{x}{r}, \qquad \tan t = \frac{y}{x},$$

where $r = \sqrt{x^2 + y^2}$ is the distance from (x, y) to the origin.

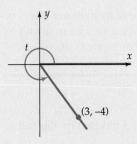

Figure 6–28

EXAMPLE 1

Figure 6–28 shows an angle of t radians in standard position. Find $\sin t$, $\cos t$, and $\tan t$.

SOLUTION Apply the facts in the box above with $(x, y) = (3, -4)$ and

$$r = \sqrt{x^2 + y^2} = \sqrt{3^2 + (-4)^2} = \sqrt{25} = 5.$$

Then we have

$$\sin t = \frac{y}{r} = \frac{-4}{5} = -\frac{4}{5}. \qquad \cos t = \frac{x}{r} = \frac{3}{5}, \qquad \tan t = \frac{y}{x} = \frac{-4}{3} = -\frac{4}{3}. \quad \blacksquare$$

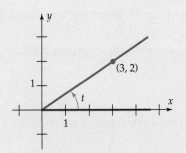

Figure 6–29

EXAMPLE 2

The terminal side of a first-quadrant angle of t radians in standard position lies on the line with equation $2x - 3y = 0$. Evaluate the three trigonometric functions at t.

SOLUTION Verify that the point $(3, 2)$ satisfies the equation and hence lies on the terminal side of the angle (Figure 6–29). Now we have $(x, y) = (3, 2)$ and $r = \sqrt{x^2 + y^2} = \sqrt{3^2 + 2^2} = \sqrt{13}$. Therefore,

$$\sin t = \frac{y}{r} = \frac{2}{\sqrt{13}}, \qquad \cos t = \frac{x}{r} = \frac{3}{\sqrt{13}}, \qquad \tan t = \frac{y}{x} = \frac{2/\sqrt{13}}{3/\sqrt{13}} = \frac{2}{3}. \quad \blacksquare$$

For a few numbers, we can use our knowledge of special angles to evaluate the trigonometric functions exactly.

EXAMPLE 3

Find the exact value of each of the following.

(a) $\sin \dfrac{\pi}{6}$ and $\cos \dfrac{\pi}{6}$ (b) $\sin \dfrac{\pi}{4}$ and $\cos \dfrac{\pi}{4}$

(c) $\tan \dfrac{3\pi}{4}$

SOLUTION

(a) An angle of $\pi/6$ radians is the same as an angle of $30°$. From Example 4(a) of Section 8.1 we know the values of sine and cosine at $30°$:

$$\sin \frac{\pi}{6} = \sin 30° = \frac{1}{2} \qquad \text{and} \qquad \cos \frac{\pi}{6} = \cos 30° = \frac{\sqrt{3}}{2}.$$

(b) An angle of $\pi/4$ radians is the same as an angle of 45°. By Example 5 of Section 8.1,

$$\sin \frac{\pi}{4} = \sin 45° = \frac{\sqrt{2}}{2} \qquad \text{and} \qquad \cos \frac{\pi}{4} = \cos 45° = \frac{\sqrt{2}}{2}.$$

(c) An angle of $3\pi/4$ radians is the same as an angle of 135°, so from Example 12 of Alternate Section 8.1, we now have

$$\tan \frac{3\pi}{4} = \tan 135° = -1. \qquad \blacksquare$$

In most cases, evaluating trigonometric functions is not as simple as it was in the preceding examples. Usually, you must use a calculator (in radian mode) to approximate the values of these functions, as illustrated in Figure 6–30.

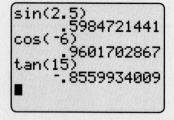

Figure 6–30

EXAMPLE 4

When a baseball is hit by a bat, the horizontal distance d traveled by the ball is approximated by

$$d = \frac{v^2 \sin t \cos t}{16},$$

where the ball leaves the bat at an angle of t radians and has initial velocity v feet per second, as shown in Figure 6–31.

(a) How far does the ball travel when the initial velocity is 90 feet per second and $t = .7$?

(b) If the initial velocity is 105 feet per second and $t = 1$, how far does the ball travel?

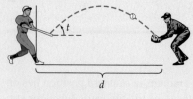

Figure 6–31

SOLUTION

(a) Let $v = 90$ and $t = .7$ in the formula for d. Then a calculator (in radian mode) shows that

$$d = \frac{v^2 \sin t \cos t}{16} = \frac{90^2 \sin .7 \cos .7}{16} \approx 249.44 \text{ feet.}$$

(b) Now let $v = 105$ and $t = 1$. Then

$$d = \frac{v^2 \sin t \cos t}{16} = \frac{105^2 \sin 1 \cos 1}{16} \approx 313.28. \qquad \blacksquare$$

THE UNIT CIRCLE DESCRIPTION

We now develop a description of sine, cosine, and tangent that is based on the **unit circle,** which is the circle of radius 1 with center at the origin, whose equation is $x^2 + y^2 = 1$. Let t be any real number and construct an angle of t radians in standard position. Let $P = (x, y)$ be the point where the terminal side of this angle intersects the unit circle, as shown in Figure 6–32.

The distance from (x, y) to the origin is 1 unit because the radius of the unit circle is 1. Using the point (x, y) and $r = 1$, we see that

$$\cos t = \frac{x}{r} = \frac{x}{1} = x \qquad \text{and} \qquad \sin t = \frac{y}{r} = \frac{y}{1} = y.$$

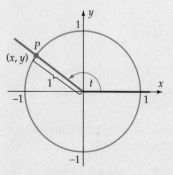

Figure 6–32

In other words,

Sine and Cosine

If P is the point where the terminal side of an angle of t radians in standard position meets the unit circle, then

$$P \text{ has coordinates } (\cos t, \sin t).$$

This description is often used as a definition of sine and cosine. To get a better feel for it, do the following Graphing Exploration.

GRAPHING EXPLORATION

With your calculator in radian mode and parametric graphing mode, set the range values as follows:

$$0 \le t \le 2\pi, \quad -1.8 \le x \le 1.8, \quad -1.2 \le y \le 1.2^*$$

Then graph the curve given by the parametric equations.

$$x = \cos t \quad \text{and} \quad y = \sin t.$$

The graph is the unit circle. Use the trace to move around the circle. At each point, the screen will display three numbers: the values of t, x, and y. For each t, the cursor is on the point where the terminal side of an angle of t radians meets the unit circle, so the corresponding x is the number $\cos t$ and the corresponding y is the number $\sin t$.

Suppose an angle of t radians in standard position intersects the unit circle at the point (x, y), as in Figure 6–32. From our earlier definition, $\tan t = y/x$ and by the unit circle description, we have $x = \cos t$ and $y = \sin t$. Therefore, we have another description of the tangent function:

$$\tan t = \frac{\sin t}{\cos t}.$$

EXAMPLE 5

Use the unit circle description and the preceding equation to find the exact values of $\sin t$, $\cos t$, and $\tan t$ when

(a) $t = \pi$ (b) $t = \pi/2$.

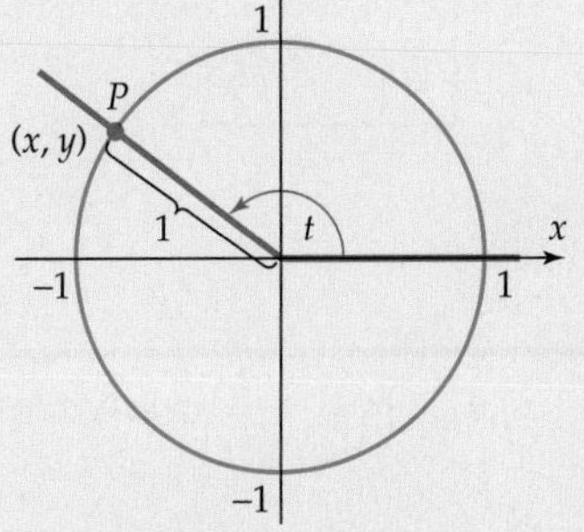

SOLUTION

(a) Construct an angle of π radians, as in Figure 6–33. Its terminal side lies on the negative x-axis and intersects the unit circle at $P = (-1, 0)$. Hence,

$$\sin \pi = y\text{-coordinate of } P = 0$$

$$\cos \pi = x\text{-coordinate of } P = -1$$

$$\tan \pi = \frac{\sin \pi}{\cos \pi} = \frac{0}{-1} = 0.$$

Figure 6–33

---

*Parametric graphing is explained in *Special Topics* 3.3.A. These settings give a square viewing window on calculators with a screen measuring approximately 95 by 63 pixels (such as TI-84+), and hence the unit circle will look like a circle. For wider screens, adjust the x range settings to obtain a square window.

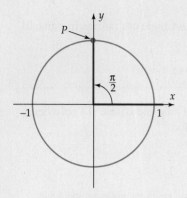

Figure 6-34

(b) An angle of $\pi/2$ radians (Figure 6-34) has its terminal side on the positive y-axis and intersects the unit circle at $P = (0, 1)$.

$$\cos\frac{\pi}{2} = x\text{-coordinate of } P = 0$$

$$\sin\frac{\pi}{2} = y\text{-coordinate of } P = 1$$

$$\tan\frac{\pi}{2} = \frac{\sin(\pi/2)}{\cos(\pi/2)} = \frac{1}{0} \; undefined. \quad \blacksquare$$

The definitions of sine and cosine show that $\sin t$ and $\cos t$ are defined for every real number t. Example 5(b), however, shows that $\tan t$ is *not* defined when the x-coordinate of the point P is 0. This occurs when P has coordinates $(0, 1)$ or $(0, -1)$, that is, when $t = \pm\pi/2, \pm3\pi/2, \pm5\pi/2$, etc. Consequently, the domain (set of inputs) of each trigonometric function is as follows.

Function	Domain
$f(t) = \sin t$	All real numbers
$g(t) = \cos t$	All real numbers
$h(t) = \tan t$	All real numbers except $\pm\pi/2, \pm3\pi/2, \pm5\pi/2, \ldots$

SPECIAL VALUES

The trigonometric functions can be evaluated exactly at $t = \pi/3$, $t = \pi/4$, $t = \pi/6$, and any integer multiples of these numbers by using the following facts (which are explained in Examples 2–4 of the Geometry Review Appendix).*

A right triangle with hypotenuse 1 and angles of $\pi/6$ and $\pi/3$ radians has sides of lengths $1/2$ (opposite the angle of $\pi/6$) and $\sqrt{3}/2$ (opposite the angle of $\pi/3$).

A right triangle with hypotenuse 1 and two angles of $\pi/4$ radians has two sides of length $\sqrt{2}/2$.

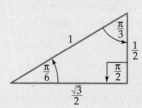
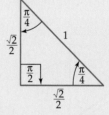

Figure 6-35

EXAMPLE 6

Evaluate the trigonometric functions at $-5\pi/4$.

SOLUTION   Construct an angle of $-5\pi/4$ radians in standard position and let P be the point where the terminal side intersects the unit circle. Draw a vertical

---

*Angles in the Geometry Review are given in degree measure: 60°, 45°, 30° instead of radian measure $\pi/3$, $\pi/4$, $\pi/6$, as is done here.

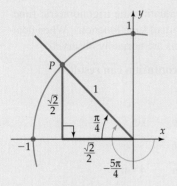

Figure 6–36

line from P to the x-axis, as shown in Figure 6–36, forming a right triangle that matches the second triangle in Figure 6–35. The sides of the triangle in Figure 6–35 show that P has coordinates $(-\sqrt{2}/2, \sqrt{2}/2)$. Hence,

$$\sin \frac{-5\pi}{4} = y\text{-coordinate of } P = \frac{\sqrt{2}}{2}$$

$$\cos \frac{-5\pi}{4} = x\text{-coordinate of } P = -\frac{\sqrt{2}}{2}$$

$$\tan \frac{-5\pi}{4} = \frac{\sin t}{\cos t} = \frac{\sqrt{2}/2}{-\sqrt{2}/2} = -1.$$ ∎

EXAMPLE 7

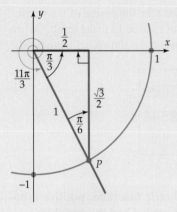

Figure 6–37

Evaluate the trigonometric functions at $11\pi/3$.

SOLUTION   Construct an angle of $11\pi/3$ radians in standard position and draw a vertical line from the x-axis to the point P where the terminal side of the angle meets the unit circle, as shown in Figure 6–37. The right triangle formed in this way matches the first triangle in Figure 6–35. The sides of the triangle in Figure 6–36, show that the coordinates of P are $(1/2, -\sqrt{3}/2)$. Therefore,

$$\sin \frac{11\pi}{3} = y\text{-coordinate of } P = -\frac{\sqrt{3}}{2}$$

$$\cos \frac{11\pi}{3} = x\text{-coordinate of } P = \frac{1}{2}$$

$$\tan \frac{11\pi}{3} = \frac{(\sin 11\pi/3)}{(\cos 11\pi/3)} = \frac{-\sqrt{3}/2}{1/2} = -\sqrt{3}.$$ ∎

EXERCISES ALTERNATE 6.2

Use the exercises for Section 6.2 on page 449.

6.3   Algebra and Identities

Section Objectives

- Learn how functional notation is used with trigonometric functions.
- Apply the rules of algebra to trigonometric functions.
- Use the Pythagorean identity to evaluate and simplify trigonometric expressions.
- Use the periodic identities to evaluate trigonometric expressions.
- Use the negative angle identities to evaluate and simplify trigonometric expressions.

Roadmap

Section 8.1 may be covered at this point by instructors who prefer to introduce right triangle trigonometry early.

In the previous section, we concentrated on evaluating the trigonometric functions. In this section, the emphasis is on the algebra of such functions. When dealing with trigonometric functions, two conventions are normally observed:

1. **Parentheses are omitted whenever no confusion can result.**

   For example,

$\sin(t)$	is written	$\sin t$
$-(\cos(5t))$	is written	$-\cos 5t$
$4(\tan t)$	is written	$4 \tan t.$

   On the other hand, parentheses *are* needed to distinguish

   $$\cos(t + 3) \quad \text{from} \quad \cos t + 3,$$

   because the first one says, "Add 3 to t and take the cosine of the result," but the second one says, "Take the cosine of t and add 3 to the result." When $t = 5$, for example, these are different numbers, as shown in Figure 6–38.*

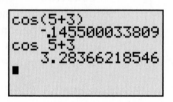

Figure 6–38

TECHNOLOGY TIP

TI-84+ and HP-39gs automatically insert an opening parenthesis when the COS key is pushed. The display COS(5 + 3 is interpreted as COS(5 + 3). If you want cos 5 + 3, you must insert a parenthesis after the 5: COS(5) + 3.

2. **When dealing with powers of trigonometric functions, positive exponents are written between the function symbol and the variable.**

   For example,

$(\cos t)^3$	is written	$\cos^3 t$
$(\sin t)^4(\tan 7t)^2$	is written	$\sin^4 t \tan^2 7t.$

   Furthermore,

   $$\sin t^3 \quad \text{means} \quad \sin(t^3) \quad [not\ (\sin t)^3 \quad \text{or} \quad \sin^3 t]$$

   For instance, when $t = 4$, we have Figure 6–39.

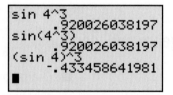

Figure 6–39

CAUTION

Convention 2 is not used when the exponent is -1. For example, $\sin^{-1} t$ does *not* mean $(\sin t)^{-1}$ or $1/(\sin t)$; it has an entirely different meaning that will be discussed in Section 7.4. Similar remarks apply to $\cos^{-1} t$ and $\tan^{-1} t$.

TECHNOLOGY TIP

Calculators do not use convention 2. To obtain $\sin^3 4$, you must enter $(\sin 4)^3$.

    Except for these two conventions and the Caution in the margin, the algebra of trigonometric functions is just like the algebra of other functions. They may be added, subtracted, multiplied, composed, etc.

---

*Figures 6–38 and 6–39 show a TI-86 screen.

EXAMPLE 1

If $f(t) = \sin^2 t + \tan t$ and $g(t) = \tan^3 t + 5$, then the product function fg is given by the rule

$$(fg)(t) = f(t)g(t) = (\sin^2 t + \tan t)(\tan^3 t + 5)$$
$$= \sin^2 t \tan^3 t + 5 \sin^2 t + \tan^4 t + 5 \tan t. \qquad ■$$

EXAMPLE 2

Factor $2 \cos^2 t - 5 \cos t - 3$.

SOLUTION   You can do this directly, but it may be easier to understand if you make a substitution. Let $u = \cos t$, then

$$2 \cos^2 t - 5 \cos t - 3 = 2(\cos t)^2 - 5 \cos t - 3$$
$$= 2u^2 - 5u - 3$$
$$= (2u + 1)(u - 3)$$
$$= (2 \cos t + 1)(\cos t - 3). \qquad ■$$

EXAMPLE 3

If $f(t) = \cos^2 t - 9$ and $g(t) = \cos t + 3$, then the quotient function f/g is given by the rule

$$\left(\frac{f}{g}\right)(t) = \frac{f(t)}{g(t)} = \frac{\cos^2 t - 9}{\cos t + 3} = \frac{(\cos t + 3)(\cos t - 3)}{\cos t + 3} = \cos t - 3. \qquad ■$$

CAUTION

You are dealing with *functional notation* here, so the symbol sin t is a *single entity*, as are cos t and tan t. Don't try some nonsensical "canceling" operation, such as

$$\frac{\sin t}{\cos t} = \frac{\sin}{\cos} \qquad \text{or} \qquad \frac{\cos t^2}{\cos t} = \frac{\cos t}{\cos} = t.$$

EXAMPLE 4

If $f(t) = \sin t$ and $g(t) = t^2 + 3$, then the composite function $g \circ f$ is given by the rule

$$(g \circ f)(t) = g(f(t)) = g(\sin t) = \sin^2 t + 3.$$

The composite function $f \circ g$ is given by the rule

$$(f \circ g)(t) = f(g(t)) = f(t^2 + 3) = \sin(t^2 + 3).$$

The parentheses are crucial here because $\sin(t^2 + 3)$ is *not* the same function as $\sin t^2 + 3$. For instance, a calculator in radian mode shows that for $t = 5$,

$$\sin(5^2 + 3) = \sin(25 + 3) = \sin 28 \approx .2709,$$

whereas

$$\sin 5^2 + 3 = \sin 25 + 3 \approx (-.1324) + 3 = 2.8676.$$ ∎

 THE PYTHAGOREAN IDENTITY

Trigonometric functions have numerous interrelationships that are usually expressed as *identities*. An **identity** is an equation with this property: For every value of the variable for which both sides of the equation are defined, the equation is true. Here is one of the most important trigonometric identities.

Pythagorean Identity

> For every real number t,
>
> $$\sin^2 t + \cos^2 t = 1.$$

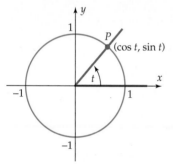

Figure 6–40

Proof   For each real number t, the point P, where the terminal side of an angle of t radians intersects the unit circle, has coordinates ($\cos t$, $\sin t$), as in Figure 6–40. Since P lies on the unit circle, its coordinates must satisfy the equation of the unit circle: $x^2 + y^2 = 1$, that is, $\cos^2 t + \sin^2 t = 1$. ∎

GRAPHING EXPLORATION

Recall that the graph of $y = 1$ is a horizontal line through (0, 1). Verify the Pythagorean identity by graphing the equation

$$y = (\sin x)^2 + (\cos x)^2$$

in the window with $-10 \le x \le 10$ and $-3 \le y \le 3$ and using the trace feature.

EXAMPLE 5

If $\pi/2 < t < \pi$ and $\sin t = 2/3$, find $\cos t$ and $\tan t$.

SOLUTION   By the Pythagorean identity,

$$\cos^2 t = 1 - \sin^2 t = 1 - \left(\frac{2}{3}\right)^2 = 1 - \frac{4}{9} = \frac{5}{9}.$$

So there are two possibilities:

$$\cos t = \sqrt{5/9} = \sqrt{5}/3 \qquad \text{or} \qquad \cos t = -\sqrt{5/9} = -\sqrt{5}/3.$$

Since $\pi/2 < t < \pi$, $\cos t$ is negative (see Exercise 63 on page 451). Therefore, $\cos t = -\sqrt{5}/3$, and

$$\tan t = \frac{\sin t}{\cos t} = \frac{2/3}{-\sqrt{5}/3} = \frac{-2}{\sqrt{5}} = \frac{-2\sqrt{5}}{5}.$$ ∎

EXAMPLE 6

The Pythagorean identity is valid for *any* number t. For instance, if $t = 3k + 7$, then $\sin^2(3k + 7) + \cos^2(3k + 7) = 1$. ∎

EXAMPLE 7

To simplify the expression $\tan^2 t \cos^2 t + \cos^2 t$, we use the definition of tangent and the Pythagorean identity:

$$\tan^2 t \cos^2 t + \cos^2 t = \frac{\sin^2 t}{\cos^2 t} \cos^2 t + \cos^2 t$$

$$= \sin^2 t + \cos^2 t = 1. \quad \blacksquare$$

For every real number t, the point $(\cos t, \sin t)$ is on the unit circle, as illustrated in Figure 6–40. Since the coordinates of any point on the unit circle are between -1 and 1, we have this useful fact:

Range of Sine
and Cosine

For every real number t

$$-1 \le \sin t \le 1$$

and

$$-1 \le \cos t \le 1.$$

As we shall see in Section 6.4,

The range of the tangent function consists of all real numbers.

You can confirm this fact by doing the following Exploration.

CALCULATOR EXPLORATION

Use the table feature to evaluate $\tan\left(\dfrac{\pi}{2} + x\right)$ when $x = .01, .001, .0001$, and so on.

What does this suggest about the outputs of the tangent function? Now evaluate when $x = -.01, -.001, -.0001$, and so on, and answer the same question.

PERIODICITY IDENTITIES

Let t be any real number and construct two angles in standard position of measure t and $t + 2\pi$ radians, respectively, as shown in Figure 6–41. As we saw in Section 6.1, both of these angles have the same terminal side. Therefore, the point P where the terminal side meets the unit circle is the *same* in both cases.

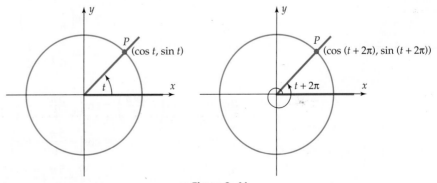

Figure 6–41

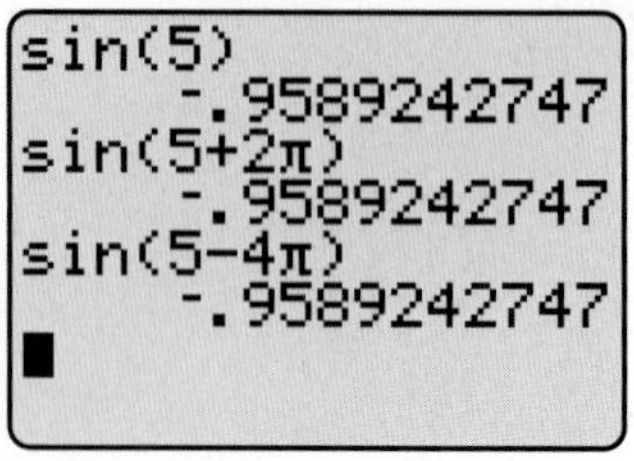

```
sin(5)
       -.9589242747
sin(5+2π)
       -.9589242747
sin(5-4π)
       -.9589242747
■
```

Figure 6–42

Therefore, the coordinates of P are the same, that is,

$$\sin t = \sin(t + 2\pi) \qquad \text{and} \qquad \cos t = \cos(t + 2\pi).$$

Furthermore, since an angle of t radians has the same terminal side as angles of radian measure $t \pm 2\pi, t \pm 4\pi, t \pm 6\pi$, and so forth, the same argument shows that

$$\sin t = \sin(t \pm 2\pi) = \sin(t \pm 4\pi) = \sin(t \pm 6\pi) = \cdots$$

$$\cos t = \cos(t \pm 2\pi) = \cos(t \pm 4\pi) = \cos(t \pm 6\pi) = \cdots$$

as illustrated (for $t = 5$) in Figure 6–42.

There is a special name for functions that repeat their values at regular intervals. A function f is said to be **periodic** if there is a positive constant k such that $f(t) = f(t + k)$ for every number t in the domain of f. There will be more than one constant k with this property; the smallest one is called the **period** of the function f. We have just seen that sine and cosine are periodic with $k = 2\pi$. Exercises 69 and 70 show that 2π is the smallest such positive constant k. Therefore, we have the following.

Period of Sine and Cosine

> The sine and cosine functions are periodic with period 2π: For every real number t,
>
> $$\sin t = \sin(t \pm 2\pi) = \sin(t \pm 4\pi) = \sin(t \pm 6\pi) = \cdots$$
>
> and
>
> $$\cos t = \cos(t \pm 2\pi) = \cos(t \pm 4\pi) = \cos(t \pm 6\pi) = \cdots$$

EXAMPLE 8

As we saw in Examples 3 and 5 of Section 6.2, $\sin \dfrac{\pi}{6} = \dfrac{1}{2}$ and $\cos\left(-\dfrac{5\pi}{4}\right) = -\dfrac{\sqrt{2}}{2}$. Use these facts to find

(a) $\sin \dfrac{13\pi}{6}$ \qquad\qquad (b) $\cos\left(-\dfrac{29\pi}{4}\right)$.

SOLUTION The key is to write the given number as a sum in which one summand is an even multiple of π, and then apply a periodicity identity.

(a)
$$\sin\frac{13\pi}{6} = \sin\left(\frac{\pi}{6} + \frac{12\pi}{6}\right)$$

$$= \sin\left(\frac{\pi}{6} + 2\pi\right)$$

$$= \sin\left(\frac{\pi}{6}\right) = \frac{1}{2} \qquad \text{[Periodicity Identity]}$$

(b)
$$\cos\left(-\frac{29\pi}{4}\right) = \cos\left(-\frac{5\pi}{4} - \frac{24\pi}{4}\right)$$

$$= \cos\left(-\frac{5\pi}{4} - 6\pi\right)$$

$$= \cos\left(-\frac{5\pi}{4}\right) = -\frac{\sqrt{2}}{2} \qquad \text{[Periodicity Identity]} \qquad ■$$

The tangent function is also periodic (see Exercise 36), but its period is π rather than 2π, that is,

$$\tan(t + \pi) = \tan t \quad \text{for every real number } t,$$

as we shall see in Section 6.4.

NEGATIVE ANGLE IDENTITIES

GRAPHING EXPLORATION

(a) In a viewing window with $-2\pi \le x \le 2\pi$, graph $y_1 = \sin x$ and $y_2 = \sin(-x)$ on the same screen. Use trace to move along $y_1 = \sin x$. Stop at a point and note its y-coordinate. Use the up or down arrow to move vertically to the graph of $y_2 = \sin(-x)$. The x-coordinate remains the same, but the y-coordinate is different. How are the two y-coordinates related? Is one the negative of the other? Repeat the procedure for other points. Are the results the same?

(b) Now graph $y_1 = \cos x$ and $y_2 = \cos(-x)$ on the same screen. How do the graphs compare?

(c) Repeat part (a) for $y_1 = \tan x$ and $y_2 = \tan(-x)$. Are the results similar to those for sine?

The preceding Graphing Exploration suggests the truth of the following statement.

Negative Angle Identities

For every real number t,

$$\sin(-t) = -\sin t$$

$$\cos(-t) = \cos t$$

$$\tan(-t) = -\tan t.$$

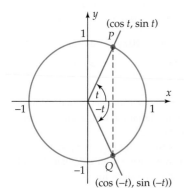

Figure 6–43

Proof   Consider angles of t radians and $-t$ radians in standard position, as in Figure 6–43. By the definition of sine and cosine, P has coordinates $(\cos t, \sin t)$, and Q has coordinates $(\cos(-t), \sin(-t))$. As Figure 6–43 suggests, P and Q lie on the same vertical line. Therefore, they have the same first coordinate, that is, $\cos(-t) = \cos t$. As the figure also suggests, P and Q lie at equal distances from the x-axis.* So the y-coordinate of Q must be the negative of the y-coordinate of P, that is, $\sin(-t) = -\sin t$. Finally, by the definition of the tangent function and the two identities just proved, we have

$$\tan(-t) = \frac{\sin(-t)}{\cos(-t)} = \frac{-\sin t}{\cos t} = -\frac{\sin t}{\cos t} = -\tan t. \qquad \blacksquare$$

EXAMPLE 9

In Example 3 of Section 6.2, we showed that

$$\sin\frac{\pi}{6} = \frac{1}{2}, \qquad \cos\frac{\pi}{6} = \frac{\sqrt{3}}{2}, \qquad \tan\frac{\pi}{6} = \frac{\sqrt{3}}{3}.$$

---

*These facts can be proved by using congruent triangles. (See Exercise 13 in the Geometry Review Appendix.)

Using the negative angle identities, we have

$$\sin\left(-\frac{\pi}{6}\right) = -\sin\frac{\pi}{6} = -\frac{1}{2}, \qquad \cos\left(-\frac{\pi}{6}\right) = \cos\frac{\pi}{6} = \frac{\sqrt{3}}{2},$$

$$\tan\left(-\frac{\pi}{6}\right) = -\tan\frac{\pi}{6} = -\frac{\sqrt{3}}{3}.$$ ∎

EXAMPLE 10

To simplify $(1 + \sin t)(1 + \sin(-t))$, we use the negative angle identity and the Pythagorean identity.

$$(1 + \sin t)(1 + \sin(-t)) = (1 + \sin t)(1 - \sin t)$$
$$= 1 - \sin^2 t$$
$$= \cos^2 t.$$ ∎

EXERCISES 6.3

In Exercises 1–4, find the rule of the product function fg.

1. $f(t) = 3 \sin t$; $\quad g(t) = \sin t + 2 \cos t$

2. $f(t) = 5 \tan t$; $\quad g(t) = \tan^3 t - 1$

3. $f(t) = 3 \sin^2 t$; $\quad g(t) = \sin t + \tan t$

4. $f(t) = \sin 2t + \cos^4 t$; $\quad g(t) = \cos 2t + \cos^2 t$

In Exercises 5–14, factor the given expression.

5. $\cos^2 t - 4$

6. $25 - \tan^2 t$

7. $\sin^2 t - \cos^2 t$

8. $\sin^3 t - \sin t$

9. $\tan^2 t + 6 \tan t + 9$

10. $\cos^2 t - \cos t - 2$

11. $6 \sin^2 t - \sin t - 1$

12. $\tan t \cos t + \cos^2 t$

13. $\cos^4 t + 4 \cos^2 t - 5$

14. $3 \tan^2 t + 5 \tan t - 2$

In Exercises 15–18, find the rules of the composite functions $f \circ g$ and $g \circ f$.

15. $f(t) = \cos t$; $\quad g(t) = 2t + 4$

16. $f(t) = \sin t + 2$; $\quad g(t) = t^2$

17. $f(t) = \tan(t + 3)$; $\quad g(t) = t^2 - 1$

18. $f(t) = \cos^2(t - 2)$; $\quad g(t) = 5t + 2$

In Exercises 19–24, determine if it is possible for a number t to satisfy the given conditions. [Hint: Think Pythagorean.]

19. $\sin t = 5/13$ and $\cos t = 12/13$

20. $\sin t = -2$ and $\cos t = 1$

21. $\sin t = -1$ and $\cos t = 1$

22. $\sin t = 1/\sqrt{2}$ and $\cos t = -1/\sqrt{2}$

23. $\sin t = 1$ and $\tan t = 1$

24. $\cos t = 8/17$ and $\tan t = 15/8$

In Exercises 25–28, use the Pythagorean identity to find sin t.

25. $\cos t = -.5$ $\quad$ and $\quad$ $\pi < t < 3\pi/2$

26. $\cos t = -3/\sqrt{10}$ $\quad$ and $\quad$ $\pi/2 < t < \pi$

27. $\cos t = 1/2$ $\quad$ and $\quad$ $0 < t < \pi/2$

28. $\cos t = 2/\sqrt{5}$ $\quad$ and $\quad$ $3\pi/2 < t < 2\pi$

In Exercises 29–35, assume that sin t = 3/5 and $0 < t < \pi/2$. Use identities in the text to find the number.

29. $\sin(-t)$ $\qquad$ **30.** $\sin(t + 10\pi)$ $\qquad$ **31.** $\sin(2\pi - t)$

32. $\cos t$ $\qquad$ **33.** $\tan t$ $\qquad$ **34.** $\cos(-t)$

35. $\tan(2\pi - t)$

36. (a) Show that $\tan(t + 2\pi) = \tan t$ for every t in the domain of $\tan t$. [*Hint:* Use the definition of tangent and some identities proved in the text.]

(b) Verify that it appears true that $\tan(x + \pi) = \tan x$ for every t in the domain by using your calculator's table feature to make a table of values for $y_1 = \tan(x + \pi)$ and $y_2 = \tan x$.

In Exercises 37–42, assume that

$$\cos t = -2/5 \quad and \quad \pi < t < 3\pi/2.$$

Use identities to find the number.

37. $\sin t$ $\qquad$ **38.** $\tan t$ $\qquad$ **39.** $\cos(2\pi - t)$

40. $\cos(-t)$ $\qquad$ **41.** $\sin(4\pi + t)$ $\qquad$ **42.** $\tan(4\pi - t)$

In Exercises 43–46, assume that

$$\sin(\pi/8) = \frac{\sqrt{2 - \sqrt{2}}}{2}$$

and use identities to find the exact functional value.

43. $\cos(\pi/8)$ **44.** $\tan(\pi/8)$

45. $\sin(17\pi/8)$ **46.** $\tan(-15\pi/8)$

In Exercises 47–58, use algebra and identities in the text to simplify the expression. Assume all denominators are nonzero.

47. $(\sin t + \cos t)(\sin t - \cos t)$

48. $(\sin t - \cos t)^2$ **49.** $\tan t \cos t$

50. $(\sin t)/(\tan t)$ **51.** $\sqrt{\sin^3 t \cos t} \sqrt{\cos t}$

52. $(\tan t + 2)(\tan t - 3) - (6 - \tan t) + 2 \tan t$

53. $\left(\dfrac{4 \cos^2 t}{\sin^2 t}\right)\left(\dfrac{\sin t}{4 \cos t}\right)^2$

54. $\dfrac{5 \cos t}{\sin^2 t} \cdot \dfrac{\sin^2 t - \sin t \cos t}{\sin^2 t - \cos^2 t}$

55. $\dfrac{\cos^2 t + 4 \cos t + 4}{\cos t + 2}$

56. $\dfrac{\sin^2 t - 2 \sin t + 1}{\sin t - 1}$

57. $\dfrac{1}{\cos t} - \sin t \tan t$

58. $\dfrac{1 - \tan^2 t}{1 + \tan^2 t} + 2 \sin^2 t$

59. The average monthly temperature in Cleveland, Ohio is approximated by

$$f(t) = 22.7 \sin(.52x - 2.18) + 49.6,$$

where $t = 1$ corresponds to January, $t = 2$ to February, and so on.

(a) Construct a table of values ($t = 1, 2, \ldots, 12$) for the function $f(t)$ and another table for $f(t + 12.083)$.

(b) Based on these tables would you say that the function f is (approximately) periodic? If so, what is the period? Is this reasonable?

60. A typical healthy person's blood pressure can be modeled by the periodic function

$$f(t) = 22 \cos(2.5\pi t) + 95,$$

where t is time (in seconds) and $f(t)$ is in millimeters of mercury. Which one of .5, .8, or 1 appears to be the period of this function?

61. The percentage of the face of the moon that is illuminated (as seen from earth) on day t of the lunar month is given by

$$g(t) = .5\left(1 - \cos \frac{2\pi t}{29.5}\right).$$

(a) What percentage of the face of the moon is illuminated on day 0? Day 10? Day 22?

(b) Construct appropriate tables to confirm that g is a periodic function with period 29.5 days.

(c) When does a full moon occur ($g(t) = 1$)?

In Exercises 62–67, show that the given function is periodic with period less than 2π. [Hint: Find a positive number k with $k < 2\pi$ such that $f(t + k) = f(t)$ for every t in the domain of f.]

62. $f(t) = \sin 2t$

63. $f(t) = \cos 3t$

64. $f(t) = \sin 4t$

65. $f(t) = \sin(\pi t)$

66. $f(t) = \cos(3\pi t/2)$

67. $f(t) = \tan 2t$

68. Fill the blanks with "even" or "odd" so that the resulting statement is true. Then prove the statement by using an appropriate identity. [*Hint: Special Topics* 3.4.A may be helpful.]

(a) $f(t) = \sin t$ is an ____ function.

(b) $g(t) = \cos t$ is an ____ function.

(c) $h(t) = \tan t$ is an ____ function.

(d) $f(t) = t \sin t$ is an ____ function.

(e) $g(t) = t + \tan t$ is an ____ function.

69. Here is a proof that the cosine function has period 2π. We saw in the text that $\cos(t + 2\pi) = \cos t$ for every t. We must show that there is no positive number smaller than 2π with this property. Do this as follows:

(a) Find all numbers k such that $0 < k < 2\pi$ and $\cos k = 1$. [*Hint:* Draw a picture and use the definition of the cosine function.]

(b) Suppose k is a number such that $\cos(t + k) = \cos t$ for every number t. Show that $\cos k = 1$. [*Hint:* Consider $t = 0$.]

(c) Use parts (a) and (b) to show that there is no positive number k less than 2π with the property that $\cos(t + k) = \cos t$ for *every* number t. Therefore, $k = 2\pi$ is the smallest such number, and the cosine function has period 2π.

70. Here is proof that the sine function has period 2π. We saw in the text that $\sin(t + 2\pi) = \sin t$ for every t. We must show that there is no positive number smaller than 2π with this property. Do this as follows:

(a) Find a number t such that $\sin(t + \pi) \neq \sin t$.

(b) Find all numbers k such that $0 < k < 2\pi$ and $\sin k = 0$. [*Hint:* Draw a picture and use the definition of the sine function.]

(c) Suppose k is a number such that $\sin(t + k) = \sin t$ for every number t. Show that $\sin k = 0$. [*Hint:* Consider $t = 0$.]

(d) Use parts (a)–(c) to show that there is no positive number k less than 2π with the property that $\sin(t + k) = \sin t$ for *every* number t. Therefore, $k = 2\pi$ is the smallest such number, and the sine function has period 2π.

6.4   Basic Graphs

Section Objectives
- Analyze the graphs of the sine, cosine, and tangent functions.
- Derive the graphs of other trigonometric functions from the graphs of sine, cosine, and tangent.
- Explore trigonometric identities graphically.

Although a graphing calculator will quickly sketch the graphs of the sine, cosine, and tangent functions, it will not give you much insight into why these graphs have the shapes they do and why these shapes are important. So the emphasis here is on the connection between the definition of these functions and their graphs.

As t Increases	The Point P Moves	The y-coordinate of $P (= \sin t)$	Rough Sketch of the Graph
from 0 to $\frac{\pi}{2}$	from $(1, 0)$ to $(0, 1)$	increases from 0 to 1	
from $\frac{\pi}{2}$ to π	from $(0, 1)$ to $(-1, 0)$	decreases from 1 to 0	
from π to $\frac{3\pi}{2}$	from $(-1, 0)$ to $(0, -1)$	decreases from 0 to -1	
from $\frac{3\pi}{2}$ to 2π	from $(0, -1)$ to $(1, 0)$	increases from -1 to 0	

If P is the point where the unit circle meets the terminal side of an angle of t radians, then the y-coordinate of P is the number sin t. As shown in the chart on the facing page, we can get a rough sketch of the graph of $f(t) = \sin t$ by watching the y-coordinate of P.

GRAPHING EXPLORATION

Your calculator can provide a dynamic simulation of this process. Put it in parametric graphing mode and set the range values as follows:

$$0 \le t \le 6.28 \qquad -1 \le x \le 6.28 \qquad -2.5 \le y \le 2.5.$$

On the same screen, graph the two functions given by

$$x_1 = \cos t, \quad y_1 = \sin t \quad \text{and} \quad x_2 = t, \quad y_2 = \sin t.$$

Using the trace feature, move the cursor along the first graph (the unit circle). Stop at a point on the circle, and note the value of t and the y-coordinate of the point. Then switch the trace to the second graph (the sine function) by using the up or down cursor arrows. The value of t remains the same. What are the x- and y-coordinates of the new point? How does the y-coordinate of the new point compare with the y-coordinate of the original point on the unit circle?

To complete the graph of the sine function, note that as t goes from 2π to 4π, the point P on the unit circle *retraces* the path it took from 0 to 2π, so *the same wave shape will repeat* on the graph. The same thing happens when t goes from 4π to 6π, or from -2π to 0, and so on. This repetition of the same pattern is simply the graphical expression of the fact that the sine function has period 2π: For any number t, the points

$$(t, \sin t) \qquad \text{and} \qquad (t + 2\pi, \sin(t + 2\pi))$$

on the graph have the same second coordinate.

A graphing calculator or some point plotting with an ordinary calculator now produces the graph of $f(t) = \sin t$ (Figure 6–44).

TECHNOLOGY TIP

Calculators have built-in windows for trigonometric functions, in which the x-axis tick marks are at intervals of $\pi/2$.
Choose TRIG or ZTRIG in this menu:

TI: ZOOM

HP-39gs: VIEWS

Casio: V-WINDOW

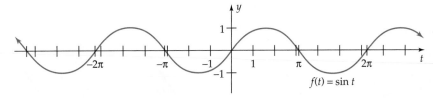

Figure 6–44

NOTE

Throughout this chapter, we use t as the variable for trigonometric functions to avoid any confusion with the x's and y's that are part of the definition of these functions. For calculator graphing in "function mode," however, you must use x as the variable: $f(x) = \sin x$, $g(x) = \cos x$, etc.

The graph of the sine function and the techniques of Section 3.4 can be used to graph other trigonometric functions.

EXAMPLE 1

The graph of $h(t) = 3 \sin t$ is the graph of $f(t) = \sin t$ stretched away from the horizontal axis by a factor of 3, as shown in Figure 6–45. ■

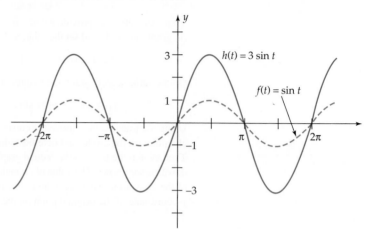

Figure 6–45

EXAMPLE 2

The graph of $k(t) = -\frac{1}{2} \sin t$ is the graph of $f(t) = \sin t$ shrunk by a factor of $1/2$ toward the horizontal axis and then reflected in the horizontal axis, as shown in Figure 6–46. ■

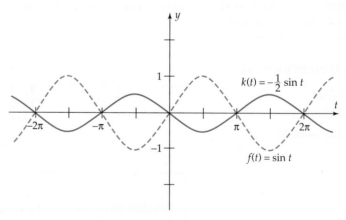

Figure 6–46

GRAPH OF THE COSINE FUNCTION

To obtain the graph of $g(t) = \cos t$, we follow the same procedure as with sine, except that we now watch the x-coordinate of P (which is $\cos t$).

As t Increases	The Point P Moves	The x-coordinate of P ($= \cos t$)	Rough Sketch of the Graph
from 0 to $\frac{\pi}{2}$	from $(1, 0)$ to $(0, 1)$	decreases from 1 to 0	
from $\frac{\pi}{2}$ to π	from $(0, 1)$ to $(-1, 0)$	decreases from 0 to -1	
from π to $\frac{3\pi}{2}$	from $(-1, 0)$ to $(0, -1)$	increases from -1 to 0	
from $\frac{3\pi}{2}$ to 2π	from $(0, -1)$ to $(1, 0)$	increases from 0 to 1	

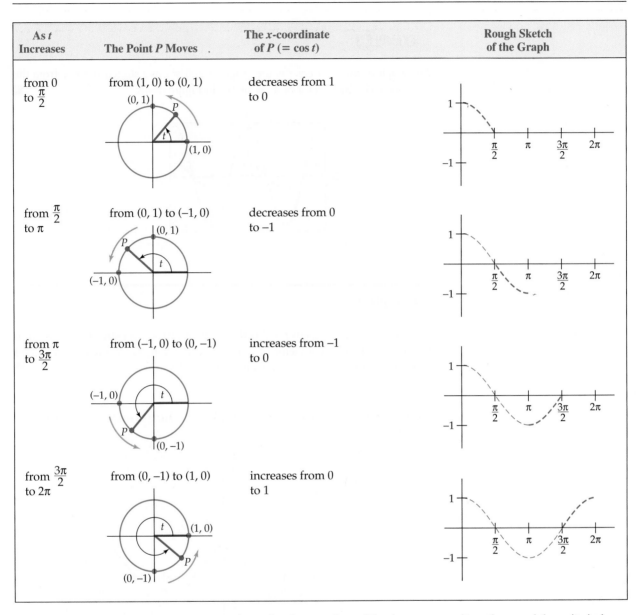

As t takes larger values, P begins to retrace its path around the unit circle, so the graph of $g(t) = \cos t$ repeats the same wave pattern, and similarly for negative values of t. So the graph looks like Figure 6–47.

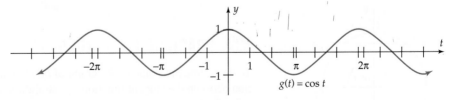

Figure 6–47

For a dynamic simulation of the cosine graphing process described above, see Exercise 69.

The techniques of Section 3.4 can be used to graph variations of the cosine function.

EXAMPLE 3

The graph of $h(t) = 4 \cos t$ is the graph of $g(t) = \cos t$ stretched away from the horizontal axis by a factor of 4, as shown in Figure 6–48. ∎

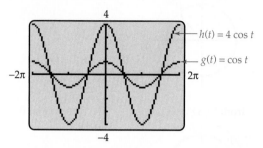

Figure 6–48

EXAMPLE 4

The graph of $k(t) = -2 \cos t + 3$ is the graph of $g(t) = \cos t$ stretched away from the horizontal axis by a factor of 2, reflected in the horizontal axis, and shifted vertically 3 units upward as shown in Figure 6–49. ∎

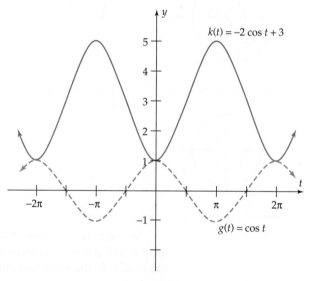

Figure 6–49

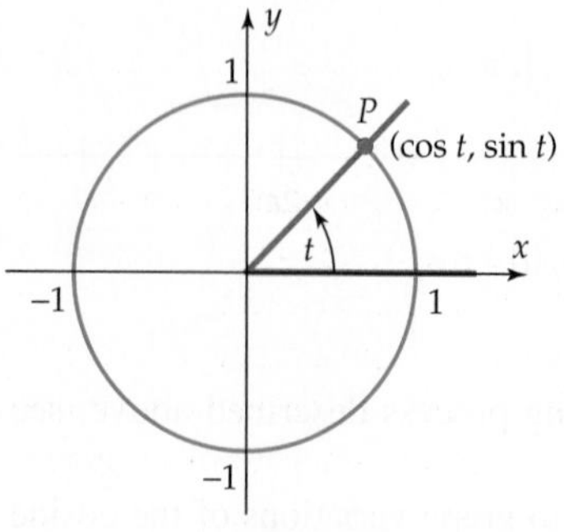

GRAPH OF THE TANGENT FUNCTION

To determine the shape of the graph of $h(t) = \tan t$, we use an interesting connection between the tangent function and straight lines. As shown in Figure 6–50, the point P where the terminal side of an angle of t radians in standard position meets the unit circle has coordinates $(\cos t, \sin t)$. We can use this point and the point $(0, 0)$ to compute the *slope* of the terminal side.

$$\text{slope} = \frac{\sin t - 0}{\cos t - 0} = \frac{\sin t}{\cos t} = \tan t$$

Figure 6–50

Therefore, we have the following.

Slope and Tangent

The slope of the terminal side of an angle of t radians in standard position is the number $\tan t$.

The graph of $h(t) = \tan t$ can now be sketched by watching the slope of the terminal side of an angle of t radians, as t takes different values. Recall that the more steeply a line rises from left to right, the larger its slope. Similarly, lines that fall from left to right have negative slopes that increase in absolute value as the line falls more steeply.

As t Changes	The Terminal Side of the Angle Moves	Its Slope ($\tan t$)	Rough Sketch of the Graph
from 0 to $\frac{\pi}{2}$	from horizontal upward toward vertical	increases from 0 in the positive direction and keeps getting larger	
from 0 to $-\frac{\pi}{2}$	from horizontal downward toward vertical	decreases from 0 in the negative direction and keeps getting larger in absolute value	

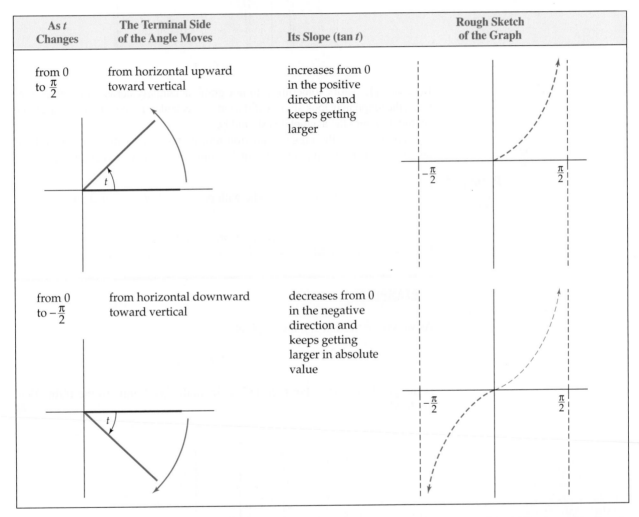

When $t = \pm\pi/2$, the terminal side of the angle is vertical, and hence its slope is not defined. This corresponds to the fact that the tangent function is not defined when $t = \pm\pi/2$. The vertical lines through $\pm\pi/2$ are vertical asymptotes of the graph: It gets closer and closer to these lines but never touches them.

As t goes from $\pi/2$ to $3\pi/2$, the terminal side goes from almost vertical with negative slope to horizontal to almost vertical with positive slope (draw a picture), exactly as it does between $-\pi/2$ and $\pi/2$. So the graph repeats the same pattern. The same thing happens between $3\pi/2$ and $5\pi/2$, between $-3\pi/2$ and $-\pi/2$, etc. Therefore, the entire graph looks like Figure 6–51 on the next page.

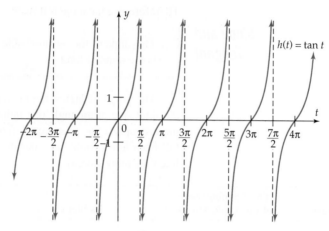

Figure 6–51

Because calculators sometimes do not graph accurately across vertical asymptotes, the graph may look slightly different on a calculator screen (with vertical line segments where the asymptotes should be).

The graph of the tangent function repeats the same pattern at intervals of length π. This means that the tangent function repeats its values at intervals of π.

Period of
Tangent

> The tangent function is periodic with period π: For every real number t in its domain,
>
> $$\tan(t \pm \pi) = \tan t.$$

EXAMPLE 5

As we saw in Section 3.4, the graph of

$$k(t) = \tan\left(t - \frac{\pi}{2}\right)$$

is the graph of $h(t) = \tan t$ shifted horizontally $\pi/2$ units to the right (Figure 6–52). ∎

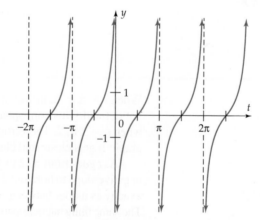

Figure 6–52

GRAPHS AND IDENTITIES

Graphing calculators can be used to identify equations that could possibly be identities. A calculator cannot *prove* that such an equation is an identity; but it can provide evidence that it *might* be one. On the other hand, a calculator *can* prove that a particular equation is *not* an identity.

EXAMPLE 6

Which of the following equations could possibly be an identity?

(a) $\cos\left(\dfrac{\pi}{2} + t\right) = \sin t$ (b) $\cos\left(\dfrac{\pi}{2} - t\right) = \sin t$

SOLUTION

(a) Consider the functions $f(t) = \cos\left(\dfrac{\pi}{2} + t\right)$ and $g(t) = \sin t$, whose rules are given by the two sides of the equation

$$\cos\left(\frac{\pi}{2} + t\right) = \sin t.$$

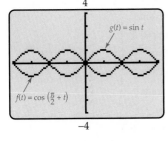

Figure 6–53

If this equation is an identity, then $f(t) = g(t)$ for every real number t, and hence, f and g have the same graph. But the graphs of f and g on the interval $[-2\pi, 2\pi]$ (Figure 6–53) are obviously different. Therefore, this equation is *not* an identity.

(b) We can test this equation in the same manner. The graph of the left side, that is, the graph of

$$h(t) = \cos\left(\frac{\pi}{2} - t\right),$$

in Figure 6–54 appears to be the same as the graph of $g(t) = \sin t$ on the interval $[-2\pi, 2\pi]$ (Figure 6–53). To check this, do the Graphing Exploration in the margin.

GRAPHING EXPLORATION

Graph $h(t) = \cos\left(\dfrac{\pi}{2} - t\right)$ and $g(t) = \sin t$ on the same screen and use the trace feature to confirm that the graphs appear to be identical.

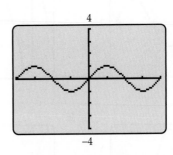

Figure 6–54

The fact that the graphs appear to be identical means that the two functions have the same value at every number t that the calculator computed in making the graphs (at least 95 numbers). This evidence strongly suggests that the equation

$\cos\left(\dfrac{\pi}{2} - t\right) = \sin t$ is an identity, but does not prove it. All we can say at this point is that the equation possibly is an identity.   ∎

CAUTION

Do not assume that two graphs that look the same on a calculator screen actually are the same. Depending on the viewing window, two graphs that are actually quite different may appear to be identical. See Exercises 61, 62, and 64–67 for some examples.

EXERCISES 6.4

In Exercises 1–6, use the graphs of the sine and cosine functions to find all the solutions of the equation.

1. $\sin t = 0$ **2.** $\cos t = 0$ **3.** $\sin t = 1$

4. $\sin t = -1$ **5.** $\cos t = -1$ **6.** $\cos t = 1$

In Exercises 7–10, find $\tan t$, where the terminal side of an angle of t radians lies on the given line.

7. $y = 11x$ **8.** $y = 1.5x$ **9.** $y = 1.4x$ **10.** $y = .32x$

In Exercises 11–22, list the transformations needed to change the graph of $f(t)$ into the graph of $g(t)$. [See Section 3.4.]

11. $f(t) = \sin t$; $g(t) = \sin t + 3$

12. $f(t) = \cos t$; $g(t) = \cos t - 2$

13. $f(t) = \cos t$; $g(t) = -\cos t$

14. $f(x) = \sin t$; $g(t) = -3 \sin t$

15. $f(t) = \tan t$; $g(t) = \tan t + 5$

16. $f(t) = \tan t$; $g(t) = -\tan t$

17. $f(t) = \cos t$; $g(t) = 3 \cos t$

18. $f(t) = \sin t$; $g(t) = -2 \sin t$

19. $f(t) = \sin t$; $g(t) = 3 \sin t + 2$

20. $f(t) = \cos t$; $g(t) = 5 \cos t + 3$

21. $f(t) = \sin t$; $g(t) = \sin(t - 2)$

22. $f(t) = \cos t$; $g(t) = 3 \cos(t + 2) - 3$

In Exercises 23–30 match the function with its graph, which is one of A–J below. [Note: the tangent graphs have erroneous vertical lines where the vertical asymptotes should be.]

23. $f(t) = 4 \sin t$ **24.** $g(t) = -\tan t$ **25.** $h(t) = 3 \tan t$ **26.** $k(t) = 2 - \sin t$

27. $f(t) = 2 \cos t$ **28.** $g(t) = -2 \sin t$ **29.** $h(t) = \cos t + 2$ **30.** $k(t) = 2 - 2 \sin t$

A.
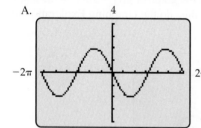

B.
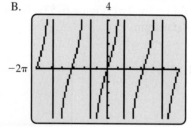

C.

D.
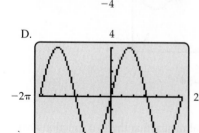

E.
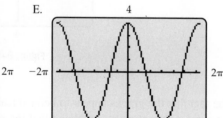

F.
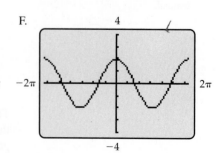

G.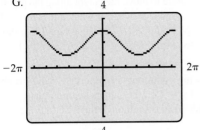

H.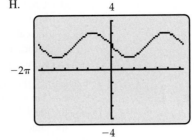

I.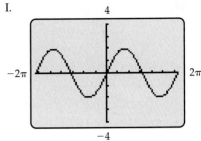

J.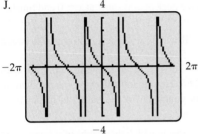

In Exercises 31–38, use the graphs of the trigonometric functions to determine the number of solutions of the equation between 0 and 2π.

31. $\sin t = 3/5$ [*Hint:* How many points on the graph of $f(t) = \sin t$ between $t = 0$ and $t = 2\pi$ have second coordinate 3/5?]

32. $\cos t = -1/4$ **33.** $\tan t = 4$

34. $\cos t = 2/3$ **35.** $\sin t = -1/2$

36. $\sin t = k$, where k is a nonzero constant such that $-1 < k < 1$.

37. $\cos t = k$, where k is a constant such that $-1 < k < 1$.

38. $\tan t = k$, where k is any constant.

In Exercises 39–50, use graphs to determine whether the equation could possibly be an identity or definitely is not an identity.

39. $\sin(-t) = -\sin t$

40. $\cos(-t) = \cos t$

41. $\sin^2 t + \cos^2 t = 1$

42. $\sin(t + \pi) = -\sin t$

43. $\sin t = \cos(t - \pi/2)$

44. $\sin^2 t - \tan^2 t = -(\sin^2 t)(\tan^2 t)$

45. $\dfrac{\sin t}{1 + \cos t} = \tan t$

46. $\dfrac{\cos t}{1 - \sin t} = \dfrac{1}{\cos t} + \tan t$

47. $\cos\left(\dfrac{\pi}{2} + t\right) = -\sin t$

48. $\sin\left(\dfrac{\pi}{2} + t\right) = -\cos t$

49. $(1 + \tan t)^2 = \dfrac{1}{\cos t}$

50. $(\cos^2 t - 1)(\tan^2 t + 1) = -\tan^2 t$

In Exercises 51–54, determine if the graph appears to be the graph of a periodic function. If it is, state the period.

51.

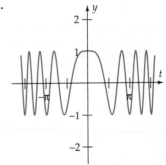

52.

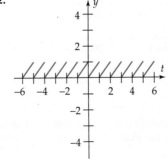

53.

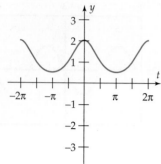

54.

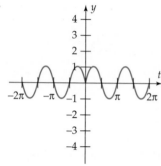

In Exercises 55–60, graph the function. Does the function appear to be periodic? If so, what is the period?

55. $f(t) = \cos |t|$

56. $f(t) = |\cos t|$

57. $g(t) = \sin |t|$

58. $g(t) = |\sin t|$

59. $h(t) = \tan |t|$

60. $h(t) = |\tan t|$

THINKERS

Exercises 61–64, explore various ways in which a calculator can produce inaccurate graphs of trigonometric functions. These exercises also provide examples of two functions, with different graphs, whose graphs appear identical in certain viewing windows.

61. Choose a viewing window with $-3 \le y \le 3$ and $0 \le x \le k$, where k is chosen as follows.

Width of Screen	k
95 pixels (TI-83/84+)	188π
127 pixels (TI-86, Casio)	252π
131 pixels (HP-39gs)	260π
159 pixels (TI-89)	316π

(a) Graph $y = \cos x$ and the constant function $y = 1$ on the same screen. Do the graphs look identical? Are the functions the same?

(b) Use the trace feature to move the cursor along the graph of $y = \cos x$, starting at $x = 0$. For what values of x did the calculator plot points? [*Hint:* $2\pi \approx 6.28$.] Use this information to explain why the two graphs look identical.

62. Using the viewing window in Exercise 61, graph $y = \tan x + 2$ and $y = 2$ on the same screen. Explain why the graphs look identical even though the functions are not the same.

63. The graph of $g(x) = \cos x$ is a series of repeated waves (see Figure 6–47). A full wave (from the peak, down to the trough, and up to the peak again) starts at $x = 0$ and finishes at $x = 2\pi$.

(a) How many full waves will the graph make between $x = 0$ and $x = 502.65$ ($\approx 80 \cdot 2\pi$)?

(b) Graph $g(t) = \cos t$ in a viewing window with $0 \le t \le 502.65$. How many full waves are shown on the graph? Is your answer the same as in part (a)? What's going on?

64. Find a viewing window in which the graphs of $y = \cos x$ and $y = .54$ appear identical. [*Hint:* See the chart in Exercise 61 and note that $\cos 1 \approx .54$.]

Exercises 65–68 provide further examples of functions with different graphs, whose graphs appear identical in certain viewing windows.

65. *Approximating trigonometric functions by polynomials.* For each odd positive integer n, let f_n be the function whose rule is

$$f_n(t) = t - \frac{t^3}{3!} + \frac{t^5}{5!} - \frac{t^7}{7!} + \cdots - \frac{t^n}{n!}.$$

Since the signs alternate, the sign of the last term might be $+$ instead of $-$, depending on what n is. Recall that $n!$ is the product of all integers from 1 to n; for instance, $5! = 1 \cdot 2 \cdot 3 \cdot 4 \cdot 5 = 120$.

(a) Graph $f_7(t)$ and $g(t) = \sin t$ on the same screen in a viewing window with $-2\pi \le t \le 2\pi$. For what values of t does f_7 appear to be a good approximation of g?

(b) What is the smallest value of n for which the graphs of f_n and g appear to coincide in this window? In this case, determine how accurate the approximation is by finding $f_n(2)$ and $g(2)$.

66. For each even positive integer n, let f_n be the function whose rule is

$$f_n(t) = 1 - \frac{t^2}{2!} + \frac{t^4}{4!} - \frac{t^6}{6!} + \frac{t^8}{8!} - \cdots + \frac{t^n}{n!}.$$

(The sign of the last term may be $-$ instead of $+$, depending on what n is.)

(a) In a viewing window with $-2\pi \le t \le 2\pi$, graph f_6, f_{10}, and f_{12}.

(b) Find a value of n for which the graph of f_n appears to coincide (in this window) with the graph of a well-known trigonometric function. What is the function?

67. Find a rational function whose graph appears to coincide with the graph of $h(t) = \tan t$ when

$$-2\pi \le t \le 2\pi.$$

[*Hint:* Exercises 65 and 66.]

68. Find a periodic function whose graph consists of "square waves." [*Hint:* Consider the sum

$$\sin \pi t + \frac{1}{3} \sin 3\pi t + \frac{1}{5} \sin 5\pi t + \frac{1}{7} \sin 7\pi t + \cdots .]$$

69. With your calculator in parametric graphing mode and the range values

$$0 \le t \le 6.28 \qquad -1 \le x \le 6.28 \qquad -2.5 \le y \le 2.5,$$

graph the following two functions on the same screen:

$$x_1 = \cos t,\, y_1 = \sin t \qquad \text{and} \qquad x_2 = t,\, y_2 = \cos t.$$

Using the trace feature, move the cursor along the first graph (the unit circle). Stop at a point on the circle, note the value of t and the x-coordinate of the point. Then switch the trace to the second graph (the cosine function) by using the up or down cursor arrows. The value of t remains the same. How does the y-coordinate of the new point compare with the x-coordinate of the original point on the unit circle? Explain what's going on.

70. (a) Judging from their graphs, which of the functions $f(t) = \sin t$, $g(t) = \cos t$, and $h(t) = \tan t$ appear to be even functions? Which appear to be odd functions?

(b) Confirm your answers in part (a) algebraically by using appropriate identities from Section 6.3.

6.5   Periodic Graphs and Simple Harmonic Motion

Section Objectives

■ Identify the period, amplitude, and phase shift of the functions $f(t) = A \sin(bt + c)$ and $g(t) = A \cos(bt + c)$

■ Explore simple harmonic motion.

We now analyze functions whose rule is of the form

$$f(t) = A \sin(bt + c) \qquad \text{or} \qquad g(t) = A \cos(bt + c),$$

where A, b, and c are constants. Many periodic phenomena can be modeled by such functions, as we shall see below.

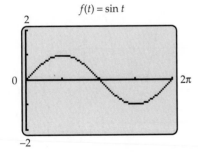

Figure 6–55

PERIOD

The functions $f(t) = \sin t$ and $g(t) = \cos t$ have period 2π, so each of their graphs makes one full wave between 0 and 2π. The sine wave begins on the horizontal axis, rises to height 1, falls to -1, and returns to the axis (Figure 6–55). The cosine wave between 0 and 2π begins at height 1, falls to -1, and rises to height 1 again (Figure 6–56).

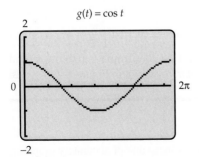

Figure 6–56

GRAPHING EXPLORATION

Graph the following functions, one at a time, in a viewing window with $0 \le t \le 2\pi$. Determine the number of complete waves in each graph and the period of the function (the length of one wave).

$$f(t) = \sin 2t, \qquad g(t) = \cos 3t, \qquad h(x) = \sin 4t, \qquad k(x) = \cos 5t.$$

This exploration suggests the following.

Period

> If $b > 0$, then the graph of either
>
> $$f(t) = \sin bt \qquad \text{or} \qquad g(t) = \cos bt$$
>
> makes b complete waves between 0 and 2π. Hence, each function has period $2\pi/b$.

Although we arrived at this statement by generalizing from several graphs, it can also be explained algebraically.

EXAMPLE 1

The graph of $g(t) = \cos t$ makes one complete wave as t takes values from 0 to 2π. Similarly, the graph of $k(t) = \cos 3t$ will complete one wave as the quantity $3t$ takes values from 0 to 2π. However,

$$3t = 0 \text{ when } t = 0 \qquad \text{and} \qquad 3t = 2\pi \text{ when } t = 2\pi/3.$$

So the graph of $k(t) = \cos 3t$ makes one complete wave between $t = 0$ and $t = 2\pi/3$, as shown in Figure 6–57, and hence k has period $2\pi/3$. Similarly, the graph makes a complete wave from $t = 2\pi/3$ to $t = 4\pi/3$ and another one from $t = 4\pi/3$ to $t = 2\pi$, as shown in Figure 6–57. ∎

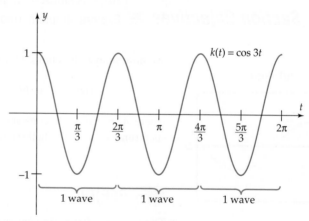

Figure 6–57

EXAMPLE 2

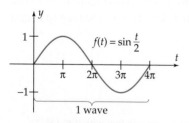

Figure 6–58

According to the box above, the function $f(t) = \sin \frac{1}{2}t$ has period $\dfrac{2\pi}{1/2} = 4\pi$. Its graph makes *half* a wave from $t = 0$ to $t = 2\pi$ (just as $\sin t$ does from $t = 0$ to $t = \pi$) and the other half of the wave from $t = 2\pi$ to $t = 4\pi$, as shown in Figure 6–58. ∎

EXAMPLE 3

Except over *very* tiny intervals, your calculator is incapable of accurately graphing $f(t) = \sin bt$ or $g(t) = \cos bt$ when b is large. For instance, we know that the graph of

$$f(t) = \sin 500t$$

should show 500 complete waves between 0 and 2π. Depending on the model, however, your calculator will produce either garbage (Figure 6–59) or a graph with far fewer than 500 waves (Figure 6–60). For the reason why, see Exercises 59 and 60. ∎

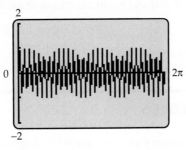

Figure 6–59   Figure 6–60

AMPLITUDE

As we saw in Section 3.4, multiplying the rule of a function by a positive constant has the effect of stretching its graph away from or shrinking it toward the horizontal axis.

EXAMPLE 4

The function $g(t) = 7\cos 3t$ is just the function $k(t) = \cos 3t$ multiplied by 7. Consequently, the graph of g is just the graph of k (which was obtained in Example 1) stretched away from the horizontal axis by a factor of 7, as shown in Figure 6–61.

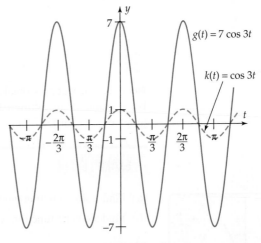

Figure 6–61

Stretching the graph affects only the height of the waves, not the period of the function: Both graphs have period $2\pi/3$, and each full wave has length $2\pi/3$. ∎

The waves of the graph of $g(t) = 7\cos 3t$ in Figure 6–61 rise 7 units above the t-axis and drop 7 units below the axis. More generally, the waves of the graph of $f(t) = A \sin bt$ or $g(t) = A \cos bt$ move a distance of $|A|$ units above and below the t-axis, and we say that these functions have **amplitude** $|A|$. In summary, we have the following.

Amplitude and Period

If $A \neq 0$ and $b > 0$, then each of the functions

$$f(t) = A \sin bt \qquad \text{or} \qquad g(t) = A \cos bt$$

has amplitude $|A|$ and period $2\pi/b$.

EXAMPLE 5

The function $f(t) = -2 \sin 4t$ has amplitude $|-2| = 2$ and period $2\pi/4 = \pi/2$. So the graph consists of waves of length $\pi/2$ that rise and fall between -2 and 2. But be careful: The waves in the graph of $2 \sin 4t$ (like the waves of $\sin t$) begin at height 0, rise, and then fall. But the graph of $f(t) = -2 \sin 4t$ is the graph of $2 \sin 4t$ reflected in the horizontal axis (see page 184). So its waves start at height 0, move *downward,* and then rise, as shown in Figure 6–62. ■

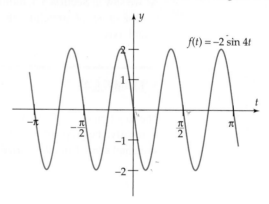

Figure 6–62

PHASE SHIFT

Next, we consider horizontal shifts. As we saw in Section 3.4, the graph of $\sin(t - 3)$ is the graph of $\sin t$ shifted 3 units to the right, and the graph of $\sin(t + 3)$ is the graph of $\sin t$ shifted 3 units to the left.

EXAMPLE 6

(a) Find a sine function whose graph looks like Figure 6–63.

(b) Find a cosine function whose graph looks like Figure 6–63.

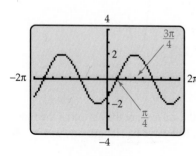

Figure 6–63

SOLUTION

(a) Since each wave has height 2, Figure 6–63 looks like the graph of $2 \sin t$ shifted $\pi/4$ units to the right (so that a sine wave starts at $t = \pi/4$). Since the graph of $2 \sin(t - \pi/4)$ is the graph of $2 \sin t$ shifted $\pi/4$ units to the right (see page 181), we conclude that Figure 6–63 closely resembles the graph of $f(t) = 2 \sin(t - \pi/4)$.

(b) Figure 6–63 also looks like the graph of 2 cos t shifted $3\pi/4$ units to the right (so that a cosine wave starts at $t = 3\pi/4$). Hence, Figure 6–63 could also be the graph of $g(t) = 2 \cos(t - 3\pi/4)$. ∎

EXAMPLE 7

(a) Find the amplitude and the period of

$$f(t) = 3 \sin(2t + 5).$$

(b) Do the same for the function $f(t) = A \sin(bt + c)$, where A, b, c are constants.

SOLUTION   The analysis of $f(t) = 3 \sin(2t + 5)$ is in the left-hand column below, and the analysis of the general case $f(t) = A \sin(bt + c)$ is in the right-hand column. Observe that exactly the same procedure is used in both cases: Just change 3 to A, 2 to b, and 5 to c.

(a) Rewrite the rule of $f(t) = 3 \sin(2t + 5)$ as

$$f(t) = 3 \sin(2t + 5) = 3 \sin\left(2\left(t + \frac{5}{2}\right)\right).$$

Thus, the rule of f can be obtained from the rule of the function $k(t) = 3 \sin 2t$ by replacing t with $t + \frac{5}{2}$. Therefore, the graph of f is just the graph of k shifted horizontally $5/2$ units to the left, as shown in Figure 6–64.

Hence, $f(t) = 3 \sin(2t + 5)$ has the same amplitude as $k(t) = 3 \sin 2t$, namely, 3, and the same period, namely, $2\pi/2 = \pi$.

On the graph of $k(t) = 3 \sin 2t$, a wave begins when $t = 0$. On the graph of

$$f(t) = 3 \sin 2\left(t + \frac{5}{2}\right),$$

the shifted wave begins when $t + 5/2 = 0$, that is, when $t = -5/2$.

(b) Rewrite the rule of $f(t) = A \sin(bt + c)$ as

$$f(t) = A \sin(bt + c) = A \sin\left(b\left(t + \frac{c}{b}\right)\right).$$

Thus, the rule of f can be obtained from the rule of the function $k(t) = A \sin bt$ by replacing t with $t + \frac{c}{b}$. Therefore, the graph of f is just the graph of k shifted horizontally by c/b units.

Hence, $f(t) = A \sin(bt + c)$ has the same amplitude as $k(t) = A \sin bt$, namely, $|A|$, and the same period, namely, $2\pi/b$.

On the graph of $k(t) = A \sin bt$, a wave begins when $t = 0$. On the graph of

$$f(t) = A \sin b\left(t + \frac{c}{b}\right),$$

the shifted wave begins when $t + c/b = 0$, that is, when $t = -c/b$. ∎

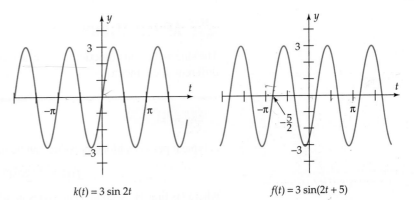

$k(t) = 3 \sin 2t$ $f(t) = 3 \sin(2t + 5)$

Figure 6–64

We say that the function $f(t) = A \sin(bt + c)$ has **phase shift** $-c/b$. A similar analysis applies to the function $g(t) = \cos(bt + c)$ and leads to this conclusion.

Amplitude, Period, and Phase Shift

> If $A \neq 0$ and $b > 0$, then each of the functions
>
> $$f(t) = A \sin(bt + c) \qquad \text{and} \qquad g(t) = A \cos(bt + c)$$
>
> has
>
> $$\text{amplitude } |A|, \qquad \text{period } 2\pi/b, \qquad \text{phase shift } -c/b.$$
>
> A wave of the graph begins at $t = -c/b$.

EXAMPLE 8

Describe the graph of $g(t) = 2 \cos(3t - 4)$.

SOLUTION   The rule of g can be rewritten as

$$g(t) = 2 \cos(3t + (-4)).$$

This is the case described in the preceding box with $A = 2$, $b = 3$, and $c = -4$. Therefore, the function g has

$$\text{amplitude } |A| = |2| = 2, \qquad \text{period } \frac{2\pi}{b} = \frac{2\pi}{3},$$

$$\text{phase shift } -\frac{c}{b} = -\frac{-4}{3} = \frac{4}{3}.$$

Hence, the graph of g consists of waves of length of $2\pi/3$ that run vertically between 2 and -2. A wave begins at $t = 4/3$.

GRAPHING EXPLORATION

Verify the accuracy of this analysis by graphing $y = 2 \cos(3t - 4)$ in the viewing window with $-2\pi \leq t \leq 2\pi$ and $-3 \leq y \leq 3$.

∎

Many other types of trigonometric graphs, including those consisting of waves of varying height and length, are considered in Special Topics 6.5.A.

 **APPLICATIONS**

The sine and cosine functions, or variations of them, can be used to describe many different phenomena.

EXAMPLE 9

A typical person's blood pressure can be modeled by the function

$$f(t) = 22 \cos(2.5\pi t) + 95,$$

where t is time (in seconds) and $f(t)$ is in millimeters of mercury. The highest pressure (systolic) occurs when the heart beats, and the lowest pressure (diastolic)

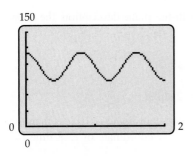

Figure 6–65

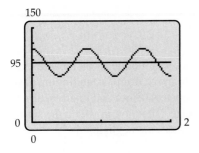

Figure 6–66

occurs when the heart is at rest between beats. The blood pressure is the ratio systolic/diastolic.

(a) Graph the blood pressure function over a period of two seconds and determine the person's blood pressure.

(b) Find the person's pulse rate (number of heartbeats per minute).

SOLUTION

(a) The graph of f is shown in Figure 6–65. The systolic pressure occurs at each local maximum of the graph and the diastolic pressure at each local minimum. Their heights can be determined by using our knowledge of periodic functions. The graph of f is the graph of $22 \cos(2.5\pi t)$ shifted upward by 95 units (as explained in Section 3.4). Since the amplitude of $22 \cos(2.5\pi t)$ is 22, its graph rises 22 units above and falls 22 units below the x-axis. When this graph is shifted 95 units upward, it rises and falls 22 units above and below the horizontal line $y = 95$ (see Figure 6–66), that is

$$\text{from a high of } 95 + 22 = 117 \text{ to a low of } 95 - 22 = 73.$$

In other words, the systolic pressure is 117 and the diastolic pressure is 73. So the person's blood pressure is 117/73.

GRAPHING EXPLORATION

Use a maximum/minimum finder to confirm that the local maxima of the graph in Figure 6–65 occur when $y = 117$ and the local minima when $y = 73$.

(b) The time between heartbeats is the horizontal distance between peaks of the graph, that is, the period of the function. The period of $\cos(2.5\pi t)$ is

$$\frac{2\pi}{b} = \frac{2\pi}{2.5\pi} = .8 \text{ second.}$$

Since one minute is 60 seconds, the number of beats per minute (pulse rate) is

$$\frac{60}{.8} = 75. \qquad \blacksquare$$

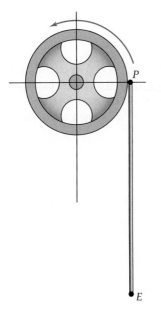

Figure 6–67

EXAMPLE 10

A wheel of radius 2 centimeters is rotating counterclockwise at 3 radians per second. A free-hanging rod 10 centimeters long is connected to the edge of the wheel at point P and remains vertical as the wheel rotates (Figure 6–67). Assuming that the center of the wheel is at the origin and that P is at $(2, 0)$ at time $t = 0$, find a function that describes the y-coordinate of the tip E of the rod at time t.

SOLUTION   The wheel is rotating at 3 radians per second, so after t seconds, the point P has moved through an angle of $3t$ radians and is 2 units from the

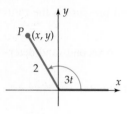

Figure 6–68

origin, as shown in Figure 6–68. By the point-in-the-plane description, the coordinates (x, y) of P satisfy

$$\frac{x}{2} = \cos 3t \qquad \frac{y}{2} = \sin 3t$$

$$x = 2 \cos 3t \qquad y = 2 \sin 3t.$$

Since E lies 10 centimeters directly below P, its y-coordinate is 10 less than the y-coordinate of P. Hence, the function giving the y-coordinate of E at time t is

$$f(t) = y - 10 = 2 \sin 3t - 10. \qquad \blacksquare$$

EXAMPLE 11

Suppose that a weight hanging from a spring is set in motion by an upward push (Figure 6–69) and that it takes 5 seconds for it to move from its equilibrium position to 8 centimeters above, then drop to 8 centimeters below, and finally return to its equilibrium position. [We consider an idealized situation in which the spring has perfect elasticity and friction, air resistance, etc., are negligible.]

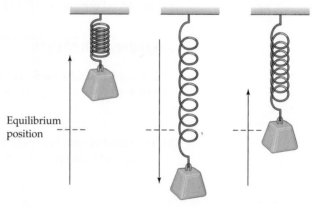

Figure 6–69

Let $h(t)$ denote the distance of the weight above $(+)$ or below $(-)$ its equilibrium position at time t. Then $h(t)$ is 0 when $t = 0$. As t runs from 0 to 5, $h(t)$ increases from 0 to 8, decreases to -8, and increases again to 0. In the next 5 seconds, it repeats the same pattern, and so on. Thus, the graph of h has some kind of wave shape. Two possibilities are shown in Figure 6–70.

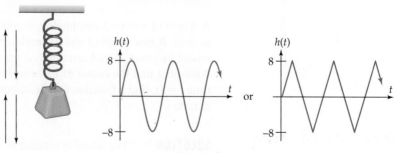

Figure 6–70

Careful physical experimentation suggests that the left-hand curve in Figure 6–70, which resembles the sine graphs studied earlier, is a reasonably accurate model of this process. Facts from physics, calculus, and differential equations show that the rule of the function h is the form $h(t) = A \sin(bt + c)$ for some constants A, b, c. Since the amplitude of h is 8, its period is 5, and its phase shift is 0, the constants A, b, and c must satisfy

$$A = 8, \qquad \frac{2\pi}{b} = 5, \qquad -\frac{c}{b} = 0$$

or, equivalently,

$$A = 8 \qquad b = \frac{2\pi}{5}, \qquad c = 0.$$

Therefore, the motion of the moving spring can be described by the function

$$h(t) = A \sin(bt + c) = 8 \sin\left(\frac{2\pi}{5}t + 0\right) = 8 \sin\frac{2\pi t}{5}. \qquad \blacksquare$$

Motion that can be described by a function of the form $f(t) = A \sin(bt + c)$ or $f(t) = A \cos(bt + c)$ is called **simple harmonic motion.** Many kinds of physical motion are simple harmonic motions. Other periodic phenomena, such as sound waves, are more complicated to describe. Their graphs consist of waves of varying amplitude. Such graphs are discussed in Special Topics 6.5.A.

EXAMPLE 12*

The table shows the average monthly temperature in Cleveland, OH, based on 30 years of data from the National Climatic Data Center. Since average temperatures are not likely to vary much from year to year, the data essentially repeats the same pattern in subsequent years. So a periodic model is appropriate.

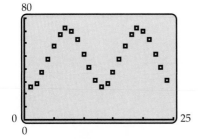

Figure 6–71

Month	Temperature (°F)	Month	Temperature (°F)
Jan	25.7	Jul	71.9
Feb	28.4	Aug	70.2
Mar	37.5	Sep	63.3
Apr	47.6	Oct	52.2
May	58.5	Nov	41.8
Jun	67.5	Dec	31.1

The data for a two-year period is plotted in Figure 6–71 (with $x = 1$ corresponding to January, $x = 2$ to February, and so on).[†] The sine regression feature on a calculator produces this model from the 24 data points:

$$y = 22.7 \sin(.5219x - 2.1842) + 49.5731.$$

The period of this function is $2\pi/.5219 \approx 12.04$ slightly off from the 12-month period we would expect. However, its graph in Figure 6–72 appears to fit the data well. $\qquad \blacksquare$

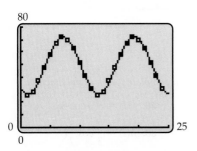

Figure 6–72

---

*Skip this example if you haven't read Sections 2.5 and 5.5 on regression.
[†]The reasons why a two-year period is used are considered in Exercises 62 and 63.

EXERCISES 6.5

In Exercises 1–7, state the amplitude, period, and phase shift of the function.

1. $g(t) = 3 \sin(2t - \pi)$ **2.** $h(t) = -6 \cos(4t - \pi/4)$

3. $q(t) = -5 \sin(5t + 1/5)$

4. $g(t) = 97 \cos(14t + 5)$

5. $f(t) = \cos 2\pi t$ **6.** $k(t) = \cos(2\pi t/3)$

7. $p(t) = 6 \cos(3\pi t + 1)$

8. (a) What is the period of $f(t) = \sin 2\pi t$?
   (b) For what values of t (with $0 \le t \le 2\pi$) is $f(t) = 0$?
   (c) For what values of t (with $0 \le t \le 2\pi$) is $f(t) = 1$? or $f(t) = -1$?

In Exercises 9–14, give the rule of a periodic function with the given numbers as amplitude, period, and phase shift (in this order).

9. $3, \pi/4, \pi/5$ **10.** $4, 5, 0$ **11.** $3/4, 2, 0$

12. $4/5, 3, 1$ **13.** $7, 5/3, -\pi/2$ **14.** $18, 3, -6$

In Exercises 15–18, state the rule of a function of the form $f(t) = A \sin bt$ or $g(t) = A \cos bt$ whose graph appears to be identical to the given graph.

15.

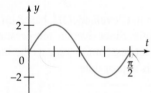

16.

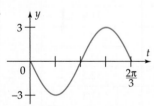

17.

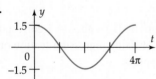

18.

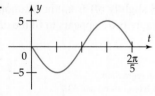

In Exercises 19–22,
(a) *State the period of the function.*
(b) *Describe the graph of the function between 0 and 2π.*
(c) *Find a viewing window that accurately shows exactly four complete waves of the graph.*

19. $f(t) = \sin(200t)$ **20.** $f(t) = \sin 600t$

21. $g(t) = \cos 900t$ **22.** $g(t) = \cos 575t$

In Exercises 23–26,
(a) *State the rule of a function of the form*
$$f(t) = A \sin(bt + c)$$
whose graph appears to be identical with the given graph.
(b) *State the rule of a function of the form*
$$g(t) = A \cos(bt + c)$$
whose graph appears to be identical with the given graph.

23.

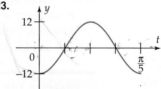

24.

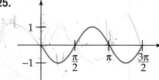

25.

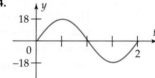

26.

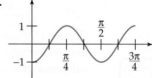

In Exercises 27–32, sketch a complete graph of the function.

27. $k(t) = -3 \sin t$ **28.** $y(t) = -2 \cos 3t$

29. $p(t) = -\dfrac{1}{2} \sin 2t$ **30.** $q(t) = \dfrac{2}{3} \cos \dfrac{3}{2}t$

31. $h(t) = 3 \sin(2t + \pi/2)$ **32.** $p(t) = 3 \cos(3t - \pi)$

In Exercises 33–36, graph the function over the interval $[0, 2\pi)$ *and determine the location of all local maxima and minima. [This can be done either graphically or algebraically.]*

33. $f(t) = \dfrac{1}{2} \sin\left(t - \dfrac{\pi}{3}\right)$

34. $g(t) = 2 \sin(2t/3 - \pi/9)$

35. $f(t) = -2 \sin(3t - \pi)$

36. $h(t) = \dfrac{1}{2} \cos\left(\dfrac{\pi}{2}t - \dfrac{\pi}{8}\right) + 1$

In Exercises 37–40, graph f(t) in a viewing window with $-2\pi \le t \le 2\pi$. *Use a maximum finder and a root finder to determine constants A, b, c such that the graph of f(t) appears to coincide with the graph of* $g(t) = A \sin(bt + c)$.

37. $f(t) = 3 \sin t + 2 \cos t$

38. $f(t) = -5 \sin t + 3 \cos t$

39. $f(t) = 3 \sin(4t + 2) + 2 \cos(4t - 1)$

40. $f(t) = 2 \sin(3t - 5) - 3 \cos(3t + 2)$

In Exercises 41 and 42, explain why there could not possibly be constants A, b, and c such that the graph of $g(t) = A \sin(bt + c)$ *coincides with the graph of f(t).*

41. $f(t) = \sin 2t + \cos 3t$

42. $f(t) = 2 \sin(3t - 1) + 3 \cos(4t + 1)$

43. Do parts (a) and (b) of Example 9 for a person whose blood pressure is given by

$$g(t) = 21 \cos(2.5\pi t) + 113.$$

According to current guidelines, someone with systolic pressure above 140 or diastolic pressure above 90 has high blood pressure and should see a doctor about it. What would you advise the person in this case?

44. Find the function in Example 10 if the wheel has a radius of 13 centimeters and the rod is 18 centimeters long.

45. The volume $V(t)$ of air (in cubic inches) in an adult's lungs t seconds after exhaling is approximately

$$V(t) = 55 + 24.5 \sin\left(\dfrac{\pi x}{2} - \dfrac{\pi}{2}\right).$$

(a) Find the maximum and minimum amount of air in the lungs.

(b) How often does the person exhale?

(c) How many breaths per minute does the person take?

46. The brightness of the binary star Beta Lyrae (as seen from the earth) varies. Its visual magnitude $M(t)$ after t days is approximately

$$M(t) = .55 \cos(.97t) + 3.85.$$

The visual magnitude scale is reversed from what you would expect: The lower the number, the brighter the star. With this in mind, answer the following questions.

(a) Graph the function M when $0 \le t \le 21$.

(b) What is the visual magnitude when the star is brightest? When it is dimmest?

(c) What is the period of the magnitude (the interval between its brightest times)?

47. The current generated by an AM radio transmitter is given by a function of the form $f(t) = A \sin 2000\pi m t$, where $550 \le m \le 1600$ is the location on the broadcast dial and t is measured in seconds. For example, a station at 980 on the AM dial has a function of the form

$$f(t) = A \sin 2000\pi(980)t = A \sin 1{,}960{,}000\pi t.$$

Sound information is added to this signal by varying (modulating) A, that is, by changing the amplitude of the waves being transmitted. (*AM* means "amplitude modulation.") For a station at 980 on the dial, what is the period of function f? What is the frequency (number of complete waves per second)?

48. The number of hours of daylight in Winnipeg, Manitoba, can be approximated by

$$d(t) = 4.15 \sin(.0172t - 1.377) + 12,$$

where t is measured in days, with $t = 1$ being January 1.

(a) On what day is there the most daylight? The least? How much daylight is there on these days?

(b) On which days are there 11 hours or more of daylight?

(c) What do you think the period of this function is? Why?

49. The original Ferris wheel, built by George Ferris for the Columbian Exposition of 1893, was much larger and slower than its modern counterparts: It had a diameter of 250 feet and contained 36 cars, each of which held 60 people; it made one revolution every 10 minutes. Imagine that the Ferris wheel revolves counterclockwise in the x-y plane with its center at the origin. A car had coordinates (125, 0) at time $t = 0$. Find the rule of a function that gives the y-coordinate of the car at time t.

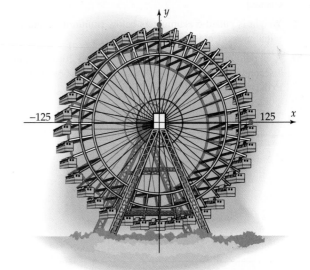

50. Do Exercise 49 if the wheel turns at 2 radians per minute and the car is at $(0, -125)$ at time $t = 0$.

51. A circular wheel of radius 1 foot rotates counterclockwise. A 4-foot-long rod has one end attached to the edge of this wheel and the other end to the base of a piston (see the figure). It transfers the rotary motion of the wheel into a back-and-forth linear motion of the piston. If the wheel is rotating at 10 revolutions per second, point W is at $(1, 0)$ at time $t = 0$, and point P is always on the x-axis, find the rule of a function that gives the x-coordinate of P at time t.

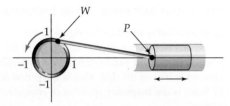

52. Do Exercise 51 if the wheel has a radius of 2 feet, rotates at 50 revolutions per second, and is at $(2, 0)$ when $t = 0$.

In Exercises 53–56, suppose there is a weight hanging from a spring (under the same idealized conditions as described in Example 11). The weight is given a push to start it moving. At any time t, let h(t) be the height (or depth) of the weight above (or below) its equilibrium point. Assume that the maximum distance the weight moves in either direction from the equilibrium point is 6 centimeters and that it moves through a complete cycle every 4 seconds. Express h(t) in terms of the sine or cosine function under the stated conditions.

53. Initial push is *upward* from the equilibrium point.

54. Initial push is *downward* from the equilibrium point. [*Hint:* What does the graph of $A \sin bt$ look like when $A < 0$?]

55. Weight is pulled 6 centimeters above equilibrium, and the initial movement (at $t = 0$) is downward. [*Hint:* Think cosine.]

56. Weight is pulled 6 centimeters below equilibrium, and the initial movement is upward.

57. A pendulum swings uniformly back and forth, taking 2 seconds to move from the position directly above point A to the position directly above point B.

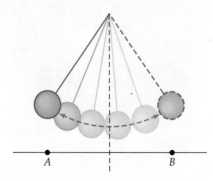

The distance from A to B is 20 centimeters. Let $d(t)$ be the horizontal distance from the pendulum to the (dashed) center line at time t seconds (with distances to the right of the line measured by positive numbers and distances to the left by negative ones). Assume that the pendulum is on the center line at time $t = 0$ and moving to the right. Assume that the motion of the pendulum is simple harmonic motion. Find the rule of the function $d(t)$.

58. The diagram shows a merry-go-round that is turning counterclockwise at a constant rate, making 2 revolutions in 1 minute. On the merry-go-round are horses A, B, C, and D at 4 meters from the center and horses E, F, and G at 8 meters from the center. There is a function $a(t)$ that gives the distance the horse A is from the y-axis (this is the x-coordinate of the position A is in) as a function of time t (measured in minutes). Similarly, $b(t)$ gives the x-coordinate for B as a function of time, and so on. Assume that the diagram shows the situation at time $t = 0$.

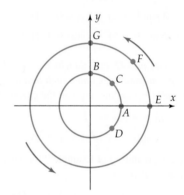

(a) Which of the following functions does $a(t)$ equal?

$$4 \cos t, \quad 4 \cos \pi t, \quad 4 \cos 2t, \quad 4 \cos 2\pi t,$$
$$4 \cos \left(\tfrac{1}{2}t\right), \quad 4 \cos \left((\pi/2)t\right), \quad 4 \cos 4\pi t$$

Explain.

(b) Describe the functions $b(t)$, $c(t)$, $d(t)$, and so on using the cosine function:

$$b(t) = \underline{\quad}, c(t) = \underline{\quad}, d(t) = \underline{\quad}.$$
$$e(t) = \underline{\quad}, f(t) = \underline{\quad}, g(t) = \underline{\quad}.$$

(c) Suppose the x-coordinate of a horse S is given by the function $4 \cos(4\pi t - (5\pi/6))$ and the x-coordinate of another horse T is given by $8 \cos(4\pi t - (\pi/3))$. Where are these horses located in relation to the rest of the horses? Mark the positions of T and S at $t = 0$ into the figure.

Exercises 59–60 explore various ways in which a calculator can produce inaccurate or misleading graphs of trigonometric functions.

59. (a) If you were going to draw a rough picture of a full wave of the sine function by plotting some points and connecting them with straight-line segments, approximately how many points would you have to plot?

(b) If you were drawing a rough sketch of the graph of $f(t) = \sin 100t$ when $0 \le t \le 2\pi$, according to the method in part (a), approximately how many points would have to be plotted?

(c) How wide (in pixels) is your calculator screen? Your answer to this question is the maximum number of points that your calculator plots when graphing any function.

(d) Use parts (a)–(c) to explain why your calculator cannot possibly produce an accurate graph of $f(t) = \sin 100t$ in any viewing window with $0 \le t \le 2\pi$.

60. (a) Using a viewing window with $0 \le t \le 2\pi$, use the trace feature to move the cursor along the horizontal axis. [On some calculators, it may be necessary to graph $y = 0$ to do this.] What is the distance between one pixel and the next (to the nearest hundredth)?

(b) What is the period of $f(t) = \sin 300t$? Since the period is the length of one full wave of the graph, approximately how many waves should there be between two adjacent pixels? What does this say about the possibility of your calculator's producing an accurate graph of this function between 0 and 2π?

61. The table below shows the number of unemployed people in the labor force (in millions) for 1984–2005.*

(a) Sketch a scatter plot of the data, with $x = 0$ corresponding to 1980.

(b) Does the data appear to be periodic? If so, find an appropriate model.

(c) Do you think this model is likely to be accurate much beyond 2005? Why?

Year	Unemployed	Year	Unemployed
1984	8.539	1995	7.404
1985	8.312	1996	7.236
1986	8.237	1997	6.739
1987	7.425	1998	6.210
1988	6.701	1999	5.880
1989	6.528	2000	5.692
1990	7.047	2001	6.801
1991	8.628	2002	8.378
1992	9.613	2003	8.774
1993	8.940	2004	8.149
1994	7.996	2005	7.591

In Exercises 62 and 63, do the following.

(a) Use 12 data points (with $x = 1$ corresponding to January) to find a periodic model of the data.

*U.S. Bureau of Labor Statistics.

(b) What is the period of the function found in part (a)? Is this reasonable?

(c) Plot 24 data points (two years) and graph the function from part (a) on the same screen. Is the function a good model in the second year?

(d) Use the 24 data points in part (c) to find another periodic model for the data.

(e) What is the period of the function in part (d)? Does its graph fit the data well?

62. The table shows the average monthly temperature in Chicago, IL, based on data from 1971 to 2000.*

Month	Temperature (°F)
Jan	22.0
Feb	27.0
Mar	37.3
Apr	47.8
May	58.7
Jun	68.2
Jul	73.3
Aug	71.7
Sep	63.8
Oct	52.1
Nov	39.3
Dec	27.4

63. The table shows the average monthly precipitation (in inches) in San Francisco, CA, based on data from 1971 to 2000.†

Month	Precipitation
Jan	4.45
Feb	4.01
Mar	3.26
Apr	1.17
May	.38
Jun	.11
Jul	.03
Aug	.07
Sep	.2
Oct	1.04
Nov	2.49
Dec	2.89

*National Climatic Data Center.
†National Climatic Data Center.

THINKERS

64. On the basis of the results of Exercises 37–42, under what conditions on the constants a, k, h, d, r, s does it appear that the graph of

$$f(t) = a \sin(kt + h) + d \cos(rt + s)$$

coincides with the graph of the function

$$g(t) = A \sin(bt + c)?$$

65. A grandfather clock has a pendulum length of k meters and its swing is given (as in Exercise 57) by the function

$f(t) = .25 \sin(\omega t)$, where

$$\omega = \sqrt{\frac{9.8}{k}}.$$

(a) Find k such that the period of the pendulum is 2 seconds.

(b) The temperature in the summer months causes the pendulum to increase its length by .01%. How much time will the clock lose in June, July, and August? [*Hint:* These three months have a total of 92 days (7,948,800 seconds). If k is increased by .01%, what is $f(2)$?]

6.5.A *SPECIAL TOPICS* Other Trigonometric Graphs

Section Objectives
- Explore the behavior of sinusoidal functions and their graphs.
- Explore the graphs of damped and compressed trigonometric functions.

A graphing calculator or computer enables you to explore with ease a wide variety of trigonometric functions.

GRAPHING EXPLORATION

Graph

$$g(t) = \cos t \qquad \text{and} \qquad f(t) = \sin(t + \pi/2)$$

on the same screen. Is there any apparent difference between the two graphs?

This exploration suggests that the equation $\cos t = \sin(t + \pi/2)$ is an identity and hence that the graph of the cosine function can be obtained by horizontally shifting the graph of the sine function. This is indeed the case, as will be proved in Section 7.2. Consequently, every graph in Section 6.5 is actually the graph of a function of the form $f(t) = A \sin(bt + c)$. In fact, considerably more is true.

EXAMPLE 1

Show that the graph of

$$g(t) = -2 \sin(t + 7) + 3 \cos(t + 2)$$

appears identical to the graph of a function of the form $f(t) = A \sin(bt + c)$ for suitable constants A, b, and c.

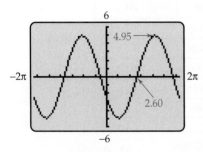

Figure 6–73

SOLUTION The function $g(t)$ has period 2π because this is the period of both $\sin(t + 7)$ and $\cos(t + 2)$. Its graph in Figure 6–73 consists of repeating waves of uniform height. By using a maximum finder and a root finder, we see that the maximum height of a wave is approximately 4.95 and that a wave similar to a sine wave begins at approximately $t = 2.60$, as indicated in Figure 6–73. Thus, the graph looks very much like a sine wave with amplitude 4.95 and phase shift 2.60. As we saw in Section 6.5, the function

$$f(t) = 4.95 \sin(t - 2.60)$$

has amplitude of 4.95, period $2\pi/1 = 2\pi$, and phase shift $-(-2.60)/1 = 2.60$.

GRAPHING EXPLORATION

Graph

$$g(t) = -2 \sin(t + 7) + 3 \cos(t + 2)$$

and

$$f(t) = 4.95 \sin(t - 2.60)$$

on the same screen. Do the graphs look identical?

Example 1 is an illustration (but not a proof) of the following fact.

Sinusoidal Graphs

If b, D, E, r, s are constants, then the graph of the function

$$g(t) = D \sin(bt + r) + E \cos(bt + s)$$

is a sine curve: There exist constants A and c such that

$$D \sin(bt + r) + E \cos(bt + s) = A \sin(bt + c).$$

EXAMPLE 2

Estimate the constants A, b, c such that

$$A \sin(bt + c) = 4 \sin(3t + 2) + 2 \cos(3t - 4).$$

SOLUTION   The function $g(t) = 4 \sin(3t + 2) + 2 \cos(3t - 4)$ has period $2\pi/3$ because this is the period of both $\sin(3t + 2)$ and $\cos(3t - 4)$. The function $f(t) = A \sin(bt + c)$ has period $2\pi/b$. So we must have

$$\frac{2\pi}{b} = \frac{2\pi}{3}, \qquad \text{or equivalently,} \qquad b = 3.$$

Using a maximum finder and a root finder on the graph of

$$g(t) = 4 \sin(3t + 2) + 2 \cos(3t - 4)$$

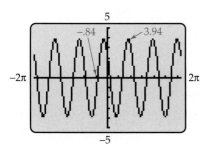

Figure 6–74

in Figure 6–74, we see that the maximum height (amplitude) of a wave is approximately 3.94 and that a sine wave begins at approximately $t = -.84$. Therefore, the graph has (approximate) amplitude 3.94 and phase shift $-.84$. Since $b = 3$ and $f(t) = A \sin(bt + c)$ has amplitude $|A|$ and phase shift $-c/b = -c/3$, we have $A \approx 3.94$ and

$$-\frac{c}{3} \approx -.84 \qquad \text{or, equivalently,} \qquad c \approx 3(.84) = 2.52.$$

Therefore,

$$3.94 \sin(3t + 2.52) \approx 4 \sin(3t + 2) + 2 \cos(3t - 4).$$

GRAPHING EXPLORATION

Graphically confirm this fact by graphing

$$f(t) = 3.94 \sin(3t + 2.52)$$

and

$$g(t) = 4 \sin(3t + 2) + 2 \cos(3t - 4)$$

on the same screen. Do the graphs appear identical?

In the preceding examples, the variable t had the same coefficient in both the sine and cosine term of the function's rule. When this is not the case, the graph will consist of waves of varying size and shape, as you can readily illustrate.

GRAPHING EXPLORATION

Graph each of the following functions separately in the viewing window with $-2\pi \le t \le 2\pi$ and $-6 \le y \le 6$.

$$f(t) = \sin 3t + \cos 2t, \qquad g(t) = -2 \sin(3t + 5) + 4 \cos(t + 2),$$

$$h(t) = 2 \sin 2t - 3 \cos 3t.$$

EXAMPLE 3

Find a complete graph of

$$f(t) = 4 \sin 100\pi t + 2 \cos 40\pi t.$$

SOLUTION   If you graph f in a window with $-2\pi \le t \le 2\pi$, you will get garbage on the screen (try it!). Trial and error might lead to a viewing window that shows a readable graph, but the graph might not be accurate. A better procedure is to note that this is a periodic function. Hence, we need only graph it over one period to have a complete graph.

To find the period of $f(t)$, we must first find the periods of its components, $\sin 100\pi t$ and of $\cos 40 \pi t$.

$$\text{Period of } \sin 100 \, \pi t = \frac{2\pi}{100\pi} = \frac{1}{50} = .02$$

$$\text{Period of } \cos 40 \, \pi t = \frac{2\pi}{40\pi} = \frac{1}{20} = .05$$

The period of $f(t) = 4 \sin 100 \, \pi t + 2 \cos 40 \, \pi t$ is the least common integer multiple of .02 and .05. The integer multiples of .02 are:

$$.02, \; 2(.02), \; 3(.02), \; 4(.02), \; 5(.02), \; 6(.02), \ldots$$

that is,

$$.02, \; .04, \; .06, \; .08, \; .10, \; .12, \ldots$$

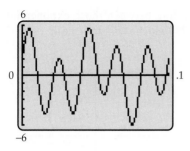

Figure 6–75

Similarly, the integer multiples of .05 are

$$.05, \ 2(.05) = .10, \ldots$$

We need go no further, since we now see that .1 is the smallest number that is on both lists. Hence, the period of $f(t)$ is .1. By graphing f in the viewing window with $0 \le t \le .1$ and $-6 \le y \le 6$, we obtain the complete graph in Figure 6–75. ■

DAMPED AND COMPRESSED TRIGONOMETRIC GRAPHS

Many physical situations can be described by functions whose graphs consist of waves of different heights. Other situations (for instance, sound waves in FM radio transmission) are modeled by functions whose graphs consist of waves of uniform height and varying frequency. Here are some examples of such functions.

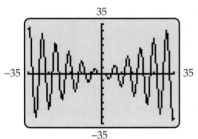

Figure 6–76

EXAMPLE 4

Explain why the graph of $f(t) = t \cos t$ in Figure 6–76 has the shape it does.

SOLUTION   The graph appears to consist of waves that get larger and larger as you move away from the origin. To explain this situation, we analyze the situation algebraically. We know that

$$-1 \le \cos t \le 1 \quad \text{for every } t.$$

If we multiply each term of this inequality by t and remember the rules for changing the direction of inequalities when multiplying by negatives, we see that

$$-t \le t \cos t \le t \quad \text{when } t \ge 0$$

and

$$-t \ge t \cos t \ge t \quad \text{when } t < 0.$$

In graphical terms, this means that the graph of $f(x) = t \cos t$ lies between the straight lines $y = t$ and $y = -t$, with the waves growing larger or smaller to fill this space. The graph touches the lines $y = \pm t$ exactly when $t \cos t = \pm t$, that is, when $\cos t = \pm 1$. This occurs when $t = 0, \pm \pi, \pm 2\pi, \pm 3\pi, \ldots$.

GRAPHING EXPLORATION

Illustrate this analysis by graphing $f(t) = t \cos t$, $y = t$, and $y = -t$ on the same screen.

■

EXAMPLE 5

No single viewing window gives a completely readable graph of $g(t) = .5^t \sin t$ (try some). To the left of the y-axis, the graph gets quite large, but to the right, it almost coincides with the horizontal axis. To get a better mental picture, note that $.5^t > 0$ for every t. Multiplying each term of the known inequality $-1 \le \sin t \le 1$ by $.5^t$, we see that

$$-.5^t \le .5^t \sin t \le .5^t \quad \text{for every } t.$$

Hence, the graph of g lies between the graphs of the exponential functions $y = -.5^t$ and $y = .5^t$, which are shown in Figure 6–77 on the next page. The graph of g will consist of sine waves rising and falling between those exponential graphs, as indicated in the sketch in Figure 6–78 (which is not to scale).

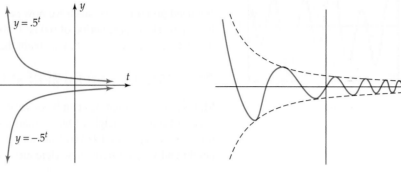

Figure 6–77   Figure 6–78

The best you can do with a calculator is to look at various viewing windows in which a portion of the graph is readable.

GRAPHING EXPLORATION

Find viewing windows that clearly show the graph of $g(t) = .5^t \sin t$ in each of these ranges.

$$-2\pi \le t \le 0, \qquad 0 \le t \le 2\pi, \qquad 2\pi \le t \le 4\pi.$$

EXAMPLE 6

If you graph $f(t) = \sin(\pi/t)$ in a wide viewing window such as Figure 6–79, it is clear that the horizontal axis is an asymptote of the graph.* Near the origin, however, the graph is not very readable, even in a very narrow viewing window like Figure 6–80.

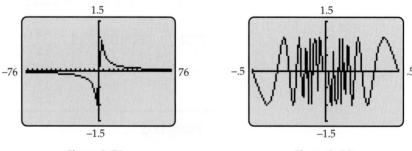

Figure 6–79   Figure 6–80

To understand the behavior of f near the origin, consider what happens as you move left from $t = 1/2$ to $t = 0$:

As t goes from $\dfrac{1}{2}$ to $\dfrac{1}{4}$, then $\dfrac{\pi}{t}$ goes from $\dfrac{\pi}{1/2} = 2\pi$ to $\dfrac{\pi}{1/4} = 4\pi$.

---

*This can also be demonstrated algebraically: When t is very large in absolute value, then π/t is very close to 0 by the Big-Little Principle, and hence, $\sin(\pi/t)$ is very close to 0 as well.

As π/t takes all values from 2π to 4π, the graph of $f(t) = \sin(\pi/t)$ makes one complete sine wave. Similarly,

$$\text{As } t \text{ goes from } \frac{1}{4} \text{ to } \frac{1}{6}, \text{ then } \frac{\pi}{t} \text{ goes from } \frac{\pi}{1/4} = 4\pi \text{ to } \frac{\pi}{1/6} = 6\pi.$$

As π/t takes all values from 4π to 6π, the graph of $f(t) = \sin(\pi/t)$ makes another complete sine wave. The same pattern continues, so the graph of f makes a complete wave from $t = 1/2$ to $t = 1/4$, another from $t = 1/4$ to $t = 1/6$, another from $t = 1/6$ to $t = 1/8$, and so on. A similar phenomenon occurs as t takes values between $-1/2$ and 0. Consequently, the graph of f near 0 oscillates infinitely often between -1 and 1, with the waves becoming more and more compressed as t gets closer to 0, as indicated in Figure 6–81. Since the function is not defined at $t = 0$, the left and right halves of the graph are not connected. ■

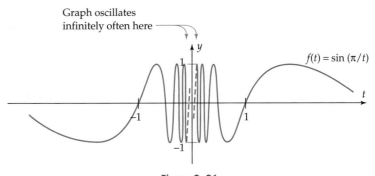

Figure 6–81

EXAMPLE 7

Describe the graph of $g(t) = \cos e^t$.

SOLUTION   When t is negative, then e^t is very close to 0 (why?), and hence, $\cos e^t$ is very close to 1. Therefore, the horizontal line $y = 1$ is an asymptote of the half of the graph to the left of the origin. As t takes increasing positive values, the corresponding values of e^t increase at a much faster rate (remember exponential growth). For instance, as t goes from 0 to 2π,

$$e^t \text{ goes from } e^0 = 1 \text{ to } e^{2\pi} \approx 535.5 \approx 170\pi = 85(2\pi).$$

Consequently, $\cos e^t$ runs through 85 periods, that is, the graph of g makes 85 full waves between 0 and 2π. As t gets larger, the graph of g makes waves at a faster and faster rate.

GRAPHING EXPLORATION

To see how compressed the waves become, graph $g(t)$ in three viewing windows, with

$$0 \leq t \leq 3.5, \qquad 4.5 \leq t \leq 5, \qquad 6 \leq t \leq 6.2,$$

and note how the number of waves increases in each succeeding window, even though the widths of the windows are getting smaller.

■

EXERCISES 6.5.A

In Exercises 1–6, estimate constants A, b, c such that

$$f(t) = A \sin(bt + c).$$

1. $f(t) = \sin t + 2 \cos t$

2. $f(t) = 3 \sin t + 2 \cos t$

3. $f(t) = 2 \sin 4t - 5 \cos 4t$

4. $f(t) = 3 \sin(2t - 1) + 4 \cos(2t + 3)$

5. $f(t) = -5 \sin(3t + 2) + 2 \cos(3t - 1)$

6. $f(t) = .3 \sin(2t + 4) - .4 \cos(2t - 3)$

In Exercises 7–16, find a viewing window that shows a complete graph of the function.

7. $g(t) = (5 \sin 2t)(\cos 5t)$ **8.** $h(t) = e^{\sin t}$

9. $f(t) = t/2 + \cos 2t$

10. $g(t) = \sin\left(\dfrac{t}{3} - 2\right) + 2 \cos\left(\dfrac{t}{4} - 2\right)$

11. $h(t) = \sin 100t + \cos 50t$

12. $f(t) = 3 \sin(200t + 1) - 2 \cos(300t + 2)$

13. $g(t) = -6 \sin(250\pi t + 5) + 3 \cos(400\pi t - 7)$

14. $h(t) = 4 \sin(600\pi t + 3) - 6 \cos(500\pi t - 3)$

15. $f(t) = 4 \sin .2\pi t - 8 \cos .1\pi t$

16. $g(t) = 9 \sin .05\pi t + 5 \cos .04\pi t$

In Exercises 17–24, describe the graph of the function verbally (including such features as asymptotes, undefined points, amplitude and number of waves between 0 and 2π, etc.) as in Examples 4–6. Find viewing windows that illustrate the main features of the graph.

17. $g(t) = \sin e^t$

18. $h(t) = \dfrac{\cos 2t}{1 + t^2}$

19. $f(t) = \sqrt{|t|} \cos t$

20. $g(t) = e^{-t^2/8} \sin 2\pi t$

21. $h(t) = \dfrac{1}{t} \sin t$

22. $f(t) = t \sin \dfrac{1}{t}$

23. $g(t) = \ln |\cos t|$

24. $h(t) = \ln |\sin t + 1|$

25. At a beach in Maui, Hawaii, the level of the tides is approximated by

$$f(t) = -.7 \sin(.52t - 1.3728) + .73 \cos(.26t - .6864) + 1.4,$$

where t is measured in hours and $f(t)$ in feet.

(a) Graph the tide function over a three-day period.

(b) At approximately what times during the day does the highest tide occur? The lowest?

(c) What is the period of this function?

6.6   Other Trigonometric Functions

Section Objectives
- Define and graph the cotangent, secant and cosecant functions.
- Use the point-in-the-plane description of these functions.
- Apply the periodicity and Pythagorean identities for these functions.

This section introduces three more trigonometric functions. It is divided into three parts, each of which may be covered earlier, as shown in the table.

Subsection of Section 6.6	May be covered at the end of
Part I	Section 6.2
Part II	Section 6.3
Part III	Section 6.4

PART I: Definitions and Descriptions

The three remaining trigonometric functions are defined in terms of sine and cosine, as follows.

Definition of Cotangent,
Secant, and
Cosecant Functions

Name of Function	Value of Function at t Is Denoted	Rule of Function
contangent	$\cot t$	$\cot t = \dfrac{\cos t}{\sin t}$
secant	$\sec t$	$\sec t = \dfrac{1}{\cos t}$
cosecant	$\csc t$	$\csc t = \dfrac{1}{\sin t}$

The domain of each function consists of all real numbers for which the denominator is not 0. The graphs of the sine and cosine function in Section 6.4 show that $\sin t = 0$ only when $t = 0, \pm\pi, \pm2\pi, \pm3\pi, \ldots$ and $\cos t = 0$ only when $t = \pm\pi/2, \pm3\pi/2, \pm5\pi/2, \ldots$. So the domains of cotangent, secant, and cosecant are as follows.

Function	Domain
$f(t) = \cot t$	All real numbers except $0, \pm\pi, \pm2\pi, \pm3\pi, \ldots$
$g(t) = \sec t$	All real numbers except $\pm\pi/2, \pm3\pi/2, \pm5\pi/2, \ldots$
$h(t) = \csc t$	All real numbers except $0, \pm\pi, \pm2\pi, \pm3\pi, \ldots$

The values of these functions may be approximated on a calculator by using the SIN and COS keys. For instance,

$$\cot(-3.1) = \frac{\cos(-3.1)}{\sin(-3.1)} \approx 24.0288, \qquad \sec 7 = \frac{1}{\cos 7} \approx 1.3264,$$

$$\csc 18.5 = \frac{1}{\sin 18.5} \approx -2.9199.$$

The cotangent function can also be evaluated with the TAN key, by using this fact:

$$\cot t = \frac{\cos t}{\sin t} = \frac{1}{\dfrac{\sin t}{\cos t}} = \frac{1}{\tan t}.^*$$

For example,

$$\cot(-5) = \frac{1}{\tan(-5)} \approx .2958.$$

---

*This identity is valid except for $t = \pm\pi/2, \pm3\pi/2, \pm5\pi/2, \ldots$. At these values, $\cos t = 0$ and $\sin t = \pm1$, so $\cot t = 0$, but $\tan t$ is not defined.

EXAMPLE 1

A batter hits a baseball. The ball is three feet above the ground and leaves the bat with an initial velocity of 100 feet per second at an angle of t radians from the horizontal. According to physics, the ball reaches a maximum height of

$$\frac{156.25 \tan^2 t}{\sec^2 t} + 3 \text{ feet.*}$$

What is the maximum height of the ball when it leaves the bat at an angle of .6 radians?

SOLUTION   Use a calculator to evaluate the formula for $t = .6$. The maximum height is

$$\frac{156.25 \tan^2 .6}{\sec^2 .6} + 3 = \frac{156.25 \tan^2 .6}{\dfrac{1}{\cos^2 .6}} + 3 = 156.25(\tan .6)^2 (\cos .6)^2 + 3$$

$$\approx 52.816 \text{ feet.} \qquad \blacksquare$$

These new trigonometric functions may be evaluated exactly at any integer multiple of $\pi/3$, $\pi/4$, or $\pi/6$.

EXAMPLE 2

Evaluate the cotangent, secant, and cosecant functions at $t = \pi/3$.

SOLUTION   Let P be the point where the terminal side of an angle of $\pi/3$ radians in standard position meets the unit circle (Figure 6–82). Draw the vertical line from P to the x-axis, forming a right triangle with hypotenuse 1, angles of $\pi/3$ and $\pi/6$ radians, and sides of lengths of $1/2$ and $\sqrt{3}/2$ as explained on page 445. Then P has coordinates $(1/2, \sqrt{3}/2)$, and by definition,

$$\sin \frac{\pi}{3} = y\text{-coordinate of } P = \sqrt{3}/2,$$

$$\cos \frac{\pi}{3} = x\text{-coordinate of } P = 1/2.$$

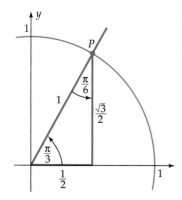

Figure 6–82

Therefore,

$$\csc \frac{\pi}{3} = \frac{1}{\sin(\pi/3)} = \frac{1}{\sqrt{3}/2} = \frac{2}{\sqrt{3}} = \frac{2\sqrt{3}}{3},$$

$$\sec \frac{\pi}{3} = \frac{1}{\cos(\pi/3)} = \frac{1}{1/2} = 2,$$

$$\cot \frac{\pi}{3} = \frac{\cos(\pi/3)}{\sin(\pi/3)} = \frac{1/2}{\sqrt{3}/2} = \frac{1}{\sqrt{3}} = \frac{\sqrt{3}}{3}. \qquad \blacksquare$$

ALTERNATE DESCRIPTIONS

The point-in-the-plane description of sine, cosine, and tangent readily extends to these new functions.

---

*Wind resistance is ignored here.

Point-in-the-Plane
Description

Let t be a real number and (x, y) any point (except the origin) on the terminal side of an angle of t radians in standard position. Let

$$r = \sqrt{x^2 + y^2}.$$

Then,

$$\cot t = \frac{x}{y}, \qquad \sec t = \frac{r}{x}, \qquad \csc t = \frac{r}{y}$$

for each number t in the domain of the given function.

These statements are proved by using the similar descriptions of sine and cosine. For instance,

$$\cot t = \frac{\cos t}{\sin t} = \frac{x/r}{y/r} = \frac{x}{y}.$$

The proofs of the other statements are similar.

EXAMPLE 3

Evaluate all six trigonometric functions at $t = 3\pi/4$.

SOLUTION The terminal side of an angle of $3\pi/4$ radians in standard position lies on the line $y = -x$, as shown in Figure 6–83. We shall use the point $(-1, 1)$ on this line to compute the function values. In this case,

$$r = \sqrt{x^2 + y^2} = \sqrt{(-1)^2 + 1^2} = \sqrt{2}.$$

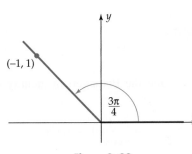

(−1, 1)

$\frac{3\pi}{4}$

Figure 6–83

Therefore,

$$\sin \frac{3\pi}{4} = \frac{y}{r} = \frac{1}{\sqrt{2}} = \frac{\sqrt{2}}{2} \qquad \cos \frac{3\pi}{4} = \frac{x}{r} = \frac{-1}{\sqrt{2}} = \frac{-\sqrt{2}}{2}$$

$$\tan \frac{3\pi}{4} = \frac{y}{x} = \frac{1}{-1} = -1 \qquad \csc \frac{3\pi}{4} = \frac{r}{y} = \frac{\sqrt{2}}{1} = \sqrt{2}$$

$$\sec \frac{3\pi}{4} = \frac{r}{x} = \frac{\sqrt{2}}{-1} = -\sqrt{2} \qquad \cot \frac{3\pi}{4} = \frac{x}{y} = \frac{-1}{1} = -1. \qquad \blacksquare$$

PART II: Algebra and Identities

We begin by noting the relationship between the cotangent and tangent functions.

Reciprocal
Identities

The cotangent and tangent functions are reciprocals; that is,

$$\cot t = \frac{1}{\tan t} \qquad \text{and} \qquad \tan t = \frac{1}{\cot t}$$

for every number t in the domain of both functions.

The first of these identities was proved on page 497, and the second is proved similarly (Exercise 49).

Period of Secant,
Cosecant, Cotangent

> The secant and cosecant functions are periodic with period 2π and the cotangent function is periodic with period π. In symbols,
>
> $$\sec(t + 2\pi) = \sec t, \qquad \csc(t + 2\pi) = \csc t,$$
>
> $$\cot(t + \pi) = \cot t$$
>
> for every number t in the domain of the given function.

The proof of these statements uses the fact that each of these functions is the reciprocal of a function whose period is known. For instance,

$$\csc(t + 2\pi) = \frac{1}{\sin(t + 2\pi)} = \frac{1}{\sin t} = \csc t,$$

$$\cot(t + \pi) = \frac{1}{\tan(t + \pi)} = \frac{1}{\tan t} = \cot t.$$

The other details are left as an exercise.

Pythagorean
Identities

> For every number t in the domain of both functions,
>
> $$1 + \tan^2 t = \sec^2 t$$
>
> and
>
> $$1 + \cot^2 t = \csc^2 t.$$

Proof   By the definitions of the functions and the Pythagorean identity $(\sin^2 t + \cos^2 t = 1)$, we have

$$1 + \tan^2 t = 1 + \frac{\sin^2 t}{\cos^2 t} = \frac{\cos^2 t + \sin^2 t}{\cos^2 t} = \frac{1}{\cos^2 t} = \left(\frac{1}{\cos t}\right)^2 = \sec^2 t.$$

The second identity is proved similarly.   ■

EXAMPLE 4

Simplify the expression $\dfrac{30\cos^3 t \sin t}{6\sin^2 t \cos t}$, assuming that $\sin t \neq 0$, $\cos t \neq 0$.

SOLUTION

$$\frac{30\cos^3 t \sin t}{6\sin^2 t \cos t} = \frac{5\cos^3 t \sin t}{\cos t \sin^2 t} = \frac{5\cos^2 t}{\sin t} = 5\frac{\cos t}{\sin t}\cos t = 5\cot t \cos t. \qquad ■$$

EXAMPLE 5

Assume that $\cos t \neq 0$ and simplify $\cos^2 t + \cos^2 t \tan^2 t$.

SOLUTION

$$\cos^2 t + \cos^2 t \tan^2 t = \cos^2 t(1 + \tan^2 t) = \cos^2 t \sec^2 t = \cos^2 t \cdot \frac{1}{\cos^2 t} = 1. \quad \blacksquare$$

EXAMPLE 6

If $\tan t = 3/4$ and $\sin t < 0$, find $\cot t$, $\cos t$, $\sin t$, $\sec t$, and $\csc t$.

SOLUTION   First we have $\cot t = 1/\tan t = 1/(3/4) = 4/3$. Next we use the Pythagorean identity to obtain

$$\sec^2 t = 1 + \tan^2 t = 1 + \left(\frac{3}{4}\right)^2 = 1 + \frac{9}{16} = \frac{25}{16}$$

$$\sec t = \pm\sqrt{\frac{25}{16}} = \pm\frac{5}{4}$$

$$\frac{1}{\cos t} = \pm\frac{5}{4} \qquad \text{or, equivalently,} \qquad \cos t = \pm\frac{4}{5}.$$

Since $\sin t$ is given as negative and $\tan t = \sin t/\cos t$ is positive, $\cos t$ must be negative. Hence, $\cos t = -4/5$. Consequently,

$$\frac{3}{4} = \tan t = \frac{\sin t}{\cos t} = \frac{\sin t}{(-4/5)}$$

so

$$\sin t = \left(-\frac{4}{5}\right)\left(\frac{3}{4}\right) = -\frac{3}{5}.$$

Therefore,

$$\sec t = \frac{1}{\cos t} = \frac{1}{(-4/5)} = -\frac{5}{4} \quad \text{and} \quad \csc t = \frac{1}{\sin t} = \frac{1}{(-3/5)} = -\frac{5}{3}. \quad \blacksquare$$

PART III: Graphs

The graph of the secant function is shown in red in Figure 6–84.

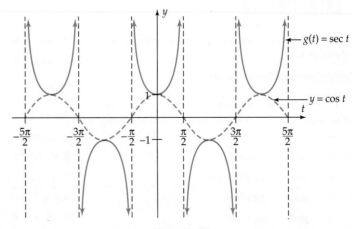

Figure 6–84

The shape of the secant graph can be understood by looking at the graph of cosine (blue in Figure 6–84) and noting these facts:

1. $\sec t = 1/\cos t$ is not defined when $\cos t = 0$, that is, when $t = \pm\pi/2$, $\pm3\pi/2$, $\pm5\pi/2$, and so on.

2. The graph of $\sec t$ has a vertical asymptote at $t = \pm\pi/2$, $\pm3\pi/2$, $\pm5\pi/2$, . . . The reason is that when $\cos t$ is close to 0 (graph close to t-axis), then $\sec t = 1/\cos t$ is very large in absolute value,* so its graph is far from the axis.

3. When $\cos t$ is near 1 or -1 (that is, when t is near 0, $\pm\pi$, $\pm2\pi$, $\pm3\pi$, . . .), then so is $\sec t = 1/\cos t$.

The graphs of $h(t) = \csc t = 1/\sin t$ and $f(t) = \cot t = 1/\tan t$ can be obtained in a similar fashion (Figure 6–85).

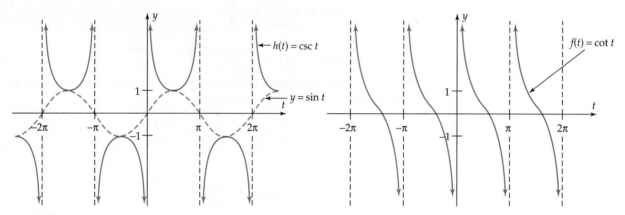

Figure 6–85

---

*See the Big-Little Principle on page 288.

EXERCISES 6.6

Note: The arrangement of the exercises corresponds to the sub-sections of this section.

Part I: Definitions and Descriptions

In Exercises 1–6, determine the quadrant containing the terminal side of an angle of t radians in standard position under the given conditions.

1. $\cos t > 0$ and $\sin t < 0$

2. $\sin t < 0$ and $\tan t > 0$

3. $\sec t < 0$ and $\cot t < 0$

4. $\csc t < 0$ and $\sec t > 0$

5. $\sec t > 0$ and $\cot t < 0$

6. $\sin t > 0$ and $\sec t < 0$

In Exercises 7–16, evaluate all six trigonometric functions at t, where the given point lies on the terminal side of an angle of t radians in standard position.

7. $(3, 4)$

8. $(0, 6)$

9. $(-5, 12)$

10. $(-2, -3)$

11. $(-1/5, 1)$

12. $(4/5, -3/5)$

13. $(\sqrt{2}, \sqrt{3})$

14. $(-2\sqrt{3}, \sqrt{3})$

15. $(1 + \sqrt{2}, 3)$

16. $(1 + \sqrt{3}, 1 - \sqrt{3})$

17. Suppose the batter in Example 1 hits a popup (the ball leaves the bat at an angle of 1.4 radians). What is the maximum height of the ball?

Exercises 18–20 deal with the path of a projectile (such as a baseball, a rocket, or an arrow). If the projectile is fired with an initial velocity of v feet per second at angle of t radians and its initial height is k feet, then the path of the projectile is given by

$$y = \left(\frac{-16}{v^2}\sec^2 t\right)x^2 + (\tan t)x + k. *$$

You can think of the projectile as being fired in the direction of the x-axis from the point (0, k) on the y-axis.

18. (a) Find a viewing window that shows the path of a projectile that is fired from a 20-foot high platform at an initial velocity of 120 feet per second at an angle of .8 radians.
   (b) What is the maximum height reached by the projectile?
   (c) How far down range does the projectile hit the ground?

19. Do Exercise 18 for a projectile that is fired from ground level at an initial velocity of 80 feet per second at an angle of .4 radians.

20. Do Exercise 18 for a projectile that is fired from a 40-foot high platform at an initial velocity of 125 feet per second at an angle of 1.2 radians.

In Exercises 21–25, evaluate all six trigonometric functions at the given number without using a calculator.

21. $\dfrac{4\pi}{3}$ **22.** $-\dfrac{7\pi}{6}$ **23.** $\dfrac{7\pi}{4}$

24. $\dfrac{11\pi}{3}$ **25.** $\dfrac{-11\pi}{4}$

26. Fill in the missing entries in the following table. Give exact answers, not decimal approximations.

27. Find the average rate of change of $f(t) = \cot t$ from $t = 1$ to $t = 3$.

28. Find the average rate of change of $g(t) = \csc t$ from $t = 2$ to $t = 3$.

29. (a) Find the average rate of change of $f(t) = \tan t$ from $t = 2$ to $t = 2 + h$, for each of these values of h: .01, .001, .0001, and .00001.
   (b) Compare your answers in part (a) with the number $(\sec 2)^2$. What would you guess that the instantaneous rate of change of $f(t) = \tan t$ is at $t = 2$?

Part II: Algebra and Identities

In Exercises 30–36, perform the indicated operations, then simplify your answers by using appropriate definitions and identities.

30. $\tan t\,(\cos t - \csc t)$ **31.** $\cos t \sin t\,(\csc t + \sec t)$

32. $(1 + \cot t)^2$ **33.** $(1 - \sec t)^2$

34. $(\sin t - \csc t)^2$

35. $(\cot t - \tan t)(\cot^2 t + 1 + \tan^2 t)$

36. $(\sin t + \csc t)(\sin^2 t + \csc^2 t - 1)$

In Exercises 37–42, factor and simplify the given expression.

37. $\sec t \csc t - \csc^2 t$ **38.** $\tan^2 t - \cot^2 t$

39. $\tan^4 t - \sec^4 t$ **40.** $4\sec^2 t + 8\sec t + 4$

41. $\cos^3 t - \sec^3 t$ **42.** $\csc^4 t + 4\csc^2 t - 5$

In Exercises 43–48, simplify the given expression. Assume that all denominators are nonzero and all quantities under radicals are nonnegative.

43. $\dfrac{\cos^2 t \sin t}{\sin^2 t \cos t}$ **44.** $\dfrac{\sec^2 t + 2\sec t + 1}{\sec t}$

45. $\dfrac{4\tan t \sec t + 2\sec t}{6\sin t \sec t + 2\sec t}$ **46.** $\dfrac{\sec^2 t \csc t}{\csc^2 t \sec t}$

47. $(2 + \sqrt{\tan t})(2 - \sqrt{\tan t})$

t	0	$\dfrac{\pi}{6}$	$\dfrac{\pi}{4}$	$\dfrac{\pi}{3}$	$\dfrac{\pi}{2}$	$\dfrac{2\pi}{3}$	$\dfrac{3\pi}{4}$	$\dfrac{5\pi}{6}$	π	$\dfrac{3\pi}{2}$
$\sin t$										
$\cos t$										
$\tan t$					—					—
$\cot t$	—								—	
$\sec t$					—					—
$\csc t$	—								—	

*Wind resistance is ignored in this equation.

48. $\dfrac{6 \tan t \sin t - 3 \sin t}{9 \sin^2 t + 3 \sin t}$

In Exercises 49–54, prove the given identity.

49. $\tan t = \dfrac{1}{\cot t}$ [*Hint:* See page 497.]

50. $\sec(t + 2\pi) = \sec t$ [*Hint:* See page 500.]

51. $1 + \cot^2 t = \csc^2 t$ [*Hint:* Look at the proof of the similar identity on page 500]

52. $\cot(-t) = -\cot t$ [*Hint:* Express the left side in terms of sine and cosine; then use the negative angle identities and express the result in terms of cotangent.]

53. $\sec(-t) = \sec t$ [Adapt the hint for Exercise 52.]

54. $\csc(-t) = -\csc t$

In Exercises 55–60, find the values of all six trigonometric functions at t if the given conditions are true.

55. $\cos t = -1/2$ and $\sin t > 0$
[*Hint:* $\sin^2 t + \cos^2 t = 1.$]

56. $\cos t = \dfrac{1}{2}$ and $\sin t < 0$

57. $\cos t = 0$ and $\sin t = 1$

58. $\sin t = -2/3$ and $\sec t > 0$

59. $\sec t = -13/5$ and $\tan t < 0$

60. $\csc t = 8$ and $\cos t < 0$

Part III: Graphs

In Exercises 61–64, use graphs to determine whether the equation could possibly be an identity or is definitely not an identity.

61. $\tan t = \cot\left(\dfrac{\pi}{2} - t\right)$ **62.** $\dfrac{\cos t}{\cos(t - \pi/2)} = \cot t$

63. $\dfrac{\sin t}{1 - \cos t} = \cot t$

64. $\dfrac{\sec t + \csc t}{1 + \tan t} = \csc t$

65. Show graphically that the equation $\sec t = t$ has infinitely many solutions, but none between $-\pi/2$ and $\pi/2$.

THINKERS

66. In the diagram of the unit circle in the figure, find six line segments whose respective lengths are $\sin t$, $\cos t$, $\tan t$, $\cot t$, $\sec t$, $\csc t$. [*Hint:* $\sin t =$ length CA. Why? Note that OC has length 1 and various right triangles in the figure are similar.]

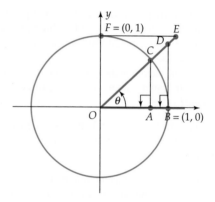

67. In the figure for Exercise 66, find the following areas in terms of θ.

(a) triangle OCA
(b) triangle ODB
(c) circular segment OCB

Chapter 6 Review

IMPORTANT CONCEPTS

IMPORTANT FACTS & FORMULAS

- *Conversion Rules:* To convert radians to degrees, multiply by $180/\pi$. To convert degrees to radians, multiply by $\pi/180$.

- *Definition of Trigonometric Functions:* If P is the point where the terminal side of an angle of t radians in standard position meets the unit circle, then

$$\sin t = y\text{-coordinate of } P,$$

$$\cos t = x\text{-coordinate of } P,$$

$$\tan t = \frac{\sin t}{\cos t}, \qquad \cot t = \frac{\cos t}{\sin t}, \qquad \sec t = \frac{1}{\cos t}, \qquad \csc t = \frac{1}{\sin t}.$$

- *Point-in-the-Plane Description:* If (x, y) is any point other than the origin on the terminal side of an angle of t radians in standard position and $r = \sqrt{x^2 + y^2}$, then

$$\sin t = \frac{y}{r}, \qquad \cos t = \frac{x}{r},$$

$$\tan t = \frac{y}{x}, \qquad \cot t = \frac{x}{y},$$

$$\sec t = \frac{r}{x}, \qquad \csc t = \frac{r}{y}.$$

■ *Basic Identities:*

$$\sin^2 t + \cos^2 t = 1 \qquad \sin(-t) = -\sin t$$

$$1 + \tan^2 t = \sec^2 t \qquad \cos(-t) = \cos t$$

$$1 + \cot^2 t = \csc^2 t \qquad \tan(-t) = -\tan t$$

$$\tan t = \frac{1}{\cot t} \qquad \cot t = \frac{1}{\tan t}$$

$$\sin(t \pm 2\pi) = \sin t \qquad \csc(t \pm 2\pi) = \csc t$$

$$\cos(t \pm 2\pi) = \cos t \qquad \sec(t \pm 2\pi) = \sec t$$

$$\tan(t \pm \pi) = \tan t \qquad \cot(t \pm \pi) = \cot t$$

■ If $A \neq 0$ and $b > 0$, then each of the $f(t) = A \sin(bt + c)$ and $g(t) = A \cos(bt + c)$ has

amplitude $|A|$,   period $2\pi/b$,   phase shift $-c/b$.

CATALOG OF BASIC FUNCTIONS—PART 4

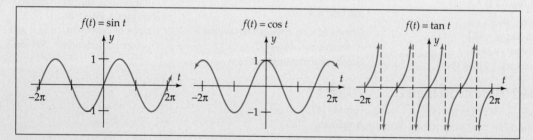

REVIEW QUESTIONS

1. Find a number t between 0 and 2π such that an angle of t radians in standard position is coterminal with an angle of $-23\pi/3$ radians in standard position.

2. Through how many radians does the second hand of a clock move in 2 minutes and 40 seconds?

3. $\dfrac{9\pi}{5}$ radians = _____ degrees.

4. 36 degrees = _____ radians.

5. $220° =$ _____ radians.

6. $\dfrac{17\pi}{12}$ radians = _____ degrees.

7. $-\dfrac{11\pi}{4}$ radians = _____ degrees.

8. $-135° =$ _____ radians.

9. If an angle of v radians has its terminal side in the second quadrant and $\sin v = \sqrt{8/9}$, then find $\cos v$.

10. $\cos \dfrac{47\pi}{2} = ?$

11. $\sin(-13\pi) = ?$

12. Simplify: $\dfrac{\tan(t + \pi)}{\sin(t + 2\pi)}$

Use the figure on the next page in Questions 13–23.

13. $\cos\left(\dfrac{\pi}{5}\right) = ?$ **14.** $\sin\left(\dfrac{7\pi}{6}\right) = ?$

15. $\tan\left(\dfrac{-5\pi}{6}\right) = ?$ **16.** $\tan\left(\dfrac{16\pi}{6}\right) = ?$

17. $\sin\left(\dfrac{-4\pi}{3}\right) = ?$ **18.** $\cos\left(\dfrac{-9\pi}{5}\right) = ?$

19. $\csc\left(\dfrac{\pi}{5}\right) = ?$ **20.** $\sec\left(\dfrac{7\pi}{6}\right) = ?$

21. $\cot\left(\dfrac{-4\pi}{3}\right) = ?$ **22.** $\csc\left(\dfrac{-9\pi}{5}\right) = ?$

23. $\sec\left(\dfrac{-4\pi}{3}\right) = ?$

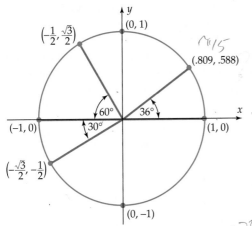

$$24. \left[3 \sin\left(\frac{\pi}{5^{500}}\right)\right]^2 + \left[3 \cos\left(\frac{\pi}{5^{500}}\right)\right]^2 = ?$$

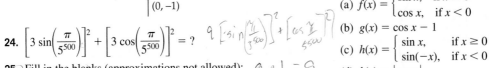

25. Fill in the blanks (approximations not allowed):

t	0	$\frac{\pi}{6}$	$\frac{\pi}{4}$	$\frac{\pi}{3}$	$\frac{\pi}{2}$
sin t					
cos t					

26. Express as a single real number:

$$\cos\frac{3\pi}{4}\sin\frac{5\pi}{6} - \sin\frac{3\pi}{4}\cos\frac{5\pi}{6}$$

27. $\left(\sin\frac{\pi}{6} + 1\right)^2 = ?$

28. $\sin(\pi/2) + \sin 0 + \cos 0 = ?$

29. If $f(x) = \log_{10} x$ and $g(t) = -\cos t$, then $(f \circ g)(\pi) = ?$

30. Cos t is negative when the terminal side of an angle of t radians in standard position lies in which quadrants?

31. If $\sin t = 1/\sqrt{3}$ and the terminal side of an angle of t radians in standard position lies in the second quadrant, then $\cos t = ?$

32. Which of the following could possibly be a true statement about a real number t?

(a) $\sin t = -2$ and $\cos t = 1$
(b) $\sin t = 1/2$ and $\cos t = \sqrt{2}/2$
(c) $\sin t = -1$ and $\cos t = 1$
(d) $\sin t = \pi/2$ and $\cos t = 1 - (\pi/2)$
(e) $\sin t = 3/5$ and $\cos t = 4/5$

33. If $\sin t = -4/5$ and the terminal side of an angle of t radians in standard position lies in the third quadrant, then $\cos t = \_\_\_\_$.

34. If $\sin(-101\pi/2) = -1$, then $\sin(-105\pi/2) = ?$

35. If $\pi/2 < t < \pi$ and $\sin t = 5/13$, then $\cos t = ?$

36. $\cos\left(-\frac{\pi}{6}\right) = ?$

37. $\cos\left(\frac{2\pi}{3}\right) = ?$

38. $\sin\left(-\frac{11\pi}{6}\right) = ?$

39. $\sin\left(\frac{\pi}{3}\right) = ?$

40. $\tan(5\pi/3) = ?$

41. Which of the following is *not* true about the graph of $f(t) = \sin t$?

(a) It has no sharp corners.
(b) It crosses the horizontal axis more than once.
(c) It rises higher and higher as t gets larger.
(d) It is periodic.
(e) It has no vertical asymptotes.

42. Which of the following functions has the graph in the figure between $-\pi$ and π?

(a) $f(x) = \begin{cases} \sin x, & \text{if } x \geq 0 \\ \cos x, & \text{if } x < 0 \end{cases}$

(b) $g(x) = \cos x - 1$

(c) $h(x) = \begin{cases} \sin x, & \text{if } x \geq 0 \\ \sin(-x), & \text{if } x < 0 \end{cases}$

(d) $k(x) = |\cos x|$

(e) $p(x) = \sqrt{1 - \sin^2 x}$

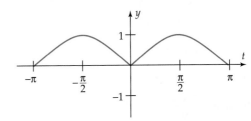

The point $(-3/\sqrt{50}, 7/\sqrt{50})$ *lies on the terminal side of an angle of t radians (in standard position). Find:*

43. $\sin t$

44. $\cos t$

45. $\tan t$

46. $\csc t$

47. $\sec t$

48. $\cot t$

49. Find the equation of the straight line containing the terminal side of an angle of $5\pi/3$ radians (in standard position).

Suppose that an angle of w radians has its terminal side in the fourth quadrant and $\cos w = 2/\sqrt{13}$. *Find:*

50. $\sin w$

51. $\tan w$

52. $\cot w$

53. $\cos(-w)$

54. $\sec(w)$

55. $\csc(-w)$

56. Fill in the blanks (approximations not allowed):

t	sin t	tan t	sec t
$\pi/4$			
$2\pi/3$			
$5\pi/6$			

57. Sketch the graphs of $f(t) = \sin t$ and $h(t) = \csc t$ on the same set of coordinate axes $(-2\pi \leq t \leq 2\pi)$.

58. *Let* θ be the angle shown in the figure. Which of the following statements is true?

(a) $\sin\theta = \dfrac{\sqrt{2}}{2}$

(b) $\cos\theta = \dfrac{\sqrt{2}}{2}$

(c) $\tan\theta = 1$

(d) $\cos\theta = \sqrt{2}$

(e) $\tan\theta = -1$

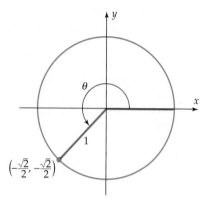

59. Let θ be as indicated in the figure below. Which of the statements (i)–(iii) are *true*?

(i) $\cos\theta = -\dfrac{1}{3}$

(ii) $\tan\theta = \dfrac{2\sqrt{2}}{9}$

(iii) $\sin\theta = -\dfrac{2\sqrt{2}}{3}$

(a) only ii

(b) only ii and iii

(c) all of them

(d) only i and iii

(e) none of them

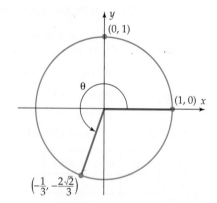

60. Between (and including) 0 and 2π, the function $h(t) = \tan t$ has

(a) 3 roots and is undefined at 2 places.

(b) 2 roots and is undefined at 3 places.

(c) 2 roots and is undefined at 2 places.

(d) 3 roots and is defined everywhere.

(e) no roots and is undefined at 3 places.

61. If the terminal side of an angle of θ radians in standard position passes through the point $(-2, 3)$, then $\tan\theta =$ ___.

62. Which of the statements (i)–(iii) are true?

(i) $\sin(-x) = -\sin x$

(ii) $\cos(-x) = -\cos x$

(iii) $\tan(-x) = -\tan x$

(a) (i) and (ii) only

(b) (ii) only

(c) (i) and (iii) only

(d) all of them

(e) none of them

63. If $\sec x = 1$ and $-\pi/2 < x < \pi/2$, then $x = ?$

64. If $\tan t = 4/3$ and $0 < t < \pi$, what is $\cos t$?

65. Which of the following is true about $\sec t$?

(a) $\sec(0) = 0$

(b) $\sec t = 1/\sin t$

(c) Its graph has no asymptotes.

(d) It is a periodic function.

(e) It is never negative.

66. If $\cot t = 0$ and $0 < t \le \pi$, then $t =$ ___.

67. What is $\cot\left(\dfrac{2\pi}{3}\right)$?

68. Which of the following functions has the graph in the figure?

(a) $f(t) = \tan t$

(b) $g(t) = \tan\left(t + \dfrac{\pi}{2}\right)$

(c) $h(t) = 1 + \tan t$

(d) $k(t) = 3\tan t$

(e) $p(t) = -\tan t$

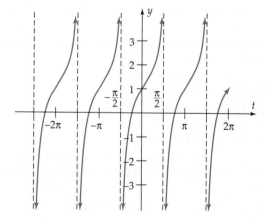

69. Let $f(t) = \sqrt{\dfrac{3}{2}}\sin 5t.$

(a) What is the largest possible value of $f(x)$?

(b) Find the smallest positive number t such that $f(t) = 0$.

70. Sketch the graph of $g(t) = -2 \cos t$.

71. Sketch the graph of $f(t) = -\frac{1}{2} \sin 2t \, (-2\pi \le t \le 2\pi)$.

72. Sketch the graph of $f(t) = \sin 4t \, (0 \le t \le 2\pi)$.

In Questions 73–78, determine graphically whether the given equation could possibly be an identity.

73. $\cos t = \sin\left(t - \dfrac{\pi}{2}\right)$

74. $\tan \dfrac{t}{2} = \dfrac{\sin t}{1 + \cos t}$

75. $\dfrac{\sin t - \sin 3t}{\cos t + \cos 3t} = -\tan t$

76. $\cos 2t = \dfrac{1}{1 - 2\sin^2 t}$

77. $\sec^2 x + \csc^2 x = (\sec^2 x)(\csc^2 x)$

78. $\tan x = \sqrt{\sec^2 x - 1}$

79. What is the period of the function $g(t) = \sin 4\pi t$?

80. What are the amplitude, period, and phase shift of the function
$$h(t) = 13 \cos(14t + 15)?$$

81. The number of hours of daylight in Boston on day t of the year is approximated by
$$d(t) = 3.1 \sin(.0172t - 1.377) + 12.$$

   (a) On the day with the most daylight, how many hours of daylight are there? On the day with the least daylight, how many hours are there?

   (b) On what days are there less than 11 hours of daylight?

82. A certain person's blood pressure $P(t)$ at time t seconds is approximately
$$P(t) = 22 \cos(2.5\pi t) + 98.$$

   (a) The period of this function is time between heartbeats. What is this person's pulse rate (heartbeats per minute)?

   (b) What is this person's blood pressure [systolic pressure (maximum) over diastolic pressure (minimum)]?

83. State the rule of a periodic function whose graph from $t = 0$ to $t = 2\pi$ closely resembles the one in the figure.

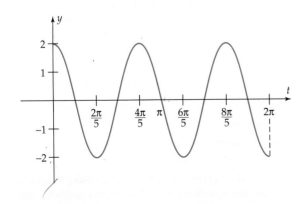

84. State the rule of a periodic function with amplitude 3, period π, and phase shift $\pi/3$.

85. State the rule of a periodic function with amplitude 8, period 5, and phase shift 14.

86. If $g(t) = 20 \sin(200t)$, for how many values of t, with $0 \le t \le 2\pi$, is it true that $g(t) = 1$?

In Exercises 87 and 88, estimate constants A, b, c such that $f(t) = A \sin(bt + c)$.

87. $f(t) = 6 \sin(4t + 7) - 5 \cos(4t + 8)$

88. $f(t) = -5 \sin(5t - 3) + 2 \cos(5t + 2)$

In Exercises 89 and 90, find a viewing window that shows a complete graph of the function.

89. $f(t) = 3 \sin(300t + 5) - 2 \cos(500t + 8)$

90. $g(t) = -5 \sin(400\pi t + 1) + 2 \cos(150\pi t - 6)$

91. The average monthly temperatures in St. Louis, Missouri, are shown in the table.*

Month	Temperature (°F)
Jan.	30
Feb.	35
March	46
April	57
May	67
June	77
July	80
Aug.	78
Sept.	70
Oct.	58
Nov.	45
Dec.	34

   (a) Let $x = 1$ correspond to January. Plot 24 data points (two years).

   (b) Use regression to find a sine function that models this data.

   (c) What is the period of the function in part (b)? Does it fit the data well?

---

*Based on data from the National Climatic Data Center.

Chapter 6 Test

Sections 6.1–6.3; Special Topics 6.1.A

1. Find the radian measure of an angle in standard position formed by rotating the terminal side $\frac{1}{72}$ of a circle.

2. Find the exact values:

   (a) $\sin\left(-\frac{13\pi}{3}\right)$ (b) $\cos\left(-\frac{13\pi}{3}\right)$

   (c) $\tan\left(-\frac{13\pi}{3}\right)$

3. Factor: $\sin^3 t - \sin t$

4. Assume that a wheel of a car has radius 36 cm. Find the angle (in radians) through which the wheel turns when the car travels 5 kilometers.

5. Match each angle in column 1 with an angle in column II that is coterminal with it.

I	II
(a) $\frac{8\pi}{7}$	(i) $\frac{25\pi}{7}$
(b) $-\frac{3\pi}{7}$	(ii) $-\frac{4\pi}{7}$
(c) $\frac{9\pi}{7}$	(iii) $-\frac{6\pi}{7}$
	(iv) $\frac{23\pi}{7}$

6. The terminal side of an angle of t radians in standard position passes through the point $(-3, 2)$. Find:

   (a) $\sin t$ (b) $\cos t$
   (c) $\tan t$

7. Find the exact value of $\sin t$ if $\cos t = \frac{1}{\sqrt{17}}$ and $\frac{3\pi}{2} < t < 2\pi$.

8. If the radius of the circle is 4 cm and angle θ measures 2 radians, find the area of the shaded sector.

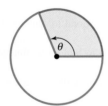

9. Do not use a calculator and show your work.

   (a) $480° = $ _____ radians

   (b) $\frac{\pi}{40}$ radians = _____ °.

10. The terminal side of an angle of t radians in standard position lies in quadrant I on the straight line that passes through the origin and is parallel to the line $7y - 2x = -4$. Find

    (a) $\sin t$ (b) $\cos t$ (c) $\tan t$

11. If $\sin t = 3/5$ and $0 < t < \frac{\pi}{2}$, find $\tan(5\pi - t)$.

12. The second hand on a clock is 5 inches long. Find the linear speed (in inches per minute) and the angular speed (in radians per minute) of the tip of the second hand as it moves on the face of the clock.

13. Simplify the expression (assume all denominators are nonzero).

$$\frac{8\cos t}{\sin^2 t} \cdot \frac{\sin^2 t - \sin t \cos t}{\sin^2 t - \cos^2 t}$$

14. Find the exact average rate of change of $f(t) = \tan t$ from $t = \frac{\pi}{6}$ to $t = \frac{\pi}{4}$. Do not give decimal approximations.

15. Find the exact value of $\sin\left(-\frac{17\pi}{4}\right)$.

16. Simplify the expression (assume all denominators are nonzero).

$$\frac{4\tan t \sin t - 2\sin t}{6\sin^2 t + 2\sin t}$$

Sections 6.4–6.6; Special Topics 6.5.A

17. If the terminal side of an angle of t radians lies on the line $y = 6.2t$, then $\tan t = $ _____.

18. Let $f(t) = \cos(2\pi t)$. Find the values of t (with $0 \le t \le 2\pi$) for which $f(t) = -1$.

19. Find the exact values:

    (a) $\sin\left(-\frac{11\pi}{6}\right)$ (b) $\cos\left(-\frac{11\pi}{6}\right)$

    (c) $\tan\left(-\frac{11\pi}{6}\right)$ (d) $\cot\left(-\frac{11\pi}{6}\right)$

    (e) $\csc\left(-\frac{11\pi}{6}\right)$ (f) $\sec\left(-\frac{11\pi}{6}\right)$

20. Sketch the graph of a periodic function with period 2 that is *not* a trigonometric function.

21. Find a viewing window that accurately shows exactly six complete waves of the graph of $f(t) = \sin(1100t)$.

22. Multiply and simplify your answer:

$$(\sin t - \csc t)(\sin^2 t + \csc^2 t + 1).$$

23. Use graphs to determine the *number* of solutions of the equation $\sin t = -\frac{2}{9}$ between 0 and 4π. You need not solve the equation.

24. The table on the next page shows the average monthly temperature in a western city.

    (a) Use 12 data points (with $x = 1$ corresponding to January) and sine regression to find a function that models this data.

(b) Use 24 data points (two years of data) to find another model for this data.

(c) Which model is likely to be more accurate? [*Hint:* what is the period of each model?]

Month	Temperature (°F)
Jan	22.2
Feb	27.2
Mar	37.5
Apr	48.0
May	58.9
Jun	68.4
Jul	73.5
Aug	71.9
Sep	64.0
Oct	52.3
Nov	39.5
Dec	27.6

25. Prove the identity: $1 + \tan^2 x = \sec^2 x$.

26. Use graphs to determine whether this equation could possibly be an identity or is definitely not an identity:

$$(\cos^2 t + 1)(\tan^2 t - 1) = -\tan^2 t.$$

27. (a) State the amplitude, period, and phase shift of the function $f(t) = -3 \cos(4t + 7)$.

(b) Write the rule of a periodic function that has amplitude 6, period 2, and phase shift -1.

28. Assume that t is a number such that

$$\sin t = -\frac{6}{7} \text{ and } \sec t > 0.$$

Find the exact value of:

(a) $\cos t$ (b) $\tan t$
(c) $\cot t$ (d) $\csc t$
(e) $\sec t$.

29. Let $f(t) = 2 \sin(6t)$. Use algebra (not a calculator) to find all the local maxima of this function in the interval $[0, \pi]$ (that is, the coordinates of the highest points on the graph over this interval).

30. Find constants A, b, and c (rounded to four decimal places) such that

$$A \cos(bt + c) \approx 2 \sin(3t - 5) + 2 \cos(3t + 2).$$

DISCOVERY PROJECT 6   Pistons and Flywheels

A common and well-proven piece of technology is the piston and flywheel combination. It is clearly visible in photographs of steam locomotives from the middle of the nineteenth century. The structure involves a wheel (or crankshaft) connected to a sliding plug in a cylinder by a rigid arm. The axis of rotation of the wheel is perpendicular to the central axis of the cylinder. The sliding plug, the piston, moves in one dimension, in and out of the cylinder. The motion of the piston is periodic, like the basic trigonometric functions, but doesn't have the same elegant symmetry.

It is quite easy to superimpose a coordinate plane on the flywheel-piston system so that the center of the flywheel is the origin, the flywheel rotates counterclockwise, and the piston moves along the x-axis. As you can see in the diagram, the flywheel typically has a larger radius than the radial distance from the center to the point where the arm attaches. In this particular figure, the radius of the flywheel is 50 centimeters, and the radial distance to the attachment point Q is 45 centimeters. The arm is 150 centimeters long measured from the base of the piston to the attachment point.

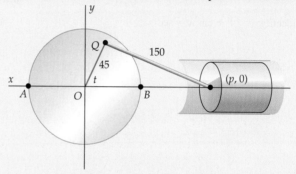

1. What are the coordinates of A and B in the figure? How close does the base of the piston come to the flywheel? What is the length of the piston stroke?

2. Show that Q has coordinates $(45 \cos t, 45 \sin t)$, where t is the radian measure of angle BOQ. [*Hint:* If Q has coordinates (x, y), use the point-in-the-plane description to compute $\cos t$ and solve for x; find y similarly.]

3. Let p be the x-coordinate of the center point of the base of the piston. Express p as a function of t. [*Hint:* Use the distance formula to express the distance from Q to $(p, 0)$ in terms of p and t. Set this expression equal to 150 (why?) and solve for p.]

4. Let $p(t)$ be the function found in Question 3 (that is, $p(t)$ = the x-coordinate of the base of the piston when angle BOQ measures t radians). What is the range of this function? How does this relate to Question 1?

5. Approximate the values of $p(0)$, $p(\pi/2)$, and $p(\pi)$. Note that $p(\pi/2)$ is not halfway between $p(0)$ and $p(\pi)$. Which side of the halfway point is it on? Does this mean that the piston moves faster on the average when it is to the right or left of the halfway point? Find the value of t that places the base of the piston at the halfway point of its motion.

6. If the flywheel spins at a constant speed, does the piston move back and forth at a constant speed? How do you know?

TRIGONOMETRIC IDENTITIES AND EQUATIONS

Beam me up, Scotty!

When a light beam passes from one medium to another (for instance, from air to water), it changes both its speed and its direction. If you know what some of these numbers are (say, the speed of light in air or the angle at which a light beam hits the water), then you can determine the unknown ones by solving a trigonometric equation. See Exercises 39–42 on page 564.

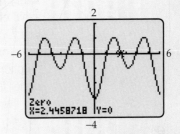

© Stefano Torrione/Getty Images

Chapter Outline

Until now, the variable t has been used for trigonometric functions, to avoid confusion with the x's and y's that appear in their definitions. Now that you are comfortable with these functions, we shall usually use the letter x (or occasionally y) for the variable. Unless stated, all trigonometric functions in this chapter are considered to be functions of real numbers, rather than functions of angles in degree measure.

Two kinds of trigonometric equations are considered here. *Identities* (Sections 7.1–7.3) are equations that are valid for all values of the variable for which the equation is defined, such as

$$\sin^2 x + \cos^2 x = 1 \qquad \text{and} \qquad \cot x = \frac{1}{\tan x}.$$

Identities can be used for simplifying expressions, rewriting the rule of a trigonometric function, performing numerical computations, and in other ways. *Conditional equations* (Section 7.5) are valid only for certain values of the variable, such as

$$\sin x = 0 \qquad \text{(true only when x is an integer multiple of π).}$$

Inverse trigonometric functions, which have a number of uses, are considered in Section 7.4.

7.1 Basic Identities and Proofs

Section Objectives
- Examine possible identities graphically.
- Use basic identities to simplify trigonometric expressions.
- Learn strategies for proving identities algebraically.

When you suspect that an equation might be an identity, it's a good idea to see whether there is any graphical evidence to support this conclusion.

EXAMPLE 1

Is either of the following equations an identity?

(a) $2 \sin^2 x - \cos x = 2 \cos^2 x + \sin x$

(b) $\dfrac{1 + \sin x - \sin^2 x}{\cos x} = \cos x + \tan x$

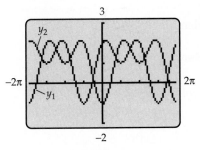

Figure 7–1

SOLUTION

(a) Test the equation by graphing these two equations on the same screen:

$$y_1 = 2\sin^2 x - \cos x \qquad \text{[left side of equation being tested]}$$

$$y_2 = 2\cos^2 x + \sin x \qquad \text{[right side of equation being tested]}$$

If the given equation is an identity (meaning that $y_1 = y_2$), then the two graphs will be identical. As Figure 7–1 shows, however, the graphs are quite different. Hence, this equation is not an identity.

(b) Test the second equation graphically.

GRAPHING EXPLORATION

Graph the functions

$$f(x) = \frac{1 + \sin x - \sin^2 x}{\cos x} \qquad \text{and} \qquad g(x) = \cos x + \tan x$$

on the same screen, using a viewing window with $-2\pi \leq x \leq 2\pi$. Do the graphs appear to be identical?

The exploration suggests that the equation *may* be an identity because the two graphs *appear* to be identical. However, this graphical evidence is *not* a proof—that must be done algebraically. ∎

Any equation can be graphically tested, as in Example 1, to see whether it might be an identity. If the left-side and right-side graphs are different, then it definitely is *not* an identity. If the graphs appear to be the same, then it is *possible,* but not certain, that the equation is an identity.

The fact that two graphs appear identical on a calculator screen does *not* prove that they are actually the same, as the following Graphing Exploration shows.

GRAPHING EXPLORATION

In the viewing window with $-\pi \leq x \leq \pi$ and $-2 \leq y \leq 2$, graph both sides of the equation

$$\cos x = 1 - \frac{x^2}{2} + \frac{x^4}{24} - \frac{x^6}{720} + \frac{x^8}{40,320}.$$

Do the graphs appear to be identical? Now change the viewing window so that $-2\pi \leq x \leq 2\pi$. Is the equation an identity?

PROVING IDENTITIES

The phrases "prove the identity" and "verify the identity" mean "prove that the given equation is an identity." We shall assume the elementary identities that were proved in Chapter 6 and are summarized here.

Basic Trigonometric Identities

Reciprocal Identities

$$\sec x = \frac{1}{\cos x} \qquad \csc x = \frac{1}{\sin x}$$

$$\tan x = \frac{\sin x}{\cos x} \qquad \cot x = \frac{\cos x}{\sin x}$$

$$\tan x = \frac{1}{\cot x} \qquad \cot x = \frac{1}{\tan x}$$

Periodicity Identities

$$\sin(x + 2\pi) = \sin x \qquad \cos(x + 2\pi) = \cos x$$

$$\sec(x + 2\pi) = \sec x \qquad \csc(x + 2\pi) = \csc x$$

$$\tan(x + \pi) = \tan x \qquad \cot(x + \pi) = \cot x$$

Pythagorean Identities

$$\sin^2 x + \cos^2 x = 1 \qquad 1 + \tan^2 x = \sec^2 x \qquad 1 + \cot^2 x = \csc^2 x$$

Negative Angle Identities

$$\sin(-x) = -\sin x \qquad \cos(-x) = \cos x \qquad \tan(-x) = -\tan x$$

There are no cut-and-dried rules for simplifying trigonometric expressions or proving identities, but there are some common strategies that are often helpful. Six of these strategies are illustrated in the following examples. There are often a variety of ways to proceed, and it will take some practice before you can easily decide which strategies are likely to be the most efficient in a particular case.

Strategy 1

> Express everything in terms of sine and cosine.

EXAMPLE 2

Simplify $(\csc x + \cot x)(1 - \cos x)$.

SOLUTION   Using Strategy 1, we have

$$(\csc x + \cot x)(1 - \cos x) = \left(\frac{1}{\sin x} + \frac{\cos x}{\sin x}\right)(1 - \cos x) \qquad \text{[Reciprocal identities]}$$

$$= \frac{(1 + \cos x)}{\sin x}(1 - \cos x)$$

$$= \frac{(1 + \cos x)(1 - \cos x)}{\sin x}$$

$$= \frac{1 - \cos^2 x}{\sin x} = \frac{\sin^2 x}{\sin x} \qquad \text{[Pythagorean identity]}$$

$$= \sin x.$$

■

Strategy 2

> Use algebra and identities to transform the expression on one side of the equal sign into the expression on the other side.*

EXAMPLE 3

In Example 1, we verified graphically that the equation

$$\frac{1 + \sin x - \sin^2 x}{\cos x} = \cos x + \tan x$$

might be an identity. Prove that it is.

SOLUTION   We use Strategy 2, beginning with the left side of the equation:

$$\frac{1 + \sin x - \sin^2 x}{\cos x} = \frac{(1 - \sin^2 x) + \sin x}{\cos x}$$

$$= \frac{\cos^2 x + \sin x}{\cos x} \qquad \text{[Pythagorean identity]}$$

$$= \frac{\cos^2 x}{\cos x} + \frac{\sin x}{\cos x}$$

$$= \cos x + \frac{\sin x}{\cos x}$$

$$= \cos x + \tan x. \qquad \blacksquare$$

Strategy 3

> Deal separately with each side of the equation $A = B$. First use identities and algebra to transform A into some expression C (so that $A = C$). Then use (possibly different) identities and algebra to transform B into the *same* expression C (so that $B = C$). Conclude that $A = B$.

EXAMPLE 4

Prove that

$$(1 + \cos x)(\sec x - 1) = \cos x \tan^2 x.$$

SOLUTION   We shall use Strategy 3. We begin by multiplying out the left side and simplifying the result.

$$(1 + \cos x)(\sec x - 1) = \sec x - 1 + \cos x \sec x - \cos x$$

$$= \sec x - 1 + \cos x \cdot \frac{1}{\cos x} - \cos x \qquad \text{[Reciprocal identity]}$$

$$= \sec x - 1 + 1 - \cos x$$

$$(1 + \cos x)(\sec x - 1) = \sec x - \cos x \qquad (*)$$

---

*That is, start with expression A on one side and use identities and algebra to produce a sequence of equalities $A = B, B = C, C = D, D = E$, where E is the other side of the identity to be proved; conclude that $A = E$.

Since the right side of (∗) is reasonably simple, we now work on the right side of the alleged identity.

$$\cos x \tan^2 x = \cos x \, (\sec^2 x - 1) \qquad \text{[Pythagorean identity]}$$

$$= \cos x \sec^2 x - \cos x$$

$$= \cos x \left(\frac{1}{\cos x}\right)^2 - \cos x \qquad \text{[Reciprocal identity]}$$

$$= \cos x \, \frac{1}{\cos^2 x} - \cos x$$

$$= \frac{1}{\cos x} - \cos x$$

$$\cos x \tan^2 x = \sec x - \cos x \qquad \text{[Reciprocal identity]} \qquad (\ast\ast)$$

Equations (∗) and (∗∗) show that $(1 + \cos x)(\sec x - 1)$ and $\cos x \tan^2 x$ are each equal to $\sec x - \cos x$. Therefore, they are equal to each other:

$$(1 + \cos x)(\sec x - 1) = \cos x \tan^2 x. \qquad \blacksquare$$

There are several useful strategies for dealing with fractions.

Strategy 4

> Combine the sum or difference of two fractions into a single fraction.

EXAMPLE 5

Is this equation an identity?

$$\frac{1 + \sin x}{\cos x} + \frac{\cos x}{1 + \sin x} = 2 \sec x$$

SOLUTION   We first check to see if the equation *might* be an identity.

GRAPHING EXPLORATION

On the same screen, graph

$$y = \frac{1 + \sin x}{\cos x} + \frac{\cos x}{1 + \sin x} \quad \text{and} \quad y = 2 \sec x = \frac{2}{\cos x}.$$

If the two graphs appear to be identical, the equation could possibly be an identity. Do they?

The exploration suggests that this equation might be an identity. So we try to prove it, beginning with Strategy 4. Express the fractions on the left side in terms of a common denominator and add them. Then (using strategy 2) we attempt to transform this result into $2 \sec x$.

$$\frac{1 + \sin x}{\cos x} + \frac{\cos x}{1 + \sin x} = \frac{(1 + \sin x)(1 + \sin x)}{\cos x(1 + \sin x)} + \frac{(\cos x)(\cos x)}{\cos x(1 + \sin x)}$$

[Common denominator]

$$= \frac{(1 + \sin x)^2 + \cos^2 x}{\cos x(1 + \sin x)}$$

[Add fractions]

$$= \frac{1 + 2\sin x + \sin^2 x + \cos^2 x}{\cos x (1 + \sin x)} \qquad \text{[Expand numerator]}$$

$$= \frac{1 + 2\sin x + 1}{\cos x (1 + \sin x)} \qquad \text{[Pythagorean identity]}$$

$$= \frac{2 + 2\sin x}{\cos x (1 + \sin x)} = \frac{2(1 + \sin x)}{\cos x (1 + \sin x)}$$

$$= \frac{2}{\cos x} = 2\sec x \qquad \text{[Reciprocal identity]} \quad \blacksquare$$

Strategy 5

> Rewrite a fraction in an equivalent form by multiplying its numerator and denominator by the same quantity.

EXAMPLE 6

Prove that

$$\frac{\sin x}{1 + \cos x} = \frac{1 - \cos x}{\sin x}.$$

SOLUTION   We shall use Strategy 2 and transform the left side into the right side. We apply strategy 5 by multiplying the numerator and denominator of the left side by $1 - \cos x$:*

$$\frac{\sin x}{1 + \cos x} = \frac{\sin x}{1 + \cos x} \cdot \frac{1 - \cos x}{1 - \cos x} = \frac{\sin x(1 - \cos x)}{(1 + \cos x)(1 - \cos x)}$$

$$= \frac{\sin x(1 - \cos x)}{1 - \cos^2 x}$$

$$= \frac{\sin x(1 - \cos x)}{\sin^2 x} \qquad \text{[Pythagorean identity]}$$

$$= \frac{1 - \cos x}{\sin x}.$$

ALTERNATE SOLUTION   The numerators of the given equation look similar to the Pythagorean identity—with the squares missing. So we begin with the left side and introduce some squares by multiplying it by $\dfrac{\sin x}{\sin x} = 1$.

$$\frac{\sin x}{1 + \cos x} = 1 \cdot \frac{\sin x}{1 + \cos x} = \frac{\sin x}{\sin x} \cdot \frac{\sin x}{1 + \cos x} = \frac{\sin^2 x}{\sin x(1 + \cos x)}$$

$$= \frac{1 - \cos^2 x}{\sin x(1 + \cos x)} \qquad \text{[Pythagorean identity]}$$

$$= \frac{(1 - \cos x)(1 + \cos x)}{\sin x(1 + \cos x)} \qquad \text{[Factor numerator]}$$

$$= \frac{1 - \cos x}{\sin x}. \qquad \qquad \blacksquare$$

---

*This is analogous to the process used to rationalize the denominator of a fraction by multiplying its numerator and denominator by the conjugate of the denominator, as in the example:

$$\frac{1}{3 + \sqrt{2}} = \frac{1}{3 + \sqrt{2}} \cdot \frac{3 - \sqrt{2}}{3 - \sqrt{2}} = \frac{3 - \sqrt{2}}{3^2 - (\sqrt{2})^2} = \frac{3 - \sqrt{2}}{7}.$$

Proving identities involving fractions can sometimes be quite complicated. It often helps to approach a fractional identity indirectly, as in the following example.

EXAMPLE 7

Prove these identities.

(a) $\sec x(\sec x - \cos x) = \tan^2 x$ (b) $\dfrac{\sec x}{\tan x} = \dfrac{\tan x}{\sec x - \cos x}$.

SOLUTION

(a) Beginning with the left side (Strategy 2), we have

$$\sec x(\sec x - \cos x) = \sec^2 x - \sec x \cos x$$

$$= \sec^2 x - \frac{1}{\cos x}\cos x \qquad \text{[Reciprocal identity]}$$

$$= \sec^2 x - 1$$

$$= \tan^2 x. \qquad \text{[Pythagorean identity]}$$

Therefore, $\sec x(\sec x - \cos x) = \tan^2 x$.

(b) By part (a), we know that

$$\sec x(\sec x - \cos x) = \tan x \tan x.$$

Dividing both sides of this equation by $\tan x(\sec x - \cos x)$ shows that

$$\frac{\sec x(\sec x - \cos x)}{\tan x(\sec x - \cos x)} = \frac{\tan x \tan x}{\tan x(\sec x - \cos x)}$$

$$\frac{\sec x}{\tan x} = \frac{\tan x}{\sec x - \cos x}.$$ ∎

Look carefully at how identity (b) was proved in Example 7. We first proved identity (a):

$$\sec x\underbrace{(\sec x - \cos x)}_{} = \tan x \cdot \tan x.$$

$$AD \qquad = BC$$

Then we divided both sides of $AD = BC$ by $BD = \tan x(\sec x - \cos x)$ and cancelled factors to obtain identity (b):

$$\frac{A\cancel{D}}{B\cancel{D}} = \frac{\cancel{B}C}{\cancel{B}D}$$

$$\frac{\sec x\cancel{(\sec x - \cos x)}}{\tan x\cancel{(\sec x - \cos x)}} = \frac{\cancel{\tan x}\, \tan x}{\cancel{\tan x}(\sec x - \cos x)}$$

$$\frac{\sec x}{\tan x} = \frac{\tan x}{\sec x - \cos x}$$

$$\frac{A}{B} = \frac{C}{D}.$$

The same argument works in the general case and provides the following strategy for dealing with identities involving fractions.

Strategy 6

> If you can prove that $AD = BC$, with $B \neq 0$ and $D \neq 0$, then you can conclude that
>
> $$\frac{A}{B} = \frac{C}{D}.$$

Many students misunderstand Strategy 6: It does *not* say that you begin with a fractional equation $A/B = C/D$ and cross-multiply to eliminate the fractions. If you did that, you would be assuming what has to be proved. What the strategy says is that to prove an identity involving fractions, you need only prove a different identity that does not involve fractions. In other words, if you prove that $AD = BC$ whenever $B \neq 0$ and $D \neq 0$, then you can conclude that $A/B = C/D$. Note that you do not *assume* that $AD = BC$; you use Strategy 2 or 3 or some other means to *prove* this statement.

EXAMPLE 8

Prove that $\dfrac{\cot x - 1}{\cot x + 1} = \dfrac{1 - \tan x}{1 + \tan x}$.

SOLUTION We use Strategy 6, with $A = \cot x - 1$, $B = \cot x + 1$, $C = 1 - \tan x$, and $D = 1 + \tan x$. We must prove that this equation is an identity:

$$AD = BC$$

$(***)$ $\qquad (\cot x - 1)(1 + \tan x) = (\cot x + 1)(1 - \tan x).$

Strategy 3 will be used. Multiplying out the left side shows that

$$(\cot x - 1)(1 + \tan x) = \cot x - 1 + \cot x \tan x - \tan x$$

$$= \cot x - 1 + \frac{1}{\tan x}\tan x - \tan x$$

$$= \cot x - 1 + 1 - \tan x$$

$$= \cot x - \tan x.$$

Similarly, on the right side of $(***)$,

$$(\cot x + 1)(1 - \tan x) = \cot x + 1 - \cot x \tan x - \tan x$$

$$= \cot x + 1 - 1 - \tan x$$

$$= \cot x - \tan x.$$

TECHNOLOGY TIP

Using SOLVE in the TI-89 ALGEBRA menu to solve an equation that might be an identity produces one of three responses. "True" means the equation *probably* is an identity [algebraic proof is required for certainty]. "False" means the equation is not an identity. A numerical answer is inconclusive [the equation may or may not be an identity].

Since the left and right sides are equal to the same expression, we have proved that $(***)$ is an identity. Therefore, by Strategy 4, we conclude that

$$\frac{\cot x - 1}{\cot x + 1} = \frac{1 - \tan x}{1 + \tan x}$$

is also an identity. ∎

It takes a good deal of practice, as well as *much* trial and error, to become proficient in proving identities. The more practice you have, the easier it will get. Since there are many correct methods, your proofs may be quite different from those of your instructor or the text answers.

If you don't see what to do immediately, try something and see where it leads: Multiply out or factor or multiply numerator and denominator by the same nonzero quantity. Even if this doesn't lead anywhere, it might give you some ideas on other things to try. When you do obtain a proof, check to see whether it can be done more efficiently. In your final proof, don't include the side trips that may have given you some ideas but aren't themselves part of the proof.

EXERCISES 7.1

In Exercises 1–4, test the equation graphically to determine whether it might be an identity. You need not prove those equations that seem to be identities.

1. $\dfrac{\sec x - \cos x}{\sec x} = \sin^2 x$

2. $\tan x + \cot x = (\sin x)(\cos x)$

3. $\dfrac{1 - \cos(2x)}{2} = \sin^2 x$

4. $\dfrac{\tan x + \cot x}{\csc x} = \sec x$

In Exercises 5–8, insert one of A–F on the right of the equal sign so that the resulting equation appears to be an identity when you test it graphically. You need not prove the identity.

A. $\cos x$ **B.** $\sec x$ **C.** $\sin^2 x$

D. $\sec^2 x$ **E.** $\sin x - \cos x$ **F.** $\dfrac{1}{\sin x \cos x}$

5. $\csc x \tan x =$ ____

6. $\dfrac{\sin x}{\tan x} =$ ____

7. $\dfrac{\sin^4 x - \cos^4 x}{\sin x + \cos x} =$ ____

8. $\tan^2(-x) - \dfrac{\sin(-x)}{\sin x} =$ ____

In Exercises 9–24, prove the identity.

9. $\tan x \cos x = \sin x$
10. $\cot x \sin x = \cos x$
11. $\cos x \sec x = 1$
12. $\sin x \csc x = 1$
13. $\tan x \csc x = \sec x$
14. $\sec x \cot x = \csc x$
15. $\dfrac{\tan x}{\sec x} = \sin x$
16. $\dfrac{\cot x}{\csc x} = \cos x$
17. $(1 + \cos x)(1 - \cos x) = \sin^2 x$
18. $(\csc x - 1)(\csc x + 1) = \cot^2 x$
19. $\cot x \sec x \sin x = 1$
20. $\sin x \tan x + \cos x = \sec x$
21. $\tan x + \cot x = \sec x \csc x$
22. $\tan x(\cos x + \csc x) = \sin x + \sec x$
23. $\cos x - \cos x \sin^2 x = \cos^3 x$
24. $\cos x \sec x - \sin^2 x = \cos^2 x$

In Exercises 25–64, state whether or not the equation is an identity. If it is an identity, prove it.

25. $\sin x = \sqrt{1 - \cos^2 x}$

26. $\cot x = \dfrac{\csc x}{\sec x}$

27. $\dfrac{\sin(-x)}{\cos(-x)} = -\tan x$

28. $\tan x = \sqrt{\sec^2 x - 1}$

29. $\cot(-x) = -\cot x$

30. $\sec(-x) = \sec x$

31. $1 + \sec^2 x = \tan^2 x$

32. $\sec^4 x - \tan^4 x = 1 + 2 \tan^2 x$

33. $\sec^2 x - \csc^2 x = \tan^2 x - \cot^2 x$

34. $\sec^2 x + \csc^2 x = \sec^2 x \csc^2 x$

35. $\sin^2 x(\cot x + 1)^2 = \cos^2 x(\tan x + 1)^2$

36. $\cos^2 x(\sec x + 1)^2 = (1 + \cos x)^2$

37. $\sin^2 x - \tan^2 x = -\sin^2 x \tan^2 x$

38. $\cot^2 x - 1 = \csc^2 x$

39. $(\cos^2 x - 1)(\tan^2 x + 1) = -\tan^2 x$

40. $(1 - \cos^2 x)\csc x = \sin x$

41. $\tan x = \dfrac{\sec x}{\csc x}$

42. $\dfrac{\cos(-x)}{\sin(-x)} = -\cot x$

43. $\cos^4 x - \sin^4 x = \cos^2 x - \sin^2 x$

44. $\cot^2 x - \cos^2 x = \cos^2 x \cot^2 x$

45. $(\sin x + \cos x)^2 = \sin^2 x + \cos^2 x$

46. $(1 + \tan x)^2 = \sec^2 x$

47. $\dfrac{1 + \sin x}{\sin x} = \dfrac{\cot^2 x}{\csc x - 1}$

48. $\dfrac{1}{1 - \sin x} - \dfrac{1}{1 + \sin x} = 2 \tan x \sec x$

49. $\dfrac{1 + \sin x}{1 - \sin x} = \dfrac{\sec x + \tan x}{\sec x - \tan x}$

50. $\dfrac{\sin x}{\cos x} + \dfrac{\cos x}{1 + \sin x} = \sec x$

51. $\dfrac{\tan x + \sin x}{1 + \cos x} = \tan x$

52. $\dfrac{1}{1 - \sin x} + \dfrac{1}{1 + \sin x} = 2 \sec^2 x$

53. $\dfrac{1 - \sin x}{\cos x} = \dfrac{\cos x}{1 + \sin x}$

54. $\dfrac{1 + \tan x}{1 + \cot x} = \tan x$

55. $\dfrac{\sec^2 x - 1}{\sec^2 x} = \sin^2 x$

56. $\dfrac{\csc^2 x - 1}{\csc^2 x} = \cos^2 x$

57. $\dfrac{\sec x}{\csc x} + \dfrac{\sin x}{\cos x} = 2 \tan x$

58. $\dfrac{1 + \cos x}{\sin x} + \dfrac{\sin x}{1 + \cos x} = 2 \csc x$

59. $\dfrac{\sec x + \csc x}{1 + \tan x} = \csc x$ **60.** $\dfrac{\cot x - 1}{1 - \tan x} = \dfrac{\csc x}{\sec x}$

61. $\dfrac{1}{\csc x - \sin x} = \sec x \tan x$

62. $\dfrac{1 + \csc x}{\csc x} = \dfrac{\cos^2 x}{1 - \sin x}$

63. $\dfrac{\sin x - \cos x}{\tan x} = \dfrac{\tan x}{\sin x + \cos x}$

64. $\dfrac{\cot x}{\csc x - 1} = \dfrac{\csc x + 1}{\cot x}$

In Exercises 65–68, half of an identity is given. Graph this half in a viewing window with $-2\pi \le x \le 2\pi$ and make a conjecture as to what the right side of the identity is. Then prove your conjecture.

65. $1 - \dfrac{\sin^2 x}{1 + \cos x} = ?$ [*Hint:* What familiar function has a graph that looks like this?]

66. $\dfrac{1 + \cos x - \cos^2 x}{\sin x} - \cot x = ?$

67. $(\sin x + \cos x)(\sec x + \csc x) - \cot x - 2 = ?$

68. $\cos^3 x (1 - \tan^4 x + \sec^4 x) = ?$

In Exercises 69–82, prove the identity.

69. $\dfrac{1 - \sin x}{\sec x} = \dfrac{\cos^3 x}{1 + \sin x}$

70. $\dfrac{\sin x}{1 - \cot x} + \dfrac{\cos x}{1 - \tan x} = \cos x + \sin x$

71. $\dfrac{\cos x}{1 - \sin x} = \sec x + \tan x$

72. $\dfrac{1 + \sec x}{\tan x + \sin x} = \csc x$

73. $\dfrac{\cos x \cot x}{\cot x - \cos x} = \dfrac{\cot x + \cos x}{\cos x \cot x}$

74. $\dfrac{\cos^3 x - \sin^3 x}{\cos x - \sin x} = 1 + \sin x \cos x$

75. $\log_{10}(\cot x) = -\log_{10}(\tan x)$

76. $\log_{10}(\sec x) = -\log_{10}(\cos x)$

77. $\log_{10}(\csc x + \cot x) = -\log_{10}(\csc x - \cot x)$

78. $\log_{10}(\sec x + \tan x) = -\log_{10}(\sec x - \tan x)$

79. $\tan x - \tan y = -\tan x \tan y (\cot x - \cot y)$

80. $\dfrac{\tan x - \tan y}{\cot x - \cot y} = -\tan x \tan y$

81. $\dfrac{\cos x - \sin y}{\cos y - \sin x} = \dfrac{\cos y + \sin x}{\cos x + \sin y}$

82. $\dfrac{\tan x + \tan y}{\cot x + \cot y} = \dfrac{\tan x \tan y - 1}{1 - \cot x \cot y}$

7.2   Addition and Subtraction Identities

Section Objectives
- Use the addition and subtraction identities to evaluate trigonometric functions.
- Use the addition and subtraction identities to prove other identities.
- Use the cofunction identities to prove other identities.

A common student ERROR is to write

$$\sin\left(x + \frac{\pi}{6}\right) = \sin x + \sin \frac{\pi}{6} = \sin x + \frac{1}{2}.$$

GRAPHING EXPLORATION

Verify graphically that the equation above is NOT an identity by graphing $y = \sin(x + \pi/6)$ and $y = \sin x + 1/2$ on the same screen.

The exploration shows that "$\sin(x + y) = \sin x + \sin y$" is NOT an identity (because it's false when $y = \pi/6$). There is an identity that enables us to express $\sin(x + y)$, but it is a bit more complicated, as we now see.

Addition and Subtraction Identities

$$\sin(x + y) = \sin x \cos y + \cos x \sin y$$

$$\sin(x - y) = \sin x \cos y - \cos x \sin y$$

$$\cos(x + y) = \cos x \cos y - \sin x \sin y$$

$$\cos(x - y) = \cos x \cos y + \sin x \sin y$$

The addition and subtraction identities are probably the most important of all the trigonometric identities. Before reading their proofs at the end of this section, you should become familiar with the examples and special cases below.

EXAMPLE 1

Use the addition and subtraction identities to find the *exact* values of

(a) $\sin\left(\dfrac{\pi}{12}\right)$ (b) $\cos\left(\dfrac{7\pi}{12}\right)$.

SOLUTION   The key here is to write $\pi/12$ and $7\pi/12$ as a sum or difference of two numbers whose sine and cosine are known.

(a) Note that

$$\frac{\pi}{12} = \frac{4\pi}{12} - \frac{3\pi}{12} = \frac{\pi}{3} - \frac{\pi}{4}.$$

we apply the subtraction identity for sine with $x = \pi/3$ and $y = \pi/4$.

$$\sin\left(\frac{\pi}{12}\right) = \sin\left(\frac{\pi}{3} - \frac{\pi}{4}\right)$$

$$= \sin\frac{\pi}{3}\cos\frac{\pi}{4} - \cos\frac{\pi}{3}\sin\frac{\pi}{4}$$

$$= \frac{\sqrt{3}}{2} \cdot \frac{\sqrt{2}}{2} - \frac{1}{2} \cdot \frac{\sqrt{2}}{2}$$

$$= \frac{\sqrt{3}\sqrt{2}}{4} - \frac{\sqrt{2}}{4} = \frac{\sqrt{6}}{4} - \frac{\sqrt{2}}{4} = \frac{\sqrt{6} - \sqrt{2}}{4}$$

(b) In this case, we see that $\dfrac{7\pi}{12} = \dfrac{4\pi}{12} + \dfrac{3\pi}{12} = \dfrac{\pi}{3} + \dfrac{\pi}{4}$. So we apply the addition identity for cosine with $x = \pi/3$ and $y = \pi/4$.

$$\cos\left(\frac{7\pi}{12}\right) = \cos\left(\frac{\pi}{3} + \frac{\pi}{4}\right)$$

$$= \cos\frac{\pi}{3}\cos\frac{\pi}{4} - \sin\frac{\pi}{3}\sin\frac{\pi}{4}$$

$$= \frac{1}{2} \cdot \frac{\sqrt{2}}{2} - \frac{\sqrt{3}}{2} \cdot \frac{\sqrt{2}}{2}$$

$$= \frac{\sqrt{2}}{4} - \frac{\sqrt{3}\sqrt{2}}{4} = \frac{\sqrt{2}}{4} - \frac{\sqrt{6}}{4} = \frac{\sqrt{2} - \sqrt{6}}{4}.$$ ∎

EXAMPLE 2

Find $\sin(\pi - y)$.

SOLUTION Apply the subtraction identity for sine with $x = \pi$.

$$\sin(\pi - y) = \sin \pi \cos y - \cos \pi \sin y$$
$$= (0)(\cos y) - (-1)(\sin y)$$
$$= \sin y. \qquad \blacksquare$$

EXAMPLE 3

Show that the difference quotient of the function $f(x) = \sin x$ is given by:

$$\frac{f(x + h) - f(x)}{h} = \sin x \left(\frac{\cos h - 1}{h} \right) + \cos x \left(\frac{\sin h}{h} \right).$$

[This fact is needed in calculus.]

SOLUTION Use the addition identity for $\sin(x + y)$ with $y = h$.

$$\frac{f(x + h) - f(x)}{h} = \frac{\sin(x + h) - \sin x}{h}$$

$$= \frac{\sin x \cos h + \cos x \sin h - \sin x}{h}$$

$$= \frac{\sin x(\cos h - 1) + \cos x \sin h}{h}$$

$$= \sin x \left(\frac{\cos h - 1}{h} \right) + \cos x \left(\frac{\sin h}{h} \right). \qquad \blacksquare$$

EXAMPLE 4

Prove the identity:

$$\frac{\cos(x + y)}{\cos x \cos y} = 1 - \tan x \tan y.$$

SOLUTION Begin with the left side and apply the addition identity for cosine to the numerator.

$$\frac{\cos(x + y)}{\cos x \cos y} = \frac{\cos x \cos y - \sin x \sin y}{\cos x \cos y}$$

$$= \frac{\cos x \cos y}{\cos x \cos y} - \frac{\sin x \sin y}{\cos x \cos y}$$

$$= 1 - \frac{\sin x}{\cos x} \cdot \frac{\sin y}{\cos y} = 1 - \tan x \tan y. \qquad \blacksquare$$

EXAMPLE 5

Prove that

$$\cos x \cos y = \frac{1}{2}[\cos(x + y) + \cos(x - y)].$$

SOLUTION   We begin with the more complicated right side and use the addition and subtraction identities for cosine to transform it into the left side.

$$\frac{1}{2}[\cos(x + y) + \cos(x - y)] = \frac{1}{2}[(\cos x \cos y - \sin x \sin y)$$
$$+ (\cos x \cos y + \sin x \sin y)]$$

$$= \frac{1}{2}(\cos x \cos y + \cos x \cos y)$$

$$= \frac{1}{2}(2 \cos x \cos y) = \cos x \cos y. \qquad \blacksquare$$

The addition and subtraction identities for sine and cosine can be used to obtain the following identities, as outlined in Exercise 38.

Addition and Subtraction Identities for Tangent

$$\tan(x + y) = \frac{\tan x + \tan y}{1 - \tan x \tan y}$$

$$\tan(x - y) = \frac{\tan x - \tan y}{1 + \tan x \tan y}$$

It is sometimes convenient to say that x is a number *in the first quadrant* if $0 < x < \pi/2$, that x is a number *in the second quadrant* if $\pi/2 < x < \pi$, and so on.

EXAMPLE 6

Suppose x is a number in the first quadrant and y is a number in the third quadrant. If $\sin x = 3/4$ and $\cos y = -1/3$, find the exact values of $\sin(x + y)$ and $\tan(x + y)$ and determine in which quadrant $x + y$ lies.

SOLUTION   We want to apply the addition identities for sine and tangent. To do so we must first find $\cos x$, $\tan x$, $\sin y$ and $\tan y$. Using the Pythagorean identity and the fact that $\cos x$ and $\tan x$ are positive when $0 < x < \pi/2$, we have

$$\cos x = \sqrt{1 - \sin^2 x} = \sqrt{1 - \left(\frac{3}{4}\right)^2} = \sqrt{1 - \frac{9}{16}} = \sqrt{\frac{7}{16}} = \frac{\sqrt{7}}{4},$$

$$\tan x = \frac{\sin x}{\cos x} = \frac{3/4}{\sqrt{7}/4} = \frac{3}{4} \cdot \frac{4}{\sqrt{7}} = \frac{3}{\sqrt{7}} = \frac{3\sqrt{7}}{7}.$$

Since y lies between π and $3\pi/2$, its sine is negative; hence,

$$\sin y = -\sqrt{1 - \cos^2 y} = -\sqrt{1 - \left(-\frac{1}{3}\right)^2} = -\sqrt{\frac{8}{9}} = -\frac{\sqrt{8}}{3} = -\frac{2\sqrt{2}}{3},$$

$$\tan y = \frac{\sin y}{\cos y} = \frac{-2\sqrt{2}/3}{-1/3} = \frac{-2\sqrt{2}}{3} \cdot \frac{3}{-1} = 2\sqrt{2}.$$

The addition identities for sine and tangent now show that

$$\sin(x + y) = \sin x \cos y + \cos x \sin y$$

$$= \frac{3}{4} \cdot \frac{-1}{3} + \frac{\sqrt{7}}{4} \cdot \frac{-2\sqrt{2}}{3} = \frac{-3}{12} - \frac{2\sqrt{14}}{12} = \frac{-3 - 2\sqrt{14}}{12},$$

$$\tan(x + y) = \frac{\tan x + \tan y}{1 - \tan x \tan y}$$

$$= \frac{\dfrac{3\sqrt{7}}{7} + 2\sqrt{2}}{1 - \left(\dfrac{3\sqrt{7}}{7}\right)(2\sqrt{2})} = \frac{\dfrac{3\sqrt{7} + 14\sqrt{2}}{7}}{\dfrac{7 - 6\sqrt{14}}{7}} = \frac{3\sqrt{7} + 14\sqrt{2}}{7 - 6\sqrt{14}}.$$

The numerator of $\sin(x + y)$ is negative and the denominator of $\tan(x + y)$ is negative, as you can easily verify, so the sine and tangent of $x + y$ are negative numbers. The fourth quadrant is the only one in which both sine and tangent are negative (see the sign chart in Exercise 63 on page 451). Hence, $x + y$ must be in the fourth quadrant, that is, in the interval $(3\pi/2, 2\pi)$. ■

COFUNCTION IDENTITIES

Other special cases of the addition and subtraction identities are the cofunction identities:

Cofunction Identities

$$\sin x = \cos\left(\frac{\pi}{2} - x\right) \qquad \cos x = \sin\left(\frac{\pi}{2} - x\right)$$

$$\tan x = \cot\left(\frac{\pi}{2} - x\right) \qquad \cot x = \tan\left(\frac{\pi}{2} - x\right)$$

$$\sec x = \csc\left(\frac{\pi}{2} - x\right) \qquad \csc x = \sec\left(\frac{\pi}{2} - x\right)$$

The first confunction identity is proved by using the identity for $\cos(x - y)$ with $\pi/2$ in place of x and x in place of y.

$$\cos\left(\frac{\pi}{2} - x\right) = \cos\frac{\pi}{2}\cos x + \sin\frac{\pi}{2}\sin x = (0)(\cos x) + (1)(\sin x) = \sin x.$$

Since the first cofunction identity is valid for *every* number x, it is also valid with the number $\pi/2 - x$ in place of x.

$$\sin\left(\frac{\pi}{2} - x\right) = \cos\left[\frac{\pi}{2} - \left(\frac{\pi}{2} - x\right)\right] = \cos x.$$

Thus, we have proved the second cofunction identity. The others now follow from these two. For instance,

$$\tan\left(\frac{\pi}{2} - x\right) = \frac{\sin[(\pi/2) - x]}{\cos[(\pi/2) - x]} = \frac{\cos x}{\sin x} = \cot x.$$

EXAMPLE 7

Verify that $\dfrac{\cos(x - \pi/2)}{\cos x} = \tan x.$

SOLUTION   Beginning on the left side, we see that the term $\cos(x - \pi/2)$ looks almost, but not quite, like the term $\cos(\pi/2 - x)$ in the cofunction identity. But note that $-(x - \pi/2) = \pi/2 - x$. Therefore,

$$\frac{\cos\left(x - \dfrac{\pi}{2}\right)}{\cos x} = \frac{\cos\left[-\left(x - \dfrac{\pi}{2}\right)\right]}{\cos x} \qquad \text{[Negative angle identity with } x - \dfrac{\pi}{2} \text{ in place of } x]$$

$$= \frac{\cos\left(\dfrac{\pi}{2} - x\right)}{\cos x}$$

$$= \frac{\sin x}{\cos x} \qquad \text{[Cofunction identity]}$$

$$= \tan x. \qquad \text{[Reciprocal identity]} \qquad \blacksquare$$

PROOF OF THE ADDITION AND SUBTRACTION IDENTITIES

We first prove the subtraction identity for cosine:

$$\cos(x - y) = \cos x \cos y + \sin x \sin y.$$

If $x = y$, then this is true by the Pythagorean identity:

$$\cos(x - x) = \cos 0 = 1 = \cos^2 x + \sin^2 x = \cos x \cos x + \sin x \sin x.$$

Next we prove the identity in the case when $x > y$. Let P be the point where the terminal side of an angle of x radians in standard position meets the unit circle and let Q be the point where the terminal side of an angle of y radians in standard position meets the circle, as shown in Figure 7–2. According to the definitions of sine and cosine, P has coordinates $(\cos x, \sin x)$ and Q has coordinates $(\cos y, \sin y)$.

Using the distance formula, we have

Distance from P to Q

$$= \sqrt{(\cos x - \cos y)^2 + (\sin x - \sin y)^2}$$

$$= \sqrt{\cos^2 x - 2\cos x \cos y + \cos^2 y + \sin^2 x - 2\sin x \sin y + \sin^2 y}$$

$$= \sqrt{(\cos^2 x + \sin^2 x) + (\cos^2 y + \sin^2 y) - 2\cos x \cos y - 2\sin x \sin y}$$

$$= \sqrt{1 + 1 - 2\cos x \cos y - 2\sin x \sin y}$$

$$= \sqrt{2 - 2\cos x \cos y - 2\sin x \sin y}.$$

The angle QOP formed by the two terminal sides has radian measure $x - y$ (Figure 7.2). Rotate this angle clockwise until side OQ lies on the horizontal axis, as shown in Figure 7–3. Angle QOP is now in standard position, and its terminal side meets the unit circle at P. Since angle QOP has radian measure $x - y$, the

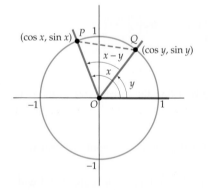

Figure 7–2

definitions of sine and cosine show that the point P, in this new location, has coordinates $(\cos(x - y), \sin(x - y))$. Q now has coordinates $(1, 0)$.

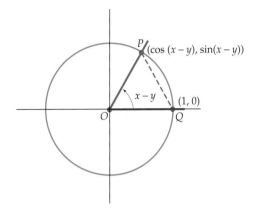

Figure 7–3

Using the coordinates of P and Q *after* the angle is rotated shows that

$$\text{Distance from } P \text{ to } Q = \sqrt{[\cos(x - y) - 1]^2 + [\sin(x - y) - 0]^2}$$

$$= \sqrt{\cos^2(x - y) - 2\cos(x - y) + 1 + \sin^2(x - y)}$$

$$= \sqrt{\cos^2(x - y) + \sin^2(x - y) - 2\cos(x - y) + 1}$$

$$= \sqrt{1 - 2\cos(x - y) + 1} \quad \text{[Pythagorean identity]}$$

$$= \sqrt{2 - 2\cos(x - y)}.$$

The two expressions for the distance from P to Q must be equal. Hence,

$$\sqrt{2 - 2\cos(x - y)} = \sqrt{2 - 2\cos x \cos y - 2\sin x \sin y}.$$

Squaring both sides of this equation and simplifying the result yields

$$2 - 2\cos(x - y) = 2 - 2\cos x \cos y - 2\sin x \sin y$$

$$-2\cos(x - y) = -2(\cos x \cos y + \sin x \sin y)$$

$$\cos(x - y) = \cos x \cos y + \sin x \sin y.$$

This completes the proof of the subtraction identity for cosine when $x > y$. If $y > x$, then the proof just given is valid with the roles of x and y interchanged; it shows that

$$\cos(y - x) = \cos y \cos x + \sin y \sin x$$

$$= \cos x \cos y + \sin x \sin y.$$

The negative angle identity with $x - y$ in place of x shows that

$$\cos(x - y) = \cos[-(x - y)] = \cos(y - x).$$

Combining this fact with the previous one shows that

$$\cos(x - y) = \cos x \cos y + \sin x \sin y$$

in this case also. Therefore, the subtraction identity for cosine is proved.

Next, we prove the addition identity for cosine:

$$\mathbf{\cos(x + y) = \cos x \cos y - \sin x \sin y.}$$

The proof uses the subtraction identity for cosine just proved and the fact that $x + y = x - (-y)$.

$$\cos(x + y) = \cos[x - (-y)]$$

$$= \cos x \cos(-y) + \sin x \sin(-y) \quad \text{[Subtraction identity for cosine]}$$

$$= \cos x \cos y + \sin x(-\sin y) \quad \text{[Negative angle identities]}$$

$$= \cos x \cos y - \sin x \sin y.$$

The proofs of the addition and subtraction identities for sine are in Exercises 36 and 37.

EXERCISES 7.2

In Exercises 1–12, find the exact value.

1. $\cos \dfrac{\pi}{12}$ **2.** $\tan \dfrac{\pi}{12}$ **3.** $\sin \dfrac{5\pi}{12}$

4. $\cos \dfrac{5\pi}{12}$ **5.** $\cot \dfrac{5\pi}{12}$ **6.** $\sin \dfrac{7\pi}{12}$

7. $\tan \dfrac{7\pi}{12}$ **8.** $\sin \dfrac{11\pi}{12}$ **9.** $\cos \dfrac{11\pi}{12}$

10. $\sin 75°$ [*Hint:* $75° = 45° + 30°$.]*

11. $\sin 105°$*   **12.** $\cos 165°$*

In Exercises 13–18, rewrite the given expression in terms of $\sin x$ *and* $\cos x$.

13. $\sin\left(\dfrac{\pi}{2} + x\right)$ **14.** $\cos\left(x + \dfrac{\pi}{2}\right)$ **15.** $\cos\left(x - \dfrac{3\pi}{2}\right)$

16. $\csc\left(x + \dfrac{\pi}{2}\right)$ **17.** $\sec(x - \pi)$ **18.** $\cot(x + \pi)$

In Exercises 19–24, simplify the given expression.

19. $\sin 3 \cos 5 - \cos 3 \sin 5$

20. $\sin 37° \sin 53° - \cos 37° \cos 53°$*

21. $\cos(x + y) \cos y + \sin(x + y) \sin y$

22. $\sin(x - y) \cos y + \cos(x - y) \sin y$

23. $\cos(x + y) - \cos(x - y)$

24. $\sin(x + y) - \sin(x - y)$

25. If $\sin x = \dfrac{1}{3}$ and $0 < x < \dfrac{\pi}{2}$, then $\sin\left(\dfrac{\pi}{4} + x\right) = ?$

26. If $\cos x = -\dfrac{1}{4}$ and $\dfrac{\pi}{2} < x < \pi$, then $\cos\left(\dfrac{\pi}{6} - x\right) = ?$

27. If $\cos x = -\dfrac{1}{5}$ and $\pi < x < \dfrac{3\pi}{2}$, then $\sin\left(\dfrac{\pi}{3} - x\right) = ?$

28. If $\sin x = -\dfrac{3}{4}$ and $\dfrac{3\pi}{2} < x < 2\pi$, then $\cos\left(\dfrac{\pi}{4} + x\right) = ?$

In Exercises 29–32, assume that $\sin x = .8$ *and* $\sin y = \sqrt{.75}$ *and that* x *and* y *lie between* 0 *and* $\pi/2$. *Evaluate the given expressions.*

29. $\sin(x + y)$ **30.** $\cos(x - y)$.

31. $\sin(x - y)$ **32.** $\tan(x + y)$

33. The figure shows an angle of t radians. Prove that for any number x,

$$5 \sin(x + t) = 3 \sin x + 4 \cos x.$$

34. The figure shows an angle of t radians. Prove that for any number y,

$$13 \cos(t - y) = 12 \cos y + 5 \sin y.$$

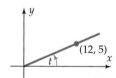

35. If $f(x) = \cos x$ and h is a fixed nonzero number, prove that:

$$\dfrac{f(x + h) - f(x)}{h} = \cos x \left(\dfrac{\cos h - 1}{h}\right) - \sin x \left(\dfrac{\sin h}{h}\right).$$

36. Prove the subtraction identity for sine:

$$\sin(x - y) = \sin x \cos y - \cos x \sin y.$$

[*Hint:* Use the first cofunction identity*

$$\sin(x - y) = \cos\left[\dfrac{\pi}{2} - (x - y)\right] = \cos\left[\left(\dfrac{\pi}{2} - x\right) + y\right]$$

and the addition identity for cosine.]

---

*Skip Exercises 10–12 and 20 if you haven't read Section 8.1.

*The cofunction identity may be validly used here because its proof on page 527 depends only on the subtraction identity for cosine which was proved in the text.

37. Prove the addition identity for sine:

$$\sin(x + y) = \sin x \cos y + \cos x \sin y.$$

[*Hint:* You may assume Exercise 36. Use the same method by which the addition identity for cosine was obtained from the subtraction identity for cosine in the text.]

38. Prove the addition and subtraction identities for the tangent function (page 526). [*Hint:*

$$\tan (x + y) = \frac{\sin(x + y)}{\cos(x + y)}.$$

Use the addition identities on the numerator and denominator; then divide both numerator and denominator by $\cos x \cos y$ and simplify.]

In Exercises 39–44, prove the identity.

39. $\dfrac{\cos(x - y)}{\cos x \cos y} = 1 + \tan x \tan y$

40. $\dfrac{\sin(x + y)}{\sin x \sin y} = \cot x + \cot y$

41. $\dfrac{\sin(x - y)}{\sin x \sin y} = \cot y - \cot x$

42. $\dfrac{\cos(x - y)}{\sin x \sin y} = 1 + \cot x \cot y$

43. $\dfrac{\sin(x + y)}{\sin x \cos y} = 1 + \cot x \tan y$

44. $\dfrac{\sin(x - y)}{\sin x \cos y} = 1 - \cot x \tan y$

45. If x is in the first and y is in the second quadrant, $\sin x = 24/25$, and $\sin y = 4/5$, find the exact value of $\sin(x + y)$ and $\tan(x + y)$ and the quadrant in which $x + y$ lies.

46. If x and y are in the second quadrant, $\sin x = 1/3$, and $\cos y = -3/4$, find the exact value of $\sin(x + y)$, $\cos(x + y)$, $\tan(x + y)$, and find the quadrant in which $x + y$ lies.

47. If x is in the first and y is in the second quadrant, $\sin x = 4/5$, and $\cos y = -12/13$, find the exact value of $\cos(x + y)$ and $\tan(x + y)$ and the quadrant in which $x + y$ lies.

48. If x is in the fourth and y is in the first quadrant, $\cos x = 1/3$, and $\cos y = 2/3$, find the exact value of $\sin(x - y)$ and $\tan(x - y)$ and the quadrant in which $x - y$ lies.

49. Express $\sin(u + v + w)$ in terms of sines and cosines of u, v, and w. [*Hint:* First apply the addition identity with $x = u + v$ and $y = w$.]

50. Express $\cos(x + y + z)$ in terms of sines and cosines of x, y, and z.

51. If $x + y = \pi/2$, show that $\sin^2 x + \sin^2 y = 1$.

52. Prove that $\cot(x + y) = \dfrac{\cot x \cot y - 1}{\cot x + \cot y}$.

In Exercises 53–64, prove the identity.

53. $\sin(x - \pi) = -\sin x$

54. $\cos(x - \pi) = -\cos x$

55. $\cos(\pi - x) = -\cos x$

56. $\tan(\pi - x) = -\tan x$

57. $\sin(x + \pi) = -\sin x$

58. $\cos(x + \pi) = -\cos x$

59. $\tan(x + \pi) = \tan x$

60. $\sin x \cos y = \frac{1}{2}[\sin(x + y) + \sin(x - y)]$

61. $\sin x \sin y = \frac{1}{2}[\cos(x - y) - \cos(x + y)]$

62. $\cos x \sin y = \frac{1}{2}[\sin(x + y) - \sin(x - y)]$

63. $\cos(x + y) \cos(x - y) = \cos^2 x \cos^2 y - \sin^2 x \sin^2 y$

64. $\sin(x + y) \sin(x - y) = \sin^2 x \cos^2 y - \cos^2 x \sin^2 y$

In Exercises 65–74, determine graphically whether the equation could possibly be an identity (by choosing a numerical value for y and graphing both sides). If it could, prove that it is.

65. $\dfrac{\cos(x - y)}{\sin x \cos y} = \cot x + \tan y$

66. $\dfrac{\cos(x + y)}{\sin x \cos y} = \cot x - \tan y$

67. $\sin(x - y) = \sin x - \sin y$

68. $\cos(x + y) = \cos x + \cos y$

69. $\dfrac{\sin(x + y)}{\sin(x - y)} = \dfrac{\tan x + \tan y}{\tan x - \tan y}$

70. $\dfrac{\sin(x + y)}{\sin(x - y)} = \dfrac{\cot y + \cot x}{\cot y - \cot x}$

71. $\dfrac{\cos(x + y)}{\cos(x - y)} = \dfrac{\cot x + \tan y}{\cot x - \tan y}$

72. $\dfrac{\cos(x - y)}{\cos(x + y)} = \dfrac{\cot y + \tan x}{\cot y - \tan x}$

73. $\tan(x + y) = \tan x + \tan y$

74. $\cot(x - y) = \cot x - \cot y$

7.2.A *SPECIAL TOPICS* Lines and Angles

Section Objective ■ Find the angle between two lines.

If L is a nonhorizontal straight line, the **angle of inclination** of L is the angle θ formed by the part of L above the x-axis and the x-axis in the positive direction, as shown in Figure 7–4.

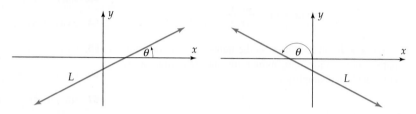

Figure 7–4

The angle of inclination of a horizontal line is defined to be $\theta = 0$. Thus, the radian measure of the angle of inclination of any line satisfies $0 \le \theta < \pi$. Furthermore,

***Angle of
Inclination***

> If L is a nonvertical line with an angle of inclination of θ radians, then
>
> $$\tan \theta = \text{slope of } L.$$

Proof   First, suppose L is horizontal. Then L has slope 0 and angle of inclination $\theta = 0$. Hence,

$$\tan \theta = \tan 0 = 0 = \text{slope } L.$$

Next, suppose L is not horizontal. L is parallel to a line M through the origin, as shown in Figure 7–5.*

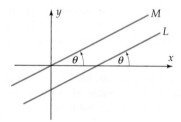

Figure 7–5

Basic facts about parallel lines show that M has the same angle of inclination θ as L. Furthermore, M lies on the terminal side of an angle of θ radians in standard position. Therefore, as we proved in Section 6.4,

$$\text{slope } M = \tan \theta.$$

Since parallel lines have the same slope, we have

$$\text{slope } L = \text{slope } M = \tan \theta.$$ ■

---

*Figure 7–5 illustrates the case when θ is acute and L lies to the right of M. The pictures are different in the other possible cases, but the argument is the same.

ANGLES BETWEEN TWO LINES

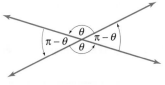

Figure 7–6

If two lines intersect, then they determine four angles with vertex at the point of intersection, as shown in Figure 7–6. If one of these angles measures θ radians, then each of the two angles adjacent to it measures $\pi - \theta$ radians. (Why?) The fourth angle also measures θ radians by the vertical angle theorem from plane geometry.

The angles between intersecting lines can be determined from the angles of inclination of the lines. Suppose L and M have angles of inclination α and β, respectively, such that $\beta \geq \alpha$. Basic facts about parallel lines, as illustrated in Figure 7–7, show that $\beta - \alpha$ is one angle between L and M and $\pi - (\beta - \alpha)$ is the other one.

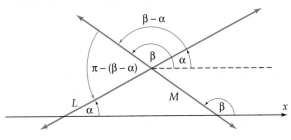

Figure 7–7

The angle between two lines can also be found from their slopes by using this fact.

Angle Between Two Lines

> If two nonvertical, nonperpendicular lines have slopes m and k, then one angle θ between them satisfies
>
> $$\tan \theta = \left| \frac{m - k}{1 + mk} \right|.$$

Proof   Suppose the line with slope k has angle of inclination α and the line with slope m has angle of inclination β. Then

$$\tan \alpha = k \qquad \text{and} \qquad \tan \beta = m.$$

If $\beta \geq \alpha$, then $\theta = \beta - \alpha$ is one angle between the lines. By the subtraction identity for tangent,

$$\tan \theta = \tan(\beta - \alpha) = \frac{\tan \beta - \tan \alpha}{1 + \tan \beta \tan \alpha} = \frac{m - k}{1 + mk}.$$

The other angle between the lines is $\pi - \theta$, and

$$\tan(\pi - \theta) = \tan(-\theta) \qquad \text{[tangent has period } \pi]$$

$$= -\tan \theta \qquad \text{[negative angle identity]}$$

$$= -\frac{m - k}{1 + mk}.$$

By the definition of absolute value,

$$\left| \frac{m - k}{1 + mk} \right| = \pm \frac{m - k}{1 + mk},$$

whichever is positive. Thus, the tangent of one of the angles θ or $\pi - \theta$ is

$$\left| \frac{m - k}{1 + mk} \right|.$$

This completes the proof when $\beta \geq \alpha$. The proof when $\alpha \geq \beta$ is similar. ■

EXAMPLE 1

Figure 7–8

Find the angle between a line L with slope 8 and a line M of slope -3.

SOLUTION   One angle θ between the lines satisfies

$$\tan \theta = \left| \frac{8 - (-3)}{1 + 8(-3)} \right| = \left| \frac{11}{-23} \right| = \frac{11}{23}.$$

Although you could solve the equation $\tan \theta = 11/23$ graphically, it is easier to use the TAN^{-1} key on your calculator. When you key in $\text{TAN}^{-1}(11/23)$, the calculator displays an angle between 0 and $\pi/2$ radians such that $\tan \theta = 11/23$, as in Figure 7–8.* Therefore, $\theta \approx .4461$ radians. ■

We can now prove the following fact, which was first presented in Section 1.4.

Slope Theorem for Perpendicular Lines

Let L be a line with slope k and M a line with slope m. Then

L and M are perpendicular exactly when $km = -1$.

Proof   First, suppose L and M are perpendicular. We must show that $km = -1$. If α and β (with $\beta \geq \alpha$) are the angles of inclination of L and M, then $\beta - \alpha$ is the angle between L and M, so $\beta - \alpha = \pi/2$, or, equivalently, $\beta = \alpha + \pi/2$. Therefore, by the addition identities for sine and cosine,

$$m = \tan \beta = \tan\left(\alpha + \frac{\pi}{2} \right) = \frac{\sin[\alpha + (\pi/2)]}{\cos[\alpha + (\pi/2)]}$$

$$= \frac{\sin \alpha \cos(\pi/2) + \cos \alpha \sin(\pi/2)}{\cos \alpha \cos(\pi/2) - \sin \alpha \sin(\pi/2)}$$

$$= \frac{\sin \alpha(0) + \cos \alpha(1)}{\cos \alpha(0) - \sin \alpha(1)}$$

$$= -\frac{\cos \alpha}{\sin \alpha} = -\cot \alpha = \frac{-1}{\tan \alpha} = \frac{-1}{k}.$$

Thus, $m = -1/k$, and hence, $mk = -1$.

Now suppose that $mk = -1$. We must show that L and M are perpendicular. If L and M are *not* perpendicular, then neither of the angles between them is $\pi/2$.

---

*$\text{Tan}^{-1}(x)$ denotes the inverse tangent function, which is explained in Section 7.4. In this context, using the TAN^{-1} key is the electronic equivalent of searching through a table of tangent values until you find a number whose tangent is $11/23$.

In this case, if θ is either of the angles between L and M, then $\tan \theta$ is a well-defined real number. But we know that one of these angles must satisfy

$$\tan \theta = \left| \frac{m-k}{1+mk} \right| = \left| \frac{m-k}{1+(-1)} \right| = \frac{|m-k|}{0},$$

which is *not* defined. This contradiction shows that L and M must be perpendicular. ■

EXERCISES 7.2.A

In Exercises 1–6, find $\tan \theta$, *where* θ *is the angle of inclination of the line through the given points.*

1. $(-1, 2), (3, 5)$ **2.** $(0, 4), (5, -1)$

3. $(1, 4), (6, 0)$ **4.** $(4, 2), (-3, -2)$

5. $(3, -7), (3, 5)$ **6.** $(0, 0), (-4, -5)$

In Exercises 7–13, find one of the angles between the straight lines L and M.

7. L has slope $3/2$ and M has slope -1.

8. L has slope 1 and M has slope 3.

9. L has slope -1 and M has slope 0.

10. L has slope -2 and M has slope -3.

11. $(3, 2)$ and $(5, 6)$ are on L; $(0, 3)$ and $(4, 0)$ are on M.

12. $(-1, 2)$ and $(3, -3)$ are on L; $(3, -3)$ and $(6, 1)$ are on M.

13. L is parallel to the line with equation $y = 3x - 2$ and M is perpendicular to the line with equation $y = -.5x + 1$.

14. If θ is an angle between two nonperpendicular lines with slopes m and k, respectively, and $\tan \theta = \left| \frac{m-k}{1+mk} \right|$, explain why θ is an acute angle. [*Hint:* For what values of θ is $\tan \theta$ positive?]

7.3 Other Identities

Section Objectives

■ Use the double-angle, power-reducing, and half-angle identities to evaluate and simplify trigonometric functions.

■ Use the product-to-sum and sum-to-product identities to prove other identities.

We now present a variety of identities that are special cases of the addition and subtraction identities of Section 7.2, beginning with

Double-Angle Identities

$$\sin 2x = 2 \sin x \cos x$$
$$\cos 2x = \cos^2 x - \sin^2 x$$
$$\tan 2x = \frac{2 \tan x}{1 - \tan^2 x}$$

Proof   Let $x = y$ in the addition identities:

$$\sin 2x = \sin(x + x) = \sin x \cos x + \cos x \sin x = 2 \sin x \cos x,$$
$$\cos 2x = \cos(x + x) = \cos x \cos x - \sin x \sin x = \cos^2 x - \sin^2 x,$$
$$\tan 2x = \tan(x + x) = \frac{\tan x + \tan x}{1 - \tan x \tan x} = \frac{2 \tan x}{1 - \tan^2 x}.$$ ■

EXAMPLE 1

If $\pi < x < 3\pi/2$ and $\cos x = -8/17$, find $\sin 2x$ and $\cos 2x$, and show that $5\pi/2 < 2x < 3\pi$.

SOLUTION   To use the double-angle identities, we first must determine $\sin x$. It can be found by using the Pythagorean identities.

$$\sin^2 x = 1 - \cos^2 x = 1 - \left(-\frac{8}{17}\right)^2 = 1 - \frac{64}{289} = \frac{225}{289}.$$

Since $\pi < x < 3\pi/2$, we know $\sin x$ is negative. Therefore,

$$\sin x = -\sqrt{\frac{225}{289}} = -\frac{15}{17}.$$

We now substitute these values in the double-angle identities.

$$\sin 2x = 2 \sin x \cos x = 2\left(-\frac{15}{17}\right)\left(-\frac{8}{17}\right) = \frac{240}{289} \approx .83$$

$$\cos 2x = \cos^2 x - \sin^2 x = \left(-\frac{8}{17}\right)^2 - \left(-\frac{15}{17}\right)^2$$

$$= \frac{64}{289} - \frac{225}{289} = -\frac{161}{289} \approx -.56.$$

Since $\pi < x < 3\pi/2$, we know that $2\pi < 2x < 3\pi$. The calculations above show that at $2x$, sine is positive and cosine is negative. This can occur only if $2x$ lies between $5\pi/2$ and 3π. ■

EXAMPLE 2

Express the rule of the function $f(x) = \sin 3x$ in terms of $\sin x$ and constants.

SOLUTION   We first use the addition identity for $\sin(x + y)$ with $y = 2x$.

$$f(x) = \sin 3x = \sin(x + 2x) = \sin x \cos 2x + \cos x \sin 2x.$$

Next apply the double-angle identities for $\cos 2x$ and $\sin 2x$.

$$f(x) = \sin 3x = \sin x \cos 2x + \cos x \sin 2x$$

$$= \sin x(\cos^2 x - \sin^2 x) + \cos x(2 \sin x \cos x)$$

$$= \sin x \cos^2 x - \sin^3 x + 2 \sin x \cos^2 x$$

$$= 3 \sin x \cos^2 x - \sin^3 x.$$

Finally, use the Pythagorean identity.

$$f(x) = \sin 3x = 3 \sin x \cos^2 x - \sin^3 x = 3 \sin x(1 - \sin^2 x) - \sin^3 x$$

$$= 3 \sin x - 3 \sin^3 x - \sin^3 x = 3 \sin x - 4 \sin^3 x. \quad ■$$

The double-angle identity for $\cos 2x$ can be rewritten in several useful ways. For instance, we can use the Pythagorean identity in the form of $\cos^2 x = 1 - \sin^2 x$ to obtain:

$$\cos 2x = \cos^2 x - \sin^2 x = (1 - \sin^2 x) - \sin^2 x = 1 - 2 \sin^2 x.$$

Similarly, using the Pythagorean identity in the form $\sin^2 x = 1 - \cos^2 x$, we have:

$$\cos 2x = \cos^2 x - \sin^2 x = \cos^2 x - (1 - \cos^2 x) = 2\cos^2 x - 1.$$

In summary:

More Double-Angle Identities

$$\cos 2x = 1 - 2\sin^2 x$$

$$\cos 2x = 2\cos^2 x - 1$$

EXAMPLE 3

Prove that

$$\frac{1 - \cos 2x}{\sin 2x} = \tan x.$$

SOLUTION   The first identity in the preceding box and the double-angle identity for sine show that

$$\frac{1 - \cos 2x}{\sin 2x} = \frac{1 - (1 - 2\sin^2 x)}{2\sin x \cos x} = \frac{2\sin^2 x}{2\sin x \cos x} = \frac{\sin x}{\cos x} = \tan x. \quad\blacksquare$$

If we solve the first equation in the preceding box for $\sin^2 x$ and the second one for $\cos^2 x$, we obtain a useful alternate form for these identities.

Power-Reducing Identities

$$\sin^2 x = \frac{1 - \cos 2x}{2}$$

$$\cos^2 x = \frac{1 + \cos 2x}{2}$$

EXAMPLE 4

Express the rule of the function $f(x) = \sin^4 x$ in terms of constants and first powers of the cosine function.

SOLUTION   We begin by applying the power-reducing identity.

$$f(x) = \sin^4 x = \sin^2 x \, \sin^2 x = \frac{1 - \cos 2x}{2} \cdot \frac{1 - \cos 2x}{2}$$

$$= \frac{1 - 2\cos 2x + \cos^2 2x}{4}.$$

Next we apply the power-reducing identity for cosine to $\cos^2 2x$. Note that this means using $2x$ in place of x in the identity.

$$\cos^2 2x = \frac{1 + \cos 2(2x)}{2} = \frac{1 + \cos 4x}{2}.$$

Finally, we substitute this last result in the expression for $\sin^4 x$ above.

$$f(x) = \sin^4 x = \frac{1 - 2\cos 2x + \cos^2 2x}{4} = \frac{1 - 2\cos 2x + \dfrac{1 + \cos 4x}{2}}{4}$$

$$= \frac{1}{4} - \frac{1}{2}\cos 2x + \frac{1}{8}(1 + \cos 4x)$$

$$= \frac{3}{8} - \frac{1}{2}\cos 2x + \frac{1}{8}\cos 4x. \qquad \blacksquare$$

HALF-ANGLE IDENTITIES

If we use the power-reducing identity with $x/2$ in place of x, we obtain

$$\sin^2\left(\frac{x}{2}\right) = \frac{1 - \cos 2\left(\dfrac{x}{2}\right)}{2} = \frac{1 - \cos x}{2}.$$

Consequently, we must have

$$\sin\left(\frac{x}{2}\right) = \pm\sqrt{\frac{1 - \cos x}{2}}.$$

This proves the first of the half-angle identities.

Half-Angle Identities

$$\sin\frac{x}{2} = \pm\sqrt{\frac{1 - \cos x}{2}} \qquad \cos\frac{x}{2} = \pm\sqrt{\frac{1 + \cos x}{2}}$$

$$\tan\frac{x}{2} = \pm\sqrt{\frac{1 - \cos x}{1 + \cos x}}$$

The half-angle identity for cosine is derived from a power-reducing identity, as was the half-angle identity for sine. The half-angle identity for tangent then follows immediately since $\tan(x/2) = \sin(x/2)/\cos(x/2)$. In all cases, *the sign in front of the radical depends on the quadrant in which $x/2$ lies.*

EXAMPLE 5

Find the exact value of

(a) $\cos\dfrac{5\pi}{8}$ (b) $\sin\dfrac{\pi}{12}$.

SOLUTION

(a) Since $\dfrac{5\pi}{8} = \dfrac{1}{2}\left(\dfrac{5\pi}{4}\right) = \dfrac{5\pi/4}{2}$, we use the half-angle identity with $x = 5\pi/4$ and the fact that $\cos(5\pi/4) = -\sqrt{2}/2$. The sign chart in Exercise 63 on page 451 shows that $\cos(5\pi/8)$ is negative because $5\pi/8$ is in the second quadrant. So we use the negative sign in front of the radical.

$$\cos\frac{5\pi}{8} = \cos\frac{5\pi/4}{2} = -\sqrt{\frac{1+\cos(5\pi/4)}{2}}$$

$$= -\sqrt{\frac{1+(-\sqrt{2}/2)}{2}} = -\sqrt{\frac{(2-\sqrt{2})/2}{2}}$$

$$= -\sqrt{\frac{2-\sqrt{2}}{4}}$$

$$= \frac{-\sqrt{2-\sqrt{2}}}{2}.$$

(b) Since $\dfrac{\pi}{12} = \dfrac{1}{2}\left(\dfrac{\pi}{6}\right) = \dfrac{\pi/6}{2}$ and $\pi/12$ is in the first quadrant, where sine is positive, we have

$$\sin\frac{\pi}{12} = \sin\frac{\pi/6}{2} = \sqrt{\frac{1-\cos(\pi/6)}{2}}$$

$$= \sqrt{\frac{1-\sqrt{3}/2}{2}} = \sqrt{\frac{(2-\sqrt{3})/2}{2}} = \sqrt{\frac{2-\sqrt{3}}{4}}$$

$$= \frac{\sqrt{2-\sqrt{3}}}{2} \qquad\blacksquare$$

Example 5(b) shows that

$$\sin\frac{\pi}{12} = \frac{\sqrt{2-\sqrt{3}}}{2}.$$

On the other hand, in Example 1(a) of Section 7.2 we proved that

$$\sin\frac{\pi}{12} = \frac{\sqrt{6}-\sqrt{2}}{4}.$$

So we can conclude that

$$\frac{\sqrt{2-\sqrt{3}}}{2} = \frac{\sqrt{6}-\sqrt{2}}{4},$$

a fact that can readily be confirmed by a calculator (Figure 7–9). The moral here is that there may be several correct ways to express the exact value of a trigonometric function.

The problem of determining signs in the half-angle formulas can be eliminated with tangent by using these identities.

```
√(2-√(3))/2
        .2588190451
(√(6)-√(2))/4
        .2588190451
```

Figure 7–9

Half-Angle Identities for Tangent

$$\tan\frac{x}{2} = \frac{1-\cos x}{\sin x}$$

$$\tan\frac{x}{2} = \frac{\sin x}{1+\cos x}$$

Proof In the identity

$$\tan x = \frac{1-\cos 2x}{\sin 2x},$$

which was proved in Example 3, replace x by $x/2$.

$$\tan\left(\frac{x}{2}\right) = \frac{1 - \cos 2(x/2)}{\sin 2(x/2)} = \frac{1 - \cos x}{\sin x}.$$

The second identity in the box is proved in Exercise 89. ∎

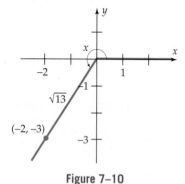

Figure 7–10

EXAMPLE 6

If $\tan x = \dfrac{3}{2}$ and $\pi < x < \dfrac{3\pi}{2}$, find $\tan \dfrac{x}{2}$.

SOLUTION   The terminal side of an angle of x radians in standard position lies in the third quadrant, as shown in Figure 7–10. The tangent of the angle in standard position whose terminal side passes through the point $(-2, -3)$ is $\dfrac{-3}{-2} = \dfrac{3}{2}$. Since there is only one angle in the third quadrant with tangent $3/2$, the point $(-2, -3)$ must lie on the terminal side of the angle of x radians.
 Since the distance from $(-2, -3)$ to the origin is

$$\sqrt{(-2 - 0)^2 + (-3 - 0)^2} = \sqrt{13},$$

we have

$$\sin x = \frac{-3}{\sqrt{13}} \qquad \text{and} \qquad \cos x = \frac{-2}{\sqrt{13}}.$$

Therefore, by the first of the half-angle identities for tangent

$$\tan \frac{x}{2} = \frac{1 - \cos x}{\sin x} = \frac{1 - \left(\dfrac{-2}{\sqrt{13}}\right)}{\dfrac{-3}{\sqrt{13}}} = \frac{\dfrac{\sqrt{13} + 2}{\sqrt{13}}}{\dfrac{-3}{\sqrt{13}}} = -\frac{\sqrt{13} + 2}{3}. \quad ∎$$

SUM/PRODUCT IDENTITIES

The following identities were proved in Example 5 and Exercises 60–62 of Section 7.2.

Product to Sum Identities

$$\sin x \cos y = \frac{1}{2}[\sin(x + y) + \sin(x - y)]$$

$$\sin x \sin y = \frac{1}{2}[\cos(x - y) - \cos(x + y)]$$

$$\cos x \cos y = \frac{1}{2}[\cos(x + y) + \cos(x - y)]$$

$$\cos x \sin y = \frac{1}{2}[\sin(x + y) - \sin(x - y)]$$

EXAMPLE 7

Express $\sin(3x)\cos(5x)$ as a sum or difference of trigonometric functions.

SOLUTION   We use the first product to sum identity, with $3x$ in place of x and $5x$ in place of y.

$$\sin(3x)\cos(5x) = \frac{1}{2}[\sin(3x+5x) + \sin(3x-5x)]$$

$$= \frac{1}{2}[\sin(8x) + \sin(-2x)]$$

$$= \frac{1}{2}[\sin(8x) - \sin(2x)] \qquad \text{[Negative angle identity]}$$

$$= \frac{1}{2}\sin(8x) - \frac{1}{2}\sin(2x) \qquad \blacksquare$$

If we use the first product to sum identity with $\frac{1}{2}(x+y)$ in place of x and $\frac{1}{2}(x-y)$ in place of y, we obtain

$$\sin\left[\frac{1}{2}(x+y)\right]\cos\left[\frac{1}{2}(x-y)\right] = \frac{1}{2}\left[\sin\left(\frac{1}{2}(x+y) + \frac{1}{2}(x-y)\right) + \sin\left(\frac{1}{2}(x+y) - \frac{1}{2}(x-y)\right)\right]$$

$$= \frac{1}{2}(\sin x + \sin y).$$

Multiplying both sides of the last equation by 2 produces the first of the following identities.

Sum to Product Identities

$$\sin x + \sin y = 2\sin\left(\frac{x+y}{2}\right)\cos\left(\frac{x-y}{2}\right)$$

$$\sin x - \sin y = 2\cos\left(\frac{x+y}{2}\right)\sin\left(\frac{x-y}{2}\right)$$

$$\cos x + \cos y = 2\cos\left(\frac{x+y}{2}\right)\cos\left(\frac{x-y}{2}\right)$$

$$\cos x - \cos y = -2\sin\left(\frac{x+y}{2}\right)\sin\left(\frac{x-y}{2}\right)$$

The last three sum to product identities are proved in the same way as the first. (See Exercises 59–61.)

EXAMPLE 8

Express $\cos(7x) + \cos(3x)$ as a product of trigonometric functions.

SOLUTION   We use the third sum to product identity with $7x$ in place of x and $3x$ in place of y.

$$\cos(7x) + \cos(3x) = 2\cos\left(\frac{7x+3x}{2}\right)\cos\left(\frac{7x-3x}{2}\right)$$

$$= 2\cos\left(\frac{10x}{2}\right)\cos\left(\frac{4x}{2}\right)$$

$$= 2\cos(5x)\cos(2x) \qquad \blacksquare$$

EXAMPLE 9

Prove the identity

$$\frac{\sin t + \sin 3t}{\cos t + \cos 3t} = \tan 2t.$$

SOLUTION   Using the first factoring identity with $x = t$ and $y = 3t$ yields

$$\sin t + \sin 3t = 2 \sin\left(\frac{t + 3t}{2}\right)\cos\left(\frac{t - 3t}{2}\right) = 2 \sin 2t \cos(-t).$$

Similarly,

$$\cos t + \cos 3t = 2 \cos\left(\frac{t + 3t}{2}\right)\cos\left(\frac{t - 3t}{2}\right) = 2 \cos 2t \cos(-t),$$

so

$$\frac{\sin t + \sin 3t}{\cos t + \cos 3t} = \frac{2 \sin 2t \cos(-t)}{2 \cos 2t \cos(-t)} = \frac{\sin 2t}{\cos 2t} = \tan 2t. \qquad\blacksquare$$

EXERCISES 7.3

In Exercises 1–7, find $\sin 2x$, $\cos 2x$, *and* $\tan 2x$ *under the given conditions.*

1. $\sin x = \dfrac{5}{13}$ $\left(0 < x < \dfrac{\pi}{2}\right)$

2. $\sin x = -\dfrac{4}{5}$ $\left(\pi < x < \dfrac{3\pi}{2}\right)$

3. $\cos x = -\dfrac{3}{5}$ $\left(\pi < x < \dfrac{3\pi}{2}\right)$

4. $\cos x = -\dfrac{1}{3}$ $\left(\dfrac{\pi}{2} < x < \pi\right)$

5. $\tan x = \dfrac{3}{4}$ $\left(\pi < x < \dfrac{3\pi}{2}\right)$

6. $\tan x = -\dfrac{3}{2}$ $\left(\dfrac{\pi}{2} < x < \pi\right)$

7. $\csc x = 4$ $\left(0 < x < \dfrac{\pi}{2}\right)$

8. A batter hits a baseball that is caught by a fielder. If the ball leaves the bat at an angle of θ radians to the horizontal, with an initial velocity of v feet per second, then the approximate horizontal distance d traveled by the ball is given by

$$d = \frac{v^2 \sin \theta \cos \theta}{16}.$$

(a) Use an identity to show that

$$d = \frac{v^2 \sin 2\theta}{32}.$$

(b) If the initial velocity is 115 ft/second, what angle θ will produce the maximum distance? [*Hint:* Use part (a). For what value of θ is $\sin 2\theta$ as large as possible?]

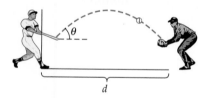

9. A rectangle is inscribed in a semicircle of radius 3 inches and the radius to the corner makes an angle of t radians with the horizontal, as shown in the figure.

(a) Express the horizontal length, vertical height, and area of the rectangle in terms of x and y.

(b) Express x and y in terms of sine and cosine.

(c) Use parts (a) and (b) and suitable identities to show that the area A of the rectangle is given by

$$A = 9 \sin 2t.$$

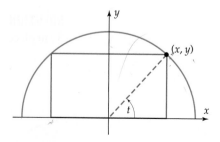

10. In Exercise 9, what angle will produce a rectangle with largest possible area? What is this maximum area?

In Exercises 11–26, use the half-angle identities to evaluate the given expression exactly.

11. $\cos \dfrac{\pi}{8}$ 12. $\tan \dfrac{\pi}{8}$ 13. $\sin \dfrac{3\pi}{8}$ 14. $\cos \dfrac{3\pi}{8}$

15. $\tan \dfrac{\pi}{12}$ 16. $\sin \dfrac{5\pi}{8}$ 17. $\cos \dfrac{\pi}{12}$ 18. $\tan \dfrac{5\pi}{8}$

19. $\sin \dfrac{7\pi}{8}$ 20. $\cos \dfrac{7\pi}{8}$ 21. $\tan \dfrac{7\pi}{8}$ 22. $\cot \dfrac{\pi}{8}$

23. $\cos \dfrac{\pi}{16}$ [*Hint:* Exercise 11]   24. $\sin \dfrac{\pi}{16}$

25. $\sin \dfrac{\pi}{24}$ [*Hint:* Exercise 17]   26. $\cos \dfrac{\pi}{24}$

In Exercises 27–32, find $\sin \dfrac{x}{2}$, $\cos \dfrac{x}{2}$, and $\tan \dfrac{x}{2}$ under the given conditions.

27. $\cos x = .4$ $\left(0 < x < \dfrac{\pi}{2}\right)$

28. $\sin x = .6$ $\left(\dfrac{\pi}{2} < x < \pi\right)$

29. $\sin x = -\dfrac{3}{5}$ $\left(\dfrac{3\pi}{2} < x < 2\pi\right)$

30. $\cos x = .8$ $\left(\dfrac{3\pi}{2} < x < 2\pi\right)$

31. $\tan x = \dfrac{1}{2}$ $\left(\pi < x < \dfrac{3\pi}{2}\right)$

32. $\cot x = 1$ $\left(-\pi < x < -\dfrac{\pi}{2}\right)$

In Exercises 33–38, write each expression as a sum or difference.

33. $\sin 4x \cos 6x$ 34. $\sin 5x \sin 7x$

35. $\cos 2x \cos 4x$ 36. $\sin 3x \cos 5x$

37. $\sin 17x \sin(-3x)$ 38. $\cos 13x \cos(-5x)$

In Exercises 39–44, write each expression as a product.

39. $\sin 3x + \sin 5x$ 40. $\cos 2x + \cos 6x$

41. $\sin 9x - \sin 5x$ 42. $\cos 5x - \cos 7x$

43. $\cos 2x + \cos 5x$ 44. $\sin 4x + \sin 3x$

In Exercises 45–50, assume $\sin x = .6$ and $0 < x < \pi/2$ and evaluate the given expression.

45. $\sin 2x$ 46. $\cos 4x$ 47. $\cos 2x$ 48. $\sin 4x$

49. $\sin \dfrac{x}{2}$ 50. $\cos \dfrac{x}{2}$

51. Express $\cos 3x$ in terms of $\cos x$.

52. (a) Express the rule of the function $f(x) = \cos^3 x$ in terms of constants and first powers of the cosine function as in Example 4.
    (b) Do the same for $f(x) = \cos^4 x$.

In Exercises 53–58, simplify the given expression.

53. $\dfrac{\sin 2x}{2 \sin x}$ 54. $1 - 2\sin^2\left(\dfrac{x}{2}\right)$

55. $2\cos 2y \sin 2y$ (Think!)

56. $\cos^2\left(\dfrac{x}{2}\right) - \sin^2\left(\dfrac{x}{2}\right)$ 57. $(\sin x + \cos x)^2 - \sin 2x$

58. $2\sin x \cos^3 x - 2\sin^3 x \cos x$

In Exercises 59–61, prove the given sum to product identity. [Hint: See the proof on page 541.]

59. $\sin x - \sin y = 2\cos\left(\dfrac{x+y}{2}\right)\sin\left(\dfrac{x-y}{2}\right)$

60. $\cos x + \cos y = 2\cos\left(\dfrac{x+y}{2}\right)\cos\left(\dfrac{x-y}{2}\right)$

61. $\cos x - \cos y = -2\sin\left(\dfrac{x+y}{2}\right)\sin\left(\dfrac{x-y}{2}\right)$

62. When you press a key on a touch-tone phone, the key emits two tones that combine to produce the sound wave
$$f(t) = \sin(2\pi Lt) + \sin(2\pi Ht),$$
Where t is in seconds, L is the low frequency tone for the row the key is in, and H is the high frequency tone for the column the key is in, as shown in the diagram below. For example, pressing 2 produces the sound wave $f(t) = \sin[2\pi(697)t] + \sin[2\pi(1336)t]$.

High frequency

1209   1336   1477 Hz

1	2	3	← 697 Hz
4	5	6	← 770 Hz
7	8	9	← 852 Hz
*	0	#	← 941 Hz

Low frequency

(a) Write the function that gives the sound wave produced by pressing the 6 key.
(b) Express the 6 key function in part (a) as the product of a sine and a cosine function.

In Exercises 63–76, determine graphically whether the equation could possibly be an identity. If it could, prove that it is.

63. $\sin 16x = 2\sin 8x \cos 8x$ 64. $\cos 8x = \cos^2 4x - \sin^2 4x$

65. $\cos^4 x - \sin^4 x = \cos 2x$ 66. $\sec 2x = \dfrac{1}{1 - 2\sin^2 x}$

67. $\cos 4x = 2\cos 2x - 1$ **68.** $\sin^2 x = \cos^2 x - 2\sin x$

69. $\dfrac{1 + \cos 2x}{\sin 2x} = \cot x$ **70.** $\sin 2x = \dfrac{2\cot x}{\csc^2 x}$

71. $\sin 3x = (\sin x)(3 - 4\sin^2 x)$

72. $\sin 4x = (4\cos x \sin x)(1 - 2\sin^2 x)$

$2\cos x$
$\sin x$

73. $\cos 2x = \dfrac{2\tan x}{\sec^2 x}$

74. $\cos 3x = (\cos x)(3 - 4\cos^2 x)$

75. $\csc^2\left(\dfrac{x}{2}\right) = \dfrac{2}{1 - \cos x}$

76. $\sec^2\left(\dfrac{x}{2}\right) = \dfrac{2}{1 + \cos x}$

In Exercises 77 and 78, the graph of the left side of the expression is shown. Fill the blank on the right side with a simple trigonometric expression and prove that the resulting equation is an identity. [Hint: What trigonometric function has a graph that closely resembles the given one?]

77. $\dfrac{\sin 5x - \sin 3x}{2\cos 4x} = $ _____

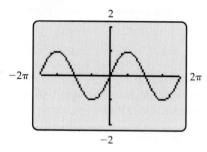

78. $\dfrac{\sin 5x + \sin 3x}{2\sin 4x} = $ _____

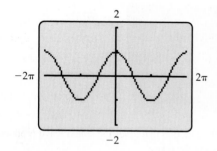

In Exercises 79–82, fill the blank on the right side with a simple trigonometric expression and prove that the resulting equation is an identity. [Hint: Exercises 77 and 78.]

79. $\dfrac{\cos x + \cos 3x}{2\cos 2x} = $ _____   **80.** $\dfrac{\cos 4x - \cos 6x}{\sin 4x + \sin 6x} = $ _____

81. $\dfrac{\sin 3x - \sin x}{\cos x + \cos 3x} = $ _____   **82.** $\dfrac{\cos x + \cos 3x}{\sin x - \sin 3x} = $ _____

In Exercises 83–88, prove the identity.

83. $\dfrac{\sin x - \sin 3x}{\cos x + \cos 3x} = -\tan x$ **84.** $\dfrac{\sin x - \sin 3x}{\cos x - \cos 3x} = -\cot 2x$

85. $\dfrac{\sin 4x + \sin 6x}{\cos 4x - \cos 6x} = \cot x$

86. $\dfrac{\cos 8x + \cos 4x}{\cos 8x - \cos 4x} = -\cot 6x \cot 2x$

87. $\dfrac{\sin x + \sin y}{\cos x - \cos y} = -\cot\left(\dfrac{x - y}{2}\right)$

88. $\dfrac{\sin x - \sin y}{\cos x + \cos y} = \tan\left(\dfrac{x - y}{2}\right)$

89. (a) Prove that $\dfrac{1 - \cos x}{\sin x} = \dfrac{\sin x}{1 + \cos x}$.

(b) Use part (a) and the half-angle identity proved in the text to prove that

$$\tan\dfrac{x}{2} = \dfrac{\sin x}{1 + \cos x}.$$

90. To avoid a steep hill, a road is being built in straight segments from P to Q and from Q to R; it makes a turn of t radians at Q, as shown in the figure. The distance from P to S is 40 miles, and the distance from R to S is 10 miles. Use suitable trigonometric functions to express:

(a) c in terms of b and t [*Hint:* Place the figure on a coordinate plane with P and Q on the x-axis, with Q at the origin. Then what are the coordinates of R?]

(b) b in terms of t

(c) a in terms of t [*Hint:* $a = 40 - c$; use parts (a) and (b).]

(d) Use parts (b) and (c) and a suitable identity to show that the length $a + b$ of the road is

$$40 + 10\tan\dfrac{t}{2}.$$

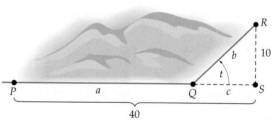

91. Find the exact value of $\cos\dfrac{\pi}{32}$. [*Hint:* Exercise 23.]

92. (a) List the exact values of $\cos\dfrac{\pi}{4}$, $\cos\dfrac{\pi}{8}$, $\cos\dfrac{\pi}{16}$, and $\cos\dfrac{\pi}{32}$. [*Hint:* Exercises 11, 23 and 91.]

(b) Based on the pattern you see in the answers to part (a) make a conjecture about the exact value of $\cos\dfrac{\pi}{64}$. Use a calculator to support your answer.

(c) Make a conjecture about the exact value of $\cos\dfrac{\pi}{128}$ and support the truth of your conjecture with a calculator.

(d) What do you think the exact value of $\cos\dfrac{\pi}{256}$ is?

7.4 Inverse Trigonometric Functions

Section Objectives
- Evaluate the inverse sine, cosine, and tangent functions.
- Investigate the properties of the inverse trigonometric functions.
- Prove identities involving inverse trigonometric functions.
- Use inverse trigonometric functions to to solve applied problems.

Before reading this section, you should review the concept of an inverse function (Section 3.7). As explained there, a function f has an inverse function only when its graph passes the Horizontal Line Test:

No horizontal line intersects the graph of f more than once.

The graphs of the sine, cosine, and tangent functions certainly do not have this property. However, functions that are closely related to them (same rules but smaller domains) *do* have inverse functions.

The **restricted sine function** is defined as follows:

Domain: $[-\pi/2, \pi/2]$ Rule: $f(x) = \sin x$.

Its graph in Figure 7–11 shows that for each number v between -1 and 1, there is exactly one number u between $-\pi/2$ and $\pi/2$ such that $\sin u = v$.

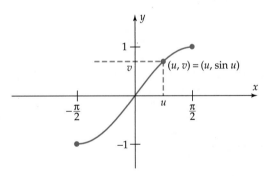

Figure 7–11

Since the graph of the restricted sine function passes the horizontal line test, we know that it has an inverse function. This inverse function is called the **inverse sine** (or **arcsine**) **function** and is denoted by $g(x) = \sin^{-1}x$ or $g(x) = \arcsin x$. The domain of the inverse sine function is the interval $[-1, 1]$, and its rule is as follows.

Inverse Sine Function

For each v with $-1 \le v \le 1$,

$\sin^{-1}v$ = the unique number u between $-\pi/2$ and $\pi/2$ whose sine is v;

that is,

$$\sin^{-1}v = u \quad \text{exactly when} \quad \sin u = v.$$

EXAMPLE 1

Find

(a) $\sin^{-1}(1/2)$

(b) $\sin^{-1}(-\sqrt{2}/2)$.

SOLUTION

(a) $\sin^{-1}(1/2)$ is the one number between $-\pi/2$ and $\pi/2$ whose sine is $1/2$. From our study of special values, we know that $\sin \pi/6 = 1/2$, and $\pi/6$ is between $-\pi/2$ and $\pi/2$. Hence, $\sin^{-1}(1/2) = \pi/6$.

(b) $\sin^{-1}(-\sqrt{2}/2) = -\pi/4$ because $\sin(-\pi/4) = -\sqrt{2}/2$ and $-\pi/4$ is between $-\pi/2$ and $\pi/2$. ■

EXAMPLE 2

Except for special values (as in Example 1), you should use the SIN^{-1} key (labeled ASIN on some calculators) in *radian mode* to evaluate the inverse sine function. For instance,

$$\sin^{-1}(-.67) \approx -.7342 \quad \text{and} \quad \sin^{-1}(.42) \approx .4334. \quad ■$$

EXAMPLE 3

If you key in SIN^{-1} 2 ENTER, you will get an error message, because 2 is not in the domain of the inverse sine function.* ■

CAUTION

The notation $\sin^{-1}x$ is *not* exponential notation. It does *not* mean either $(\sin x)^{-1}$ or $\dfrac{1}{\sin x}$. For instance, Example 1 shows that

$$\sin^{-1}(1/2) = \pi/6 \approx .5236,$$

but

$$\left(\sin \frac{1}{2}\right)^{-1} = \frac{1}{\sin \frac{1}{2}} \approx \frac{1}{.4794} \approx 2.0858.$$

Suppose $-1 \le v \le 1$ and $\sin^{-1}v = u$. Then by the definition of the inverse sine function, we know that $-\pi/2 \le u \le \pi/2$ and $\sin u = v$. Therefore,

$$\sin^{-1}(\sin u) = \sin^{-1}(v) = u \quad \text{and} \quad \sin(\sin^{-1}v) = \sin u = v.$$

This shows that the restricted sine function and the inverse sine function have the usual "round-trip properties" of inverse functions. In summary,

Properties of Inverse Sine

$$\sin^{-1}(\sin u) = u \quad \text{if} \quad -\frac{\pi}{2} \le u \le \frac{\pi}{2}$$

$$\sin(\sin^{-1}v) = v \quad \text{if} \quad -1 \le v \le 1$$

---

*TI-85/86 and HP-39gs display the complex number $(1.5707\cdots, -1.3169\cdots)$ for $\sin^{-1}(2)$. For our purposes, this is equivalent to an error message, since we deal only with functions whose values are real numbers.

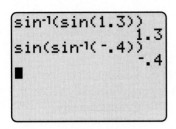

Figure 7–12

A calculator can illustrate the identities in the preceding box, as shown in Figure 7–12. Nevertheless, when special values are involved, you should be able to deal with them by hand.

EXAMPLE 4

Find   (a) $\sin^{-1}(\sin \pi/6)$ (b) $\sin^{-1}(\sin 5\pi/6)$.

SOLUTION

(a) We know that $\sin \pi/6 = 1/2$. Hence,

$$\sin^{-1}\left(\sin \frac{\pi}{6}\right) = \sin^{-1}\left(\frac{1}{2}\right) = \frac{\pi}{6}$$

because $\pi/6$ is the number between $-\pi/2$ and $\pi/2$ whose sine is $1/2$.

(b) We also have $\sin 5\pi/6 = 1/2$, so the expression $\sin^{-1}(\sin 5\pi/6)$ is defined. However,

$$\sin^{-1}\left(\sin \frac{5\pi}{6}\right) \qquad \text{is NOT equal to} \qquad \frac{5\pi}{6}$$

because the identity in the box on page 546 is valid only when u is between $-\pi/2$ and $\pi/2$. Using the result of part (a), we see that

$$\sin^{-1}\left(\sin \frac{5\pi}{6}\right) = \sin^{-1}\left(\frac{1}{2}\right) = \frac{\pi}{6}. \qquad \blacksquare$$

EXAMPLE 5

Find the exact value of $\tan\left[\sin^{-1}\left(\frac{5}{9}\right)\right]$.

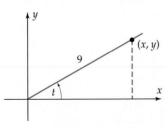

Figure 7–13

SOLUTION   Let $\sin^{-1}\left(\frac{5}{9}\right) = t$. We must find $\tan t$, so we construct an angle of t radians in standard position (Figure 7–13). Let (x, y) be the point on the terminal side of the angle that is 9 units from the origin. By the point-in-the plane description, we have

$$\sin t = \frac{y}{9} \quad \text{and} \quad \tan t = \frac{y}{x}.$$

Consequently,

$$\frac{y}{9} = \sin t = \sin\left[\sin^{-1}\left(\frac{5}{9}\right)\right] = \frac{5}{9}.$$

The first and last terms of this equation show that $y = 5$. Applying the Pythagorean Theorem to the right triangle in Figure 7–13, we see that

$$x^2 + 5^2 = 9^2$$

$$x^2 + 25 = 81$$

$$x^2 = 56$$

$$x = \sqrt{56}.$$

Therefore,

$$\tan\left[\sin^{-1}\left(\frac{5}{9}\right)\right] = \tan t = \frac{y}{x} = \frac{5}{\sqrt{56}}.$$ ∎

Recall that the graph of the inverse function of f can be obtained in two ways: Reverse the coordinates of each point on the graph of f or, equivalently, reflect the graph of f in the line $y = x$, as explained on pages 224–225. In summary,

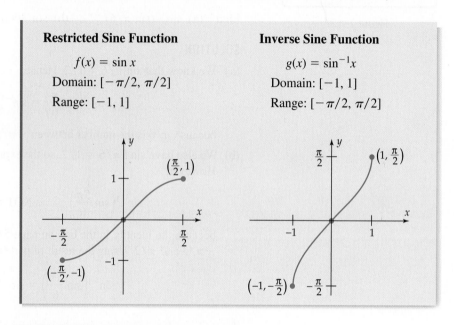

Restricted Sine Function

$f(x) = \sin x$

Domain: $[-\pi/2, \pi/2]$

Range: $[-1, 1]$

Inverse Sine Function

$g(x) = \sin^{-1}x$

Domain: $[-1, 1]$

Range: $[-\pi/2, \pi/2]$

THE INVERSE COSINE FUNCTION

The **restricted cosine function** is defined as follows:

Domain: $[0, \pi]$ Rule: $f(x) = \cos x.$

Its graph in Figure 7–14 shows that for each number v between -1 and 1, there is exactly one number u between 0 and π such that $\cos u = v$.

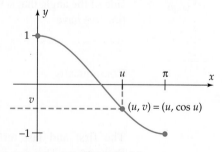

Figure 7–14

Since the graph of the restricted cosine function passes the Horizontal Line Test, we know that it has an inverse function. This inverse function is called the **inverse cosine (or arccosine) function** and is denoted by $h(x) = \cos^{-1}x$ or $h(x) = \arccos x$. The domain of the inverse cosine function is the interval $[-1, 1]$, and its rule is as follows.

Inverse Cosine
Function

For each v with $-1 \leq v \leq 1$,

$\cos^{-1}v =$ the unique number u between 0 and π whose cosine is v;

that is,

$$\cos^{-1}v = u \qquad \text{exactly when} \qquad \cos u = v.$$

The inverse cosine function has these properties:

$$\cos^{-1}(\cos u) = u \qquad \text{if} \qquad 0 \leq u \leq \pi;$$

$$\cos(\cos^{-1}v) = v \qquad \text{if} \qquad -1 \leq v \leq 1.$$

CAUTION

$\cos^{-1}x$ does *not* mean
$(\cos x)^{-1}$, or $1/\cos x$.

EXAMPLE 6

Find   (a) $\cos^{-1}(1/2)$ (b) $\cos^{-1}(0)$ (c) $\cos^{-1}(-.63)$.

SOLUTION

(a) $\text{Cos}^{-1}(1/2) = \pi/3$ since $\pi/3$ is the unique number between 0 and π whose cosine is $1/2$.

(b) $\text{Cos}^{-1}(0) = \pi/2$ because $\cos \pi/2 = 0$ and $0 \leq \pi/2 \leq \pi$.

(c) The COS^{-1} key on a calculator in *radian mode* shows that $\cos^{-1}(-.63) \approx 2.2523.$ ∎

EXAMPLE 7

Write $\sin(\cos^{-1}v)$ as an algebraic expression in v.

SOLUTION $\text{Cos}^{-1}v = u$, where $\cos u = v$ and $0 \leq u \leq \pi$. Hence, $\sin u$ is non-negative, and by the Pythagorean identity, $\sin u = \sqrt{\sin^2 u} = \sqrt{1 - \cos^2 u}$. Also, $\cos^2 u = v^2$. Therefore,

$$\sin(\cos^{-1}v) = \sin u = \sqrt{1 - \cos^2 u} = \sqrt{1 - v^2}.$$ ∎

EXAMPLE 8

Prove the identity $\sin^{-1}x + \cos^{-1}x = \pi/2$.

SOLUTION   Suppose $\sin^{-1}x = u$, with $-\pi/2 \leq u \leq \pi/2$. Verify that $0 \leq \pi/2 - u \leq \pi$ (Exercise 24). Then we have

$$\sin u = x \qquad \text{[Definition of inverse sine]}$$

$$\cos\left(\frac{\pi}{2} - u\right) = x \qquad \text{[Cofunction identity]}$$

$$\cos^{-1}x = \frac{\pi}{2} - u. \qquad \text{[Definition of inverse cosine]}$$

Therefore,

$$\sin^{-1}x + \cos^{-1}x = u + \left(\frac{\pi}{2} - u\right) = \frac{\pi}{2}. \qquad \blacksquare$$

EXAMPLE 9

Prove that

$$\sin(\cos^{-1}x) = \cos(\sin^{-1}x).$$

SOLUTION   By the identity in Example 8,

$$\sin(\cos^{-1}x) = \sin\left(\frac{\pi}{2} - \sin^{-1}x\right)$$

$$= \cos(\sin^{-1}x) \qquad \text{[Cofunction identity]} \qquad \blacksquare$$

The graph of the inverse cosine function is the reflection of the graph of the restricted cosine function in the line $y = x$, as shown below.

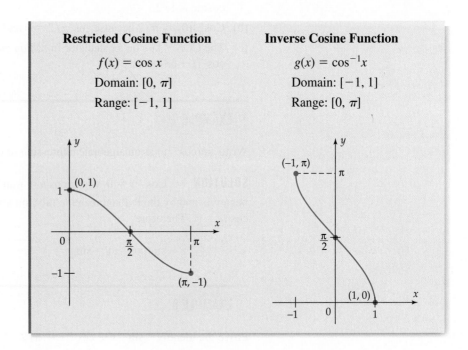

Restricted Cosine Function	Inverse Cosine Function
$f(x) = \cos x$	$g(x) = \cos^{-1}x$
Domain: $[0, \pi]$	Domain: $[-1, 1]$
Range: $[-1, 1]$	Range: $[0, \pi]$

THE INVERSE TANGENT FUNCTION

The **restricted tangent function** is defined as follows:

$$\text{Domain: } (-\pi/2,\ \pi/2) \qquad \text{Rule: } f(x) = \tan x.$$

Its graph in Figure 7–15 shows that for every real number v, there is exactly one number u between $-\pi/2$ and $\pi/2$ such that $\tan u = v$.

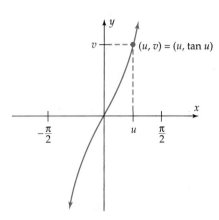

Figure 7–15

Since the graph of the restricted tangent function passes the horizontal line test, we know that it has an inverse function. This inverse function is called the **inverse tangent** (or **arctangent**) **function** and is denoted by $g(x) =$ **tan⁻¹x** or $g(x) =$ **arctan x.** The domain of the inverse tangent function is the set of all real numbers, and its rule is as follows.

Inverse Tangent Function

For each real number v,

$\tan^{-1}v =$ the unique number u between $-\pi/2$ and $\pi/2$ whose tangent is v;

that is,

$$\tan^{-1}v = u \qquad \text{exactly when} \qquad \tan u = v.$$

The inverse tangent function has these properties:

$$\tan^{-1}(\tan u) = u \qquad \text{if} \qquad -\frac{\pi}{2} < u < \frac{\pi}{2};$$

$$\tan(\tan^{-1}v) = v \qquad \text{for every number } v.$$

CAUTION

$\tan^{-1}x$ does *not* mean $(\tan x)^{-1}$, or $1/\tan x$.

EXAMPLE 10

$\text{Tan}^{-1}1 = \pi/4$ because $\pi/4$ is the unique number between $-\pi/2$ and $\pi/2$ such that $\tan \pi/4 = 1$. A calculator in *radian mode* shows that $\tan^{-1}(136) \approx 1.5634$.

∎

EXAMPLE 11

Find the exact value of $\cos[\tan^{-1}(\sqrt{5}/2)]$.

SOLUTION Consider an angle of u radians in standard position whose terminal side passes through $(2, \sqrt{5})$, as in Figure 7–16. By the point-in-the-plane description,

$$\tan u = \sqrt{5}/2.$$

Since u is between $-\pi/2$ and $\pi/2$ and $\tan u = \sqrt{5}/2$, we must have

$$u = \tan^{-1}(\sqrt{5}/2).$$

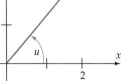

Figure 7–16

Furthermore, the distance from $(2, \sqrt{5})$ to the origin is

$$\sqrt{(2-0)^2 + (\sqrt{5}-0)^2} = \sqrt{4+5} = 3,$$

so

$$\cos u = 2/3.$$

Therefore,

$$\cos[\tan^{-1}(\sqrt{5}/2)] = \cos u = 2/3. \qquad \blacksquare$$

The graph of the inverse tangent function is the reflection of the graph of the restricted tangent function in the line $y = x$, as shown below.

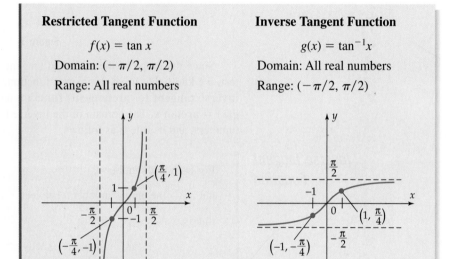

Restricted Tangent Function	**Inverse Tangent Function**
$f(x) = \tan x$	$g(x) = \tan^{-1}x$
Domain: $(-\pi/2, \pi/2)$	Domain: All real numbers
Range: All real numbers	Range: $(-\pi/2, \pi/2)$

EXAMPLE 12

A 26-foot high movie screen is located 12 feet above the ground, as shown in Figure 7–17. Assume that when you are seated your eye is 4 feet above the ground.

(a) Express the angle of t radians as a function of your distance x from the wall holding the screen.

(b) How far should you be from the wall to make t as large as possible?

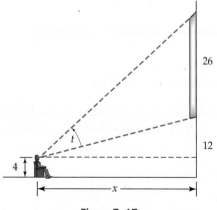

Figure 7–17

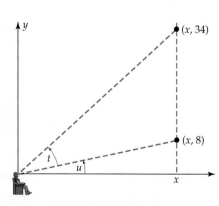

Figure 7–18

SOLUTION

(a) Imagine that the *x*-axis is at your eye level with your eye at the origin. Since your eye is 4 feet above the ground, the bottom of the screen is 8 feet above your eye level and the situation looks like Figure 7–18. The point-in-the-plane description shows that

$$\tan u = \frac{8}{x} \qquad \text{and} \qquad \tan(u + t) = \frac{34}{x}.$$

Hence,

$$u = \tan^{-1}\frac{8}{x} \qquad \text{and} \qquad u + t = \tan^{-1}\frac{34}{x},$$

so that

$$t = (u + t) - u = \tan^{-1}\frac{34}{x} - \tan^{-1}\frac{8}{x}.$$

(b) We graph the function $t = \tan^{-1}\dfrac{34}{x} - \tan^{-1}\dfrac{8}{x}$ and use a maximum finder to determine that *t* is largest when $x \approx 16.5$ feet (Figure 7–19). ∎

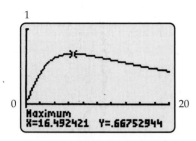

Figure 7–19

EXERCISES 7.4

In Exercises 1–14, find the exact functional value without using a calculator.

1. $\sin^{-1}1$ **2.** $\cos^{-1}0$ **3.** $\tan^{-1}(-1)$

4. $\sin^{-1}(-1)$ **5.** $\cos^{-1}1$ **6.** $\tan^{-1}1$

7. $\tan^{-1}(\sqrt{3}/3)$ **8.** $\cos^{-1}(\sqrt{3}/2)$

9. $\sin^{-1}(-\sqrt{2}/2)$ **10.** $\sin^{-1}(\sqrt{3}/2)$

11. $\tan^{-1}(-\sqrt{3})$ **12.** $\cos^{-1}(-\sqrt{2}/2)$

13. $\cos^{-1}\left(-\dfrac{1}{2}\right)$ **14.** $\sin^{-1}\left(-\dfrac{1}{2}\right)$

In Exercises 15–23, use a calculator in radian mode to approximate the functional value.

15. $\sin^{-1}.35$ **16.** $\cos^{-1}.76$

17. $\tan^{-1}(-3.256)$ **18.** $\sin^{-1}(-.795)$

19. $\sin^{-1}(\sin 7)$ [The answer is *not* 7.]

20. $\cos^{-1}(\cos 3.5)$ **21.** $\tan^{-1}[\tan(-4)]$

22. $\sin^{-1}[\sin(-2)]$ **23.** $\cos^{-1}[\cos(-8.5)]$

24. Let *u* be a number such that

$$-\frac{\pi}{2} \le u \le \frac{\pi}{2}.$$

Prove that

$$0 \le \frac{\pi}{2} - u \le \pi.$$

25. Given that $u = \sin^{-1}(-\sqrt{3}/2)$, find the exact value of $\cos u$ and $\tan u$.

26. Given that $u = \tan^{-1}(4/3)$, find the exact value of $\sin u$ and $\sec u$.

In Exercises 27–48, find the exact functional value without using a calculator.

27. $\sin^{-1}(\cos 0)$ **28.** $\cos^{-1}(\sin \pi/6)$

29. $\cos^{-1}(\sin 4\pi/3)$ **30.** $\tan^{-1}(\cos \pi)$

31. $\sin^{-1}(\cos 7\pi/6)$ **32.** $\cos^{-1}(\tan 7\pi/4)$

33. $\sin^{-1}(\sin 2\pi/3)$ (See Exercise 19.)

34. $\cos^{-1}(\cos 5\pi/4)$ **35.** $\cos^{-1}[\cos(-\pi/6)]$

36. $\tan^{-1}[\tan(-4\pi/3)]$

37. $\sin[\cos^{-1}(3/5)]$ (See Example 11.)

38. $\tan[\sin^{-1}(3/5)]$ **39.** $\cos[\tan^{-1}(-3/4)]$

40. $\cos[\sin^{-1}(12/13)]$ **41.** $\tan[\cos^{-1}(5/13)]$

42. $\sin[\tan^{-1}(12/5)]$ **43.** $\cos[\sin^{-1}(\sqrt{3}/5)]$

44. $\tan[\sin^{-1}(\sqrt{7}/12)]$ **45.** $\sin[\cos^{-1}(3/\sqrt{13})]$

46. $\tan[\cos^{-1}(8/9)]$ **47.** $\sin[\tan^{-1}(\sqrt{5}/10)]$

48. $\cos[\tan^{-1}(3/7)]$

In Exercises 49–55, write the expression as an algebraic expression in v, as in Example 7.

49. $\cos(\sin^{-1}v)$ **50.** $\tan(\cos^{-1}v)$

51. $\tan(\sin^{-1}v)$ **52.** $\sin(\tan^{-1}v)$

53. $\cos(\tan^{-1}v)$ **54.** $\sin(2\sin^{-1}v)$

55. $\sin(2\cos^{-1}v)$

In Exercises 56–58, prove the identity.

56. $\tan(\sin^{-1}v) = \cot(\cos^{-1}v)$

57. $\tan(\cos^{-1}v) = \cot(\sin^{-1}v)$

58. $\sec(\sin^{-1}v) = \csc(\cos^{-1}v)$

In Exercises 59–62, graph the function.

59. $f(x) = \cos^{-1}(x + 1)$ **60.** $g(x) = \tan^{-1}x + \pi$

61. $h(x) = \sin^{-1}(\sin x)$ **62.** $k(x) = \sin(\sin^{-1}x)$

63. In an alternating current circuit, the voltage is given by the formula

$$V = V_{\max} \cdot \sin(2\pi ft + \phi),$$

where $V_{\max}$ is the maximum voltage, f is the frequency (in cycles per second), t is the time in seconds, and ϕ is the phase angle.

(a) If the phase angle is 0, solve the voltage equation for t.
(b) If $\phi = 0$, $V_{\max} = 20$, $V = 8.5$, and $f = 120$, find the smallest positive value of t.

64. Calculus can be used to show that the area A between the x-axis and the graph of $y = \dfrac{1}{x^2 + 1}$ from $x = a$ to $x = b$ is given by

$$A = \tan^{-1} b - \tan^{-1} a.$$

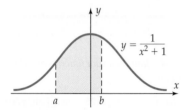

Find the area A when
(a) $a = 0$ and $b = 1$ (b) $a = -1$ and $b = 2$
(c) $a = -2.5$ and $b = -.5$.

Note: *Example 12 may be helpful for Exercises 65–71.*

65. A model plane 40 feet above the ground is flying away from an observer.

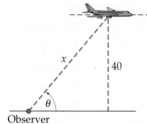

(a) Express the angle of elevation θ of the plane as a function of the distance x from the observer to the plane.
(b) What is θ when the plane is 250 feet away from the observer?

66. Suppose that another model plane is flying while attached to the ground by a 100 foot long wire that is always kept taut. Let h denote the height of the plane above the ground and θ the radian measure of the angle the wire makes with the ground. (The figure for Exercise 65 is the case when $x = 100$ and $h = 40$.)

(a) Express θ as a function of the height h.
(b) What is θ when the plane is 55 feet above the ground?
(c) When $\theta = 1$ radian, how high is the plane?

67. A rocket is fired straight up. The line of sight from an observer 4 miles away makes an angle of t radians with the horizontal.

(a) Express t as a function of the height h of the rocket.
(b) Find t when the rocket is .25 mile, 1 mile, and 2 miles high respectively.
(c) When $t = .4$ radian, how high is the rocket?

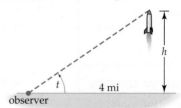

68. A cable from the top of a 60-foot high tower is to be attached to the ground x feet from the base of the tower.

(a) If the cable makes an angle of t radians with the ground when attached, express t as a function of x. [*Hint:* Select a coordinate system in which both x and t are positive, or use Section 8.1.]
(b) What is t when the distance $x = 40$ feet? When $x = 70$ feet? When $x = 100$ feet?
(c) If $t = \pi/5$, how far is the end of the cable from the base of the tower?

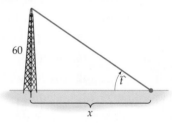

69. Suppose that the movie screen in Example 12 is 24 feet high and is 10 feet above the ground and that the eye level of the watcher is 4 feet above the ground.

(a) Express the angle t at the watcher's eye as a function of her distance x from the wall holding the screen.
(b) At what distance from the screen is the angle t as large as possible?

70. Section 8.1 is a prerequisite for this exercise. A camera on a 5-foot-high tripod is placed in front of a 6-foot-high picture that is mounted 3 feet above the floor.

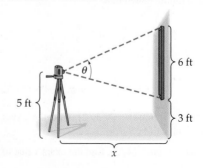

(a) Express angle θ as a function of the distance x from the camera to the wall.

(b) The photographer wants to use a particular lens, for which $\theta = 36°$ ($\pi/5$ radians). How far should she place the camera from the wall to be sure that the entire picture will show in the photograph?

71. A 15-foot-wide highway sign is placed 10 feet from a road, perpendicular to the road (see figure). A spotlight at the edge of the road is aimed at the sign.

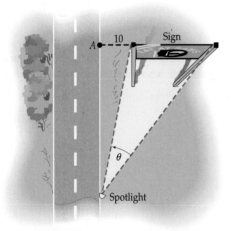

(a) Express θ as a function of the distance x from point A to the spotlight.

(b) How far from point A should the spotlight be placed so that the angle θ is as large as possible?

72. Show that the restricted secant function, whose domain consists of all numbers x such that $0 \leq x \leq \pi$ and $x \neq \pi/2$, has an inverse function. Sketch its graph.

73. Show that the restricted cosecant function, whose domain consists of all numbers x such that $-\pi/2 \leq x \leq \pi/2$ and $x \neq 0$, has an inverse function. Sketch its graph.

74. Show that the restricted cotangent function, whose domain is the interval $(0, \pi)$, has an inverse function. Sketch its graph.

75. Show that the inverse cosine function actually has the two properties listed in the box on page 549.

76. Show that the inverse tangent function actually has the two properties listed in the box on page 551.

In Exercises 77–84, prove the identity.

77. $\sin^{-1}(-x) = -\sin^{-1}x$ [*Hint:* Let $u = \sin^{-1}(-x)$ and show that $\sin^{-1}x = -u$.]

78. $\tan^{-1}(-x) = -\tan^{-1}x$

79. $\cos^{-1}(-x) = \pi - \cos^{-1}x$ [*Hint:* Let $u = \cos^{-1}(-x)$ and show that $0 \leq \pi - u \leq \pi$; use the identity

$$\cos(\pi - u) = -\cos u.]$$

80. $\sin^{-1}(\cos x) = \pi/2 - x$ $(0 \leq x \leq \pi)$

81. $\tan^{-1}(\cot x) = \pi/2 - x$ $(0 < x < \pi)$

82. $\tan^{-1}x + \tan^{-1}\left(\dfrac{1}{x}\right) = \dfrac{\pi}{2}$

83. $\sin^{-1}x = \tan^{-1}\left(\dfrac{x}{\sqrt{1 - x^2}}\right)$ $(-1 < x < 1)$

[*Hint:* Let $u = \sin^{-1}x$ and show that $\tan u = x/\sqrt{1 - x^2}$. Since $\sin u = x$, $\cos u = \pm\sqrt{1 - x^2}$. Show that in this case, $\cos u = \sqrt{1 - x^2}$.]

84. $\cos^{-1}x = \dfrac{\pi}{2} - \tan^{-1}\left(\dfrac{x}{\sqrt{1 - x^2}}\right)$ $(-1 < x < 1)$

[*Hint:* See Example 8 and Exercise 83.]

85. Is it true that $\tan^{-1}x = \dfrac{\sin^{-1}x}{\cos^{-1}x}$? Justify your answer.

86. Using the viewing window with $-2\pi \leq x \leq 2\pi$ and $-4 \leq y \leq 4$ graph the functions $f(x) = \cos(\cos^{-1}x)$ and $g(x) = \cos^{-1}(\cos x)$. How do you explain the shapes of the two graphs?

7.5 Trigonometric Equations

Section Objectives ■ Solve basic trigonometric equations.
■ Solve other trigonometric equations.

Any equation that involves trigonometric functions can be solved graphically, and many can be solved algebraically. Unlike the equations solved previously, trigonometric equations typically have an infinite number of solutions. In most cases, these solutions can be systematically determined by using periodicity, as we now see.

BASIC EQUATIONS

We begin with **basic equations,** such as

$$\sin x = .39, \qquad \cos x = .2, \qquad \tan x = -3.$$

Basic equations can be solved by the methods illustrated in Examples 1–3.

EXAMPLE 1

Solve $\tan x = 2$.

SOLUTION The equation can be solved graphically by graphing $y = \tan x$ and $y = 2$ on the same coordinate axes and finding the intersection points. The x-coordinate of every such point is a number whose tangent is 2, that is, a solution of the equation.

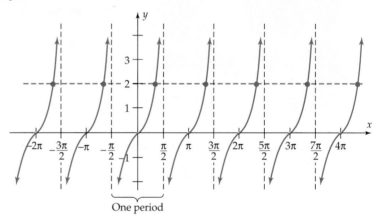

One period

Figure 7–20

Figure 7–20 shows that there is exactly one solution in each period of $\tan x$. The solution between $-\pi/2$ and $\pi/2$ could be found graphically, but it's faster to compute $\tan^{-1} 2$ on a calculator.* The calculator then displays the number between $-\pi/2$ and $\pi/2$ whose tangent is 2, namely, $x = 1.1071$, as shown in Figure 7–21.† Since the tangent graph repeats its pattern with period π, all the other solutions differ from this one by an integer multiple of π. Thus, all the solutions are

$$1.1071, \qquad 1.1071 \pm \pi, \qquad 1.1071 \pm 2\pi, \qquad 1.1071 \pm 3\pi, \qquad \text{etc.}$$

These solutions are customarily written like this:

$$x = 1.1071 + k\pi \quad (k = 0, \pm 1, \pm 2, \pm 3, \ldots). \qquad \blacksquare$$

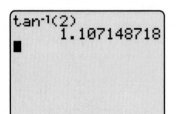

Figure 7–21

EXAMPLE 2

Solve $\tan^{-1}x = 1$.

SOLUTION We can construct a picture of the situation by replacing the horizontal blue line through 2 in Figure 7–20 by a horizontal line through 1. As in

---

*Unless stated otherwise, radian mode is used throughout this section.
†You need not have read about inverse trigonometric functions in Section 7.4 to understand this section. Here the calculator's $\tan^{-1}$ key is used only to produce one number with the given tangent, and similarly for the $\sin^{-1}$ and $\cos^{-1}$ keys.

Example 1, there is just one solution between $-\pi/2$ and $\pi/2$. Our knowledge of special values tells us that this solution is $x = \pi/4$ (because $\tan(\pi/4) = 1$ by Example 4 of Section 6.2). Since the tangent function has period π, the other solutions differ from $x = \pi/4$ by integer multiples of π. So all solutions are given by

$$x = \frac{\pi}{4} + k\pi \quad (k = 0, \pm1, \pm2, \pm3, \ldots).$$ ∎

The techniques illustrated in Examples 1 and 2 apply in the general case.

Solving
tan x = c

If c is any real number, then the equation

$$\tan x = c$$

can be solved as follows.

1. Find one solution u by using your knowledge of special values or by computing $\tan^{-1}c$ on a calculator.

2. Then all solutions are given by
$$x = u + k\pi \, (k = 0, \pm1, \pm2, \pm3, \ldots).$$

Solving basic sine equations is similar to solving basic tangent equations, but involves one additional step, as illustrated in the next example.

EXAMPLE 3

Solve $\sin x = -.75$

SOLUTION   The solutions are the x-coordinates of the points where the graphs of $y = \sin x$ and $y = -.75$ intersect (why?). Note that there are exactly two solutions in every period of $\sin x$ (for instance, between $-\pi/2$ and $3\pi/2$).

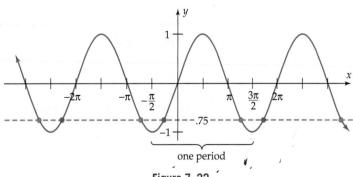

Figure 7–22

Figure 7–22 shows that there is one solution between $-\pi/2$ and $\pi/2$. It can be found by computing $\sin^{-1}(-.75)$ on a calculator. The calculator displays the number between $-\pi/2$ and $\pi/2$ whose sine is $-.75$, namely $x = -.8481$, as shown in Figure 7–23. Since the sine graph repeats is pattern with period 2π, all of the following numbers are also solutions:

$$-.8481, \quad -.8481 \pm 2\pi, \quad -.8481 \pm 4\pi, \quad -.8481 \pm 6\pi, \quad \text{etc.}$$

Figure 7–23

These solutions correspond to the red intersection points in Figure 7–22, each of which is 2π units from the next red point. As you can see, there are still more solutions (corresponding to the blue points). One of them can be found by using the identity that was proved in Example 2 of Section 7.2:

$$\sin(\pi - x) = \sin x.$$

Applying this identity with the solution $x = -.8481$ shows that

$$\sin[\pi - (-.8481] = \sin(-.8481) = -.75.$$

In other words, $\pi - (-.8481) = 3.9897$ is also a solution of $\sin x = -.75$. The other solutions of the equation are

$$3.9897, \quad 3.9897 \pm 2\pi, \quad 3.9897 \pm 4\pi, \quad 3.9897 \pm 6\pi, \quad \text{etc.,}$$

corresponding to the blue intersection points in Figure 7–22, each of which is 2π units from the next blue point. Therefore, all the solutions of $\sin x = -0.75$ are

$$x = -.8481 + 2k\pi \quad \text{and} \quad x = 3.9897 + 2k\pi$$

$$(k = 0, \pm 1, \pm 2, \pm 3, \ldots).\quad\blacksquare$$

The solution methods of Example 3 extend to the general case.

Solving
sin x = c

If c is a number between -1 and 1, then the equation

$$\sin x = c$$

can be solved as follows.*

1. Find one solution u by using your knowledge of special values or by computing $\sin^{-1}c$ on a calculator.

2. A second solution is $\pi - u$.

3. All solutions are given by

$$x = u + 2k\pi \quad \text{and} \quad x = (\pi - u) + 2k\pi \quad (k = 0, \pm 1, \pm 2, \pm 3, \ldots).$$

EXAMPLE 4

Solve $\sin v = \sqrt{2}/2$ without using a calculator.

SOLUTION   Our knowledge of special values shows that $v = \pi/4$ is one solution (see Example 4 of Section 6.2). Hence, a second solution is

$$\pi - v = \pi - \frac{\pi}{4} = \frac{3\pi}{4},$$

and all solutions are

$$v = \frac{\pi}{4} + 2k\pi \quad \text{and} \quad v = \frac{3\pi}{4} + 2k\pi \quad (k = 0, \pm 1, \pm 2, \pm 3, \ldots).\quad\blacksquare$$

---

*Equations of the form $\sin x = c$, with $|c| > 1$, have no solutions because the values of sine are always between -1 and 1, as we saw in Chapter 6.

Solving basic cosine equations is similar to solving basic sine equations, except that a different identity must be used to find the second solution.

EXAMPLE 5

Solve $\cos x = \sqrt{3}/2$.

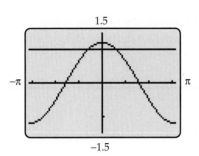

1.5

−π   π

−1.5

Figure 7–24

SOLUTION By graphing $y = \cos x$ and $y = \sqrt{3}/2$ on the same screen (Figure 7–24), we see that there are two solutions of the equation between $-\pi$ and π (one full period of cosine). The positive solution could be approximated by computing $\cos^{-1}(\sqrt{3}/2)$ on a calculator. However, our knowledge of special values provides an exact solution. Example 3 of Section 6.2 shows that $\cos(\pi/6) = \sqrt{3}/2$. So one solution of the equation is $x = \pi/6$. The negative angle identity $\cos(-x) = \cos x$ shows that the second solution is $x = -\pi/6$, because

$$\cos\left(-\frac{\pi}{6}\right) = \cos\left(\frac{\pi}{6}\right) = \frac{\sqrt{3}}{2}.$$

Since the interval $[-\pi, \pi]$ is one full period of cosine, all the solutions of the equation are

$$x = \frac{\pi}{6} + 2k\pi \quad \text{and} \quad x = -\frac{\pi}{6} + 2k\pi \quad (k = 0, \pm1, \pm2, \pm3, \ldots). \quad \blacksquare$$

In the general case, we have the following result.

Solving
COS X = C

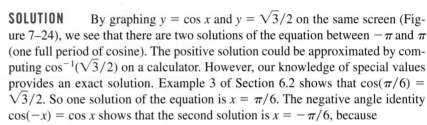

If c is a number between -1 and 1, then the equation

$$\cos x = c$$

can be solved as follows.*

1. Find one solution u by using your knowledge of special values or by computing $\cos^{-1}c$ on a calculator.

2. A second solution is $-u$.

3. All solutions are given by

$$x = u + 2k\pi \quad \text{and} \quad x = -u + 2k\pi \quad (k = 0, \pm1, \pm2, \pm3, \ldots).$$

EXAMPLE 6

Find all solutions of $\sec x = 8$ in the interval $[0, 2\pi)$.

SOLUTION Note that $\sec x = 8$ exactly when

$$\frac{1}{\cos x} = 8 \quad \text{or, equivalently,} \quad \cos x = \frac{1}{8} = .125.$$

Since $\cos^{-1}(.125) = 1.4455$, the solutions of $\cos x = .125$, and hence of $\sec x = 8$, are

$$x = 1.4455 + 2k\pi \quad \text{and} \quad x = -1.4455 + 2k\pi$$
$$(k = 0, \pm1, \pm2, \pm3, \ldots).$$

---

*Equations of the form $\cos x = c$, with $|c| > 1$, have no solutions because the values of cosine are always between -1 and 1, as we saw in Chapter 6.

Of these solutions, the two between 0 and 2π are

$$x = 1.4455 \qquad \text{and} \qquad x = -1.4455 + 2\pi = 4.8377. \qquad \blacksquare$$

 ## ALGEBRAIC SOLUTION OF OTHER TRIGONOMETRIC EQUATIONS

Many trigonometric equations can be solved algebraically by using substitution, factoring, the quadratic formula, and identities to reduce the problem to an equivalent one that involves only basic equations.

EXAMPLE 7

Solve exactly: $\sin 2x = \sqrt{2}/2$.

SOLUTION First, let $v = 2x$ and solve the basic equation $\sin v = \sqrt{2}/2$. As we saw in Example 4, the solutions are

$$v = \frac{\pi}{4} + 2k\pi \qquad \text{and} \qquad v = \frac{3\pi}{4} + 2k\pi \qquad (k = 0, \pm1, \pm2, \pm3, \ldots).$$

Since $v = 2x$, each of these solutions leads to a solution of the original equation.

$$2x = v = \frac{\pi}{4} + 2k\pi \qquad \text{or, equivalently,} \qquad x = \frac{1}{2}\left(\frac{\pi}{4} + 2k\pi\right) = \frac{\pi}{8} + k\pi.$$

Similarly,

$$2x = v = \frac{3\pi}{4} + 2k\pi \qquad \text{or, equivalently,} \qquad x = \frac{1}{2}\left(\frac{3\pi}{4} + 2k\pi\right) = \frac{3\pi}{8} + k\pi.$$

Therefore, all solutions of $\sin 2x = \sqrt{2}/2$ are given by

$$x = \frac{\pi}{8} + k\pi \qquad \text{and} \qquad x = \frac{3\pi}{8} + k\pi \qquad (k = 0, \pm1, \pm2, \pm3, \ldots).$$

The fact that the solutions are obtained by adding multiples of π rather than 2π is a reflection of the fact that the period of $\sin 2x$ is π. $\qquad \blacksquare$

EXAMPLE 8

Solve $-10\cos^2 x - 3\sin x + 9 = 0$.

SOLUTION We first use the Pythagorean identity to rewrite the equation in terms of the sine function.

$$-10\cos^2 x - 3\sin x + 9 = 0$$

$$-10(1 - \sin^2 x) - 3\sin x + 9 = 0$$

$$-10 + 10\sin^2 x - 3\sin x + 9 = 0$$

$$10\sin^2 x - 3\sin x - 1 = 0.$$

Now factor the left side:*

$$(2 \sin x - 1)(5 \sin x + 1) = 0$$

$$2 \sin x - 1 = 0 \qquad \text{or} \qquad 5 \sin x + 1 = 0$$

$$2 \sin x = 1 \qquad\qquad\qquad 5 \sin x = -1$$

$$\sin x = 1/2 \qquad\qquad\qquad \sin x = -1/5 = -.2.$$

Each of these basic equations is readily solved. We note that $\sin(\pi/6) = 1/2$, so $x = \pi/6$ and $x = \pi - \pi/6 = 5\pi/6$ are solutions of the first one. Since $\sin^{-1}(-.2) = -.2014$, both $x = -.2014$ and $x = \pi - (-.2014) = 3.3430$ are solutions of the second equation. Therefore, all solutions of the original equation are given by

$$x = \frac{\pi}{6} + 2k\pi, \qquad x = \frac{5\pi}{6} + 2k\pi,$$

$$x = -.2014 + 2k\pi, \qquad x = 3.3430 + 2k\pi,$$

where $k = 0, \pm 1, \pm 2, \pm 3, \ldots$. ∎

EXAMPLE 9

Solve $\sec^2 x + 5 \tan x = -2$.

SOLUTION   We use the Pythagorean identity $\sec^2 x = 1 + \tan^2 x$ to obtain an equivalent equation.

$$\sec^2 x + 5 \tan x = -2$$

$$\sec^2 x + 5 \tan x + 2 = 0$$

$$(1 + \tan^2 x) + 5 \tan x + 2 = 0$$

$$\tan^2 x + 5 \tan x + 3 = 0.$$

If we let $u = \tan x$, this last equation becomes $u^2 + 5u + 3 = 0$. Since the left side does not readily factor, we use the quadratic formula to solve the equation.

$$u = \frac{-5 \pm \sqrt{5^2 - 4 \cdot 1 \cdot 3}}{2} = \frac{-5 \pm \sqrt{13}}{2}.$$

Since $u = \tan x$, the original equation is equivalent to

$$\tan x = \frac{-5 + \sqrt{13}}{2} \approx -.6972 \qquad \text{or} \qquad \tan x = \frac{-5 - \sqrt{13}}{2} \approx -4.3028.$$

Solving these basic equations as above, we find that $x = -.6089$ is a solution of the first and $x = -1.3424$ is a solution of the second. Hence, the solutions of the original equation are

$$x = -.6089 + k\pi \qquad \text{and} \qquad x = -1.3424 + k\pi$$

$$(k = 0, \pm 1, \pm 2, \pm 3, \ldots).$$ ∎

---

*The factorization may be easier to see if you first substitute v for $\sin x$, so that $10 \sin^2 x - 3 \sin x - 1$ becomes $10v^2 - 3v - 1 = (2v - 1)(5v + 1)$.

EXAMPLE 10

Solve $5 \cos x + 3 \cos 2x = 3$.

SOLUTION   We use the double-angle identity: $\cos 2x = 2 \cos^2 x - 1$ as follows.

$$5 \cos x + 3 \cos 2x = 3$$

Use double-angle identity: $$5 \cos x + 3(2 \cos^2 x - 1) = 3$$

Multiply out left side: $$5 \cos x + 6 \cos^2 x - 3 = 3$$

Rearrange terms: $$6 \cos^2 x + 5 \cos x - 6 = 0$$

Factor left side: $$(2 \cos x + 3)(3 \cos x - 2) = 0$$

$$2 \cos x + 3 = 0 \quad \text{or} \quad 3 \cos x - 2 = 0$$

$$2 \cos x = -3 \qquad\qquad 3 \cos x = 2$$

$$\cos x = -\frac{3}{2} \qquad\qquad \cos x = \frac{2}{3}.$$

The equation $\cos x = -3/2$ has no solutions because $\cos x$ always lies between -1 and 1. A calculator shows that the solutions of $\cos x = 2/3$ are

$$x = .8411 + 2k\pi \quad \text{and} \quad x = -.8411 + 2k\pi$$

$$(k = 0, \pm 1, \pm 2, \pm 3, \ldots). \qquad\blacksquare$$

 ## GRAPHICAL SOLUTION METHOD

When the techniques of the preceding examples are inadequate, trigonometric equations may be solved by the following graphical procedure.

Graphical Method
for Solving
Trigonometric Equations

1. Write the equation in the form $f(x) = 0$.

2. Determine the period of p of $f(x)$.

3. Graph $f(x)$ over an interval of length p.

4. Use a graphical root finder to determine the x-intercepts of the graph in this interval.

5. For each x-intercept u, all of the numbers

   $$u + kp \qquad (k = 0, \pm 1, \pm 2, \pm 3, \ldots)$$

   are solutions of the equation.

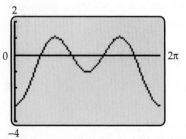

Figure 7–25

EXAMPLE 11

Solve $3 \sin^2 x - \cos x - 2 = 0$.

SOLUTION   Both sine and cosine have period 2π, so $f(x) = 3 \sin^2 x - \cos x - 2$ also has period 2π. Figure 7–25 shows one full period of the graph of f. A graphical root finder shows that the four x-intercepts (solutions of the equation) in this window are

$$x = 1.1216, \qquad x = 2.4459, \qquad x = 3.8373, \qquad x = 5.1616.$$

Since the graph repeats its pattern to the left and right, the other x-intercepts (solutions) will differ from these four by multiples of 2π. For instance, in addition to the solution $x = 1.1216$, each of the following is a solution.

$$x = 1.1216 \pm 2\pi, \qquad x = 1.1216 \pm 4\pi, \qquad x = 1.1216 \pm 6\pi, \text{ etc.}$$

A similar analysis applies to the other solutions between 0 and 2π. Hence, all solutions of the equation are given by

$$x = 1.1216 + 2k\pi, \qquad x = 2.4459 + 2k\pi, \qquad x = 3.8373 + 2k\pi,$$

$$x = 5.1616 + 2k\pi, \qquad \text{where } k = 0, \pm 1, \pm 2, \pm 3, \dots. \qquad ■$$

EXAMPLE 12

Solve $\tan x = 3 \sin 2x$.

SOLUTION   We first rewrite the equation as

$$\tan x - 3 \sin 2x = 0.$$

Both $\tan x$ and $\sin 2x$ have period π (see pages 472 and 478). Hence, the function given by the left side of the equation, $f(x) = \tan x - 3 \sin 2x$, also has period π. The graph of f on the interval $[0, \pi)$ (Figure 7–26) shows an erroneous vertical line segment at $x = \pi/2$, where tangent is not defined, as well as x-intercepts at the endpoints of the interval. Consequently, we use the more easily read graph f in Figure 7–27, which uses the interval $(-\pi/2, \pi/2)$.

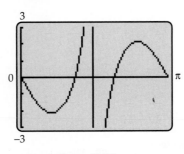

Figure 7–26

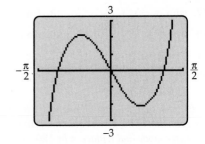

Figure 7–27

Even without the graph, we can verify that there is an x-intercept at the origin because

$$f(0) = \tan 0 - 3 \sin(2 \cdot 0) = 0.$$

A root finder shows that the other two x-intercepts in Figure 7–22 are

$$x = -1.1503 \qquad \text{and} \qquad x = 1.1503.$$

Since $f(x)$ has period π, all solutions of the equation are given by

$$x = -1.1503 + k\pi, \qquad x = 0 + k\pi, \qquad x = 1.1503 + k\pi$$

$$(k = 0, \pm 1, \pm 2, \pm 3, \dots). \qquad ■$$

EXERCISES 7.5

In all exercises, find exact solutions if possible (as in Examples 2, 4, 5, and 7) and approximate ones otherwise. When a calculator is used, round your answers (but not any intermediate results) to four decimal places.

In Exercises 1–10, find all solutions of the equation.

1. $\sin x = .465$

2. $\sin x = .682$

3. $\cos x = -.564$

4. $\cos x = -.371$

5. $\tan x = -.354$

6. $\tan x = 10$

7. $\cot x = 2.3$ [*Remember:* $\cot x = 1/\tan x$.]

8. $\cot x = -3.5$ 9. $\sec x = -1.6$

10. $\csc x = 6.4$

In Exercises 11–14, approximate all solutions in $[0, 2\pi)$ of the given equation.

11. $\sin x = .119$ 12. $\cos x = .958$

13. $\tan x = 4$ 14. $\tan x = 18$

In Exercises 15–24, use your knowledge of special values to find the exact solutions of the equation.

15. $\sin x = \sqrt{3}/2$ 16. $2 \cos x = \sqrt{2}$

17. $\tan x = -\sqrt{3}$ 18. $\tan x = 1$

19. $2 \cos x = -\sqrt{3}$ 20. $\sin x = 0$

21. $2 \sin x + 1 = 0$ 22. $\csc x = \sqrt{2}$

23. $\csc x = 2$ 24. $-2 \sec x = 4$

In Exercises 25–34, find all angles θ with $0° \le \theta < 360°$ that are solutions of the given equation. [Hint: Put your calculator in degree mode and replace π by $180°$ in the solution algorithms for basic equations.]

25. $\tan \theta = 7.95$ 26. $\tan \theta = 69.4$

27. $\cos \theta = -.42$ 28. $\cot \theta = -2.4$

29. $2 \sin^2\theta + 3 \sin \theta + 1 = 0$

30. $4 \cos^2\theta + 4 \cos \theta - 3 = 0$

31. $\tan^2\theta - 3 = 0$

32. $2 \sin^2\theta = 1$

33. $4 \cos^2\theta + 4 \cos \theta + 1 = 0$

34. $\sin^2\theta - 3 \sin \theta = 10$

At the instant you hear a sonic boom from an airplane overhead, your angle of elevation α to the plane is given by the equation $\sin \alpha = 1/m$, where m is the Mach number for the speed of the plane (Mach 1 is the speed of sound, Mach 2.5 is 2.5 times the speed of sound, etc.). In Exercises 35–38, find the angle of elevation (in degrees) for the given Mach number. Remember that an angle of elevation must be between $0°$ and $90°$.

35. $m = 1.1$ 36. $m = 1.6$

37. $m = 2$ 38. $m = 2.4$

When a light beam passes from one medium to another (for instance, from air to water), it changes both its speed and direction. According to Snell's Law of Refraction,

$$\frac{\sin \theta_1}{\sin \theta_2} = \frac{v_1}{v_2},$$

where v_1 is the speed of light in the first medium, v_2 its speed in the second medium, θ_1 the angle of incidence, and θ_2 the angle of refraction, as shown in the figure. The number v_1/v_2 is called the index of refraction. *Use this information to do Exercises 39–42.*

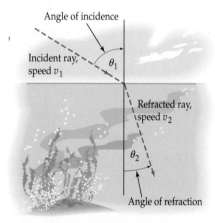

39. The index of refraction of light passing from air to water is 1.33. If the angle of incidence is $38°$, find the angle of refraction.

40. The index of refraction of light passing from air to ordinary glass is 1.52. If the angle of incidence is $17°$, find the angle of refraction.

41. The index of refraction of light passing from air to dense glass is 1.66. If the angle of incidence is $24°$, find the angle of refraction.

42. The index of refraction of light passing from air to quartz is 1.46. If the angle of incidence is $50°$, find the angle of refraction.

In Exercises 43–52, use an appropriate substitution (as in Example 7) to find all solutions of the equation.

43. $\sin 2x = -\sqrt{3}/2$ **44.** $\cos 2x = \sqrt{2}/2$

45. $2 \cos \dfrac{x}{2} = \sqrt{2}$ **46.** $2 \sin \dfrac{x}{3} = 1$

47. $\tan 3x = -\sqrt{3}$ **48.** $5 \sin 2x = 2$

49. $5 \cos 3x = -3$ **50.** $2 \tan 4x = 16$

51. $4 \tan \dfrac{x}{2} = 8$ **52.** $5 \sin \dfrac{x}{4} = 4$

Exercises 53–60, deal with a circle of radius r and a central angle of t radians, $(0 < t < \pi)$, as shown in the figure. The length L of the chord determined by the angle and the area A of the shaded segment are given by

$$L = 2r \sin \frac{t}{2} \quad and \quad A = \frac{r^2}{2}(t - \sin t).$$

(See Exercise 108 for a proof of the first of these formulas.)

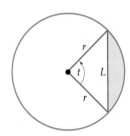

In Exercises 53–56, find the radian measure of the angle and the area of the segment under the given conditions.

53. $r = 5$ and $L = 8$ **54.** $r = 8$ and $L = 5$

55. $r = 1$ and $L = 1.5$ **56.** $r = 10$ and $L = 12$.

In Exercises 57–60, find the radian measure of the angle and the length L of the chord under the given conditions.

57. $r = 10$ and $A = 50$ **58.** $r = 1$ and $A = .5$

59. $r = 8$ and $A = 20$ **60.** $r = 5$ and $A = 2$

Exercises 61 and 62 deal with a rectangle inscribed in the segment of the graph of $f(x) = 2 \cos 2x$ shown in the figure.

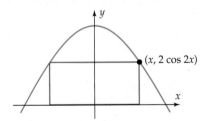

61. (a) Find a formula for the area of the rectangle in terms of x. [*Hint:* The length is $2x$.]

(b) For what values of x does the rectangle have an area of 1 square unit?

62. Use the formula in Exercise 61(a) to determine the value of x that determines the rectangle with the largest possible area. What is this maximum area?

In Exercises 63–88, use factoring, the quadratic formula, or identities to solve the equation. Find all solutions in the interval $[0, 2\pi)$.

63. $3 \sin^2 x - 8 \sin x - 3 = 0$

64. $5 \cos^2 x + 6 \cos x = 8$

65. $2 \tan^2 x + 7 \tan x + 5 = 0$

66. $3 \sin^2 x + 2 \sin x = 5$

67. $\cot x \cos x = \cos x$ [Be careful; see Exercise 109.]

68. $\tan x \cos x = \cos x$ **69.** $\cos x \csc x = 2 \cos x$

70. $\tan x \sec x + 3 \tan x = 0$

71. $4 \sin x \tan x - 3 \tan x + 20 \sin x - 15 = 0$
[*Hint:* One factor is $\tan x + 5$.]

72. $25 \sin x \cos x - 5 \sin x + 20 \cos x = 4$

73. $\sin^2 x + 2 \sin x - 2 = 0$ **74.** $\cos^2 x + 5 \cos x = 1$

75. $\tan^2 x + 1 = 3 \tan x$ **76.** $4 \cos^2 x - 2 \cos x = 1$

77. $2 \tan^2 x - 1 = 3 \tan x$ **78.** $6 \sin^2 x + 4 \sin x = 1$

79. $\sin^2 x + 3 \cos^2 x = 0$ **80.** $\sec^2 x - 2 \tan^2 x = 0$

81. $\sin 2x + \cos x = 0$ **82.** $\cos 2x - \sin x = 1$

83. $9 - 12 \sin x = 4 \cos^2 x$ **84.** $\sec^2 x + \tan x = 3$

85. $\cos^2 x - \sin^2 x + \sin x = 0$

86. $2 \tan^2 x + \tan x = 5 - \sec^2 x$

87. $\sin \dfrac{x}{2} = 1 - \cos x$ **88.** $4 \sin^2 \left(\dfrac{x}{2}\right) + \cos^2 x = 2$

In Exercises 89–100, solve the equation graphically.

89. $4 \sin 2x - 3 \cos 2x = 2$

90. $5 \sin 3x + 6 \cos 3x = 1$

91. $3 \sin^3 2x = 2 \cos x$

92. $\sin^2 2x - 3 \cos 2x + 2 = 0$

93. $\tan x + 5 \sin x = 1$

94. $2 \cos^2 x + \sin x + 1 = 0$

95. $\cos^3 x - 3 \cos x + 1 = 0$ **96.** $\tan x = 3 \cos x$

97. $\cos^4 x - 3 \cos^3 x + \cos x = 1$

98. $\sec x + \tan x = 3$

99. $\sin^3 x + 2 \sin^2 x - 3 \cos x + 2 = 0$

100. $\csc^2 x + \sec x = 1$

101. The number of hours of daylight in Detroit on day t of a non–leap year (with $t = 0$ being January 1) is given by the function

$$d(t) = 3 \sin\left[\frac{2\pi}{365}(t - 80)\right] + 12.$$

 (a) On what days of the year are there exactly 11 hours of daylight?

 (b) What day has the maximum amount of daylight?

102. A weight hanging from a spring is set into motion (see Figure 6–69 on page 484), moving up and down. Its distance (in centimeters) above or below the equilibrium point at time t seconds is given by

$$d = 5(\sin 6t - 4 \cos 6t).$$

 At what times during the first 2 seconds is the weight at the equilibrium position ($d = 0$)?

In Exercises 103–106, use the following fact: When a projectile (such as a ball or a bullet) leaves its starting point at angle of elevation θ with velocity v, the horizontal distance d it travels is given by the equation

$$d = \frac{v^2}{32} \sin 2\theta,$$

where d is measured in feet and v in feet per second. Note that the horizontal distance traveled may be the same for two different angles of elevation, so some of these exercises may have more than one correct answer.

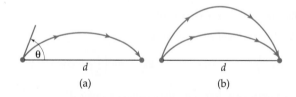

 (a)   (b)

103. If muzzle velocity of a rifle is 300 feet per second, at what angle of elevation (in radians) should it be aimed for the bullet to hit a target 2500 feet away?

104. Is it possible for the rifle in Exercise 103 to hit a target that is 3000 feet away? [At what angle of elevation would it have to be aimed?]

105. A fly ball leaves the bat at a velocity of 98 mph and is caught by an outfielder 288 feet away. At what angle of elevation (in degrees) did the ball leave that bat?

106. An outfielder throws the ball at a speed of 75 mph to the catcher who is 200 feet away. At what angle of elevation was the ball thrown?

THINKERS

107. Under what conditions (on the constant) does a basic equation involving the sine and cosine function have *no* solutions?

108. Prove the formula $L = 2r \sin \dfrac{t}{2}$ used in Exercises 53–60 as follows.

 (a) Construct the perpendicular line from the center of the circle to the chord PQ, as shown in the figure. Verify that triangles OCP and OCQ are congruent. [*Hint:* Angles P and Q are equal by the Isosceles Triangle Theorem,* and in each triangle, angle C is a right angle (why?). Use the Congruent Triangles Theorem.*]

 (b) Use part (a) to explain why angle POC measures $t/2$ radians.

 (c) Show that the length of PC is $r \sin \dfrac{t}{2}$.

 (d) Use the fact that PC and QC have the same length to conclude that the length L of PC is

$$L = 2r \sin \frac{t}{2}.$$

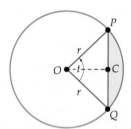

109. What is wrong with this so-called solution?

$$\sin x \tan x = \sin x$$

$$\tan x = 1$$

$$x = \frac{\pi}{4} \quad \text{or} \quad \frac{5\pi}{4}.$$

 [*Hint:* Solve the original equation by moving all terms to one side and factoring. Compare your answers with the ones above.]

110. Let n be a fixed positive integer. Describe *all* solutions of the equation $\sin nx = 1/2$. [*Hint:* See Exercises 43–52.]

_____

*See the Geometry Review Appendix.

Chapter 7 Review

IMPORTANT CONCEPTS

IMPORTANT FACTS & FORMULAS

- All identities in the Chapter 6 Review
- *Addition and Subtraction Identities:*

$$\sin(x + y) = \sin x \cos y + \cos x \sin y$$

$$\sin(x - y) = \sin x \cos y - \cos x \sin y$$

$$\cos(x + y) = \cos x \cos y - \sin x \sin y$$

$$\cos(x - y) = \cos x \cos y + \sin x \sin y$$

$$\tan(x + y) = \frac{\tan x + \tan y}{1 - \tan x \tan y}$$

$$\tan(x - y) = \frac{\tan x - \tan y}{1 + \tan x \tan y}$$

- *Cofunction Identities:*

$$\sin x = \cos\left(\frac{\pi}{2} - x\right) \qquad \cos x = \sin\left(\frac{\pi}{2} - x\right)$$

$$\tan x = \cot\left(\frac{\pi}{2} - x\right) \qquad \cot x = \tan\left(\frac{\pi}{2} - x\right)$$

$$\sec x = \csc\left(\frac{\pi}{2} - x\right) \qquad \csc x = \sec\left(\frac{\pi}{2} - x\right)$$

- *Double-Angle Identities:*

$$\sin 2x = 2 \sin x \cos x$$

$$\cos 2x = \cos^2 x - \sin^2 x$$

$$\cos 2x = 2 \cos^2 x - 1 \qquad \cos 2x = 1 - 2 \sin^2 x$$

$$\tan 2x = \frac{2 \tan x}{1 - \tan^2 x}$$

■ *Half-Angle Identities:*

$$\sin \frac{x}{2} = \pm \sqrt{\frac{1 - \cos x}{2}}$$

$$\cos \frac{x}{2} = \pm \sqrt{\frac{1 + \cos x}{2}}$$

$$\tan \frac{x}{2} = \frac{1 - \cos x}{\sin x} \qquad \tan \frac{x}{2} = \frac{\sin x}{1 + \cos x} \qquad \tan \frac{x}{2} = \pm \sqrt{\frac{1 - \cos x}{1 + \cos x}}$$

■ $\sin^{-1}v = u$ exactly when $\sin u = v$ $\left(-\frac{\pi}{2} \leq u \leq \frac{\pi}{2}, -1 \leq v \leq 1\right)$

■ $\cos^{-1}v = u$ exactly when $\cos u = v$ $(0 \leq u \leq \pi, -1 \leq v \leq 1)$

■ $\tan^{-1}v = u$ exactly when $\tan u = v$ $\left(-\frac{\pi}{2} < u < \frac{\pi}{2}, \text{any } v\right)$

REVIEW QUESTIONS

In Questions 1–4, simplify the given expression.

1. $\dfrac{\sin^2 t + (\tan^2 t + 2 \tan t - 4) + \cos^2 t}{3 \tan^2 t - 3 \tan t}$

2. $\dfrac{\sec^2 t \csc t}{\csc^2 t \sec t}$

3. $\dfrac{\tan^2 x - \sin^2 x}{\sec^2 x}$

4. $\dfrac{(\sin x + \cos x)(\sin x - \cos x) + 1}{\sin^2 x}$

In Questions 5–12, determine graphically whether the equation could possibly be an identity. If it could, prove that it is.

5. $\sin^4 t - \cos^4 t = 2 \sin^2 t - 1$

6. $1 + 2 \cos^2 t + \cos^4 t = \sin^4 t$

7. $\dfrac{\sin t}{1 - \cos t} = \dfrac{1 + \cos t}{\sin t}$

8. $\dfrac{\sin^2 t}{\cos^2 t} + 1 = \dfrac{1}{\cos^2 t}$

9. $\dfrac{\cos^2(\pi + t)}{\sin^2(\pi + t)} - 1 = \dfrac{1}{\sin^2 t}$

10. $\tan x + \cot x = \sec x \csc x$

11. $(\sin x + \cos x)^2 - \sin 2x = 1$

12. $\dfrac{1 - \cos 2x}{\tan x} = \sin 2x$

In Questions 13–22, prove the given identity.

13. $\dfrac{\tan x - \sin x}{2 \tan x} = \sin^2\left(\dfrac{x}{2}\right)$

14. $2 \cos x - 2 \cos^3 x = \sin x \sin 2x$

15. $\cos(x + y)\cos(x - y) = \cos^2 x - \sin^2 y$

16. $\dfrac{\cos(x - y)}{\cos x \cos y} = 1 + \tan x \tan y$

17. $\dfrac{\sec x + 1}{\tan x} = \dfrac{\tan x}{\sec x - 1}$

18. $\dfrac{\cos^4 x - \sin^4 x}{1 - \tan^4 x} = \cos^4 x$

19. $\dfrac{1 + \tan^2 x}{\tan^2 x} = \csc^2 x$

20. $\sec x - \cos x = \sin x \tan x$

21. $\tan^2 x - \sec^2 x = \cot^2 x - \csc^2 x$

22. $\sin 2x = \dfrac{1}{\tan x + \cot 2x}$

23. If $\tan x = 5/12$ and $\sin x > 0$, find $\sin 2x$.

24. If $\cos x = 15/17$ and $0 < x < \pi/2$, find $\sin(x/2)$.

25. If $\tan x = 4/3$ with $\pi < x < 3\pi/2$, and $\cot y = -5/12$ with $3\pi/2 < y < 2\pi$, find $\sin(x - y)$.

26. If $\sin x = -12/13$ with $\pi < x < 3\pi/2$, and $\sec y = 13/12$ with $3\pi/2 < y < 2\pi$, find $\cos(x + y)$.

27. If $\sin x = 1/4$ and $0 < x < \pi/2$, then $\sin(\pi/3 + x) = $?

28. If $\sin x = -2/5$ and $3\pi/2 < x < 2\pi$, then $\cos(\pi/4 + x) = $?

29. If $\sin x = 0$, is it true that $\sin 2x = 0$? Justify your answer.

30. If $\cos x = 0$, is it true that $\cos 2x = 0$? Justify your answer.

31. Show that

$$\sqrt{2 + \sqrt{3}} = \dfrac{\sqrt{2} + \sqrt{6}}{2}$$

by computing $\cos(\pi/12)$ in two ways, using the half-angle identity and the subtraction identity for cosine.

32. True or false: $2 \sin x = \sin 2x$. Justify your answer.

33. Compute $\sin(5\pi/12)$ exactly.

34. Express $\sec(x - \pi)$ in terms of $\sin x$ and $\cos x$.

35. $\sqrt{\dfrac{1 - \cos^2 x}{1 - \sin^2 x}} = \underline{\qquad}$.

(a) $|\tan x|$ (b) $|\cot x|$

(c) $\sqrt{\dfrac{1 - \sin^2 x}{1 - \cos^2 x}}$ (d) $\sec x$

(e) undefined

36. $\dfrac{1}{(\csc x)(\sec^2 x)} = \underline{\qquad}$.

(a) $\dfrac{1}{(\sin x)(\cos^2 x)}$ (b) $\sin x - \sin^3 x$

(c) $\dfrac{1}{(\sin x)(1 + \tan^2 x)}$ (d) $\sin x - \dfrac{1}{1 + \tan^2 x}$

(e) $1 + \tan^3 x$

37. If $\sin x = .6$ and $0 < x < \pi/2$, find $\sin 2x$.

38. If $\sin x = .6$ and $0 < x < \pi/2$, find $\sin(x/2)$.

39. If $\dfrac{\cos\left(\dfrac{u}{2} + \dfrac{v}{2}\right)}{\cos\dfrac{u}{2}} = 1.9$ and $v = \dfrac{\pi}{3}$, find u.

40. A coil of wire rotating in a magnetic field induces voltage k, where

$$k = 15 \sin\left(\frac{\pi t}{4} - \frac{\pi}{2}\right).$$

Use an appropriate identity to express k in terms of $\cos\dfrac{\pi t}{4}$.

41. Find the angle of inclination of the straight line through the points $(2, 6)$ and $(-2, 2)$.

42. Find one of the angles between the line L through the points $(-3, 2)$ and $(5, 1)$ and the line M, which has slope 2.

In Questions 43–52, exact answers are required.

43. $\cos^{-1}(\sqrt{2}/2) = ?$ **44.** $\cos^{-1}(\sqrt{3}/2) = ?$

45. $\tan^{-1}\sqrt{3}/3 = ?$ **46.** $\sin^{-1}(\cos 11\pi/6) = ?$

47. $\cos^{-1}(\sin 5\pi/3) = ?$ **48.** $\tan^{-1}(\cos 7\pi/2) = ?$

49. $\sin^{-1}(\sin .75) = ?$ **50.** $\cos^{-1}(\cos 2) = ?$

51. $\sin^{-1}(\sin 8\pi/3) = ?$ **52.** $\cos^{-1}(\cos 13\pi/4) = ?$

53. Sketch the graph of $f(x) = \tan^{-1}x - \pi$.

54. Sketch the graph of $g(x) = \sin^{-1}(x - 2)$.

55. Find the exact value of $\sin[\cos^{-1}(1/4)]$.

56. Find the exact value of $\sin[\tan^{-1}(1/2) - \cos^{-1}(4/5)]$.

In Questions 57–76, solve the equation by any means. Find exact solutions when possible and approximate ones otherwise.

57. $2 \sin x = 1$ **58.** $\sin x = -\sqrt{3}/2$

59. $\cos x = \sqrt{2}/2$ **60.** $\cos x = .5$

61. $\tan x = -1$ **62.** $\tan x = \sqrt{3}$

63. $\sin 3x = -\sqrt{3}/2$ **64.** $\cos 4x = .5$

65. $\sin x = .4$ **66.** $\cos x = -.7$

67. $\tan x = 15$ **68.** $\cot x = .3$

69. $2 \sin^2 x + 5 \sin x = 3$ **70.** $4 \cos^2 x - 2 = 0$

71. $2 \sin^2 x - 3 \sin x = 2$

72. $\cos 2x = \cos x$ [*Hint:* First use an identity.]

73. $\sec^2 x + 3 \tan^2 x = 13$ **74.** $\sec^2 x = 4 \tan x - 2$

75. $2 \sin^2 x + \sin x - 2 = 0$ **76.** $\cos^2 x - 3 \cos x - 2 = 0$

In Questions 77–80, solve the equation graphically.

77. $5 \tan x = 2 \sin 2x$ **78.** $\sin^3 x + \cos^2 x - \tan x = 2$

79. $\sin x + \sec^2 x = 3$

80. $\cos^2 x - \csc^2 x + \tan(x - \pi/2) + 5 = 0$

81. Find all angles θ with $0° \le \theta \le 360°$ such that $\sin \theta = -.7133$.

82. Find all angles θ with $0° \le \theta \le 360°$ such that $\tan \theta = 3.7321$.

83. A cannon has a muzzle velocity of 600 feet per second. At what angle of elevation should it be fired in order to hit a target 3500 feet away? [*Hint:* Use the projectile equation preceding Exercise 103 of Section 7.5.]

84. A weight hanging from a spring is set into motion (see Figure 6–69 on page 484), moving up and down. Its distance (in centimeters) above or below the equilibrium point at time t seconds is given by $d = 5 \sin 3t - 3 \cos 3t$. At what times during the first 2 seconds is the weight at the equilibrium position ($d = 0$)?

Chapter 7 Test

Sections 7.1–7.3   (1–18)

1. Prove: $1 - \sin x \cos x \tan x = \cos^2 x$

2. Find the exact value of $\cos\left(-\dfrac{5\pi}{12}\right)$.

3. Find the exact value of $\tan\dfrac{3\pi}{8}$.

4. Prove: $\cos^2 x - \cot^2 x = -\cos^2 x \cot^2 x$

5. Simplify: $\sin(x - y) \sin x + \cos(x - y) \cos x$

6. If $\sin x = -\dfrac{40}{41}$ and $\pi < x < \dfrac{3\pi}{2}$, find the exact value of

(a) $\cos 2x$ (b) $\sin 2x$ (c) $\tan 2x$

7. Prove: $(\sin^2 x - 1)(\cot^2 x + 1) = -\cot^2 x$

8. If $\cos x = -\dfrac{1}{4}$ and $\pi < x < \dfrac{3\pi}{2}$, then find the exact value of $\cos\left(\dfrac{\pi}{6} - x\right)$.

9. Simplify: $2\cos^2\left(\dfrac{x}{2}\right) - 1$

10. Prove: $\dfrac{1 - \tan x}{\cot x - 1} = \tan x$

11. Prove: $\sin x \sin(\pi - x) = 1 - \cos^2 x$

12. If $\sin x = .4$ and $\dfrac{\pi}{2} < x < \pi$, find the exact values of

    (a) $\sin \dfrac{x}{2}$ (b) $\cos \dfrac{x}{2}$.

Sections 7.4 and 7.5

13. Find the exact value of $\cos^{-1}\left(-\dfrac{\sqrt{3}}{2}\right)$.

14. Find all solutions of the equation $7\cos x = 5$.

15. Find the exact value of $\cos^{-1}\left(\sin \dfrac{\pi}{4}\right)$.

16. Find all solutions in $[0, 2\pi)$ of the equation $\cos^2 x + 7\cos x = 7$.

17. A rocket is fired straight up. The line of sight from an observer 3 miles away makes an angle of t radians with the horizontal. Find t when the rocket is 4 miles high.

18. Find all solutions of $2\sin \dfrac{x}{2} = \sqrt{2}$. Exact answers are required.

19. Assume that $-\dfrac{\pi}{2} < x < \dfrac{\pi}{2}$. Prove that $\cos^{-1}(\sin x) = \dfrac{\pi}{2} - x$.

20. When a projectile (such as a ball) leaves its starting point at an angle of elevation of θ radians with velocity v, the horizontal distance d it travels is given by

$$d = \dfrac{v^2}{32}\sin 2\theta,$$

where d is measured in feet and v in feet per second. Suppose an outfielder throws a baseball at a speed of 76 mph to the catcher who is 170 feet away. At what angle of elevation (in radians) was the ball thrown?

DISCOVERY PROJECT 7   The Sun and the Moon

Jack Hollingsworth/Getty Images

It has long been known that the cycles of the sun and the moon are periodic; that is, the moon is full at regular intervals and new at regular intervals. The same is true of the sun; the interval between the summer and winter solstices is also regular. It is therefore quite natural to use the sun and the moon to keep track of time, and to study the interaction between the two to predict events such as full moons, new moons, solstices, equinoxes, and eclipses. Indeed, these solar and lunar events have consequences on the earth, including the succession of the seasons and the severity of tides.

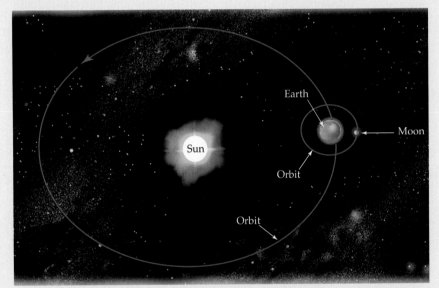

1. Thep following is a list of the days in 1999 when the moon was full. The length of the lunar month is the length of time between full moons. Use the data to approximate the length of the lunar month.

January 2	May 30	October 24
January 31	June 28	November 23
March 2	July 28	December 22
March 31	August 26	
April 30	September 25	

2. Write a function that has value 1 when the moon is full and 0 when the moon is new. Measure the independent variable in days with January 2, 1999, set as time 0. Use a function of the form

$$m(x) = \frac{\cos kt + 1}{2}$$

with period equal to the length of the lunar month.

571

3. Use your function to predict the date of the first full moon of the 21st century (in January of the year 2001). Does your function agree with the actual date of January 9? If not, what could have caused the discrepancy?

4. The solar year is approximately 365.24 days long. Write a function $s(x)$ with period equal to the length of the solar year so that on the date of the summer solstice, $s(x) = 1$ and on the day of the winter solstice, $s(x) = 0$. The summer solstice in 1999 was June 21 and the winter solstice falls midway between summer solstices.

5. If your functions $s(x)$ and $m(x)$ were accurate, when would you expect to see the next full moon on the summer solstice?

6. How would you go about making your models $s(x)$ and $m(x)$ more accurate?

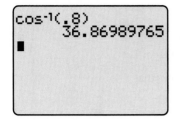

Figure 8–13

A faster, more accurate method of finding θ is to use the COS^{-1} key on your calculator (labeled ACOS on some models). When you key in COS^{-1} .8, as in Figure 8–13, the calculator produces an acute angle whose cosine is .8, namely, $\theta \approx 36.8699°$. Thus, the COS^{-1} key provides the electronic equivalent of searching the cosine table, without actually having to construct the table. ■

NOTE

In this chapter, we shall use the COS^{-1} key, and the analogous keys SIN^{-1} and TAN^{-1}, as they were used in the preceding example: as a way to find an angle θ in a triangle, when $\sin \theta$ or $\cos \theta$ or $\tan \theta$ is known. The other uses of these keys are discussed in Section 7.4, which deals with the inverse functions of sine, cosine, and tangent.

EXAMPLE 9

Without using the Pythagorean Theorem, find angles α and β and side c of the triangle in Figure 8–14. Make your answers as accurate as your technology allows.

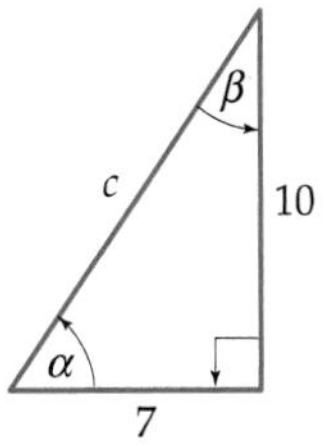

SOLUTION   We first note that

$$\tan \alpha = \frac{\text{opposite}}{\text{adjacent}} = \frac{10}{7}.$$

We use the TAN^{-1} key on a calculator to approximate α and store the result

$$\alpha \approx 55.0079798°$$

in memory A (Figure 8–15).* Since the sum of the angles of a triangle is 180°, we have

$$\alpha + \beta + 90° = 180°$$
$$\beta = 180° - 90° - \alpha.$$

Using a calculator and the stored value of α, we see that $\beta \approx 34.9920202°$ (Figure 8–16).

Figure 8–14

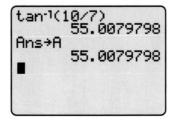

Figure 8–15

Figure 8–16

---

*We are using a TI-84+. Other calculators and computer programs may give answers with a different degree of accuracy. For instance, a TI-89 gives $\alpha \approx 55.0079798014°$.

Finally, we use the fact that

$$\sin \alpha = \frac{\text{opposite}}{\text{hypotenuse}} = \frac{10}{c}.$$

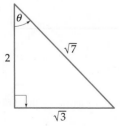

Multiplying both ends of this equation by c shows that

$$c \sin \alpha = 10$$

$$c = \frac{10}{\sin \alpha}.$$

Using a calculator and the stored value of α, we find that $c \approx 12.20655562$ (Figure 8–17). ■

Figure 8–17

EXERCISES 8.1

Directions: *When solving triangles here, all decimal approximations should be rounded off to one decimal place at the end of the computation.*

In Exercises 1–6, *evaluate the trigonometric functions at the angle (in standard position) whose terminal side contains the given point.*

1. $(2, 3)$ **2.** $(4, -2)$ **3.** $(-3, 7)$

4. $(\sqrt{2}, \sqrt{3})$ **5.** $(-3, -\sqrt{2})$ **6.** $(3, -5)$

In Exercises 7–12, *find* $\sin \theta$, $\cos \theta$, $\tan \theta$.

7.

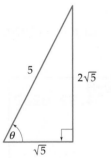

8.

9.

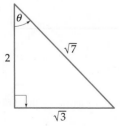

10.

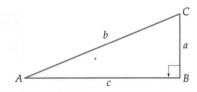

11.

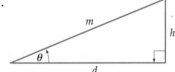

12.

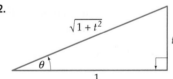

In Exercises 13–18, *find side c of the right triangle in the figure under the given conditions.*

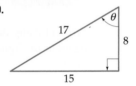

13. $\cos A = 12/13$ and $b = 39$

14. $\sin C = 3/4$ and $b = 12$

15. $\tan A = 5/12$ and $a = 15$

16. $\sec A = 2$ and $b = 8$

17. $\cot A = 6$ and $a = 1.4$

18. $\csc C = 1.5$ and $b = 4.5$

In Exercises 19–24, find the length h of the side of the right triangle, without using a calculator.

19.

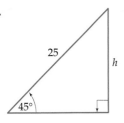

20.

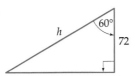

21.

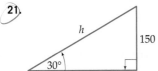

22.

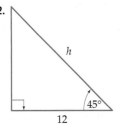

23.

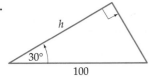

24.

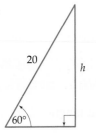

In Exercises 25–28, find the required side without using a calculator.

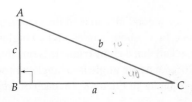

25. $a = 4$ and angle A measures $60°$; find c.

26. $c = 5$ and angle A measures $60°$; find a.

27. $c = 10$ and angle A measures $30°$; find a.

28. $a = 12$ and angle A measures $30°$; find c.

In Exercises 29–36, use the figure for Exercises 25–28. Solve the right triangle under the given conditions.

29. $b = 10$ and ⊾$C = 40°$

30. $c = 12$ and ⊾$C = 37°$

31. $a = 16$ and ⊾$A = 14°$

32. $a = 8$ and ⊾$A = 40°$

33. $c = 5$ and ⊾$A = 65°$

34. $c = 4$ and ⊾$C = 28°$

35. $b = 3.5$ and ⊾$A = 72°$

36. $a = 4.2$ and ⊾$C = 33°$

In Exercises 37–40, find angle θ.

37.

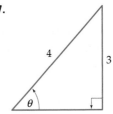

38.

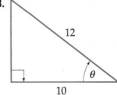

39.

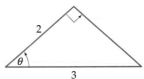

40.

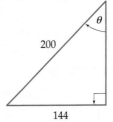

In Exercises 41–48, use the figure for Exercises 25–28 to find angles A and C under the given conditions.

41. $a = 8$ and $c = 15$

42. $b = 14$ and $c = 5$

43. $a = 7$ and $b = 10$

44. $a = 7$ and $c = 3$

45. $b = 18$ and $c = 12$

46. $a = 4$ and $b = 9$

47. $a = 2.5$ and $c = 1.4$

48. $b = 3.7$ and $c = 2.2$

49. Let θ be an acute angle with sides a and b in a triangle, as in the figure below.

    (a) Find the area of the triangle (in terms of h and a).

    (b) Find $\sin \theta$.

    (c) Use part (b) to show that $h = b \sin \theta$.

    (d) Use parts (a) and (c) to show that the area A of a triangle in which an acute angle θ has sides a and b is

$$A = \frac{1}{2} ab \sin \theta.$$

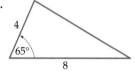

In Exercises 50–54, use the result of Exercise 49 to find the area of the given triangle.

50.

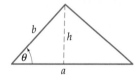

51.

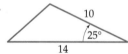

52.

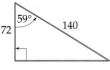

53.

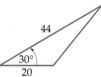

54.

55. Let θ and α be acute angles of a right triangle, as shown in the figure.

    (a) Find $\sin \theta$ and $\cos \alpha$.

    (b) Explain why $\theta + \alpha = 90°$.

    (c) Use parts (a) and (b) to conclude that for any acute angle θ,

$$\cos(90° - \theta) = \sin \theta.$$

56. (a) Using the figure for Exercise 55, find $\cos \theta$ and $\sin \alpha$.

    (b) Use part (a) and part (b) of Exercise 55 to show that for any acute angle θ,

$$\sin(90° - \theta) = \cos \theta.$$

This equation and the one in Exercise 55(c) are called *cofunction identities.*

 8.1   *ALTERNATE*   Trigonometric Functions of Angles

Section Objectives
- ■ Use right triangles to evaluate the trigonometric functions of acute angles.
- ■ Use the point-in-the-plane description to evaluate trigonometric functions of any angle.

NOTE

If you have read Chapter 6, omit this section. If you have not read Chapter 6, use this section in place of Section 8.1.

Before reading this section, it might be a good idea to read the Geometry Review Appendix, which presents the basic facts about angles and triangles that frequently are used here. In particular, recall that a **right triangle** is one that contains a **right angle,** that is, an angle of 90°. An **acute angle** is an angle whose measure is less than 90°. Consider the right triangles in Figure 8–18, each of which has an acute angle of θ degrees.

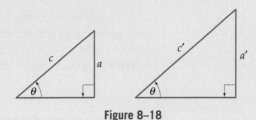

Figure 8–18

Since the sum of the angles of any triangle is 180°, we see that the third angle in each of these triangles has the same measure, namely, $180° - 90° - \theta$. Thus, both triangles have equal corresponding angles and, therefore, are similar. Consequently, by the Ratios Theorem of the Geometry Review Appendix, we know that the ratio of corresponding sides is the same, that is

$$\frac{a}{c} = \frac{a'}{c'}.$$

Each of these fractions is the ratio

$$\frac{\text{length of the side opposite angle } \theta}{\text{length of the hypotenuse}},$$

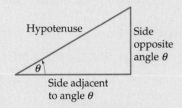

Figure 8–19

as indicated in Figure 8–19. Consequently, *this ratio depends only on the angle θ* and *not* on the size of the triangle. Similar remarks apply to the ratios of other sides of the triangle in Figure 8–19 and make it possible to define three new functions. For each function, the input is an acute angle θ, and the corresponding output is a ratio of sides in any right triangle containing angle θ, as summarized here.

Trigonometric Functions of Acute Angles

Name of Function	Abbre- viation	Rule of Function
sine	sin	$\sin \theta = \dfrac{\text{length of side opposite angle } \theta}{\text{length of hypotenuse}}$
cosine	cos	$\cos \theta = \dfrac{\text{length of side adjacent to angle } \theta}{\text{length of hypotenuse}}$
tangent	tan	$\tan \theta = \dfrac{\text{length of side opposite angle } \theta}{\text{length of side adjacent to angle } \theta}$

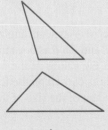

NOTE

Now turn to page 577 and begin reading at Example 2. When you have finished Example 9 on page 582, return here and continue reading below. [If you prefer, you can delay reading the material below until Section 8.3, where it will first be used.]

Figure 8–20

The preceding discussion applies only to right triangles. The next step is to learn how to solve other triangles, such as those in Figure 8–20. To do this, we must find a description of the trigonometric functions that is not limited to acute angles of a right triangle. First, we introduce some terminology.

An angle in the coordinate plane is said to be in **standard position** if its vertex is at the origin and one of its sides (which we call the **initial side**) is on the positive x-axis, as shown in Figure 8–21. The other side of the angle will be called its **terminal side.**

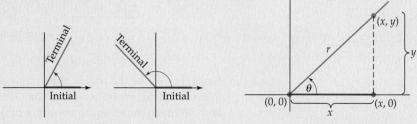

Figure 8–21   Figure 8–22

Let θ be an acute angle in standard position. Choose any point (x, y) on the terminal side of θ (except the origin) and consider the right triangle with vertices $(0, 0)$, $(x, 0)$, and (x, y) shown in Figure 8–22.

The legs of this triangle have lengths x and y, respectively. The distance formula shows that the hypotenuse has length

$$\sqrt{(x - 0)^2 + (y - 0)^2} = \sqrt{x^2 + y^2},$$

which we denote by r. The triangle shows that

$$\sin \theta = \frac{y}{r} \qquad \cos \theta = \frac{x}{r} \qquad \tan \theta = \frac{y}{x}.$$

We now have a description of the trigonometric functions in terms of the coordinate plane rather than right triangles. Furthermore, this description makes sense for *any* angle. Consequently, we make the following definition.

Point-in-the-Plane Description

Let θ be an angle in standard position and (x, y) any point (except the origin) on its terminal side. Then the trigonometric functions are defined by these rules:

$$\sin \theta = \frac{y}{r}, \qquad \cos \theta = \frac{x}{r}, \qquad \tan \theta = \frac{y}{x} \quad (x \neq 0),$$

where $r = \sqrt{x^2 + y^2}$ is the distance from (x, y) to the origin.

The discussion preceding the box shows that when θ is an acute angle, these definitions produce the same numbers for $\sin \theta$, $\cos \theta$, and $\tan \theta$ as does the right triangle definition on page 585. It can be shown that the values of the trigonometric functions of θ are independent of the point that is chosen on the terminal side (just as the definition for acute angles is independent of the size of the right triangle).

EXAMPLE 10

Find $\sin \theta$, $\cos \theta$, and $\tan \theta$ for the angle θ shown in Figure 8–23.

SOLUTION   Since $(-5, 7)$ is on the terminal side of θ, we apply the definitions in the preceding box with $(x, y) = (-5, 7)$ and

$$r = \sqrt{x^2 + y^2} = \sqrt{(-5)^2 + 7^2} = \sqrt{74}:$$

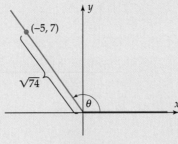

$$\sin \theta = \frac{y}{r} = \frac{7}{\sqrt{74}}$$

$$\cos \theta = \frac{x}{r} = \frac{-5}{\sqrt{74}}$$

$$\tan \theta = \frac{y}{x} = \frac{7}{-5} = -\frac{7}{5}.$$ ∎

Figure 8–23

For most angles, we use a calculator in degree mode to approximate the values of the trigonometric functions, as in Figure 8–24. But there are a few angles for which we can compute the exact values of the trigonometric functions.

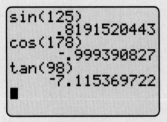

```
sin(125)
        .8191520443
cos(178)
        -.999390827
tan(98)
        -7.115369722
■
```

Figure 8–24

EXAMPLE 11

Find the exact values of the trigonometric functions at $\theta = 90°$.

SOLUTION   We use the point $(0, 1)$ on the terminal side of an angle of $90°$ in standard position (Figure 8–25). In this case, $r = \sqrt{x^2 + y^2} = \sqrt{0^2 + 1^2} = 1$, so

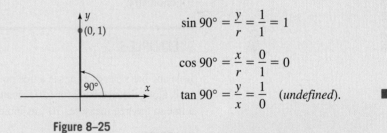

$$\sin 90° = \frac{y}{r} = \frac{1}{1} = 1$$

$$\cos 90° = \frac{x}{r} = \frac{0}{1} = 0$$

$$\tan 90° = \frac{y}{x} = \frac{1}{0} \quad (undefined).$$ ∎

Figure 8–25

The last part of Example 11 shows that the domain of the tangent function excludes $90°$, whereas the domains of sine and cosine include all angles.

EXAMPLE 12

Find the exact values of $\sin 135°$, $\cos 135°$, and $\tan 135°$.

SOLUTION   Construct an angle of $135°$ in standard position and let P be the point on the terminal side that is 1 unit from the origin (Figure 8–26 on the next page). Draw a vertical line from P to the x-axis, forming a right triangle with hypotenuse 1 and two angles of $45°$. Each side of this triangle has length $\sqrt{2}/2$, as explained in Example 2 of the Geometry Review Appendix. Therefore, the coordinates of P are $(-\sqrt{2}/2, \sqrt{2}/2)$, and we have

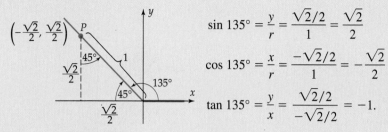

$$\sin 135° = \frac{y}{r} = \frac{\sqrt{2}/2}{1} = \frac{\sqrt{2}}{2}$$

$$\cos 135° = \frac{x}{r} = \frac{-\sqrt{2}/2}{1} = -\frac{\sqrt{2}}{2}$$

$$\tan 135° = \frac{y}{x} = \frac{\sqrt{2}/2}{-\sqrt{2}/2} = -1.$$

Figure 8–26

EXERCISES ALTERNATE 8.1

Use the exercises for Section 8.1 on page 582.

8.2   Applications of Right Triangle Trigonometry

Section Objective ■ Use right triangle trigonometry to solve applied problems.

The following examples illustrate a variety of practical applications of triangle trigonometry.

EXAMPLE 1

Lola and her sister Harper see a tree on the river's edge directly opposite them. They walk along the riverbank for 120 feet and note that the angle formed by their path and a line to the tree measures 70°, as indicated in Figure 8–27. How wide is the river?

SOLUTION   Using the right triangle whose one leg is the width w of the river and whose other leg is the 120-foot path on this side of the river, we see that

$$\tan 70° = \frac{\text{opposite}}{\text{adjacent}} = \frac{w}{120}.$$

Solving this equation for w, we have

$$w = 120 \tan 70° \approx 329.7.$$

So the river is about 330 feet wide.   ■

Figure 8–27

EXAMPLE 2

A plane takes off, making an angle of 18° with the ground. After the plane travels three miles along this flight path, how high (in feet) is it above the ground?

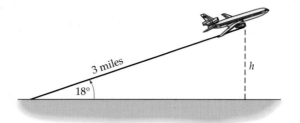

Figure 8–28

SOLUTION Figure 8–28 shows that

$$\sin 18° = \frac{\text{opposite}}{\text{hypotenuse}} = \frac{h}{3}$$

Multiply both sides by 3: $\qquad h = 3 \sin 18° \approx .92705 \text{ miles.}$

Since there are 5280 feet in a mile, the height of the plane in feet is

$$h = .92705 \cdot 5280 \approx 4894.8 \text{ feet.} \qquad \blacksquare$$

EXAMPLE 3

According to the safety sticker on an extension ladder, the distance from the foot of the ladder to the base of the wall on which it leans should be one-fourth of the length of the ladder. If the ladder is in this position, what angle does it make with the ground?

SOLUTION Let c be the distance from the foot of the ladder to the base of the wall. Then the ladder's length is $4c$, as shown in Figure 8–29. If θ is the angle the ladder makes with the ground, then

$$\cos \theta = \frac{\text{adjacent}}{\text{hypotenuse}} = \frac{c}{4c} = \frac{1}{4}.$$

Using the COS⁻¹ key, we find that $\theta \approx 75.52°$, as shown in Figure 8–30. $\qquad \blacksquare$

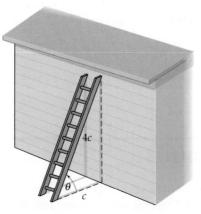

Figure 8–29

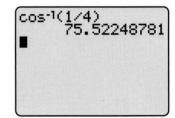

Figure 8–30

In many practical applications, one uses the angle between the horizontal and some other line (for instance, the line of sight from an observer to a distant object). This angle is called the **angle of elevation** or the **angle of depression,** depending on whether the line is above or below the horizontal, as shown in Figure 8–31.

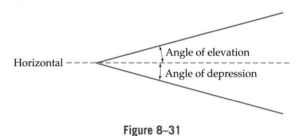

Figure 8–31

EXAMPLE 4

A surveyor stands on one edge of a ravine. By using the method in Example 1, she determines that the ravine is 125 feet wide. She then determines that the angle of depression from the edge where she is standing to a point on the bottom of the ravine is 57.5°, as shown in Figure 8–32 (which is not to scale). How deep is the ravine?

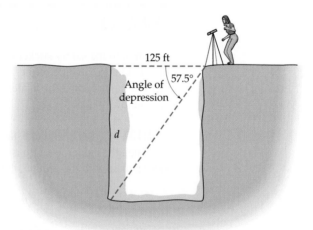

Figure 8–32

SOLUTION   From Figure 8–32, we see that

$$\tan 57.5° = \frac{\text{opposite}}{\text{adjacent}} = \frac{d}{125}$$

Multiply both sides by 125: $d = 125 \tan 57.5°$

$d \approx 196.21 \text{ feet}$ ■

EXAMPLE 5

A wire is to be stretched from the top of a 10-meter-high building to a point on the ground. From the top of the building, the angle of depression to the ground point is 22°. How long must the wire be?

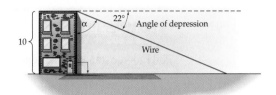

Figure 8–33

SOLUTION Figure 8–33 shows that the sum of the angle of depression and the angle α is 90°. Hence, α measures 90° − 22° = 68°. We know the length of the side of the triangle adjacent to the angle α and must find the hypotenuse w (the length of the wire). Using the cosine function, we see that

$$\cos 68° = \frac{\text{adjacent}}{\text{hypotenuse}} = \frac{10}{w}$$

$$w = \frac{10}{\cos 68°} \approx 26.7 \text{ meters.} \quad \blacksquare$$

EXAMPLE 6

A large American flag flies from a pole on top of the Terminal Tower in Cleveland (Figure 8–34). At a point 203 feet from the base of the tower, the angle of elevation to the bottom of the flag pole is 74°, and the angle of elevation to the top of the pole is 75.285°. To the nearest foot, how long is the flagpole?

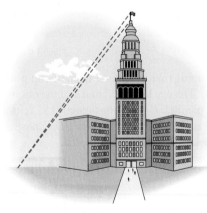

Figure 8–34

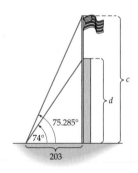

Figure 8–35

SOLUTION By abstracting the given information, we see that there are two right triangles, as shown in Figure 8–35 (which is not to scale). The length of the flagpole is $c - d$. We can use the two triangles to find c and d.

Larger Triangle

$$\frac{c}{203} = \frac{\text{opposite}}{\text{adjacent}} = \tan 75.285°$$

$$c = 203 \tan 75.285° \approx 773$$

Smaller Triangle

$$\frac{d}{203} = \frac{\text{opposite}}{\text{adjacent}} = \tan 74°$$

$$d = 203 \tan 74° \approx 708$$

As shown in Figure 8–36, the length of the flagpole is

$$c - d \approx 773 - 708 = 65 \text{ feet.} \quad \blacksquare$$

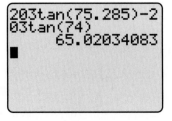

Figure 8–36

EXAMPLE 7

Phil Embree stands on the edge of one bank of a canal and observes a lamp post on the edge of the other bank of the canal. His eye level is 152 centimeters above the ground (approximately 5 feet). The angle of elevation from eye level to the top of the lamp post is 12°, and the angle of depression from eye level to the bottom of the lamp post is 7°, as shown in Figure 8–37. How wide is the canal? How high is the lamp post?

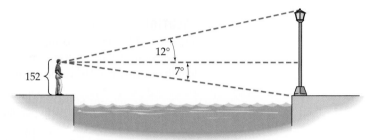

Figure 8–37

SOLUTION   Abstracting the essential information, we obtain the diagram in Figure 8–38.

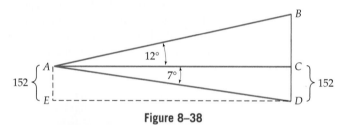

Figure 8–38

We must find the height of the lamp post *BD* and the width of the canal *AC* (or *ED*). The eye level height *AE* of the observer is 152 centimeters. Since *AC* and *ED* are parallel, *CD* also has length 152 centimeters. In right triangle *ACD*, we know the angle of 7° and the side *CD* opposite it. We must find the adjacent side *AC*. The tangent function is needed.

$$\tan 7° = \frac{\text{opposite}}{\text{adjacent}} = \frac{152}{AC} \quad \text{or, equivalently,} \quad AC = \frac{152}{\tan 7°}$$

$$AC = \frac{152}{\tan 7°} \approx 1237.94 \text{ centimeters.}$$

So the canal is approximately 12.3794 meters* wide (about 40.6 feet). Now using right triangle *ACB*, we see that

$$\tan 12° = \frac{\text{opposite}}{\text{adjacent}} = \frac{BC}{AC} \approx \frac{BC}{1237.94}$$

or, equivalently,

$$BC \approx 1237.94(\tan 12°) \approx 263.13 \text{ centimeters.}$$

Therefore, the height of the lamp post *BD* is *BC* + *CD* ≈ 263.13 + 152 = 415.13 centimeters or, equivalently, 4.1513 meters.   ■

---

*Remember, 100 centimeters = 1 meter.

EXERCISES 8.2

In Exercises 1–4, solve the right triangle.

1.

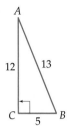

2.

3.

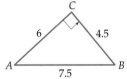

4.

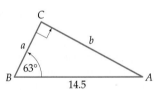

5. What is the width of the river in Example 1 if the angle to the tree is 40° (and all the other information is the same)?

6. Suppose the plane in Example 2 takes off at an angle of 5° and travels along this path for one mile. How high (in feet) is the plane above the ground?

7. Suppose you have a 24-foot-long ladder and you ignore the safety advice in Example 3 by placing the foot of the ladder 9 feet from the base of the wall. What angle does the ladder make with the ground?

8. The surveyor in Example 4 stands at the edge of another ravine, which is known to be 115 feet wide. She notes that the angle of depression from the edge she is standing on to the bottom of the oposite side is 64.3°. How deep is this ravine?

9. How long a wire is needed in Example 5 if the angle of depression is 25.8°?

10. Suppose that the flagpole on the Terminal Tower (Example 6) has been replaced. Now from a point 240 feet from the base of the tower, the angle of elevation to the bottom of the flagpole is 71.3°, and the angle of elevation to the top of the pole is 72.9°. To the nearest foot, how long is the new flagpole?

11. A 20-foot-long ladder leans on a wall of a building. The foot of the ladder makes an angle of 50° with the ground.

How far above the ground does the top of the ladder touch the wall?

12. A pilot flying at an altitude of 14,500 feet notes that his angle of depression to the control tower of a nearby airport is 15°. If the plane continues flying at this altitude toward the control tower, how far must it travel before it is directly over the tower?

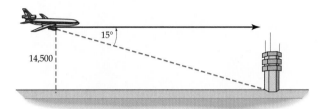

13. A straight road leads from an ocean beach into the nearby hills. The road has a constant upward grade of 3°. After taking this road for one mile, how high above sea level (in feet) are you?

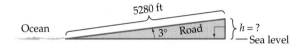

14. If you travel the road in Exercise 13 for a mile and a half, how high above sea level are you?

15. A powerful searchlight projects a beam of light vertically upward so that it shines on the bottom of a cloud. A clinometer, 600 feet from the searchlight, measures the angle θ, as shown in the figure. If θ measures 80°, how high is the cloud?

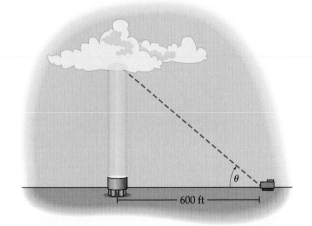

16. A wire from the top of a TV tower to the ground makes an angle of 49.5° with the ground and touches ground 225 feet from the base of the tower. How high is the tower?

17. Find the distance across the pond (from B to C) if AC is 110 feet and angle A measures $38°$.

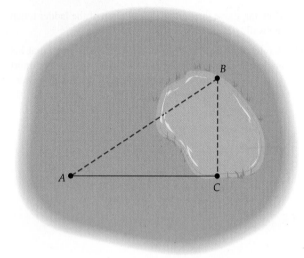

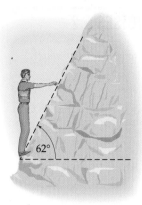

18. The Seattle Space Needle casts a 225-foot-long shadow. If the angle of elevation from the tip of the shadow to the top of the Space Needle is $69.6°$, how high is the Space Needle?

19. Batman is on the edge of a 200-foot-deep chasm and wants to jump to the other side. A tree on the edge of the chasm is directly across from him. He walks 20 feet to his right and notes that the angle to the tree is $54°$. His jet belt enables him to jump a maximum of 24 feet. How wide is the chasm, and is it safe for Batman to jump?

20. From the top of a 130-foot-high lighthouse, the angle of depression to a boat in Lake Erie is $2.5°$. How far is the boat from the lighthouse?

21. If you stand upright on a mountainside that makes a $62°$ angle with the horizontal and stretch your arm straight out at shoulder height, you may be able to touch the mountain (as shown in the figure). Can a person with an arm reach of 27 inches, whose shoulder is five feet above the ground, touch the mountain?

22. Alice is flying a kite. Her hand is three feet above ground level and is holding the end of a 300-foot-long kite string, which makes an angle of $57°$ with the horizontal. How high is the kite above the ground?

23. It is claimed that the Ohio Turnpike never has an uphill grade of more than $3°$. How long must a straight uphill segment of the road be to allow a vertical rise of 450 feet?

24. A swimming pool is three feet deep in the shallow end. The bottom of the pool has a steady downward drop of $12°$. If the pool is 50 feet long, how deep is it at the deep end?

25. Consider a 16-foot-long drawbridge on a medieval castle, as shown in the figure. The royal army is engaged in ignominious retreat. The king would like to raise the end of the drawbridge 8 feet off the ground so that Sir Rodney can jump onto the drawbridge and scramble into the castle while the enemy's cavalry are held at bay. Through how much of an angle must the drawbridge be raised for the end of it to be 8 feet off the ground?

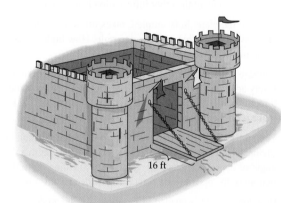

26. Through what angle must the drawbridge in Exercise 25 be raised in order that its end be directly above the center of the moat?

27. A buoy in the ocean is observed from the top of a 40-meter-high radar tower on shore. The angle of depression from the top of the tower to the base of the buoy is $6.5°$. How far is the buoy from the base of the radar tower?

28. A 150-foot-long ramp connects a ground-level parking lot with the entrance of a building. If the entrance is 8 feet above the ground, what angle does the ramp make with the ground?

29. A plane flies a straight course. On the ground directly below the flight path, observers two miles apart spot the plane at the same time. The plane's angle of elevation is 46° from one observation point and 71° from the other. How high is the plane?

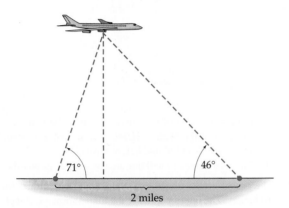

30. A man stands 20 feet from a statue. The angle of elevation from his eye level to the top of the statue is 30°, and the angle of depression to the base of the statue is 15°. How tall is the statue?

31. Two boats lie on a straight line with the base of a lighthouse. From the top of the lighthouse (21 meters above water level), it is observed that the angle of depression of the nearer boat is 53° and the angle of depression of the farther boat is 27°. How far apart are the boats?

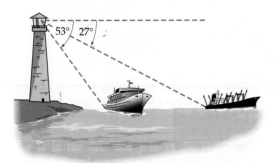

32. A rocket shoots straight up from the launchpad. Five seconds after liftoff, an observer two miles away notes that the rocket's angle of elevation is 3.5°. Four seconds later, the angle of elevation is 41°. How far did the rocket rise during those four seconds?

33. From a 35-meter-high window, the angle of depression to the top of a nearby streetlight is 55°. The angle of depression to the base of the streetlight is 57.8°. How high is the streetlight?

34. A plane takes off at an angle of 6° traveling at the rate of 200 feet/second. If it continues on this flight path at the same speed, how many minutes will it take to reach an altitude of 8000 feet?

35. A car on a straight road passes under a bridge. Two seconds later an observer on the bridge, 20 feet above the road, notes that the angle of depression to the car is 7.4°. How fast (in miles per hour) is the car traveling? [*Note:* 60 mph is equivalent to 88 feet/second.]

36. A plane passes directly over your head at an altitude of 500 feet. Two seconds later, you observe that its angle of elevation is 42°. How far did the plane travel during those two seconds?

37. Laura Bernett is 5 ft-4 inches tall. She stands 10 feet from a streetlight and casts a 4-foot-long shadow. How tall is the streetlight? What is angle θ?

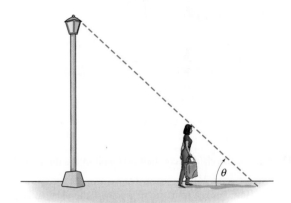

38. One plane flies straight east at an altitude of 31,000 feet. A second plane is flying west at an altitude of 14,000 feet on a course that lies directly below that of the first plane and directly above the straight road from Thomasville to Johnsburg. As the first plane passes over Thomasville, the second is passing over Johnsburg. At that instant, both planes spot a beacon next to the road between Thomasville to Johnsburg. The angle of depression from the first plane to the beacon is 61°, and the angle of depression from the second plane to the beacon is 34°. How far is Thomasville from Johnsburg?

39. A schematic diagram of a pedestrian overpass is shown in the figure. If you walk on the overpass from one end to the other, how far have you walked?

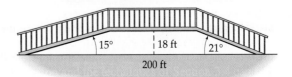

40. A 5-inch-high plastic beverage glass has a 2.5-inch-diameter base. Its sides slope outward at a 4° angle as shown. What is the diameter of the top of the glass?

41. In aerial navigation, directions are given in degrees clockwise from north. Thus, east is 90°, south is 180°, and so on, as shown in the figure. A plane travels from an airport for 200 miles in the direction 300°. How far west of the airport is the plane then?

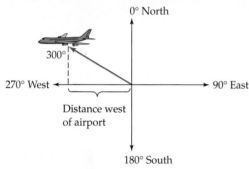

42. A plane travels at a constant 300 mph in the direction 65° (see Exercise 41).

(a) How far east of its starting point is the plane after half an hour?

(b) How far north of its starting point is the plane after 2 hours and 24 minutes?

43. A closed 60-foot-long drawbridge is 24 feet above water level. When open, the bridge makes an angle of 33° with the horizontal.

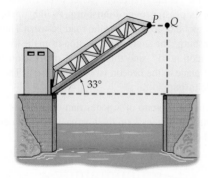

(a) How high is the tip P of the open bridge above the water?

(b) When the bridge is open, what is the distance from P to Q?

THINKERS

44. A gutter is to be made from a strip of metal 24 inches wide by bending up the sides to form a trapezoid.

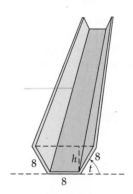

(a) Express the area of the cross section of the gutter as a function of the angle t. [*Hint:* The area of a trapezoid with bases b and b' and height h is $h(b + b')/2$.]

(b) For what value of t will this area be as large as possible?

45. The cross section of a tunnel is a semicircle with radius 10 meters. The interior walls of the tunnel form a rectangle.

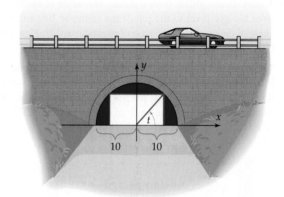

(a) Express the area of the rectangular cross section of the tunnel opening as a function of angle t.

(b) For what value of t is the cross-sectional area of the tunnel opening as large as possible? What are the dimensions of the tunnel opening in this case?

46. A spy plane on a practice run over the Midwest takes a picture that shows Cleveland, Ohio, on the eastern horizon and St. Louis, Missouri, 520 miles away, on the western horizon (the figure is not to scale). Assuming that the radius of the earth is 3950 miles, how high was the plane when the picture was taken? [*Hint:* The sight lines from the plane to the horizons are tangent to the earth, and a tangent line to a circle is perpendicular to the radius at that point. The arc of the earth between St. Louis and Cleveland is 520 miles long. Use this fact and the arc length formula to find angle θ (your answers will be in radians). Note that $\alpha = \theta/2$ (why?).]

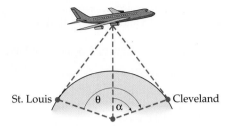

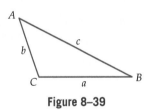

St. Louis θ α Cleveland

47. A 50-foot-high flagpole stands on top of a building. From a point on the ground, the angle of elevation of the top of the pole is 43°, and the angle of elevation of the bottom of the pole is 40°. How high is the building?

48. Two points on level ground are 500 meters apart. The angles of elevation from these points to the top of a nearby hill are 52° and 67°, respectively. The two points and the ground level point directly below the top of the hill lie on a straight line. How high is the hill?

8.3 The Law of Cosines

Section Objectives
- Use the Law of Cosines to solve oblique triangles.
- Use the Law of Cosines to solve applied problems.

Figure 8–39

We now consider the solution of *oblique* triangles (ones that don't contain a right angle). We shall use **standard notation** for triangles: Each vertex is labeled with a capital letter, and the length of the side opposite that vertex is denoted by the same letter in lower case, as shown in Figure 8–39. The letter A will also be used to label the *angle* at vertex A and similarly for B and C. So we shall make statements such as $A = 37°$ or $\cos B = .326$.

The first fact needed to solve oblique triangles is the Law of Cosines, whose proof is given at the end of this section.

Law of Cosines

In any triangle ABC, with sides of lengths a, b, c, as in Figure 8–39,

$$a^2 = b^2 + c^2 - 2bc \cos A$$
$$b^2 = a^2 + c^2 - 2ac \cos B$$
$$c^2 = a^2 + b^2 - 2ab \cos C$$

You need only memorize one of these equations since each of them provides essentially the same information: a description of one side of a triangle in terms of the angle opposite it and the other two sides.

NOTE

When C is a right angle, then c is the hypotenuse and

$$\cos C = \cos 90° = 0.$$

In this case, the third equation in the Law of Cosines becomes the Pythagorean Theorem:

$$c^2 = a^2 + b^2.$$

So the Pythagorean Theorem is a special case of the Law of Cosines.

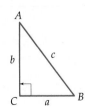

EXAMPLE 1

If the triangle in Figure 8–39 has $a = 7$, $c = 15$ and $B = 60°$, find b.

SOLUTION   Using the second equation in the Law of Cosines, we have

$$b^2 = a^2 + c^2 - 2ac \cos B$$
$$= 7^2 + 15^2 - 2 \cdot 7 \cdot 15 \cos 60°$$
$$= 49 + 225 - 210 \cos 60°$$
$$= 274 - 210 \cdot \frac{1}{2}$$
$$b^2 = 169$$

Hence, $b = \sqrt{169} = 13$. ∎

Sometimes it is more convenient to use the Law of Cosines in a slightly different form. To do this we solve the first equation in the Law of Cosines for $\cos A$.

$$a^2 = b^2 + c^2 - 2bc \cos A$$

Add $2bc \cos A$ to both sides: $2bc \cos A + a^2 = b^2 + c^2$

Subtract a^2 from both sides: $2bc \cos A = b^2 + c^2 - a^2$

Divide both sides by $2bc$: $\cos A = \dfrac{b^2 + c^2 - a^2}{2bc}$

So we have the following result.

Law of Cosines: Alternate Form

> In any triangle ABC, with sides of lengths a, b, c, as in Figure 8–39,
>
> $$\cos A = \frac{b^2 + c^2 - a^2}{2bc}.$$

The other two equations can be similarly rewritten. In this form, the Law of Cosines provides a description of each angle of a triangle in terms of the three sides. Consequently, the Law of Cosines can be used to solve triangles in these cases:

1. Two sides and the angle between them are known (SAS).
2. Three sides are known (SSS).

EXAMPLE 2

SAS   Solve triangle ABC in Figure 8–40.

SOLUTION   We have $a = 16$, $b = 10$, and $C = 110°$. The right side of the third equation in the Law of Cosines involves only these known quantities. Hence,

$$c^2 = a^2 + b^2 - 2ab \cos C$$
$$c^2 = 16^2 + 10^2 - 2 \cdot 16 \cdot 10 \cos 110°$$
$$c^2 \approx 256 + 100 - 320(-.342) \approx 465.4.^*$$

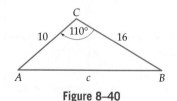

Figure 8–40

---

*Throughout this chapter, all decimals are printed in rounded-off form for reading convenience, but no rounding is done in the actual computation until the final answer is obtained.

Therefore, $c \approx \sqrt{465.4} \approx 21.6$. Now use the alternate form of the Law of Cosines.

$$\cos A = \frac{b^2 + c^2 - a^2}{2bc}$$

$$\approx \frac{10^2 + (21.6)^2 - 16^2}{2 \cdot 10 \cdot 21.6} \approx .7172.$$

A calculator (in degree mode) shows that $\cos^{-1}(.7172) \approx 44.2°$. So $A \approx 44.2°$ is an angle with cosine .7172. Hence,

$$B = 180° - A - C$$

$$\approx 180° - 44.2° - 110° = 25.8°.$$ ■

EXAMPLE 3

SSS   Find the angles of triangle ABC in Figure 8–41.

SOLUTION   In this case, $a = 20$, $b = 15$, and $c = 8.3$. By the alternate form of the Law of Cosines,

$$\cos A = \frac{b^2 + c^2 - a^2}{2bc}$$

$$= \frac{15^2 + 8.3^2 - 20^2}{2 \cdot 15 \cdot 8.3} = \frac{-106.11}{249} \approx -.4261.$$

The $\cos^{-1}$ key shows that $A \approx 115.2°$. Similarly, the alternate form of the Law of Cosines yields

$$\cos B = \frac{a^2 + c^2 - b^2}{2ac}$$

$$\cos B = \frac{20^2 + 8.3^2 - 15^2}{2 \cdot 20 \cdot 8.3} = \frac{243.89}{332} \approx .7346$$

$$B \approx 42.7°.$$

Therefore, $C \approx 180° - 115.2° - 42.7° = 22.1°.$ ■

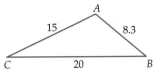

A
15 8.3
C 20 B

Figure 8–41

EXAMPLE 4

Two trains leave a station on different tracks. The tracks make an angle of 125° with the station as vertex. The first train travels at an average speed of 100 kilometers per hour, and the second travels at an average speed of 65 kilometers per hour. How far apart are the trains after 2 hours?

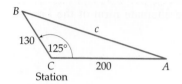

Figure 8–42

SOLUTION   The first train A traveling at 100 kilometers per hour for 2 hours goes a distance of $100 \times 2 = 200$ kilometers. The second train B travels a distance of $65 \times 2 = 130$ kilometers. So we have the situation shown in Figure 8–42.

By the Law of Cosines,

$$c^2 = a^2 + b^2 - 2ab \cos C$$
$$= 130^2 + 200^2 - 2 \cdot 130 \cdot 200 \cos 125°$$
$$= 56,900 - 52,000 \cos 125° \approx 86,725.97$$
$$c \approx \sqrt{86,725.97} = 294.5 \text{ kilometers.}$$

The trains are 294.5 kilometers apart after 2 hours.   ∎

EXAMPLE 5

A small powerboat leaves Chicago and sails 35 miles due east on Lake Michigan. It then changes course 59° northward, heading for Grand Haven, Michigan, as shown in Figure 8–43. After traveling 60 miles on this course, how far is the boat from Chicago?

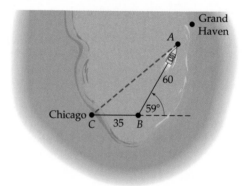

Figure 8–43

SOLUTION   Figure 8–43 shows that

$$B + 59° = 180°$$
$$B = 180° - 59° = 121°.$$

We must find the side of the triangle opposite angle B. By the Law of Cosines,

$$b^2 = a^2 + c^2 - 2ac \cos B$$
$$b^2 = 35^2 + 60^2 - 2 \cdot 35 \cdot 60 \cos 121°$$
$$b^2 \approx 6988.1599$$
$$b \approx \sqrt{6988.1599} \approx 83.5952.$$

So the boat is about 83.6 miles from Chicago.   ∎

EXAMPLE 6

A sculpture is being placed in front of a new office building. The sculpture consists of two steel beams of lengths 10 and 12 feet, respectively, and a 15.2-foot cable, as shown in Figure 8–44. If the 10-foot beam makes an angle of 50° with the ground, what angle does the 12-foot beam make with the ground?

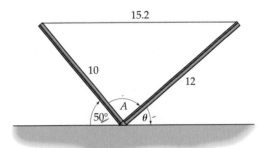

15.2

10

12

A

50°

θ

Figure 8–44

SOLUTION   We must find the measure of angle θ. As you can see in Figure 8–44,

$$50° + A + θ = 180°.$$

We first find the measure of angle A and then solve this equation for $θ$. The triangle formed by the beams and cable has sides of lengths 10, 12, and 15.2, with angle A opposite the 15.2-foot side. By the alternate form of the Law of Cosines,

$$\cos A = \frac{10^2 + 12^2 - 15.2^2}{2 \cdot 10 \cdot 12} = .054.$$

Hence, the measure of angle A is about 86.9° (Figure 8–45). Therefore,

$$50° + A + θ = 180°$$
$$50° + 86.9° + θ = 180°$$
$$θ = 180° - 50° - 86.9° = 43.1°. \qquad ■$$

cos⁻¹(.054)
           86.90452226

Figure 8–45

EXAMPLE 7

A 100-foot-tall antenna tower is to be placed on a hillside that makes an angle of 12° with the horizontal. It is to be anchored by two cables from the top of the tower to points 85 feet uphill and 95 feet downhill from the base. How much cable is needed?

SOLUTION   The situation is shown in Figure 8–46 on the next page.

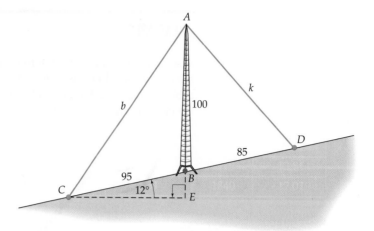

Figure 8–46

In triangle BEC, angle E is a right angle and by hypothesis, angle C measures $12°$. Since the sum of the angles of a triangle is $180°$, we must have

$$\angle CBE = 180° - (90° + 12°) = 78°.$$

As shown in the figure, the sum of angles CBE and CBA is a straight angle ($180°$). Hence,

$$\angle CBA = 180° - 78° = 102°.$$

Apply the Law of Cosines to triangle ABC.

$$b^2 = a^2 + c^2 - 2ac \cos B$$

$$b^2 = 95^2 + 100^2 - 2 \cdot 95 \cdot 100 \cos 102°$$

$$= 9025 + 10{,}000 - 19{,}000 \cos 102°$$

$$\approx 22{,}975.32.$$

Therefore, the length of the downhill cable is $b \approx \sqrt{22{,}975.32} \approx 151.58$ feet.

To find the length of the uphill cable, note that the sum of angles CBA and DBA is a straight angle, so

$$\angle DBA = 180° - \angle CBA = 180° - 102° = 78°.$$

Applying the Law of Cosines to triangle DBA, we have

$$k^2 = 85^2 + 100^2 - 2 \cdot 85 \cdot 100 \cos 78°$$

$$= 7225 + 10{,}000 - 17{,}000 \cos 78° \approx 13{,}690.50.$$

Hence, the length of the uphill cable is $k = \sqrt{13{,}690.50} \approx 117.01$ feet. ■

PROOF OF THE LAW OF COSINES

Given triangle ABC, position it on a coordinate plane so that angle A is in standard position with initial side c and terminal side b. Depending on the size of angle A, there are two possibilities, as shown in Figure 8–47.

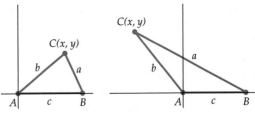

Figure 8–47

The coordinates of B are $(c, 0)$. Let (x, y) be the coordinates of C. Now C is a point on the terminal side of angle A, and the distance from C to the origin A is obviously b. Therefore, according to the point-in-the-plane description of sine and cosine, we have

$$\frac{x}{b} = \cos A \qquad \text{or, equivalently,} \qquad x = b \cos A,$$

$$\frac{y}{b} = \sin A \qquad \text{or, equivalently,} \qquad y = b \sin A.$$

Using the distance formula on the coordinates of B and C, we have

$$a = \text{distance from } C \text{ to } B$$

$$= \sqrt{(x - c)^2 + (y - 0)^2} = \sqrt{(b \cos A - c)^2 + (b \sin A - 0)^2}.$$

Squaring both sides of this last equation and simplifying, using the Pythagorean identity, yields

$$a^2 = (b \cos A - c)^2 + (b \sin A)^2$$

$$a^2 = b^2 \cos^2 A - 2bc \cos A + c^2 + b^2 \sin^2 A$$

$$a^2 = b^2(\sin^2 A + \cos^2 A) + c^2 - 2bc \cos A$$

$$a^2 = b^2 + c^2 - 2bc \cos A.$$

This proves the first equation in the Law of Cosines. Similar arguments beginning with angle B or C in standard position prove the other two equations.

EXERCISES 8.3

Directions: Standard notation for triangle ABC is used throughout. Use a calculator and round off your answers to one decimal place at the end of the computation.

In Exercises 1–16, solve the triangle ABC under the given conditions.

1. $A = 40°, b = 10, c = 7$

2. $B = 40°, a = 12, c = 20$

3. $C = 118°, a = 6, b = 12$

4. $C = 52.5°, a = 6.5, b = 9$

5. $A = 140°, b = 12, c = 14$

6. $B = 25.4°, a = 6.8, c = 10.5$

7. $C = 78.6°, a = 12.1, b = 20.3$

8. $A = 118.2°, b = 16.5, c = 10.7$

9. $a = 7, b = 3, c = 5$

10. $a = 8, b = 5, c = 10$

11. $a = 16, b = 30, c = 32$

12. $a = 5.3, b = 7.2, c = 10$

13. $a = 7.2, b = 6.5, c = 11$

14. $a = 6.8, b = 12.4, c = 15.1$

15. $a = 12, b = 16.5, c = 20.6$

16. $a = 5.7, b = 20.4, c = 16.8$

17. Find the angles of the triangle whose vertices are $(0, 0)$, $(5, -2)$, $(1, -4)$.

18. Find the angles of the triangle whose vertices are $(-3, 4)$, $(6, 1)$, $(2, -1)$.

19. In Example 4, suppose that the angle between the two tracks is $112°$ and that the average speeds are 90 kilometers per

hour for the first train and 55 kilometers per hour for the second train. How far apart are the trains after two hours and 45 minutes?

20. Suppose that the boat in Example 5 goes 25 miles due east and then changes course 56° northward. After traveling 50 miles on this course, how far is the boat from Chicago?

21. The sculptor builds a smaller version of the sculpture in Example 6, in which the beams are six feet and nine feet long, respectively, and the cable is 10.4 feet long. If the six-foot beam makes an angle of 40° with the ground, what angle does the nine-foot beam make with the ground?

22. Suppose that the tower in Example 7 is 175 feet high and that the cable on the downhill side is 120 feet from the base of the tower. How long is that cable?

23. The pitcher's mound on a standard baseball diamond (which is actually a square) is 60.5 feet from home plate (see the figure). How far is the pitcher's mound from first base?

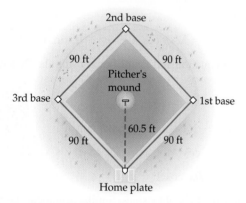

24. At Wrigley Field in Chicago, the straight-line distance from home plate over second base to the center field wall is 400 feet. How far is it from first base to the same point at the center field wall? [*Hint:* Adapt and extend the figure from Exercise 23.]

25. A stake is located 10.8 feet from the end of a closed gate that is 8 feet long. The gate swings open, and its end hits the stake. Through what angle did the gate swing?

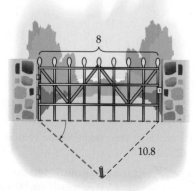

26. The distance from Chicago to St. Louis is 440 kilometers, that from St. Louis to Atlanta 795 kilometers, and that from Atlanta to Chicago 950 kilometers. What are the angles in the triangle with these three cities as vertices?

27. A satellite is placed in an orbit such that the satellite remains stationary 24,000 miles over a fixed point on the surface of the earth. The angle *CES*, where *C* is Cape Canaveral, *E* is the center of the earth, and *S* is the satellite, measures 60°. Assuming that the radius of the earth is 3960 miles, how far is the satellite from Cape Canaveral?

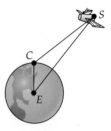

28. One plane flies west from Cleveland at 350 mph. A second plane leaves Cleveland at the same time and flies southeast at 200 mph. How far apart are the planes after 1 hour and 36 minutes?

29. A weight is hung by two cables from a beam. What angles do the cables make with the beam?

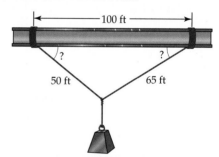

30. Two ships leave port, one traveling in a straight course at 22 mph and the other traveling a straight course at 31 mph. Their courses diverge by 38°. How far apart are they after 3 hours?

31. A boat runs in a straight line for 3 kilometers, then makes a 45° turn and goes for another 6 kilometers (see the figure). How far is the boat from its starting point?

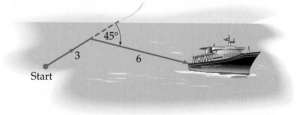

32. A plane flies in a straight line at 400 mph for 1 hour and 12 minutes. It makes a 15° turn and flies at 375 mph for 2 hours and 27 minutes. How far is it from its starting point?

33. A surveyor wants to measure the width *CD* of a sinkhole. So he places a stake *B* and determines the measurements shown in the figure. How wide is the sinkhole?

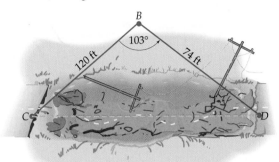

34. A straight tunnel is to be dug through a hill. Two people stand on opposite sides of the hill where the tunnel entrances are to be located. Both can see a stake located 530 meters from the first person and 755 meters from the second. The angle determined by the two people and the stake (vertex) is 77°. How long must the tunnel be?

35. A 400-foot-high tower stands on level ground, anchored by two cables on the west side. The end of the cable closest to the tower makes an angle of 70° with the horizontal. The two cable ends are 100 feet apart, as shown in the figure. How long are the cables?

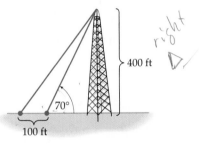

36. One diagonal of a parallelogram is 6 centimeters long, and the other is 13 centimeters long. They form an angle of 42° with each other. How long are the sides of the parallelogram? [*Hint:* The diagonals of a parallelogram bisect each other.]

37. A ship is traveling at 18 mph from Corsica to Barcelona, a distance of 350 miles. To avoid bad weather, the ship leaves Corsica on a route 22° south of the direct route (see the figure). After 7 hours, the bad weather has been bypassed. Through what angle should the ship now turn to head directly to Barcelona?

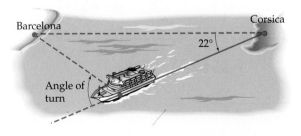

38. A plane leaves South Bend for Buffalo, 400 miles away, intending to fly a straight course in the direction 70° (aerial navigation is explained in Exercise 41 of Section 8.2). After flying 180 miles, the pilot realizes that an error has been made and that he has actually been flying in the direction 55°.

(a) At that time, how far is the plane from Buffalo?
(b) In what direction should the plane now go to reach Buffalo?

39. Assume that the earth is a sphere of radius 3960 miles. A satellite travels in a circular orbit around the earth, 900 miles above the equator, making one full orbit every 6 hours. If it passes directly over a tracking station at 2 P.M., what is the distance from the satellite to the tracking station at 2:05 P.M.?

40. A surveyor has determined the distance and angles in the figure. He wants you to find the straight-line distance from *A* to *B*. Do so.

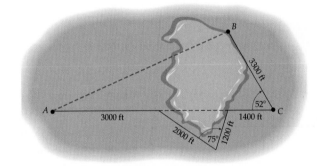

41. A parallelogram has diagonals of lengths 12 and 15 inches that intersect at an angle of 63.7°. How long are the sides of the parallelogram? [See the hint for Exercise 36.]

42. Two planes at the same altitude approach an airport. One plane is 16 miles from the control tower and the other is 22 miles from the tower. The angle determined by the planes and the tower, with the tower as vertex, is 11°. How far apart are the planes?

43. Assuming that the circles in the figure are mutually tangent, find the lengths of the sides and the measures of the angles in triangle *ABC*.

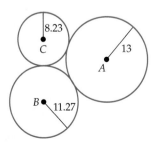

44. Assuming that the circles in the figure are mutually tangent, find the lengths of the sides and the measures of the angles in triangle *ABC*.

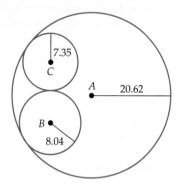

THINKERS

45. A rope is attached at points *A* and *B* and taut around a pulley whose center is at *C*, as shown in the figure (in which *AC*

has length 8 and *BC* length 7). The rope lies on the pulley from *D* to *E* and the radius of the pulley is 1 meter. How long is the rope?

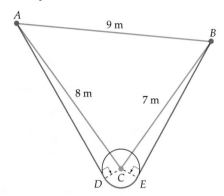

46. Use the Law of Cosines to prove that the sum of the squares of the lengths of the two diagonals of a parallelogram equals the sum of the squares of the lengths of the four sides.

8.4   The Law of Sines

Section Objectives
- Use the Law of Sines to solve oblique triangles.
- Use the Law of Sines to solve applied problems.

To solve oblique triangles in cases in which the Law of Cosines cannot be used, we need this fact.

AAS
SSA

cant us
path. th. on
non right △

Law of Sines	In any triangle *ABC* (in standard notation), $$\frac{\sin A}{a} = \frac{\sin B}{b} = \frac{\sin C}{c}.^*$$

Proof   Position triangle *ABC* on a coordinate plane so that angle *C* is in standard position, with initial side *b* and terminal side *a*, as shown in Figure 8–48.

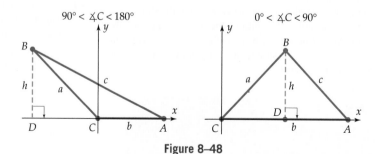

Figure 8–48

*An equality of the form $u = v = w$ is shorthand for the statement $u = v$ and $v = w$ and $w = u$.

In each case, we can compute sin C by using the point B on the terminal side of angle C. The second coordinate of B is h, and the distance from B to the origin is a. Therefore, by the point-in-the-plane description of sine,

$$\sin C = \frac{h}{a} \qquad \text{or, equivalently,} \qquad h = a \sin C.$$

In each case, right triangle ADB shows that

$$\sin A = \frac{\text{opposite}}{\text{hypotenuse}} = \frac{h}{c} \qquad \text{or, equivalently,} \qquad h = c \sin A.$$

Combining this with the fact that $h = a \sin C$, we have

$$c \sin A = a \sin C.$$

Dividing both sides of the last equation by ac yields

$$\frac{\sin A}{a} = \frac{\sin C}{c}.$$

This proves one equation in the Law of Sines. Similar arguments beginning with angles A or B in standard position prove the other equations. ■

The Law of Sines can be used to solve triangles in these cases:

1. Two angles and one side are known (AAS).

2. Two sides and the angle opposite one of them are known (SSA).

EXAMPLE 1

AAS   If $B = 20°$, $C = 31°$, and $b = 210$ in Figure 8–49, find the other angles and sides.

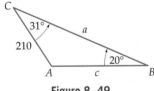

Figure 8–49

SOLUTION   Since the sum of the angles of a triangle is $180°$,

$$A = 180° - (20° + 31°) = 180° - 51° = 129°.$$

To find side c, we observe that we know three of the four quantities in one of the equations given by the Law of Sines.

$$\frac{\sin B}{b} = \frac{\sin C}{c}$$

Substitute known quantities: $\dfrac{\sin 20°}{210} = \dfrac{\sin 31°}{c}$

Multiply both sides by $210c$: $c \sin 20° = 210 \sin 31°$

Divide both sides by $\sin 20°$: $c = \dfrac{210 \sin 31°}{\sin 20°} \approx 316.2.$

Side a is found similarly. Beginning with an equation of the Law of Sines involving a and three known quantities, we have:

$$\frac{\sin B}{b} = \frac{\sin A}{a}$$

Substitute known quantities: $\dfrac{\sin 20°}{210} = \dfrac{\sin 129°}{a}$

Multiply both sides by $210a$: $a \sin 20° = 210 \sin 129°$

Divide both sides by $\sin 20°$: $a = \dfrac{210 \sin 129°}{\sin 20°} \approx 477.2.$ ■

THE AMBIGUOUS CASE (SSA)

In the AAS case, there is exactly one triangle that satisfies the given data.* But when two sides of a triangle and the angle opposite one of them are known (SSA), there may be one, two, or no triangles that satisfy the given data. Figure 8–50 shows some of the possibilities when sides a and b and angle A are given.

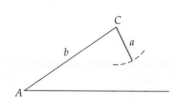

No Solution

(side a is too short)

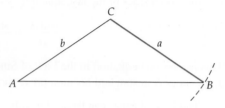

One Solution

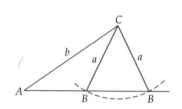

Two Solutions

Figure 8–50

Determining the situation geometrically may require careful measurement and drawing. So it will be easier to use an analytic approach for solving SSA triangles, as illustrated in the next four examples.

EXAMPLE 2

SSA   Solve the triangle ABC when $A = 65°$, $a = 6$, and $b = 7$.

SOLUTION   To find angle B, we use an equation from the Law of Sines that involves B and three known quantities.

$$\frac{\sin B}{b} = \frac{\sin A}{a}$$

Substitute given values:
$$\frac{\sin B}{7} = \frac{\sin 65°}{6}$$

Multiply both sides by $6 \cdot 7$:
$$6 \sin B = 7 \sin 65°$$

Divide both sides by 6:
$$\sin B = \frac{7 \sin 65°}{6} \approx 1.06$$

There is no angle B whose sine is greater than 1. Therefore, there is no triangle satisfying the given data.   ■

---

*Once you know two angles, you know all three (their sum must be 180°). Hence, you know two angles and the included side. Any two triangles satisfying these conditions will be congruent by the ASA Theorem of plane geometry.

When there is no solution for an SSA problem, that fact will become apparent as it did in Example 2, with an impossible value for sine. In other cases, you should use the following identity to determine whether there are one or two solutions, as illustrated in Examples 3 and 4.*

Supplementary Angle Identity

If $0° \leq \theta \leq 90°$, then

$$\sin \theta = \sin(180° - \theta).$$

EXAMPLE 3

SSA   Solve triangle ABC when $B = 50°$, $b = 12$ and $c = 11$.

SOLUTION   A rough picture of the situation is in Figure 8–51. We must find angles A and C and side a. We begin with an equation from the Law of Sines that involves the three known quantities.

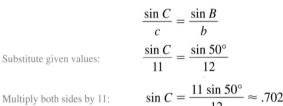

Figure 8–51

$$\frac{\sin C}{c} = \frac{\sin B}{b}$$

Substitute given values: $\dfrac{\sin C}{11} = \dfrac{\sin 50°}{12}$

Multiply both sides by 11: $\sin C = \dfrac{11 \sin 50°}{12} \approx .7022$

A calculator shows that one possibility for C is $\sin^{-1}(.7022) \approx 44.6°$. According to the supplementary angle identity,

$$\sin(180° - 44.6°) = \sin 44.6° = .7022.$$

So another possibility is $C = 180° - 44.6° = 135.4°$. If $C = 135.4°$, however, then $B + C = 50° + 135.4° = 185.4°$, which is impossible in a triangle. So the only solution here is $C = 44.6°$. Consequently,

$$A = 180° - B - C = 180° - 50° - 44.6° = 85.4°.$$

Finally, we use the Law of Sines to find a.

$$\frac{\sin B}{b} = \frac{\sin A}{a}$$

Substitute known values: $\dfrac{\sin 50°}{12} = \dfrac{\sin 85.4°}{a}$

Multiply both sides by $12a$: $a \sin 50° = 12 \sin 85.4°$

Divide both sides by $\sin 50°$: $a = \dfrac{12 \sin 85.4°}{\sin 50°} \approx 15.6$ ∎

*The identity was proved for all angles in Example 2 of Section 7.2. An alternate proof that does not depend on Chapter 7 is in Exercise 50.

EXAMPLE 4

SSA   Solve triangle ABC when $a = 7.5$, $b = 12$, and $A = 35°$.

SOLUTION   The Law of Sines shows that

$$\frac{\sin B}{b} = \frac{\sin A}{a}$$

Substitute given values: $$\frac{\sin B}{12} = \frac{\sin 35°}{7.5}$$

Multiply both sides by 12: $$\sin B = \frac{12 \sin 35°}{7.5} \approx .9177$$

The SIN^{-1} key shows that $66.6°$ is a solution of $\sin B = .9177$. Therefore, $180° - 66.6° = 113.4°$ is also a solution of $\sin B = .9177$ by the supplementary angle identity. In each case the sum of angles A and B is less than $180°$:

Case 1. $A + B = 35° + 66.6° = 101.6°$

Case 2. $A + B = 35° + 113.4° = 148.4°.$

So there are two triangles ABC satisfying the given data, as shown in Figure 8–52.

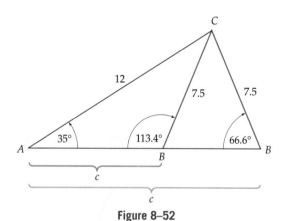

Figure 8–52

Case 1. $A = 35°$ and $B = 66.6°$. Then

$$C = 180° - A - B = 180° - 35° - 66.6° = 78.4°.$$

By the Law of Sines,

$$\frac{\sin A}{a} = \frac{\sin C}{c}$$

Substitute known values: $$\frac{\sin 35°}{7.5} = \frac{\sin 78.4°}{c}$$

Multiply both sides by 7.5c: $$c \sin 35° = 7.5 \sin 78.4°$$

Divide both sides by sin 35°: $$c = \frac{7.5 \sin 78.4°}{\sin 35°} \approx 12.8.$$

Case 2. $A = 35°$ and $B = 113.4°$. Then

$$C = 180° - A - B = 180° - 35° - 113.4° = 31.6°.$$

In Exercises 9 and 10, find the area of the triangle with the given vertices.

9. $(0, 0), (2, -5), (-3, 1)$ **10.** $(-4, 2), (5, 7), (3, 0)$

In Exercises 11 and 12, find the area of the polygonal region. [Hint: Divide the region into triangles.]

11.

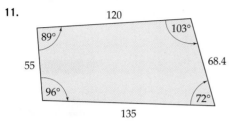

12.

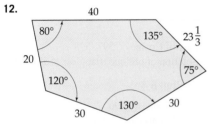

13. A triangular banner is hung from a window along the side of a building. The edges that touch the window are 20 and 24 feet long respectively. The third side is parallel to the ground. The angle between the 20-foot side and the third side is 44°. What is the area of the banner?

14. A triangular lot has sides of length 120 feet and 160 feet. The angle between these sides is 42°. Adjacent to this lot is a rectangular lot whose longest side has length 200 feet and whose shortest side is the same length as the shortest side of the triangular lot. What is the total area of both lots?

15. If a gallon of paint covers 400 square feet, how many gallons are needed to paint a triangular deck with sides of lengths 65 feet, 72 feet, and 88 feet?

16. Find the volume of the prism in the figure. The volume is given by the formula $V = \frac{1}{3} Bh$, where B is the area of the base and h is the height.

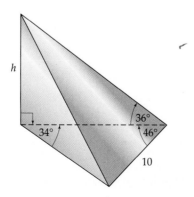

17. A rigid plastic triangle ABC rests on three vertical rods, as shown in the figure. What is its area?

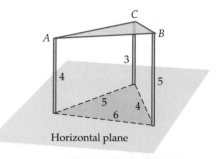

18. Prove that the area of triangle ABC (standard notation) is given by $\frac{a^2 \sin B \sin C}{2 \sin A}$.

19. A *regular polygon* has n equal sides and n equal angles formed by the sides. For example, a regular polygon of three sides is an equilateral triangle and a regular polygon of 4 sides is a square. If a regular polygon of n sides is inscribed in a circle of radius r, then Exercise 20 shows that its area is

$$\frac{1}{2}nr^2 \sin\left(\frac{360}{n}\right)^°.$$

Find the area of

(a) A square inscribed in a circle of radius 5.
(b) A regular hexagon (6 sides) inscribed in a circle of radius 2.

20. Prove the area formula in Exercise 19 as follows.

(a) Draw lines from each vertex of the polygon to the center of the circle, as shown in the figure for $n = 6$.

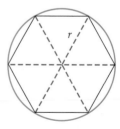

(b) Explain why all the triangles are congruent and hence, all the angles at the center of the circle have equal measure. [*Hint:* SSS.]
(c) Explain why the measure of each center angle is $360°/n$.
(d) Find the area of each triangle. [*Hint:* See the first box on page 617.]
(e) The area of the polygon is the sum of the areas of these n triangles. Use this fact and part (d) to obtain the area formula in Exercise 19.

21. What is the area of a triangle whose sides have lengths 12, 20, and 36? (If your answer turns out strangely, try drawing a picture.)

22. Complete the proof of Heron's Formula as follows. Let $s = \frac{1}{2}(a + b + c)$.

(a) Show that

$$\frac{1}{2}ab(1 + \cos C) = \frac{(a + b)^2 - c^2}{4}$$

$$= \frac{(a + b) + c}{2} \cdot \frac{(a + b) - c}{2}$$

$$= s(s - c).$$

[*Hint:* Use the Law of Cosines to express $\cos C$ in terms of a, b, c; then simplify.]

(b) Show that

$$\frac{1}{2}ab(1 - \cos C) = \frac{c^2 - (a - b)^2}{4}$$

$$= \frac{c - (a - b)}{2} \cdot \frac{c + (a - b)}{2}$$

$$= (s - a)(s - b).$$

Chapter 8 Review

IMPORTANT CONCEPTS

IMPORTANT FACTS & FORMULAS

■ *Right Triangle Description:* In a right triangle containing an angle θ,

$$\sin \theta = \frac{\text{opposite}}{\text{hypotenuse}} \qquad \cos \theta = \frac{\text{adjacent}}{\text{hypotenuse}}$$

$$\tan \theta = \frac{\text{opposite}}{\text{adjacent}} \qquad \cot \theta = \frac{\text{adjacent}}{\text{opposite}}$$

$$\sec \theta = \frac{\text{hypotenuse}}{\text{adjacent}} \qquad \csc \theta = \frac{\text{hypotenuse}}{\text{opposite}}$$

■ *Law of Cosines:* $a^2 = b^2 + c^2 - 2bc \cos A$

■ *Law of Cosines—Alternate Form:* $\cos A = \dfrac{b^2 + c^2 - a^2}{2bc}$

■ *Law of Sines:* $\dfrac{\sin A}{a} = \dfrac{\sin B}{b} = \dfrac{\sin C}{c}$

■ $\sin D = \sin(180° - D)$

■ Area of triangle $ABC = \dfrac{ab \sin C}{2}$

■ *Heron's Formula:*

$$\text{Area of triangle } ABC = \sqrt{s(s - a)(s - b)(s - c)}$$

where $s = \frac{1}{2}(a + b + c)$.

REVIEW QUESTIONS

Note: *Standard notation is used for triangles.*

In Questions 1 and 2, find sin θ, cos θ, and tan θ.

1.

2.

3. Which of the following statements about the angle θ is true?

   (a) sin θ = 3/4
   (b) cos θ = 5/4
   (c) tan θ = 3/5
   (d) sin θ = 4/5
   (e) sin θ = 4/3

   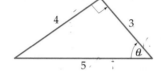

4. Suppose θ is a real number. Consider the right triangle with sides as shown in the figure. Then

   (a) x = 1
   (b) x = 2
   (c) x = 4
   (d) x = 2 (cos θ + sin θ)
   (e) none of the above

   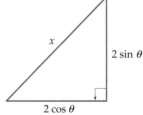

5. Use the right triangle in the figure to find sec θ.

   (a) sec θ = 7/4
   (b) sec θ = 4/√65
   (c) sec θ = 7/√65
   (d) sec θ = √65/7
   (e) sec θ = √65/4

   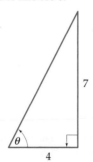

6. Find the length of side h in the triangle, where angle A measures 40° and the distance from C to A is 25.

   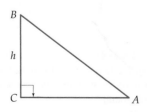

In Questions 7–10, angle B is a right angle. Solve triangle ABC.

7. a = 10, c = 14

8. A = 40°, b = 10

9. C = 35°, a = 12

10. A = 56°, a = 11

11. From a point on level ground 145 feet from the base of a tower, the angle of elevation to the top of the tower is 57.3°. How high is the tower?

12. A pilot in a plane at an altitude of 22,000 feet observes that the angle of depression to a nearby airport is 26°. How many *miles* is the airport from a point on the ground directly below the plane?

13. A road rises 140 feet per horizontal mile. What angle does the road make with the horizontal?

14. A lighthouse keeper 100 feet above the water sees a boat sailing in a straight line directly toward her. As she watches, the angle of depression to the boat changes from 25° to 40°. How far has the boat traveled during this time?

In Questions 15–18, use the Law of Cosines to solve triangle ABC.

15. a = 12, b = 8, c = 14

16. a = 7.5, b = 3.2, c = 6.4

17. a = 10, c = 14, B = 115°

18. a = 7, b = 8.6, C = 72.4°

19. Two trains depart simultaneously from the same station. The angle between the two tracks on which they leave is 120°. One train travels at an average speed of 45 mph and the other at 70 mph. How far apart are the trains after 3 hours?

20. A 40-foot-high flagpole sits on the side of a hill. The hillside makes a 17° angle with the horizontal. How long is a wire that runs from the top of the pole to a point 72 feet downhill from the base of the pole?

In Questions 21–26, use the Law of Sines to solve triangle ABC.

21. B = 124°, C = 40°, c = 3.5

22. A = 96°, B = 44°, b = 12

23. a = 75, c = 95, C = 62°

24. a = 5, c = 2.5, C = 30°

25. a = 3.5, b = 4, A = 60°

26. a = 3.8, c = 2.8, C = 41°

27. Find the area of triangle ABC if b = 24, c = 15, and A = 55°.

28. Find the area of triangle ABC if a = 10, c = 14, and B = 75°.

29. A boat travels for 8 kilometers in a straight line from the dock. It is then sighted from a lighthouse that is 6.5 kilometers from the dock. The angle determined by the dock, the lighthouse (vertex), and the boat is 25°. How far is the boat from the lighthouse?

30. A pole tilts 12° from the vertical, away from the sun, and casts a 34-foot-long shadow on level ground. The angle of elevation from the end of the shadow to the top of the pole is 64°. How long is the pole?

In Questions 31–34, solve triangle ABC.

31. $A = 48°, B = 75°, b = 50$

32. $A = 67°, c = 125, a = 100$

33. $a = 12, c = 6, B = 76°$

34. $a = 90, b = 70, c = 40$

35. Two surveyors, Joe and Alice, are 240 meters apart on a riverbank. Each sights a flagpole on the opposite bank. The angle from the pole to Joe (vertex) to Alice is 63°. The angle from the pole to Alice (vertex) to Joe is 54°. How far are Joe and Alice from the pole?

36. A surveyor stakes out points A and B on opposite sides of a building. Point C on the side of the building is 300 feet from A and 440 feet from B. Angle ACB measures 38°. What is the distance from A to B?

37. A woman on the top of a 448-foot-high building spots a small plane. As she views the plane, its angle of elevation is 62°. At the same instant, a man at the ground-level entrance to the building sees the plane and notes that its angle of elevation is 65°.

   (a) How far is the woman from the plane?
   (b) How far is the man from the plane?
   (c) How high is the plane?

38. A straight road slopes at an angle of 10° with the horizontal. When the angle of elevation of the sun (from horizontal) is 62.5°, a telephone pole at the side of the road casts a 15-foot shadow downhill, parallel to the road. How high is the telephone pole?

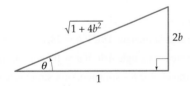

39. Find angle *ABC*.

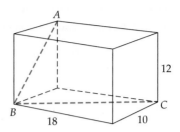

40. Use the Law of Sines to prove Engelsohn's equations: For any triangle *ABC* (standard notation),

$$\frac{a + b}{c} = \frac{\sin A + \sin B}{\sin C}$$

and

$$\frac{a - b}{c} = \frac{\sin A - \sin B}{\sin C}.$$

In Questions 41–44, find the area of triangle ABC under the given conditions.

41. There is an angle of 30°, the sides of which have lengths 5 and 8.

42. There is an angle of 40°, the sides of which have lengths 3 and 12.

43. The sides have lengths 7, 11, and 14.

44. The sides have lengths 4, 8, and 10.

Chapter 8 Test

Sections 8.1 and 8.2

1. In the right triangle shown below, find
   (a) $\sin \theta$ (b) $\cos \theta$ (c) $\tan \theta$

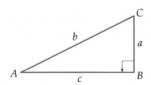

2. A 160-ft long ramp connects a ground-level parking lot with the entrance of a building. If the entrance is 7 feet above the ground, what angle does the ramp make with the ground?

3. Find side c of the right triangle shown below, if $\csc C = 3.5$ and $b = 31.5$.

4. Ruth is flying a kite. Her hand is 3 feet above ground level and is holding the end of a 310-ft long kite string, which makes an angle of 55° with the horizontal. How high is the kite above the ground?

5. In the right triangle shown here, find the exact length of side c (no decimal approximations).

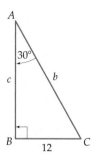

6. A wire from the top of a TV tower to the ground makes and angle of 42° with the ground and touches the ground 200 feet from the base of the tower. How high is the tower?

7. Solve the right triangle below.

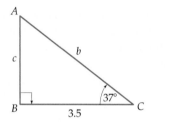

8. A plane passes directly over your head at an altitude of 5500 feet. Two seconds later, you observe that its angle of elevation is 72°. How far did the plane travel during those two seconds?

9. Find angle θ in the right triangle shown here.

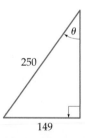

10. Find angles A and C in the triangle shown here.

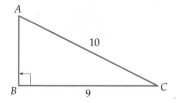

Sections 8.3 and 8.4; Special Topics 8.4.A

11. The side of a hill makes an angle of 17° with the horizontal. A wire is to be run from the top of a 180-ft tower on the top of the hill to a stake located 120 feet down the hillside from the base of the tower. How long a wire is needed?

In Questions 12–17, solve triangle ABC (standard notation) under the given conditions.

12. $B = 93.5°$, $C = 49.5°$, and $b = 6$

13. $B = 26.6°$, $a = 6.9$, and $c = 16.5$

14. $a = 22$, $c = 15.3$, and $B = 74°$

15. $a = 5.4$, $b = 7.2$, and $c = 12$.

16. $b = 15.2$, $c = 19.3$, and $A = 44°$

17. $b = 12.3$, $c = 20.2$, and $B = 37.2°$

18. Let A, B and C be the points in the plane with coordinates $(-8, 5)$, $(5, 1)$ and $(1, -1)$ respectively. Find the angles of triangle ABC.

19. A plane flies in a straight line at 425 mph for 1 hour and 42 minutes. It makes a 25° turn and flies at 375 mph for 2 hours and 33 minutes. How far is it from its starting point?

20. A straight path makes an angle of 7° with the horizontal. A statue at the higher end of the path casts a 9-meter long shadow straight down the path. The angle of elevation from the end of the shadow to the top of the statue is 33°. How tall is the statue?

21. Each of two observers 300 feet apart measures the angle of elevation to the top of a tree that sits on the straight line between them. These angles are 54° and 67° respectively. How tall is the tree?

22. If ABC is a triangle (standard notation), find its area when

   (a) $a = 4$, $b = 6$, and $c = 9$.
   (b) $a = 7$, $b = 9$, and $C = 43°$.

DISCOVERY PROJECT 8   Life on a Sphere

Although we experience the surface of the earth as a level surface, it is not. The earth, of course, is a sphere whose radius is about 6370 kilometers. When you stand on a sphere, strange things happen, especially when the surface is particularly smooth.

Granted, the horizon is often interfered with by hills, forests, buildings, and such. However, if you are out in an area of flat land like that found in eastern Kansas or northern Ontario, the hills are very slight, and the ground is very much like a sphere. The same is true of large lakes and oceans.

Generally speaking, you are not as interested in seeing the horizon as you are in objects that the horizon might hide. On the ocean, for instance, you might want to see another ship. On land, you might be looking for a particular building or vehicle.

1. Suppose that a person is standing on a highway and a 3.5-meter-high truck passes by. Assume also that an average human has eyes that are 1.55 meters off the ground. If the road is flat and straight, how far away is the truck when it disappears below the horizon?

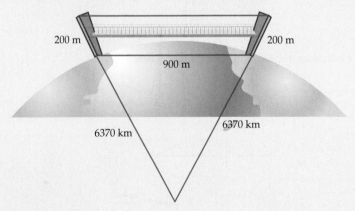

2. Engineers building large bridges must also account for the curvature of the earth. In the figure, a bridge with towers 200 meters tall and 900 meters apart (straight-line distance) at the base is constructed. How much farther apart are the tops of the towers than the bases?

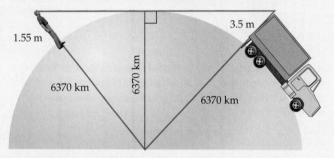

3. How much difference does it make if the distance between the towers is measured along the curve of the earth?

Russell Illig / Getty Images

Chapter 9

APPLICATIONS OF TRIGONOMETRY

Is this bridge safe?

When planning a bridge or a building, architects and engineers must determine the stress on cables and other parts of the structure to be sure that all parts are adequately supported. Problems like these can be modeled and solved by using vectors. See Exercise 73 on page 652.

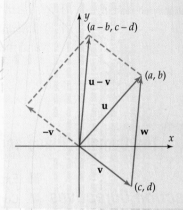

© Jeff Spielman/Getty Images

Chapter Outline

Trigonometry has a variety of useful applications in geometry, algebra, and the physical sciences, several of which are discussed in this chapter.

9.1   The Complex Plane and Polar Form for Complex Numbers*

Section Objectives

■ Explore the complex plane.
■ Convert a complex number from rectangular to polar form.
■ Multiply and divide complex numbers in polar form.

The complex number system can be represented geometrically by the coordinate plane:

The complex number $a + bi$ corresponds to the point (a, b) in the plane.

For example, the point $(2, 3)$ in Figure 9–1 is labeled by $2 + 3i$, and similarly for the other points shown:

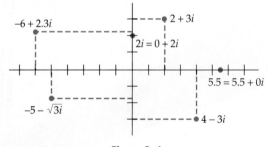

Figure 9–1

When the coordinate plane is labeled by complex numbers in this way, it is called the **complex plane.** Each real number $a = a + 0i$ corresponds to the point $(a, 0)$ on the horizontal axis; so this axis is called the **real axis.** The vertical axis is called the **imaginary axis** because every imaginary number $bi = 0 + bi$ corresponds to the point $(0, b)$ on the vertical axis.

The absolute value of a real number c is the distance from c to 0 on the number line (see page 12). So we define the **absolute value** (or **modulus**) of the

TECHNOLOGY TIP

To do complex arithmetic on TI-86 and HP-39gs, enter $a + bi$ as (a, b). On other calculators, use the special *i* key whose location is:

TI-84+/89: keyboard

Casio 9850: OPTN/CPLX

---

*Section 4.7 is a prerequisite for this section.

complex number $a + bi$ to be the distance from $a + bi$ to the origin in the complex plane:

$$|a + bi| = \text{distance from } (a, b) \text{ to } (0, 0) = \sqrt{(a - 0)^2 + (b - 0)^2}.$$

Therefore, we have the following.

Absolute Value

> The **absolute value** (or **modulus**) of the complex number $a + bi$ is
> $$|a + bi| = \sqrt{a^2 + b^2}.$$

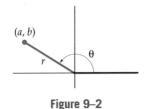

Figure 9–2

EXAMPLE 1

Find the modulus of each of the following complex numbers:
(a) $3 + 2i$ (b) $4 - 5i$ (c) $-3i$

SOLUTION

(a) $|3 + 2i| = \sqrt{3^2 + 2^2} = \sqrt{13}$.

(b) $|4 - 5i| = \sqrt{4^2 + (-5)^2} = \sqrt{41}$.

(c) $-3i = 0 - 3i$, so

$$|-3i| = |0 - 3i| = \sqrt{0^2 + (-3)^2} = \sqrt{9} = 3. \quad\blacksquare$$

Let $a + bi$ be a nonzero complex number, and denote $|a + bi|$ by r. Then r is the length of the line segment joining (a, b) and $(0, 0)$ in the plane. Let θ be the angle in standard position with this line segment as its terminal side (Figure 9–2).

According to the point-in-the-plane description of sine and cosine,

$$\cos \theta = \frac{a}{r} \quad \text{and} \quad \sin \theta = \frac{b}{r},$$

so

$$a = r \cos \theta \quad \text{and} \quad b = r \sin \theta.$$

Consequently,

$$a + bi = r \cos \theta + (r \sin \theta)i = r(\cos \theta + i \sin \theta).^*$$

When a complex number $a + bi$ is written in this way, it is said to be in **polar form** or **trigonometric form**. The angle θ is called the **argument** and is usually expressed in radian measure. The number 0 can also be written in polar notation by letting $r = 0$ and θ be any angle. Thus, we have the following.

Polar Form

> Every complex number $a + bi$ can be written in polar form:
> $$r(\cos \theta + i \sin \theta),$$
> where $r = |a + bi| = \sqrt{a^2 + b^2}$, $a = r \cos \theta$, and $b = r \sin \theta$.

---

*It is customary to place i in front of $\sin \theta$ rather than after it. Some books abbreviate $r(\cos \theta + i \sin \theta)$ as $r \operatorname{cis} \theta$.

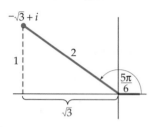

Figure 9–3

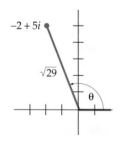

Figure 9–4

When a complex number is written in polar form, the argument θ is not uniquely determined, since θ, $\theta \pm 2\pi$, $\theta \pm 4\pi$, etc., all satisfy the conditions in the box.

EXAMPLE 2

Express $-\sqrt{3} + i$ in polar form.

SOLUTION   Here $a = -\sqrt{3}$ and $b = 1$, so

$$r = \sqrt{a^2 + b^2} = \sqrt{(-\sqrt{3})^2 + 1^2} = \sqrt{3 + 1} = 2.$$

The angle θ must satisfy

$$\cos \theta = \frac{a}{r} = \frac{-\sqrt{3}}{2} \qquad \text{and} \qquad \sin \theta = \frac{b}{r} = \frac{1}{2}.$$

Since $-\sqrt{3} + i$ lies in the second quadrant (Figure 9–3), θ must be a second-quadrant angle. Our knowledge of special angles and Figure 9–3 show that $\theta = 5\pi/6$ satisfies these conditions. Hence,

$$-\sqrt{3} + i = 2\left(\cos \frac{5\pi}{6} + i \sin \frac{5\pi}{6}\right).$$ ∎

EXAMPLE 3*

Express $-2 + 5i$ in polar form.

SOLUTION   Since $a = -2$ and $b = 5$, $r = \sqrt{(-2)^2 + 5^2} = \sqrt{29}$. The angle θ must satisfy

$$\cos \theta = \frac{a}{r} = \frac{-2}{\sqrt{29}} \qquad \text{and} \qquad \sin \theta = \frac{b}{r} = \frac{5}{\sqrt{29}},$$

so

$$\tan \theta = \frac{\sin \theta}{\cos \theta} = \frac{5/\sqrt{29}}{-2/\sqrt{29}} = -\frac{5}{2} = -2.5.$$

Since $-2 + 5i$ lies in the second quadrant (Figure 9–4), θ lies between $\pi/2$ and π. As we saw in Section 7.5, the only solution of the equation $\tan \theta = -2.5$ that lies between $\pi/2$ and π is $\theta \approx -1.1903 + \pi = 1.9513$. Therefore,

$$-2 + 5i \approx \sqrt{29}(\cos 1.9513 + i \sin 1.9513).$$ ∎

Multiplication and division of complex numbers in polar form are done by the following rules, which are proved at the end of the section.

---

*Omit this example if you haven't read Section 7.5.

Polar Multiplication and Division Rules

If $z_1 = r_1(\cos \theta_1 + i \sin \theta_1)$ and $z_2 = r_2(\cos \theta_2 + i \sin \theta_2)$ are any two complex numbers, then

$$z_1 z_2 = r_1 r_2 [\cos(\theta_1 + \theta_2) + i \sin(\theta_1 + \theta_2)]$$

and

$$\frac{z_1}{z_2} = \frac{r_1}{r_2} [\cos(\theta_1 - \theta_2) + i \sin(\theta_1 - \theta_2)] \quad (z_2 \neq 0).$$

In other words, to multiply two numbers in polar form, just *multiply the moduli and add the arguments.* To divide, just *divide the moduli and subtract the arguments.* Before proving the statements in the box, we will illustrate them with some examples.

EXAMPLE 4

Find $z_1 z_2$, when

$$z_1 = 2[\cos(5\pi/6) + i \sin(5\pi/6)] \quad \text{and} \quad z_2 = 3[\cos(7\pi/4) + i \sin(7\pi/4)].$$

TECHNOLOGY TIP

Complex arithmetic can be done with numbers in polar form on TI calculators. Some answers may be expressed in rectangular form.

SOLUTION   Here r_1 is the number 2, and $\theta_1 = 5\pi/6$; similarly, $r_2 = 3$, and $\theta_2 = 7\pi/4$, and we have

$$z_1 z_2 = r_1 r_2 [\cos(\theta_1 + \theta_2) + i \sin(\theta_1 + \theta_2)]$$

$$= 2 \cdot 3 \left[\cos\left(\frac{5\pi}{6} + \frac{7\pi}{4}\right) + i \sin\left(\frac{5\pi}{6} + \frac{7\pi}{4}\right) \right]$$

$$= 6 \left[\cos\left(\frac{10\pi}{12} + \frac{21\pi}{12}\right) + i \sin\left(\frac{10\pi}{12} + \frac{21\pi}{12}\right) \right]$$

$$= 6 \left(\cos\frac{31\pi}{12} + i \sin\frac{31\pi}{12} \right). \qquad \blacksquare$$

EXAMPLE 5

Find z_1/z_2, where

$$z_1 = 10[\cos(\pi/3) + i \sin(\pi/3)] \quad \text{and} \quad z_2 = 2[\cos(\pi/4) + i \sin(\pi/4)].$$

SOLUTION

$$\frac{z_1}{z_2} = \frac{10\left(\cos\frac{\pi}{3} + i \sin\frac{\pi}{3} \right)}{2\left(\cos\frac{\pi}{4} + i \sin\frac{\pi}{4} \right)} = \frac{10}{2}\left[\cos\left(\frac{\pi}{3} - \frac{\pi}{4}\right) + i \sin\left(\frac{\pi}{3} - \frac{\pi}{4}\right) \right]$$

$$= 5\left(\cos\frac{\pi}{12} + i \sin\frac{\pi}{12} \right). \qquad \blacksquare$$

PROOF OF THE POLAR MULTIPLICATION RULE

If $z_1 = r_1(\cos\theta_1 + i\sin\theta_1)$ and $z_2 = r_2(\cos\theta_2 + i\sin\theta_2)$, then

$$z_1 z_2 = r_1(\cos\theta_1 + i\sin\theta_1)r_2(\cos\theta_2 + i\sin\theta_2)$$

$$= r_1 r_2(\cos\theta_1 + i\sin\theta_1)(\cos\theta_2 + i\sin\theta_2)$$

$$= r_1 r_2(\cos\theta_1\cos\theta_2 + i\sin\theta_1\cos\theta_2 + i\cos\theta_1\sin\theta_2 + i^2\sin\theta_1\sin\theta_2)$$

$$= r_1 r_2[(\cos\theta_1\cos\theta_2 - \sin\theta_1\sin\theta_2) + i(\sin\theta_1\cos\theta_2 + \cos\theta_1\sin\theta_2)].$$

But the addition identities for sine and cosine (page 524) show that

$$\cos\theta_1\cos\theta_2 - \sin\theta_1\sin\theta_2 = \cos(\theta_1 + \theta_2)$$

$$\sin\theta_1\cos\theta_2 + \cos\theta_1\sin\theta_2 = \sin(\theta_1 + \theta_2).$$

Therefore,

$$z_1 z_2 = r_1 r_2[(\cos\theta_1\cos\theta_2 - \sin\theta_1\sin\theta_2) + i(\sin\theta_1\cos\theta_2 + \cos\theta_1\sin\theta_2)]$$

$$= r_1 r_2[\cos(\theta_1 + \theta_2) + i\sin(\theta_1 + \theta_2)].$$

This completes the proof of the multiplication rule. The division rule is proved similarly (Exercise 77).

EXERCISES 9.1

In Exercises 1–8, plot the point in the complex plane corresponding to the number.

1. $3 + 2i$

2. $-7 + 6i$

3. $-\dfrac{8}{3} - \dfrac{5}{3}i$

4. $\sqrt{2} - 7i$

5. $(1 + i)(1 - i)$

6. $(2 + i)(1 - 2i)$

7. $2i\left(3 - \dfrac{5}{2}i\right)$

8. $\dfrac{4i}{3}(-6 - 3i)$

In Exercises 9–14, find the absolute value.

9. $|5 - 12i|$ **10.** $|2i|$ **11.** $|1 + \sqrt{2}i|$

12. $|2 - 3i|$ **13.** $|-12i|$.   **14.** $|i^7|$

15. Give an example of complex numbers z and w such that $|z + w| \neq |z| + |w|$.

16. If $z = 3 - 4i$, find $|z|^2$ and $z\bar{z}$, where $\bar{z}$ is the conjugate of z (see page 323).

In Exercises 17–24, sketch the graph of the equation in the complex plane (z denotes a complex number of the form $a + bi$).

17. $|z| = 4$ [*Hint:* The graph consists of all points that lie 4 units from the origin.]

18. $|z| = 1$

19. $|z - 1| = 10$ [*Hint:* 1 corresponds to $(1, 0)$ in the complex plane. What does the equation say about the distance from z to 1?]

20. $|z + 3| = 1$ **21.** $|z - 2i| = 4$

22. $|z - 3i + 2| = 9$ [*Hint:* Rewrite it as $|z - (-2 + 3i)| = 9.$]

23. $\text{Re}(z) = 2$ [The **real part** of the complex number $z = a + bi$ is defined to be the number a and is denoted $\text{Re}(z)$.]

24. $\text{Im}(z) = -5/2$ [The **imaginary part** of $z = a + bi$ is defined to be the number b (*not bi*) and is denoted $\text{Im}(z)$.]

In Exercises 25–36, express the number in the form $a + bi$.

25. $2\left(\cos\dfrac{\pi}{4} + i\sin\dfrac{\pi}{4}\right)$ **26.** $3\left(\cos\dfrac{\pi}{3} + i\sin\dfrac{\pi}{3}\right)$

27. $\cos\dfrac{\pi}{2} + i\sin\dfrac{\pi}{2}$ **28.** $4\left(\cos\dfrac{3\pi}{4} + i\sin\dfrac{3\pi}{4}\right)$

29. $5\left(\cos\dfrac{2\pi}{3} + i\sin\dfrac{2\pi}{3}\right)$ **30.** $2\left(\cos\dfrac{7\pi}{6} + i\sin\dfrac{7\pi}{6}\right)$

31. $1.5\left(\cos\dfrac{\pi}{6} + i\sin\dfrac{\pi}{6}\right)$ **32.** $5(\cos 3 + i\sin 3)$

33. $2(\cos 4 + i\sin 4)$ **34.** $3(\cos 5 + i\sin 5)$

35. $4(\cos 2 + i\sin 2)$ **36.** $2(\cos 1.5 + i\sin 1.5)$

In Exercises 37–52, express the number in polar form.

37. $3 + 3i$ **38.** $5 - 5i$ **39.** $2 + 2\sqrt{3}i$

40. $5\sqrt{3} + 5i$ **41.** $3\sqrt{3} - 3i$ **42.** $-4 - 4\sqrt{3}i$

43. $-\sqrt{3} - \sqrt{3}i$

44. $2\sqrt{5} - 2\sqrt{5}i$

45. $3 + 4i$

46. $-4 + 3i$

47. $5 - 12i$

48. $-\sqrt{7} - 3i$

49. $1 + 2i$

50. $3 - 5i$

51. $-\dfrac{5}{2} + \dfrac{7}{2}i$

52. $\sqrt{5} + \sqrt{11}i$

In Exercises 53–64, perform the indicated multiplication or division. Express your answer in both polar form $r(\cos\theta + i\sin\theta)$ and rectangular form $a + bi$.

53. $\left(\cos\dfrac{\pi}{2} + i\sin\dfrac{\pi}{2}\right) \cdot (\cos\pi + i\sin\pi)$

54. $2\left(\cos\dfrac{\pi}{6} + i\sin\dfrac{\pi}{6}\right) \cdot 5\left(\cos\dfrac{\pi}{3} + i\sin\dfrac{\pi}{3}\right)$

55. $4\left(\cos\dfrac{\pi}{4} + i\sin\dfrac{\pi}{4}\right) \cdot 3\left(\cos\dfrac{\pi}{12} + i\sin\dfrac{\pi}{12}\right)$

56. $\left(\cos\dfrac{\pi}{12} + i\sin\dfrac{\pi}{12}\right) \cdot 2\left(\cos\dfrac{7\pi}{12} + i\sin\dfrac{7\pi}{12}\right)$

57. $3\left(\cos\dfrac{\pi}{8} + i\sin\dfrac{\pi}{8}\right) \cdot 12\left(\cos\dfrac{3\pi}{8} + i\sin\dfrac{3\pi}{8}\right)$

58. $12\left(\cos\dfrac{11\pi}{12} + i\sin\dfrac{11\pi}{12}\right) \cdot \dfrac{7}{2}\left(\cos\dfrac{\pi}{4} + i\sin\dfrac{\pi}{4}\right)$

59. $\dfrac{\cos\pi + i\sin\pi}{\cos\dfrac{2\pi}{3} + i\sin\dfrac{2\pi}{3}}$

60. $\dfrac{\cos\dfrac{3\pi}{4} + i\sin\dfrac{3\pi}{4}}{\cos\dfrac{\pi}{4} + i\sin\dfrac{\pi}{4}}$

61. $\dfrac{8\left(\cos\dfrac{4\pi}{3} + i\sin\dfrac{4\pi}{3}\right)}{4\left(\cos\dfrac{7\pi}{6} + i\sin\dfrac{7\pi}{6}\right)}$

62. $\dfrac{8\left(\cos\dfrac{5\pi}{18} + i\sin\dfrac{5\pi}{18}\right)}{4\left(\cos\dfrac{\pi}{9} + i\sin\dfrac{\pi}{9}\right)}$

63. $\dfrac{6\left(\cos\dfrac{7\pi}{20} + i\sin\dfrac{7\pi}{20}\right)}{4\left(\cos\dfrac{\pi}{10} + i\sin\dfrac{\pi}{10}\right)}$

64. $\dfrac{\sqrt{54}\left(\cos\dfrac{9\pi}{4} + i\sin\dfrac{9\pi}{4}\right)}{\sqrt{6}\left(\cos\dfrac{7\pi}{12} + i\sin\dfrac{7\pi}{12}\right)}$

In Exercises 65–72, convert to polar form and then multiply or divide. Express your answer in polar form.

65. $(1 + i)(1 + \sqrt{3}i)$

66. $(1 - i)(3 - 3i)$

67. $\dfrac{1 + i}{1 - i}$

68. $\dfrac{2 - 2i}{-1 - i}$

69. $3i(2\sqrt{3} + 2i)$

70. $\dfrac{-4i}{\sqrt{3} + i}$

71. $i(i + 1)(-\sqrt{3} + i)$

72. $(1 - i)(2\sqrt{3} - 2i)(-4 - 4\sqrt{3}i)$

73. Explain what is meant by saying that multiplying a complex number $z = r(\cos\theta + i\sin\theta)$ by i amounts to rotating z $90°$ counterclockwise around the origin. [*Hint:* Express i and iz in polar form. What are their relative positions in the complex plane?]

74. Describe what happens geometrically when you multiply a complex number by 2.

THINKERS

75. The sum of two distinct complex numbers, $a + bi$ and $c + di$, can be found geometrically by means of the so-called **parallelogram rule:** Plot the points $a + bi$ and $c + di$ in the complex plane, and form the parallelogram, three of whose vertices are 0, $a + bi$, and $c + di$, as in the figure. Then the fourth vertex of the parallelogram is the point whose coordinate is the sum

$$(a + bi) + (c + di) = (a + c) + (b + d)i.$$

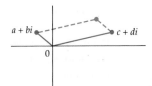 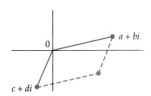

Complete the following *proof* of the parallelogram rule when $a \neq 0$ and $c \neq 0$.

(a) Find the *slope* of the line K from 0 to $a + bi$. [*Hint:* K contains the points $(0, 0)$ and (a, b).]

(b) Find the *slope* of the line N from 0 to $c + di$.

(c) Find the *equation* of the line L through $a + bi$ and parallel to line N of part (b). [*Hint:* The point (a, b) is on L; find the slope of L by using part (b) and facts about the slope of parallel lines.]

(d) Find the *equation* of the line M through $c + di$ and parallel to line K of part (a).

(e) Label the lines K, L, M, and N in the figure.

(f) Show by using substitution that the point $(a + c, b + d)$ satisfies both the equation of line L and the equation of line M. Therefore, $(a + c, b + d)$ lies on both L and M. Since the only point on both L and M is the fourth vertex of the parallelogram (see the figure), this vertex must be $(a + c, b + d)$. Hence, this vertex has coordinate

$$(a + c) + (b + d)i = (a + bi) + (c + di).$$

76. Let $z = a + bi$ be a complex number and denote its conjugate $a - bi$ by $\bar{z}$. Prove that $|z|^2 = z\bar{z}$.

77. *Proof of the polar division rule.* Let $z_1 = r_1(\cos \theta_1 + i \sin \theta_1)$ and $z_2 = r_2(\cos \theta_2 + i \sin \theta_2)$. Then

$$\frac{z_1}{z_2} = \frac{r_1(\cos \theta_1 + i \sin \theta_1)}{r_2(\cos \theta_2 + i \sin \theta_2)}$$

$$= \frac{r_1(\cos \theta_1 + i \sin \theta_1)}{r_2(\cos \theta_2 + i \sin \theta_2)} \cdot \frac{\cos \theta_2 - i \sin \theta_2}{\cos \theta_2 - i \sin \theta_2}.$$

(a) Multiply out the denominator on the right side and use the Pythagorean identity to show that it is just the number r_2.

(b) Multiply out the numerator on the right side; use the subtraction identities for sine and cosine (page 524) to show that it is

$$r_1[\cos(\theta_1 - \theta_2) + i \sin(\theta_1 - \theta_2)].$$

Therefore,

$$\frac{z_1}{z_2} = \left(\frac{r_1}{r_2}\right)[\cos(\theta_1 - \theta_2) + i \sin(\theta_1 - \theta_2)].$$

78. (a) If $s(\cos \beta + i \sin \beta) = r(\cos \theta + i \sin \theta)$, (with $r > 0$, $s > 0$), explain why we must have $s = r$. [*Hint:* Think distance.]

(b) If $r(\cos \beta + i \sin \beta) = r(\cos \theta + i \sin \theta)$, explain why $\cos \beta = \cos \theta$ and $\sin \beta = \sin \theta$. [*Hint:* See property 5 of the complex numbers on page 322.]

(c) If $\cos \beta = \cos \theta$ and $\sin \beta = \sin \theta$, show that angles β and θ in standard position have the same terminal side. [*Hint:* $(\cos \beta, \sin \beta)$ and $(\cos \theta, \sin \theta)$ are points on the unit circle.]

(d) Use parts (a)–(c) to prove this **equality rule for polar form:**

$$s(\cos \beta + i \sin \beta) = r(\cos \theta + i \sin \theta)$$

exactly when $s = r$ and $\beta = \theta + 2k\pi$ for some integer k. [*Hint:* Angles with the same terminal side must differ by an integer multiple of 2π.]

9.2   DeMoivre's Theorem and *n*th Roots of Complex Numbers

Section Objectives
- Use DeMoivre's Theorem to compute powers of complex numbers.
- Find the *n*th roots of a complex number.
- Find the *n*th roots of unity algebraically and geometrically.

Polar form provides a convenient way to calculate both powers and roots of complex numbers. If $z = r(\cos \theta + i \sin \theta)$, then the multiplication formula on page 629 shows that

$$z^2 = z \cdot z = r \cdot r[\cos(\theta + \theta) + i \sin(\theta + \theta)]$$

$$= r^2(\cos 2\theta + i \sin 2\theta).$$

Similarly,

$$z^3 = z^2 \cdot z = r^2 \cdot r[\cos(2\theta + \theta) + i \sin(2\theta + \theta)]$$

$$= r^3(\cos 3\theta + i \sin 3\theta).$$

Repeated application of the multiplication formula proves the following theorem.

DeMoivre's Theorem

For any complex number $z = r(\cos \theta + i \sin \theta)$ and any positive integer n,

$$z^n = r^n(\cos n\theta + i \sin n\theta).$$

EXAMPLE 1

Compute $(-\sqrt{3} + i)^5$.

SOLUTION   We first express $-\sqrt{3} + i$ in polar form (as in Example 2 on page 628):

$$-\sqrt{3} + i = 2\left(\cos \frac{5\pi}{6} + i \sin \frac{5\pi}{6}\right).$$

By DeMoivre's Theorem,

$$(-\sqrt{3} + i)^5 = 2^5\left[\cos\left(5 \cdot \frac{5\pi}{6}\right) + i\sin\left(5 \cdot \frac{5\pi}{6}\right)\right] = 32\left(\cos\frac{25\pi}{6} + i\sin\frac{25\pi}{6}\right).$$

Since $25\pi/6 = (\pi/6) + (24\pi/6) = (\pi/6) + 4\pi$, we have

$$(-\sqrt{3} + i)^5 = 32\left(\cos\frac{25\pi}{6} + i\sin\frac{25\pi}{6}\right) = 32\left(\cos\frac{\pi}{6} + i\sin\frac{\pi}{6}\right)$$

$$= 32\left(\frac{\sqrt{3}}{2} + \frac{1}{2}i\right) = 16\sqrt{3} + 16i. \qquad \blacksquare$$

EXAMPLE 2

Find $(1 + i)^{10}$.

SOLUTION   First verify that the polar form of $1 + i$ is

$$1 + i = \sqrt{2}\left(\cos\frac{\pi}{4} + i\sin\frac{\pi}{4}\right).$$

Therefore, by DeMoivre's Theorem,

$$(1 + i)^{10} = (\sqrt{2})^{10}\left(\cos\frac{10\pi}{4} + i\sin\frac{10\pi}{4}\right)$$

$$= (2^{1/2})^{10}\left(\cos\frac{5\pi}{2} + i\sin\frac{5\pi}{2}\right) = 2^5(0 + i \cdot 1) = 32i.$$

A calculator that can do complex arithmetic confirms this result (Figure 9–5). $\blacksquare$

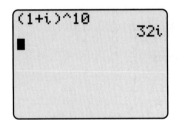

Figure 9–5

*n*TH ROOTS

Recall that for a real number c, we called a solution of the equation $x^n = c$ an *n*th root of c. Similarly, if $a + bi$ is a complex number, then any solution of the equation

$$z^n = a + bi$$

is called an ***n*th root** of $a + bi$. In this context, the radical symbol will be used only for nonnegative real numbers and will have the same meaning as before: If r is a nonnegative real number, then $\sqrt[n]{r}$ denotes the unique nonnegative real number whose *n*th power is r.

All *n*th roots of a complex number $a + bi$ can easily be found if $a + bi$ is written in polar form, as illustrated in the next example.

EXAMPLE 3

Find the fourth roots of $-8 + 8\sqrt{3}i$.

SOLUTION   To solve $z^4 = -8 + 8\sqrt{3}i$, first verify that the polar form of $-8 + 8\sqrt{3}i$ is $16\left(\cos\frac{2\pi}{3} + i\sin\frac{2\pi}{3}\right)$. We must find numbers s and β such that

$$[s(\cos\beta + i\sin\beta)]^4 = 16\left(\cos\frac{2\pi}{3} + i\sin\frac{2\pi}{3}\right).$$

By DeMoivre's Theorem, we must have

$$s^4(\cos 4\beta + i \sin 4\beta) = 16\left(\cos \frac{2\pi}{3} + i \sin \frac{2\pi}{3}\right).$$

The equality rules for complex numbers in polar form (Exercise 78 in Section 9.1) show that this can happen only when

$$s^4 = 16 \qquad \text{and} \qquad 4\beta = \frac{2\pi}{3} + 2k\pi \quad (k \text{ an integer})$$

$$s = \sqrt[4]{16} = 2 \qquad\qquad \beta = \frac{2\pi/3 + 2k\pi}{4}.$$

Substituting these values in $s(\cos \beta + i \sin \beta)$ shows that the solutions of

$$z^4 = 16\left(\cos \frac{2\pi}{3} + i \sin \frac{2\pi}{3}\right) \text{ are}$$

$$z = 2\left(\cos \frac{2\pi/3 + 2k\pi}{4} + i \sin \frac{2\pi/3 + 2k\pi}{4}\right) \quad (k = 0, \pm 1, \pm 2, \pm 3, \ldots).$$

which can be simplified as

$$z = 2\left[\cos\left(\frac{\pi}{6} + \frac{k\pi}{2}\right) + i \sin\left(\frac{\pi}{6} + \frac{k\pi}{2}\right)\right] \quad (k = 0, \pm 1, \pm 2, \pm 3, \ldots).$$

Letting $k = 0, 1, 2, 3$, produces four distinct solutions:

$k = 0$: $\quad z = 2\left(\cos \frac{\pi}{6} + i \sin \frac{\pi}{6}\right) = \sqrt{3} + i.$

$k = 1$: $\quad z = 2\left[\cos\left(\frac{\pi}{6} + \frac{\pi}{2}\right) + i \sin\left(\frac{\pi}{6} + \frac{\pi}{2}\right)\right] = 2\left(\cos \frac{2\pi}{3} + i \sin \frac{2\pi}{3}\right)$

$\qquad\qquad = -1 + \sqrt{3}i.$

$k = 2$: $\quad z = 2\left[\cos\left(\frac{\pi}{6} + \pi\right) + i \sin\left(\frac{\pi}{6} + \pi\right)\right] = 2\left(\cos \frac{7\pi}{6} + i \sin \frac{7\pi}{6}\right)$

$\qquad\qquad = -\sqrt{3} - i.$

$k = 3$: $\quad z = 2\left[\cos\left(\frac{\pi}{6} + \frac{3\pi}{2}\right) + i \sin\left(\frac{\pi}{6} + \frac{3\pi}{2}\right)\right] = 2\left(\cos \frac{5\pi}{3} + i \sin \frac{5\pi}{3}\right).$

$\qquad\qquad = 1 - \sqrt{3}i.$

Any other value of k produces an angle β with the same terminal side as one of the four angles used above, and hence leads to the same solution. For instance, when $k = 4$, then $\beta = \frac{\pi}{6} + \frac{4\pi}{2} = \frac{\pi}{6} + 2\pi$ and β has the same terminal side as $\pi/6$. Therefore, we have found *all* the solutions—the four fourth roots of $-8 + 8\sqrt{3}i$.* ∎

The general equation $z^n = r(\cos \theta + i \sin \theta)$ can be solved by exactly the same method used in the preceding example—just substitute n for 4, r for 16, and θ for $2\pi/3$, as follows. A solution is a number $s(\cos \beta + i \sin \beta)$ such that

$$[s(\cos \beta + i \sin \beta)]^n = r(\cos \theta + i \sin \theta)$$

$$s^n(\cos n\beta + i \sin n\beta) = r(\cos \theta + i \sin \theta).$$

---

*Alternatively, page 330 shows that a fourth-degree equation, such as $z^4 = -8 + 8\sqrt{3}i$, has at most four distinct solutions.

Therefore,

$$s^n = r \qquad \text{and} \qquad n\beta = \theta + 2k\pi \quad (k \text{ any integer})$$

$$s = \sqrt[n]{r} \qquad\qquad \beta = \frac{\theta + 2k\pi}{n}$$

Taking $k = 0, 1, 2, \ldots, n - 1$ produces n distinct angles β. Any other value of k leads to an angle β with the same terminal side as one of these. Hence,

Formula for nth Roots

For each positive integer n, nonzero complex number

$$r(\cos \theta + i \sin \theta)$$

has exactly n distinct nth roots. They are given by

$$\sqrt[n]{r}\left[\cos\left(\frac{\theta + 2k\pi}{n}\right) + i \sin\left(\frac{\theta + 2k\pi}{n}\right)\right],$$

where $k = 0, 1, 2, 3, \ldots n - 1$.

TECHNOLOGY TIP

The polynomial solvers on TI-86 and HP-39gs can solve

$$z^5 = 4 + 4i$$

and similar equations.
 On TI-89, use cSOLVE in the COMPLEX submenu of the ALGEBRA menu.

EXAMPLE 4

Find the fifth roots of $4 + 4i$.

SOLUTION First write $4 + 4i$ in polar form as $4\sqrt{2}\left(\cos\dfrac{\pi}{4} + i \sin\dfrac{\pi}{4}\right)$. Now apply the root formula with $n = 5$, $r = 4\sqrt{2}$, $\theta = \pi/4$, and $k = 0, 1, 2, 3, 4$. Note that

$$\sqrt[5]{r} = \sqrt[5]{4\sqrt{2}} = (4\sqrt{2})^{1/5} = (2^2 2^{1/2})^{1/5} = (2^{5/2})^{1/5} = 2^{5/10} = 2^{1/2} = \sqrt{2}.$$

Therefore, the fifth roots are

$$\sqrt{2}\left[\cos\left(\frac{\pi/4 + 2k\pi}{5}\right) + i \sin\left(\frac{\pi/4 + 2k\pi}{5}\right)\right] \quad k = 0, 1, 2, 3, 4,$$

that is,

$$k = 0: \sqrt{2}\left[\cos\left(\frac{\pi/4 + 0}{5}\right) + i \sin\left(\frac{\pi/4 + 0}{5}\right)\right] = \sqrt{2}\left(\cos\frac{\pi}{20} + i \sin\frac{\pi}{20}\right),$$

$$k = 1: \sqrt{2}\left[\cos\left(\frac{\pi/4 + 2\pi}{5}\right) + i \sin\left(\frac{\pi/4 + 2\pi}{5}\right)\right] = \sqrt{2}\left(\cos\frac{9\pi}{20} + i \sin\frac{9\pi}{20}\right),$$

$$k = 2: \sqrt{2}\left[\cos\left(\frac{\pi/4 + 4\pi}{5}\right) + i \sin\left(\frac{\pi/4 + 4\pi}{5}\right)\right] = \sqrt{2}\left(\cos\frac{17\pi}{20} + i \sin\frac{17\pi}{20}\right),$$

$$k = 3: \sqrt{2}\left[\cos\left(\frac{\pi/4 + 6\pi}{5}\right) + i \sin\left(\frac{\pi/4 + 6\pi}{5}\right)\right] = \sqrt{2}\left(\cos\frac{25\pi}{20} + i \sin\frac{25\pi}{20}\right),$$

$$k = 4: \sqrt{2}\left[\cos\left(\frac{\pi/4 + 8\pi}{5}\right) + i \sin\left(\frac{\pi/4 + 8\pi}{5}\right)\right] = \sqrt{2}\left(\cos\frac{33\pi}{20} + i \sin\frac{33\pi}{20}\right).$$

ROOTS OF UNITY

The n distinct nth roots of 1 (the solutions of $z^n = 1$) are called the **nth roots of unity.** Since $\cos 0 = 1$ and $\sin 0 = 0$, the polar form of the number 1 is $\cos 0 + i \sin 0$. Applying the root formula with $r = 1$ and $\theta = 0$ shows that

Roots of Unity

> For each positive integer n, there are n distinct nth roots of unity:
>
> $$\cos \frac{2k\pi}{n} + i \sin \frac{2k\pi}{n} \quad (k = 0, 1, 2, \ldots, n - 1).$$

EXAMPLE 5

Find the cube roots of unity.

SOLUTION   Apply the formula with $n = 3$ and $k = 0, 1, 2$:

$$k = 0: \qquad\qquad \cos 0 + i \sin 0 = 1,$$

$$k = 1: \qquad\qquad \cos \frac{2\pi}{3} + i \sin \frac{2\pi}{3} = -\frac{1}{2} + \frac{\sqrt{3}}{2} i,$$

$$k = 2: \qquad\qquad \cos \frac{4\pi}{3} + i \sin \frac{4\pi}{3} = -\frac{1}{2} - \frac{\sqrt{3}}{2} i. \qquad\blacksquare$$

Denote by ω the first complex cube root of unity obtained in Example 5:

$$\omega = \cos \frac{2\pi}{3} + i \sin \frac{2\pi}{3}.$$

If we use DeMoivre's Theorem to find ω^2 and ω^3, we see that these numbers are the other two cube roots of unity found in Example 5:

$$\omega^2 = \left(\cos \frac{2\pi}{3} + i \sin \frac{2\pi}{3} \right)^2 = \cos \frac{4\pi}{3} + i \sin \frac{4\pi}{3},$$

$$\omega^3 = \left(\cos \frac{2\pi}{3} + i \sin \frac{2\pi}{3} \right)^3 = \cos \frac{6\pi}{3} + i \sin \frac{6\pi}{3} = \cos 2\pi + i \sin 2\pi$$

$$= 1 + 0 \cdot i = 1.$$

In other words, all the cube roots of unity are powers of ω. The same thing is true in the general case.

Roots of Unity

> Let n be a positive integer with $n > 1$. Then the number
>
> $$z = \cos \frac{2\pi}{n} + i \sin \frac{2\pi}{n}$$
>
> is an nth root of unity and all the nth roots of unity are
>
> $$z, z^2, z^3, z^4, \ldots z^{n-1}, z^n = 1.$$

The *n*th roots of unity have an interesting geometric interpretation. Every *n*th root of unity has absolute value 1 by the Pythagorean identity:

$$\left| \cos \frac{2k\pi}{n} + i \sin \frac{2k\pi}{n} \right| = \left(\cos \frac{2k\pi}{n} \right)^2 + \left(\sin \frac{2k\pi}{n} \right)^2$$

$$= \cos^2 \left(\frac{2k\pi}{n} \right) + \sin^2 \left(\frac{2k\pi}{n} \right) = 1.$$

Therefore, in the complex plane, every *n*th root of unity is exactly 1 unit from the origin. In other words, the *n*th roots of unity all lie on the unit circle.

EXAMPLE 6

Find the fifth roots of unity.

SOLUTION   They are

$$\cos \frac{2k\pi}{5} + i \sin \frac{2k\pi}{5} \quad (k = 0, 1, 2, 3, 4),$$

that is,

$$\cos 0 + i \sin 0 = 1, \qquad \cos \frac{2\pi}{5} + i \sin \frac{2\pi}{5}, \qquad \cos \frac{4\pi}{5} + i \sin \frac{4\pi}{5},$$

$$\cos \frac{6\pi}{5} + i \sin \frac{6\pi}{5}, \qquad \cos \frac{8\pi}{5} + i \sin \frac{8\pi}{5}.$$

These five roots can be plotted in the complex plane by starting at $1 = 1 + 0i$ and moving counterclockwise around the unit circle, moving through an angle of $2\pi/5$ at each step, as shown in Figure 9–6. If you connect these five roots, they form the vertices of a regular pentagon (Figure 9–7). ■

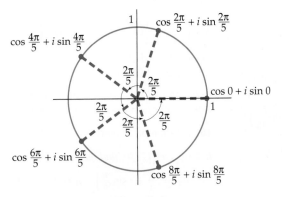

Figure 9–6

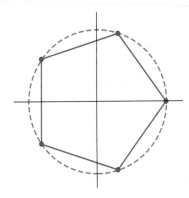

Figure 9–7

GRAPHING EXPLORATION

With your calculator in parametric graphing mode, set these range values:

$$0 \le t \le 2\pi, \qquad t\text{-step} \approx .067,$$

$$-1.5 \le x \le 1.5, \qquad -1 \le y \le 1$$

and graph the unit circle, whose parametric equations are

$$x = \cos t \qquad \text{and} \qquad y = \sin t.^*$$

Reset the t-step to be $2\pi/5$ and graph again. Your screen now looks exactly like the red lines in Figure 9–7 because the calculator plotted only the five points corresponding to $t = 0, 2\pi/5, 4\pi/5, 6\pi/5, 8\pi/5^{\dagger}$ and connected them with the shortest possible segments. Use the trace feature to move along the graph. The cursor will jump from vertex to vertex, that is, it will move from one fifth root of unity to the next.

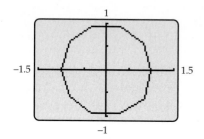

Figure 9–8

EXAMPLE 7

Find the tenth roots of unity graphically.

SOLUTION   Graph the unit circle as in the preceding exploration, but use $2\pi/10$ as the t-step. The result (Figure 9–8) is a regular decagon whose vertices are the tenth roots of unity. By using the trace feature, you can approximate each of them.

GRAPHING EXPLORATION

Verify that the two tenth roots of unity in the first quadrant are (approximately) $.8090 + .5878i$ and $.3090 + .9511i$.

---

*On wide-screen calculators, use $-2 \le x \le 2$ or $-1.7 \le x \le 1.7$ so that the unit circle looks like a circle.
†The point corresponding to $t = 10\pi/5 = 2\pi$ is the same as the one corresponding to $t = 0$.

EXERCISES 9.2

In Exercises 1–6, calculate the given product and express your answer in the form $a + bi$.

1. $\left(\cos \dfrac{\pi}{12} + i \sin \dfrac{\pi}{12}\right)^6$ **2.** $\left(\cos \dfrac{\pi}{5} + i \sin \dfrac{\pi}{5}\right)^{20}$

3. $\left[2\left(\cos \dfrac{\pi}{24} + i \sin \dfrac{\pi}{24}\right)\right]^8$ **4.** $\left[\sqrt{2}\left(\cos \dfrac{\pi}{60} + i \sin \dfrac{\pi}{60}\right)\right]^{10}$

5. $\left[3\left(\cos \dfrac{7\pi}{30} + i \sin \dfrac{7\pi}{30}\right)\right]^5$ **6.** $\left[\sqrt[3]{4}\left(\cos \dfrac{7\pi}{36} + i \sin \dfrac{7\pi}{36}\right)\right]^{12}$

In Exercises 7–14, calculate the product by expressing the number in polar form and using DeMoivre's Theorem. Express your answer in the form $a + bi$.

7. $\left(\dfrac{1}{2} + \dfrac{\sqrt{3}}{2}i\right)^3$ **8.** $\left(-\dfrac{\sqrt{2}}{2} + \dfrac{\sqrt{2}}{2}i\right)^4$

9. $(1 - i)^{12}$ **10.** $(2 + 2i)^8$

11. $\left(\dfrac{\sqrt{3}}{2} + \dfrac{1}{2}i\right)^{10}$ **12.** $\left(-\dfrac{1}{2} + \dfrac{\sqrt{3}}{2}i\right)^{20}$

13. $\left(\dfrac{-1}{\sqrt{2}} + \dfrac{i}{\sqrt{2}}\right)^{14}$ **14.** $(-1 + \sqrt{3}i)^8$

In Exercises 15 and 16, find the indicated roots of unity and express your answers in the form $a + bi$.

15. Fourth roots of unity   **16.** Sixth roots of unity

In Exercises 17–30, find the nth roots in polar form.

17. $36\left(\cos \dfrac{\pi}{3} + i \sin \dfrac{\pi}{3}\right); \quad n = 2$

18. $64\left(\cos \dfrac{\pi}{4} + i \sin \dfrac{\pi}{4}\right); \quad n = 2$

19. $64\left(\cos\dfrac{\pi}{5} + i\sin\dfrac{\pi}{5}\right)$; $n = 3$

20. $8\left(\cos\dfrac{\pi}{10} + i\sin\dfrac{\pi}{10}\right)$; $n = 3$

21. $81\left(\cos\dfrac{\pi}{12} + i\sin\dfrac{\pi}{12}\right)$; $n = 4$

22. $16\left(\cos\dfrac{\pi}{7} + i\sin\dfrac{\pi}{7}\right)$; $n = 5$

23. -1; $n = 5$ **24.** 1; $n = 7$

25. i; $n = 5$ **26.** $-i$; $n = 6$

27. $1 + i$; $n = 2$ **28.** $1 - \sqrt{3}i$; $n = 3$

29. $8\sqrt{3} + 8i$; $n = 4$

30. $-16\sqrt{2} - 16\sqrt{2}i$; $n = 5$

In Exercises 31–40, solve the given equation in the complex number system.

31. $x^6 = -1$ **32.** $x^6 + 64 = 0$

33. $x^3 = i$ **34.** $x^4 = i$

35. $x^3 + 27i = 0$ **36.** $x^6 + 729 = 0$

37. $x^5 - 243i = 0$ **38.** $x^7 = 1 - i$

39. $x^4 = -1 + \sqrt{3}i$ **40.** $x^4 = -8 - 8\sqrt{3}i$

In Exercises 41–46, represent the roots of unity graphically. Then use the trace feature to obtain approximations of the form a + bi for each root (round to four places).

41. Seventh roots of unity   **42.** Fifth roots of unity

43. Eighth roots of unity   **44.** Twelfth roots of unity

45. Ninth roots of unity   **46.** Tenth roots of unity

47. Solve the equation $x^3 + x^2 + x + 1 = 0$. [*Hint:* First find the quotient when $x^4 - 1$ is divided by $x - 1$ and then consider solutions of $x^4 - 1 = 0$.]

48. Solve the equation $x^4 + x^3 + x^2 + x + 1 = 0$. [*Hint:* Consider $x^5 - 1$ and $x - 1$ and see Exercise 47.]

49. Solve $x^5 + x^4 + x^3 + x^2 + x + 1 = 0$. [*Hint:* Consider $x^6 - 1$ and $x - 1$ and see Exercise 47.]

50. What do you think are the solutions of $x^{n-1} + x^{n-2} + \cdots + x^3 + x^2 + x + 1 = 0$? (See Exercises 47–49.)

THINKERS

51. In the complex plane, identify each point with its complex number label. The unit circle consists of all points (numbers) z such that $|z| = 1$. Suppose v and w are two points (numbers) that move around the unit circle in such a way that $v = w^{12}$ at all times. When w has made one complete trip around the circle, how many trips has v made? [*Hint:* Think polar and DeMoivre.]

52. Suppose u is an nth root of unity. Show that $1/u$ is also an nth root of unity. [*Hint:* Use the definition, *not* polar form.]

53. Let $u_1, u_2, \ldots, u_n$ be the distinct nth roots of unity and suppose v is a nonzero solution of the equation

$$z^n = r(\cos\theta + i\sin\theta).$$

Show that $vu_1, vu_2, \ldots, vu_n$ are n distinct solutions of the equation. [*Remember:* Each u_i is a solution of $x^n = 1$.]

54. Use the formula for nth roots and the identities

$$\cos(x + \pi) = -\cos x \qquad \sin(x + \pi) = -\sin x$$

to show that the nonzero complex number $r(\cos\theta + i\sin\theta)$ has two square roots and that these square roots are negatives of each other.

9.3 Vectors in the Plane

Section Objectives
- Find the components and magnitude of a vector.
- Use scalar multiplication, vector addition, and vector subtraction.
- Find a unit vector with the same direction as a given vector, **v**.
- Find the direction angle of a vector.
- Use vectors to solve applied problems.

Once a unit of measure has been agreed upon, quantities such as area, length, time, and temperature can be described by a single number. Other quantities, such as an east wind of 10 mph, require two numbers to describe them because they involve both *magnitude* and *direction.* Such quantities are called **vectors** and are represented geometrically by a directed line segment or arrow, as in Figure 9–9.

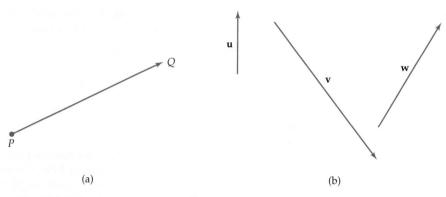

(a)   (b)

Figure 9–9

When a vector extends from a point P to a point Q, as in Figure 9–9(a), P is called the **initial point** of the vector and Q is called the **terminal point,** and the vector is written $\overrightarrow{PQ}$. Its **length** is denoted by $\|\overrightarrow{PQ}\|$. When the endpoints are not specified, as in Figure 9–9(b), vectors are denoted by boldface letters such as **u, v,** and **w.** The length of a vector **u** is denoted by $\|\mathbf{u}\|$ and is called the **magnitude** of **u.**

If **u** and **v** are vectors with the same magnitude and direction, we say that **u** and **v** are **equivalent** and write **u = v.** Some examples are shown in Figure 9–10.

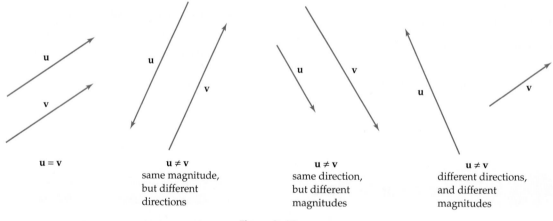

| **u = v** | **u ≠ v**
same magnitude,
but different
directions | **u ≠ v**
same direction,
but different
magnitudes | **u ≠ v**
different directions,
and different
magnitudes |

Figure 9–10

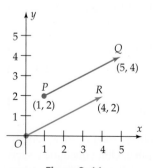

Figure 9–11

EXAMPLE 1

Let $P = (1, 2)$, $Q = (5, 4)$, $O = (0, 0)$, and $R = (4, 2)$, as in Figure 9–11. Show that $\overrightarrow{PQ} = \overrightarrow{OR}$.

SOLUTION   The distance formula shows that $\overrightarrow{PQ}$ and $\overrightarrow{OR}$ have the *same length:*

$$\|\overrightarrow{PQ}\| = \sqrt{(5-1)^2 + (4-2)^2} = \sqrt{4^2 + 2^2} = \sqrt{20}.$$
$$\|\overrightarrow{OR}\| = \sqrt{(4-0)^2 + (2-0)^2} = \sqrt{4^2 + 2^2} = \sqrt{20}.$$

Furthermore, the lines through PQ and OR have the same slope:

$$\text{slope } PQ = \frac{4-2}{5-1} = \frac{2}{4} = \frac{1}{2}, \qquad \text{slope } OR = \frac{2-0}{4-0} = \frac{2}{4} = \frac{1}{2}.$$

Since $\overrightarrow{PQ}$ and $\overrightarrow{OR}$ both point to the upper right on lines of the same slope, $\overrightarrow{PQ}$ and $\overrightarrow{OR}$ have the *same direction*. Therefore, $\overrightarrow{PQ} = \overrightarrow{OR}$. ∎

According to the definition of equivalence, a vector may be moved from one location to another, provided that its magnitude and direction are not changed. Consequently, we have the following.

Equivalent Vectors

> Every vector $\overrightarrow{PQ}$ is equivalent to a vector $\overrightarrow{OR}$ with initial point at the origin: If $P = (x_1, y_1)$ and $Q = (x_2, y_2)$, then
>
> $$\overrightarrow{PQ} = \overrightarrow{OR}, \qquad \text{where } R = (x_2 - x_1, y_2 - y_1).$$

Proof The proof is similar to the one used in Example 1. It follows from the fact that $\overrightarrow{PQ}$ and $\overrightarrow{OR}$ have the same length,

$$\|\overrightarrow{OR}\| = \sqrt{[(x_2 - x_1) - 0]^2 + [(y_2 - y_1) - 0]^2}$$
$$= \sqrt{(x_2 - x_1)^2 + (y_2 - y_1)^2} = \|\overrightarrow{PQ}\|,$$

and that either the line segments PQ and OR are both vertical or they have the same slope,

$$\text{slope } OR = \frac{(y_2 - y_1) - 0}{(x_2 - x_1) - 0} = \frac{y_2 - y_1}{x_2 - x_1} = \text{slope } PQ,$$

as shown in Figure 9–12. ∎

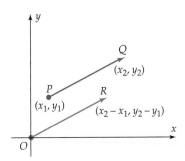

Figure 9–12

The magnitude and direction of a vector with the origin as initial point are completely determined by the coordinates of its terminal point. Consequently, we denote the vector with initial point $(0, 0)$ and terminal point (a, b) by $\langle a, b \rangle$. The numbers a and b are called the **components** of the vector $\langle a, b \rangle$.

Since the length of the vector $\langle a, b \rangle$ is the distance from $(0, 0)$ to (a, b), the distance formula shows that

Magnitude

> The **magnitude** (or **norm**) of the vector $\mathbf{v} = \langle a, b \rangle$ is
>
> $$\|\mathbf{v}\| = \sqrt{a^2 + b^2}.$$

EXAMPLE 2

Find the components and the magnitude of the vector with initial point $P = (-2, 6)$ and terminal point $Q = (4, -3)$.

SOLUTION According to the fact in the first box on page 641 (with $x_1 = -2$, $y_1 = 6$, $x_2 = 4$, $y_2 = -3$):

$$\overrightarrow{PQ} = \overrightarrow{OR}, \qquad \text{where } R = (4 - (-2), -3 - 6) = (6, -9)$$

that is,

$$\overrightarrow{PQ} = \overrightarrow{OR} = \langle 6, -9 \rangle.$$

Therefore,

$$\|\overrightarrow{PQ}\| = \|\overrightarrow{OR}\| = \sqrt{6^2 + (-9)^2} = \sqrt{36 + 81} = \sqrt{117}. \qquad \blacksquare$$

VECTOR ARITHMETIC

When dealing with vectors, it is customary to refer to ordinary real numbers as **scalars. Scalar multiplication** is an operation in which a scalar k is "multiplied" by a vector $\mathbf{v}$ to produce another *vector* denoted by $k\mathbf{v}$. Here is the formal definition.

Scalar Multiplication

If k is a real number and $\mathbf{v} = \langle a, b \rangle$ is a vector, then

$$k\mathbf{v} \text{ is the vector } \langle ka, kb \rangle.$$

The vector $k\mathbf{v}$ is called a **scalar multiple** of $\mathbf{v}$.

EXAMPLE 3

If $\mathbf{v} = \langle 3, 1 \rangle$, then

$$3\mathbf{v} = 3\langle 3, 1 \rangle = \langle 3 \cdot 3, 3 \cdot 1 \rangle = \langle 9, 3 \rangle,$$

$$-2\mathbf{v} = -2\langle 3, 1 \rangle = \langle -2 \cdot 3, -2 \cdot 1 \rangle = \langle -6, -2 \rangle,$$

as shown in Figure 9–13:

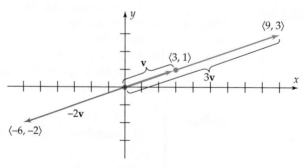

Figure 9–13

Figure 9–13 shows that $3\mathbf{v}$ has the same direction as $\mathbf{v}$, while $-2\mathbf{v}$ has the opposite direction. Also note that

$$\|\mathbf{v}\| = \|\langle 3, 1\rangle\| = \sqrt{3^2 + 1^2} = \sqrt{10}$$
$$\|-2\mathbf{v}\| = \|\langle -6, -2\rangle\| = \sqrt{(-6)^2 + (-2)^2} = \sqrt{40} = 2\sqrt{10}.$$

Therefore,

$$\|-2\mathbf{v}\| = 2\sqrt{10} = 2\|\mathbf{v}\| = |-2| \cdot \|\mathbf{v}\|.$$

Similarly, you can verify that $\|3\mathbf{v}\| = |3| \cdot \|\mathbf{v}\| = 3\|\mathbf{v}\|$. ■

Example 3 is an illustration of the following facts.

Geometric Interpretation of Scalar Multiplication

> The *magnitude* of the vector $k\mathbf{v}$ is $|k|$ times the length of $\mathbf{v}$, that is,
>
> $$\|k\mathbf{v}\| = |k| \cdot \|\mathbf{v}\|.$$
>
> The *direction of* $k\mathbf{v}$ is the same as that of $\mathbf{v}$ when k is positive and opposite that of $\mathbf{v}$ when k is negative.

See Exercise 77 for a proof of this statement.

Vector addition is an operation in which two vectors $\mathbf{u}$ and $\mathbf{v}$ are added to produce a new vector denoted $\mathbf{u} + \mathbf{v}$. Formally, we have the following.

Vector Addition

> If $\mathbf{u} = \langle a, b\rangle$ and $\mathbf{v} = \langle c, d\rangle$, then
>
> $$\mathbf{u} + \mathbf{v} = \langle a + c, b + d\rangle.$$

EXAMPLE 4

If $\mathbf{u} = \langle -5, 2\rangle$ and $\mathbf{v} = \langle 3, 1\rangle$, find $\mathbf{u} + \mathbf{v}$.

TECHNOLOGY TIP

Vector arithmetic and other vector operations can be done on TI-86/89 and HP-39gs.

SOLUTION

$$\mathbf{u} + \mathbf{v} = \langle -5, 2\rangle + \langle 3, 1\rangle = \langle -5 + 3, 2 + 1\rangle = \langle -2, 3\rangle$$

as shown in Figure 9–14. ■

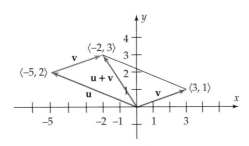

Figure 9–14

Example 4 is an illustration of these facts.

Geometric Interpretations
of Vector Addition

1. If **u** and **v** are vectors with the same initial point P, then **u** + **v** is the vector $\overrightarrow{PQ}$, where $\overrightarrow{PQ}$ is the diagonal of the parallelogram with adjacent sides **u** and **v**.

2. If the vector **v** is moved (without changing its magnitude or direction) so that its initial point lies on the endpoint of the vector **u**, then **u** + **v** is the vector with the same initial point P as **u** and the same terminal point Q as **v**.

See Exercise 78 for a proof of these statements.

The **negative** of a vector $\mathbf{v} = \langle c, d \rangle$ is defined to be the vector $(-1)\mathbf{v} = (-1)\langle c, d \rangle = \langle -c, -d \rangle$ and is denoted $-\mathbf{v}$. **Vector subtraction** is then defined as follows.

Vector
Subtraction

If $\mathbf{u} = \langle a, b \rangle$ and $\mathbf{v} = \langle c, d \rangle$, then $\mathbf{u} - \mathbf{v}$ is the vector

$$\mathbf{u} + (-\mathbf{v}) = \langle a, b \rangle + \langle -c, -d \rangle$$

$$= \langle a - c, b - d \rangle.$$

A geometric interpretation of vector subtraction is given in Exercise 79.

EXAMPLE 5

If $\mathbf{u} = \langle 2, 5 \rangle$ and $\mathbf{v} = \langle 6, 1 \rangle$, find $\mathbf{u} - \mathbf{v}$.

SOLUTION

$$\mathbf{u} - \mathbf{v} = \langle 2, 5 \rangle - \langle 6, 1 \rangle = \langle 2 - 6, 5 - 1 \rangle = \langle -4, 4 \rangle,$$

as shown in Figure 9–15. ■

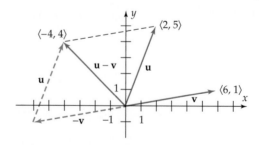

Figure 9–15

The vector $\langle 0, 0 \rangle$ is called the **zero vector** and is denoted **0**.

EXAMPLE 6

If $\mathbf{u} = \langle -1, 6 \rangle$, $\mathbf{v} = \langle 2/3, -4 \rangle$, and $\mathbf{w} = \langle 2, 5/2 \rangle$, find $2\mathbf{u} + 3\mathbf{v}$ and $4\mathbf{w} - 2\mathbf{u}$.

SOLUTION

$$2\mathbf{u} + 3\mathbf{v} = 2\langle -1, 6 \rangle + 3\left\langle \frac{2}{3}, -4 \right\rangle = \langle -2, 12 \rangle + \langle 2, -12 \rangle$$

$$= \langle 0, 0 \rangle = \mathbf{0},$$

and

$$4\mathbf{w} - 2\mathbf{u} = 4\left\langle 2, \frac{5}{2} \right\rangle - 2\langle -1, 6 \rangle$$

$$= \langle 8, 10 \rangle - \langle -2, 12 \rangle$$

$$= \langle 8 - (-2), 10 - 12 \rangle = \langle 10, -2 \rangle. \quad \blacksquare$$

Operations on vectors share many of the same properties as arithmetical operations on numbers.

Properties of Vector Addition and Scalar Multiplication

For any vectors $\mathbf{u}$, $\mathbf{v}$, and $\mathbf{w}$ and any scalars r and s,

1. $\mathbf{u} + (\mathbf{v} + \mathbf{w}) = (\mathbf{u} + \mathbf{v}) + \mathbf{w}$

2. $\mathbf{u} + \mathbf{v} = \mathbf{v} + \mathbf{u}$

3. $\mathbf{v} + \mathbf{0} = \mathbf{v} = \mathbf{0} + \mathbf{v}$

4. $\mathbf{v} + (-\mathbf{v}) = \mathbf{0}$

5. $r(\mathbf{u} + \mathbf{v}) = r\mathbf{u} + r\mathbf{v}$

6. $(r + s)\mathbf{v} = r\mathbf{v} + s\mathbf{v}$

7. $(rs)\mathbf{v} = r(s\mathbf{v}) = s(r\mathbf{v})$

8. $1\mathbf{v} = \mathbf{v}$

9. $0\mathbf{v} = \mathbf{0}$ and $r\mathbf{0} = \mathbf{0}$

Proof   If $\mathbf{u} = \langle a, b \rangle$ and $\mathbf{v} = \langle c, d \rangle$, then because addition of real numbers is commutative, we have

$$\mathbf{u} + \mathbf{v} = \langle a, b \rangle + \langle c, d \rangle = \langle a + c, b + d \rangle$$

$$= \langle c + a, d + b \rangle = \langle c, d \rangle + \langle a, b \rangle = \mathbf{v} + \mathbf{u}.$$

The other properties are proved similarly; see Exercises 53–58. $\blacksquare$

UNIT VECTORS

A vector with length 1 is called a **unit vector.** For instance, $\langle 3/5, 4/5 \rangle$ is a unit vector, since

$$\left\| \left\langle \frac{3}{5}, \frac{4}{5} \right\rangle \right\| = \sqrt{\left(\frac{3}{5} \right)^2 + \left(\frac{4}{5} \right)^2} = \sqrt{\frac{9}{25} + \frac{16}{25}} = \sqrt{\frac{25}{25}} = 1.$$

EXAMPLE 7

Find a unit vector that has the same direction as the vector $\mathbf{v} = \langle 5, 12 \rangle$.

SOLUTION The length of $\mathbf{v}$ is

$$\|\mathbf{v}\| = \|\langle 5, 12 \rangle\| = \sqrt{5^2 + 12^2} = \sqrt{169} = 13.$$

The vector

$$\mathbf{u} = \frac{1}{13}\mathbf{v} = \left\langle \frac{5}{13}, \frac{12}{13} \right\rangle$$

has the same direction as $\mathbf{v}$ (since it is a scalar multiple by a positive number), and $\mathbf{u}$ is a unit vector because

$$\|\mathbf{u}\| = \left\| \frac{1}{13}\mathbf{v} \right\| = \left| \frac{1}{13} \right| \cdot \|\mathbf{v}\| = \frac{1}{13} \cdot 13 = 1. \qquad \blacksquare$$

The procedure used in Example 7 (multiplying a vector by the reciprocal of its length) works in the general case.

TECHNOLOGY TIP

To find a unit vector in the same direction as **v**, use UNITV in this menu/submenu:

TI-86: VECTOR/MATH

TI-89: MATH/MATRIX/
VECTOR OPS

Unit Vectors

> If $\mathbf{v}$ is a nonzero vector, then $\dfrac{1}{\|\mathbf{v}\|}\mathbf{v}$ is a unit vector with the same direction as $\mathbf{v}$.

You can easily verify that the vectors $\mathbf{i} = \langle 1, 0 \rangle$ and $\mathbf{j} = \langle 0, 1 \rangle$ are unit vectors. The vectors $\mathbf{i}$ and $\mathbf{j}$ play a special role because they lead to a useful alternate notation for vectors. For example, if $\mathbf{u} = \langle 5, -7 \rangle$ then

$$\mathbf{u} = \langle 5, 0 \rangle + \langle 0, -7 \rangle = 5\langle 1, 0 \rangle - 7\langle 0, 1 \rangle = 5\mathbf{i} - 7\mathbf{j}.$$

Similarly, if $\mathbf{v} = \langle a, b \rangle$ is any vector, then

$$\mathbf{v} = \langle a, b \rangle = \langle a, 0 \rangle + \langle 0, b \rangle = a\langle 1, 0 \rangle + b\langle 0, 1 \rangle = a\mathbf{i} + b\mathbf{j}.$$

The vector $\mathbf{v}$ is said to be a **linear combination** of $\mathbf{i}$ and $\mathbf{j}$. When vectors are written as linear combinations of $\mathbf{i}$ and $\mathbf{j}$, then the properties in the box on page 645 can be used to write the rules for vector addition and scalar multiplication in this form.

$$(a\mathbf{i} + b\mathbf{j}) + (c\mathbf{i} + d\mathbf{j}) = (a + c)\mathbf{i} + (b + d)\mathbf{j}$$

and

$$c(a\mathbf{i} + b\mathbf{j}) = ca\mathbf{i} + cb\mathbf{j}.$$

EXAMPLE 8

If $\mathbf{u} = 2\mathbf{i} - 6\mathbf{j}$ and $\mathbf{v} = -5\mathbf{i} + 2\mathbf{j}$, find $3\mathbf{u} - 2\mathbf{v}$.

SOLUTION

$$3\mathbf{u} - 2\mathbf{v} = 3(2\mathbf{i} - 6\mathbf{j}) - 2(-5\mathbf{i} + 2\mathbf{j}) = 6\mathbf{i} - 18\mathbf{j} + 10\mathbf{i} - 4\mathbf{j}$$

$$= 16\mathbf{i} - 22\mathbf{j}. \qquad \blacksquare$$

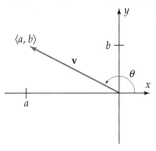

Figure 9–16

DIRECTION ANGLES

If $\mathbf{v} = \langle a, b \rangle = a\mathbf{i} + b\mathbf{j}$ is a vector, then the direction of $\mathbf{v}$ is completely determined by the standard position angle θ between 0° and 360° whose terminal side is $\mathbf{v}$, as shown in Figure 9–16. The angle θ is called the **direction angle** of the vector $\mathbf{v}$. According to the point-in-the-plane description of the trigonometric functions,

$$\cos\theta = \frac{a}{\|\mathbf{v}\|} \quad \text{and} \quad \sin\theta = \frac{b}{\|\mathbf{v}\|}.$$

Rewriting each of these equations produces the following result.

Components of the
Direction Angle

If $\mathbf{v} = \langle a, b \rangle = a\mathbf{i} + b\mathbf{j}$, then

$$a = \|\mathbf{v}\| \cos\theta \quad \text{and} \quad b = \|\mathbf{v}\| \sin\theta$$

where θ is the direction angle of $\mathbf{v}$.

EXAMPLE 9

Find the component form of the vector that represents the velocity of an airplane at the instant its wheels leave the ground if the plane is going 60 mph and the body of the plane makes a 7° angle with the horizontal.

SOLUTION   The velocity vector $\mathbf{v} = a\mathbf{i} + b\mathbf{j}$ has magnitude 60 and direction angle $\theta = 7°$, as shown in Figure 9–17. Hence,

$$\mathbf{v} = (\|\mathbf{v}\| \cos\theta)\mathbf{i} + (\|\mathbf{v}\| \sin\theta)\mathbf{j}$$
$$= (60 \cos 7°)\mathbf{i} + (60 \sin 7°)\mathbf{j}$$
$$\approx (60 \cdot .9925)\mathbf{i} + (60 \cdot .1219)\mathbf{j}$$
$$\approx 59.55\mathbf{i} + 7.31\mathbf{j} = \langle 59.55, 7.31 \rangle. \qquad \blacksquare$$

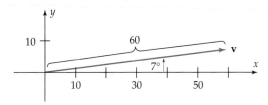

Figure 9–17

If $\mathbf{v} = a\mathbf{i} + b\mathbf{j}$ is a nonzero vector with direction angle θ, then

$$\tan\theta = \frac{\sin\theta}{\cos\theta} = \frac{b/\|\mathbf{v}\|}{a/\|\mathbf{v}\|} = \frac{b}{a}.$$

This fact provides a convenient way to find the direction angle of a vector.

EXAMPLE 10

Find the direction angle of

(a) **u** = 5**i** + 13**j**   (b) **v** = −10**i** + 7**j**.

SOLUTION

(a) The direction angle θ of **u** satisfies $\tan \theta = b/a = 13/5 = 2.6$. Using the TAN^{-1} key on a calculator, we find that $\theta \approx 68.96°$, as shown in Figure 9–18(a).

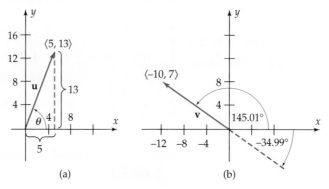

(a)   (b)

Figure 9–18

(b) The direction angle of **v** satisfies $\tan \theta = -7/10 = -.7$. Since **v** lies in the second quadrant, θ must be between $90°$ and $180°$. A calculator shows that $-34.99°$ is an angle with tangent (approximately) $-.7$. Since tangent has period $\pi (= 180°)$, we know that $\tan t = \tan(t + 180°)$ for every t. Therefore, $\theta = -34.99° + 180° = 145.01°$ is the angle between $90°$ and $180°$ such that $\tan \theta \approx -.7$. See Figure 9–18(b). ∎

EXAMPLE 11

An object at the origin is acted upon by two forces. A 150-pound force makes an angle of $20°$ with the positive x-axis, and the other force of 100 pounds makes an angle of $70°$, as shown in Figure 9–19. Find the direction and magnitude of the resultant force.

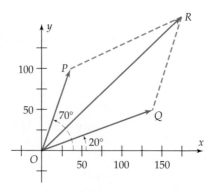

Figure 9–19

SOLUTION   The forces acting on the object are

$$\overrightarrow{OP} = (100 \cos 70°)\mathbf{i} + (100 \sin 70°)\mathbf{j},$$

$$\overrightarrow{OQ} = (150 \cos 20°)\mathbf{i} + (150 \sin 20°)\mathbf{j}.$$

The resultant force $\overrightarrow{OR}$ is the sum of $\overrightarrow{OP}$ and $\overrightarrow{OQ}$. Hence,

$$\overrightarrow{OR} = (100 \cos 70° + 150° \cos 20°)\mathbf{i} + (100 \sin 70° + 150 \sin 20°)\mathbf{j}$$

$$\approx 175.16\mathbf{i} + 145.27\mathbf{j}.$$

Therefore, the magnitude of the resultant force is

$$\|\overrightarrow{OR}\| \approx \sqrt{(175.16)^2 + (145.27)^2} \approx 227.56.$$

The direction angle θ of the resultant force satisfies

$$\tan \theta \approx 145.27/175.16 \approx .8294.$$

A calculator shows that $\theta \approx 39.67°$.   ∎

APPLICATIONS

EXAMPLE 12

A 200-pound box lies on a ramp that makes an angle of 24° with the horizontal. A rope is tied to the box from a post at the top of the ramp to keep it in position. Ignoring friction, how much force is being exerted on the rope by the box?

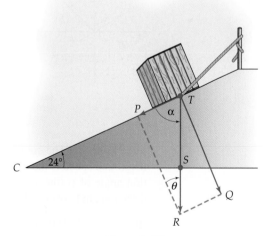

Figure 9–20

SOLUTION   Because of gravity, the box exerts a 200-pound weight straight down (vector $\overrightarrow{TR}$). As Figure 9–20 shows, $\overrightarrow{TR}$ is the sum of $\overrightarrow{TP}$ and $\overrightarrow{TQ}$. The force on the rope is exerted by $\overrightarrow{TP}$, the vector of the force pulling the box down the ramp, so we must find $\|\overrightarrow{TP}\|$. In right triangle TSC, $\alpha + 24° = 90°$, and in right triangle TPR, $\alpha + \theta = 90°$. Hence,

$$\alpha + \theta = \alpha + 24°; \qquad \text{hence,} \qquad \theta = 24°.$$

Therefore,

$$\frac{\|\overrightarrow{TP}\|}{\|\overrightarrow{TR}\|} = \sin \theta$$

$$\frac{\|\overrightarrow{TP}\|}{200} = \sin 24°$$

$$\|\overrightarrow{TP}\| = 200 \sin 24° \approx 81.35.$$

So the force on the rope is 81.35 pounds. ■

In aerial navigation, directions are given in terms of the angle measured in degrees clockwise from true north. Thus, north is 0°, east is 90°, and so on.

EXAMPLE 13

An airplane is traveling in the direction 50° with an air speed of 300 mph, and there is a 35-mph wind from the direction 120°, as represented by the vectors **p** and **w** in Figure 9–21(a). Find the *course* and *ground speed* of the plane (that is, its direction and speed relative to the ground).

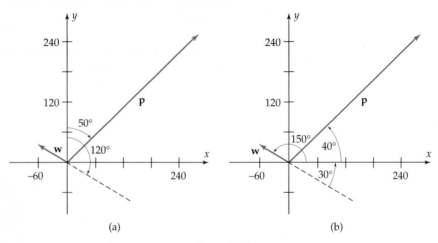

(a)   (b)

Figure 9–21

SOLUTION   The course of the plane is the direction of the vector **p** + **w**, and its ground speed is the magnitude of **p** + **w**. Figure 9–21(b) shows that the direction angle of **p** (the angle it makes with the positive *x*-axis) is 40° and that the direction angle of **w** is 150°. Therefore,

$$\mathbf{p} + \mathbf{w} = [(300 \cos 40°)\mathbf{i} + (300 \sin 40°)\mathbf{j}] + [(35 \cos 150°)\mathbf{i} + (35 \sin 150°)\mathbf{j}]$$

$$= (300 \cos 40° + 35 \cos 150°)\mathbf{i} + (300 \sin 40° + 35 \sin 150°)\mathbf{j}$$

$$\approx 199.50\mathbf{i} + 210.34\mathbf{j}.$$

The direction angle of **p** + **w** satisfies $\tan \theta = 210.34/199.50 \approx 1.0543$, and a calculator shows that $\theta \approx 46.5°$. This is the angle **p** + **w** makes with the positive *x*-axis; hence, the course of the plane (the angle between true north and **p** + **w**) is $90° - 46.5° = 43.5°$. The ground speed of the plane is

$$\|\mathbf{p} + \mathbf{w}\| \approx \sqrt{(199.5)^2 + (210.34)^2} \approx 289.9 \text{ mph.}$$ ■

EXERCISES 9.3

In Exercises 1–4, find the magnitude of the vector $\overrightarrow{PQ}$.

1. $P = (2, 3)$, $Q = (5, 9)$

2. $P = (-3, 5)$, $Q = (7, -11)$

3. $P = (-7, 0)$, $Q = (-4, -5)$

4. $P = (30, 12)$, $Q = (25, 5)$

In Exercises 5–10, find a vector with the origin as initial point that is equivalent to the vector $\overrightarrow{PQ}$.

5. $P = (1, 5)$, $Q = (7, 11)$

6. $P = (2, 7)$, $Q = (-2, 9)$

7. $P = (-4, -8)$, $Q = (-10, 2)$

8. $P = (-5, 6)$, $Q = (-7, -9)$

9. $P = \left(\frac{4}{5}, -2\right)$, $Q = \left(\frac{17}{5}, -\frac{12}{5}\right)$

10. $P = (\sqrt{2}, 4)$, $Q = (\sqrt{3}, -1)$

In Exercises 11–20, find $u + v$, $v - u$, and $2u - 3v$.

11. $u = \langle -2, 4 \rangle$, $v = \langle 6, 1 \rangle$

12. $u = \langle 4, 0 \rangle$, $v = \langle 1, -3 \rangle$

13. $u = \langle 3, 3\sqrt{2} \rangle$, $v = \langle 4\sqrt{2}, 1 \rangle$

14. $u = \left\langle \frac{2}{3}, 4 \right\rangle$, $v = \left\langle -7, \frac{19}{3} \right\rangle$

15. $u = 2\langle -2, 5 \rangle$, $v = \frac{1}{4}\langle -7, 12 \rangle$

16. $u = i - j$, $v = 2i + j$

17. $u = 8i$, $v = 2(3i - 2j)$

18. $u = -4(-i + j)$, $v = -3i$

19. $u = -\left(2i + \frac{3}{2}j\right)$, $v = \frac{3}{4}i$

20. $u = \sqrt{2}j$, $v = \sqrt{3}i$

In Exercises 21–26, find the components of the given vector, where $u = i - 2j$, $v = 3i + j$, $w = -4i + j$.

21. $u + 2w$

22. $\frac{1}{2}(3v + w)$

23. $\frac{1}{2}w$

24. $-2u + 3v$

25. $\frac{1}{4}(8u + 4v - w)$

26. $3(u - 2v) - 6w$

In Exercises 27–34, find the component form of the vector v whose magnitude and direction angle θ are given.

27. $\|v\| = 4$, $\theta = 0°$

28. $\|v\| = 5$, $\theta = 30°$

29. $\|v\| = 10$, $\theta = 225°$

30. $\|v\| = 20$, $\theta = 120°$

31. $\|v\| = 6$, $\theta = 40°$

32. $\|v\| = 8$, $\theta = 160°$

33. $\|v\| = 1/2$, $\theta = 250°$

34. $\|v\| = 3$, $\theta = 310°$

In Exercises 35–42, find the magnitude and direction angle of the vector v.

35. $v = \langle 4, 4 \rangle$

36. $v = \langle 5, 5\sqrt{3} \rangle$

37. $v = \langle -8, 0 \rangle$

38. $v = \langle 4, 5 \rangle$

39. $v = 6j$

40. $v = 4i - 8j$

41. $v = -2i + 8j$

42. $v = -15i - 10j$

In Exercises 43–46, find a unit vector that has the same direction as v.

43. $\langle 4, -5 \rangle$

44. $-7i + 8j$

45. $5i + 10j$

46. $-3i - 9j$

In Exercises 47–50, an object at the origin is acted upon by two forces, u and v, with direction angle θ_u and θ_v, respectively. Find the direction and magnitude of the resultant force.

47. $u = 30$ pounds, $\theta_u = 0°$; $v = 90$ pounds, $\theta_v = 60°$

48. $u = 6$ pounds, $\theta_u = 45°$; $v = 6$ pounds, $\theta_v = 120°$

49. $u = 12$ newtons, $\theta_u = 130°$; $v = 20$ newtons $\theta_v = 250°$

50. $u = 30$ newtons, $\theta_u = 300°$; $v = 80$ newtons, $\theta_v = 40°$

If forces $u_1, u_2, \ldots, u_k$ act on an object at the origin, the result-ant force is the sum $u_1 + u_2 + \cdots + u_k$. The forces are said to be in equilibrium if their resultant force is 0. In Exercises 51 and 52, find the resultant force and find an additional force v that, if added to the system, produces equilibrium.

51. $u_1 = \langle 2, 5 \rangle$, $u_2 = \langle -6, 1 \rangle$, $u_3 = \langle -4, -8 \rangle$

52. $u_1 = \langle 3, 7 \rangle$, $u_2 = \langle 8, -2 \rangle$, $u_3 = \langle -9, 0 \rangle$, $u_4 = \langle -5, 4 \rangle$

In Exercises 53–58, let $u = \langle a, b \rangle$ and $v = \langle c, d \rangle$, and let r and s be scalars. Prove that the stated property holds by calculat-ing the vector on each side of the equal sign.

53. $v + 0 = v = 0 + v$

54. $v + (-v) = 0$

55. $r(u + v) = ru + rv$

56. $(r + s)v = rv + sv$

57. $(rs)v = r(sv) = s(rv)$

58. $1v = v$ and $0v = 0$

59. Two ropes are tied to a wagon. A child pulls one with a force of 20 pounds while another child pulls the other with a force of 30 pounds (see figure on the next page). If the angle be-tween the two ropes is 28°, how much force must be exerted by a third child, standing behind the wagon, to keep the wagon from moving? [*Hint:* Assume that the wagon is at the

origin and one rope runs along the positive *x*-axis. Proceed as in Example 11 to find the resultant force on the wagon from the ropes. The third child must use the same amount in the opposite direction.]

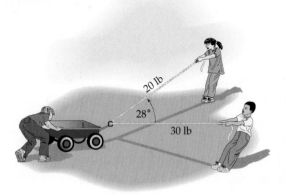

60. Two circus elephants, Bessie and Maybelle, are dragging a large wagon, as shown in the figure. If Bessie pulls with a force of 2200 pounds and Maybelle with a force of 1500 pounds and the wagon moves along the dashed line, what is angle θ?

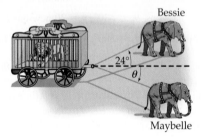

Exercises 61–64 deal with an object on an inclined plane. The situation is similar to that in Figure 9–20 of Example 12, where $\|\overrightarrow{TP}\|$ is the component of the weight of the object parallel to the plane and $\|\overrightarrow{TQ}\|$ is the component of the weight perpendicular to the plane.

61. An object weighing 50 pounds lies on an inclined plane that makes a 40° angle with the horizontal. Find the components of the weight parallel and perpendicular to the plane. [*Hint:* Solve an appropriate triangle.]

62. Do Exercise 61 when the object weighs 200 pounds and the inclined plane makes a 20° angle with the horizontal.

63. If an object on an inclined plane weighs 150 pounds and the component of the weight perpendicular to the plane is 60 pounds, what angle does the plane make with the horizontal?

64. A force of 500 pounds is needed to pull a cart up a ramp that makes a 15° angle with the ground. Assume that no friction is involved, and find the weight of the cart. [*Hint:* Draw a picture similar to Figure 9–20; the 500-pound force is parallel to the ramp.]

In Exercises 65–68, find the course and ground speed of the plane under the given conditions. (See Example 13.)

65. Air speed 250 mph in the direction 60°; wind speed 40 mph from the direction 330°.

66. Air speed 400 mph in the direction 150°; wind speed 30 mph from the direction 60°.

67. Air speed 300 mph in the direction 300°; wind speed 50 mph in (*not* from) the direction 30°.

68. Air speed 500 mph in the direction 180°; wind speed 70 mph in the direction 40°.

69. The course and ground speed of a plane are 70° and 400 mph, respectively. There is a 60-mph wind blowing south. Find the (approximate) direction and air speed of the plane.

70. A plane is flying in the direction 200° with an air speed of 500 mph. Its course and ground speed are 210° and 450 mph, respectively. What are the direction and speed of the wind?

71. A river flows from east to west. A swimmer on the south bank wants to swim to a point on the opposite shore directly north of her starting point. She can swim at 2.8 mph, and there is a 1-mph current in the river. In what direction should she head so as to travel directly north (that is, what angle should her path make with the south bank of the river)?

72. A river flows from west to east. A swimmer on the north bank swims at 3.1 mph along a straight course that makes a 75° angle with the north bank of the river and reaches the south bank at a point directly south of his starting point. How fast is the current in the river?

73. A 400-pound weight is suspended by two cables (see figure). What is the force (tension) on each cable? [*Hint:* Imagine that the weight is at the origin and that the dashed line is the *x*-axis. Then cable **v** is represented by the vector $(c \cos 65°)\mathbf{i} + (c \sin 65°)\mathbf{j}$, which has magnitude *c* (why?). Represent cable **u** similarly, denoting its magnitude by *d*. Use the fact that $\mathbf{u} + \mathbf{v} = 0\mathbf{i} + 400\mathbf{j}$ (why?) to set up a system of two equations in the unknowns *c* and *d*.]

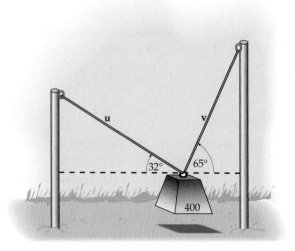

74. An 800-pound weight is suspended from two cables, as shown in the figure. What is the tension (force) on each cable? [See the Hint for Exercise 73.]

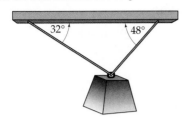

75. Do Exercise 74 when the weight is 600 pounds and the angles are 28° and 38°.

76. A 175-pound high-wire artist stands balanced on a tightrope, which sags slightly at the point where he is standing. The rope in front of him makes a 6° angle with the horizontal, and the rope behind him makes a 4° angle with the horizontal. Find the force on each end of the rope. [*Hint:* Use a picture and procedure similar to that in Exercise 73.]

77. Let **v** be the vector with initial point (x_1, y_1) and terminal point (x_2, y_2), and let k be any real number.

(a) Find the component form of **v** and k**v**.
(b) Calculate $\|\mathbf{v}\|$ and $\|k\mathbf{v}\|$.
(c) Use the fact that $\sqrt{k^2} = |k|$ to verify that $\|k\mathbf{v}\| = |k| \cdot \|\mathbf{v}\|$.
(d) Show that $\tan \theta = \tan \beta$, where θ is the direction angle of **v** and β is the direction angle of k**v**. Use the fact that $\tan t = \tan(t + 180°)$ to conclude that **v** and k**v** have either the same or opposite directions.
(e) Use the fact that (c, d) and $(-c, -d)$ lie on the same straight line on opposite sides of the origin (Exercise 85 in Section 1.3) to verify that **v** and k**v** have the same direction if $k > 0$ and opposite directions if $k < 0$.

78. Let $\mathbf{u} = \langle a, b \rangle$, $\mathbf{v} = \langle c, d \rangle$. Verify the accuracy of the two geometric interpretations of vector addition given on page 644 as follows.

(a) Show that the distance from (a, b) to $(a + c, b + d)$ is the same as $\|\mathbf{v}\|$.
(b) Show that the distance from (c, d) to $(a + c, b + d)$ is the same as $\|\mathbf{u}\|$.
(c) Show that the line through (a, b) and $(a + c, b + d)$ is parallel to **v** by showing they have the same slope.
(d) Show that the line through (c, d) and $(a + c, b + d)$ is parallel to **u**.

79. Let $\mathbf{u} = \langle a, b \rangle$ and $\mathbf{v} = \langle c, d \rangle$. Show that $\mathbf{u} - \mathbf{v}$ is equivalent to the vector **w** with initial point (c, d) and terminal point (a, b) as follows. (See the figure.)

(a) Show that $\|\mathbf{u} - \mathbf{v}\| = \|\mathbf{w}\|$.
(b) Show that $\mathbf{u} - \mathbf{v}$ and **w** have the same direction.

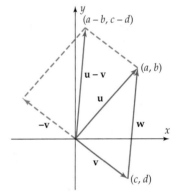

9.4 The Dot Product

Section Objectives
- ■ Find the dot product of two vectors.
- ■ Find the angle between two vectors.
- ■ Determine when two vectors are parallel or orthogonal.
- ■ Find the projections and components of a pair of vectors.
- ■ Use vectors to solve application problems.

When two vectors are added, their sum is another vector, but the situation is different with products. The dot product of two *vectors* is the *real number* defined as follows.

Dot Product

> The **dot product** of vectors $\mathbf{u} = \langle a, b \rangle = a\mathbf{i} + b\mathbf{j}$ and $\mathbf{v} = \langle c, d \rangle = c\mathbf{i} + d\mathbf{j}$ is denoted $\mathbf{u} \cdot \mathbf{v}$ and is defined to be the real number $ac + bd$. Thus,
>
> $$\mathbf{u} \cdot \mathbf{v} = ac + bd.$$

TECHNOLOGY TIP

To compute dot products use DOT or DOTP in this menu/submenu:

TI-86: VECTOR/MATH
TI-89: MATH/MATRIX/VECTOR OPS
HP-39gs: MATH/MATRIX

EXAMPLE 1

(a) If $\mathbf{u} = \langle 5, 3\rangle$ and $\mathbf{v} = \langle -2, 6\rangle$, then

$$\mathbf{u} \cdot \mathbf{v} = 5(-2) + 3 \cdot 6 = 8.$$

(b) If $\mathbf{u} = 4\mathbf{i} - 2\mathbf{j}$ and $\mathbf{v} = 3\mathbf{i} - \mathbf{j}$, then

$$\mathbf{u} \cdot \mathbf{v} = 4 \cdot 3 + (-2)(-1) = 14.$$

(c) $\langle 2, -4\rangle \cdot \langle 6, 3\rangle = 2 \cdot 6 + (-4)3 = 0.$ ■

The dot product has a number of useful properties.

Properties of the Dot Product

If $\mathbf{u}$, $\mathbf{v}$, $\mathbf{w}$ are vectors and k is a real number, then

1. $\mathbf{u} \cdot \mathbf{u} = \|\mathbf{u}\|^2$.
2. $\mathbf{u} \cdot \mathbf{v} = \mathbf{v} \cdot \mathbf{u}$.
3. $\mathbf{u} \cdot (\mathbf{v} + \mathbf{w}) = \mathbf{u} \cdot \mathbf{v} + \mathbf{u} \cdot \mathbf{w}$.
4. $k\mathbf{u} \cdot \mathbf{v} = k(\mathbf{u} \cdot \mathbf{v}) = \mathbf{u} \cdot k\mathbf{v}$.
5. $\mathbf{0} \cdot \mathbf{u} = 0$.

Proof

1. If $\mathbf{u} = \langle a, b\rangle$, then

$$\|\mathbf{u}\| = \sqrt{a^2 + b^2}.$$

Hence,

$$\mathbf{u} \cdot \mathbf{u} = \langle a, b\rangle \cdot \langle a, b\rangle = a \cdot a + b \cdot b = a^2 + b^2 = (\sqrt{a^2 + b^2})^2 = \|\mathbf{u}\|^2.$$

2. If $\mathbf{u} = \langle a, b\rangle$ and $\mathbf{v} = \langle c, d\rangle$, then

$$\mathbf{u} \cdot \mathbf{v} = \langle a, b\rangle \cdot \langle c, d\rangle = ac + bd = ca + db = \langle c, d\rangle \cdot \langle a, b\rangle = \mathbf{v} \cdot \mathbf{u}.$$

The last three statements are proved similarly (Exercises 37–39.) ■

ANGLES

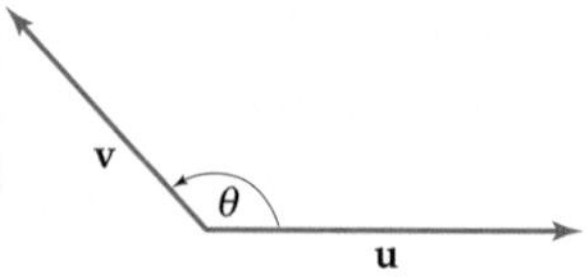

Figure 9–22

If $\mathbf{u} = \langle a, b\rangle$ and $\mathbf{v} = \langle c, d\rangle$ are nonzero vectors, then the **angle between u and v** is the smallest angle θ formed by these two line segments, as shown in Figure 9–22. We ignore clockwise or counterclockwise rotation and consider the angle between $\mathbf{v}$ and $\mathbf{u}$ to be the same as the angle between $\mathbf{u}$ and $\mathbf{v}$. Thus, the radian measure of θ ranges from 0 to π.

Nonzero vectors $\mathbf{u}$ and $\mathbf{v}$ are said to be **parallel** if the angle between them is either 0 or π radians (that is, $\mathbf{u}$ and $\mathbf{v}$ lie on the same straight line through the origin and have either the same or opposite directions). The zero vector $\mathbf{0}$ is considered to be parallel to every vector.

Any scalar multiple of $\mathbf{u}$ is parallel to $\mathbf{u}$, since it lies on the same straight line as $\mathbf{u}$ (see Example 3 in Section 9.3). Conversely, if $\mathbf{v}$ is parallel to $\mathbf{u}$, it is easy to show that $\mathbf{v}$ must be a scalar multiple of $\mathbf{u}$ (Exercise 40). Hence, we have the following.

Parallel Vectors

> Vectors **u** and **v** are parallel exactly when
>
> $$\mathbf{v} = k\mathbf{u} \text{ for some real number } k.$$

EXAMPLE 2

The vectors $\langle 2, 3 \rangle$ and $\langle 8, 12 \rangle$ are parallel because $\langle 8, 12 \rangle = 4\langle 2, 3 \rangle$.   ∎

The angle between nonzero vectors **u** and **v** is closely related to their dot product.

Angle Theorem

> If θ is the angle between the nonzero vectors **u** and **v**, then
>
> $$\mathbf{u} \cdot \mathbf{v} = \|\mathbf{u}\| \, \|\mathbf{v}\| \cos \theta$$
>
> or, equivalently,
>
> $$\cos \theta = \frac{\mathbf{u} \cdot \mathbf{v}}{\|\mathbf{u}\| \, \|\mathbf{v}\|}.$$

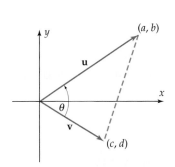

Figure 9–23

Proof   If $\mathbf{u} = \langle a, b \rangle$, $\mathbf{v} = \langle c, d \rangle$, and the angle θ is not 0 or π, then **u** and **v** form two sides of a triangle, as shown in Figure 9–23.

The lengths of two sides of the triangle are $\|\mathbf{u}\| = \sqrt{a^2 + b^2}$ and $\|\mathbf{v}\| = \sqrt{c^2 + d^2}$. The distance formula shows that the length of the third side (opposite angle θ) is $\sqrt{(a - c)^2 + (b - d)^2}$. Therefore, by the Law of Cosines,

$$[\sqrt{(a - c)^2 + (b - d)^2}]^2 = \|\mathbf{u}\|^2 + \|\mathbf{v}\|^2 - 2\|\mathbf{u}\| \, \|\mathbf{v}\| \cos \theta$$

$$(a - c)^2 + (b - d)^2 = (a^2 + b^2) + (c^2 + d^2) - 2\|\mathbf{u}\| \, \|\mathbf{v}\| \cos \theta$$

$$a^2 - 2ac + c^2 + b^2 - 2bd + d^2 = (a^2 + c^2) + (b^2 + d^2) - 2\|\mathbf{u}\| \, \|\mathbf{v}\| \cos \theta$$

$$-2ac - 2bd = -2\|\mathbf{u}\| \, \|\mathbf{v}\| \cos \theta.$$

Dividing both sides by -2 shows that

$$ac + bd = \|\mathbf{u}\| \, \|\mathbf{v}\| \cos \theta.$$

Since the left side of this equation is precisely $\mathbf{u} \cdot \mathbf{v}$, the proof is complete in this case. The proof when θ is 0 or π is left to the reader (Exercise 41).   ∎

EXAMPLE 3

Find the angle θ between the vectors $\langle -3, 1 \rangle$ and $\langle 5, 2 \rangle$ shown in Figure 9–24.

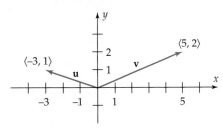

Figure 9–24

SOLUTION   Apply the formula in the box above with $\mathbf{u} = \langle -3, 1 \rangle$ and $\mathbf{v} = \langle 5, 2 \rangle$.

$$\cos \theta = \frac{\mathbf{u} \cdot \mathbf{v}}{\|\mathbf{u}\| \, \|\mathbf{v}\|} = \frac{(-3)5 + 1 \cdot 2}{\sqrt{(-3)^2 + 1^2} \, \sqrt{5^2 + 2^2}} = \frac{-13}{\sqrt{10} \, \sqrt{29}} = \frac{-13}{\sqrt{290}}.$$

Using the $\cos^{-1}$ key, we see that

$$\theta \approx 2.4393 \text{ radians } (\approx 139.76°).$$

The Angle Theorem has several useful consequences. For instance, by taking absolute values on both sides of $\mathbf{u} \cdot \mathbf{v} = \|\mathbf{u}\| \, \|\mathbf{v}\| \cos \theta$ and using the fact that $|\|\mathbf{u}\| \, \|\mathbf{v}\|| = \|\mathbf{u}\| \, \|\mathbf{v}\|$ (because $\|\mathbf{u}\| \, \|\mathbf{v}\| \geq 0$), we see that

$$|\mathbf{u} \cdot \mathbf{v}| = |\|\mathbf{u}\| \, \|\mathbf{v}\| \cos \theta| = |\|\mathbf{u}\| \, \|\mathbf{v}\|| \, |\cos \theta| = \|\mathbf{u}\| \, \|\mathbf{v}\| \, |\cos \theta|.$$

But for any angle θ, $|\cos \theta| \leq 1$, so

$$|\mathbf{u} \cdot \mathbf{v}| = \|\mathbf{u}\| \, \|\mathbf{v}\| \, |\cos \theta| \leq \|\mathbf{u}\| \, \|\mathbf{v}\|.$$

This proves the Schwarz inequality.

Schwarz Inequality

> For any vectors $\mathbf{u}$ and $\mathbf{v}$,
>
> $$|\mathbf{u} \cdot \mathbf{v}| \leq \|\mathbf{u}\| \, \|\mathbf{v}\|.$$

Vectors $\mathbf{u}$ and $\mathbf{v}$ are said to be **orthogonal** (or **perpendicular**) if the angle between them is $\pi/2$ radians (90°), or if at least one of them is $\mathbf{0}$. Here is the key fact about orthogonal vectors.

Orthogonal Vectors

> Let $\mathbf{u}$ and $\mathbf{v}$ be vectors. Then
>
> $\mathbf{u}$ and $\mathbf{v}$ are orthogonal exactly when $\mathbf{u} \cdot \mathbf{v} = 0$.

Proof   If $\mathbf{u}$ or $\mathbf{v}$ is $\mathbf{0}$, then $\mathbf{u} \cdot \mathbf{v} = 0$, and if $\mathbf{u}$ and $\mathbf{v}$ are nonzero orthogonal vectors, then by the Angle Theorem,

$$\mathbf{u} \cdot \mathbf{v} = \|\mathbf{u}\| \, \|\mathbf{v}\| \cos \theta = \|\mathbf{u}\| \, \|\mathbf{v}\| \cos(\pi/2) = \|\mathbf{u}\| \, \|\mathbf{v}\| \, (0) = 0.$$

Conversely, if $\mathbf{u}$ and $\mathbf{v}$ are vectors such that $\mathbf{u} \cdot \mathbf{v} = 0$, then Exercise 42 shows that $\mathbf{u}$ and $\mathbf{v}$ are orthogonal.

EXAMPLE 4

(a) The vectors $\mathbf{u} = \langle 2, -6 \rangle$ and $\mathbf{v} = \langle 9, 3 \rangle$ are orthogonal because

$$\mathbf{u} \cdot \mathbf{v} = \langle 2, -6 \rangle \cdot \langle 9, 3 \rangle = 2 \cdot 9 + (-6)3 = 18 - 18 = 0.$$

(b) The vectors $\frac{1}{2}\mathbf{i} + 5\mathbf{j}$ and $10\mathbf{i} - \mathbf{j}$ are orthogonal, since

$$\left(\frac{1}{2}\mathbf{i} + 5\mathbf{j} \right) \cdot (10\mathbf{i} - \mathbf{j}) = \frac{1}{2}(10) + 5(-1) = 5 - 5 = 0.$$

PROJECTIONS AND COMPONENTS

If **u** and **v** are nonzero vectors and θ is the angle between them, construct the perpendicular line segment from the terminal point P of **u** to the straight line on which **v** lies. This perpendicular segment intersects the line at a point Q, as shown in Figure 9–25.

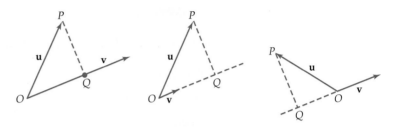

Figure 9–25

The vector $\overrightarrow{OQ}$ is called the **projection of u on v** and is denoted $\text{proj}_v\mathbf{u}$. Here is a useful description of $\text{proj}_v\mathbf{u}$.

Projection of u on v

If **u** and **v** are nonzero vectors, then the projection of **u** on **v** is the vector

$$\text{proj}_v\mathbf{u} = \left(\frac{\mathbf{u} \cdot \mathbf{v}}{\|\mathbf{v}\|^2}\right)\mathbf{v}.$$

Proof   Since $\text{proj}_v\mathbf{u}$ and **v** lie on the same straight line, they are parallel, and hence, $\text{proj}_v\mathbf{u} = k\mathbf{v}$ for some real number k. Let **w** be the vector with initial point at the origin and the same length and direction as $\overrightarrow{QP}$, as in the two cases shown in Figure 9–26.

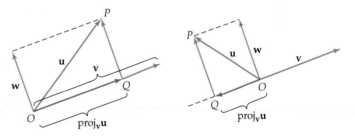

Figure 9–26

Note that **w** is parallel to $\overrightarrow{QP}$ and hence is orthogonal to **v**. As is shown in Figure 9–26, $\mathbf{u} = \text{proj}_v\mathbf{u} + \mathbf{w} = k\mathbf{v} + \mathbf{w}$. Consequently, by the properties of the dot product,

$$\mathbf{u} \cdot \mathbf{v} = (k\mathbf{v} + \mathbf{w}) \cdot \mathbf{v} = (k\mathbf{v}) \cdot \mathbf{v} + \mathbf{w} \cdot \mathbf{v}$$

$$= k(\mathbf{v} \cdot \mathbf{v}) + \mathbf{w} \cdot \mathbf{v} = k\|\mathbf{v}\|^2 + \mathbf{w} \cdot \mathbf{v}.$$

But $\mathbf{w} \cdot \mathbf{v} = 0$ because **w** and **v** are orthogonal. Hence,

$$\mathbf{u} \cdot \mathbf{v} = k\|\mathbf{v}\|^2 \qquad \text{or, equivalently,} \qquad k = \frac{\mathbf{u} \cdot \mathbf{v}}{\|\mathbf{v}\|^2}.$$

Therefore,

$$\text{proj}_\mathbf{v}\mathbf{u} = k\mathbf{v} = \left(\frac{\mathbf{u} \cdot \mathbf{v}}{\|\mathbf{v}\|^2} \right)\mathbf{v},$$

and the proof is complete. ■

EXAMPLE 5

If $\mathbf{u} = 8\mathbf{i} + 3\mathbf{j}$ and $\mathbf{v} = 4\mathbf{i} - 2\mathbf{j}$, find $\text{proj}_\mathbf{v}\mathbf{u}$ and $\text{proj}_\mathbf{u}\mathbf{v}$.

SOLUTION

$$\mathbf{u} \cdot \mathbf{v} = 8 \cdot 4 + 3(-2) = 26, \quad \text{and} \quad \|\mathbf{v}\|^2 = \mathbf{v} \cdot \mathbf{v} = 4^2 + (-2)^2 = 20.$$

Therefore,

$$\text{proj}_\mathbf{v}\mathbf{u} = \left(\frac{\mathbf{u} \cdot \mathbf{v}}{\|\mathbf{v}\|^2} \right)\mathbf{v} = \frac{26}{20}(4\mathbf{i} - 2\mathbf{j}) = \frac{26}{5}\mathbf{i} - \frac{13}{5}\mathbf{j},$$

as is shown in Figure 9–27. We can find the projection of $\mathbf{v}$ on $\mathbf{u}$ by noting that $\|\mathbf{u}\|^2 = \mathbf{u} \cdot \mathbf{u} = 8^2 + 3^2 = 73$, and hence,

$$\text{proj}_\mathbf{u}\mathbf{v} = \left(\frac{\mathbf{v} \cdot \mathbf{u}}{\|\mathbf{u}\|^2} \right)\mathbf{u} = \frac{26}{73}(8\mathbf{i} + 3\mathbf{j}) = \frac{208}{73}\mathbf{i} + \frac{78}{73}\mathbf{j}. \qquad ■$$

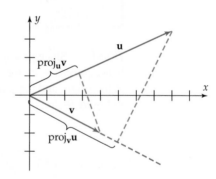

Figure 9–27

Recall that $\frac{1}{\|\mathbf{v}\|}\mathbf{v}$ is a unit vector in the direction of $\mathbf{v}$ (see page 646). We can express $\text{proj}_\mathbf{v}\mathbf{u}$ as a scalar multiple of this unit vector as follows.

$$\text{proj}_\mathbf{v}\mathbf{u} = \left(\frac{\mathbf{u} \cdot \mathbf{v}}{\|\mathbf{v}\|^2} \right)\mathbf{v} = \left(\frac{\mathbf{u} \cdot \mathbf{v}}{\|\mathbf{v}\|} \right)\left(\frac{1}{\|\mathbf{v}\|}\mathbf{v} \right).$$

The scalar $\frac{\mathbf{u} \cdot \mathbf{v}}{\|\mathbf{v}\|}$ is called the **component of u along v** and is denoted $\text{comp}_\mathbf{v}\mathbf{u}$.

Thus,

$$\text{proj}_\mathbf{v}\mathbf{u} = \left(\frac{\mathbf{u} \cdot \mathbf{v}}{\|\mathbf{v}\|} \right)\left(\frac{1}{\|\mathbf{v}\|}\mathbf{v} \right) = \text{comp}_\mathbf{v}\mathbf{u}\left(\frac{1}{\|\mathbf{v}\|}\mathbf{v} \right).$$

Since $\frac{1}{\|\mathbf{v}\|}\mathbf{v}$ is a unit vector, the length of $\text{proj}_\mathbf{v}\mathbf{u}$ is

$$\|\text{proj}_\mathbf{v}\mathbf{u}\| = \left\| \text{comp}_\mathbf{v}\mathbf{u}\left(\frac{1}{\|\mathbf{v}\|}\mathbf{v} \right) \right\| = |\text{comp}_\mathbf{v}\mathbf{u}| \left\| \frac{1}{\|\mathbf{v}\|}\mathbf{v} \right\| = |\text{comp}_\mathbf{v}\mathbf{u}|.$$

Furthermore, since $\mathbf{u} \cdot \mathbf{v} = \|\mathbf{u}\|\,\|\mathbf{v}\| \cos \theta$, where θ is the angle between $\mathbf{u}$ and $\mathbf{v}$, we have

$$\text{comp}_\mathbf{v}\mathbf{u} = \frac{\mathbf{u} \cdot \mathbf{v}}{\|\mathbf{v}\|} = \frac{\|\mathbf{u}\|\,\|\mathbf{v}\| \cos \theta}{\|\mathbf{v}\|}.$$

Cancelling $\|\mathbf{v}\|$ on the right side produces this result.

Projections and Components

If $\mathbf{u}$ and $\mathbf{v}$ are nonzero vectors and θ is the angle between them, then

$$\text{comp}_\mathbf{v}\mathbf{u} = \frac{\mathbf{u} \cdot \mathbf{v}}{\|\mathbf{v}\|} = \|\mathbf{u}\| \cos \theta$$

and

$$\|\text{proj}_\mathbf{v}\mathbf{u}\| = |\text{comp}_\mathbf{v}\mathbf{u}|.$$

EXAMPLE 6

If $\mathbf{u} = 2\mathbf{i} + 3\mathbf{j}$ and $\mathbf{v} = -5\mathbf{i} + 2\mathbf{j}$, find $\text{comp}_\mathbf{v}\mathbf{u}$ and $\text{comp}_\mathbf{u}\mathbf{v}$.

SOLUTION

$$\text{comp}_\mathbf{v}\mathbf{u} = \frac{\mathbf{u} \cdot \mathbf{v}}{\|\mathbf{v}\|} = \frac{2(-5) + 3 \cdot 2}{\sqrt{(-5)^2 + 2^2}} = \frac{-4}{\sqrt{29}}.$$

$$\text{comp}_\mathbf{u}\mathbf{v} = \frac{\mathbf{v} \cdot \mathbf{u}}{\|\mathbf{u}\|} = \frac{-4}{\sqrt{2^2 + 3^2}} = \frac{-4}{\sqrt{13}}.$$

 APPLICATIONS

Vectors and the dot product can be used to solve a variety of physical problems.

EXAMPLE 7

A 4000-pound automobile is on an inclined ramp that makes a 15° angle with the horizontal. Find the force required to keep it from rolling down the ramp, assuming that the only force that must be overcome is that due to gravity.

SOLUTION The situation is shown in Figure 9–28, where the coordinate system is chosen so that the car is at the origin, the vector $\mathbf{F}$ representing the downward force of gravity is on the y-axis, and $\mathbf{v}$ is a unit vector from the origin down the ramp. Since the car weighs 4000 pounds, $\mathbf{F} = -4000\mathbf{j}$. Figure 9–28 shows that the angle between $\mathbf{v}$ and $\mathbf{F}$ is 75°. The vector $\text{proj}_\mathbf{v}\mathbf{F}$ is the force pulling the car down the ramp, so a force of the same magnitude in the opposite direction is needed to keep the car motionless. As we saw in the preceding box,

$$\|\text{proj}_\mathbf{v}\mathbf{F}\| = |\text{comp}_\mathbf{v}\mathbf{F}| = |\,\|\mathbf{F}\| \cos 75°\,|$$

$$\approx 4000(.25882) \approx 1035.3.$$

Therefore, a force of 1035.3 pounds is required to hold the car in place.

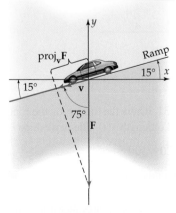

Figure 9–28

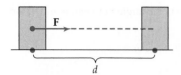

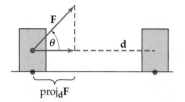

Figure 9–29

Figure 9–30

If a constant force **F** is applied to an object, pushing or pulling it a distance d in the direction of the force as shown in Figure 9–29, the amount of **work** done by the force is defined to be the product

$$W = (\text{magnitude of force})(\text{distance}) = \|\mathbf{F}\| \cdot d.$$

If the magnitude of **F** is measured in pounds and d in feet, then the units for W are foot-pounds. For example, if you push a car for 35 feet along a level driveway by exerting a constant force of 110 pounds, the amount of work done is $110 \cdot 35 = 3850$ foot-pounds.

When a force **F** moves an object in the direction of a vector **d** rather than in the direction of **F**, as shown in Figure 9–30, then the motion of the object can be considered as the result of the vector $\text{proj}_\mathbf{d}\mathbf{F}$, which is a force in the same direction as **d**.

Therefore, the amount of work done by **F** is the same as the amount of work done by $\text{proj}_\mathbf{d}\mathbf{F}$, namely,

$$W = (\text{magnitude of } \text{proj}_\mathbf{d}\mathbf{F})(\text{length of } \mathbf{d}) = \|\text{proj}_\mathbf{d}\mathbf{F}\| \cdot \|\mathbf{d}\|.$$

The box on page 659 and the Angle Theorem (page 655) show that

$$W = \|\text{proj}_\mathbf{d}\mathbf{F}\| \cdot \|\mathbf{d}\| = |\text{comp}_\mathbf{d}\mathbf{F}| \cdot \|\mathbf{d}\|$$

$$= \|\mathbf{F}\|(\cos \theta)\|\mathbf{d}\|$$

$$= \mathbf{F} \cdot \mathbf{d}.*$$

Consequently, we have these descriptions of work.

Work

> The work W done by a constant force F as its point of application moves along the vector **d** is
>
> $$W = |\text{comp}_\mathbf{u}\mathbf{F}| \cdot \|\mathbf{d}\| \qquad \text{or, equivalently,} \qquad W = \mathbf{F} \cdot \mathbf{d}.$$

EXAMPLE 8

How much work is done by a child who pulls a sled 100 feet over level ground by exerting a constant 20-pound force on a rope that makes a 45° angle with the ground?

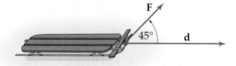

Figure 9–31

SOLUTION   The situation is shown in Figure 9–31, where the force **F** on the rope has magnitude 20 and the sled moves along vector **d** of length 100. The work done is

$$W = \mathbf{F} \cdot \mathbf{d} = \|\mathbf{F}\| \, \|\mathbf{d}\| \cdot \cos \theta = 20 \cdot 100 \cdot \frac{\sqrt{2}}{2}$$

$$= 1000 \sqrt{2} \approx 1414.2 \text{ foot-pounds.} \qquad \blacksquare$$

---

*This formula reduces to the previous one when **F** and **d** have the same direction because in that case, $\cos \theta = \cos 0 = 1$, so $W = \|\mathbf{F}\| \cdot \|\mathbf{d}\| = (\text{magnitude of force})(\text{distance moved})$.

EXERCISES 9.4

In Exercises 1–6, find $u \cdot v$, $u \cdot u$, and $v \cdot v$.

1. $\mathbf{u} = \langle 3, 4 \rangle$, $\mathbf{v} = \langle -5, 2 \rangle$

2. $\mathbf{u} = \langle -1, 6 \rangle$, $\mathbf{v} = \langle -4, 1/3 \rangle$

3. $\mathbf{u} = 2\mathbf{i} + \mathbf{j}$, $\mathbf{v} = 3\mathbf{i}$

4. $\mathbf{u} = \mathbf{i} - \mathbf{j}$, $\mathbf{v} = 5\mathbf{j}$

5. $\mathbf{u} = 3\mathbf{i} + 2\mathbf{j}$, $\mathbf{v} = 2\mathbf{i} + 3\mathbf{j}$

6. $\mathbf{u} = 4\mathbf{i} - \mathbf{j}$, $\mathbf{v} = -\mathbf{i} + 2\mathbf{j}$

In Exercises 7–12, find the dot product when $u = \langle 4, 3 \rangle$, $v = \langle -5, 2 \rangle$, and $w = \langle 4, -1 \rangle$.

7. $\mathbf{u} \cdot (\mathbf{v} + \mathbf{w})$

8. $\mathbf{u} \cdot (\mathbf{v} - \mathbf{w})$

9. $(\mathbf{u} + \mathbf{v}) \cdot (\mathbf{v} + \mathbf{w})$

10. $(\mathbf{u} + \mathbf{v}) \cdot (\mathbf{u} - \mathbf{v})$

11. $(3\mathbf{u} + \mathbf{v}) \cdot (2\mathbf{w})$

12. $(\mathbf{u} + 4\mathbf{v}) \cdot (2\mathbf{u} + \mathbf{w})$

In Exercises 13–18, find the angle between the two vectors.

13. $\langle 4, -3 \rangle$, $\langle 1, 2 \rangle$

14. $\langle 2, 4 \rangle$, $\langle 0, -5 \rangle$

15. $2\mathbf{i} - 3\mathbf{j}$, $-\mathbf{i}$

16. $2\mathbf{j}$, $4\mathbf{i} + \mathbf{j}$

17. $\sqrt{2}\mathbf{i} + \sqrt{2}\mathbf{j}$, $\mathbf{i} - \mathbf{j}$

18. $3\mathbf{i} - 5\mathbf{j}$, $-2\mathbf{i} + 3\mathbf{j}$

In Exercises 19–24, determine whether the given vectors are parallel, orthogonal, or neither.

19. $\langle 2, 6 \rangle$, $\langle 3, -1 \rangle$

20. $\langle -5, 3 \rangle$, $\langle 2, 6 \rangle$

21. $\langle 9, -6 \rangle$, $\langle -6, 4 \rangle$

22. $-\mathbf{i} + 2\mathbf{j}$, $2\mathbf{i} - 4\mathbf{j}$

23. $2\mathbf{i} - 2\mathbf{j}$, $5\mathbf{i} + 8\mathbf{j}$

24. $6\mathbf{i} - 4\mathbf{j}$, $2\mathbf{i} + 3\mathbf{j}$

In Exercises 25–28, find a real number k such that the two vectors are orthogonal.

25. $2\mathbf{i} + 3\mathbf{j}$, $3\mathbf{i} - k\mathbf{j}$

26. $-3\mathbf{i} + \mathbf{j}$, $2k\mathbf{i} - 4\mathbf{j}$

27. $\mathbf{i} - \mathbf{j}$, $k\mathbf{i} + \sqrt{2}\mathbf{j}$

28. $-4\mathbf{i} + 5\mathbf{j}$, $2\mathbf{i} + 2k\mathbf{j}$

In Exercises 29–32, find $proj_u v$ and $proj_v u$.

29. $\mathbf{u} = 3\mathbf{i} - 5\mathbf{j}$, $\mathbf{v} = 6\mathbf{i} + 2\mathbf{j}$

30. $\mathbf{u} = 2\mathbf{i} - 3\mathbf{j}$, $\mathbf{v} = \mathbf{i} + 2\mathbf{j}$

31. $\mathbf{u} = \mathbf{i} + \mathbf{j}$, $\mathbf{v} = \mathbf{i} - \mathbf{j}$

32. $\mathbf{u} = 5\mathbf{i} + \mathbf{j}$, $\mathbf{v} = -2\mathbf{i} + 3\mathbf{j}$

In Exercises 33–36, find $comp_v u$.

33. $\mathbf{u} = 10\mathbf{i} + 4\mathbf{j}$, $\mathbf{v} = 3\mathbf{i} - 2\mathbf{j}$

34. $\mathbf{u} = \mathbf{i} - 2\mathbf{j}$, $\mathbf{v} = 3\mathbf{i} + \mathbf{j}$

35. $\mathbf{u} = 3\mathbf{i} + 2\mathbf{j}$, $\mathbf{v} = -\mathbf{i} + 3\mathbf{j}$

36. $\mathbf{u} = \mathbf{i} + \mathbf{j}$, $\mathbf{v} = -3\mathbf{i} - 2\mathbf{j}$

In Exercises 37–39, let $u = \langle a, b \rangle$, $v = \langle c, d \rangle$, and $w = \langle r, s \rangle$. Verify that the given property of dot products is valid by calculating the quantities on each side of the equal sign.

37. $\mathbf{u} \cdot (\mathbf{v} + \mathbf{w}) = \mathbf{u} \cdot \mathbf{v} + \mathbf{u} \cdot \mathbf{w}$

38. $k\mathbf{u} \cdot \mathbf{v} = k(\mathbf{u} \cdot \mathbf{v}) = \mathbf{u} \cdot k\mathbf{v}$

39. $\mathbf{0} \cdot \mathbf{u} = 0$

40. Suppose $\mathbf{u} = \langle a, b \rangle$ and $\mathbf{v} = \langle c, d \rangle$ are nonzero parallel vectors.

    (a) If $c \neq 0$, show that $\mathbf{u}$ and $\mathbf{v}$ lie on the same nonvertical straight line through the origin.

    (b) If $c \neq 0$, show that $\mathbf{v} = \dfrac{a}{c}\mathbf{u}$ (that is, $\mathbf{v}$ is a scalar multiple of $\mathbf{u}$). [*Hint:* The equation of the line on which $\mathbf{u}$ and $\mathbf{v}$ lie is $y = mx$ for some constant m (why?), which implies that $b = ma$ and $d = mc$.]

    (c) If $c = 0$, show that $\mathbf{v}$ is a scalar multiple of $\mathbf{u}$. [*Hint:* If $c = 0$, then $a = 0$ (why?), and hence, $b \neq 0$ (otherwise, $\mathbf{u} = \mathbf{0}$).]

41. Prove the Angle Theorem in the case when θ is 0 or π.

42. If $\mathbf{u}$ and $\mathbf{v}$ are nonzero vectors such that $\mathbf{u} \cdot \mathbf{v} = 0$, show that $\mathbf{u}$ and $\mathbf{v}$ are orthogonal. [*Hint:* If θ is the angle between $\mathbf{u}$ and $\mathbf{v}$, what is $\cos \theta$ and what does this say about θ?]

43. Show that $(1, 2)$, $(3, 4)$, $(5, 2)$ are the vertices of a right triangle by considering the sides of the triangle as vectors.

44. Find a number x such that the angle between the vectors $\langle 1, 1 \rangle$ and $\langle x, 1 \rangle$ is $\pi/4$ radians.

45. Find nonzero vectors $\mathbf{u}$, $\mathbf{v}$, and $\mathbf{w}$ such that $\mathbf{u} \cdot \mathbf{v} = \mathbf{u} \cdot \mathbf{w}$ and $\mathbf{v} \neq \mathbf{w}$ and neither $\mathbf{v}$ nor $\mathbf{w}$ is orthogonal to $\mathbf{u}$.

46. If $\mathbf{u}$ and $\mathbf{v}$ are nonzero vectors, show that the vectors $\|\mathbf{u}\|\mathbf{v} + \|\mathbf{v}\|\mathbf{u}$ and $\|\mathbf{u}\|\mathbf{v} - \|\mathbf{v}\|\mathbf{u}$ are orthogonal.

47. A 600-pound trailer is on an inclined ramp that makes a 30° angle with the horizontal. Find the force required to keep it from rolling down the ramp, assuming that the only force that must be overcome is that due to gravity.

48. In Example 7, find the vector that represents the force necessary to keep the car motionless.

*In Exercises 49–52, find the work done by a constant force **F** as the point of application of **F** moves along the vector $\overrightarrow{PQ}$.*

49. $\mathbf{F} = 2\mathbf{i} + 5\mathbf{j}$, $P = (0, 0)$, $Q = (4, 1)$

50. $\mathbf{F} = \mathbf{i} - 2\mathbf{j}$, $P = (0, 0)$, $Q = (-5, 2)$

51. $\mathbf{F} = 2\mathbf{i} + 3\mathbf{j}$, $P = (2, 3)$, $Q = (5, 9)$ [*Hint:* Find the component form of $\overrightarrow{PQ}$.]

52. $\mathbf{F} = 5\mathbf{i} + \mathbf{j}$, $P = (-1, 2)$, $Q = (4, -3)$

53. A lawn mower handle makes an angle of 60° with the ground. A woman pushes on the handle with a force of 30 pounds. How much work is done in moving the lawn mower a distance of 75 feet on level ground?

54. A child pulls a wagon along a level sidewalk by exerting a force of 18 pounds on the wagon handle, which makes an angle of 25° with the horizontal. How much work is done in pulling the wagon 200 feet?

55. A 40-pound cart is pushed 100 feet up a ramp that makes a 20° angle with the horizontal (see the figure). How much work is done against gravity? [*Hint:* The amount of work done against gravity is the negative of the amount of work

done *by* gravity. Coordinatize the situation so that the cart is at the origin. Then the cart moves along vector $\mathbf{d} = (100 \cos 20°)\mathbf{i} + (100 \sin 20°)\mathbf{j}$, and the downward force of gravity is $\mathbf{F} = 0\mathbf{i} - 40\mathbf{j}$.]

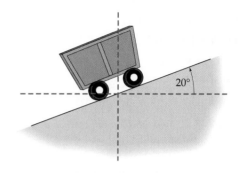

56. Suppose the child in Exercise 54 is pulling the wagon up a hill that makes an angle of 20° with the horizontal and all other facts remain the same. How much work is done in pulling the wagon 150 feet?

Chapter 9 Review

IMPORTANT CONCEPTS

Section 9.1
Complex plane   626
Real axis   626
Imaginary axis   626
Absolute value   627
Modulus   627
Argument   627
Polar form   627
Multiplication and division in polar
  form   629

Section 9.2
DeMoivre's Theorem   632
Formula for nth roots   635

Roots of unity   636

Section 9.3
Vector   640
Magnitude   640–641
Components   641
Scalar multiplication   642
Vector addition and subtraction
  643–645
Unit vector   645–646
Linear combination of $\mathbf{i}$ and $\mathbf{j}$ 646
Direction angle of a vector
  647

Section 9.4
Dot product   653
Angle between vectors   654
Parallel vectors   655
Angle Theorem   655
Schwarz inequality   656
Orthogonal vectors   656
Projection of $\mathbf{u}$ on $\mathbf{v}$ 657
Component of $\mathbf{u}$ along $\mathbf{v}$ 658–659
Work   660

IMPORTANT FACTS & FORMULAS

- $|a + bi| = \sqrt{a^2 + b^2}$
- $a + bi = r(\cos \theta + i \sin \theta)$, where

$$r = \sqrt{a^2 + b^2}, \qquad a = r \cos \theta, \qquad b = r \sin \theta.$$

- $r_1(\cos \theta_1 + i \sin \theta_1) \cdot r_2(\cos \theta_2 + i \sin \theta_2) = r_1 r_2 [\cos(\theta_1 + \theta_2) + i \sin(\theta_1 + \theta_2)]$

- $\dfrac{r_1(\cos \theta_1 + i \sin \theta_1)}{r_2(\cos \theta_2 + i \sin \theta_2)} = \dfrac{r_1}{r_2}[\cos(\theta_1 - \theta_2) + i \sin(\theta_1 - \theta_2)]$

- *DeMoivre's Theorem:*

$$[r(\cos \theta + i \sin \theta)]^n = r^n[\cos(n\theta) + i \sin(n\theta)]$$

- The distinct nth roots of $r(\cos \theta + i \sin \theta)$ are

$$\sqrt[n]{r}\left[\cos\left(\frac{\theta + 2k\pi}{n}\right) + i \sin\left(\frac{\theta + 2k\pi}{n}\right)\right] \quad (k = 0, 1, 2, \ldots, n-1).$$

- The distinct nth roots of unity are

$$\cos \frac{2k\pi}{n} + i \sin \frac{2k\pi}{n} \quad (k = 0, 1, 2, \ldots, n-1).$$

- If $P = (x_1, y_1)$ and $Q = (x_2, y_2)$, then $\overrightarrow{PQ} = \langle x_2 - x_1, y_2 - y_1 \rangle$.
- $\|\langle a, b \rangle\| = \sqrt{a^2 + b^2}$
- If $\mathbf{u} = \langle a, b \rangle$ and k is a scalar, then $k\mathbf{u} = \langle ka, kb \rangle$.
- If $\mathbf{u} = \langle a, b \rangle$ and $\mathbf{v} = \langle c, d \rangle$, then

$$\mathbf{u} + \mathbf{v} = \langle a + c, b + d \rangle \qquad \text{and} \qquad \mathbf{u} - \mathbf{v} = \langle a - c, b - d \rangle.$$

- *Properties of Vector Addition and Scalar Multiplication:* For any vectors $\mathbf{u}$, $\mathbf{v}$, and $\mathbf{w}$ and any scalars r and s,
    1. $\mathbf{u} + (\mathbf{v} + \mathbf{w}) = (\mathbf{u} + \mathbf{v}) + \mathbf{w}$
    2. $\mathbf{u} + \mathbf{v} = \mathbf{v} + \mathbf{u}$
    3. $\mathbf{v} + \mathbf{0} = \mathbf{v} = \mathbf{0} + \mathbf{v}$
    4. $\mathbf{v} + (-\mathbf{v}) = \mathbf{0}$
    5. $r(\mathbf{u} + \mathbf{v}) = r\mathbf{u} + r\mathbf{v}$
    6. $(r + s)\mathbf{v} = r\mathbf{v} + s\mathbf{v}$
    7. $(rs)\mathbf{v} = r(s\mathbf{v}) = s(r\mathbf{v})$
    8. $1\mathbf{v} = \mathbf{v}$
    9. $0\mathbf{v} = \mathbf{0} = r\mathbf{0}$

- If $\mathbf{v} = \langle a, b \rangle = a\mathbf{i} + b\mathbf{j}$, then

$$a = \|\mathbf{v}\| \cos \theta \qquad \text{and} \qquad b = \|\mathbf{v}\| \sin \theta,$$

where θ is the direction angle of $\mathbf{v}$.

- If $\mathbf{u} = \langle a, b \rangle = a\mathbf{i} + b\mathbf{j}$ and $\mathbf{v} = \langle c, d \rangle = c\mathbf{i} + d\mathbf{j}$, then

$$\mathbf{u} \cdot \mathbf{v} = ac + bd.$$

- If θ is the angle between nonzero vectors $\mathbf{u}$ and $\mathbf{v}$, then

$$\mathbf{u} \cdot \mathbf{v} = \|\mathbf{u}\|\,\|\mathbf{v}\| \cos \theta.$$

- *Schwarz Inequality:* $|\mathbf{u} \cdot \mathbf{v}| \le \|\mathbf{u}\|\,\|\mathbf{v}\|$.
- Vectors $\mathbf{u}$ and $\mathbf{v}$ are orthogonal exactly when $\mathbf{u} \cdot \mathbf{v} = 0$.
- $\text{proj}_{\mathbf{v}}\mathbf{u} = \left(\dfrac{\mathbf{u} \cdot \mathbf{v}}{\|\mathbf{v}\|^2}\right)\mathbf{v}$

- $\text{comp}_{\mathbf{v}}\mathbf{u} = \dfrac{\mathbf{u} \cdot \mathbf{v}}{\|\mathbf{v}\|} = \|\mathbf{u}\| \cos \theta$, where θ is the angle between $\mathbf{u}$ and $\mathbf{v}$.

REVIEW QUESTIONS

1. Simplify: $|i(4 + 2i)| + |3 - i|$.

2. Simplify: $|3 + 2i| - |1 - 2i|$.

3. Graph the equation $|z| = 2$ in the complex plane.

4. Graph the equation $|z - 3| = 1$ in the complex plane.

5. Express in polar form: $1 + \sqrt{3}i$.

6. Express in polar form: $4 - 5i$.

In Questions 7–11, multiply or divide, and express the answer in the form $a + bi$.

7. $2\left(\cos\dfrac{\pi}{12} + i\sin\dfrac{\pi}{12}\right) \cdot 4\left(\cos\dfrac{\pi}{6} + i\sin\dfrac{\pi}{6}\right)$

8. $3\left(\cos\dfrac{\pi}{8} + i\sin\dfrac{\pi}{8}\right) \cdot 2\left(\cos\dfrac{3\pi}{8} + i\sin\dfrac{3\pi}{8}\right)$

9. $\dfrac{12\left(\cos\dfrac{7\pi}{12} + i\sin\dfrac{7\pi}{12}\right)}{3\left(\cos\dfrac{5\pi}{12} + i\sin\dfrac{5\pi}{12}\right)}$

10. $\left(\cos\dfrac{\pi}{12} + i\sin\dfrac{\pi}{12}\right)^{18}$

11. $\left[\sqrt[3]{3}\left(\cos\dfrac{5\pi}{36} + i\sin\dfrac{5\pi}{36}\right)\right]^{12}$

In Questions 12–16, solve the given equation in the complex number system, and express your answers in polar form.

12. $x^3 = i$

13. $x^6 = 1$

14. $x^8 = -\sqrt{3} - 3i$

15. $x^4 = i$

16. $x^3 = 1 - i$

In Questions 17–20, let $u = \langle 2, -5 \rangle$ and $v = \langle 6, 1 \rangle$. Find

17. $u + v$

18. $\|-3v\|$

19. $\|2v - 4u\|$

20. $3u - \dfrac{1}{2}v$

In Questions 21–24, let $u = -3i + j$ and $v = 2i - 5j$. Find

21. $4u - v$

22. $u + 2v$

23. $\|u + v\|$

24. $\|u\| + \|v\|$

25. Find the components of the vector v such that $\|v\| = 5$ and the direction angle of v is $45°$.

26. Find the magnitude and direction angle of $3i + 4j$.

27. Find a unit vector whose direction is *opposite* the direction of $3i - 6j$.

28. An object at the origin is acted upon by a 10-pound force with direction angle $90°$ and a 20-pound force with direction angle $30°$. Find the magnitude and direction of the resultant force.

29. A plane flies in the direction $120°$ with an air speed of 300 mph. The wind is blowing from north to south at 40 mph. Find the course and ground speed of the plane.

30. An object weighing 40 pounds lies on an inclined plane that makes a $30°$ angle with the horizontal. Find the components of the weight parallel and perpendicular to the plane.

In Questions 31–34, $u = \langle 4, -3 \rangle$, $v = \langle -1, 6 \rangle$, and $w = \langle 5, 0 \rangle$. Find

31. $u \cdot v$

32. $u \cdot u - v \cdot v$

33. $(u + v) \cdot w$

34. $(u + w) \cdot (w - 3v)$

35. What is the angle between the vectors $5i - 2j$ and $3i + j$?

36. Is $3i - 2j$ orthogonal to $4i + 6j$?

In Questions 37 and 38, $u = 4i - 3j$ and $v = 2i + j$. Find

37. $\text{proj}_v u$

38. $\text{comp}_u v$

39. If u and v have the same magnitude, show that $u + v$ and $u - v$ are orthogonal.

40. If u and v are nonzero vectors, show that the vector $u - kv$ is orthogonal to v, where $k = \dfrac{u \cdot v}{\|v\|^2}$.

41. A 3500-pound automobile is on an inclined ramp that makes a $30°$ angle with the horizontal. Find the force required to keep it from rolling down the ramp, assuming that the only force that must be overcome is that due to gravity.

42. A sled is pulled along level ground by a rope that makes a $50°$ angle with the horizontal. If a force of 40 pounds is used to pull the sled, how much work is done in pulling it 100 feet?

Chapter 9 Test

Sections 9.1 and 9.2

Note: If the directions say "show your work," it is *not* acceptable to respond "I used my calculator".

1. Plot the number $\frac{3i}{2}(6 + 2i)$ in the complex plane.

2. Calculate the product $(3 + 3i)^6$ and express your answer in the form $a + bi$. Show your work.

3. Find the exact value of $|3 + 2i|$. Show your work.

4. Find the cube roots (in polar form) of $64\left(\cos \frac{\pi}{16} + i \sin \frac{\pi}{16}\right)$. No decimal approximations allowed.

5. Let $z = 5 - 3i$. Find each of the following and show your work.
   (a) $|z^2|$
   (b) $z\bar{z}$, where $\bar{z}$ is the complex conjugate of z.

6. Find the cube roots (in polar form) of $1 + \sqrt{3}i$.

7. Sketch the graph of the equation $|z - 3| = 5$ in the complex plane.

8. Solve: $x^4 = -i$. Express your answers in polar form.

9. Compute $3\left(\cos \frac{3\pi}{8} + i \sin \frac{3\pi}{8}\right) \cdot 13\left(\cos \frac{\pi}{8} + i \sin \frac{\pi}{8}\right)$ and express your answer in the form $a + bi$. Show your work.

10. Solve: $x^4 = -648 + 648\sqrt{3}i$. Express your answers in the form $a + bi$.

11. Convert to polar form and divide: $\frac{-10i}{\sqrt{3} + i}$. Express the exact answer in polar form.

12. Solve: $x^3 + x^2 + x + 1 = 0$. [*Hint:* First, multiply both sides by $x - 1$.]

Sections 9.3 and 9.4

13. Find a vector with initial point at the origin that is equivalent to $\overrightarrow{PQ}$, where $P = (-1, 4)$ and $Q = (-8, -9)$.

14. If $\mathbf{u} = \langle 3, 5 \rangle$, $\mathbf{v} = \langle -2, 1 \rangle$ and $\mathbf{w} = \langle 3, -1 \rangle$, find $(\mathbf{u} + 2\mathbf{v}) \cdot (3\mathbf{u} + \mathbf{w})$.

15. If $\mathbf{u} = \sqrt{5}\mathbf{j}$ and $\mathbf{v} = \sqrt{11}\mathbf{i}$, find
    (a) $\mathbf{u} + \mathbf{v}$ (b) $\mathbf{u} - \mathbf{v}$ (c) $6\mathbf{u} - 5\mathbf{v}$

16. Find the angle (in radians) between the vectors $\langle 9, 3 \rangle$ and $\langle 0, -2 \rangle$. Round your answer to four decimal places.

17. Find the component form of the vector $\mathbf{v}$ whose magnitude is 3 and whose direction angle is $210°$.

18. Find a real number k such that $3k\mathbf{i} - 6\mathbf{j}$ is orthogonal to $-7\mathbf{i} + \mathbf{j}$.

19. Find a unit vector that has the same direction as $-7\mathbf{i} - 6\mathbf{j}$.

20. Find $\text{comp}_\mathbf{v}\mathbf{u}$, where $\mathbf{u} = \mathbf{i} - 2\mathbf{j}$ and $\mathbf{v} = 3\mathbf{i} + \mathbf{j}$.

21. If vectors $\mathbf{u}_1, \mathbf{u}_2, \mathbf{u}_3, \ldots, \mathbf{u}_k$ act on an object at the origin, the *resultant force* is defined to be $\mathbf{u}_1 + \mathbf{u}_2 + \mathbf{u}_3 + \cdots + \mathbf{u}_k$.
    (a) Find the resultant force when $\mathbf{u}_1 = \langle 1, 6 \rangle$, $\mathbf{u}_2 = \langle 7, -3 \rangle$, $\mathbf{u}_3 = \langle -2, 0 \rangle$, and $\mathbf{u}_4 = \langle -9, 8 \rangle$.
    (b) Find a vector $\mathbf{v}$ such that the resultant force of $\mathbf{u}_1, \mathbf{u}_2, \mathbf{u}_3, \mathbf{u}_4, \mathbf{v}$ is $\mathbf{0}$.

22. Find the work done by a constant force $\mathbf{F} = 8\mathbf{i} + \mathbf{j}$ as the point of application of $\mathbf{F}$ moves along the vector $\overrightarrow{PQ}$, where $P = \langle -1, 3 \rangle$ and $Q = \langle 5, -3 \rangle$.

23. A river flows from west to east. A swimmer on the north bank swims at 3.9 mph along a straight course that makes a $75°$ angle with the north bank of the river. He reaches the south bank at a point directly south of his starting point. How fast is the current in the river? Round your answer to two decimal places.

24. A child pulls a wagon along a level sidewalk by exerting a force of 17 pounds on the wagon handle, which makes an angle of $20°$ with the horizontal. How much work is done in pulling the wagon 200 feet? Round your answer to two decimal places.

You might ask, how far apart are two points *A* and *B*? Under ordinary circumstances, the easiest thing to do would be to measure the distance. However, sometimes it is impractical to make such a direct measurement because of intervening obstacles. Two historical methods for measuring the distance between the two mutually invisible points *A* and *B* are given below.

1. Find the distance from *A* to *B*, using the classic surveyor's method shown in the figure below. A transit is used to measure the angle between two distant objects*, and a reasonably short baseline is measured. Angles are reported in degrees. Triangles are drawn, and the law of sines (page 606) is used to calculate the length of unknown edges. To find the distance from *A* to *B*, draw an additional triangle that has *AB* as one of the sides. Caution: The picture is not drawn to scale.

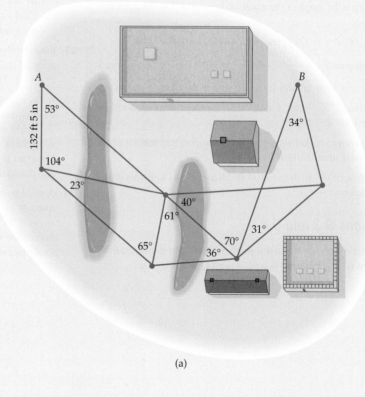

(a)

Arthur S. Aubry / Getty Images

*A transit is a small telescope on a tripod that can be used to measure angles precisely. (See the photograph at the left.)

2. Find the distance from A to B, using a modern laser range finder. Laser technology removes the necessity of drawing triangles; the tool provides a vector (angle and distance) from point to point, so the distance from A to B can be found by using vector arithmetic in place of repeated application of the law of sines. Angles are taken from the internal compass and are reported in degrees measured clockwise from North (0°), so you have to convert compass direction angles to angles in standard position. For example, the angle from point A (straight South) is reported as 180°, but in standard position, you would list the angle as 270°.

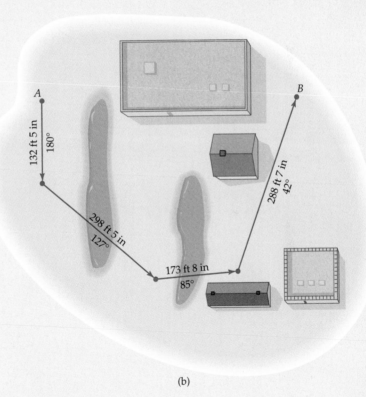

(b)

Chapter 10

ANALYTIC GEOMETRY

Calling all ships!

The planets travel in elliptical orbits around the sun, and satellites travel in elliptical orbits around the earth. Parabolic reflectors are used in spotlights, radar antennas, and satellite dishes. The long-range navigation system (LORAN) uses hyperbolas to enable a ship to determine its exact location. See Exercise 54 on page 699.

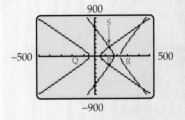

© Digital Vision/Getty Images

Chapter Outline

When a right circular cone is cut by a plane, the intersection is a curve called a **conic section,** as shown in Figure 10–1. Conic sections were first studied by the ancient Greeks and over the centuries have appeared time and again in astronomy, warfare, medicine, economics, and other branches of science. For instance, planets travel in elliptical orbits, thrown objects trace out parabolas, and certain atomic particles follow hyperbolic paths.

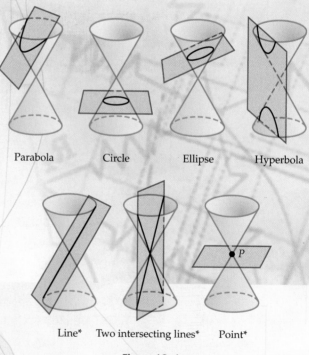

Parabola   Circle   Ellipse   Hyperbola

Line*   Two intersecting lines*   Point*

Figure 10–1

---

*A point, a line, or two intersecting lines are sometimes called **degenerate** conic sections.

Although the Greeks studied conic sections from a purely geometric point of view, the modern approach is to describe them in terms of the coordinate plane and distance or as the graphs of certain types of equations. This was done for circles in Section 1.3 and will be done in Sections 10.1–10.3 for ellipses, hyperbolas, and parabolas.

Sections 10.5–10.7 present a more thorough treatment of parametric graphing and an alternate method of coordinatizing the plane and graphing.

10.1   Circles and Ellipses

Section Objectives

■ Find the center and radius of a circle from its equation.
■ Use the standard equation of an ellipse to sketch its graph.
■ Set up and solve applied problems involving ellipses.

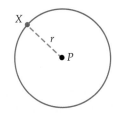

Figure 10–2

Let P be a point in the plane, and let r be a positive number. Then the **circle** with center P and radius r consists of all points X in the plane such that

$$\text{Distance from } X \text{ to } P = r,$$

as shown in Figure 10–2. In Section 1.3, we proved the following result.

Equation of a Circle

The circle with center $(0, 0)$ and radius r is the graph of the equation

$$x^2 + y^2 = r^2.$$

The circle with center (h, k) and radius r is the graph of

$$(x - h)^2 + (y - k)^2 = r^2.$$

EXAMPLE 1

Find the center and radius of the circle whose equation is given.

(a) $(x - 2)^2 + (y - 7)^2 = 25$
(b) $(x - 4)^2 + (y + 3)^2 = 36$
(c) $2x^2 + 2y^2 - 4x - 12y + 12 = 0$

SOLUTION

(a) The right side of the equation is 5^2. So the center is at $(2, 7)$, and the radius is 5.

(b) First, we rewrite the equation to match the form in the box above:

$$(x - 4)^2 + \quad (y + 3)^2 = 36$$

$$(x - 4)^2 + [y - (-3)]^2 = 6^2.$$

Hence, the circle has center $(4, -3)$ and radius 6.

(c) We begin by rewriting the equation.

$$2x^2 + 2y^2 - 4x - 12y + 12 = 0$$

Divide both sides by 2: $$x^2 + y^2 - 2x - 6y + 6 = 0$$

Rearrange terms: $$(x^2 - 2x) + (y^2 - 6y) = -6$$

Now we complete the square in both expressions in parentheses by adding 1 to the x-expression and 9 to the y-expression, and adding the same amounts to the right side of the equation.

$$(x^2 - 2x + 1) + (y^2 - 6y + 9) = -6 + 1 + 9$$

Factor: $$(x - 1)^2 + (y - 3)^2 = 4$$

$$(x - 1)^2 + (y - 3)^2 = 2^2$$

The circle has center $(1, 3)$ and radius 2.   ∎

The preceding discussion of circles provides the model for our discussion of the other conic sections. In each case, the conic is defined in terms of points and distances, and its Cartesian equation is determined. The standard form of the equation of a conic includes the key information necessary for a rough sketch of its graph, just as the standard form of the equation of a circle tells you its center and radius.

ELLIPSES

Definition.   Let P and Q be points in the plane, and let r be a number greater than the distance from P to Q. The **ellipse** with **foci*** P and Q is the set of all points X such that

(Distance from X to P) + (Distance from X to Q) = r.

To draw this ellipse, take a piece of string of length r and pin its ends on P and Q. Put your pencil point against the string and move it, keeping the string taut. You will trace out the ellipse, as shown in Figure 10–3.

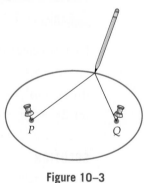

Figure 10–3

---

*"Foci" is the plural of "focus."

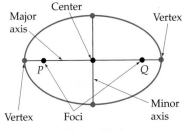

Figure 10–4

The midpoint of the line segment from P to Q is the **center** of the ellipse. The points where the straight line through the foci intersects the ellipse are its **vertices.** The **major axis** of the ellipse is the line segment joining the vertices; its **minor axis** is the line segment through the center, perpendicular to the major axis, as shown in Figure 10–4.

Equation.   Suppose that the foci P and Q are on the x-axis, with coordinates

$$P = (-c, 0) \qquad \text{and} \qquad Q = (c, 0) \quad \text{for some } c > 0.$$

Let $a = r/2$, so that $2a = r$. Then the point (x, y) is on the ellipse exactly when

[Distance from (x, y) to P] + [Distance from (x, y) to Q] $= r$

$$\sqrt{(x + c)^2 + (y - 0)^2} + \sqrt{(x - c)^2 + (y - 0)^2} = 2a$$

$$\sqrt{(x + c)^2 + y^2} = 2a - \sqrt{(x - c)^2 + y^2}.$$

Squaring both sides and simplifying (Exercise 76), we obtain

$$a\sqrt{(x - c)^2 + y^2} = a^2 - cx.$$

Again squaring both sides and simplifying, we have

$$(a^2 - c^2)x^2 + a^2 y^2 = a^2(a^2 - c^2).$$

To simplify the form of this equation, let $b = \sqrt{a^2 - c^2}$* so that $b^2 = a^2 - c^2$ and the equation becomes

$$b^2 x^2 + a^2 y^2 = a^2 b^2.$$

Dividing both sides by $a^2 b^2$ shows that the coordinates of every point on the ellipse satisfy the equation

$$\frac{x^2}{a^2} + \frac{y^2}{b^2} = 1.$$

Conversely, it can be shown that every point whose coordinates satisfy this equation is on the ellipse. When the equation is in this form the x- and y-intercepts of the graph are easily found. For instance, to find the x-intercepts, we set $y = 0$ and solve

$$\frac{x^2}{a^2} + \frac{0^2}{b^2} = 1$$

$$x^2 = a^2$$

$$x = \pm a.$$

Similarly, we find the y-intercepts by setting $x = 0$ and solving for y:

$$\frac{0^2}{a^2} + \frac{y^2}{b^2} = 1$$

$$y^2 = b^2$$

$$y = \pm b.$$

An analogous argument applies when the foci are on the y-axis and leads to this conclusion.

---

*The distance between the foci is $2c$. Since $r = 2a$ and $r > 2c$ by definition, we have $2a > 2c$, and hence, $a > c$. Therefore, $a^2 - c^2$ is a positive number and has a real square root.

*Standard Equations
of Ellipses Centered
at the Origin*

Let a and b be real numbers with $a > b > 0$. Then the graph of each of the following equations is an ellipse centered at the origin:

$$\frac{x^2}{a^2} + \frac{y^2}{b^2} = 1 \begin{cases} x\text{-intercepts: } \pm a \qquad y\text{-intercepts: } \pm b \\ \text{major axis on the } x\text{-axis, with vertices } (a, 0) \text{ and } (-a, 0) \\ \text{foci: } (c, 0) \text{ and } (-c, 0), \text{ where } c = \sqrt{a^2 - b^2} \end{cases}$$

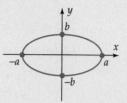

$$\frac{x^2}{b^2} + \frac{y^2}{a^2} = 1 \begin{cases} x\text{-intercepts: } \pm b \qquad y\text{-intercepts: } \pm a \\ \text{major axis on the } y\text{-axis, with vertices } (0, a) \text{ and } (0, -a) \\ \text{foci: } (0, c) \text{ and } (0, -c), \text{ where } c = \sqrt{a^2 - b^2} \end{cases}$$

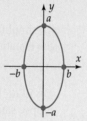

In the preceding box $a > b$, but don't let all the letters confuse you: When the equation is in standard form, the denominator of the x term tells you the x-intercepts, the denominator of the y term tells you the y-intercepts, and the major axis is the longer one, as illustrated in the following examples.

EXAMPLE 2

Identify the intercepts, axes, vertices and foci of the ellipse

$$\frac{x^2}{9} + \frac{y^2}{25} = 1$$

and sketch its graph.

SOLUTION   The equation can be written as

$$\frac{x^2}{3^2} + \frac{y^2}{5^2} = 1.$$

Its x-intercepts are ± 3 and its y-intercepts are ± 5. Since the larger denominator 5^2 is in the y-term of the equation, the major axis lies on the y-axis. So the vertices are determined by the y-intercepts: $(0, 5)$ and $(0, -5)$. The foci have coordinates $(0, c)$ and $(0, -c)$, where c is the square root of the difference of the larger and smaller denominators:

$$c = \sqrt{5^2 - 3^2} = \sqrt{16} = 4.$$

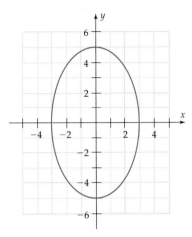

Figure 10–5

Thus, the foci are $(0, 4)$ and $(0, -4)$. The minor axis (the shorter one, corresponding to the smaller denominator in the equation) lies on the x-axis.

By plotting the four intercepts and a few more points, we obtain the graph of the ellipse in Figure 10–5. ■

EXAMPLE 3

Identify and sketch the graph of the equation $4x^2 + 9y^2 = 36$.

SOLUTION   To identify the graph, we put the equation in standard form.

$$4x^2 + 9y^2 = 36$$

Divide both sides by 36:
$$\frac{4x^2}{36} + \frac{9y^2}{36} = \frac{36}{36}$$

Simplify:
$$\frac{x^2}{9} + \frac{y^2}{4} = 1$$

$$\frac{x^2}{3^2} + \frac{y^2}{2^2} = 1.$$

This form of the equation shows that the graph is an ellipse with

x-intercepts: ± 3 and   y-intercepts: ± 2.

The major axis is the x-axis (because the larger denominator 3^2 is in the x-term). The foci are on the major axis, so they have the form $(c, 0)$ and $(-c, 0)$, where

$$c = \sqrt{3^2 - 2^2} = \sqrt{5}.$$

A hand-sketched graph is shown in Figure 10–6. To graph this ellipse on a calculator, we first solve its equation for y:

$$4x^2 + 9y^2 = 36$$

Subtract $4x^2$ from both sides:
$$9y^2 = 36 - 4x^2$$

Divide both sides by 9:
$$y^2 = \frac{36 - 4x^2}{9}.$$

Taking square roots on both sides, we see that

$$y = \sqrt{\frac{36 - 4x^2}{9}} \qquad \text{or} \qquad y = -\sqrt{\frac{36 - 4x^2}{9}}.$$

Graphing both of these equations on the same screen, we obtain Figure 10–7. ■

TECHNOLOGY TIP

On most calculators, you can graph both equations in Example 3 simultaneously by keying in

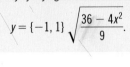

$$y = \{-1, 1\}\sqrt{\frac{36 - 4x^2}{9}}.$$

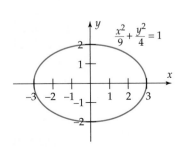

Figure 10–6

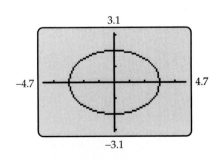

Figure 10–7

EXAMPLE 4

Find the equation of the ellipse with vertices $(0, \pm 6)$ and foci $(0, \pm 2\sqrt{6})$, and sketch its graph.

SOLUTION   Since the foci are $(0, 2\sqrt{6})$ and $(0, -2\sqrt{6})$, the center of the ellipse is $(0, 0)$, and its major axis lies on the y-axis. Hence, its equation is of the form

$$\frac{x^2}{b^2} + \frac{y^2}{a^2} = 1 \quad (a > b).$$

From the box on page 674, we see that $a = 6$ and $c = 2\sqrt{6}$. Since $c = \sqrt{a^2 - b^2}$, we have $c^2 = a^2 - b^2$, so

$$b^2 = a^2 - c^2 = 6^2 - (2\sqrt{6})^2 = 36 - 4 \cdot 6 = 12.$$

Hence, $b = \sqrt{12}$, and the equation of the ellipse is

$$\frac{x^2}{(\sqrt{12})^2} + \frac{y^2}{6^2} = 1 \quad \text{or, equivalently,} \quad \frac{x^2}{12} + \frac{y^2}{36} = 1.$$

The graph has x-intercepts $\pm\sqrt{12} \approx \pm 3.46$ and y-intercepts ± 6, as sketched in Figure 10–8. ∎

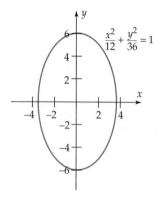

$$\frac{x^2}{12} + \frac{y^2}{36} = 1$$

Figure 10–8

VERTICAL AND HORIZONTAL SHIFTS

The circle with equation

$$x^2 + y^2 = 4$$

has radius 2 and center $(0, 0)$. If we replace x with $x - 5$ and replace y with $y - 3$ in the equation, we obtain

$$(x - 5)^2 + (y - 3)^2 = 4,$$

which is the equation of a circle with radius 2 and center $(5, 3)$. So shifting the center of the first circle 5 units to the right and 3 units upward, from $(0, 0)$ to $(5, 3)$, produces the second circle. Equivalently, the second circle can be obtained from the first by shifting each point on the first circle 5 units horizontally to the right and 3 units vertically upward, as shown in Figure 10–9.

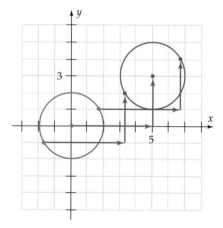

Figure 10–9

The same thing is true for an ellipse centered at the origin.

Vertical and Horizontal Shifts

Consider an ellipse centered at the origin with equation

$$\frac{x^2}{a^2} + \frac{y^2}{b^2} = 1.$$

Let h and k be constants. Replacing x with $x - h$ and replacing y with $y - k$ in this equation produces the equation

$$\frac{(x - h)^2}{a^2} + \frac{(y - k)^2}{b^2} = 1,$$

whose graph is the original ellipse shifted horizontally and vertically so that its center is (h, k).

EXAMPLE 5

Identify and sketch the graph of

$$\frac{(x - 2)^2}{9} + \frac{(y - 6)^2}{4} = 1$$

SOLUTION The equation $\dfrac{(x - 2)^2}{9} + \dfrac{(y - 6)^2}{4} = 1$ was obtained from $\dfrac{x^2}{9} + \dfrac{y^2}{4} = 1$ by replacing x with $x - 2$ and replacing y with $y - 6$. This is the situation described in the preceding box, with $h = 2$ and $k = 6$. The graph is an ellipse with center $(2, 6)$ that can be obtained from the ellipse $\dfrac{x^2}{9} + \dfrac{y^2}{4} = 1$ (shown in Figure 10–6) by shifting it 2 units to the right and 6 units upward, as shown in Figure 10–10. Its major axis lies on the horizontal line $y = 6$, as do its foci. Its minor axis lies on the vertical line $x = 2$. ∎

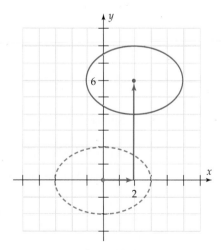

Figure 10–10

EXAMPLE 6

Identify and sketch the graph of

$$\frac{(x-5)^2}{9} + \frac{(y+4)^2}{36} = 1.$$

SOLUTION   The preceding box deals with differences ($y - k$, not $y + k$). To use it here, we must rewrite $y + 4$ as $y - (-4)$, so that the given equation becomes $\dfrac{(x-5)^2}{9} + \dfrac{(y-(-4))^2}{36} = 1$. This equation can be obtained from the equation $\dfrac{x^2}{9} + \dfrac{y^2}{36} = 1$ by replacing x with $x - 5$ and replacing y with $y - (-4)$. Therefore, the graph of $\dfrac{(x-5)^2}{9} + \dfrac{(y-(-4))^2}{36} = 1$ is an ellipse with center $(5, -4)$. Its graph is the ellipse $\dfrac{x^2}{9} + \dfrac{y^2}{36} = 1$ shifted 5 units to the right and 4 units downward, as shown in Figure 10–11. Its major axis lies on the vertical line $x = 5$, as do its foci, and its minor axis lies on the horizontal line $y = -4$. ∎

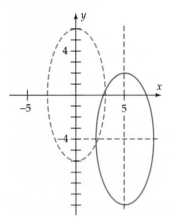

Figure 10–11

EXAMPLE 7

Find the center, vertices and foci of the ellipse

$$\frac{(x+5)^2}{16} + \frac{(y-2)^2}{9} = 1.$$

SOLUTION   By writing $x + 5$ as $x - (-5)$, we see that the center of the ellipse is $(-5, 2)$. It is obtained from the ellipse $\dfrac{x^2}{16} + \dfrac{y^2}{9} = 1$ by shifting it 5 units to the left and 2 units upward. As shown in the box on page 674, the vertices of $\dfrac{x^2}{16} + \dfrac{y^2}{9} = 1$ are $(-4, 0)$ and $(4, 0)$. So the vertices of $\dfrac{(x-(-5))^2}{16} + \dfrac{(y-2)^2}{9} = 1$ are obtained as follows:

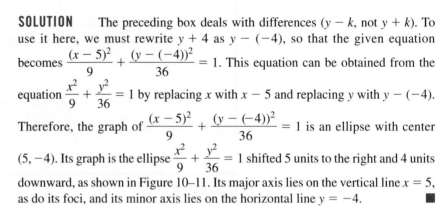

$$
\begin{array}{c}
\text{Shift 5 units left and} \\
\text{2 units upward} \\
(-4, 0) \text{------------------->} (-4 - 5, 0 + 2) = (-9, 2) \\
(4, 0) \text{------------------->} (4 - 5, 0 + 2) = (-1, 2).
\end{array}
$$

Similarly, the box on page 674 shows that the foci of $\dfrac{x^2}{16} + \dfrac{y^2}{9} = 1$ are $(-c, 0)$ and $(c, 0)$, where $c = \sqrt{16 - 9} = \sqrt{7}$. The foci of $\dfrac{(x-(-5))^2}{16} + \dfrac{(y-2)^2}{9} = 1$ are found in the same way the vertices were:

$$
\begin{array}{c}
\text{Shift 5 units left and} \\
\text{2 units upward} \\
(-\sqrt{7}, 0) \text{------------------->} (-\sqrt{7} - 5, 0 + 2) = (-\sqrt{7} - 5, 2) \\
(\sqrt{7}, 0) \text{------------------->} (\sqrt{7} - 5, 0 + 2) = (\sqrt{7} - 5, 2).
\end{array}
$$
∎

Examples 5–7 illustrate the following facts.

Standard Equations of Ellipses with Center at (h, k)

Let (h, k) be any point in the plane. If a and b are real numbers with $a > b > 0$, then the graph of each of the following equations is an ellipse with center (h, k).

$$\frac{(x - h)^2}{a^2} + \frac{(y - k)^2}{b^2} = 1 \begin{cases} \text{major axis on the horizontal line } y = k \\ \text{minor axis on the vertical line } x = h \\ \text{foci: } (-c + h, k) \text{ and } (c + h, k), \text{ where} \\ \quad c = \sqrt{a^2 - b^2} \end{cases}$$

$$\frac{(x - h)^2}{b^2} + \frac{(y - k)^2}{a^2} = 1 \begin{cases} \text{major axis on the vertical line } x = h \\ \text{minor axis on the horizontal line } y = k \\ \text{foci: } (h, -c + k) \text{ and } (h, c + k), \text{ where} \\ \quad c = \sqrt{a^2 - b^2} \end{cases}$$

 GRAPHING TECHNIQUES

In order to graph an ellipse by hand, you must first put its equation in standard form, as illustrated in the next example.

EXAMPLE 8

Identify and sketch the graph of

$$4x^2 + 9y^2 - 32x - 90y + 253 = 0.$$

SOLUTION   We want to put the equation in standard form. We begin by rewriting it as follows.

$$4x^2 + 9y^2 - 32x - 90y + 253 = 0$$

Rearrange terms: $(4x^2 - 32x) + (9y^2 - 90y) = -253$

Factor out 4 and 9: $4(x^2 - 8x) + 9(y^2 - 10y) = -253$

CAUTION

Completing the square works only when the coefficient of x^2 is 1. In an expression such as $4x^2 - 32x$, you must first factor out the 4,

$$4(x^2 - 8x),$$

and then complete the square on the expression in parentheses.

The factoring in the last equation was done in preparation for completing the square (see the Caution in the margin). To complete the square on $x^2 - 8x$, we add 16 (the square of half the coefficient of x), and to complete the square on $y^2 - 10y$, we add 25 (the square of half the coefficient of y):

$$4(x^2 - 8x + 16) + 9(y^2 - 10y + 25) = -253 + ? + ?.$$

Be careful here: On the left side, we haven't just added 16 and 25. When the left side is multiplied out, we have actually added $4 \cdot 16 = 64$ and $9 \cdot 25 = 225$. To leave the equation unchanged, we must add these numbers on the right:

$$4(x^2 - 8x + 16) + 9(y^2 - 10y + 25) = -253 + 64 + 225$$

Factor and simplify: $4(x - 4)^2 + 9(y - 5)^2 = 36$

Divide both sides by 36: $\dfrac{4(x - 4)^2}{36} + \dfrac{9(y - 5)^2}{36} = \dfrac{36}{36}$

Simplify: $\dfrac{(x - 4)^2}{9} + \dfrac{(y - 5)^2}{4} = 1.$

The graph of this equation is the ellipse $\dfrac{x^2}{9} + \dfrac{y^2}{4} = 1$ shifted 4 units to the right and 5 units upward. Its center is at (4, 5). Its major axis lies on the horizontal line $y = 5$, and its minor axis lies on the vertical line $x = 4$, as shown in Figure 10–12. ∎

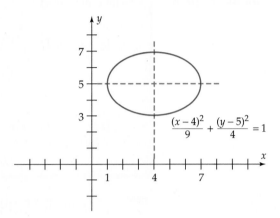

$$\frac{(x-4)^2}{9} + \frac{(y-5)^2}{4} = 1$$

Figure 10–12

EXAMPLE 9

Find an appropriate viewing window and use technology to graph the ellipse

$$\frac{(x-3)^2}{40} + \frac{(y+2)^2}{120} = 1.$$

SOLUTION To graph the ellipse, we first solve its equation for y.

Subtract $\dfrac{(x-3)^2}{40}$ from both sides:
$$\frac{(y+2)^2}{120} = 1 - \frac{(x-3)^2}{40}$$

Multiply both sides by 120:
$$(y+2)^2 = 120\left[1 - \frac{(x-3)^2}{40}\right]$$

Multiply out right side:
$$(y+2)^2 = 120 - 3(x-3)^2$$

Take square roots on both sides:
$$y + 2 = \pm\sqrt{120 - 3(x-3)^2}$$

$$y = \sqrt{120 - 3(x-3)^2} - 2 \qquad \text{or} \qquad y = -\sqrt{120 - 3(x-3)^2} - 2$$

So we should graph both of these last two equations on the same screen.

By trial and error we found the window in Figure 10–13, in which the graph looks longer horizontally than vertically. However, from the original form of the equation

$$\frac{(x-3)^2}{40} + \frac{(y+2)^2}{120} = 1,$$

we know that the major axis of the ellipse (the longer one) should be vertical, because the larger constant 120 is in the denominator of the y-term. So we change to a square window and obtain the more accurate graph in Figure 10–14.*

*Figure 10–14 shows a square window for a TI-84+. On wide-screen calculators, a longer x-axis is needed for a square window.

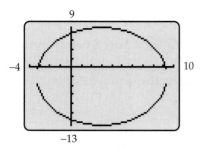

Figure 10–13

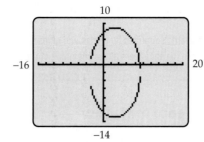

Figure 10–14

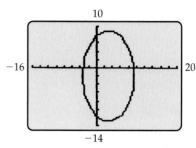

Figure 10–15

TECHNOLOGY TIP

TI-84+ and Casio 9850 have conic section graphers for equations in standard form. They are located in the Casio main menu or the TI APPS menu (if it's not there, you can download it free from TI).

The graphs in Figures 10–13 and 10–14 have gaps that should not be there (an ellipse is a connected figure). The gaps can be eliminated by using a conic section grapher (see the Technology Tip in the margin) because it uses a slightly different program than does the regular equation grapher, as Figure 10–15 illustrates. ■

If you only want the graph of an ellipse and don't need to know its center, vertices, or foci, it is not necessary to put its equation in standard form.

EXAMPLE 10

Graph the equation $x^2 + 8y^2 + 6x + 9y + 4 = 0$ without first putting it in standard form.

SOLUTION   Rewrite it like this:

$$8y^2 + 9y + (x^2 + 6x + 4) = 0.$$

This is a quadratic equation of the form $ay^2 + by + c = 0$, with

$$a = 8, \qquad b = 9, \qquad c = x^2 + 6x + 4$$

and hence can be solved by using the quadratic formula.

$$y = \frac{-b \pm \sqrt{b^2 - 4ac}}{2a}$$

$$y = \frac{-9 \pm \sqrt{9^2 - 4 \cdot 8 \cdot (x^2 + 6x + 4)}}{2 \cdot 8}$$

$$y = \frac{-9 \pm \sqrt{81 - 32(x^2 + 6x + 4)}}{16}$$

TECHNOLOGY TIP

TI users can save keystrokes by entering the first equation in Example 10 as y_1 and then using the RCL key to copy the text of y_1 to y_2. [On TI-89, use COPY and PASTE in place of RCL.] Then only one sign needs to be changed to make y_2 into the second equation of Example 10.

GRAPHING EXPLORATION

Find a complete graph of the original equation by graphing both of the following functions on the same screen. The Technology Tip in the margin may be helpful.

$$y = \frac{-9 + \sqrt{81 - 32(x^2 + 6x + 4)}}{16}$$

$$y = \frac{-9 - \sqrt{81 - 32(x^2 + 6x + 4)}}{16}.$$

■

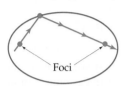

Figure 10–16

> **NOTE**
> Parametric equations for circles and ellipses are discussed in Special Topics 10.3.A and summarized in the endpapers at the beginning of the book.

 APPLICATIONS

Elliptical surfaces have interesting reflective properties. If a sound or light ray passes through one focus and reflects off an ellipse, the ray will pass through the other focus, as shown in Figure 10–16. Exactly this situation occurs under the elliptical dome of the U.S. Capitol. A person who stands at one focus and whispers can be clearly heard by anyone at the other focus. Before this fact was widely known, when Congress used to sit under the dome, several political secrets were inadvertently revealed by congressmen to members of the other party.

The planets and many comets have elliptical orbits, with the sun as one focus. The moon travels in an elliptical orbit with the earth as one focus. Satellites are usually put into elliptical orbits around the earth.

EXAMPLE 11

The earth's orbit around the sun is an ellipse that is almost a circle. The sun is one focus, and the major and minor axes have lengths 186,000,000 miles and 185,974,062 miles, respectively. What are the minimum and maximum distances from the earth to the sun?

SOLUTION   The orbit is shown in Figure 10–17. If we use a coordinate system with the major axis on the x-axis and the sun having coordinates $(c, 0)$, then we obtain Figure 10–18.

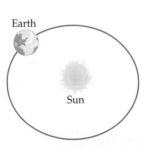

Figure 10–17

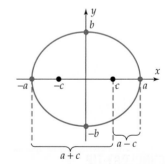

Figure 10–18

The length of the major axis is $2a = 186{,}000{,}000$, so that $a = 93{,}000{,}000$. Similarly, $2b = 185{,}974{,}062$, so $b = 92{,}987{,}031$. As was shown earlier, the equation of the orbit is

$$\frac{x^2}{a^2} + \frac{y^2}{b^2} = 1, \quad \text{where}$$

$$c = \sqrt{a^2 - b^2} = \sqrt{(93{,}000{,}000)^2 - (92{,}987{,}031)^2} \approx 1{,}553{,}083.$$

Figure 10–18 suggests a fact that can also be proven algebraically: The minimum and maximum distances from a point on the ellipse to the focus $(c, 0)$ occur at the endpoints of the major axis:

$$\text{Minimum distance} = a - c \approx 93,000,000 - 1,553,083 = 91,446,917 \text{ miles}$$

$$\text{Maximum distance} = a + c \approx 93,000,000 + 1,553,083 = 94,553,083 \text{ miles.} \quad \blacksquare$$

EXERCISES 10.1

In Exercises 1–6, determine which of the following equations could possibly have the given graph.

$2x^2 + y^2 = 12,$ $(x - 4)^2 + (y - 3)^2 = 4,$

$x^2 + 6y^2 = 18,$ $(x + 3)^2 + (y + 4)^2 = 6,$

$(x + 3)^2 + y^2 = 9,$ $x^2 + (y - 3)^2 = 4,$

$(x - 3)^2 + (y + 4)^2 = 5,$ $(x + 2)^2 + (y - 3)^2 = 2$

1.

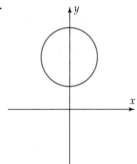

2.

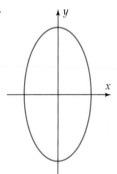

3.

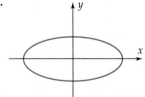

4.

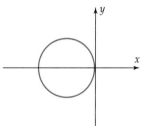

5.

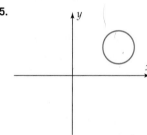

6.

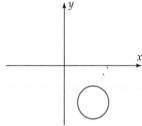

In Exercises 7–12, find the center and radius of the circle whose equation is given.

7. $x^2 + y^2 + 8x - 6y - 15 = 0$

8. $15x^2 + 15y^2 = 10$

9. $x^2 + y^2 + 6x - 4y - 15 = 0$

10. $x^2 + y^2 + 10x - 75 = 0$

11. $x^2 + y^2 + 25x + 10y = -12$

12. $3x^2 + 3y^2 + 12x + 12 = 18y$

In Exercises 13–20, identify the conic section whose equation is given and find its graph. If it is a circle, list its center and radius. If it is an ellipse, list its center, vertices, and foci.

13. $\dfrac{x^2}{25} + \dfrac{y^2}{4} = 1$ **14.** $\dfrac{x^2}{6} + \dfrac{y^2}{16} = 1$

15. $4x^2 + 3y^2 = 12$ **16.** $9x^2 + 4y^2 = 72$

17. $\dfrac{y^2}{49} + \dfrac{x^2}{81} = 1$ **18.** $\dfrac{x^2}{10} - 1 = \dfrac{-y^2}{36}$

19. $4x^2 + 4y^2 = 1$ **20.** $x^2 + 4y^2 = 1$

In Exercises 21–26, find the equation of the ellipse that satisfies the given conditions.

21. Center $(0, 0)$; foci on x-axis; x-intercepts ± 7; y-intercepts ± 2.

22. Center $(0, 0)$; foci on y-axis; x-intercepts ± 1; y-intercepts ± 8.

23. Center $(0, 0)$; foci on x-axis; major axis of length 12; minor axis of length 8.

24. Center $(0, 0)$; foci on y-axis; major axis of length 20; minor axis of length 18.

25. Center $(0, 0)$; endpoints of major and minor axes: $(0, -7)$, $(0, 7)$, $(-3, 0)$, $(3, 0)$.

26. Center $(0, 0)$; vertices $(8, 0)$ and $(-8, 0)$; minor axis of length 8.

Calculus can be used to show that the area of the ellipse with equation $\dfrac{x^2}{a^2} + \dfrac{y^2}{b^2} = 1$ is πab. Use this fact to find the area of each ellipse in Exercises 27–32.

27. $\dfrac{x^2}{16} + \dfrac{y^2}{4} = 1$ **28.** $\dfrac{x^2}{9} + \dfrac{y^2}{5} = 1$

29. $3x^2 + 4y^2 = 12$ **30.** $7x^2 + 5y^2 = 35$

31. $6x^2 + 2y^2 = 14$ **32.** $5x^2 + y^2 = 5$

In Exercises 33–38, identify the conic section whose equation is given, and find its graph. If it is a circle, list its center and radius. If it is an ellipse, list its center, vertices, and foci.

33. $\dfrac{(x - 1)^2}{4} + \dfrac{(y - 5)^2}{9} = 1$

34. $\dfrac{(x - 2)^2}{16} + \dfrac{(y + 3)^2}{12} = 1$

35. $\dfrac{(x + 1)^2}{16} + \dfrac{(y - 4)^2}{8} = 1$

36. $\dfrac{(x + 5)^2}{4} + \dfrac{(y + 2)^2}{12} = 1$

37. $9x^2 + 4y^2 + 54x - 8y + 49 = 0$

38. $4x^2 + 5y^2 - 8x + 30y + 29 = 0$

In Exercises 39–44, identify the conic section and use technology to graph it.

39. $x^2 + y^2 + 6x - 8y + 5 = 0$

40. $x^2 + y^2 - 4x + 2y - 7 = 0$

41. $4x^2 + y^2 + 24x - 4y + 36 = 0$

42. $9x^2 + y^2 - 36x + 10y + 52 = 0$

43. $9x^2 + 25y^2 - 18x + 50y = 191$

44. $25x^2 + 16y^2 + 50x + 96y = 231$

In Exercises 45–50, find the equation of the ellipse that satisfies the given conditions.

45. Center $(2, 3)$; endpoints of major and minor axes: $(2, -1)$, $(0, 3)$, $(2, 7)$, $(4, 3)$.

46. Center $(-5, 2)$; endpoints of major and minor axes: $(0, 2)$, $(-5, 17)$, $(-10, 2)$, $(-5, -13)$.

47. Center $(7, -4)$; foci on the line $x = 7$; major axis of length 12; minor axis of length 5.

48. Center $(-3, -9)$; foci on the line $y = -9$; major axis of length 15; minor axis of length 7.

49. Center $(3, -2)$; passing through $(3, -6)$ and $(9, -2)$.

50. Center $(2, 5)$; passing through $(2, 4)$ and $(-3, 5)$.

In Exercises 51 and 52, find the equations of two distinct ellipses satisfying the given conditions.

51. Center at $(-5, 3)$; major axis of length 14; minor axis of length 8.

52. Center at $(2, -6)$; major axis of length 15; minor axis of length 6.

In Exercises 53–58 all viewing windows are square. Determine which of the following equations could possibly have the given graph.

$$\dfrac{(x + 3)^2}{4} + \dfrac{(y + 3)^2}{8} = 1, \qquad \dfrac{(x - 3)^2}{9} + \dfrac{(y + 4)^2}{4} = 1,$$

$$2x^2 + 2y^2 - 8 = 0, \qquad 4x^2 + 2y^2 - 8 = 0,$$

$$2x^2 + y^2 - 8x - 6y + 9 = 0,$$

$$x^2 + 3y^2 + 6x - 12y + 17 = 0.$$

53.

54.

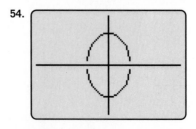

55.

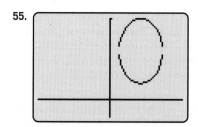

56.

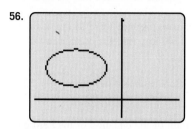

57.

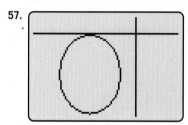

58.

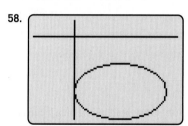

59. A circle is inscribed in the ellipse with equation

$$x^2 + 4y^2 = 64.$$

A diameter of the circle lies on the minor axis of the ellipse. Find the equation of the circle.

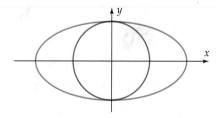

60. If (c, d) is a point on the ellipse in Exercise 59, prove that $(c/2, d)$ is a point on the circle.

61. The orbit of the moon around the earth is an ellipse with the earth as one focus. If the length of the major axis of the orbit is 477,736 miles and the length of the minor axis is 477,078 miles, find the minimum and maximum distances from the earth to the moon.

62. Halley's Comet has an elliptical orbit with the sun as one focus and a major axis that is 1,636,484,848 miles long. The closest the comet comes to the sun is 54,004,000 miles. What is the maximum distance from the comet to the sun?

63. A cross section of the ceiling of a "whispering room" at a museum is half of an ellipse. Two people stand so that their heads are approximately at the foci of the ellipse. If one whispers upward, the sound waves are reflected off the elliptical ceiling to the other person, as indicated in the figure. (For the reason why, see Figure 10–16 and the accompanying text.) How far apart are the two people standing? [*Hint:* Use the rectangular coordinate system suggested by the blue lines in the figure to find the equation of the ellipse; then find the foci.]

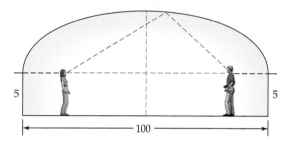

64. Suppose that the whispering room in Exercise 63 is 100 feet long and that the two people at the foci are 80 feet apart. How high is the center of the roof?

65. The bottom of a bridge is shaped like half an ellipse and is 20 feet above the river at the center, as shown in the figure. Find the height of the bridge bottom over a point on the river 25 feet from the center of the river. [*Hint:* Use a coordinate system, with the *x*-axis on the surface of the water and the *y*-axis running through the center of the bridge, to find an equation for the ellipse.]

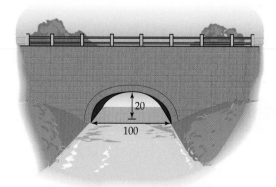

66. The stained glass window in the figure on the next page is shaped like the top half of an ellipse. The window is 10 feet wide, and the two figures in the window are located 3 feet from the center. If the figures are 1.6 feet high, find the height of the window at the center. [*Hint:* Use a coordinate system, with the bottom of the window as the *x*-axis and the vertical line through the center of the window as the *y*-axis.]

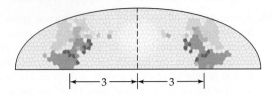

If $a > b > 0$, then the **eccentricity** of the ellipse

$$\frac{(x-h)^2}{a^2} + \frac{(y-k)^2}{b^2} = 1 \quad \text{or} \quad \frac{(x-h)^2}{b^2} + \frac{(y-k)^2}{a^2} = 1$$

is the number $\dfrac{\sqrt{a^2-b^2}}{a}$. In Exercises 67–70, find the eccentricity of the ellipse whose equation is given.

67. $\dfrac{x^2}{100} + \dfrac{y^2}{99} = 1$ **68.** $\dfrac{x^2}{18} + \dfrac{y^2}{25} = 1$

69. $\dfrac{(x-3)^2}{10} + \dfrac{(y-9)^2}{40} = 1$

70. $\dfrac{(x+5)^2}{12} + \dfrac{(y-4)^2}{8} = 1$

71. On the basis of your answers to Exercises 67–70, how is the eccentricity of an ellipse related to its graph? [*Hint:* What is the shape of the graph when the eccentricity is close to 0? When it is close to 1?]

72. Assuming that these viewing windows are square, which of these ellipses has the larger eccentricity?

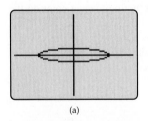

(a)

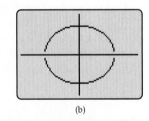

(b)

73. A satellite is to be placed in an elliptical orbit, with the center of the earth as one focus. The satellite's maximum distance from the surface of the earth is to be 22,380 km, and its minimum distance is to be 6540 km. Assume that the radius of the earth is 6400 km, and find the eccentricity of the satellite's orbit.

74. The first step in landing Apollo 11 on the moon was to place the spacecraft in an elliptical orbit such that the minimum distance from the *surface* of the moon to the spacecraft was 110 km and the maximum distance was 314 km. If the radius of the moon is 1740 km, find the eccentricity of the Apollo 11 orbit.

75. Consider the ellipse whose equation is $\dfrac{x^2}{a^2} + \dfrac{y^2}{b^2} = 1$. Show that if $a = b$, then the graph is actually a circle.

76. Complete the derivation of the equation of the ellipse on page 673 as follows.

(a) By squaring both sides, show that the equation

$$\sqrt{(x+c)^2 + y^2} = 2a - \sqrt{(x-c)^2 + y^2}$$

may be simplified as

$$a\sqrt{(x-c)^2 + y^2} = a^2 - cx.$$

(b) Show that the last equation in part (a) may be further simplified as

$$(a^2 - c^2)x^2 + a^2y^2 = a^2(a^2 - c^2).$$

THINKER

77. The punch bowl and a table holding the punch cups are placed 50 feet apart at a garden party. A portable fence is then set up so that any guest inside the fence can walk straight to the table, then to the punch bowl, and then return to his or her starting point without traveling more than 150 feet. Describe the longest possible such fence that encloses the largest possible area.

10.2 Hyperbolas

Section Objectives
■ Find the vertices, foci and asymptotes of a hyperbola and sketch its graph.
■ Set up and solve applied problems involving hyperbolas.

Definition. Let *P* and *Q* be points in the plane, and let *r* be a positive number. The set of all points *X* such that

$$\big|(\text{Distance from } P \text{ to } X) - (\text{Distance from } Q \text{ to } X)\big| = r$$

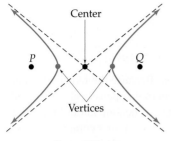

Center

Vertices

Figure 10–19

Standard Equations of Hyperbolas Centered at the Origin

is the **hyperbola** with **foci** P and Q; r will be called the **distance difference**. Every hyperbola has the general shape shown by the red curve in Figure 10–19. The dotted straight lines are the **asymptotes** of the hyperbola; it gets closer and closer to the asymptotes, but never touches them. The asymptotes intersect at the midpoint of the line segment from P to Q; this point is called the **center** of the hyperbola. The **vertices** of the hyperbola are the points where it intersects the line segment from P to Q. The line through P and Q is called the **focal axis.**

Equation. A complicated exercise in the use of the distance formula, which will be omitted here, leads to the following algebraic description.

Let a and b be positive real numbers. Then the graph of each of the following equations is a hyperbola centered at the origin:

$$\frac{x^2}{a^2} - \frac{y^2}{b^2} = 1 \begin{cases} \text{x-intercepts: } \pm a \qquad \text{y-intercepts: none} \\ \text{focal axis on the x-axis, with vertices } (a, 0) \text{ and } (-a, 0) \\ \text{foci: } (c, 0) \text{ and } (-c, 0), \text{ where } c = \sqrt{a^2 + b^2}. \\ \text{asymptotes: } y = \frac{b}{a}x \text{ and } y = -\frac{b}{a}x \end{cases}$$

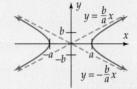

$$\frac{y^2}{a^2} - \frac{x^2}{b^2} = 1 \begin{cases} \text{x-intercepts: none} \qquad \text{y-intercepts: } \pm a \\ \text{focal axis on the y-axis, with vertices } (0, a) \text{ and } (0, -a) \\ \text{foci: } (0, c) \text{ and } (0, -c), \text{ where } c = \sqrt{a^2 + b^2}. \\ \text{asymptotes: } y = \frac{a}{b}x \text{ and } y = -\frac{a}{b}x \end{cases}$$

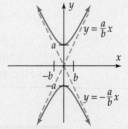

Once again, don't worry about all the letters in the box. The standard form of the equation gives all the necessary information, as explained in the following example.

EXAMPLE 1

List the vertices, foci, and asymptotes of these hyperbolas.

(a) $\dfrac{x^2}{25} - \dfrac{y^2}{16} = 1$ (b) $\dfrac{y^2}{4} - \dfrac{x^2}{16} = 1$

SOLUTION

(a) We first rewrite the equation in standard form.

$$\frac{x^2}{5^2} - \frac{y^2}{4^2} = 1$$

Now we can read off the required information. Because the hyperbola equation is of the form "x-term $-$ y-term," the hyperbola opens from side to side and its vertices (x-intercepts) are determined by the denominator of the x-term of the equation: $(-5, 0)$ and $(5, 0)$. To find the foci, we first compute c, which is the square root of the sum of the denominators in the hyperbola's equation:

$$c = \sqrt{a^2 + b^2}$$
$$= \sqrt{5^2 + 4^2}$$
$$= \sqrt{25 + 16} = \sqrt{41}.$$

Hence, the foci are $(-\sqrt{41}, 0)$ and $(\sqrt{41}, 0)$. Finally, to remember the correct asymptote coefficients, think of them as

$$\pm \frac{\text{square root of the hyperbola's } y\text{-term denominator}}{\text{square root of the hyperbola's } x\text{-term denominator}} = \pm \frac{4}{5}.$$

So the asymptotes are

$$y = \frac{4}{5}x \qquad \text{and} \qquad y = -\frac{4}{5}x.$$

(b) In standard form the equation is

$$\frac{y^2}{2^2} - \frac{x^2}{4^2} = 1.$$

Because the hyperbola equation is of the form "y-term $-$ x-term," the hyperbola opens up and down and its vertices (y-intercepts) are determined by the denominator of the y-term of the equation: $(0, -2)$ and $(0, 2)$. The foci are $(0, c)$ and $(0, -c)$, where

$$c = \sqrt{2^2 + 4^2}$$
$$= \sqrt{4 + 16} = \sqrt{20} = 2\sqrt{5}.$$

Once again, the asymptote coefficients are

$$\pm \frac{\text{square root of the hyperbola's } y\text{-term denominator}}{\text{square root of the hyperbola's } x\text{-term denominator}} = \pm \frac{2}{4} = \frac{1}{2},$$

so the asymptotes are

$$y = \frac{1}{2}x \qquad \text{and} \qquad y = -\frac{1}{2}x. \qquad \blacksquare$$

EXAMPLE 2

Identify and sketch the graph of the equation $9x^2 - 4y^2 = 36$.

SOLUTION   We first put the equation in standard form:

$$9x^2 - 4y^2 = 36$$

Divide both sides by 36:
$$\frac{9x^2}{36} - \frac{4y^2}{36} = \frac{36}{36}$$

Simplify:
$$\frac{x^2}{4} - \frac{y^2}{9} = 1$$

$$\frac{x^2}{2^2} - \frac{y^2}{3^2} = 1.$$

Applying the techniques of Example 1 or the facts in the preceding box (with $a = 2$ and $b = 3$) shows that the graph is a hyperbola with vertices $(2, 0)$ and $(-2, 0)$ and asymptotes $y = \frac{3}{2}x$ and $y = -\frac{3}{2}x$. We first plot the vertices and sketch the rectangle determined by the vertical lines $x = \pm 2$ and the horizontal lines $y = \pm 3$. The asymptotes go through the origin and the corners of this rectangle, as shown on the left in Figure 10–20. It is then easy to sketch the hyperbola. ■

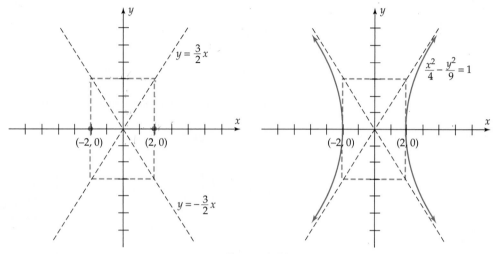

Figure 10–20

EXAMPLE 3

Find the equation of the hyperbola with vertices $(0, 1)$ and $(0, -1)$ that passes through the point $(3, \sqrt{2})$. Then sketch its graph.

SOLUTION   The vertices are on the y-axis, and the equation is of the form

$$\frac{y^2}{a^2} - \frac{x^2}{b^2} = 1$$

with $a = 1$. Since $(3, \sqrt{2})$ is on the graph, we have

$$\frac{(\sqrt{2})^2}{1^2} - \frac{3^2}{b^2} = 1$$

Simplify:
$$2 - \frac{9}{b^2} = 1$$

Subtract 2 from both sides:
$$-\frac{9}{b^2} = -1$$

Multiply both sides by $-b^2$:
$$9 = b^2.$$

Therefore, $b = 3$, and the equation is

$$\frac{y^2}{1^2} - \frac{x^2}{3^2} = 1 \qquad \text{or, equivalently,} \qquad y^2 - \frac{x^2}{9} = 1.$$

The asymptotes of the hyperbola are the lines $y = \pm\frac{1}{3}x$.

There are several ways to graph $\dfrac{y^2}{1} - \dfrac{x^2}{3^2} = 1$. You can graph by hand using the technique of Example 2: Sketch the rectangle determined by the horizontal lines $y = \pm 1$ and the vertical lines $x = \pm 3$. The asymptotes run through the origin and the corners of this rectangle. Once you have these, it's easy to plot a few points and obtain the graph in Figure 10–21.

You can also use a calculator to graph the equation. A conic section grapher, if your calculator has one, is the easiest way (Figure 10–22).* Otherwise, solve the hyperbola equation for y:

$$y^2 - \frac{x^2}{9} = 1$$

Add $\frac{x^2}{9}$ to both sides:

$$y^2 = 1 + \frac{x^2}{9}$$

Take square roots on both sides: $\quad y = \sqrt{1 + \frac{x^2}{9}} \quad$ or $\quad y = -\sqrt{1 + \frac{x^2}{9}}.$

Graphing these last two equations on the same screen, also produces Figure 10–22. ∎

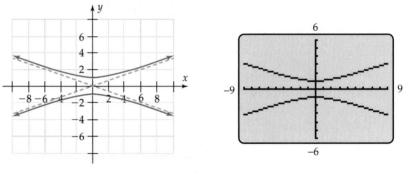

Figure 10–21   Figure 10–22

VERTICAL AND HORIZONTAL SHIFTS

In Section 10.1 we saw that replacing x with $x - h$ and y with $y - k$ in the equation of an ellipse, shifts the ellipse vertically and horizontally. The same thing is true for hyperbolas.

Vertical and Horizontal Shifts

> Consider a hyperbola centered at the origin with equation
>
> $$\frac{x^2}{a^2} - \frac{y^2}{b^2} = 1.$$
>
> Let h and k be constants. Replacing x with $x - h$ and replacing y with $y - k$ in this equation produces the equation
>
> $$\frac{(x - h)^2}{a^2} - \frac{(y - k)^2}{b^2} = 1,$$
>
> whose graph is the original hyperbola shifted horizontally and vertically so that its center is (h, k).
>
> Analogous facts are true for the hyperbola with equation $\frac{y^2}{a^2} - \frac{x^2}{b^2} = 1.$

*See the Technology Tip on page 681.

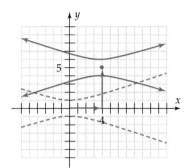

Figure 10–23

EXAMPLE 4

Identify and sketch the graph of

$$\frac{(y-5)^2}{1} - \frac{(x-4)^2}{9} = 1.$$

SOLUTION   This equation can be obtained from $\dfrac{y^2}{1} - \dfrac{x^2}{9} = 1$ by replacing x with $x-4$ and replacing y with $y-5$. This is the situation described in the preceding box with $h = 4$ and $k = 5$. So the graph of $\dfrac{(y-5)^2}{1} - \dfrac{(x-4)^2}{9} = 1$ is a hyperbola with center $(4, 5)$. That can be obtained from the hyperbola $\dfrac{y^2}{1} - \dfrac{x^2}{9} = 1$ (shown in Figure 10–21) by shifting it 5 units upward and 4 units to the right, as shown in Figure 10–23. Its vertices, foci and focal axis lie on the vertical line $x = 4$. ∎

EXAMPLE 5

Identify and sketch the graph of

$$\frac{(x-3)^2}{4} - \frac{(y+2)^2}{9} = 1.$$

SOLUTION   If we rewrite the equation as

$$\frac{(x-3)^2}{2^2} - \frac{(y-(-2))^2}{3^2} = 1,$$

then it has the form in the preceding box with $h = 3$ and $k = -2$. Its graph is a hyperbola with center $(3, -2)$. There are several ways to obtain the graph.

Method 1.   The equation of this hyperbola can be obtained from

$$\frac{x^2}{4} - \frac{y^2}{9} = 1$$

by replacing x with $x-3$ and y with $y+2 = y-(-2)$. So its graph is the hyperbola $\dfrac{x^2}{4} - \dfrac{y^2}{9} = 1$ (see Figure 10–20) shifted 3 units to the right and 2 units downward, as shown in Figure 10–24. The vertices, foci, and focal axis lie on the horizontal line $y = -2$.

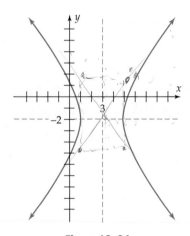

Figure 10–24

Method 2.   If your calculator has a conic section grapher, you can insert the appropriate values ($a = 2$, $b = 3$, $h = 3$, and $k = -2$) and obtain Figure 10–25.*

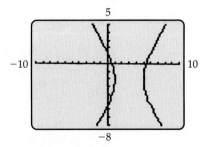

Figure 10–25

*See the Technology Tip on page 681.

Method 3.   Solve the original equation for y:

$$\frac{(x-3)^2}{4} - \frac{(y+2)^2}{9} = 1$$

Multiply both sides by 36: $9(x-3)^2 - 4(y+2)^2 = 36$

Rearrange terms: $4(y+2)^2 = 9(x-3)^2 - 36$

Divide both sides by 4: $(y+2)^2 = \dfrac{9(x-3)^2 - 36}{4}$

Take square roots of both sides: $y + 2 = \pm\sqrt{\dfrac{9(x-3)^2 - 36}{4}}$

$$y = -2 + \sqrt{\frac{9(x-3)^2 - 36}{4}} \quad \text{or} \quad y = -2 - \sqrt{\frac{9(x-3)^2 - 36}{4}}.$$

GRAPHING EXPLORATION

Graph these last two equations in viewing window of Figure 10–24. How does your graph compare with Figure 10–24? What is the explanation for the gaps?

■

EXAMPLE 6

Find the center, vertices, foci, and asymptotes of the hyperbola

$$\frac{(x+2)^2}{25} - \frac{(y-3)^2}{16} = 1.$$

SOLUTION   By writing $x + 2$ as $x - (-2)$, we see that the center of the hyperbola is $(-2, 3)$. Its graph is obtained from the hyperbola $\dfrac{x^2}{25} - \dfrac{y^2}{16} = 1$ by shifting it 2 units to the left and 3 units upward. As shown in the box on page 687, the vertices of $\dfrac{x^2}{25} - \dfrac{y^2}{16} = 1$ are $(-5, 0)$ and $(5, 0)$. So the vertices of $\dfrac{(x-(-2))^2}{25} - \dfrac{(y-3)^2}{16} = 1$ are obtained as follows:

Shift 2 units left and
3 units upward

$(-5, 0)$---------------------->$(-5 - 2, 0 + 3) = (-7, 3)$

$(5, 0)$------------------------>$(5 - 2, 0 + 3) = (3, 3)$.

Similarly, the box on page 687 shows that the foci of $\dfrac{x^2}{25} - \dfrac{y^2}{16} = 1$ are $(-c, 0)$ and $(c, 0)$, where $c = \sqrt{25 + 16} = \sqrt{41}$. So the foci of $\dfrac{(x-(-2))^2}{25} - \dfrac{(y-3)^2}{16} = 1$ are given by:

Shift 2 units left and
3 units upward

$(-\sqrt{41}, 0)$---------------------->$(-\sqrt{41} - 2, 0 + 3) = (-\sqrt{41} - 2, 3)$

$(\sqrt{41}, 0)$------------------------>$(\sqrt{41} - 2, 0 + 3) = (\sqrt{41} - 2, 3)$.

The asymptotes of $\dfrac{x^2}{25} - \dfrac{y^2}{16} = 1$ are $y = \pm\dfrac{4}{5}x$. So the asymptotes of $\dfrac{(x-(-2))^2}{25} - \dfrac{(y-3)^2}{16} = 1$ are

$$y - 3 = \frac{4}{5}(x - (-2)) \qquad \text{and} \qquad y - 3 = -\frac{4}{5}(x - (-2))$$

$$y = \frac{4}{5}(x - (-2)) + 3 \qquad\qquad y = -\frac{4}{5}(x - (-2)) + 3$$

$$y = \frac{4}{5}(x + 2) + 3 \qquad\qquad y = -\frac{4}{5}(x + 2) + 3. \qquad \blacksquare$$

Examples 4–6 illustrate the following facts.

Standard Equations of Hyperbolas with Center at (h, k)

If a and b are positive real numbers, then the graph of each of the following equations is a hyperbola with center (h, k).

$$\frac{(x-h)^2}{a^2} - \frac{(y-k)^2}{b^2} = 1 \begin{cases} \text{focal axis on the horizontal line } y = k \\ \text{foci: } (-c + h, k) \text{ and } (c + h, k), \text{ where} \\ \quad c = \sqrt{a^2 + b^2} \\ \text{vertices: } (-a + h, k) \text{ and } (a + h, k) \\ \text{asymptotes: } y = \pm\dfrac{b}{a}(x - h) + k \end{cases}$$

$$\frac{(y-k)^2}{a^2} - \frac{(x-h)^2}{b^2} = 1 \begin{cases} \text{focal axis on the vertical line } x = h \\ \text{foci: } (h, -c + k) \text{ and } (h, c + k), \text{ where} \\ \quad c = \sqrt{a^2 + b^2} \\ \text{vertices: } (h, -a + k) \text{ and } (h, a + k) \\ \text{asymptotes: } y = \pm\dfrac{a}{b}(x - h) + k \end{cases}$$

GRAPHING TECHNIQUES

When the equation of a hyperbola is in standard form, it can be graphed relatively easily, either by hand or with technology, as illustrated in Examples 4 and 5. However, an equation can be graphed directly on a calculator or computer without first putting it in standard form.

EXAMPLE 7

Sketch the graph of $6y^2 - 8x^2 - 24y - 48x - 96 = 0$.

SOLUTION We first solve the equation for y. Begin by rewriting it as

$$6y^2 - 24y + (-8x^2 - 48x - 96) = 0.$$

This is a quadratic equation of the form $ay^2 + by + c = 0$, with

$$a = 6, \qquad b = -24, \qquad c = -8x^2 - 48x - 96.$$

We use the quadratic formula to solve it.

$$y = \frac{-b \pm \sqrt{b^2 - 4ac}}{2a}$$

$$y = \frac{-(-24) \pm \sqrt{(-24)^2 - 4 \cdot 6(-8x^2 - 48x - 96)}}{2 \cdot 6}$$

$$y = \frac{24 \pm \sqrt{576 - 24(-8x^2 - 48x - 96)}}{12}.$$

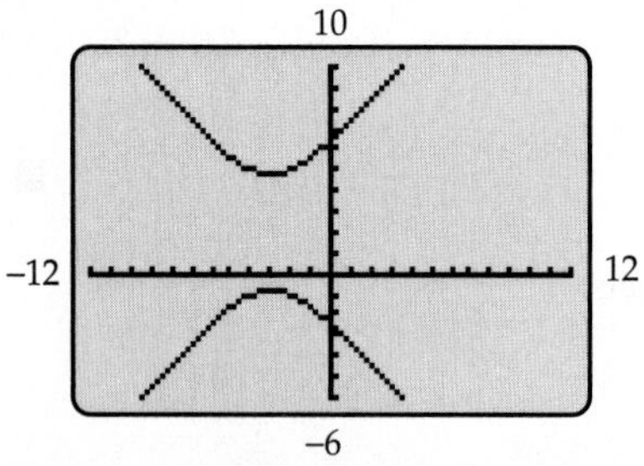

Figure 10–26

Now we graph both

$$y = \frac{24 + \sqrt{576 - 24(-8x^2 - 48x - 96)}}{12} \quad \text{and}$$

$$y = \frac{24 - \sqrt{576 - 24(-8x^2 - 48x - 96)}}{12}$$

on the same screen to obtain the hyperbola in Figure 10–26. ■

> **NOTE**
> Parametric equations for hyperbolas are discussed in Special Topics 10.3.A and summarized in the endpapers at the beginning of the book.

Although the graph in Figure 10–26 shows you what the hyperbola looks like, it does not provide all the pertinent information about the hyperbola. To get that information, you do need to put the equation in standard form.

EXAMPLE 8

Find the center of the hyperbola in Example 7.

> **CAUTION**
> Completing the square only works when the coefficient of y^2 is 1. In an expression such as $6y^2 - 24y$, you must first factor out the 6,
> $$6(y^2 - 4y),$$
> and then complete the square on the expression in parentheses.

SOLUTION Begin by rearranging the equation.

$$6y^2 - 8x^2 - 24y - 48x - 96 = 0$$

Add 96 to both sides: $$6y^2 - 24y - 8x^2 - 48x = 96$$

Group x- and y-terms: $$(6y^2 - 24y) - (8x^2 + 48x) = 96$$

Factor out coefficients of y^2 and x^2: $$6(y^2 - 4y) - 8(x^2 + 6x) = 96.$$

Complete the square in the expression $y^2 - 4y$ by adding 4 (the square of half the coefficient of y), and complete the square in $x^2 + 6x$ by adding 9 (the square of half the coefficient of x).

$$6(y^2 - 4y + 4) - 8(x^2 + 6x + 9) = 96 + ? + ?$$

On the left side, we have actually added $6 \cdot 4 = 24$ and $-8 \cdot 9 = -72$, so we must add these numbers on the right to keep the equation unchanged.

$$6(y^2 - 4y + 4) - 8(x^2 + 6x + 9) = 96 + 24 - 72$$

Factor and simplify: $$6(y - 2)^2 - 8(x + 3)^2 = 48$$

Divide both sides by 48: $$\frac{(y - 2)^2}{8} - \frac{(x + 3)^2}{6} = 1$$

$$\frac{(y - 2)^2}{8} - \frac{(x - (-3))^2}{6} = 1.$$

In this form, we can see that the graph is a hyperbola with center at $(-3, 2)$. ■

APPLICATIONS

The reflective properties of hyperbolas are used in the design of camera and telescope lenses. If a light ray passes through one focus of a hyperbola and reflects off the hyperbola at a point P, then the reflected ray moves along the straight line determined by P and the other focus, as shown in Figure 10–27.

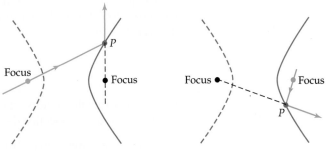

Figure 10–27

Hyperbolas are also the basis of the long-range navigation system (LORAN), which enables a ship to determine its exact location by radio, as illustrated in the next example.

EXAMPLE 9

Three LORAN transmitters Q, P, and R are located 200 miles apart along a straight shoreline and simultaneously transmit signals at regular intervals. These signals travel at a speed of 980 feet per microsecond. A ship S receives a signal from P and 305 microseconds later a signal from R. It also receives a signal from Q 528 microseconds after the one from P. Determine the ship's location.

SOLUTION   Take the line through the LORAN stations as the x-axis, with the origin located midway between Q and P, so that the situation looks like Figure 10–28.

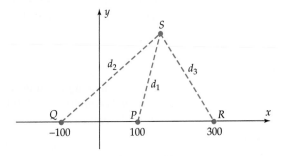

Figure 10–28

If the signal takes t microseconds to go from P to S, then

$$d_1 = 980t \quad \text{and} \quad d_2 = 980(t + 528),$$

so

$$|d_1 - d_2| = |980t - 980(t + 528)| = 980 \cdot 528 = 517{,}440 \text{ feet.}$$

Since 1 mile is 5280 feet, this means that

$$|d_1 - d_2| = 517,440/5,280 \text{ miles} = 98 \text{ miles}.$$

In other words,

$$|(\text{Distance from } P \text{ to } S) - (\text{Distance from } Q \text{ to } S)| = |d_1 - d_2| = 98.$$

This is precisely the situation described in the definition of "hyperbola" on pages 686–687: S is on the hyperbola with foci $P = (100, 0)$, $Q = (-100, 0)$, and distance difference $r = 98$. This hyperbola has an equation of the form

$$\frac{x^2}{a^2} - \frac{y^2}{b^2} = 1,$$

where $(\pm a, 0)$ are the vertices, $(\pm c, 0) = (\pm 100, 0)$ are the foci, and $c^2 = a^2 + b^2$. Figure 10–29 and the fact that the vertex $(a, 0)$ is on the hyperbola show that

$$|[\text{Distance from } P \text{ to } (a, 0)] - [\text{Distance from } Q \text{ to } (a, 0)]| = r = 98$$

$$|(100 - a) - (100 + a)| = 98$$

$$|-2a| = 98$$

$$|a| = 49.$$

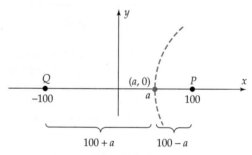

Figure 10–29

Consequently, $a^2 = 49^2 = 2401$, and hence, $b^2 = c^2 - a^2 = 100^2 - 49^2 = 7599$. Thus, the ship lies on the hyperbola

(*)
$$\frac{x^2}{2401} - \frac{y^2}{7599} = 1.$$

A similar argument using P and R as foci shows that the ship also lies on the hyperbola with foci $P = (100, 0)$ and $R = (300, 0)$ and center $(200, 0)$, whose distance difference r is

$$|d_1 - d_3| = 980 \cdot 305 = 298,900 \text{ feet} \approx 56.61 \text{ miles}.$$

As before, you can verify that $a = 56.61/2 = 28.305$, and hence, $a^2 = 28.305^2 = 801.17$. This hyperbola has center $(200, 0)$, and its foci are $(200 - c, k) = (100, 0)$ and $(200 + c, k) = (300, 0)$, which implies that $c = 100$. Hence, $b^2 = c^2 - a^2 = 100^2 - 801.17 = 9198.83$, and the ship also lies on the hyperbola

(**)
$$\frac{(x - 200)^2}{801.17} - \frac{y^2}{9198.83} = 1.$$

900

S

−500   Q   D   R   500

−900

Figure 10–30

Since the ship lies on both hyperbolas, its coordinates are solutions of both the equations (∗) and (∗∗). They can be found algebraically by solving each of the equations for y^2, setting the results equal, and solving for x. They can be found geometrically by graphing both hyperbolas and finding the intersection point. As shown in Figure 10–30, there are actually four points of intersection. However, the two below the x-axis represent points on land in our situation. Furthermore, since the signal from P was received first, the ship is closest to P. So it is located at the point S in Figure 10–30. A graphical intersection finder shows that this point is approximately (130.48, 215.14), where the coordinates are in miles from the origin. ∎

EXERCISES 10.2

In Exercises 1–6, determine which of the following equations could possibly have the given graph.

$3x^2 + 3y^2 = 12$, $6y^2 − x^2 = 6$,

$x^2 + 4y^2 = 1$, $4x^2 + 4(y + 2)^2 = 12$,

$4(x + 4)^2 + 4y^2 = 12$, $6x^2 + 2y^2 = 18$

$2x^2 − y^2 = 8$, $3x^2 − y = 6$

1.

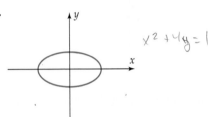

$x^2 + 4y = 1$

2.

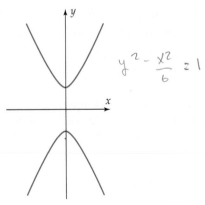

$y^2 - \dfrac{x^2}{6} = 1$

3.

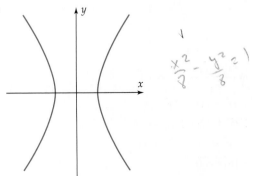

$\dfrac{x^2}{8} - \dfrac{y^2}{8} = 1$

4.

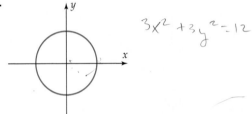

$3x^2 + 3y^2 = 12$

5.

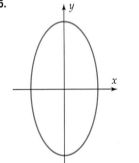

$6x^2 + 2y^2 = 18$

6.

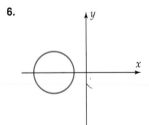

$4x^2 + 4(y + 2)^2 = 12$

In Exercises 7–14, identify the conic section whose equation is given and find its graph. List its vertices, foci, and asymptotes.

7. $\dfrac{x^2}{4} − y^2 = 1$

8. $\dfrac{x^2}{6} − \dfrac{y^2}{16} = 1$

9. $3y^2 − 5x^2 = 15$

10. $4x^2 − y^2 = 16$

11. $\dfrac{y^2}{9} − \dfrac{x^2}{16} = 1$

12. $\dfrac{x^2}{10} − \dfrac{y^2}{36} = 1$

13. $x^2 − 4y^2 = 1$

14. $2x^2 − y^2 = 4$

In Exercises 15–20, find the equation of the hyperbola.

15.

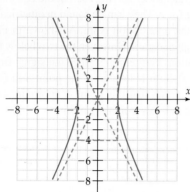

16.

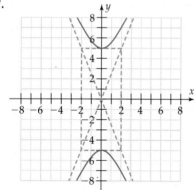

17.

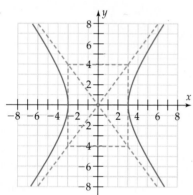

18.

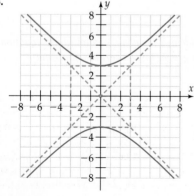

19.

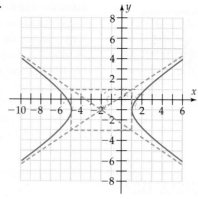

20.

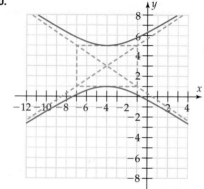

In Exercises 21–24, find the equation of the hyperbola that satisfies the given conditions.

21. Center $(0, 0)$; x-intercepts ± 3; asymptote $y = 2x$.

22. Center $(0, 0)$; y-intercepts ± 12; asymptote $y = 3x/2$.

23. Center $(0, 0)$; vertex $(2, 0)$; passing through $(4, \sqrt{3})$.

24. Center $(0, 0)$; vertex $(0, \sqrt{12})$; passing through $(2\sqrt{3}, 6)$.

In Exercises 25–32, identify the conic whose equation is given and find its graph. If it is an ellipse, list its center, vertices, and foci. If it is a hyperbola, list its center, vertices, foci, and asymptotes.

25. $\dfrac{(y + 3)^2}{25} - \dfrac{(x + 1)^2}{16} = 1$

26. $\dfrac{(y + 1)^2}{9} - \dfrac{(x - 1)^2}{25} = 1$

27. $\dfrac{(x + 3)^2}{1} - \dfrac{(y - 2)^2}{4} = 1$

28. $\dfrac{(y + 5)^2}{9} - \dfrac{(x - 2)^2}{1} = 1$

29. $(y + 4)^2 - 8(x - 1)^2 = 8$

30. $(x - 3)^2 + 12(y - 2)^2 = 24$

31. $4y^2 - x^2 + 6x - 24y + 11 = 0$

32. $x^2 - 16y^2 = 0$

In Exercises 33–38, identify the conic section whose equation is given and use technology to graph it.

33. $2x^2 + 2y^2 - 12x - 16y + 26 = 0$

34. $3x^2 + 3y^2 + 12x + 6y = 0$

35. $2x^2 + 3y^2 - 12x - 24y + 54 = 0$

36. $x^2 + 2y^2 + 4x - 4y = 8$

37. $x^2 - 3y^2 + 4x + 12y = 20$

38. $2x^2 + 16x = y^2 - 6y - 55$

In Exercises 39–42, find the equation of the hyperbola that satisfies the given conditions.

39. Center $(-2, 3)$; vertex $(-2, 1)$; passing through $(-2 + 3\sqrt{10}, 11)$.

40. Center $(-5, 1)$; vertex $(-3, 1)$; passing through $(-1, 1 - 4\sqrt{3})$.

41. Center $(4, 2)$; vertex $(7, 2)$; asymptote $3y = 4x - 10$.

42. Center $(-3, -5)$; vertex $(-3, 0)$; asymptote $6y = 5x - 15$.

In Exercises 43–48, determine which of the following equations could possibly have the given graph.

$$\frac{(y-2)^2}{4} - \frac{(x-3)^2}{9} = 1, \qquad \frac{(x+3)^2}{3} - \frac{(y+3)^2}{4} = 1,$$

$$4x^2 - 2y^2 = 8, \qquad 9(y-2)^2 = 36 + 4(x+3)^2,$$

$$3(y+3)^2 = 4(x-3)^2 - 12, \qquad y^2 - 2x^2 = 6.$$

43.

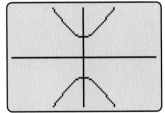

44.

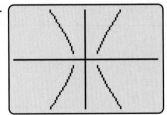

45.

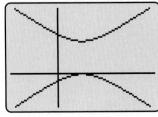

46.

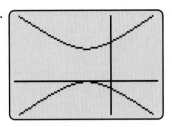

47.

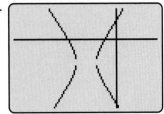

48.

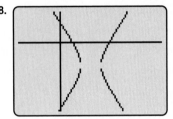

49. Sketch the graph of $\dfrac{y^2}{4} - \dfrac{x^2}{b^2} = 1$ for $b = 2$, $b = 4$, $b = 8$, $b = 12$, and $b = 20$. What happens to the hyperbola as b takes larger and larger values? Could the graph ever degenerate into a pair of horizontal lines?

50. Find a number k such that $(-2, 1)$ is on the graph of $3x^2 + ky^2 = 4$. Then graph the equation.

51. Show that the asymptotes of the hyperbola $\dfrac{x^2}{a^2} - \dfrac{y^2}{a^2} = 1$ are perpendicular to each other.

52. Find the approximate coordinates of the points where these hyperbolas intersect:

$$\frac{(x-1)^2}{4} - \frac{(y+1)^2}{8} = 1 \qquad \text{and} \qquad 4y^2 - x^2 = 1.$$

53. Two listening stations that are 1 mile apart record an explosion. One microphone receives the sound 2 seconds after the other does. Use the line through the microphones as the x-axis, with the origin midway between the microphones, and the fact that sound travels at 1100 feet per second to find the equation of a hyperbola on which the explosion is located. Can you determine the exact location of the explosion?

54. Two transmission stations P and Q are located 200 miles apart on a straight shoreline. A ship 50 miles from shore is moving parallel to the shoreline. A signal from Q reaches the ship 400 microseconds after a signal from P. If the signals travel at 980 feet per microsecond, find the location of the ship (in terms of miles) in the coordinate system with x-axis through P and Q and origin midway between them.

Exercises 55 and 56 deal with an experiment conducted by the physicist Ernest Rutherford in 1911. Rutherford discovered that when alpha particles are directed toward the nucleus of a gold atom, the particles follow a hyperbolic path, as shown in the figure, in which the nucleus is at the origin and the dashed lines are the asymptotes of the particles path.

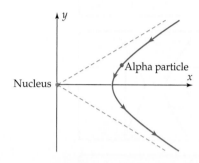

55. If the asymptotes of the hyperbolic path of the particle are given by $y = \pm\dfrac{3}{4}x$ and the closest the particle comes to the nucleus is 5 units, find the equation of the particle's path.

56. Do Exercise 55 when the asymptotes are $y = \pm\dfrac{2}{3}x$ and the minimum distance from the particle to the nucleus is 2 units.

*If $a > 0$ and $b > 0$, then the **eccentricity** of the hyperbola*

$$\frac{(x-h)^2}{a^2} - \frac{(y-k)^2}{b^2} = 1 \quad \text{or} \quad \frac{(y-k)^2}{a^2} - \frac{(x-h)^2}{b^2} = 1$$

is the number $\dfrac{\sqrt{a^2+b^2}}{a}$. In Exercises 57–61, find the eccentricity of the hyperbola whose equation is given.

57. $\dfrac{(x-6)^2}{10} - \dfrac{y^2}{40} = 1$

58. $\dfrac{y^2}{18} - \dfrac{x^2}{25} = 1$

59. $6(y-2)^2 = 18 + 3(x+2)^2$

60. $16x^2 - 9y^2 - 32x + 36y + 124 = 0$

61. $4x^2 - 5y^2 - 16x - 50y + 71 = 0$

62. (a) Graph these hyperbolas (on the same screen if possible).

$$\frac{y^2}{4} - \frac{x^2}{1} = 1 \qquad \frac{y^2}{4} - \frac{x^2}{12} = 1 \qquad \frac{y^2}{4} - \frac{x^2}{96} = 1$$

   (b) Compute the eccentricity of each hyperbola in part (a).
   (c) On the basis of parts (a) and (b), how is the shape of a hyperbola related to its eccentricity?

10.3   Parabolas

Section Objectives
- ■ Identify the focus, directrix, and standard equation of a parabola and sketch its graph.
- ■ Set up and solve applied problems involving parabolas.

Definition.   Parabolas appeared in Section 4.1 as the graphs of quadratic functions. Parabolas of this kind are a special case of the following more general definition. Let L be a line in the plane, and let P be a point not on L. If X is any point not on L, the distance from X to L is defined to be the length of the perpendicular line segment from X to L. The **parabola** with **focus** P and **directrix** L is the set of all points X such that

$$\text{Distance from } X \text{ to } P = \text{Distance from } X \text{ to } L$$

as shown in Figure 10–31.
   The line through P perpendicular to L is called the **axis.** The intersection of the axis with the parabola (the midpoint of the segment of the axis from P to L) is the **vertex** of the parabola, as illustrated in Figure 10–31. The parabola is symmetric with respect to its axis.

Equation.   Suppose that the focus is on the y-axis at the point $(0, p)$, where p is a nonzero constant, and that the directrix is the horizontal line $y = -p$. If (x, y) is any point on the parabola, then the distance from (x, y) to the horizontal line

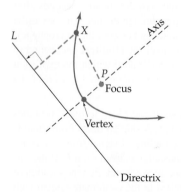

Figure 10–31

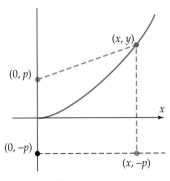

Figure 10–32

$y = -p$ is the length of the vertical segment from (x, y) to $(x, -p)$ as shown in Figure 10–32.

By the definition of the parabola,

Distance from (x, y) to $(0, p)$ = Distance from (x, y) to line $y = -p$

Distance from (x, y) to $(0, p)$ = Distance from (x, y) to $(x, -p)$

$$\sqrt{(x - 0)^2 + (y - p)^2} = \sqrt{(x - x)^2 + [y - (-p)]^2}.$$

Squaring both sides and simplifying, we have

$$(x - 0)^2 + (y - p)^2 = (x - x)^2 + (y + p)^2$$
$$x^2 + y^2 - 2py + p^2 = 0^2 + y^2 + 2py + p^2$$
$$x^2 = 4py.$$

Conversely, it can be shown that every point whose coordinates satisfy this equation is on the parabola.

A similar argument works for the parabola with focus $(p, 0)$ on the x-axis and directrix the vertical line $x = -p$. Furthermore, these arguments work for both positive and negative p, and leads to this conclusion.

Standard Equations of Parabolas with Vertex at the Origin

Let p be a nonzero real number. Then the graph of each of the following equations is a parabola with vertex at the origin.

$$x^2 = 4py$$

| focus: $(0, p)$ | directrix: $y = -p$ | axis: y-axis |

| $p > 0$ | $p < 0$ |
| opens upward | opens downward |

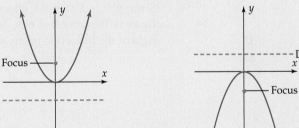

$$y^2 = 4px$$

| focus: $(p, 0)$ | directrix: $x = -p$ | axis: x-axis |

| $p > 0$ | $p < 0$ |
| opens right | opens left |

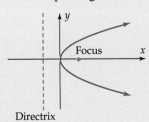
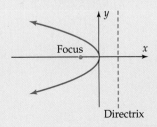

EXAMPLE 1

Find the equation of the parabola with vertex (0, 0) and focus (4, 0), and sketch its graph.

SOLUTION   Since the focus is $(p, 0) = (4, 0)$, we see that $p = 4$ and that the equation is

$$y^2 = 4px$$
$$= 4 \cdot 4x = 16x$$

Because $p = 4 > 0$, the parabola opens to the right, as in the lower left-hand figure in the preceding box. To get a reasonably accurate graph by hand we find the points directly above and below the focus—that is, the points on the parabola with $x = 4$.

$$y^2 = 16x$$

Let $x = 4$: $$y^2 = 16 \cdot 4 = 64$$

Take square roots: $$y = \pm 8$$

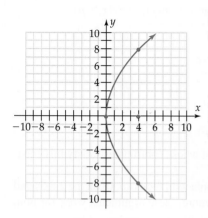

Figure 10–33

So we plot the points $(4, -8)$ and $(4, 8)$, and sketch the graph in Figure 10–33.   ∎

EXAMPLE 2

Find the equation of the parabola with vertex (0, 0) and focus (0, −3), and sketch its graph.

SOLUTION   The focus is $(0, p) = (0, -3)$, so $p = -3$ and the equation is

$$x^2 = 4py = 4(-3)y = -12y.$$

Since $p = -3 < 0$, the parabola opens downward, as in the upper right-hand figure in the preceding box. To sketch its graph we find the points to the left and right of the focus (the points with $y = -3$).

$$x^2 = -12y$$

Let $y = -3$: $$x^2 = -12(-3) = 36$$

Take square roots: $$x = \pm 6$$

We plot the points $(-6, -3)$ and $(6, -3)$, and sketch the graph in Figure 10–34.

∎

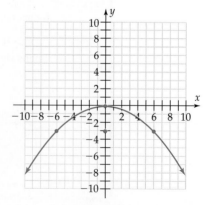

Figure 10–34

The line segment through the focus and perpendicular to the axis of a parabola, with endpoints on the parabola, is called the **latus rectum.** In both Example 1 and Example 2, the points we plotted to graph the parabola were the endpoints of the latus rectum, as shown in Figure 10–35.

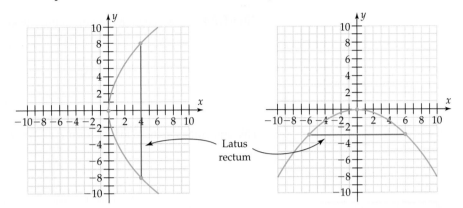

Figure 10–35

When graphing, the latus rectum can be thought of as indicating the "width" of a parabola. Exercise 74 shows that the latus rectum is $4|p|$ units long when the equation of the parabola is $x^2 = 4py$ or $y^2 = 4px$.

EXAMPLE 3

Show that the graph of $x^2 + 8y = 0$ is a parabola. Find its focus, vertex, and directrix, and sketch its graph.

SOLUTION   We first rewrite the equation in standard form.

$$x^2 + 8y = 0$$

Subtract 8y from both sides: $x^2 = -8y$

The last equation is in the standard form $x^2 = 4py$, with

$$4p = -8$$

Divide both sides by 4: $p = -2$

Hence, the graph is a downward-opening parabola with focus $(0, -2)$ and vertex $(0, 0)$. The directrix is the horizontal line $y = -p = -(-2) = 2$. The graph can be sketched by hand (Figure 10–36) or by rewriting the equation and using a calculator (Figure 10–37). ■

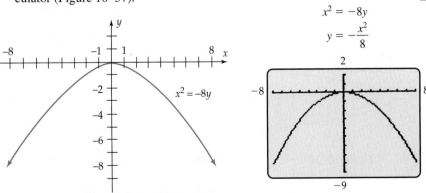

Figure 10–36   **Figure 10–37**

EXAMPLE 4

Identify the graph of $3y^2 = x$ and use technology to graph the equation.

SOLUTION   We begin by rewriting the equation in standard form.

$$3y^2 = x$$

Divide both sides by 3: $$y^2 = \frac{x}{3}$$

$$y^2 = \frac{1}{3}x$$

This equation is in the standard form $y^2 = 4px$ with

$$4p = \frac{1}{3}$$

Multiply both sides by $\frac{1}{4}$: $$p = \frac{1}{12}$$

So the graph is a parabola with focus $(1/12, 0)$ and directrix $x = -1/12$ that opens to the right.

To sketch its graph, you can either use a conic section grapher*, or you can solve the equation $y^2 = x/3$ for y and graph the two resulting equations

$$y = \sqrt{\frac{x}{3}} \qquad \text{and} \qquad y = -\sqrt{\frac{x}{3}}$$

on the same screen. Both methods produce Figure 10–38.   ∎

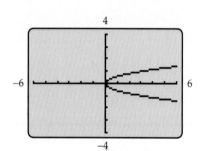

Figure 10–38

EXAMPLE 5

Find the focus, directrix, and equation of the parabola that passes through the point $(8, 2)$, has vertex $(0, 0)$ and focus on the x-axis.

SOLUTION   Since the vertex is $(0, 0)$ and the focus is on the x-axis, the equation is of the form $y^2 = 4px$. Since $(8, 2)$ is on the graph, we have

$$y^2 = 4px$$

Let $x = 8$ and $y = 2$: $$2^2 = 4p \cdot 8$$

$$4 = 32p$$

Divide both sides by 32: $$p = \frac{4}{32} = \frac{1}{8}.$$

Therefore, the equation of the parabola is

$$y^2 = 4px$$

$$y^2 = 4\left(\frac{1}{8}\right)x$$

$$y^2 = \frac{1}{2}x.$$

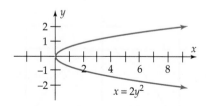

Figure 10–39

Its graph is sketched in Figure 10–39.   ∎

---

*See the Technology Tip on page 681.

VERTICAL AND HORIZONTAL SHIFTS

We have seen that replacing x with $x - h$ and y with $y - k$ in the equation of an ellipse or hyperbola shifts the graph vertically and horizontally. The same thing is true for parabolas.

Vertical and Horizontal Shifts

> Consider a parabola with equation
>
> $$x^2 = 4py \qquad \text{or} \qquad y^2 = 4px.$$
>
> Let h and k be constants. Replacing x with $x - h$ and replacing y with $y - k$ in one of these equations produces the equation
>
> $$(x - h)^2 = 4p(y - k) \qquad \text{or} \qquad (y - k)^2 = 4p(x - h),$$
>
> whose graph is the original parabola shifted vertically and horizontally so that its vertex is (h, k).

EXAMPLE 6

Identify and sketch the graph of $(x - 3)^2 = -8(y - 4)$.

SOLUTION This equation can be obtained from the equation $x^2 = -8y$ by replacing x with $x - 3$ and replacing y with $y - 4$. This is the situation described in the preceding box, with $h = 3$ and $k = 4$. So the graph of $(x - 3)^2 = -8(y - 4)$ is a parabola with vertex $(3, 4)$ that can be obtained from the parabola $x^2 = -8y$ (shown in Figure 10–36) by shifting it 3 units to the right and 4 units upward, as shown in Figure 10–40. Its focus lies on the vertical line $x = 3$. ■

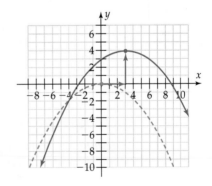

Figure 10–40

EXAMPLE 7

Identify and sketch the graph of $(y + 3)^2 = \dfrac{x + 4}{3}$.

SOLUTION To use the information in the preceding box, we must rewrite the equation in the form given there ($x - h$, not $x + h$, and similarly for the y term), namely,

$$[y - (-3)]^2 = \frac{1}{3}[x - (-4)].$$

This is the situation in the box, with $h = -4$ and $k = -3$. So the graph is a parabola with vertex $(-4, -3)$. Its focus is on the horizontal line $y = -3$. There are several ways to graph this parabola.

Method 1. The equation of this parabola can be obtained from the equation $y^2 = \dfrac{1}{3}x$ by replacing x with $x + 4 = x - (-4)$ and y with $y + 3 = y - (-3)$. So its graph is the parabola $y^2 = \dfrac{1}{3}x$ (Figure 10–38) shifted 4 units to the left and 3 units downward, as shown in Figure 10–41.

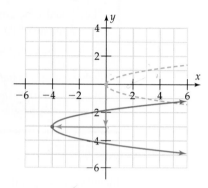

Figure 10–41

Method 2. To use technology for the graph, we first solve the equation for y:

$$(y + 3)^2 = \frac{x + 4}{3}$$

Take square roots on both sides: $\quad y + 3 = \pm\sqrt{\dfrac{x + 4}{3}}$

Subtract 3 from both sides: $\quad y = \pm\sqrt{\dfrac{x + 4}{3}} - 3.$

Graphing

$$y = \sqrt{\frac{x + 4}{3}} - 3 \qquad \text{and} \qquad y = -\sqrt{\frac{x + 4}{3}} - 3$$

on the same screen produces the graph in Figure 10–42.

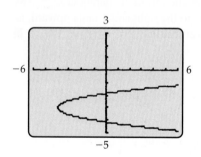

Figure 10–42

Method 3. To use a conic section grapher*, you must have the equation in the form given in the preceding box and find the value of p:

$$(y - k)^2 = 4p(x - h)$$

$$[y - (-3)]^2 = \frac{1}{3}[x - (-4)].$$

Therefore,

$$4p = \frac{1}{3}$$

Multiply both sides by $\dfrac{1}{4}$: $\quad p = \dfrac{1}{12}.$

Now insert the appropriate values ($h = -4$, $k = -3$, and $p = 1/12$) in the conic section grapher to obtain Figure 10–42. ∎

---

*See the Technology Tip on page 681.

EXAMPLE 6

Use parametric equations to graph $y^2 + 13 = 8y + x$.

SOLUTION   We first solve the equation for x:

$$x = y^2 - 8y + 13.$$

This is the equation of a parabola. Furthermore, the equation shows that x is a function of y (each value of y produces a unique value of x). Consequently, we can graph this equation by using the parametric equations

$$x = t^2 - 8t + 13$$

$$y = t \qquad (t \text{ any real number}),$$

as explained in Special Topics 3.3.A on page 178. The graph is shown in Figure 10–50. ■

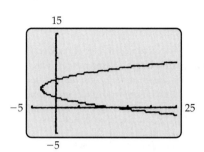

Figure 10–50

The parametric graphing technique used in Example 6 can be applied to the parabola with equation $(y - k)^2 = 4p(x - h)$ by solving the equation for x and letting $y = t$.

Parametric Equations for Parabolas

> The parabola with vertex (h, k) and equation
>
> $$(y - k)^2 = 4p(x - h)$$
>
> is given by the parametric equations
>
> $$x = \frac{(t - k)^2}{4p} + h \qquad \text{and} \qquad y = t \quad (t \text{ any real number}).$$

NOTE

The rectangular and parametric equations for conic sections are summarized in the endpapers at the beginning of the book.

EXERCISES 10.3.A

In Exercises 1–28, find parametric equations for the conic section whose rectangular equation is given, and use them to find a complete graph of the conic.

1. $x^2 + y^2 = 25$

2. $x^2 + y^2 = 40$

3. $(x - 4)^2 + (y + 2)^2 = 9$

4. $(x - 1)^2 = 1 - (y - 1)^2$

5. $(x + 1)^2 = 5 - y^2$

6. $x^2 - 4x + 4 = 9 - y^2 - 6y - 9$

7. $\dfrac{x^2}{10} - 1 = \dfrac{-y^2}{36}$

8. $\dfrac{y^2}{49} + \dfrac{x^2}{81} = 1$

9. $4x^2 + 4y^2 = 1$

10. $x^2 + 4y^2 = 1$

11. $\dfrac{(x - 1)^2}{4} + \dfrac{(y - 5)^2}{9} = 1$

12. $\dfrac{(x - 2)^2}{16} + \dfrac{(y + 3)^2}{12} = 1$

13. $\dfrac{(x + 1)^2}{16} + \dfrac{(y - 4)^2}{8} = 1$

14. $\dfrac{(x + 5)^2}{4} + \dfrac{(y + 2)^2}{12} = 1$

15. $\dfrac{x^2}{10} - \dfrac{y^2}{36} = 1$

16. $\dfrac{y^2}{9} - \dfrac{x^2}{16} = 1$

17. $x^2 - 4y^2 = 1$

18. $2x^2 - y^2 = 4$

19. $\dfrac{(y+3)^2}{25} - \dfrac{(x+1)^2}{16} = 1$ **20.** $\dfrac{(y+1)^2}{9} - \dfrac{(x-1)^2}{25} = 1$

21. $\dfrac{(x+3)^2}{1} - \dfrac{(y-2)^2}{4} = 1$ **22.** $\dfrac{(y+5)^2}{9} - \dfrac{(x-2)^2}{1} = 1$

23. $8x = 2y^2$ **24.** $4y = x^2$

25. $y = 4(x-1)^2 + 2$ **26.** $y = 3(x-2)^2 - 3$

27. $x = 2(y-2)^2$ **28.** $x = -3(y-1)^2 - 2$

In Exercises 29–38, identify the conic section whose parametric equations are given without graphing. *For circles, list the center and radius. For ellipses and hyperbolas, list the center. For parabolas, list the vertex.*

29. $x = 3 \cos t$ and $y = 3 \sin t - 5$ $(0 \le t \le 2\pi)$

30. $x = 7 \cos t - 4$ and $y = 7 \sin t + 3$ $(0 \le t \le 2\pi)$

31. $x = 3 \cos t + 4$ and $y = 5 \sin t$ $(0 \le t \le 2\pi)$

32. $x = 6 \cos t - 4$ and $y = 3 \sin t - 4$ $(0 \le t \le 2\pi)$

33. $x = \dfrac{2}{\cos t} + 2$ and $y = 4 \tan t + 4$ $(0 \le t \le 2\pi)$

34. $x = \tan t - 3$ and $y = \dfrac{7}{\cos t} + 5$ $(0 \le t \le 2)$

35. $x = 4 \tan t$ and $y = \dfrac{3}{\cos t} - 3$ $(0 \le t \le 2\pi)$

36. $x = \dfrac{1}{\cos t} - 1$ and $y = 3 \tan t + 2$ $(0 \le t \le 2\pi)$

37. $x = \dfrac{(t-4)^2}{4} - 3$ and $y = t$ (*t* any real number)

38. $x = \dfrac{(t+2)^2}{2} + 2$ and $y = t$ (*t* any real number)

THINKERS

39. (a) Verify that the curve with parametric equations

$$x = \cos(.5t) \text{ and } y = \sin(.5t) \quad (0 \le t \le 2\pi)$$

lies on the circle $x^2 + y^2 = 1$. [*Hint:* Use the argument in Example 1.]

(b) Verify that the curve with parametric equations

$$x = \cos(.5t) \text{ and } y = -\sin(.5t) \qquad (0 \le t \le 2\pi)$$

lies on the circle $x^2 + y^2 = 1$.

(c) Explain why neither of the curves in parts (a) and (b) is a complete circle. [*Hint:* What are the periods of cos(.5t) and sin(.5t)?]

In Exercises 40–41, use parametric equations (and trial and error) to draw a face on your calculator screen that closely resembles the one shown. [Hint: Use a square viewing window. Let the head be a circle with center at the origin and radius 3. Let the eyes be smaller circles with appropriate centers and radii. Let the mouth be a half circle (see Exercise 39). Finally, turn off the axes on your calculator screen.*]

40.   **41.**

In Exercises 42–43, use parametric equations (and trial and error) to draw a face on your calculator screen that closely resembles the one shown. [Hint: Adapt the hint for Exercises 40–41, using ellipses in place of circles.]

42.   **43.**

---

*In the FORMAT menu of TI-84+; in the GRAPH SET-UP menu of Casio 9850; on the second page of the PLOT SET-UP menu of HP-39gs.

 10.4   **Rotations and Second-Degree Equations**

Section Objective ■ Use the discriminant to identify the graph of a second-degree equation.

A **second-degree equation** in x and y is one that can be written in the form

$$Ax^2 + Bxy + Cy^2 + Dx + Ey + F = 0$$

for some constants A, B, C, D, E, F, with at least one of A, B, C nonzero.

EXAMPLE 1

Show that each of the following conic sections is the graph of a second-degree equation.

(a) Ellipse: $\dfrac{x^2}{6} + \dfrac{y^2}{5} = 1$ (b) Hyperbola: $\dfrac{(x+1)^2}{4} - \dfrac{(y-3)^2}{6} = 1$

SOLUTION   We need only show that each of these equations is in fact a second-degree equation. In each case, eliminate denominators, multiply out all terms, and gather them on one side of the equal sign.

(a)

$$\frac{x^2}{6} + \frac{y^2}{5} = 1$$

Multiply both sides by 30: $5x^2 + 6y^2 = 30$

Rearrange terms: $5x^2 + 6y^2 - 30 = 0.$

This equation is a second-degree equation because it has the form

$$Ax^2 + Bxy + Cy^2 + Dx + Ey + F = 0$$

with $A = 5$, $B = 0$, $C = 6$, $D = 0$, $E = 0$, and $F = -30$.

(b)

$$\frac{(x+1)^2}{4} - \frac{(y-3)^2}{6} = 1$$

Multiply both sides by 12: $3(x+1)^2 - 2(y-3)^2 = 12$

Multiply out left side: $3(x^2 + 2x + 1) - 2(y^2 - 6y + 9) = 12$

$3x^2 + 6x + 3 - 2y^2 + 12y - 18 = 12$

Rearrange terms: $3x^2 - 2y^2 + 6x + 12y - 27 = 0.$

This is a second-degree equation with $A = 3$, $B = 0$, $C = -2$, $D = 6$, $E = -12$, and $F = -27$. ∎

Calculations like those in Example 1 can be used on the equation of any conic section to show that it is the graph of a second-degree equation. Conversely, it can be shown that

The graph of every second-degree equation is a conic section

(possibly degenerate). When the second-degree equation has no xy-term (that is, $B = 0$), as was the case in Example 1, the graph is a conic section in standard position (axis or axes parallel to the coordinate axes). When $B \neq 0$, however, the conic is rotated from standard position, so its axis or axes are not parallel to the coordinate axes.

EXAMPLE 2

Graph the equation

$$3x^2 + 6xy + y^2 + x - 2y + 7 = 0.$$

SOLUTION   We first rewrite it as

$$y^2 + 6xy - 2y + 3x^2 + x + 7 = 0$$

$$y^2 + (6x - 2)y + (3x^2 + x + 7) = 0.$$

This equation has the form $ay^2 + by + c = 0$, with $a = 1$, $b = 6x - 2$, and $c = 3x^2 + x + 7$. It can be solved by using the quadratic formula

$$y = \frac{-b \pm \sqrt{b^2 - 4ac}}{2a} = \frac{-(6x - 2) \pm \sqrt{(6x - 2)^2 - 4 \cdot 1 \cdot (3x^2 + x + 7)}}{2 \cdot 1}.$$

The top half of the graph is obtained by graphing

$$y = \frac{-6x + 2 + \sqrt{(6x - 2)^2 - 4(3x^2 + x + 7)}}{2},$$

and the bottom half is obtained by graphing

$$y = \frac{-6x + 2 - \sqrt{(6x - 2)^2 - 4(3x^2 + x + 7)}}{2}.$$

The graph is a hyperbola whose focal axis tilts upward to the left, as shown in Figure 10–51.   ■

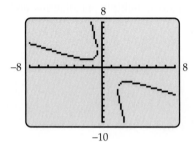

Figure 10–51

THE DISCRIMINANT

The following fact, whose proof is outlined in Exercise 15 of Special Topics 10.4.A, makes it easy to identify the graphs of second-degree equations without graphing them.

Graphs of
Second-Degree
Equations

> If the equation
>
> $$Ax^2 + Bxy + Cy^2 + Dx + Ey + F = 0 \quad (A, B, C \text{ not all } 0)$$
>
> has a graph, then that graph is
>
> A circle or an ellipse (or a point), if $B^2 - 4AC < 0$;
>
> A parabola (or a line or two parallel lines), if $B^2 - 4AC = 0$;
>
> A hyperbola (or two intersecting lines), if $B^2 - 4AC > 0$.

The expression $B^2 - 4AC$ is called the **discriminant** of the equation.

EXAMPLE 3

Identify the graph of

$$2x^2 - 4xy + 3y^2 + 5x + 6y - 8 = 0$$

and sketch the graph.

SOLUTION   We compute the discriminant with $A = 2$, $B = -4$, and $C = 3$.

$$B^2 - 4AC = (-4)^2 - 4 \cdot 2 \cdot 3 = 16 - 24 = -8.$$

Hence, the graph is an ellipse (possibly a circle or a single point). To find the graph, we rewrite the equation as

$$3y^2 - 4xy + 6y + 2x^2 + 5x - 8 = 0$$

$$3y^2 + (-4x + 6)y + (2x^2 + 5x - 8) = 0.$$

The equation has the form $ay^2 + by + c = 0$ and can be solved by the quadratic formula.

$$y = \frac{-b \pm \sqrt{b^2 - 4ac}}{2a}$$

$$= \frac{-(-4x + 6) \pm \sqrt{(-4x + 6)^2 - 4 \cdot 3 \cdot (2x^2 + 5x - 8)}}{2 \cdot 3}.$$

The graph can now be found by graphing the last two equations on the same screen.

GRAPHING EXPLORATION

Find a viewing window that shows a complete graph of the equation. In what direction does the major axis run?

■

EXAMPLE 4

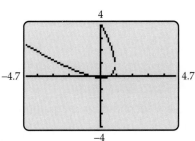

Figure 10–52

Does Figure 10–52 show a complete graph of

$$3x^2 + 5xy + 2y^2 - 8y - 1 = 0?$$

SOLUTION   Although the graph in the figure looks like a parabola, appearances are deceiving. The discriminant of the equation is

$$B^2 - 4AC = 5^2 - 4 \cdot 3 \cdot 2 = 1,$$

which means that the graph is a hyperbola. So Figure 10–52 cannot possibly be a complete graph of the equation.*

GRAPHING EXPLORATION

Solve the equation for y, as in Example 3. Then find a viewing window that clearly shows both branches of the hyperbola.

■

EXAMPLE 5

GRAPHING EXPLORATION

Find a viewing window that shows a complete graph of the equation.

Sketch the graph of

$$3x^2 + 6xy + 3y^2 + 13x + 9y + 53 = 0.$$

SOLUTION   The discriminant is $B^2 - 4AC = 6^2 - 4 \cdot 3 \cdot 3 = 0$. Hence, the graph is a parabola (or a line or parallel lines in the degenerate case).   ■

---

*When one of the authors asked students to graph this equation on an examination, many of them did not bother to compute the discriminant and produced something similar to Figure 10–52. They didn't get much credit for this answer. *Moral:* Use the discriminant to identify the conic so that you know the shape of the graph you are looking for.

EXERCISES 10.4

In Exercises 1–6, assume that the graph of the equation is a nondegenerate conic section. Without graphing, determine whether the graph an ellipse, hyperbola, or parabola.

1. $x^2 - 2xy + 3y^2 - 1 = 0$

2. $xy - 1 = 0$

3. $x^2 + 2xy + y^2 + 2\sqrt{2}x - 2\sqrt{2}y = 0$

4. $2x^2 - 4xy + 5y^2 - 6 = 0$

5. $17x^2 - 48xy + 31y^2 + 50 = 0$

6. $2x^2 - 4xy - 2y^2 + 3x + 5y - 10 = 0$

In Exercises 7–24, use the discriminant to identify the conic section whose equation is given, and find a viewing window that shows a complete graph.

7. $9x^2 + 4y^2 + 54x - 8y + 49 = 0$

8. $4x^2 + 5y^2 - 8x + 30y + 29 = 0$

9. $4y^2 - x^2 + 6x - 24y + 11 = 0$

10. $x^2 - 16y^2 = 0$

11. $3y^2 - x - 2y + 1 = 0$

12. $x^2 - 6x + y + 5 = 0$

13. $41x^2 - 24xy + 34y^2 - 25 = 0$

14. $x^2 + 2\sqrt{3}xy + 3y^2 + 8\sqrt{3}x - 8y + 32 = 0$

15. $17x^2 - 48xy + 31y^2 + 49 = 0$

16. $52x^2 - 72xy + 73y^2 = 200$

17. $9x^2 + 24xy + 16y^2 + 90x - 130y = 0$

18. $x^2 + 10xy + y^2 + 1 = 0$

19. $23x^2 + 26\sqrt{3}xy - 3y^2 - 16x + 16\sqrt{3}y + 128 = 0$

20. $x^2 + 2xy + y^2 + 12\sqrt{2}x - 12\sqrt{2}y = 0$

21. $17x^2 - 12xy + 8y^2 - 80 = 0$

22. $11x^2 - 24xy + 4y^2 + 30x + 40y - 45 = 0$

23. $3x^2 + 2\sqrt{3}xy + y^2 + 4x - 4\sqrt{3}y - 16 = 0$

24. $3x^2 + 2\sqrt{2}xy + 2y^2 - 12 = 0$

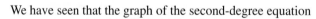

10.4.A *SPECIAL TOPICS* Rotation of Axes

Section Objectives

- Use the rotation equations to express a second-degree equation in *x* and *y* as an equation in *u* and *v*.
- Determine the angle of rotation needed to write a second-degree equation in *x* and *y* as an equation in *u* and *v* that has no *uv* term.

We have seen that the graph of the second-degree equation

$$Ax^2 + Bxy + Cy^2 + Dx + Ey + F = 0 \quad (B \neq 0)$$

is a conic section whose axes are not parallel to the coordinate axes, as in Figure 10–53. Although we can graph the equation on a calculator (as in Section 10.4), we cannot read off useful information about the center, vertices, etc. from the equation, as we can from an equation in standard form.

The key is to replace the *xy* coordinate system by a new coordinate system, as indicated by the blue *u*- and *v*-axes in Figure 10–53. The conic is not rotated in the new coordinate system, so it has an equation (in *u* and *v*) in standard form that will provide the desired information.

To use this approach, we must first determine the relationship between the *xy* coordinates of a point and its coordinates in the *uv* system. Suppose the *uv* coordinate system is obtained by rotating the *xy* axes about the origin, counterclockwise through an angle θ.* If a point *P* has coordinates (x, y) in the *xy* system, we can find its coordinates (u, v) in the rotated coordinate system by using Figure 10–54.

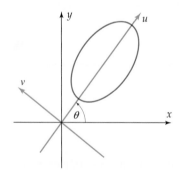

Figure 10–53

---

*All rotations in this section are counterclockwise about the origin, with $0° < \theta < 90°$.

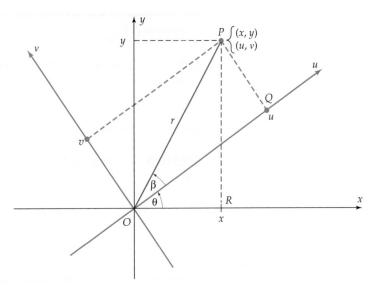

Figure 10–54

Triangle OPQ shows that

$$\cos \beta = \frac{OQ}{OP} = \frac{u}{r} \qquad \text{and} \qquad \sin \beta = \frac{PQ}{OP} = \frac{v}{r}.$$

Therefore,

$$u = r \cos \beta \qquad \text{and} \qquad v = r \sin \beta.$$

Similarly, triangle OPR shows that

$$\cos(\theta + \beta) = \frac{OR}{OP} = \frac{x}{r} \qquad \text{and} \qquad \sin(\theta + \beta) = \frac{PR}{OP} = \frac{y}{r},$$

so

$$x = r \cos(\theta + \beta) \qquad \text{and} \qquad y = r \sin(\theta + \beta).$$

Applying the addition identity for cosine (page 524) shows that

$$
\begin{aligned}
x &= r \cos(\theta + \beta) \\
&= r(\cos \theta \cos \beta - \sin \theta \sin \beta) \\
&= (r \cos \beta) \cos \theta - (r \sin \beta) \sin \theta \\
&= u \cos \theta - v \sin \theta.
\end{aligned}
$$

A similar argument with $y = r \sin(\theta + \beta)$ and the addition identity for sine leads to this result.

**The Rotation
Equations**

> If the xy coordinate axes are rotated through an angle θ to produce the uv coordinate axes, then the coordinates (x, y) and (u, v) of a point are related by these equations:
>
> $$x = u \cos \theta - v \sin \theta,$$
>
> $$y = u \sin \theta + v \cos \theta.$$

EXAMPLE 1

If the xy axes are rotated 30°, find the equation relative to the uv axes of the graph of

$$3x^2 + 2\sqrt{3}xy + y^2 + x - \sqrt{3}y = 0,$$

and graph the equation.

SOLUTION   Since $\sin 30° = 1/2$ and $\cos 30° = \sqrt{3}/2$, the rotation equations are

$$x = u \cos 30° - v \sin 30° = \frac{\sqrt{3}}{2}u - \frac{1}{2}v,$$

$$y = u \sin 30° + v \cos 30° = \frac{1}{2}u + \frac{\sqrt{3}}{2}v.$$

Substitute these expressions in the original equation.

$$3x^2 + 2\sqrt{3}xy + y^2 + x - \sqrt{3}y = 0$$

$$3\left(\frac{\sqrt{3}}{2}u - \frac{1}{2}v\right)^2 + 2\sqrt{3}\left(\frac{\sqrt{3}}{2}u - \frac{1}{2}v\right)\left(\frac{1}{2}u + \frac{\sqrt{3}}{2}v\right)$$

$$+ \left(\frac{1}{2}u + \frac{\sqrt{3}}{2}v\right)^2 + \left(\frac{\sqrt{3}}{2}u - \frac{1}{2}v\right) - \sqrt{3}\left(\frac{1}{2}u + \frac{\sqrt{3}}{2}v\right) = 0$$

$$3\left(\frac{3}{4}u^2 - \frac{\sqrt{3}}{2}uv + \frac{1}{4}v^2\right) + 2\sqrt{3}\left(\frac{\sqrt{3}}{4}u^2 + \frac{1}{2}uv - \frac{\sqrt{3}}{4}v^2\right)$$

$$+ \left(\frac{1}{4}u^2 + \frac{\sqrt{3}}{2}uv + \frac{3}{4}v^2\right) + \left(\frac{\sqrt{3}}{2}u - \frac{1}{2}v\right) - \sqrt{3}\left(\frac{1}{2}u + \frac{\sqrt{3}}{2}v\right) = 0.$$

Verify that the last equation simplifies as

$$4u^2 - 2v = 0 \qquad \text{or, equivalently,} \qquad v = 2u^2.$$

In the uv system, $v = 2u^2$ is the equation of an upward-opening parabola with vertex at $(0, 0)$ as shown in Figure 10–55. ■

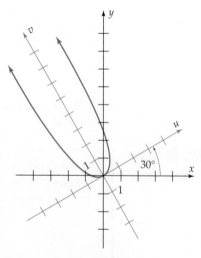

Figure 10–55

Rotating the axes in the preceding example changed the original equation, which included an xy term, to an equation that had no uv term. That enabled us to identify the graph of the new equation as a conic section. This can be done for any second-degree equation by choosing an angle of rotation that will eliminate the xy term. The choice of the angle is determined by this fact, which is proved in Exercise 13.

Rotation Angle

> The equation $Ax^2 + Bxy + Cy^2 + Dx + Ey + F = 0$ $(B \neq 0)$ can be rewritten as $A'u^2 + C'v^2 + D'u + E'v + F' = 0$ by rotating the xy axes through an angle θ such that
>
> $$\cot 2\theta = \frac{A - C}{B} \quad (0° < \theta < 90°)$$
>
> and using the rotation equations.

EXAMPLE 2

What angle of rotation will eliminate the xy term in the equation

$$153x^2 + 192xy + 97y^2 - 1710x - 1470y + 5625 = 0,$$

and what are the rotation equations in this case?

SOLUTION   According to the fact in the box with $A = 153$, $B = 192$, and $C = 97$, we should rotate through an angle of θ, where

$$\cot 2\theta = \frac{153 - 97}{192} = \frac{56}{192} = \frac{7}{24}.$$

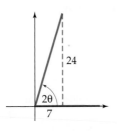

Figure 10–56

Since $0° < 2\theta < 180°$ and $\cot 2\theta$ is positive, the terminal side of the angle 2θ lies in the first quadrant, as shown in Figure 10–56. The hypotenuse of this triangle has length $\sqrt{7^2 + 24^2} = \sqrt{625} = 25$. Hence, $\cos 2\theta = 7/25$. The half-angle identities with $x = 2\theta$ (page 538) show that

$$\sin \theta = \sqrt{\frac{1 - \cos 2\theta}{2}} = \sqrt{\frac{1 - 7/25}{2}} = \sqrt{\frac{9}{25}} = \frac{3}{5},$$

$$\cos \theta = \sqrt{\frac{1 + \cos 2\theta}{2}} = \sqrt{\frac{1 + 7/25}{2}} = \sqrt{\frac{16}{25}} = \frac{4}{5}.$$

Using $\sin \theta = 3/5$ and the SIN^{-1} key on a calculator, we find that the angle θ of rotation is approximately $36.87°$. The rotation equations are

$$x = u \cos \theta - v \sin \theta = \frac{4}{5}u - \frac{3}{5}v,$$

$$y = u \sin \theta + v \cos \theta = \frac{3}{5}u + \frac{4}{5}v.$$

∎

EXAMPLE 3

Graph the equation without using a calculator.

$$153x^2 + 192xy + 97y^2 - 1710x - 1470y + 5625 = 0.$$

SOLUTION   The angle θ and the rotation equations for eliminating the xy term were found in the preceding example. Substitute the rotation equations in the given equation.

$$153x^2 + 192xy + 97y^2 - 1710x - 1470y + 5625 = 0$$

$$153\left(\frac{4}{5}u - \frac{3}{5}v\right)^2 + 192\left(\frac{4}{5}u - \frac{3}{5}v\right)\left(\frac{3}{5}u + \frac{4}{5}v\right)$$

$$+ 97\left(\frac{3}{5}u + \frac{4}{5}v\right)^2 - 1710\left(\frac{4}{5}u - \frac{3}{5}v\right) - 1470\left(\frac{3}{5}u + \frac{4}{5}v\right) + 5625 = 0$$

$$153\left(\frac{16}{25}u^2 - \frac{24}{25}uv + \frac{9}{25}v^2\right) + 192\left(\frac{12}{25}u^2 + \frac{7}{25}uv - \frac{12}{25}v^2\right)$$

$$+ 97\left(\frac{9}{25}u^2 + \frac{24}{25}uv + \frac{16}{25}v^2\right) - 2250u - 150v + 5625 = 0$$

$$225u^2 + 25v^2 - 2250u - 150v + 5625 = 0$$

$$9u^2 + v^2 - 90u - 6v + 225 = 0$$

$$9(u^2 - 10u) + (v^2 - 6v) = -225.$$

Finally, complete the square in u and v (adding the appropriate amounts to the right side so as not to change the equation).

$$9(u^2 - 10u + 25) + (v^2 - 6v + 9) = -225 + 9 \cdot 25 + 9$$

$$9(u - 5)^2 + (v - 3)^2 = 9$$

$$\frac{(u - 5)^2}{1} + \frac{(v - 3)^2}{9} = 1.$$

Therefore the graph is an ellipse centered at (5, 3) in the uv coordinate system, as shown in Figure 10–57. ■

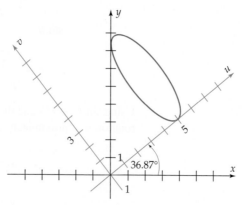

Figure 10–57

EXERCISES 10.4.A

In Exercises 1–4, find the new coordinates of the point when the coordinate axes are rotated through the given angle.

1. $(3, 2); \theta = 45°$

2. $(-2, 4); \theta = 60°$

3. $(1, 0); \theta = 30°$

4. $(3, 3); \sin \theta = 5/13$

In Exercises 5–8, rotate the axes through the given angle to form the uv coordinate system. Express the given equation in terms of the uv coordinate system.

5. $\theta = 45°; xy = 1$

6. $\theta = 45°; 13x^2 + 10xy + 13y^2 = 72$

7. $\theta = 30°; 7x^2 - 6\sqrt{3}xy + 13y^2 - 16 = 0$

8. $\sin \theta = 1/\sqrt{5}; x^2 - 4xy + 4y^2 + 5\sqrt{5}y + 1 = 0$

In Exercises 9–12, find the angle of rotation that will eliminate the xy term of the equation and list the rotation equations in this case.

9. $41x^2 - 24xy + 34y^2 - 25 = 0$

10. $x^2 + 2\sqrt{3}xy + 3y^2 + 8\sqrt{3}x - 8y + 32 = 0$

11. $17x^2 - 48xy + 31y^2 + 49 = 0$

12. $52x^2 - 72xy + 73y^2 = 200$

THINKERS

13. (a) Given an equation
$Ax^2 + Bxy + Cy^2 + Dx + Ey + F = 0$, with $B \neq 0$. and an angle θ, use the rotation equations in the box on page 723 to rewrite the equation in the form
$A'u^2 + B'uv + C'v^2 + D'u + E'v + F' = 0,$

where $A', \ldots, F'$ are expressions involving $\sin \theta$, $\cos \theta$, and the constants $A, \ldots, F$.

(b) Verify that $B' = 2(C - A) \sin \theta \cos \theta + B(\cos^2\theta - \sin^2\theta)$.

(c) Use the double-angle identities to show that
$B' = (C - A) \sin 2\theta + B \cos 2\theta$.

(d) If θ is chosen so that $\cot 2\theta = (A - C)/B$, show that $B' = 0$. This proves the statement in the box on page 725.

14. Assume that the graph of $A'u^2 + C'v^2 + D'u + E'v + F' = 0$ (with at least one of A', C' nonzero) in the *uv* coordinate system is a nondegenerate conic. Show that its graph is an ellipse if $A'C' > 0$ (A' and C' have the same sign), a hyperbola if $A'C' < 0$ (A' and C' have opposite signs), or a parabola if $A'C' = 0$.

15. Assume the graph of $Ax^2 + Bxy + Cy^2 + Dx + Ey + F = 0$ is a nondegenerate conic section. Prove the statement in the box on page 720 as follows.

(a) In Exercise 13(a), show that $(B')^2 - 4A'C' = B^2 - 4AC$.

(b) Assume that θ has been chosen so that $B' = 0$. Use Exercise 14 to show that the graph of the original equation is an ellipse if $B^2 - 4AC < 0$, a parabola if $B^2 - 4AC = 0$, and a hyperbola if $B^2 - 4AC > 0$.

10.5 Plane Curves and Parametric Equations*

Section Objectives
■ Graph a curve given by parametric equations.
■ Express a parametric curve as part of the graph of an equation in *x* and *y*.
■ Find a parametric representation for the graph of an equation in *x* and *y*.

There are many curves in the plane that cannot be represented as the graph of a function $y = f(x)$. Parametric graphing enables us to represent such curves in terms of functions and also provides a formal definition of a curve in the plane.

Consider, for example, an object moving in the plane during a particular time interval. To describe both the path of the object and its location at any particular

---

*Parametric graphing was introduced in Special Topics 3.3.A, which is *not* a prerequisite for this section.

time, three variables are needed: the time t and the coordinates (x, y) of the object at time t. For instance, the coordinates x and y might be given by

$$x = 4 \cos t + 5 \cos(3t) \qquad \text{and} \qquad y = \sin(3t) + t.$$

From $t = 0$ to $t = 12.5$, the object traces out the curve shown in Figure 10–58. The points marked on the graph show the location of the object at various times. Note that the object may be at the same location at different times (the points where the graph crosses itself).

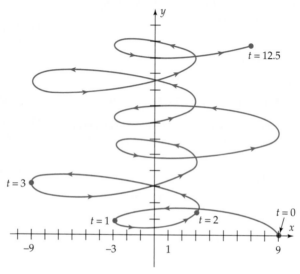

Figure 10–58

In the preceding example, both x and y were determined by continuous functions of t, with t taking values in the interval $[0, 12.5]$.* The example suggests the following definition.

Definition of Plane Curve

Let f and g be continuous functions of t on an interval I. The set of all points (x, y) where

$$x = f(t) \qquad \text{and} \qquad y = g(t)$$

is called a **plane curve.** The variable t is called a **parameter,** and the equations defining x and y are **parametric equations.**

In this general definition of "curve," the variable t need not represent time. As the following examples illustrate, different pairs of parametric equations may produce the same curve. Each such pair of parametric equations is called a **parameterization** of the curve.

A curve given by parametric equations can be graphed by hand, as in the next example.

---

*Intuitively, "continuous" means that the graph of the function that determines x, namely, $f(t) = 4 \cos t + 5 \cos 3t$, is a connected curve with no gaps or holes and similarly for the function that determines y. Continuous functions are defined more precisely in Chapter 13.

EXAMPLE 1

Sketch the graph of the curve whose parametric equations are

$$x = t^3 + 1 \qquad \text{and} \qquad y = 2t \qquad (-2 \le t \le 2).$$

SOLUTION We choose values of t between -2 and 2, and compute the corresponding values of x and y.

t	$x = t^3 + 1$	$y = 2t$
-2	-7	-4
-1.5	-2.375	-3
-1	0	-2
0	1	0
1	2	2
1.5	4.375	3
2	9	4

Next we plot the points (x, y) given by the table: $(-7, -4)$, $(-2.375, -3)$, $(0, -2)$, etc. (Figure 10–59). The plotted points suggest the graph in Figure 10–60. ■

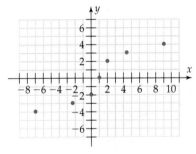

y
x
6
4
2
-2
-4
-6
-8 -6 -4 -2
2 4 6 8 10

Figure 10–59

y
x
6
4
2
-2
-4
-6
-8 -6 -4 -2
2 4 6 8 10

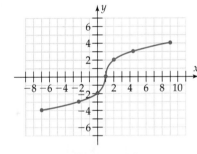

Figure 10–60

Graphing more complicated curves is difficult to do by hand, but quite easy when technology is used. For example, to graph the curve shown in Figure 10–58, which is given by the parametric equations

$$x = 4\cos t + 5\cos(3t) \qquad \text{and} \qquad y = \sin(3t) + t,$$

put your calculator or computer in parametric graphing mode. Then enter the equations (Figure 10–61) and the viewing window, as partially shown in Figure 10–62 (scroll down to see the rest).

TECHNOLOGY TIP

To change to parametric graphing mode, select PAR(AMETRIC) in the following menu/submenu:

TI: MODE

Casio: GRAPH/TYPE

HP-39gs: APLET

Plot1 Plot2 Plot3
\X1T=4cos(T)+5co
s(3T)
Y1T=sin(3T)+T
\X2T=
Y2T=
\X3T=
Y3T=

Figure 10–61

WINDOW
Tmin=0
Tmax=12.5
Tstep=.1
Xmin=-10
Xmax=10
Xscl=1
↓Ymin=0

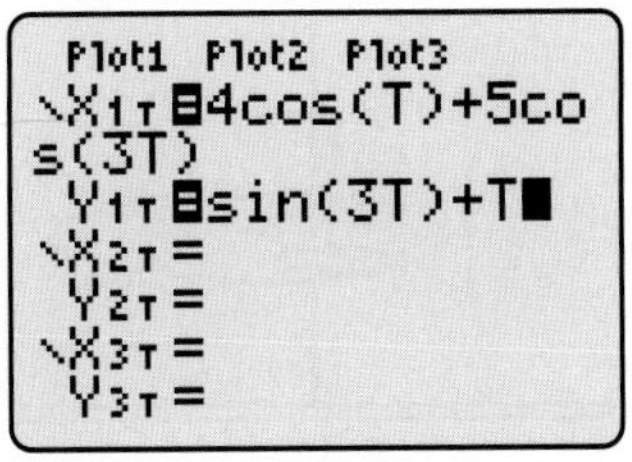

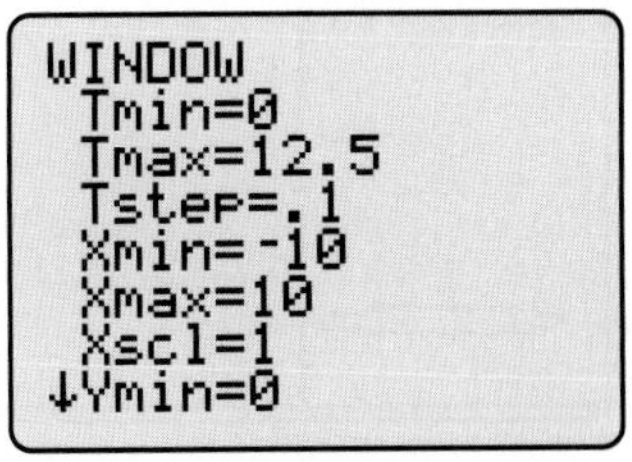

Figure 10–62

GRAPHING EXPLORATION

Enter the equations in Figure 10–61 in your calculator and set the viewing window to match Figure 10–62 (with $0 \le y \le 15$). Then graph the curve. How does your graph compare with Figure 10–58? Use the trace feature to see which points correspond to $t = 0, 1, 2, 3$, and 12.5.

In parametric graphing, failure to make appropriate choices for the range of t-values and the t-step may result in an inaccurate graph, as the next example illustrates.

EXAMPLE 2

In the window with $-10 \le x \le 10$ and $0 \le y \le 15$, graph the curve given by

$$x = 4 \cos t + 5 \cos(3t) \qquad \text{and} \qquad y = \sin(3t) + t$$

(a) with $0 \le t \le 12.5$ and each of these t-steps: 2, 1, and .5;

(b) with t-step $= .1$ and each of these ranges: $2.4 \le t \le 3.9$, $4.6 \le t \le 8$, and $0 \le t \le 12.5$.

How do these graphs compare with Figure 10–58?

SOLUTION

(a) The graphs are shown in Figure 10–63. Only the last one looks even remotely like Figure 10–58.

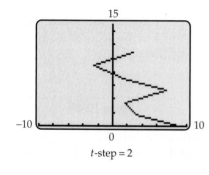

t-step $= 2$

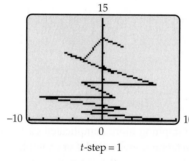

t-step $= 1$

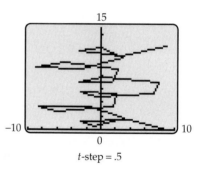

t-step $= .5$

Figure 10–63

(b) The graphs are shown in Figure 10–64. The first two show only a small portion of Figure 10–58, but the last one is a complete graph that closely resembles Figure 10–58, as you verified in the preceding Graphing Exploration. ∎

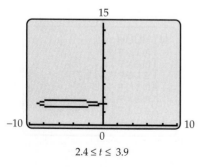

$2.4 \le t \le 3.9$

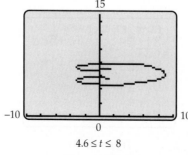

$4.6 \le t \le 8$

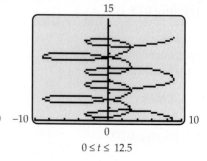

$0 \le t \le 12.5$

Figure 10–64

Example 2 suggests that

A *t*-step between .05 and 1.5 usually produces a reasonably smooth graph.

Larger *t*-steps may produce jagged or totally inaccurate graphs. Smaller *t*-steps may involve long graphing times and usually don't improve smoothness very much.

EXAMPLE 3

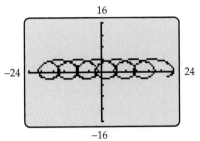

Figure 10–65

Graph the curve given by

$$x = t + 5 \cos t \qquad \text{and} \qquad y = 1 - 3 \sin t$$

in the viewing window with

$$-24 \le x \le 24, \qquad -16 \le y \le 16, \qquad -20 \le t \le 20.$$

SOLUTION See Figure 10–65. Reproduce this graph yourself to see how the curve spirals along the *x*-axis. ∎

ELIMINATING THE PARAMETER

In some cases a curve given by parametric equations can also be described by a rectangular equation in *x* and *y*. The process of finding such a rectangular equation is called **eliminating the parameter.** Here is one method for eliminating the parameter:

> **Solve one of the parametric equations for *t* and substitute this solution in the other parametric equation. The result is an equation in *x* and *y* whose graph includes the parametric curve.**

EXAMPLE 4

Consider the curve given by

$$x = -2t \qquad \text{and} \qquad y = 4t^2 - 4 \qquad (-1 \le t \le 2).$$

(a) Graph the curve.

(b) Eliminate the parameter and find an equation in *x* and *y* whose graph includes the graph in part (a).

SOLUTION

(a) The graph is shown in Figure 10–66.

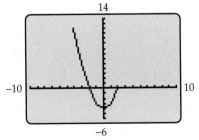

Figure 10–66

GRAPHING EXPLORATION

Graph this curve on your calculator, using the same viewing window and range of *t* values as in Figure 10–66. In what direction is the curve traced out (that is, what is the first point graphed (corresponding to $t = -1$) and what is the last point graphed (corresponding to $t = 2$))?

(b) We first solve one of the parametric equations for t. We'll solve the x-equation since it is simpler.

$$x = -2t$$

Divide both sides by -2: $$t = -\frac{x}{2}$$

Now substitute this result in the y-equation.

$$y = 4t^2 - 4$$

Let $t = -\frac{x}{2}$: $$y = 4\left(-\frac{x}{2}\right)^2 - 4$$

$$y = 4\left(\frac{x^2}{4}\right) - 4$$

$$y = x^2 - 4.$$

Therefore, the coordinates of every point on the parametric curve satisfy the equation $y = x^2 - 4$. So every point on the curve lies on the graph of $y = x^2 - 4$.

From Section 3.4 and 4.1 we know that the graph of $y = x^2 - 4$ is the parabola in Figure 10–67. However, the curve given by the parametric equations is *not* the entire parabola, but only the part of it shown in red—the part from $(2, 0)$ to $(-4, 12)$, which corresponds to the minimum and maximum values of t, namely, $t = -1$ and $t = 2$.   ∎

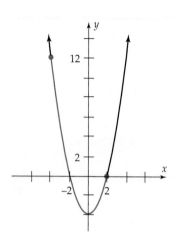

Figure 10–67

When neither parametric equation can be solved for t, other methods are needed to eliminate the parameter. Appropriate trigonometric identities sometimes do the trick.

EXAMPLE 5

Find a rectangular equation for the *witch of Agnesi,* which is the curve given by

$$x = \cot t \quad\text{and}\quad y = \sin^2 t \quad (0 < t < \pi).^*$$

SOLUTION   We have

$$y = \sin^2 t$$

$$y = \frac{1}{\csc^2 t} \qquad \text{[Reciprocal identity]}$$

$$y = \frac{1}{\cot^2 t + 1} \qquad \text{[Pythagorean identity]}$$

$$y = \frac{1}{x^2 + 1} \qquad \text{[}x = \cot t\text{]}$$

In this case the parametric curve is the entire rectangular graph because, as t goes from 0 to π, $\cot t$ (which is x here) takes all real values (see Figure 6–85).   ∎

---

*This curve is a special case of one discussed in a textbook by Maria Agnesi in 1748. Because of its shape, the curve was called "la versiera" in Italian (meaning "rope that turns into a sail"). When Agnesi's book was translated into English, however, the translator mistook "la versiera" for "l' aversiera", which means "the witch". See Exercises 51–52.

FINDING PARAMETRIC EQUATIONS FOR CURVES

Having seen how to find a rectangular equation that describes a curve given parametrically, we now consider the reverse problem: finding a parametric respresentation of an equation in x and y. In two cases, this is easy:

1. If the equation defines y as a function of x, such as

$$y = x^3 + 5x^2 - 3x + 4,$$

a parametric description can be found by changing the variable:

$$x = t \qquad \text{and} \qquad y = t^3 + 5t^2 - 3t + 4.$$

2. If the equation defines x as a function of y, such as

$$x = y^2 + 3,$$

a parametric description is given by

$$x = t^2 + 3 \qquad \text{and} \qquad y = t.$$

EXAMPLE 6

Find a parameterization for the graph of $y^3 - 4y - x + 5 = 0$ and use it to sketch the graph.

SOLUTION   We rewrite the equation as

$$x = y^3 - 4y + 5,$$

and see that x is a function of y (each value of y leads to a unique value of x). So we use the parameterization

$$x = t^3 - 4t + 5 \qquad \text{and} \qquad y = t.$$

The graph is shown in Figure 10–68.   ■

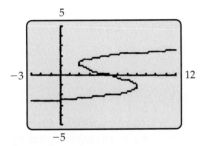

Figure 10–68

EXAMPLE 7

Find three parameterizations of the straight line through $(1, -3)$ with slope -2.

SOLUTION   The point-slope form of the equation of this line is

$$y - (-3) = -2(x - 1) \qquad \text{or, equivalently,} \qquad y = -2x - 1.$$

Its graph is shown in Figure 10–69. Since this equation defines y as a function of x, one parameterization is

$$x = t \qquad \text{and} \qquad y = -2t - 1 \qquad (t \text{ any real number}).$$

A second parameterization is given by letting $x = t + 1$; then

$$y = -2x - 1 = -2(t + 1) - 1 = -2t - 3 \qquad (t \text{ any real number}).$$

A third parameterization can be obtained by letting

$$x = \tan t \qquad \text{and} \qquad y = -2x - 1 = -2 \tan t - 1 \qquad (-\pi/2 < t < \pi/2).$$

When t runs from $-\pi/2$ to $\pi/2$, then $x = \tan t$ takes all possible real number values, and hence so does y.   ■

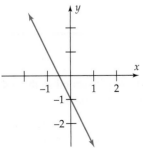

Figure 10–69

CAUTION

Some substitutions in an equation $y = f(x)$ do *not* lead to a parameterization of the entire graph. For instance, in Example 7, letting $x = t^2$ and substituting in the equation $y = -2x - 1$ lead to

$$x = t^2 \quad \text{and} \quad y = -2t^2 - 1 \quad \text{(any real number } t\text{)}.$$

Thus, x is always nonnegative, and y is always negative. So the parameterization produces only the half of the line $y = -2x - 1$ to the right of the y-axis in Figure 10–69.

Parameterizations of conic sections, such as

$$(x - 3)^2 + (y + 7)^2 = 16, \quad \frac{x^2}{25} + \frac{y^2}{4} = 1, \quad \text{and} \quad \frac{(x - 3)^2}{4} - \frac{(y + 5)^2}{9} = 1,$$

are discussed in Special Topics 10.3.A and are summarized in the endpapers at the beginning of this book. As with lines, there are many ways to parameterize conic sections.

EXAMPLE 8

In Special Topics 10.3.A we saw that a parameterization of the circle

$$(x - 4)^2 + (y - 1)^2 = 9$$

is given by

$$(*) \qquad x = 3 \cos t + 4 \quad \text{and} \quad y = 3 \sin t + 1 \quad (0 \le t \le 2\pi).$$

With this parameterization, the circle is traced out in a counterclockwise direction from the point $(7, 1)$, as shown in Figure 10–70. Another parameterization is given by

$$x = 3 \cos 2t + 4 \quad \text{and} \quad y = -3 \sin 2t + 1 \quad (0 \le t \le \pi).$$

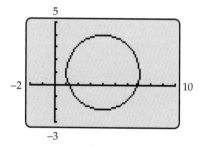

Figure 10–70

GRAPHING EXPLORATION

Verify that this parameterization traces out the circle in a clockwise direction, twice as fast as the parameterization given by $(*)$, since t runs from 0 to π rather than 2π.

APPLICATIONS

In the following applications, we ignore air resistance and assume some facts about gravity that are proved in physics.

EXAMPLE 9

Bob Lahr hits a golf ball with an initial velocity of 140 feet per second so that its path as it leaves the ground makes an angle of 31° with the horizontal.

(a) When does the ball hit the ground?

(b) How far from its starting point does it land?

(c) What is the maximum height of the ball during its flight?

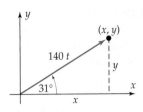

Figure 10–71

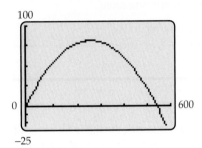

Figure 10–72

SOLUTION   Imagine that the golf ball starts at the origin and travels in the direction of the positive x-axis. If there were no gravity, the distance traveled by the ball in t seconds would be $140t$ feet. As shown in Figure 10–71, the coordinates (x, y) of the ball would satisfy

$$\frac{x}{140t} = \cos 31° \qquad \frac{y}{140t} = \sin 31°,$$

$$x = (140 \cos 31°)t \qquad y = (140 \sin 31°)t.$$

However, there *is* gravity, and at time t, it exerts a force of $16t^2$ downward (that is, in the negative direction on the y-coordinate). Consequently, the coordinates of the golf ball at time t are

$$x = (140 \cos 31°)t \qquad \text{and} \qquad y = (140 \sin 31°)t - 16t^2.$$

The path given by these parametric equations is graphed in Figure 10–72.*

(a) The ball is on the ground when $y = 0$, that is, at the x-intercepts of the graph. They can be found geometrically by using trace and zoom-in (the graphical root finder does not operate in parametric mode), but this is very time-consuming. To find the intercepts algebraically we need only set $y = 0$ and solve for t.

$$(140 \sin 31°)t - 16t^2 = 0$$

$$t(140 \sin 31° - 16t) = 0$$

$$t = 0 \qquad \text{or} \qquad 140 \sin 31° - 16t = 0$$

$$t = \frac{140 \sin 31°}{16} \approx 4.5066.$$

Thus, the ball hits the ground after approximately 4.5066 seconds.

(b) The horizontal distance traveled by the ball is given by the x-coordinate of the intercept. The x-coordinate when $t \approx 4.5066$ is

$$x = (140 \cos 31°)(4.5066) \approx 540.81 \text{ feet.}$$

(c) The graph in Figure 10–72 looks like a parabola, and it is, as you can verify by eliminating the parameter t (Exercise 56). The y-coordinate of the vertex is the maximum height of the ball. It can be found geometrically by using trace and zoom-in (the maximum finder doesn't work in parametric mode) or algebraically as follows. The vertex occurs halfway between its two x-intercepts ($x = 0$ and $x \approx 540.81$), that is, when $x \approx 270.405$. Hence,

$$(140 \cos 31°)t = x = 270.405,$$

so

$$t = \frac{270.405}{140 \cos 31°} \approx 2.2533.$$

Therefore, the y-coordinate of the vertex (the maximum height of the ball) is

$$y = (140 \sin 31°)(2.2533) - 16(2.2533)^2 \approx 81.237 \text{ feet.} \qquad ■$$

---

*Only the part of the graph on or above the x-axis represents the ball's path, since the ball does not go underground after it lands.

The argument used in Example 9 also applies when the initial position of the golf ball is k feet above the ground (for instance, if the golfer were on a platform at a driving range). In that case, the ball begins at $(0, k)$ instead of $(0, 0)$, and its position at time t is k feet higher than before, so its coordinates are

$$x = (140 \cos 31°)t \qquad \text{and} \qquad y = (140 \sin 31°)t - 16t^2 + k.$$

Then replacing 140 with v and 31° with θ in Example 9 leads to this conclusion.

Projectile Motion

> When a projectile is fired from the position $(0, k)$ on the positive y-axis at an angle θ with the horizontal, in the direction of the positive x-axis, with initial velocity v feet per second, with negligible air resistance, then its position at time t seconds is given by the parametric equations
>
> $$x = (v \cos \theta)t \qquad \text{and} \qquad y = (v \sin \theta)t - 16t^2 + k.$$

EXAMPLE 10

A batter hits a ball that is 3 feet above the ground. It leaves the bat at an angle of 26° with the horizontal and is headed toward a 25-foot high fence that is 400 feet away. Will the ball go over the fence if its initial velocity is

(a) 138 feet per second?   (b) 135.5 feet per second?

SOLUTION

(a) According to the preceding box (with $v = 138$, $\theta = 26°$, and $k = 3$), the path of the ball is given by the parametric equations

$$x = (138 \cos 26°)t \qquad \text{and} \qquad y = (138 \sin 26°)t - 16t^2 + 3.$$

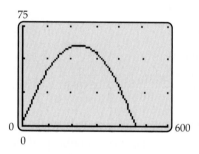

Figure 10–73

The graph of the ball's path in Figure 10–73 was made with the grid-on feature (in the TI FORMAT menu). Vertical tick marks are 25 units apart and it is easy to see that when the x-coordinate of the ball is 400, its y-coordinate is larger than 25. So the ball goes over the fence.

(b) In this case, the ball's path is given by

$$x = (135.5 \cos 26°)t \qquad \text{and} \qquad y = (135.5 \sin 26°)t - 16t^2 + 3.$$

GRAPHING EXPLORATION

Find the graph of the ball's path, using degree mode and the viewing window of Figure 10–73 (with $0 \le t \le 4$ and t-step = .1). If the graph is hard to read, try one or more of the following:

1. Use the trace feature.

2. Change the t-step to .01 and regraph.

3. Use the table feature with Δ Tbl set very small.

Does the ball clear the wall?

■

Our final example is a curve that has several interesting applications.

EXAMPLE 11

Choose a point P on a circle of radius 3, and find a parametric description of the curve that is traced out by P as the circle rolls along the x-axis, as shown in Figure 10–74.

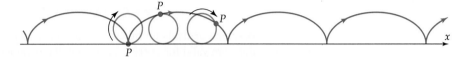

Figure 10–74

SOLUTION   This curve is called a **cycloid.** Begin with P at the origin and the center C of the circle at $(0, 3)$. As the circle rolls along the x-axis, the line segment CP moves from vertical through an angle of t radians, as shown in Figure 10–75.

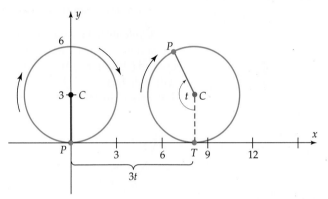

Figure 10–75

The distance from point T to the origin is the length of arc of the circle from T to P. As shown on page 435, this arc has length $3t$. Therefore the center C has coordinates $(3t, 3)$. When $0 < t < \pi/2$, the situation looks like Figure 10–76.

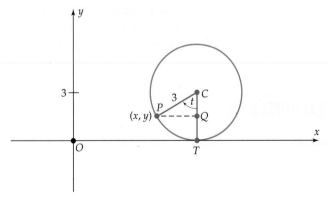

Figure 10–76

Right triangle PQC shows that

$$\sin t = \frac{PQ}{3} \qquad \text{or, equivalently,} \qquad PQ = 3 \sin t$$

and

$$\cos t = \frac{CQ}{3} \qquad \text{or, equivalently,} \qquad CQ = 3 \cos t.$$

In Figure 10–76, P has coordinates (x, y) and we have

$$x = OT - PQ = 3t - 3 \sin t = 3(t - \sin t),$$

$$y = CT - CQ = 3 - 3 \cos t = 3(1 - \cos t).$$

A similar analysis for other values of t (Exercises 65–67) shows that these equations are valid for every t. Therefore, the parametric equations of this cycloid are

$$x = 3(t - \sin t) \qquad \text{and} \qquad y = 3(1 - \cos t) \quad (t \text{ any real number}). \qquad \blacksquare$$

If a cycloid is traced out by a circle of radius r, then the argument given in Example 11, with r in place of 3, shows that the parametric equations of the cycloid are

$$x = r(t - \sin t) \qquad \text{and} \qquad y = r(1 - \cos t) \quad (t \text{ any real number}).$$

Cycloids have a number of interesting applications. For example, among all the possible paths joining points P and Q in Figure 10–77, an arch of an inverted cycloid (shown in red) is the curve along which a particle (subject only to gravity) will slide from P to Q in the shortest possible time. This fact was first proved by J. Bernoulli in 1696.

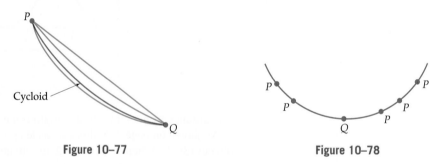

Figure 10–77   **Figure 10–78**

The Dutch physicist Christiaan Huygens (who invented the pendulum clock) proved that a particle takes the same time to slide to the bottom point Q of an inverted cycloid arch (as in Figure 10–78) from *any* point P on the curve.

EXERCISES 10.5

In Exercises 1–18, find a viewing window that shows a complete graph of the curve.

1. $x = t^2 - 4, \qquad y = t/2, \qquad -2 \leq t \leq 3$

2. $x = 3t^2, \qquad y = 2 + 5t, \quad 0 \leq t \leq 2$

3. $x = 2t, \qquad y = t^2 - 1, \quad -1 \leq t \leq 2$

4. $x = t - 1, \qquad y = \dfrac{t + 1}{t - 1}, \quad t \geq 1$

5. $x = \cos 4t, \qquad y = \cos 3t, \quad 0 \leq t \leq \pi$

6. $x = \sin 4t, \qquad y = \sin 3t, \quad 0 \leq t \leq 2\pi$

7. $x = \cos 3t + t, \qquad y = \cos 4t + t, \quad -\pi \leq t \leq \pi$

8. $x = 3 \sin t, \qquad y = \ln t, \quad 0 \leq t \leq 4\pi$

9. $x = 4 \sin 2t + 9, \qquad y = 6 \cos t - 8, \quad 0 \leq t \leq 2\pi$

10. $x = t^3 - 3t - 8, \qquad y = 3t^2 - 15, \quad -4 \leq t \leq 4$

11. $x = 6 \cos t + 12 \cos^2 t, \quad y = 8 \sin t + 8 \sin t \cos t,$ $\quad 0 \leq t \leq 2\pi$

12. $x = 12 \cos t, \qquad y = 12 \sin 2t, \quad 0 \leq t \leq 2\pi$

10.6 Polar Coordinates

Section Objectives
- ■ Convert from rectangular to polar coordinates and vice versa.
- ■ Graph polar coordinate equations.

In the past, we used a rectangular coordinate system in the plane, based on two perpendicular coordinate axes. Now we introduce another coordinate system for the plane, based on angles.

Choose a point O in the plane (called the **origin** or **pole**) and a half-line extending from O (called the **polar axis**). As shown in Figure 10–79, a point P is given **polar coordinates** (r, θ), where

$r = $ Distance from P to O

$\theta = $ Angle with polar axis as initial
side and OP as terminal side.

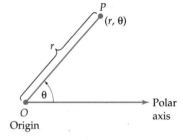

Figure 10–79

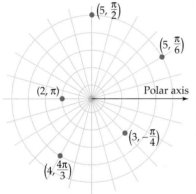

Figure 10–80

We shall usually measure the angle θ in radians; it may be either positive or negative, depending on whether it is generated by a clockwise or counterclockwise rotation. Some typical points are shown in Figure 10–80, which also illustrates the "circular grid" that a polar coordinate system imposes on the plane.

The polar coordinates of a point P are *not* unique. The angle θ may be replaced by any angle that has the same terminal side as θ, such as $\theta \pm 2\pi$. For instance, the coordinates $(2, \pi/3)$, $(2, 7\pi/3)$, and $(2, -5\pi/3)$ all represent the same point, as shown in Figure 10–81.

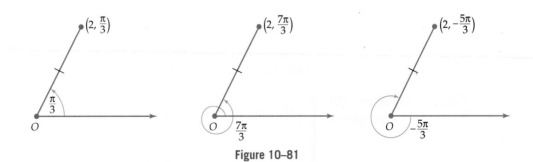

Figure 10–81

We shall consider the coordinates of the origin to be $(0, \theta)$, where θ is *any* angle.

Negative values for the first coordinate will be allowed according to this convention: For each positive r, the point $(-r, \theta)$ lies on the straight line containing the terminal side of θ, at distance r from the origin, on the *opposite* side of the origin from the point (r, θ), as shown in Figure 10–82 on the next page.

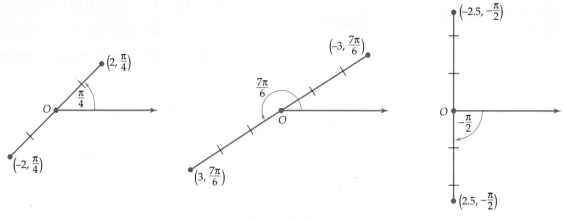

Figure 10–82

It is sometimes convenient to use both a rectangular and a polar coordinate system in the plane, with the polar axis coinciding with the positive x-axis. Then the y-axis is the polar line $\theta = \pi/2$. Suppose P has rectangular coordinates (x, y) and polar coordinates (r, θ), with $r > 0$, as in Figure 10–83.

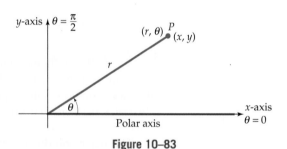

Figure 10–83

Since r is the distance from (x, y) to $(0, 0)$, the distance formula shows that $r = \sqrt{x^2 + y^2}$, and hence, $r^2 = x^2 + y^2$. The point-in-the-plane description of the trigonometric functions shows that

$$\cos \theta = \frac{x}{r}, \qquad \sin \theta = \frac{y}{r}, \qquad \tan \theta = \frac{y}{x}.$$

Solving the first two equations for x and y, we obtain the relationship between polar and rectangular coordinates.*

Coordinate Conversion Formulas

If a point has polar coordinates (r, θ), then its rectangular coordinates (x, y) are

$$x = r \cos \theta \qquad \text{and} \qquad y = r \sin \theta.$$

If a point has rectangular coordinates (x, y), with $x \neq 0$, then its polar coordinates (r, θ) satisfy

$$r^2 = x^2 + y^2 \qquad \text{and} \qquad \tan \theta = \frac{y}{x}.$$

*The conclusions in the next box are also true when $r < 0$ (Exercise 86).

EXAMPLE 1

Convert each of the following points in polar coordinates to rectangular coordinates.

(a) $(2, \pi/6)$ (b) $(3, 4)$

SOLUTION

(a) Apply the first set of equations in the box with $r = 2$ and $\theta = \pi/6$.

$$x = 2 \cos \frac{\pi}{6} = 2 \cdot \frac{\sqrt{3}}{2} = \sqrt{3} \qquad \text{and} \qquad y = 2 \sin \frac{\pi}{6} = 2 \cdot \frac{1}{2} = 1$$

So the rectangular coordinates are $(\sqrt{3}, 1)$.

(b) The point with polar coordinates $(3, 4)$ has $r = 3$ and $\theta = 4$ radians. Therefore, its rectangular coordinates are

$$(r \cos \theta, r \sin \theta) = (3 \cos 4, 3 \sin 4) \approx (-1.9609, -2.2704). \qquad \blacksquare$$

TECHNOLOGY TIP

Keys to convert from rectangular to polar coordinates, or vice versa, are in this menu/submenu:

TI-84+:	ANGLE
TI-86:	VECTOR/OPS
TI-89:	MATH/ANGLE
Casio:	OPTN/ANGLE

Conversion programs for HP-39gs are in the Program Appendix.

EXAMPLE 2

Find the polar coordinates of the point with rectangular coordinates $(2, -2)$.

SOLUTION The second set of equations in the box, with $x = 2$, $y = -2$, shows that

$$r = \sqrt{2^2 + (-2)^2} = \sqrt{8} = 2\sqrt{2} \qquad \text{and} \qquad \tan \theta = -2/2 = -1.$$

We must find an angle θ whose terminal side passes through $(2, -2)$ and whose tangent is -1. Figure 10–84 shows that two of the many possibilities are

$$\theta = -\frac{\pi}{4} \qquad \text{and} \qquad \theta = \frac{7\pi}{4}.$$

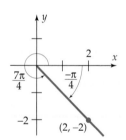

Figure 10–84

So one pair of polar coordinates is $\left(2\sqrt{2}, -\frac{\pi}{4} \right)$, and another is $\left(2\sqrt{2}, \frac{7\pi}{4} \right)$. $\blacksquare$

Rectangular to polar conversion is relatively easy when special angles are involved, as in Example 2. In other cases technology may be necessary.

EXAMPLE 3

Find the polar coordinates of the points whose rectangular coordinates are

(a) $(3, 5)$ (b) $(-2, 4)$

SOLUTION

(a) Applying the second set of equations in the box, with $x = 3$, $y = 5$, we have

$$r = \sqrt{3^2 + 5^2} = \sqrt{34} \qquad \text{and} \qquad \tan \theta = 5/3.$$

The TAN^{-1} key on a calculator shows that $\theta \approx 1.0304$ radians is an angle between 0 and $\pi/2$ with tangent 5/3. Since $(3, 5)$ is in the first quadrant, one pair of (approximate) polar coordinates is $(\sqrt{34}, 1.0304)$.

(b) In this case,

$$r = \sqrt{(-2)^2 + 4^2} = \sqrt{20} = 2\sqrt{5} \qquad \text{and} \qquad \tan\theta = \frac{4}{-2} = -2.$$

Using the TAN^{-1} key, we find that $\theta \approx -1.1071$ is an angle between $-\pi/2$ and 0 with tangent -2. However, we want an angle between $\pi/2$ and π because $(-2, 4)$ is in the second quadrant. Since tangent has period π,

$$\tan(-1.1071 + \pi) = \tan(-1.1071) = -2.$$

Thus, $-1.1071 + \pi \approx 2.0344$ is an angle between $\pi/2$ and π whose tangent is -2. Therefore, one pair of polar coordinates is $(2\sqrt{5}, 2.0344)$. ∎

The technique used in Example 3 may be summarized as follows.

Rectangular to Polar Conversion

If the rectangular coordinates of a point are (x, y), let $r = \sqrt{x^2 + y^2}$. If (x, y) lies in the first or fourth quadrant, then its polar coordinates are

$$\left(r, \tan^{-1}\frac{y}{x} \right).$$

If (x, y) lies in the second or third quadrant, then its polar coordinates are

$$\left(r, \tan^{-1}\left(\frac{y}{x}\right) + \pi \right).$$

EQUATION CONVERSION

When an equation in rectangular coordinates is given, it can be converted to polar coordinates by making the substitutions $x = r\cos\theta$ and $y = r\cos\theta$. Converting a polar equation to rectangular coordinates, however, is a bit trickier.

EXAMPLE 4

Find an equivalent rectangular equation for the given polar equation, and use it to identify the shape of the graph.

(a) $r = 4\cos\theta$ (b) $r = \dfrac{1}{1 - \sin\theta}$

SOLUTION

(a) Rewrite the equation $r = 4\cos\theta$ as follows.

Multiply both sides by r: $\qquad\qquad\qquad\qquad\qquad\qquad r^2 = 4r\cos\theta$

Substitute $r^2 = x^2 + y^2$ and $r\cos\theta = x$: $\qquad\qquad x^2 + y^2 = 4x$

$$x^2 - 4x + y^2 = 0$$

Add 4 to both sides: $\qquad\qquad\qquad\qquad (x^2 - 4x + 4) + y^2 = 4$

Factor: $\qquad\qquad\qquad\qquad\qquad\qquad (x - 2)^2 + y^2 = 2^2$

As we saw in Section 1.1, the graph of this equation is the circle with center $(2, 0)$ and radius 2.

(b) Begin by eliminating fractions.

$$r = \frac{1}{1 - \sin\theta}$$

$$r(1 - \sin\theta) = 1$$

$$r - r\sin\theta = 1$$

Substitute $r = \sqrt{x^2 + y^2}$ and $y = r\sin\theta$: $\sqrt{x^2 + y^2} - y = 1$

Rearrange terms: $\sqrt{x^2 + y^2} = y + 1$

Square both sides: $x^2 + y^2 = (y + 1)^2$

Simplify: $x^2 + y^2 = y^2 + 2y + 1$

$$x^2 = 2y + 1$$

$$y = \frac{1}{2}(x^2 - 1)$$

As we saw in Section 4.1, the graph is an upward-opening parabola. ■

POLAR GRAPHS

The graphs of a few polar coordinate equations can be easily determined from the appropriate definitions.

EXAMPLE 5

Graph the equations

(a) $r = 3$*   (b) $\theta = \pi/6$.*

SOLUTION

(a) The graph consists of all points (r, θ) with first coordinate 3, that is, all points whose distance from the origin is 3. So the graph is a circle with center O and radius 3, as shown in Figure 10–85.

(b) The graph consists of all points $(r, \pi/6)$. If $r \geq 0$, then $(r, \pi/6)$ lies on the terminal side of an angle of $\pi/6$ radians, whose initial side is the polar axis. If $r < 0$, then $(r, \pi/6)$ lies on the extension of this terminal side across the origin. So the graph is the straight line in Figure 10–86. ■

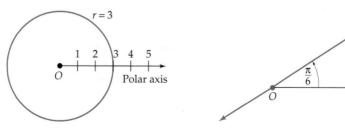

Figure 10–85   Figure 10–86

---

*Every equation is understood to involve two variables, but one may have coefficient 0, as is the case here: $r = 3 + 0 \cdot \theta$ and $\theta = 0 \cdot r + \pi/6$. This is analogous to equations such as $y = 5$ and $x = 2$ in rectangular coordinates.

Some polar graphs can be sketched by hand by using basic facts about trigonometric functions.

EXAMPLE 6

Graph $r = 1 + \sin \theta$.

SOLUTION   Remember the behavior of $\sin \theta$ between 0 and 2π:

As θ increases from 0 to $\pi/2$, $\sin \theta$ increases from 0 to 1. So $r = 1 + \sin \theta$ increases from 1 to 2.

As θ increases from $\pi/2$ to π, $\sin \theta$ decreases from 1 to 0. So $r = 1 + \sin \theta$ decreases from 2 to 1.

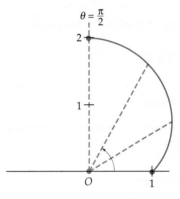

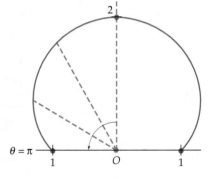

As θ increases from π to $3\pi/2$, $\sin \theta$ decreases from 0 to -1. So $r = 1 + \sin \theta$ decreases from 1 to 0.

As θ increases from $3\pi/2$ to 2π, $\sin \theta$ increases from -1 to 0. So $r = 1 + \sin \theta$ increases from 0 to 1.

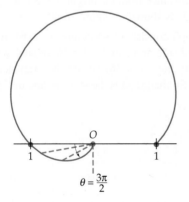

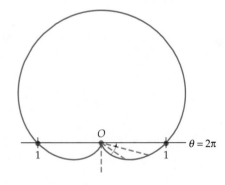

Figure 10–87

As θ takes values larger than 2π, $\sin \theta$ repeats the same pattern, and hence, so does $r = 1 + \sin \theta$. The same is true for negative values of θ. The full graph (called a **cardioid**) is at the lower right in Figure 10–87.   ∎

The easiest way to graph polar equation $r = f(\theta)$ is to use a calculator in polar graphing mode. A second way is to use parametric graphing mode, with the coordinate converison formulas as a parameterization.

$$x = r \cos \theta = f(\theta) \cos \theta,$$

$$y = r \sin \theta = f(\theta) \sin \theta.$$

EXAMPLE 7

Graph $r = 2 + 4 \cos \theta$.

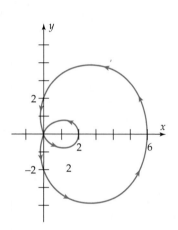

Figure 10–88

SOLUTION   ***Polar Method:***   Put your calculator in polar graphing mode and enter $r = 2 + 4 \cos \theta$ in the function memory. Set the viewing window by entering minimum and maximum values for x, y, and θ. Since cosine has period 2π, a complete graph can be obtained by taking $0 \le \theta \le 2\pi$. You must also set the θ step (or θ pitch), which determines how many values of θ the calculator uses to plot the graph. With an appropriate θ step, the graph should look like Figure 10–88.

Parametric Method: Put your calculator in parametric graphing mode. The parametric equations for $r = 2 + 4 \cos \theta$ are as follows (using t as the variable instead of θ with $0 \le t \le 2\pi$):

$$x = r \cos t = (2 + 4 \cos t) \cos t = 2 \cos t + 4 \cos^2 t$$

$$y = r \sin t = (2 + 4 \cos t) \sin t = 2 \sin t + 4 \sin t \cos t.$$

They also produce the graph in Figure 10–88.   ■

EXAMPLE 8

The graph of $r = \sin 2\theta$ in Figure 10–89 can be obtained either by graphing directly in polar mode or by using parametric mode and the equations

$$x = r \cos t = \sin 2t \cos t \quad \text{and} \quad y = r \sin t = \sin 2t \sin t \quad (0 \le t \le 2\pi). \quad ■$$

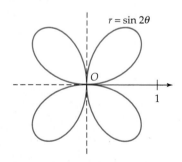

Figure 10–89

Here is a summary of commonly encountered polar graphs (in each case, a and b are constants).

Equation	Name of Graph	Shape of Graph*
$r = a\theta \ (\theta \geq 0)$ $r = a\theta \ (\theta \leq 0)$	Archimedean spiral	 $r = a\theta \ (\theta \geq 0)$ $r = a\theta \ (\theta \leq 0)$
$r = a(1 \pm \sin \theta)$ $r = a(1 \pm \cos \theta)$	cardioid	 $r = a(1 + \cos \theta)$ $r = a(1 - \sin \theta)$
$r = a \sin n\theta$ $r = a \cos n\theta$ $(n \geq 2)$	rose (There are n petals when n is odd and $2n$ petals when n is even.)	 $r = a \cos n\theta$ $r = a \sin n\theta$

*Depending on the plus or minus sign and whether sine or cosine is involved, the basic shape of a specific graph may differ from those shown by a rotation, reversal, or horizontal or vertical shift.

Equation	Name of Graph	Shape of Graph		
$r = a \sin \theta$ $r = a \cos \theta$	circle	 $r = a \cos \theta$	 $r = a \sin \theta$	
$r^2 = \pm a^2 \sin 2\theta$ $r^2 = \pm a^2 \cos 2\theta$	lemniscate	 $r^2 = a^2 \sin 2\theta$	 $r^2 = a^2 \cos 2\theta$	
$r = a \pm b \sin \theta$ $r = a \pm b \cos \theta$ $(a, b > 0; a \neq b)$	limaçon	 $a < b$ $r = a + b \cos \theta$	 $b < a < 2b$ $r = a + b \sin \theta$	 $a \geq 2b$ $r = a - b \sin \theta$

EXERCISES 10.6

1. What are the polar coordinates of the points P, Q, R, S, T, U, V in the figure?

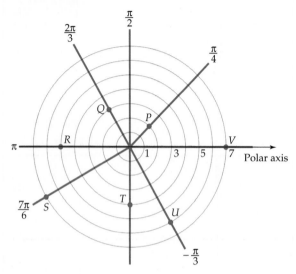

In Exercises 2–6, plot the point whose polar coordinates are given.

2. $(1, \pi/4)$ 3. $(2, -3\pi/4)$ 4. $(-2, 2\pi/3)$

5. $(-3, -5\pi/3)$ 6. $(3, \pi/6)$

In Exercises 7–12, list four other pairs of polar coordinates for the given point, each with a different combination of signs (that is, $r > 0$, $\theta > 0$; $r > 0$, $\theta < 0$; $r < 0$, $\theta > 0$; $r < 0$, $\theta < 0$).

7. $(3, \pi/3)$ 8. $(-5, \pi)$ 9. $(2, -2\pi/3)$

10. $(-1, -\pi/6)$ 11. $(\sqrt{3}, 3\pi/4)$ 12. $(-3, 7\pi/6)$

In Exercises 13–20, convert the polar coordinates to rectangular coordinates.

13. $(3, \pi/3)$ 14. $(-2, \pi/4)$ 15. $(-1, 5\pi/6)$

16. $(2, 0)$ 17. $(1.5, 5)$ → 18. $(2.2, -2.2)$

19. $(-4, -\pi/7)$ 20. $(-1, 1)$

In Exercises 21–34, convert the rectangular coordinates to polar coordinates.

21. $(3\sqrt{3}, -3)$ 22. $(2\sqrt{3}, 2)$ 23. $(1, 1)$

24. $(\sqrt{2}, -\sqrt{2})$ 25. $(3, 3\sqrt{3})$ 26. $(-\sqrt{2}, \sqrt{6})$

27. $(2, 4)$ 28. $(3, -2)$ 29. $(-5, 2.5)$

30. $(-6.2, -3)$ 31. $(0, -2)$ 32. $(.5, 3.5)$

33. $(-2, 4)$ 34. $(\sqrt{5}, \sqrt{10})$

In Exercises 35–40, find a polar equation that is equivalent to the given rectangular equation.

35. $x^2 + y^2 = 25$ 36. $4xy = 1$ 37. $x = 12$

38. $y = 4$ 39. $y = 2x + 1$ 40. $y = x - 2$

In Exercises 41–52, find a rectangular equation that is equivalent to the given polar equation.

41. $r = 3$ [Hint: Square both sides, then substitute.]

42. $r = 5$

43. $\theta = \pi/6$ {Hint: Take the tangent of both sides, then substitute.]

44. $\theta = -\pi/4$

45. $r = \sec \theta$ [Hint: Express the right side in terms of cosine.]

46. $r = \csc \theta$ 47. $r^2 = \tan \theta$ 48. $r^2 = \sin \theta$

49. $r = 2 \sin \theta$ 50. $r = 3 \cos \theta$

51. $r = \dfrac{4}{1 + \sin \theta}$ 52. $r = \dfrac{6}{1 - \cos \theta}$

In Exercises 53–58, sketch the graph of the equation without using a calculator.

53. $r = 4$ 54. $r = -1$ 55. $\theta = -\pi/3$

56. $\theta = 5\pi/6$ 57. $\theta = 1$ 58. $\theta = -4$

In Exercises 59–82, sketch the graph of the equation.

59. $r = \theta$ $(\theta \le 0)$ 60. $r = 3\theta$ $(\theta \ge 0)$

61. $r = 1 - \sin \theta$ 62. $r = 3 - 3 \cos \theta$

63. $r = -2 \cos \theta$ 64. $r = -6 \sin \theta$

65. $r = \cos 2\theta$ 66. $r = \cos 3\theta$

67. $r = \sin 3\theta$ 68. $r = \sin 4\theta$

69. $r^2 = 4 \cos 2\theta$ 70. $r^2 = \sin 2\theta$

71. $r = 2 + 4 \cos \theta$ 72. $r = 1 + 2 \cos \theta$

73. $r = \sin \theta + \cos \theta$ 74. $r = 4 \cos \theta + 4 \sin \theta$

75. $r = \sin (\theta/2)$ 76. $r = 4 \tan \theta$

77. $r = \sin \theta \tan \theta$ (cissoid)

78. $r = 4 + 2 \sec \theta$ (conchoid)

79. $r = e^\theta$ (logarithmic spiral)

80. $r^2 = 1/\theta$ 81. $r = 1/\theta$ $(\theta > 0)$ 82. $r^2 = \theta$

83. (a) Find a complete graph of $r = 1 - 2 \sin 3\theta$.
    (b) Predict what the graph of $r = 1 - 2 \sin 4\theta$ will look like. Then check your prediction with a calculator.
    (c) Predict what the graph of $r = 1 - 2 \sin 5\theta$ will look like. Then check your prediction with a calculator.

84. (a) Find a complete graph of $r = 1 - 3 \sin 2\theta$.
    (b) Predict what the graph of $r = 1 - 3 \sin 3\theta$ will look like. Then check your prediction with a calculator.
    (c) Predict what the graph of $r = 1 - 3 \sin 4\theta$ will look like. Then check your prediction with a calculator.

85. If a, b are nonzero constants, show that the graph of $r = a \sin \theta + b \cos \theta$ is a circle. [Hint: Multiply both sides by r and convert to rectangular coordinates.]

86. Prove that the coordinate conversion formulas are valid when $r < 0$. [Hint: If P has coordinates (x, y) and (r, θ), with $r < 0$,

verify that the point Q with rectangular coordinates $(-x, -y)$ has polar coordinates $(-r, \theta)$. Since $r < 0$, $-r$ is positive and the conversion formulas proved in the text apply to Q. For instance, $-x = -r \cos \theta$, which implies that $x = r \cos \theta$.]

87. ***Distance Formula for Polar Coordinates:*** Prove that the distance from (r, θ) to (s, β) is

$$\sqrt{r^2 + s^2 - 2rs \cos(\theta - \beta)}$$

[*Hint:* If $r > 0$, $s > 0$, and $\theta > \beta$, then the triangle with vertices (r, θ), (s, β), $(0, 0)$ has an angle of $\theta - \beta$, whose sides have lengths r and s. Use the Law of Cosines.]

88. Explain why the following symmetry tests for the graphs of polar equations are valid.

(a) If replacing θ by $-\theta$ produces an equivalent equation, then the graph is symmetric with respect to the line $\theta = 0$ (the x-axis).

(b) If replacing θ by $\pi - \theta$ produces an equivalent equation, then the graph is symmetric with respect to the line $\theta = \pi/2$ (the y-axis).

(c) If replacing r by $-r$ produces an equivalent equation, then the graph is symmetric with respect to the origin.

10.7 Polar Equations of Conics

Section Objectives
- Find the eccentricity of a conic section.
- Learn the polar form for equations of conic sections.
- Find the polar equation of a conic section.

In a rectangular coordinate system, each type of conic section has a different definition. By using polar coordinates, it is possible to give a unified treatment of conics and their equations. Before doing this, we must first introduce a concept that will play a key role in the development.

Recall that both ellipses and hyperbolas are defined in terms of two foci and both have two vertices that lie on the line through the foci (see pages 672–673 and 686–687). The **eccentricity** of an ellipse or a hyperbola is denoted e and is defined to be the ratio

$$e = \frac{\text{distance between the foci}}{\text{distance between the vertices}}.$$

For conics centered at the origin, with foci on the x-axis, the situation is as follows.

Ellipse

$$\frac{x^2}{a^2} + \frac{y^2}{b^2} = 1 \quad (a > b)$$

foci: $(\pm c, 0)$ vertices: $(\pm a, 0)$

$$c = \sqrt{a^2 - b^2}$$

Hyperbola

$$\frac{x^2}{a^2} - \frac{y^2}{b^2} = 1$$

foci: $(\pm c, 0)$ vertices: $(\pm a, 0)$

$$c = \sqrt{a^2 + b^2}$$

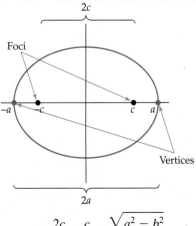

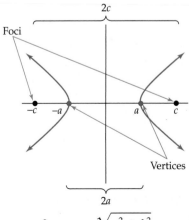

$$e = \frac{2c}{2a} = \frac{c}{a} = \frac{\sqrt{a^2 - b^2}}{a} \qquad\qquad e = \frac{2c}{2a} = \frac{c}{a} = \frac{\sqrt{a^2 + b^2}}{a}$$

A similar analysis shows that the formulas for e are also valid for conics whose equations are of the form

$$\frac{x^2}{b^2} + \frac{y^2}{a^2} = 1 \quad (a > b) \qquad \text{or} \qquad \frac{y^2}{a^2} - \frac{x^2}{b^2} = 1.$$

These formulas can be used to compute the eccentricity of any ellipse or hyperbola whose equation can be put in standard form.

EXAMPLE 1

Find the eccentricity of the conic with equation

(a) $\dfrac{y^2}{4} - \dfrac{x^2}{21} = 1$; (b) $4x^2 + 9y^2 - 32x - 90y + 253 = 0.$

SOLUTION

(a) In this case, $a^2 = 4$ (so $a = 2$) and $b^2 = 21$. Hence, the eccentricity is

$$e = \frac{\sqrt{a^2 + b^2}}{a} = \frac{\sqrt{4 + 21}}{2} = \frac{\sqrt{25}}{2} = \frac{5}{2} = 2.5.$$

(b) In Example 8 of Section 10.1, we saw that the equation can be put into this standard form:

$$\frac{(x - 4)^2}{9} + \frac{(y - 5)^2}{4} = 1.$$

Hence, its graph is just the ellipse

$$\frac{x^2}{9} + \frac{y^2}{4} = 1$$

shifted vertically and horizontally. Since the shifting does not change the distances between foci or vertices, both ellipses have the same eccentricity, which can be computed by using $a^2 = 9$ and $b^2 = 4$.

$$e = \frac{\sqrt{a^2 - b^2}}{a} = \frac{\sqrt{9 - 4}}{3} = \frac{\sqrt{5}}{3} \approx .745. \qquad \blacksquare$$

Example 1 and the preceding pictures illustrate the following fact. For ellipses, the distance between the foci (numerator of e) is less than the distance between the vertices (denominator), so $e < 1$. For hyperbolas, however, $e > 1$ because the distance between the foci is greater than that between the vertices.

The eccentricity of an ellipse measures its "roundness." An ellipse whose eccentricity is close to 0 is almost circular (Exercise 19). The eccentricity of a hyperbola measures how "flat" its branches are. The branches of a hyperbola with large eccentricity look almost like parallel lines (Exercise 20).

CONICS AND POLAR EQUATIONS

The polar analogues of the standard equations of ellipses, parabolas, and hyperbolas are given in the following chart. The proof of these statements is given at the end of the section. In the chart, e and d are constants, with $e > 0$. Remember that in a rectangular coordinate system whose positive x-axis coincides with the polar axis, a point with polar coordinates (r, θ) is on the x-axis when $\theta = 0$ or π and on the y-axis when $\theta = \pi/2$ or $3\pi/2$.

Polar Equations
for Conic Sections

Equation	Graph	
$r = \dfrac{ed}{1 + e \cos \theta}$ or $r = \dfrac{ed}{1 - e \cos \theta}$	$0 < e < 1$	*Ellipse* with eccentricity e One of the foci: $(0, 0)$ Vertices at $\theta = 0$ and $\theta = \pi$
	$e = 1$	*Parabola* with focus $(0, 0)$ Vertex at $\theta = 0$ or $\theta = \pi$; (r is not defined for the other value of θ)
	$e > 1$	*Hyperbola* with eccentricity e One of the foci: $(0, 0)$ Vertices at $\theta = 0$ and $\theta = \pi$
$r = \dfrac{ed}{1 + e \sin \theta}$ or $r = \dfrac{ed}{1 - e \sin \theta}$	$0 < e < 1$	*Ellipse* with eccentricity e One of the foci: $(0, 0)$ Vertices at $\theta = \pi/2$ and $\theta = 3\pi/2$
	$e = 1$	*Parabola* with focus $(0, 0)$ Vertex at $\theta = \pi/2$ or $\theta = 3\pi/2$; (r is not defined for the other value of θ)
	$e > 1$	*Hyperbola* with eccentricity e One of the foci: $(0, 0)$ Vertices at $\theta = \pi/2$ and $\theta = 3\pi/2$

EXAMPLE 2

Find a complete graph of

$$r = \frac{3e}{1 + e \cos \theta}$$

when

(a) $e = .7$ (b) $e = 1$ (c) $e = 2$.

SOLUTION   From the first equation in the preceding chart (with $d = 3$), we know that the graphs are an ellipse, a parabola, and a hyperbola, respectively, as shown in Figure 10–90.   ■

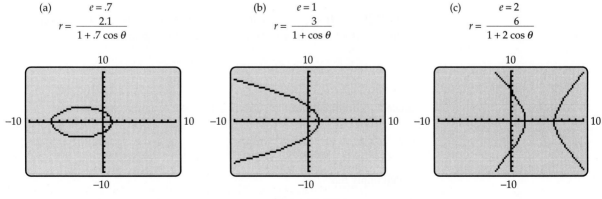

(a) $e = .7$
$r = \dfrac{2.1}{1 + .7 \cos \theta}$

(b) $e = 1$
$r = \dfrac{3}{1 + \cos \theta}$

(c) $e = 2$
$r = \dfrac{6}{1 + 2 \cos \theta}$

Figure 10–90

EXAMPLE 3

Identify the conic section that is the graph of

$$r = \frac{20}{4 - 10 \sin \theta},$$

and find its vertices and eccentricity.

SOLUTION First, rewrite the equation in one of the forms listed in the preceding box.

$$r = \frac{20}{4 - 10 \sin \theta} = \frac{20}{4\left(1 - \frac{10}{4} \sin \theta\right)} = \frac{5}{1 - 2.5 \sin \theta}.$$

This is one such form, with $e = 2.5$ and $ed = 5$ (so $d = 2$). Consequently, the graph is a hyperbola with eccentricity $e = 2.5$ whose vertices are at

$$\theta = \frac{\pi}{2}, \qquad r = \frac{20}{4 - 10 \sin \frac{\pi}{2}} = \frac{20}{4 - 10 \cdot 1} = -\frac{20}{6} = -\frac{10}{3}$$

and

$$\theta = \frac{3\pi}{2}, \qquad r = \frac{20}{4 - 10 \sin \frac{3\pi}{2}} = \frac{20}{4 - 10(-1)} = \frac{20}{14} = \frac{10}{7}.$$

GRAPHING EXPLORATION

Find a viewing window that shows a complete graph of this hyperbola.

EXAMPLE 4

Find a polar equation of the ellipse with $(0, 0)$ as a focus and vertices $(3, 0)$ and $(6, \pi)$.

SOLUTION Because of the location of the vertices, the polar equation is of the form $r = ed/(1 \pm e \cos \theta)$. We first consider the equation

$$r = \frac{ed}{1 + e \cos \theta}.$$

Since the coordinates of the vertices satisfy the equation, we must have

$$3 = \frac{ed}{1 + e \cos 0} = \frac{ed}{1 + e} \qquad \text{and} \qquad 6 = \frac{ed}{1 + e \cos \pi} = \frac{ed}{1 - e},$$

which imply that

$$3(1 + e) = ed \qquad \text{and} \qquad 6(1 - e) = ed.$$

Therefore,

$$3(1 + e) = 6(1 - e)$$

$$3 + 3e = 6 - 6e$$

$$9e = 3$$

$$e = 1/3.$$

Substituting $e = 1/3$ in either of the original equations shows that $d = 12$. So an equation of the ellipse is

$$r = \frac{ed}{1 + e \cos \theta} = \frac{\frac{1}{3} \cdot 12}{1 + \frac{1}{3} \cos \theta} = \frac{12}{3 + \cos \theta}.$$

If we had started instead with the equation $r = ed/(1 - e \cos \theta)$ and solved for e as above, we would have obtained $e = -1/3$, which is impossible, since $e > 0$.

ALTERNATE SOLUTION   Verify that the vertex $(3, 0)$ also has polar coordinates $(-3, \pi)$. Similarly, $(6, \pi)$ also has polar coordinates $(-6, 0)$. If you begin with the equation $r = ed/(1 - e \cos \theta)$ and the vertices $(-3, \pi)$ and $(-6, 0)$ and proceed as before to find e and d, you obtain the equation

$$r = \frac{-12}{3 - \cos \theta}. \qquad \blacksquare$$

ALTERNATE DEFINITION OF CONICS

The theorem stated in the following box is sometimes used as a definition of the conic sections because it provides a unified approach instead of the variety of descriptions given in Sections 10.1–10.3. Its proof also provides a proof of the statements in the box on page 755.

The basic idea is to describe every conic in terms of a straight line L (the **directrix**) and a point P not on L (the **focus**), in much the same way that parabolas were defined in Section 10.3. The number e in the theorem turns out to be the eccentricity of the conic.

Conic Section Theorem

> Let L be a straight line, P a point not on L, and e a positive constant. The set of all points X in the plane such that
>
> $$\frac{\text{distance from } X \text{ to } P}{\text{distance from } X \text{ to } L} = e$$
>
> is a conic section with P as one of the foci.* The conic is an ellipse if $0 < e < 1$, a parabola if $e = 1$,[†] and a hyperbola if $e > 1$.

---

*The distance from X to L is measured along the line through X that is perpendicular to L.
[†]When $e = 1$, the given condition is equivalent to

$$\text{distance from } X \text{ to } P = \text{distance from } X \text{ to } L$$

which is the definition of a parabola given in Section 10.3.

Proof   Coordinatize the plane so that the pole is the point P, the polar axis is horizontal, and the directrix L is a vertical line to the left of the pole, as in Figure 10–91. Let d be the distance from P to L, and let (r, θ) be the polar coordinates of X.

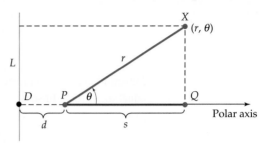

Figure 10–91

If X satisfies the condition

$$\frac{\text{distance from } X \text{ to } P}{\text{distance from } X \text{ to } L} = e,$$

then

(∗)   distance from X to $P = e(\text{distance from } X \text{ to } L)$.

Figure 10–91 shows that r is the distance from P to X and that the distance from X to L is the same as that from D to Q, namely, $d + s$. Furthermore, $\cos\theta = s/r$, so $s = r\cos\theta$. Consequently, equation (∗) can be written in polar coordinates as follows.

$$\text{distance from } X \text{ to } P = e(\text{distance from } X \text{ to } L)$$
$$r = e(d + s)$$
$$r = e(d + r\cos\theta)$$
$$r - er\cos\theta = ed$$
$$r(1 - e\cos\theta) = ed$$
$$r = \frac{ed}{1 - e\cos\theta}.$$

To show that this is actually the equation of a conic, we translate it into rectangular coordinates using the conversion formulas from Section 10.6.

$$r^2 = x^2 + y^2 \qquad \text{and} \qquad \cos\theta = \frac{x}{r} = \frac{x}{\pm\sqrt{x^2 + y^2}}.$$

Then the polar coordinate equation becomes

$$\pm\sqrt{x^2 + y^2} = \frac{ed}{1 - e\left(\dfrac{x}{\pm\sqrt{x^2 + y^2}}\right)}$$

$$\pm\sqrt{x^2 + y^2}\left(1 - \frac{ex}{\pm\sqrt{x^2 + y^2}}\right) = ed$$

$$\pm\sqrt{x^2 + y^2} - ex = ed$$

$$\pm\sqrt{x^2 + y^2} = ed + ex.$$

Squaring both sides and rearranging terms, we have

$$x^2 + y^2 = e^2d^2 + 2de^2x + e^2x^2$$

(∗∗) $$(1 - e^2)x^2 - 2de^2x + y^2 = e^2d^2.$$

Now we consider the two possibilities $e = 1$ and $e \neq 1$.

5. Sewing machines use similar devices, called feed dogs, to move the cloth beneath the presser foot. The motion of the feed dogs can be simulated using functions of the type $\dfrac{|\sin kt|}{\sin kt}$ in place of sines and cosines. Use this to design a set of parametric equations that would produce stitches like the ones shown below.

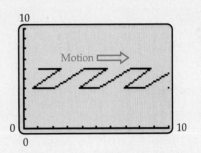

6. How would you change your answer to question 5 if the width of the stitches, measured perpendicularly to the direction of motion, needs to be 5 times as wide but the length when measured from peak to peak in the direction of motion remains the same?

Chapter 11

SYSTEMS OF EQUATIONS

Is this a diamond in the rough?

The structure of certain crystals can be described by a large system of linear equations (more than one hundred equations and variables). A variety of resource allocation problems involving many variables can be handled by solving an appropriate system of equations. The fastest solution methods involve matrices and are easily implemented on a computer or calculator. See Exercise 51 on page 804.

```
rref C
          [[1 0 0 3 ]
           [0 1 0 -1]
■          [0 0 1 4 ]]
```

Chapter Outline

This chapter deals with *systems of equations,* such as

$$2x - 5y + 3z = 1 \qquad 2x + 5y + z + w = 0 \qquad x^2 + y^2 = 25$$
$$x + 2y - z = 2 \qquad 2y - 4z + 41w = 5 \qquad x^2 - y = 7$$
$$3x + y + 2z = 11 \qquad 3x + 7y + 5z - 8w = -6$$

Three equations in three variables   Three equations in four variables   Two equations in two variables

Sections 11.1–11.3 deal with systems of linear equations (such as the first two shown above). Systems involving nonlinear equations are considered in Special Topics 11.1.A.

A **solution of a system** is a solution that satisfies *all* the equations in the system. For instance, in the first system of equations above, $x = 1$, $y = 2$, $z = 3$ is a solution of all three equations (check it) and hence is a solution of the system. On the other hand, $x = 0$, $y = 7$, $z = 12$ is a solution of the first two equations but not of the third (check it). So $x = 0$, $y = 7$, $z = 12$ is not a solution of the system.

11.1   Systems of Linear Equations in Two Variables

Section Objectives

■ Solve systems of linear equations graphically.

■ Use the substitution method to solve systems of linear equations.

■ Use the elimination method to solve systems of linear equations.

■ Identify inconsistent and dependent systems.

■ Use systems of linear equations to solve applied problems.

Systems of linear equations in two variables may be solved graphically or algebraically. The geometric method is similar to what we have done previously.

EXAMPLE 1

Solve this system graphically.

$$2x - y = 1$$
$$3x + 2y = 4.$$

SOLUTION   First, we solve each equation for y.

$$2x - y = 1 \qquad\qquad 3x + 2y = 4$$

$$-y = -2x + 1 \qquad\qquad 2y = -3x + 4$$

$$y = 2x - 1 \qquad\qquad y = \frac{-3x + 4}{2}.$$

Next, we graph both equations on the same screen (Figure 11–1). As we saw in Section 1.4, each graph is a straight line, and every point on the graph represents a solution of the equation. Therefore, the solution of the system is given by the coordinates of the point that lies on both lines. An intersection finder (Figure 11–2) shows that the approximate coordinates of this point are

$$x \approx .85714286 \qquad \text{and} \qquad y \approx .71428571. \qquad\blacksquare$$

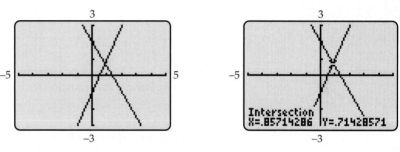

Figure 11–1   Figure 11–2

As is shown in Example 1, the solutions of a system of linear equations are determined by the points where their graphs intersect. There are exactly three geometric possibilities for two lines in the plane: They are parallel, they intersect at a single point, or they coincide, as illustrated in Figure 11–3. Each of these possibilities leads to a different number of solutions for the system.

Number of Solutions
of a System

A system of two linear equations in two variables must have

No solutions *or*

Exactly one solution *or*

An infinite number of solutions.

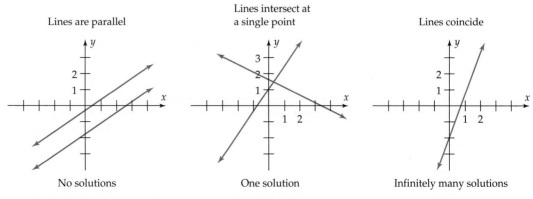

Figure 11–3

THE SUBSTITUTION METHOD

When you use a calculator to solve systems graphically, you may have to settle for an approximate solution, as we did in Example 1. Algebraic methods, however, produce exact solutions. Furthermore, algebraic methods are often as easy to implement as graphical ones, so we shall use them, whenever practical, to obtain exact solutions. One algebraic method is **substitution,** which is explained in the next example.

EXAMPLE 2

Use substitution to find the exact solution of the system from Example 1.

$$2x - y = 1$$
$$3x + 2y = 4.$$

SOLUTION   Any solution of this system must satisfy the first equation, $2x - y = 1$. Solving this equation for y, as in Example 1, shows that

$$y = 2x - 1.$$

Substituting this expression for y in the second equation, we have

$$3x + 2y = 4$$
$$3x + 2(2x - 1) = 4$$
$$3x + 4x - 2 = 4$$
$$7x = 6$$
$$x = 6/7.$$

Therefore, every solution of the original system must have $x = 6/7$. But when $x = 6/7$, we see from the first equation that

$$2x - y = 1$$
$$2\left(\frac{6}{7}\right) - y = 1$$
$$\frac{12}{7} - y = 1$$
$$-y = -\frac{12}{7} + 1$$
$$y = \frac{12}{7} - 1 = \frac{5}{7}.$$

(We would also have found that $y = 5/7$ if we had substituted $x = 6/7$ in the second equation.) Consequently, the exact solution of the original system is $x = 6/7, y = 5/7$. ∎

CAUTION

To guard against arithmetic mistakes, you should always *check your answers* by substituting them into *all* the equations of the original system. We have in fact checked the answers in all the examples, but these checks are omitted to save space.

When using the substitution method, you may solve either of the given equations for either one of the variables and then substitute that result in the other equation. In Example 2, we solved for y in the first equation because that avoided the fractional expression that would have occurred if we had solved for x or had solved the second equation for x or y.

 ## THE ELIMINATION METHOD

The **elimination method** of solving systems of linear equations is often more convenient than substitution. It depends on this fact.

> **Multiplying both sides of an equation by a nonzero constant does not change the solutions of the equation.**

For example, the equation $x + 3 = 5$ has the same solution as $2x + 6 = 10$ (the first equation multiplied by 2). The elimination method also uses this fact from basic algebra:

$$\text{If } A = B \text{ and } C = D, \text{ then } A + C = B + D \quad \text{and} \quad A - C = B - D.$$

EXAMPLE 3

Solve this system using the elimination method:

$$x - 3y = 4$$
$$2x + y = 1.$$

SOLUTION We replace the first equation by an equivalent one (that is, one with the same solutions).

$$-2x + 6y = -8 \quad \text{[First equation multiplied by } -2]$$
$$2x + y = 1.$$

The multiplier -2 was chosen so that the coefficients of x in the two equations would be negatives of each other. Any solution of this last system must also be a solution of the sum of the two equations.

$$-2x + 6y = -8$$
$$\underline{2x + y = 1}$$
$$7y = -7. \quad \text{[The first variable has been eliminated]}$$

Solving this last equation, we see that $y = -1$. Substituting this value in the first of the original equations shows that

$$x - 3(-1) = 4$$
$$x = 1.$$

Therefore, $x = 1, y = -1$ is the solution of the original system. ∎

EXAMPLE 4

Solve the following system:

$$5x - 3y = 3$$
$$3x - 2y = 1.$$

SOLUTION Any solution of the above system must also be a solution of this system:

$$10x - 6y = 6 \qquad \text{[First equation multiplied by 2]}$$

$$-9x + 6y = -3. \qquad \text{[Second equation multiplied by } -3\text{]}$$

The multipliers 2 and -3 were chosen so that the coefficients of y in the new equations would be negatives of each other. Any solution of this last system must also be a solution of the equation obtained by adding these two equations.

$$\begin{aligned} 10x - 6y &= 6 \\ \underline{-9x + 6y} &= \underline{-3} \\ x &= 3. \qquad \text{[The second variable has been eliminated]} \end{aligned}$$

Substituting $x = 3$ in the first of the original equations shows that

$$5(3) - 3y = 3$$

$$-3y = -12$$

$$y = 4.$$

Therefore, the solution of the original system is $x = 3$, $y = 4$. ∎

EXAMPLE 5

To solve the system

$$2x - 3y = 5$$

$$4x - 6y = 1,$$

we multiply the first equation by -2 and add.

$$\begin{aligned} -4x + 6y &= -10 \\ \underline{4x - 6y} &= \underline{1} \\ 0 &= -9. \end{aligned}$$

Since $0 = -9$ is always false, the original system cannot possibly have any solutions. A system with no solution is said to be **inconsistent.** ∎

GRAPHING EXPLORATION

Confirm the result of Example 5 geometrically by graphing the two equations in the system. Do these lines intersect, or are they parallel?

EXAMPLE 6

To solve the system

$$3x - y = 2$$

$$6x - 2y = 4$$

we multiply the first equation by 2 to obtain the following system.

$$6x - 2y = 4$$
$$6x - 2y = 4.$$

The two equations are identical. So the solutions of this system are the same as the solutions of the single equation $6x - 2y = 4$, which can be rewritten as

$$2y = 6x - 4$$
$$y = 3x - 2.$$

This equation, and hence the original system, has infinitely many solutions. They can be described as follows: Choose any real number for x, say, $x = b$. Then $y = 3x - 2 = 3b - 2$. So the solutions of the system are all pairs of numbers of the form

$$x = b, \qquad y = 3b - 2, \quad \text{where } b \text{ is any real number.}$$

A system such as this is said to be **dependent.**   ∎

Some nonlinear systems can be solved by replacing them with equivalent linear systems.

EXAMPLE 7

Solve the following system:

$$\frac{1}{x} + \frac{3}{y} = -1$$
$$\frac{2}{x} - \frac{1}{y} = 5.$$

SOLUTION   We wish to transform this system into a linear one. We let $u = 1/x$ and $v = 1/y$ so that the system becomes

$$u + 3v = -1$$
$$2u - v = 5.$$

We can solve this system by multiplying the first equation by -2 and adding it to the second equation.

$$-2u - 6v = 2$$
$$\underline{2u - v = 5}$$
$$-7v = 7$$
$$v = -1.$$

Substituting $v = -1$ in the equation $u + 3v = -1$, we see that

$$u = -3v - 1 = -3(-1) - 1 = 2.$$

Consequently, the possible solution of the original system is

$$x = \frac{1}{u} = \frac{1}{2} \quad \text{and} \quad y = \frac{1}{v} = \frac{1}{(-1)} = -1.$$

You should substitute this possible solution in both equations of the original system to check that it is actually a solution of the system.   ∎

APPLICATIONS

Producers are happy to supply items at a high price, but if they do, the consumers' demand for them may be low. At lower prices, more items will be demanded, but if the price is too low, producers may not want to supply the items. The **equilibrium price** is the price at which the number of items demanded by consumers is the same as the number supplied by producers. The number of items demanded and supplied at the equilibrium price is the **equilibrium quantity.**

EXAMPLE 8

The consumer demand for a certain type of floor tile is related to its price by the equation $p = 60 - .75x$, where p is the price (in dollars) at which x thousand boxes of tile will be demanded. The supply of these tiles is related to their price by the equation $p = .8x + 5.75$, where p is the price at which x thousand boxes will be supplied by the producer. Find the equilibrium quantity and the equilibrium price.

SOLUTION   We must find the values of x and p that satisfy both the supply and demand equations. In other words, we must solve this system:

$$p = 60 - .75x$$
$$p = .8x + 5.75.$$

Algebraic Method.   Since both equations are already solved for p, we use substitution. Substituting the value of p given by the first equation into the second, we obtain

$$60 - .75x = .8x + 5.75$$

Subtract .8x from both sides: $60 - 1.55x = 5.75$

Subtract 60 from both sides: $-1.55x = -54.25$

Divide both sides by -1.55: $x = \dfrac{-54.25}{-1.55} = 35.$

We can determine p by substituting $x = 35$ in either equation, say, the first one.

$$p = 60 - .75x$$
$$p = 60 - .75(35) = 33.75.$$

Therefore, the equilibrium quantity is 35,000 boxes (x is measured in thousands), and the equilibrium price is \$33.75 per box.

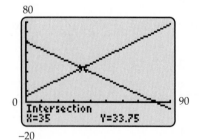

Figure 11–4

Graphical Method.   We graph the two equations in the form $y = 60 - .75x$ and $y = .8x + 5.75$ on the same screen and find their intersection point (Figure 11–4). The intersection point (35, 33.75) is called the **equilibrium point.** Its first coordinate is the equilibrium quantity, and its second coordinate is the equilibrium price.   ∎

EXAMPLE 9

575 people attend a ball game, and total ticket sales are \$2575. If adult tickets cost \$5 and children's tickets cost \$3, how many adults attended the game? How many children?

SOLUTION Let x be the number of adults, and let y be the number of children. Then,

$$\text{Number of adults} + \text{Number of children} = \text{Total attendance}$$

$$x + y = 575.$$

We can obtain a second equation by using the information about ticket sales.

$$\text{Adult ticket sales} + \text{Children ticket sales} = \text{Total ticket sales}$$

$$\begin{pmatrix} \text{Price} \\ \text{per} \\ \text{ticket} \end{pmatrix} \times \begin{pmatrix} \text{Number} \\ \text{of} \\ \text{adults} \end{pmatrix} + \begin{pmatrix} \text{Price} \\ \text{per} \\ \text{ticket} \end{pmatrix} \times \begin{pmatrix} \text{Number} \\ \text{of} \\ \text{children} \end{pmatrix} = 2575$$

$$5x + 3y = 2575.$$

To find x and y, we need only solve this system of equations:

$$x + y = 575$$
$$5x + 3y = 2575.$$

Multiplying the first equation by -3 and adding, we have

$$\begin{array}{rcl} -3x - 3y &=& -1725 \\ 5x + 3y &=& 2575 \\ \hline 2x &=& 850 \\ x &=& 425. \end{array}$$

So 425 adults attended the game. The number of children was

$$y = 575 - x = 575 - 425 = 150. \qquad \blacksquare$$

EXAMPLE 10

A plane flies 3000 miles from San Francisco to Boston in 5 hours, with a tailwind all the way. The return trip on the same route, now with a headwind, takes 6 hours. Assuming that both remain constant, find the speed of the plane and the speed of the wind.

SOLUTION Let x be the plane's speed, and let y be the wind speed (both in miles per hour). Then on the trip to Boston with a tailwind,

$x + y = $ actual speed of the plane (wind and plane go in same direction).

On the return trip against a headwind,

$x - y = $ actual speed of plane (wind and plane go in opposite directions).

Using the basic rate/distance equation, we have

Trip to Boston	**Return Trip**

time × rate = distance   time × rate = distance

$$5(x + y) = 3000 \qquad\qquad 6(x - y) = 3000$$

$$5x + 5y = 3000 \qquad\qquad 6x - 6y = 3000$$

$$x + y = 600 \qquad\qquad x - y = 500$$

Thus, we need only solve this system of equations:

$$x + y = 600$$
$$x - y = 500.$$

Adding the two equations shows that

$$2x = 1100$$
$$x = 550.$$

Substituting this result in the first equation, we have

$$550 + y = 600$$
$$y = 50.$$

Thus, the plane's speed is 550 mph, and the wind speed is 50 mph. ∎

EXAMPLE 11

How many pounds of tin and how many pounds of copper should be added to 1000 pounds of an alloy that is 10% tin and 30% copper to produce a new alloy that is 27.5% tin and 35% copper?

SOLUTION   Let x be the number of pounds of tin and y the number of pounds of copper to be added to the 1000 pounds of the old alloy. Then there will be $1000 + x + y$ pounds of the new alloy. We first find the *amounts* of tin and copper in the new alloy.

	Pounds in old alloy	+ Pounds added =	Pounds in new alloy
Tin	10% of 1000 +	x =	$100 + x$
Copper	30% of 1000 +	y =	$300 + y$

Now consider the *percentages* of tin and copper in the new alloy.

	Percentage in new alloy	×	Total weight of new alloy	= Pounds in new alloy
Tin	27.5%	of	$1000 + x + y$	= $.275(1000 + x + y)$
Copper	35%	of	$1000 + x + y$	= $.35(1000 + x + y)$

The two ways of computing the weight of each metal in the alloy must produce the same result, that is,

$$100 + x = .275(1000 + x + y) \quad \text{[pounds of tin]}$$

$$300 + y = .35(1000 + x + y). \quad \text{[pounds of copper]}$$

Multiplying out the right sides and rearranging terms produces this system of equations:

$$.725x - .275y = 175$$

$$-.35x + .65y = 50.$$

Multiplying the first equation by .65 and the second by .275 and adding the results, we have

$$.47125x - .17875y = 113.75$$

$$\underline{-.09625x + .17875y = 13.75}$$

$$.37500x = 127.50$$

$$x = 340.$$

Substituting this in the first equation and solving for y shows that $y = 260$. Therefore, 340 pounds of tin and 260 pounds of copper should be added. ■

EXERCISES 11.1

In Exercises 1–6, determine whether the given values of x, y, and z are a solution of the system of equations.

1. $x = -1, y = 3$

$2x + y = 1$

$-3x + 2y = 9$

2. $x = -2, y = 5$

$2x + 3y = 11$

$x - 2y = -12$

3. $x = 2, y = -1$

$\dfrac{1}{3}x + \dfrac{1}{2}y = \dfrac{1}{6}$

$\dfrac{1}{2}x + \dfrac{1}{3}y = \dfrac{2}{3}$

4. $x = .3, y = .7$

$4x - 1.2y = .36$

$3.1x + 2y = 4.7$

5. $x = \dfrac{1}{3}, y = 2, z = -1$

$3x - y + 2z = 1$

$4y - 3z = 5$

$3z = 3$

6. $x = 3, y = -\dfrac{5}{4}, z = \dfrac{3}{4}$

$4x - 6y + 10z = 27$

$4x + 8y - 28z = -19$

$x + 4y - 8z = -8$

In Exercises 7–14, use substitution to solve the system.

7. $x + 3y = 6$

$3x - y = 2$

8. $2x - 3y = 6$

$5x + 7y = 2$

9. $3x - 2y = 4$

$2x + y = -1$

10. $5x - 3y = 1$

$5x - 2y = -7$

11. $r + s = 0$

$r - s = 5$

12. $t = 7u - 4$

$t = 5u + 6$

13. $x + y = c + d$ (where c, d are constants)

$x - y = 2c - d$

14. $x - 2y = 2c + 5d$ (where c, d are constants)

$3x - y = c - 2d$

In Exercises 15–30, use the elimination method to solve the system.

15. $2x - 2y = 12$

$-2x + 3y = 10$

16. $4x - 3y = 10$

$7x + 4y = -1$

17. $x + 3y = -1$

$2x - y = 5$

18. $3x - 2y = -12$

$x + 5y = 13$

19. $2x + 3y = 15$

$8x + 12y = 40$

20. $2x - 3y = 7$

$6x - 9y = 1$

21. $4x - 5y = 2$

$12x - 15y = 6$

22. $5x - 2y = 1$

$15x - 6y = 3$

23. $\dfrac{x}{3} - \dfrac{y}{2} = -3$

$\dfrac{2x}{5} + \dfrac{y}{5} = -2$

24. $\dfrac{5x}{7} + \dfrac{3y}{7} = 1$

$x + \dfrac{2y}{3} = 1$

25. $\dfrac{x+y}{4} - \dfrac{x-y}{3} = 1$

$\dfrac{x+y}{4} + \dfrac{x-y}{2} = 9$

26. $\dfrac{x-y}{3} + \dfrac{x+y}{5} = \dfrac{2}{3}$

$\dfrac{x-3y}{4} - \dfrac{2x-y}{5} = \dfrac{3}{8}$

27. $\dfrac{1}{x} - \dfrac{3}{y} = 2$

$\dfrac{2}{x} + \dfrac{1}{y} = 3$

28. $\dfrac{3}{x} + \dfrac{2}{y} = -7$

$\dfrac{1}{x} - \dfrac{4}{y} = -14$

29. $\dfrac{2}{x} + \dfrac{3}{y} = 8$

$\dfrac{3}{x} - \dfrac{1}{y} = 1$

30. $\dfrac{3}{x-7} + \dfrac{2}{y-4} = 7$

$\dfrac{5}{x-7} - \dfrac{1}{y-4} = 3$

$\left[\text{Hint: Let } u = \dfrac{1}{x-7} \text{ and }\right.$

$\left. v = \dfrac{1}{y-4}.\right]$

31. The population y in year x of Philadelphia and Houston is approximated by these equations:

> Philadelphia: $7.96x + y = 1588.47$
>
> Houston: $-26.67x + y = 1644.64,$

where $x = 0$ corresponds to 1990 and y is in thousands.*

(a) How can you tell from the equation whether a city's population was increasing or decreasing since 1990?

(b) According to this model, in what year did the two cities have the same population?

32. The cost of bread, y, in dollars per pound, is approximated by these equations:*

> White Bread: $-.0071t + y = .982$
>
> Whole Wheat: $.014t + y = 1.360,$

where t is the number of years since 2000. According to this model, in what year will white bread and wheat bread cost the same?

33. On the basis of data from 2000–2004, the median income y in year x for men and women is approximated by these equations:

> Men: $135x + y = 31,065$
>
> Women: $-29.31x + 3y = 42,908$

where $x = 0$ corresponds to 2000 and y is in constant 2004 dollars.* If the equations remain valid in the future, when will the median income of men and women be the same?

34. The death rate per 100,000 population y in year x for heart disease and cancer is approximated by these equations:

> Heart Disease: $6.9x + 2y = 728.4$
>
> Cancer: $-1.3x + y = 167.5,$

where $x = 0$ corresponds to 1970.* If the equations remain accurate, when will the death rates for heart disease and cancer be the same?

In Exercises 35 and 36, solve the system of equations.

35. $ax + by = r$ (where a, b, c, d, r, s are

$cx + dy = s$ constants and $ad - bc \neq 0$)

36. $ax + by = ab$ (where a, b are nonzero constants)

$bx - ay = ab$

37. Let c be any real number. Show that this system has exactly one solution.

$$x + 2y = c$$
$$6x - 3y = 4.$$

38. (a) Find the values of c for which this system has an infinite number of solutions.

$$2x - 4y = 6$$
$$-3x + 6y = c.$$

(b) Find the values of c for which the system in part (a) has no solutions.

In Exercises 39 and 40, find the values of c and d for which both given points lie on the given straight line.

39. $cx + dy = 2$; $(0, 4)$ and $(2, 16)$

40. $cx + dy = -6$; $(1, 3)$ and $(-2, 12)$

In Exercises 41–44, find the equilibrium quantity and the equilibrium price. In the supply and demand equations, p is price (in dollars) and x is quantity (in thousands).

41. Supply: $p = .85x$
Demand: $p = 40 - 1.15x$

42. Supply: $p = 1.4x - .6$
Demand: $p = -2x + 3.2$

43. Supply: $p = 300 - 30x$
Demand: $p = 80 + 25x$

44. Supply: $p = 181 - .01x$
Demand: $p = 52 + .02x$

45. A 200-seat theater charges $8 for adults and $5 for children. If all seats were filled and the total ticket income was $1,435, how many adults and how many children were in the audience?

46. A theater charges $50 for main floor seats and $15 for balcony seats. If all seats are sold, the ticket income is $27,000. At one show, 25% of the main floor seats and 40% of the balcony seats were sold, and ticket income was $8,550. How many seats are on the main floor and how many are in the balcony?

---

47. A boat made a 4-mile trip upstream against a constant current in 15 minutes. The return trip at the same constant speed with the same current took 12 minutes. What is the speed of the boat and what is the speed of the current?

48. A plane flying into a headwind travels 2000 miles in 4 hours and 24 minutes. The return flight along the same route with a tailwind takes 4 hours. Find the wind speed and the plane's speed (assuming that both are constant).

49. At a certain store, cashews cost \$4.40/pound and peanuts cost \$1.20/pound. If you want to buy exactly 6 pounds of nuts for \$12 dollars, how many pounds of each kind of nuts should you buy?

50. How many cubic centimeters (cm^3) of a solution that is 20% acid and of another solution that is 45% acid should be mixed to produce 100 cm^3 of a solution that is 30% acid?

51. How many grams of a 50%-silver alloy should be mixed with a 75%-silver alloy to obtain 40 grams of a 60%-silver alloy?

52. A winemaker has two large casks of wine. One wine is 8% alcohol, and the other is 18% alcohol. How many liters of each wine should be mixed to produce 30 liters of wine that is 12% alcohol?

53. Bill and Ann plan to install a heating system for their swimming pool. Since gas is not available, they have a choice of electric or solar heat. They have gathered the following cost information.

System	Installation Costs	Monthly Operational Cost
Electric	\$2,000	\$80
Solar	\$14,000	\$9.50

(a) Ignoring changes in fuel prices, write a linear equation for each heating system that expresses its total cost y in terms of the number of *years x* of operation.

(b) What is the five-year total cost of electric heat? Of solar heat?

(c) In what year will the total cost of the two heating systems be the same? Which is the cheaper system before that time? After that time?

54. A toy company makes Boomie Babies, as well as collector cases for each Boomie Baby. To make x cases cost the company \$5000 in fixed overhead, plus \$7.50 per case. An outside supplier has offered to produce any desired volume of cases for \$8.20 per case.

(a) Write an equation that expresses the company's cost to make x cases itself.

(b) Write an equation that expresses the cost of buying x cases from the outside supplier.

(c) Graph both equations on the same axes and determine when the two costs are the same.

(d) When should the company make the cases themselves and when should they buy them from the outside supplier?

55. When Neil Simon planned to open his play *London Suite,* his producer, Emanuel Azenberg, made the following cost and revenue estimates for opening on Broadway or off Broadway.*

	On Broadway	Off Broadway
Initial cost to open[†]	\$1,295,000	\$440,000
Weekly costs[†]	\$206,500	\$82,000
Weekly revenue	\$250,500	\$109,000

For a production on Broadway that runs x weeks, find a linear equation that gives the

(a) total revenue R;
(b) total cost C;
(c) total profit P.
(d) Do parts (a)–(c) for an off-Broadway production.
(e) After how many weeks will the on-Broadway profit equal the off-Broadway profit? What is the profit then?
(f) When would it be better to open off Broadway rather than on Broadway?

56. A store sells their 80-GB iPods for \$350, and their 30-GB iPods for \$250. Their total iPod inventory would sell for \$15,250. During a recent month, the store actually sold half of the 80-GB iPods and two-thirds of their 30-GB iPods, taking in a total of \$9,000. How many of each kind did they have at the beginning of the month?

*Exercises 57–60 deal with the **break-even point,** which is the point at which revenue equals cost.*

57. The cost C of making x hedge trimmers is given by $C = 45x + 6000$. Each hedge trimmer can be sold for \$60.

(a) Find an equation that expresses the revenue R from selling x hedge trimmers.
(b) How many hedge trimmers must be sold for the company to break even?

58. The cost C of making x cases of pet food is given by $C = 100x + 2600$. Each case of 100 boxes sells for \$120.

(a) Find an equation that expresses the revenue from selling x cases.
(b) How many cases must be sold for the company to break even?

*Albert Goetz, "Basic Economics: Calculating against Theatrical Disaster," *Mathematics Teacher* 89, no. 1 (January 1996) © 1996. Reprinted with permission of the National Councils of Teachers of Mathematics. All rights reserved.
[†]Initial cost includes sets, costumes, rehearsals, etc. Weekly expenses include theater rent, salaries, advertising, etc.

59. A firm is developing a new product. The marketing department estimates that no more than 450 units can be sold. Costs for producing x units are expected to be given by $C = 80x + 7500$, and revenues are given by $R = 95x$. You are the manager who must determine whether or not to produce the product. What is your decision? Why?

60. Do Exercise 59 when costs are given by

$$C = 1750x + 96{,}175,$$

and revenue is given by $R = 1975x$; no more than 800 units can be produced.

61. An investor has part of her money in an account that pays 9% annual interest and the rest in an account that pays 11% annual interest. If she has $8000 less in the higher-paying account than in the lower-paying one and her total annual interest income is $2010, how much does she have invested in each account?

62. Joyce has money in two investment funds. Last year, the first fund paid a dividend of 8%, the second paid a dividend of 2%, and Joyce received a total of $780. This year, the first fund paid a 10% dividend, the second paid only 1%, and Joyce received $810. How much money does she have invested in each fund?

63. The table shows the percentages of men and women in the labor force who are college graduates in selected years.*

Year	1992	1995	1997	1998	1999	2000	2005	2006
Men	27.5	29.7	29.2	29.6	30.5	30.9	32.1	32.5
Women	25.0	26.6	28.0	28.6	29.5	29.8	32.8	33.5

(a) Use linear regression (with $x = 0$ corresponding to 1990) to find an equation that gives the percentage of men in the labor force in year x who are college graduates.

(b) Do part (a) for women.

(c) When did the percentage of women equal that of men?

64. The table shows the number of passenger cars (in thousands) imported into the United States from Japan and Canada in selected years.*

Year	1995	1996	1997	1998	1999	2000
Japan	1114.4	1190.9	1387.8	1456.1	1707.3	1839.1
Canada	1552.7	1690.7	1731.2	1837.6	2170.4	2138.8

(a) Use linear regression to find an equation that approximates the number y of cars imported from Japan in year x, with $x = 5$ corresponding to 1995.

(b) Do part (a) for Canada.

(c) If these models remain accurate, in what year will the imports from Japan and Canada be the same? Approximately how many cars will be imported that year?

65. A machine in a pottery factory takes 3 minutes to form a bowl and 2 minutes to form a plate. The material for a bowl costs $.25, and the material for a plate costs $.20. If the machine runs for 8 hours straight and exactly $44 is spent for material, how many bowls and plates can be produced?

66. Because Chevrolet and Saturn produce cars in the same price range, Chevrolet's sales are a function not only of Chevy prices (x), but of Saturn prices (y) as well. Saturn prices are related similarly to both Saturn and Chevy prices. Suppose General Motors forecasts the demand z_1 for Chevrolets and the demand z_2 for Saturns to be given by

$$z_1 = 68{,}000 - 6x + 4y \quad \text{and} \quad z_2 = 42{,}000 + 3x - 3y.$$

Solve this system of equations and express

(a) the price x of Chevrolets as a function of z_1 and z_2;

(b) the price y of Saturns as a function of z_1 and z_2.

---

*U.S. Bureau of Labor Statistics.

*U.S. Census Bureau, Foreign Trade Division.

11.1.A *SPECIAL TOPICS*   Systems of Nonlinear Equations

Section Objectives
- Solve systems of nonlinear equations algebraically and graphically.
- Use systems of nonlinear equations to solve applied problems.

Some systems that include nonlinear equations can be solved algebraically.

EXAMPLE 1

Solve the system

$$-2x + y = -1$$
$$xy = 3.$$

SOLUTION   Solve the first equation for y:

$$y = 2x - 1$$

and substitute this into the second equation:

$$xy = 3$$
$$x(2x - 1) = 3$$
$$2x^2 - x = 3$$
$$2x^2 - x - 3 = 0$$
$$(2x - 3)(x + 1) = 0$$
$$2x - 3 = 0 \quad \text{or} \quad x + 1 = 0$$
$$x = 3/2 \qquad\qquad x = -1.$$

Using the equation $y = 2x - 1$ to find the corresponding values of y, we see that

$$\text{If } x = \frac{3}{2}, \quad \text{then } y = 2\left(\frac{3}{2}\right) - 1 = 2.$$

$$\text{If } x = -1, \quad \text{then } y = 2(-1) - 1 = -3.$$

Therefore, the solutions of the system are $x = 3/2, y = 2$, and $x = -1, y = -3$. ∎

EXAMPLE 2

Solve the system

$$x^2 + y^2 = 8$$
$$x^2 - y = 6.$$

SOLUTION   Solve the second equation for y, obtaining $y = x^2 - 6$, and substitute this into the first equation.

$$x^2 + y^2 = 8$$
$$x^2 + (x^2 - 6)^2 = 8$$
$$x^2 + x^4 - 12x^2 + 36 = 8$$
$$x^4 - 11x^2 + 28 = 0$$
$$(x^2 - 4)(x^2 - 7) = 0$$
$$x^2 - 4 = 0 \quad \text{or} \quad x^2 - 7 = 0$$
$$x^2 = 4 \qquad\qquad x^2 = 7$$
$$x = \pm 2 \qquad\qquad x = \pm\sqrt{7}.$$

Using the equation $y = x^2 - 6$ to find the corresponding values of y, we find that the solutions of the system are

$$x = 2, y = -2; \quad x = -2, y = -2; \quad x = \sqrt{7}, y = 1;$$
$$x = -\sqrt{7}, y = 1.$$
∎

Algebraic techniques were successful in Examples 1 and 2 because substitution led to equations whose solutions could be found exactly. When this is not the case, graphical methods are needed. The solutions of a system of equations are the points that are on the graphs of all the equations in the system. They can be approximated with a graphical intersection finder.

EXAMPLE 3

Solve the system

$$y = x^4 - 4x^3 + 9x - 1$$
$$y = 3x^2 - 3x - 7.$$

SOLUTION   If you try substitution on this system, say, by substituting the expression for y from the first equation into the second, you obtain

$$x^4 - 4x^3 + 9x - 1 = 3x^2 - 3x - 7$$
$$x^4 - 4x^3 - 3x^2 + 12x + 6 = 0.$$

This fourth-degree equation cannot be readily solved algebraically, so a graphical approach is appropriate.

Graph both equations of the original system on the same screen. In the viewing window of Figure 11–5, the graphs intersect at three points. However, the graphs seem to be getting closer together as they run off the screen at the top right, which suggests that there may be another intersection point.

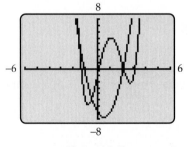

Figure 11–5

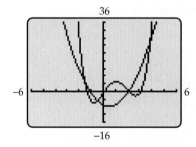

Figure 11–6

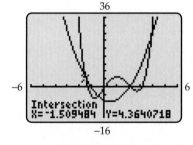

Figure 11–7

The larger window in Figure 11–6 shows four intersection points. There cannot be any more because, as we saw in the previous paragraph, the intersection points (solutions of the system) correspond to the solutions of a fourth-degree polynomial equation, which has a maximum of four solutions. An intersection finder (Figure 11–7) shows that one of the approximate solutions of the system is

$$x = -1.509484, \qquad y = 4.3640718.$$

GRAPHING EXPLORATION

Graph the two equations in the viewing window of Figure 11–6 and use your intersection finder to approximate the other three solutions of this system.

SYSTEMS WITH SECOND-DEGREE EQUATIONS

Solving systems of equations graphically depends on our ability to graph each equation in the system. With some equations of higher degree, this may require special techniques.

EXAMPLE 4

Solve this system graphically.

$$x^2 - 4x - y + 1 = 0$$

$$10x^2 + 25y^2 = 100.$$

SOLUTION It's easy to graph the first equation, since it can be rewritten as $y = x^2 - 4x + 1$. To graph the second equation, we must first solve for y.

$$10x^2 + 25y^2 = 100$$

$$25y^2 = 100 - 10x^2$$

$$y^2 = \frac{100 - 10x^2}{25}$$

$$y = \sqrt{\frac{100 - 10x^2}{25}} \quad \text{or} \quad y = -\sqrt{\frac{100 - 10x^2}{25}}.$$

TECHNOLOGY TIP

On most calculators, you can graph both of these functions at once by keying in

$$y = \{1, -1\}\sqrt{\frac{100 - 10x^2}{25}}.$$

By graphing both these functions on the same screen (see the Technology Tip), we obtain the complete graph of the equation $10x^2 + 25y^2 = 100$, as shown in Figure 11–8. The graphs of both equations in the system are shown in Figure 11–9.

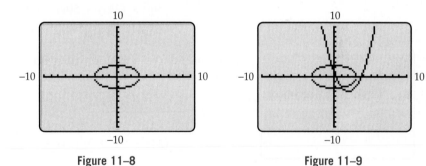

Figure 11–8   Figure 11–9

The two intersection points (solutions of the system) in Figure 11–9 can now be determined by an intersection finder.

$$x = -.2348, y = 1.9945 \quad \text{and} \quad x = .9544, y = -1.9067. \qquad \blacksquare$$

EXAMPLE 5

To solve the system

$$-4x^2 + 24xy + 3y^2 - 48 = 0$$

$$16x^2 + 24xy + 9y^2 + 100x + 50y + 100 = 0,$$

we must express each equation in terms of functions in order to graph it. The first equation may be rewritten as

$$3y^2 + (24x)y + (-4x^2 - 48) = 0.$$

This is a quadratic equation of the form $ay^2 + by + c = 0$, with

$$a = 3, \qquad b = 24x, \qquad c = -4x^2 - 48,$$

and hence can be solved by using the quadratic formula.

$$y = \frac{-b \pm \sqrt{b^2 - 4ac}}{2a}$$

$$y = \frac{-24x \pm \sqrt{(24x)^2 - 4 \cdot 3 \cdot (-4x^2 - 48)}}{2 \cdot 3}$$

$$= \frac{-24x \pm \sqrt{(24x)^2 - 12(-4x^2 - 48)}}{6}.$$

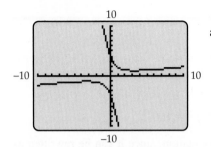

Figure 11–10

Consequently, the graph of the first equation can be obtained by graphing both of these functions on the same screen (Figure 11–10).

(∗)
$$y = \frac{-24x + \sqrt{(24x)^2 - 12(-4x^2 - 48)}}{6}.$$

$$y = \frac{-24x - \sqrt{(24x)^2 - 12(-4x^2 - 48)}}{6}.$$

TECHNOLOGY TIP

TI-84+/86 users can save keystrokes by entering the first equation of (∗) as y_1 and then using the RCL key to copy the text of y_1 to y_2. On TI-89, use COPY and PASTE in place of RCL. Then only one sign needs to be changed to make y_2 into the second equation to be graphed.

The second equation can also be solved for y by rewriting it as follows.

$$16x^2 + 24xy + 9y^2 + 100x + 50y + 100 = 0$$

$$9y^2 + (24xy + 50y) + (16x^2 + 100x + 100) = 0$$

$$9y^2 + (24x + 50)y + (16x^2 + 100x + 100) = 0.$$

Now apply the quadratic formula with $a = 9$, $b = 24x + 50$, and $c = 16x^2 + 100x + 100$.

$$y = \frac{-b \pm \sqrt{b^2 - 4ac}}{2a}$$

$$= \frac{-(24x + 50) \pm \sqrt{(24x + 50)^2 - 4 \cdot 9(16x^2 + 100x + 100)}}{2 \cdot 9}.$$

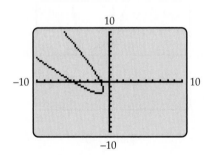

Figure 11–11

Thus, the graph of the second equation (Figure 11–11) consists of the graphs of these two functions.

(∗∗)
$$y = \frac{-(24x + 50) + \sqrt{(24x + 50)^2 - 36(16x^2 + 100x + 100)}}{18}.$$

$$y = \frac{-(24x + 50) - \sqrt{(24x + 50)^2 - 36(16x^2 + 100x + 100)}}{18}.$$

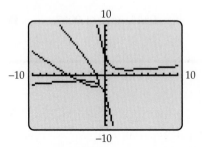

Figure 11–12

By graphing both equations (that is, all four functions shown in (∗) and (∗∗)), we obtain Figure 11–12. Then an intersection finder shows that the solutions of the system (points of intersection) are

$$x = -3.623, y = -1.113 \qquad \text{and} \qquad x = -.943, y = -1.833. \quad \blacksquare$$

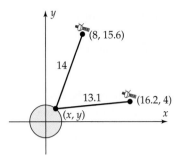

Figure 11–13

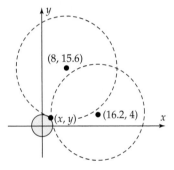

Figure 11–14

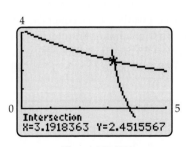

Figure 11–15

![icon] **APPLICATIONS**

EXAMPLE 6

Suppose the earth is a flat, circular disc of radius 4000 miles, centered at the origin of a two-dimensional coordinate system in which 1 unit represents 1000 miles. Figure 11–13 (which is not to scale) shows a GPS device at point (x, y) and two satellites at $(8, 15.6)$ and $(16.2, 4)$. Radio signals show that the distances of the satellites to (x, y) are 14 thousand and 13.1 thousand miles respectively, as shown in the figure. What are the coordinates of the device?

SOLUTION   The distance from (x, y) to $(8, 15.6)$ is 14. By the distance formula,

$$\sqrt{(x - 8)^2 + (y - 15.6)^2} = 14$$

Square both sides: $(x - 8)^2 + (y - 15.6)^2 = 14^2.$

Thus, (x, y) lies on the circle with center $(8, 15.6)$ and radius 14. Similarly, since the distance from (x, y) to $(16.2, 4)$ is 13.1,

$$\sqrt{(x - 16.2)^2 + (y - 4)^2} = 13.1$$

$$(x - 16.2)^2 + (y - 4)^2 = 13.1^2.$$

So (x, y) also lies on the circle with center $(16.2, 4)$ and radius 13.1.

Therefore, (x, y) is one of the intersection points of these two circles, as shown in Figure 11–14*. We must find this intersection point, that is, solve the system

$$(x - 8)^2 + (y - 15.6)^2 = 14^2$$

$$(x - 16.2)^2 + (y - 4)^2 = 13.1^2.$$

To solve this system graphically, we first solve each equation for y.

$$(x - 8)^2 + (y - 15.6)^2 = 14^2$$

$$(y - 15.6)^2 = 14^2 - (x - 8)^2$$

$$y - 15.6 = \pm\sqrt{14^2 - (x - 8)^2}$$

$$y = \sqrt{14^2 - (x - 8)^2} + 15.6 \quad \text{or} \quad y = -\sqrt{14^2 - (x - 8)^2} + 15.6.$$

A similar computation with the equation $(x - 16.2)^2 + (y - 4)^2 = 13.1^2$ shows

$$y = \sqrt{13.1^2 - (x - 16.2)^2} + 4 \quad \text{or} \quad y = -\sqrt{13.1^2 - (x - 16.2)^2} + 4.$$

Graphing the four preceding equations should produce something similar to Figure 11–14. Because of limited resolution, however, a graphing calculator will show only parts of the circles (try it!). Fortunately, we are interested only in the intersection point closest to the origin. Using an intersection finder in the partial graph in Figure 11–15, shows that this point is

$$(x, y) \approx (3.1918, 2.4516).$$ ∎

---

*In our actual three-dimensional world, the GPS location would be an intersection point of three or more spheres, i.e., a solution of a system of three or more second-degree equations in three variables.

EXAMPLE 7

A 52-foot-long piece of wire is to be cut into three pieces, two of which are the same length. The two equal pieces are to be bent into squares and the third piece into a circle. What should the length of each piece be if the total area enclosed by the two squares and the circle is 100 square feet?

SOLUTION   Let x be the length of each piece of wire that is to be bent into a square, and let y be the length of the piece that is to be bent into a circle. Since the original wire is 52 feet long,

$$x + x + y = 52 \qquad \text{or, equivalently,} \qquad y = 52 - 2x.$$

If a piece of wire of length x is bent into a square, the side of the square will have length $x/4$ and hence the area of the square will be $(x/4)^2 = x^2/16$. The remaining piece of wire will be made into a circle of circumference (length) y. Since the circumference is 2π times the radius (that is, $y = 2\pi r$), the circle has radius $r = y/2\pi$. Therefore, the area of the circle is

$$\pi r^2 = \pi\left(\frac{y}{2\pi}\right)^2 = \frac{\pi y^2}{4\pi^2} = \frac{y^2}{4\pi}.$$

The sum of the areas of the two squares and the circle is 100, that is,

$$\frac{x^2}{16} + \frac{x^2}{16} + \frac{y^2}{4\pi} = 100$$

$$\frac{y^2}{4\pi} = 100 - \frac{2x^2}{16}$$

$$y^2 = 4\pi\left(100 - \frac{x^2}{8}\right).$$

Therefore, the lengths x and y are solutions of this system of equations.

$$y = 52 - 2x$$

$$y^2 = 4\pi\left(100 - \frac{x^2}{8}\right).$$

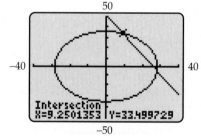

50

−40   40

Intersection
X=9.2501353 Y=33.499729

−50

Figure 11–16

The system may be solved either algebraically or graphically, using Figure 11–16. Since x and y are lengths, both must be positive. Consequently, we need only consider the intersection point in the first quadrant. An intersection finder shows that its coordinates are $x \approx 9.25$, $y \approx 33.50$. Therefore, the wire should be cut into two 9.25-foot pieces and one 33.5-foot piece. ■

EXERCISES 11.1.A

In Exercises 1–12, solve the system algebraically.

1. $x^2 - y = 0$
$-2x + y = 3$

2. $x^2 - y = 0$
$3x + y = 18$

3. $x^2 - y = 0$
$x + 2y = 5$

4. $x^2 - y = 0$
$x + 5y = 7$

5. $x + y = 10$
$xy = 21$

6. $2x + y = 5$
$xy = 3$

7. $xy + y^2 = 9$
$x - 2y = 6$

8. $xy + 3x^2 = -2$
$4x + y = 1$

9. $x^2 + y^2 - 4x - 4y = -4$
$x - y = 2$

10. $x^2 + y^2 - 4x - y = -1$
$x + 3y = 2$

11. $x^2 + y^2 = 25$
$x^2 + y = 19$

12. $x^2 + y^2 = 4$
$x^2 - y = 10$

In Exercises 13–28, solve the system by any means.

13. $y = x^3 - 3x^2 + 4$
$y = -.5x^2 + 3x - 2$

14. $y = -x^3 + 5x^2 + x - 4$
$y = 3x^2 - 2$

15. $y = x^3 - 3x + 2$
$y = \dfrac{3}{x^2 + 3}$

16. $y = .5x^4 + x^2 - 5$
$y = x^3 + 2x^2 - 3x + 2$

17. $y = x^3 + x + 1$
$y = \sin x$

18. $y = x^3 + 2x^2 - 1$
$y = \cos x$

19. $25x^2 - 16y^2 = 400$
$-9x^2 + 4y^2 = -36$

20. $5x^2 + 8y^2 = 100$
$2x^2 - 4y^2 = 1$

21. $3x^2 + 4y^2 - 18x + 14y = 1$
$x - y = 3$

22. $3x^2 + 6y^2 = 40$
$2x - y = 3$

23. $x^2 + 4xy + 4y^2 - 30x - 90y + 450 = 0$
$x^2 + x - y + 1 = 0$

24. $4x^2 + xy - 4y^2 - 2x + 3y + 50 = 0$
$x^2 - 3x - y = 0$

25. $4x^2 - 6xy + 2y^2 - 3x + 10y = 6$
$4x^2 + y^2 = 64$

26. $4x^2 + xy + y^2 - x + 2y - 3 = 0$
$3x^2 + 5xy + y^2 + 3x - 4y + 6 = 0$

27. $x^2 + 3xy + y^2 = 2$
$3x^2 - 5xy + 3y^2 = 7$

28. $2x^2 - 8xy + 8y^2 + 2x - 5 = 0$
$16x^2 - 24xy + 9y^2 + 100x - 200y + 100 = 0$

29. Suppose that the first satellite in Example 6 is located at (2.7, 15.9) and is 12.7 thousand miles from the GPS device. If the second satellite is at (13.9, 9.9) and is 13.3 thousand miles from the device, what are the coordinates of the device?

30. Internet sales of apparel (in billions of dollars) are projected to be given by

$$f(x) = .38x^2 - .35x + 4.85.$$

where $x = 2$ corresponds to 2002.* Similarly, sales of computer-related items are projected to be given by

$$g(x) = -.17x^2 + 2.73x + 2.71.$$

In what year were Internet sales of apparel and computer-related items at the same level?

31. A rectangular box (including top) with square ends and a volume of 16 cubic meters is to be constructed from 40 square meters of cardboard. What should its dimensions be?

32. A rectangular sheet of metal is to be rolled into a circular tube. If the tube is to have a surface area (excluding ends) of 210 square inches and a volume of 252 cubic inches, what size sheet of metal should be used?

33. Find two real numbers whose sum is -17 and whose product is 52.

34. Find two real numbers whose sum is 34.5 and whose product is 297.

35. Find two real numbers whose difference is 25.75 and whose product is 127.5.

36. Find two real numbers whose sum is 3 such that the sum of their squares is 369.

37. Find two real numbers whose sum is 2 such that the difference of their squares is 60.

38. Find the dimensions of a rectangular room whose perimeter is 58 feet and whose area is 204 square feet.

39. Find the dimensions of a rectangular room whose perimeter is 53 feet and whose area is 165 square feet.

40. A rectangle has area 120 square inches and a diagonal of length 17 inches. What are its dimensions?

41. A right triangle has area 225 square centimeters and a hypotenuse of length 35 centimeters. To the nearest tenth of a centimeter, how long are the legs of the triangle?

42. Find the equation of the straight line that intersects the parabola $y = x^2$ *only* at the point (3, 9). [*Hint:* What condition on the discriminant guarantees that a quadratic equation has exactly one real solution?]

---

*Based on data from Forrester Research.

11.2   Large Systems of Linear Equations

Section Objectives

■ Use Gaussian elimination to solve large systems of linear equations.

■ Use matrix methods to solve systems of linear equations.

■ Identify matrices that are in row echelon or reduced row echelon form.

■ Use the Gauss-Jordan method to solve systems of linear equations.

■ Identify inconsistent and dependent systems.

■ Use systems of linear equations to solve applied problems.

Systems of linear equations in three variables can be interpreted geometrically as the intersection of planes. However, algebraic methods are the only practical means to solve such systems or ones with more variables. Large systems can be solved by **Gaussian elimination,*** which is an extension of the elimination method used in Section 11.1.

Two systems of equations are said to be **equivalent** if they have the same solutions. The basic idea of Gaussian elimination is to transform a given system into an equivalent system that can easily be solved. There are several operations on a system of equations that leave the solutions to the system unchanged and, hence, produce an equivalent system. The first one is to

Interchange any two equations in the system,

which obviously won't affect the solutions of the system. The second is to

Multiply an equation in the system by a nonzero constant.

This does not change the solutions of the equation and therefore does not change the solutions of the system. To understand how the next operation works, we shall examine an earlier example from a different viewpoint.

EXAMPLE 1

In Example 3 of Section 11.1, we solved the system

$$x - 3y = 4$$
$$2x + y = 1$$

by multiplying the first equation by -2 and adding it to the second to eliminate the variable x.

$$
\begin{array}{ll}
-2x + 6y = -8 & \text{[-2 times first equation]} \\
\underline{2x + y = 1} & \text{[Second equation]} \\
 7y = -7. & \text{[Sum of second equation and -2 times first equation]}
\end{array}
$$

---

*Named after the German mathematician K. F. Gauss (1777–1855).

We then solved this last equation for y and substituted the answer, $y = -1$, in the original first equation to find that $x = 1$. What we did, in effect, was

> **Replace the original system by the following system, in which x has been eliminated from the second equation; then solve this new system.**

$(*)$
$$x - 3y = 4 \qquad \text{[First equation]}$$
$$7y = -7. \qquad \text{[Sum of second equation and } -2 \text{ times first equation]}$$

The solution of system $(*)$ is easily seen to be $y = -1$, $x = 1$, and you can readily verify that this is the solution of the original system. So the two systems are equivalent. [*Note:* We are not claiming that the second equations in the two systems have the same solutions—they don't—but only that the two *systems* have the same solutions.] ∎

Example 1 is an illustration of the third of the following operations.

Elementary Operations

> Performing any of the following operations on a system of equations produces an equivalent system.
>
> 1. Interchange any two equations in the system.
>
> 2. Replace an equation in the system by a nonzero constant multiple of itself.
>
> 3. Replace an equation in the system by the sum of itself and a constant multiple of another equation in the system.

The next example shows how elementary operations can be used to transform a system into an equivalent system that can be solved.

EXAMPLE 2

Solve the system

$$x + 4y - 3z = 1 \qquad \text{[Equation A]}$$
$$-3x - 6y + z = 3 \qquad \text{[Equation B]}$$
$$2x + 11y - 5z = 0. \qquad \text{[Equation C]}$$

SOLUTION We first use elementary operations to produce an equivalent system in which the variable x has been eliminated from the second and third equations.

To eliminate x from equation B, replace equation B by the sum of itself and 3 times equation A.

$$\begin{array}{rl} \text{[3 times } A\text{]} & 3x + 12y - 9z = 3 \\ \text{[}B\text{]} & -3x - 6y + z = 3 \\ \hline & 6y - 8z = 6 \end{array}$$

$$x + 4y - 3z = 1 \qquad \text{[}A\text{]}$$
$$6y - 8z = 6 \qquad \text{[Sum of } B \text{ and 3 times } A\text{]}$$
$$2x + 11y - 5z = 0. \qquad \text{[}C\text{]}$$

To eliminate x from equation C, we replace equation C by the sum of itself and -2 times equation A.

$$[-2 \text{ times } A] \quad -2x - 8y + 6z = -2$$
$$[C] \quad \underline{2x + 11y - 5z = \quad 0}$$
$$3y + \quad z = -2$$

$$x + 4y - 3z = \quad 1$$
$$6y - 8z = \quad 6$$
$$3y + \quad z = -2. \qquad [\text{Sum of } C \text{ and } -2 \text{ times } A]$$

The next step is to eliminate the y term in one of the last two equations. This can be done by replacing the second equation by the sum of itself and -2 times the third equation.

$$x + 4y - \quad 3z = \quad 1$$
$$-10z = \quad 10 \qquad \begin{array}{l}[\text{Sum of second equation and} \\ -2 \text{ times third equation}]\end{array}$$
$$3y + \quad z = -2.$$

Finally, interchange the last two equations.

$$(*) \qquad \begin{aligned} x + 4y - 3z &= \quad 1 \\ 3y + \quad z &= -2 \\ -10z &= \quad 10. \end{aligned}$$

This last system, which is equivalent to the original one, is easily solved. The last equation shows that

$$-10z = 10 \qquad \text{or, equivalently,} \qquad z = -1.$$

Substituting $z = -1$ in the second equation shows that

$$3y + z = -2$$
$$3y + (-1) = -2$$
$$3y = -1$$
$$y = -\frac{1}{3}.$$

Substituting $y = -1/3$ and $z = -1$ in the first equation yields

$$x + 4y - 3z = 1$$
$$x + 4\left(-\frac{1}{3}\right) - 3(-1) = 1$$
$$x = 1 + \frac{4}{3} - 3 = -\frac{2}{3}.$$

Therefore, the original system has just one solution: $x = -2/3$, $y = -1/3$, $z = -1$. ■

The process used to solve the final system $(*)$ in Example 2 is called **back substitution** because you begin with the last equation and work back to the first. It works because system $(*)$ is in **triangular form:** The first variable in the first equation, x, does not appear in any subsequent equation; the first variable in the second equation, y, does not appear in any subsequent equation, and so on. It can be shown that the procedure in Example 2 works in every case.

Gaussian Elimination

> Any system of linear equations can be transformed into an equivalent system in triangular form by using a finite number of elementary operations. If the system has solutions, they can then be found by back substitution in the triangular form system.

TECHNOLOGY TIP

To solve a system of linear equations on a calculator, use:

   TI-84+: POLYSMLT (APPS menu*)

   TI-86: SIMULT (Keyboard)

   Casio: EQUA (Main menu)

The TI-84+ PolySmlt solver can handle all systems (but see the Tip on page 800). Other solvers can handle only systems with the same number of equations as variables (when they display an error message, the system may have no solutions or it may have infinitely many solutions).

Most people prefer to use a calculator or computer to solve large systems of equations. However, the system solvers on some calculators are limited (see the Technology Tip in the margin). So we now develop a version of Gaussian elimination that works with all systems and is easily implemented on a calculator.

MATRIX METHODS

When solving systems by hand, a lot of time is wasted copying the x's, y's, z's, and so on. This fact suggests a shorthand system for representing a system of equations. For example, the system

$$
\begin{aligned}
x + 2y + 3z &= -2 \\
2x + 6y + z &= 2 \\
3x + 3y + 10z &= -2
\end{aligned}
$$

(*)

can be represented by the following rectangular array of numbers, consisting of the coefficients of the variables and the constants on the right of the equal sign, arranged in the same order in which they appear in the system.

$$
\begin{pmatrix}
1 & 2 & 3 & \vdots & -2 \\
2 & 6 & 1 & \vdots & 2 \\
3 & 3 & 10 & \vdots & -2
\end{pmatrix}
$$

TECHNOLOGY TIP

To enter and store a matrix in the matrix editor, use MAT(RIX), which is located here:

   TI: Keyboard

   HP-39gs: Keyboard

   Casio 9850: main menu.

This array is called the **augmented matrix** of the system. It has three horizontal **rows** and four vertical **columns.**

EXAMPLE 3

Use the matrix form of the preceding system (*) to solve the system.

SOLUTION   To solve the system in its original equation form, we would use elementary operations to eliminate the x terms from the last two equations, and then eliminate the y term from the last equation. With matrices, we do essentially the same thing, with the elementary operations on equations being replaced by the corresponding **row operations** on the augmented matrix in order to make certain

---

*If it's not in the APPS menu, it can be downloaded from TI.

entries in the first and second columns 0. Here is a side-by-side development of the two solution methods.

Equation Method	**Matrix Method**

Equation Method

Replace the second equation by the sum of itself and -2 times the first equation:

$$x + 2y + 3z = -2$$
$$2y - 5z = 6$$
$$3x + 3y + 10z = -2$$

Replace the third equation by the sum of itself and -3 times the first equation:

$$x + 2y + 3z = -2$$
$$2y - 5z = 6$$
$$-3y + z = 4$$

Multiply the second equation by $1/2$ (so that y has coefficient 1):

$$x + 2y + 3z = -2$$
$$y - \frac{5}{2}z = 3$$
$$-3y + z = 4$$

Replace the third equation by the sum of itself and 3 times the second equation:

$$x + 2y + 3z = -2$$
$$y - \frac{5}{2}z = 3$$
$$-\frac{13}{2}z = 13$$

Finally, multiply the last equation by $-2/13$:*

$$x + 2y + 3z = -2$$
(**) $$y - \frac{5}{2}z = 3$$
$$z = -2$$

Matrix Method

Replace the second row by the sum of itself and -2 times the first row:

$$\begin{pmatrix} 1 & 2 & 3 & | & -2 \\ 0 & 2 & -5 & | & 6 \\ 3 & 3 & 10 & | & -2 \end{pmatrix}$$

Replace the third row by the sum of itself and -3 times the first row:

$$\begin{pmatrix} 1 & 2 & 3 & | & -2 \\ 0 & 2 & -5 & | & 6 \\ 0 & -3 & 1 & | & 4 \end{pmatrix}$$

Multiply the second row by $1/2$:

$$\begin{pmatrix} 1 & 2 & 3 & | & -2 \\ 0 & 1 & -\frac{5}{2} & | & 3 \\ 0 & -3 & 1 & | & 4 \end{pmatrix}$$

Replace the third row by the sum of itself and 3 times the second row:

$$\begin{pmatrix} 1 & 2 & 3 & | & -2 \\ 0 & 1 & -\frac{5}{2} & | & 3 \\ 0 & 0 & -\frac{13}{2} & | & 13 \end{pmatrix}$$

Finally, multiply the last row by $-2/13$:

$$\begin{pmatrix} 1 & 2 & 3 & | & -2 \\ 0 & 1 & -\frac{5}{2} & | & 3 \\ 0 & 0 & 1 & | & -2 \end{pmatrix}$$

---

*This step isn't necessary, but it is often convenient to have 1 as the coefficient of the first variable in each equation.

System (∗∗) is easily solved. The third equation shows that $z = -2$, and substituting this in the second equation shows that

$$y - \frac{5}{2}(-2) = 3$$
$$y = 3 - 5 = -2.$$

Substituting $y = -2$ and $z = -2$ in the first equation yields

$$x + 2(-2) + 3(-2) = -2$$
$$x = -2 + 4 + 6 = 8.$$

Therefore, the only solution of the original system is $x = 8$, $y = -2$, $z = -2$. ∎

When using matrix notation, row operations replace elementary operations on equations, as shown in Example 3. The solution process ends when you reach a matrix, (and corresponding system), such as the final matrix in Example 1:

$$(\ast\ast) \qquad \begin{pmatrix} 1 & 2 & 3 & -2 \\ 0 & 1 & -\frac{5}{2} & 3 \\ 0 & 0 & 1 & -2 \end{pmatrix} \qquad \begin{array}{l} x + 2y + 3z = -2 \\ y - \frac{5}{2}z = 3 \\ z = -2. \end{array}$$

The final matrix should satisfy these conditions:

All rows consisting entirely of zeros (if any) are at the bottom.

The first nonzero entry in each nonzero row is a 1 (called a **leading 1**).

Each leading 1 appears to the right of leading 1's in any preceding rows.

Such a matrix is said to be in **row echelon form.**

Most calculators have a key that uses row operations to put a given matrix into row echelon form (see the Technology Tip in the margin). For example, using the TI-84+ REF key on the first matrix in Example 3 produced this row echelon matrix and corresponding system of equations.

$$\begin{pmatrix} 1 & 1 & \frac{10}{3} & -\frac{2}{3} \\ 0 & 1 & -\frac{17}{12} & \frac{5}{6} \\ 0 & 0 & 1 & -2 \end{pmatrix} \qquad \begin{array}{l} x + y + \frac{10}{3}z = -\frac{2}{3} \\ y - \frac{17}{12}z = \frac{5}{6} \\ z = -2. \end{array}$$

TECHNOLOGY TIP

To put a matrix in row echelon form, use REF in this menu/submenu:

TI-84+: MATRIX/MATH

TI-86: MATRIX/OPS

TI-89: MATH/MATRIX

Casio and HP-39gs users should consult the Technology Tip after Example 5.

Because the calculator used a different sequence of row operations than was used in Example 3, it produced a row echelon matrix (and corresponding system) that differs from matrix (∗∗) above. You can easily verify, however, that the preceding system has the same solutions as system (∗∗), namely,

$$x = 8, \quad y = -2, \quad z = -2.$$

EXAMPLE 4

Solve the system

$$\begin{aligned} x + y + 2z &= 1 \\ 2x + 4y + 5z &= 2 \\ 3x + 5y + 7z &= 4. \end{aligned}$$

SOLUTION   If you try to use a systems equation solver on a calculator, you may get an error message. So we form the augmented matrix and reduce it to row echelon form either by hand or by using the REF key. A TI-86 produced the row echelon matrix in Figure 11–17.

```
ref A▶Frac
   [[1 5/3 7/3 4/3]
    [0 1  1/2 1/2]
    [0 0  0   1  ]]
```

Figure 11–17

Look at the last row of the matrix in Figure 11–17; it represents the equation

$$0x + 0y + 0z = 1.$$

Since this equation has no solutions (the left side is always 0 and the right side is always 1), neither does the original system. Such a system is said to be **inconsistent.** ■

THE GAUSS-JORDAN METHOD

Gaussian elimination on a calculator is an efficient method of solving systems of equations but may involve some messy calculations when you solve the final triangular form system by hand. Most hand computations can be eliminated by using a slight variation, known as the **Gauss-Jordan method,** * which is illustrated in the next example.

EXAMPLE 5

Use the Gauss-Jordan method to solve this system.

$$x - y + 5z = -6$$
$$3x + 3y - z = 10$$
$$x - 5y + 8z = -17$$
$$x + 3y + 2z = 5.$$

SOLUTION   The augmented matrix of the system is shown in Figure 11–18, and an equivalent row echelon matrix (obtained by using the REF and FRAC keys) is shown in Figure 11–19.

$$\begin{pmatrix} 1 & -1 & 5 & -6 \\ 3 & 3 & -1 & 10 \\ 1 & -5 & 8 & -17 \\ 1 & 3 & 2 & 5 \end{pmatrix}$$

Figure 11–18

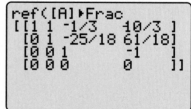

Figure 11–19

*This method was developed by the German engineer Wilhelm Jordan (1842–1899).

At this point in Gaussian elimination, we would use back substitution to solve the triangular form system represented by the last matrix in Figure 11–19. In the Gauss-Jordan method, however, additional elimination of variables replaces back substitution. Look at the leading 1 in the third row (shown in color).

$$\begin{pmatrix} 1 & 1 & -\dfrac{1}{3} & \dfrac{10}{3} \\ 0 & 1 & -\dfrac{25}{18} & \dfrac{61}{18} \\ 0 & 0 & 1 & -1 \\ 0 & 0 & 0 & 0 \end{pmatrix}.$$

In Gauss-Jordan elimination, we make the entries above this leading 1 into 0's.

Replace the second row by the sum of itself and 25/18 times the third row:
$$\begin{pmatrix} 1 & 1 & -\dfrac{1}{3} & \dfrac{10}{3} \\ 0 & 1 & 0 & 2 \\ 0 & 0 & 1 & -1 \\ 0 & 0 & 0 & 0 \end{pmatrix}$$

Replace the first row by the sum of itself and 1/3 times the third row:
$$\begin{pmatrix} 1 & 1 & 0 & 3 \\ 0 & 1 & 0 & 2 \\ 0 & 0 & 1 & -1 \\ 0 & 0 & 0 & 0 \end{pmatrix}$$

Now consider the leading 1 in the second row (shown in color), and make the entry above it 0.

Replace the first row by the sum of itself and -1 times the second row:
$$\begin{pmatrix} 1 & 0 & 0 & 1 \\ 0 & 1 & 0 & 2 \\ 0 & 0 & 1 & -1 \\ 0 & 0 & 0 & 0 \end{pmatrix}$$

This last matrix represents the following system, whose solution is obvious:

$$\begin{aligned} x &= 1 \\ y &= 2 \\ z &= -1. \end{aligned}$$

∎

TECHNOLOGY TIP

To put a matrix in reduced row echelon form, use RREF in this menu/submenu:

TI-84+: MATRIX/MATH

TI-86: MATRIX/OPS

TI-89: MATH/MATRIX

HP-39gs: MATH/MATRIX

RREF programs for Casio are in the Program Appendix.

A row echelon form matrix, such as the last one in Example 5, in which any column containing a leading 1 has 0's in all other positions, is said to be in **reduced row echelon form.** The goal in the Gauss-Jordan method is to use row operations to put a given augmented matrix into reduced row echelon form (from which the solutions can be read immediately, as in Example 5).

As a general rule, Gaussian elimination (matrix version) is the method of choice when working by hand, (the additional row operations needed to put a matrix in reduced row echelon form are usually more time-consuming—and error-prone—than back substitution). With a calculator or computer, however, it's better to find a reduced row echelon matrix for the system. You can do this in one step on a calculator by using the RREF key (see the Technology Tip in the margin).

EXAMPLE 6

Solve this system.

$$2x + 5y + z + 3w = 0$$
$$2y - 4z + 6w = 0$$
$$2x + 17y - 23z + 40w = 0.$$

SOLUTION   A system such as this, in which all the constants on the right side are zero, is called a **homogeneous system.** Every homogeneous system has at least one solution, namely, $x = 0$, $y = 0$, $z = 0$, $w = 0$, which is called the **trivial solution.** The issue with homogeneous systems is whether or not they have any nonzero solutions. The augmented matrix of the system is shown in Figure 11–20, and an equivalent reduced row echelon form matrix is shown in Figure 11–21.*

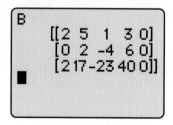

Figure 11–20   **Figure 11–21**

The system corresponding to the reduced echelon form matrix in Figure 11–21 is

$$x + \frac{11}{2}z = 0$$
$$y - 2z = 0$$
$$w = 0.$$

The second equation shows that

$$y = 2z.$$

This equation has an infinite number of solutions, for instance,

$$z = 1, y = 2 \quad \text{or} \quad z = 3, y = 6 \quad \text{or} \quad z = -2.5, y = -5.$$

In fact, for each real number t, there is a solution: $z = t$, $y = 2t$. Substituting $z = t$ into the first equation shows that

$$x + \frac{11}{2}t = 0$$
$$x = -\frac{11}{2}t.$$

Therefore, this system, and hence the original one, has an infinite number of solutions, one for each real number t.

$$x = -\frac{11}{2}t, \qquad y = 2t, \qquad z = t, \qquad w = 0.$$

A system with infinitely many solutions, such as this one, is said to be **dependent.**

■

TECHNOLOGY TIP

If the TI-84+ PolySmlt solver displays the message

    "No Solutions Found",

press the RREF key at the bottom of the screen to display the reduced row echelon matrix of the system, from which you can determine the solutions, if any.

*When dealing with homogeneous systems, it's not really necessary to include the last column of zeros, as is done here, because row operations do not change this column.

NOTE

Every system that has more variables than equations (as in Example 6) is dependent, but other systems may be dependent as well.

APPLICATIONS

In calculus, it is sometimes necessary to write a complicated rational expression as the sum of simpler ones. One technique for doing this involves systems of equations.

EXAMPLE 7

Find constants A, B, and C such that

$$\frac{2x^2 + 15x + 10}{(x - 1)(x + 2)^2} = \frac{A}{x - 1} + \frac{B}{x + 2} + \frac{C}{(x + 2)^2}.$$

SOLUTION   Multiply both sides of the equation by the common denominator $(x - 1)(x + 2)^2$ and collect like terms on the right side.

$$\begin{aligned}
2x^2 + 15x + 10 &= A(x + 2)^2 + B(x - 1)(x + 2) + C(x - 1) \\
&= A(x^2 + 4x + 4) + B(x^2 + x - 2) + C(x - 1) \\
&= Ax^2 + 4Ax + 4A + Bx^2 + Bx - 2B + Cx - C \\
&= (A + B)x^2 + (4A + B + C)x + (4A - 2B - C).
\end{aligned}$$

Since the polynomials on the left and right sides of the last equation are equal, their coefficients must be equal term by term, that is,

$$
\begin{aligned}
A + B \quad\quad &= 2 \quad\quad \text{[Coefficients of } x^2\text{]}\\
4A + B + C &= 15 \quad\quad \text{[Coefficients of } x\text{]}\\
4A - 2B - C &= 10. \quad\quad \text{[Constant terms]}
\end{aligned}
$$

We can consider this as a system of equations with unknowns A, B, C. The augmented matrix of the system is shown in Figure 11–22, and an equivalent reduced row echelon form matrix is shown in Figure 11–23.

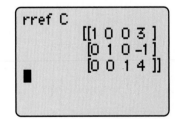

Figure 11–22   Figure 11–23

The solutions of the system can be read from the reduced row echelon form matrix in Figure 11–23.

$$A = 3, \quad\quad B = -1, \quad\quad C = 4.$$

Therefore,

$$\frac{2x^2 + 15x + 10}{(x - 1)(x + 2)^2} = \frac{3}{x - 1} + \frac{-1}{x + 2} + \frac{4}{(x + 2)^2}.$$

The right side of this equation is called the **partial fraction decomposition** of the fraction on the left side. ∎

EXAMPLE 8

Charlie is starting a small business and borrows $10,000 on three different credit cards, with annual interest rates of 18%, 15%, and 9%, respectively. He borrows three times as much on the 15% card as on the 18% card, and his total annual interest on all three cards is $1244.25. How much did he borrow on each credit card?

SOLUTION   Let x be the amount on the 18% card, let y be the amount on the 15% card, and let z be the amount on the 9% card. Then $x + y + z = 10,000$. Furthermore,

$$\underset{\text{18\% card}}{\text{Interest on}} + \underset{\text{15\% card}}{\text{Interest on}} + \underset{\text{9\% card}}{\text{Interest on}} = \underset{\text{interest}}{\text{Total}}$$

$$.18x \quad + \quad .15y \quad + \quad .09z \quad = 1244.25.$$

Finally, we have

$$\underset{\text{15\% card}}{\text{Amount on}} = \underset{\text{on 18\% card}}{\text{3 times amount}}$$

$$y = 3x,$$

which is equivalent to $3x - y = 0$. Therefore, we must solve this system of equations.

$$x + \quad y + \quad z = 10,000$$
$$.18x + .15y + .09z = \quad 1,244.25$$
$$3x - \quad y \qquad = \qquad 0,$$

whose augmented matrix is

$$\begin{pmatrix} 1 & 1 & 1 & 10,000 \\ .18 & .15 & .09 & 1,244.25 \\ 3 & -1 & 0 & 0 \end{pmatrix}.$$

CALCULATOR EXPLORATION

Enter this matrix in your calculator. Use row operations or the RREF key to put it in reduced row echelon form. Read the solutions of the system from this last matrix.

The Calculator Exploration shows that Charlie borrowed $1275 on the 18% card, $3825 on the 15% card, and $4900 on the 9% card.   ∎

The preceding examples illustrate the following fact, whose proof is omitted.

Number of Solutions of a System

> Any system of linear equations must have
>
> No solutions (an inconsistent system) *or*
>
> Exactly one solution *or*
>
> An infinite number of solutions (a dependent system).

EXERCISES 11.2

In Exercises 1–4, write the augmented matrix of the system.

1. $2x - 3y + 4z = 1$
$\quad x + 2y - 6z = 0$
$\quad 3x - 7y + 4z = -3$

2. $4x + y + z + 7w = 4$
$\quad x - 4y \quad - 3w = 0$
$\quad 5x \quad - 5z + 10w = -3$

3. $2x - \dfrac{5}{2}y + \dfrac{2}{3}z = 0$

$\quad x - \dfrac{1}{4}y + 4z = 0$

$\quad - 3y + \dfrac{1}{2}z = 0$

4. $x + 7y - \dfrac{2}{5}z + \dfrac{5}{6}w = 0$

$\quad \dfrac{1}{8}x - y - 8z \qquad = 1$

$\quad \dfrac{2}{3}y - 5z + \quad w = -2$

$\quad \dfrac{1}{6}x + 4y + \dfrac{2}{7}z \qquad = 3$

In Exercises 5–8, the augmented matrix of a system of equations is given. Express the system in equation notation.

5. $\begin{pmatrix} 3 & -5 & 4 \\ 9 & 7 & 2 \end{pmatrix}$

6. $\begin{pmatrix} 2 & -3 & 5 & 0 \\ 7 & 0 & -1 & 5 \end{pmatrix}$

7. $\begin{pmatrix} 1 & 0 & 1 & 0 & 1 \\ 1 & -1 & 4 & -2 & 3 \\ 4 & 2 & 5 & 0 & 2 \end{pmatrix}$

8. $\begin{pmatrix} -1 & 0 & 2 & 6 \\ 0 & 5 & -4 & 1 \\ 8 & -2 & 3 & 4 \end{pmatrix}$

In Exercises 9–12, the reduced row echelon form of the augmented matrix of a system of equations is given. Find the solutions of the system.

9. $\begin{pmatrix} 1 & 0 & 0 & 0 & 3/2 \\ 0 & 1 & 0 & 0 & 5 \\ 0 & 0 & 1 & 0 & -2 \\ 0 & 0 & 0 & 1 & 0 \end{pmatrix}$

10. $\begin{pmatrix} 1 & 0 & 0 & 0 & -1 \\ 0 & 1 & 0 & 0 & 3 \\ 0 & 0 & 1 & 0 & 1 \\ 0 & 0 & 0 & 0 & 1 \end{pmatrix}$

11. $\begin{pmatrix} 1 & 0 & 0 & 2 & 3 \\ 0 & 1 & 0 & 3 & 5 \\ 0 & 0 & 1 & 0 & 2 \\ 0 & 0 & 0 & 0 & 0 \end{pmatrix}$

12. $\begin{pmatrix} 1 & 0 & 0 & -2 \\ 0 & 1 & 0 & 3 \\ 0 & 0 & 1 & 0 \\ 0 & 0 & 0 & 0 \end{pmatrix}$

In Exercises 13–16, use Gaussian elimination to solve the system.

13. $-x + 3y + 2z = 0$
$\quad 2x - y - z = 3$
$\quad x + 2y + 3z = 0$

14. $2x - 3y + 2z = 8$
$\quad 3x + 2y + z = 3$
$\quad x + 2y + 3z = 1$

15. $x + y + z = 1$
$\quad x - 2y + 2z = 4$
$\quad 2x - y + 3z = 5$

16. $2x \quad - z = 3$
$\quad 8x + y + 4z = -1$
$\quad 4x + y + 6z = -7$

In Exercises 17–20, use the Gauss-Jordan method to solve the system.

17. $x - 2y + 4z = 6$
$\quad x + y + 13z = 6$
$\quad -2x + 6y - z = -10$

18. $5x - 2y - 3z = 31$
$\quad 2x + y - 7z = -10$
$\quad x + y + 2z = 3$

19. $x + y + z = 200$
$\quad x - 2y = 0$
$\quad 2x + 3y + 5z = 600$
$\quad 2x - y + z = 200$

20. $2x - 4y + z = 2$
$\quad x + y - 5z = 3$

In Exercises 21–36, solve the system.

21. $11x + 10y + 9z = 5$
$\quad x + 2y + 3z = 1$
$\quad 3x + 2y + z = 1$

22. $2x + y = -1$
$\quad x - 3y = 5$
$\quad 3x + 5y = -7$

23. $5x - y = 7$
$\quad x + y = 5$
$\quad 4x - 2y = 2$

24. $6x + 2y - z = 4$
$\quad 3x \quad - 2z = 0$
$\quad 3x - 8y - 14z = -1$

25. $x - 4y - 13z = 4$
$\quad x - 2y - 3z = 2$
$\quad -3x + 5y + 4z = 2$

26. $3x - y + 4z = -2$
$\quad 4x - 2y = 2$
$\quad 2x - y + 8z = 2$

27. $4x + y + 3z = 7$
$\quad x - y + 2z = 3$
$\quad 3x + 2y + z = 4$

28. $2x + 5y + 3z = -5$
$\quad 5x - 8y - 2z = -2$
$\quad x - 18y - 8z = 8$

29. $x + y + z = 0$
$\quad 3x - y + z = 0$
$\quad -5x - y + z = 0$

30. $3x - y + 4z = 0$
$\quad -x - y - 3z = 0$
$\quad 2x + y + 5z = 0$

31.
$$2x + y + 3z - 2w = -6$$
$$4x + 3y + z - w = -2$$
$$x + y + z + w = -5$$
$$-2x - 2y + 2z + 2w = -10$$

32.
$$x + 7y - z + 2w = 24$$
$$5x - 3y - 8z = 7$$
$$x + 4y + 7z + w = 6$$
$$3y + 4z - w = -2$$

33.
$$x - 2y - z - 3w = -3$$
$$-x + y + z = 0$$
$$4y + 3z - 2w = -1$$
$$2x - 2y + w = 1$$

34.
$$2x - 7y - 2z + 2w = -6$$
$$3x + 2y - z = 14$$
$$x + 4y + 7z + 3w = 4$$
$$x - 3y - 4z - w = 1$$

35.
$$\frac{3}{x} - \frac{1}{y} + \frac{4}{z} = -13$$
$$\frac{1}{x} + \frac{2}{y} - \frac{1}{z} = 12$$
$$\frac{4}{x} - \frac{1}{y} + \frac{3}{z} = -7$$
[*Hint:* Let $u = 1/x$, $v = 1/y$, $w = 1/z$.]

36.
$$\frac{1}{x+1} - \frac{2}{y-3} + \frac{3}{z-2} = 4$$
$$\frac{5}{y-3} - \frac{10}{z-2} = -5$$
$$\frac{-3}{x+1} + \frac{4}{y-3} - \frac{1}{z-2} = -2$$
[*Hint:* Let $u = 1/(x+1)$, $v = 1/(y-3)$, $w = 1/(z-2)$.]

Exercises 37–40, solve the system. [Note: *The REF and RREF keys on some calculators produce an error message when there are more rows than columns in a matrix, in which case you will have to solve the system by some other means.*]

37.
$$2x - y = 1$$
$$3x + y = 2$$
$$4x - 2y = 2$$
$$5x + 5y = 4$$

38.
$$x + y = 3$$
$$-x + 2y = 3$$
$$5x - y = 3$$
$$-7x + 5y = 3$$

39.
$$x + 2y = 3$$
$$2x + 3y = 4$$
$$3x + 4y = 5$$
$$4x + 5y = 6$$

40.
$$x - y = 2$$
$$x + y = 4$$
$$2x + 3y = 9$$
$$3x - 2y = 6$$

In Exercises 41–46, find the constants A, B, and C.

41. $\dfrac{4x}{(x-1)(x+3)} = \dfrac{A}{x-1} + \dfrac{B}{x+3}$

42. $\dfrac{1}{(x+1)(x-1)} = \dfrac{A}{x+1} + \dfrac{B}{x-1}$

43. $\dfrac{2x+1}{(x+2)(x-3)^2} = \dfrac{A}{x+2} + \dfrac{B}{x-3} + \dfrac{C}{(x-3)^2}$

44. $\dfrac{x^2 - x - 21}{(2x-1)(x^2+4)} = \dfrac{A}{2x-1} + \dfrac{Bx+C}{x^2+4}$

45. $\dfrac{5x^2 + 1}{(x+1)(x^2-x+1)} = \dfrac{A}{x+1} + \dfrac{Bx+C}{x^2-x+1}$

46. $\dfrac{x-2}{(x+4)(x^2+2x+2)} = \dfrac{A}{x+4} + \dfrac{Bx+C}{x^2+2x+2}$

47. Lillian borrows $10,000. She borrows some from her friend at 8% annual interest, twice as much as that from her bank at 9%, and the remainder from her insurance company at 5%. She pays a total of $830 in interest for the first year. How much did she borrow from each source?

48. An investor puts a total of $25,000 into three very speculative stocks. She invests some of it in Crystalcomp and $2000 more than one-half that amount in Flyboys. The remainder is invested in Zumcorp. Crystalcomp rises 16% in value, Flyboys rises 20%, and Zumcorp rises 18%. Her investment in the three stocks is now worth $29,440. How much was originally invested in each stock?

49. An investor has $70,000 invested in a mutual fund, bonds, and a fast food franchise. She has twice as much invested in bonds as in the mutual fund. Last year the mutual fund paid a 2% dividend, the bonds paid 10%, and the fast food franchise paid 6%; her dividend income was $4800. How much is invested in each of the three investments?

50. Tickets to a band concert cost $2 for children, $3 for teenagers, and $5 for adults. 570 people attended the concert and total ticket receipts were $1950. Three-fourths as many teenagers as children attended. How many children, adults, and teenagers attended?

51. The table shows the calories, sodium, and protein in one cup of various kinds of soup.

	Progresso Roasted Chicken Rotini	Healthy Choice Hearty Chicken	Campbell's Chunky Chicken Noodle
Calories	100	130	130
Sodium (mg)	970	480	880
Protein (g)	6	8	8

How many cups of each kind of soup should be mixed together to produce ten servings of soup, each of which provides 203 calories, 1190 milligrams of sodium, and

12.4 grams of protein? What is the serving size (in cups)? (*Hint:* In ten servings, there must be 2030 calories, 11,900 milligrams of sodium, and 124 grams of protein.)

52. The table shows the calories, sodium, and fat in one ounce of various snack foods (all produced by Planters).

	Cashews	Dry Roasted Honey Peanuts	Cajun Crunch Trail Mix
Calories	170	150	160
Sodium (mg)	115	110	270
Fat (g)	14	12	11

How many ounces of each kind of snack should be combined to produce ten servings, each of which provides 220 calories, 188 milligrams of sodium, and 17.4 grams of fat? What is the serving size?

53. Comfort Systems, Inc., sells three models of humidifiers. The bedroom model weighs 10 pounds and comes in an 8-cubic-foot box; the living-room model weighs 20 pounds and comes in an 8-cubic-foot box; the whole-house model weighs 60 pounds and comes in a 28-cubic-foot box. Each of the company's delivery vans has 248 cubic feet of space and can hold a maximum of 440 pounds. For a van to be as fully loaded as possible, how many of each model should it carry?

54. Peanuts cost $3 per pound, almonds cost $4 per pound, and cashews costs $8 per pound. How many pounds of each should be used to produce 140 pounds of a mixture costing $6 per pound, in which there are twice as many peanuts as almonds?

Exercises 55 and 56 deal with computer-aided tomography (CAT) scanners that take X-rays of body parts from different directions to create a picture of a cross section of the body.* The amount by which the X-ray energy decreases (measured in linear-attenuation units) indicates whether the X-ray has passed through healthy tissue, tumorous tissue, or bone, according to the following table.

Tissue Type	Linear-Attenuation Units
Healthy	.1625–.2977
Tumorous	.2679–.3930
Bone	.3857–.5108

**Exercises 55 and 56 are based on D. Jabon, G. Nord, B. W. Wilson, and P. Coffman, "Medical Applications of Linear Equations," *Mathematics Teacher 89*, no. 5 (May 1996).*

The body part being scanned is divided into cells. The total linear-attenuation value is the sum of the values for each cell the X-ray passes through. In the figure for Exercise 55, for example, let a, b, and c be the values for cells A, B, and C, respectively; then the attenuation value for X-ray 3 is b + c.

55. (a) In the figure, find the linear-attenuation values for X-rays 1 and 2.
   (b) If the total linear-attenuation values for X-rays 1, 2, and 3 are .75, .60, and .54, respectively, set up and solve a system of three equations in *a, b, c*.
   (c) What kind of tissue are cells *A, B,* and *C*?

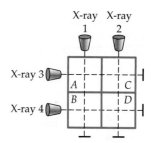

56. Four X-ray beams are aimed at four cells, as shown in the figure.

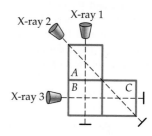

   (a) If the total linear-attenuation values for X-rays 1, 2, 3, and 4 are .60, .75, .65, and .70, respectively, is there enough information to determine the values of *a, b, c,* and *d*? Explain.
   (b) If an additional X-ray beam is added, with a linear-attenuation value of .85, as shown in the figure below, can the values of *a, b, c,* and *d* be determined? If so, what are they? What can be said about cells *A, B, C,* and *D*?

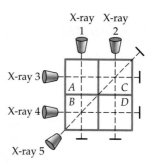

57. A furniture manufacturer has 1950 machine hours available each week in the cutting department, 1490 hours in the assembly department, and 2160 in the finishing department. Manufacturing a chair requires .2 hours of cutting, .3 hours of assembly, and .1 hours of finishing. A chest requires .5 hours of cutting, .4 hours of assembly, and .6 hours of finishing. A table requires .3 hours of cutting, .1 hours of assembly, and .4 hours of finishing. How many chairs, chests, and tables should be produced to use all the available production capacity?

58. A stereo equipment manufacturer produces three models of speakers, R, S, and T, and has three kinds of delivery vehicles: trucks, vans, and station wagons. A truck holds two boxes of model R, one of model S, and three of model T. A van holds one box of model R, three of model S, and two of model T. A station wagon holds one box of model R, three of model S, and one of model T. If 15 boxes of model R, 20 boxes of model S, and 22 boxes of model T are to be delivered, how many vehicles of each type should be used so that all operate at full capacity?

59. The diagram shows the traffic flow at four intersections during a typical one-hour period. The streets are all one-way, as indicated by the arrows. To adjust the traffic lights to avoid congestion, engineers must determine the possible values of x, y, z, and t.

(a) Write a system of linear equations that describes congestion-free traffic flow. [*Hint:* 600 cars per hour come down Euclid to intersection A, and 400 come down 4th Avenue to intersection A. Also, x cars leave intersection A on Euclid, and t cars leave on 4th Avenue. To avoid congestion, the number of cars leaving the intersection must be the same as the number entering, that is, $x + t = 600 + 400$. Use intersections B, C, and D to find three more equations.]

(b) Solve the system in part (a), which is dependent. Express your answers in terms of the variable t.

(c) Find the largest and smallest number of cars that can leave the given intersection on the given street: A on 4th Avenue, A on Euclid, C on 5th Avenue, and C on Chester.

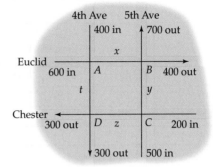

60. The diagram shows the traffic flow at four intersections during rush hour, as in Exercise 59.

(a) What are the possible values of x, y, z, and t in order to avoid any congestion? [Express your answers in terms of t.]

(b) What are the possible values of t?

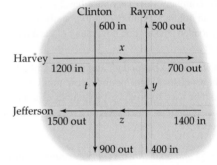

11.3 Matrix Methods for Square Systems

Section Objectives

■ Perform matrix multiplication.

■ Find the inverse of an invertible matrix.

■ Write certain systems of equations in matrix form and use matrix inverses to solve them.

■ Find the equation of a parabola, given three points on it.

Matrices were used in Section 11.2 as a convenient shorthand for solving systems of linear equations. We now consider matrices in a more general setting and show how the algebra of matrices provides an alternative method for solving systems of equations that are not dependent and have the same number of equations as variables.

Let m and n be positive integers. An $m \times n$ **matrix** (read "m by n matrix") is a rectangular array of numbers, with m horizontal rows and n vertical columns. For example,

$$\begin{pmatrix} 3 & 2 & -5 \\ 6 & 1 & 7 \\ -2 & 5 & 0 \end{pmatrix} \qquad \begin{pmatrix} -3 & 4 \\ 2 & 0 \\ 0 & 1 \\ 7 & 3 \\ 1 & -6 \end{pmatrix} \qquad \begin{pmatrix} 3 & 0 & 1 & 0 \\ \sqrt{2} & -\frac{1}{2} & 4 & \frac{8}{3} \\ 10 & 2 & -\frac{3}{4} & 12 \end{pmatrix} \qquad \begin{pmatrix} \sqrt{3} \\ 2 \\ 0 \\ 11 \end{pmatrix}$$

3×3 matrix	5×2 matrix	3×4 matrix	4×1 matrix
3 rows	5 rows	3 rows	4 rows
3 columns	2 columns	4 columns	1 column

In a matrix, the rows are horizontal and are numbered from top to bottom. The columns are vertical and are numbered from left to right. For example,

$$\begin{array}{l} \text{Row } 1 \rightarrow \\ \text{Row } 2 \rightarrow \\ \text{Row } 3 \rightarrow \end{array} \begin{pmatrix} 11 & 3 & 14 \\ -2 & 0 & -5 \\ \frac{1}{3} & 6 & 7 \end{pmatrix}$$

$$\begin{array}{ccc} \uparrow & \uparrow & \uparrow \\ \text{Column 1} & \text{Column 2} & \text{Column 3} \end{array}$$

Each entry in a matrix can be located by stating the row and column in which it appears. For instance, in the preceding 3×3 matrix, 14 is the entry in row 1, column 3, and 0 is the entry in row 2, column 2. When you enter a matrix on a calculator, the words "row" and "column" won't be displayed, but the row numbers will always be listed before the column number. Thus, a display such as "$A[3, 2]$," or simply "3, 2," indicates the entry in row 3, column 2.

Two matrices are said to be **equal** if they have the same size (same number of rows and columns) and the corresponding entries are equal. For example,

$$\begin{pmatrix} 3 & (-1)^2 \\ 6 & 12 \end{pmatrix} = \begin{pmatrix} 3 & 1 \\ \sqrt{36} & 12 \end{pmatrix}, \qquad \text{but} \qquad \begin{pmatrix} 6 & 4 \\ 5 & 1 \end{pmatrix} \neq \begin{pmatrix} 6 & 5 \\ 4 & 1 \end{pmatrix}.$$

MATRIX MULTIPLICATION

Although there is an extensive arithmetic of matrices, we shall need only matrix multiplication. The simplest case is the product of a matrix with a single row and a matrix with a single column, where the row and column have the same number of entries. This is done by multiplying corresponding entries (first by first, second by second, and so on) and then adding the results. An example is shown in Figure 11–24.

$$(3 \quad 1 \quad 2) \begin{pmatrix} 2 \\ 0 \\ 1 \end{pmatrix} = 3 \cdot 2 + 1 \cdot 0 + 2 \cdot 1 = 8$$

$$\begin{array}{ccc} \uparrow & \uparrow & \uparrow \\ \text{First} & \text{Second} & \text{Third} \\ \text{Terms} & \text{Terms} & \text{Terms} \end{array}$$

Figure 11–24

Note that the product of a row and a column is a single number.

Now let A be an $m \times n$ matrix, and let B be an $n \times p$ matrix, so that the number of columns of A is the same as the number of rows of B (namely, n). The product matrix AB is defined to be an $m \times p$ matrix (same number of rows as A and same number of columns as B). The product AB is defined as follows.

Matrix Multiplication

If A is an $m \times n$ and B is an $n \times p$ matrix, then AB is the $m \times p$ matrix whose entry in the ith row and jth column is

the product of the ith row of A and the jth column of B.

EXAMPLE 1

If it is defined, find the product AB, where

$$A = \begin{pmatrix} 3 & 1 & 2 \\ -1 & 0 & 4 \end{pmatrix} \quad \text{and} \quad B = \begin{pmatrix} 2 & -3 & 0 & 1 \\ 0 & 5 & 2 & 7 \\ 1 & 8 & -4 & 1 \end{pmatrix}.$$

SOLUTION A has 3 columns, and B has 3 rows. So the product matrix AB is defined. AB has 2 rows (same as A) and 4 columns (same as B). Its entries are calculated as follows. The entry in row 1, column 1 of AB is the product of row 1 of A and column 1 of B, which is the number 8, as shown in Figure 11–24 and indicated at the right here below.

row 1, column 1 $\begin{pmatrix} 3 & 1 & 2 \\ -1 & 0 & 4 \end{pmatrix} \begin{pmatrix} 2 & -3 & 0 & 1 \\ 0 & 5 & 2 & 7 \\ 1 & 8 & -4 & 1 \end{pmatrix} = \begin{pmatrix} 8 & & & \\ & & & \end{pmatrix}$ $3 \cdot 2 + 1 \cdot 0 + 2 \cdot 1 = 8$

The other entries in AB are obtained similarly.

row 1, column 2 $\begin{pmatrix} 3 & 1 & 2 \\ -1 & 0 & 4 \end{pmatrix} \begin{pmatrix} 2 & -3 & 0 & 1 \\ 0 & 5 & 2 & 7 \\ 1 & 8 & -4 & 1 \end{pmatrix} = \begin{pmatrix} 8 & 12 & & \\ & & & \end{pmatrix}$ $3(-3) + 1 \cdot 5 + 2 \cdot 8 = 12$

row 1, column 3 $\begin{pmatrix} 3 & 1 & 2 \\ -1 & 0 & 4 \end{pmatrix} \begin{pmatrix} 2 & -3 & 0 & 1 \\ 0 & 5 & 2 & 7 \\ 1 & 8 & -4 & 1 \end{pmatrix} = \begin{pmatrix} 8 & 12 & -6 & \\ & & & \end{pmatrix}$ $3 \cdot 0 + 1 \cdot 2 + 2(-4) = -6$

row 1, column 4 $\begin{pmatrix} 3 & 1 & 2 \\ -1 & 0 & 4 \end{pmatrix} \begin{pmatrix} 2 & -3 & 0 & 1 \\ 0 & 5 & 2 & 7 \\ 1 & 8 & -4 & 1 \end{pmatrix} = \begin{pmatrix} 8 & 12 & -6 & 12 \\ & & & \end{pmatrix}$ $3 \cdot 1 + 1 \cdot 7 + 2 \cdot 1 = 12$

row 2, column 1 $\begin{pmatrix} 3 & 1 & 2 \\ -1 & 0 & 4 \end{pmatrix} \begin{pmatrix} 2 & -3 & 0 & 1 \\ 0 & 5 & 2 & 7 \\ 1 & 8 & -4 & 1 \end{pmatrix} = \begin{pmatrix} 8 & 12 & -6 & 12 \\ 2 & & & \end{pmatrix}$ $(-1)2 + 0 \cdot 0 + 4 \cdot 1 = 2$

row 2, column 2 $\begin{pmatrix} 3 & 1 & 2 \\ -1 & 0 & 4 \end{pmatrix} \begin{pmatrix} 2 & -3 & 0 & 1 \\ 0 & 5 & 2 & 7 \\ 1 & 8 & -4 & 1 \end{pmatrix} = \begin{pmatrix} 8 & 12 & -6 & 12 \\ 2 & 35 & & \end{pmatrix}$ $(-1)(-3) + 0 \cdot 5 + 4 \cdot 8 = 35$

row 2, column 3 $\begin{pmatrix} 3 & 1 & 2 \\ -1 & 0 & 4 \end{pmatrix} \begin{pmatrix} 2 & -3 & 0 & 1 \\ 0 & 5 & 2 & 7 \\ 1 & 8 & -4 & 1 \end{pmatrix} = \begin{pmatrix} 8 & 12 & -6 & 12 \\ 2 & 35 & -16 & \end{pmatrix}$ $(-1)0 + 0 \cdot 2 + 4(-4) = -16$

row 2, column 4 $\begin{pmatrix} 3 & 1 & 2 \\ -1 & 0 & 4 \end{pmatrix} \begin{pmatrix} 2 & -3 & 0 & 1 \\ 0 & 5 & 2 & 7 \\ 1 & 8 & -4 & 1 \end{pmatrix} = \begin{pmatrix} 8 & 12 & -6 & 12 \\ 2 & 35 & -16 & 3 \end{pmatrix}$ $(-1)1 + 0 \cdot 7 + 4 \cdot 1 = 3$

The last matrix on the right is the product AB.   ∎

EXAMPLE 2

Let A, B, C, and D be the following matrices.

$$A = \begin{pmatrix} 3 & 2 & 1 \\ -2 & 0 & 4 \\ 1 & -2 & 5 \end{pmatrix}, \quad B = \begin{pmatrix} 5 & -2 \\ 1 & -1 \\ 4 & 2 \end{pmatrix}, \quad C = \begin{pmatrix} 4 & -2 & 7 \\ 6 & 3 & 1 \\ 2 & 1 & 4 \end{pmatrix},$$

$$D = \begin{pmatrix} 1 & 0 & -5 \\ 2 & 3 & -4 \\ 3 & 7 & 2 \end{pmatrix}$$

Find each of the following matrices, if possible.

(a) AB (b) BC (c) CD and DC

SOLUTION

(a) Following the same procedure as in Example 1, we have

$$AB = \begin{pmatrix} 3 & 2 & 1 \\ -2 & 0 & 4 \\ 1 & -2 & 5 \end{pmatrix} \begin{pmatrix} 5 & -2 \\ 1 & -1 \\ 4 & 2 \end{pmatrix}$$

$$= \begin{pmatrix} 3 \cdot 5 + 2 \cdot 1 + 1 \cdot 4 & 3(-2) + 2(-1) + 1 \cdot 2 \\ (-2) \cdot 5 + 0 \cdot 1 + 4 \cdot 4 & (-2)(-2) + 0(-1) + 4 \cdot 2 \\ 1 \cdot 5 + (-2) \cdot 1 + 5 \cdot 4 & 1(-2) + (-2)(-1) + 5 \cdot 2 \end{pmatrix}$$

$$= \begin{pmatrix} 21 & -6 \\ 6 & 12 \\ 23 & 10 \end{pmatrix}.$$

(b) Matrix B is 3×2, and C is 3×3. Since the number of columns in B is different from the number of rows in C, the product is not defined.

(c) We use the matrix editor of a calculator to enter the matrices (Figure 11–25).

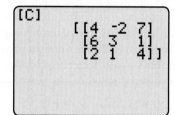
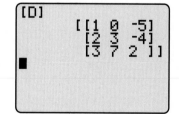

Figure 11–25

We then use the calculator to compute both products (Figure 11–26).

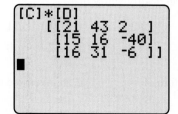
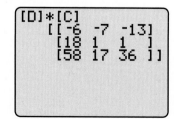

Figure 11–26

Note that DC is not equal to CD.

It can be shown that matrix multiplication is associative, meaning that $A(BC) = (AB)C$ for all matrices A, B, C for which the products are defined. As we saw in Example 2, however, matrix multiplication is *not* commutative, that is, AB might not be equal to BA, even when both products are defined.

TECHNOLOGY TIP

To display an $n \times n$ identity matrix, use IDENT(ITY)n or IDENMATn in this menu/submenu:

TI-84+: MATRIX/MATH

TI-86: MATRIX/OPS

TI-89: MATH/MATRIX

Casio: OPTN/MAT

HP-39gs: MATH/MATRIX

IDENTITY MATRICES AND INVERSES

The $n \times n$ **identity matrix** I_n is the matrix with 1's on the diagonal from upper left to lower right and 0's everywhere else; for example,

$$I_2 = \begin{pmatrix} 1 & 0 \\ 0 & 1 \end{pmatrix}, \qquad I_3 = \begin{pmatrix} 1 & 0 & 0 \\ 0 & 1 & 0 \\ 0 & 0 & 1 \end{pmatrix}, \qquad I_4 = \begin{pmatrix} 1 & 0 & 0 & 0 \\ 0 & 1 & 0 & 0 \\ 0 & 0 & 1 & 0 \\ 0 & 0 & 0 & 1 \end{pmatrix}.$$

The number 1 is the multiplicative identity of the number system because $a \cdot 1 = a = 1 \cdot a$ for every number a. The identity matrix I_n is the multiplicative identity for $n \times n$ matrices.

Identity Matrix

For any $n \times n$ matrix A,

$$AI_n = A = I_nA.$$

For example, in the 2×2 case,

$$\begin{pmatrix} a & b \\ c & d \end{pmatrix}\begin{pmatrix} 1 & 0 \\ 0 & 1 \end{pmatrix} = \begin{pmatrix} a \cdot 1 + b \cdot 0 & a \cdot 0 + b \cdot 1 \\ c \cdot 1 + d \cdot 0 & c \cdot 0 + d \cdot 1 \end{pmatrix} = \begin{pmatrix} a & b \\ c & d \end{pmatrix}.$$

Verify that the same answer results if you reverse the order of multiplication.

Every nonzero number c has a multiplicative inverse $c^{-1} = 1/c$ with the property that $cc^{-1} = 1$. The analogous statement for matrix multiplication does not always hold, and special terminology is used when it does. An $n \times n$ matrix A is said to be **invertible** (or **nonsingular**) if there is an $n \times n$ matrix B such that $AB = I_n$. In this case, it can be proved that $BA = I_n$ also. The matrix B is called the **inverse** of A and is sometimes denoted A^{-1}.

EXAMPLE 3

You can readily verify that

$$\begin{pmatrix} 2 & 1 \\ 3 & 1 \end{pmatrix}\begin{pmatrix} -1 & 1 \\ 3 & -2 \end{pmatrix} = \begin{pmatrix} 1 & 0 \\ 0 & 1 \end{pmatrix} = \begin{pmatrix} -1 & 1 \\ 3 & -2 \end{pmatrix}\begin{pmatrix} 2 & 1 \\ 3 & 1 \end{pmatrix}.$$

Therefore, $A = \begin{pmatrix} 2 & 1 \\ 3 & 1 \end{pmatrix}$ is an invertible matrix with inverse

$$A^{-1} = \begin{pmatrix} -1 & 1 \\ 3 & -2 \end{pmatrix}.$$ ∎

EXAMPLE 4

Find the inverse of the matrix $\begin{pmatrix} 2 & 6 \\ 1 & 4 \end{pmatrix}$.

SOLUTION   We must find numbers x, y, u, v such that

$$\begin{pmatrix} 2 & 6 \\ 1 & 4 \end{pmatrix}\begin{pmatrix} x & u \\ y & v \end{pmatrix} = \begin{pmatrix} 1 & 0 \\ 0 & 1 \end{pmatrix},$$

which is the same as

$$\begin{pmatrix} 2x + 6y & 2u + 6v \\ x + 4y & u + 4v \end{pmatrix} = \begin{pmatrix} 1 & 0 \\ 0 & 1 \end{pmatrix}.$$

Since corresponding entries in these last two matrices are equal, finding x, y, u, v amounts to solving these systems of equations:

$$\begin{array}{ccc} 2x + 6y = 1 & & 2u + 6v = 0 \\ x + 4y = 0 & \text{and} & u + 4v = 1. \end{array}$$

We shall solve the systems by the Gauss-Jordan method of Section 11.2. The augmented matrices for the two systems are

$$A = \begin{pmatrix} 2 & 6 & \vdots & 1 \\ 1 & 4 & \vdots & 0 \end{pmatrix} \quad \text{and} \quad B = \begin{pmatrix} 2 & 6 & \vdots & 0 \\ 1 & 4 & \vdots & 1 \end{pmatrix}.$$

Note that the row operations that are needed for both A and B will be the same (because the first two columns are the same in both A and B). Consequently, we can save space and time by combining both of these matrix into this single matrix.

$$\begin{pmatrix} 2 & 6 & \vdots & 1 & 0 \\ 1 & 4 & \vdots & 0 & 1 \end{pmatrix}.$$

The first three columns of the last matrix are matrix A, and the first two and last columns are matrix B. Performing row operations on this matrix amounts to doing the operations simultaneously on both A and B.

Multiply row 1 by $1/2$: $\begin{pmatrix} 1 & 3 & \vdots & \frac{1}{2} & 0 \\ 1 & 4 & \vdots & 0 & 1 \end{pmatrix}.$

Replace row 2 by the sum of itself and $+1$ times row 1: $\begin{pmatrix} 1 & 3 & \vdots & \frac{1}{2} & 0 \\ 0 & 1 & \vdots & -\frac{1}{2} & 1 \end{pmatrix}$

Replace row 1 by the sum of itself and -3 times row 2: $\begin{pmatrix} 1 & 0 & \vdots & 2 & -3 \\ 0 & 1 & \vdots & -\frac{1}{2} & 1 \end{pmatrix}$

The first three columns of the last matrix show that $x = 2$ and $y = -1/2$. Similarly, the first two and last columns show that $u = -3$ and $v = 1$. Therefore,

$$A^{-1} = \begin{pmatrix} 2 & -3 \\ -\frac{1}{2} & 1 \end{pmatrix}.$$

Observe that A^{-1} is just the right half of the final form of the preceding augmented matrix and that the left half is the identity matrix I_2. ∎

TECHNOLOGY TIP

On calculators other than TI-89, you can find the inverse A^{-1} of matrix A by keying in A (or Mat A) and using the x^{-1} key. Using the $\wedge$ key and -1 produces A^{-1} on TI-89 and HP-39gs but leads to an error message on other calculators.

Although the technique in Example 4 can be used to find the inverse of any matrix that has one, it's quicker to use a calculator. Any calculator with matrix capabilities can find the inverse of an invertible matrix (see the Technology Tip in the margin).

CAUTION

A calculator should produce an error message when asked for the inverse of a matrix A that does not have one. However, because of round-off errors, it may sometimes display a matrix that it says is A^{-1}. As an accuracy check when finding inverses, multiply A by A^{-1} to see whether the product is the identity matrix. If it isn't, A does not have an inverse.

 INVERSE MATRICES AND SYSTEMS OF EQUATIONS

Any system of linear equations can be expressed in matrix form, as shown in the next example.

EXAMPLE 5

Use matrix multiplication to express this system of equations in matrix form.

$$\begin{aligned} x + y + z &= 2 \\ 2x + 3y &= 5 \\ x + 2y + z &= -1. \end{aligned}$$

SOLUTION   Let A be the 3×3 matrix of coefficients on the left side of the equations, let B be the column matrix of constants on the right side, and let X be the column matrix of unknowns.

$$A = \begin{pmatrix} 1 & 1 & 1 \\ 2 & 3 & 0 \\ 1 & 2 & 1 \end{pmatrix}, \qquad X = \begin{pmatrix} x \\ y \\ z \end{pmatrix}, \qquad B = \begin{pmatrix} 2 \\ 5 \\ -1 \end{pmatrix}.$$

Then AX is a matrix with three rows and one column, as is B.

$$AX = \begin{pmatrix} 1 & 1 & 1 \\ 2 & 3 & 0 \\ 1 & 2 & 1 \end{pmatrix}\begin{pmatrix} x \\ y \\ z \end{pmatrix} = \begin{pmatrix} x + y + z \\ 2x + 3y + 0z \\ x + 2y + z \end{pmatrix} \quad \text{and} \quad B = \begin{pmatrix} 2 \\ 5 \\ -1 \end{pmatrix}.$$

The entries in AX are just the left sides of the equations of the system, and the entries in B are the constants on the right sides. Therefore, the system can be expressed as the matrix equation $AX = B$. ∎

Suppose a system of equations is written in matrix form $AX = B$ and that the matrix A has an inverse. Then we can solve $AX = B$ by multiplying both sides by A^{-1}.

$$A^{-1}(AX) = A^{-1}B$$

$$(A^{-1}A)X = A^{-1}B \qquad \text{[Matrix multiplication is associative]}$$

$$I_nX = A^{-1}B \qquad \text{[$A^{-1}A$ is the identity matrix]}$$

$$X = A^{-1}B \qquad \text{[Product of identity matrix and X is X]}$$

The next example shows how this works in practice.

EXAMPLE 6

Solve the system

$$x + y + z = 2$$
$$2x + 3y \quad\ \ = 5$$
$$x + 2y + z = -1.$$

SOLUTION As we saw in Example 5, this system is equivalent to the matrix equation

$$AX = B$$

$$\begin{pmatrix} 1 & 1 & 1 \\ 2 & 3 & 0 \\ 1 & 2 & 1 \end{pmatrix} \begin{pmatrix} x \\ y \\ z \end{pmatrix} = \begin{pmatrix} 2 \\ 5 \\ -1 \end{pmatrix}.$$

Use a calculator to find the inverse of the coefficient matrix A and multiply both sides of the equation by A^{-1}.

$$A^{-1}AX = A^{-1}B$$

$$\begin{pmatrix} 1.5 & .5 & -1.5 \\ -1 & 0 & 1 \\ .5 & -.5 & .5 \end{pmatrix} \begin{pmatrix} 1 & 1 & 1 \\ 2 & 3 & 0 \\ 1 & 2 & 1 \end{pmatrix} \begin{pmatrix} x \\ y \\ z \end{pmatrix} = \begin{pmatrix} 1.5 & .5 & -1.5 \\ -1 & 0 & 1 \\ .5 & -.5 & .5 \end{pmatrix} \begin{pmatrix} 2 \\ 5 \\ -1 \end{pmatrix}$$

$$\begin{pmatrix} 1 & 0 & 0 \\ 0 & 1 & 0 \\ 0 & 0 & 1 \end{pmatrix} \begin{pmatrix} x \\ y \\ z \end{pmatrix} = \begin{pmatrix} 1.5 & .5 & -1.5 \\ -1 & 0 & 1 \\ .5 & -.5 & .5 \end{pmatrix} \begin{pmatrix} 2 \\ 5 \\ -1 \end{pmatrix} \qquad \text{[Since } A^{-1}A = I_3\text{]}$$

$$\begin{pmatrix} x \\ y \\ z \end{pmatrix} = \begin{pmatrix} 7 \\ -3 \\ -2 \end{pmatrix} \qquad \text{[Since } I_3X = X\text{]}$$

Therefore, the solution of the original system is $x = 7$, $y = -3$, $z = -2$. ∎

Only a matrix with the same number of rows as columns can possibly have an inverse. Consequently, the method of Example 6 can be tried only when the system has the same number of equations as unknowns. In this case, you should use your calculator to verify that the coefficient matrix actually has an inverse (see the Caution on page 812). If it does not, other methods must be used. Here is a summary of the possibilities.

Matrix Solution of a System of Equations

Suppose a system with the same number of equations as unknowns is written in matrix form as $AX = B$.

If the matrix A has an inverse, then the unique solution of the system is

$$X = A^{-1}B.$$

If A does not have an inverse, then the system either has no solutions or has infinitely many solutions. Its solutions (if any) may be found by using Gauss-Jordan elimination (Section 11.2).

EXAMPLE 7

Solve the system

$$2x + y - z = 2$$
$$x + 3y + 2z = 1$$
$$x + y + z = 2.$$

SOLUTION   Since there are the same number of equations as unknowns, we can try the method of Example 6. In this case, we have

$$A = \begin{pmatrix} 2 & 1 & -1 \\ 1 & 3 & 2 \\ 1 & 1 & 1 \end{pmatrix} \quad \text{and} \quad B = \begin{pmatrix} 2 \\ 1 \\ 2 \end{pmatrix}.$$

CALCULATOR EXPLORATION

Verify that the matrix A does have an inverse. Show that the solutions of the system are $x = 2, y = -1, z = 1$ by computing $A^{-1}B$.

APPLICATIONS

Just as two points determine a unique line, three points (that aren't on the same line) determine a unique parabola, as the next example demonstrates.

EXAMPLE 8

Find the equation of the parabola that passes through the points $(-1, 6)$, $(3, -2)$, and $(4, 1)$.

SOLUTION   As we saw in Section 4.1, a parabola is the graph of an equation of the form

$$y = ax^2 + bx + c$$

for some constants a, b, and c. Since $(-1, 6)$ is on the graph, we know that when $x = -1$, then $y = 6$.

$$ax^2 + bx + c = y$$

Let $x = -1$ and $y = 6$: $a(-1)^2 + b(-1) + c = 6$

$$a - b + c = 6. \qquad \text{[Equation 1]}$$

Similarly, since $(3, -2)$ is on the graph, we have

$$ax^2 + bx + c = y$$

Let $x = 3$ and $y = -2$: $a(3^2) + b(3) + c = -2$

$$9a + 3b + c = -2. \qquad \text{[Equation 2]}$$

Finally, since (4, 1) is on the graph,

$$ax^2 + bx + c = y$$

Let $x = 4$ and $y = 1$: $$a(4^2) + b(4) + c = 1$$

$$16a + 4b + c = 1. \quad \text{[Equation 3]}$$

We can determine a, b, and c by solving the system determined by Equations 1–3.

$$\begin{aligned} a - b + c &= 6 \\ 9a + 3b + c &= -2 \\ 16a + 4b + c &= 1 \end{aligned}$$

or, in matrix form,

$$\begin{pmatrix} 1 & -1 & 1 \\ 9 & 3 & 1 \\ 16 & 4 & 1 \end{pmatrix} \begin{pmatrix} a \\ b \\ c \end{pmatrix} = \begin{pmatrix} 6 \\ -2 \\ 1 \end{pmatrix}.$$

A calculator shows that the solution is

$$\begin{pmatrix} a \\ b \\ c \end{pmatrix} = \begin{pmatrix} 1 & -1 & 1 \\ 9 & 3 & 1 \\ 16 & 4 & 1 \end{pmatrix}^{-1} \begin{pmatrix} 6 \\ -2 \\ 1 \end{pmatrix} = \begin{pmatrix} 1 \\ -4 \\ 1 \end{pmatrix}.$$

Using these values for a, b, and c, we obtain the equation of the parabola.

$$y = ax^2 + bx + c$$

Let $a = 1$, $b = -4$, and $c = 1$: $$y = x^2 - 4x + 1.$$ ■

EXAMPLE 9

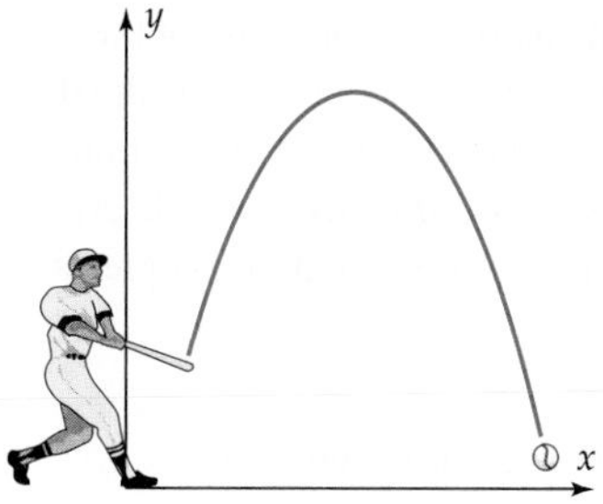

Figure 11–27

Matt Mahoney hits a baseball, and special measuring devices locate its position at various times during its flight. If the path of the ball is drawn on a coordinate plane, with the batter at the origin, it looks like Figure 11–27. According to the measuring devices, the ball passes through the points (7, 9), (47, 38), and (136, 70).

(a) What is the equation of the path of the ball?

(b) How far from Matt does the ball hit the ground?

SOLUTION

(a) The path of the ball appears to be part of a parabola (a fact that will not be proved here) and hence has an equation of the form

$$y = ax^2 + bx + c.$$

As in Example 8, each of the points (7, 9), (47, 38) and (136, 70) determines an equation.

$$\begin{aligned} &(7, 9) && a(7^2) + b(7) + c = 9 \\ &(47, 38): && a(47^2) + b(47) + c = 38 \\ &(136, 70): && a(136^2) + b(136) + c = 70. \end{aligned}$$

We must solve the resulting system.

$$49a + 7b + c = 9$$
$$2209a + 47b + c = 38$$
$$18{,}496a + 136b + c = 70$$

whose matrix form is

$$\begin{pmatrix} 49 & 7 & 1 \\ 2209 & 47 & 1 \\ 18{,}496 & 136 & 1 \end{pmatrix} \begin{pmatrix} a \\ b \\ c \end{pmatrix} = \begin{pmatrix} 9 \\ 38 \\ 70 \end{pmatrix}.$$

A calculator shows that the solution is

$$\begin{pmatrix} a \\ b \\ c \end{pmatrix} = \begin{pmatrix} 49 & 7 & 1 \\ 2209 & 47 & 1 \\ 18{,}496 & 136 & 1 \end{pmatrix}^{-1} \begin{pmatrix} 9 \\ 38 \\ 70 \end{pmatrix} \approx \begin{pmatrix} -.00283 \\ .87798 \\ 2.99296 \end{pmatrix}.$$

Therefore, the approximate equation for the ball's path is

$$y = -.00283x^2 + .87798x + 2.99296.$$

(b) The ball hits the ground at a point where its height y is 0, that is, when

$$-.00283x^2 + .87798x + 2.99296 = 0.$$

Use the quadratic formula or an equation solver to verify that the solutions of this equation are $x \approx -3.37$ (which is not applicable here) and $x \approx 313.61$. Therefore, the ball hits the ground approximately 314 feet from Matt.   ■

Given any three points not on a straight line, the method in Examples 8 and 9 can be used to find the unique parabola that passes through these points. This parabola can also be found by quadratic regression.*

To see why regression produces the parabola that actually passes through the three points, recall that the error in a quadratic model is measured by taking the difference between the y-coordinate of each data point and the y-coordinate of the corresponding point on the model and summing the squares of these errors. If the data points actually lie on a parabola (which is always the case with three points that are not on a line), then the error for that parabola will be 0 (the smallest possible error). Hence, it will be the parabola produced by the least squares quadratic regression procedure.

CALCULATOR EXPLORATION

Use quadratic regression on the three points in Example 9 and verify that the equation obtained is the same as the one in Example 9.

---

*If you have not read Section 2.5 and Special Topics 4.4.A, you may skip this discussion.

EXERCISES 11.3

In Exercises 1–6, determine whether the product AB or BA is defined. If a product is defined, state its size (number of rows and columns). Do not actually calculate any products.

1. $A = \begin{pmatrix} 3 & 6 & 7 \\ 8 & 0 & 1 \end{pmatrix}$, $B = \begin{pmatrix} 2 & 5 & 9 & 1 \\ 7 & 0 & 0 & 6 \\ -1 & 3 & 8 & 7 \end{pmatrix}$

2. $A = \begin{pmatrix} 0 & 5 & 9 \\ 0 & 5 & 3 \\ 0 & 0 & 0 \end{pmatrix}$, $B = \begin{pmatrix} -8 & 7 & 1 \\ 0 & -3 & 4 \end{pmatrix}$

3. $A = \begin{pmatrix} 1 & 0 \\ 1 & 1 \\ 0 & 1 \end{pmatrix}$, $B = \begin{pmatrix} 5 & 6 & 11 \\ 7 & 8 & 15 \end{pmatrix}$

4. $A = \begin{pmatrix} -1 & 0 \\ 6 & 8 \end{pmatrix}$, $B = \begin{pmatrix} 4 & 0 \\ 1 & -7 \end{pmatrix}$

5. $A = \begin{pmatrix} -4 & 15 \\ 3 & -7 \\ 2 & 10 \end{pmatrix}$, $B = \begin{pmatrix} 1 & 2 \\ 3 & 4 \end{pmatrix}$

6. $A = (-1 \quad 5 \quad 6)$, $\quad B = \begin{pmatrix} 7 \\ 8 \\ -1 \end{pmatrix}$

In Exercises 7–12, find AB.

7. $A = \begin{pmatrix} 1 & 2 \\ -4 & 3 \end{pmatrix}$, $\quad B = \begin{pmatrix} -1 & 2 & 3 \\ 7 & 1 & 0 \end{pmatrix}$

8. $A = \begin{pmatrix} 1 & 5 & 9 \\ -2 & 3 & 3 \\ 1 & 7 & 0 \end{pmatrix}$, $\quad B = \begin{pmatrix} 2 & -3 & 0 \\ 4 & 5 & -1 \\ 0 & -1 & 1 \end{pmatrix}$

9. $A = \begin{pmatrix} 1 & 0 & -2 \\ 0 & 3 & -1 \\ 2 & 4 & 0 \end{pmatrix}$, $\quad B = \begin{pmatrix} 1 & 1 \\ 1 & 0 \\ 0 & 1 \end{pmatrix}$

10. $A = \begin{pmatrix} -1 & 7 & 1 \\ -5 & 3 & 2 \\ 0 & 1 & 5 \\ -3 & 6 & 7 \end{pmatrix}$, $\quad B = \begin{pmatrix} 7 & -2 & 6 & 2 \\ -2 & 8 & 4 & 1 \\ 0 & 7 & 0 & -5 \end{pmatrix}$

11. $A = \begin{pmatrix} 2 & 0 & -1 \\ 1 & 1 & 2 \\ 0 & 2 & -3 \\ 2 & 3 & 0 \end{pmatrix}$, $\quad B = \begin{pmatrix} 1 & 0 & 1 & 1 \\ 1 & 1 & 0 & 1 \\ 1 & 1 & 1 & 0 \end{pmatrix}$

12. $A = \begin{pmatrix} 1 & 2 & 6 \\ 0 & 3 & -1 \\ 1 & 4 & -8 \end{pmatrix}$, $\quad B = \begin{pmatrix} -1 & 0 & 8 \\ 11 & -4 & 3 \end{pmatrix}$

In Exercises 13–16, show that AB is not equal to BA by computing both products.

13. $A = \begin{pmatrix} 2 & 3 \\ 1 & 5 \end{pmatrix}$, $\quad B = \begin{pmatrix} -1 & 1 \\ 3 & 2 \end{pmatrix}$

14. $A = \begin{pmatrix} 5/4 & -2 \\ -3/4 & 1 \end{pmatrix}$, $\quad B = \begin{pmatrix} 9/4 & 5/4 \\ -2 & 3 \end{pmatrix}$

15. $A = \begin{pmatrix} 4 & 2 & -1 \\ 0 & 1 & 2 \\ -3 & 0 & 1 \end{pmatrix}$, $\quad B = \begin{pmatrix} 1 & 7 & -5 \\ 2 & -2 & 6 \\ 0 & 0 & 0 \end{pmatrix}$

16. $A = \begin{pmatrix} -1 & 3 & 2 & -4 \\ 8 & 0 & 5 & 6 \\ -1 & 0 & 6 & 3 \\ 2 & 3 & -2 & 5 \end{pmatrix}$, $\quad B = \begin{pmatrix} 4 & -2 & -2 & 0 \\ 1 & -7 & 4 & 1 \\ 0 & 5 & 7 & -2 \\ 1 & 4 & 0 & 5 \end{pmatrix}$

In Exercises 17–24, find the inverse of the matrix, if it exists.

17. $\begin{pmatrix} 1 & 2 \\ 3 & 4 \end{pmatrix}$

18. $\begin{pmatrix} 3 & 5 \\ -1 & 2 \end{pmatrix}$

19. $\begin{pmatrix} 2 & -4 \\ -3 & 6 \end{pmatrix}$

20. $\begin{pmatrix} 1 & -1 & 4 \\ 0 & 1 & 3 \\ 2 & -3 & 4 \end{pmatrix}$

21. $\begin{pmatrix} 1 & 2 & 0 \\ 3 & -1 & 2 \\ -2 & 3 & -2 \end{pmatrix}$

22. $\begin{pmatrix} -1 & 3 & 1 \\ -2 & 1 & -3 \\ 1 & -3 & -2 \end{pmatrix}$

23. $\begin{pmatrix} 5 & 0 & 2 \\ 2 & 2 & 1 \\ -3 & 1 & -1 \end{pmatrix}$

24. $\begin{pmatrix} -2 & 5 & 1 \\ 4 & -7 & 0 \\ 8 & -17 & -2 \end{pmatrix}$

In Exercises 25–28, solve the system of equations by using the method of Example 6.

25.
$$\begin{aligned} -x + y \phantom{{}+ z} &= 1 \\ -x \phantom{{}+ y} + z &= -2 \\ 6x - 2y - 3z &= 3 \end{aligned}$$

26.
$$\begin{aligned} 8x - 3y + 4z &= 47 \\ 4x - 5y \phantom{{}+ 2z} &= 33 \\ x + 2y + 2z &= 0 \end{aligned}$$

27.
$$\begin{aligned} 2x + y \phantom{{}+ 3z} &= 0 \\ -4x - y - 3z &= 1 \\ 3x + y + 2z &= 2 \end{aligned}$$

28.
$$\begin{aligned} 7x - 3y + z &= 24 \\ 2x \phantom{{}- 3y} + z &= 2 \\ 3x - 2y - z &= 21 \end{aligned}$$

In Exercises 29–39, solve the system by any method.

29.
$$\begin{aligned} x + y + 2w &= 3 \\ 2x - y + z - w &= 5 \\ 3x + 3y + 2z - 2w &= 0 \\ x + 2y + z \phantom{{}- 2w} &= 2 \end{aligned}$$

30.
$$\begin{aligned} x + 7y - z + 5w &= 70 \\ 2x - y - 8z \phantom{{}+ 5w} &= 6 \\ 3y - 7z - w &= 20 \\ 3x - 2y + 5z \phantom{{}- w} &= -11 \end{aligned}$$

31.
$$\begin{aligned} x + y + 6z + 2v \phantom{{}- 3w} &= 1.5 \\ x + 5z + 2v - 3w &= 2 \\ 3x + 2y + 17z + 6v - 4w &= 2.5 \\ 4x + 3y + 21z + 7v - 2w &= 3 \\ -6x - 5y - 36z - 12v + 3w &= 3.5 \end{aligned}$$

32.
$$\begin{aligned} x + 1.5y - 4.5z + 2.5v - w &= 1 \\ x - 2.5y + 2z - 3v - 2w &= -20 \\ 2x - y - 8.5z \phantom{{}+ 3v} + 3.5w &= -51 \\ y - 6z + 1.5v - w &= -20 \\ x \phantom{{}+ y} + 3.5z - 4v \phantom{{}- w} &= -9 \end{aligned}$$

33.
$$\begin{aligned} x + 2y + 2z - 2w &= -23 \\ 4x + 4y - z + 5w &= 7 \\ -2x + 5y + 6z + 4w &= 0 \\ 5x + 13y + 7z + 12w &= -7 \end{aligned}$$

34.
$$\begin{aligned} 2x - 2y - z + 5w &= 0 \\ 3x - y + 5z - 8w &= 1 \\ y + 3z - w &= 3 \\ x - 2y + 5z + 4w &= -2 \end{aligned}$$

35. $x + 2y + 5z - 2v + 4w = 0$

$2x - 4y + 6z + v + 4w = 0$

$5x + 2y - 3z + 2v + 3w = 0$

$6x - 5y - 2z + 5v + 3w = 0$

$x + 2y - z - 2v + 4w = 0$

36. $2x + y - 2z - u + v - 3w = 1$

$3x - y + z + u - 2v + w = 2$

$x - y + 3z + u + 5v - w = 5$

$5x + y - 4z + u + 2v - 2w = -1$

$x + y + z - 3u - 7v + w = 3$

$x - y + 3u + 4w = 0$

37. $x + 2y + 3z = 1$

$3x + 2y + 4z = -1$

$2x + 6y + 8z + w = 3$

$2x + 2z - 2w = 3$

38. $2x + 6y - z = 4$

$5x - z = 4$

$4x - 18y + z = -4$

39. $x + 3w = -2$

$x - 4y - z + 3w = -7$

$4y + z = 5$

$-x + 12y + 3z - 3w = 17$

In Exercises 40–42, find constants a, b, c such that the three given points lie on the parabola $y = ax^2 + bx + c$. See Example 8.

40. $(-3, 2), (1, 1), (2, -1)$

41. $(-3, 15), (1, -7), (5, 111)$

42. $(1, -2), (3, 1), (4, -1)$

43. Concentrations of the greenhouse gas carbon dioxide (CO_2) have increased quadratically over the past half-century.* So the concentration y of CO_2 (in parts per million) in year x is given by an equation of the form

$$y = ax^2 + bx + c.$$

(a) Let $x = 0$ correspond to 1958 and use the following data to find a, b, and c.

Year	1958	1979	2001
CO_2 Concentration	315	337	371

(b) Use this equation to estimate the CO_2 concentration in the years 1983, 1993, and 2003. [For comparison purposes, the actual concentrations in 1983 and 1993 were 343 ppm and 357 ppm respectively.]

44. The table shows the per capita consumption of chicken (in pounds) in selected years.*

Year	1990	1999	2004
Chicken Consumption	42.4	54.2	59.2

(a) Let $x = 0$ correspond to 1990. Find a quadratic equation that models this data.

(b) Use the equation to estimate the consumption of chicken in 1995 and 2015. [The actual consumption in 1995 was 49.0 pounds.]

45. The table shows the per capita sales (in dollars) of electronics and appliances in selected years.†

Year	1992	1996	2000
Sales	$169	$260	$320

(a) Find a quadratic equation that models this data, with $x = 2$ corresponding to 1992.

(b) Use the equation to estimate sales of electronics and appliances in 1993, 1998, and 2002.

46. The table shows the percentage y of total consumer spending on sound recordings that is spent on pop music in year x.‡

Year	1998	2000	2002	2004	2006
% Pop	10	11	9	10	7.1

(a) Assume that the data can be modeled by an equation of the form

$$y = ax^4 + bx^3 + cx^2 + dx + k.$$

(Let $x = 0$ correspond to 1998.) Use the same method used in Exercises 43–45 (with five equations instead of three) to find the constants a, b, c, d, and k.

(b) Use the equation in part (a) to estimate the percentage of consumer spending spent on pop music in 2005.

47. Find constants a, b, c such that the points $(0, -2)$, $(\ln 2, 1)$, and $(\ln 4, 4)$ lie on the graph of $f(x) = ae^x + be^{-x} + c$. [*Hint:* Proceed as in Example 8.]

48. Find constants a, b, c such that the points $(0, -1)$, $(\ln 2, 4)$, and $(\ln 3, 7)$ lie on the graph of $f(x) = ae^x + be^{-x} + c$.

49. New Army Stores, a national chain of casual clothing stores, recently sent shipments of jeans, jackets, sweaters, and shirts to its stores in various cities. The number of items shipped to each city and their total wholesale cost are shown in the table on the next page. Find the wholesale price of each of the following: one pair of jeans, one jacket, one sweater, and one shirt.

---

*C. D. Keeling and T. P. Whorf, Scripps Institution of Oceanography.

*U.S. Department of Agriculture.
†U.S. Census Bureau.
‡Recording Industry Association of America, Inc.

City	Jeans	Jackets	Sweaters	Shirts	Total Cost
Cleveland	3000	3000	2200	4200	$507,650
St. Louis	2700	2500	2100	4300	459,075
Seattle	5000	2000	1400	7500	541,225
Phoenix	7000	1800	600	8000	571,500

50. A 100-bed nursing home provides two levels of long-term care: regular and maximum. Patients at each level have a choice of a private room or a less expensive semiprivate room. The table below shows the number of patients in each category at various times last year and the total daily cost for these patients. Find the daily cost of each of the following: a private room (regular care), a private room (maximum care), a semiprivate room (regular care), and a semiprivate room (maximum care).

Month	Regular Care Patients Semi-private	Regular Care Patients Private	Maximum Care Patients Semi-private	Maximum Care Patients Private	Daily Cost
January	22	8	60	10	$18,824
April	26	8	54	12	18,738
July	24	14	56	6	18,606
October	20	10	62	8	18,824

51. A candy company produces three types of gift boxes: *A*, *B*, and *C*. A box of variety *A* contains .6 pound of chocolates and .4 pound of mints. A box of variety *B* contains .3 pound of chocolates, .4 pound of mints, and .3 pound of caramels. A box of variety *C* contains .5 pound of chocolates, .3 pound of mints, and .2 pound of caramels. The company has 41,400 pounds of chocolates, 29,400 pounds of mints, and 16,200 pounds of caramels in stock. How many boxes of each variety should be made to use up all the stock?

52. Certain circus animals are fed the same three food mixes: *R*, *S*, and *T*. Lions receive 1.1 units of *R*, 2.4 units of *S*, and 3.7 units of *T* each day. Horses receive 8.1 units of *R*, 2.9 units of *S*, and 5.1 units of *T* each day. Bears receive 1.3 units of *R*, 1.3 units of *S*, and 2.3 units of *T* each day. If 16,000 units of *R*, 28,000 units of *S*, and 44,000 units of *T* are available each day, how many of each type of animal can be supported?

Chapter 11 Review

IMPORTANT CONCEPTS

REVIEW QUESTIONS

In Questions 1–4, solve the system of linear equations by any means you want.

1. $-5x + 3y = 4$
$\quad 2x - y = -3$

2. $3x - y = 6$
$\quad 2x + 3y = 7$

3. $3x + 4y = 7$
$\quad 2x - 3y = -1$

4. $\frac{1}{4}x - \frac{1}{3}y = -\frac{1}{4}$
$\quad \frac{1}{10}x + \frac{2}{5}y = \frac{2}{5}$

5. The number of days y in year x in which acceptable air quality standards were not met in two metropolitan areas are approximated by these equations (in which $x = 4$ corresponds to 1994).*

Riverside–San Bernadino, CA: $\quad 10.89x + y = 184.92$

Atlanta, GA: $\quad 16.06x - 2y = 32.04$

In what year is the number of unacceptable days the same in both areas? How many such days are there in that year?

6. The sum of one number and three times a second number is -20. The sum of the second number and two times the first number is 55. Find the two numbers.

7. An alloy containing 40% gold and an alloy containing 80% gold are to be mixed to produce 50 pounds of an alloy containing 75% gold. How much of each alloy is needed?

8. The table shows the population (in thousands) of the cities of Miami and Cleveland in selected years.[†]

Year	1950	1960	1970	1980	1990	2000	2005
Miami	249.3	291.7	335.0	347.0	358.5	362.5	386.4
Cleveland	914.8	876.1	751.0	574.0	505.6	478.4	452.2

(a) Use linear regression to find an equation that gives the population y of Miami in year x, with $x = 0$ corresponding to 1950.
(b) Do part (a) for Cleveland.
(c) If these models remain accurate, when do the two cities have the same population and what is that population?

In Questions 9–12, solve the nonlinear system.

9. $x^2 - y = 0$
$\quad y - 2x = 3$

10. $x^2 + 2y^2 = 29$
$\quad x^2 + 5y = 17$

11. $x^2 + y^2 = 16$
$\quad x + y = 2$

12. $2x^2 - 3xy + y^2 = 63$
$\quad x^2 - 2xy + 4y^2 = 124$

13. Find the augmented matrix corresponding to this system.
$$3x - 5y + 2z = -9$$
$$3x + 2y = 0$$
$$x + y + 3z = 19$$

14. Use matrix methods to solve the system in Question 13.

15. Find the augmented matrix corresponding to this system.
$$x + 7y - z + 2w = 13$$
$$3x - 5y - z - w = -15$$
$$2x - 2y - 3z = -19$$

16. Use matrix methods to solve the system in Question 15.

In Questions 17–22, solve the system.

17. $2x + 5y + z = 8$
$\quad x + 2z = 9$
$\quad -2x - 5y - 2z = -13$

18. $x - 3y + 2z = 6$
$\quad 2x + 3y - 7z = -1$
$\quad 4x - 3y - 3z = 11$

19. $4x + 3y - 3z = 2$
$\quad 5x - 3y + 2z = 10$
$\quad 2x - 2y + 3z = 14$

20. $2x + 6y - z = 4$
$\quad 2x + y + 3z = 0$
$\quad x - 2y = -1$

21. $x - 2y - 3z = 1$
$\quad 5y + 10z = 0$
$\quad 8x - 6y - 4z = 8$

22. $2x - \frac{1}{2}y - z = 1$
$\quad 3x + 4y - z = -8$
$\quad x - 5y - z = -3$

23. Let L be the line with equation $4x - 2y = 6$, and let M be the line with equation $-10x + 5y = -15$. Which of the following statements is true?

(a) L and M do not intersect.
(b) L and M intersect at a single point.
(c) L and M are the same line.
(d) All of the above are true.
(e) None of the above are true.

24. Which of the following statements about this system of equations are *false*?

$$x + z = 2$$
$$6x + 4y + 14z = 24$$
$$2x + y + 4z = 7$$

(a) $x = 2, y = 3, z = 0$ is a solution.
(b) $x = 1, y = 1, z = 1$ is a solution.
(c) $x = 1, y = -3, z = 3$ is a solution.
(d) The system has an infinite number of solutions.
(e) $x = 2, y = 5, z = -1$ is not a solution.

In Questions 25 and 26, find the constants A, B, C that make the statement true.

25. $\dfrac{4x - 7}{x^2 - x - 6} = \dfrac{A}{x - 3} + \dfrac{B}{x + 2}$

26. $\dfrac{6x^2 + 6x - 6}{(x^2 - 1)(x + 2)} = \dfrac{A}{x + 1} + \dfrac{B}{x - 1} + \dfrac{C}{x + 2}$

In Questions 27–30, perform the indicated matrix multiplication or state that the product is not defined. Use these matrices.

$A = \begin{pmatrix} -1 & 0 \\ 0 & -1 \end{pmatrix}$, $B = \begin{pmatrix} 2 & -3 \\ 4 & 1 \end{pmatrix}$, $C = \begin{pmatrix} 3 & 2 \\ 2 & 4 \end{pmatrix}$,

$D = \begin{pmatrix} -3 & 1 & 2 \\ 1 & 0 & 4 \end{pmatrix}$, $E = \begin{pmatrix} 1 & 2 \\ -3 & 4 \\ 0 & 5 \end{pmatrix}$, $F = \begin{pmatrix} 2 & 3 \\ 6 & 3 \\ 6 & 1 \end{pmatrix}$

27. AB **28.** CD **29.** AE **30.** DF

In Questions 31–34, find the inverse of the matrix, if it exists.

31. $\begin{pmatrix} -5 & 2 \\ 7 & -3 \end{pmatrix}$ **32.** $\begin{pmatrix} 2 & 6 \\ 1 & 3 \end{pmatrix}$

33. $\begin{pmatrix} 3 & 2 & 6 \\ 1 & 1 & 2 \\ 2 & 2 & 5 \end{pmatrix}$ **34.** $\begin{pmatrix} 1 & -1 & 1 \\ 2 & -3 & 2 \\ -4 & 6 & 1 \end{pmatrix}$

In Questions 35 and 36, use matrix inverses to solve the system.

35. $x + 2z + 6w = 2$
  $3x + 4y - 2z - w = 0$
  $5x + 2z - 5w = -4$
  $4x - 4y + 2z + 3w = 1$

36. $2x + y + 2z + u = 2$
  $x + 3y - 4z - 2u + 2v = -2$
  $2x + 3y + 5z - 4u + v = 1$
  $x - 2z + 4v = 4$
  $2x + 6z - 5v = 0$

In Questions 37 and 38, find the equation of the parabola passing through the given points.

37. $(-3, 52), (2, 17), (8, 305)$

38. $(-2, -18), (2, 6), (4, -12)$

39. The table shows the number of hours spent per person per year on home video games.*

Year	1999	2002	2004
Hours	58	70	77

*Data from *Statistical Abstract of the United States: 2007*.

(a) Find a quadratic equation that models this data, with $x = 0$ corresponding to 1990.
(b) Use the equation to estimate the number of hours spent in 2000 and 2015.

40. The table shows the per capita consumption of eggs in selected years.*

Year	1980	1990	1999
Egg Consumption	271	234	255

(a) Find a quadratic equation that models this data, with $x = 0$ corresponding to 1980.
(b) Use the equation to estimate egg consumption in 1985, 1995, and 2005.

41. Tickets to a lecture cost $1 for students, $1.50 for faculty, and $2 for others. Total attendance at the lecture was 460, and the total income from tickets was $570. Three times as many students as faculty attended. How many faculty members attended the lecture?

42. If Andrew, Laura, and Ryan work together, they can paint a large room in 4 hours. When only Laura and Ryan work together, it takes 8 hours to paint the room. Andrew and Laura, working together, take 6 hours to paint the room. How long would it take each of them to paint the room alone? [*Hint:* If x is the amount of the room painted in 1 hour by Andrew, y is the amount painted by Laura, and z is the amount painted by Ryan, then $x + y + z = 1/4$.]

43. An animal feed is to be made from corn, soybeans, and meat by-products. One bag is to supply 1800 units of fiber, 2800 units of fat, and 2200 units of protein. Each pound of corn has 10 units of fiber, 30 units of fat, and 20 units of protein. Each pound of soybeans has 20 units of fiber, 20 units of fat, and 40 units of protein. Each pound of by-products has 30 units of fiber, 40 units of fat, and 25 units of protein. How many pounds of corn, soybeans, and by-products should each bag contain?

44. A company produces three camera models: A, B, and C. Each model A requires 3 hours of lens polishing, 2 hours of assembly time, and 2 hours of finishing time. Each model B requires 2 hours of lens polishing, 2 hours of assembly time, and 1 hour of finishing time. Each model C requires 1, 3, and 1 hours of lens polishing, assembly, and finishing time, respectively. There are 100 hours available for lens polishing, 100 hours for assembly, and 65 hours for finishing each week. How many of each model should be produced if all available time is to be used?

*U.S. Department of Agriculture.

Chapter 11 Test

1. I paid $35.25 for 15 boxes of cereal. I bought some Basho Bites, which cost $2.25 per box, and some Health Nuggets, which cost $2.50 per box. How many boxes of each kind of cereal did I buy?

2. Find the constants A, B, and C so that the following is an identity:

$$\frac{5x^2 - 10 - 8x}{x^3 - 4x^2 + 3x - 12} = \frac{A}{x - 4} + \frac{7Bx + C}{x^2 + 3}$$

3. Find a, b, c so the function $f(x) = ax^2 + bx + c$ conforms to the following table of values.

x	$f(x)$
0	3
1	4
2	3

4. Solve the following system of equations geometrically.

$$x + y = 8$$
$$3x - 5y = 0$$

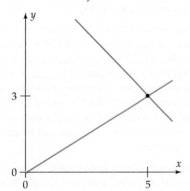

5. Solve the following systems of equations:

(a) $x + 2y = 1$
     $3x + 4y = 4$

(b) $x - y = -5.0$
     $3x + 3y = 7.8$

(c) $x^2 + 3 \ln y = -2$
     $y = e^x$

6. Determine the number of solutions of the following systems:

(a) $x + y = 2$
     $x - y = 8$

(b) $5x - 3y = 6$
     $15x - 9y = 18$

(c) $2x + y = 3$
     $-x - \frac{1}{2}y = 4$

(d) $y = \sin x$
     $y = \frac{x^2}{40}$

7. Find two real numbers whose sum is 100 and whose product is 2343.75.

8. Solve the following systems:

(a) $6x - y - 4z = -4$
                $y + z = 0$
                    $6z = 12$

(b) $x + y + z = 8$
     $10x - 3y - 5z = 4$
     $-2x - 4y + 2z = 7$

9. For the following matrices A, B, compute AB if it is defined, and then compute BA if it is defined.

(a) $A = \begin{pmatrix} 1 & 2 & -1 \\ 0 & 2 & 1 \\ -1 & 2 & 1 \end{pmatrix}$ $B = \begin{pmatrix} 3 & 0 & 0 \\ 1 & 2 & 1 \\ -1 & -1 & 1 \end{pmatrix}$

(b) $A = \begin{pmatrix} 1 & 2 & 0 \\ -1 & 3 & 0 \\ -2 & 1 & 1 \end{pmatrix}$ $B = \begin{pmatrix} 1 & 2 & 1 \\ -4 & 2 & 3 \end{pmatrix}$

(c) $A = \begin{pmatrix} 3 & 5 \\ 1 & 2 \end{pmatrix}$ $B = \begin{pmatrix} 2 & -5 \\ -1 & 3 \end{pmatrix}$

DISCOVERY PROJECT 11   Input-Output Analysis

Wassily Leontief won the Nobel Prize in economics in 1973 for his method of input-output analysis of the economies of industrialized nations. This method has become a permanent part of production planning and forecasting by both national governments and private corporations. During the Arab oil boycott in 1973, for example, General Electric used input-output analysis on 184 sectors of the economy (such as energy, agriculture, and transportation) to predict the effect of the energy crisis on public demand for its products. The key to Leontief's method is knowing how much each sector of the economy needs from other sectors in order to do its job.

A simple economic model will illustrate the basic ideas behind input-output analysis. Suppose the country Hypothetica has just two sectors in its economy: agriculture and manufacturing. The production of a ton of agricultural products requires the use of .1 ton of agricultural products and .1 ton of manufactured products. Similarly, the production of a ton of manufactured products consumes .1 ton of agricultural products and .3 ton of manufactured products. The key economic question is: How many tons of each sector must Hypothetica produce in order to have enough surplus to export 10,000 tons of agricultural products and 10,000 tons of manufactured goods?

DISCOVERY PROJECT 11

Let A be the total amount of agricultural goods, and let M be the total amount of manufactured goods produced. The total agricultural production A is the sum of the amount needed for producing the agricultural products plus the amount needed for producing manufactured products plus the amount targeted for export. In other words,

agricultural needs		manufacturing needs		export needs		total agricultural production
$.1A$	$+$	$.1M$	$+$	$100{,}000$	$=$	$A.$

1. Write a similar equation for manufactured goods.

2. Solve the system of equations given by the agricultural equation above and the manufacturing equation of Problem 1. What level of production of agricultural and manufactured goods should Hypothetica work toward to reach its export goals?

3. The same kind of analysis can be used to determine the surplus available for exports in an economy. Suppose that Hypothetica can produce a total of 120,000 tons of agricultural products and 28,000 tons of manufactured goods. How much of each is available for export?

4. Input-output analysis can also be used for allocating resources in smaller-scale situations. Suppose a horse outfitter is hired by the State of Washington Department of Fish and Wildlife to haul 20 salt blocks into a game area for winter feeding. Each horse can carry a 200-pound load. A single person can handle three horses. Each person requires one horse to ride and one-half horse to carry his or her personal gear. A single horse can carry the feed and equipment for five horses. How many people and horses must go on the trip? [*Note:* Each horse load of salt must consist of an even number of blocks. Each salt block weighs 50 pounds. Also, fractional numbers of horses and people are not allowed. Make sure to adjust your answer to integer values, and be sure that the answer still provides sufficient carrying capacity for the salt.]

Chapter 12

DISCRETE ALGEBRA

What's next?

CD and CD-ROM players, fax machines, cameras, and other devices incorporate digital technology, which uses sequences of 0's and 1's to send signals. Determining the monthly payment on a car loan involves the sum of a geometric sequence. Problems involving the action of bouncing balls, vacuum pumps, and other devices can sometimes be solved by using sequences. See Exercises 55, 58, and 62 on page 851.

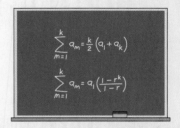

$$\sum_{m=1}^{k} a_m = \frac{k}{2}\left(a_1 + a_k\right)$$

$$\sum_{m=1}^{k} a_m = a_1\left(\frac{1-r^k}{1-r}\right)$$

© Terry Oakley/Alamy

Chapter Outline

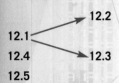
This chapter deals with a variety of subjects involving counting processes and the nonnegative integers 0, 1, 2, 3, ...

12.1   Sequences and Sums

Section Objectives

- Find terms of a sequence.
- Write the formula for a sequence, given a few of its terms.
- Find the formula for a recursively defined sequence.
- Set up and solve applied problems using sequences.
- Use summation notation.
- Find partial sums of a sequence.

A **sequence** is an ordered list of numbers, such as

$$2, 4, 6, 8, 10, 12, \ldots$$

$$1, -3, 5, -7, 9, -11, 13, \ldots$$

$$1, 0, 1, 0, 1, 0, 1, 0, \ldots$$

$$2, 1, \frac{2}{3}, \frac{3}{4}, \frac{4}{5}, \frac{5}{6}, \frac{6}{7}, \ldots,$$

where the dots indicate that the same pattern continues forever.* Each number on the list is called a **term** of the sequence.

$$
\begin{array}{cccccc}
2, & 1, & 0, & 1, & 2, & 3, \quad 2, 1, 0, 1, 2, 3, 2, \ldots. \\
\uparrow & \uparrow & \uparrow & \uparrow & \uparrow & \uparrow \\
\text{1st} & \text{2nd} & \text{3rd} & \text{4th} & \text{5th} & \text{6th} \\
\text{term} & \text{term} & \text{term} & \text{term} & \text{term} & \text{term}
\end{array}
$$

When the pattern isn't obvious, as in the preceding examples, sequences are usually described in terms of a formula.

---

*Such a list defines a function f whose domain is the set of positive integers. The rule is $f(1) =$ first number on the list, $f(2) =$ second number on the list, and so on. Conversely, any function g whose domain is the set of positive integers leads to an ordered list of numbers, namely, $g(1), g(2), g(3), \ldots.$ So a sequence is formally defined to be a function whose domain is the set of positive integers.

EXAMPLE 1

(a) Find the first three terms of the sequence $a_1, a_2, a_3, \ldots, a_n, \ldots$ where a_n is given by the formula

$$a_n = \frac{n^2 - 3n + 1}{2n + 5}.$$

(b) Find a_{39}.

SOLUTION

(a) To find a_1, we substitute $n = 1$ in the formula for a_n; to find a_2, we substitute $n = 2$ in the formula; and so on.

$$a_1 = \frac{1^2 - 3 \cdot 1 + 1}{2 \cdot 1 + 5} = -\frac{1}{7},$$

$$a_2 = \frac{2^2 - 3 \cdot 2 + 1}{2 \cdot 2 + 5} = -\frac{1}{9},$$

$$a_3 = \frac{3^2 - 3 \cdot 3 + 1}{2 \cdot 3 + 5} = \frac{1}{11}.$$

Thus, the sequence begins $-1/7, -1/9, 1/11, \ldots$.

(b) The 39th term is

$$a_{39} = \frac{39^2 - 3 \cdot 39 + 1}{2 \cdot 39 + 5} = \frac{1405}{83}.$$ ∎

The subscript notation for sequences is sometimes abbreviated by writing $\{a_n\}$ in place of $a_1, a_2, a_3, \ldots$.

EXAMPLE 2

Find the first three terms, the 41st term, and the 206th term of the sequence

$$\left\{ \frac{(-1)^n}{n + 2} \right\}.$$

SOLUTION   The formula is

$$a_n = \frac{(-1)^n}{n + 2}.$$

Substituting $n = 1$, $n = 2$, and $n = 3$ shows that

$$a_1 = \frac{(-1)^1}{1 + 2} = -\frac{1}{3}, \qquad a_2 = \frac{(-1)^2}{2 + 2} = \frac{1}{4}, \qquad a_3 = \frac{(-1)^3}{3 + 2} = -\frac{1}{5}.$$

Similarly,

$$a_{41} = \frac{(-1)^{41}}{41 + 2} = -\frac{1}{43} \qquad \text{and} \qquad a_{206} = \frac{(-1)^{206}}{206 + 2} = \frac{1}{208}.$$ ∎

EXAMPLE 3

Here are some other sequences whose nth term can be described by a formula.

Sequence	nth Term	First 5 Terms
$a_1, a_2, a_3, \ldots$	$a_n = n^2 + 1$	$2, 5, 10, 17, 26$
$b_1, b_2, b_3, \ldots$	$b_n = \dfrac{1}{n}$	$1, \dfrac{1}{2}, \dfrac{1}{3}, \dfrac{1}{4}, \dfrac{1}{5}$
$c_1, c_2, c_3, \ldots$	$c_n = \dfrac{(-1)^{n+1}2n}{(n+1)(n+2)}$	$\dfrac{1}{3}, -\dfrac{1}{3}, \dfrac{3}{10}, -\dfrac{4}{15}, \dfrac{5}{21}$
$x_1, x_2, x_3, \ldots$	$x_n = 3 + \dfrac{1}{10^n}$	$3.1, 3.01, 3.001, 3.0001, 3.00001.$
$a_1, a_2, a_3, \ldots$	$a_n = 7$	$7, 7, 7, 7, 7.$ ■

A sequence in which every term is the same, such as the last one in Example 3, is called a **constant sequence.** A calculator is often useful for computing and displaying the terms of more complicated sequences.

EXAMPLE 4

(a) Display the first five terms of the sequence $\{a_n = n^2 - n - 3\}$ on your calculator screen.

(b) Display the first, fifth, ninth, and thirteenth terms of this sequence.

SOLUTION

Method 1:   Enter the sequence in the function memory as $y_1 = x^2 - x - 3$. In the table set-up screen, begin the table at $x = 1$ and set the increment at 1 and display a table of values (Figure 12–1). To display the first, fifth, ninth, and thirteenth terms, set the table increment at 4 (Figure 12–2).

| Figure 12–1 | Figure 12–2 |

TECHNOLOGY TIP

SEQ (or MAKELIST on HP-39gs) is in this menu/submenu:

TI-84+/86: LIST/*Ops*

TI-89: MATH/*List*

Casio: OPTN/*List*

HP-39gs: MATH/*List*

Method 2:   Using the Technology Tip in the margin, enter the following, which produces Figure 12–3 on the next page.

$$\text{SEQ}(x^2 - x - 3, x, 1, 5, 1).$$

To display the first, fifth, ninth and thirteenth terms, enter

$$\text{SEQ}(x^2 - x - 3, x, 1, 13, 4),$$

which tells the calculator to look at every fourth term from 1 to 13 and produces Figure 12–4.

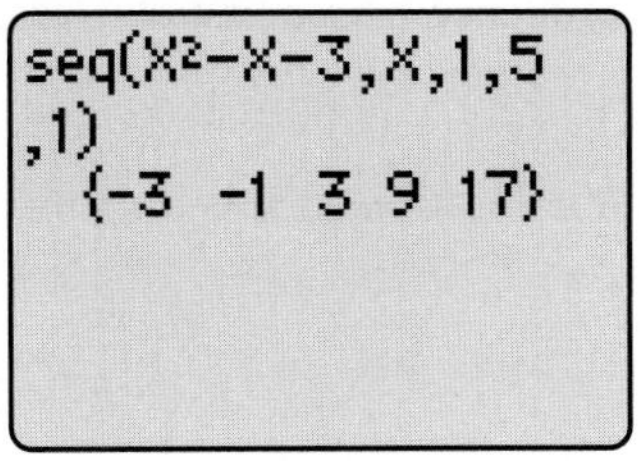

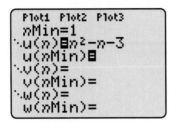

Figure 12–3

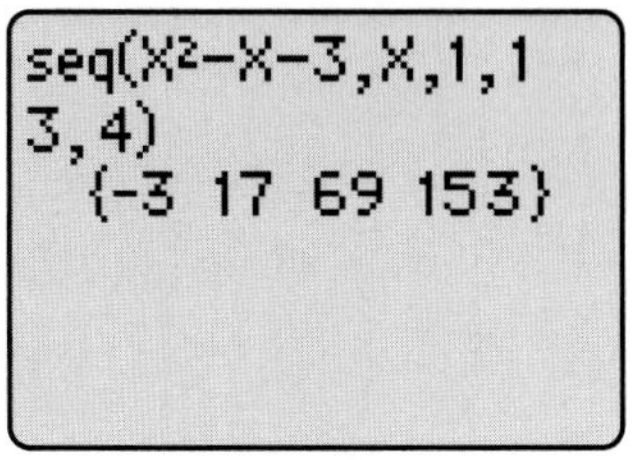

Figure 12–4

TECHNOLOGY TIP

To change to sequence graphing mode, choose SEQUENCE in this menu:

TI-84+/89: MODE

HP-39gs: APLET

On Casio, choose RECUR in the main menu.

Method 3: If possible, put your calculator in sequence graphing mode (see the Technology Tip in the margin). Enter the formula in the equation memory (Figure 12–5). Now you can either construct a table (as in *Method 1*) or graph the sequence and use the trace feature to determine its terms, as in Figure 12–6.* ■

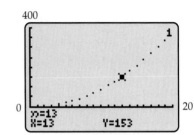

Figure 12–5

Figure 12–6

EXAMPLE 5

Display the first ten terms of the sequence $\left\{\dfrac{n}{n+1}\right\}$ on your calculator screen in fractional form, if possible.

SOLUTION Creating a table always produces decimal approximations, as you can easily verify. The same is usually true of the SEQ key, unless you take special steps. On HP-39gs, change the number format mode to "fraction" (MODE menu). On TI calculators (other than TI-89), either use the *Frac* key after obtaining the sequence (Figure 12–7) or use parentheses and the *Frac* key as part of the function: Entering

$$\text{SEQ}(x/(x+1) \blacktriangleright \textit{Frac}, x, 1, 10, 1)$$

produces Figure 12–8. In each figure, you must use the arrow key to scroll to the right to see all the terms. ■

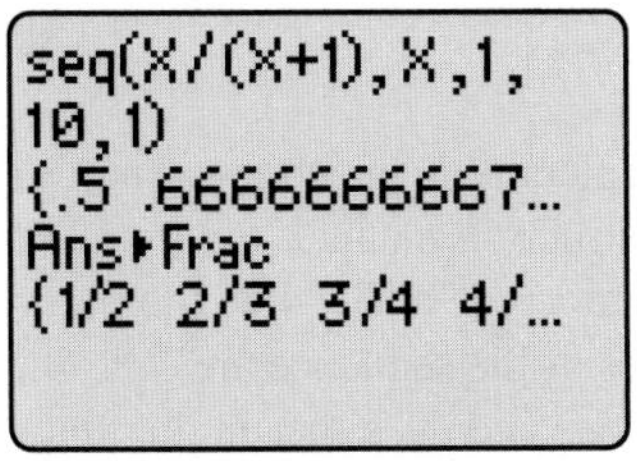

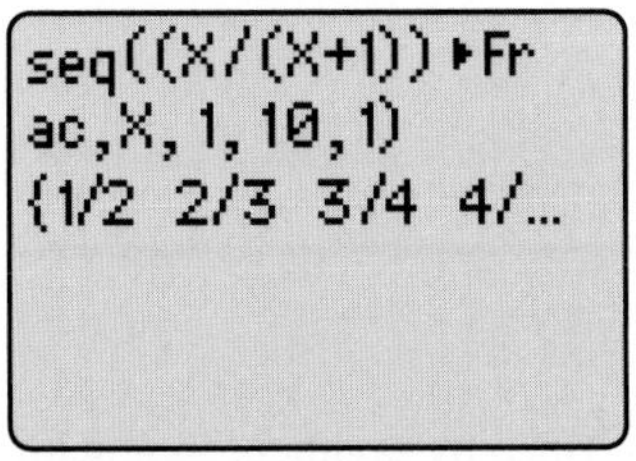

Figure 12–7

Figure 12–8

A sequence is said to be defined **recursively** (or **inductively**) if the first term is given (or the first several terms) and there is a rule for determining the nth term by using the terms that precede it.

*Set TI calculators for DOT instead of CONNECTED graphing in the MODE menu. This is not necessary on Casio and not available on HP-39gs when it is in sequence mode.

EXAMPLE 6

Consider the sequence whose first two terms are

$$a_1 = 1 \quad \text{and} \quad a_2 = 1$$

and whose nth term (for $n \geq 3$) is the sum of the two preceding terms.

$$a_3 = a_2 + a_1 = 1 + 1 = 2,$$
$$a_4 = a_3 + a_2 = 2 + 1 = 3,$$
$$a_5 = a_4 + a_3 = 3 + 2 = 5.$$

For each integer n, the two preceding integers are $n - 1$ and $n - 2$. So

$$a_n = a_{n-1} + a_{n-2} \quad (n \geq 3).$$

This sequence 1, 1, 2, 3, 5, 8, 13, . . . is called the **Fibonacci sequence,** and the numbers that appear in it are called **Fibonacci numbers.** Fibonacci numbers have many surprising and interesting properties, and are often found in nature. See Exercises 76–82 for details. ∎

EXAMPLE 7

The sequence given by

$$a_1 = -7 \quad \text{and} \quad a_n = a_{n-1} + 3 \quad \text{for } n \geq 2$$

is defined recursively. Its first three terms are

$$a_1 = -7, \quad a_2 = a_1 + 3 = -7 + 3 = -4,$$
$$a_3 = a_2 + 3 = -4 + 3 = -1.$$ ∎

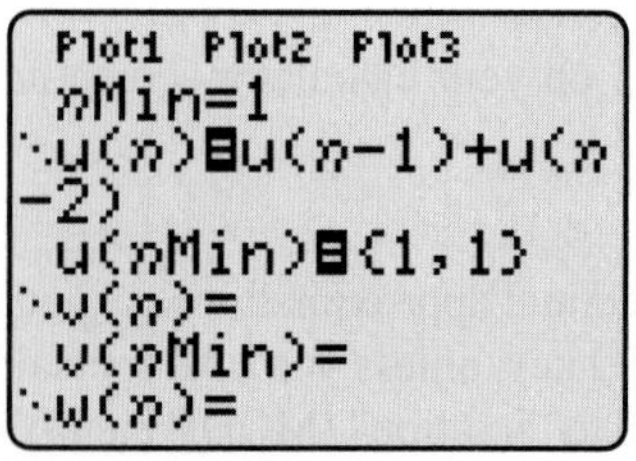

```
Plot1 Plot2 Plot3
nMin=1
\u(n)Bu(n-1)+u(n
-2)
u(nMin)B(1,1)
\v(n)=
v(nMin)=
\w(n)=
```

Figure 12–9

Entering a recursively defined sequence in a calculator (in sequence mode) may require the use of special keys; check your instruction manual. Figure 12–9 shows the Fibonacci sequence in the function memory of a TI-84+; the entry "u(nMin) = {1, 1}" indicates that the first two terms of the sequence are 1, 1. On other calculators, these terms are entered directly as a_1 and a_2, either in the function memory (HP-39gs) or in the RANG menu (Casio).

Sometimes, it is convenient or more natural to begin numbering the terms of a sequence with a number other than 1. So we may consider sequences such as

$$b_4, b_5, b_6, \ldots \quad \text{or} \quad c_0, c_1, c_2, \ldots.$$

EXAMPLE 8

The sequence 4, 5, 6, 7, . . . can be conveniently described by saying $b_n = n$, with $n \geq 4$. In the brackets notation, we write $\{n\}_{n \geq 4}$. Similarly, the sequence

$$2^0, 2^1, 2^2, 2^3, \ldots$$

may be described as $\{2^n\}_{n \geq 0}$ or by saying $c_n = 2^n$, with $n \geq 0$. ∎

EXAMPLE 9

To buy a car, Leslie borrows $14,000 at 7% annual interest. Her monthly payment is $277.22 for 60 months.

(a) Find a formula for a recursively defined sequence $\{u_n\}$ such that u_n is the balance due on the loan after the nth payment.

(b) Find the balance after 30 months.

SOLUTION   With car loans and home mortgages, interest is computed monthly on the unpaid balance. The monthly interest rate is understood to be one-twelfth of the annual rate, that is, $.07/12$.

(a) Let $u_0 = 14,000$, the balance after 0 payments. When the first payment is made, the loan balance is

$$\$14,000 + \text{one month's interest} = \$14,000 + \frac{.07}{12}(14,000).$$

Subtracting the first loan payment gives the balance after one month (rounded to the nearest penny).

$$u_1 = \$14,000 + \frac{.07}{12}(14,000) - 277.22 = \$13,804.45.$$

Similarly,

$$u_n = u_{n-1} + \frac{.07}{12}u_{n-1} - 277.22,$$

| Balance after | Balance after | Interest | nth payment |
| n payments | $n-1$ payments | on u_{n-1} | |

which can be written as

$$u_n = \left(1 + \frac{.07}{12}\right)u_{n-1} - 277.22.$$

(b) Using the table feature of a calculator in sequence mode, we find that $u_{30} = \$7609.07$, as shown in Figure 12–10. ∎

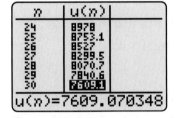

n	$u(n)$
24	8978
25	8753.1
26	8527
27	8299.5
28	8070.7
29	7840.6
30	7609.1

$u(n)=7609.070348$

Figure 12–10

SUMMATION NOTATION

It is sometimes necessary to find the sum of various terms in a sequence. For instance, we might want to find the sum of the first nine terms of the sequence $\{a_n\}$. Mathematicians often use the Greek letter sigma (Σ) to abbreviate such a sum:*

$$\sum_{k=1}^{9} a_k = a_1 + a_2 + a_3 + a_4 + a_5 + a_6 + a_7 + a_8 + a_9.$$

Similarly, for any positive integer m and numbers $a_1, a_2, \ldots, a_m$, we have the following.

Summation Notation

$$\sum_{k=1}^{m} a_k \quad \text{means} \quad a_1 + a_2 + a_3 + \cdots + a_m.$$

---

*Σ is the letter S in the Greek alphabet, the first letter in *Sum*.

An example of this situation is the sum $2^1 + 2^2 + 2^3 + 2^4 + 2^5 + 2^6$. To express this sum in summation notation, you could let $a_1 = 2^1$, $a_2 = 2^2$, $a_3 = 2^3$, $a_4 = 2^4$, $a_5 = 2^5$, and $a_6 = 2^6$, and write $\sum_{k=1}^{6} a_k$. However, since $a_k = 2^k$ for each k, it is more efficient to express this sum directly (without mentioning any a's) as $\sum_{k=1}^{6} 2^k$.

Conversely, if you are given a sum such as $\sum_{k=1}^{5} k^2$, you find the sum by successively substituting $k = 1, 2, 3, 4, 5$ for k in the expression k^2 and adding up the result:

$$\sum_{k=1}^{5} k^2 = \underset{\underset{k=1}{\uparrow}}{1^2} + \underset{\underset{k=2}{\uparrow}}{2^2} + \underset{\underset{k=3}{\uparrow}}{3^2} + \underset{\underset{k=4}{\uparrow}}{4^2} + \underset{\underset{k=5}{\uparrow}}{5^2} = 55.$$

EXAMPLE 10

Compute each of these sums.

(a) $\displaystyle\sum_{k=1}^{4} k^2(k-2)$

(b) $\displaystyle\sum_{k=1}^{6} (-1)^k k$.

SOLUTION

(a) Successively substituting 1, 2, 3, 4 for k in $k^2(k-2)$ and adding the results, we have

$$\sum_{k=1}^{4} k^2(k-2) = 1^2(1-2) + 2^2(2-2) + 3^2(3-2) + 4^2(4-2)$$
$$= 1(-1) + 4(0) + 9(1) + 16(2) = 40.$$

(b) $\displaystyle\sum_{k=1}^{6} (-1)^k k =$
$$(-1)^1 \cdot 1 + (-1)^2 \cdot 2 + (-1)^3 \cdot 3 + (-1)^4 \cdot 4 + (-1)^5 \cdot 5 + (-1)^6 \cdot 6$$
$$= -1 + 2 - 3 + 4 - 5 + 6 = 3. \qquad \blacksquare$$

In sums such as $\sum_{k=1}^{5} k^2$ and $\sum_{k=1}^{6} (-1)^k k$, The letter k is called the **summation index.** Any letter may be used for the summation index, just as the rule of a function f may be denoted by $f(x)$ or $f(t)$ or $f(k)$. For example, $\sum_{n=1}^{5} n^2$ means: Take the sum of the terms n^2 as n takes values from 1 to 5. In other words,

$$\sum_{n=1}^{5} n^2 = \sum_{k=1}^{5} k^2.$$

Similarly,

$$\sum_{k=1}^{4} k^2(k-2) = \sum_{j=1}^{4} j^2(j-2) = \sum_{n=1}^{4} n^2(n-2).$$

The Σ notation for sums can also be used for sums that don't begin with $k = 1$. For instance,

$$\sum_{k=4}^{10} k^2 = 4^2 + 5^2 + 6^2 + 7^2 + 8^2 + 9^2 + 10^2 = 371$$

$$\sum_{j=0}^{3} j^2(2j+5) = 0^2(2 \cdot 0 + 5) + 1^2(2 \cdot 1 + 5) + 2^2(2 \cdot 2 + 5) + 3^2(2 \cdot 3 + 5).$$

EXAMPLE 11

Use a calculator to compute these sums.

(a) $\displaystyle\sum_{k=1}^{50} k^2$ (b) $\displaystyle\sum_{k=38}^{75} k^2.$

TECHNOLOGY TIP

SUM (or ΣLIST on HP-39gs) is in this menu/submenu:

    TI-84+: LIST/*Math*

    TI-86: LIST/*Ops*

    TI-89: MATH/*List*

    Casio: OPTN/*List*

    HP-39gs: MATH/*List*

SOLUTION   In each case, use SUM together with SEQ (or ΣLIST and MAKE-LIST on HP-39gs), with the same syntax for SEQ as in Example 4:

$$\sum_{k=1}^{50} k^2 = \text{SUM SEQ}(x^2, x, 1, 50, 1) = 42{,}925$$

and

$$\sum_{k=38}^{75} k^2 = \text{SUM SEQ}(x^2, x, 38, 75, 1) = 125{,}875,$$

as shown in Figure 12–11. ∎

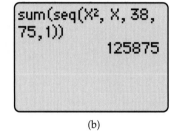

(a)   (b)

Figure 12–11

EXAMPLE 12

Express the following sum in Σ notation in two ways:

$$\frac{1}{2 \ln 2} + \frac{1}{3 \ln 4} + \frac{1}{4 \ln 8} + \frac{1}{5 \ln 16} + \frac{1}{6 \ln 32}.$$

SOLUTION   If we examine the pattern of the terms, we see that 2, 4, 8, 16, and 32 are powers of 2 and that the sum can be written as

$$\frac{1}{2 \ln 2^1} + \frac{1}{3 \ln 2^2} + \frac{1}{4 \ln 2^3} + \frac{1}{5 \ln 2^4} + \frac{1}{6 \ln 2^5}.$$

In each denominator, the exponent of 2 is one less than the first term of the denominator. Thus, when the first denominator term is k, the exponent is $k - 1$ and each term of the sum has the form $\dfrac{1}{k \ln 2^{k-1}}$. Since the first term begins with $k = 2$ and $k - 1 = 1$, we can write the sum as

$$\frac{1}{2 \ln 2^1} + \frac{1}{3 \ln 2^2} + \frac{1}{4 \ln 2^3} + \frac{1}{5 \ln 2^4} + \frac{1}{6 \ln 2^5} = \sum_{k=2}^{6} \frac{1}{k \ln 2^{k-1}}.$$

For a second way to express this sum, we first write it as

$$\frac{1}{(1+1) \ln 2^1} + \frac{1}{(2+1) \ln 2^2} + \frac{1}{(3+1) \ln 2^3} + \frac{1}{(4+1) \ln 2^4} + \frac{1}{(5+1) \ln 2^5}.$$

Now each term is of the form $\dfrac{1}{(k+1)\ln 2^k}$, with $k = 1$ corresponding to the first term. So the sum can also be written as

$$\frac{1}{2\ln 2^1} + \frac{1}{3\ln 2^2} + \frac{1}{4\ln 2^3} + \frac{1}{5\ln 2^4} + \frac{1}{6\ln 2^5} = \sum_{k=1}^{5} \frac{1}{(k+1)\ln 2^k}. \quad \blacksquare$$

PARTIAL SUMS

Suppose $\{a_n\}$ is a sequence and k is a positive integer. The sum of the first k terms of the sequence is called the **kth partial sum** of the sequence. Thus, we have the following.

Partial Sums

> The **kth partial sum** of $\{a_n\}$ is $\displaystyle\sum_{n=1}^{k} a_n = a_1 + a_2 + a_3 + \cdots + a_k$.

EXAMPLE 13

Here are some partial sums of the sequence $\{n^3\}$:

First partial sum: $\displaystyle\sum_{n=1}^{1} n^3 = 1^3 = 1,$

Second partial sum: $\displaystyle\sum_{n=1}^{2} n^3 = 1^3 + 2^3 = 9,$

Sixth partial sum: $\displaystyle\sum_{n=1}^{6} n^3 = 1^3 + 2^3 + 3^3 + 4^3 + 5^3 + 6^3 = 441. \quad \blacksquare$

EXAMPLE 14

The sequence $\{2^n\}_{n \geq 0}$ begins with the 0th term, so the fourth partial sum (the sum of the first four terms) is

$$2^0 + 2^1 + 2^2 + 2^3 = \sum_{n=0}^{3} 2^n.$$

Similarly, the fifth partial sum of the sequence $\left\{\dfrac{1}{n(n-2)}\right\}_{n \geq 3}$ is the sum of the first five terms.

$$\frac{1}{3(3-2)} + \frac{1}{4(4-2)} + \frac{1}{5(5-2)} + \frac{1}{6(6-2)} + \frac{1}{7(7-2)} = \sum_{n=3}^{7} \frac{1}{n(n-2)}. \quad \blacksquare$$

Certain calculations can be written very compactly in summation notation. For example, the distributive law shows that

$$ca_1 + ca_2 + ca_3 + \cdots + ca_r = c(a_1 + a_2 + a_3 + \cdots + a_r).$$

In summation notation, this becomes

$$\sum_{n=1}^{r} ca_n = c\left(\sum_{n=1}^{r} a_n\right).$$

This proves the first of the following statements.

Properties
of Sums

1. $\sum_{n=1}^{r} ca_n = c\left(\sum_{n=1}^{r} a_n\right)$ for any number c.

2. $\sum_{n=1}^{r} (a_n + b_n) = \sum_{n=1}^{r} a_n + \sum_{n=1}^{r} b_n$

3. $\sum_{n=1}^{r} (a_n - b_n) = \sum_{n=1}^{r} a_n - \sum_{n=1}^{r} b_n$

To prove statement 2, use the commutative and associative laws repeatedly to show that

$$(a_1 + b_1) + (a_2 + b_2) + (a_3 + b_3) + \cdots + (a_r + b_r)$$
$$= (a_1 + a_2 + a_3 + \cdots + a_r) + (b_1 + b_2 + b_3 + \cdots + b_r),$$

which can be written in summation notation as

$$\sum_{n=1}^{r} (a_n + b_n) = \sum_{n=1}^{r} a_n + \sum_{n=1}^{r} b_n.$$

The last statement is proved similarly.

EXERCISES 12.1

In Exercises 1–14, find the first five terms of the sequence
$\{a_n\}$.

1. $a_n = 2n + 6$

2. $a_n = 2^n - 7$

3. $a_n = \dfrac{1}{n^3}$

4. $a_n = \dfrac{1}{(n+3)(n+1)}$

5. $a_n = \dfrac{n}{2^n}$

6. $a_n = \sqrt{n^2 + 1}$

7. $a_n = (-1)^n \sqrt{n+2}$

8. $a_n = (-1)^{n+1} n(n-1)$

9. $a_n = 4 + (-.1)^n$

10. $a_n = 5 - (.1)^n$

11. $a_n = (-1)^n + 3n$

12. $a_n = (-1)^{n+2} - (n+1)$

13. a_n is the nth digit in the decimal expansion of π.

14. a_n is the nth digit in the decimal expansion of $1/13$.

In Exercises 15–24, find a formula for the nth term of the
sequence whose first few terms are given.

15. $-1, 1, -1, 1, -1, 1, \ldots$

16. $2, -2, 2, -2, 2, -2, \ldots$

17. $\dfrac{1}{2}, \dfrac{2}{3}, \dfrac{3}{4}, \dfrac{4}{5}, \dfrac{5}{6}, \ldots$

18. $\dfrac{1}{2 \cdot 3}, \dfrac{1}{3 \cdot 4}, \dfrac{1}{4 \cdot 5}, \dfrac{1}{5 \cdot 6}, \dfrac{1}{6 \cdot 7}, \ldots$

19. $2, 7, 12, 17, 22, 27, \ldots$

20. $8, 5, 2, -1, -4, \ldots$

21. $3, 6, 12, 24, 48, \ldots$

22. $-\dfrac{1}{8}, -\dfrac{1}{2}, -2, -8, -32, \ldots$

23. $4, \sqrt{32}, \sqrt{48}, 8, \sqrt{80}, \ldots$

24. $8, -5, 2, -11, -4, -17, -10, \ldots$

In Exercises 25–34, find the first five terms of the recursively
defined sequence.

25. $a_1 = 4$ and $a_n = 2a_{n-1} + 3$ for $n \geq 2$

26. $a_1 = 0$ and $a_n = 3a_{n-1} - 2$ for $n \geq 2$

27. $a_1 = -16$ and $a_n = \dfrac{a_{n-1}}{2}$ for $n \geq 2$

28. $a_1 = 3$ and $a_n = n + 2a_{n-1}$ for $n \geq 2$

29. $a_1 = 2$ and $a_n = n - a_{n-1}$ for $n \geq 2$

30. $a_1 = -3$ and $a_n = (-1)^n 4a_{n-1} - 5$ for $n \geq 2$

31. $a_1 = 1, a_2 = -2, a_3 = 3,$ and
$a_n = a_{n-1} + a_{n-2} + a_{n-3}$ for $n \geq 4$

32. $a_1 = 1, a_2 = 3,$ and $a_n = 2a_{n-1} + 3a_{n-2}$ for $n \geq 3$

33. $a_0 = 2, a_1 = 3,$ and $a_n = (a_{n-1})\left(\dfrac{1}{2} a_{n-2}\right)$ for $n \geq 2$

34. $a_0 = 1, a_1 = 1,$ and $a_n = na_{n-1}$ for $n \geq 2$

In Exercises 35–40, express the sum in Σ notation.

35. $1 + 2 + 3 + 4 + 5 + 6 + 7 + 8 + 9 + 10 + 11$

36. $1^1 + 2^2 + 3^3 + 4^4 + 5^5$

37. $\dfrac{1}{2^7} + \dfrac{1}{2^8} + \dfrac{1}{2^9} + \dfrac{1}{2^{10}} + \dfrac{1}{2^{11}} + \dfrac{1}{2^{12}} + \dfrac{1}{2^{13}}$

38. $(-6)^{11} + (-6)^{12} + (-6)^{13} + (-6)^{14} + (-6)^{15}$

39. $2 + \dfrac{2^2}{2} + \dfrac{2^3}{3} + \dfrac{2^4}{4} + \dfrac{2^5}{5} + \dfrac{2^6}{6} + \dfrac{2^7}{7}$

40. $2 + \dfrac{3}{2} + \dfrac{4}{3} + \dfrac{5}{4} + \dfrac{6}{5} + \dfrac{7}{6} + \dfrac{8}{7} + \dfrac{9}{8} + \dfrac{10}{9}$

In Exercises 41–50, find the sum.

41. $\displaystyle\sum_{k=1}^{7} k$

42. $\displaystyle\sum_{k=1}^{5} (-k)$

43. $\displaystyle\sum_{i=1}^{4} (i^2 + 1)$

44. $\displaystyle\sum_{i=1}^{10} 6$

45. $\displaystyle\sum_{i=1}^{5} 3i$

46. $\displaystyle\sum_{i=1}^{4} \dfrac{1}{2^i}$

47. $\displaystyle\sum_{n=1}^{16} (2n - 3)$

48. $\displaystyle\sum_{n=1}^{75} (-1)^n(3n + 1)$

49. $\displaystyle\sum_{n=15}^{36} (n^2 - 8)$

50. $\displaystyle\sum_{k=0}^{25} (2k^2 - 5k + 1)$

In Exercises 51–54, find the third and the sixth partial sums of the sequence.

51. $\{n^2 - 5n + 2\}$

52. $\{(2n - 3n^2)^2\}$

53. $\{(-1)^{n+1}5\}$

54. $\{2^n(2 - n^2)\}_{n \geq 0}$

In Exercises 55–58, express the given sum in Σ notation and find the sum.

55. $\dfrac{1}{3} + \dfrac{1}{5} + \dfrac{1}{7} + \dfrac{1}{9} + \dfrac{1}{11} + \dfrac{1}{13}$

56. $2 + 1 + \dfrac{4}{5} + \dfrac{5}{7} + \dfrac{2}{3} + \dfrac{7}{11} + \dfrac{8}{13}$

57. $\dfrac{1}{8} - \dfrac{2}{9} + \dfrac{3}{10} - \dfrac{4}{11} + \dfrac{5}{12}$

58. $\dfrac{2}{3 \cdot 5} + \dfrac{4}{5 \cdot 7} + \dfrac{8}{7 \cdot 9} + \dfrac{16}{9 \cdot 11}$

$\qquad\qquad + \dfrac{32}{11 \cdot 13} + \dfrac{64}{13 \cdot 15} + \dfrac{128}{15 \cdot 17}$

In Exercises 59–64, use a calculator to approximate the required term or sum.

59. a_{12} where $a_n = \left(1 + \dfrac{1}{n}\right)^n$

60. a_{50} where $a_n = \dfrac{\ln n}{n^2}$

61. a_{102} where $a_n = \dfrac{n^3 - n^2 + 5n}{3n^2 + 2n - 1}$

62. a_{125} where $a_n = \sqrt[n]{n}$

63. $\displaystyle\sum_{k=1}^{14} \dfrac{1}{k^2}$

64. $\displaystyle\sum_{n=8}^{22} \dfrac{1}{n}$

In Exercises 65–66, use a recursively defined sequence to find the answer, as in Example 9.

65. Suppose you take out a car loan for $9500 at 5% annual interest for 48 months. Your monthly payment is $218.78. How much do you owe after one year (12 payments)? After three years?

66. Lisa Chow is buying a condo. She takes out an $80,000 mortgage for 30 years at 6% annual interest. Her monthly payment is $479.64.

   (a) How much does she owe after one year (12 payments)?
   (b) How much interest does Lisa pay during the first five years? [*Hint:* The interest paid is the difference between her total payments and the amount of the loan paid off after five years

   ($80,000 − remaining balance).]

   (c) After 15 years, Lisa sells the condo and pays off the remaining mortgage balance. How much does she pay?

67. The number of bachelor's degrees (in thousands) awarded in year n is approximated by the sequence $\{a_n\}$, where $a_n = 25.3n + 1250$ and $n = 0$ corresponds to 2000.*

   (a) Approximately how many bachelor's degrees were awarded in 2004 and in 2007?
   (b) Approximately how many bachelor's degrees will be awarded between 2004 and 2009 (inclusive)?

68. The number of hours the average person spends listening to satellite radio in year n is approximated by the sequence $\{b_n\}$, where $n = 1$ corresponds to 2001 and $b_n = 945.54 + 73.57 \ln n$.* Round your final answers (not your calculations) to the following questions to the nearest hour.

   (a) For how many hours did the average person listen to satellite radio in 2002 and 2006?
   (b) How much total time did the average person spend listening to satellite radio from 2001 to 2005 (inclusive)?

69. Consumer spending per person (in dollars) on video games in year n is approximately $c_n = .16n^2 + 2.7n + 25.1$, where $n = 0$ corresponds to 2000.*

   (a) What was per person spending in 2005 and in 2007?
   (b) How much was spent per person from 2000 to 2007 (inclusive)?

70. Book sales (in billions of dollars) in the United States are approximated by the sequence $\{d_n\}$, where $d_n = 1.89n + 42.09$ and $n = 4$ corresponds to 2004.†

   (a) What were the sales in 2007?
   (b) What was the total spent on books from 2001 to 2005?

---

*Based on data and projections from the *Statistical Abstract of the United States: 2007.*

†Based on data and projections from the Book Industry Study Group, Inc.

THINKERS

*Exercises 71–75 deal with prime numbers. A positive integer greater than 1 is **prime** if its only positive integer factors are itself and 1. For example, 7 is prime because its only factors are 7 and 1, but 15 is not prime because it has factors other than 15 and 1 (namely, 3 and 5).*

71. (a) Let $\{a_n\}$ be the sequence of prime integers in their usual ordering. Verify that the first ten terms are 2, 3, 5, 7, 11, 13, 17, 19, 23, 29.

(b) Find $a_{17}, a_{18}, a_{19}, a_{20}$.

In Exercises 72–75 find the first five terms of the sequence.

72. a_n is the nth prime integer larger than 10. [*Hint:* $a_1 = 11$.]

73. a_n is the square of the nth prime integer.

74. a_n is the number of prime integers less than n.

75. a_n is the largest prime integer less than $5n$.

Exercises 76–82 deal with the Fibonacci sequence $\{a_n\}$ that was discussed in Example 6.

76. Leonardo Fibonacci discovered the sequence in the thirteenth century in connection with this problem: A rabbit colony begins with one pair of adult rabbits (one male, one female). Each adult pair produces one pair of babies (one male, one female) every month. Each pair of baby rabbits becomes adult and produces the first offspring at age two months. Assuming that no rabbits die, how many adult pairs of rabbits are in the colony at the end of n months ($n = 1, 2, 3, \ldots$)? [*Hint:* It may be helpful to make up a chart listing for each month the number of adult pairs, the number of one-month-old pairs, and the number of baby pairs.]

77. (a) List the first 10 terms of the Fibonacci sequence.

(b) List the first 10 partial sums of the sequence.

(c) Do the partial sums follow an identifiable pattern?

78. Verify that every positive integer less than or equal to 15 can be written as a sum of Fibonacci numbers, with none used more than once. For example, $12 = 1 + 3 + 8$ and 1, 3, and 8 are Fibonacci numbers.

79. Verify that $5(a_n)^2 + 4(-1)^n$ is always a perfect square for $n = 1, 2, \ldots, 10$.

80. Verify that $(a_n)^2 = a_{n+1}a_{n-1} + (-1)^{n-1}$ for $n = 2, \ldots, 10$.

81. Show that $\displaystyle\sum_{n=1}^{k} a_n = a_{k+2} - 1$. [*Hint:* $a_1 = a_3 - a_2$; $a_2 = a_4 - a_3$; etc.]

82. Show that $\displaystyle\sum_{n=1}^{k} a_{2n-1} = a_{2k}$, that is, the sum of the first k odd-numbered terms is the kth even-numbered term. [*Hint:* $a_3 = a_4 - a_2$; $a_5 = a_6 - a_4$; etc.]

12.2 Arithmetic Sequences

Section Objectives
- ■ Identify arithmetic sequences.
- ■ Find terms and a formula for an arithmetic sequence.
- ■ Find partial sums of arithmetic sequences.

An **arithmetic sequence** (sometimes called an **arithmetic progression**) is a sequence in which the difference between each term and the preceding one is always the same constant.

EXAMPLE 1

In the sequence 3, 8, 13, 18, 23, 28, . . . the difference between each term and the preceding one is always 5. So this is an arithmetic sequence. ■

EXAMPLE 2

Which of the following sequences are arithmetic?

(a) $14, 10, 6, 2, -2, -6, -10, -14, \ldots$

(b) $7, 7^2, 7^3, 7^4, 7^5, \ldots$

(c) $\log 5, \log 5^2, \log 5^3, \log 5^4, \log 5^5, \ldots$

SOLUTION

(a) The difference between each term and the preceding one is -4.

$$10 - 14 = -4 \qquad\qquad -2 - 2 = -4,$$
$$6 - 10 = -4 \qquad\qquad -6 - (-2) = -4,$$
$$2 - 6 = -4 \qquad\qquad -10 - (-6) = -4,$$

and so on. Hence, the sequence is arithmetic.

(b) Note that

$$7^2 - 7 = 49 - 7 = 42, \qquad \text{but} \qquad 7^3 - 7^2 = 343 - 49 = 294.$$

Since the difference between consecutive terms is not always the same, this sequence is not arithmetic.

(c) Using the Power Law for logarithms (Section 5.4), we see that

$$\log 5^2 - \log 5 = 2 \log 5 - \log 5 = \log 5,$$
$$\log 5^3 - \log 5^2 = 3 \log 5 - 2 \log 5 = \log 5,$$
$$\log 5^4 - \log 5^3 = 4 \log 5 - 3 \log 5 = \log 5,$$

and so on. The difference between each term and the preceding one is always $\log 5 \approx .69897$, so the sequence is arithmetic. ∎

If $\{a_n\}$ is an arithmetic sequence, then for each $n \geq 2$, the term preceding a_n is a_{n-1}, and the difference $a_n - a_{n-1}$ is some constant—call it d. Therefore, $a_n - a_{n-1} = d$ or, equivalently,

Arithmetic Sequences

> In an arithmetic sequence $\{a_n\}$,
>
> $$a_n = a_{n-1} + d$$
>
> for some constant d and all $n \geq 2$.

The number d is called the **common difference** of the arithmetic sequence.

EXAMPLE 3

If $\{a_n\}$ is an arithmetic sequence with $a_1 = 3$ and $a_2 = 4.5$, find the common difference and list the first eight terms.

SOLUTION   The common difference is

$$d = a_2 - a_1 = 4.5 - 3 = 1.5.$$

Therefore, the sequence begins $3, 4.5, 6, 7.5, 9, 10.5, 12, 13.5, \ldots$ ∎

EXAMPLE 4

Show that the sequence $\{-7 + 4n\}$ is arithmetic and find the common difference.

SOLUTION   The sequence is arithmetic because for each $n \geq 2$,

$$a_n - a_{n-1} = (-7 + 4n) - [-7 + 4(n-1)]$$
$$= (-7 + 4n) - (-7 + 4n - 4) = 4.$$

Therefore, the common difference is $d = 4$.   ■

If $\{a_n\}$ is an arithmetic sequence with common difference d, then for each $n \geq 2$, we know that $a_n = a_{n-1} + d$. Applying this fact repeatedly shows that

$$a_2 = a_1 + d$$
$$a_3 = a_2 + d = (a_1 + d) + d = a_1 + 2d$$
$$a_4 = a_3 + d = (a_1 + 2d) + d = a_1 + 3d$$
$$a_5 = a_4 + d = (a_1 + 3d) + d = a_1 + 4d,$$

and in general, we have the following.

nth Term of an
Arithmetic Sequence

> In an arithmetic sequence $\{a_n\}$ with common difference d,
>
> $$a_n = a_1 + (n-1)d \quad \text{for every } n \geq 1.$$

EXAMPLE 5

Find the nth term of the arithmetic sequence with first term -5 and common difference 3.

SOLUTION   Since $a_1 = -5$ and $d = 3$, the formula in the preceding box shows that

$$a_n = a_1 + (n-1)d = -5 + (n-1)3 = 3n - 8.$$ ■

EXAMPLE 6

What is the 45th term of the arithmetic sequence whose first three terms are 5, 9, 13?

SOLUTION   The first three terms show that $a_1 = 5$ and that the common difference d is 4. Applying the formula in the box with $n = 45$, we have

$$a_{45} = a_1 + (45 - 1)d = 5 + (44)4 = 181.$$ ■

EXAMPLE 7

If $\{a_n\}$ is an arithmetic sequence with $a_6 = 57$ and $a_{10} = 93$, find a_1 and a formula for a_n.

SOLUTION   Apply the formula $a_n = a_1 + (n-1)d$ with $n = 6$ and $n = 10$.

$$a_6 = a_1 + (6-1)d \quad \text{and} \quad a_{10} = a_1 + (10-1)d$$
$$57 = a_1 + 5d \qquad\qquad\quad 93 = a_1 + 9d$$
$$a_1 = 57 - 5d \qquad\qquad\quad a_1 = 93 - 9d.$$

These two expressions for a_1 must be equal.

$$57 - 5d = 93 - 9d$$

$$4d = 36$$

$$d = 9.$$

Substituting $d = 9$ in either of the equations above shows that $a_1 = 12$.

$$a_1 = 57 - 5(9) = 12 \qquad \text{or} \qquad a_1 = 93 - 9(9) = 12.$$

So the formula for a_n is

$$a_n = a_1 + (n - 1)d = 12 + (n - 1)9 = 9n + 3. \qquad \blacksquare$$

 PARTIAL SUMS

It's easy to compute partial sums of arithmetic sequences by using the following formulas.

Partial Sums of an Arithmetic Sequence

If $\{a_n\}$ is an arithmetic sequence with common difference d, then for each positive integer k, the kth partial sum can be found by using *either* of these formulas.

1. $\displaystyle\sum_{n=1}^{k} a_n = \frac{k}{2}(a_1 + a_k)$ or

2. $\displaystyle\sum_{n=1}^{k} a_n = ka_1 + \frac{k(k-1)}{2}d.$

Proof   Let S denote the kth partial sum $a_1 + a_2 + \cdots + a_k$. For reasons that will become apparent later, we shall calculate the number $2S$.

$$2S = S + S = (a_1 + a_2 + \cdots + a_k) + (a_1 + a_2 + \cdots + a_k).$$

Now we rearrange the terms on the right by grouping the first and last terms together, then the first and last of the remaining terms, and so on.

$$2S = (a_1 + a_k) + (a_2 + a_{k-1}) + (a_3 + 2_{k-2}) + \cdots + (a_k + a_1).$$

Since adjacent terms of the sequence differ by d, we have

$$a_2 + a_{k-1} = (a_1 + d) + (a_k - d) = a_1 + a_k.$$

Using this fact, we have

$$a_3 + a_{k-2} = (a_2 + d) + (a_{k-1} - d) = a_2 + a_{k-1} = a_1 + a_k.$$

Continuing in this manner, we see that every pair in the sum for $2S$ is equal to $a_1 + a_k$. Therefore,

$$2S = (a_1 + a_k) + (a_2 + a_{k-1}) + (a_3 + a_{k-2}) + \cdots + (a_k + a_1)$$

$$= (a_1 + a_k) + (a_1 + a_k) + (a_1 + a_k) + \cdots + (a_1 + a_k) \quad (k \text{ terms})$$

$$= k(a_1 + a_k).$$

Dividing both sides of this last equation by 2 shows that $S = \dfrac{k}{2}(a_1 + a_k)$. This proves the first formula. To obtain the second one, note that

$$a_1 + a_k = a_1 + [a_1 + (k-1)d] = 2a_1 + (k-1)d.$$

Substituting the right side of this equation in the first formula for S shows that

$$S = \frac{k}{2}(a_1 + a_k) = \frac{k}{2}[2a_1 + (k-1)d] = ka_1 + \frac{k(k-1)}{2}d.$$

This proves the second formula. ■

EXAMPLE 8

Find the 12th partial sum of the arithmetic sequence that begins $-8, -3, 2, 7, \ldots$.

SOLUTION   We first note that the common difference d is 5. Since $a_1 = -8$ and $d = 5$, the second formula in the box with $k = 12$ shows that

$$\sum_{n=1}^{12} a_n = 12(-8) + \frac{12(11)}{2}5 = -96 + 330 = 234.$$ ■

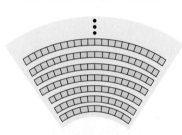

Figure 12–12

EXAMPLE 9

A corner section of a stadium has 28 rows of seats. There are 10 seats in the first row, 12 in the second row, 14 in the third row, and so on, as indicated in Figure 12–12.

(a) How many seats are in the 20th row?

(b) How many seats are in the entire section?

SOLUTION

(a) Let a_n be the number of seats in row n. Then $a_1 = 10$, $a_2 = 12$, $a_3 = 14$, and so on. Thus, we have an arithmetic sequence, with common difference $d = 2$. According to the box before Example 5, the nth term of the sequence is

$$a_n = a_1 + (n-1)d$$
$$a_n = 10 + (n-1)2 = 8 + 2n.$$

The number of seats in the 20th row is

$$a_{20} = 8 + 2(20) = 48.$$

(b) The total number of seats in all 28 rows is the 28th the partial sum of the sequence $\{a_n\}$, which is given by the second formula in the box on page 840.

$$\sum_{n=1}^{k} a_n = ka_1 + \frac{k(k-1)}{2}d.$$

Here $k = 28$, $a_1 = 10$, and $d = 2$, so

$$\sum_{n=1}^{28} a_n = 28(10) + \frac{28(28-1)}{2}2$$

$$= 280 + 28(27) = 1036 \text{ seats.}$$ ■

EXAMPLE 10

Find the sum of all multiples of 3 from 3 to 333.

SOLUTION   The answer can be found in two ways.

Algebraic Method: This sum is a partial sum of the arithmetic sequence 3, 6, 9, 12, ... Since this sequence can be written in the form

$$3 \cdot 1, 3 \cdot 2, 3 \cdot 3, 3 \cdot 4, 3 \cdot 5, 3 \cdot 6, \ldots,$$

we see that $333 = 3 \cdot 111$ is the 111th term. The 111th partial sum of this sequence can be found by using the first formula in the box with $k = 111$, and $a_1 = 3$, and $a_{111} = 333$.

$$\sum_{n=1}^{111} a_n = \frac{111}{2}(3 + 333) = \frac{111}{2}(336) = 18{,}648.$$

Calculator Method: Note that the nth term of the sequence 3, 6, 9, . . . is $3n$. Consequently, we can compute the 111th partial sum as in Figure 12–13.   ∎

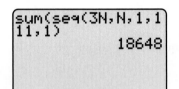

```
sum(seq(3N,N,1,1
11,1)
                18648
```

Figure 12–13

EXAMPLE 11

If the starting salary for a job is $20,000 and you get a $2000 raise at the beginning of each subsequent year, what will your salary be during the tenth year? How much will you earn during the first ten years?

SOLUTION   Your yearly salary rates form a sequence: 20,000, 22,000, 24,000, 26,000, and so on. It is an arithmetic sequence with $a_1 = 20{,}000$ and $d = 2000$. Your tenth-year salary is

$$a_{10} = a_1 + (10 - 1)d = 20{,}000 + 9 \cdot 2000 = \$38{,}000.$$

Your ten-year total earnings are the tenth partial sum of the sequence.

$$\frac{10}{2}(a_1 + a_{10}) + \frac{10}{2}(20{,}000 + 38{,}000) = 5(58{,}000) = \$290{,}000.$$ ∎

EXERCISES 12.2

In Exercises 1–8, determine whether the sequence is arithmetic or not. If it is, find the common difference.

1. $1, 3, 5, 7, 9, \ldots$

2. $\dfrac{1}{3}, \dfrac{2}{3}, \dfrac{3}{3}, \dfrac{4}{3}, \dfrac{5}{3}, \ldots$

3. $1, 2, 4, 8, 16, \ldots$

4. $-9, -6, -3, 0, \ldots$

5. $\log 1, \log 2, \log 4, \log 8, \log 16, \ldots$

6. $\log 3, \log 6, \log 9, \log 12, \log, 15, \ldots$

7. $\dfrac{1}{3}, -\dfrac{4}{6}, -\dfrac{15}{9}, -\dfrac{32}{12}, -\dfrac{55}{15}, -\dfrac{84}{18}, \ldots$

8. $\dfrac{1}{2}, \dfrac{3}{4}, 1, \dfrac{3}{4}, \dfrac{1}{2}, \dfrac{1}{4}, \ldots$

In Exercises 9–16, write the first five terms of the sequence whose nth term is given. Use them to decide whether the sequence is arithmetic. If it is, list the common difference.

9. $a_n = 5 + 4n$

10. $b_n = n - \dfrac{5}{4}$

11. $c_n = (-1)^n$

12. $d_n = 2n + \dfrac{1}{n}$

13. $a_n = 2 + (-1)^n n$

14. $c_n = 1 + \dfrac{n}{3}$

15. $a_n = e^{\ln n}$

16. $b_n = 3\sqrt{7n^2}$

In Exercises 17–24, show that the sequence is arithmetic and find its common difference.

17. $\{3 - 2n\}$ **18.** $\{1.5 + 1.5n\}$

19. $\left\{4 + \dfrac{n}{3}\right\}$ **20.** $\left\{-3 - \dfrac{n}{2}\right\}$

21. $\left\{\dfrac{5 + 3n}{2}\right\}$ **22.** $\left\{\dfrac{\pi - n}{2}\right\}$

23. $\{c + 2n\}$ (*c* constant)

24. $\{2b + 3nc\}$ (*b*, *c* constants)

In Exercises 25–32, the first term a_1 and the common difference d of an arithmetic sequence are given. Find the fifth term and the formula for the nth term.

25. $a_1 = 5, d = 2$ **26.** $a_1 = -4, d = 5$

27. $a_1 = 4, d = \dfrac{1}{4}$ **28.** $a_1 = -6, d = \dfrac{2}{3}$

29. $a_1 = 10, d = -\dfrac{1}{2}$ **30.** $a_1 = \pi, d = \dfrac{1}{5}$

31. $a_1 = 8, d = .1$ **32.** $a_1 = -.1, d = -8$

In Exercises 33–40, use the given information about the arithmetic sequence with common difference d to find a_1 and a formula for a_n.

33. $a_4 = 12, d = 2$ **34.** $a_7 = -8, d = 3$

35. $a_3 = 3, d = 5$ **36.** $a_4 = -5, d = -5$

37. $a_2 = 4, a_6 = 32$ **38.** $a_7 = 6, a_{12} = -4$

39. $a_5 = 0, a_9 = 6$ **40.** $a_5 = -3, a_9 = -18$

In Exercises 41–48, find the kth partial sum of the arithmetic sequence $\{a_n\}$ with common difference d.

41. $k = 6, a_1 = 2, d = 5$ **42.** $k = 8, a_1 = \dfrac{2}{3}, d = -\dfrac{4}{3}$

43. $k = 7, a_1 = \dfrac{3}{4}, d = -\dfrac{1}{2}$ **44.** $k = 9, a_1 = -4, d = \dfrac{1}{2}$

45. $k = 6, a_1 = -4, a_6 = 14$ **46.** $k = 10, a_1 = 0, a_{10} = 30$

47. $k = 9, a_1 = 6, a_9 = -24$ **48.** $k = 8, a_1 = -6, a_8 = 13$

In Exercises 49–54, find the sum.

49. $\displaystyle\sum_{n=1}^{20}(3n + 4)$ **50.** $\displaystyle\sum_{n=1}^{25}\left(\dfrac{n}{4} + 5\right)$

51. $\displaystyle\sum_{n=1}^{30}\left(3 - \dfrac{n}{2}\right)$ **52.** $\displaystyle\sum_{n=1}^{35}\left(\dfrac{2n + 4}{8}\right)$

53. $\displaystyle\sum_{n=1}^{40}\dfrac{n + 3}{6}$ **54.** $\displaystyle\sum_{n=1}^{30}\dfrac{4 - 6n}{3}$

55. The sequence in which

a_n = per capita amount spent on health services and supplies in year n,

with $n = 1$ corresponding to 1999, is approximately arithmetic.*

(a) If the per capita amount was \$4154 in 1999 and \$5864 in 2004, find a formula for a_n.

(b) Use the sequence to estimate the per capita amount in 2005 and 2008.

56. The sequence in which

b_n = remaining life expectancy of a man at age n

is approximately arithmetic.†

(a) Use the fact that a man's remaining life expectancy is 60.6 years at age 15 and 51.2 years at age 25 to find a formula for b_n.

(b) Determine the remaining life expectancy of a man at these ages: 20, 22, 30 and 40.

57. A lecture hall has six seats in the first row, eight in the second, ten in the third, and so on, through row 12. Rows 12 through 20 (the last row) all have the same number of seats. Find the number of seats in the lecture hall.

58. A monument is constructed by first laying a row of 60 bricks at ground level. A second row, with two fewer bricks, is centered on that; a third row, with two fewer bricks, is centered on the second; and so on. The top row contains 10 bricks. How many bricks are there in the monument?

59. A ladder with nine rungs is to be built, with the bottom rung 24 inches wide and the top rung 18 inches wide. If the lengths of the rungs decrease uniformly from bottom to top, how long should each of the seven intermediate rungs be?

60. Find the first eight numbers in an arithmetic sequence in which the sum of the first and seventh term is 40 and the product of the first and fourth terms is 160.

61. Find the sum of all the even integers from 2 to 100.

62. Find the sum of all the integer multiples of 7 from 7 to 700.

63. Find the sum of the first 200 positive integers.

64. Find the sum of the positive integers from 101 to 200 (inclusive). [*Hint:* What's the sum from 1 to 100? Use it and Exercise 63.]

65. A business makes a \$10,000 profit during its first year. If the yearly profit increases by \$7500 in each subsequent year, what will the profit be in the tenth year and what will the total profit for the first 10 years be?

66. If a man's starting salary is \$24,000 and he receives a \$1000 increase every six months, what will his salary be during the last six months of the sixth year? How much will he earn during the first six years?

---

*U.S. Centers for Medicare and Medicaid Services
†National Center for Health Statistics

67. The sequence with

$$a_n = \begin{array}{l} \text{holiday retail sales in year } n \\ \text{(in billions of dollars),} \end{array}$$

with $n = 1$ corresponding to 1994, is approximately arithmetic.*

(a) If sales were \$136.4 billion in 1994 and \$217.4 billion in 2003, find the formula for a_n.
(b) Estimate holiday sales in 2005.
(c) Find the holiday sales from 2000 to 2005.

68. Let $\{b_n\}$ be the sequence that gives U.S. personal income in year n (in trillions of dollars), where $n = 1$ corresponds to 2000. This sequence is approximately arithmetic.†

(a) If personal income was \$8.4 trillion in 2000 and \$9.7 trillion in 2004, find a formula for b_n.
(b) Estimate personal income in 2005.
(c) Find the total personal income from 2000 to 2005.

---

*Based on data from the U.S. Department of Commerce.
†Based on data from the U.S. Bureau of Economic Analysis.

69. Let $n = 1$ correspond to 1981. The arithmetic sequence in which

$$c_n = \begin{array}{l} \text{average yearly earnings of a production} \\ \text{worker in manufacturing in year } n \end{array}$$

has $c_1 = \$15,828.80$ and $c_{19} = \$28,932.80$.*

(a) Find a formula for c_n.
(b) Find the total earnings of an average production worker from 1981 to 2005.

70. Let $n = 1$ correspond to 2001. The arithmetic sequence in which

$$d_n = \text{hours per person spent on the Internet in year } n$$

has $d_1 = 136$ and $d_6 = 205$.†

(a) Find a formula for d_n.
(b) How many hours per person were spent on the Internet in 2007?
(c) Find the total number of hours per person on the Internet from 2001 to 2005.

---

*Based on data from the U.S. Bureau of Labor Statistics.
†Based on data and projections in *Statistical Abstract of the United States: 2006.*

12.3   Geometric Sequences

Section Objectives
- Identify geometric sequences.
- Find terms and a formula for a geometric sequence.
- Find partial sums of geometric sequences.
- Use geometric sequences to solve applied problems.

A **geometric sequence** (sometimes called a **geometric progression**) is a sequence in which the quotient of each term and the preceding one is the same constant r. This constant r is called the **common ratio** of the geometric sequence.

EXAMPLE 1

The sequence $3, 9, 27, \ldots 3^n, \ldots$ is geometric with common ratio 3. For instance, $a_2/a_1 = 9/3 = 3$ and $a_3/a_2 = 27/9 = 3$. If 3^n is any term ($n \geq 2$), then the preceding term is 3^{n-1}, and

$$\frac{3^n}{3^{n-1}} = \frac{3 \cdot 3^{n-1}}{3^{n-1}} = 3. \qquad \blacksquare$$

EXAMPLE 2

Which of the following sequences are geometric?

(a) $\dfrac{1}{2}, \dfrac{2}{3}, \dfrac{3}{4}, \dfrac{4}{5}, \dfrac{5}{6}, \ldots$ (b) $e, -e^3, e^5, -e^7, e^9, \ldots$ (c) $\left\{\dfrac{5}{2^n}\right\}$

SOLUTION

(a) Note that

$$\frac{a_2}{a_1} = \frac{2/3}{1/2} = \frac{2}{3} \cdot \frac{2}{1} = \frac{4}{3}, \qquad \text{but} \qquad \frac{a_3}{a_2} = \frac{3/4}{2/3} = \frac{3}{4} \cdot \frac{3}{2} = \frac{9}{8}.$$

Since the quotient of consecutive terms is not always the same, this sequence is not geometric.

(b) Compute the quotient of each term and the preceding one.

$$\frac{a_2}{a_1} = \frac{-e^3}{e} = -e^2, \qquad \frac{a_4}{a_3} = \frac{-e^7}{e^5} = -e^2,$$

$$\frac{a_3}{a_2} = \frac{e^5}{-e^3} = -e^2, \qquad \frac{a_5}{a_4} = \frac{e^9}{-e^7} = -e^2,$$

and so on. The consecutive quotients are always the number $-e^2$, so the sequence is geometric with common ratio $-e^2$.

(c) For each $n \geq 2$, the term preceding a_n is a_{n-1}. Here,

$$a_n = \frac{5}{2^n} \qquad \text{and} \qquad a_{n-1} = \frac{5}{2^{n-1}},$$

so

$$\frac{a_n}{a_{n-1}} = \frac{5/2^n}{5/2^{n-1}} = \frac{5}{2^n} \cdot \frac{2^{n-1}}{5} = \frac{2^{n-1}}{2^n} = \frac{1}{2}.$$

Since the quotient of each term and the preceding one is always $1/2$, the sequence is geometric with common ratio $1/2$. ∎

Suppose $\{a_n\}$ is a geometric sequence with common ratio r. For each $n \geq 2$, the quotient of the term a_n and the preceding one is $\dfrac{a_n}{a_{n-1}} = r$. Multiplying both sides of the equation by a_{n-1} produces this fact.

Geometric Sequences

> In a geometric sequence $\{a_n\}$ with common ratio r,
>
> $$a_n = ra_{n-1} \qquad \text{for each } n \geq 2.$$

Applying this last formula for $n = 2, 3, 4, \ldots$, we have

$$a_2 = ra_1,$$
$$a_3 = ra_2 = r(ra_1) = r^2 a_1,$$
$$a_4 = ra_3 = r(r^2 a_1) = r^3 a_1,$$
$$a_5 = ra_4 = r(r^3 a_1) = r^4 a_1,$$

and in general we have the following.

nth Term of a Geometric Sequence

> If $\{a_n\}$ is a geometric sequence with common ratio r, then for all $n \geq 1$,
>
> $$a_n = r^{n-1} a_1.$$

EXAMPLE 3

Find a formula for the nth term of the geometric sequence $\{a_n\}$ that satisfies the given conditions.

(a) $a_1 = 7$ and $r = 2$ (b) $a_3 = 8$ and $r = 1/2$

SOLUTION

(a) The equation in the preceding box shows that

$$a_n = r^{n-1}a_1 = 2^{n-1} \cdot 7.$$

(b) We must first find a_1.

$$a_n = r^{n-1}a_1$$

Let $n = 3$: $\qquad a_3 = r^{3-1}a_1$

Substitute $a_3 = 8$ and $r = 1/2$: $\quad 8 = \left(\dfrac{1}{2}\right)^2 a_1$

$$8 = \frac{1}{4}a_1$$

Multiply both sides by 4: $\qquad 32 = a_1.$

Therefore,

$$a_n = r^{n-1}a_1 = \left(\frac{1}{2}\right)^{n-1} \cdot 32. \qquad \blacksquare$$

EXAMPLE 4

Find a formula for the nth term of the geometric sequence whose first two terms are 2 and $-2/5$.

SOLUTION The common ratio is

$$r = \frac{a_2}{a_1} = \frac{-2/5}{2} = \frac{-2}{5} \cdot \frac{1}{2} = -\frac{1}{5}.$$

Using the equation in the box, we now see that the formula for the nth term is

$$a_n = r^{n-1}a_1 = \left(-\frac{1}{5}\right)^{n-1}(2) = \frac{(1)^{n-1}}{(-5)^{n-1}}(2) = \frac{2}{(-5)^{n-1}}.$$

So, the sequence begins $2, -2/5, 2/5^2, -2/5^3, 2/5^4, \ldots$. $\qquad \blacksquare$

EXAMPLE 5

Find a formula for the nth term of the geometric sequence $\{a_n\}$ in which $a_2 = 20/9$ and $a_5 = 160/243$.

SOLUTION By the equation in the box above, we have

$$\frac{160/243}{20/9} = \frac{a_5}{a_2} = \frac{r^4 a_1}{r a_1} = r^3.$$

Consequently,

$$r = \sqrt[3]{\frac{160/243}{20/9}} = \sqrt[3]{\frac{160}{243} \cdot \frac{9}{20}} = \sqrt[3]{\frac{8 \cdot 9}{243}} = \sqrt[3]{\frac{8}{27}} = \frac{2}{3}.$$

Since $a_2 = ra_1$, we see that

$$a_1 = \frac{a_2}{r} = \frac{20/9}{2/3} = \frac{20}{9} \cdot \frac{3}{2} = \frac{10}{3}.$$

Therefore,

$$a_n = r^{n-1}a_1 = \left(\frac{2}{3}\right)^{n-1} \cdot \frac{10}{3} = \frac{2^{n-1} \cdot 2 \cdot 5}{3^{n-1} \cdot 3} = \frac{2^n \cdot 5}{3^n} = 5\left(\frac{2}{3}\right)^n. \qquad \blacksquare$$

PARTIAL SUMS

If the common ratio r of a geometric sequence is the number 1, then we have

$$a_n = 1^{n-1}a_1 \quad \text{for every } n \geq 1.$$

Therefore, the sequence is just the constant sequence $a_1, a_1, a_1, \ldots$. For any positive integer k, the kth partial sum of this constant sequence is

$$\underbrace{a_1 + a_1 + \cdots + a_1}_{k \text{ terms}} = ka_1$$

In other words, the kth partial sum of a constant sequence is just k times the constant. If a geometric sequence is not constant (that is, $r \neq 1$), then its partial sums are given by the following formula.

Partial Sums of a Geometric Sequence

> The kth partial sum of the geometric sequence $\{a_n\}$ with common ratio $r \neq 1$ is
>
> $$\sum_{n=1}^{k} a_n = a_1\left(\frac{1 - r^k}{1 - r}\right).$$

Proof If S denotes the kth partial sum, then the formula for the nth term of a geometric sequence shows that

$$S = a_1 + a_2 + \cdots + a_k = a_1 + a_1 r + a_1 r^2 + a_1 r^3 + \cdots + a_1 r^{k-1}.$$

Use this equation to compute $S - rS$.

$$S = a_1 + a_1 r + a_1 r^2 + a_1 r^3 + \cdots + a_1 r^{k-1}$$
$$\underline{rS = \qquad a_1 r + a_1 r^2 + a_1 r^3 + \cdots + a_1 r^{k-1} + a_1 r^k}$$
$$S - rS = a_1 \qquad\qquad\qquad\qquad\qquad\qquad\qquad - a_1 r^k$$

$$(1 - r)S = a_1(1 - r^k).$$

Since $r \neq 1$, we can divide both sides of this last equation by $1 - r$ to complete the proof.

$$S = \frac{a_1(1 - r^k)}{1 - r} = a_1\left(\frac{1 - r^k}{1 - r}\right). \qquad \blacksquare$$

EXAMPLE 6

Find the sum

$$-\frac{3}{2}+\frac{3}{4}-\frac{3}{8}+\frac{3}{16}-\frac{3}{32}+\frac{3}{64}-\frac{3}{128}+\frac{3}{256}-\frac{3}{512}.$$

SOLUTION Note that this is the ninth partial sum of the geometric sequence $\left\{3\left(\dfrac{-1}{2}\right)^n\right\}$. The common ratio is $r = -1/2$. The formula in the box shows that

$$\sum_{n=1}^{9} 3\left(\frac{-1}{2}\right)^n = a_1\left(\frac{1-r^9}{1-r}\right) = \left(\frac{-3}{2}\right)\left[\frac{1-(-1/2)^9}{1-(-1/2)}\right]$$

$$= \left(\frac{-3}{2}\right)\left(\frac{1+1/2^9}{3/2}\right) = \left(\frac{-3}{2}\right)\left(\frac{2}{3}\right)\left(1+\frac{1}{2^9}\right)$$

$$= -1 - \frac{1}{2^9} = -1 - \frac{1}{512} = -\frac{513}{512}.$$

Figure 12–14

The sum can also be found with a calculator (Figure 12–14) and expressed in fractional form.*

EXAMPLE 7

A superball is dropped from a height of 9 feet. It hits the ground and bounces to a height of 6 feet. It continues to bounce up and down. On each bounce, it rises to 2/3 of the height of the previous bounce. How far has the ball traveled (both up and down) when it hits the ground for the seventh time?

SOLUTION We first consider how far the ball travels on each bounce. On the first bounce, it rises 6 feet and falls 6 feet for a total of 12 feet. On the second bounce, it rises and falls 2/3 of the previous height and hence travels 2/3 of 12 feet. If a_n denotes the distance traveled on the nth bounce, then

$$a_1 = 12, \qquad a_2 = \left(\frac{2}{3}\right)a_1, \qquad a_3 = \left(\frac{2}{3}\right)a_2 = \left(\frac{2}{3}\right)^2 a_1,$$

and in general

$$a_n = \left(\frac{2}{3}\right)a_{n-1} = \left(\frac{2}{3}\right)^{n-1} a_1.$$

So $\{a_n\}$ is a geometric sequence with common ratio $r = 2/3$. When the ball hits the ground for the seventh time, it has completed six bounces. Therefore, the total distance it has traveled is the distance it was originally dropped (9 feet) plus the distance traveled in six bounces, namely,

$$9 + a_1 + a_2 + a_3 + a_4 + a_5 + a_6 = 9 + \sum_{n=1}^{6} a_n = 9 + a_1\left(\frac{1-r^6}{1-r}\right)$$

$$= 9 + 12\left[\frac{1-(2/3)^6}{1-(2/3)}\right] \approx 41.84 \text{ feet.}$$

---

*For fractional form on TI, use the *Frac* key, as in Figure 12–14. On HP-39gs, change the number format in the MODE menu to "fraction." On Casio 9850, use the *Frac* program in the Program Appendix.

EXAMPLE 8*

When a patient is given a 20 mg dose of aminophylline, the amount of the drug in the bloodstream t hours later is given by

$$C(t) = 20e^{-.1155t}.^\dagger$$

Suppose that a 20 mg dose is given every four hours. Let $\{b_n\}$ be the sequence with

b_n = the total amount of aminophylline in the bloodstream after the nth dose.

(a) Find a formula for the nth term of the sequence $\{b_n\}$.

(b) Find the total amount of aminophylline in the bloodstream after 7 doses.

SOLUTION

(a) First, consider what happens to a single dose. The initial amount is $C(0) = 20$ mg. The amount after 4 hours is

$$C(4) = 20e^{-.1155(4)} = 20e^{-.462} \approx 12.60 \text{ mg}.$$

Similarly, the amounts after 8, 12, and 16 hours, respectively, are

$$C(8) = C(4 \cdot 2) = 20e^{-.1155(4)(2)} = 20[e^{-.1155(4)}]^2 = 20[e^{-.462}]^2 \approx 7.94 \text{ mg}$$
$$C(12) = C(4 \cdot 3) = 20e^{-.1155(4)(3)} = 20[e^{-.1155(4)}]^3 = 20[e^{-.462}]^3 \approx 5.00 \text{ mg}$$
$$C(16) = C(4 \cdot 4) = 20e^{-.1155(4)(4)} = 20[e^{-.1155(4)}]^4 = 20[e^{-.462}]^4 \approx 3.15 \text{ mg}.$$

So the amount of the drug in the bloodstream from a single dose decreases slowly over time.

When a dose is given every four hours, the amount of the drug in the bloodstream after each new dose is 20 mg from the new dose plus the amounts remaining from all of the earlier doses. The following chart describes the situation.

Amount of Drug in bloodstream at time t from each dose

Time	Dose 1	Dose 2	Dose 3	Dose 4 · · ·
$t = 0$	20			
$t = 4$	$20e^{-.462}$	20		
$t = 4 \cdot 2$	$20[e^{-.462}]^2$	$20e^{-.462}$	20	
$t = 4 \cdot 3$	$20[e^{-.462}]^3$	$20[e^{-.462}]^2$	$20e^{-.462}$	20
⋮	⋮	⋮	⋮	⋮

Each column shows the amount of a particular dose that remains at the given time. The sum of each row is the total amount in the bloodstream at the given time. Summing each row from right to left, we see that

$b_1 = 20$ [Amount after 1st dose $(t = 0)$]

$b_2 = 20 + 20e^{-.462}$ [Amount after 2nd dose $(t = 4 \cdot 1)$]

$b_3 = 20 + 20e^{-.462} + 20[e^{-.462}]^2$ [Amount after 3rd dose $(t = 4 \cdot 2)$]

$b_4 = 20 + 20e^{-.462} + 20[e^{-.462}]^2 + 20[e^{-.462}]^3$ [Amount after 4th dose $(t = 4 \cdot 3)$]

---

*Our thanks to Theresa Laurent of the St. Louis College of Pharmacy for providing data and other helpful information for this example and several related exercises.

†We assume that the drug is administered intravenously and that the entire dose enters the bloodstream immediately.

The same pattern continues, so the amount after the nth dose [when $t = 4 \cdot (n - 1)$] is

$$b_n = 20 + 20e^{-.462} + 20[e^{-.462}]^2 + 20[e^{-.462}]^3 + \cdots + 20[e^{-.462}]^{n-1}$$

If you look carefully at b_n, you see that it is the nth partial sum of the geometric series with $a_1 = 20$ and $r = e^{-.462}$. By the partial sum formula,

$$b_n = a_1\left(\frac{1 - r^n}{1 - r}\right) = 20\left(\frac{1 - (e^{-.462})^n}{1 - e^{-.462}}\right) = 20\left(\frac{1 - e^{-.462n}}{1 - e^{-.462}}\right).$$

(b) To find the amount in the bloodstream after 7 doses, we use the preceding formula to compute b_7.

$$b_n = 20\left(\frac{1 - e^{-.462n}}{1 - e^{-.462}}\right)$$

Let $n = 7$: $b_7 = 20\left(\frac{1 - e^{-.462(7)}}{1 - e^{-.462}}\right) \approx 51.927$ mg.   ■

EXERCISES 12.3

In Exercises 1–12, determine whether the sequence is arithmetic, geometric, or neither.

1. $2, 7, 12, 17, 22, \ldots$ 2. $2, 6, 18, 54, 162, \ldots$

3. $13, 13/2, 13/4, 13/8, \ldots$

4. $-1, -\frac{1}{2}, 0, \frac{1}{2}, \ldots$ 5. $50, 48, 46, 44, \ldots$

6. $2, -3, 9/2, -27/4, -81/8, \ldots$

7. $3, -3/2, 3/4, -3/8, 3/16, \ldots$

8. $-6, -3.7, -1.4, .9, 3.2, \ldots$

9. $3, 3\sqrt{2}, 6, 6\sqrt{2}, 12, 12\sqrt{2}, \ldots$

10. $\ln e, \ln e^2, \ln e^3, \ln e^4, \ln e^5, \ldots$

11. $6, 6, 6, 6, 6, \ldots$

12. $1, \sqrt[3]{3}, \sqrt[3]{9}, 3, 3\sqrt[3]{3}, 3\sqrt[3]{9}, 9, \ldots$

In Exercises 13–22, one term and the common ratio r of a geometric sequence are given. Find the sixth term and a formula for the nth term.

13. $a_1 = 5, r = 2$ 14. $a_1 = 1, r = -2$

15. $a_1 = 4, r = \frac{1}{4}$ 16. $a_1 = -6, r = \frac{2}{3}$

17. $a_1 = 10, r = -\frac{1}{2}$ 18. $a_1 = \pi, r = \frac{1}{5}$

19. $a_2 = 12, r = 1/3$ 20. $a_3 = 1/2, r = 3$

21. $a_4 = -4/5, r = 2/5$ 22. $a_5 = 2/3, r = -1/3$

In Exercises 23–30, show that the given sequence is geometric and find the common ratio.

23. $\left\{\left(-\frac{1}{2}\right)^n\right\}$ 24. $\{2^{3n}\}$

25. $\{5^{n+2}\}$ 26. $\{3^{n/2}\}$

27. $\{(\sqrt{5})^n\}$ 28. $\{4^{n-4}\}$

29. $\{10e^{.4n}\}$ 30. $\{5e^{-.5n}\}$

In Exercises 31–38, use the given information about the geometric sequence $\{a_n\}$ to find a_5 and a formula for a_n.

31. $a_1 = 256, a_2 = -64$ 32. $a_1 = 1/6, a_2 = -1/18$

33. $a_1 = 1/2, a_2 = 5$ 34. $a_1 = \sqrt{7}, a_2 = \sqrt{42}$

35. $a_3 = 4, a_6 = 1/16$ 36. $a_3 = 4, a_6 = -32$

37. $a_1 = 5, a_7 = 20$ (assume that $r > 0$)

38. $a_2 = 6, a_7 = 192$

In Exercises 39–42, find the kth partial sum of the geometric sequence $\{a_n\}$ with common ratio r.

39. $k = 6, a_1 = 5, r = \frac{1}{2}$ 40. $k = 8, a_1 = 9, r = \frac{1}{3}$

41. $k = 7, a_2 = 6, r = 2$ 42. $k = 9, a_2 = 6, r = \frac{1}{4}$

In Exercises 43–48, find the sum.

43. $\sum_{n=1}^{7} 2^n$ 44. $\sum_{k=1}^{6} 3\left(\frac{1}{2}\right)^k$

45. $\sum_{n=1}^{9} \left(-\frac{1}{3}\right)^n$ 46. $\sum_{n=1}^{5} 5 \cdot 3^{n-1}$

47. $\sum_{j=1}^{6} 4\left(\frac{3}{2}\right)^{j-1}$ 48. $\sum_{t=1}^{8} 6(.9)^{t-1}$

In Exercises 49–54, you are asked to find a geometric sequence. In each case, round the common ratio r to four decimal places.

49. The number of students per computer in U.S. schools in year n (with $n = 1$ corresponding to 1997) can be approximated by a geometric sequence whose first two terms are $a_1 = 5.912$ and $a_2 = 5.687$.*

   (a) Find a formula for a_n.
   (b) What is the number of students per computer in 2007?
   (c) In what year will there first be fewer than 3 students per computer?

50. According to data from the U.S. Census Bureau, the population of the United States (in millions) in year n can be approximated by a geometric sequence $\{b_n\}$, where $n = 1$ corresponds to 2001.

   (a) If $b_1 = 285.108$ and $b_3 = 290.850$, find a formula for b_n.
   (b) Estimate the U.S. population in 2008 and 2011.

51. The value of all group life insurance (in billions of dollars) in year n can be approximated by a geometric sequence $\{c_n\}$, where $n = 1$ corresponds to 1991.[†]

   (a) If there was $3.9631 billion in effect in 1991 and $4.1672 billion in 1992, find a formula for c_n.
   (b) How much group life insurance is in effect in 2000? In 2004? In 2008?

52. Data from the U.S. Centers for Disease Control and Prevention indicate that the number of newly reported cases of AIDs each year can be approximated by a geometric sequence $\{a_n\}$, where $n = 1$ corresponds to 2000.

   (a) If there were 40,758 cases reported in 2000 and 41,573 cases reported in 2001, find a formula for a_n.
   (b) About how many cases were reported in 2004?
   (c) Find the total number of cases reported from 2000 to 2007 (inclusive).

53. The amount spent per person per year on cable and satellite TV can be approximated by a geometric sequence $\{b_n\}$, where $n = 1$ corresponds to 2001.[‡]

   (a) If $204.74 was spent in 2001 and $232.22 was spent in 2003, find a formula for b_n.
   (b) Find the total that will be spent per person from 2001 to 2009 (inclusive).

54. According to data from the U.S. National Center for Education Statistics, the number of bachelor's degrees earned by women can be approximated by a geometric sequence $\{c_n\}$, where $n = 1$ corresponds to 1996.

   (a) If 642,000 degrees were earned in 1996 and 659,334 in 1997 find a formula for c_n.
   (b) How many degrees were earned in 2000? In 2002? In 2005?

   (c) Find the total number of degrees earned from 1996 to 2005.

55. A ball is dropped from a height of 8 feet. On each bounce, it rises to half its previous height. When the ball hits the ground for the seventh time, how far has it traveled?

56. A ball is dropped from a height of 10 feet. On each bounce, it rises to 45% of its previous height. When it hits the ground for the tenth time, how far has it traveled?

57. In Example 8, suppose a 20 mg dose of aminophylline is given to a patient every 5 hours. Let c_n denote the total amount of aminophylline in the patient's bloodstream after the nth dose.

   (a) Find a formula for the nth term of the sequence $\{c_n\}$.
   (b) How much aminophylline is in the bloodstream after the 9th dose?

58. When a patient is given a 300 mg dose of the drug cimetidine, the amount of the drug remaining in the bloodstream t hours later is given by $C(t) = 300\, e^{-.3466t}$. Suppose the patient receives a 300 mg dose every three hours and d_n denotes the total amount of cimetidine in the bloodstream after n doses.

   (a) Find a formula for the nth term of the sequence $\{d_n\}$.
   (b) How much cimetidine is in the patient's bloodstream after the 8th dose?

59. If you are paid a salary of 1¢ on the first day of March and 2¢ on the second day, and your salary continues to double each day, how much will you earn in the month of March?

60. Starting with your parents, how many ancestors do you have for the preceding ten generations?

61. A car that sold for $8000 depreciates in value 25% each year. What is it worth after five years?

62. A vacuum pump removes 60% of the air in a container at each stroke. What percentage of the original amount of air remains after six strokes?

THINKERS

63. Suppose $\{a_n\}$ is a geometric sequence with common ratio $r > 0$ and each $a_n > 0$. Show that the sequence $\{\log a_n\}$ is an arithmetic sequence with common difference $\log r$.

64. Suppose $\{a_n\}$ is an arithmetic sequence with common difference d. Let C be any positive number. Show that the sequence $\{C^{a_n}\}$ is a geometric sequence with common ratio C^d.

65. In the geometric sequence $1, 2, 4, 8, 16, \ldots$, show that each term is 1 plus the sum of all preceding terms.

66. In the geometric sequence $2, 6, 18, 54, \ldots$, show that each term is twice the sum of 1 and all preceding terms.

67. The minimum monthly payment for a certain bank credit card is the larger of $1/25$ of the outstanding balance or $5. If the balance is less than $5, the entire balance is due. If you make only the minimum payment each month, how long will it take to pay off a balance of $200 (excluding any interest that might be due)?

---

*Based on data from Quality Education Data, Inc.
[†]Based on data from the American Council of Life Insurance.
[‡]Based on data and projections in the *Statistical Abstract of the United States: 2007.*

12.3.A *SPECIAL TOPICS* Infinite Series

Section Objectives
- Find the sum of an infinite geometric series.
- Use geometric series to set up and solve applied problems.

An **infinite series** (or simply **series**) is an expression of the form

$$a_1 + a_2 + a_3 + a_4 + a_5 + \cdots,$$

where each a_n is a real number. This notation suggests the "sum" of the infinite sequence $a_1, a_2, a_3, a_4, a_5, \ldots$ However, ordinary addition is defined only for a finite list of numbers. So we must *define* what is meant by the sum of an infinite series. Consider, for example, the infinite series

$$\frac{3}{10} + \frac{3}{10^2} + \frac{3}{10^3} + \frac{3}{10^4} + \cdots.$$

We begin with the *sequence* $3/10, 3/10^2, 3/10^3, 3/10^4, \ldots$ and compute its partial sums.

$$S_1 = \frac{3}{10},$$

$$S_2 = \frac{3}{10} + \frac{3}{10^2} = \frac{33}{100},$$

$$S_3 = \frac{3}{10} + \frac{3}{10^2} + \frac{3}{10^3} = \frac{333}{1000},$$

$$S_4 = \frac{3}{10} + \frac{3}{10^2} + \frac{3}{10^3} + \frac{3}{10^4} = \frac{3333}{10,000}.$$

These partial sums $S_1, S_2, S_3, S_4, \ldots$ themselves form a sequence.

$$\frac{3}{10}, \frac{33}{100}, \frac{333}{1000}, \frac{3333}{10,000}, \ldots.$$

The terms in the sequence of partial sums appear to be getting closer and closer to $1/3$. In other words, as k gets larger and larger, the corresponding partial sum S_k gets closer and closer to $1/3$. So we say that $1/3$ is the *sum* of the infinite series

$$\frac{3}{10} + \frac{3}{10^2} + \frac{3}{10^3} + \frac{3}{10^4} + \cdots,$$

or that this series *converges* to $1/3$, and we write

$$\frac{3}{10} + \frac{3}{10^2} + \frac{3}{10^3} + \frac{3}{10^4} + \cdots = \frac{1}{3}.$$

The general case is handled similarly. If $a_1 + a_2 + a_3 + a_4 + \cdots$ is an infinite series, then its **partial sums** are the numbers

$$S_1 = a_1,$$

$$S_2 = a_1 + a_2,$$

$$S_3 = a_1 + a_2 + a_3,$$

and in general, for any $k \geq 1$,

$$S_k = a_1 + a_2 + a_3 + a_4 + \cdots + a_k.$$

If it happens that the terms $S_1, S_2, S_3, S_4, \ldots$ of the *sequence* of partial sums get closer and closer to a particular real number S in such a way that the partial sum S_k is arbitrarily close to S when k is large enough, then we say that the series **converges** and that S is the **sum of the convergent series.**

Not every series has a sum. For instance, the partial sums of the series

$$1 + 2 + 3 + 4 + \cdots$$

get larger and larger and never get closer and closer to a single real number. So this series is not convergent and does not have a sum.

EXAMPLE 1

Although no proof will be given here, it is clear that every infinite decimal may be thought of as the sum of a convergent series. For instance,

$$\pi = 3.1415926 \cdots = 3 + .1 + .04 + .001 + .0005 + .00009 + \cdots.$$

Note that the third partial sum is $3 + .1 + .04 = 3.14$, which is π to two decimal places. Similarly, the kth partial sum of this series is just π to $k - 1$ decimal places. ∎

INFINITE GEOMETRIC SERIES

If $\{a_n\}$ is a geometric sequence with common ratio r, then the corresponding infinite series

$$a_1 + a_2 + a_3 + a_4 + a_5 + \cdots$$

is called an **infinite geometric series.** By using the formula for the nth term of a geometric sequence, we can also express the corresponding geometric series in the form

$$a_1 + ra_1 + r^2a_1 + r^3a_1 + r^4a_1 + \cdots.$$

Under certain circumstances, an infinite geometric series is convergent and has a sum.

Sum of an Infinite Geometric Series

If $|r| < 1$, then the infinite geometric series

$$a_1 + ra_1 + r^2a_1 + r^3a_1 + r^4a_1 + \cdots$$

converges, and its sum is

$$\frac{a_1}{1 - r}.$$

Although we cannot prove this fact rigorously here, we can make it highly plausible both geometrically and algebraically.

EXAMPLE 2

$\dfrac{8}{5} + \dfrac{8}{5^2} + \dfrac{8}{5^3} + \cdots$ is an infinite geometric series with $a_1 = 8/5$ and $r = 1/5$.

The kth partial sum of this series is the same as the kth partial sum of the sequence $\{8/5^n\}$, and hence, from the box on p. 847, we know that

$$S_k = a_1\left(\frac{1 - r^k}{1 - r}\right) = \frac{8}{5}\left[\frac{1 - \left(\frac{1}{5}\right)^k}{1 - \frac{1}{5}}\right]$$

$$= \frac{8}{5}\left(\frac{1 - \frac{1}{5^k}}{\frac{4}{5}}\right) = \frac{8}{5} \cdot \frac{5}{4}\left(1 - \frac{1}{5^k}\right)$$

$$= 2\left(1 - \frac{1}{5^k}\right) = 2 - \frac{2}{5^k}.$$

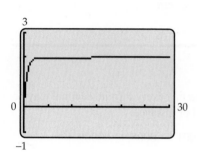

Figure 12–15

The function $f(x) = 2 - 2/5^x$ is defined for all real numbers. When $x = k$ is a positive integer, then $f(k)$ is S_k, the kth partial sum of the series. Using a calculator, we obtain the graph of $f(x)$ in Figure 12–15.

GRAPHING EXPLORATION

Graph $f(x)$ in the same viewing window as in Figure 12–15. Use the trace feature to move the cursor along the graph. As x gets larger, what is the apparent value of $f(x)$ (that is, the value of the partial sum)?

Your calculator will probably tell you that every partial sum is 2, once you move beyond approximately $x = 15$. Actually, the partial sums are slightly smaller than 2 but are rounded to 2 by the calculator. In any case, the horizontal line through 2 is a horizontal asymptote of the graph (meaning that the graph gets very close to the line as x gets larger), so it is very plausible that the sequence converges to the number 2. But 2 is exactly what the preceding box says the sum should be.

$$\frac{a_1}{1 - r} = \frac{\frac{8}{5}}{1 - \frac{1}{5}} = \frac{\frac{8}{5}}{\frac{4}{5}} = \frac{8}{4} = 2. \qquad \blacksquare$$

Example 2 is typical of the general case, as can be seen algebraically. Consider the geometric series $a_1 + a_2 + a_3 + \cdots$ with common ratio r such that $|r| < 1$. The kth partial sum S_k is the same as the kth partial sum of the geometric sequence $\{a_n\}$, and hence,

$$S_k = a_1\left(\frac{1 - r^k}{1 - r}\right).$$

As k gets larger and larger, the number r^k gets very close to 0 because $|r| < 1$ (for instance, $(-.6)^{20} \approx .0000366$ and $.2^9 \approx .000000512$). Consequently, when k is very large, $1 - r^k$ is very close to $1 - 0$ so that

$$S_k = a_1\left(\frac{1 - r^k}{1 - r}\right) \qquad \text{is very close to} \qquad a_1\left(\frac{1 - 0}{1 - r}\right) = \frac{a_1}{1 - r}.$$

EXAMPLE 3

$-\dfrac{1}{2} + \dfrac{1}{4} - \dfrac{1}{8} + \dfrac{1}{16} + \cdots$ is an infinite geometric series with $a_1 = -1/2$ and $r = -1/2$. Since $|r| < 1$, this series converges, and its sum is

$$\frac{a_1}{1-r} = \frac{-\dfrac{1}{2}}{1 - \left(-\dfrac{1}{2}\right)} = \frac{-\dfrac{1}{2}}{\dfrac{3}{2}} = -\frac{1}{3}. \qquad \blacksquare$$

Infinite geometric series provide another way of writing an infinite repeating decimal as a rational number.

EXAMPLE 4

To express $6.8573573573 \cdots$ as a rational number, we first write it as $6.8 + .0573573573 \cdots$. Consider $.0573573573 \cdots$ as an infinite series.

$$.0573 + .0000573 + .0000000573 + .0000000000573 + \cdots,$$

which is the same as

$$.0573 + (.001)(.0573) + (.001)^2(.0573) + (.001)^3(.0573) + \cdots.$$

This is a convergent geometric series with $a_1 = .0573$ and $r = .001$. Its sum is

$$\frac{a_1}{1-r} = \frac{.0573}{1 - .001} = \frac{.0573}{.999} = \frac{573}{9990}.$$

Therefore,

$$6.8573573573 \cdots = 6.8 + [.0573 + .0000573 + \cdots]$$

$$= 6.8 + \frac{573}{9990}$$

$$= \frac{68}{10} + \frac{573}{9990}$$

$$= \frac{68{,}505}{9990} = \frac{4567}{666}. \qquad \blacksquare$$

EXAMPLE 5

Suppose a patient is given a 20 mg dose of aminophylline every four hours, as in Example 8 of Section 12.3. Does the total amount of aminophylline in the bloodstream eventually level off? If so, what is this *steady state* amount?

SOLUTION   As we saw in Example 8(a), the total amount of aminophylline in the patient's bloodstream after n doses is

$$b_n = 20 + 20e^{-.462} + 20[e^{-.462}]^2 + 20[e^{-.462}]^3 + \cdots + 20[e^{-.462}]^{n-1}.$$

Each new dose adds another term to this sum. So if the doses are continued for a long time, the total amount will be approximately given by the geometric *series*

$$20 + 20e^{-.462} + 20[e^{-.462}]^2 + 20[e^{-.462}]^3 + \cdots + 20[e^{-.462}]^{n-1} + \cdots,$$

which has $a_1 = 20$ and $r = e^{-.462}$. Since $e^{-.462} < 1$, the series converges and its sum is

$$\frac{a_1}{1 - r} = \frac{20}{1 - e^{-.462}} \approx 54.0573 \text{ mg.}$$

So the amount levels off at approximately 54 mg. ■

GRAPHING EXPLORATION

In Example 8(a) of Section 12.3, we saw that b_n (the total amount in the bloodstream after the nth dose) is given by

$$b_n = 20\left(\frac{1 - e^{-.462n}}{1 - e^{-.462}}\right).$$

Confirm the answer in Example 5 by graphing the function $f(x) = 20\left(\frac{1 - e^{-.462x}}{1 - e^{-.462}}\right)$

in the viewing window with $0 \le x \le 35$ and $0 \le y \le 70$. Use the trace feature to show that as x gets larger and larger (more and more doses), y (total amount of aminophylline) gets very close to 54.0573.

We close this section by mentioning an alternative notation for infinite series. The series

$$a_1 + a_2 + a_3 + a_4 + \cdots \qquad \text{is denoted by} \qquad \sum_{n=1}^{\infty} a_n.$$

For example, $\sum_{n=1}^{\infty} 2(.6^n)$ denotes the series

$$2(.6) + 2(.6^2) + 2(.6^3) + 2(.6^4) + \cdots,$$

and $\sum_{n=1}^{\infty} \left(\frac{-1}{2}\right)^n$ denotes

$$-\frac{1}{2} + \left(\frac{-1}{2}\right)^2 + \left(\frac{-1}{2}\right)^3 + \left(\frac{-1}{2}\right)^4 + \cdots = -\frac{1}{2} + \frac{1}{4} - \frac{1}{8} + \frac{1}{16} - \frac{1}{32} + \cdots.$$

EXERCISES 12.3.A

In Exercises 1–8, find the sum of the infinite series, if it has one.

1. $\sum_{n=1}^{\infty} \frac{1}{2^n}$ 2. $\sum_{n=1}^{\infty} \left(-\frac{3}{4}\right)^n$ 3. $\sum_{n=1}^{\infty} (.06)^n$

4. $1 - .5 + .25 - .125 + .0625 - \cdots$

5. $500 + 200 + 80 + 32 + \cdots$

6. $9 - 3\sqrt{3} + 3 - \sqrt{3} + 1 - \frac{1}{\sqrt{3}} + \cdots$

7. $2 + \sqrt{2} + 1 + \frac{1}{\sqrt{2}} + \frac{1}{2} + \cdots$

8. $\sum_{n=1}^{\infty} \left(\frac{1}{2^n} - \frac{1}{3^n}\right)$

In Exercises 9–15, express the repeating decimal as a rational number.

9. $.22222 \cdots$ 10. $.37373737 \cdots$

11. $5.4272727 \cdots$ 12. $85.131313 \cdots$

13. $2.1425425425 \cdots$ 14. $3.7165165165 \cdots$

15. $1.74241241241 \cdots$

16. If $\{a_n\}$ is an arithmetic sequence with common difference $d > 0$ and each $a_i > 0$, explain why the infinite series $a_1 + a_2 + a_3 + a_4 + \cdots$ is not convergent.

17. (a) Verify that $\sum_{n=1}^{\infty} 2(1.5)^n$ is a geometric series with $a_1 = 3$ and $r = 1.5$.

(b) Find the kth partial sum of the series and use this expression to define a function f, as in Example 2.

(c) Graph the function f in a viewing window with $0 \leq x \leq 30$. As x gets very large, what happens to the corresponding value of $f(x)$? Does the graph get closer and closer to some horizontal line, as in Example 2? What does this say about the convergence of the series?

18. Use the graphical approach illustrated in Example 2 to find the sum of the series in Example 3. Does the graph get very close to the horizontal line through $-1/3$? What's going on?

In Exercises 19 and 20, use your calculator to compute partial sums and estimate the sum of the convergent infinite series.

19. $\dfrac{\pi}{2} - \dfrac{(\pi/2)^3}{3!} + \dfrac{(\pi/2)^5}{5!} - \dfrac{(\pi/2)^7}{7!} + \cdots$, where $n!$ is the product $1 \cdot 2 \cdot 3 \cdot 4 \cdot \cdots n$.*

_____

*See the *n Factorial* box below and the sentence immediately following it.

20. $1 + 1 + \dfrac{1}{2!} + \dfrac{1}{3!} + \dfrac{1}{4!} + \cdots$ [*Hint:* What is the decimal expansion of e?]

21. Find the steady state amount of aminophylline in the bloodstream of the patient in Exercise 57 of Section 12.3. [See Example 5 of this section.]

22. Find the steady state amount of cimetidine in the bloodstream of the patient in Exercise 58 of Section 12.3.

12.4   The Binomial Theorem

Section Objectives
- Use the Binomial Theorem to expand binomial expressions.
- Find terms of a binomial expansion.

The Binomial Theorem provides a formula for calculating the product $(x + y)^n$ for any positive integer n. Before we state the theorem, some preliminaries are needed.

Let n be a positive integer. The symbol $n!$ (read **n factorial**) denotes the product of all the integers from 1 to n. For example,

$$2! = 1 \cdot 2 = 2, \qquad 3! = 1 \cdot 2 \cdot 3 = 6, \qquad 4! = 1 \cdot 2 \cdot 3 \cdot 4 = 24,$$

$$5! = 1 \cdot 2 \cdot 3 \cdot 4 \cdot 5 = 120,$$

$$10! = 1 \cdot 2 \cdot 3 \cdot 4 \cdot 5 \cdot 6 \cdot 7 \cdot 8 \cdot 9 \cdot 10 = 3{,}628{,}800.$$

In general, we have this result.

n Factorial

Let n be a positive integer. Then

$$n! = 1 \cdot 2 \cdot 3 \cdot 4 \cdots (n - 2)(n - 1)n.$$

0! is defined to be the number 1.

Learn to use your calculator to compute factorials. You will find ! in the *Prob* (or *Prb*) submenu of the MATH or OPTN menu.

CALCULATOR EXPLORATION

15! is such a large number your calculator will switch to scientific notation to express it. What is this approximation? Many calculators cannot compute factorials larger than 69! If yours does compute larger ones, how large a one can you compute without getting an error message [or on HP-39gs, getting the number $9.9999 \cdots \text{E}499$]?

If r and n are integers with $0 \le r \le n$, then we have the following.

Binomial Coefficients

> Either of the symbols $\binom{n}{r}$ or $_nC_r$ denotes the number $\dfrac{n!}{r!(n-r)!}$.
>
> $\binom{n}{r}$ is called a **binomial coefficient**.

For example,

$$_5C_3 = \binom{5}{3} = \frac{5!}{3!(5-3)!} = \frac{5}{3!2!} = \frac{1 \cdot 2 \cdot 3 \cdot 4 \cdot 5}{(1 \cdot 2 \cdot 3)(1 \cdot 2)} = \frac{4 \cdot 5}{2} = 10,$$

$$_4C_2 = \binom{4}{2} = \frac{4!}{2!(4-2)!} = \frac{4!}{2!2!} = \frac{1 \cdot 2 \cdot 3 \cdot 4}{(1 \cdot 2)(1 \cdot 2)} = \frac{3 \cdot 4}{2} = 6.$$

Binomial coefficients can be computed on a calculator by using $_nC_r$ or Comb in the *Prob* (or *Prb*) submenu of the MATH or OPTN menu.

CALCULATOR EXPLORATION

Although calculators cannot compute 475!, they can compute many binomial coefficients, such as $\binom{475}{400}$, because most of the factors cancel out (as in the previous example). Check yours. Will it also compute $\binom{475}{100}$?

The preceding examples illustrate a fact whose proof will be omitted: *Every binomial coefficient is an integer.* Furthermore, for every nonnegative integer n,

$$\binom{n}{0} = 1 \qquad \text{and} \qquad \binom{n}{n} = 1$$

because

$$\binom{n}{0} = \frac{n!}{0!(n-0)!} + \frac{n!}{0!n!} = \frac{n!}{n!} = 1 \qquad \text{and}$$

$$\binom{n}{n} = \frac{n!}{n!(n-n)!} = \frac{n!}{n!0!} = \frac{n!}{n!} = 1.$$

If we list the binomial coefficients for each value of n in this manner:

$n = 0$ $\binom{0}{0}$

$n = 1$ $\binom{1}{0}$ $\binom{1}{1}$

$n = 2$ $\binom{2}{0}$ $\binom{2}{1}$ $\binom{2}{2}$

$n = 3$ $\binom{3}{0}$ $\binom{3}{1}$ $\binom{3}{2}$ $\binom{3}{3}$

$n = 4$ $\binom{4}{0}$ $\binom{4}{1}$ $\binom{4}{2}$ $\binom{4}{3}$ $\binom{4}{4}$

and then calculate each of them, we obtain the following array of numbers.

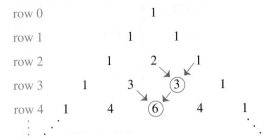

row 0   1

row 1   1   1

row 2   1   2   1

row 3   1   3   ③   1

row 4   1   4   ⑥   4   1

This array is called **Pascal's triangle.** It has an interesting property:

*Each interior entry of Pascal's triangle is the sum of the
two closest entries in the row above it.*

Two examples of this are shown above by the colored circles and arrows. Using this fact we can easily construct row 5 of the triangle from row 4.

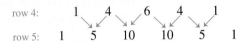

row 4:   1   4   6   4   1

row 5:   1   5   10   10   5   1

This property of Pacal's triangle is proved in Exercise 69.

THE BINOMIAL THEOREM

To develop a formula for calculating $(x + y)^n$, we first calculate these products for small values of n to see whether we can find some kind of pattern.

$$n = 0 \qquad (x + y)^0 = \qquad\qquad\qquad 1$$

$$n = 1 \qquad (x + y)^1 = \qquad\qquad\qquad 1x + 1y$$

(∗) $\qquad n = 2 \qquad (x + y)^2 = \qquad\qquad 1x^2 + 2xy + 1y^2$

$$n = 3 \qquad (x + y)^3 = \qquad 1x^3 + 3x^2y + 3xy^2 + 1y^3$$

$$n = 4 \qquad (x + y)^4 = 1x^4 + 4x^3y + 6x^2y^2 + 4xy^3 + 1y^4$$

One pattern is immediately obvious: The coefficients shown in color above are the top part of Pascal's triangle! In the case $n = 4$, for example, this means that the coefficients are the numbers

$$\begin{matrix} 1 & 4 & 6 & 4 & 1 \end{matrix}$$
$$\binom{4}{0}, \ \binom{4}{1}, \ \binom{4}{2}, \ \binom{4}{3}, \ \binom{4}{4}.$$

If this pattern holds for larger n, then the coefficients in the expansion of $(x + y)^n$ are

$$\binom{n}{0}, \binom{n}{1}, \binom{n}{2}, \binom{n}{3}, \ \ldots, \ \binom{n}{n-1}, \binom{n}{n}.$$

As for the xy-terms associated with each of these coefficients, look at the pattern in (∗) above: The exponent of x goes down by 1 and the exponent of y goes up by 1 as you go from term to term, which suggests that the terms of the expansion of $(x + y)^n$ (without the coefficients) are

$$x^n, \quad x^{n-1}y, \quad x^{n-2}y^2, \quad x^{n-3}y^3, \quad \ldots, \quad xy^{n-1}, \quad y^n.$$

Combining the patterns of coefficients and xy-terms suggests that the following result is true about the expansion of $(x + y)^n$.

The Binomial Theorem

For each positive integer n,

$$(x + y)^n = x^n + \binom{n}{1}x^{n-1}y + \binom{n}{2}x^{n-2}y^2 +$$

$$\binom{n}{3}x^{n-3}y^3 + \cdots + \binom{n}{n-1}xy^{n-1} + y^n.$$

Using summation notation and the fact that

$$\binom{n}{0} = 1 = \binom{n}{n},$$

we can write the Binomial Theorem compactly as

$$(x + y)^n = \sum_{j=0}^{n}\binom{n}{j}x^{n-j}y^j.$$

The Binomial Theorem will be proved in Section 12.5 by means of mathematical induction. We shall assume its truth for now and illustrate some of its uses.

EXAMPLE 1

Expand $(x + y)^8$.

SOLUTION   We apply the Binomial Theorem in the case $n = 8$.

$$(x + y)^8 = x^8 + \binom{8}{1}x^7y + \binom{8}{2}x^6y^2 + \binom{8}{3}x^5y^3$$

$$+ \binom{8}{4}x^4y^4 + \binom{8}{5}x^3y^5 + \binom{8}{6}x^2y^6 + \binom{8}{7}xy^7 + y^8.$$

The coefficients can be computed individually by hand or by using $_nC_r$ (or COMB) on a calculator; for instance,

$$_8C_2 = \binom{8}{2} = \frac{8!}{2!6!} = 28 \qquad \text{and} \qquad _8C_3 = \binom{8}{3} = \frac{8!}{3!5!} = 56.$$

Alternatively, you can display all the coefficients at once by making a table of values for the function $f(x) = {}_8C_x$, as shown in Figure 12–16.*

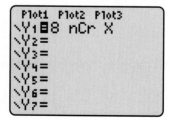

Figure 12–16

Substituting these values in the preceding expansion, we have

$$(x + y)^8 = x^8 + 8x^7y + 28x^6y^2$$

$$+ 56x^5y^3 + 70x^4y^4 + 56x^3y^5 + 28x^2y^6 + 8xy^7 + y^8. \qquad \blacksquare$$

*Thanks to Nick Goodbody for suggesting this method.

EXAMPLE 2

Expand $(1 - z)^6$.

SOLUTION   Note that $1 - z = 1 + (-z)$ and apply the Binomial Theorem with $x = 1$, $y = -z$, and $n = 6$.

$$(1 - z)^6 = 1^6 + \binom{6}{1}1^5(-z) + \binom{6}{2}1^4(-z)^2 + \binom{6}{3}1^3(-z)^3$$

$$+ \binom{6}{4}1^2(-z)^4 + \binom{6}{5}1(-z)^5 + (-z)^6$$

$$= 1 - \binom{6}{1}z + \binom{6}{2}z^2 - \binom{6}{3}z^3 + \binom{6}{4}z^4 - \binom{6}{5}z^5 + z^6$$

$$= 1 - 6z + 15z^2 - 20z^3 + 15z^4 - 6z^5 + z^6. \qquad \blacksquare$$

EXAMPLE 3

If $f(x) = x^4 + 2x^3 + 5$ and $g(x) = f(x + 1)$, expand and simplify the rule of g.

SOLUTION   The rule of g is found by replacing x by $x + 1$ in the rule of f.

$$f(x) = x^4 + 2x^3 + 5$$

$$g(x) = f(x + 1) = (x + 1)^4 + 2(x + 1)^3 + 5.$$

Now we apply the Binomial Theorem to the terms $(x + 1)^4$ and $(x + 1)^3$.

$$g(x) = (x + 1)^4 + 2(x + 1)^3 + 5$$

$$= (x^4 + 4x^3 + 6x^2 + 4x + 1) + 2(x^3 + 3x^2 + 3x + 1) + 5$$

$$= x^4 + 4x^3 + 6x^2 + 4x + 1 + 2x^3 + 6x^2 + 6x + 2 + 5$$

$$= x^4 + 6x^3 + 12x^2 + 10x + 8. \qquad \blacksquare$$

TECHNOLOGY TIP

Binomial expansions, such as those in Examples 1–4 can be done on TI–89. Use *Expand* in the ALGEBRA menu.

EXAMPLE 4

Expand $(x^2 + x^{-1})^4$.

SOLUTION   Use the Binomial Theorem with x^2 in place of x and x^{-1} in place of y.

$$(x^2 + x^{-1})^4 = (x^2)^4 + \binom{4}{1}(x^2)^3(x^{-1}) + \binom{4}{2}(x^2)^2(x^{-1})^2 + \binom{4}{3}(x^2)(x^{-1})^3 + (x^{-1})^4$$

$$= x^8 + 4x^6x^{-1} + 6x^4x^{-2} + 4x^2x^{-3} + x^{-4}$$

$$= x^8 + 4x^5 + 6x^2 + 4x^{-1} + x^{-4}. \qquad \blacksquare$$

Sometimes we need to know only one term in the expansion of $(x + y)^n$. If you examine the expansion given by the Binomial Theorem, you will see that in the second term y has exponent 1, in the third term y has exponent 2, and so on. Thus, we have the following.

Properties of the Binomial Expansion

In the binomial expansion of $(x + y)^n$,

The exponent of y is always one less than the number of the term.

Furthermore, in each of the middle terms of the expansion,

The coefficient of the term containing y^r is $\dbinom{n}{r}$.

The sum of the x exponent and the y exponent is n.

For instance, in the *ninth* term of the expansion of $(x + y)^{13}$, y has exponent 8, the coefficient is $\dbinom{13}{8}$, and x must have exponent 5 (since $8 + 5 = 13$). Thus, the ninth term is $\dbinom{13}{8}x^5y^8$.

EXAMPLE 5

Find the ninth term of the expansion of $\left(2x^2 + \dfrac{\sqrt[4]{y}}{\sqrt{6}}\right)^{13}$.

SOLUTION   We use the Binomial Theorem with $n = 13$ and with $2x^2$ in place of x and $\sqrt[4]{y}/\sqrt{6}$ in place of y. The remarks above show that the ninth term is

$$\binom{13}{8}(2x^2)^5\left(\frac{\sqrt[4]{y}}{\sqrt{6}}\right)^8.$$

Since $\sqrt[4]{y} = y^{1/4}$ and $\sqrt{6} = \sqrt{3}\sqrt{2} = 3^{1/2}2^{1/2}$, we can simplify as follows.

$$\binom{13}{8}(2x^2)^5\left(\frac{\sqrt[4]{y}}{\sqrt{6}}\right)^8 = \binom{13}{8}2^5(x^2)^5\frac{(y^{1/4})^8}{(3^{1/2})^8(2^{1/2})^8} = \binom{13}{8}2^5x^{10}\frac{y^2}{3^4 \cdot 2^4}$$

$$= \binom{13}{8}\frac{2}{3^4}x^{10}y^2 = \frac{13 \cdot 12 \cdot 11 \cdot 10 \cdot 9}{5 \cdot 4 \cdot 3 \cdot 2} \cdot \frac{2}{3^4}x^{10}y^2$$

$$= \frac{286}{9}x^{10}y^2. \qquad \blacksquare$$

EXERCISES 12.4

In Exercises 1–18, evaluate the expression.

1. $6!$

2. $10!$

3. $\dfrac{8!}{6!}$

4. $\dfrac{11!}{8!}$

5. $\dfrac{12!}{9!3!}$

6. $\dfrac{9! - 8!}{7!}$

7. $\dbinom{6}{2}$

8. $\dbinom{8}{4}$

9. $\dbinom{100}{99}$

10. $\dbinom{200}{198}$

11. $\dbinom{4}{1}\dbinom{5}{3}$

12. $\dbinom{5}{3}\dbinom{4}{2}$

13. $\dbinom{5}{3} + \dbinom{5}{2} - \dbinom{6}{3}$

14. $\dbinom{12}{11} - \dbinom{11}{10} + \dbinom{7}{0}$

15. $\dbinom{6}{0} + \dbinom{6}{1} + \dbinom{6}{2} + \dbinom{6}{3} + \dbinom{6}{4} + \dbinom{6}{5} + \dbinom{6}{6}$

16. $\binom{6}{0} - \binom{6}{1} + \binom{6}{2} - \binom{6}{3} + \binom{6}{4} - \binom{6}{5} + \binom{6}{6}$

17. $\binom{100}{96}$ **18.** $\binom{75}{72}$

In Exercises 19–38, expand and (where possible) simplify the expression.

19. $(x + 1)^4$ **20.** $(x - 1)^5$ **21.** $(x + 2)^4$

22. $(x - 2)^6$ **23.** $(x + y)^5$ **24.** $(a + b)^7$

25. $(a - b)^5$ **26.** $(c - d)^8$ **27.** $(2x + y^2)^5$

28. $(3u - v^3)^6$ **29.** $(\sqrt{x} + 1)^6$ **30.** $(2 - \sqrt{y})^5$

31. $(1 - c)^{10}$ **32.** $\left(\sqrt{c} + \frac{1}{\sqrt{c}}\right)^7$

33. $(x^{-3} + x)^4$ **34.** $(3x^{-2} - x^2)^6$

35. $(1 + \sqrt{3})^4 + (1 - \sqrt{3})^4$

36. $(\sqrt{3} + 1)^6 - (\sqrt{3} - 1)^6$

37. $(1 + i)^6$, where $i^2 = -1$

38. $(\sqrt{2} - i)^4$, where $i^2 = -1$

In Exercises 39–42, expand and simplify the rule of the function g.

39. $f(x) = x^3 - 2x^2 + x - 4$ and $g(x) = f(x + 1)$

40. $f(x) = 2x^3 + 3x^2 + x$ and $g(x) = f(x - 1)$

41. $f(x) = x^4 + 3x + 5$ and $g(x) = f(x - 2)$

42. $f(x) = 2x^3 - x + 4$ and $g(x) = f(x + 3)$

In Exercises 43–48, find the indicated term of the expansion of the given expression.

43. third, $(x + y)^5$ **44.** fourth, $(a + b)^6$

45. fifth, $(c - d)^7$ **46.** third, $(a + 2)^8$

47. fourth, $\left(u^{-2} + \frac{u}{2}\right)^7$ **48.** fifth, $(\sqrt{x} - \sqrt{2})^7$

49. Find the coefficient of x^5y^8 in the expansion of $(2x - y^2)^9$.

50. Find the coefficient of $x^{12}y^6$ in the expansion of $(x^3 - 3y)^{10}$.

51. Find the coefficient of $1/x^3$ in the expansion of $\left(2x + \frac{1}{x^2}\right)^6$.

52. Find the constant term in the expansion of $\left(y - \frac{1}{2y}\right)^{10}$.

In Exercises 53–56, use the Binomial Theorem to factor the expression.

53. $x^5 + 5x^4 + 10x^3 + 10x^2 + 5x + 1$

54. $x^4 - 4x^3 + 6x^2 - 4x + 1$

55. $16z^4 + 32z^3 + 24z^2 + 8z + 1$

56. $27y^3 - 27y^2 + 9y - 1$

57. (a) Verify that $\binom{9}{1} = 9$ and $\binom{9}{8} = 9$.

(b) Prove that for each positive integer n, $\binom{n}{1} = n$ and $\binom{n}{n-1} = n$. [*Note:* Part (a) is the case when $n = 9$ and $n - 1 = 8$.]

58. (a) Verify that $\binom{7}{2} = \binom{7}{5}$.

(b) Let r and n be integers with $0 \le r \le n$. Prove that $\binom{n}{r} = \binom{n}{n-r}$. [*Note:* Part (a) is the case when $n = 7$ and $r = 2$.]

59. Prove that for any positive integer n,
$$2^n = \binom{n}{0} + \binom{n}{1} + \binom{n}{2} + \cdots + \binom{n}{n}.$$
[*Hint:* $2 = 1 + 1$.]

60. Prove that for any positive integer n,
$$\binom{n}{0} - \binom{n}{1} + \binom{n}{2} - \binom{n}{3} + \binom{n}{4} - \cdots$$
$$+ (-1)^k\binom{n}{k} + \cdots + (-1)^n\binom{n}{n} = 0.$$

61. Use the Binomial Theorem with $y = i \sin \theta$ and $x = \cos \theta$ to find $(\cos \theta + i \sin \theta)^4$, where $i^2 = -1$.

62. (a) Use DeMoivre's Theorem to find
$$(\cos \theta + i \sin \theta)^4.$$

(b) Use the fact that the two expressions obtained in part (a) and in Exercise 61 must be equal to express $\cos 4\theta$ and $\sin 4\theta$ in terms of $\sin \theta$ and $\cos \theta$.

63. (a) Let f be the function given by $f(x) = x^5$. Let h be a nonzero number and compute $f(x + h) - f(x)$ (but leave all binomial coefficients in the form $\binom{5}{r}$ here and below).

(b) Use part (a) to show that h is a factor of $f(x + h) - f(x)$ and find $\dfrac{f(x + h) - f(x)}{h}$.

(c) If h is *very* close to 0, find a simple approximation of the quantity $\dfrac{f(x + h) - f(x)}{h}$. [See part (b).]

64. Do Exercise 63 with $f(x) = x^8$ in place of $f(x) = x^5$.

65. Do Exercise 63 with $f(x) = x^{12}$ in place of $f(x) = x^5$.

66. Let n be a fixed positive integer. Do Exercise 63 with $f(x) = x^n$ in place of $f(x) = x^5$.

67. Total credit card debt (in billions of dollars) in the United States is approximated by
$$f(x) = -.052x^3 + .7x^2 + 42.14x + 213.4 \quad (0 \le x \le 19)$$

where $x = 0$ corresponds to 1990.* Suppose we want to describe this situation by a function g, for which $x = 0$ corresponds to 2000 instead of 1990.

(a) Explain why the function $g(x) = f(x + 10)$ will work.
(b) How is the graph of $g(x)$ related to the graph of $f(x)$? [*Hint:* See Section 3.4.]
(c) Compute and simplify the rule of $g(x)$.

68. According to data and projections in an article in *USA Today,* the number of Internet phone calls (in millions) is approximated by

$$f(x) = .075x^3 - .86x^2 + 3.66x - 4 \qquad (3 \le x \le 7),$$

where $x = 3$ corresponds to 2003.† Write the rule of a function $g(x)$, which provides the same information as f, but has $x = 0$ corresponding to 2003. [*Hint:* See Exercise 67.]

THINKERS

69. Let r and n be integers such that $0 \le r \le n$.

(a) Verify that $(n - r)! = (n - r)[n - (r + 1)]!$
(b) Verify that $(n - r)! = [(n + 1) - (r + 1)]!$

(c) Prove that $\binom{n}{r+1} + \binom{n}{r} = \binom{n+1}{r+1}$ for any $r \le n - 1$.
[*Hint:* Write out the terms on the left side and use parts (a) and (b) to express each of them as a fraction with denominator $(r + 1)!(n - r)!$. Then add these two fractions, simplify the numerator, and compare the result with $\binom{n+1}{r+1}$.]

(d) Use part (c) to explain why each entry in Pascal's triangle (except the 1's at the beginning or end of a row) is the sum of the two closest entries in the row above it.

70. (a) Find these numbers and write them one *below* the next: $11^0, 11^1, 11^2, 11^3, 11^4$.
(b) Compare the list in part (a) with rows 0 to 4 of Pascal's triangle. What's the explanation?
(c) What can be said about 11^5 and row 5 of Pascal's triangle?
(d) Calculate all integer powers of 101 from 101^0 to 101^8, list the results one under the other, and compare the list with rows 0 to 8 of Pascal's triangle. What's the explanation? What happens with 101^9?

71. Use the Binomial Theorem to show that $1.001^{1000} > 2$. [*Hint:* Write 1.001 as a sum.]

12.5 Mathematical Induction

Section Objective ■ Use mathematical induction.

Mathematical induction is a method of proof that can be used to prove a wide variety of mathematical facts, including the Binomial Theorem, DeMoivre's Theorem, and statements such as the following.

The sum of the first n positive integers is the number $\dfrac{n(n + 1)}{2}$.

$2^n > n$ for every positive integer n.

For each positive integer n, 4 is a factor of $7^n - 3^n$.

All of the preceding statements have a common property. For example, a statement such as

The sum of the first n positive integers is the number $\dfrac{n(n + 1)}{2}$

or, in symbols,

$$1 + 2 + 3 + \cdots + n = \frac{n(n + 1)}{2}$$

is really an infinite sequence of statements, one for each possible value of n.

$$n = 1: \qquad 1 = \frac{1(2)}{2},$$

$$n = 2: \qquad 1 + 2 = \frac{2(3)}{2},$$

$$n = 3: \qquad 1 + 2 + 3 = \frac{3(4)}{2},$$

and so on. Obviously, there isn't time enough to verify every one of the statements on this list, one at a time. But we can find a workable method of proof by examining how each statement on the list is *related* to the *next* statement on the list.

For example, for $n = 50$, the statement is

$$1 + 2 + 3 + \cdots + 50 = \frac{50(51)}{2}.$$

At the moment, we don't know whether or not this statement is true. But just *suppose* that it were true. What could then be said about the next statement, the one for $n = 51$?

$$1 + 2 + 3 + \cdots + 50 + 51 = \frac{51(52)}{2}.$$

Well, *if* it is true that

$$1 + 2 + 3 + \cdots + 50 = \frac{50(51)}{2},$$

then adding 51 to both sides and simplifying the right side would yield these equalities.

$$1 + 2 + 3 + \cdots + 50 + 51 = \frac{50(51)}{2} + 51,$$

$$1 + 2 + 3 + \cdots + 50 + 51 = \frac{50(51)}{2} + \frac{2(51)}{2} = \frac{50(51) + 2(51)}{2},$$

$$1 + 2 + 3 + \cdots + 50 + 51 = \frac{(50 + 2)51}{2},$$

$$1 + 2 + 3 + \cdots + 50 + 51 = \frac{51(52)}{2}.$$

Since this last equality is just the original statement for $n = 51$, we conclude that

If the statement is true for $n = 50$, *then* it is also true for $n = 51$.

We have *not* proved that the statement actually *is* true for $n = 50$, but only that *if* it is, then it is also true for $n = 51$.

We claim that this same conditional relationship holds for any two consecutive values of n. In other words, we claim that for any positive integer k,

① **If the statement is true for $n = k$, *then* it is also true for $n = k + 1$.**

The proof of this claim is the same argument used earlier (with k and $k + 1$ in place of 50 and 51): *If* it is true that

$$1 + 2 + 3 + \cdots + k = \frac{k(k + 1)}{2}, \qquad \text{[Original statement for } n = k]$$

then adding $k + 1$ to both sides and simplifying the right side produces these equalities.

$$1 + 2 + 3 + \cdots + k + (k + 1) = \frac{k(k + 1)}{2} + (k + 1),$$

$$1 + 2 + 3 + \cdots + k + (k + 1)$$
$$= \frac{k(k + 1)}{2} + \frac{2(k + 1)}{2} = \frac{k(k + 1) + 2(k + 1)}{2},$$

$$1 + 2 + 3 + \cdots + k + (k + 1) = \frac{(k + 2)(k + 1)}{2},$$

$$1 + 2 + 3 + \cdots + k + (k + 1) = \frac{(k + 1)[(k + 1) + 1]}{2}.$$

[Original statement for $n = k + 1$]

We have proved that claim ① is valid for each positive integer k. We have *not* proved that the original statement is true for any value of n, but only that *if* it is true for $n = k$, then it is also true for $n = k + 1$. Applying this fact when $k = 1, 2, 3, \ldots$, we see that

② $\begin{cases} \textit{If} \text{ the statement is true for } n = 1, \quad \textit{then} \text{ it is also true for } \\ n = 1 + 1 = 2; \\ \\ \textit{If} \text{ the statement is true for } n = 2, \quad \textit{then} \text{ it is also true for } \\ n = 2 + 1 = 3; \\ \\ \textit{If} \text{ the statement is true for } n = 3, \quad \textit{then} \text{ it is also true for } \\ n = 3 + 1 = 4; \\ \quad \vdots \\ \\ \textit{If} \text{ the statement is true for } n = 50, \quad \textit{then} \text{ it is also true for } \\ n = 50 + 1 = 51; \\ \\ \textit{If} \text{ the statement is true for } n = 51, \quad \textit{then} \text{ it is also true for } \\ n = 51 + 1 = 52; \\ \quad \vdots \end{cases}$

and so on.

We are finally in a position to *prove* the original statement:

$$1 + 2 + 3 + \cdots + n = n(n + 1)/2.$$

Obviously, it *is true* for $n = 1$, since $1 = 1(2)/2$. Now apply in turn each of the propositions on list ②. Since the statement *is* true for $n = 1$, it must also be true for $n = 2$, and hence for $n = 3$, and hence for $n = 4$, and so on, for every value of n. Therefore, the original statement is true for *every* positive integer n.

The preceding proof is an illustration of the following principle.

Principle of Mathematical Induction

Suppose there is given a statement involving the positive integer n and that

(i) The statement is true for $n = 1$.

(ii) If the statement is true for $n = k$ (where k is any positive integer), then the statement is also true for $n = k + 1$.

Then the statement is true for every positive integer n.

Property (i) is simply a statement of fact. To verify that it holds, you must prove the given statement is true for $n = 1$. This is usually easy, as in the preceding example.

Property (ii) is a *conditional* property. It does not assert that the given statement *is* true for $n = k$, but only that *if* it is true for $n = k$, then it is also true for $n = k + 1$. So to verify that property (ii) holds, you need only prove this conditional proposition.

If the statement is true for $n = k$, *then* it is also true for $n = k + 1$.

To prove this or any conditional proposition, you must proceed as in the previous example: Assume the "if" part and use this assumption to prove the "then" part. As we saw earlier, the same argument will usually work for any possible k. Once this conditional proposition has been proved, you can use it *together* with property (i) to conclude that the given statement is necessarily true for every n, just as in the preceding example.

Thus proof by mathematical induction reduces to two steps.

Step 1   Prove that the given statement is true for $n = 1$.

Step 2   Let k be a positive integer. Assume that the given statement is true for $n = k$. Use this assumption to prove that the statement is true for $n = k + 1$.

Step 2 may be performed before step 1 if you wish. Step 2 is sometimes referred to as the **inductive step.** The assumption that the given statement is true for $n = k$ in this inductive step is called the **induction hypothesis.**

EXAMPLE 1

Prove that $2^n > n$ for every positive integer n.

SOLUTION   Here, the statement involving n is $2^n > n$.

Step 1   When $n = 1$, we have the statement $2^1 > 1$. This is obviously true.

Step 2   Let k be any positive integer. We assume that the statement is true for $n = k$, that is, we assume that $2^k > k$. We shall use this assumption to prove that the statement is true for $n = k + 1$, that is, that $2^{k+1} > k + 1$.

We begin with the induction hypothesis: $2^k > k$.* Multiplying both sides of this inequality by 2 yields

$$2 \cdot 2^k > 2k$$

(3)

$$2^{k+1} > 2k.$$

Since k is a positive integer, we know that $k \geq 1$. Adding k to each side of the inequality $k \geq 1$, we have

$$k + k \geq k + 1$$

$$2k \geq k + 1.$$

Combining this result with inequality ③, we see that

$$2^{k+1} > 2k \geq k + 1.$$

The first and last terms of this inequality show that $2^{k+1} > k + 1$. Therefore, the statement is true for $n = k + 1$. This argument works for any positive integer k. Thus, we have completed the inductive step. By the Principle of Mathematical Induction, we conclude that $2^n > n$ for every positive integer n. ∎

EXAMPLE 2

The first n positive even integers are 2, 4, 6, 8, ..., $2n$. Prove that their sum is $n^2 + n$.

SOLUTION   The statement to be proved is this: for each positive integer n,

$$2 + 4 + 6 + 8 + \cdots + 2n = n^2 + n.$$

---

*This is the point at which you usually must do some work. Remember that what follows is the "finished proof." It does not include all the thought, scratch work, false starts, and so on that were done before this proof was actually found.

Step 1 When $n = 1$, the statement is $2 = 1^2 + 1$, which is certainly true.

Step 2 Let k be a positive integer and assume that the statement is true for $n = k$. In other words, the induction hypothesis is

$(*)$ $$2 + 4 + 6 + \cdots + 2k = k^2 + k.$$

We must use this to show that the statement is true for $n = k + 1$, that is, we must show that

$(**)$ $$2 + 4 + 6 + \cdots + 2k + 2(k + 1) = (k + 1)^2 + (k + 1).$$

We begin with the left side of equation $(**)$ and use algebra and the induction hypothesis to show that it equals the right side.

$$
\begin{aligned}
2 + 4 + 6 + \cdots + 2k + 2(k + 1) & \\
= (2 + 4 + 6 + \cdots + 2k) + 2(k + 1) \quad & \text{[Regroup.]} \\
= \qquad (k^2 + k) \qquad + 2(k + 1) \quad & \text{[Induction hypothesis]} \\
= \qquad k^2 + k + 2k + 2 \quad & \text{[Expand.]}^\dagger \\
= (k^2 + 2k + 1) + (k + 1) \quad & \text{[Rearrange terms.]} \\
= \qquad (k + 1)^2 + (k + 1) \quad & \text{[Factor first expression.]}
\end{aligned}
$$

The first and last lines of the preceding computation form equation $(**)$, so we have proved that if the original statement is true for $n = k$, then it is also true for $n = k + 1$. By the Principle of Mathematical Induction we conclude that

$$2 + 4 + 6 + 8 + \cdots + 2n = n^2 + n \text{ for every positive integer } n. \quad \blacksquare$$

EXAMPLE 3

Simple arithmetic shows that

$$7^2 - 3^2 = 49 - 9 = 40 = 4 \cdot 10$$

and

$$7^3 - 3^3 = 343 - 27 = 316 = 4 \cdot 79.$$

In each case, 4 is a factor. These examples suggest that

For each positive integer n, 4 is a factor of $7^n - 3^n$.

This conjecture can be proved by induction as follows.

Step 1 When $n = 1$, the statement is "4 is a factor of $7^1 - 3^1$." Since $7^1 - 3^1 = 4 = 4 \cdot 1$, the statement is true for $n = 1$.

Step 2 Let k be a positive integer and assume that the statement is true for $n = k$, that is, that 4 is a factor of $7^k - 3^k$. Let us denote the other factor by D, so that the induction hypothesis is: $7^k - 3^k = 4D$. We must

---

†At this point, some algebraic trial and error (which is omitted) was used to find the next step in showing that this expression is the same as the right side of equation $(**)$. We were guided by the fact that $(k + 1)^2 = k^2 + 2k + 1$.

use this assumption to prove that the statement is true for $n = k + 1$, that is, that 4 is a factor of $7^{k+1} - 3^{k+1}$. Here is the proof.

$$7^{k+1} - 3^{k+1} = 7^{k+1} - 7 \cdot 3^k + 7 \cdot 3^k - 3^{k+1} \qquad [\text{Since } -7 \cdot 3^k + 7 \cdot 3^k = 0]^\dagger$$

$$= 7(7^k - 3^k) + (7 - 3)3^k \qquad [\text{Factor}]$$

$$= 7(4D) + (7 - 3)3^k \qquad [\text{Induction hypothesis}]$$

$$= 7(4D) + 4 \cdot 3^k \qquad [7 - 3 = 4]$$

$$= 4(7D + 3^k). \qquad [\text{Factor out 4}]$$

From this last line, we see that 4 is a factor of $7^{k+1} - 3^{k+1}$. Thus, the statement is true for $n = k + 1$, and the inductive step is complete. Therefore, by the Principle of Mathematical Induction, the conjecture is actually true for every positive integer n. ∎

Another example of mathematical induction, the proof of the Binomial Theorem, is given at the end of this section.

Sometimes a statement involving the integer n may be false for $n = 1$ and (possibly) other small values of n but true for all values of n beyond a particular number. For instance, the statement $2^n > n^2$ is false for $n = 1, 2, 3$, and 4. But it is true for $n = 5$ and all larger values of n. A variation on the Principle of Mathematical Induction can be used to prove this fact and similar statements. See Exercise 28 for details.

A COMMON MISTAKE WITH INDUCTION

It is sometimes tempting to omit step 2 of an inductive proof when the given statement can easily be verified for small values of n, especially if a clear pattern seems to be developing. As the next example shows, however, *omitting step 2 may lead to error.*

EXAMPLE 4

An integer (> 1) is said to be *prime* if its only positive integer factors are itself and 1. For instance, 11 is prime, since its only positive integer factors are 11 and 1. But 15 is not prime because it has factors other than 15 and 1 (namely, 3 and 5). For each positive integer n, consider the number

$$f(n) = n^2 - n + 11.$$

You can readily verify that

$$f(1) = 11, \qquad f(2) = 13, \qquad f(3) = 17, \qquad f(4) = 23, \qquad f(5) = 31$$

and that *each of these numbers is prime.* Furthermore, there is a clear pattern: The first two numbers (11 and 13) differ by 2; the next two (13 and 17) differ by 4; the next two (17 and 23) differ by 6; and so on. On the basis of this evidence, we might conjecture that

For each positive integer n, the number $f(n) = n^2 - n + 11$ is prime.

---

†Adding and subtracting the same quantity is often a useful technique to rewrite an expression without changing its value.

We have seen that this conjecture is true for $n = 1, 2, 3, 4, 5$. Unfortunately, however, it is *false* for some values of n. For instance, when $n = 11$,

$$f(11) = 11^2 - 11 + 11 = 11^2 = 121.$$

But 121 is obviously *not* prime, since it has a factor other than 121 and 1, namely, 11. You can verify that the statement is also false for $n = 12$ but true for $n = 13$. ■

In the preceding example, the proposition

If the statement is true for $n = k$, then it is true for $n = k + 1$

is false when $k = 10$ and $k + 1 = 11$. If you were not aware of this and tried to complete step 2 of an inductive proof, you would not have been able to find a valid proof for it. Of course, the fact that you can't find a proof of a proposition doesn't always mean that no proof exists. But when you are unable to complete step 2, you are warned that there is a possibility that the given statement may be false for some values of n. This warning should prevent you from drawing any wrong conclusions.

 PROOF OF THE BINOMIAL THEOREM

We shall use induction to prove that for every positive integer n,

$$(x + y)^n = x^n + \binom{n}{1}x^{n-1}y$$

$$+ \binom{n}{2}x^{n-2}y^2 + \binom{n}{3}x^{n-3}y^3 + \cdots + \binom{n}{n-1}xy^{n-1} + y^n.$$

This theorem was discussed and its notation explained in Section 12.4.

Step 1   When $n = 1$, there are only two terms on the right side of the preceding equation, and the statement reads $(x + y)^1 = x^1 + y^1$. This is certainly true.

Step 2   Let k be any positive integer and assume that the theorem is true for $n = k$, that is, that

$$(x + y)^k = x^k + \binom{k}{1}x^{k-1}y + \binom{k}{2}x^{k-2}y^2 + \cdots$$

$$+ \binom{k}{r}x^{k-r}y^r + \cdots + \binom{k}{k-1}xy^{k-1} + y^k.$$

[On the right side of this equation, we have included a typical middle term $\binom{k}{r}x^{k-r}y^r$. The sum of the exponents is k, and the bottom part of the binomial coefficient is the same as the y exponent.] We shall use this assumption to prove that the theorem is true for $n = k + 1$, that is, that

$$(x + y)^{k+1} = x^{k+1} + \binom{k+1}{1}x^k y + \binom{k+1}{2}x^{k-1}y^2 + \cdots$$

$$+ \binom{k+1}{r+1}x^{k-r}y^{r+1} + \cdots + \binom{k+1}{k}xy^k + y^{k+1}.$$

We have simplified some of the terms on the right side; for instance, $(k + 1) - 1 = k$ and $(k + 1) - (r + 1) = k - r$. But this is the correct statement for $n = k + 1$: The coefficients of the middle terms are $\binom{k+1}{1}$, $\binom{k+1}{2}$, $\binom{k+1}{3}$, and so on; the sum of the exponents of each middle term is $k + 1$, and the bottom part of each binomial coefficient is the same as the y exponent.

To prove the theorem for $n = k + 1$, we shall need this fact about binomial coefficients: For any integers r and k with $0 \le r < k$,

(4)
$$\binom{k}{r+1} + \binom{k}{r} = \binom{k+1}{r+1}.$$

A proof of this fact is outlined in Exercise 69 on page 864.

To prove the theorem for $n = k + 1$, we first note that

$$(x + y)^{k+1} = (x + y)(x + y)^k.$$

Applying the induction hypothesis to $(x + y)^k$, we see that

$$(x + y)^{k+1} = (x + y)\left[x^k + \binom{k}{1}x^{k-1}y + \binom{k}{2}x^{k-2}y^2 + \cdots + \binom{k}{r}x^{k-r}y^r \right.$$
$$\left. + \binom{k}{r+1}x^{k-(r+1)}y^{r+1} + \cdots + \binom{k}{k-1}xy^{k-1} + y^k \right]$$
$$= x\left[x^k + \binom{k}{1}x^{k-1}y + \cdots + y^k \right] + y\left[x^k + \binom{k}{1}x^{k-1}y + \cdots + y^k \right].$$

Next we multiply out the right side. Remember that multiplying by x increases the x exponent by 1 and multiplying by y increases the y exponent by 1.

$$(x + y)^{k+1} = \left[x^{k+1} + \binom{k}{1}x^k y + \binom{k}{2}x^{k-1}y^2 + \cdots + \binom{k}{r}x^{k-r+1}y^r \right.$$
$$\left. + \binom{k}{r+1}x^{k-r}y^{r+1} + \cdots + \binom{k}{k-1}x^2y^{k-1} + xy^k \right]$$
$$+ \left[x^k y + \binom{k}{1}x^{k-1}y^2 + \binom{k}{2}x^{k-2}y^3 + \cdots + \binom{k}{r}x^{k-r}y^{r+1} \right.$$
$$\left. + \binom{k}{r+1}x^{k-(r+1)}y^{r+2} + \cdots + \binom{k}{k-1}xy^k + y^{k+1} \right]$$
$$= x^{k+1} + \left[\binom{k}{1} + 1 \right]x^k y + \left[\binom{k}{2} + \binom{k}{1} \right]x^{k-1}y^2 + \cdots$$
$$+ \left[\binom{k}{r+1} + \binom{k}{r} \right]x^{k-r}y^{r+1} + \cdots + \left[1 + \binom{k}{k-1} \right]xy^k + y^{k+1}.$$

Now apply statement (4) to each of the coefficients of the middle terms. For instance, with $r = 1$, statement (4) shows that

$$\binom{k}{2} + \binom{k}{1} = \binom{k+1}{2}.$$

Similarly, with $r = 0$,

$$\binom{k}{1} + 1 = \binom{k}{1} + \binom{k}{0} = \binom{k+1}{1},$$

and so on. Then the expression above for $(x + y)^{k+1}$ becomes

$$(x + y)^{k+1} = x^{k+1} + \binom{k+1}{1}x^k y + \binom{k+1}{2}x^{k-1}y^2 + \cdots$$

$$+ \binom{k+1}{r+1}x^{k-r}y^{r+1} + \cdots + \binom{k+1}{k}xy^k + y^{k+1}.$$

Since this last statement says that the theorem is true for $n = k + 1$, the inductive step is complete. By the Principle of Mathematical Induction, the theorem is true for every positive integer n.

 EXERCISES 12.5

In Exercises 1–18, use mathematical induction to prove that each of the given statements is true for every positive integer n.

1. $1 + 2 + 2^2 + 2^3 + 2^4 + \cdots + 2^{n-1} = 2^n - 1$

2. $1 + 3 + 3^2 + 3^3 + 3^4 + \cdots + 3^{n-1} = \dfrac{3^n - 1}{2}$

3. $1 + 3 + 5 + 7 + \cdots + (2n - 1) = n^2$

4. $1^2 + 2^2 + 3^2 + \cdots + n^2 = \dfrac{n(n + 1)(2n + 1)}{6}$

5. $\dfrac{1}{2} + \dfrac{1}{4} + \dfrac{1}{8} + \cdots + \dfrac{1}{2^n} = 1 - \dfrac{1}{2^n}$

6. $\dfrac{1}{3} + \dfrac{1}{9} + \dfrac{1}{27} + \cdots + \dfrac{1}{3^n} = \dfrac{1}{2} - \dfrac{1}{2 \cdot 3^n}$

7. $\dfrac{1}{1 \cdot 2} + \dfrac{1}{2 \cdot 3} + \dfrac{1}{3 \cdot 4} + \cdots + \dfrac{1}{n(n + 1)} = \dfrac{n}{n + 1}$

8. $\left(1 + \dfrac{1}{1}\right)\left(1 + \dfrac{1}{2}\right)\left(1 + \dfrac{1}{3}\right) \cdots \left(1 + \dfrac{1}{n}\right) = n + 1$

9. $n + 2 > n$ **10.** $2n + 2 > n$

11. $3^n \geq 3n$ **12.** $3^n \geq 1 + 2n$

13. $3n > n + 1$ **14.** $\left(\dfrac{3}{2}\right)^n > n$

15. 3 is a factor of $2^{2n+1} + 1$

16. 5 is a factor of $2^{4n-2} + 1$

17. 64 is a factor of $3^{2n+2} - 8n - 9$

18. 64 is a factor of $9^n - 8n - 1$

19. Let c and d be fixed real numbers. Prove that

$$c + (c + d) + (c + 2d) + (c + 3d) + \cdots$$

$$+ [c + (n - 1)d] = \dfrac{n[2c + (n - 1)d]}{2}.$$

20. Let r be a fixed real number with $r \neq 1$. Prove that

$$1 + r + r^2 + r^3 + \cdots + r^{n-1} = \dfrac{r^n - 1}{r - 1}.$$

[*Remember:* $1 = r^0$; so when $n = 1$, the left side reduces to $r^0 = 1$.]

21. (a) Write *each* of $x^2 - y^2$, $x^3 - y^3$, and $x^4 - y^4$ as a product of $x - y$ and another factor.
    (b) Make a conjecture as to how $x^n - y^n$ can be written as a product of $x - y$ and another factor. Use induction to prove your conjecture.

22. Let $x_1 = \sqrt{2}$; $x_2 = \sqrt{2 + \sqrt{2}}$; $x_3 = \sqrt{2 + \sqrt{2 + \sqrt{2}}}$; and so on. Prove that $x_n < 2$ for every positive integer n.

In Exercises 23–27, if the given statement is true, prove it. If it is false, give a counterexample.

23. Every odd positive integer is prime.

24. The number $n^2 + n + 17$ is prime for every positive integer n.

25. $(n + 1)^2 > n^2 + 1$ for every positive integer n.

26. 3 is a factor of the number $n^3 - n + 3$ for every positive integer n.

27. 4 is a factor of the number $n^4 - n + 4$ for every positive integer n.

28. Let q be a *fixed* integer. Suppose a statement involving the integer n has these two properties:

    (i) The statement is true for $n = q$.
    (ii) *If* the statement is true for $n = k$ (where k is any integer with $k \geq q$), then the statement is also true for $n = k + 1$.

Then we claim that the statement is true for every integer n greater than or equal to q.

    (a) Give an informal explanation that shows why this claim should be valid. Note that when $q = 1$, this claim is precisely the Principle of Mathematical Induction.
    (b) The claim made before part (a) will be called the *Extended Principle of Mathematical Induction*. State the two steps necessary to use this principle to prove that a given statement is true for all $n \geq q$. (See the discussion on pages 866–867.)

In Exercises 29–34, use the Extended Principle of Mathematical Induction (Exercise 28) to prove the given statement.

29. $2n - 4 > n$ for every $n \geq 5$. (Use 5 for q here.)

30. Let r be a fixed real number with $r > 1$. Then $(1 + r)^n > 1 + nr$ for every integer $n \geq 2$. (Use 2 for q here.)

31. $n^2 > n$ for all $n \geq 2$ **32.** $2^n > n^2$ for all $n \geq 5$

33. $3^n > 2^n + 10n$ for all $n \geq 4$

34. $2n < n!$ for all $n \geq 4$

THINKERS

35. Let n be a positive integer. Suppose that there are three pegs and on one of them, n rings are stacked, each ring being smaller in diameter than the one below it (see the figure). We want to transfer the stack of rings to another peg according to these rules: (i) Only one ring may be moved at a time; (ii) a ring can be moved to any peg, provided that it is never placed on top of a smaller ring; (iii) the final order of the rings on the new peg must be the same as the original order on the first peg.

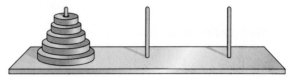

(a) What is the smallest possible number of moves when $n = 2$? $n = 3$? $n = 4$?

(b) Make a conjecture as to the smallest possible number of moves required for any n. Prove your conjecture by induction.

36. The basic formula for compound interest $T(x) = P(1 + r)^x$ was discussed on page 369. Prove by induction that the formula is valid whenever x is a positive integer. [*Note:* P and r are assumed to be constant.]

37. Use induction to prove DeMoivre's Theorem: For any complex number $z = r(\cos \theta + i \sin \theta)$ and any positive integer n,
$$z^n = r^n[\cos(n\theta) + i \sin(n\theta)].$$

Chapter 12 Review

IMPORTANT CONCEPTS

Section 12.1

Sequence   826
Term   826
Constant sequence   828
Recursively defined sequence   829
Fibonacci sequence   830
Summation notation   831
Summation index   832
Partial sum   834

Section 12.2

Arithmetic sequence   837
Common difference   838

nth term formula   838–839
Partial sum formulas   840

Section 12.3

Geometric sequence   844
Common ratio   844
nth term formula   845
Partial sum formula   847

Special Topics: 12.3.A

Infinite series   852
Infinite geometric series   853

Section 12.4

n factorial ($n!$)   857
Binomial coefficient   858
Pascal's triangle   859
Binomial Theorem   860

Section 12.5

Principle of Mathematical Induction   866
Inductive step   867
Induction hypothesis   867

IMPORTANT FACTS & FORMULAS

■ In an arithmetic sequence $\{a_n\}$ with common difference d,

$$a_n = a_1 + (n - 1)d, \qquad \sum_{n=1}^{k} a_n = \frac{k}{2}(a_1 + a_k),$$

$$\sum_{n=1}^{k} a_n = ka_1 + \frac{k(k - 1)}{2} d.$$

- In a geometric sequence $\{a_n\}$ with common ratio $r \neq 1$,

$$a_n = r^{n-1}a_1, \qquad \sum_{n=1}^{k} a_n = a_1\left(\frac{1-r^k}{1-r}\right).$$

- The sum of the infinite geometric series $a_1 + ra_1 + r^2a_1 + \cdots$ (with $|r| < 1$) is

$$\frac{a_1}{1-r}.$$

- $n! = 1 \cdot 2 \cdot 3 \cdots (n-2)(n-1)n$

- $\displaystyle \binom{n}{r} = \frac{n!}{r!(n-r)!} = {}_nC_r$

- *The Binomial Theorem:*

$$(x+y)^n = x^n + \binom{n}{1}x^{n-1}y + \binom{n}{2}x^{n-2}y^2 + \binom{n}{3}x^{n-3}y^3 + \cdots + \binom{n}{n-1}xy^{n-1} + y^n$$

$$= \sum_{j=0}^{n} \binom{n}{j}x^{n-j}y^j$$

REVIEW QUESTIONS

In Questions 1–4, find the first four terms of the sequence $\{a_n\}$.

1. $a_n = 2n - 5$

2. $a_n = 3^n - 27$

3. $a_n = \left(\dfrac{-1}{n}\right)^2$

4. $a_n = (-1)^{n+1}(n-1)$

In Questions 5 and 6, find a formula for the nth term of the sequence.

5. $\dfrac{2}{5}, \dfrac{4}{25}, \dfrac{8}{125}, \dfrac{16}{625}, \cdots$

6. $\sqrt{4}, \sqrt{10}, \sqrt{28}, \sqrt{82}, \cdots$ [*Hint:* $4 = 3 + 1$ and $10 = 9 + 1$.]

7. The Pension Benefit Guaranty Corporation (PBGC) is the government agency that insures pensions. Data and projections from the Center on Federal Financial Institutions show that PBGC's annual cash payouts (in billions of dollars) are approximated by the sequence $\{a_n\}$, where

$$a_n = -.0035x^2 + .5x + 2.5 \qquad (n \geq 4),$$

and $n = 4$ corresponds to 2004.

(a) How much did PBGC pay out in 2004? In 2006? What will it pay out in 2010?

(b) Approximately when will the yearly cash payout exceed \$9.6 billion?

In Questions 8 and 9, express the given sum in summation notation. Do not find the sum.

8. $\dfrac{1}{\ln 2} + \dfrac{1}{\ln 4} + \dfrac{1}{\ln 8} + \dfrac{1}{\ln 16} + \cdots + \dfrac{1}{\ln 256}$

9. $\dfrac{1}{2} + \dfrac{2}{3} + \dfrac{3}{4} + \dfrac{4}{5} + \cdots + \dfrac{9}{10}$

In Questions 10–12, find the sum.

10. $\displaystyle \sum_{n=0}^{4} 2^n(n+1) = ?$

11. $\displaystyle \sum_{n=2}^{4} (3n^2 - n + 1) = ?$

12. $\displaystyle \sum_{n=1}^{4} n^2 + \sum_{n=5}^{9} (n+1)^2$

13. Find the fourth partial sum of the sequence $\{a_n\}$, where $a_1 = 1/9$ and $a_n = 3a_{n-1}$.

14. Find the fifth partial sum of the sequence $\{a_n\}$, where $a_1 = -4$ and $a_n = 3a_{n-1} + 2$.

15. Average tuition and fees each year in private colleges can be approximated by the sequence given by

$$b_n = 11{,}561.7 \cdot e^{.05642n},$$

where $n = 0$ corresponds to the 1990–1991 school year.*

(a) What were the average tuition and fees in 2004–2005? What are they this year?

(b) If you started at a private college with average tuition and fees in Fall 1990 and graduated after four years, what was the total that you (or your parents) spent on tuition and fees?

(c) Do part (b), assuming that you started in Fall 2004.

16. Find the total cash payout by PBGC in Question 7 for the years 2004–2008 (inclusive).

In Questions 17–20, find a formula for a_n; assume that the sequence is arithmetic.

17. $a_1 = 3$ and the common difference is -6.

18. $a_2 = 4$ and the common difference is 3.

19. $a_1 = -5$ and $a_3 = 7$.

20. $a_3 = 2$ and $a_7 = -1$.

21. Find the 11th partial sum of the arithmetic sequence with $a_1 = 5$ and common difference -2.

22. Find the 12th partial sum of the arithmetic sequence with $a_1 = -3$ and $a_{12} = 16$.

---

*Based on data from the U.S. Center for Educational Statistics.

In Questions 23–26, find a formula for a_n; assume that the sequence is geometric.

23. $a_1 = 2$ and the common ratio is 3.

24. $a_1 = 5$ and the common ratio is $-1/2$.

25. $a_2 = 192$ and $a_7 = 6$.

26. $a_3 = 9/2$ and $a_6 = -243/16$.

27. Find the fifth partial sum of the geometric sequence with $a_1 = 1/4$ and common ratio 3.

28. Find the sixth partial sum of the geometric sequence with $a_1 = 5$ and common ratio $1/2$.

29. Find numbers b, c, d such that $4, b, c, d, 23$ are the first five terms of an arithmetic sequence.

30. Find numbers c and d such that $8, c, d, 27$ are the first four terms of a geometric sequence.

31. The net revenue in year n (in billions of dollars) of Texas Instruments is approximated by a geometric series $\{a_n\}$ in which $n = 1$ corresponds to 2001.*

(a) If net revenues were \$7.898 billion in 2001 and \$8.972 billion in 2002, find a formula for a_n in which the common ratio r is rounded to three decimal places.

(b) What were net revenues in 2005?

(c) What were the total net revenues from 2001 to 2006 (inclusive)?

32. Retail expenditures on boating (in billions of dollars) is approximated by an arithmetic sequence $\{b_n\}$ in which $n = 1$ corresponds to 2000.†

(a) If expenditures were \$27.217 billion in 2000 and \$28.923 billion in 2001, find the expenditures in 2006.

(b) What were the total expenditures from 2000 to 2005 (inclusive)?

33. When a patient is given a 100 mg dose of furocimide, the amount of the drug remaining in the bloodstream t hours later is given by $C(t) = 100\, e^{-1.3863t}$. Suppose the patient receives a 100 mg dose every two hours and let b_n denote the total amount of furocimide in the bloodstream after n doses.

(a) Find a formula for the nth term of the sequence $\{b_n\}$.

(b) How much furocimide is in the patient's bloodstream after the 5th dose?

(c) Find the approximate steady state amount of furocimide when the doses are given every two hours for a long time.

In Questions 34–36, find the sums, if they exist.

34. $\displaystyle\sum_{n=1}^{\infty} \left(\frac{3}{2}\right)^{n-1}$

35. $\displaystyle\sum_{n=1}^{\infty} \frac{1}{2^{n-1}}$

36. $\displaystyle\sum_{n=1}^{\infty} \left(\frac{-1}{4^n}\right)$

37. $\dbinom{15}{12} = ?$

38. $\dbinom{18}{3} = ?$

39. $\dfrac{20!5!}{6!17!} = ?$

40. $\dfrac{7! - 5!}{4!} = ?$

41. Let n be a positive integer. Simplify $\dbinom{n+1}{n}$.

42. Use the Binomial Theorem to expand $(\sqrt{x} + 1)^5$. Simplify your answer.

43. If $f(x) = x^3 - 2x^2 + 3x + 1$ and $g(x) = f(x - 1)$, find and simplify the rule of $g(x)$.

44. Factor $x^5 - 5x^4 + 10x^3 - 10x^2 + 5x - 1$. [*Hint:* Think Binomial.]

45. Find the coefficient of $x^2 y^4$ in the expansion of $(2y + x^2)^5$.

46. Prove that for every positive integer n,

$$1^3 + 2^3 + 3^3 + \cdots + n^3 = \frac{n^2(n+1)^2}{4}.$$

47. Prove that for every positive integer n,

$$1 + 5 + 5^2 + 5^3 + \cdots + 5^{n-1} = \frac{5^n - 1}{4}.$$

48. Prove that $2^n \geq 2n$ for every positive integer n.

49. If x is a real number with $|x| < 1$, then prove that $|x^n| < 1$ for all $n \geq 1$.

50. Prove that for any positive integer n,

$$1 + 5 + 9 + \cdots + (4n - 3) = n(2n - 1).$$

51. Prove that for any positive integer n,

$$1 + 4 + 4^2 + 4^3 + \cdots + 4^{n-1} = \frac{1}{3}(4^n - 1).$$

52. Prove that $3n < n!$ for every $n \geq 4$.

53. Prove that for every positive integer n, 8 is a factor of $9^n - 8n - 1$.

---

*Based on data from TI annual reports.
†Based on data from the *Statistical Abstracts of the United States 2007*.

Chapter **12** Test

Sections 12.1–12.3; Special topics 12.3.A

1. Find the first five terms of the sequence $\{a_n\}$, where $a_n = (-1)^{n+2} - (n+7)$.

2. The average cost of a dormitory room for one academic year at a four-year public university is approximated by the arithmetic sequence $\{b_n\}$, where $n = 1$ corresponds to the 2004–2005 school year and

$$b_n = 185n + 3227.*$$

   (a) What was the cost in 2006–2007?
   (b) If you begin attending a four-year public university in 2004–2005 and stay for *six* years (not uncommon), what will the total cost of your dorm room be?

3. In the geometric sequence $\{c_n\}$, $c_5 = 2/3$ and the common ratio $r = -1/3$.

   (a) Find c_6.
   (b) Find the rule for c_n.

4. Find the sum: $\displaystyle\sum_{k=1}^{25} (2k^2 - 4k + 3)$.

5. Show that the sequence log 4, log 8, log 16, log 32, log 64, . . . is arithmetic and find its common difference.

6. Show that the sequence $\{4^{n-4}\}$ is geometric and find its common ratio.

7. Find the first five terms of the sequence with

$$a_1 = 1, \quad a_2 = 3, \quad \text{and} \quad a_n = 2a_{n-1} + 3a_{n-2} \quad \text{for } n \ge 3.$$

8. Suppose b and c are constants. Show that the sequence $\{9b + 8cn\}$ is arithmetic and find its common difference.

9. In the geometric sequence $\{a_n\}$, $a_2 = 10$ and $a_7 = 320$.

   (a) Find a formula for a_n.
   (b) Find a_5.

10. Find the third and sixth partial sum of the sequence $\{(2n - 3n^2)^2\}$.

11. The arithmetic sequence $\{b_n\}$ has $b_1 = 0$ and $b_8 = 34$. Find the 8th partial sum of the sequence.

12. The geometric sequence $\{c_n\}$ has $c_2 = 9$ and common ratio $r = 1/2$. Find the 7th partial sum and round your answer to the nearest integer.

13. Express this sum in Σ notation:

$$(-2)^{12} + (-2)^{13} + (-2)^{14} + (-2)^{15} + (-2)^{16}.$$

14. Find the sum: $\displaystyle\sum_{n=1}^{40} \frac{7 - 8n}{4}$.

15. A ball is dropped from a height of 10 feet. On each bounce, it rises to 44% of the previous height. When it hits the ground for the tenth time, how far has it traveled? Round your answer to two decimal places.

16. The number of people living with AIDs in a given year is approximated by the sequence $\{a_n\}$, where $n = 1$ corresponds to 1998 and $a_n = 258{,}205(1.07)^n.*$

   (a) How many people are living with AIDs in the years 2004 and 2007?
   (b) If the sequence remains accurate, how many will be living with AIDs in 2010?

17. Find the sum of all the integer multiples of 8 from 8 to 800.

18. Find the sum of the geometric series

$$\frac{1}{2} + \frac{1}{10} + \frac{1}{50} + \frac{1}{250} + \cdots.$$

Sections 12.4–12.5

19. $\dbinom{5}{0} - \dbinom{5}{1} + \dbinom{5}{2} - \dbinom{5}{3} + \dbinom{5}{4} - \dbinom{5}{5} = ?$

20. If $f(x) = x^6 + x + 1$ and $g(x) = f(x + 1)$, find and simplify the rule of $g(x)$.

21. Use mathematical induction to prove that for every positive integer n,

$$10 + 12 + 14 + \cdots + 2(n + 4) = n^2 + 9n.$$

22. Expand $(4u - v^3)^6$.

23. Find the fifth term in the expansion of $(\sqrt{x} - \sqrt{10})^7$.

24. Prove that for every positive integer n, 4 is a factor of $5^n + 3$.

25. Prove that for each positive integer n,

$$\dbinom{n}{1} = n \quad \text{and} \quad \dbinom{n}{n-1} = n.$$

26. Prove that for every positive integer n with $n \ge 3$,

$$n^2 > 2n + 1.$$

---

*Based on data from the U.S. National Center for Education Statistics.

*Based on estimates from the U.S. Centers for Disease Control and Prevention.

DISCOVERY PROJECT 12   Taking Your Chances

Games of chance have been played in many societies throughout history, so it should not be surprising that as mathematics becomes more sophisticated, its principles are applied to the study of games of chance. Many common games can be analyzed by using the techniques of discrete mathematics, particularly the Binomial Theorem. The Binomial Theorem comes into play because of the nature of gaming; each play is either a win or a loss, one of two choices. Consider the simple game of flipping a coin. What does a sequence of two games look like?

The tree diagram indicates four paths through the sequence of two games: head-head, head-tail, tail-head, and tail-tail.

```
          H   HH
      H  <
  .       T   HT
   \
    \     H   TH
      T  <
          T
              TT
```

If you think of heads and tails as the variables H and T, then you have three final results: H^2, T^2, and two HT's if you consider $HT = TH$. Indeed, you can think of the last column of the tree as the polynomial of two variables

$$H^2 + 2HT + T^2.$$

The coefficients of the terms are the binomial coefficients you learned about in the Binomial Theorem. You can use the theorem to help you describe many such sequential games.

1. Write a polynomial that describes the results of a game in which you flip a coin eight times in a row.

The variables H and T can also be used to represent the probability that you will get heads or tails, respectively, when you flip a coin. If the coin you are using is a fair one, then the probability of heads or tails is equally likely; that is, you would expect to see heads half the time and tails half the time. These values of $\frac{1}{2}$ are the probabilities that heads or tails will appear on a given flip.

2. What is the probability that you will get exactly four heads and four tails when you flip a coin eight times? Answer this by evaluating the H^4T^4 term of the probability polynomial using the values $H = \frac{1}{2}$ and $T = \frac{1}{2}$.

3. Show that the probability polynomial for flipping eight coins adds to exactly 1.

DISCOVERY PROJECT 12

There is no magic to the values of $\frac{1}{2}$ for winning or losing a game. The only constraint is that the probability values for winning and losing must add to 1.

4. Suppose that you play a game in which the probability that you win is 0.47 and the probability that you lose is 0.53. What is the probability that you win exactly four out of seven games? Evaluate the $W^4 L^3$ term of the probability polynomial $(W + L)^7$.

LIMITS AND CONTINUITY

Where did that number come from?

Doctors use a formula to determine when a dialysis patient has been adequately dialyzed. Engineers used a function to determine the shape of the Gateway Arch in St. Louis. Economists and scientists estimate the capacity of the earth to support its population. Geologists and physicists compute the amount of energy released by an earthquake. What do these disparate formulas, functions, estimates, and computations have in common? They all involve *e*, a number that is given by a limit. See Exercise 50 on page 889 and Exercise 51 on page 923.

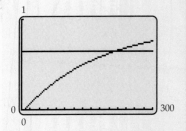

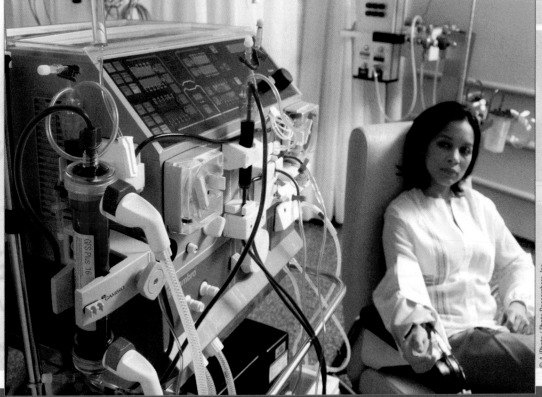

Chapter Outline

In earlier chapters, we dealt with *static* situations, such as the following:

What is the size of the angle?

How far did the rock fall?

How old is the skeleton?

Calculus, by contrast, deals with *dynamic* questions:

At what rate is the angle changing size?

How fast is the rock falling at time $t = 2$ seconds?

What production rate will increase profits the most?

The key to dealing with such dynamic problems is the concept of limit, which is introduced in this chapter.

13.1   Limits of Functions

Section Objectives
- ■ Understand the informal definition of limit.
- ■ Use tables and graphs to find limits.
- ■ Identify situations in which a limit does not exist.
- ■ Find one-sided limits.

You have often dealt with questions such as this:

What is the value of the function $f(x)$ when $x = 4$?

Underlying the idea of "limit," however, is the question:

How does the function $f(x)$ behave near $x = 4$?

EXAMPLE 1

The function

$$f(x) = \frac{.1x^4 - .8x^3 + 1.6x^2 + 2x - 8}{x - 4}$$

is not defined at $x = 4$ (why?). What happens to the values of $f(x)$ when x is *very close* to 4?

SOLUTION   We evaluate f at several numbers that are very close to 4, including numbers that are smaller than 4 and numbers that are larger than 4, that is, numbers that lie to the left of 4 and to the right of 4 on the number line, as shown below.

	x gets closer and closer to 4 from the left					*x* gets closer and closer to 4 from the right			
x	3.9	3.99	3.999	3.9999	4	4.0001	4.001	4.01	4.1
f(*x*)	1.8479	1.9841	1.9984	1.9998	*	2.0002	2.0016	2.0161	2.1681

The table suggests the following.

> As *x* gets very close to 4 (from either side), the corresponding values of *f*(*x*) get very close to 2.

Furthermore, by taking *x* close enough to 4, the corresponding values of *f*(*x*) can be made *as close as you want* to 2. For instance,

$$f(3.999999) = 1.999998 \quad \text{and} \quad f(4.000001) = 2.000002.$$

This situation is usually expressed by saying that "the *limit* of *f*(*x*) as *x* approaches 4 is the number 2," which is written symbolically as

$$\lim_{x \to 4} f(x) = 2 \quad \text{or} \quad \lim_{x \to 4} \frac{.1x^4 - .8x^3 + 1.6x^2 + 2x - 8}{x - 4} = 2. \quad ■$$

Example 1 is an illustration of the following definition of "limit." Now *f* is any function, and *c* and *L* are fixed real numbers (in Example 1, *c* = 4 and *L* = 2). The phrase "arbitrarily close" means "as close as you want."

Informal Definition of Limit

Let *f* be a function, and let *c* be a real number such that *f*(*x*) is defined for all values of *x* near *x* = *c*, except possibly at *x* = *c* itself. Suppose that

> as *x* takes values very close (but not equal) to *c* (on both sides of *c*), the corresponding values of *f*(*x*) are very close (and possibly equal) to the real number *L*

and that

> the values of *f*(*x*) can be made arbitrarily close to *L* for all values of *x* that are sufficiently close (but not equal) to *c*.

Then we say that

> the **limit** of the function *f*(*x*) as *x* approaches *c* is the number *L*,

which is written

$$\lim_{x \to c} f(x) = L.$$

EXAMPLE 2

Find $\lim_{x \to 2} 3x^2$.

SOLUTION   In this case, *f*(*x*) = 3*x*². When *x* is a number very close to 2, then *x*² is very close to 2 · 2 = 4. For instance,

$$x = 1.9995 \text{ is very close to 2} \quad \text{and} \quad x^2 = 1.9995^2 = 3.99800025.$$

X	Y2
1.99	11.88
1.999	11.988
1.9999	11.999
2	12
2.0001	12.001
2.001	12.012
2.01	12.12

Y2=3X²

Figure 13–1

Similarly, when x^2 is very close to 4, then $3x^2$ is very close to $3 \cdot 4 = 12$. This reasoning suggests that

$$\lim_{x \to 2} 3x^2 = 12.$$

The table in Figure 13–1 supports this conclusion (values of x close to 2 produce values of $f(x) = 3x^2$ close to 12). Furthermore, the *limit* of f as x approaches 2 is the *value* of f at $x = 2$ because $f(2) = 3(2^2) = 12$. ∎

EXAMPLE 3

Find $\lim_{x \to 2} f(x)$, where f is the function given by this two-part rule:

$$f(x) = \begin{cases} 0 \text{ if } x \text{ is an integer,} \\ 1 \text{ if } x \text{ is not an integer.} \end{cases}$$

SOLUTION   A calculator is not much help here, but the function is easily graphed by hand, as shown in Figure 13–2.

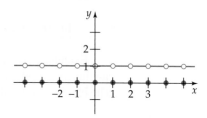

Figure 13–2

When x is a number very close (but not equal) to 2 (on either side of 2), the corresponding value of $f(x)$ is the number 1, and this is true no matter how close x is to 2. Thus,

$$\lim_{x \to 2} f(x) = 1.$$

Since $f(2) = 0$ by definition, we see that the limit of f as x approaches 2 is *not* the same as the value of the function f at $x = 2$. ∎

Examples 1–3 illustrate the following facts.

Limits and
Functional Values

If the function $f(x)$ has a limit as x approaches c, then there are three possibilities:

1. $f(c)$ is not defined, but $\lim_{x \to c} f(x)$ is defined.   [Example 1]

2. $f(c)$ is defined and $\lim_{x \to c} f(x) = f(c)$.   [Example 2]

3. $f(c)$ is defined, but $\lim_{x \to c} f(x) \neq f(c)$.   [Example 3]

EXAMPLE 4

The graph of $f(x) = \dfrac{\sin x}{x}$ is shown in Figure 13–3. Find $\lim_{x \to 0} f(x)$.

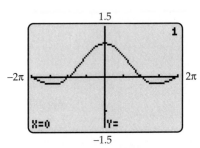

Figure 13–3

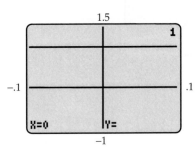

Figure 13–4

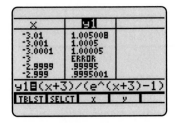

Figure 13–5

TECHNOLOGY TIP

To compute $\lim\limits_{x \to c} f(x)$ on TI-89, use this syntax:

$$\text{limit}(f(x), x, c)$$

LIMIT is in the CALC menu on the TI-89 home screen.

SOLUTION The function is not defined at $x = 0$ (which is indicated by the trace feature in Figure 13–3). To see what's going on very close to $x = 0$, we graph f in the narrow viewing window of Figure 13–4, in which the graph looks like a straight line.

GRAPHING EXPLORATION

Graph f in the viewing window of Figure 13–4. Use the trace feature to move closer and closer to the y-axis from both the left and the right, and note the values of $f(x)$ as x gets closer and closer to 0.

The Graphing Exploration strongly suggests that

$$\lim_{x \to 0} f(x) = 1 \qquad \text{or, equivalently,} \qquad \lim_{x \to 0} \frac{\sin x}{x} = 1.$$

This is indeed the case, as will be proved in calculus. ∎

EXAMPLE 5

Use numerical methods to find

$$\lim_{x \to -3} \frac{x + 3}{e^{x+3} - 1}.$$

Confirm your answer graphically.

SOLUTION Create a table of values for $f(x) = \dfrac{x + 3}{e^{x+3} - 1}$ when x is very close to -3, as in Figure 13–5 (which shows a TI-86 screen). The table indicates that f is not defined when $x = -3$. Nevertheless, the table suggests that

$$\lim_{x \to -3} \frac{x + 3}{e^{x+3} - 1} = 1.$$

We can confirm this conclusion by graphing f in a very narrow window around $x = -3$ and using the trace feature to examine the values of $f(x)$ when x is very close to -3. Figure 13–6 also suggests that the limit is 1. ∎

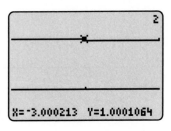 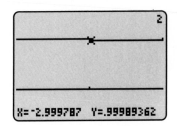

Figure 13–6

NONEXISTENCE OF LIMITS

Not every function has a limit at every number. Limits can fail to exist for a variety of reasons, some of which are illustrated in the following examples.

EXAMPLE 6

Figure 13–7 shows the graph of

$$f(x) = \frac{1}{(x+1)^2}.$$

Find $\lim\limits_{x \to -1} f(x)$, if it exists.

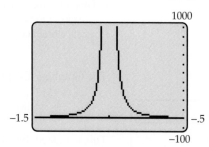

Figure 13–7

SOLUTION

GRAPHING EXPLORATION

Graph $f(x)$ in the viewing window of Figure 13–7. Use the trace feature to show that as x approaches -1 from the left or right, the corresponding values of $f(x)$ become larger and larger, without bound, rather than getting closer and closer to one particular number.

Therefore, the limit of $f(x)$ as x approaches -1 does not exist. ■

EXAMPLE 7

As was explained in Example 6 on page 494, the graph of $f(x) = \sin(\pi/x)$ oscillates infinitely often between -1 and 1 as x approaches 0, as shown in Figure 13–8.

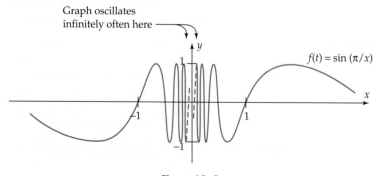

Figure 13–8

As x approaches 0, the function takes every value between -1 and 1 infinitely many times. In particular, $f(x)$ does not get closer and closer to one particular number. Therefore, the limit of $f(x) = \sin(\pi/x)$ as x approaches 0 does not exist. ■

EXAMPLE 8

Find $\lim\limits_{x \to 0} \dfrac{|x|}{x}$ if it exists.

SOLUTION   The function $f(x) = |x|/x$ is not defined when $x = 0$. According to the definition of absolute value, $|x| = x$ when $x > 0$ and $|x| = -x$ when $x < 0$. Therefore,

$$\text{When } x > 0, \text{ then } f(x) = \frac{|x|}{x} = \frac{x}{x} = 1;$$

$$\text{When } x < 0, \text{ then } f(x) = \frac{|x|}{x} = \frac{-x}{x} = -1.$$

Consequently, the graph of f looks like Figure 13–9, as you can verify with a calculator.

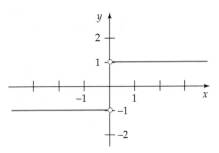

Figure 13–9

If x takes positive values closer and closer to 0, then the corresponding value of $f(x)$ is always 1. If x takes negative values closer and closer to 0, however, the corresponding value of $f(x)$ is always -1. Thus, as x takes values close to 0 on *both* sides of 0, as required by the definition of limit, the corresponding values of $f(x)$ do not get closer and closer to the same real number. Therefore, the limit does not exist. ■

ONE-SIDED LIMITS

Although the limit does not exist in Example 8, we can adapt the limit concept to describe the behavior of the function $f(x) = \dfrac{|x|}{x}$ near $x = 0$. If you consider only *positive* values of x, then the graph in Figure 13–9 shows that

> as x takes values very close to 0 (with $x > 0$), the corresponding values of $f(x)$ are very close (in fact, equal) to 1.

We express this fact by writing

$$\lim_{x \to 0^+} f(x) = 1$$

and saying

> the limit of $f(x)$ as x approaches 0 from the right is 1.

Similarly, if you consider only *negative* values of x, then

> as x takes values very close to 0 (with $x < 0$), the corresponding values of $f(x)$ are very close (in fact, equal) to -1.

We express this fact by writing

$$\lim_{x \to 0^-} f(x) = -1$$

and saying

the limit of $f(x)$ as x approaches 0 from the left is -1.

The same idea carries over to the general case, in which f is any function and c and L are real numbers.

One-Sided Limits

Suppose that

as x takes values very close to c (with $x > c$), the corresponding values of $f(x)$ are very close (and possibly equal) to the real number L

and that

the values of $f(x)$ can be made arbitrarily close to L for all values of x (with $x > c$) that are sufficiently close to c.

Then we say that

the limit of $f(x)$ as x approaches c from the right is L,

which is written

$$\lim_{x \to c^+} f(x) = L.$$

The symbol

$$\lim_{x \to c^-} f(x) = L,$$

which is read

the limit of $f(x)$ as x approaches c from the left is L,

is defined analogously (just replace "$x > c$" by "$x < c$" in the preceding definition).

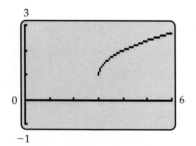

Figure 13–10

EXAMPLE 9

The function $f(x) = \sqrt{x - 3} + 1$ is defined only when $x \geq 3$, that is, for values of x to the right of 3. Find the limit of $f(x)$ as x approaches 3 from the right.

SOLUTION The graph of f in Figure 13–10 suggests that the limit is 1. You can confirm this by using a very narrow window.

GRAPHING EXPLORATION

Graph f in the viewing window with $2.9999 \leq x \leq 3.0001$ and $.98 \leq y \leq 1.02$. Use the trace feature to move along the graph. Verify that as x approaches 3 from the right, the corresponding values of $f(x)$ get closer and closer to 1.

Therefore, we conclude that $\lim_{x \to 3^+} (\sqrt{x - 3} + 1) = 1$. ∎

We now have three kinds of limits: the left-hand limits and right-hand limits defined on the preceding page and the two-sided limits defined on page 881. Example 9 exhibits a function that has a right-hand limit at $x = 3$ but no left-hand or two-sided limit. The function in Example 8 has both a left- and a right-hand limit but no two-sided limit.

When a function has a two-sided limit L at $x = c$, then it automatically has L as both the left- and right-hand limit at $x = c$. Example 5 illustrates this clearly. Figures 13–5 and 13–6 show that as x gets closer and closer to -3 (on *both* sides of -3), the corresponding values of

$$f(x) = \frac{x + 3}{e^{x+3} - 1}$$

get closer and closer to 1. So it is certainly true that when x takes values to the right of -3 (that is, $x > -3$), the corresponding values of $f(x)$ get closer and closer to 1—and similarly, when x takes values to the left of -3. A similar argument applies in the general case.

Two-Sided Limits

Let f be a function and let c and L be real numbers. Then

$$\lim_{x \to c} f(x) = L \quad \text{exactly when} \quad \lim_{x \to c^-} f(x) = L \quad \text{and} \quad \lim_{x \to c^+} f(x) = L.$$

EXAMPLE 10

The graph of the function f in Figure 13–11 shows that $f(2) = 1$ and that

$$\lim_{x \to 2} f(x) = 3, \qquad \lim_{x \to 2^-} f(x) = 3, \qquad \lim_{x \to 2^+} f(x) = 3.$$

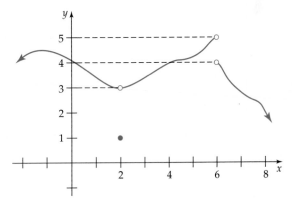

Figure 13–11

CAUTION

Whenever a calculator was used in the preceding examples, we were careful to say that the information it provided *suggested* that the limit was a particular number. Although calculators can provide strong numerical and graphical evidence for the existence of a limit, they do *not* provide a proof of this fact. In fact, calculators can occasionally be misleading (see Exercise 63).

The graph also shows that

$$\lim_{x \to 6^-} f(x) = 5 \quad \text{and} \quad \lim_{x \to 6^+} f(x) = 4,$$

so the two-sided limit, $\lim_{x \to 6} f(x)$, is not defined. ∎

EXERCISES 13.1

Note: In the following exercises, "find the limit" means "find the exact limit, if possible, and otherwise the best possible approximation."

In Exercises 1–16, use the table feature of your calculator to find the limit.

1. $\lim\limits_{x\to-1}\dfrac{x^6-1}{x^4-1}$

2. $\lim\limits_{x\to2}\dfrac{x^5-32}{x^3-8}$

3. $\lim\limits_{x\to0}\dfrac{\tan x-x}{x^3}$

4. $\lim\limits_{x\to0}\dfrac{\tan x}{x+\sin x}$

5. $\lim\limits_{x\to0}\dfrac{x-\tan x}{x-\sin x}$

6. $\lim\limits_{x\to0}\dfrac{x+\sin 2x}{x-\sin 2x}$

7. $\lim\limits_{x\to0}\dfrac{x}{\ln|x|}$

8. $\lim\limits_{x\to0}\dfrac{e^{2x}-1}{x}$

9. $\lim\limits_{x\to0}\dfrac{e^x-1}{\sin x}$

10. $\lim\limits_{x\to0}\left(x\sin\dfrac{1}{x}\right)$

11. $\lim\limits_{x\to\pi^-}\dfrac{\sin x}{1-\cos x}$

12. $\lim\limits_{x\to\frac{\pi}{2}}(\sec x-\tan x)$

13. $\lim\limits_{x\to0^+}\sqrt{x}\,(\ln x)$

14. $\lim\limits_{x\to0^+}\left(\dfrac{1}{x}\right)^x$

15. $\lim\limits_{x\to0^-}\dfrac{\sin(6x)}{x}$

16. $\lim\limits_{x\to0^+}\dfrac{\sin(3x)}{1+\sin(4x)}$

In Exercises 17–22, use the graph of the function f to determine the following limits.

$$\lim_{x\to-3}f(x)\qquad \lim_{x\to0}f(x),\qquad \lim_{x\to2}f(x).$$

17.

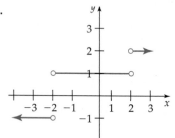

18.

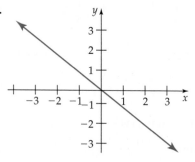

19.

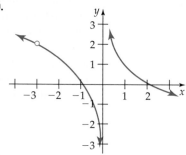

20.

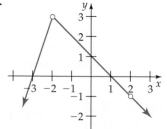

21.

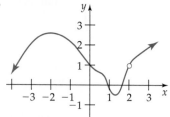

22.

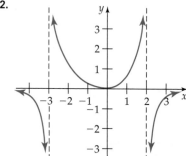

In Exercises 23–26, use the graph of the function f to determine

(a) $\lim\limits_{x\to-2^-}f(x)$

(b) $\lim\limits_{x\to0^+}f(x)$

(c) $\lim\limits_{x\to3^-}f(x)$

(d) $\lim\limits_{x\to3^+}f(x)$

23.

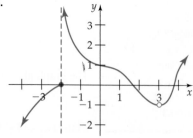

24.

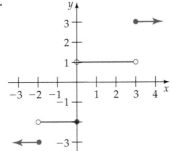

25.

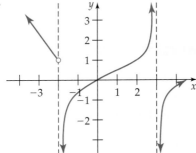

26.

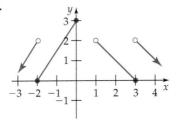

In Exercises 27–49, use numerical or graphical means to find the limit, if it exists. If the limit of f as x approaches c does exist, answer this question: Is it equal to f(c)?

27. $\lim\limits_{x\to 3} \dfrac{5x-4}{2x-1}$

28. $\lim\limits_{x\to -2} \dfrac{2x+3}{2x-5}$

29. $\lim\limits_{x\to 2} \dfrac{x^2-5x+6}{x^2-6x+8}$

30. $\lim\limits_{x\to -2} \dfrac{x^2+3x+2}{x^2-x-6}$

31. $\lim\limits_{x\to 4} \dfrac{x-6}{x-4}$

32. $\lim\limits_{x\to -3} \dfrac{(x-3)(x+3)(x+4)}{(x-4)(x+3)(x+8)}$

33. $\lim\limits_{x\to 2^+} \sqrt{x^2-4}$

34. $\lim\limits_{x\to 3^-} \sqrt{9-x^2}$

35. $\lim\limits_{x\to 4} \dfrac{-6}{(x-4)^2}$

36. $\lim\limits_{x\to -2} \dfrac{2x}{(x+2)^3}$

37. $\lim\limits_{x\to 0} \dfrac{x}{e^x-1}$

38. $\lim\limits_{x\to 0} \dfrac{e^{2x}+e^x-2}{e^x-1}$

39. $\lim\limits_{x\to 1} \dfrac{\ln x}{x-1}$

40. $\lim\limits_{x\to 5} \dfrac{\ln x - \ln 5}{x-5}$

41. $\lim\limits_{x\to 0} (x \ln |x|)$

42. $\lim\limits_{x\to \pi/2} x \sin x$

43. $\lim\limits_{x\to \pi/2} x \cos x$

44. $\lim\limits_{x\to 1} \tan \pi x$

45. $\lim\limits_{x\to 1^-} \dfrac{|x-1|}{x-1}$

46. $\lim\limits_{x\to -5} \dfrac{|x+5|}{x+5}$

47. $\lim\limits_{x\to 0} \cos\left(\dfrac{\pi}{x}\right)$

48. $\lim\limits_{x\to 0} \tan\left(\dfrac{\pi}{2}-x\right)$

49. $\lim\limits_{x\to 1} e^x \sin \dfrac{\pi x}{2}$

50. (a) Approximate $\lim\limits_{x\to 0} (1+x)^{1/x}$ to seven decimal places. (Evaluate the function at numbers closer and closer to 0 until successive approximations agree in the first seven decimal places.)

(b) Find the decimal expansion of e to at least nine decimal places.

(c) On the basis of the results of parts (a) and (b), what do you think is the exact value of $\lim\limits_{x\to 0} (1+x)^{1/x}$?

51. (a) Graph the function f whose rule is

$$f(x) = \begin{cases} 3-x & \text{if } x < -2, \\ x+2 & \text{if } -2 \le x < 2, \\ 1 & \text{if } x = 2, \\ 4-x & \text{if } x > 2. \end{cases}$$

Use the graph in part (a) to evaluate these limits:

(b) $\lim\limits_{x\to -2} f(x)$ **(c)** $\lim\limits_{x\to 1} f(x)$ **(d)** $\lim\limits_{x\to 2} f(x)$

52. In Exercise 51, find

(a) $\lim\limits_{x\to -2^-} f(x)$ **(b)** $\lim\limits_{x\to -2^+} f(x)$

(c) $\lim\limits_{x\to 2^-} f(x)$ **(d)** $\lim\limits_{x\to 2^+} f(x)$

53. (a) Graph the function g whose rule is

$$g(x) = \begin{cases} x^2 & \text{if } x < -1, \\ x+2 & \text{if } -1 \le x < 1, \\ 3-x & \text{if } x \ge 1. \end{cases}$$

Use the graph in part (a) to evaluate these limits:

(b) $\lim\limits_{x\to -1} g(x)$ **(c)** $\lim\limits_{x\to 0} g(x)$ **(d)** $\lim\limits_{x\to 1} g(x)$

54. In Exercise 53, find

(a) $\lim\limits_{x\to -1^-} g(x)$ **(b)** $\lim\limits_{x\to -1^+} g(x)$

(c) $\lim\limits_{x\to 1^-} g(x)$ **(d)** $\lim\limits_{x\to 1^+} g(x)$

In Exercises 55 and 56, use a unit circle diagram to explain why the given statement is true.

55. $\lim\limits_{t\to \pi/2} \sin t = 1$

56. $\lim\limits_{t\to \pi/2} \cos t = 0$

Exercises 57–62 involve the greatest integer function $f(x) = [\![x]\!]$, which was defined in Example 7 on page 145. You may use your calculator as an aid in analyzing these problems.

57. Let $h(x) = [\![x]\!] + [\![-x]\!]$; find $\lim\limits_{x\to 2} h(x)$, if this limit exists.

58. Let $g(x) = x - [\![-x]\!]$; find $\lim\limits_{x\to 2} g(x)$, if this limit exists.

59. Find $\lim\limits_{x\to 2^+} [\![x]\!]$ and $\lim\limits_{x\to 2^-} [\![x]\!]$.

60. Find $\lim\limits_{x\to 3^-} (x - [\![x]\!])$ and $\lim\limits_{x\to 3^+} (x - [\![x]\!])$.

61. Let $r(x) = \dfrac{[\![x]\!] + [\![-x]\!]}{x}$; find $\lim\limits_{x\to 3} r(x)$, if this limit exists.

62. Let $k(x) = \dfrac{x}{[\![x]\!] + [\![-x]\!]}$; find $\lim\limits_{x \to 1} k(x)$, if this limit exists.

63. If $f(x) = \dfrac{1 - \cos(x^6)}{x^{12}}$, then calculus shows that $\lim\limits_{x \to 0} f(x) = 1/2$. A calculator or computer, however, may indicate otherwise. Graph $f(x)$ in a viewing window with

$$-.1 \le x \le .1,$$

and use the trace feature to determine the values of $f(x)$ when x is very close to 0. What does this suggest that the limit is?

64. Consider the function t whose rule is

$$t(x) = \begin{cases} 0 & \text{if } x \text{ is rational,} \\ 1 & \text{if } x \text{ is irrational.} \end{cases}$$

Explain why $\lim\limits_{x \to 4} t(x)$ does not exist.

13.2   Properties of Limits

Section Objectives
- Learn the algebraic properties of limits.
- Find limits of polynomial and rational functions.
- Find the limit of a difference quotient when h approaches 0.

We now consider a number of facts that greatly simplify the computation of limits.

Properties of Limits

If f and g are functions and c, L, M are numbers such that

$$\lim_{x \to c} f(x) = L \qquad \text{and} \qquad \lim_{x \to c} g(x) = M,$$

then

1. $\lim\limits_{x \to c} (f(x) + g(x)) = L + M$

2. $\lim\limits_{x \to c} (f(x) - g(x)) = L - M$

3. $\lim\limits_{x \to c} (f(x) \cdot g(x)) = L \cdot M$

4. $\lim\limits_{x \to c} \left(\dfrac{f(x)}{g(x)} \right) = \dfrac{L}{M}$ (provided that $M \neq 0$)

5. $\lim\limits_{x \to c} \sqrt{f(x)} = \sqrt{L}$ (provided that $f(x) \geq 0$ for all x near c)

All of these properties remain valid for one-sided limits (that is, when "$x \to c$" is replaced throughout either by "$x \to c^{+}$" or by "$x \to c^{-}$").

Although formal proofs of these properties won't be given here, the central idea is easily understood.* Consider, for example, Properties 1 and 3. We are given that

$$\lim_{x \to c} f(x) = L \qquad \text{and} \qquad \lim_{x \to c} g(x) = M.$$

Consequently, as x gets very close to c, the corresponding values of $f(x)$ are very close to L, and the corresponding values of $g(x)$ are very close to M. So it seems plausible that $f(x) + g(x)$ is very close to $L + M$ and $f(x) g(x)$ is very close to LM.

---

*A formal proof requires a rigorous definition of "limit." This rigorous definition and proofs of several of these properties are discussed in *Special Topics* 13.2.A.

These properties are often stated somewhat differently. Since $\lim\limits_{x \to c} f(x) = L$ and $\lim\limits_{x \to c} g(x) = M$, Properties 1–5 can be written as follows.

1. $\lim\limits_{x \to c} [f(x) + g(x)] = \lim\limits_{x \to c} f(x) + \lim\limits_{x \to c} g(x)$

2. $\lim\limits_{x \to c} [f(x) - g(x)] = \lim\limits_{x \to c} f(x) - \lim\limits_{x \to c} g(x)$

3. $\lim\limits_{x \to c} [f(x) \cdot g(x)] = \lim\limits_{x \to c} f(x) \cdot \lim\limits_{x \to c} g(x)$

4. $\lim\limits_{x \to c} \left(\dfrac{f(x)}{g(x)} \right) = \dfrac{\lim\limits_{x \to c} f(x)}{\lim\limits_{x \to c} g(x)}$ (provided that $\lim\limits_{x \to c} g(x) \neq 0$)

5. $\lim\limits_{x \to c} \sqrt{f(x)} = \sqrt{\lim\limits_{x \to c} f(x)}$ (provided that $f(x) \geq 0$ for all x near c)

LIMITS OF POLYNOMIAL FUNCTIONS

We begin with the most simple polynomial functions, the constant functions, such as

$$f(x) = 5.$$

To compute $\lim\limits_{x \to c} f(x)$, for example, you must ask, "When x is very close to c, what is the value of $f(x)$?" The answer is easy, because

no matter what x is, the value of $f(x)$ is *always* the number 5.

So as x gets closer and closer to c, the value of $f(x)$ is always 5. Hence,

$$\lim\limits_{x \to c} f(x) = 5, \qquad \text{which is usually written} \qquad \lim\limits_{x \to c} 5 = 5.$$

The same thing is true for any constant function.

Constant Limits

> If d is a constant, then $\lim\limits_{x \to c} d = d$.

Now consider the *identity function*, whose rule is $f(x) = x$, and the limit $\lim\limits_{x \to c} f(x)$. When x is very close to c, the corresponding value of $f(x)$ (namely, x itself) obviously is very close to c. So we have this conclusion.

Identity Function Limit

> For every real number c, $\lim\limits_{x \to c} x = c$.

The preceding facts, together with Limit Properties 1–3, now make it easy to find the limit of any polynomial function.

EXAMPLE 1

If $f(x) = x^2 - 2x + 3$, find $\lim\limits_{x \to -4} f(x)$.

SOLUTION By Properties 1 and 2,

$$\lim\limits_{x \to -4} f(x) = \lim\limits_{x \to -4} (x^2 - 2x + 3)$$
$$= \lim\limits_{x \to -4} x^2 - \lim\limits_{x \to -4} 2x + \lim\limits_{x \to -4} 3.$$

Consequently, by Property 3,

$$\lim\limits_{x \to -4} f(x) = (\lim\limits_{x \to -4} x)(\lim\limits_{x \to -4} x) - (\lim\limits_{x \to -4} 2)(\lim\limits_{x \to -4} x) + \lim\limits_{x \to -4} 3.$$

Using the facts about the limits of constant functions and the identity function in the preceding boxes, we see that

$$\lim_{x \to -4} f(x) = (\lim_{x \to -4} x)(\lim_{x \to -4} x) - (\lim_{x \to -4} 2)(\lim_{x \to -4} x) + \lim_{x \to -4} 3$$

$$= (\lim_{x \to -4} x)(\lim_{x \to -4} x) - 2(\lim_{x \to -4} x) + 3$$

$$= (-4)(-4) - 2(-4) + 3 = 27.$$

Note that the limit of $f(x) = x^2 - 2x + 3$ at $x = -4$ is the same as the value of the function at $x = -4$, namely, $f(-4) = 27$. ∎

Since any polynomial function consists of sums and products of constants and x, the argument used in Example 1 works for any polynomial function and leads to the following conclusion.

Polynomial
Limits

> If $f(x)$ is a polynomial function and c is any real number, then
>
> $$\lim_{x \to c} f(x) = f(c).$$
>
> In other words, the limit is the value of the polynomial function at $x = c$.

This result also applies to one-sided limits because the left- and right-hand limits at $x = c$ must be the same as the two-sided limit (see the box on page 887).

EXAMPLE 2

The function $g(x) = \sqrt{9 - x^2}$ is defined only when $-3 \le x \le 3$ (why?). Find $\lim_{x \to 3^-} \sqrt{9 - x^2}$.

SOLUTION   By Property 5,

$$\lim_{x \to 3^-} \sqrt{9 - x^2} = \sqrt{\lim_{x \to 3^-} (9 - x^2)}.$$

The limit under the right-side radical is the limit of the polynomial function $f(x) = 9 - x^2$. According to the preceding box (and the remark after it), the limit can be found by evaluating $f(x)$ at 3. Therefore,

$$\lim_{x \to 3^-} \sqrt{9 - x^2} = \sqrt{\lim_{x \to 3^-} (9 - x^2)} = \sqrt{9 - 3^2} = 0.$$ ∎

 LIMITS OF RATIONAL FUNCTIONS

Property 4 and the fact that polynomial limits can be found by evaluation make it easy to compute the limits of rational functions.

EXAMPLE 3

If

$$f(x) = \frac{x^3 - 3x^2 + 10}{x^2 - 6x + 1},$$

find $\lim_{x \to 2} f(x)$.

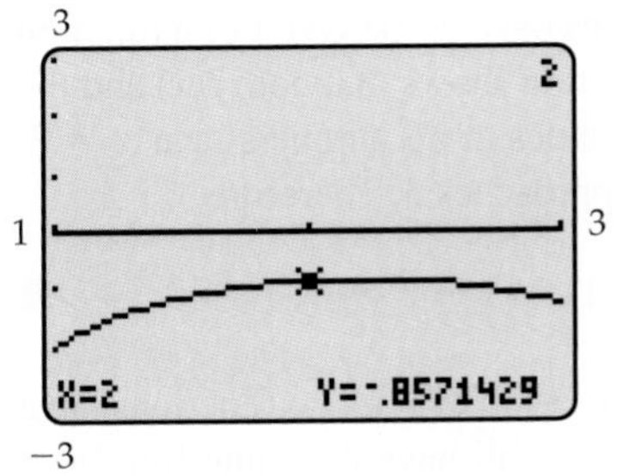

Figure 13–12

SOLUTION   The graph of $f(x)$ near $x = 2$ (Figure 13–12) suggests that the limit is approximately $-.86$. We can determine the limit exactly by noting that $f(x)$ is the quotient of two polynomial functions

$$g(x) = x^3 - 3x^2 + 10 \quad \text{and} \quad h(x) = x^2 - 6x + 1,$$

each of whose limits, as x approaches 2, can be found by evaluation of the functions at $x = 2$. Therefore,

$$\lim_{x \to 2} f(x) = \lim_{x \to 2} \frac{x^3 - 3x^2 + 10}{x^2 - 6x + 1}$$

$$= \frac{\lim_{x \to 2} (x^3 - 3x^2 + 10)}{\lim_{x \to 2} (x^2 - 6x + 1)} \qquad \text{[Property 4]}$$

$$= \frac{2^3 - 3 \cdot 2^2 + 10}{2^2 - 6 \cdot 2 + 1} = \frac{6}{-7} = -\frac{6}{7} \qquad \text{[Limits of Polynomial Functions]}$$

Note that the limit of $f(x)$, as x approaches 2, is the number $f(2)$. ∎

The procedure in Example 3 works for other rational functions as well.

Rational Limits

> Let $f(x)$ be a rational function, and let c be a real number such that $f(c)$ is defined. Then
>
> $$\lim_{x \to c} f(x) = f(c).$$
>
> In other words, the limit is the value of the function at $x = c$.

This result is also valid for one-sided limits (see the box on page 887). When a rational function is not defined at a number, different techniques must be used to find the limit (if there is one).

EXAMPLE 4

If

$$f(x) = \frac{x^2 - 2x - 3}{x - 3},$$

find $\lim_{x \to 3} f(x)$.

SOLUTION   Since $f(x)$ is not defined when $x = 3$, we cannot find the limit by evaluation. Nor can we use Property 4 because the limit of the denominator as x approaches 3 is 0. To find the limit, we first factor the numerator.

$$\frac{x^2 - 2x - 3}{x - 3} = \frac{(x + 1)(x - 3)}{x - 3}.$$

The factor $x - 3$ on the right side may be canceled *provided that $x - 3 \neq 0$*, that is, provided that $x \neq 3$. In other words,

$$\frac{x^2 - 2x - 3}{x - 3} = \frac{(x + 1)(x - 3)}{x - 3} = x + 1 \qquad \text{for all } x \neq 3.$$

The definition of limit, as x approaches 3, involves only the behavior of a function *near* $x = 3$ and not *at* $x = 3$. The preceding equation shows that both $f(x)$ and the function $g(x) = x + 1$ have exactly the same values at all numbers near $x = 3$. Therefore, they must have the same limit as x approaches 3. Therefore,

$$\lim_{x \to 3} \left(\frac{x^2 - 2x - 3}{x - 3} \right) = \lim_{x \to 3} (x + 1) = 3 + 1 = 4.$$ ■

The technique in Example 4 applies in many cases. When two functions have identical behavior, except possibly as $x = c$, they will have the same limit as x approaches c. More precisely,

*Limit
Theorem*

If f and g are functions that have limits as x approaches c and

$$f(x) = g(x) \qquad \text{for all } x \neq c,$$

then

$$\lim_{x \to c} f(x) = \lim_{x \to c} g(x).$$

EXAMPLE 5

Find $\lim\limits_{x \to 4} \dfrac{\sqrt{x} - 2}{x - 4}$.

SOLUTION   Property 4 does not apply here since the limit of the denominator is 0 when x approaches 4. However, the denominator can be factored.

$$x - 4 = (\sqrt{x} + 2)(\sqrt{x} - 2).$$

Consequently,

$$\frac{\sqrt{x} - 2}{x - 4} = \frac{\sqrt{x} - 2}{(\sqrt{x} + 2)(\sqrt{x} - 2)} = \frac{1}{\sqrt{x} + 2} \cdot \frac{\sqrt{x} - 2}{\sqrt{x} - 2}.$$

When $x \neq 4$, then $\sqrt{x} - 2 \neq 0$ and $\dfrac{\sqrt{x} - 2}{\sqrt{x} - 2} = 1$, so

$$\frac{\sqrt{x} - 2}{x - 4} = \frac{1}{\sqrt{x} + 2} \cdot \frac{\sqrt{x} - 2}{\sqrt{x} - 2} = \frac{1}{\sqrt{x} + 2} \qquad \text{for all } x \neq 4.$$

By the Limit Theorem,

$$\lim_{x \to 4} \frac{\sqrt{x} - 2}{x - 4} = \lim_{x \to 4} \frac{1}{\sqrt{x} + 2}$$

$$= \frac{\lim\limits_{x \to 4} 1}{\lim\limits_{x \to 4} (\sqrt{x} + 2)} \qquad \text{[Property 4]}$$

$$= \frac{\lim\limits_{x \to 4} 1}{\lim\limits_{x \to 4} \sqrt{x} + \lim\limits_{x \to 4} 2} \qquad \text{[Property 1]}$$

$$= \frac{\lim\limits_{x \to 4} 1}{\sqrt{\lim\limits_{x \to 4} x} + \lim\limits_{x \to 4} 2} \qquad \text{[Property 5]}$$

$$= \frac{1}{\sqrt{4} + 2} = \frac{1}{4}. \qquad \text{[Constant Limits; Polynomial Limit]}$$ ■

LIMITS OF DIFFERENCE QUOTIENTS

Limits involving the difference quotient of a function play an important role in calculus. Recall from Section 3.2 that the difference quotient of a function f is the quantity

$$\frac{f(x + h) - f(x)}{h}.$$

In limit computations with difference quotients, the variable is the quantity h and x is treated as a constant.

EXAMPLE 6

Suppose $f(x) = x^2$.

(a) Find and simplify the difference quotient of f, when $x = 5$.

(b) Find the limit of this difference quotient as h approaches 0.

SOLUTION

(a) $\dfrac{f(5 + h) - f(5)}{h} = \dfrac{(5 + h)^2 - 5^2}{h} = \dfrac{(25 + 10h + h^2) - 25}{h}$

$$= \frac{10h + h^2}{h}.$$

(b) $\displaystyle\lim_{h \to 0} \frac{f(5 + h) - f(5)}{h} = \lim_{h \to 0} \frac{10h + h^2}{h}$

$$= \lim_{h \to 0} \frac{h(10 + h)}{h} \qquad \text{[Factor numerator]}$$

$$= \lim_{h \to 0} (10 + h) \qquad \text{[Limit Theorem]}$$

$$= 10 + 0 = 10. \qquad \text{[Limit of a polynomial function]}$$ ∎

EXAMPLE 7

Find the limit of the difference quotient of $f(x) = \sqrt{x}$ as h approaches 7.

SOLUTION The difference quotient is

$$\frac{f(x + h) - f(x)}{h} = \frac{\sqrt{x + h} - \sqrt{x}}{h}.$$

Rationalizing the numerator of the last fraction (as was done in Example 9 of Section 5.1) shows that the fraction on the right above is equal to

$$\frac{1}{\sqrt{x + h} + \sqrt{x}}.$$

Therefore,

$$\lim_{h \to 0} \frac{f(x+h) - f(x)}{h} = \lim_{h \to 0} \frac{\sqrt{x+h} - \sqrt{x}}{h}$$

$$= \lim_{h \to 0} \frac{1}{\sqrt{x+h} + \sqrt{x}}$$

$$= \frac{\lim_{h \to 0} 1}{\lim_{h \to 0} \sqrt{x+h} + \lim_{h \to 0} \sqrt{x}} \qquad \text{[Property 4]}$$

$$= \frac{1}{\lim_{h \to 0} \sqrt{x+h} + \sqrt{x}} \qquad \text{[Constant Limits*]}$$

$$= \frac{1}{\sqrt{\lim_{h \to 0} (x+h)} + \sqrt{x}} \qquad \text{[Property 5]}$$

$$= \frac{1}{\sqrt{x} + \sqrt{x}} = \frac{1}{2\sqrt{x}}. \qquad \text{[Polynomial Limit*]} \quad \blacksquare$$

Limits such as the one in Example 7 play a central role in calculus and have a special name. The **derivative** of the function f, denoted f', is defined to be the function whose rule is

$$f'(x) = \lim_{h \to 0} \frac{f(x+h) - f(x)}{h}.$$

Example 7 shows that the derivative of $f(x) = \sqrt{x}$ is the function f' given by

$$f'(x) = \frac{1}{2\sqrt{x}}.$$

---

*Remember that x is treated as a constant here. Hence, $x + h$ is a polynomial in h with constant term x.

EXERCISES 13.2

In Exercises 1–8, use the following facts about the functions f, g, and h to find the required limit.

$$\lim_{x \to 4} f(x) = 5 \qquad \lim_{x \to 4} g(x) = 0 \qquad \lim_{x \to 4} h(x) = -2$$

1. $\lim\limits_{x \to 4} (f(x) + g(x))$

2. $\lim\limits_{x \to 4} (g(x) - h(x))$

3. $\lim\limits_{x \to 4} \dfrac{f(x)}{g(x)}$

4. $\lim\limits_{x \to 4} \dfrac{g(x)}{h(x)}$

5. $\lim\limits_{x \to 4} f(x)g(x)$

6. $\lim\limits_{x \to 4} h(x)^2$

7. $\lim\limits_{x \to 4} \dfrac{3h(x)}{2f(x) + g(x)}$

8. $\lim\limits_{x \to 4} \dfrac{f(x) - 2g(x)}{4h(x)}$

In Exercises 9–40, find the limit if it exists. If the limit does not exist, explain why.

9. $\lim\limits_{x \to 2} (6x^3 - 2x^2 + 5x - 3)$

10. $\lim\limits_{x \to -1} (x^7 + 2x^5 - x^4 + 3x + 4)$

11. $\lim\limits_{x \to -2} \dfrac{3x - 1}{2x + 3}$

12. $\lim\limits_{x \to 3} \dfrac{x^2 + x + 1}{x^2 - 2x}$

13. $\lim\limits_{x \to 3} \dfrac{x^2 - x - 6}{x^2 - 2x - 3}$

14. $\lim\limits_{x \to 1} \dfrac{x^2 - 1}{x^2 + x - 2}$

15. $\lim\limits_{x \to 1} \dfrac{x^3 - 1}{x^2 - 1}$

16. $\lim\limits_{x \to -2} \dfrac{x^2 + 5x + 6}{x^2 - x - 6}$

17. $\lim\limits_{x \to 4^-} \dfrac{x - 4}{x^2 - 16}$

18. $\lim\limits_{x \to 2^+} \dfrac{|x - 2|}{x - 2}$

19. $\lim\limits_{x \to 3^+} \dfrac{3}{x^2 - 9}$

20. $\lim\limits_{x \to 2^-} \dfrac{x + 1}{x^2 - x - 2}$

21. $\lim\limits_{x \to 1} \sqrt{x^3 + 6x^2 + 2x + 5}$

22. $\lim\limits_{x \to 2} \sqrt{x^2 + x + 3}$

23. $\lim\limits_{x \to 1^+} (\sqrt{x - 1} + 3)$

24. $\lim\limits_{x \to 3^-} \sqrt{-3 - x}$

25. $\lim\limits_{x \to -2.5^+} (\sqrt{5 + 2x} + x)$

26. $\lim\limits_{x \to 3^+} (\sqrt{x - 3} + \sqrt{3x})$

27. $\lim\limits_{x \to 3} \dfrac{\sqrt{x} - \sqrt{3}}{x - 3}$

28. $\lim\limits_{x\to 25} \dfrac{\sqrt{x}-5}{x-25}$

29. $\lim\limits_{x\to 0}\left(\dfrac{1/(x+5)-1/5}{x}\right)$ [*Hint:* Write the expression in parentheses as a single fraction.]

30. $\lim\limits_{x\to 0}\left(\dfrac{2/(x+6)-1/3}{x}\right)$ **31.** $\lim\limits_{x\to 1}\left[\dfrac{1}{x-1}-\dfrac{2}{x^2-1}\right]$

32. $\lim\limits_{x\to -2}\left[\dfrac{x^2}{x+2}+\dfrac{2x}{x+2}\right]$ **33.** $\lim\limits_{x\to 0}\dfrac{x^2}{|x|}$

34. $\lim\limits_{x\to -2}|x+2|$

35. $\lim\limits_{x\to -3^+}\left(\dfrac{|x+3|}{x+3}+\sqrt{x+3}+1\right)$

36. $\lim\limits_{x\to -4^-}(\sqrt{|x|-4}+x^2)$

37. $\lim\limits_{x\to 0}\dfrac{\sqrt{2-x}-\sqrt{2}}{x}$ [*Hint:* Rationalize the numerator; see Section 5.1.]

38. $\lim\limits_{x\to 0}\left[\dfrac{1}{x\sqrt{x+1}}-\dfrac{1}{x}\right]$ **39.** $\lim\limits_{x\to 0}\left[\dfrac{|x|}{x}-\dfrac{x}{|x|}\right]$

40. $\lim\limits_{x\to -3}\dfrac{|x+3|}{x+3}$

In Exercises 41–44, find
$$\lim\limits_{h\to 0}\dfrac{f(2+h)-f(2)}{h}.$$

41. $f(x)=x^2$ **42.** $f(x)=x^3$

43. $f(x)=x^2+x+1$ **44.** $f(x)=\sqrt{x}$

In Exercises 45 and 46, find $\lim\limits_{h\to 0}\dfrac{f(0+h)-f(0)}{h}$, *if it exists.*

45. $f(x)=|x|$ **46.** $f(x)=x|x|$

In Exercises 47–50, find the rule of the derivative of the function f. [See Example 7 and the remarks following it.]

47. $f(x)=2x+3$ **48.** $f(x)=3x-5$

49. $f(x)=x^2+x$ **50.** $f(x)=x^2-x+1$

In Exercises 51–54, find the rule of the derivative of the function f. [The difference quotients of these functions were found and simplified in Exercises 57–60 of Section 5.1.]

51. $f(x)=\sqrt{x+1}$ **52.** $f(x)=2\sqrt{x+3}$

53. $f(x)=\sqrt{x^2+1}$ **54.** $f(x)=\sqrt{x^2-x}$

55. Give an example of functions f and g and a number c such that neither $\lim\limits_{x\to c}f(x)$ nor $\lim\limits_{x\to c}g(x)$ exists, but
$$\lim\limits_{x\to c}(f(x)+g(x))\text{ does exist.}$$

56. Give an example of functions f and g and a number c such that neither $\lim\limits_{x\to c}f(x)$ nor $\lim\limits_{x\to c}g(x)$ exists, but $\lim\limits_{x\to c}f(x)g(x)$ does exist.

13.2.A SPECIAL TOPICS   The Formal Definition of Limit

Section Objectives
■ Understand the formal definition of limit.
■ Prove statements about limits.

The informal definition of limit in Section 13.1 (or one very much like it) was used for more than a century and played a crucial role in the development of calculus. To avoid error, however, mathematical concepts such as limit must be based on precise definitions and theorems. In this section, we take the first step in building this rigorous foundation by developing the formal definition of limit.

To keep the discussion as concrete as possible, suppose we have a function f such that $\lim\limits_{x\to 5}f(x)=12$. You don't need to know the rule of f or anything else about it, except the informal definition of limit, to understand the following discussion.

The informal definition of the statement "$\lim\limits_{x\to 5}f(x)=12$" that was given in Section 13.1 has two components:

A. As x takes values very close (but not equal) to 5, the corresponding values of $f(x)$ are very close (and possibly equal) to 12.

B. The value of $f(x)$ can be made arbitrarily close (as close as you want) to 12 for all x sufficiently close (but not equal) to 5.

Strictly speaking, only component B is necessary, since it implies component A: If the values of $f(x)$ can be made *arbitrarily close* to 12 by taking x close enough to 5, then it is certainly true that when x is *very* close to 5, $f(x)$ must be very close to 12.* Consequently, we begin with component B of the informal definition.

①

> $\lim_{x \to 5} f(x) = 12$ means that the values of $f(x)$ can be made as close as you want to 12 for all x sufficiently close to 5.

We shall modify this definition step by step until we reach the desired formal definition.

Definition ① says, in effect, that a two-step process is involved: Whenever you tell us how close you want $f(x)$ to 12, we can tell you how close x must be to 5 to guarantee this. So it can be restated like this.

②

> $\lim_{x \to 5} f(x) = 12$ means that whenever you specify how close $f(x)$ should be to 12, we can tell you how close x must be to 5 to guarantee this.

For example, if you want $f(x)$ to be within .01 of 12 (that is, between 11.99 and 12.01), we must tell you how close x must be to 5 to guarantee that the corresponding values of $f(x)$ are between 11.99 and 12.01. But "arbitrarily close" implies much more. We must be able to answer your demands, regardless of how close you want $f(x)$ to be to 12. If you want $f(x)$ to be within .002 of 12, or within .0001 of 12, or within any distance of 12, we must be able to tell you how close x must be to 5 in each case to accomplish this. So Definition ② can be restated as follows.

③

> $\lim_{x \to 5} f(x) = 12$ means that no matter what positive number you specify (measuring how close you want $f(x)$ to 12), we can tell you how close x must be to 5 to guarantee that $f(x)$ is that close to 12.

Hereafter, whatever small positive number you specify (measuring how close $f(x)$ should be to 12) will be denoted by the Greek letter ϵ (epsilon).[†] When we tell you how close x should be to 5 to accomplish this, we must give some small positive number that measures "how close" (for instance, x must be within .0003 of 5, or x must be within .00002 of 5). We denote the number that measures the closeness of x to 5 by the Greek letter δ (delta).[‡] Presumably the number δ, which measures how close x must be to 5, will depend on the number ϵ, which measures how close you want $f(x)$ to be to 12. In this language, Definition ③ becomes the following.

---

*Component A was included in the informal definition because it describes the procedures that are actually used to estimate a limit. In this case, we would compute values of $f(x)$ when $x = 4.9, 4.99, 4.999, 5.1, 5.01, 5.001$, etc. to see whether they are very close to 12.

[†]Mathematicians have used Greek letters in this context since the formal definition was first developed in the nineteenth century. Epsilon is the Greek letter e, the first letter in the word "error." You can think of ϵ as measuring the allowable degree of error—the amount by which you will permit $f(x)$ to differ from 12.

[‡]Delta is the Greek letter d, the first letter of "difference." We must state the difference between x and 5 that will guarantee that $f(x)$ is within ϵ of 12.

④

$\lim\limits_{x \to 5} f(x) = 12$ means that for every positive number ϵ, there is a positive number δ (depending on ϵ) with this property:

If x is within δ of 5 (but not equal to 5), then $f(x)$ is within ϵ of 12 (and possibly equal to 12).

Although Definition ④ is essentially the formal definition, somewhat briefer notation is usually used, as we now explain.

If we think of $f(x)$ and 12 as numbers on the number line, then the statement

$$f(x) \text{ is within } \epsilon \text{ of } 12$$

means that

the distance from $f(x)$ to 12 is less than ϵ.

Since distance on the number line is measured by absolute value (see Section 1.1), this last statement means

$$\left| f(x) - 12 \right| < \epsilon.$$

Similarly, saying that x is within δ of 5 and not equal to 5 means that the distance from x to 5 is less than δ but greater than 0, that is,

$$0 < |x - 5| < \delta.$$

In this notation, Definition ④ becomes the desired formal definition.

⑤

$\lim\limits_{x \to 5} f(x) = 12$ means that for each positive number ϵ, there is a positive number δ (depending on ϵ) with this property:

If $\quad 0 < |x - 5| < \delta, \quad$ then $\quad \left| f(x) - 12 \right| < \epsilon.$

Definition ⑤ is sufficiently rigorous because the imprecise terms such as "arbitrarily close" and "sufficiently close" in the informal definition have been replaced by a precise statement about inequalities that can be verified in specific cases, as will be shown in the examples below. There is nothing special about 5 and 12 in the preceding discussion; the entire analysis applies equally well in the general case and leads to this formal definition of limit (which is just Definition ⑤ with c in place of 5, L in place of 12, and f any function).

Definition of Limit

Let f be a function and let c be a real number such that $f(x)$ is defined for all x (except possibly $x = c$) in some open interval containing c.* We say "the limit of $f(x)$, as x approaches c, is L" and write

$$\lim_{x \to c} f(x) = L,$$

provided that for each positive number ϵ, there is a positive number δ (depending on ϵ) with this property:

If $\quad 0 < |x - c| < \delta, \quad$ then $\quad \left| f(x) - L \right| < \epsilon.$

---

*This means that there are numbers a and b with $a < c < b$ such that $f(x)$ is defined for all x with $a < x < b$, except possibly $x = c$.

EXAMPLE 1

Let $f(x) = 4x - 8$, and prove that $\lim_{x \to 5} f(x) = 12$.

SOLUTION   We must apply the definition of limit with $c = 5$, $L = 12$, and $f(x) = 4x - 8$. Suppose that ϵ is any positive number. We must find a positive number δ with this property.

$$\text{If} \quad 0 < |x - 5| < \delta, \quad \text{then} \quad |f(x) - 12| < \epsilon.$$

Let δ be the number $\epsilon/4$; we claim that this δ will work. For now, don't worry about how we figured out that $\delta = \epsilon/4$ will work; just verify that the following argument is valid: If $|x - 5| < \delta$, then

$$|x - 5| < \epsilon/4 \qquad \text{[Since } \delta = \epsilon/4]$$

$$4|x - 5| < \epsilon \qquad \text{[Both sides multiplied by 4]}$$

$$|4||x - 5| < \epsilon \qquad \text{[Since } 4 = |4|]$$

$$|4(x - 5)| < \epsilon. \qquad \text{[Property 4 of absolute value (page 10)]}$$

$$|4x - 20| < \epsilon$$

$$|(4x - 8) - 12| < \epsilon \qquad \text{[Rewrite } -20 \text{ as } -8 - 12]$$

$$|f(x) - 12| < \epsilon. \qquad \text{[Because } f(x) = 4x - 8]$$

We have verified that for each positive ϵ, there is a positive δ (namely, $\epsilon/4$) with this property:

$$\text{If} \quad 0 < |x - 5| < \delta, \quad \text{then} \quad |f(x) - 12| < \epsilon.$$

This completes the proof that $\lim_{x \to 5} f(x) = 12$. ∎

Proofs like the one in Example 1 often seem mysterious to beginners. Although they can follow the argument after the appropriate δ has been announced, they don't see how anyone found that particular δ in the first place. So here's an example that gives a somewhat fuller picture of the mental processes that are used in proving statements about limits.

EXAMPLE 2

Prove that $\lim_{x \to 1} (2x + 7) = 9$.

SOLUTION   *Scratch Work*   In this case, $f(x) = 2x + 7$, $c = 1$, and $L = 9$. Let ϵ be any positive number. We must find a δ with this property.

$$\text{If} \quad 0 < |x - 1| < \delta, \quad \text{then} \quad |(2x + 7) - 9| < \epsilon.$$

To get some idea which δ might have this property, we *work backwards*. The conclusion we want to reach, namely,

$$|(2x + 7) - 9| < \epsilon,$$

is equivalent to

$$|2x - 2| < \epsilon,$$

which in turn is equivalent to each of these statements.

$$|2(x - 1)| < \epsilon$$

$$|2||x - 1| < \epsilon$$

$$2|x - 1| < \epsilon$$

$$|x - 1| < \epsilon/2.$$

When the conclusion is written this way, it suggests that number $\epsilon/2$ would be a good choice for δ.

Everything up to here has been "scratch work" (the similar scratch work was not included in Example 1). Now we must give the actual proof.

Proof   Given a positive number ϵ, let δ be the number $\epsilon/2$. If $0 < |x - 1| < \delta$, then

$$|x - 1| < \epsilon/2 \quad \text{[Since } \delta = \epsilon/2\text{]}$$

$$2|x - 1| < \epsilon \quad \text{[Both sides multiplied by 2]}$$

$$|2||x - 1| < \epsilon \quad \text{[Since } 2 = |2|\text{]}$$

$$|2(x - 1)| < \epsilon \quad \text{[Property of absolute value]}$$

$$|2x - 2| < \epsilon$$

$$|(2x + 7) - 9| < \epsilon \quad \text{[Rewrite } -2 \text{ as } 7 - 9\text{]}$$

$$|f(x) - 9| < \epsilon. \quad \text{[Since } f(x) = 2x + 7\text{]}$$

Therefore, $\delta = \epsilon/2$ has the required property, and the proof is complete. ∎

Once we did some algebraic scratch work in Examples 1 and 2, the limit proofs were relatively easy. In most cases, however, a more involved argument is required. In fact, it can be quite difficult to prove directly from the definition, for example, that

$$\lim_{x \to 3} \frac{x^2 - 4x + 1}{x^3 - 2x^2 - x} = -\frac{1}{3}.$$

Fortunately, such complicated calculations can often be avoided by using the various limit properties given in Section 13.2. Of course, these properties must first be proved by using the definition. Surprisingly, these proofs are comparatively easy.

EXAMPLE 3

Let f and g be functions such that

$$\lim_{x \to c} f(x) = L \quad \text{and} \quad \lim_{x \to c} g(x) = M.$$

Prove that

$$\lim_{x \to c} (f(x) + g(x)) = L + M.$$

SOLUTION   *Scratch Work*   If ϵ is any positive number, we must find a positive δ with this property:

If $0 < |x - c| < \delta$,   then $|(f(x) + g(x)) - (L + M)| < \epsilon.$

We first note that by the triangle inequality (page 11),

$$|(f(x) + g(x)) - (L + M)| = |(f(x) - L) + (g(x) - M)|$$
$$\leq |f(x) - L| + |g(x) - M|.$$

If we can find a δ with this property:

$$\text{If} \qquad 0 < |x - c| < \delta, \qquad \text{then} \qquad |f(x) - L| + |g(x) - M| < \epsilon,$$

then the smaller quantity $|(f(x) + g(x)) - (L + M)|$ will also be less than ϵ when $0 < |x - c| < \delta$. We can find such a δ as follows.

Proof   Let ϵ be any positive number. Apply the definition of $\lim_{x \to c} f(x) = L$ with $\epsilon/2$ in place of ϵ. There is a positive number δ_1 with the following property.

$$\text{If} \qquad 0 < |x - c| < \delta_1, \qquad \text{then} \qquad |f(x) - L| < \epsilon/2.$$

Similarly, by the definition of $\lim_{x \to c} g(x) = M$ with $\epsilon/2$ in place of ϵ, there is a positive number δ_2 with the following property.

$$\text{If} \qquad 0 < |x - c| < \delta_2, \qquad \text{then} \qquad |g(x) - M| < \epsilon/2.$$

Now let δ be the smaller of the two numbers δ_1 and δ_2, so that $\delta \leq \delta_1$ and $\delta \leq \delta_2$. Then if $0 < |x - c| < \delta$, we must have

$$0 < |x - c| < \delta_1 \qquad \text{and} \qquad 0 < |x - c| < \delta_2,$$

and therefore,

$$|f(x) - L| < \epsilon/2 \qquad \text{and} \qquad |g(x) - M| < \epsilon/2.$$

Consequently, if $0 < |x - c| < \delta$, then

$$|(f(x) + g(x)) - (L + M)| = |(f(x) - L) + (g(x)) - M)|$$
$$\leq |f(x) - L| + |g(x) - M|$$
$$< \frac{\epsilon}{2} + \frac{\epsilon}{2} = \epsilon.$$

Thus, we have shown that for any $\epsilon > 0$, there is a $\delta > 0$ with the following property.

$$\text{If} \qquad 0 < |x - c| < \delta, \qquad \text{then} \qquad |(f(x) + g(x)) - (L + M)| < \epsilon.$$

Therefore, $\lim_{x \to c} (f(x) + g(x)) = L + M$.   ∎

The proofs of the other limit properties and theorems of Section 13.2 are dealt with in Exercise 15 and in calculus.

Finally, the formal definition of limit may easily be carried over to one-sided limits by using the following fact (see Special Topics 4.6.A).

$$|x - c| < \delta \qquad \text{exactly when} \qquad -\delta < x - c < \delta,$$

which is equivalent to

$$|x - c| < \delta \qquad \text{exactly when} \qquad c - \delta < x < c + \delta.$$

Thus, the numbers between c and $c + \delta$ lie to the right of c, within distance δ of c, and the numbers between $c - \delta$ and c lie to the left of c, within distance δ of c.

Consequently, the formal definition of right-hand limits can be obtained from the definition above by replacing the phrase "if $0 < |x - c| < \delta$" by "$c < x < c + \delta$." For a formal definition of left-hand limits, replace the phrase "if $0 < |x - c| < \delta$" by "$c - \delta < x < c$."

EXERCISES 13.2.A

In Exercises 1–12, use the formal definition of limit to prove the statement, as in Examples 1 and 2.

1. $\lim_{x\to 3} (3x - 2) = 7$ **2.** $\lim_{x\to 1} (4x + 2) = 6$

3. $\lim_{x\to 5} x = 5$ **4.** $\lim_{x\to 0} (x + 2) = 2$

5. $\lim_{x\to 2} (6x + 3) = 15$ **6.** $\lim_{x\to 7} (-2x + 19) = 5$

7. $\lim_{x\to 1} 4 = 4$ **8.** $\lim_{x\to 2} \pi = \pi$

9. $\lim_{x\to 4} (x - 6) = -2$ **10.** $\lim_{x\to 1} (2x - 7) = -5$

11. $\lim_{x\to -2} (2x + 5) = 1$ **12.** $\lim_{x\to -3} (2x - 4) = -10$

In Exercises 13 and 14, use the formal definition of limit to prove the statement.

13. $\lim_{x\to 0} x^2 = 0$ **14.** $\lim_{x\to 0} x^3 = 0$

In Exercises 15 and 16, let f and g be functions such that

$$\lim_{x\to c} f(x) = L \quad \text{and} \quad \lim_{x\to c} g(x) = M.$$

15. Prove that $\lim_{x\to c} (f(x) - g(x)) = L - M$.

16. If k is a constant, prove that $\lim_{x\to c} kf(x) = kL$.

17. (a) Describe the graph of $g(x) = x \sin\left(\dfrac{\pi}{x}\right)$ near the origin.

(b) Use the formal definition of limit to prove that
$$\lim_{x\to 0} x \sin\left(\frac{\pi}{x}\right) = 0.$$

18. Let f be the function defined by
$$f(x) = \begin{cases} 0 & \text{if } x \text{ is irrational} \\ |x| & \text{if } x \text{ is rational} \end{cases}.$$

Use the formal definition of limit to prove that
$$\lim_{x\to 0} f(x) = 0.$$

13.3 Continuity

Section Objectives
- Understand the definition of continuity.
- Show that a function is continuous at a point.
- Explore continuity on an interval.
- Learn the properties of continuous functions.

Let c be a real number in the domain of a function f. Intuitively speaking, the function f is **continuous** at $x = c$ if you can draw the graph of f at and near the point $(c, f(c))$ without lifting your pencil from the paper. For example, each of the four graphs in Figure 13–13 is the graph of a function that is continuous at $x = c$.

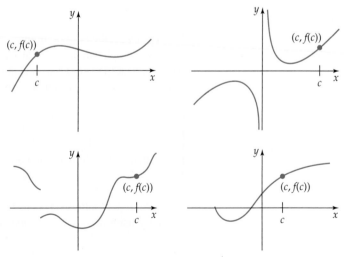

Figure 13–13

On the other hand, none of the functions whose graphs are shown in Figure 13–14 is continuous at $x = c$. If you don't believe it, just try to draw one of these graphs near $x = c$ without lifting your pencil from the paper.

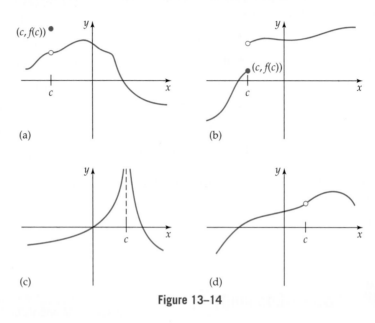

Figure 13–14

Our goal is to find an analytical description of continuity at a point that does not depend on having the graph given in advance. So let us consider the two possibilities when a function is continuous at $x = c$, as shown in Figure 13–15.

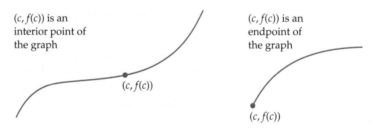

Figure 13–15

We begin with the case of an interior point.*

If the graph can be drawn around the point $(c, f(c))$ without lifting pencil from paper, then at the very least, $f(x)$ must be defined for $x = c$ and for $x = t$ when t is any number near c. We can describe this more precisely by saying the following.

① $f(x)$ is defined for all x in some open interval containing c. In other words, there are numbers a and b with $a < c < b$ such that $f(x)$ is defined for all x with $a < x < b$. In particular, $f(c)$ is defined.

Although condition ① is necessary for f to be continuous at $x = c$, this condition by itself does not *guarantee* continuity. For instance, the functions whose

---

*The endpoint case is considered on page 907.

graphs are shown in Figure 13–14(a) and 13–14(b) on the facing page, are defined for all values of x near c but are *not* continuous at $x = c$.

Consider Figure 13–16, which shows a typical function f that is continuous at $x = c$. The point $(x, f(x))$ represents the pencil point as it moves along the graph near the point $(c, f(c))$.

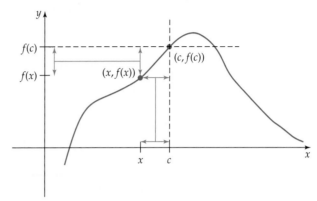

Figure 13–16

Note that

②
> As the pencil point $(x, f(x))$ gets closer to the vertical line through $x = c$ (from the left or the right), it also gets closer and closer to the horizontal line through $y = f(c)$.

Furthermore,

③
> The pencil point $(x, f(x))$ can be brought as close as you want to the horizontal line by moving it close enough to the vertical line.

As shown in Figure 13–16, the distance from the point $(x, f(x))$ to the dashed vertical line is the same as the distance from x to c on the x-axis. Similarly, the distance from the point $(x, f(x))$ to the dashed horizontal line is the same as the distance from $f(x)$ to $f(c)$ on the y-axis. Consequently, statements ② and ③ mean that

> as x takes values closer and closer to c (on both sides of c),
> the corresponding values of $f(x)$ get closer and closer to $f(c)$

and that

> the values of $f(x)$ can be made as close as you want
> to $f(c)$ for all x close enough to c.

This last statement, together with statement ① above, looks suspiciously like the definition of limit (with $f(c)$ in place of L) and suggests that the idea of "not lifting the pencil" can be expressed mathematically by saying

④
$$\lim_{x \to c} f(x) = f(c).$$

The preceding analysis leads to the following.

Definition of Continuity

Let f be a function that is defined for all x in some open interval containing c. Then f is said to be **continuous** at $x = c$ provided that

(i) $f(c)$ is defined;

(ii) $\lim_{x \to c} f(x)$ exists; and

(iii) $\lim_{x \to c} f(x) = f(c)$.

Strictly speaking, it isn't necessary to list all three of the conditions in the definition, since condition (iii) implies that the other two conditions hold.

EXAMPLE 1

Without graphing, show that the function

$$f(x) = \frac{\sqrt{x^2 - x + 1}}{x - 5}$$

is continuous at $x = -3$.

SOLUTION

$$f(-3) = \frac{\sqrt{(-3)^2 - (-3) + 1}}{(-3) - 5} = -\frac{\sqrt{13}}{8},$$

and by the properties of limits (page 890),

$$\lim_{x \to -3} f(x) = \lim_{x \to -3} \frac{\sqrt{x^2 - x + 1}}{x - 5} = \frac{\lim_{x \to -3} \sqrt{x^2 - x + 1}}{\lim_{x \to -3} (x - 5)}$$

$$= \frac{\sqrt{\lim_{x \to -3} (x^2 - x + 1)}}{\lim_{x \to -3} (x - 5)} = \frac{\sqrt{(-3)^2 - (-3) + 1}}{(-3) - 5}$$

$$= -\frac{\sqrt{13}}{8}.$$

Therefore, $\lim_{x \to -3} f(x) = f(-3)$, and hence, f is continuous at $x = -3$. ∎

The properties of limits (Section 13.2) and the definition of continuity provide justification for assumptions made about various graphs in earlier chapters. For example, in Section 13.2, we saw that for any polynomial or rational function f and any number c such that $f(c)$ is defined,

$$\lim_{x \to c} f(x) = f(c).$$

Thus, every polynomial and every rational function is continuous wherever it is defined. That proves the first two of the following statements.

Continuous Functions

Each of the following functions is continuous at every number in its domain.

1. Polynomial functions

2. Rational functions

3. The exponential functions
$$f(x) = 10^x, \qquad g(x) = e^x, \qquad h(x) = b^x$$

4. The logarithmic functions
$$f(x) = \log x, \qquad g(x) = \ln x, \qquad h(x) = \log_b x$$

5. The trigonometric functions
$$f(x) = \sin x, \qquad g(x) = \cos x, \qquad h(x) = \tan x,$$
$$k(x) = \csc x, \qquad r(x) = \sec x, \qquad s(x) = \cot x$$

6. The inverse trigonometric functions
$$f(x) = \sin^{-1}x, \qquad g(x) = \cos^{-1}x, \qquad h(x) = \tan^{-1}x$$

The facts in the preceding box justify the graphing techniques used in this book. Plotting points (a few when graphing by hand, many when graphing with technology) and connecting them with an unbroken curve is a reasonable procedure when graphing continuous functions.

CONTINUITY ON AN INTERVAL

Now let us consider continuity at an endpoint of the graph of a function f, such as $(a, f(a))$ or $(b, f(b))$ in Figure 13–17.

$(a, f(a))$ $(b, f(b))$

Figure 13–17

The intuitive idea of continuity at $(a, f(a))$ is that we can draw the graph of f at, and to the right of, the point $(a, f(a))$ without lifting the pencil from the paper. Essentially the same analysis that was given above can be made here, provided that we consider only values of x to the right of $x = a$. It shows that the intuitive idea of continuity at $x = a$ is equivalent to a statement about right-hand limits, namely, $\lim_{x \to a^+} f(x) = f(a)$. An analogous discussion (with left-hand limits) applies to the endpoint $(b, f(b))$ and leads to this definition.

Continuity from the Left and Right

A function f is **continuous from the right** at $x = a$ provided that
$$\lim_{x \to a^+} f(x) = f(a).$$

A function f is **continuous from the left** at $x = b$ provided that
$$\lim_{x \to b^-} f(x) = f(b).$$

EXAMPLE 2

The function $f(x) = \sqrt{x}$, which is not defined when $x < 0$, is continuous from the right at $x = 0$ because $f(0) = \sqrt{0} = 0$ and

$$\lim_{x \to 0^+} f(x) = \lim_{x \to 0^+} \sqrt{x} = 0 = f(0).$$ ■

The most useful functions are those that are continuous at every point in an interval. Intuitively, this means that their graphs can be drawn over the entire interval without lifting the pencil from the paper. As we saw above, most of the functions in this book are of this type. Here is a formal definition.

Continuity on
an Interval

A function f is said to be **continuous on an open interval (a, b)** provided that f is continuous at every number in the interval.

A function f is said to be **continuous on a closed interval $[a, b]$** provided that f is continuous from the right at $x = a$, continuous from the left at $x = b$, and continuous at every number in the open interval (a, b).

Analogous definitions may be given for continuity on intervals of the form $[a, b)$, $(a, b]$, (a, ∞), $[a, \infty)$, $(-\infty, b)$, and $(-\infty, b]$.

EXAMPLE 3

The function f whose graph is shown in Figure 13–18 is discontinuous at $x = -3$ and $x = 2$ but is continuous on each of the intervals $(-5, -3)$, $[-3, 2]$, and $(2, \infty)$.   ■

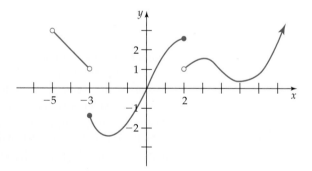

Figure 13–18

▓ PROPERTIES OF CONTINUOUS FUNCTIONS

Using the definition is not always the most convenient way to show that a particular function is continuous. It is often easier to establish continuity by using the following facts.

Properties of Continuous Functions

If the functions f and g are continuous at $x = c$, then each of the following functions is also continuous at $x = c$.

1. The sum function $f + g$

2. The difference function $f - g$

3. The product function fg

4. The quotient function f/g (provided that $g(c) \neq 0$)

Proof   **1.** By the definition of the sum function, $(f + g)(x) = f(x) + g(x)$. Since f and g are continuous at $x = c$, we know that

$$\lim_{x \to c} f(x) = f(c) \qquad \text{and} \qquad \lim_{x \to c} g(x) = g(c).$$

Therefore, by the first property of limits (see page 890),

$$\lim_{x \to c} (f + g)(x) = \lim_{x \to c} (f(x) + g(x)) = \lim_{x \to c} f(x) + \lim_{x \to c} g(x)$$

$$= f(c) + g(c)$$

$$= (f + g)(c).$$

This says that $f + g$ is continuous at $x = c$.

The remaining statements are proved similarly, using limit properties 2, 3, and 4. ∎

EXAMPLE 4

We noted above that the function $f(x) = \sin x$ is continuous at $x = 0$. We also know that the polynomial function $g(x) = x^3 + 5x - 2$ is continuous at $x = 0$. Therefore, each of the following functions is continuous at $x = 0$.

$$(f + g)(x) = \sin x + x^3 + 5x - 2$$

$$(f - g)(x) = \sin x - (x^3 + 5x - 2) = \sin x - x^3 - 5x + 2$$

$$(fg)(x) = (\sin x)(x^3 + 5x - 2)$$

$$\left(\frac{f}{g}\right)(x) = \frac{\sin x}{x^3 + 5x - 2} \qquad \text{(Note that } g(0) \neq 0.\text{)}$$ ∎

Composition of functions is often used to construct new functions from given ones.

Continuity of Composite Functions

If the function f is continuous at $x = c$ and the function g is continuous at $x = f(c)$, then the composite function $g \circ f$ is continuous at $x = c$.

EXAMPLE 5

The polynomial function $f(x) = x^3 - 3x^2 + x + 7$ is continuous at $x = 2$, and $f(2) = 2^3 - 3 \cdot 2^2 + 2 + 7 = 5$. The function $g(x) = \sqrt{x}$ is continuous at $x = 5$ because by Limit Property 5 (page 890).

$$\lim_{x \to 5} \sqrt{x} = \sqrt{\lim_{x \to 5} x} = \sqrt{5} = g(5).$$

By the box above (with $c = 2$ and $f(c) = f(2) = 5$), the composite function $g \circ f$, which is given by

$$(g \circ f)(x) = g(f(x)) = g(x^3 - 3x^2 + x + 7) = \sqrt{x^3 - 3x^2 + x + 7},$$

is also continuous at $x = 2$.   ■

THE INTERMEDIATE VALUE THEOREM

The following property of continuous functions has several important applications.

Intermediate Value Theorem

> If the function f is continuous on the closed interval $[a, b]$ and k is any number between $f(a)$ and $f(b)$, then there exists at least one number c between a and b such that $f(c) = k$.

Although we shall not prove the Intermediate Value Theorem, we can demonstrate its plausibility by a simple geometric argument. Figure 13–19 shows the graph of a function f that is continuous on the interval $[a, b]$ and a number k that is between $f(a)$ and $f(b)$.

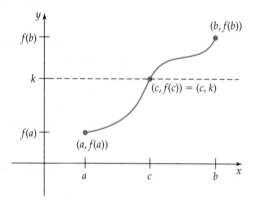

Figure 13–19

Because f is continuous on the interval, its graph can be drawn from $(a, f(a))$ to $(b, f(b))$ without lifting the pencil from the paper. As Figure 13–19 suggests, there is no way to do this unless the graph crosses the horizontal line $y = k$ at least once between $x = a$ and $x = b$. The coordinates of a point where the graph crosses the line can be described in two ways:

$(c, f(c))$ because the point is on the graph of f;

(c, k) because the point is on the line $y = k$.

Thus, this point satisfies the conclusion of the Intermediate Value Theorem:

$$a < c < b \qquad \text{and} \qquad f(c) = k.$$

The Intermediate Value Theorem further explains why the graph of a continuous function is connected and unbroken. If f is continuous on the interval $[a, b]$, then its graph cannot go from the point $(a, f(a))$ to the point $(b, f(b))$ without moving through all the y values between $f(a)$ and $f(b)$.

Here is an important special case of the Intermediate Value Theorem.

Equation Theorem

> If the function f is continuous on the interval $[a, b]$ and $f(a)$ and $f(b)$ have opposite signs, then the equation $f(x) = 0$ has a solution between a and b.

Proof Since one of $f(a)$ and $f(b)$ is positive and the other is negative, the number 0 is between them. By the Intermediate Value Theorem, there is a number c between a and b such that $f(c) = 0$. ■

Many root finders on calculators are based on the Equation Theorem. To see why, consider the equation

$$x^3 - x - 1 = 0.$$

The graph of $f(x) = x^3 - x - 1$ in Figure 13–20 shows that there is a solution between 1 and 2. So we evaluate $f(x)$, beginning at $x = 1$ and moving by tenths, as in Figure 13–21. Since $f(1.3)$ and $f(1.4)$ have opposite signs, the solution is between 1.3 and 1.4 by the Equation Theorem. Hence, the decimal expansion of the solution begins with 1.3.

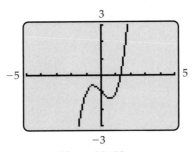

Figure 13–20

X	Y1
1	-1
1.1	-.769
1.2	-.472
1.3	-.103
1.4	.344
1.5	.875
1.6	1.496

Y1=X^3-X-1

Figure 13–21

Now evaluate $f(x)$ beginning at 1.3 and moving by hundredths, as in Figure 13–22. Since $f(1.32)$ and $f(1.33)$ have opposite signs, the solution is between 1.32 and 1.33. So its decimal expansion begins with 1.32. Next evaluate $f(x)$ beginning at 1.32 and moving by thousandths, as in Figure 13–23, which shows that the solution is between 1.324 and 1.325. Thus, its decimal approximation begins 1.324. Each time the evaluation procedure is carried out, one more decimal place in the solution is found.

X	Y1
1.3	-.103
1.31	-.0619
1.32	-.02
1.33	.02264
1.34	.0661
1.35	.11038
1.36	.15546

Y1=X^3-X-1

Figure 13–22

X	Y1
1.32	-.02
1.321	-.0158
1.322	-.0116
1.323	-.0073
1.324	-.0031
1.325	.0012
1.326	.00547

Y1=X^3-X-1

Figure 13–23

Root finders use this process (or a variation of it) to approximate solutions to as many decimal places as the calculator can handle. The Equation Theorem is also the reason that a root finder that cannot find a solution may display an error message that reads "no sign change."

EXERCISES 13.3

1. Use the graph of the function f in the figure to find all the numbers at which f is not continuous.

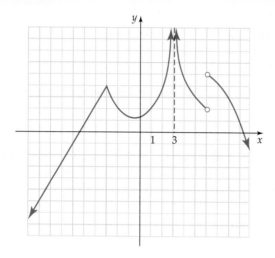

2. Use the graph of the function g in the figure to find all the numbers at which $g(x)$ is defined but not continuous.

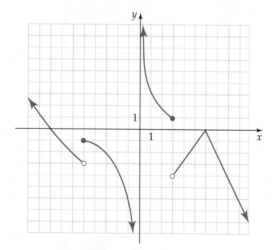

In Exercises 3–6, determine whether the function whose graph is given is continuous at $x = -2$, at $x = 0$, and at $x = 3$.

3.

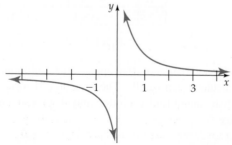

4.

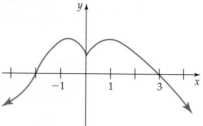

5.

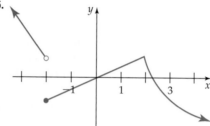

6.

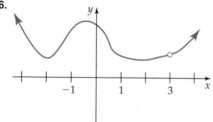

In Exercises 7–12, use the definition of continuity and the properties of limits to show that the function is continuous at the given number.

7. $f(x) = x^2 + 5(x - 2)^7$ at $x = 3$

8. $g(x) = (x^2 - 3x - 10)(x^3 + 2x^2 - 5x + 4)$ at $x = -1$

9. $f(x) = \dfrac{x^2 - 9}{(x^2 - x - 6)(x^2 + 6x + 9)}$ at $x = 2$

10. $h(x) = \dfrac{x + 3}{(x^2 - x - 1)(x^2 + 1)}$ at $x = -2$

11. $f(x) = \dfrac{x\sqrt{x}}{(x - 6)^2}$ at $x = 36$

12. $k(x) = \dfrac{\sqrt{8 - x^2}}{2x^2 - 5}$ at $x = -2$

In Exercises 13–18, explain why the function is not continuous at the given number.

13. $f(x) = 1/(x - 3)^3$ at $x = 3$

14. $h(x) = \dfrac{x^2 + 4}{x^2 - x - 2}$ at $x = 2$

15. $f(x) = \dfrac{x^2 + 4x + 3}{x^2 - x - 2}$ at $x = -1$

16. $g(x) = \begin{cases} \sin(\pi/x) & \text{if } x \neq 0 \\ 1 & \text{if } x = 0 \end{cases}$ at $x = 0$

17. $f(x) = \begin{cases} x^2 & \text{if } x \neq 0 \\ 1 & \text{if } x = 0 \end{cases}$ at $x = 0$

18. $f(x) = \dfrac{\sqrt{2+x} - \sqrt{2}}{x}$ at $x = 0$

In Exercises 19–24, determine whether or not the function is continuous at the given number.

19. $f(x) = \begin{cases} -2x + 4 & \text{if } x \leq 2 \\ 2x - 4 & \text{if } x > 2 \end{cases}$ at $x = 2$

20. $g(x) = \begin{cases} 2x + 5 & \text{if } x < -1 \\ -2x + 1 & \text{if } x \geq -1 \end{cases}$ at $x = -1$

21. $f(x) = \begin{cases} x^2 - x & \text{if } x \leq 0 \\ 2x^2 & \text{if } x > 0 \end{cases}$ at $x = 0$

22. $g(x) = \begin{cases} x^3 - x + 1 & \text{if } x < 2 \\ 3x^2 - 2x - 1 & \text{if } x \geq 2 \end{cases}$ at $x = 2$

23. $f(x) = |x - 3|$ at $x = 3$

24. $k(x) = -|x + 2| + 3$ at $x = -2$

In Exercises 25–28, determine all numbers at which the function is continuous.

25. $f(x) = \begin{cases} \dfrac{x^2 + x - 2}{x^2 - 4x + 3} & \text{if } x \neq 1 \\ -3/2 & \text{if } x = 1 \end{cases}$

26. $g(x) = \begin{cases} \dfrac{x^2 - x - 6}{x^2 - 4} & \text{if } x \neq -2 \\ 5/4 & \text{if } x = -2 \end{cases}$

27. $f(x) = \begin{cases} x^2 + 1 & \text{if } x < 0 \\ x & \text{if } 0 < x \leq 2 \\ -2x + 3 & \text{if } x > 2 \end{cases}$

28. $h(x) = \begin{cases} 1/x & \text{if } x < 1 \quad \text{and} \quad x \neq 0 \\ x^2 & \text{if } x \geq 1 \end{cases}$

In Exercises 29–32, justify your answers.

29. Taxis in New York City cost $2 plus 30 cents for each $1/5$ of a mile (or portion thereof). Let $f(x)$ be the cost of traveling x miles. Is f continuous on the interval $[0, 3]$?

30. On a four and a half hour flight from Chicago to Seattle, let $h(x)$ be the height of the plane above the ground at time x hours. Is h continuous on the interval $[0, 4.5]$?

31. The U.S. Weather Bureau at Hopkins Airport in Cleveland continuously records the temperature. If $g(x)$ is the temperature at time x hours, where $x = 0$ corresponds to midnight, at what points on the interval $[0, 24]$ is g continuous?

32. Postage on a letter from the United States to Germany is 80 cents for each ounce (or fraction thereof) for letters weighing up to 8 ounces. Let $f(x)$ be the postage for a letter weighing x ounces. At what points on the interval $(0, 8]$ is f continuous?

33. If you don't have a calculator and you don't remember how to find square roots by hand, explain how you could use the Equation Theorem (and a lot of pencil and paper

multiplication and addition) to find the decimal expansion of $\sqrt{7}$. [*Hint:* What are the solutions of $x^2 - 7 = 0$?]

34. (a) The function whose graph is shown in Figure 13–2 on page 882 is discontinuous at an infinite number of places. Where is it discontinuous?
    (b) Explain why a rational function cannot be discontinuous at an infinite number of places.

THINKERS

35. Show that the function
$$f(x) = \frac{x^4 - 5x^2 + 4}{x - 1}$$
is *not* continuous on $[-3, 3]$ but *does* satisfy the conclusion of the Intermediate Value Theorem (that is, if k is a number between $f(-3)$ and $f(3)$, there is a number c between -3 and 3 such that $f(c) = k$). [*Hint:* What can be said about f on the intervals $[-3, -2]$ and $[2, 3]$?]

36. Show that the function
$$f(x) = \frac{x^4 - 2x^3}{x - 2}$$
is not continuous on $[-3, 3]$ and does not satisfy the conclusion of the Intermediate Value Theorem (that is, there is a number k between $f(-3)$ and $f(3)$ for which there is no number c between -3 and 3 such that $f(c) = k$).

37. For what values of b is the function
$$f(x) = \begin{cases} bx + 4 & \text{if } x \leq 3 \\ bx^2 - 2 & \text{if } x > 3 \end{cases}$$
continuous at $x = 3$?

38. Show that $f(x) = \sqrt{|x|}$ is continuous at $x = 0$.

A function f that is not defined at $x = c$ is said to have a removable discontinuity at $x = c$ if there is a function g such that $g(c)$ is defined, g is continuous at $x = c$, and $g(x) = f(x)$ for $x \neq c$. In Exercises 39–43, show that the function f has a removable discontinuity by finding an appropriate function g.

39. $f(x) = \dfrac{x - 1}{x^2 - 1}$

40. $f(x) = \dfrac{x^2}{|x|}$

41. $f(x) = \dfrac{2 - \sqrt{x}}{4 - x}$

42. $f(x) = \dfrac{\sin x}{x}$ [*Hint:* See Example 4 on pages 882–883.]

43. Show that the function $f(x) = \dfrac{|x|}{x}$ has a discontinuity at $x = 0$ that is *not* removable.

44. A ranger leaves his truck at a parking lot at the trail head at 8:00 A.M. and hikes 11 miles to a fire tower, arriving there at noon. He stays overnight and starts back along the same trail at 8:00 A.M., arriving at the parking lot at noon. Show that there is a point on the trail that he passes at exactly the same time on both days. [*Hint:* Let $f(t)$ be his distance from the parking lot at time t on the first day, and let $g(t)$ be his distance from the parking lot at time t on the second day. Use an appropriate theorem to solve the equation $f(t) = g(t)$.]

13.4   Limits Involving Infinity

Section Objectives

■ Understand the definitions and properties of limits at infinity.
■ Use limits at infinity to find the horizontal asymptotes of a graph.
■ Find limits of rational functions.

In the following discussion, it is important to remember that

> ***There is no real number called "infinity," and the symbol ∞,
> which is usually read "infinity," does not represent any real number.***

Nevertheless, the word "infinity" and the symbol ∞ are often used as a convenient shorthand to describe the way some functions behave under certain circumstances. General speaking, "infinity" indicates a situation in which some numerical quantity gets larger and larger without bound, meaning that it can be made larger than any given number. Similarly, "negative infinity" (in symbols $-\infty$) indicates a situation in which a numerical quantity gets smaller and smaller without bound, meaning that it can be made smaller than any given negative number.

Consider the function f whose graph is shown in Figure 13–24.

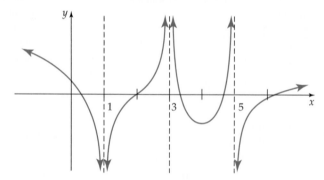

Figure 13–24

The graph shows that as x approaches 3 from the left or right, the corresponding values of $f(x)$ do not get closer and closer to a particular number. Instead, they become larger and larger without bound. Although there is no limit in the sense defined in Section 13.1, it is convenient to describe this situation symbolically by writing

$$\lim_{x \to 3} f(x) = \infty$$

which is read "the limit of $f(x)$ as x approaches 3 is infinity." Similarly, f does not have a limit as x approaches 1 (from the left or right) because the corresponding values of $f(x)$ get smaller and smaller without bound. We say that "the limit of $f(x)$ as x approaches 1 is negative infinity" and write

$$\lim_{x \to 1} f(x) = -\infty.$$

Near $x = 5$, the values of $f(x)$ get very large on the left side of 5 and very small on the right side of 5, so we write

$$\lim_{x \to 5^-} f(x) = \infty \qquad \text{and} \qquad \lim_{x \to 5^+} f(x) = -\infty$$

and say "the limit as x approaches 5 from the left is infinity" and "the limit as x approaches 5 from the right is negative infinity."

There are many cases like the ones illustrated above in which the language of limits and the word "infinity" is useful for describing the behavior of a function that actually does not have a limit in the sense of Section 13.1.

EXAMPLE 1

How does $f(x) = -5/x^4$ behave near $x = 0$?

SOLUTION

GRAPHING EXPLORATION

Graph $f(x)$ in the viewing window with $-.5 \le x \le .5$ and $-500,000 \le y \le 0$. Use the trace feature to move along the graph on both sides of $x = 0$ and confirm that

$$\lim_{x \to 0} \frac{-5}{x^4} = -\infty.$$

EXAMPLE 2

Describe the behavior of the function

$$g(x) = \frac{8}{x^2 - 2x - 8}$$

near $x = -2$.

SOLUTION

GRAPHING EXPLORATION

Graph $g(x)$ in a suitable viewing window around $x = -2$, and verify that

$$\lim_{x \to -2^-} \frac{8}{x^2 - 2x - 8} = \infty \quad \text{and} \quad \lim_{x \to -2^+} \frac{8}{x^2 - 2x - 8} = -\infty.$$

The "infinite limits" considered in Figure 13–24 and Examples 1 and 2 can be interpreted geometrically. Each such limit corresponds to a vertical asymptote of the graph. Consequently, we have this formal definition.

Vertical Asymptotes

The vertical line $x = c$ is a **vertical asymptote** of the graph of the function f if at least one of the following is true:

$$\lim_{x \to c^-} f(x) = \infty, \qquad \lim_{x \to c^+} f(x) = \infty, \qquad \lim_{x \to c} f(x) = \infty,$$

$$\lim_{x \to c^-} f(x) = -\infty, \qquad \lim_{x \to c^+} f(x) = -\infty, \qquad \lim_{x \to c} f(x) = -\infty.$$

LIMITS AT INFINITY

The word "limit" has been used thus far to describe the behavior of a function f when x is near a particular number c. Now we consider the behavior of a function when x is very large or very small.

EXAMPLE 3

Figure 13–25 shows the graph of

$$f(x) = \frac{5}{1 + 24e^{-x/4}} + 1.$$

Describe the behavior of f when x is very large and when x is very small.

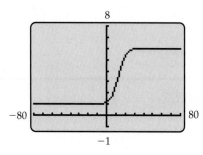
Figure 13–25

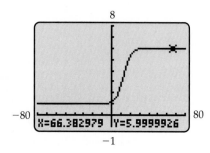
Figure 13–26

SOLUTION   As you move to the right on the graph (that is, as x gets very large), the graph gets very close to the horizontal line $y = 6$, as shown in Figure 13–26. In other words, as x gets larger and larger, the corresponding values of $f(x)$ get very close to 6. We express this fact symbolically by writing

$$\lim_{x \to \infty} f(x) = 6,$$

which is read "the limit of $f(x)$ as x approaches infinity is the number 6."

Toward the left, the graph gets very close to the horizontal line $y = 1$. As x gets smaller and smaller, the corresponding values of $f(x)$ get very close to 1, as shown in Figure 13–27.* We say that "the limit of $f(x)$ as x approaches negative infinity is 1" and write

$$\lim_{x \to -\infty} f(x) = 1.$$ ∎

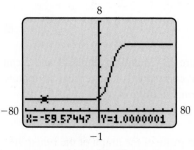
Figure 13–27

The limits illustrated in Example 3 are similar to those in Section 13.1 in that the values of the function do approach a fixed number. The definition in the general case is similar.

---

*Because of rounding off, the trace feature on most calculators will say that $f(x) = 1$ when x is smaller than approximately -60. However, the value of the function is always greater than 1 (why?).

Limits at Infinity

Let f be a function that is defined for all $x > a$, for some number a. If

as x takes larger and larger positive values, increasing without bound, the corresponding values of $f(x)$ get very close (and possibly equal) to the real number L

and

the values of $f(x)$ can be made arbitrarily close (as close as you want) to L for all large enough values of x,

then we say that

the limit of $f(x)$ as x approaches infinity is L,

which is written

$$\lim_{x \to \infty} f(x) = L.$$

Similarly, if f is defined for all $x < b$, for some number b, then

$$\lim_{x \to -\infty} f(x) = L,$$

which is read

the limit of $f(x)$ as x approaches negative infinity is L,

means that

as x takes smaller and smaller negative values, decreasing without bound, the corresponding values of $f(x)$ get very close (and possibly equal) to the real number L,

and

the values of $f(x)$ can be made arbitrarily close (as close as you want) to L for all small enough values of x.

This is an informal definition, as was the one in Section 13.1. Rigorous definitions (similar to those in *Special Topics* 13.2.A for ordinary limits) are discussed in Exercises 49 and 50.

As Example 3 shows, limits as x approaches infinity or negative infinity correspond to horizontal asymptotes of the graph of the function. Consequently, we have this formal definition.

Horizontal Asymptotes

The line $y = L$ is a **horizontal asymptote** of the graph of the function f if either

$$\lim_{x \to \infty} f(x) = L \qquad \text{or} \qquad \lim_{x \to -\infty} f(x) = L.$$

EXAMPLE 4

When x is a very large positive number, then $1/x$ is a positive number that is very close to 0. Similarly, when x is a negative number that is large in absolute value (such as $-5,000,000$), $1/x$ is a negative number that is very close to 0. These facts suggest that

$$\lim_{x \to \infty} \frac{1}{x} = 0 \qquad \text{and} \qquad \lim_{x \to -\infty} \frac{1}{x} = 0.$$

GRAPHING EXPLORATION

Confirm that fact geometrically by graphing the function $f(x) = 1/x$ and verifying that the x-axis (the horizontal line $y = 0$) is a horizontal asymptote of the graph in both directions.

No polynomial function has a limit as x approaches infinity or negative infinity, because no polynomial graph has a horizontal asymptote.

EXAMPLE 5

Describe the behavior of the polynomial function

$$f(x) = x^3 - 10x + 5$$

as x approaches infinity and as x approaches negative infinity.

SOLUTION   Look what happens when x takes larger and larger positive values.

GRAPHING EXPLORATION

Graph $f(x)$ in the viewing window with

$$0 \le x \le 5000 \text{ and } 0 \le y \le 100,000,000,000.$$

Use the trace feature to verify that as you move to the right along the graph, the values of $f(x)$ get larger and larger, increasing without bound.

Thus, $\lim_{x \to \infty} f(x)$, in the sense defined on the preceding page, does not exist. Nevertheless, we can describe this situation by writing

$$\lim_{x \to \infty} f(x) = \infty.$$

A similar situation occurs when x takes smaller and smaller negative values.

GRAPHING EXPLORATION

Graph $f(x)$ in the viewing window with $-5000 \le x \le 0$ and $-100,000,000,000 \le y \le 0$. Use the trace feature to verify that as you move to the left along the graph, the values of $f(x)$ get smaller and smaller, decreasing without bound.

We describe this situation by writing

$$\lim_{x \to -\infty} f(x) = -\infty.$$

PROPERTIES OF LIMITS AT INFINITY

The limits of constant functions are easily found. Consider, for example, the function $f(x) = 5$. As x gets larger and larger (or smaller and smaller), the corresponding value of $f(x)$ is *always* the number 5, so that $\lim_{x \to \infty} f(x) = 5$ and $\lim_{x \to -\infty} f(x) = 5$. A similar argument works for any constant function.

Limit of a Constant

If c is a constant, then

$$\lim_{x \to \infty} c = c \quad \text{and} \quad \lim_{x \to -\infty} c = c.$$

Infinite limits have the same useful properties that ordinary limits have. For instance, suppose that as x approaches infinity, the values of a function f get close to a number L and the values of a function g get close to a number M. Then it is plausible that the values of $f(x) + g(x)$ get close to $L + M$, the values of $f(x)g(x)$ get close to $L \cdot M$, and so forth. Similar remarks apply when x approaches negative infinity. More formally, we have the following.

Properties of Limits at Infinity

If f and g are functions and L and M are numbers such that

$$\lim_{x \to \infty} f(x) = L \quad \text{and} \quad \lim_{x \to \infty} g(x) = M,$$

then

1. $\lim_{x \to \infty} (f(x) + g(x)) = L + M$;
2. $\lim_{x \to \infty} (f(x) - g(x)) = L - M$;
3. $\lim_{x \to \infty} (f(x) \cdot g(x)) = L \cdot M$;
4. $\lim_{x \to \infty} \left(\dfrac{f(x)}{g(x)} \right) = \dfrac{L}{M}$;   (provided that $M \neq 0$)
5. $\lim_{x \to \infty} \sqrt{f(x)} = \sqrt{L}$.   (provided that $f(x) \geq 0$ for all large x)

Properties 1–5 also hold with $-\infty$ in place of ∞ (provided that for property 5, $f(x) \geq 0$ for all small x).

These properties are often stated somewhat differently. For example, since $\lim_{x \to \infty} f(x) = L$ and $\lim_{x \to \infty} g(x) = M$, Property 4 can be written as

$$\lim_{x \to \infty} \frac{f(x)}{g(x)} = \frac{\lim_{x \to \infty} f(x)}{\lim_{x \to \infty} g(x)},$$

and similarly for the others.

If c is a constant, then Property 3 and Example 4 show that

$$\lim_{x \to \infty} \frac{c}{x} = \lim_{x \to \infty} \left(c \cdot \frac{1}{x} \right) = \left(\lim_{x \to \infty} c \right) \left(\lim_{x \to \infty} \frac{1}{x} \right) = c \cdot 0 = 0.$$

Repeatedly using Property 3 with this fact and Example 4, we see that for any integer $n \geq 2$,

$$\lim_{x \to \infty} \frac{c}{x^n} = \lim_{x \to \infty} \left(\frac{c}{x} \cdot \frac{1}{x} \cdot \frac{1}{x} \cdots \frac{1}{x} \right)$$

$$= \left(\lim_{x \to \infty} \frac{c}{x} \right) \left(\lim_{x \to \infty} \frac{1}{x} \right) \left(\lim_{x \to \infty} \frac{1}{x} \right) \cdots \left(\lim_{x \to \infty} \frac{1}{x} \right)$$

$$= 0 \cdot 0 \cdot 0 \cdots 0 = 0.$$

A similar argument works with $-\infty$ in place of ∞ and produces this useful result.

Infinite Limit Theorem

If c is a constant, then for each positive integer n,

$$\lim_{x \to \infty} \frac{c}{x^n} = 0 \quad \text{and} \quad \lim_{x \to -\infty} \frac{c}{x^n} = 0.$$

This theorem and the limit properties above now make it possible to determine the limit (if it exists) of any rational function as x approaches infinity or negative infinity.

EXAMPLE 6

If you graph

$$f(x) = \frac{3x^2 - 2x + 1}{2x^2 + 4x - 5}$$

to the right of the y-axis, you will see that there appears to be a horizontal asymptote close to $y = 1.5$. We can confirm this fact algebraically by computing

$$\lim_{x \to \infty} \frac{3x^2 - 2x + 1}{2x^2 + 4x - 5}.$$

Property 4 cannot be used directly because neither the numerator nor the denominator has a finite limit as x approaches infinity (see Example 5). So we first divide both numerator and denominator by the highest power of x that appears, namely, x^2. When $x > 0$, this does not change the value of the fraction, and we have

$$\lim_{x \to \infty} \frac{3x^2 - 2x + 1}{2x^2 + 4x - 5} = \lim_{x \to \infty} \frac{\dfrac{3x^2 - 2x + 1}{x^2}}{\dfrac{2x^2 + 4x - 5}{x^2}}$$

$$= \lim_{x \to \infty} \frac{\dfrac{3x^2}{x^2} - \dfrac{2x}{x^2} + \dfrac{1}{x^2}}{\dfrac{2x^2}{x^2} + \dfrac{4x}{x^2} - \dfrac{5}{x^2}}$$

$$= \lim_{x \to \infty} \frac{3 - \dfrac{2}{x} + \dfrac{1}{x^2}}{2 + \dfrac{4}{x} - \dfrac{5}{x^2}}$$

$$= \frac{\displaystyle\lim_{x \to \infty}\left(3 - \frac{2}{x} + \frac{1}{x^2}\right)}{\displaystyle\lim_{x \to \infty}\left(2 + \frac{4}{x} - \frac{5}{x^2}\right)} \qquad \text{[Property 4]}$$

$$= \frac{\displaystyle\lim_{x \to \infty} 3 - \lim_{x \to \infty}\frac{2}{x} + \lim_{x \to \infty}\frac{1}{x^2}}{\displaystyle\lim_{x \to \infty} 2 + \lim_{x \to \infty}\frac{4}{x} - \lim_{x \to \infty}\frac{5}{x^2}} \qquad \text{[Properties 1, 2]}$$

$$= \frac{3 - \displaystyle\lim_{x \to \infty}\frac{2}{x} + \lim_{x \to \infty}\frac{1}{x^2}}{2 + \displaystyle\lim_{x \to \infty}\frac{4}{x} - \lim_{x \to \infty}\frac{5}{x^2}} \qquad \text{[Limit of constant]}$$

$$= \frac{3 - 0 + 0}{2 + 0 - 0} = \frac{3}{2}. \qquad \text{[Limit Theorem]} \quad \blacksquare$$

The technique in Example 6 carries over to any rational function $f(x) = g(x)/h(x)$, where g and h have the same degree, and provides the mathematical justification for the treatment of horizontal asymptotes in Section 4.5 (see Exercise 48). A slight variation of the procedure can be used to compute certain limits involving square roots.

EXAMPLE 7

Find

(a) $\displaystyle\lim_{x \to \infty} \frac{\sqrt{3x^2 + 1}}{2x + 3}$

(b) $\displaystyle\lim_{x \to -\infty} \frac{\sqrt{3x^2 + 1}}{2x + 3}$

SOLUTION

GRAPHING EXPLORATION

Graph the function in a viewing window with $-500 \le x \le 500$, and verify that there appear to be two horizontal asymptotes, one below the x-axis to the left and one above the x-axis to the right. Use the trace feature to estimate the required limits. Compare your estimates to the results that we now obtain algebraically.

(a) We need only consider positive values of x. When x is positive, $\sqrt{x^2} = x$. Therefore,

$$\lim_{x \to \infty} \frac{\sqrt{3x^2 + 1}}{2x + 3} = \lim_{x \to \infty} \frac{\dfrac{\sqrt{3x^2 + 1}}{x}}{\dfrac{2x + 3}{x}} = \lim_{x \to \infty} \frac{\dfrac{\sqrt{3x^2 + 1}}{\sqrt{x^2}}}{\dfrac{2x + 3}{x}}$$

$$= \lim_{x \to \infty} \frac{\sqrt{\dfrac{3x^2 + 1}{x^2}}}{\dfrac{2x + 3}{x}} = \lim_{x \to \infty} \frac{\sqrt{3 + \dfrac{1}{x^2}}}{2 + \dfrac{3}{x}}$$

$$= \frac{\displaystyle\lim_{x \to \infty} \sqrt{3 + \dfrac{1}{x^2}}}{\displaystyle\lim_{x \to \infty} \left(2 + \dfrac{3}{x}\right)} \qquad \text{[Property 4]}$$

$$= \frac{\sqrt{\displaystyle\lim_{x \to \infty} \left(3 + \dfrac{1}{x^2}\right)}}{\displaystyle\lim_{x \to \infty} \left(2 + \dfrac{3}{x}\right)} \qquad \text{[Property 5]}$$

$$= \frac{\sqrt{\displaystyle\lim_{x \to \infty} 3 + \lim_{x \to \infty} \dfrac{1}{x^2}}}{\displaystyle\lim_{x \to \infty} 2 + \lim_{x \to \infty} \dfrac{3}{x}} \qquad \text{[Property 1]}$$

$$= \frac{\sqrt{3 + 0}}{2 + 0} = \frac{\sqrt{3}}{2}. \qquad \text{[Constant limit and Limit Theorem]}$$

(b) To compute the limit as x approaches negative infinity, we need only consider negative values of x and use the fact that when x is negative, $x = -\sqrt{x^2}$ (for instance, $-2 = -\sqrt{(-2)^2}$). Then an argument similar to the one in part (a) shows that

$$\lim_{x \to -\infty} \frac{\sqrt{3x^2 + 1}}{2x + 3} = -\frac{\sqrt{3}}{2}.$$ ∎

EXERCISES 13.4

In Exercises 1–8, use a calculator to estimate the limit.

1. $\lim\limits_{x \to \infty} \left[\sqrt{x^2 + 1} - (x + 1) \right]$

2. $\lim\limits_{x \to -\infty} \left[\sqrt{x^2 + x + 1} + x \right]$

3. $\lim\limits_{x \to -\infty} \dfrac{x^{2/3} - x^{4/3}}{x^3}$

4. $\lim\limits_{x \to \infty} \dfrac{x^{5/4} + x}{2x - x^{5/4}}$

5. $\lim\limits_{x \to -\infty} \sin \dfrac{1}{x}$

6. $\lim\limits_{x \to \infty} \dfrac{\sin x}{x}$

7. $\lim\limits_{x \to \infty} \dfrac{\ln x}{x}$

8. $\lim\limits_{x \to \infty} \dfrac{5}{1 + (1.1)^{-x/20}}$

In Exercises 9–14, list the vertical asymptotes of the graph (if there are any). Then use the graph of the function f to find

$$\lim_{x \to \infty} f(x) \quad \text{and} \quad \lim_{x \to -\infty} f(x).$$

9.

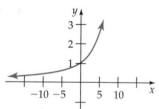

10.

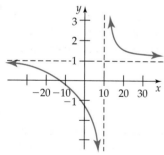

11.

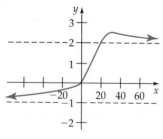

12.

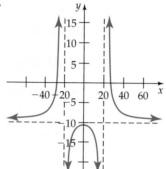

13.

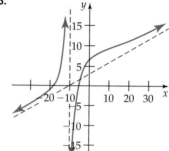

14.

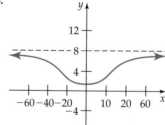

In Exercises 15–20, use the Infinite Limit Theorem and the properties of limits as in Example 6 to find the horizontal asymptotes (if any) of the graph of the given function.

15. $f(x) = \dfrac{3x^2 + 5}{4x^2 - 6x + 2}$

16. $g(x) = \dfrac{x^2}{x^2 - 2x + 1}$

17. $h(x) = \dfrac{2x^2 - 6x + 1}{2 + x - x^2}$

18. $k(x) = \dfrac{3x + x^2 - 4}{2x - x^3 + x^2}$

19. $f(x) = \dfrac{3x^4 - 2x^3 + 5x^2 - x + 1}{7x^3 - 4x^2 + 6x - 12}$

20. $g(x) = \dfrac{2x^5 - x^3 + 2x - 9}{5 - x^5}$

In Exercises 21–39, use the Infinite Limit Theorem and the properties of limits to find the limit.

21. $\lim\limits_{x \to -\infty} \dfrac{(x-3)(x+2)}{2x^2 + x + 1}$

22. $\lim\limits_{x \to \infty} \dfrac{(2x+1)(3x-2)}{3x^2 + 2x - 5}$

23. $\lim\limits_{x \to \infty} \left(3x - \dfrac{1}{x^2}\right)$

24. $\lim\limits_{x \to -\infty} (3x^2 + 1)^{-2}$

25. $\lim\limits_{x \to -\infty} \left(\dfrac{3x}{x+2} + \dfrac{2x}{x-1}\right)$

26. $\lim\limits_{x \to \infty} \left(\dfrac{x}{x^2+1} + \dfrac{2x^2}{x^3+x}\right)$

27. $\lim\limits_{x \to \infty} \dfrac{2x}{\sqrt{x^2 - 2x}}$

28. $\lim\limits_{x \to -\infty} \dfrac{x}{\sqrt{x^2 - 1}}$

29. $\lim\limits_{x \to -\infty} \dfrac{3x - 2}{\sqrt{2x^2 + 1}}$

30. $\lim\limits_{x \to \infty} \dfrac{3x - 2}{\sqrt{2x^2 + 1}}$

31. $\lim\limits_{x \to \infty} \dfrac{\sqrt{2x^2 + 1}}{3x - 5}$

32. $\lim\limits_{x \to -\infty} \dfrac{\sqrt{2x^2 + 1}}{3x - 5}$

33. $\lim\limits_{x \to -\infty} \dfrac{\sqrt{3x^2 + 3}}{x + 3}$

34. $\lim\limits_{x \to \infty} \dfrac{\sqrt{3x^2 + 2x}}{2x + 1}$

35. $\lim\limits_{x \to \infty} \dfrac{x^2 + 2x + 1}{\sqrt{x^4 + 2x}}$

36. $\lim\limits_{x \to \infty} \dfrac{\sqrt{x^6 - x^2}}{2x^3}$

37. $\lim\limits_{x \to \infty} \dfrac{1 - \sqrt{x}}{1 + \sqrt{x}}$ [*Hint:* Rationalize the denominator.]

38. $\lim\limits_{x \to \infty} \dfrac{\sqrt{x} + 2}{\sqrt{x} - 3}$

39. $\lim\limits_{x \to \infty} (\sqrt{x^2 + 1} - x)$ $\left[Hint: \text{Multiply by } \dfrac{\sqrt{x^2+1}+x}{\sqrt{x^2+1}+x}. \right]$

In Exercises 40–42, find the limit by adapting the hint for Exercise 39.

40. $\lim\limits_{x \to -\infty} (x + \sqrt{x^2 + 4})$

41. $\lim\limits_{x \to \infty} (\sqrt{x^2 - 1} - \sqrt{x^2 + 1})$

42. $\lim\limits_{x \to -\infty} (\sqrt{x^2 + 5x + 5} + x + 1)$

THINKERS

In Exercises 43–44, find the limit.

43. $\lim\limits_{x \to \infty} \dfrac{x}{|x|}$

44. $\lim\limits_{x \to -\infty} \dfrac{|x|}{|x| + 1}$

45. Let $[\![x]\!]$ denote the greatest integer function (see Example 7 on page 145) and find:

(a) $\lim\limits_{x \to \infty} \dfrac{[\![x]\!]}{x}$ (b) $\lim\limits_{x \to -\infty} \dfrac{[\![x]\!]}{x}$

46. Use the change of base formula for logarithms (*Special Topics 5.4.A*) to show that $\lim\limits_{x \to \infty} \dfrac{\ln x}{\log x} = \ln 10$.

47. Find $\lim\limits_{x \to \infty} \dfrac{\sqrt{x + \sqrt{x + \sqrt{x}}}}{\sqrt{x + 1}}$.

48. Let $f(x)$ be a nonzero polynomial with leading coefficient a, and let $g(x)$ be a nonzero polynomial with leading coefficient c. Prove that

(a) If $\deg f(x) < \deg g(x)$, then $\lim\limits_{x \to \infty} \dfrac{f(x)}{g(x)} = 0$.

(b) If $\deg f(x) = \deg g(x)$, then $\lim\limits_{x \to \infty} \dfrac{f(x)}{g(x)} = \dfrac{a}{c}$.

(c) If $\deg f(x) > \deg g(x)$, then $\lim\limits_{x \to \infty} \dfrac{f(x)}{g(x)}$ does not exist.

Formal definitions of limits at infinity and negative infinity are given in Exercises 49 and 50. Adapt the discussion in Special Topics 13.2.A to explain how these definitions are derived from the informal definitions given in this section.

49. Let f be a function, and let L be a real number. Then the statement $\lim\limits_{x \to \infty} f(x) = L$ means that for each positive number ϵ, there is a positive real number k (depending on ϵ) with this property:

If $x > k$. then $|f(x) - L| < \epsilon$.

[*Hint:* Concentrate on the second part of the informal definition. The number k measures "large enough," that is, how large the values of x must be to guarantee that $f(x)$ is as close as you want to L.]

50. Let f be a function, and let L be a real number. Then the statement $\lim\limits_{x \to -\infty} f(x) = L$ means that for each positive number ϵ, there is a negative real number n (depending on ϵ) with this property:

If $x < n$,   then $|f(x) - L| < \epsilon$.

51. (a) Approximate $\lim\limits_{x \to \infty} \left(1 + \dfrac{1}{x}\right)^x$ to seven decimal places. (Evaluate the function at larger and larger values of x until successive approximations agree in the first seven decimal places.)

(b) Find the decimal expansion of e to at least nine decimal places.

(c) On the basis of the results in parts (a) and (b), what do you think is the exact value of $\lim\limits_{x \to \infty} \left(1 + \dfrac{1}{x}\right)^x$?

(d) Compare this limit to the one in Exercise 46 of Section 13.1. How are the two related?

Chapter 13 Review

IMPORTANT CONCEPTS

Section 13.1

Limit of a function   881
Nonexistence of limits   884–885
Limit of a function from the right   886
Limit of a function from the left   886

Section 13.2

Properties of limits   890
Limits of polynomial and rational
    functions   891–893
Limit Theorem   894

Limits of difference quotients   895
Derivative   896

Special Topics 13.2.A

The formal definition of limit   899

Section 13.3

Continuity of a function at $x = c$ 906
Continuity from the left and right   907
Continuity on an interval   908
Properties of continuous functions   909

Continuity of composite functions   909
Intermediate Value Theorem   910
Equation Theorem   911

Section 13.4

Infinite limits   914–915
Vertical asymptotes   915
Limit of a function as x approaches
    infinity or negative infinity   917
Horizontal asymptotes   917
Properties of limits at infinity   919

REVIEW QUESTIONS

In Questions 1–4, use a calculator to estimate the limit.

1. $\lim\limits_{x \to 0} \dfrac{3x - \sin x}{x}$

2. $\lim\limits_{x \to \pi/2} \dfrac{1 - \sin x}{1 + \cos 2x}$

3. $\lim\limits_{x \to -1} \dfrac{10^x - .1}{x + 1}$

4. $\lim\limits_{x \to 0} \dfrac{\sqrt{x - 3} - \sqrt{3}}{x}$

In Questions 5 and 6, use the graph of the function to determine the limit.

5. $\lim\limits_{x \to 2} f(x)$

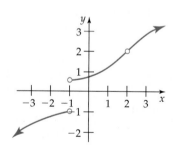

6. $\lim\limits_{x \to -1} f(x)$

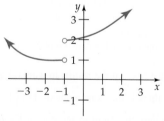

In Questions 7–10, assume that $\lim\limits_{x \to 3} f(x) = 5$ and $\lim\limits_{x \to 3} g(x) = -2$. Find the limit.

7. $\lim\limits_{x \to 3} [3f(x) - 15]$

8. $\lim\limits_{x \to 3} (f(x)[g(x) - 2])$

9. $\lim\limits_{x \to 3} \dfrac{f(x)g(x) - 2f(x)}{[g(x)]^2}$

10. $\lim\limits_{x \to 3} \dfrac{\sqrt{f(x) - 2g(x)}}{f(x) + g(x)}$

In Questions 11–20, find the limit if it exists. If the limit does not exist, explain why.

11. $\lim\limits_{x \to -2} (x^3 - 3x + 1)$

12. $\lim\limits_{x \to 4} \dfrac{2x + 1}{x - 2}$

13. $\lim\limits_{x \to 1} \dfrac{x^2 - 1}{x^2 - 3x + 2}$

14. $\lim\limits_{x \to -2} \dfrac{x^2 - x - 6}{x^2 + x - 2}$

15. $\lim\limits_{x \to 0} \dfrac{\sqrt{1 + x} - 1}{x}$

16. $\lim\limits_{x \to 2} \dfrac{x^2 - 2x - 3}{x^2 - 6x + 9}$

17. $\lim\limits_{x \to -1^+} \sqrt{9 + 8x - x^2}$

18. $\lim\limits_{x \to 8^+} \dfrac{x^2 - 64}{x - 8}$

19. $\lim\limits_{x \to -5^+} \dfrac{|x + 5|}{x + 5}$

20. $\lim\limits_{x \to 7^-} \left(\sqrt{7 - x^2 + 6x} + 2\right)$

21. If $f(x) = x^2 + 1$, find

$$\lim\limits_{h \to 0} \dfrac{f(2 + h) - f(2)}{h}.$$

22. If $f(x) = 3x - 2$ and c is a constant, find

$$\lim\limits_{h \to 0} \dfrac{f(c + h) - f(c)}{h}.$$

In Questions 23 and 24, find the rule of the derivative of the function f.

23. $f(x) = 4x - 2$

24. $f(x) = x^2$

In Questions 25 and 26, use the formal definition of limit to prove the statement.

25. $\lim\limits_{x \to 3} (2x + 1) = 7$

26. $\lim\limits_{x \to 2} \left(\dfrac{1}{2}x + 3\right) = 4$

In Questions 27 and 28, determine whether the function whose graph is given is continuous at $x = -3$ and $x = 2$.

27.

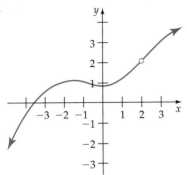

28.

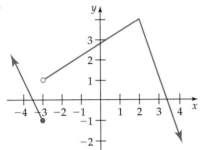

In Questions 29–32, determine if the function is continuous at the given points.

29. $f(x) = \sqrt{x^2 - 8}$; $x = 4$, $x = \sqrt{8}$

30. $g(x) = |x - 3|$; $x = 3$, $x = -3$

31. $h(x) = \dfrac{x - 7}{x + 4}$; $x = -4$, $x = 0$

32. $f(x) = \dfrac{x^2 - 1}{x + 1}$; $x = -1$, $x = 0$, $x = 1$

33. Show that $f(x) = \dfrac{x^2 - x - 6}{x^2 - 9}$ is
   (a) continuous at $x = 2$
   (b) discontinuous at $x = 3$.

34. Is the function given by

$$f(x) = \begin{cases} 3x - 2 & \text{if } x \le 3 \\ 10 - x & \text{if } x > 3 \end{cases}$$

continuous at $x = 3$? Justify your answer.

In Questions 35 and 36, find the vertical asymptotes of the graph of the given function and state whether the graph moves upward or downward on each side of each asymptote.

35. $f(x) = \dfrac{x^2 + 1}{x^2 - x - 2}$ **36.** $g(x) = \dfrac{x^2 - 1}{x^2 - 3x + 2}$

In Questions 37–44, find the limit.

37. $\lim\limits_{x \to 3} \dfrac{1}{(x - 3)^2}$ **38.** $\lim\limits_{x \to -2^-} \dfrac{x^2 + x - 2}{x - 2}$

39. $\lim\limits_{x \to \infty} \dfrac{\sin x}{x^2}$ **40.** $\lim\limits_{x \to \infty} \dfrac{x}{(\ln x)^2}$

41. $\lim\limits_{x \to \infty} \dfrac{2x^3 - 3x^2 + 5x - 1}{4x^3 + 2x^2 - x + 10}$

42. $\lim\limits_{x \to -\infty} \dfrac{4 - 3x - 2x^2}{x^3 + 2x + 5}$

43. $\lim\limits_{x \to -\infty} \left(\dfrac{2x + 1}{x - 3} + \dfrac{4x - 1}{3x} \right)$ **44.** $\lim\limits_{x \to \infty} \dfrac{\sqrt{3x^2 + 2}}{4x + 1}$

In Questions 45 and 46, find the horizontal asymptotes of the graph of the given function algebraically and verify your results graphically.

45. $f(x) = \dfrac{x^2 - x + 7}{2x^2 + 5x + 7}$ **46.** $f(x) = \dfrac{x - 9}{\sqrt{4x^2 + 3x + 2}}$

Chapter 13 Test

Sections 13.1, 13.2; Special Topics 13.2.A

1. Find each of the following limits, where f is the function whose graph is shown below.
   (a) $\lim\limits_{x \to -2^+} f(x)$ (b) $\lim\limits_{x \to 2} f(x)$ (c) $\lim\limits_{x \to 4} f(x)$ (d) $\lim\limits_{x \to 5} f(x)$

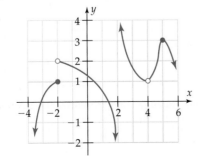

2. Find $\lim\limits_{x \to -2} \dfrac{x^2 + 3x + 2}{x^2 - 5x - 14}$. Do not use a calculator or tables. Show your work.

3. Use numerical or graphical means to find the limit, if it exists.

$$\lim\limits_{x \to 0} \dfrac{e^{8x} + e^x - 2}{e^x - 1}$$

4. Suppose that f, g and h are functions such that

$$\lim\limits_{x \to 4} f(x) = 5, \qquad \lim\limits_{x \to 4} g(x) = 0, \qquad \lim\limits_{x \to 4} h(x) = -8.$$

Find $\lim\limits_{x \to 4} \dfrac{f(x) - 2g(x)}{4h(x)}$.

5. (a) Graph the function whose rule is

$$g(x) = \begin{cases} x^2 & \text{if } x < -1 \\ 2x + 3 & \text{if } -1 \le x < 1 \\ 2 & \text{if } x = 1 \\ 3 - x & \text{if } x > 1 \end{cases}.$$

Use the graph to find the following limits.

(b) $\lim\limits_{x \to -1^-} g(x)$ **(c)** $\lim\limits_{x \to -1^+} g(x)$

(d) $\lim\limits_{x \to 1^-} g(x)$ **(e)** $\lim\limits_{x \to 1^+} g(x)$

6. Find the limit and show your work.

$$\lim_{x \to -2} \left(\frac{x^2}{x + 2} + \frac{2x}{x + 2} \right)$$

7. Find $\lim\limits_{x \to 3} \dfrac{|x - 3|}{x - 3}$.

8. If $f(x) = x^2$, find $\lim\limits_{h \to 0} \dfrac{f(3 + h) - f(3)}{h}$.

9. (a) Use graphical or numerical means to find $\lim\limits_{x \to 0} f(x)$, when
$f(x) = (e^x - 1) \ln (|x|)$.
(b) What is $f(0)$?

10. If $f(x) = x^2 - 5x + 6$, find $f'(x)$, where

$$f'(x) = \lim_{h \to 0} \frac{f(x + h) - f(x)}{h}.$$

11. Find the exact value of $\lim\limits_{x \to \sqrt{2}} \left(\dfrac{x - 2}{x - \sqrt{2}} \right)$.

12. Use the formal $(\epsilon - \delta)$ definition of "limit" to prove that

$$\lim_{x \to 2} (3x - 1) = 5.$$

Sections 13.3 and 13.4

13. Use the definition of continuity and the properties of limits to show that $k(x) = \dfrac{\sqrt{8 - x^2}}{4x^2 - 7}$ is continuous at $x = -2$.

14. Find each of the following limits, where f is the function whose graph is shown below.

(a) $\lim\limits_{x \to -2^-} f(x)$ **(b)** $\lim\limits_{x \to 0^+} f(x)$

(c) $\lim\limits_{x \to 3^-} f(x)$ **(d)** $\lim\limits_{x \to 3^+} f(x)$

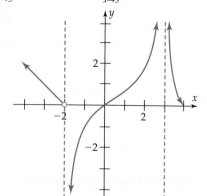

15. Determine whether the function g is continuous at $x = -3$, where

$$g(x) = \begin{cases} 4x + 17 & \text{if } x < -3 \\ -4x - 5 & \text{if } x \ge -3 \end{cases}.$$

16. The graph of a function f is shown below. Use it to find the following:

(a) The vertical asymptotes of the graph
(b) $\lim\limits_{x \to \infty} f(x)$
(c) $\lim\limits_{x \to -\infty} f(x)$

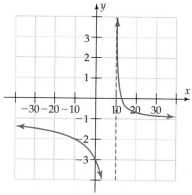

17. Determine all numbers at which the following function is continuous:

$$g(x) = \begin{cases} \dfrac{5}{x} & \text{if } x < 1 \text{ and } x \ne 0 \\ 5x^2 & \text{if } x \ge 1 \end{cases}.$$

18. Find $\lim\limits_{x \to \infty} \dfrac{4x^3 + 3x^2 - 2x + 1}{2x^3 - 3x^2 + 4x - 5}$. Show your work.

19. Taxis in a certain city cost $2.50 plus 30 cents for each 1/5 of a mile (or portion thereof). Let $f(x)$ be the cost of traveling x miles. Is f continuous on the interval

(a) $[0, .2]$?   **(b)** $[0, 1]$?

20. Find $\lim\limits_{x \to \infty} \left(\dfrac{x}{x^2 + 1} + \dfrac{2x^2}{x^3 + x} \right)$. Show your work.

21. For what values of b is the following function continuous at $x = 3$?

$$g(x) = \begin{cases} bx + 4 & \text{if } x \le 3 \\ b^2x^2 - 2 & \text{if } x > 3 \end{cases}$$

22. Find $\lim\limits_{x \to \infty} \dfrac{(2x + 5)(8x + 5)}{4x^2 + 2x - 3}$. Show your work.

23. Fill the blanks in the following table. Enter "yes," if the given function is continuous on the given interval, or "no" if it is not.

Function	Interval			
	[0, 1]	(0, 1]	[0, 1)	(0, 1)
$f(x) = \dfrac{1}{x}$				
$g(x) = \dfrac{2x^2}{x - 1}$				
$H(x) = \sqrt{x}$				
$K(x) = \dfrac{2x^2}{2x - 1}$				

24. Without graphing, find the horizontal asymptote of the graph of $f(x) = \dfrac{5x^5 - x^3 + 2x - 1}{7 - x^5}$. Show your work.

DISCOVERY PROJECT 13   Black Holes

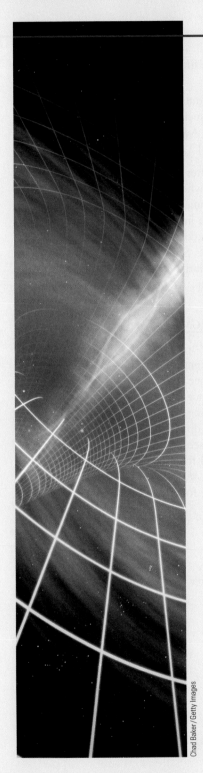

Chad Baker / Getty Images

You can't get here from there. Or, as the Eagles said in their 1976 song *Hotel California,* "You can never leave." A black hole is the ultimate "gotcha." Nothing can escape the gravitational attraction of a black hole—not even light. To understand why this is so and to understand why the entire universe isn't sucked into some black hole, we need to use a little algebra and two ideas from physics: the speed of light and escape velocity.

The **speed of light** (c in the famous equation $E = mc^2$) is approximately 3.00×10^8 meters per second. The speed of light is also the absolute maximum velocity in this universe. Nothing moves faster. The **escape velocity** is the velocity required for an object to escape the gravity of a planet or some other object.

Suppose that an object of mass m kilograms is moving with velocity v meters per second. Its **kinetic energy** (energy of motion) is given by

$$KE = \frac{1}{2} mv^2.$$

Near the surface of a planet with mass M kilograms, the gravitational **potential energy** (stored energy) of the object is given by

$$PE = \frac{GmM}{r},$$

where $G \approx 6.67 \times 10^{-11}$ is the gravitational constant and r is the distance (in meters) from the center of the planet. If the velocity is large enough, the kinetic energy equals (or exceeds) the potential energy, and the object will escape. To find the escape velocity, we solve the equation $KE = PE$.

$$\frac{1}{2} mv^2 = \frac{GmM}{r}$$

Divide both sides by m and multiply both sides by 2:

$$v^2 = \frac{2GM}{r}$$

(1)   Take the square root of both sides: $v = \sqrt{\dfrac{2GM}{r}}$ meters per second.

Notice that the escape velocity v depends only on the mass M of the planet and the distance r of the object from the planet's center. The mass of the object itself is irrelevant. For an atom or an ant or a rocket or the entire city of Atlanta, the escape velocity is the same!

Under the preceding assumptions, find the escape velocity from the following locations. Express all your answers in terms of kilograms (kg), meters (m), and seconds. Express your final answers (but not any intermediate results) in scientific notation (with the number between 1 and 10 rounded to two decimal places).

1. Surface of the earth. The mass of the earth is 5.97×10^{24} kg, and its radius is 6.38×10^6 m.

2. Surface of the moon. The mass of the moon is 7.36×10^{22} kg, and its radius is 1.74×10^6 m.

3. A satellite at an altitude of 35,786 kilometers orbits the earth at the same rate as the earth rotates. Satellites in such *geosynchronous orbits* stay "parked" above the same spot. [*Hint:* A kilometer is 10^3 meters. Add the altitude to the earth's radius to find r.]

Gravity works in all stars to pull matter toward the center. The energy released by fusion counteracts gravity. But when a star dies, gravity eventually prevails. Depending on its mass, the star becomes a white dwarf, a neutron star, or a black hole. In the case of a black hole, the mass is so large that gravity overcomes even neutron repulsion, and the mass contracts to a singularity (a point).

As a dying star contracts, it reaches its **Schwarzschild radius,** R_S, also known as the **event horizon.** At the event horizon the escape velocity is c, the speed of light. To find the Schwarzchild radius, we first solve the escape velocity equation (**1**) for r:

(**2**)
$$r = \frac{2GM}{v^2}$$

Then set $v = c$, so that $R_S = 2GM/c^2$. For points closer to the center/singularity, the escape velocity is greater than c. But since nothing moves faster than light, nothing escapes. Nothing. At greater distances, outside the imaginary sphere defined by the Schwarzschild radius, everything is pretty much normal. The gravitational attraction and escape velocity are no different than they would be for any regular star of the same mass. You just can't see what is attracting you.

Use $c = 3.00 \times 10^8$ meters per second for the speed of light, and denote the mass of the sun by M_{sun}. Assume that $M_{sun} = 1.99 \times 10^{30}$ kg. Find the Schwarzschild radius for black holes with the following masses.

4. Mass M equal to five times the mass of the sun ($5 \times M_{sun}$)

5. $M = 8.4 \times M_{sun}$

6. $M = 20 \times M_{sun}$

The figure below is a depiction of the space-time fabric in the vicinity of a black hole.

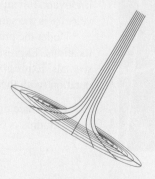

A cross-section by a perpendicular plane through the singularity will resemble the graph of $f(x) = 1/x^2$.

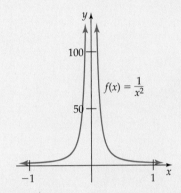

Although you may have thought that we would never make the connection, there is a relationship between the physics of black holes and the algebra of limits. As we saw in Section 13.4, the behavior of $f(x) = 1/x^2$ near $x = 0$ is described by writing

$$\lim_{x \to 0} \frac{1}{x^2} = \infty.$$

Infinite limits such as this one can be rigorously defined (as ordinary limits were in Special Topics 13.2.A) as follows.

Let f be a function and let a be a real number such that $f(x)$ is defined for all x in some open interval containing a (except possibly at $x = a$). We say that **the limit of $f(x)$, as x approaches a, is infinity,** and write

$$\lim_{x \to a} f(x) = \infty,$$

provided that for each positive number N, there is a positive number δ (depending on N) with this property:

$$\text{If } 0 < |x - a| < \delta, \qquad \text{then} \qquad f(x) > N.$$

To relate this definition to the preceding discussion, note that equation **(1)** defines the velocity v as a function of the distance r; the rule of the function is

$$f(r) = \sqrt{\frac{2GM}{r}}.$$

Finding the Schwarzschild radius amounts to finding δ for this function f when $N = c$ and a is the center/singularity of the black hole (so that $r = |x - a|$), as in Exercises 4–6.

More about this topic can be found at the following Web sites:

http://antwrp.gsfc.nasa.gov/htmltest/rjn_bht.html
http://archive.ncsa.uiuc.edu/Cyberia/NumRel/BlackHoles.html

This appendix reviews the fundamental algebraic facts that are used frequently in this book. You must be able to handle these algebraic manipulations to succeed in this course and in calculus.

1.A   Integral Exponents

Exponents provide a convenient shorthand for certain products. If c is a real number, then c^2 denotes cc, and c^3 denotes ccc. More generally, for any positive integer n,

c^n denotes the product $ccc \cdots c$ (n factors).

In this notation, c^1 is just c, so we usually omit the exponent 1.

EXAMPLE 1

$3^4 = 3 \cdot 3 \cdot 3 \cdot 3 = 81$, and

$$(-2)^5 = (-2)(-2)(-2)(-2)(-2) = -32.$$

For every positive integer n, $0^n = 0 \cdots 0 = 0$.   ∎

EXAMPLE 2

To find $(2.4)^9$, use the $\wedge$ (or a^b or x^y) key on your calculator:

$$2.4 \wedge 9 \text{ ENTER*}$$

which produces the (approximate) answer 2641.80754.   ∎

> **CAUTION**
>
> Be careful with negative bases. For instance, if you want to compute $(-12)^4$, which is a positive number, but you key in $(-)$ 12 $\wedge$ 4 ENTER, the calculator will interpret this as $-(12^4)$ and produce a negative answer. To get the correct answer, you must key in the parentheses:
>
> $$((-)\ 12) \wedge 4 \text{ ENTER.}$$

Because exponents are just shorthand for multiplication, it is easy to determine the rules they obey. For instance,

$$c^3 c^5 = (ccc)(ccccc) = c^8, \qquad \text{that is,} \qquad c^3 c^5 = c^{3+5}.$$

$$\frac{c^7}{c^4} = \frac{ccccccc}{cccc} = \frac{\cancel{cccc}ccc}{\cancel{cccc}} = ccc = c^3, \qquad \text{that is,} \qquad \frac{c^7}{c^4} = c^{7-4}.$$

Similar arguments work in the general case.

---

*The ENTER key is labeled EXE on Casio calculators.

Multiplication and Division with Exponents

To multiply c^m by c^n, add the exponents.

$$c^m \cdot c^n = c^{m+n}.$$

To divide c^m by c^n, subtract the exponents.

$$\frac{c^m}{c^n} = c^{m-n}.$$

EXAMPLE 3

$4^2 \cdot 4^7 = 4^{2+7} = 4^9$ and $2^8/2^3 = 2^{8-3} = 2^5$. ∎

The notation c^n can be extended to the cases in which n is zero or negative as follows.

Zero and Negative Exponents

If $c \neq 0$ and n is a positive integer, then

c^0 is defined to be the number 1;

c^{-n} is defined to the number $\dfrac{1}{c^n}$.

Note that 0^0 and negative powers of 0 are *not* defined (negative powers of 0 would involve division by 0). The reason for choosing these definitions is that the multiplication and division rules for exponents remain valid. For instance,

$$c^5 \cdot c^0 = c^5 \cdot 1 = c^5, \qquad \text{so} \qquad c^5 c^0 = c^{5+0}.$$
$$c^7 c^{-7} = c^7 \frac{1}{c^7} = 1 = c^0, \qquad \text{so} \qquad c^7 c^{-7} = c^{7-7}.$$

EXAMPLE 4

$6^{-3} = 1/6^3 = 1/216$ and $(-2)^{-5} = 1/(-2)^5 = -1/32$. A calculator shows that $(.287)^{-12} \approx 3{,}201{,}969.857.$* ∎

If c and d are nonzero real numbers and m and n are integers (positive, negative, or zero), then we have these laws.

Exponent Laws

1. $c^m c^n = c^{m+n}$

2. $\dfrac{c^m}{c^n} = c^{m-n}$

3. $(c^m)^n = c^{mn}$

4. $(cd)^n = c^n d^n$

5. $\left(\dfrac{c}{d}\right)^n = \dfrac{c^n}{d^n}$

6. $\dfrac{1}{c^{-n}} = c^n$

---

*$\approx$ means "approximately equal to."

EXAMPLE 5

Here are examples of the six exponent laws.

1. $\pi^{-5}\pi^2 = \pi^{-5+2} = \pi^{-3} = 1/\pi^3$.

2. $x^9/x^4 = x^{9-4} = x^5$.

3. $(5^{-3})^2 = 5^{(-3)2} = 5^{-6}$.

4. $(2x)^5 = 2^5x^5 = 32x^5$.

5. $\left(\dfrac{7}{3}\right)^{10} = \dfrac{7^{10}}{3^{10}}$.

6. $1/x^{-5} = \dfrac{1}{\left(\dfrac{1}{x^5}\right)} = x^5$.

CAUTION

$(2x)^5$ is *not* the same as $2x^5$. Part 4 of Example 5 shows that $(2x)^5 = 32x^5$ and *not* $2x^5$.

The exponent laws can often be used to simplify complicated expressions.

EXAMPLE 6

(a) $(2x^2y^3z)^4 = 2^4(x^2)^4(y^3)^4z^4 = 16x^8y^{12}z^4$.

            ↑   ↑

      Law (4)   Law (3)

(b) $(r^{-3}s^2)^{-2} = (r^{-3})^{-2}(s^2)^{-2} = r^6s^{-4} = r^6/s^4$.

           ↑   ↑

      Law (4)   Law (3)

(c) $\dfrac{x^5(y^2)^3}{(x^2y)^2} = \dfrac{x^5y^6}{(x^2y)^2} = \dfrac{x^5y^6}{(x^2)^2y^2} = \dfrac{x^5y^6}{x^4y^2} = x^{5-4}y^{6-2} = xy^4$.

      ↑   ↑   ↑ ↑

   Law (3) Law (4) Law (3) Law (2)

It is usually more efficient to use the exponent laws with the negative exponents rather than first converting to positive exponents. If positive exponents are required, the conversion can be made in the last step.

EXAMPLE 7

Simplify and express without negative exponents

$$\frac{a^{-2}(b^2c^3)^{-2}}{(a^{-3}b^{-5})^2c}.$$

SOLUTION

$$\frac{a^{-2}(b^2c^3)^{-2}}{(a^{-3}b^{-5})^2c} = \frac{a^{-2}(b^2)^{-2}(c^3)^{-2}}{(a^{-3})^2(b^{-5})^2c} = \frac{a^{-2}b^{-4}c^{-6}}{a^{-6}b^{-10}c}$$

              ↑   ↑

         Law (4)   Law (3)

$$= a^{-2-(-6)}b^{-4-(-10)}c^{-6-1} = a^4b^6c^{-7} = \frac{a^4b^6}{c^7}.$$

  ↑

Law (2)

Since $(-1)(-1) = +1$, any even power of -1, such as $(-1)^4$ or $(-1)^{12}$, will be equal to 1. Every odd power of -1 is equal to -1; for instance,

$(-1)^5 = (-1)^4(-1) = 1(-1) = -1$. Consequently, for every positive number c,

$$(-c)^n = [(-1)c]^n = (-1)^n c^n = \begin{cases} c^n & \text{if } n \text{ is even} \\ -c^n & \text{if } n \text{ is odd} \end{cases}.$$

EXAMPLE 8

$(-3)^4 = 3^4 = 81$ and $(-5)^3 = -5^3 = -125$. ∎

EXERCISES 1.A

In Exercises 1–18, evaluate the expression.

1. $(-6)^2$

2. -6^2

3. $5 + 4(3^2 + 2^3)$

4. $(-3)2^2 + 4^2 - 1$

5. $\dfrac{(-3)^2 + (-2)^4}{-2^2 - 1}$

6. $\dfrac{(-4)^2 + 2}{(-4)^2 - 7} + 1$

7. $\left(\dfrac{-5}{4}\right)^3$

8. $-\left(\dfrac{7}{4} + \dfrac{3}{4}\right)^2$

9. $\left(\dfrac{1}{3}\right)^3 + \left(\dfrac{2}{3}\right)^3$

10. $\left(\dfrac{5}{7}\right)^2 + \left(\dfrac{2}{7}\right)^2$

11. $2^4 - 2^7$

12. $3^3 - 3^{-7}$

13. $(2^{-2} + 2)^2$

14. $(3^{-1} - 3^3)^2$

15. $2^2 \cdot 3^{-3} - 3^2 \cdot 2^{-3}$

16. $4^3 \cdot 5^{-2} + 4^2 \cdot 5^{-1}$

17. $\dfrac{1}{2^3} + \dfrac{1}{2^{-4}}$

18. $3^2\left(\dfrac{1}{3} + \dfrac{1}{3^{-2}}\right)$

In Exercises 19–38, simplify the expression. Each letter represents a nonzero real number and should appear at most once in your answer.

19. $x^2 \cdot x^3 \cdot x^5$

20. $y \cdot y^4 \cdot y^6$

21. $(.03)y^2 \cdot y^7$

22. $(1.3)z^3 \cdot z^5$

23. $(2x^2)^3 3x$

24. $(3y^3)^4 5y^2$

25. $(3x^2y)^2$

26. $(2xy^3)^3$

27. $(a^2)(7a)(-3a^3)$

28. $(b^3)(-b^2)(3b)$

29. $(2w)^3(3w)(4w)^2$

30. $(3d)^4(2d)^2(5d)$

31. $a^{-2}b^3a^3$

32. $c^4 d^5 c^{-3}$

33. $(2x)^{-2}(2y)^3(4x)$

34. $(3x)^{-3}(2y)^{-2}(2x)$

35. $(-3a^4)^2(9x^3)^{-1}$

36. $(2y^3)^3(3y^2)^{-2}$

37. $(2x^2y)^0(3xy)$

38. $(3x^2y^4)^0$

In Exercises 39–42, express the given number as a power of 2.

39. $(64)^2$

40. $(1/8)^3$

41. $(2^4 \cdot 16^{-2})^3$

42. $(1/2)^{-8}(1/4)^4(1/16)^{-3}$

In Exercises 43–60, simplify and write the given expression without negative exponents. All letters represent nonzero real numbers.

43. $\dfrac{x^4(x^2)^3}{x^3}$

44. $\left(\dfrac{z^2}{t^3}\right)^4 \cdot \left(\dfrac{z^3}{t}\right)^5$

45. $\left(\dfrac{e^6}{c^4}\right)^2 \cdot \left(\dfrac{c^3}{e}\right)^3$

46. $\left(\dfrac{x^7}{y^6}\right)^2 \cdot \left(\dfrac{y^2}{x}\right)^4$

47. $\left(\dfrac{ab^2c^3d^4}{abc^2d}\right)^2$

48. $\dfrac{(3x)^2(y^2)^3x^2}{(2xy^2)^3}$

49. $\left(\dfrac{a^6}{b^{-4}}\right)^2$

50. $\left(\dfrac{x^{-2}}{y^{-2}}\right)^2$

51. $\left(\dfrac{c^5}{d^{-3}}\right)^{-2}$

52. $\left(\dfrac{x^{-1}}{2y^{-1}}\right)\left(\dfrac{2y}{x}\right)^{-2}$

53. $\left(\dfrac{3x}{y^2}\right)^{-3}\left(\dfrac{-x}{2y^3}\right)^2$

54. $\left(\dfrac{5u^2v}{2uv^2}\right)^2\left(\dfrac{-3uv}{2u^2v}\right)^{-3}$

55. $\dfrac{(a^{-3}b^2c)^{-2}}{(ab^{-2}c^3)^{-1}}$

56. $\dfrac{(-2cd^2e^{-1})^3}{(5c^{-3}de)^{-2}}$

57. $(c^{-1}d^{-2})^{-3}$

58. $[(x^2y^{-1})^2]^{-3}$

59. $a^2(a^{-1} + a^{-3})$

60. $\dfrac{a^{-2}}{b^{-2}} + \dfrac{b^2}{a^2}$

In Exercises 61–66, determine the sign of the given number without calculating the product.

61. $(-2.6)^3(-4.3)^{-2}$

62. $(4.1)^{-2}(2.5)^{-3}$

63. $(-1)^9(6.7)^5$

64. $(-4)^{12}6^9$

65. $(-3.1)^{-3}(4.6)^{-6}(7.2)^7$

66. $(45.8)^{-7}(-7.9)^{-9}(-8.5)^{-4}$

In Exercises 67–72, r, s, and t are positive integers and a, b, and c are nonzero real numbers. Simplify and write the given expression without negative exponents.

67. $\dfrac{3^{-r}}{3^{-s-r}}$

68. $\dfrac{4^{-(t+1)}}{4^{2-t}}$

69. $\left(\dfrac{a^6}{b^{-4}}\right)^t$

70. $\dfrac{c^{-t}}{(6b)^{-s}}$

71. $\dfrac{(c^{-r}b^s)^t}{(c^tb^{-s})^r}$

72. $\dfrac{(a^rb^{-s})^{-t}}{(b^tc^r)^{-s}}$

ERRORS TO AVOID

In Exercises 73–80, give an example to show that the statement may be false for some numbers.

73. $a^r + b^r = (a + b)^r$

74. $a^r a^s = a^{rs}$

75. $a^r b^s = (ab)^{r+s}$

76. $c^{-r} = -c^r$

77. $\dfrac{c^r}{c^s} = c^{r/s}$

78. $(a + 1)(b + 1) = ab + 1$

79. $(-a)^2 = -a^2$

80. $(-a)(-b) = -ab$

1.B   Arithmetic of Algebraic Expressions

Expressions such as

$$b + 3c^2, \qquad 3x^2 - 5x + 4, \qquad \sqrt{x^3 + z}, \qquad \frac{x^3 + 4xy - \pi}{x^2 + xy}$$

are called **algebraic expressions.** The letters in algebraic expressions represent numbers. A letter that represents a number whose value remains unchanged throughout the discussion is called a **constant.** Constants are usually denoted by letters near the beginning of the alphabet. A letter that may represent *any* number is called a **variable.** Letters near the end of the alphabet usually denote variables. Thus, in the expression $3x + 5 + c$, it is understood that x is a variable and c is a constant. The value of this expression depends on the value of the variable x. If $x = 4$, then

$$3x + 5 + c = 3(4) + 5 + c = 17 + c,$$

and if $x = \frac{1}{3}$, then

$$3x + 5 + c = 3\left(\tfrac{1}{3}\right) + 5 + c = 6 + c,$$

and so on.*

The usual rules of arithmetic are valid for algebraic expressions:

Commutative Laws:

$$a + b = b + a \qquad \text{and} \qquad ab = ba.$$

Associative Laws:

$$(a + b) + c = a + (b + c) \qquad \text{and} \qquad (ab)c = a(bc).$$

Distributive Laws:

$$a(b + c) = ab + ac \qquad \text{and} \qquad (b + c)a = ba + ca.$$

EXAMPLE 1

Use the distributive law to *combine like terms;* for instance,

$$3x + 5x + 4x = (3 + 5 + 4)x = 12x.$$

---

*We assume any conditions on the constants and variables necessary to guarantee that an algebraic expression does represent a real number. For instance, in $\sqrt{z}$, we assume $z \geq 0$, and in $1/c$, we assume $c \neq 0$.

In practice, you do the middle part in your head and simply write

$$3x + 5x + 4x = 12x.$$ ∎

EXAMPLE 2

In more complicated expressions, eliminate parentheses, use the commutative law to group like terms together, and then combine them.

$$(a^2b - 3\sqrt{c}) + (5ab + 7\sqrt{c}) + 7a^2b = a^2b - 3\sqrt{c} + 5ab + 7\sqrt{c} + 7a^2b$$

Regroup
$$= \underbrace{a^2b + 7a^2b} - \underbrace{3\sqrt{c} + 7\sqrt{c}} + 5ab$$

Combine like terms:
$$= \quad 8a^2b \quad + \quad 4\sqrt{c} \quad + 5ab.$$ ∎

CAUTION

Be careful when parentheses are preceded by a minus sign: $-(b + 3) = -b - 3$ and *not* $-b + 3$. Here's the reason: $-(b + 3)$ means $(-1)(b + 3)$; hence by the distributive law,

$$-(b + 3) = (-1)(b + 3) = (-1)b + (-1)3 = -b - 3.$$

Similarly, $-(7 - y) = -7 - (-y) = -7 + y$.

The examples in the Caution Box illustrate the following.

Rules for Eliminating Parentheses

> Parentheses preceded by a plus sign (or no sign) may be deleted.
>
> Parentheses preceded by a minus sign may be deleted *if* the sign of every term within the parentheses is changed.

EXAMPLE 3

Compute: $(4x^3 - 5x^2 + 7x - 2) - (2x^3 + x^2 - 6x - 8)$.

SOLUTION

Eliminate parentheses:
$$= 4x^3 - 5x^2 + 7x - 2 - 2x^3 - x^2 + 6x + 8$$

Group like terms:
$$= (4x^3 - 2x^3) + (-5x^2 - x^2) + (7x + 6x) + (-2 + 8)$$

Combine like terms:
$$= 2x^3 - 6x^2 + 13x + 6$$ ∎

To multiply algebraic expressions, use the distributive law repeatedly, as shown in the following examples. The net result is to *multiply every term in the first polynomial by every term in the second.*

EXAMPLE 4

Compute $(y - 2)(3y^2 - 7y + 4)$.

SOLUTION   We first apply the distributive law, treating $(3y^2 - 7y + 4)$ as a single number.

$$(y - 2)(3y^2 - 7y + 4) = y(3y^2 - 7y + 4) - 2(3y^2 - 7y + 4)$$

Distributive law: $= 3y^3 - 7y^2 + 4y - 6y^2 + 14y - 8$

Regroup: $= 3y^3 - \underbrace{7y^2 - 6y^2} + \underbrace{4y + 14y} - 8$

Combine like terms: $= 3y^3 - \quad 13y^2 \quad + \quad 18y \quad - 8.$ ■

EXAMPLE 5

Compute $(2x - 5y)(3x + 4y)$.

SOLUTION   Treat $3x + 4y$ as a single number, and use the distributive law.

$$(2x - 5y)(3x + 4y) = 2x(3x + 4y) - 5y(3x + 4y)$$

Distributive law: $= 2x \cdot 3x + 2x \cdot 4y + (-5y) \cdot 3x + (-5y) \cdot 4y$

Regroup $= 6x^2 + \underbrace{8xy - 15xy} - 20y^2$

Combine like terms: $= 6x^2 - \quad 7xy \quad - 20y^2.$ ■

Observe the pattern in the second equation of Example 5 and its relationship to the terms being multiplied.

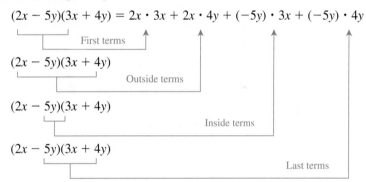

This pattern is easy to remember by using the acronym FOIL (**F**irst, **O**utside, **I**nside, **L**ast). The FOIL method makes it easy to find products such as this one mentallly, without the necessity of writing out the intermediate steps.

CAUTION

The FOIL method can be used only when multiplying two expressions that each have two terms.

EXAMPLE 6

$$(3x + 2)(x + 5) = 3x^2 + 15x + 2x + 10 = 3x^2 + 17x + 10.$$ ■

$\qquad\qquad\uparrow\qquad\quad\uparrow\qquad\uparrow\qquad\uparrow$

$\qquad$ First   Outside Inside   Last

EXERCISES 1.B

In Exercises 1–54, perform the indicated operations and simplify your answer.

1. $x + 7x$

2. $5w + 7w - 3w$

3. $6a^2b + (-8b)a^2$

4. $-6x^3 \sqrt{t} + 7x^3 \sqrt{t} + 15x^3 \sqrt{t}$

5. $(x^2 + 2x + 1) - (x^3 - 3x^2 + 4)$

6. $\left[u^4 - (-3)u^3 + \dfrac{u}{2} + 1 \right] + \left(u^4 - 2u^3 + 5 - \dfrac{u}{2} \right)$

7. $\left[u^4 - (-3)u^3 + \dfrac{u}{2} + 1 \right] - \left(u^4 - 2u^3 + 5 - \dfrac{u}{2} \right)$

8. $(6a^2b + 3a \sqrt{c} - 5ab \sqrt{c}) + (-6ab^2 - 3ab + 6ab \sqrt{c})$

9. $[4z - 6z^2w - (-2)z^3w^2] + (8 - 6z^2w - zw^3 + 4z^3w^2)$

10. $(x^5y - 2x + 3xy^3) - (-2x - x^5y + 2xy^3)$

11. $(9x - x^3 + 1) - [2x^3 + (-6)x + (-7)]$

12. $(x - \sqrt{y} - z) - (x + \sqrt{y} - z) - (\sqrt{y} + z - x)$

13. $(x^2 - 3xy) - (x + xy) - (x^2 + xy)$

14. $2x(x^2 + 2)$

15. $(-5y)(-3y^2 + 1)$

16. $x^2y(xy - 6xy^2)$

17. $3ax(4ax - 2a^2y + 2ay)$

18. $2x(x^2 - 3xy + 2y^2)$

19. $6z^3(2z + 5)$

20. $-3x^2(12x^6 - 7x^5)$

21. $3ab(4a - 6b + 2a^2b)$

22. $(-3ay)(4ay - 5y)$

23. $(x + 1)(x - 2)$

24. $(x + 2)(2x - 5)$

25. $(-2x + 4)(-x - 3)$

26. $(y - 6)(2y + 2)$

27. $(y + 3)(y + 4)$

28. $(w - 2)(3w + 1)$

29. $(3x + 7)(-2x + 5)$

30. $(ab + 1)(a - 2)$

31. $(y - 3)(3y^2 + 4)$

32. $(y + 8)(y - 8)$

33. $(x + 4)(x - 4)$

34. $(3x - y)(3x + y)$

35. $(4a + 5b)(4a - 5b)$

36. $(x + 6)^2$

37. $(y - 11)^2$

38. $(2x + 3y)^2$

39. $(5x - b)^2$

40. $(2s^2 - 9y)(2s^2 + 9y)$

41. $(4x^3 - y^4)^2$

42. $(4x^3 - 5y^2)(4x^3 + 5y^2)$

43. $(-3x^2 + 2y^4)^2$

44. $(c - 2)(2c^2 - 3c + 1)$

45. $(2y + 3)(y^2 + 3y - 1)$

46. $(x + 2y)(2x^2 - xy + y^2)$

47. $(5w + 6)(-3w^2 + 4w - 3)$

48. $(5x - 2y)(x^2 - 2xy + 3y^2)$

49. $2x(3x + 1)(4x - 2)$

50. $3y(-y + 2)(3y + 1)$

51. $(x - 1)(x - 2)(x - 3)$

52. $(y - 2)(3y + 2)(y + 2)$

53. $(x + 4y)(2y - x)(3x - y)$

54. $(2x - y)(3x + 2y)(y - x)$

In Exercises 55–64, find the coefficient of x^2 in the given product. Avoid doing any more multiplying than necessary.

55. $(x^2 + 3x + 1)(2x - 3)$

56. $(x^2 - 1)(x + 1)$

57. $(x^3 + 2x - 6)(x^2 + 1)$

58. $(\sqrt{3} + x)(\sqrt{3} - x)$

59. $(x + 2)^3$

60. $(x^2 + x + 1)(x - 1)$

61. $(x^2 + x + 1)(x^2 - x + 1)$

62. $(2x^2 + 1)(2x^2 - 1)$

63. $(2x - 1)(x^2 + 3x + 2)$

64. $(1 - 2x)(4x^2 + x - 1)$

In Exercises 65–70, perform the indicated multiplication, and simplify your answer if possible.

65. $(\sqrt{x} + 5)(\sqrt{x} - 5)$

66. $(2\sqrt{x} + \sqrt{2y})(2\sqrt{x} - \sqrt{2y})$

67. $(3 + \sqrt{y})^2$

68. $(7w - \sqrt{2x})^2$

69. $(1 + \sqrt{3}x)(x + \sqrt{3})$

70. $(2y + \sqrt{3})(\sqrt{5}y - 1)$

In Exercises 71–76, compute the product, and arrange the terms of your answer according to decreasing powers of x, with each power of x appearing at most once.

Example: $(ax + b)(4x - c) = 4ax^2 + (4b - ac)x - bc$.

71. $(ax + b)(3x + 2)$

72. $(4x - c)(dx + c)$

73. $(ax + b)(bx + a)$

74. $rx(3rx + 1)(4x - r)$

75. $(x - a)(x - b)(x - c)$

76. $(2dx - c)(3cx + d)$

In Exercises 77–79, assume that all exponents are nonnegative integers, and find the product.

Example: $2x^k(3x + x^{n+1}) = (2x^k)(3x) + (2x^k)(x^{n+1})$
$$= 6x^{k+1} + 2x^{k+n+1}.$$

77. $(2x^n)(8x^k)$

78. $(x^m + 2)(x^n - 3)$

79. $(y^r + 1)(y^s - 4)$

In Exercises 80–82, express the area of the shaded region as an algebraic expression in x.

80.

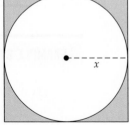

81.

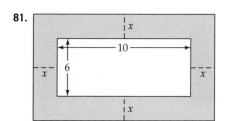

82.

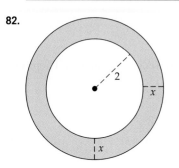

ERRORS TO AVOID

In Exercises 83–92, find a numerical example to show that the given statement is false. *Then find the mistake in the statement and correct it.*

Example: *The statement* $-(b + 2) = -b + 2$ *is false when* $b = 5$, *since* $-(5 + 2) = -7$ *but* $-5 + 2 = -3$. *The mistake is the sign on the 2. The correct statement is* $-(b + 2) = -b - 2$.

83. $3(y + 2) = 3y + 2$

84. $x - (3y + 4) = x - 3y + 4$

85. $(x + y)^2 = x + y^2$ **86.** $(2x)^3 = 2x^3$

87. $(7x)(7y) = 7xy$ **88.** $(x + y)^2 = x^2 + y^2$

89. $y + y + y = y^3$ **90.** $(a - b)^2 = a^2 - b^2$

91. $(x - 3)(x - 2) = x^2 - 5x - 6$

92. $(a + b)(a^2 + b^2) = a^3 + b^3$

THINKERS

In Exercises 93 and 94, explain algebraically why each of these parlor tricks always works.

93. Write down a nonzero number. Add 1 to it, and square the result. Subtract 1 from the original number, and square the result. Subtract this second square from the first one. Divide by the number with which you started. The answer is 4.

94. Write down a positive number. Add 4 to it. Multiply the result by the original number. Add 4 to this result, and then take the square root. Subtract the number with which you started. The answer is 2.

95. Invent a similar parlor trick in which the answer is always the number with which you started.

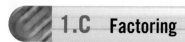

1.C Factoring

Factoring is the reverse of multiplication: We begin with a product and find the factors that multiply together to produce this product. Factoring skills are necessary to simplify expressions, to do arithmetic with fractional expressions, and to solve equations and inequalities.

Here is the first general rule for factoring.

Common Factors

> If there is a common factor in every term of the expression, factor out the common factor of highest degree.

EXAMPLE 1

In $4x^6 - 8x$, for example, each term contains a factor of $4x$, so

$$4x^6 - 8x = 4x(x^5 - 2).$$

Similarly, the common factor of highest degree in $x^3y^2 + 2xy^3 - 3x^2y^4$ is xy^2, and

$$x^3y^2 + 2xy^3 - 3x^2y^4 = xy^2(x^2 + 2y - 3xy^2). \qquad \blacksquare$$

You can greatly increase your factoring proficiency by learning to recognize multiplication patterns that appear frequently. Here are the most common ones.

Quadratic Factoring Patterns

Difference of Squares:	$u^2 - v^2 = (u + v)(u - v);$
Perfect Squares:	$u^2 + 2uv + v^2 = (u + v)^2;$
	$u^2 - 2uv + v^2 = (u - v)^2.$

EXAMPLE 2

(a) $x^2 - 9y^2$ can be written $x^2 - (3y)^2$, a difference of squares. Therefore, $x^2 - 9y^2 = (x + 3y)(x - 3y)$.

(b) $y^2 - 7 = y^2 - (\sqrt{7})^2 = (y + \sqrt{7})(y - \sqrt{7})$.*

(c) $36r^2 - 64s^2 = (6r)^2 - (8s)^2 = (6r + 8s)(6r - 8s)$
$$= 2(3r + 4s)2(3r - 4s) = 4(3r + 4s)(3r - 4s).$$ ∎

EXAMPLE 3

Factor: $4x^2 - 36x + 81$.

SOLUTION   Since the first and last terms of $4x^2 - 36x + 81$ are perfect squares, we try to use the perfect square pattern with $u = 2x$ and $v = 9$:

$$4x^2 - 36x + 81 = (2x)^2 - 36x + 9^2$$
$$= (2x)^2 - 2 \cdot 2x \cdot 9 + 9^2 = (2x - 9)^2.$$ ∎

Cubic Factoring Patterns

Difference of Cubes	$u^3 - v^3 = (u - v)(u^2 + uv + v^2)$
Sum of Cubes	$u^3 + v^3 = (u + v)(u^2 - uv + v^2)$
Perfect Cubes	$u^3 + 3u^2v + 3uv^2 + v^3 = (u + v)^3$
	$u^3 - 3u^2v + 3uv^2 - v^3 = (u - v)^3$

EXAMPLE 4

Factor:

(a) $x^3 - 125$ (b) $x^3 + 8y^3$ (c) $x^3 - 12x^2 + 48x - 64$

SOLUTION

(a) $x^3 - 125 = x^3 - 5^3 = (x - 5)(x^2 + 5x + 5^2)$
$$= (x - 5)(x^2 + 5x + 25).$$

(b) $x^3 + 8y^3 = x^3 + (2y)^3 = (x + 2y)[x^2 - x \cdot 2y + (2y)^2]$
$$= (x + 2y)(x^2 - 2xy + 4y^2).$$

---

*When a polynomial has integer coefficients, we normally look only for factors with integer coefficients. But when it is easy to find other factors, as here, we shall do so.

(c) $x^3 - 12x^2 + 48x - 64 = x^3 - 12x^2 + 48x - 4^3$

$$= x^3 - 3x^2 \cdot 4 + 3x \cdot 4^2 - 4^3$$

$$= (x - 4)^3.$$ ∎

When none of the multiplication patterns applies, other techniques must be used. The next example illustrates a method that works for quadratic polynomials with leading coefficient 1.

EXAMPLE 5

Factor: $x^2 + 9x + 18$.

SOLUTION   We must find integers b and d such that

$$x^2 + 9x + 18 = (x + b)(x + d)$$

$$= x^2 + dx + bx + bd$$

$$= x^2 + (b + d)x + bd.$$

If the polynomials are equal, then their coefficients are equal term by term, so

$$9 = b + d \qquad \text{and} \qquad 18 = bd.$$

In particular, b and d are integer factors of 18. The possibilities are summarized in this table.

Factors b, d of 18	Sum $b + d$
$1 \cdot 18$	$1 + 18 = 19$
$2 \cdot 9$	$2 + 9 = 11$
$3 \cdot 6$	$3 + 6 = 9$
$(-1)(-18)$	$-1 + (-18) = -19$
$(-2)(-9)$	$-2 + (-9) = -11$
$(-3)(-6)$	$-3 + (-6) = -9$

Note that we really didn't have to list the negative factors, since their sums are also negative and we must have $b + d = 9$. There is just one choice with sum 9, namely, 3 and 6. So we have this factorization (which you should verify by multiplying out the right-hand side):

$$x^2 + 9x + 18 = (x + 3)(x + 6).$$ ∎

EXAMPLE 6

Factor: $x^2 + 3x - 10$.

SOLUTION   We must find factors $x + b$ and $x + d$ such that $bd = -10$ and $b + d = 3$. The table shows the possibilities.

Factors b, d of -10	Sum $b + d$
$1(-10)$	$1 + (-10) = -9$
$(-1)(10)$	$-1 + 10 = 9$
$2(-5)$	$2 + (-5) = -3$
$(-2)(5)$	$-2 + 5 = 3$

The only factors with product -10 and sum 3 are -2 and 5. So the correct factorization is

$$x^2 + 3x - 10 = (x - 2)(x + 5).$$ ∎

Although we used tables in Examples 5 and 6 to summarize all the possibilities, you can often do this without writing out a table by mentally checking the various possibilities. The same idea works for factoring quadratic polynomials in which the leading coefficient is not 1, as shown in the next example.

EXAMPLE 7

Factor: $6x^2 + 11x + 4$.

SOLUTION   We must find factors $ax + b$ and $cx + d$ such that

$$
\begin{aligned}
6x^2 + 11x + 4 &= (ax + b)(cx + d) \\
&= acx^2 + adx + bcx + bd \\
&= acx^2 + (ad + bc)x + bd.
\end{aligned}
$$

Once again, the coefficients must be the same, term by term, on both sides. In particular, we must have

$$6 = ac \quad \text{and} \quad 4 = bd.$$

There are quite a few possibilities, as shown here.

$ac = 6$	a	± 1	± 2	± 3	± 6
	c	± 6	± 3	± 2	± 1

$bd = 4$	b	± 1	± 2	± 4
	d	± 4	± 2	± 1

Guided by the possibilities listed in the tables, we use trial and error to find the right factors. Typically, it might go like this.

$a = 1, c = 6$ and $b = 1, d = 4$:

$$(ax + b)(cx + d) = (x + 1)(6x + 4) = 6x^2 + 10x + 4 \qquad \text{Incorrect}$$

$a = 2, c = 3$ and $b = 4, d = 1$:

$$(ax + b)(cx + d) = (2x + 4)(3x + 1) = 6x^2 + 14x + 4 \qquad \text{Incorrect}$$

$a = 2, c = 3$ and $b = 1, d = 4$:

$$(ax + b)(cx + d) = (2x + 1)(3x + 4) = 6x^2 + 11x + 4 \qquad \text{Correct} \quad ∎$$

Substitution and appropriate factoring patterns can sometimes be used to factor expressions involving exponents larger than 2.

EXAMPLE 8

Factor: $x^4 - 2x^2 - 3$.

SOLUTION   The idea is to put the polynomial in quadratic form. Note that $x^4 = (x^2)^2$ and let $u = x^2$. Then

$$
\begin{aligned}
x^4 - 2x^2 - 3 &= (x^2)^2 - 2x^2 - 3 \\
&= u^2 - 2u - 3 = (u + 1)(u - 3) \\
&= (x^2 + 1)(x^2 - 3) \\
&= (x^2 + 1)(x + \sqrt{3})(x - \sqrt{3}).
\end{aligned}
$$ ∎

EXAMPLE 9

Factor:

(a) $x^6 - y^6$ (b) $x^8 - 1$.

SOLUTION   Once again, we want to put each polynomial in quadratic form.

(a) Note that $x^6 = (x^3)^2$ and $y^6 = (y^3)^2$. Let $u = x^3$ and $v = y^3$. Then

$$x^6 - y^6 = (x^3)^2 - (y^3)^2$$
$$= u^2 - v^2 = (u + v)(u - v) \qquad \text{[Difference of squares]}$$
$$= (x^3 + y^3)(x^3 - y^3)$$
$$= (x + y)(x^2 - xy + y^2)(x^3 - y^3) \qquad \text{[Sum of cubes]}$$
$$= (x + y)(x^2 - xy + y^2)(x - y)(x^2 + xy + y^2). \qquad \text{[Difference of cubes]}$$

(b) Note that $x^8 = (x^4)^2$ and let $u = x^4$. Then

$$x^8 - 1 = (x^4)^2 - 1 = u^2 - 1 = (u + 1)(u - 1) \qquad \text{[Difference of squares]}$$
$$= (x^4 + 1)(x^4 - 1).$$

Observe that $x^4 = (x^2)^2$ and let $v = x^2$. Then

$$x^8 - 1 = (x^4 + 1)(x^4 - 1)$$
$$= [(x^2)^2 + 1][(x^2)^2 - 1]$$
$$= (v^2 + 1)(v^2 - 1)$$
$$= (v^2 + 1)(v + 1)(v - 1) \qquad \text{[Difference of squares]}$$
$$= (x^4 + 1)(x^2 + 1)(x^2 - 1)$$
$$= (x^4 + 1)(x^2 + 1)(x + 1)(x - 1). \qquad \text{[Difference of squares]} \qquad ■$$

Occasionally, a polynomial can be factored by regrouping.

EXAMPLE 10

Factor: $3x^3 + 3x^2 + 2x + 2$.

SOLUTION   Write the polynomial as

$$(3x^3 + 3x^2) + (2x + 2)$$

and factor a common factor out of each expression in parentheses:

$$3x^2(x + 1) + 2(x + 1).$$

Now use the distributive law (considering $x + 1$ as a single quantity).

$$3x^2(x + 1) + 2(x + 1) = (3x^2 + 2)(x + 1) \qquad ■$$

EXERCISES 1.C

In Exercises 1–58, factor the expression.

1. $x^2 - 4$

2. $x^2 + 6x + 9$

3. $9y^2 - 25$

4. $y^2 - 4y + 4$

5. $81x^2 + 36x + 4$

6. $4x^2 - 12x + 9$

7. $5 - x^2$

8. $1 - 36u^2$

9. $49 + 28z + 4z^2$

10. $25u^2 - 20uv + 4v^2$

11. $x^4 - y^4$

12. $x^2 - 1/9$

13. $x^2 + x - 6$

14. $y^2 + 11y + 30$

15. $z^2 + 4z + 3$

16. $x^2 - 8x + 15$

17. $y^2 + 5y - 36$

18. $z^2 - 9z + 14$

19. $x^2 - 6x + 9$

20. $4y^2 - 81$

21. $x^2 + 7x + 10$

22. $w^2 - 6w - 16$

23. $x^2 + 11x + 18$

24. $x^2 + 3xy - 28y^2$

25. $3x^2 + 4x + 1$

26. $4y^2 + 4y + 1$

27. $2z^2 + 11z + 12$

28. $10x^2 - 17x + 3$

29. $9x^2 - 72x$

30. $4x^2 - 4x - 3$

31. $10x^2 - 8x - 2$

32. $7z^2 + 23z + 6$

33. $8u^2 + 6u - 9$

34. $2y^2 - 4y + 2$

35. $4x^2 + 20xy + 25y^2$

36. $63u^2 - 46uv + 8v^2$

37. $x^3 - 125$

38. $y^3 + 64$

39. $x^3 + 6x^2 + 12x + 8$

40. $y^3 - 3y^2 + 3y - 1$

41. $8 + x^3$

42. $z^3 - 9z^2 + 27z - 27$

43. $-x^3 + 15x^2 - 75x + 125$

44. $27 - t^3$

45. $x^3 + 1$

46. $x^3 - 1$

47. $8x^3 - y^3$

48. $(x - 1)^3 + 1$

49. $x^6 - 64$

50. $x^5 - 8x^2$

51. $y^4 + 7y^2 + 10$

52. $z^4 - 5z^2 + 6$

53. $81 - y^4$

54. $x^6 + 16x^3 + 64$

55. $z^6 - 1$

56. $y^6 + 26y^3 - 27$

57. $x^4 + 2x^2y - 3y^2$

58. $x^8 - 17x^4 + 16$

In Exercises 59–64, factor by regrouping and using the distributive law (as in Example 9).

59. $x^2 - yz + xz - xy$

60. $x^6 - 2x^4 - 8x^2 + 16$

61. $a^3 - 2b^2 + 2a^2b - ab$

62. $u^2v - 2w^2 - 2uvw + uw$

63. $x^3 + 4x^2 - 8x - 32$

64. $z^8 - 5z^7 + 2z - 10$

THINKER

65. Show that there do *not* exist real numbers c and d such that $x^2 + 1 = (x + c)(x + d)$.

1.D   Fractional Expressions

Quotients of algebraic expressions, such as

$$\frac{x^2 - 2}{x - 1}, \quad \frac{1}{x^3 + 4x^2 - x}, \quad \text{or} \quad \frac{3\sqrt{x} - 7}{x^3(x + 1)}$$

are called **fractional expressions.** Throughout this section, we assume that all denominators are nonzero.

The basic rules for dealing with fractional expressions are essentially the same as those for ordinary numerical fractions. In particular, the usual cancellation property holds.

Cancellation

If $k \neq 0$, then $\dfrac{ka}{kb} = \dfrac{a}{b}$.

EXAMPLE 1

Simplify:

(a) $\dfrac{36}{56}$ (b) $\dfrac{x^4 - 1}{x^2 + 1}$.

SOLUTION

(a) Factor numerator and denominator:
$$\frac{36}{56} = \frac{9 \cdot 4}{7 \cdot 8} = \frac{9 \cdot 4}{7 \cdot 2 \cdot 4}$$

Cancel the common factor 4:
$$= \frac{9}{7 \cdot 2} = \frac{9}{14}.$$

(b) The same procedure works for the rational expression.

Factor the numerator:
$$\frac{x^4 - 1}{x^2 + 1} = \frac{(x^2 + 1)(x^2 - 1)}{x^2 + 1}$$

Cancel the common factor $x^2 + 1$:
$$= \frac{x^2 - 1}{1} = x^2 - 1 \qquad \blacksquare$$

From arithmetic, we know that
$$\frac{2}{4} \times \frac{3}{6}.$$

Note that the cross-products are equal: $2 \cdot 6 = 4 \cdot 3$. The same thing is true in the general case.

Equality Rule

$$\frac{a}{b} = \frac{c}{d} \qquad \text{exactly when} \qquad ad = bc.$$

Proof Suppose
$$\frac{a}{b} = \frac{c}{d}.$$

Multiply both sides by bd:
$$\frac{a}{b} \cdot bd = bd \cdot \frac{c}{d}$$

Use cancellation:
$$ad = bc$$

Conversely, if $ad = bc$, then dividing both sides by bd and cancelling shows that $\frac{a}{b} = \frac{c}{d}$. $\qquad \blacksquare$

EXAMPLE 2

We know that
$$\frac{x^2 + 2x}{x^2 + x - 2} = \frac{x}{x - 1}$$

because the cross-products are equal:
$$(x^2 + 2x)(x - 1) = x^3 + x^2 - 2x \qquad \text{and} \qquad (x^2 + x - 2)x = x^3 + x^2 - 2x. \qquad \blacksquare$$

A fraction is in **lowest terms** if its numerator and denominator have no common factors except ± 1.

EXAMPLE 3

Express $\dfrac{x^2 + x - 6}{x^2 - 3x + 2}$ in lowest terms.

SOLUTION   Proceed as we did in Example 1: Factor numerator and denominator, and cancel common factors.

$$\frac{x^2 + x - 6}{x^2 - 3x + 2} = \frac{(x - 2)(x + 3)}{(x - 2)(x - 1)} = \frac{x + 3}{x - 1} \qquad \blacksquare$$

To add or subtract two fractions with the same denominator, simply add or subtract the numerators.

$$\frac{a}{b} + \frac{c}{b} = \frac{a + c}{b} \qquad \text{and} \qquad \frac{a}{b} - \frac{c}{b} = \frac{a - c}{b}$$

EXAMPLE 4

$$\frac{7x^2 + 2}{x^2 + 3} - \frac{4x^2 + 2x - 5}{x^2 + 3} = \frac{(7x^2 + 2) - (4x^2 + 2x - 5)}{x^2 + 3}$$

$$= \frac{7x^2 + 2 - 4x^2 - 2x + 5}{x^2 + 3}$$

$$= \frac{3x^2 - 2x + 7}{x^2 + 3}. \qquad \blacksquare$$

To add or subtract fractions with different denominators, you must use a **common denominator,** that is, replace the given fractions by equivalent ones that have the same denominator. For instance, to add 1/3 and 1/4, we use the common denominator 12 (the product of 3 and 4):

$$\frac{1}{3} = \frac{4}{12} \qquad \text{and} \qquad \frac{1}{4} = \frac{3}{12},$$

so that

$$\frac{1}{3} + \frac{1}{4} = \frac{4}{12} + \frac{3}{12} = \frac{4 + 3}{12} = \frac{7}{12}.$$

EXAMPLE 5

Compute: $\dfrac{2x + 1}{3x} - \dfrac{x^2 - 2}{x - 1}.$

SOLUTION   We use the product of the denominators $3x(x - 1)$ as the common denominator. To rewrite the first fraction in terms of this denominator, we must find the right-side numerator here.

$$\frac{2x + 1}{3x} = \frac{?}{3x(x - 1)}$$

Note that to get from the left-side denominator to the right-side one, you multiply by $x - 1$. To keep the fractions equal, you must multiply the left-hand numerator by $x - 1$ also:

$$\frac{2x + 1}{3x} = \frac{(2x + 1)(x - 1)}{3x(x - 1)}$$

A similar procedure works with the other fraction.

$$\frac{x^2 - 2}{x - 1} = \frac{?}{3x(x - 1)} \qquad \frac{x^2 - 2}{x - 1} = \frac{3x(x^2 - 2)}{3x(x - 1)}$$

Multiply by $3x$

Now that the fractions have the same denominator, it's easy to subtract them.

$$\frac{2x + 1}{3x} - \frac{x^2 - 2}{x - 1} = \frac{(2x + 1)(x - 1)}{3x(x - 1)} - \frac{3x(x^2 - 2)}{3x(x - 1)}$$

$$= \frac{(2x + 1)(x - 1) - 3x(x^2 - 2)}{3x(x - 1)}$$

$$= \frac{2x^2 - x - 1 - 3x^3 + 6x}{3x^2 - 3x}$$

$$= \frac{-3x^3 + 2x^2 + 5x - 1}{3x^2 - 3x} \qquad \blacksquare$$

Although the product of the denominators can always be used as a common denominator, it's usually more efficient to use the **least common denominator,** that is, the smallest quantity that can be the denominator of both fractions. The next example illustrates an easy way to find the least common denominator when dealing with numbers.

EXAMPLE 6

Find the least common denominator for $\dfrac{1}{100}$ and $\dfrac{1}{120}$.

SOLUTION The two denominators can be factored as

$$100 = 2^2 \cdot 5^2 \qquad \text{and} \qquad 120 = 2^3 \cdot 3 \cdot 5.$$

The distinct factors that appear in both numbers are 2, 3, and 5. For each factor, take the highest power to which it appears in either number (namely, 2^3, 3^1, and 5^2). The product of these highest powers is the least common denominator.

$$2^3 \cdot 3 \cdot 5^2 = 600$$

Note that $600 = 100 \cdot 6$ and $600 = 120 \cdot 5$, so

$$\frac{1}{100} = \frac{1 \cdot 6}{100 \cdot 6} = \frac{6}{600} \qquad \text{and} \qquad \frac{1}{120} = \frac{1 \cdot 5}{120 \cdot 5} = \frac{5}{600}. \qquad \blacksquare$$

The technique of Example 6 can also be used to find the least common denominator of rational expressions.

EXAMPLE 7

Find the least common denominator for

$$\frac{1}{x^2 + 2x + 1}, \quad \frac{5x}{x^2 - x}, \quad \text{and} \quad \frac{3x - 7}{x^4 + x^3}$$

and express each fraction in terms of this common denominator.

SOLUTION   Begin by factoring each denominator completely.

$$x^2 + 2x + 1 = (x + 1)^2 \qquad x^2 - x = x(x - 1) \qquad x^4 + x^3 = x^3(x + 1)$$

The distinct factors are x, $x + 1$, and $x - 1$. The least common denominator is

$$x^3(x + 1)^2(x - 1).$$

Therefore,

$$\frac{1}{(x + 1)^2} = \frac{1}{(x + 1)^2} \cdot \frac{x^3(x - 1)}{x^3(x - 1)} = \frac{x^3(x - 1)}{x^3(x + 1)^2(x - 1)}$$

$$\frac{5x}{x(x - 1)} = \frac{5x}{x(x - 1)} \cdot \frac{x^2(x + 1)^2}{x^2(x + 1)^2} = \frac{5x^3(x + 1)^2}{x^3(x + 1)^2(x - 1)}$$

$$\frac{3x - 7}{x^3(x + 1)} = \frac{3x - 7}{x^3(x + 1)} \cdot \frac{(x + 1)(x - 1)}{(x + 1)(x - 1)} = \frac{(3x - 7)(x + 1)(x - 1)}{x^3(x + 1)^2(x - 1)}. \quad \blacksquare$$

EXAMPLE 8

Compute $\dfrac{1}{z} + \dfrac{3z}{z + 1} - \dfrac{z^2}{(z + 1)^2}$.

SOLUTION   We use the least common denominator $z(z + 1)^2$ and express each fraction in terms of this denominator.

$$\frac{1}{z} = \frac{1}{z} \cdot \frac{(z + 1)^2}{(z + 1)^2} = \frac{(z + 1)^2}{z(z + 1)^2}$$

$$\frac{3z}{z + 1} = \frac{3z}{z + 1} \cdot \frac{z(z + 1)}{z(z + 1)} = \frac{3z^2(z + 1)}{z(z + 1)^2}$$

$$\frac{z^2}{(z + 1)^2} = \frac{z^2}{(z + 1)^2} \cdot \frac{z}{z} = \frac{z^3}{z(z + 1)^2}$$

Then

$$\frac{1}{z} + \frac{3z}{z + 1} - \frac{z^2}{(z + 1)^2} = \frac{(z + 1)^2}{z(z + 1)^2} + \frac{3z^2(z + 1)}{z(z + 1)^2} - \frac{z^3}{z(z + 1)^2}$$

$$= \frac{(z + 1)^2 + 3z^2(z + 1) - z^3}{z(z + 1)^2}$$

$$= \frac{z^2 + 2z + 1 + 3z^3 + 3z^2 - z^3}{z(z + 1)^2}$$

$$= \frac{2z^3 + 4z^2 + 2z + 1}{z(z + 1)^2}. \quad \blacksquare$$

Multiplication of fractional expressions is done in the same way as multiplication of numerical fractions.

Multiplication of Fractional Expressions

If $a, b, c,$ and d are algebraic expressions, then

$$\frac{a}{b} \cdot \frac{c}{d} = \frac{ac}{bd}.$$

In other words, to multiply fractional expressions, multiply the corresponding numerators and denominators.

EXAMPLE 9

Multiply $\dfrac{x^2-1}{x^2+2} \cdot \dfrac{3x-4}{x+1}$ and simplify your answer.

SOLUTION

$$\frac{x^2-1}{x^2+2} \cdot \frac{3x-4}{x+1} = \frac{(x^2-1)(3x-4)}{(x^2+2)(x+1)}$$

$$= \frac{(x-1)\cancel{(x+1)}(3x-4)}{(x^2+2)\cancel{(x+1)}} = \frac{(x-1)(3x-4)}{x^2+2}. \qquad \blacksquare$$

In Example 9, we multiplied and then factored and canceled. You can also factor and cancel before multiplying, as illustrated in the next example.

EXAMPLE 10

$$\frac{x^2+6x+8}{x+3} \cdot \frac{x^2+x-6}{x^2+3x+2} = \frac{\cancel{(x+2)}(x+4)}{\cancel{x+3}} \cdot \frac{\cancel{(x+3)}(x-2)}{\cancel{(x+2)}(x+1)}$$

$$= \frac{(x+4)(x-2)}{x+1}$$

$$= \frac{x^2+2x-8}{x+1} \qquad \blacksquare$$

Division of fractional expressions is also done in the same manner as division of numerical fractions.

Division of Fractional Expressions

> If a, b, c, and d are algebraic expressions, then
>
> $$\frac{a}{b} \div \frac{c}{d} = \frac{a}{b} \cdot \frac{d}{c} = \frac{ad}{bc}.$$
>
> That is, to divide fractional expressions, invert the divisor and multiply.

EXAMPLE 11

$$\frac{x^2+x-2}{x^2-6x+9} \div \frac{x^2-1}{x-3} = \frac{x^2+x-2}{x^2-6x+9} \cdot \frac{x-3}{x^2-1}$$

$$= \frac{(x+2)\cancel{(x-1)}}{(x-3)\cancel{(x-3)}} \cdot \frac{\cancel{x-3}}{\cancel{(x-1)}(x+1)}$$

$$= \frac{x+2}{(x-3)(x+1)}. \qquad \blacksquare$$

Division problems can also be written in fractional form. For instance, $8/2$ means $8 \div 2$. Similarly, a compound fraction such as

$$\frac{\dfrac{x+1}{x^2-4}}{\dfrac{3x+2}{x+2}}$$

means

$$\frac{x+1}{x^2-4} \div \frac{3x+2}{x+2}, \quad \text{which is equivalent to} \quad \frac{x+1}{x^2-4} \cdot \frac{x+2}{3x+2}.$$

Consequently, the basic rule for simplifying a compound fraction is:

Invert the denominator and multiply it by the numerator.

EXAMPLE 12

Compute:

(a) $\dfrac{\dfrac{16y^2z}{8yz^2}}{\dfrac{yz}{6y^3z^3}}$ (b) $\dfrac{\dfrac{y^2}{y+2}}{y^3+y}$

SOLUTION

(a) $\dfrac{16y^2z/8yz^2}{yz/6y^3z^3} = \dfrac{16y^2z}{8yz^2} \cdot \dfrac{6y^3z^3}{yz} = \dfrac{\overset{2}{\cancel{16}} \cdot 6 \cdot y^5z^4}{\cancel{8y^2z^3}}$

$$= 2 \cdot 6 \cdot y^{5-2}z^{4-3} = 12y^3z$$

(b) $\dfrac{\dfrac{y^2}{y+2}}{y^3+y} = \dfrac{y^2}{y+2} \cdot \dfrac{1}{y^3+y} = \dfrac{y^2}{(y+2)(y^3+y)}$

$$= \dfrac{y^2}{(y+2)y(y^2+1)}$$

$$= \dfrac{y}{(y+2)(y^2+1)} \qquad \blacksquare$$

EXAMPLE 13

Simplify the compound fraction

$$\dfrac{\dfrac{1}{x+h} - \dfrac{1}{x}}{h}.$$

SOLUTION First write the numerator as a single fraction, then invert and multiply.

$$\dfrac{\dfrac{1}{x+h} - \dfrac{1}{x}}{h} = \dfrac{\dfrac{x}{x(x+h)} - \dfrac{x+h}{x(x+h)}}{h}$$

$$= \dfrac{\dfrac{x-(x+h)}{x(x+h)}}{h} = \dfrac{\dfrac{-h}{x(x+h)}}{h}$$

$$= \dfrac{-h}{x(x+h)} \cdot \dfrac{1}{h} = \dfrac{-1}{x(x+h)} \qquad \blacksquare$$

EXERCISES 1.D

In Exercises 1–10, express the fraction in lowest terms.

1. $\dfrac{63}{49}$

2. $\dfrac{121}{33}$

3. $\dfrac{13 \cdot 27 \cdot 22 \cdot 10}{6 \cdot 4 \cdot 11 \cdot 12}$

4. $\dfrac{x^2 - 4}{x + 2}$

5. $\dfrac{x^2 - x - 2}{x^2 + 2x + 1}$

6. $\dfrac{z + 1}{z^3 + 1}$

7. $\dfrac{a^2 - b^2}{a^3 - b^3}$

8. $\dfrac{x^4 - 3x^2}{x^3}$

9. $\dfrac{(x + c)(x^2 - cx + c^2)}{x^4 + c^3 x}$

10. $\dfrac{x^4 - y^4}{(x^2 + y^2)(x^2 - xy)}$

In Exercises 11–28, perform the indicated operations.

11. $\dfrac{3}{7} + \dfrac{2}{5}$

12. $\dfrac{7}{8} - \dfrac{5}{6}$

13. $\left(\dfrac{19}{7} + \dfrac{1}{2} \right) - \dfrac{1}{3}$

14. $\dfrac{1}{a} - \dfrac{2a}{b}$

15. $\dfrac{c}{d} + \dfrac{3c}{e}$

16. $\dfrac{r}{s} + \dfrac{s}{t} + \dfrac{t}{r}$

17. $\dfrac{b}{c} - \dfrac{c}{b}$

18. $\dfrac{a}{b} + \dfrac{2a}{b^2} + \dfrac{3a}{b^3}$

19. $\dfrac{1}{x + 1} - \dfrac{1}{x}$

20. $\dfrac{1}{2x + 1} + \dfrac{1}{2x - 1}$

21. $\dfrac{1}{x + 4} + \dfrac{2}{(x + 4)^2} - \dfrac{3}{x^2 + 8x + 16}$

22. $\dfrac{1}{x} + \dfrac{1}{xy} + \dfrac{1}{xy^2}$

23. $\dfrac{1}{x} - \dfrac{1}{3x - 4}$

24. $\dfrac{3}{x - 1} + \dfrac{4}{x + 1}$

25. $\dfrac{1}{x + y} + \dfrac{x + y}{x^3 + y^3}$

26. $\dfrac{6}{5(x - 1)(x - 2)^2} + \dfrac{x}{3(x - 1)^2(x - 2)}$

27. $\dfrac{1}{4x(x + 1)(x + 2)^3} - \dfrac{6x + 2}{4(x + 1)^3}$

28. $\dfrac{x + y}{(x^2 - xy)(x - y)^2} - \dfrac{2}{(x^2 - y^2)^2}$

In Exercises 29–42, express in lowest terms.

29. $\dfrac{3}{4} \cdot \dfrac{12}{5} \cdot \dfrac{10}{9}$

30. $\dfrac{10}{45} \cdot \dfrac{6}{14} \cdot \dfrac{1}{2}$

31. $\dfrac{3a^2 c}{4ac} \cdot \dfrac{8ac^3}{9a^2 c^4}$

32. $\dfrac{6x^2 y}{2x} \cdot \dfrac{y}{21xy}$

33. $\dfrac{7x}{11y} \cdot \dfrac{66y^2}{14x^3}$

34. $\dfrac{ab}{c^2} \cdot \dfrac{cd}{a^2 b} \cdot \dfrac{ad}{bc^2}$

35. $\dfrac{3x + 9}{2x} \cdot \dfrac{8x^2}{x^2 - 9}$

36. $\dfrac{4x + 16}{3x + 15} \cdot \dfrac{2x + 10}{x + 4}$

37. $\dfrac{5y - 25}{3} \cdot \dfrac{y^2}{y^2 - 25}$

38. $\dfrac{6x - 12}{6x} \cdot \dfrac{8x^2}{x - 2}$

39. $\dfrac{u}{u - 1} \cdot \dfrac{u^2 - 1}{u^2}$

40. $\dfrac{t^2 - t - 6}{t^2 - 6t + 9} \cdot \dfrac{t^2 + 4t - 5}{t^2 - 25}$

41. $\dfrac{2u^2 + uv - v^2}{4u^2 - 4uv + v^2} \cdot \dfrac{8u^2 + 6uv - 9v^2}{4u^2 - 9v^2}$

42. $\dfrac{2x^2 - 3xy - 2y^2}{6x^2 - 5xy - 4y^2} \cdot \dfrac{6x^2 + 6xy}{x^2 - xy - 2y^2}$

In Exercises 43–60, compute the quotient and express in lowest terms.

43. $\dfrac{5}{12} \div \dfrac{4}{14}$

44. $\dfrac{\frac{100}{52}}{\frac{27}{26}}$

45. $\dfrac{uv}{v^2 w} \div \dfrac{uv}{u^2 v}$

46. $\dfrac{3x^2 y}{(xy)^2} \div \dfrac{3xyz}{x^2 y}$

47. $\dfrac{\frac{x + 3}{x + 4}}{\frac{2x}{x + 4}}$

48. $\dfrac{\frac{(x + 2)^2}{(x - 2)^2}}{\frac{x^2 + 2x}{x^2 - 4}}$

49. $\dfrac{x + y}{x + 2y} \div \left(\dfrac{x + y}{xy} \right)^2$

50. $\dfrac{\frac{u^3 + v^3}{u^2 - v^2}}{\frac{u^2 - uv + v^2}{u + v}}$

51. $\dfrac{\frac{(c + d)^2}{c^2 - d^2}}{cd}$

52. $\dfrac{\frac{1}{x} - \frac{3}{2}}{\frac{2}{x - 2} + \frac{5}{x}}$

53. $\dfrac{\frac{1}{x^2} - \frac{1}{y^2}}{\frac{1}{x} + \frac{1}{y}}$

54. $\dfrac{\frac{x}{x + 1} + \frac{1}{x}}{\frac{1}{x} + \frac{1}{x + 1}}$

55. $\dfrac{\frac{6}{y} - 3}{1 - \frac{1}{y - 1}}$

56. $\dfrac{\frac{1}{3x} - \frac{1}{4y}}{\frac{5}{6x^2} + \frac{1}{y}}$

57. $\dfrac{\frac{1}{2x + 2h} - \frac{1}{2x}}{h}$

58. $\dfrac{\frac{1}{(x + h)^2} - \frac{1}{x^2}}{h}$

59. $(x^{-1} + y^{-1})^{-1}$

60. $\dfrac{(x + y)^{-1}}{x^{-1} + y^{-1}}$

In Exercises 61–64, a dartboard is shown. Assuming that every dart actually hits the board, find the probability that the dart will land in the shaded area. This probability is the quotient.

$$\dfrac{\text{Area of shaded region}}{\text{Area of dartboard}}.$$

61.

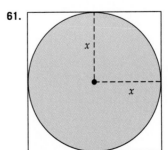

62.

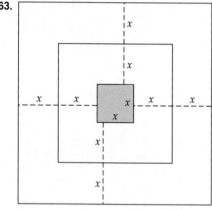

63.

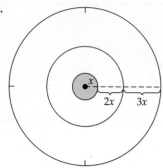

64.

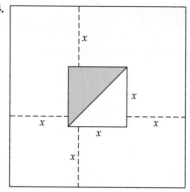

ERRORS TO AVOID

In Exercises 65–71, find a numerical example to show that the given statement is false. *Then find the mistake in the statement and correct it.*

65. $\dfrac{1}{a} + \dfrac{1}{b} = \dfrac{1}{a+b}$

66. $\dfrac{x^2}{x^2 + x^6} = 1 + x^3$

67. $\left(\dfrac{1}{\sqrt{a} + \sqrt{b}}\right)^2 = \dfrac{1}{a+b}$

68. $\dfrac{r+s}{r+t} = 1 + \dfrac{s}{t}$

69. $\dfrac{u}{v} + \dfrac{v}{u} = 1$

70. $\dfrac{\dfrac{1}{x}}{\dfrac{1}{y}} = \dfrac{1}{xy}$

71. $(\sqrt{x} + \sqrt{y})\, \dfrac{1}{\sqrt{x} + \sqrt{y}} = x + y$

An **angle** consists of two half-lines that begin at the same point P, as in Figure 1. The point P is called the **vertex** of the angle, and the half-lines are called the **sides** of the angle.

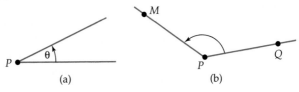

Figure 1

An angle may be labeled by a Greek letter, such as angle θ in Figure 1(a), or by listing three points (a point on one side, the vertex, a point on the other side), such as angle QPM in Figure 1(b).

To measure the size of an angle, we must assign a number to each angle. Here is the classical method for doing this.

1. Construct a circle whose center is the vertex of the angle.

2. Divide the circumference of the circle into 360 equal parts (called **degrees**) by marking 360 points on the circumference, beginning with the point where one side of the angle intersects the circle. Label these points $0°, 1°, 2°, 3°$, and so on.

3. The label of the point where the second side of the angle intersects the circle is the degree measure of the angle.

For example, Figure 2 shows an angle θ of measure 25 degrees (in symbols, $25°$) and an angle β of measure $135°$.

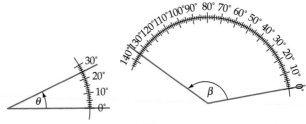

Figure 2

An **acute angle** is an angle whose measure is strictly between $0°$ and $90°$, such as angle θ in Figure 2. A **right angle** is an angle that measures $90°$. An **obtuse angle** is an angle whose measure is strictly between $90°$ and $180°$, such as angle β in Figure 2.

TRIANGLES

A **triangle** has three sides (straight line segments) and three angles, formed at the points where the various sides meet. When angles are measured in degrees,

the sum of the measures of all three angles of a triangle is *always* 180°.

For instance, see Figure 3. This fact is proved in Exercise 37.

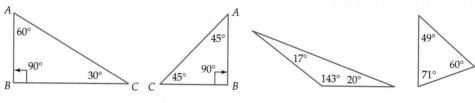

Figure 3

A **right triangle** is a triangle, one of whose angles is a right angle, such as the first two triangles shown in Figure 3. The side of a right triangle that lies opposite the right angle is called the **hypotenuse.** In each of the right triangles in Figure 3, side AC is the hypotenuse.

Pythagorean Theorem

If the sides of a right triangle have lengths a and b and the hypotenuse has length c, then

$$c^2 = a^2 + b^2.$$

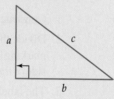

See Exercise 38 for a proof of the Pythagorean Theorem.

EXAMPLE 1

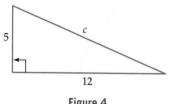

Figure 4

Consider the right triangle with sides of lengths 5 and 12, as shown in Figure 4.
According to the Pythagorean Theorem, the length c of the hypotenuse satisfies the equation $c^2 = 5^2 + 12^2 = 25 + 144 = 169$. Since $169 = 13^2$, we see that c must be 13. ∎

Isosceles Triangle Theorem

If two angles of a triangle are equal, then the two sides opposite these angles have the same length.

EXAMPLE 2

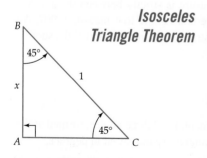

Figure 5

Suppose the hypotenuse of the right triangle shown in Figure 5 has length 1 and that angles B and C measure 45° each.

Answers

Chapter 1

Section 1.1, page 13

1.

3. Positive **5.** Positive **7.** Negative

9. $\dfrac{2040}{523}, \dfrac{189}{37}, \sqrt{27}, \dfrac{4587}{691}, 6.735, \sqrt{47}$

11. $-4 > -8$ **13.** $\pi < 100$ **15.** $z \geq 4$

17. $d \leq 7$ **19.** $z \geq -17$ **21.** $5 > -3$

23. $>$ **25.** 3 **27.** 5

29. 0 **31.** $b + c = a$

33. a lies to the right of b. **35.** $a < b$

37.

39.

41.

43. $[5, 10]$ **45.** $(-3, 14)$ **47.** $[-9, \infty)$

49. 6.506×10^9 **51.** $5.91 \cdot 10^{12}$ **53.** $2 \cdot 10^{-9}$ m

55. 150,000,000,000 m

57. .00000000000000000016726 kg

59. (a) $8.365 \times 10^{12}; 2.984 \times 10^8$ (b) \$28,032.84

61. 4 **63.** 6/5 **65.** 2

67. u^2 **69.** 11 **71.** 0

73. 169 **75.** $\pi - \sqrt{2}$ **77.** π

79. $<$ **81.** $>$ **83.** $<$

85. 7 **87.** 14.5 **89.** $\pi - 3$

91. $\sqrt{3} - \sqrt{2}$ **93.** $\dfrac{4 \cdot \pi \cdot \sqrt{805}}{161} \approx 2.21$

95. $|x - 143| \leq 21; |x - 163| \leq 26$

97. 15 **99.** 30 **101.** t^2

103. $b - 3$ **105.** $-(c - d) = d - c$

107. 0

109. $|(c - d)^2| = (c - d)^2 = c^2 - 2cd + d^2$

111. $|x - 5| < 4$ **113.** $|x + 4| \leq 17$

115. $|c| < |b|$

117. The distance from x to 3 is less than 2 units.

119. The distance from x to -7 is at most 3 units.

121. (a) iii (b) i (c) ii (d) v (e) iv

123. $x = 1$ or -1 **125.** $x = 1$ or 3

127. $x = -\pi + 4$ or $-\pi - 4$ **129.** $-7 < x < 7$

131. $3 < x < 7$ **133.** $x \leq -5$ or $x \geq 1$

135. Since $|a| \geq 0, |b| \geq 0$, and $|c| \geq 0$, the sum $|a| + |b| + |c|$ is positive only when one or more of $|a|, |b|, |c|$ is positive. But $|a|$ is positive only when $a \neq 0$; similarly for b, c.

Special Topics 1.1.A, page 17

1. $.7777 \cdots$ **3.** $.8181 \cdots$

5. $3.142857142857 \cdots$ **7.** $\dfrac{37}{99}$

9. $\dfrac{758,679}{9900} = \dfrac{252,893}{3300}$ **11.** $\dfrac{5}{37}$

13. $\dfrac{517,896}{9900} = \dfrac{14,386}{275}$ **15.** No

17. Yes **19.** No **21.** Yes

23. If $d = .74999 \cdots$, then $1000d - 100d = (749.999 \cdots) - (74.999 \cdots) = 675$. Hence $900d = 675$ so that $d = \dfrac{675}{900} = \dfrac{3}{4}$. Also $.75000 \cdots = .75 = \dfrac{75}{100} = \dfrac{3}{4}$.

25. .0588235294117647058823

27. $\dfrac{1}{29} = .0344827586206896551724137931034 \cdots$

29. $\dfrac{283}{47} =$

$6.0212765957446808510638297872340425531914893617021\overline{2} \cdots$

31. All of these numbers are *approximations* of $1/17$, but *not* equal to $1/17$. If you subtract $1/17$ from any one of them, the answer is not 0.

Section 1.2, page 26

1. $x = 8$ **3.** $x = \dfrac{-5}{6}$ **5.** $y = -32$

7. $y = \dfrac{x + 5}{3}$ **9.** $b = \dfrac{2A - hc}{h}$ $(h \neq 0)$

11. $h = \dfrac{4V}{\pi d^2}$ $(d \neq 0)$ **13.** $x = 3$ or 5

15. $x = -2$ or 7

17. $y = \dfrac{1}{2}$ or -3

19. $t = -2$ or $\dfrac{-1}{4}$

21. $u = 2$ or $\dfrac{-2}{3}$

23. $x = 1 \pm \sqrt{13}$

25. $x = \dfrac{1 + \sqrt{5}}{2}$ or $\dfrac{1 - \sqrt{5}}{2}$

27. 2   **29.** 2   **31.** 1

33. $x = 2 \pm \sqrt{3}$

35. $x = -3 \pm \sqrt{2}$

37. No real solutions

39. $x = \dfrac{-1}{2} \pm \sqrt{2}$

41. $x = \dfrac{-1 \pm \sqrt{2}}{2}$

43. $x = -3$ or -6

45. $x = \dfrac{-1 \pm \sqrt{2}}{2}$

47. $x = \dfrac{-3}{2}$ or 5

49. No real solutions

51. No real solutions

53. $x \approx 1.824$ or 0.470

55. $x = 13.79$

57. $y = \pm 1$ or $\pm\sqrt{6}$

59. $x = \pm\sqrt{7}$

61. $y = \pm 2$ or $\dfrac{\pm 1}{\sqrt{2}}$

63. $x = \pm\dfrac{\sqrt{5}}{5}$

65. No real solution

67. 1

69. $x = \dfrac{2}{5}$

71. $x = \dfrac{-5 \pm \sqrt{57}}{8}$

73. $x = \dfrac{3}{4}$ or -2

75. 2011

77. 2006

79. 1500

81. About 6.32 sec

83. (a) About 4.38 sec   (b) 50 sec

85. (a) $670.5 \dfrac{\text{lb}}{\text{ft}^2}$ (b) 14,400 ft

87. 1960

89. (a) 2003   (b) 2007   (c) mid-2010

91. (a) About 32 and 62   (b) About 17 and 77

93. (a) 73.53%   (b) 84.75%   (c) 91.74%

95. $k = 10$ or -10 **97.** $k = 16$

99. Yes   **101.** No   **103.** $k = 4$

105. (a) $x = \dfrac{-5 \pm \sqrt{17}}{2}$
(b) Answers vary
(c) $x \approx -.3068$ or $.6959$ or -1.1183 or 1.3959

Special Topics 1.2.A, page 33

1. $x = 3$ or -6 **3.** $x = \dfrac{3}{2}$ **5.** $x = 2$

7. $x = \dfrac{3}{2}$ **9.** $x = -5, -3, -1, 1$

11. $x = \dfrac{5 \pm \sqrt{33}}{2}, 1, 4$

13. Upper control limit: 0.0497;
Lower control limit: -0.0097

Special Topics 1.2.B, page 37

1. A varies directly as the square of r; the constant of variation is π.

3. A varies jointly as l and w; the constant of variation is 1.

5. V varies jointly as the square of r and h; the constant of variation is $\dfrac{\pi}{3}$.

7. $a = \dfrac{k}{b}$ **9.** $z = kxyw$ **11.** $d = k\sqrt{h}$

13. $v = ku$; $k = 4$ **15.** $v = k/u$; $k = 16$

17. $t = krs$; $k = 4$ **19.** $w = kxy^2$; $k = 2$

21. $T = kpv^3/u^2$; $k = 16$ **23.** $r = 4$

25. $b = \dfrac{9}{4}$ **27.** $u = 50$

29. $r = 3$ **31.** \$1521.00   **33.** 12 lb

35. 80 kg/cm^2 **37.** 400 ft   **39.** 30 in.

41. 3750 kg

43. (a) 2400 lb   (b) No more than 4 ft

Section 1.3, page 48

1. $A(-3, 3)$; $B(-1.5, 3)$; $C(-2.5, 0)$; $D(-1.5, -3)$; $E(0, 2)$; $F(0, 0)$; $G(2, 0)$; $H(3, 1)$; $I(3, -1)$

3. $(-6, 3)$ **5.** $(4, 2)$

7.

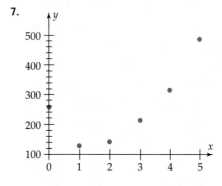

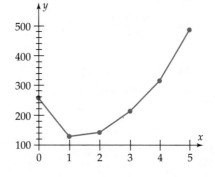

9. (a) Quadrant IV   (b) Quadrants III or IV

11. (a)

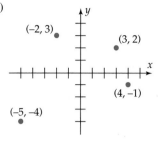

(b)

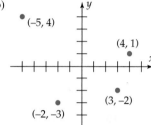

(c) They are mirror images of each other, with the *x*-axis being the mirror. In other words, they lie on the same vertical line, on opposite sides of the *x*-axis, the same distance from the axis.

13. $13; \left(\dfrac{-1}{2}, -1\right)$ **15.** $\sqrt{10}; \left(\dfrac{-3}{2}, \dfrac{7}{2}\right)$

17. $\sqrt{6 - 2\sqrt{6}} \approx 1.05; \dfrac{\sqrt{2} + \sqrt{3}}{2}, \dfrac{3}{2}$

19. $\sqrt{2} \cdot |a - b|; \left(\dfrac{a + b}{2}, \dfrac{a + b}{2}\right)$

21. $(-3.5, 4.6)$ **23.** $13 + 2\sqrt{2} + \sqrt{13}$

25. 15 square units

27. Hypotenuse from $(1, 1)$ to $(2, -2)$ has length $\sqrt{10}$; other sides have lengths $\sqrt{2}$ and $\sqrt{8}$. Since $(\sqrt{2})^2 + (\sqrt{8})^2 = (\sqrt{10})^2$, this is a right triangle.

29. The distances are $(1, 4)$ to $(5, 2)$: $\sqrt{20}$; $(5, 2)$ to $(3, -2)$: $\sqrt{20}$; $(3, -2)$ to $(1, 4)$: $\sqrt{40}$. But $(\sqrt{20})^2 + (\sqrt{20})^2 = 40 = (\sqrt{40})^2$, so the triangle is a right triangle with hypotenuse $\sqrt{40}$.

31. (a) About 45 yd

(b) $\left(34\dfrac{1}{6}, 27\dfrac{1}{2}\right)$

33. (a)

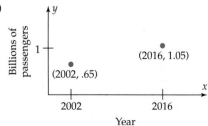

(b) $(2009, .85)$

(c) If linear growth is assumed, the midpoint suggests that there will be 850 million passengers in 2009.

35. No   **37.** Yes   **39.** No

41. $x = 1, 5; y = -5$

43. $x = 5, -1; y = \sqrt{5}, -\sqrt{5}$

45. $x = 0, -10; y = 0, 8$

47. (a) Approximately 7.3 million.
(b) Approximately 2007; Approximately 7.8 million.
(c) 2000–2003, 2011–.

49. (a) Under 56;   (b) Age 56;
(c) Retiring at age 60: about \$30,000; retiring at age 65: About \$35,000.

51. B   **53.** C

55. $(x + 3)^2 + (y - 4)^2 = 4$

57. $x^2 + y^2 = 3$

59. $(x - 5)^2 + (y + 2)^2 = 5$

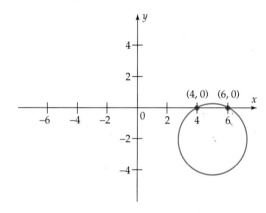

61. $(x + 1)^2 + (y - 3)^2 = 9$

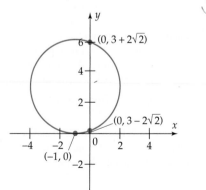

63. Center: $(-4, 3)$; radius: $2\sqrt{10}$

65. Center: $(-3, 2)$; radius: $2\sqrt{7}$

67. Center: $\left(-\dfrac{25}{2}, -5\right)$; radius: $\dfrac{\sqrt{677}}{2}$

69. (a) On   (b) Outside   (c) Inside
(d) Outside   (e) On

71. $(x - 3)^2 + (y - 3)^2 = 18$

73. $(x - 1)^2 + (y - 2)^2 = 8$

75. $(x + 5)^2 + (y - 4)^2 = 16$

77. $x^2 + y^2 - 4x - 2y = 0$

79. $(-3, -4)$ and $(2, 1)$

81. Assume $k > d$. The other two vertices of one possible square are $(c + k - d, d), (c + k - d, k)$; those of another square are $(c - (k - d), d), (c - (k - d), k)$; those of a third square are $\left(c + \dfrac{k - d}{2}, \dfrac{k + d}{2}\right), \left(c - \dfrac{k - d}{2}, \dfrac{k + d}{2}\right)$.

83. $(0, 0), (6, 0)$

85. $(3, -5 + \sqrt{11}), (3, -5 - \sqrt{11})$

87. $x = 6$

89. M has coordinates $(s/2, r/2)$ by the midpoint formula. Hence the distance from M to $(0, 0)$ is
$$\sqrt{\left(\frac{s}{2} - 0\right)^2 + \left(\frac{r}{2} - 0\right)^2} = \sqrt{\frac{s^2}{4} + \frac{r^2}{4}},$$
and the distance from M to $(0, r)$ is the same:
$$\sqrt{\left(\frac{s}{2} - 0\right)^2 + \left(\frac{r}{2} - r\right)^2} = \sqrt{\left(\frac{s}{2}\right)^2 + \left(-\frac{r}{2}\right)^2}$$
$$= \sqrt{\frac{s^2}{4} + \frac{r^2}{4}}$$
as is the distance from M to $(s, 0)$:
$$\sqrt{\left(\frac{s}{2} - s\right)^2 + \left(\frac{r}{2} - 0\right)^2} = \sqrt{\left(-\frac{s}{2}\right)^2 + \left(\frac{r}{s}\right)^2}$$
$$= \sqrt{\frac{s^2}{4} + \frac{r^2}{4}}.$$

91. Place one vertex of the rectangle at the origin, with one side on the positive x-axis and another on the positive y-axis. Let $(a, 0)$ be the coordinates of the vertex on the x-axis and $(0, b)$ the coordinates of the vertex on the y-axis. Then the fourth vertex has coordinates (a, b) (draw a picture!). One diagonal has endpoints $(0, b)$ and $(a, 0)$, so that its length is $\sqrt{(0 - a)^2 + (b - 0)^2} = \sqrt{a^2 + b^2}$. The other diagonal has endpoints $(0, 0)$ and (a, b) and hence has the same length: $\sqrt{(0 - a)^2 + (0 - b)^2} = \sqrt{a^2 + b^2}$.

93. The circle $(x - k)^2 + y^2 = k^2$ has center $(k, 0)$ and radius $|k|$ (the distance from $(k, 0)$ to $(0, 0)$). So the family consists of every circle that is tangent to the y-axis *and* has center on the x-axis.

95. The points are on opposite sides of the origin because one first coordinate is positive and one is negative. They are equidistant from the origin because the midpoint on the line segment joining them is
$$\left(\frac{c + (-c)}{2}, \frac{d + (-d)}{2}\right) = (0, 0).$$

Section 1.4, page 64

1. (a) C (b) B (c) B (d) D

3. $\dfrac{5}{2}$ **5.** 4   **7.** $t = 22$ **9.** $t = \dfrac{12}{5}$

11.

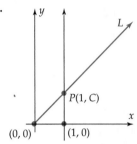

Slope of $L = \dfrac{C - 0}{1 - 0} = C$

13. (d)   **15.** (a)

17. $y = 4x + 5$ **19.** $y = -2.3x + 1.5$

21. $y = -\dfrac{2}{3}x + 2$ **23.** $y = \dfrac{3}{4}x - 3$

25. Slope 2; y-intercept 5

27. Slope $\dfrac{-3}{7}$; y-intercept $\dfrac{-11}{7}$

29. $y = x + 3$ **31.** $y = -x + 8$

33. $y = -x - 5$ **35.** $y = \dfrac{-12}{5}x + \dfrac{87}{25}$

37.

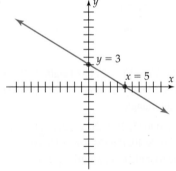

39.

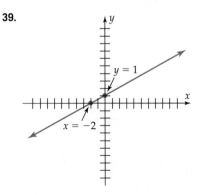

41.

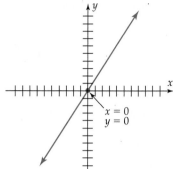

$x = 0$
$y = 0$

43. Perpendicular   **45.** Parallel

47. Parallel   **49.** Perpendicular

51. Yes   **53.** $y = -\dfrac{1}{2}x + 6$

55. $y = \dfrac{-1}{5}x + \dfrac{13}{5}$ **57.** $y = 3x + 7$

59. $y = \dfrac{3}{2}x$ **61.** $y = x - 5$

63. $y = -x + 2$ **65.** $k = \dfrac{-11}{3}$

67. $y = \dfrac{-3}{4}x + \dfrac{25}{4}$ **69.** $y = \dfrac{-x}{2} + 6$

71. Both have slope $\dfrac{-A}{B}$.

73. (a) $(0, 60), (5, 66)$ (b) $y = 1.2x + 60$
(c) $64,800,000$ (d) 2010

75. (a) $y = .03x$ (b) About 158.33 ft

77. (a) $y = \dfrac{-1}{550}x + 212$

(b) $211°$ (c) $209.6°$ (d) $206.3°$ (e) $199.5°$

79. (a) $y = .23x + 3.2$ (b) $7,800,000$ (c) 2020

81. (a) $y = \dfrac{2.75x + 26,000}{x}$ (b) $104,000$

83. (a) $r = 1.4x$ (b) $24,167$ items

85. (a) $y = 60 - 2x$ (b) $y = 80 - 2x$ (c) $y = 160 - 4x$

87. (a) $F = \dfrac{9}{5}C + 32$ (b) $C = \dfrac{5}{9}(F - 32)$ (c) $-40°$

89. (a) 12.5 gpm; 8.33 gpm; 25 gpm.   (b) $y = -10x + 75$

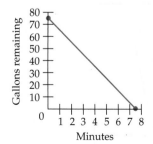

91. (a) $y = -1360x + 15,350$ (b) $\$1360$ per year
(c) $\$7190$

93. Let $y = mx + b$ and $y = mx + c$ be equations of lines with same slope m, and $b \neq c$. Suppose (x_1, y_1) is an arbitrary point lying on both lines. Then, $y_1 = mx_1 + b$ and $y_1 = mx_1 + c$. So, $mx_1 + b = mx_1 + c$ and, $b = c$, a contradiction. Thus, the lines share no point in common so must be parallel.

95.

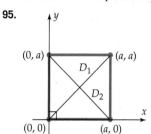

Equation of D_1; $y = x$, with slope 1
Equation of D_2; $y = -x + a$, with slope -1
(slope D_1)(slope D_2) = $(1)(-1) = -1$

Chapter 1 Review, page 70

1. (a) $>$ (b) $<$ (c) $<$ (d) $>$ (e) $=$

3. (a) $-10 < y < 0$
(b) $0 \leq x \leq 10$

5. (a) $(-8, \infty)$
(b) $(-\infty, 5]$

7. (a) 1.232×10^{16}
(b) 7.89×10^{-11}

9. (a) $|x + 7| < 3$
(b) $|y| > |x - 3|$

11. $x = 2$ or 8 **13.** $x = \dfrac{-1}{2}$ or $\dfrac{-11}{2}$

15. $x < -4$ or $x > 0$

17. (a) $7 - \pi$
(b) $\sqrt{23} - \sqrt{3}$

19. $28/99$ **21.** $x = 44/7$

23. No real solutions   **25.** $z = \dfrac{-3 \pm 2\sqrt{11}}{5}$

27. 2.25 times as large   **29.** 2

31. $x = 3$ or -1 **33.** 6

35. 1

37. $x = 3$ or -3 or $\sqrt{2}$ or $-\sqrt{2}$

39. $x = -1$ or $5/3$

41. $\sqrt{58}$ **43.** $\sqrt{c^2 + d^2}$

45. $\left(d, \dfrac{c + 2d}{2}\right)$

47. (a) $\sqrt{17}$ (b) $(x - 2)^2 + (y + 3)^2 = 17$

49.

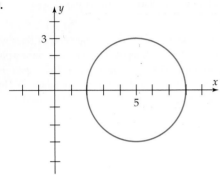

51. b and d   **53.** c

55.

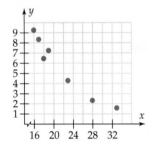

57. (a) 1 (b) 4/5   **59.** $y = 3x - 7$

61. $y = -2x + 1$ **63.** $x - 5y = -29$

65. 25,000 ft   **67.** False   **69.** False   **71.** False

73. False   **75.** (d)   **77.** (e)   **79.** (c)

81. (a) $y = .155x + 74.7$ (b) About 74.5 yrs
(c) 2019

83. (c)   **85.** (d)

Chapter 1 Test, page 73

1. (a)

-3   0   2

          (b) $(-\infty, -2]$

2. 49

3. (a) 740.76 pounds per square ft   (b) 14,112 ft

4. (a) 5.622×10^{12} (b) 2.811×10^8
(c) $20,000

5. $t = -1/2$ **6.** $x = -7$ or $1/2$

7. 14/99

8. $b = \dfrac{2E - ch}{h}$ **9.** $x = \dfrac{3 \pm \sqrt{29}}{4}$

10. (a) All numbers that are less than 6 units from 16 on the number line.
(b) $|c - 16| > 5$

11. $k = 8$ or -8

12. (a) $(y - 1)^2$ (b) $x^2 - 1$

13. $x = -7/4$ or $17/4$ **14.** $(x + 1)^2 + (y + 6)^2 = 37$

15. Neither. Reason: one line has slope -4 and the other has slope -2.

16. (a) (1/2, 1)   (b) 5

17. $\dfrac{2}{\sqrt{5} - 5}$

18. y-intercept is -2; x-intercepts are -1 and 2/3.

19. (a) $425/year   (b) $17,609   (c) $21,859

20. (a) (1, 1)   (b) $y = -\dfrac{1}{2}x + 4$

21. Center $(-1, 4)$; radius 3

22.

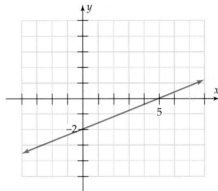

23. $12 + 2\sqrt{8}$

24. (a) $y = -7.82x + 412.1$
(b) About 247.88 per 100,000
(c) About 185.32 per 100,000
(d) No; it predicts a negative death rate in 2033.

25. $r = 9/4$ **26.** Neither

Chapter 2

Section 2.1, page 89

1.

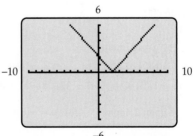

3.

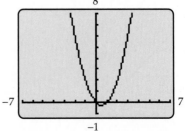

5.

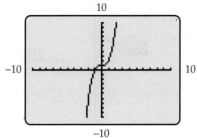

7.

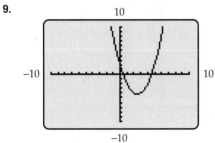

9.

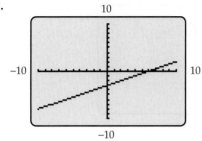

11.

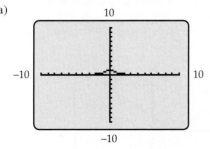

13. (a)

(b) Yes. For *x* larger than about 2.2 in absolute value, the value of the function is less than the height of one pixel.

(c)

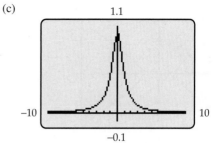

There are many possible square windows for Exercises 15–24; these were made on a TI-84+. To have a square window with the same y-axis shown here on wider screen calculators (such as TI-86/89), the x-axis should be longer than the one shown here. Because of the calculator's limited resolution, some graphs that should be connected (such as circles) may appear to have breaks in them.

15.

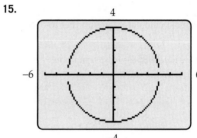

17.

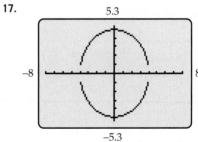

19.

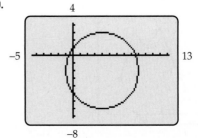

21.

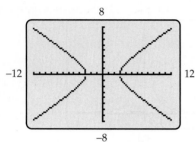

23.

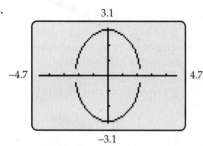

25. Answers vary, but most calculators are correct to 6–8 decimal places.

27. Highest at about (.7612, 3.9358); lowest at about (4.0812, −3.3356)

29. in late 2003 ($x \approx 3.9$); 5.17%

31. (c)   **33.** (d)

35. (e)

37. $-5 \leq x \leq 5$ and $-100 \leq y \leq 100$

39. $-10 \leq x \leq 10$ and $-2 \leq y \leq 20$

41. $-6 \leq x \leq 12$ and $-100 \leq y \leq 250$

43.

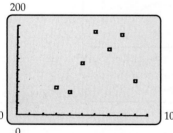

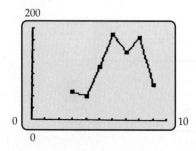

45.

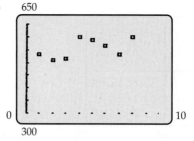

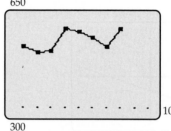

47. (a)

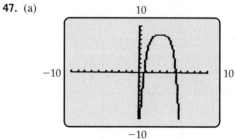

(b)

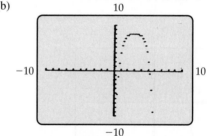

(c) Because the *vertical* distance between points on the graph is very small at the top of the graph, the graphed pixels make a solid line on the screen. When the vertical distance between adjacent pixels is larger, more "space" shows and the individual points appear isolated.

49. No

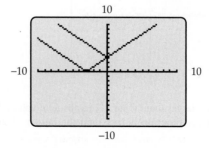

51. Yes

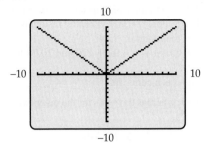

53. No

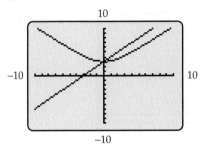

55. (a)

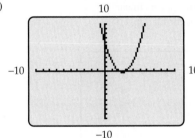

(b)

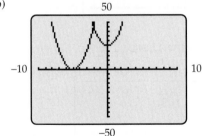

57. Possibly true

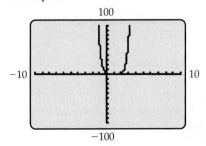

59. (c)

61. An appropriate viewing window should be large enough to contain the points $(0, 4000)$, $(20, 104000)$, and $(25.080869, 0)$, and should not show negative values.

63. x should range from 0 to 48, and y should range from 0 to at least 31.5.

65. (a) 1917; 1929   (b) 1925; 104,000

67. (a) After about 6.3 hours   (b) After about 18.7 hours

69. All four graphs have the same shape. Graph (b) is graph (a) shifted 5 units vertically upward, graph (c) is graph (a) shifted 5 units vertically downward, and graph (d) is graph (a) shifted 2 units vertically downward.

71. Graph (b) is graph (a) stretched by a factor of 2 vertically away from the x-axis; similarly for graph (c), except the stretching factor is 3. Graph (d) is graph (a) shrunk vertically toward the x-axis by a factor of $\frac{1}{2}$.

73. The graphs are "mirror images" of each other, with the straight line $y = x$ being the mirror.

75. Same answer as Exercise 73

Section 2.2, page 100

1. 3   **3.** 3   **5.** 2

7. $x = -2.42645$ **9.** -1.4526 **11.** -1.4751

13. 1.1921235   **15.** -1.379414 **17.** 1.3289

19. -2.1149 **21.** $x = 2.1017$ **23.** $x = -1.7521$

25. $x = .9505$ **27.** $x = 0$ or 2.2074

29. $x = 2.3901454$

31. $x = -.6513878188$ or 1.151387819

33. $x = 7.033393042$ **35.** $x = 2/3$

37. $x = 1/12$ **39.** $x = \sqrt{3}$

41. $x = 1.4528$ or -112.00 **43.** 3.00242

45. 1.7388   **47.** 2009

49. (a) late 2003
(b) mid 2010

51. about 96 ft   **53.** late 2012

Section 2.3, page 111

1. (a) Let x represent the student's score on the fourth exam.
(b) x, $88 + 62 + 79 + x$, $(88 + 62 + 79 + x)/4$

3.

English Language	Mathematical Language
Length of rectangle	x
Width of rectangle	y
Perimeter is 45	$x + y + x + y = 45$ or $2x + 2y = 45$
Area is 112.5	$xy = 112.5$

5. $x + .08x = 2619$

7. The circle has radius $r = 16/2 = 8$, so its area is $\pi r^2 = \pi \cdot 8^2 = 64\pi$. Let x be the amount by which the radius is to be reduced. Then $r = 8 - x$ and the new area is $\pi(8 - x)^2$, which must be 48π less than the original area, that is, $\pi(8 - x)^2 = 64\pi - 48\pi$, or equivalently, $\pi(8 - x)^2 = 16\pi$.

9. $366.67 at 12% and $733.33 at 6%

11. $2\frac{2}{3}$ qt   **13.** 60 mph   **15.** 44, 38.25

17. 65 mph   **19.** About 1.753 ft   **21.** 2 meters

23. About 132.7 ft

25. Red Riding Hood, 54 mph; wolf, 48 mph

27. 2.234 in. × 2.234 in.   **29.** 4.658 in.   **31.** 8.02 in.

33. 11.47 ft or 29.91 ft   **35.** 6.205 miles

37. 157.6 mph   **39.** 2.2 × 4.4 × 4 ft

Section 2.4, page 118

1. $(-1, 8)$ **3.** $(1.142855, -2.0625)$

5. $(-.3409, .0003222)$

7. (a) $(-1, 4)$ (b) $(-1, 4)$ and $(2, 4)$ (c) $(3, 20)$

9. 52.86 mph   **11.** 21 ft

13. 450 ft × 900 ft   **15.** $8800

17. 34.2 cm × 34.2 cm × 17.1 cm

19. (a) 4.4267 by 4.4267 in.   (b) 10/3 by 10/3 in.

21. (a) Approximately 206
 (b) Approximately 269; approximately $577

23. (a) 600   (b) 958

25. $x = 9.306$; area ≈ 220.18 ft²

27. Approximately $(1.871, 1.5)$ **29.** 12 times

Section 2.5, page 128

1.

MODEL: $y = x$			
Data Point	**Model Point**	**Residual**	**Squared Residual**
(1, 2)	(1, 1)	1	1
(2, 2)	(2, 2)	0	0
(3, 3)	(3, 3)	0	0
(4, 3)	(4, 4)	−1	1
(5, 5)	(5, 5)	0	0
		Sum: 0	Sum: 2

MODEL: $y = .5x + 1.5$			
Data Point	**Model Point**	**Residual**	**Squared Residual**
(1, 2)	(1, 2)	0	0
(2, 2)	(2, 2.5)	−0.5	0.25
(3, 3)	(3, 3)	0	0
(4, 3)	(4, 3.5)	−0.5	0.25
(5, 5)	(5, 4)	1	1
		Sum: 0	Sum: 1.5

The model $y = .5x + 1.5$ is better.

3. (a) Data points are (0, 171.3), (2, 179.8), (4, 188), (6, 201.5). For $y = 4.9x + 170$, the residuals are 1.3, 0, −1.6, 2.1; their sum is 1.8. For $y = 5x + 171$, the residuals are .3, −1.2, −3, .5; their sum is −3.4.
 (b) The sum is 8.66 for $y = 4.9x + 170$ and 10.78 for $y = 5x + 171$.
 (c) The first model is a better fit.

5. The sum is 1.9055, whereas the sums for the other two models are 2.8 and 2.75.

7. Negative   **9.** Negative

11.

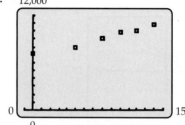

(a) The data appears linear.
(b) Positive correlation

13.

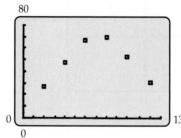

(a) The data is not linear.

15.

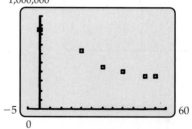

(a) The data appears linear.
(b) Negative correlation

17. (a) $y = 14.9x + 2822$
 (b) 5057, 6994, 9080; the estimates are quite close to the actual values.
 (c) 6323.5; 6500

19. (a) (6, 1.8), (7, 2.3), (8, 2.5), (9, 3.1), (10, 3.9), (11, 3.8), (12, 4), (13, 4.4), (14, 4.8), (15, 5.1)
 (b) $y = .3594x - .2036$
 (c) About $6.98 billion

21. (a) $y = 9.9429x + 600.8095$ and
$\quad y = 20.2286x + 900.4286$
(b) Earnings are increasing at the approximate rate of $9.63, $16.29, $9.94, and $20.23 per week respectively.

23. (a) $y = .2179x + 52.2805$
(b) $y = .1447x + 65.4246$
(c) Yes; in 2079

25. (a) $y = .03119x + .5635$
(b) About .938 billion; about 1.06 billion

Chapter 2 Review, page 132

1. (a) a, d
(b) b, c do not show peaks and valleys near the origin; e crowds the graph onto the y-axis.
(c) a or $-4 \le x \le 6$ and $-10 \le y \le 10$

3. (a) None of them
(b) a, b, d do not show any peaks or valleys; c does not show the valleys; e shows only one point on the graph (which can't be distinguished because it's on the y-axis).
(c) $-7 \le x \le 11$ and $-1000 \le y \le 500$

5. (a) b, c
(b) a shows no peaks or valleys; d is too crowded horizontally; e shows only one point.
(c) $-10 \le x \le 10$ and $-150 \le y \le 150$

7. As x moves left or right from the origin, x^2 grows larger and larger, and hence so does $x^2 - 10$. Therefore, the graph always rises as it moves away from the y-axis and is complete.

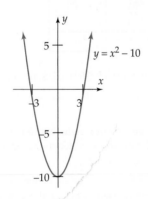

9. Note that y is defined only when $x \ge 5$ (why?). Also, $x - 5$ grows larger as x gets larger (that is, as you move to the right) and hence the same is true of $\sqrt{x - 5}$. Therefore, the graph always rises as you move to the right and is complete.

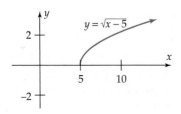

11.

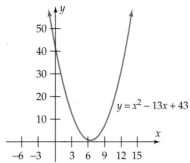

13.

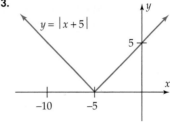

15. $x = 2.7644$ **17.** $x = 3.2678$

19. $x = -3.2843$ **21.** $x = 1.6511$

23. Gold, $\frac{3}{11}$ oz; silver, $\frac{8}{11}$ oz

25. $2\frac{2}{9}$ hrs   **27.** 9.6 ft

29. 4 ft   **31.** 25

33. 20 yd by 30 yd (interior fence is 20 yd)

35. $x = \sqrt{3} \approx 1.732$

37. (a)

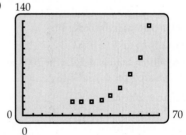

(b) Nonlinear

39. (a) $y = 1.5868x + .2473$
(b) 89 and 108 years respectively   (c) About 2.5 ft

41. (a) $y = 14.8998x + 59.0163$
(b) $193.1 billion in 1999; $297.4 billion in 2008
(c) in 2011

Chapter 2 Test, page 135

1. The standard viewing window clearly shows the parts of the graph near the origin. To see the rest of the graph a larger window is needed, such as $-4 \le x \le 14$ and $-212 \le y \le 10$.

2. $x = -17$ **3.** (8.1510, 95.8884)

4. (a) 5   (b) any k such that $k \le -4.43$ or $k \ge 4.43$

5. $-12 \le x \le 12$ and $-8 \le y \le 8$ on a standard screen (such as TI-84+); on a wide-screen calculator, $-6 \le y \le 6$ produces an approximately square window.

6. $x = -2$, but $x = 2$ is *not* a solution (why?).

7. (a) $0 \le x \le 48$ and $0 \le y \le 30$
(b) $t = 7.937$ hours; 26.457 mg per liter

8. $x \approx -1.3865$ or $-.4258$ **9.** $(35 - x)(40 - x) = 785$

10. 8.347 in by 7.4 in   **11.** 1.14 qt

12. (a) $(1, 1.1)$ (b) $(-2, 1.1)$

13. 27 cm by 28 cm by 29 cm

14. About 1942   **15.** 2.3 ft by 4.6 ft by 4 ft high

16. (a) 2.937 in by 2.937 in   (b) 2 in by 2 in

17. (a)

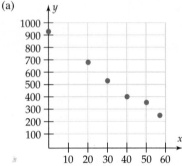

(b) Yes; negative   (c) $y = -11.823x + 912.576$

18. (a) Residuals: .4, $-.5$, .3, .8; sum of residuals; 1; sum of squares of residuals: 1.14
(b) Residuals: $-.5$, -1.8, -1.2, $-.9$; sum of residuals; -4.4; sum of squares of residuals: 5.74
(c) The model in part (a) is the better fit.

19. (a) $y = .3272x + 5.0259$
(b) About 7.6 million barrels per day in 2008 and about 10.9 million barrels per day in 2018

20. (a) $y = 19.2286x + 206.0952$
(b) About 340,700 vehicles
(c) About 19,200 cars per year

Chapter 3

Section 3.1, page 148

1. This is a function because, for every input there is a unique output.

3. This could not be a table of values of a function, because two output values are associated with the input -5.

5. 6   **7.** -2 **9.** -17

11. This defines y as a function of x.

13. This defines x as a function of y.

15. This defines both y as a function of x and x as a function of y.

17. Neither

19.

X	Y1
-2	-2
-1.5	-3.25
-1	-4
-.5	-4.25
0	-4
.5	-3.25
1	-2

X=-2

X	Y1
1.5	-.25
2	2
2.5	4.75
3	8
3.5	11.75
4	16
4.5	20.75

X=4.5

21.

X	Y2
-8	59
-6	31
-4	11
-2	1
0	5
2	1
4	11

X=-8

X	Y2
6	31
8	59
10	95
12	139
14	191
16	251
18	319

X=12

23. $8.40, $31.69, $693.75, $521.25, $262.50, $2150.17

25. A function may assign the same output to many different input.

27. Postage is a function of weight, but weight is not a function of postage; for example, all letters less than one ounce use the same postage amount.

29. This could not be the rule of a function, since, for example, it would assign two numbers (2 and -2) to the input 4.

31. (a) Jan. 2000, 8½%. Jan. 2001, 9½%. Mid-2005, 6%.
(b) After Jan. 2002 until Dec. 2004.
(c) The prime rate is a function of time, but time is not a function of the prime rate.

33. (a) $A = x^2$ (b) $A = \dfrac{d^2}{2}$

35. $S = 10\pi r^2$ **37.** $C = 125x + 26,000$

39. (a) $y = 2.3107x + 20.0821$ (b) $40.9 million

41. Output for -2 is -2. The output for -1 is 0. The output for 0 is approximately 1¼ and the output for 1 is approximately 2¾.

43. Output for -2 is -1. The output for 0 is 3. The output for 1 is 2.

The output for 2.5 is -1.

The output for -1.5 is 0.

45. Output for -2 is 1. The output for -1 is -2.9. The output for 0 is -1. The output for ½ is 1. The output for 1 is 1½.

47. (a) $x > 0$
(b) $-1 \le y \le 1$ where $y = \cos(\ln x)$

49. (a) For a nonnegative number, the part of the number to the left of the decimal point (the integer part) is the closest integer to the left of the number on the number line.
(b) the negative integers.
(c) negative numbers that are not integers.

51. (a)

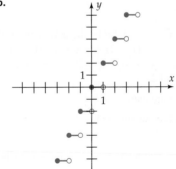

```
   X     Y1
  -5    -112
 -11   -1306
  8     499
 7.2   361.85
-.44   3.7948

Y1◼X^3-2X+3
```

(b) $10^3 - 2(10) + 3 = 983$

Section 3.2, page 158

1. (a) $-.8$ (b) $-.75$ (c) $-.4$ (d) $-.125$ (e) 0

3. $\sqrt{3} + 1$

5. $\sqrt{\sqrt{2} + 3} - \sqrt{2} + 1$

7. 4

9. 17.75

11. $\pi^2 + 2\pi + 3 + \dfrac{1}{\pi + 1}$

13. $(a + k)^2 + \dfrac{1}{a + k} + 2$

15. $x^2 - \dfrac{1}{x} + 2$

17. $x^2 - 6x + 11 + \dfrac{1}{x - 3}$

19. $s^2 + 2s$

21. $t^2 - 1$

23. $\sqrt{11} - 7$

25. 1

27. 3

29. $1 - 2x - h$

31. $\dfrac{1}{\sqrt{x + h} + \sqrt{x}}$

33. $2x + h$

35. (a) $f(a) = a^2, f(b) = b^2, f(a + b) = (a + b)^2 = a^2 + 2ab + b^2$, so $(a + b)^2 \ne a^2 + b^2$, that is, $f(a + b) \ne f(a) + f(b)$
(b) $f(a) = 3a, f(b) = 3b, f(a + b) = 3(a + b) = 3a + 3b = f(a) + f(b)$.
(c) $f(a) = 5, f(b) = 5, f(a + b) = 5, f(a + b) \ne f(a) + f(b)$

37. $c = -2$

39. (a) $-3 \le x \le 4$ (b) $-2 \le y \le 3$ (c) -2
(d) $1/2$ (e) 1   (f) -1

41. $|x|$

43. (a) $x \le 20$ (b) 3   (c) -1 (d) 1   (e) 2

45. all real numbers   **47.** all real numbers

49. $x \ge 0$ **51.** all real numbers except 0

53. all real numbers

55. all real numbers except 3 and -2

57. $6 \le x \le 12$

59. Many examples including $f(x) = x^2$ and $g(x) = x^4$.

61. Many examples including $g(x) = x^3$.

63. $3/2, -4$

65. $\dfrac{-3 \pm \sqrt{21}}{2}$

67. $f(x) = 8 - 3x^2$

69. $P(x) = .7x - 1800$

71. (a) $38 + .72y$
(b) $2 + .72x$

73. $d = \begin{cases} 55t & 0 \le t \le 2 \\ 20 + 45t & t > 2 \end{cases}$

75. (a) $y = 108 - 4x$
(b) $V = x^2(108 - 4x) = 108x^2 - 4x^3$

77. $C = 2t^2 + \dfrac{20}{t}$

79. (a) $f(x) = 18.3x + 244.4$
(b) $g(8) = 391, g(11) = 446$. These are not that accurate, but in the right ballpark
(c) $g(10) = 427$
(d) $g(21) = 629$

Section 3.3, page 171

1. Yes, $f(3) = 0$ **3.** No

5.

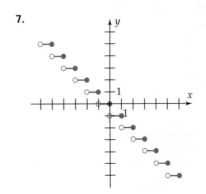

7.

9.

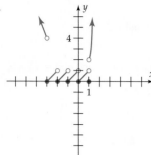

11. The graph fails the vertical line test; hence, it cannot be the graph of a function. For example, $x = 3$ corresponds to both 5 and 9.

13. $[-3, 5]$ **15.** $[-3, 3)$ **17.** 3.5

19. 4.5   **21.** 1 and 5

23. (a) $h(x) = \begin{cases} \dfrac{x}{2} - 2 & x \geq 0 \\ -\dfrac{x}{2} - 2 & x < 0 \end{cases}$

(b)

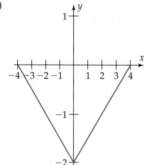

25. If $0 \leq x \leq 2$, then $x - 2 \leq 0$. So by the definition of absolute value, $|x - 2| = -(x - 2) = -x + 2$. Also, since $x \geq 0$, $|x| = x$. Hence, $f(x) = |x| + |x - 2| = x - x + 2 = 2$ for all x between 0 and 2.

27. Maxima at $(\pm 4, 0)$, minimum at $(0, -4)$

29. Maximum at $(-1, 3)$, minimum at $(1, -1)$

31. None

33. Decreasing when $x < -5.8$ and $x > .46$, increasing when $-5.8 < x < .46$

35. Only decreasing in $(-\infty, 0) \cup (0, \infty)$

37. (a) $2x + 2z = 100$
(b) $A(x) = x(50 - x)$
(c)

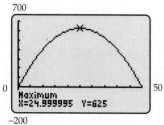

$x = 25$ inches, $z = 25$ inches

39. (a) $S = 2x^2 + 4xh$
(b) $x^2 h = 867$
(c) $S = 2x^2 + \dfrac{3468}{x}$
(d)

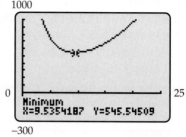

$x = 9.5354$ inches, $h = 9.5354$ inches

41. (a) (iv)   (b) (i)   (c) (v)   (d) (iii)   (e) (ii)

43.

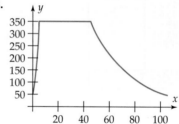

Domain is $x \geq 0$, range is $50 \leq y \leq 350$.

45.

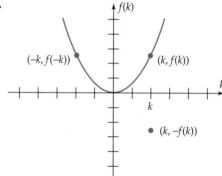

47. (a) 9%, 16%, 8%
(b) Lowest: 2003. Highest: 1990
(c) 1980–1983; steepness of curve

49. (a) False
(b) False
(c) False

51.

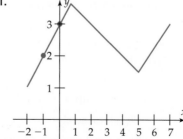

53. (a)

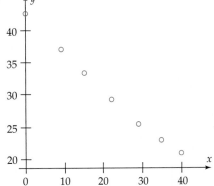

(b) $y = -.545x + 41.8$ (c) 27.7%, 15.7%   (d) 2014
(e) It will disappear completely in 2042 according to this model.

55.

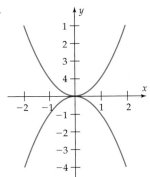

57. (a) $(-\infty, 4]$ (b) $[2, \infty)$
(c)

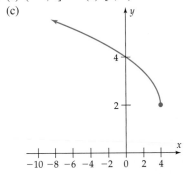

59. After 15 minutes she took a break then picked up her pace for 10 minutes. After 30 minutes she jogged back home at a constant rate for a total jog of 55 minutes.

Section 3.3.A, page 178

1. $-6 \le x \le 44, 0 \le y \le 16$

3. $-2 \le x \le 32, -60 \le y \le 60$

5. $-16 \le x \le 2, -60 \le y \le 60$

7. $-50 \le x \le 50, -7 \le y \le 3, -7 \le t \le 3$ (answers will vary)

9. $-8 \le x \le 5, -5 \le y \le 5, -5 \le t \le 5$ (answers will vary)

11. $-10 \le x \le -7, 0 \le y \le 2, 0 \le t \le 2$ (answers will vary)

13.

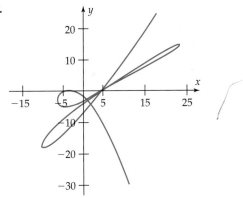

It crosses itself 6 times

Section 3.4, page 186

1. H   **3.** F   **5.** K   **7.** C

9.

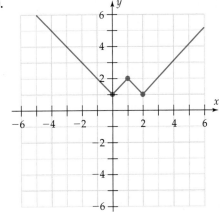

11.

t	$f(t)$	$g(t) = f(t) - 3$	$h(t) = 4f(-t)$	$i(t) = f(t - 1) - 2$
-2	3	0	20	Not possible
-1	6	3	0	1
0	8	5	32	4
1	0	-3	24	6
2	5	2	12	-2

13.

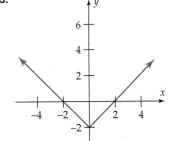

15.

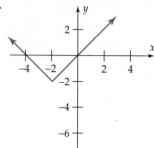

17. $-15 \leq x \leq 15$ and $-12 \leq y \leq 10$

19. $-5 \leq x \leq 7$ and $-10 \leq y \leq 10$

21.

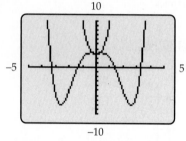

23. Shift 2 units to the left then 5 units up.

25. Reflect across the x-axis, then stretch vertically by a factor of 2, and then shift 10 units up.

27. $g(x) = -x^2 + x + 2$ **29.** $f(x) = -\frac{1}{2}\sqrt{-(x+3)}$

31. (a) $g(x) = x^2 - 2x + 6$
    (b) difference quotient for f: $2x + h$; difference quotient for
      g: $2x + h - 2$
    (c) $2x + h - 2 = 2(x - 1) + h = d(x - 1)$

33.

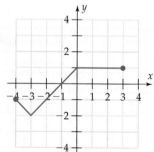

35.

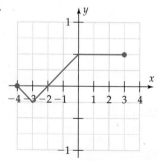

37.

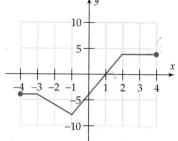

39.

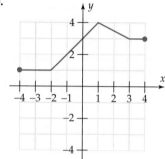

41.

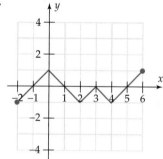

43.

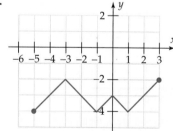

45.

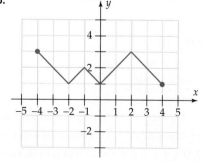

47. (a) The graph shifts 35 units up.
    (b) The graph stretches in the y-direction.

49.

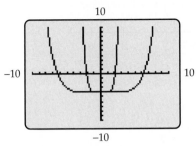

51. The graph of $f(cx)$, with $c > 1$, is the graph of $f(x)$ contracted toward the y-axis by a factor of c.

53.

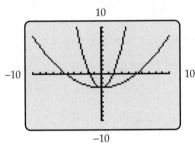

55.

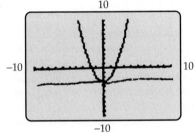

57.

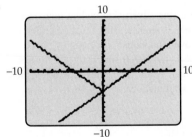

59.

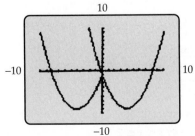

61. For $x > 0$, $f(x)$ and $g(x)$ are the same. For $x < 0$, $g(x) = f(-x)$.

63.

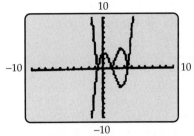

65.

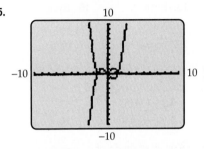

67. (a)

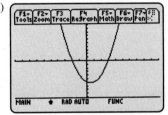

    $h(x)$ will look just like $f(x)$, shifted 1000 units to the right
    (b) $990 < x < 1010$, $-10 < y < 10$
    (c) The problem is difficult because we have to include x values that are 1000 units apart, which makes seeing any details of the graphs impossible.
    (d) $g(x)$ should look like $f(x)$, only stretched by a factor of 1000
    (e) $-10 < x < 10$, $-10000 < y < 10000$. While we can display both f and g, any window that shows g in detail will not show f in detail.

Section 3.4.A, page 193

1.

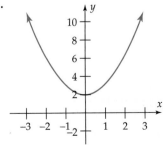

Graph has y-axis symmetry

3.

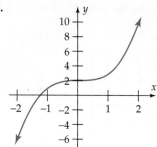

5. Odd   **7.** Even   **9.** Even

11. Odd   **13.** Even   **15.** Even

17. Yes   **19.** No   **21.** Origin

23. Origin   **25.** y-axis   **27.** None

29.

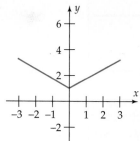

31.

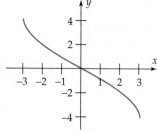

33. (a)

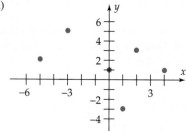

(b)

35. (a)

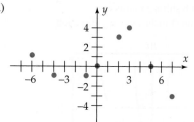

(b)

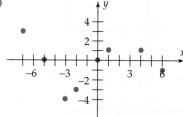

(c) Many correct answers, including this one.

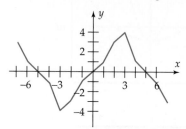

37. If replacing x by $-x$ does not change the graph and replacing y by $-y$ does not change the graph, then doing both does not change the graph. Thus, x-axis plus y-axis symmetry guarantees origin symmetry. If replacing x by $-x$ does not change the graph and replacing y by $-y$ *does* change the graph, then doing both must change the graph. Thus, y-axis symmetry and origin symmetry without x-axis symmetry is impossible. The argument is analogous for x-axis and origin symmetry.

39. (a) No because as the advertising budget increases, so do the sales which is an increasing function which is not a property of even functions.
  (b) Yes, because as the advertising budget increases, so do the sales which is an increasing function which is a property of odd functions.

Section 3.5, page 202

1. $x^3 - 3x + 2$; $-x^3 - 3x + 2$; $x^3 + 3x - 2$

3. $x^2 + 2x - 5 + \dfrac{1}{x}$; $\dfrac{1}{x} - x^2 - 2x + 5$; $x^2 + 2x - 5 - \dfrac{1}{x}$

5. $-3x^4 + 2x^3$; $\dfrac{-3x + 2}{x^3}$; $\dfrac{x^3}{-3x + 2}$

7. $x^2 - 25$, $\dfrac{x + 5}{x - 5}$, $\dfrac{x - 5}{x + 5}$

9. All real numbers except 0; all real numbers except 0

11. $-4/3 \le x \le 2$; $-4/3 < x \le 2$

13. 0

15. 30

17. 49; 1; −8

19. −3; −3; 0

21. $-3x^3 + 2$, all real numbers; $(-3x + 2)^3$, all real numbers

23. $\dfrac{1}{2x^2 - 1}$, all real numbers except $\dfrac{\pm\sqrt{2}}{2}$; $\dfrac{-4x^2 - 4x}{4x^2 + 4x + 1}$, all real numbers except $-1/2$

25. x^6; x^9

27. $\dfrac{1}{x^2}$; x

29. $(f \circ g)(x) = f(g(x)) = 9g(x) + 8 = 9\left(\dfrac{x - 8}{9}\right) + 8 =$
$x - 8 + 8 = x$; $(g \circ f)(x) = g(f(x)) = \dfrac{f(x) - 8}{9} =$
$\dfrac{9x + 8 - 8}{9} = \dfrac{9x}{9} = x$

31. $(f \circ g)(x) = f(g(x)) = \sqrt[3]{g(x)} + 2 = \sqrt[3]{(x - 2)^3} + 2 =$
$x - 2 + 2 = x$; $(g \circ f)(x) = g(f(x)) = (f(x) - 2)^3 =$
$(\sqrt[3]{x} + 2 - 2)^3 = (\sqrt[3]{x})^3 = x$

33.

x	$f(x)$	$g(x) = f(f(x))$
−4	−3	−1
−3	−1	1/2
−2	0	1
−1	1/2	5/4
0	1	3/2
1	3/2	2
2	1	3/2
3	−2	0
4	−2	0

35.

x	$(g \circ f)(x)$
1	4
2	2
3	5
4	4
5	4

37.

x	$(f \circ f)(x)$
1	1
2	3
3	3
4	5
5	1

There may be correct answers to Exercises 39–44, other than those given here.

39. Let $g(x) = x^2 + 2$ and $h(t) = \sqrt[3]{t}$. Then $(h \circ g)(x) = h(g(x)) = \sqrt[3]{g(x)} = \sqrt[3]{x^2 + 2}$.

41. Let $g(x) = 7x^3 - 10x + 17$ and $f(t) = t^7$. Then $(f \circ g)(x) = f(g(x)) = (7x^3 - 10x + 17)^7$.

43. Let $h(x) = x^2 + 2x$
$k(t) = t + 1$

45. Domain of $f \circ g : x \geq 0$ Domain of $g \circ f : x \geq 0$

47. Domain of $f \circ g : x \geq -2$ Domain of $g \circ f : x \geq -10$

49. (a) $2x^6 + 5x^2 - 1$
(b) $4x^6 + 20x^4 - 4x^3 + 25x^2 - 10x + 1$
(c) The answers are not the same. We conclude $f(x^2) \neq (f(x))^2$ in general.

51.

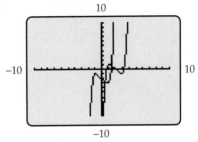

Functions are not the same.

53. $x^2 + 6x + 10$; $2x + h + 6$

55. $\dfrac{2}{x}$; $\dfrac{-2}{x(x + h)}$

57. (a) about 1.22×10^{-4} square inches, about 2.5×10^{-6} square inches, about 1.36×10^{-7} square inches
(b) no, no, over a certain time period

59. (a) $A = \dfrac{\pi}{4}d^2 = \dfrac{\pi}{4}\left(6 - \dfrac{50}{t^2 + 10}\right)^2$
(b) about .7854 square inches, about 22.265 square inches
(c) about 11.4 weeks

61. $V(t) = \dfrac{256\pi t^3}{3}$, about 17,157 cm^3

63. $s(t) = \dfrac{10t}{3}$

65. $f(x) = |x|$

67. (a) As the composition is applied you will stabilize at $-.1708$
(b) yes.
(c) Now we wind up oscillating between $-.8873$ and $-.1127$ (starting at either 0 or 1)
(d) We wind up cycling between four numbers this time.

Section 3.6, page 214

1. (a) 14 ft per sec
(b) 54 ft per sec
(c) 112 ft per sec
(d) 93.3 ft per sec

3. (a) 9954.5 million subscribers/year
(b) 19218.6 million subscribers/year

5. (a) Decreasing at 291,000 per year
(b) Increasing at 541,800 per year
(c) Increasing at 353,500 per year
(d) Increasing at 179,778 per year
(e) Fastest from 1985 to 1995, slowest from 2005 to 2014

7. (a) 29.18   (b) 51.8
(c) 24.08   (d) 33.5
(e) 1999 to 2001

9. (a) $5000 per page
(b) $1875 per page
(c) $625 per page
(d) $1750 per page
(e) No, no, no

11. 4   **13.** -7 **15.** .371

17. .417   **19.** 7   **21.** $2x + 3 + h$

23. $3x^2 + 3xh + h^2$

25. $-\dfrac{5}{p(p + h)}$

27. (a) 49.2
(b) 48.1
(c) 48
(d) 48

29. Decreasing at $-\dfrac{1}{500}$.

31. (a) They started at the same profit and ended at the same profit.
(b) Dec 2008

33. (a) 10
(b) -15
(c) 5

35. (a) $y = 13.58x + 184.95$
(b) 13.58
(c) 12.8, 16.8. They are similar.
(d) 2010

Section 3.7, page 226

1. No   **3.** Yes   **5.** Yes

7. Yes   **9.** $f^{-1}(x) = -x$ **11.** $f^{-1}(x) = \dfrac{x + 4}{5}$

13. $f^{-1}(x) = \sqrt[3]{\dfrac{-x + 5}{2}}$ **15.** $f^{-1}(x) = \dfrac{x^2 + 7}{4}$

17. $f^{-1}(x) = \dfrac{1}{x}$ **19.** $f^{-1}(x) = \dfrac{1 - x}{2x}$

21. $f^{-1}(x) = \sqrt[3]{\dfrac{-5x - 1}{x - 1}}$

23. $f(g(x)) = x$ and $g(f(x)) = x$

25. $f(g(x)) = x$ and $g(f(x)) = x$

27. $f(g(x)) = x$ and $g(f(x)) = x$

29. $f(f(x)) = x$

31. (a) 48
(b) 2
(c) 5
(d) 12
(e) Not enough information
(f) Not enough information
(g) Not enough information

33.

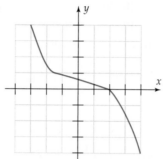

35.

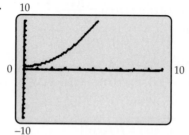

37.

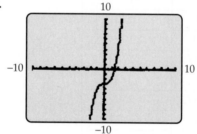

39.

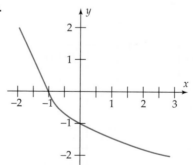

41. $x \geq 0$ **43.** $x \geq 0$

45. (a) $f^{-1}(x) = \dfrac{x - 2}{3}$
(b) $f^{-1}(1) = \dfrac{-1}{3}, \dfrac{1}{f(1)} = 1/5$, therefore not equal

47. $g(x) = \dfrac{x - b}{m}$

49. (a) -1
(b) 1
(c) since the slopes are negative reciprocal of each other, the lines are perpendicular to each other as well as symmetric with respect to the $y = x$ axis.

51. $f(g(x)) = x$ and $g(f(x)) = x$

53. True. Rotating the graph of an increasing function over $y = x$ gives an increasing function.

Chapter 3 Review, page 229

1. (a) -3 (b) 1755   (c) 2   (d) -14

3.

x	0	1	2	-4	t	k	$b - 1$	$1 - b$	$6 - 2u$
$f(x)$	7	5	3	15	$7 - 2t$	$7 - 2k$	$9 - 2b$	$5 + 2b$	$4u - 5$

5. Many possible answers, including these.

    (a) If $f(x) = x + 1$, $a = 1$, $b = 2$, then $f(a + b) = f(1 + 2) = f(3) = 4$, but $f(a) + f(b) = f(1) + f(2) = 2 + 3 = 5$, so the statement is false.

    (b) If $f(x) = x + 1$, $a = 1$, $b = 2$, then $f(ab) = f(1 \cdot 2) = f(2) = 3$, but $f(a) \cdot f(b) = f(1) \cdot f(2) = 2 \cdot 3 = 6$, so the statement is false.

7. $r \geq 4$

9. $t^2 + t - 2$

11. $\dfrac{x^3 + 2x + 4}{4}$

13. (a) $f(t) = 50 \sqrt{t}$
    (b) $g(t) = 2500\pi t$
    (c) radius is 150 meters, area is $22{,}500\pi$ square meters
    (d) after approximately 12.7 hours

15.

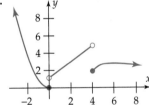

17. (b)

19. No local maximum, local minimum at $-.5$, increasing when $x > -.5$, decreasing when $x < -.5$

21. Local maximum at -5.0704, local minimum at $-.263$, increasing when $x < -5.0704$ and when $x > -.263$, decreasing when $-5.0704 < x < -.263$.

23.

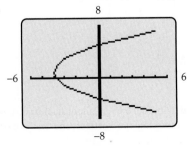

25. Here is one possibility:

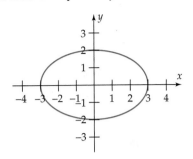

27. Graph has symmetry with respect to x-axis, y-axis, and the origin.

29. Even   **31.** Odd

33. Symmetric with respect to y-axis.

35.

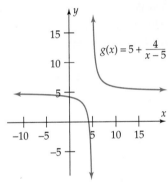

37. $-3 \leq y \leq 4$

39. Many correct answers, including: any number from 2 to 3.5 or from 4.5 to 6

41. 1   **43.** -3 **45.** True

47. 4   **49.** $x \leq 3$ **51.** $x < 3$

53. (a) King Richard
    (b) King Richard
    (c) Fireball Bob

55. Shrink vertically by a factor of .25, then shift 2 units up.

57. Shift 7 units right, then stretch vertically by a factor of 3, then reflect in the x-axis, and then shift 2 units up.

59. (e)

61. (a) $1/3$
    (b) $(x - 1) \sqrt{x^2 + 5}$ $(x \neq 1)$
    (c) $\dfrac{\sqrt{(c + 1)^2 + 5}}{c}$ $(c \neq 0)$

63.

x	-4	-3	-2	-1	0	1	2	3	4
$g(x)$	1	4	3	1	-1	-3	-2	-4	-3
$h(x) = g(g(x))$	-3	-3	-4	-3	1	4	3	1	4

65. $82/27$

67. $\dfrac{1}{x^3} + 3$

69. $1/4$

71. (a) $\dfrac{1}{x^2 - 1}$

    (b) $\dfrac{1}{x^2} - 1$

73. $x \geq 0, x \neq 1$

75. (a) $-1/3$ (b) $5/8$

77. 6   **79.** 3

81. $2x + h$

83. (a) \$290 per ton
    (b) \$230 per ton
    (c) \$212 per ton

85. (a) approximately 45 to 50
    (b) approximately 25 to 35
    (c) approximately 30 to 44

87. (a) \$5338; \$9900; \$6194 dollars per year
    (b) \$234,100

89. $g(x) = 5 - (x - 7)^2 = -x^2 + 14x - 44;\quad x \geq 7$

91.

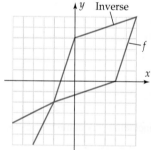

93. The graph of f passes the horizontal line test and hence has an inverse function. It is easy to verify either geometrically [by reflecting the graph of f in the line $y = x$] or algebraically [by calculating $f(f(x))$] that f is its own inverse function.

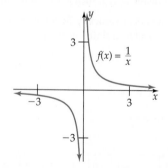

95. There is no inverse function because the graph of f fails the horizontal line test (use the viewing window with $-10 \leq x \leq 20$ and $-200 \leq y \leq 100$).

Chapter 3 Test, page 234

1. $A(r) = \pi r^2$

2.

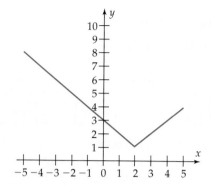

$f(x) - 3$

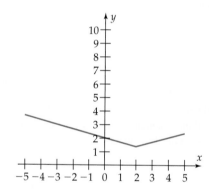

$\dfrac{1}{3} f(x)$

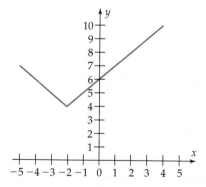

$f(-x)$

3. (a) Domain: $-12 \leq x \leq 12$ or $\mathbb{R}$
    (b) $-2 \leq y < 1, 0 \leq y \leq 6$
    (c) 6

4. It is a function of x, because every input has exactly one output. Answers invoking the vertical line test are also acceptable.

5. $\dfrac{f(x + h) - f(x)}{h} = \dfrac{\dfrac{2}{x + h} + 3 - \dfrac{2}{x} - 3}{h}$

$= \dfrac{\dfrac{2}{x + h} - \dfrac{2}{x}}{h}$

$= \dfrac{\dfrac{2x}{x(x + h)} - \dfrac{2(x + h)}{x(x + h)}}{h}$

$= \dfrac{2x - 2x - 2h}{x(x + h)h}$

$= \dfrac{-2h}{x(x + h)h}$

$= -\dfrac{2}{x(x + h)}$

6.

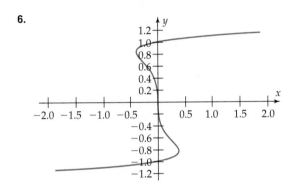

7. (a) $s(0) = 25$ If the child has read no books, the score will be 25.
$s^{-1}(35) = 30$ If the child wants a score of 35, s/he must read 30 books.
(b) 80 books

8. Determine whether the given graph is symmetric with respect to the y-axis, the x-axis, the origin, or none of these:

(a) The origin
(b) The x axis
(c) None of these

9. Describe a sequence of transformations that will transform the graph of the function f into the graph of the function g
f: Shift up by 2
g: Shift left by 5, and up by 9

10. The height in feet of a dropped ball after t seconds is given by $h(t) = -16t^2 + 500$.

(a) 97.6 feet/second
(b) 962 feet/second
(c) 96 feet/second

11. -0.63667 million shares/month

12.

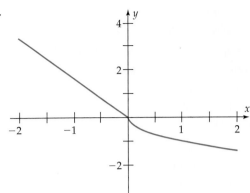

13. $f \circ g(x) = f(g(x))$
$= f(x^5 + 1)$
$= \sqrt[5]{(x^5 + 1) - 1}$
$= \sqrt[5]{x^5}$
$= x$

and
$g \circ f(x) = g(f(x))$
$= g\left(\sqrt[5]{x - 1}\right)$
$= \left(\sqrt[5]{x - 1}\right)^5 + 1$
$= x - 1 + 1$
$= x$

Chapter 4

Section 4.1, page 247

1. Graph **I**

3. Graph **K**

5. Graph **J**

7. Graph **F**

9. Vertex: $(5, 2)$. Since $a = 3$, opens upward.

11. Vertex: $\left(\sqrt{2}, \pi\right)$. Since $a = -1$, opens downward.

13. Vertex: $(4, -3)$. Since $a = 2$, opens upward.

15. Vertex: $\left(\dfrac{3}{2}, \dfrac{-5}{4}\right)$. Since $a = 1$, opens upward.

17. Vertex: $(1, 3)$. Since $a = -4$, opens downward.

19. Vertex: $(0, 3)$. Since $a = 2$, opens upward.

21. (a) $-6x - 3h + 1$
(b) Vertex: $\left(\dfrac{1}{6}, \dfrac{1}{12}\right)$
(c) $-3h$

23. (a) $-4x - 2h + 2$
(b) Vertex: $\left(\dfrac{1}{2}, -\dfrac{1}{2}\right)$
(c) $-2h$

25. Rule: $g(x) = 2x^2 - 3$. Vertex: $(0, -3)$.

27. $h(x) = 2(x - 5)^2 + 4$
Vertex: $(5, 4)$

29. The rule is $f(x) = 3x^2$.

31. The rule is $f(x) = 2(x - 3)^2 + 4$.

33. $f(x) = (x - 1)^2 + 4$ **35.** $f(x) = x^2 + 6$

37. $b = 0$ **39.** $b = -4, c = 8$

41. $x = \dfrac{111}{2}, y = \dfrac{111}{2}$

43. (a) $20,000
$100,000
(b) 200 units. Then
$p(200) = \$110,000$
(c) Approximately 365 units

45. The distance will be approximately 6,886 feet when the height is 842 feet. The shell hits the water 13,987 feet, or 2.65 miles, away.

47. (a) $B(30) = 30$ meters; $B(100) = 170$ meters
(b) 50 kilometers per hour

49. 196 ft at 2.5 sec   **51.** 6.9 feet

53. $b = 15, h = 15$

55. When $x = 50$ feet, then $y = 100$ feet.

57. Maximum area is 704.2 square feet.

59. (a) $y = \dfrac{1}{1152}x^2$
(b) Thus $x \approx 34$ inches

61. $4.00   **63.** $7.08

65. Minimum area: $A = \dfrac{10,000}{\pi} \approx 3183.1$

Section 4.2, page 257

1. This is a polynomial with leading coefficient 1, constant term 1, and degree 3.

3. This is a polynomial with leading coefficient 1, constant term -1, and degree 3.

5. This is a polynomial with leading coefficient 1, constant term -3, and degree 2.

7. This is not a polynomial.

9. This is not a polynomial.

11. Quotient: $3x^3 - 3x^2 + 11x - 17$; remainder: 18

13. Quotient: $x^2 + 2x - 6$; remainder: $-7x + 7$

15. Quotient: $x^2 + 2x + 3$; remainder: 0

17. Quotient: $5x^2 + 5x + 5$; remainder: 0

19. No   **21.** Yes

23. $x = \{2, -5\}$ **25.** $x = \{2\sqrt{2}, -1\}$

27. $x = \{-3, 0\}$ **29.** The remainder is 2.

31. The remainder is 31.4375.   **33.** The remainder is -36.

35. The remainder is 183,424.

37. The remainder is $5,935,832\pi$.

39. No   **41.** No

43. Yes   **45.** Yes

47. $(x + 4)(2x - 7)(3x - 5)$

49. $(x - 3)(x + 3)(2x + 1)^2$

51. $f(x) = x^5 - 3x^4 - 5x^3 + 15x^2 + 4x - 12$

53. $f(x) = x^5 - 5x^4 + 5x^3 + 5x^2 - 6x$

55. $f(x) = x^3 - 4x^2 - 25x + 28$

57. $f(x) = (x - 1)(x + 1)$

59. $f(x) = (x - 1)^2(x - 2)^2(x - \pi)^2$

61. $f(x) = .25(x - 8)(x - 5)x$

63. $k = -9$ **65.** $k = 1$

67. If $x - c$ were a factor, by the Factor Theorem, c would be a root. Then $c^4 + c^2 + 1 = 0$. This equation has discriminant given by $1^2 - 4(1)(1) = -3$, hence it has no real roots. Thus $x - c$ cannot be a factor for any real c.

69. (a) $(x + 2)$ is not a factor of $x^3 - 2^3$.
(b) $(-c)^n + c^n = -c^n + c^n = 0$.

71. $k = 5$

73. $-3, 2 - 4\sqrt{5}, 2 + 4\sqrt{5}$

Section 4.2.A, page 262

1. Quotient: $3x^3 - 2x^2 - 4x + 1$; remainder: 10

3. Quotient: $2x^3 + x^2 - 3x + 7$; remainder: -29

5. Quotient: $5x^3 + 35x^2 + 242x + 1690$; remainder: 11836

7. Quotient: $x^5 + x^4 + x^3 + x^2 + x + 1$; remainder: 0

9. Quotient: $3x^3 + \dfrac{3}{4}x^2 - \dfrac{29}{16}x - \dfrac{29}{64}$; remainder: $\dfrac{483}{256}$

11. Quotient: $x^2 + \sqrt{2}x - x - \sqrt{2}$; remainder: 0

13. Quotient: $x^2 - x - 30$; remainder: 0

15. $\left(x - \dfrac{1}{2}\right)(2x^4 - 6x^3 + 12x^2 - 10)$

17. Quotient: $x^2 - \dfrac{43}{20}x + 4$; remainder: 2.25

19. $c = -4$

21. (a) $x^4 + x^3 + 2x^2 + 6x + 21$; remainder: 64
(b) When dividing by $(x - a)$ whenever $a > 3$, the coefficients will be positive. When $a > 3$, $a - 2$ is positive and $a(a - 2)$ is greater than 1.
(c) For a root a to be greater than three, the remainder when dividing by $(x - a)$ would have to be zero. But in part b we've shown it has to be a positive number.

Section 4.3, page 268

1. $x = -1, 1, 3$ **3.** $x = -2, 1, 4$

5. $x = -\dfrac{2}{3}, \dfrac{-1}{2}, 3$ **7.** $x = -2, 0, \dfrac{1}{2}, 3$

9. $x = -3, 2$ **11.** $x = -1, \dfrac{3}{2}, 2$

13. $x = -5, 2, 3$

15. $x = 1$

17. $(x - 1)(2x^2 + 3)$

19. $x^3(x - 4)(x^2 + 3)$

21. $(x - 2)^3(x^2 - 7)$

23. Lower bound: -5; upper bound: 2

25. Lower bound: -1; no upper bound less than 3

27. Lower bound: -8; upper bound 4

29. $x = \left\{-2, \dfrac{-1}{2}, 3\right\}$

31. $x = \left\{-\dfrac{1}{3}, \dfrac{1}{2}, 2\right\}$

33. $x = \left\{2, \dfrac{\sqrt{37} - 5}{2}, \dfrac{-(\sqrt{37} + 5)}{2}\right\}$

35. $x = \left\{\dfrac{1}{2}, -\sqrt{3}, -\sqrt{2}, \sqrt{3}, -\sqrt{2}\right\}$

37. $x = \{-1, 4, \sqrt{3}, -\sqrt{3}\}$

39. $x = 50, x = -2.24698, x = -0.5549581,$
$x = 0.80193774$

41. (a) The only *possible* rational roots of $x^2 - 2$ are $1, -1,$ $2, -2$. None of those numbers are roots of the polynomial, so the square root of two is irrational.
(b) The only *possible* rational roots of $x^2 - 3$ are $1, -1,$ $3, -3$. None of those numbers are roots of the polynomial, so the square root of three is irrational.
(c) The only *possible* rational roots of $x^2 - 4$ are $1, -1,$ $2, -2, 4, -4$. We try them all and find that 2 and -2 both are roots of the polynomial.

43. (a) 5.78 per 100,000 and 5.62 per 100,000
(b) Middle of 1997
(c) 1995
(d) 2004
(e) 2002–2004

45. The sides should be 2 inches.

47. (a) 6 degrees per day, 6.6435 degrees per day
(b) $t = 2.033$ and $t = 10.7069$ (days)
(c) $t = 5.0768$ and $t = 9.6126$
(d) $t = 4$

49. (a) The carrying capacity is an upper bound on the population of bunnies, while the threshold population is a lower bound.
(b) The population is increasing
(c) It is decreasing
(d) It is decreasing
(e) $kx(x - T)(C - x)$
(f) $x = T, x = C, x = 0$

Section 4.4, page 278

1. Yes

3. No

5. No

7. This could be the graph of a polynomial function of degree at least 3 or 5.

9. Not the graph of a polynomial function.

11. This could be the graph of a polynomial function of degree at least 5.

13. In a very large window like $-50 \le x \le 50$, $-100,000 \le y \le 1,000,000$ the graph looks like the graph of $y = x^4$.

15. Roots $-2, 1, 3$; each has (odd) multiplicity 1.

17. Root -2 has multiplicity 1, root -1 has multiplicity 1, and root 2 has (even) multiplicity 2 (or possibly higher).

19. (e)

21. (f)

23. (c)

25. Since this is a polynomial function of degree 3, the complete graph must have another x-intercept.

27. Since this is a polynomial function of degree 4 with positive leading coefficient, both of the ends should point up; here they are pointing down.

29. $-5 \le x \le 5, -50 \le y \le 30$

31. $-6 \le x \le 6, -60 \le y \le 320$

33. $-3 \le x \le 4, -35 \le y \le 20$

35. $-33 \le x \le -2, -50,000 \le y \le 260,000$ then $-2 \le x \le 3,$ $-20 \le y \le 30$

37. (a) The graph of a cubic polynomial can have no more than two local extrema. If only one, the ends would go in the same direction. Hence it can have two or none.
(b) When the end behavior and the number of extrema are both accounted for, these four shapes are the only possible ones.

39. (a) The ends go in opposite directions, so the degree is odd.
(b) Since if x is large and positive, so is $f(x)$, the leading coefficient is positive.
(c) Root -2 has multiplicity 1, root 0 has multiplicity at least 2, root 4 has multiplicity 1, root 6 has multiplicity 1.
(d) Adding the multiplicities from **c**, the polynomial must have degree at least 5.

41.

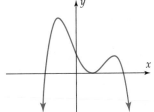

43. x intercepts: $1, 1 \pm \sqrt{3}$;
local max: $(2, 2)$;
local min: $(0, -2)$

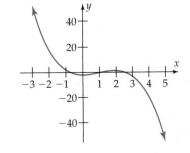

45. x intercepts: 2, 3;

local min: $\left(\dfrac{11}{4}, -\dfrac{27}{256}\right)$

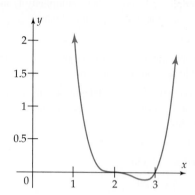

47. x intercepts: $\dfrac{\pm 1 \pm \sqrt{5}}{2}$, 0; local max: $(-1.30, 1.58)$

$(.345, .227)$; local min: $(-.345, -.227)(1.30, -1.58)$

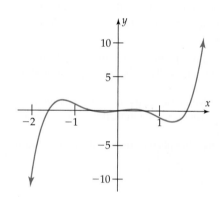

49. (a)

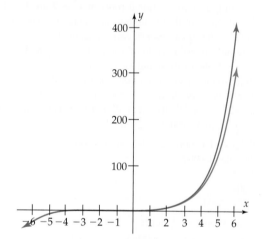

   (b) 4950, 3750. The rate of increase is decreasing.

   (c) After this point, you get less and less benefits for a unit of expenditure.

51. 6 additional trees per acre

53. $r = 3.046$, $h = 12.18$

55. It is a good approximation for $-3 < x < 3$.

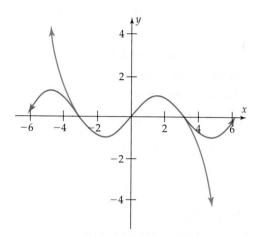

57. It is a good approximation for $-3 < x < 2.25$.

59. (a)

If the graph had the horizontal portion appearing in the window, then the equation $g(x) = 4$ would have infinitely many solutions, which is impossible, since a polynomial equation of degree 3 can have at most three solutions.

   (b) $1 \le x \le 3$ and $3.99 \le y \le 4.01$

   (c) If the graph of a polynomial of degree n had a horizontal portion appearing in the window, it would be a portion also of the line $y = k$. Then the equation $f(x) = k$ would have infinitely many solutions, which is impossible, since a polynomial equation of degree n can have at most n solutions.

61. (a)

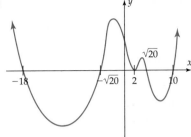

(b)

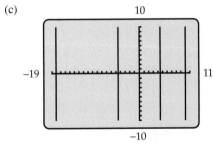

(c)

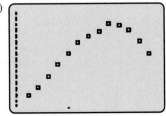

(d) Different windows would be needed for different portions of the graph. $-20 \le x \le -3$ and $-5 \times 10^6 \le y \le 10^6$; then $-3 \le x \le 2$ and $-5000 \le y \le 60,000$; then $1 \le x \le 5$ and $-5000 \le y \le 5000$; finally $5 \le x \le 11$ and $-100,000 \le y \le 100,000$.

Section 4.4.A, page 286

1. A cubic model seems best.

3. A quadratic model seems best.

5. (a) $f(x) = 1.239x^3 - 26.221x^2 + 19.339x + 4944.881$
(b) 4055.7, 3633.7
(c) Very accurate
(d) The model predicts a large surge in crime after 2005. The data do not support this conclusion. So the model should not be used to predict data after 2005.

7. (a)

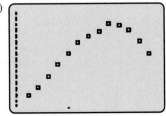

(b)

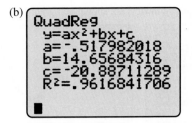

(c) Evaluating the function in **(a)** gives:
noon ($x = 12$) 80.4°, 9 A.M. ($x = 9$) 69.1°, 2 P.M. ($x = 14$) 82.8°.

9. (a)

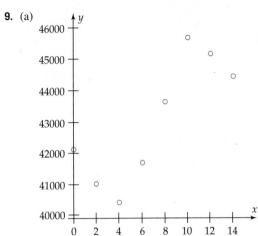

(b) cubic model $f(x) = -14.847x^3 + 321.847x^2 - 1444.238x + 42266.864$
(c) \$8,181
(d) The model predicts that the dropoff in median income that began in 2002 will continue, and get worse and worse. The reasonability of the model depends on your opinion of this prediction.

11. (a)

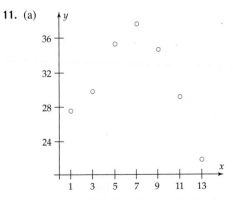

(b) $f(x) = -.340x^2 + 4.410x + 22.092$,
$f(x) = -.0156x^3 - .0115x^2 + 2.55x + 24.39$
$f(x) = .00746x^4 - .224x^3 + 1.896x^2 - 3.684x + 29.531$
(c) The quartic model fits the data best, but given the downward trend in smoking, the cubic model is probably best for the future.

13. (a) Cubic and quartic models are shown and graphed:

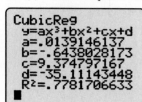

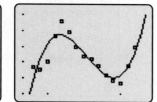

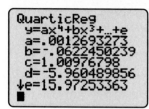

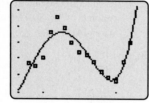

(b) Both models seem to rise disturbingly steeply and only a pessimist would expect them to be accurate very far into the future.

Section 4.5, page 300

1. $\left(-\infty, \dfrac{4}{3}\right) \cup \left(\dfrac{4}{3}, \infty\right)$ i.e. $x \neq \dfrac{4}{3}$ **3.** All real numbers

5. $(-\infty, -3) \cup (-3, -1) \cup (-1, 3) \cup (3, \infty)$

7. $f(x) = \dfrac{2x^2}{x^2 - 9}$

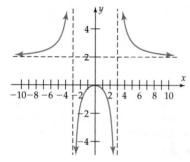

9. $f(x) = \dfrac{-x^4 + 36}{(x + 3)(x - 1)(x + 1)(x - 3)} = \dfrac{-x^4 + 36}{x^4 - 10x^2 + 9}$

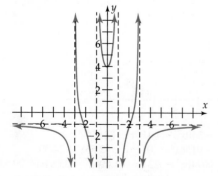

11. F   **13.** A

15. Vertical asymptotes at $x = 0$ and $x = 1$

17. Vertical asymptotes at $x = \dfrac{-7 \pm \sqrt{41}}{2}$

19. Horizontal asymptote: $y = 0$. All viewing windows will work.

21. Horizontal asymptote: none

23. Horizontal asymptote: $y = \dfrac{2}{3}$. $-25 < x < 25$, $-5 < y < 5$

25. Vertical asymptote: $x = 2$
Horizontal asymptote: $y = 0$
y-intercept: $-\dfrac{1}{2}$

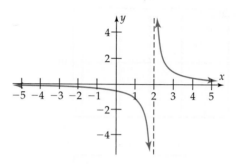

27. Vertical asymptote: $x = -1$
Horizontal asymptote: $y = 2$
y-intercept: 0

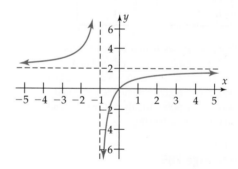

29. Vertical asymptotes: $x = 2$, $x = 3$
Horizontal asymptote: $y = 0$
Hole: $\left(0, \dfrac{1}{6}\right)$
y-intercept: none

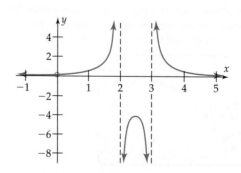

31. Horizontal asymptote: $y = 0$
y-intercept: 2

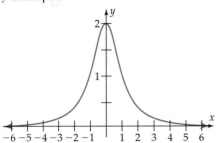

33. Vertical asymptotes: $x = -5, x = 1$
Horizontal asymptotes: $y = 0$
y-intercept: $-\dfrac{1}{5}$

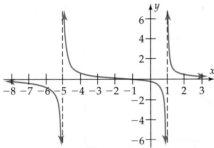

35. Vertical asymptotes: $x = 2, x = -1$
Horizontal asymptotes: $y = 2$
Holes: $\left(-2, \dfrac{9}{4}\right)$
y-intercept: $\dfrac{1}{2}$

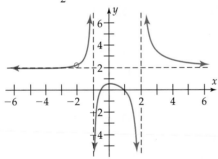

37. Three windows are needed: $-5 \le x \le 4.4$ and $-8 \le y \le 4$; then $-2 \le x \le 2$ and $-.5 \le y \le 5$; finally, $-15 \le x \le -3$ and $-.07 \le y \le .02$

39. $-9.4 \le x \le 9.4$ and $-4 \le y \le 4$

41. $-5 < x < 15, -2 < y < 2$

43. $\dfrac{-1}{x(x+h)}$

45. $-\dfrac{3}{(x-2)(x+h-2)}$

47. $-3\left(\dfrac{2x+h}{x^2(x+h)^2}\right)$

49. (a) $-1/4$ (b) $-1/9$ (c) They are identical.

51. (a) $g(x) = \dfrac{x-1}{x-2}$ (b) $g(x) = \dfrac{x-1}{-x-2}$

(c)

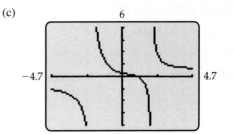

53. 8.4343 in. by 8.4343 in. by 14.057 in.

55. (a) 10
(b) 17
(c) Number of figures $= \dfrac{20 + 5t}{2 + .25t}$

(d)

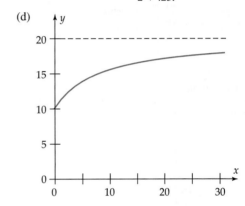

(e) He will be able to buy 18 in the year 2012. He will never be able to buy 21.

57. (a) $x + \dfrac{400}{x}$
(b) $30 - 10\sqrt{5} < x < 30 + 10\sqrt{5}$ or $(7.6 < x < 52.4)$
(c) $20, 20 \times 10$

59. (a) $h_1 = h - 2$
(b) $h_1 = 150/\pi r^2 - 2$
(c) $\pi(r-1)^2(150/\pi r^2 - 2)$
(d) Otherwise $(r - 1)$ would be negative.
(c) $r = 2.879, h = 5.759$

61. (a) 9.801 m/s^2
(b)

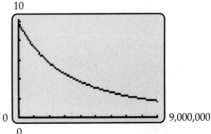

(c) No

Special Topics 4.5.A, page 307

1. $y = x$; $-14 \le x \le 14$ and $-14 \le y \le 14$

3. $y = 2x + 3$; $-15 < x < 17$ and $-30 < y < 30$

5. Vertical asymptote: $x = 2$
Nonvertical asymptote: $y = x + 1$,

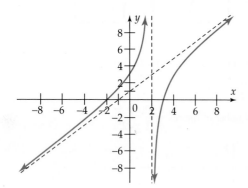

7. Vertical asymptote: $x = \dfrac{5}{2}$

Nonvertical asymptote: $y = 2x + 1$

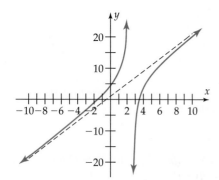

9.

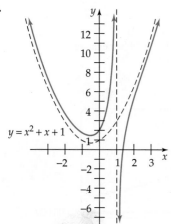

Vertical asymptote: $x = 1$
Parabolic asymptote: $y = x^2 + x + 1$

11. Vertical asymptotes: $1, 2, -1$
Nonvertical asymptote: $y = x^2 + 5x$

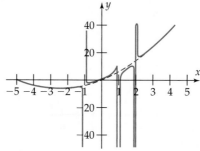

13. $-15.5 \le x \le 8.5$ and $-16 \le y \le 8$

15. $7 \le x \le 14$ and $-30 \le y \le 30$

17. Overall: $-13 \le x \le 7$ and $-20 \le y \le 20$; hidden area near the origin: $-2.5 \le x \le 1$ and $-.02 \le y \le .02$

19. (a)

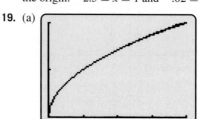

(b) Approx. $0.06 \le x \le 2.78$

21. $y = rx + b$

Section 4.6, page 315

1. $(-\infty, 3/2)$ **3.** $(-2, \infty)$

5. $(-\infty, -8/5]$ **7.** $(1, \infty)$

9. $(2, 4)$ **11.** $[-3, 5/2]$

13. $(-4/45, \infty)$ **15.** $[-7/17, \infty)$

17. $[-1, 1/8)$ **19.** $[5, \infty)$

21. $x < \dfrac{b + c}{a}$ **23.** $c < x < a + c$

25. $1 \le x \le 3$

27. $x \le \dfrac{1 - \sqrt{33}}{2}$ or $x \ge \dfrac{1 + \sqrt{33}}{2}$

29. $-1 \le x \le 0$ or $x \ge 1$

31. $-1 < x < 1$ and $x < -3$

33. $-2 < x < -1$ or $1 < x < 2$

35. $.5 < x < .84$ **37.** $x < -1/3$ or $x > 2$

39. $x > 1$ **41.** $-3 < x < 1$ or $x \ge 5$

43. $-\sqrt{7} < x < \sqrt{7}$ or $x > 5.34$

45. $x < -3$ or $1/2 < x < 5$ **47.** $x \le 0$ or $x \ge 1$

49. $x > -1.43$ **51.** $x \le -3.79$ or $x \ge .79$

53. Approximately 8.608 cents per kwh

55. More than \$37,500

57. (a) They would have to talk more than 6 months.
  (b) They are saving between 42 and 60 dollars per month, not counting the up-front fees to switch.

59. $1 < x < 19$ and $y = 20 - x$

61. $10 < x < 35$ **63.** $1 \le t \le 4$ **65.** $2 < t < 2.25$

67. (a) $x^2 < x$ when $0 < x < 1$ and $x^2 > x$ when $x < 0$ or $x > 1$.
  (b) If c is nonzero and $|c| < 1$, then either $0 < c < 1$ or $-1 < c < 0$ (which is equivalent to $1 > -c > 0$). If $0 < c < 1$, then $|c| = c$ and c is a solution of $x^2 < x$ by part (a), so that $c^2 < c = |c|$. If $1 > -c > 0$, then $|c| = -c$, which is a solution of $x^2 < x$ by part (a), so that $c^2 = (-c)^2 < (-c) = |c|$.
  (c) If $|c| > 1$, then either $c < -1$ or $c > 1$. In either case, c is a solution of $x^2 > x$ by part (a).

69. No solutions

71. $(-\infty, \infty)$, all real numbers are solutions.

73. $x = -2, x = 3$ **75.** No solutions

77. $x = \dfrac{5}{2}$

Special Topics 4.6.A, page 320

1. $-4/3 \le x \le 0$ **3.** $7/6 < x < 11/6$

5. $x \le -11/20$ or $x \ge -1/4$

7. $x < -53/40$ or $x > -43/40$

9. $x \le -7/2$ or $x \ge -5/4$ **11.** $-1/7 < x < 3$

13. $-\sqrt{3} < x < -1$ or $1 < x < \sqrt{3}$

15. $x < -\sqrt{6}$ or $x > \sqrt{6}$

17. $x \le -2$ or $-1 \le x \le 0$ or $x \ge 1$

19. $-1.43 < x < 1.24$ **21.** $x < -.89$ or $x > 1.56$

23. $x \le 2$ or $x \ge 14/3$

25. $-1.13 < x < 1.35$ or $1.35 < x < 1.67$

27. $(-\infty, \infty)$, all real numbers are solutions.

29. $x = 1, x = 2$

31. If $|x - 3| < E/5$, then multiplying both sides by 5 shows that $5|x - 3| < E$. But
$5|x - 3| = |5| \cdot |x - 3| = |5(x - 3)|$
$= |5x - 15| = |(5x - 4) - 11|$. Thus,
$|(5x - 4) - 11| < E$.

33. (a) $2.999 < x < 3.001$
  (b) $|\delta - l| < .001$ where δ is the desired length, and l is the length of the manufactured rod.

Section 4.7, page 326

1. $8 + 2i$ **3.** $-2 - 10i$

5. $-1/2 - 2i$ **7.** $\dfrac{\sqrt{2} - \sqrt{3}}{2} + 2i$

9. $1 + 13i$ **11.** $-30i$

13. $-21 - 20i$ **15.** 4

17. $-i$ **19.** i

21. i **23.** $\dfrac{3 - 2i}{13}$

25. $-\dfrac{4}{3}i$ **27.** $\dfrac{12}{41} - \dfrac{15}{41}i$

29. $-\dfrac{5}{41} - \dfrac{4}{41}i$ **31.** $\dfrac{10}{17} - \dfrac{11}{17}i$

33. $\dfrac{7}{10} + \dfrac{11}{10}i$ **35.** $-\dfrac{113}{170} + \dfrac{41}{170}i$

37. $6i$ **39.** $\sqrt{14}i$

41. $-4i$ **43.** $11i$

45. $(\sqrt{15} - 3\sqrt{2})i$ **47.** $2/3$

49. $-41 - i$

51. $(2 + 5\sqrt{2}) + (\sqrt{5} - 2\sqrt{10})i$

53. $\dfrac{1}{6} - \dfrac{1}{6}i\sqrt{5}$ **55.** $x = 2, y = -2$

57. $x = -3/4, y = 3/2$ **59.** $\dfrac{1}{3} + \dfrac{\sqrt{14}}{3}i, \dfrac{1}{3} - \dfrac{\sqrt{14}}{3}i$

61. $-2 + 0i, -3 + 0i$ **63.** $\dfrac{1}{4} + \dfrac{\sqrt{31}}{4}i, \dfrac{1}{4} - \dfrac{\sqrt{31}}{4}i$

65. $-42 - 2.5i, -42 + 2.5i$

67. $2 + 0i, -1 + \sqrt{3}i, -1 - \sqrt{3}i$

69. $1 + 0i, -1 + 0i, 0 + i, 0 - i$

71. -1

73. (a) (i) 5   (ii) 25   (iii) 8
      (iv) 8   (v) 8
  (b) $\text{mod}(5 + 12i)$

75. $\overline{zw} = \overline{(a + bi)(c + di)}$
    $= \overline{(ac - bd) + (bc + ad)i}$
    $= (ac - bd) - (bc + ad)i$
  $\bar{z} \cdot \bar{w} = \overline{(a + bi)} \cdot \overline{(c + di)}$
    $= (a - bi)(c - di)$
    $= (ac - bd) - (bc + ad)i$
  Thus $\overline{zw} = \bar{z} \cdot \bar{w}$

77. $\bar{\bar{z}} = \overline{\overline{a + bi}} = \overline{a - bi} = a + bi = z$

79. (a) $\dfrac{z + \bar{z}}{2} = \dfrac{(a + bi) + (a - bi)}{2} = \dfrac{2a}{a} = a$

  (b) $\dfrac{z - \bar{z}}{2i} = \dfrac{(a + bi) - (a - bi)}{2i} = \dfrac{2bi}{2i} = b$

81. (a) (i) $(a, b) + (c, d) = (a + c, b + d) = (c + a, d + b)$
      $= (c, d) + (a, b)$
    (ii) $[(a, b) + (c, d)] + (e, f) = (a + c, b + d) + (e, f)$
      $= ((a + c) + e, (b + d) + f)$
      $= (a + (c + e), b + (d + f))$
      $= (a, b) + [(c, d) + (e, f)]$
    (iii) $(a, b) + (0, 0) = (a + 0, b + 0) = (a, b)$
    (iv) $(a, b) + (-a, -b) = (a + (-a), b + (-b)) = (0, 0)$

(b) (i) $(a, b)(c, d) = (ac - bd, bc, + ad)$
$= (ca - db, cb + da) = (c, d)(a, b)$
(ii) $[(a, b)(c, d)](e, f) = (ac - bd, bc + ad)(e, f)$
$= (ace - adf - bcf - bde,$
$acf + ade + bce - bdf)$
$= (a, b)(ce - df, cf + de)$
$= (a, b)[(c, d)(e, f)]$
(iii) $(a, b)(1, 0) = (a \cdot 1 - b \cdot 0, a \cdot 0 + b \cdot 1) = (a, b)$
(iv) $(a, b)(0, 0) = (a \cdot 0 - b \cdot 0, a \cdot 0 + b \cdot 0) = (0, 0)$
(c) (i) $(a, 0) + (c, 0) = (a + c, 0 + 0) = (a + c, 0)$
(ii) $(a, 0)(c, 0) = (a \cdot c - 0 \cdot 0, a \cdot 0 + c \cdot 0) = (ac, 0)$
(d) (i) $(0, 1)(0, 1) = (0 \cdot 0 - 1 \cdot 1, 0 \cdot 1 + 0 \cdot 1) = (-1, 0)$
(ii) $(b, 0)(0, 1) = (b \cdot 0 - 0 \cdot 1, b \cdot 1 + 0 \cdot 0) = (0, b)$
(iii) $(a, 0) + (b, 0)(0, 1) = (a, 0) + (0, b) = (a + 0, 0 + b)$
$= (a, b)$

Section 4.8, page 332

1. 3 **3.** 0 **5.** Yes **7.** Yes

9. 0: multiplicity 54, $-4/5$: multiplicity 1

11. 0: multiplicity 15; π: multiplicity 14; $\pi + 1$: multiplicity 13

13. $1 \pm 2i$, $[x - (1 + 2i)][x - (1 - 2i)]$

15. -3 with multiplicity 2; $3(x + 3)(x + 3)$

17. $3, \dfrac{-3}{2} + \dfrac{3\sqrt{3}}{2}i, \dfrac{-3}{2} - \dfrac{3\sqrt{3}}{2}i,$
$(x - 3)\left(x - \dfrac{-3 + 3\sqrt{3}i}{2}\right)\left(x - \dfrac{-3 - 3\sqrt{3}i}{2}\right)$

19. $-2, 1 + \sqrt{3}i, 1 - \sqrt{3}i,$
$(x + 2)[x - (1 + \sqrt{3}i)][x - (1 - \sqrt{3}i)]$

21. $\pm 1, \pm i, (x - 1)(x + 1)(x - i)(x + i)$

23. $\pm 1, -1 \pm i, (x - 1)(x + 1)(x + 1 - i)(x + 1 + i)$

In Exercises 25–42, there may be correct answers other than the ones shown here.

25. $(x - 1)(x - 7)(x + 4)$ **27.** $(x - 1)^2(x - 2)^2(x - \pi)^2$

29. $f(x) = 2x(x + 3)(x - 4)$ **31.** $f(x) = x^2 - 4x + 5$

33. $f(x) = x^5 - x^4 - x^3 + 19x^2 - 32x + 30$

35. $f(x) = x^2 - 2x + 5$

37. $f(x) = x^4 - 14x^3 + 74x^2 - 176x + 160$

39. $f(x) = x^6 - 5x^5 + 8x^4 - 6x^3$

41. $f(x) = 3x^2 - 6x + 6$ **43.** $x^2 + (-1 + i)x + (2 + i)$

45. $x^3 - 5x^2 + (7 + 2i)x + (-3 - 6i)$

47. $3, -\dfrac{1}{2} + \dfrac{\sqrt{3}}{2}i, -\dfrac{1}{2} - \dfrac{\sqrt{3}}{2}i$

49. $i, -i, -1, -2$ **51.** $2 - i, 2 + i, i, -i$

53. (a) $\overline{z + w} = \overline{(a + bi) + (c + di)}$
$= \overline{(a + c) + (b + d)i}$
$= (a + c) - (b + d)i$
$= (a - bi) + (c - di)$
$= \bar{z} + \bar{w}$

(b) $\overline{z \cdot w} = \overline{(a + bi) \cdot (c + di)}$
$= \overline{(ac - bd) + (bc + ad)i}$
$= (ac - bd) - (bc + ad)i$
$= (a - bi) \cdot (c - di)$
$= \bar{z} \cdot \bar{w}$

55. (a) If z is a root of $f(x)$, $f(z) = 0$. Then $\overline{f(z)} = \bar{0} = 0$. Also
$$\overline{f(z)} = \overline{az^3 + bz^2 + cz + d}$$
$$= \overline{az^3} + \overline{bz^2} + \overline{cz} + \bar{d}$$
$$= \bar{a}\overline{z^3} + \bar{b}\overline{z^2} + \bar{c}\bar{z} + \bar{d}$$
$$= a\bar{z}^3 + b\bar{z}^2 + c\bar{z} + d = f(\bar{z})$$

(b) Since $\overline{f(z)} = 0$, $a\bar{z}^3 + b\bar{z}^2 + c\bar{z} + d = 0$; therefore, $\bar{z}$ is a root of $f(x)$.

57. For each nonreal complex root z, there must be two factors of the polynomial: $(x - z)$ and $(x - \bar{z})$. This yields an even number of factors. There will remain at least one factor, hence at least one root, which must be real.

Chapter 4 Review, page 335

1. $(2, 3)$ **3.** $(4, -4)$

5. $(1.5, -5.75)$

7. (a) $y = 260 - x$
(b) $A(x) = -x^2 + 260x - 3500$
(c) 130 ft by 130 ft

9. 30 ft perpendicular to the building by 60 ft parallel to the building

11. (a), (c), (d), and (f) **13.** 0

15.
$$
\begin{array}{r|rrrrrr}
2 & 1 & -5 & 8 & 1 & -17 & 16 & -4 \\
  &   & 2 & -6 & 4 & 10 & -14 & 4 \\
\hline
  & 1 & -3 & 2 & 5 & -7 & 2 & 0 \\
\end{array}
$$
the other factor is $x^5 - 3x^4 + 2x^3 + 5x^2 - 7x + 2$

17. $f(x) = x^3 - 5x^2 - x + 5$ is one of many correct answers

19. $5/3$ and -1 **21.** $\sqrt[3]{2}$

23. $0, \pm\sqrt{\dfrac{1 + \sqrt{21}}{2}}$

25. (a) $1, -1, 3, -3, \dfrac{1}{2}, -\dfrac{1}{2}, \dfrac{3}{2}, -\dfrac{3}{2}$

(b) 3 (c) $x = \dfrac{1 \pm \sqrt{3}}{2}$ and $x = 3$

27. 1

29. $3, -3, \sqrt{2}, -\sqrt{2}$

31. $(x^4 - 4x^3 + 16x - 16) \div (x - 5)$ is $x^3 + x^2 + 5x + 41$ with remainder 189. Since all coefficients and the remainder are positive, 5 is an upper bound for the roots.

33. $1, -1, 1.867, -.867$

35. $2x + h + 1$

37.

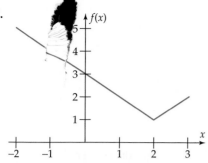

This graph could not possibly be the graph of a polynomial function because of the sharp corner. Many other graphs are possible.

39. (c)

41. $-1 \le x \le 2$ and $-10 \le y \le 60$

43. $-2 \le x \le 18$ and $-500 \le y \le 1200$

45. (a)

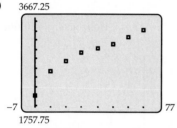

(b) $y = .00852x^3 - 1.10945x^2 + 56.2583x + 2017.2576$
(c) $C(71) - C(70) = \$26.79$
(d) About \$85.49, about \$47.84

47.

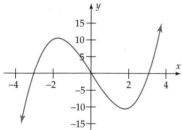

x-intercepts: $-3, 0, 3$; local maximum: $(-1.732, 10.392)$; local minimum: $(1.732, -10.392)$

49.

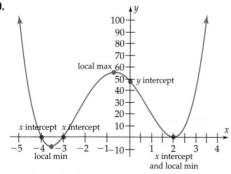

x-intercepts: $(-4, 0)$, $(-3, 0)(-2, 0)$; y-intercept: $(0, 48)$; local minima when $x \approx -3.54$ and $x = 2$; local maximum when $x \approx -.71$

51.

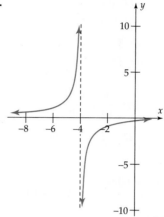

x-intercepts: none; vertical asymptote: $x = -4$; horizontal asymptote: $y = 0$

53.

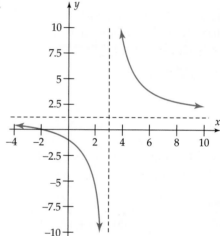

x-intercept: -2.5; vertical asymptote: $x = 3$; horizontal asymptote: $y = 4/3$

55. vertical asymptotes: $x = -2$, $x = 3$; horizontal asymptote: $y = 0$

57. $-4.7 \le x \le 4.7$ and $-5 \le y \le 5$; then $2 \le x \le 20$ and $-.2 \le y \le .1$

59. $-18.8 \le x \le 18.8$ and $-8 \le y \le 8$

61. between 400 and 2500 bags; between \$1.75 (at 400 bags) and \$0.70 (at 2500 bags)

63. (a) $0.455 < s < 5.498$
(b) 1.581
(c) 10 miles/hour. This is a very low speed limit, and the citizens would probably object.

65. $\dfrac{1}{(x + 1)(x + h + 1)}$ **67.** (d)

69. $y \le -17$ or $y \ge 13$

71. $(-\infty, -2)$ and $(-1/3, \infty)$

73. $-\sqrt{3} \le x < -1$ or $-1 < x < 1$ or $1 < x \le \sqrt{3}$

75. $x \le -1$ or $0 \le x \le 1$ **77.** (e)

79. $x \le -4/3$ or $x \ge 0$ **81.** $x \le -7$ or $x > -4$

83. $x < -2\sqrt{3}$ or $-3 < x < 2\sqrt{3}$

85. $x < \dfrac{1 - \sqrt{13}}{6}$ or $x > \dfrac{1 + \sqrt{13}}{6}$

87. $\dfrac{-3 \pm i\sqrt{31}}{2}$ **89.** $\dfrac{3}{10} \pm \dfrac{\sqrt{31}}{10}i$

91. $\pm\dfrac{\sqrt{6}}{3}, \pm i$ **93.** $-2, 1 \pm \sqrt{3}i$

95. $i, -i, 2, -1$

97. $x^4 - 2x^3 + 2x^2$ is one possibility

Chapter 4 Test, page 339

1. At $t = \dfrac{1}{\sqrt{3}} \approx .57735$ minutes.

2. (a) Quotient: $x^2 + x + 1$ Remainder: 3
   (b) Quotient: $2x^3 - x^2 - 1$ Remainder: 0
   (c) Quotient: $x^3 - 2x^2 + x - 1$ Remainder: $-4x$

3.

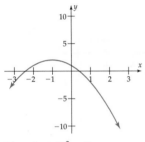

(a) $-(x + 1)^2 - 2$

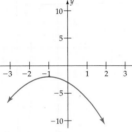

(b) $-(x + 1)^2 - 2$

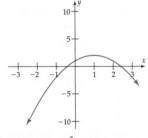

(c) $-(x - 1)^2 + 2$

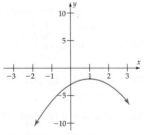

(d) $-(x - 1)^2 - 2$

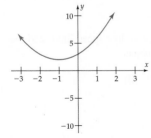

(e) $(x + 1)^2 + 2$

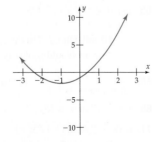

(f) $(x + 1)^2 - 2$

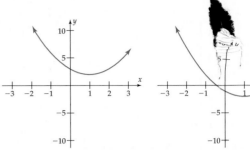

(g) $(x - 1)^2 + 2$ (h) $(x - 1)^2 - 2$

4. (a) $I = x(75 + 20(10 - x)) = -20x^2 + 275x$ (b) $6.88

5. (a) $1, -1, \dfrac{1}{3}$ (b) $-3, 3, \sqrt{2}, -\sqrt{2}$
   (c) $1, -2$ (the other two roots are complex)

6. $10 < x < 20$

7. (a) $-3 - 6i$ (b) $2 + 3i, 2 - 3i$ (c) $\dfrac{1}{2}i, -\dfrac{1}{2}i$

8. (a) $-\dfrac{1}{8}$
   (b) Vertical: $x = -4$, $x = 2$ Horizontal: $y = 0$ (the x axis)
   (c) $(-1, 0)$
   (d)

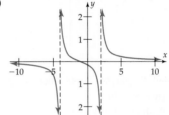

9. (a) $x \ge 2$ (b) $1 < x < 2$ (c) $x > \dfrac{2}{3}$ or $x < -2$
   (d) $-2 < x < 3$
   (e) $x \le \dfrac{-3}{2}$ or $x \ge \dfrac{9}{2}$

10. Asymptote: $y = 3x$

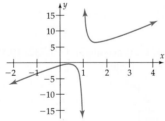

11. $(x - 1 + 2i)(x - 1 - 2i)(x - 5 + i)(x - 5 - i)x$
   $= x^5 - 12x^4 + 51x^3 - 102x^2 + 130x$

Chapter 5

Section 5.1, page 349

1. $1000k^{13/4}$ **3.** $c^{42/5} \cdot d^{10/3}$ **5.** $\dfrac{1}{3y^{2/3}}$

7. $\dfrac{a^{1/2}}{49b^{5/2}}$ **9.** a^x **11.** $x^{7/6} - x^{11/6}$

13. $x - y$ **15.** $x + y - (x + y)^{3/2}$

17. $(x^{1/3} + 3)(x^{1/3} - 2)$ **19.** $(x^{1/2} + 3)(x^{1/2} + 1)$

21. $(x^{2/5} + 9)(x^{1/5} + 3)(x^{1/5} - 3)$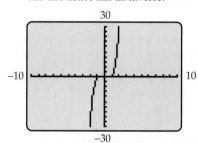

23. $x^{-1/2}$ **25.** $a(a + b)^{1/2}$

27. $4t^{27/10}$ **29.** $4\sqrt{5}$

31. $6\sqrt{2}$ **33.** $\dfrac{-2 + \sqrt{11}}{5}$

35. $-\sqrt{2}$ **37.** $15\sqrt{5}$

39. $\dfrac{4a^4}{|b|}$ **41.** $\dfrac{|d^5|}{2\sqrt{c}}$

43. $\dfrac{a^2\sqrt[3]{b^2}}{c}$ **45.** $\dfrac{3\sqrt{2}}{4}$ **47.** $\dfrac{3\sqrt{3} - 3}{4}$

49. $\dfrac{2\sqrt{x} - 4}{x - 4}$ **51.** $5\sqrt[3]{4}$ **53.** $\dfrac{\sqrt[3]{9} - \sqrt[3]{3} + 1}{4}$

55. $\dfrac{\sqrt[3]{2} + 1}{3}$ **57.** $\dfrac{1}{\sqrt{x + h + 1} + \sqrt{x + 1}}$

59. $\dfrac{2x + h}{\sqrt{(x + h)^2 + 1} + \sqrt{x^2 + 1}}$

61. About 35,863,131 mi   **63.** About 886,781,537 mi

65. About 5,312,985   **67.** About 20,003,970

69. (a) Any even power of a real number is never negative.

     (b) $\sqrt[3]{-8} = -2$, but $\sqrt[6]{(-8)^2} = 2$

71. (a) Since its graph passes the horizontal test, f is one-to-one and hence has an inverse.

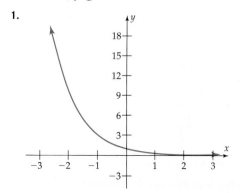

     (b) $(g \circ f)(x) = g(f(x)) = (f(x))^{1/5} = (x^5)^{1/5} = x$ and $(f \circ g)(x) = f(g(x)) = (g(x))^5 = (x^{1/5})^5 = x$

     (c) Since its graph does not pass the horizontal test, f is not one-to-one and hence has no inverse.

73. The graph of g is the graph of $f(x) = \sqrt{x}$ shifted horizontally 3 units to the left.

75. The graph of k is the graph of $f(x) = \sqrt{x}$ shifted horizontally 4 units to the left and vertically 4 units downward.

77. (a)

L	C	$Q = L^{1/4}C^{3/4}$
10	7	7.65
20	14	15.31
30	21	22.96
40	28	30.61
60	42	45.92

(b) If both labor and capital are doubled, output is doubled. If both are tripled, output is tripled.

79. (a)

L	C	$Q = L^{1/2}C^{3/4}$
10	7	13.61
20	14	32.37
30	21	53.73
40	28	76.98
60	42	127.79

(b) If both labor and capital are doubled, output is multiplied by $2^{5/4}$. If both are tripled, output is multiplied by $3^{5/4}$.

Special Topics 5.1.A, page 355

1. $x = 7$ **3.** $x = -\dfrac{9}{4}$ **5.** $x = -2$

7. $x = \pm 3$ **9.** $x = -1$ or 2 **11.** $x = 9$

13. $x = \dfrac{1}{2}$ **15.** $x = \dfrac{1}{2}$ or $x = -4$

17. $x \approx \pm .73$

19. $x \approx -1.17$ or $x \approx 2.59$ or $x = -1$

21. $x = -5$ or $x = 3$ **23.** $x = 6$

25. $x = 3$ or 7 **27.** $r = 4.658$

29. $r \approx 8.019$ **31.** $b = \dfrac{a}{\sqrt{A^2 - 1}}$

33. $x = \sqrt{1 - \dfrac{1}{y^2}}$ **35.** No solution

37. $x = 36$ **39.** $x = -1$ or -8

41. $x = 16$ **43.** $x \approx 105.236$

45. $x \approx -.283$

47. (a) 11.47 ft or 29.91 ft   (b) 21 ft

49. Approx. 6.205 mi

51. (a) 1600 feet   (b) 9.52 seconds

Section 5.2, page 365

1.

3.

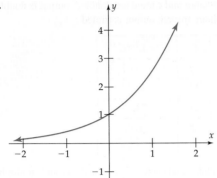

5.

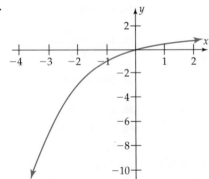

7.

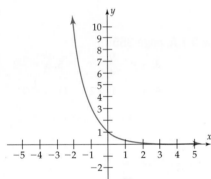

9.

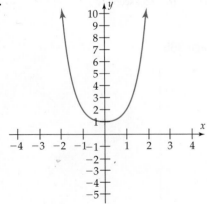

11. Shift vertically 5 units down.

13. Stretch away from the x-axis by a factor of 3.

15. Shift horizontally 2 units left, then shift vertically 5 units down.

17. Line A: $j(x)$
Line B: $g(x)$
Line C: $h(x)$
Line D: $f(x)$

19. Neither   **21.** Even   **23.** Even

25. 90   **27.** $-\dfrac{4}{27}$

29. $1000(2^{1.001} - 2) = 1.387$

31. $1000(e^{1.001} - e) = 2.720$

33. $\dfrac{10^{x+h} - 10^x}{h}$ **35.** $\dfrac{2^{x+h} + 2^{-x-h} - 2^x - 2^{-x}}{h}$

37. $-4 \le x \le 4,\ -1 \le y \le 10$

39. $-4 \le x \le 4,\ -1 \le y \le 10$

41. $-10 \le x \le 10,\ -10 \le y \le 10$

43. $-5 \le x \le 10,\ -1 \le y \le 6$

45. The negative x-axis is an asymptote; $(-1.443, -.531)$ is a local minimum.

47. No asymptote; $(0, 1)$ is a local minimum.

49. The x-axis is an asymptote; $(0, 1)$ is a local maximum.

51. (a) 100   (b) 569; 3242   (c) 13 weeks
(d) No, at some point there will not be enough space for the fruit flies and the rate the fly population grows will decrease.

53. (a) $k = 15$ (b) About 12.13 psi
(c) About .0167 psi

55. (a) About 74.1 years, about 76.3 years   (b) 1930

57. (a) 10 beavers, 149 beavers
(b) After about 9.5 years

59. (a)

Time (hours)	Number of Cells
0	1
.25	2
.5	4
.75	8
1	16

(b) $C(t) = 2^{4t}$

61. (a) $f(x) = 6(3^x)$ or $f(x) = 18(3^{x-1})$
(b) 3   (c) No; yes

63. (a)

Folds	0	1	2	3	4
Thickness	.002	.004	.008	.016	.032

(b) $f(x) = .002(2^x)$ (c) 2097.15 in. = 174.76 ft
(d) 43

65. (a) $f(x) = 1200(1.04)^x$ (b) 1349.84; 1503.57
(c) After approximately 11 years and 1 month

67. (a) $100.4(1.014)^x$ (b) 115.3 million   (c) 2016

69. (a) $32.44e^{.02216x}$
(b) 40.49 million; 56.45 million   (c) 2024

71. About 256; about 654

73. (a) $f(x) = .75^x$ (b) About 8 ft

75. (a) $M(x) = 5(0.5^{x/5730})$ (b) 3.08 g; 1.90 g
(c) After about 13,305 years

77. $f(x) = a^x$ for any nonnegative constant a.

79. (a) The graph of f is the mirror image of the graph of g
(b) $k(x) = f(x)$; see part (a).

81. (a) Not entirely   (b) $f_8(x)$
(c) Not at the right side; $f_{12}(x)$

Special Topics 5.2.A, page 374

1. (a) $1469.33
(b) $1485.95
(c) $1489.85
(d) $1491.37

3. 633.39   **5.** 674.17   **7.** 819.32

9. 686.84   **11.** $3325.29   **13.** $3359.59

15. (a) 7095.36   (b) No

17. Fund C **19.** $385.18   **21.** $1,162,003.14

23. About 5%   **25.** About 5.92%

27. (a) 9 years   (b) 9 years   (c) 9 years
(d) Investment amount and doubling time are independent.

29. 9.9 years

31. (a) 907.50
(b) CD, since the CD earns 918.06
(c) 971.27; you would make more money this way.

Section 5.3, page 383

1. 4   **3.** -2.5

5. $10^3 = 1000$ **7.** $10^{2.88} = 750$

9. $e^{1.0986} = 3$ **11.** $e^{-4.6052} = .01$

13. $e^{z+w} = x^2 + 2y$ **15.** $\log .01 = -2$

17. $\log 3 = .4771$ **19.** $\ln 25.79 = 3.25$

21. $\ln 5.5527 = 12/7$ **23.** $\ln w = 2/r$

25. $\sqrt{43}$ **27.** 15   **29.** $.5$

31. 931   **33.** $x + y$ **35.** x^2

37. $f(x) = 4e^{3.2189x}$ **39.** $g(x) = -16e^{3.4177x}$

41. $\left(\dfrac{1}{e^3}\right)^x = .0498^x$ **43.** $x > -1$

45. $x < 0$

47. (a) for all $x > 0$.
(b) According to the fourth property of logarithms,
$e^{\ln x} = x$ for every $x > 0$.

49. They are the same for $x > 0$. If $x < 0$, $g(x)$ does not exist,
while $f(x)$ is symmetric with respect to the y-axis.

51. Stretch vertically by a factor of 2 away from the x-axis.

53. Shift horizontally 4 units to the right.

55. Shift horizontally 3 units to the left, then shift vertically
4 units down.

57.

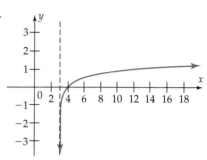

59.

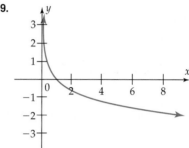

61. $0 \le x \le 20$ and $-10 \le y \le 10$; then $0 \le x \le 2$ and
$-20 \le y \le 20$

63. $-10 \le x \le 10$ and $-3 \le y \le 3$

65. $0 \le x \le 20$ and $-6 \le x \le 3$

67. $-10 \le x \le 10$ and $-10 \le y \le 10$

69. .5493   **71.** $-.2386$

73. (a) $\dfrac{\ln(2 + h) - \ln(2)}{h}$
(b) .49875415; .499875042; .499987501; .49999875
(c) .24968802; .249968755; .249996875; .24999970
(d) Answers will vary
(e) About 2.5

75. $(f \circ g)(x) = \dfrac{1}{1 + e^{-\ln(x/1-x)}} = \dfrac{1}{1 + \dfrac{1}{e^{\ln(x/1-x)}}}$

$= \dfrac{1}{1 + \dfrac{1}{\dfrac{x}{1-x}}} = \dfrac{1}{1 + \dfrac{1-x}{x}} = \dfrac{1}{\dfrac{1}{x}} = x;$

$(g \circ f)(x) = \ln\left(\dfrac{\dfrac{1}{1 + e^{-x}}}{1 - \dfrac{1}{1 + e^{-x}}}\right) = \ln\left(\dfrac{\dfrac{1}{1 + e^{-x}}}{\dfrac{e^{-x}}{1 + e^{-x}}}\right)$

$= \ln\left(\dfrac{1}{e^{-x}}\right) = \ln(e^x) = x$

77. $A = -9, B = 10$

79. About 4392 meters

81. (a) 26.55 billion pounds, 27.72 billion pounds
(b) 2022

83. (a) About 9.9 days
(b) 6986 people

85. (a) No ads: about 120 bikes; $1000: about 299 bikes;
$10,000: about 513 bikes.
(b) $1000: yes, $10,000: no.
(c) $1000: yes, $10,000: yes.

87. $n = 30$ gives an approximation with a maximum error of
.00001 when $-.7 \le x \le .7$.

89. (a) Answers will vary.
(b) (28.087, 29.087)
(c) 2529

Section 5.4, page 390

1. $\ln (x^2 y^3)$ **3.** $\log (x - 3)$ **5.** $\ln (x^{-7})$

7. $\ln (e - 1)^3$ **9.** $\log (20xy)$ **11.** $2u + 5v$

13. $\frac{1}{2}u + 2v$ **15.** $\frac{2}{3}u + \frac{1}{6}v$

17. False; the right side is not defined when $x < 0$, but the left side is.

19. True by the Power Law

21. False; the graph of the left side differs from the graph of the right side.

23. False

25. $a = 10, b = 10$, among many examples

27. Since $v = e^{\ln v}$ and $w = e^{\ln w}$, we have $\dfrac{v}{w} = \dfrac{e^{\ln v}}{e^{\ln w}} = e^{\ln v - \ln w}$
by the division property of exponents. So raising e to the
exponent $(\ln v - \ln w)$ produces $\dfrac{v}{w}$. But the exponent to which
e must be raised to produce $\dfrac{v}{w}$ is, by definition, $\ln \dfrac{v}{w}$. Therefore, $\ln \dfrac{v}{w} = \ln v - \ln w$.

29. 4.7 **31.** 3.176

33. 40 **35.** 100

37. I must be squared.

39. Following the hint: $10^{\log c} = c$ because $\log c$ is the exponent
to which 10 must be raised to produce c. Taking the natural
logarithm of both sides produces $\ln (10^{\log c}) = \ln (c)$, but
$\ln (10^{\log c}) = \log c \cdot \ln 10$ by the power law for logarithms.
Thus, $\log c \cdot \ln 10 = \ln c$, and dividing both sides of this
equation by $\ln 10$ produces the desired result.

41. If c is a positive number, then c can be written in scientific
notation as $c = b \times 10^k$, where $1 \le b < 10$ and k is an
integer. Then $\log c = \log (b \times 10^k) = \log b + \log (10^k) = \log b + k$.

43. (a)
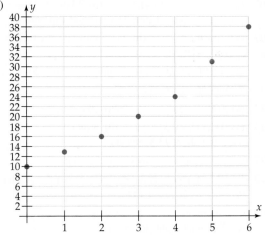
Wayland. It appears to be exponential.

(b)
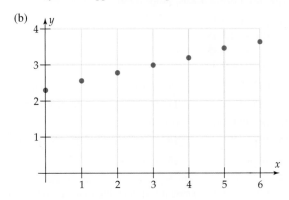

(c) Wayland. It now appears linear in part b.

Special Topics 5.4.A, page 397

1.

x	0	1	2	4
$f(x) = \log_4 x$	Not defined	0	.5	1

3.

x	1/36	1/6	1	216
$h(x) = \log_6 x$	-2	-1	0	3

5.

x	0	1/7	$\sqrt{7}$	49
$f(x) = 2 \log_7 x$	Not defined	-2	1	4

7.

x	-2.75	-1	1	29
$h(x) = 3 \log_2(x + 3)$	-6	3	6	15

9. $\log .01 = -2$ **11.** $\log \sqrt[3]{10} = 1/3$

13. $\log r = 7k$ **15.** $\log_7 5,764,801 = 8$

17. $\log_3(1/9) = -2$ **19.** $10^4 = 10,000$

21. $10^{2.88} \approx 750$ **23.** $5^3 = 125$

25. $2^{-2} = \dfrac{1}{4}$ **27.** $10^{z+w} = x^2 + 2y$

29. $\sqrt{97}$ **31.** $x^2 + y^2$

33. 1/2 **35.** 6

37. $b = 3$ **39.** $b = 20$ **41.** 5

43. 3 **45.** 4 **47.** $\log \dfrac{x^2 y^3}{z^6}$

49. $\log (x^2 - 3x)$ **51.** $\log_2 (5|c|)$

53. $\log_4 \left(\dfrac{1}{49c^2}\right)$ **55.** $\ln \left[\dfrac{(x+1)^2}{x+2}\right]$

57. $\log_2 (x)$

59. $\ln (e^2 - 2e + 1) = \ln (e-1)^2 = 2 \ln (e-1)$

61. 3.3219 **63.** .8271 **65.** 1.1115

67. 1.6199 **69.** True **71.** True

73. False **75.** 397^{398}

77. $\log_b u = \dfrac{\log_a u}{\log_a b}$ **79.** $\log_{10} u = 2 \log_{100} u$

81. $\log_b x = \dfrac{1}{2}\log_b v + 3 = \dfrac{1}{2}\log_b v + 3\log_b b = \log_b v^{1/2} + \log_b b^3 = \log_b (b^3 \sqrt{v})$, so $x = b^{\log_b x} = b^{\log_b (b^3 \sqrt{v})} = b^3 \sqrt{v}$.

83. $f(x) = g(x)$ only when $x \approx .123$, so the statement is false.

85.

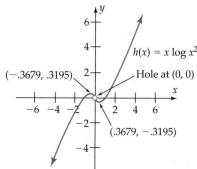

87. (a) It is greater than one google. Since $\log_5 (10^{100}) = 143.0677$, we have $5^{143.0677} = 10^{100}$. Since $5^{150} > 5^{143.0677}$, $5^{150} > 10^{100}$.
(b) 2^{3322} is slightly larger than $(10^{100})^{10} = 10^{1000}$. When we compare their logarithms we have $\log 2^{3322} = 3322 \log 2 \approx 1000.02$ and $\log 10^{1000} = 1000 \log 10 = 1000$.

Section 5.5, page 406

1. $x = 4$ **3.** $x = 1/9$ **5.** $x = \dfrac{1}{2}$ or -3

7. $x = -2$ or $-1/2$ **9.** $x = \ln 5/\ln 3 \approx 1.465$

11. $x = \ln 3/\ln 1.5 \approx 2.7095$

13. $x = \dfrac{\ln 3 - 5 \ln 5}{\ln 5 + 2 \ln 3} \approx -1.825$

15. $x = \dfrac{\ln 2 - \ln 3}{3 \ln 2 + \ln 3} \approx -.1276$

17. $x = (\ln 5)/2 \approx .805$

19. $x = (-\ln 3.5)/1.4 \approx -.895$

21. $x = 2 \ln(5/2.1)/\ln 3 \approx 1.579$

23. $x = 0$ or 1

25. $x = \ln 2 \approx .693$ or $x = \ln 3 \approx 1.099$

27. $x = \ln 3 \approx 1.099$

29. $x = \ln 2/\ln 4 = 1/2$ or $x = \ln 3/\ln 4 \approx .792$

31. $x = \ln (t + \sqrt{t^2 + 1})$

33. (a) If $\ln u = \ln v$, then $u = e^{\ln u} = e^{\ln v} = v$.
(b) No, if u and v are negative then $\ln u$ and $\ln v$ are undefined.

35. $x = 9$ **37.** $x = 5$ **39.** $x = 6$ **41.** $x = 3$

43. $x = \dfrac{-5 + \sqrt{37}}{2}$ **45.** $x = 9/(e - 1)$

47. $x = 5$ **49.** $x = \pm \sqrt{10001}$

51. $x = \sqrt{\dfrac{e + 1}{e - 1}}$ **53.** .444 billion years

55. 14.95 days **57.** Approximately 3.8132 days

59. About 3689 years old **61.** About 2534 years ago

63. About 6.99%

65. (a) About 22.5 years (b) About 22.1 years

67. $3197.05 **69.** 79.36 years

71. (a) .0146 (b) 2024

73. (a) $k \approx 21.459$ (b) $t \approx .182$

75. (a) 20 at the beginning and 2500 3 hours later
(b) $\dfrac{\ln 2}{\ln 5} \approx .43$

77. (a) 200 people; 2795 people
(b) After 6 weeks

79. (a) 16 years from this year
(b) 24 years from this year

81. (a) $c = \dfrac{250}{3}; k = -\left(\dfrac{1}{4}\right) \ln \left(\dfrac{2}{5}\right)$ (b) 12.43 weeks

Section 5.6, page 415

1. Linear **3.** Logarithmic

5. Exponential, quadratic, cubic, or power

7. Logistic or cubic **9.** Quadratic or cubic

11. Ratios are approximately 5.07, 5.06, 5.06, 5.08, and 5.05. An exponential model might be appropriate.

13. (a) As x increases, $e^{-.0216x}$ decreases from 1 toward 0. Thus, $(1 + 56.33e^{-.0216x})$ decreases toward 1, so the ratio increases toward (but never reaches) 442.1.

(b)

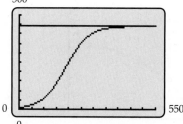

15. (a)

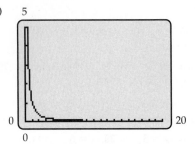

(b)

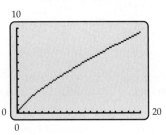

(c)

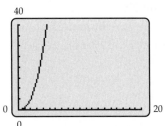

17. $\{(\ln x, \ln y)\}$ is approximately linear, so a power model is appropriate.

19. $\{(\ln x, \ln y)\}$ is approximately linear, so a power model is appropriate.

21. (a) 105,000

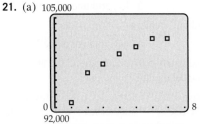

(b) 105,000

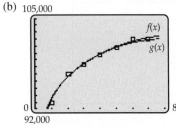

(c)

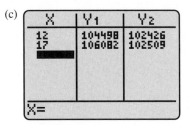

(d) Answers will vary.

23. (a)

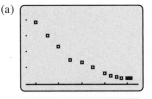

(b)

(c) An exponential model would be best:
$y = 148.55(.9700)^x$

25. (a)

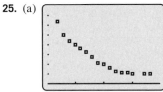

(b) $y = 73.59\,(.8572)^x$
(c) .7
(d) 2012

27. (a)

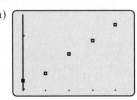

(c) $y = 1311520.25(1.02)^x$

Year	Adults on Probation	(d) Predicted Number of Adults on Probation	(b) Ratio
2000	1316333	1311520	1.01
2001	1330007	1338073	1.03
2002	1367547	1365163	1.02
2003	1392796	1392801	1.02
2004	1421911	1420999	

(e) 1,602,588

29. (a) $y = 21.34 + 12.64 \ln x$
    (b) 79.03
    (c) 2031

Chapter 5 Review, page 420

1. a^2 **3.** $x^{11/15}y^{4/3}$

5. $x^{2/3} - y^{2/3}$ **7.** $\dfrac{y^3}{2x^2}$

9. $\dfrac{3}{\sqrt{3x + 3h - 5} + \sqrt{3x - 5}}$

11. $x = 17/9$ **13.** No solutions

15. $x = -1.733$ or 5.521

17. $-3 \le x \le 3$ and $0 \le y \le 2$

19.
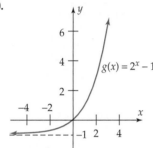

21.
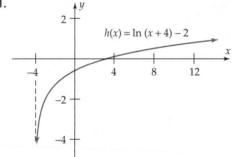

23. (a) About 2.03
    (b) About 31.97
    (c) Approximately 6 to 10 months
    (d) Never; however, at the end of 18 months about 99.6% of the program will be mastered.

25. $\ln 23 = 3.14$ **27.** $\ln (4a - b) = x - 7y$

29. $\log 177 = 2.248$ **31.** $e^{6.0014} = 404$

33. $e^{15t} = 4xy$ **35.** $6^y = 3x - 4$

37. $2/3$ **39.** $x/2$

41. $\ln x$ **43.** $\ln (3/b^4)$

45. $\ln x^{11}$ **47.** 2

49. (c)   **51.** (c)

53. $x = \pm 4$ **55.** $x = -3$ or $x = 6$

57. $x = -1/3$ **59.** $x = e^{(3b-2a)/c}$

61. $x = 3$ **63.** $x = 48.5$

65. About 1.64 mg   **67.** About 12.05 years

69. $3822.66   **71.** 7.6

73. (a) 11°F
    (b)
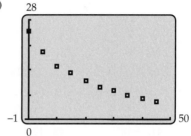

    (c) The ratios of successive entries are approximately constant.
    (d) $f(t) = 22.42e^{-.034x}$.
    (e) About 10°F

Chapter 5 Test, page 424

1. (a) $\dfrac{\sqrt{2 + h}}{2 + h}$ (b) $\dfrac{(a - b)(\sqrt{a} - \sqrt{b})}{a - b} = \sqrt{a} - \sqrt{b}$

    (c) $\dfrac{3(\sqrt{x} + 4)}{x - 16}$

2. (a) $x = 2, x = -1 + \sqrt{3}$ (b) $x = -9, x = 7$

3. (a) 905   (b) 905   (c) 1105

4. (a) k is positive, because the function approaches zero as it goes to the left, and gets larger as it goes to the right.
    (b) -3

5. (a) $2 + \sqrt{3}$ (b) $(3 - \sqrt[3]{ab})b^2$ (c) a^2b^3

6. (a) 32 million   (b) 176,777   (c) 32.768 billion

7. $3247.30

8. Compute the following
    (a) 4   (b) $x + h$ (c) 2276

9. 871 years old.

10. (a) $x^2 + y^2$ (b) $\dfrac{a}{b^2}$ (c) $-\dfrac{1}{3}$

11. $A = 3, B = 2$

12. (a) 1   (b) $\ln 2$ (c) Can't be simplified

13. A population of mayflies grows according to the equation $P = 230e^{.6t}$ where t is in days.
    (a) 230   (b) 6.29 days

14. (a) $y = 120 + 15 \ln x$, where x is in months, and $x = 1$ corresponds to January.
    (b) 130.40, 157.27

Chapter 6

Section 6.1, page 434

1. $2\pi/9$ **3.** $\pi/9$ **5.** $\pi/18$

7. About $\dfrac{\pi}{3}$ **9.** About $\dfrac{-3\pi}{4}$

11. $\dfrac{9\pi}{4}, \dfrac{17\pi}{4}, \dfrac{-7\pi}{4}, \dfrac{-15\pi}{4}$

13. $\dfrac{11\pi}{6}, \dfrac{23\pi}{6}, \dfrac{-13\pi}{6}, \dfrac{-25\pi}{6}$

15. No   **17.** Yes   **19.** $5\pi/3$

21. $3\pi/4$ **23.** $3\pi/5$ **25.** $7 - 2\pi$

27. $\pi/30$ **29.** $-\pi/15$ **31.** $5\pi/12$

33. $3\pi/4$ **35.** $-5\pi/4$ **37.** $31\pi/6$

39. $36°$ **41.** $-18°$ **43.** $135°$

45. $4°$ **47.** $-75°$ **49.** $972°$

51. $4\pi/3$ **53.** $7\pi/6$ **55.** $41\pi/6$

57. 7π **59.** 4π **61.** 42.5π

63. $2\pi k$

Special Topics 6.1.A, page 440

1. 5   **3.** 8.75   **5.** 25.13 cm

7. 2.5 rad   **9.** 2 rad   **11.** 6

13. 15   **15.** 5.56   **17.** 2000

19. 3455.8   **21.** 933.1   **23.** 12.5 sq ft

25. 76.8 sq m   **27.** 2 rad   **29.** 2.4 rad

31. $r \approx 3978.8736$ mi; $c \approx 25{,}000.0001$ mi

33. 3

35. (a) $400\pi \dfrac{\text{rad}}{\text{min}}$ (b) $2513.3 \dfrac{\text{in.}}{\text{min}}$; $209.4 \dfrac{\text{ft}}{\text{min}}$

37. (a) $5\pi \dfrac{\text{rad}}{\text{sec}}$ (b) 6.69 mph

39. 15.9 ft

Section 6.2, page 449

1. -1 **3.** 0   **5.** 0

7. 0   **9.** 0

11. $\sin t = 1/\sqrt{5}, \cos t = -2/\sqrt{5}, \tan t = -1/2$

13. $\sin t = -4/5, \cos t = -3/5, \tan t = 4/3$

15. $\sqrt{3}/2, 1/2, \sqrt{3}$ **17.** $-\sqrt{2}/2, \sqrt{2}/2, -1$

19. $\sqrt{2}/2, -\sqrt{2}/2, -1$ **21.** $1/2, -\sqrt{3}/2, \dfrac{-\sqrt{3}}{3}$

23. $\sin\left(-\dfrac{23\pi}{6}\right) = \dfrac{1}{2}, \cos\left(-\dfrac{23\pi}{6}\right) = \dfrac{\sqrt{3}}{2}, \tan\left(-\dfrac{23\pi}{6}\right) = \dfrac{\sqrt{3}}{3}$

25. $\sin\left(-\dfrac{19\pi}{3}\right) = -\dfrac{\sqrt{3}}{2}, \cos\left(-\dfrac{19\pi}{3}\right) = \dfrac{1}{2}, \tan\left(-\dfrac{19\pi}{3}\right) = -\sqrt{3}$

27. $\sin\left(-\dfrac{15\pi}{4}\right) = \dfrac{\sqrt{2}}{2}, \cos\left(-\dfrac{15\pi}{4}\right) = \dfrac{\sqrt{2}}{2}, \tan\left(-\dfrac{15\pi}{4}\right) = 1$

29. $\sin\left(-\dfrac{17\pi}{2}\right) = -1, \cos\left(-\dfrac{17\pi}{2}\right) = 0,$

   $\tan\left(-\dfrac{17\pi}{2}\right)$ is not defined

31. $-\sqrt{3}/2$ **33.** $-\sqrt{2}/2$ **35.** $\dfrac{\sqrt{2}}{4}(1 - \sqrt{3})$

37. $\sin t = 5/\sqrt{34}; \cos t = 3/\sqrt{34}; \tan t = 5/3$

39. $\sin t = -5/\sqrt{41}; \cos t = -4/\sqrt{41}; \tan t = 5/4$

41. $\sin t = -8/\sqrt{67}; \cos t = \sqrt{3/67}; \tan t = -8/\sqrt{3}$

43. (a) 467 mph   (b) 1458 mph

45. (a)

Date	Jan 1	Mar 1	May 1
Average Temperature	31.7	43.8	68.3

Date	July 1	Sept 1	Nov 1
Average Temperature	80.9	69.0	44.5

(b)

Date	June 1	June 4	June 7	June 10
Average Temperature	77.4	78.0	78.6	79.1

Date	June 13	June 16	June 19	June 22
Average Temperature	79.6	80.0	80.3	80.5

Date	June 25	June 28
Average Temperature	80.7	80.8

47. $-2/\pi$ **49.** $\dfrac{-\sqrt{3} - 1}{7\pi}$

51. $\dfrac{2\sqrt{2}}{3\pi}$ **53.** $\dfrac{4\sqrt{3}}{\pi}$

55. (a) $-.420686; -.416601; -.416192; -.416151$
    (b) $\cos 2 \approx -.416147$

57. $\sin t = -2/\sqrt{5}; \cos t = 1/\sqrt{5}; \tan t = -2$

59. $\sin t = -5/\sqrt{34}, \cos t = 3/\sqrt{34}, \tan t = -5/3$

61. $\sin t = 1/\sqrt{5}, \cos t = -2/\sqrt{5}, \tan t = -1/2$

63. Quadrant I: $\sin t(+), \cos t(+), \tan t(+)$;
    Quadrant II: $\sin t(+), \cos t(-), \tan t(-)$;
    Quadrant III: $\sin t(-), \cos t(-), \tan t(+)$;
    Quadrant IV: $\sin t(-), \cos t(+), \tan t(-)$

65. Positive since $0 < 1 < \pi/2$

67. Negative since $\pi/2 < 3 < \pi$

69. Positive since $0 < 1.5 < \pi/2$

71. $t = \dfrac{\pi}{2} + 2\pi n$, n any integer   **73.** $t = \pi n$, n any integer

75. $t = \dfrac{\pi}{2} + \pi n$, n any integer

77. $\sin(\cos 0) = \sin 1$, while $\cos(\sin 0) = \cos 0 = 1$. Since $\sin 1 < 1$ (draw a picture!), $\cos(\sin 0)$ is larger than $\sin(\cos 0)$.

79. (a) Each horse moves through an angle of 2π radians in 1 min. The angle between horses A and B is $\pi/4$ radians ($= \frac{1}{8}$ of 2π radians). It takes $\frac{1}{8}$ min for each horse to move through an angle of $\pi/4$ radians. Thus the position occupied by B at time t will be occupied by A $\frac{1}{8}$ min later, that is, at time $t + \frac{1}{8}$. Therefore, $B(t) = A\left(t + \frac{1}{8}\right)$.

(b) $C(t) = A\left(t + \frac{1}{3}\right)$.

(c) $E(t) = D\left(t + \frac{1}{8}\right)$; $F(t) = D\left(t + \frac{1}{3}\right)$.

(d) The triangles in Figure S are similar, so that $\dfrac{5}{1} = \dfrac{D(t)}{A(t)}$. Therefore, $D(t) = 5A(t)$.

(e) $E(t) = 5B(t) = 5A\left(t + \frac{1}{8}\right)$; $F(t) = 5C(t) = 5A\left(t + \frac{1}{3}\right)$.

(f) Since horse A travels through an angle of 2π radians each minute and its starting angle is 0 radians, then at the end of t min horse A will be on the terminal side of an angle of $2\pi t$ radians, at the point where it intersects the unit circle. $A(t)$ is the second coordinate of this point; hence, $A(t) = \sin(2\pi t)$.

(g) $B(t) = A\left(t + \frac{1}{8}\right) = \sin\left[2\pi\left(t + \frac{1}{8}\right)\right] = \sin(2\pi t + \pi/4)$; $C(t) = \sin(2\pi t + 2\pi/3)$

(h) $D(t) = 5\sin(2\pi t)$; $E(t) = 5\sin(2\pi t + \pi/4)$; $f(t) = 5\sin(2\pi t + 2\pi/3)$

Section 6.3, page 464

1. $(fg)(t) = 3\sin^2 t + 6\sin t \cos t$

3. $3\sin^3 t + 3\sin^2 t \tan t$ **5.** $(\cos t - 2)(\cos t + 2)$

7. $(\sin t - \cos t)(\sin t + \cos t)$

9. $(\tan t + 3)^2$ **11.** $(3\sin t + 1)(2\sin t - 1)$

13. $(\cos^2 t + 5)(\cos t + 1)(\cos t - 1)$

15. $(f \circ g)(t) = \cos(2t + 4)$, $(g \circ f)(t) = 2\cos t + 4$

17. $(f \circ g)(t) = \tan(t^2 + 2)$, $(g \circ f)(t) = \tan^2(t + 3) - 1$

19. Yes   **21.** No   **23.** No

25. $\sin t = -\sqrt{3}/2$ **27.** $\sin t = \sqrt{3}/2$

29. $-3/5$ **31.** $-3/5$ **33.** $3/4$

35. $-3/4$ **37.** $-\sqrt{21}/5$ **39.** $-2/5$

41. $-\sqrt{21}/5$ **43.** $\dfrac{\sqrt{2 + \sqrt{2}}}{2}$ **45.** $\dfrac{\sqrt{2 - \sqrt{2}}}{2}$

47. $\sin^2 t - \cos^2 t$ **49.** $\sin t$

51. $|\sin t \cos t|\sqrt{\sin t}$ **53.** $1/4$

55. $\cos t + 2$ **57.** $\cos t$

59. (a) $Y_2 = f(t) = 22.7\sin(.52x - 2.18) + 49.6$ and $Y_3 = f(x + 12.083) = 22.7\sin(.52(x + 12.083)) - 2.18)$

X	Y2
1	26.99
2	28.974
3	36.411
4	47.334
5	58.056
6	67.932
7	72.161

Y2∎22.7sin(.52X...

X	Y2
8	70.426
9	63.185
10	52.353
11	40.794
12	31.562
13	27.099
14	28.584

Y2∎22.7sin(.52X...

X	Y3
1	26.99
2	28.974
3	36.411
4	47.334
5	58.056
6	67.932
7	72.161

Y3∎Y2(X+12.083)

X	Y3
8	70.426
9	63.185
10	52.353
11	40.794
12	31.562
13	27.099
14	28.584

Y3∎Y2(X+12.083)

(b) Yes; the period is 12.083, which is a reasonable approximation of the expected period of 12 months.

61. (a) 0; about 76.5%; about 51.3%

(b) $Y_4 = g(t) = .5\left(1 - \cos\dfrac{2\pi t}{29.5}\right)$ and $Y_5 = g(t + 29.5)$ $= .5\left(1 - \cos\dfrac{2\pi(t + 29.5)}{29.5}\right)$

X	Y4	Y5
0	0	0
1	.0113	.0113
2	.04468	.04468
3	.09864	.09864
4	.17074	.17074
5	.25772	.25772
6	.35565	.35565

Y4∎.5(1-cos(2πX...

X	Y4	Y5
7	.46011	.46011
8	.56636	.56636
9	.66962	.66962
10	.76521	.76521
11	.84882	.84882
12	.91666	.91666
13	.96567	.96567

Y5∎Y4(X+29.5)

X	Y4	Y5
14	.99363	.99363
15	.99929	.99929
16	.98238	.98238
17	.94368	.94368
18	.88492	.88492
19	.80876	.80876
20	.71865	.71865

Y4∎.5(1-cos(2πX...

X	Y4	Y5
21	.61866	.61866
22	.51331	.51331
23	.40736	.40736
24	.30559	.30559
25	.21261	.21261
26	.13261	.13261
27	.06922	.06922

Y5∎Y4(X+29.5)

X	Y4	Y5
28	.0253	.0253
29	.00283	.00283
30	.00283	.00283
31	.0253	.0253
32	.06922	.06922
33	.13261	.13261
34	.21261	.21261

Y4∎.5(1-cos(2πX...

(c) Late on day 14 ($t = 14.75$)

63. $f(t + k) = \cos 3\left(t + \dfrac{2\pi}{3}\right) = \cos(3t + 2\pi) = \cos 3t = f(t)$

65. $f(t + 2) = \sin(\pi(t + 2)) = \sin(\pi t + 2\pi) = \sin(\pi t) = f(t)$

67. $f\left(t + \dfrac{\pi}{2}\right) = \tan 2\left(t + \dfrac{\pi}{2}\right) = \tan(2t + \pi) = \tan(2t) = f(t)$

69. (a) There is no such number k.
(b) If we substitute $t = 0$ in $\cos(t + k) = \cos t$, we get $\cos k = \cos 0 = 1$.
(c) If there were such a number k, then by part (b), $\cos k = 1$, which is impossible by part (a). Therefore, there is no such number k, and the period is 2π.

Section 6.4, page 474

1. $t = \ldots, -2\pi, -\pi, 0, \pi, 2\pi, \ldots$; or $t = \pi k$, where k is any integer

3. $t = \ldots, -7\pi/2, -3\pi/2, \pi/2, 5\pi/2, 9\pi/2, \ldots$; or $t = \pi/2 + 2\pi k$, where k is any integer

5. $t = \ldots, -3\pi, -\pi, \pi, 3\pi, \ldots$; or $t = \pi + 2k\pi$, where k is any integer

7. 11   **9.** 1.4

11. Shift the graph of f vertically 3 units upward.

13. Reflect the graph of f in the horizontal axis.

15. Shift the graph of f vertically 5 units upward.

17. Stretch the graph of f away from the horizontal axis by a factor of 3.

19. Stretch the graph of f away from the horizontal axis by a factor of 3, then shift the resulting graph vertically 2 units upward.

21. Shift the graph of f horizontally 2 units to the right.

23. D   **25.** B   **27.** F   **29.** G

31. 2 solutions   **33.** 2 solutions

35. 2 solutions   **37.** 2 solutions

39. Possibly an identity   **41.** Possibly an identity

43. Possibly an identity   **45.** Not an identity

47. Possibly an identity   **49.** Not an identity

51. No   **53.** Yes; period 2π

55. Yes; period 2π **57.** No

59. No

61. (a) Yes if proper value of k is used; no
(b) $0, 2\pi, 4\pi, 6\pi$, etc. So why do the graphs look identical?

63. (a) 80
(b) 14 or 15 on 96-pixel-wide screens; up to 40–50 on wider screens; quite different from part (a). Explain what's going on. [*Hint:* How many points have to be plotted in order to get even a rough approximation of one full wave? How many points is the calculator plotting for the entire graph?]

65. (a) $-\pi \leq t \leq \pi$
(b) $n = 15$; $f_{15}(2)$ and $g(2)$ are identical in the first nine decimal places and differ in the tenth, a very good approximation.

67. $r(t)/s(t)$, where $r(t) = f_{15}(t)$ in Exercise 65 and $s(t) = f_{16}(t)$ in Exercise 66.

69. The y-coordinate of the new point is the same as the x-coordinate of the point on the unit circle. To explain what's going on, look at the definition of the cosine function.

Section 6.5, page 486

1. Amplitude: 3, period: π, phase shift: $\dfrac{\pi}{2}$

3. Amplitude: 5, period: $\dfrac{2\pi}{5}$, phase shift: $\dfrac{-1}{25}$

5. Amplitude: 1, period: 1, phase shift: 0

7. Amplitude: 6, period: $\dfrac{2}{3}$, phase shift: $\dfrac{-1}{3\pi}$

9. $f(t) = 3 \sin\left(8t - \dfrac{8\pi}{5}\right)$ **11.** $f(t) = \dfrac{3}{4} \sin(\pi t)$

13. $f(t) = 7 \sin\left(\dfrac{6\pi}{5}t + \dfrac{3\pi^2}{5}\right)$

15. $f(t) = 2 \sin 4t$

17. $f(t) = 1.5 \cos \dfrac{t}{2}$

19. (a) $\pi/100$
(b) The graph makes 200 complete waves between 0 and 2π.
(c) $0 \leq x \leq \pi/25$; $-2 \leq y \leq 2$

21. (a) $\dfrac{2\pi}{900}$
(b) The graph makes 900 complete waves between 0 and 2π.
(c) $0 \leq x \leq \dfrac{2\pi}{225}$; $-2 \leq y \leq 2$

23. (a) $f(t) = -12 \sin\left(10t + \dfrac{\pi}{2}\right)$
(b) $g(t) = -12 \cos 10t$

25. (a) $f(t) = -\sin 2t$
(b) $g(t) = -\cos\left(2t - \dfrac{\pi}{2}\right)$

27.

29.

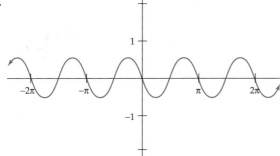

31.

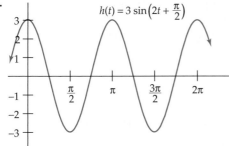

$$h(t) = 3 \sin\left(2t + \frac{\pi}{2}\right)$$

33. Local maximum at $t = 5\pi/6 \approx 2.6180$; local minimum at $t = 11\pi/6 \approx 5.7596$

35. Local maxima at $t = \pi/6 \approx .5236$, $t = 5\pi/6 \approx 2.6180$, $t = 3\pi/2 \approx 4.7124$; local minima at $t = \pi/2 \approx 1.5708$, $t = 7\pi/6 \approx 3.6652$, $t = 11\pi/6 \approx 5.7596$

37. $A \approx 3.606$; $b = 1$; $c \approx .5880$

39. $A \approx 3.8332$; $b = 4$; $c \approx 1.4572$

41. The waves do not have the same amplitude.

43. (a) The person's blood pressure is 134/92.
(b) The pulse rate is 75.

45. (a) Maximum is 79.5 cubic inches; minimum is 30.5 cubic inches.
(b) every 4 seconds
(c) 15

47. Period: $\dfrac{1}{980,000}$ seconds; frequency: 980,000

49. $f(t) = 125 \sin(\pi t/5)$

51. $f(t) = \cos 20\pi t + \sqrt{16 - \sin^2(20\pi t)}$

53. $h(t) = 6 \sin(\pi t/2)$

55. $h(t) = 6 \cos(\pi t/2)$

57. $d(t) = 10 \sin(\pi t/2)$

59. (a) At least four (starting point, high point, low point, ending point)
(b) 301
(c) Every calculator is different; the TI-84+ plots 95 points; others plot as many as 239.
(d) Obviously, 239 points (or fewer) are not enough when the absolute minimum is 301.

61. (a)

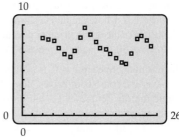

(b) The data appears to be approximately periodic. An appropriate model is $f(t) = 1.1371 \sin(.6352t - .6366) + 7.4960$.
(c) No; the model is only a fair approximation of the data and, in any case, unemployment is hard to predict.
(d) $g(x) = 24.85 \sin(.52x + (-2.15)) + 48.95$
(e) 12.05 months, yes it fits better.

63. (a) $f(t) = 2.77 \cdot \sin(.39x + 1.91) + 2.52$
(b) $f(x)$ has period ≈ 16.1. One would expect a period of 12.
(c)

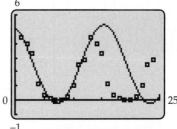

(d) $g(x) = 2.17 \sin(.51x + 1.04) + 1.71$
(e) $g(x)$ has period ≈ 12.3, and it seems to fit better than $f(x)$.

65. (a) $k = 9.8/\pi^2$
(b) When k is replaced by $(k + .01\%$ of $k)$, the value of ω changes and the period of the pendulum becomes approximately 2.000099998 sec, meaning that the clock loses .000099998 sec every 2 sec, for a total of approximately 397.43 sec (6.62 min) during the three months.

Special Topics 6.5.A, page 496

1. $A \approx 2.2361$, $b = 1$, $c \approx 1.1071$

3. $A \approx 5.3852$, $b = 4$, $c \approx -1.1903$

5. $A \approx 5.1164$, $b = 3$, $c \approx -.7442$

7. $0 \leq t \leq 2\pi$ and $-5 \leq y \leq 5$ (one period)

9. $-10 \leq t \leq 10$ and $-10 \leq y \leq 10$

11. $0 \leq t \leq \pi/25$ and $-2 \leq y \leq 2$ (one period)

13. $0 \leq t \leq .04$ and $-10 \leq y \leq 10$ (one period)

15. $0 \leq t \leq 20$ and $-11 \leq y \leq 11$ (one period)

17. To the left of the y-axis, the graph lies above the t-axis, which is a horizontal asymptote of the graph. To the right of the y-axis, the graph makes waves of amplitude 1, of shorter and shorter period.
Window: $-3 \leq t \leq 3.2$ and $-2 \leq y \leq 2$

19. The graph is symmetric with respect to the *y*-axis and consists of waves along the *t*-axis, whose amplitude slowly increases as you move farther from the origin in either direction.
Window: $-30 \leq t \leq 30$ and $-6 \leq y \leq 6$

21. The graph which is not defined at $t = 0$, is symmetric with respect to the *y*-axis and consists of waves along the *t*-axis whose amplitude rapidly decreases as you move farther from the origin in either direction.
Window: $-30 \leq t \leq 30$ and $-.3 \leq y \leq 1$

23. The function is periodic with period π. (Why?) The graph lies on or below the *t*-axis because the logarithmic function is negative for numbers between 0 and 1 and $|\cos t|$ is always between 0 and 1. The graph has vertical asymptotes when $t = \pm\pi/2, \pm3\pi/2, \pm5\pi/2, \pm7\pi/2, \ldots$ ($\cos t = 0$ at these points and ln 0 is not defined).
Window: $-2\pi \leq t \leq 2\pi$ and $-3 \leq y \leq 1$ (four periods).

25. (a)

(b) Maximum at $t = .2$, $t = 24.4$, and $t = 48.6$, that is, shortly after midnight. Minimum at $t = 17.1$, $t = 41.3$, and $t = 65.5$, that is, around 5:30 P.M.

(c) 24.2 hours

Section 6.6, page 502

1. Fourth quadrant   **3.** Second quadrant

5. Fourth quadrant

7. $\sin t = 4/5$, $\cos t = 3/5$, $\tan t = 4/3$, $\cot t = 3/4$, $\sec t = 5/3$, $\csc t = 5/4$

9. $\sin t = 12/13$, $\cos t = -5/13$, $\tan t = -12/5$, $\cot t = -5/12$, $\sec t = -13/5$, $\csc t = 13/12$

11. $\sin t = 5/\sqrt{26}$, $\cos t = -1/\sqrt{26}$, $\tan t = -5$, $\cot t = -1/5$, $\sec t = -\sqrt{26}$, $\csc t = \sqrt{26}/5$

13. $\sin t = \sqrt{3}/\sqrt{5}$, $\cos t = \sqrt{2}/\sqrt{5}$, $\tan t = \sqrt{3}/\sqrt{2}$, $\cot t = \sqrt{2}/\sqrt{3}$, $\sec t = \sqrt{5}/\sqrt{2}$, $\csc t = \sqrt{5}/\sqrt{3}$

15. $\sin t = \dfrac{3}{\sqrt{12 + 2\sqrt{2}}}$, $\cos t = \dfrac{1 + \sqrt{2}}{\sqrt{12 + 2\sqrt{2}}}$,
$\tan t = \dfrac{3}{1 + \sqrt{2}}$, $\cot t = \dfrac{1 + \sqrt{2}}{3}$,
$\sec t = \dfrac{\sqrt{12 + 2\sqrt{2}}}{1 + \sqrt{2}}$, $\csc t = \dfrac{\sqrt{12 + 2\sqrt{2}}}{3}$

17. About 154.74 ft

19. (a) $0 \leq x \leq 150$ and $-5 \leq y \leq 20$
(b) About 15.16 ft   (c) About 143 ft

21. $\sin\left(\dfrac{4\pi}{3}\right) = -\dfrac{\sqrt{3}}{2}$, $\cos\left(\dfrac{4\pi}{3}\right) = -\dfrac{1}{2}$, $\tan\left(\dfrac{4\pi}{3}\right) = \sqrt{3}$,
$\cot\left(\dfrac{4\pi}{3}\right) = \dfrac{1}{\sqrt{3}}$, $\sec\left(\dfrac{4\pi}{3}\right) = -2$, $\csc\left(\dfrac{4\pi}{3}\right) = \dfrac{-2}{\sqrt{3}}$

23. $\sin\left(\dfrac{7\pi}{4}\right) = -\dfrac{\sqrt{2}}{2}$, $\cos\left(\dfrac{7\pi}{4}\right) = \dfrac{\sqrt{2}}{2}$, $\tan\left(\dfrac{7\pi}{4}\right) = -1$,
$\cot\left(\dfrac{7\pi}{4}\right) = -1$, $\sec\left(\dfrac{7\pi}{4}\right) = \sqrt{2}$, $\csc\left(\dfrac{7\pi}{4}\right) = -\sqrt{2}$

25. $\sin\left(\dfrac{-11\pi}{4}\right) = -\dfrac{\sqrt{2}}{2}$, $\cos\left(\dfrac{-11\pi}{4}\right) = \dfrac{-\sqrt{2}}{2}$,
$\tan\left(\dfrac{-11\pi}{4}\right) = 1$, $\cot\left(\dfrac{-11\pi}{4}\right) = 1$,
$\sec\left(\dfrac{-11\pi}{4}\right) = -\sqrt{2}$, $\csc\left(\dfrac{-11\pi}{4}\right) = -\sqrt{2}$

27. -3.8287

29. (a) 5.6511; 5.7618; 5.7731; 5.7743   (b) $(\sec 2)^2$

31. $\cos t + \sin t$ **33.** $1 - 2\sec t + \sec^2 t$

35. $\cot^3 t - \tan^3 t$ **37.** $\csc t(\sec t - \csc t)$

39. $(-1)(\tan^2 t + \sec^2 t)$

41. $(\cos^2 t + 1 + \sec^2 t)(\cos t - \sec t)$

43. $\cot t$ **45.** $\dfrac{2\tan t + 1}{3\sin t + 1}$

47. $4 - \tan t$

49. $\tan t = \dfrac{\sin t}{\cos t} = \dfrac{1}{\cos t/\sin t} = \dfrac{1}{\cot t}$

51. $1 + \cot^2 t = 1 + \dfrac{\cos^2 t}{\sin^2 t} = \dfrac{\sin^2 t + \cos^2 t}{\sin^2 t} = \dfrac{1}{\sin^2 t} = \csc^2 t$

53. $\sec(-t) = \dfrac{1}{\cos(-t)} = \dfrac{1}{\cos t} = \sec t$

55. $\sin t = \dfrac{\sqrt{3}}{2}$, $\cos t = -\dfrac{1}{2}$, $\tan t = -\sqrt{3}$,
$\csc t = \dfrac{2\sqrt{3}}{3}$, $\sec t = -2$, $\cot t = -\dfrac{\sqrt{3}}{3}$

57. $\sin t = 1$, $\cos t = 0$, $\tan t$ is undefined, $\cot t = 0$, $\sec t$ is undefined, $\csc t = 1$

59. $\sin t = \dfrac{12}{13}$, $\cos t = -\dfrac{5}{13}$, $\tan t = -\dfrac{12}{5}$,
$\cot t = -\dfrac{5}{12}$, $\sec t = -\dfrac{13}{5}$, $\csc t = \dfrac{13}{12}$

61. Possibly an identity   **63.** Not an identity

65. Look at the graph of $y = \sec t$ in Figure 6–84 on page 501. If you draw in the line $y = t$, it will pass through $(-\pi/2, -\pi/2)$ and $(\pi/2, \pi/2)$, and obviously will not intersect the graph of $y = \sec t$ when $-\pi/2 \leq t \leq \pi/2$. But it will intersect each part of the graph that lies above the horizontal axis, to the right of $t = \pi/2$; it will also intersect those parts that lie below the horizontal axis, to the left of $-\pi/2$. The first coordinate of each of these infinitely many intersection points will be a solution of $\sec t = t$.

67. (a) $\dfrac{\cos\theta\sin\theta}{2}$ (b) $\dfrac{\tan\theta}{2}$ (c) $\dfrac{\theta}{2}$

Chapter 6 Review, page 506

1. $\dfrac{\pi}{3}$ **3.** $324°$ **5.** $\dfrac{11\pi}{9}$

7. $-495°$ **9.** $\cos v = -\dfrac{1}{3}$ **11.** 0

13. $.809$ **15.** $\sqrt{3}/3$ **17.** $\dfrac{\sqrt{3}}{2}$

19. 1.701 **21.** $\dfrac{-\sqrt{3}}{3}$ **23.** -2

25.

t	0	$\pi/6$	$\pi/4$	$\pi/3$	$\pi/2$
$\sin t$	0	$1/2$	$\sqrt{2}/2$	$\sqrt{3}/2$	1
$\cos t$	1	$\sqrt{3}/2$	$\sqrt{2}/2$	$1/2$	0

27. $9/4$ **29.** 0 **31.** $-\sqrt{\dfrac{2}{3}}$

33. $-\dfrac{3}{5}$ **35.** $-\dfrac{12}{13}$ **37.** $-\dfrac{1}{2}$

39. $\dfrac{\sqrt{3}}{2}$ **41.** (c)   **43.** $\dfrac{7}{\sqrt{58}}$

45. $\dfrac{-7}{3}$ **47.** $\dfrac{-\sqrt{58}}{3}$ **49.** $y = -\sqrt{3}x$

51. $\dfrac{-3}{2}$ **53.** $\dfrac{2}{\sqrt{13}}$ **55.** $\dfrac{\sqrt{13}}{3}$

57. See Figure 6–85 on page 502.

59. (d)   **61.** $-\dfrac{3}{2}$ **63.** 0

65. (d)   **67.** $-\dfrac{1}{\sqrt{3}}$

69. (a) $\dfrac{3}{2}$ (b) $t = \dfrac{\pi}{5}$

71.

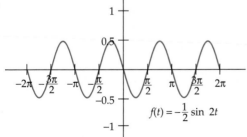

$f(t) = -\frac{1}{2}\sin 2t$

73. Not an identity   **75.** Possibly an identity

77. Possibly an identity   **79.** $\dfrac{1}{2}$

81. (a) Minimum: 8.9 hours; Maximum: 15.1 hours
(b) January 1–March 2 and October 9–December 31

83. $2\cos(5t/2)$

85. $f(t) = 8\sin\left(\dfrac{2\pi t - 28\pi}{5}\right)$ is one possibility.

87. $A \approx 10.5588$, $b = 4$, $c = .4581$

89. $0 \le t \le \pi/50$ and $-5 \le y \le 5$ (one period)

91. (a)

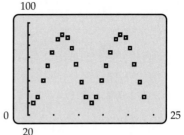

(b)

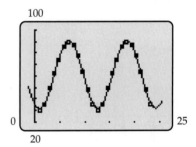

```
SinReg
y=a*sin(bx+c)+d
a=24.62950244
b=.5215404022
c=-2.086057976
d=56.32216472
```

(c) The period is 12.05 months. The graph of the function, superimposed on the data points, suggests a good fit.

Chapter 6 Test, page 510

1. $\dfrac{\pi}{36}$

2. (a) $-\dfrac{\sqrt{3}}{2}$ (b) $\dfrac{1}{2}$ (c) $-\sqrt{3}$

3. $(\sin t)(\sin t + 1)(\sin t - 1)$ **4.** About 13,888.89 radians

5. (a) (iii)   (b) (i)   (c) (iv)

6. (a) $\dfrac{2}{\sqrt{13}}$ (b) $-\dfrac{3}{\sqrt{13}}$ (c) $-\dfrac{2}{3}$

7. $\dfrac{-4}{\sqrt{17}}$ **8.** 16 sq cm

9. (a) $\dfrac{8\pi}{3}$ radians   (b) $4.5°$

10. (a) $\dfrac{2}{\sqrt{53}}$ (b) $\dfrac{7}{\sqrt{53}}$ (c) $\dfrac{2}{7}$

11. $-3/4$

12. Linear speed: 10π inches per minute; angular speed: 2π radians per minute

13. $\dfrac{8 \cos t}{(\sin t)(\cos t + \sin t)}$

14. $\dfrac{4(3 - \sqrt{3})}{\pi}$

15. $-\dfrac{\sqrt{2}}{2}$

16. $\dfrac{2 \tan t - 1}{3 \sin t + 1}$

17. 6.2

18. $t = 1/2, 3/2, 5/2, 7/2, 9/2,$ and $11/2$

19. (a) $\dfrac{1}{2}$ (b) $\dfrac{\sqrt{3}}{2}$ (c) $\dfrac{\sqrt{3}}{3}$

(d) $\sqrt{3}$ (e) 2 (f) $\dfrac{2}{\sqrt{3}}$

20. Many correct answers, including:

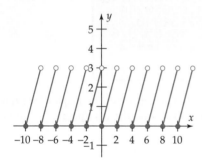

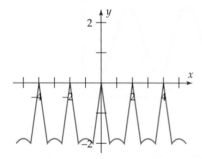

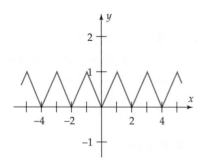

21. The function has period $\dfrac{\pi}{550}$ (why?). So any viewing window with the t-axis of length $6 \cdot \dfrac{\pi}{550} = \dfrac{3\pi}{275} \approx .03427$ (for instance, $0 \le t \le \dfrac{3\pi}{275}$) and $-1 \le y \le 1$ will show exactly six complete waves.

22. $\sin^3 t - \csc^3 t$

23. 4

24. (a) $f(t) = 25.63 \sin(.4914t - 1.9451) + 47.6516$
(b) $g(t) = 24.85 \sin(.5213t - 2.1482) + 49.1479$
(c) The model in part (b) (its period is 12.05, whereas the model in part (a) has period 12.79).

25. $1 + \tan^2 x = 1 + \left(\dfrac{\sin x}{\cos x}\right)^2 = 1 + \dfrac{\sin^2 x}{\cos^2 x} =$

$\dfrac{\cos^2 x + \sin^2 x}{\cos^2 x} = \dfrac{1}{\cos^2 x}$ (by the Pythagorean Identity)

$= \left(\dfrac{1}{\cos x}\right)^2 = \sec^2 x$

26. Definitely not an identity.

27. (a) amplitude 3, period $\pi/2$, phase shift $= -7/4$
(b) $f(t) = 6 \sin(\pi t + \pi)$ or $g(t) = 6 \cos(\pi t + \pi)$

28. (a) $\dfrac{\sqrt{13}}{7}$ (b) $\dfrac{-6}{\sqrt{13}}$ (c) $-\dfrac{\sqrt{13}}{6}$

(d) $-\dfrac{7}{6}$ (e) $\dfrac{7}{\sqrt{13}}$

29. $\left(\dfrac{\pi}{12}, 2\right), \left(\dfrac{5\pi}{12}, 2\right),$ and $\left(\dfrac{3\pi}{4}, 2\right)$

30. $A = 1.6565, b = 3, c = 2.4270$

Chapter 7

Section 7.1, page 522

1. Possibly an identity

3. Possibly an identity

5. B

7. E

9. $\tan x \cos x = \dfrac{\sin x}{\cos x} \cdot \cos x = \sin x$

11. $\cos x \sec x = \cos x \cdot \dfrac{1}{\cos x} = 1$

13. $\tan x \csc x = \dfrac{\sin x}{\cos x} \cdot \dfrac{1}{\sin x} = \dfrac{1}{\cos x} = \sec x$

15. $\dfrac{\tan x}{\sec x} = \dfrac{\sin x}{\cos x} \div \dfrac{1}{\cos x} = \dfrac{\sin x}{\cos x} \cdot \dfrac{\cos x}{1} = \sin x$

17. $(1 + \cos x)(1 - \cos x) = 1 - \cos^2 x = \sin^2 x$

19. $\cot x \sec x \sin x = \dfrac{\cos x}{\sin x} \cdot \dfrac{1}{\cos x} \cdot \sin x$

$= \dfrac{\sin x \cos x}{\sin x \cos x} = 1$

21. $\tan x + \cot x = \dfrac{\sin x}{\cos x} + \dfrac{\cos x}{\sin x}$

$= \dfrac{\sin x \sin x}{\cos x \sin x} + \dfrac{\cos x \cos x}{\cos x \sin x} = \dfrac{\sin^2 x + \cos^2 x}{\cos x \sin x}$

$= \dfrac{1}{\cos x \sin x} = \dfrac{1}{\cos x} \cdot \dfrac{1}{\sin x} = \sec x \csc x$

23. $\cos x - \cos x \sin^2 x = \cos x (1 - \sin^2 x)$

$= \cos x (\cos^2 x) = \cos^3 x$

25. This is not an identity.

67. (a) $t = \tan^{-1}\left(\dfrac{h}{4}\right)$

(b) $t \approx .0624$ radians; $t \approx .2450$ radians; $t \approx .4636$ radians

(c) about 1.6912 mi

69. (a) $t = \tan^{-1}\left(\dfrac{30}{x}\right) - \tan^{-1}\left(\dfrac{6}{x}\right)$ (b) about 13.4 ft

71. (a) $\theta = \tan^{-1}\left(\dfrac{25}{x}\right) - \tan^{-1}\left(\dfrac{10}{x}\right)$ (b) 15.8 ft

73. No horizontal line intersects the graph of $f(x) = \csc x$ more than once when $-\pi/2 \le x \le \pi/2$ (see Figure 6–85). Hence, the restricted cosecant function has an inverse function, as explained in Section 3.7.

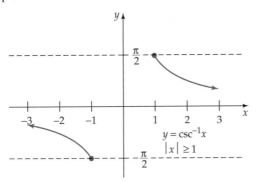

75. Let $\cos u = w$ with $0 \le u \le \pi$. Then $u = \cos^{-1} w$ and $\cos^{-1}(\cos u) = \cos^{-1} w = u$. Let $u = \cos^{-1} v$. Then $\cos u = v$ and $\cos(\cos^{-1} v) = \cos u = v$.

77. Let $u = \sin^{-1}(-x)$. Then

$$-x = \sin u \quad (-\pi/2 \le u \le \pi/2)$$
$$x = -\sin u$$
$$x = \sin(-u)$$
$$\sin^{-1} x = -u \quad (-\pi/2 \le -u \le \pi/2)$$
$$\sin^{-1} x = -\sin^{-1}(-x)$$
$$\sin^{-1}(-x) = -\sin^{-1} x$$

79. Let $u = \cos^{-1}(-x)$. Then $-x = \cos u$. Note that $0 \le u \le \pi$; hence, $-\pi \le -u \le 0$, which implies that $0 \le \pi - u \le \pi$. Hence,

$$x = -\cos u$$
$$x = \cos(\pi - u) \quad \text{(from the stated identity)}$$
$$\pi - u = \cos^{-1} x$$
$$u = \pi - \cos^{-1} x$$
$$\cos^{-1}(-x) = \pi - \cos^{-1} x$$

81. Let $u = \cot x$. Then

$$\tan(\pi/2 - x) = \cot x = u \quad (0 < u < \pi,$$
$$\text{hence, } -\pi/2 < \pi/2 - u < \pi/2)$$

$$\pi/2 - x = \tan^{-1}(u) = \tan^{-1}(\cot x)$$

83. Let $u = \sin^{-1} x$. Then $\sin u = x$, $(-\pi/2 < u < \pi/2)$

$$\tan u = \frac{\sin u}{\cos u} = \frac{\sin u}{\sqrt{1 - \sin^2 u}} = \frac{x}{\sqrt{1 - x^2}}$$

(positive square root, since u lies in quadrant I or IV)

$$u = \tan^{-1}\left(\frac{x}{\sqrt{1 - x^2}}\right)$$

$$\sin^{-1} x = \tan^{-1}\left(\frac{x}{\sqrt{1 - x^2}}\right)$$

85. The statement is false. For example, let $x = 1/2$.

$$\tan^{-1}\left(\frac{1}{2}\right) \approx .4636, \text{ but } \frac{\sin^{-1}(1/2)}{\cos^{-1}(1/2)} = \frac{\pi/6}{\pi/3} = \frac{1}{2}$$

Section 7.5, page 564

In the answers for this section $k = 0, \pm 1, \pm 2, \pm 3, \ldots$

1. $x = .4836 + 2k\pi$ or $2.6580 + 2k\pi$

3. $x = 2.1700 + 2k\pi$ or $-2.1700 + 2k\pi$

5. $x = -.3402 + k\pi$ **7.** $x = .4101 + k\pi$

9. $x = \pm 2.2459 + 2k\pi$ **11.** $x = .1193$ or 3.0223

13. $x = 1.3258$ or 4.4674

15. $x = \dfrac{\pi}{3} + 2k\pi$ or $\dfrac{2\pi}{3} + 2k\pi$

17. $x = -\dfrac{\pi}{3} + k\pi$

19. $x = 5\pi/6 + 2k\pi$ or $-5\pi/6 + 2k\pi$

21. $x = -\dfrac{\pi}{6} + 2k\pi$ or $\dfrac{7\pi}{6} + 2k\pi$

23. $x = \dfrac{\pi}{6} + 2k\pi$ or $\dfrac{5\pi}{6} + 2k\pi$

25. $\theta = 82.8°$ or $262.8°$ **27.** $\theta = 114.8°$ or $245.2°$

29. $\theta = 210°, 270°, 330°$ **31.** $\theta = 60°, 120°, 240°, 300°$

33. $\theta = 120°, 240°$ **35.** $\alpha = 65.4°$

37. $\alpha = 30°$ **39.** $\theta = 27.6°$

41. $\theta = 14.2°$

43. $x = -\dfrac{\pi}{6} + k\pi$ or $\dfrac{2\pi}{3} + k\pi$

45. $x = \pm\dfrac{\pi}{2} + 4k\pi$ **47.** $x = -\dfrac{\pi}{9} + \dfrac{k\pi}{3}$

49. $x = \pm.7381 + \dfrac{2k\pi}{3}$ **51.** $x = 2.2143 + 2k\pi$

53. $t = 1.8546; A = 11.18$ **55.** $t = 1.6961; A = .35$

57. $t = 1.9346, L = 16.47$ **59.** $t = 1.6236, L = 11.61$

61. (a) $A = 2x(2\cos 2x) = 4x\cos 2x$

(b) $x = .3050$ and $x = .5490$

63. $x = 5.9433$ and 3.4814

65. $x = \dfrac{3\pi}{4}, \dfrac{7\pi}{4}, 1.9513, 5.0929$

67. $x = \pi/2, 3\pi/2, \pi/4, 5\pi/4$

69. $x = \pi/2, 3\pi/2, \pi/6, 5\pi/6$

71. $x = .8481, 2.2935, 1.7682, 4.9098$

73. $x = .8213, 2.3203.$

75. $x = 1.2059, 4.3475, .3649,$ and 3.5065

77. $x = 1.0591, 4.2007, 2.8679,$ and 6.0095

79. No solution

81. $x = \pi/2, 3\pi/2, 7\pi/6, 11\pi/6$

83. $x = \pi/6, 5\pi/6$

85. $x = 7\pi/6, 11\pi/6, \pi/2$

87. $x = 0, \pi/3, 5\pi/3$

89. $x = .5275 + k\pi$ or $x = 1.6868 + k\pi$

91. $x = .4959, 1.2538, 1.5708, 1.8877, 2.6457,$ or $4.7124.$ Since the function is periodic with period 2π, all solutions are given by each of these six roots plus $2k\pi$.

93. $x = .1671, 1.8256, 2.8867,$ or $4.5453.$ Since the function is periodic with period 2π, all solutions are given by each of these four roots plus $2k\pi$.

95. $x = 1.2161 + 2k\pi$ or $x = 5.0671 + 2k\pi$

97. $x = 2.4620 + 2k\pi$ or $x = 3.8212 + 2k\pi$

99. $x = .5166 + 2k\pi$ or $x = 5.6766 + 2k\pi$

101. (a) days 60 and 282 (March 2 and October 10)
(b) day 171 (June 21)

103. $\theta = .5475$ or 1.0233 **105.** $13.25°$ or $76.75°$

107. The basic equations $\sin x = c$ and $\cos x = c$ have no solutions if $c > 1$ or $c < -1.$

109. The solutions $x = 0$ and $x = \pi$ are missed due to dividing by $\sin x$ (which is 0 when $x = 0$ or π).

Chapter 7 Review, page 568

1. $1/3 + \cot t$ **3.** $\sin^4 x$

5. $\sin^4 t - \cos^4 t = (\sin^2 t + \cos^2 t)(\sin^2 t - \cos^2 t)$
$= 1(\sin^2 t - (1 - \sin^2 t)) = \sin^2 t - 1$
$+ \sin^2 t = 2 \sin^2 t - 1$

7. Use Strategy 6. Prove that $\sin t \sin t = (1 - \cos t)(1 + \cos t)$ is an identity.
$\sin t \sin t = \sin^2 t = 1 - \cos^2 t = (1 - \cos t)(1 + \cos t)$

Therefore, $\dfrac{\sin t}{1 - \cos t} = \dfrac{1 + \cos t}{\sin t}.$

9. This is not an identity.

11. $(\sin x + \cos x)^2 - \sin 2x = \sin^2 x + 2 \sin x \cos x$
$+ \cos^2 x - \sin 2x = \sin^2 x + \cos^2 x + 2 \sin x \cos x$
$- 2 \sin x \cos x = 1 + 0 = 1$

13. $\dfrac{\tan x - \sin x}{2 \tan x} = \dfrac{\sin x/\cos x - \sin x}{2(\sin x/\cos x)}$

$= \dfrac{\dfrac{\sin x - \sin x \cos x}{\cos x}}{\dfrac{2 \sin x}{\cos x}} = \dfrac{\sin x - \sin x \cos x}{2 \sin x}$

$= \dfrac{\sin x(1 - \cos x)}{2 \sin x} = \dfrac{1 - \cos x}{2} = \sin^2(x/2)$

15. $\cos(x + y)\cos(x - y)$
$= (\cos x \cos y - \sin x \sin y)(\cos x \cos y + \sin x \sin y)$
$= \cos x^2 \cos^2 y - \sin^2 x \sin^2 y$
$= \cos^2 x (1 - \sin^2 y) - (1 - \cos^2 x)\sin^2 y$
$= \cos^2 x - \cos^2 x \sin^2 y - \sin^2 y + \cos^2 x \sin^2 y$
$= \cos^2 x - \sin^2 y$

17. Use Strategy 6. Prove that $(\sec x + 1)(\sec x - 1)$
$= \tan x \tan x$ is an identity.

$(\sec x + 1)(\sec x - 1) = \sec^2 x - 1 = \tan^2 x = \tan x \tan x$

Therefore, $\dfrac{\sec x + 1}{\tan x} = \dfrac{\tan x}{\sec x - 1}$ is an identity.

19. $\dfrac{1 + \tan^2 x}{\tan^2 x} = \dfrac{1}{\tan^2 x} + \dfrac{\tan^2 x}{\tan^2 x} = \cot^2 x + 1 = \csc^2 x$

21. $\tan^2 x - \sec^2 x = \sec^2 x - 1 - \sec^2 x = -1$
$= \csc^2 x - 1 - \csc^2 x = \cot^2 x - \csc^2 x$

23. $120/169$ **25.** $-\dfrac{56}{65}$ **27.** $\dfrac{\sqrt{45} + 1}{8}$ or $\dfrac{3\sqrt{5} + 1}{8}$

29. Yes. If $\sin x = 0$, $\sin 2x = 2 \sin x \cos x = 2 \cdot 0 \cdot \cos x = 0.$

31. Using the half-angle identity,

$\cos \dfrac{\pi}{12} = \cos\left[\dfrac{1}{2}\left(\dfrac{\pi}{6}\right)\right] = \sqrt{\dfrac{1 + \cos(\pi/6)}{2}} =$

$\sqrt{\dfrac{1 + \sqrt{3}/2}{2}} = \sqrt{\dfrac{2 + \sqrt{3}}{4}} = \dfrac{\sqrt{2 + \sqrt{3}}}{2}$

Using the subtraction identity for cosine,

$\cos\dfrac{\pi}{12} = \cos\left(\dfrac{\pi}{3} - \dfrac{\pi}{4}\right) = \cos\dfrac{\pi}{3}\cos\dfrac{\pi}{4} + \sin\dfrac{\pi}{3}\sin\dfrac{\pi}{4} =$

$\dfrac{1}{2} \cdot \dfrac{\sqrt{2}}{2} + \dfrac{\sqrt{3}}{2} \cdot \dfrac{\sqrt{2}}{2} = \dfrac{\sqrt{2} + \sqrt{6}}{4}$

Hence, $\dfrac{\sqrt{2 + \sqrt{3}}}{2} = \dfrac{\sqrt{2} + \sqrt{6}}{4}$ and $\sqrt{2 + \sqrt{3}} =$

$\dfrac{\sqrt{2} + \sqrt{6}}{2}.$

33. $\dfrac{\sqrt{6} + \sqrt{2}}{4}$ **35.** (a)   **37.** .96

39. $u = -2.24076 + k\pi$ $(k = 0, \pm 1, \pm 2, \ldots)$

41. $\pi/4$ **43.** $\pi/4$ **45.** $\pi/6$

47. $5\pi/6$ **49.** .75   **51.** $\pi/3$

53.

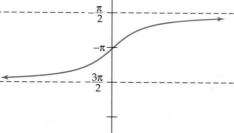
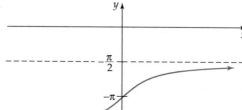

55. $\sqrt{15}/4$

In the following answers, $k = 0, \pm 1, \pm 2, \pm 3, \ldots$.

57. $x = \dfrac{\pi}{6} + 2k\pi$ or $\dfrac{5\pi}{6} + 2k\pi$

59. $x = \pm\dfrac{\pi}{4} + 2k\pi$ **61.** $x = -\pi/4 + k\pi$

63. $x = -\dfrac{\pi}{9} + \dfrac{2k\pi}{3}$ or $\dfrac{4\pi}{9} + \dfrac{2k\pi}{3}$

65. $x = .4115 + 2k\pi$ or $2.7301 + 2k\pi$

67. $x = 1.5042 + k\pi$

69. $x = \dfrac{\pi}{6} + 2k\pi$ or $\dfrac{5\pi}{6} + 2k\pi$

71. $x = -\dfrac{\pi}{6} + 2k\pi$ or $\dfrac{7\pi}{6} + 2k\pi$

73. $x = \pm\dfrac{\pi}{3} + k\pi$

75. $x = .8959 + 2k\pi$ or $2.2457 + 2k\pi$ **77.** $x = k\pi$

79. $x = .8419, 2.2997, 4.1784,$ or 5.2463. Since the function is periodic with period 2π, all solutions are given by $x = .8419 + 2k\pi$, $x = 2.2997 + 2k\pi$, and so on.

81. $\theta = 314.5°$ and $225.5°$ **83.** $\theta = 9.06°$ or $80.94°$

Chapter 7 Test, page 569

1. $1 - \sin x \cos x \tan x = 1 - \sin x \cos x \dfrac{\sin x}{\cos x} = 1 - \sin^2 x = \cos^2 x.$

2. $\dfrac{\sqrt{6} - \sqrt{2}}{4}$ **3.** $\sqrt{\dfrac{\sqrt{2} + 1}{\sqrt{2} - 1}}$

4. $\cos^2 x - \cot^2 x = \cos^2 x - \dfrac{\cos^2 x}{\sin^2 x} = \dfrac{\cos^2 x \sin^2 x - \cos^2 x}{\sin^2 x} = \dfrac{\cos^2 x (\sin^2 x - 1)}{\sin^2 x} = \dfrac{\cos^2 x}{\sin^2 x}(\sin^2 x - 1) = \cot^2 x (-\cos^2 x) = -\cos^2 x \cot^2 x.$

5. $\cos y$

6. (a) $-\dfrac{1519}{1681}$ (b) $\dfrac{720}{1681}$ (c) $-\dfrac{720}{1519}$

7. $(\sin^2 x - 1)(\cot^2 x + 1) = \sin^2 x \cot^2 x + \sin^2 x - \cot^2 x - 1 = \sin^2 x \dfrac{\cos^2 x}{\sin^2 x} + \sin^2 x - \cot^2 x - 1 = \cos^2 x + \sin^2 x - \cot^2 x - 1 = 1 - \cot^2 x - 1 = -\cot^2 x.$

8. $\dfrac{-\sqrt{3} - \sqrt{15}}{8}$ **9.** $\cos x$

10. Several possible proofs, including:

$\dfrac{1 - \tan x}{\cot x - 1} = \dfrac{1 - \dfrac{1}{\cot x}}{\cot x - 1} = \dfrac{\dfrac{\cot x - 1}{\cot x}}{\cot x - 1} = \dfrac{1}{\cot x} = \dfrac{1}{\dfrac{\cos x}{\sin x}} = \dfrac{\sin x}{\cos x} = \tan x.$

Alternatively, use Strategy 6 of Section 7.1 to prove the equivalent identity

$$(\cot x - 1)\tan x = 1 - \tan x$$

as follows:

$(\cot x - 1)\tan x = \cot x \tan x - \tan x = \dfrac{\cos x}{\sin x} \cdot \dfrac{\sin x}{\cos x} - \tan x = 1 - \tan x.$

11. $\sin x \sin(\pi - x) = \sin x (\sin \pi \cos x - \cos \pi \sin x) = \sin x (0 \cos x - (-1)\sin x) = \sin^2 x = 1 - \cos^2 x.$

12. (a) $\sqrt{\dfrac{1 + \sqrt{.84}}{2}}$ (b) $\sqrt{\dfrac{1 - \sqrt{.84}}{2}}$

13. $\dfrac{5\pi}{6}$ **14.** $x \approx \pm.7752 + 2k\pi$ **15.** $\dfrac{\pi}{4}$

16. $x \approx .4789$ and 5.8042 **17.** $t \approx .9273$

18. $x = \dfrac{\pi}{2} + 4k\pi$ and $x = \dfrac{3\pi}{2} + 4k\pi$ $(k = 0, \pm 1, \pm 2, \pm 3, \ldots)$

19. By the cofunction identity, $\sin x = \cos\left(\dfrac{\pi}{2} - x\right)$. Since $-\dfrac{\pi}{2} < x < \dfrac{\pi}{2}$, it follows that $0 < \dfrac{\pi}{2} - x < \pi$. Therefore, $\cos^{-1}(\sin x) = \cos^{-1}\left[\cos\left(\dfrac{\pi}{2} - x\right)\right] = \dfrac{\pi}{2} - x$ by the fact in the box on page 549

20. about .2266 radians (An angle of approximately 1.3442 radians is theoretically possible, but seems unlikely when throwing a baseball.)

Chapter 8

Section 8.1, page 582

	sin	cos	tan	csc	sec	cot
1.	$\dfrac{3\sqrt{13}}{13}$	$\dfrac{2\sqrt{13}}{13}$	$\dfrac{3}{2}$	$\dfrac{\sqrt{13}}{3}$	$\dfrac{\sqrt{13}}{2}$	$\dfrac{2}{3}$
3.	$\dfrac{7}{\sqrt{58}}$	$\dfrac{-3}{\sqrt{58}}$	$\dfrac{-7}{3}$	$\dfrac{\sqrt{58}}{7}$	$\dfrac{-\sqrt{58}}{3}$	$\dfrac{-3}{7}$
5.	$\dfrac{-\sqrt{22}}{11}$	$\dfrac{-3\sqrt{11}}{11}$	$\dfrac{\sqrt{2}}{3}$	$\dfrac{-\sqrt{22}}{2}$	$\dfrac{-\sqrt{11}}{3}$	$\dfrac{3\sqrt{2}}{2}$

7. $\sin \theta = \dfrac{\sqrt{2}}{\sqrt{11}}$ **9.** $\sin \theta = \dfrac{\sqrt{3}}{\sqrt{7}}$

      $\cos \theta = \dfrac{3}{\sqrt{11}}$ $\cos \theta = \dfrac{2}{\sqrt{7}}$

      $\tan \theta = \dfrac{\sqrt{2}}{3}$ $\tan \theta = \dfrac{\sqrt{3}}{2}$

11. $\sin \theta = \dfrac{h}{m}$ **13.** $c = 36$

      $\cos \theta = \dfrac{d}{m}$ **15.** $c = 36$

      $\tan \theta = \dfrac{h}{d}$ **17.** $c = 8.4$

19. $h = 25\sqrt{2}/2$ **21.** $h = 300$ **23.** $h = 50\sqrt{3}$

25. $c = 4\sqrt{3}/3$ **27.** $a = 10\sqrt{3}/3$

29. $\angle A = 50°, a = 7.7, c = 6.4$

31. $\angle C = 76°, b = 66.1, c = 64.2$

33. $\angle C = 25°, a = 10.7, b = 11.8$

35. $\angle C = 18°, a = 3.3, c = 1.1$

37. $\theta \approx 48.6°$ **39.** $\theta \approx 48.2°$

41. $\angle A = 28.1°, \angle C = 61.9°$

43. $\angle A = 44.4°, \angle C = 45.6°$

45. $\angle A = 48.2°, \angle C = 41.8°$

47. $\angle A = 60.8°, \angle C = 29.2°$

49. (a) $A = \dfrac{1}{2}ah$ (b) $\sin\theta = \dfrac{h}{b}$

  (c) $h = b\sin\theta$ follows immediately from (b).

  (d) $A = \dfrac{1}{2}ah = \dfrac{1}{2}ab\sin\theta$

51. $A = 29.6$ **53.** $A = 220$

55. (a) $\sin\theta = a/c = \cos\alpha$
  (b) Since $\theta + \alpha + 90° = 180°$, $\theta + \alpha = 90°$.
  (c) Since $\theta + \alpha = 90°$, $\alpha = 90° - \theta$; thus, $\sin\theta = \cos\alpha$
    $= \cos(90° - \theta)$

Section 8.2, page 593

1. $\angle A = 74°, c = 29.0, a = 27.9$

3. $\angle A = 36.9°, \angle B = 53.1°$

5. $w = 100.7$ ft   **7.** About $68°$ **9.** 23.0 m

11. About 15.3 ft   **13.** 276.3 ft

15. About 3402.8 ft   **17.** 85.9 ft

19. 27.5 ft. It is not safe for him to jump.

21. No   **23.** 8598.3 ft $(1.6$ mi$)$

25. $30°$ **27.** 351.1 m   **29.** 1.53 mi

31. 25.4 m   **33.** 3.52 m   **35.** 52.5 mph

37. 18.67 ft, $\theta \approx 53.13°$

39. 205.7 ft   **41.** 173.2 mi

43. (a) 56.7 ft   (b) 9.7 ft

45. (a) Area $= 200\sin t\cos t$
  (b) $t = 45°$; approximately 14.14 by 7.07 ft

47. 449.1 ft

Section 8.3, page 603

1. $a = 6.5; B = 95.9°; C = 44.1°$

3. $c = 15.7; A = 19.7°; B = 42.3°$

5. $a = 24.4; B = 18.4°; C = 21.6°$

7. $c = 21.5; A = 33.5°; B = 67.9°$

9. $A = 120°; B = 21.8°; C = 38.2°$

11. $A = 29.7°; B = 68.2°; C = 82.1°$

13. $A = 38.8°; B = 34.5°; C = 106.7°$

15. $A = 35.6°; B = 53.2°; C = 91.2°$

17. $A = 77.5°; B = 48.4°; C = 54.1°$

19. 334.9 km   **21.** $54.7°$ **23.** 63.7 ft

25. $84.9°$ **27.** $26,205$ mi

29. For the left-hand cable, $A = 34.2°$; for the right-hand cable, $B = 25.6°$

31. 8.4 km   **33.** 154.5 ft

35. 425.7 ft; 469.4 ft   **37.** $33.44°$

39. 978.7 mi   **41.** 7.2 in. and 11.5 in.

43. $a = 19.5; b = 21.23; c = 24.27.$
  $A = 50.2°; B = 56.8°; C = 73.0°.$

45. 16.99 m

Section 8.4, page 614

1. $C = 114°, b = 3.2, c = 7.9$

3. $B = 30°, b = 6.4, c = 8.2$

5. $A = 86°, a = 8.9, c = 7.1$

7. $C = 41.5°, b = 9.7, c = 10.9$

9. No solution.

11. Case 1: $A = 154.7°, C = 5.3°, c = 3.3$
  Case 2: $A = 25.3°, C = 134.7°, c = 24.9$

13. No solution.

15. Case 1: $B = 95.8°, A = 28.2°, a = 5.7$
  Case 2: $B = 84.2°, A = 39.8°, a = 7.7$

17. $b = .84, c = 5.6, C = 132.3°$

19. $a = 9.8, B = 23.3°; C = 81.7°$

21. $A = 18.6°, B = 39.6°, C = 121.8°$

23. $c = 13.9, A = 60.1°, B = 72.9°$

25. $A = 77.7°, C = 39.8°, a = 18.9$

27. No solution.

29. $a = 18.5, B = 36.7°, C = 78.3°$

31. $C = 126°, b = 193.8, c = 273.3$

33. 135.5 m   **35.** $\alpha = 5.4°$ **37.** 5.0 ft

39. $5.3°$ **41.** 30.1 km   **43.** 9.99 m

45. About 9642 ft

47. (a) Solve triangle ABC to find $\angle BAC$, the angle at A. $180° - \angle BAC = \angle EAB$: Solve triangle ABD to find $\angle ABD$, the angle at B. $180° - \angle ABD = \angle EBA$. Now solve triangle EAB using the two angles and included side to find EA.
  (b) About 94 ft

49. 13.36 m

51. (a) See Example 1 of Section 8.3.
 (b) 92.2°
 (c) 27.8°
 If you did not get the correct answer in part (b), look at Example 5 and the Note following it.

Special Topics 8.4.A, page 618

1. 9.6 square units   **3.** 38.9 square units

5. 82.3 square units   **7.** 31.4 square units

9. 6.5 square units   **11.** 7691 square units

13. 235.9 square ft   **15.** 5.8 gal

17. 11.18 square units

19. (a) 50   (b) $6\sqrt{3}$

21. Since $12 + 20 < 36$, this violates the Triangle Inequality, which states that the sum of two sides of a triangle is greater than the third side. This is not a triangle, and the area is undefined.

Chapter 8 Review, page 621

1. $\sin \theta = \dfrac{3}{\sqrt{34}}$; $\cos \theta = \dfrac{5}{\sqrt{34}}$; $\tan \theta = \dfrac{3}{5}$

3. (d)   **5.** (e)

7. $A = 35.5°$; $C = 54.5°$, $b = 17.2$

9. $b = 14.65$; $c = 8.40$; $A = 55°$

11. 225.9 ft   **13.** 1.52°

15. $A = 58.8°$, $B = 34.8°$, $C = 86.4°$

17. $b = 20.4$; $A = 26.4°$; $C = 38.6°$

19. 301 mi

21. $A = 16°$, $a = 1.5$, $b = 4.5$

23. $A = 44.2°$, $B = 73.8°$, $b = 103.3$

25. Case 1: $B = 81.8°$, $C = 38.2°$, $c = 2.5$

 Case 2: $B = 98.2°$, $C = 21.8°$, $c = 1.5$

27. 147.4 square units   **29.** 13.4 km

31. $C = 57°$, $a = 38.5$, $c = 43.4$

33. $b = 12.0$; $A = 75.1°$; $C = 28.9°$

35. Joe is 217.9 m from the flagpole. Alice is 240 m from the flagpole.

37. (a) 3617.65 ft
 (b) 4018.71 ft
 (c) 3642.19 ft

39. 71.89°

41. 10 square units

43. 37.95 square units

Chapter 8 Test, page 622

1. (a) $\dfrac{2b}{\sqrt{1 + 4b^2}}$ (b) $\dfrac{1}{\sqrt{1 + 4b^2}}$ (c) $2b$

2. 2.51°   **3.** 9

4. 256.94 ft   **5.** $12\sqrt{3}$

6. 180.08 ft

7. $A = 53°$, $b = 4.38$, $c = 2.64$

8. 1787.06 ft   **9.** 36.58°

10. $A = 64.16°$, $C = 25.84°$

11. 243.78 ft

12. $A = 37°$, $a = 3.62$, $c = 4.57$

13. $b = 10.78$, $A = 16.65°$, $C = 136.75°$

14. $A = 66.41°$, $C = 39.59°$, $b = 23.08$

15. $A = 15.29°$, $B = 20.59°$, $C = 144.11°$

16. $B = 51.61°$, $C = 84.39°$, $a = 13.47$

17. $A_1 = 45.98°$, $C_1 = 96.82°$, $a = 14.63$ or $A_2 = 59.62°$, $C_2 = 83.18°$, $a = 17.55$

18. $A = 16.59°$, $B = 43.67°$, $C = 119.74°$

19. 1639.74 mi   **20.** 4.7 m   **21.** 260.64 ft

22. (a) 9.56 Square units   (b) 21.48 Square units

Chapter 9

Section 9.1, page 630

1–7 odd.

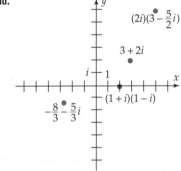

9. 13   **11.** $\sqrt{3}$ **13.** 12

15. Many correct answers, including $z = 1$, $w = i$

17.

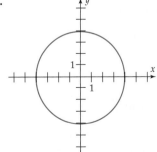

19.

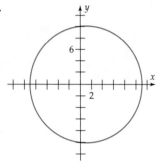

21.

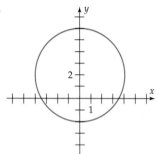

23.

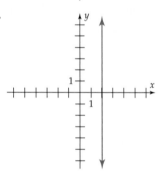

25. $\sqrt{2} + i\sqrt{2}$

27. $0 + i = i$

29. $-\dfrac{5}{2} + i\dfrac{5\sqrt{3}}{2}$

31. $\dfrac{3\sqrt{3}}{4} + \dfrac{3}{4}i$

33. $-1.3073 - 1.5136i$

35. $-1.6646 + 3.6372i$

37. $3\sqrt{2}\left(\cos\dfrac{\pi}{4} + i\sin\dfrac{\pi}{4}\right)$

39. $4\left(\cos\dfrac{\pi}{3} + i\sin\dfrac{\pi}{3}\right)$

41. $6\left(\cos\dfrac{11\pi}{6} + i\sin\dfrac{11\pi}{6}\right)$

43. $\sqrt{6}\left(\cos\dfrac{5\pi}{4} + i\sin\dfrac{5\pi}{4}\right)$

45. $5(\cos(.9273) + i\sin(.9273))$

47. $13(\cos(5.1072) + i\sin(5.1072))$

49. $\sqrt{5}(\cos(1.1071) + i\sin(1.1071))$

51. $\dfrac{\sqrt{74}}{2}(\cos(2.191) + i\sin(2.191))$

53. $\cos\dfrac{3\pi}{2} + i\sin\dfrac{3\pi}{2} = 0 + i(-1) = -i$

55. $12\left(\cos\dfrac{\pi}{3} + i\sin\dfrac{\pi}{3}\right) = 6 + 6\sqrt{3}i$

57. $36\left(\cos\dfrac{\pi}{2} + i\sin\dfrac{\pi}{2}\right) = 36i$

59. $\cos\dfrac{\pi}{3} + i\sin\dfrac{\pi}{3} = \dfrac{1}{2} + \dfrac{\sqrt{3}}{2}i$

61. $2\left(\cos\dfrac{\pi}{6} + i\sin\dfrac{\pi}{6}\right) = \sqrt{3} + i$

63. $\dfrac{3}{2}\left(\cos\dfrac{\pi}{4} + i\sin\dfrac{\pi}{4}\right) = \dfrac{3\sqrt{2}}{4} + \dfrac{3\sqrt{2}}{4}i$

65. $2\sqrt{2}\left(\cos\dfrac{7\pi}{12} + i\sin\dfrac{7\pi}{12}\right)$

67. $\cos\dfrac{\pi}{2} + i\sin\dfrac{\pi}{2}$

69. $12\left(\cos\dfrac{2\pi}{3} + i\sin\dfrac{2\pi}{3}\right)$

71. $2\sqrt{2}\left(\cos\dfrac{19\pi}{12} + i\sin\dfrac{19\pi}{12}\right)$

73. Since $z = r(\cos\theta + i\sin\theta)$ and $i = \cos\dfrac{\pi}{2} + i\sin\dfrac{\pi}{2}$ we see that $iz = r\left(\cos\left(\theta + \dfrac{\pi}{2}\right) + i\sin\left(\theta + \dfrac{\pi}{2}\right)\right)$. The argument of iz is 90° more than that of z; hence, iz is equivalent to rotating z through 90° in the complex plane.

75. (a) $\dfrac{b}{a}$

(b) $\dfrac{d}{c}$

(c) $y - b = \dfrac{d}{c}(x - a)$

(d) $y - d = \dfrac{b}{a}(x - c)$

(e)

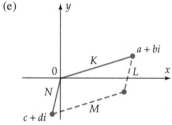

(f) Substituting $(a + c, b + d)$ in the equation of line L yields

$$y - b = \dfrac{d}{c}(x - a)$$

$$b + d - b = \dfrac{d}{c}(a + c - a)$$

$$d = d.$$

Similarly, substituting $(a + c, b + d)$ in the equation of line M yields

$$y - d = \dfrac{b}{a}(x - c)$$

$$b + d - d = \dfrac{b}{a}(a + c - c)$$

$$b = b.$$

Thus, $(a + c, b + d)$ lies on both lines L and M.

77. (a) $r_2(\cos\theta_2 + i\sin\theta_2)(\cos\theta_2 - i\sin\theta_2) =$
$r_2(\cos^2\theta_2 + \sin^2\theta_2) = r_2 \cdot 1 = r_2$

(b) $r_1(\cos\theta_1 + i\sin\theta_1)(\cos\theta_2 - i\sin\theta_2)$

$= r_1(\cos\theta_1\cos\theta_2 - i\cos\theta_1\sin\theta_2 + i\sin\theta_1\cos\theta_2$
$- i^2\sin\theta_1\sin\theta_2)$

$= r_1(\cos\theta_1\cos\theta_2 + \sin\theta_1\sin\theta_2$
$+ i(\sin\theta_1\cos\theta_2 - \cos\theta_1\sin\theta_2))$

$= r_1(\cos(\theta_1 - \theta_2) + i\sin(\theta_1 - \theta_2))$

Section 9.2, page 638

1. $0 + 1i = i$

3. $128 + 128i\sqrt{3}$

5. $-\dfrac{243\sqrt{3}}{2} - \dfrac{243}{2}i$

7. $-1 + 0i = -1$

9. $-64 + 0i = -64$

11. $\dfrac{1}{2} - \dfrac{\sqrt{3}}{2}i$

13. $0 + i1 = i$

15. $k = 0: 1; k = 1: i; k = 2: -1; k = 3: -i$

17. $k = 0: 6\left(\cos\dfrac{\pi}{6} + i\sin\dfrac{\pi}{6}\right); k = 1: 6\left(\cos\dfrac{7\pi}{6} + i\sin\dfrac{7\pi}{6}\right)$

19. $k = 0: 4\left(\cos\dfrac{\pi}{15} + i\sin\dfrac{\pi}{15}\right); k = 1: 4\left(\cos\dfrac{11\pi}{15} + i\sin\dfrac{11\pi}{15}\right);$

$k = 2: 4\left(\cos\dfrac{7\pi}{5} + i\sin\dfrac{7\pi}{5}\right)$

21. $k = 0: 3\left(\cos\dfrac{\pi}{48} + i\sin\dfrac{\pi}{48}\right); k = 1: 3\left(\cos\dfrac{25\pi}{48} + i\sin\dfrac{25\pi}{48}\right);$

$k = 2: 3\left(\cos\dfrac{49\pi}{48} + i\sin\dfrac{49\pi}{48}\right);$

$k = 3: 3\left(\cos\dfrac{73\pi}{48} + i\sin\dfrac{73\pi}{48}\right)$

23. $k = 0: \cos\dfrac{\pi}{5} + i\sin\dfrac{\pi}{5}; k = 1: \cos\dfrac{3\pi}{5} + i\sin\dfrac{3\pi}{5};$

$k = 2: \cos\pi + i\sin\pi = -1; k = 3: \cos\dfrac{7\pi}{5} + i\sin\dfrac{7\pi}{5};$

$k = 4: \cos\dfrac{9\pi}{5} + i\sin\dfrac{9\pi}{5}$

25. $k = 0: \cos\dfrac{\pi}{10} + i\sin\dfrac{\pi}{10}; k = 1: \cos\dfrac{\pi}{2} + i\sin\dfrac{\pi}{2};$

$k = 2: \cos\dfrac{9\pi}{10} + i\sin\dfrac{9\pi}{10} = i; k = 3: \cos\dfrac{13\pi}{10} + i\sin\dfrac{13\pi}{10};$

$k = 4: \cos\dfrac{17\pi}{10} + i\sin\dfrac{17\pi}{10}$

27. $k = 0: \sqrt[4]{2}\left(\cos\dfrac{\pi}{8} + i\sin\dfrac{\pi}{8}\right); k = 1: \sqrt[4]{2}\left(\cos\dfrac{9\pi}{8} + i\sin\dfrac{9\pi}{8}\right)$

29. $k = 0: 2\left(\cos\dfrac{\pi}{24} + i\sin\dfrac{\pi}{24}\right); k = 1: 2\left(\cos\dfrac{13\pi}{24} + i\sin\dfrac{13\pi}{24}\right);$

$k = 2: 2\left(\cos\dfrac{25\pi}{24} + i\sin\dfrac{25\pi}{24}\right); k = 3: 2\left(\cos\dfrac{37\pi}{24} + i\sin\dfrac{37\pi}{24}\right)$

31. $\cos\dfrac{\pi}{6} + i\sin\dfrac{\pi}{6} = \dfrac{\sqrt{3}}{2} + \dfrac{1}{2}i, \cos\dfrac{7\pi}{6} + i\sin\dfrac{7\pi}{6} =$

$-\dfrac{\sqrt{3}}{2} - \dfrac{1}{2}i, \cos\dfrac{\pi}{2} + i\sin\dfrac{\pi}{2} = i$

$\cos\dfrac{3\pi}{2} + i\sin\dfrac{3\pi}{2} = -i, \cos\dfrac{5\pi}{6} + i\sin\dfrac{5\pi}{6} = -\dfrac{\sqrt{3}}{2} + \dfrac{1}{2}i,$

$\cos\dfrac{11\pi}{6} + i\sin\dfrac{11\pi}{6} = \dfrac{\sqrt{3}}{2} - \dfrac{1}{2}i$

33. $\cos\dfrac{\pi}{6} + i\sin\dfrac{\pi}{6} = \dfrac{\sqrt{3}}{2} + \dfrac{1}{2}i, \cos\dfrac{5\pi}{6} + i\sin\dfrac{5\pi}{6} =$

$-\dfrac{\sqrt{3}}{2} + \dfrac{1}{2}i, \cos\dfrac{3\pi}{2} + i\sin\dfrac{3\pi}{2} = -i$

35. $3i, -\dfrac{3\sqrt{3}}{2} - \dfrac{3i}{2}, \dfrac{3\sqrt{3}}{2} - \dfrac{3i}{2}$

37. $3\left[\cos\left(\dfrac{\pi + 4k\pi}{10}\right) + i\sin\left(\dfrac{\pi + 4k\pi}{10}\right)\right], k = 0, 1, 2, 3, 4$

39. $\sqrt[4]{2}\left(\dfrac{\sqrt{3}}{2} + \dfrac{1}{2}i\right), \sqrt[4]{2}\left(-\dfrac{1}{2} + \dfrac{\sqrt{3}}{2}i\right), \sqrt[4]{2}\left(-\dfrac{\sqrt{3}}{2} - \dfrac{1}{2}i\right),$

$\sqrt[4]{2}\left(\dfrac{1}{2} - \dfrac{\sqrt{3}}{2}i\right)$

41. $1, .6235 \pm .7818i, -.2225 \pm .9749i, -.9010 \pm .4339i$

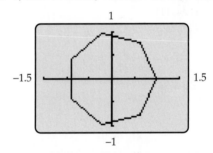

43. $\pm 1, \pm i, .7071 \pm .7071i, -.7071 \pm 7071i$

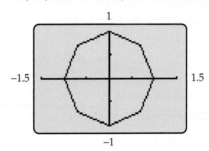

45. $1, .7660 \pm .6428i, .1736 \pm .9848i, -.5 \pm .8660i,$
$-.9397 \pm .3420i$

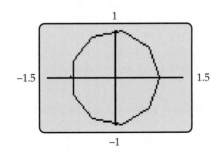

47. $-1, i, -i$

49. $(-1 \pm \sqrt{3}i)/2$, $(1 \pm \sqrt{3}i)/2$, and -1

51. 12 trips around the circle

53. Since the u_i are distinct, clearly, the vu_i are distinct. Since v is a solution of $z^n = r(\cos\theta + i\sin\theta)$, we can write $v^n = r(\cos\theta + i\sin\theta)$, and similarly $u_i^n = 1$. Then $(vu_i)^n = v^n u_i^n = r(\cos\theta + i\sin\theta) \cdot 1 = r(\cos\theta + i\sin\theta)$. Thus, the vu_i are the solutions of the equation.

Section 9.3, page 651

1. $3\sqrt{5}$ **3.** $\sqrt{34}$

5. $\langle 6, 6 \rangle$ **7.** $\langle -6, 10 \rangle$

9. $\langle 13/5, -2/5 \rangle$

11. $\mathbf{u} + \mathbf{v} = \langle 4, 5 \rangle$; $\mathbf{v} - \mathbf{u} = \langle 8, -3 \rangle$; $2\mathbf{u} - 3\mathbf{v} = \langle -22, 5 \rangle$

13. $\mathbf{u} + \mathbf{v} = \langle 3 + 4\sqrt{2}, 3\sqrt{2} + 1 \rangle$; $\mathbf{v} - \mathbf{u} = \langle 4\sqrt{2} - 3, 1 - 3\sqrt{2} \rangle$; $2\mathbf{u} - 3\mathbf{v} = \langle 6 - 12\sqrt{2}, 6\sqrt{2} - 3 \rangle$

15. $\mathbf{u} + \mathbf{v} = \langle -23/4, 13 \rangle$; $\mathbf{v} - \mathbf{u} = \langle 9/4, -7 \rangle$; $2\mathbf{u} - 3\mathbf{v} = \langle -11/4, 11 \rangle$

17. $\mathbf{u} + \mathbf{v} = 14\mathbf{i} - 4\mathbf{j}$; $\mathbf{v} - \mathbf{u} = -2\mathbf{i} - 4\mathbf{j}$; $2\mathbf{u} - 3\mathbf{v} = -2\mathbf{i} + 12\mathbf{j}$

19. $\mathbf{u} + \mathbf{v} = -\dfrac{5}{4}\mathbf{i} - \dfrac{3}{2}\mathbf{j}$; $\mathbf{v} - \mathbf{u} = \dfrac{11}{4}\mathbf{i} + \dfrac{3}{2}\mathbf{j}$; $2\mathbf{u} - 3\mathbf{v} = \dfrac{-25}{4}\mathbf{i} - 3\mathbf{j}$

21. $\langle -7, 0 \rangle$ **23.** $\langle -2, 1/2 \rangle$

25. $\langle 6, -13/4 \rangle$ **27.** $\langle 4, 0 \rangle$

29. $\langle -5\sqrt{2}, -5\sqrt{2} \rangle$ **31.** $\langle 4.5963, 3.8567 \rangle$

33. $\langle -.1710, -.4698 \rangle$ **35.** $\|\mathbf{v}\| = 4\sqrt{2}$, $\theta = 45°$

37. $\|\mathbf{v}\| = 8$, $\theta = 180°$ **39.** $\|\mathbf{v}\| = 6$, $\theta = 90°$

41. $\|\mathbf{v}\| = 2\sqrt{17}$, $\theta = 104.04°$

43. $\left\langle \dfrac{4}{\sqrt{41}}, -\dfrac{5}{\sqrt{41}} \right\rangle$ **45.** $\left\langle \dfrac{1}{\sqrt{5}}, \dfrac{2}{\sqrt{5}} \right\rangle$

47. $\|\mathbf{u} + \mathbf{v}\| = 108.2$ pounds, $\theta = 46.1°$

49. $\|\mathbf{u} + \mathbf{v}\| = 17.4356$ newtons, $\theta = 213.4132°$

51. $\langle -8, -2 \rangle$; $\mathbf{v} = \langle 8, 2 \rangle$

53. $\mathbf{v} + \mathbf{0} = \langle c, d \rangle + \langle 0, 0 \rangle = \langle c, d \rangle = \mathbf{v}$ and $\mathbf{0} + \mathbf{v} = \langle 0, 0 \rangle + \langle c, d \rangle = \langle c, d \rangle = \mathbf{v}$

55. $r(\mathbf{u} + \mathbf{v}) = r(\langle a, b \rangle + \langle c, d \rangle) = r\langle a + c, b + d \rangle = \langle ra + rc, rb + rd \rangle = \langle ra, rb \rangle + \langle rc, rd \rangle = r\langle a, b \rangle + r\langle c, d \rangle = r\mathbf{u} + r\mathbf{v}$

57. $(rs)\mathbf{v} = rs\langle c, d \rangle = \langle rsc, rsd \rangle = r\langle sc, sd \rangle = r(s\mathbf{v})$ and $\langle rsc, rsd \rangle = \langle src, srd \rangle = s\langle rc, rd \rangle = s(r\mathbf{v})$

59. 48.575 pounds

61. 32.1 pounds parallel to plane; 38.3 pounds perpendicular to plane

63. 66.4°

65. Ground speed: 253.2 mph; course: 69.1°

67. Ground speed: 304.1 mph; course: 309.5°

69. Air speed: 424.3 mph; direction: 62.4°

71. 69.08°

73. 341.77 lb on $\mathbf{v}$; 170.32 lb on $\mathbf{u}$

75. 517.55 lb on the 28° rope; 579.90 lb on the 38° rope

77. (a) $\mathbf{v} = \langle x_2 - x_1, y_2 - y_1 \rangle$; $k\mathbf{v} = \langle kx_2 - kx_1, ky_2 - ky_1 \rangle$

(b) $\|\mathbf{v}\| = \sqrt{(x_2 - x_1)^2 + (y_2 - y_1)^2}$
$\|k\mathbf{v}\| = \sqrt{(kx_2 - kx_1)^2 + (ky_2 - ky_1)^2}$

(c) $\|k\mathbf{v}\| = \sqrt{k^2(x_2 - x_1)^2 + k^2(y_2 - y_1)^2}$
$= \sqrt{k^2}\sqrt{(x_2 - x_1)^2 + (y_2 - y_1)^2}$
$= |k|\sqrt{(x_2 - x_1)^2 + (y_2 - y_1)^2} = |k|\,\|\mathbf{v}\|$

(d) $\tan\theta = \dfrac{y_2 - y_1}{x_2 - x_1} = \dfrac{k}{k}\dfrac{y_2 - y_1}{x_2 - x_1} = \dfrac{ky_2 - ky_1}{kx_2 - kx_1} = \tan\beta$
Since the angles have the same tangent, they can differ only by a multiple of π and so are parallel. They have either the same or the opposite direction.

(e) If $k > 0$, the signs of the components of $k\mathbf{v}$ are the same as those of $\mathbf{v}$, so the two vectors lie in the same quadrant. Therefore, they must have the same direction. If $k < 0$, then the components of $k\mathbf{v}$ and the components of $\mathbf{v}$ must have opposite signs, and the two vectors do not lie in the same quadrant. Therefore, they do not have the same direction and must have opposite directions.

79. (a) Since $\mathbf{u} - \mathbf{v} = \langle a - c, b - d \rangle$, $\|\mathbf{u} - \mathbf{v}\| = \sqrt{(a - c)^2 + (b - d)^2}$. The magnitude of $\mathbf{w}$ is given by the distance between the points (a, b) and (c, d). Hence, $\|\mathbf{u} - \mathbf{v}\| = \|\mathbf{w}\|$.

(b) $\mathbf{u} - \mathbf{v}$ lies on the straight line through $(0, 0)$ and $(a - c, b - d)$, which has slope $\dfrac{(b - d) - 0}{(a - c) - 0} = \dfrac{b - d}{a - c}$. $\mathbf{w}$ lies on the line joining (a, b) and (c, d). This also has slope $\dfrac{b - d}{a - c}$. Since the slopes are the same, the vectors, $\mathbf{u} - \mathbf{v}$ and $\mathbf{w}$ are parallel. Therefore, they either point in the same direction or in opposite directions. We can see that they have the same direction by considering the signs of the components of $\mathbf{u} - \mathbf{v}$. If $a - c > 0$ and $b - d > 0$, then $a > c$, $b > d$ and $\mathbf{u} - \mathbf{v}$ and $\mathbf{w}$ both point up and right. If $a - c > 0$ and $b - d < 0$, both vectors will point left and up. If $a - c < 0$ and $b - d < 0$, both vectors will point left and down. In any case, they point in the same direction.

Section 9.4, page 661

1. $\mathbf{u} \cdot \mathbf{v} = -7$, $\mathbf{u} \cdot \mathbf{u} = 25$, $\mathbf{v} \cdot \mathbf{v} = 29$

3. $\mathbf{u} \cdot \mathbf{v} = 6$, $\mathbf{u} \cdot \mathbf{u} = 5$, $\mathbf{v} \cdot \mathbf{v} = 9$

5. $\mathbf{u} \cdot \mathbf{v} = 12$, $\mathbf{u} \cdot \mathbf{u} = 13$, $\mathbf{v} \cdot \mathbf{v} = 13$

7. -1 **9.** 6

11. 34   **13.** 1.75065 radians

15. 2.1588 radians   **17.** $\pi/2$ radians

19. Orthogonal

21. Parallel

23. Neither

25. $k = 2$

27. $k = \sqrt{2}$

29. $\text{proj}_{\mathbf{u}}\mathbf{v} = \langle 12/17, -20/17 \rangle$; $\text{proj}_{\mathbf{v}}\mathbf{u} = \langle 6/5, 2/5 \rangle$

31. $\text{proj}_{\mathbf{u}}\mathbf{v} = \langle 0, 0 \rangle$; $\text{proj}_{\mathbf{v}}\mathbf{u} = \langle 0, 0 \rangle$

33. $\text{comp}_{\mathbf{v}}\mathbf{u} = 22/\sqrt{13}$ **35.** $\text{comp}_{\mathbf{v}}\mathbf{u} = 3/\sqrt{10}$

37. $\mathbf{u} \cdot (\mathbf{v} + \mathbf{w}) = \langle a, b \rangle \cdot (\langle c, d \rangle + \langle r, s \rangle)$
$= \langle a, b \rangle \cdot \langle c + r, d + s \rangle = a(c + r) + b(d + s)$
$= ac + ar + bd + bs$
$\mathbf{u} \cdot \mathbf{v} + \mathbf{u} \cdot \mathbf{w} = \langle a, b \rangle \cdot \langle c, d \rangle + \langle a, b \rangle \cdot \langle r, s \rangle$
$= (ac + bd) + (ar + bs) = ac + ar + bd + bs$

39. $\mathbf{0} \cdot \mathbf{u} = \langle 0, 0 \rangle \cdot \langle a, b \rangle = 0a + 0b = 0$

41. If $\theta = 0$ or π, then $\mathbf{u}$ and $\mathbf{v}$ are parallel, so $\mathbf{v} = k\mathbf{u}$ for some real number k. We know that $\|\mathbf{v}\| = |k| \, \|\mathbf{u}\|$ (Exercise 77, Section 9.3). If $\theta = 0$, then $\cos\theta = 1$ and $k > 0$. Since $k > 0$, $|k| = k$ and so $\|\mathbf{v}\| = k\|\mathbf{u}\|$. Therefore, $\mathbf{u} \cdot \mathbf{v} = \mathbf{u} \cdot k\mathbf{u} = k\mathbf{u} \cdot \mathbf{u} = k\|\mathbf{u}\|^2 = \|\mathbf{u}\|(k\|\mathbf{u}\|) = \|\mathbf{u}\|\|\mathbf{v}\| = \|\mathbf{u}\|\|\mathbf{v}\|\cos\theta$. On the other hand, if $\theta = \pi$, then $\cos\theta = -1$ and $k < 0$. Since $k < 0$, $|k| = -k$ and so $\|\mathbf{v}\| = -k\|\mathbf{u}\|$. Then $\mathbf{u} \cdot \mathbf{v} = \mathbf{u} \cdot k\mathbf{u} = k\mathbf{u} \cdot \mathbf{u} = k\|\mathbf{u}\|^2 = \|\mathbf{u}\|(-k\|\mathbf{u}\|) = -\|\mathbf{u}\|\|\mathbf{v}\| = \|\mathbf{u}\|\|\mathbf{v}\|\cos\theta$. In both cases we have shown $\mathbf{u} \cdot \mathbf{v} = \|\mathbf{u}\|\|\mathbf{v}\|\cos\theta$.

43. If $A = (1, 2)$, $B = (3, 4)$, and $C = (5, 2)$, then the vector $\overrightarrow{AB} = \langle 2, 2 \rangle$, $\overrightarrow{AC} = \langle 4, 0 \rangle$, and $\overrightarrow{BC} = \langle 2, -2 \rangle$. Since $\overrightarrow{AB} \cdot \overrightarrow{BC} = 0$, $\overrightarrow{AB}$ and $\overrightarrow{BC}$ are perpendicular, so the angle at vertex B is a right angle.

45. Many possible answers: One is $\mathbf{u} = \langle 1, 0 \rangle$, $\mathbf{v} = \langle 1, 1 \rangle$, and $\mathbf{w} = \langle 1, -1 \rangle$.

47. 300 lb ($= 600 \sin 30°$)   **49.** 13

51. 24

53. The force in the direction of the lawnmower's motion is $30 \cos 60° = 15$ lb. Thus, the work done is
$$15(75) = 1125 \text{ ft-lb.}$$

55. 1368 ft-lb

Chapter 9 Review, page 664

1. $\sqrt{20} + \sqrt{10}$

3.

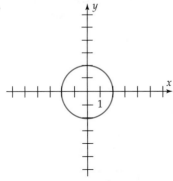

5. $2\left(\cos\dfrac{\pi}{3} + i\sin\dfrac{\pi}{3}\right)$ **7.** $4\sqrt{2} + 4\sqrt{2}i$

9. $2\sqrt{3} + 2i$ **11.** $\dfrac{81}{2} - \dfrac{81\sqrt{3}}{2}i$

13. $\cos 0 + i\sin 0, \cos\dfrac{\pi}{3} + i\sin\dfrac{\pi}{3}, \cos\dfrac{2\pi}{3} + i\sin\dfrac{2\pi}{3},$
$\cos\pi + i\sin\pi, \cos\dfrac{4\pi}{3} + i\sin\dfrac{4\pi}{3}, \cos\dfrac{5\pi}{3} + i\sin\dfrac{5\pi}{3}$

15. $\cos\dfrac{\pi}{8} + i\sin\dfrac{\pi}{8}, \cos\dfrac{5\pi}{8} + i\sin\dfrac{5\pi}{8}, \cos\dfrac{9\pi}{8} + i\sin\dfrac{9\pi}{8},$
$\cos\dfrac{13\pi}{8} + i\sin\dfrac{13\pi}{8}$

17. $\langle 8, -4 \rangle$ **19.** $10\sqrt{5}$ **21.** $\langle -14, 9 \rangle$

23. $\sqrt{17}$ **25.** $\langle 5\sqrt{2}/2, 5\sqrt{2}/2 \rangle$

27. $-\dfrac{1}{\sqrt{5}}\mathbf{i} + \dfrac{2}{\sqrt{5}}\mathbf{j}$

29. Ground speed: 321.87 mph; course: 126.18°

31. -22 **33.** 15   **35.** .70 radians

37. $\text{proj}_{\mathbf{v}}\mathbf{u} = \mathbf{v} = 2\mathbf{i} + \mathbf{j}$

39. $(\mathbf{u} + \mathbf{v}) \cdot (\mathbf{u} - \mathbf{v}) = \mathbf{u} \cdot \mathbf{u} - \mathbf{u} \cdot \mathbf{v} + \mathbf{v} \cdot \mathbf{u} - \mathbf{v} \cdot \mathbf{v} = \mathbf{u} \cdot \mathbf{u} - \mathbf{v} \cdot \mathbf{v} = \|\mathbf{u}\|^2 - \|\mathbf{v}\|^2 = 0$ since $\mathbf{u}$ and $\mathbf{v}$ have the same magnitude.

41. 1750 lb

Chapter 9 Test, page 665

1.

2. $-5832i$ **3.** $\sqrt{13}$

4. $4\left(\cos\dfrac{\pi}{48} + i\sin\dfrac{\pi}{48}\right), 4\left(\cos\dfrac{11\pi}{16} + i\sin\dfrac{11\pi}{16}\right),$
$4\left(\cos\dfrac{65\pi}{48} + i\sin\dfrac{65\pi}{48}\right)$

5. (a) 34   (b) 34

6. $\sqrt[3]{2}\left(\cos\dfrac{\pi}{9} + i\sin\dfrac{\pi}{9}\right), \sqrt[3]{2}\left(\cos\dfrac{7\pi}{9} + i\sin\dfrac{7\pi}{9}\right),$
$\sqrt[3]{2}\left(\cos\dfrac{13\pi}{9} + i\sin\dfrac{13\pi}{9}\right)$

7.

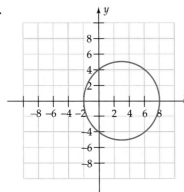

8. $\cos\dfrac{3\pi}{8} + i\sin\dfrac{3\pi}{8}, \cos\dfrac{7\pi}{8} + i\sin\dfrac{7\pi}{8},$

$\cos\dfrac{11\pi}{8} + i\sin\dfrac{11\pi}{8}, \cos\dfrac{15\pi}{8} + i\sin\dfrac{15\pi}{8}$

9. $39i$

10. $3\sqrt{3} + 3i, -3 + 3\sqrt{3}\,i, -3\sqrt{3}\,i - 3i, 3 - 3\sqrt{3}\,i$

11. $5\left(\cos\dfrac{4\pi}{3} + i\sin\dfrac{4\pi}{3}\right)$ **12.** $-1, i, -i$

13. $\langle -7, -13\rangle$ **14.** 86

15. $\sqrt{11}\,\mathbf{i} + \sqrt{5}\,\mathbf{j}, -\sqrt{11}\,\mathbf{i} + \sqrt{5}\,\mathbf{j}, -5\sqrt{11}\,\mathbf{i} + 6\sqrt{5}\,\mathbf{j}$

16. 1.8925   **17.** $\left\langle -\dfrac{3\sqrt{3}}{2}, -\dfrac{3}{2}\right\rangle$ **18.** $-\dfrac{2}{7}$

19. $-\dfrac{7}{\sqrt{85}}\,\mathbf{i} - \dfrac{6}{\sqrt{85}}\,\mathbf{j}$ **20.** $\dfrac{\sqrt{10}}{10}$

21. (a) $\langle -3, 11\rangle$ (b) $\langle 3, -11\rangle$

22. 42   **23.** 1.01   **24.** 3194.95 ft-lb

Chapter 10

Section 10.1, page 683

1. $x^2 + (y - 3)^2 = 4$ **3.** $x^2 + 6y^2 = 18$

5. $(x - 4)^2 + (y - 3)^2 = 4$

7. Center $(-4, 3)$, radius $2\sqrt{10}$

9. Center $(-3, 2)$, radius $2\sqrt{7}$

11. Center $(-25/2, -5)$, radius $\dfrac{\sqrt{677}}{2} \approx 13.01$

13. ellipse; center $(0, 0)$; vertices $(-5, 0), 5, 0)$; foci $(-\sqrt{21}, 0), (\sqrt{21}, 0)$

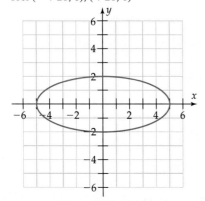

15. ellipse; center $(0, 0)$, vertices $(0, -2), (0, 2)$; foci $(0, -1),$ $(0, 1)$

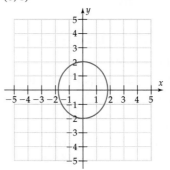

17. ellipse; center $(0, 0)$; vertices $(-9, 0), (9, 0)$; foci $(-\sqrt{32}, 0), (\sqrt{32}, 0)$

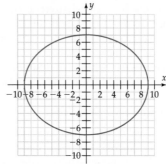

19. circle; center $(0, 0)$; radius $1/2$

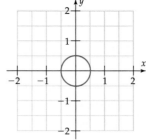

21. $\dfrac{x^2}{49} + \dfrac{y^2}{4} = 1$ **23.** $\dfrac{x^2}{36} + \dfrac{y^2}{16} = 1$

25. $\dfrac{x^2}{9} + \dfrac{y^2}{49} = 1$ **27.** 8π

29. $2\pi\sqrt{3}$ **31.** $7\pi/\sqrt{3}$

33. ellipse; center $(1, 5)$; vertices $(1, 2), (1, 8)$; foci $(0, -\sqrt{5} + 5), (0, \sqrt{5} + 5)$

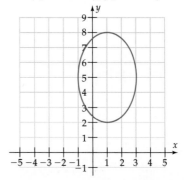

35. Ellipse, center $(-1, 4)$; vertices $(-5, 4)$, $(3, 4)$; foci $(-2\sqrt{2} - 1, 4)$, $(2\sqrt{2} - 1, 4)$

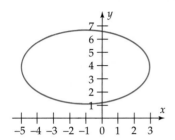

37. Ellipse, center $(-3, 1)$; vertices $(-3, 4)$, $(-3, -2)$; foci $(-3, \sqrt{5} + 1)$, $(-3, -\sqrt{5} + 1)$

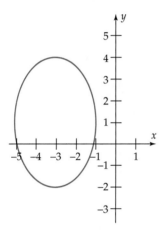

39. circle; center $(-3, 4)$

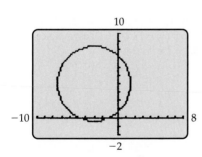

41. ellipse; center $(-3, 2)$

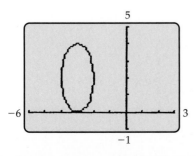

43. ellipse; center $(1, -1)$

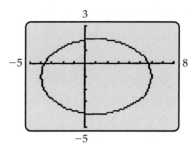

45. $\dfrac{(x - 2)^2}{4} + \dfrac{(y - 3)^2}{16} = 1$

47. $\dfrac{(x - 7)^2}{25/4} + \dfrac{(y + 4)^2}{36} = 1$

49. $\dfrac{(x - 3)^2}{36} + \dfrac{(y + 2)^2}{16} = 1$

51. $\dfrac{(x + 5)^2}{49} + \dfrac{(y - 3)^2}{16} = 1$ or $\dfrac{(x + 5)^2}{16} + \dfrac{(y - 3)^2}{49} = 1$

53. $2x^2 + 2y^2 - 8 = 0$

55. $2x^2 + y^2 - 8x - 6y + 9 = 0$

57. $\dfrac{(x + 3)^2}{4} + \dfrac{(y + 3)^2}{8} = 1$

59. The ellipse has y-intercepts $(0, \pm 4)$. Therefore, the circle, which has center at the origin, has radius 4. The equation of the circle must be $x^2 + y^2 = 16$.

61. The minimum distance is 226,335 miles. The maximum distance is 251,401 miles.

63. 80 ft   **65.** 17.3 ft

67. Eccentricity $= .1$

69. Eccentricity $= \dfrac{\sqrt{3}}{2} \approx .87$

71. The closer the eccentricity is to zero, the more the ellipse resembles a circle. The closer the eccentricity is to 1, the more elongated the ellipse becomes.

73. Eccentricity $\approx .38$.

75. If $a = b$, the equation becomes

$$\frac{x^2}{a^2} + \frac{y^2}{a^2} = 1,$$
$$x^2 + y^2 = a^2.$$

This is the equation of a circle with center at the origin, radius a.

77. The fence is an ellipse with major axis 100 ft, minor axis $50\sqrt{3} \approx 86.6$ ft.

Section 10.2, page 697

1. $x^2 + 4y^2 = 1$

3. $2x^2 - y^2 = 8$

5. $6x^2 + 2y^2 = 18$

7. hyperbola; vertices $(-2, 0)$, $(2, 0)$;
foci $(-\sqrt{5}, 0)$, $(\sqrt{5}, 0)$; asymptotes $y = \pm\dfrac{1}{2}x$

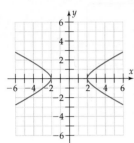

9. hyperbola; vertices $(0, -\sqrt{5})$, $(0, \sqrt{5})$;
foci $(0, -\sqrt{8})$, $(0, \sqrt{8})$; asymptotes $y = \pm\sqrt{\dfrac{5}{3}}x$

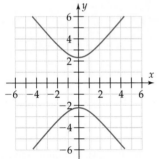

11. hyperbola; vertices $(0, -3)$, $(0, 3)$;
foci $(0, -5)$, $(0, 5)$; asymptotes $y = \pm\dfrac{3}{4}x$

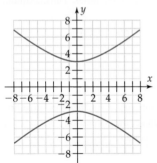

13. hyperbola; vertices $(-1, 0)$, $(1, 0)$;
foci $\left(-\dfrac{\sqrt{5}}{2}, 0\right)$, $\left(\dfrac{\sqrt{5}}{2}, 0\right)$; asymptotes $y = \pm\dfrac{1}{2}x$

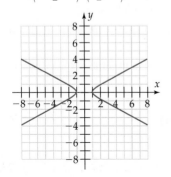

15. $\dfrac{x^2}{4} - \dfrac{y^2}{16} = 1$ **17.** $\dfrac{y^2}{25} - \dfrac{x^2}{4} = 1$

19. $\dfrac{(x+2)^2}{9} - \dfrac{(y+1)^2}{4} = 1$ **21.** $\dfrac{x^2}{9} - \dfrac{y^2}{36} = 1$

23. $\dfrac{x^2}{4} - \dfrac{y^2}{1} = 1$

25. Hyperbola, center $(-1, -3)$; vertices $(-1, -8)$, $(-1, 2)$;
foci $(-1, -3 - \sqrt{41})$, $(-1, -3 + \sqrt{41})$; asymptotes
$y + 3 = \pm\dfrac{5}{4}(x + 1)$

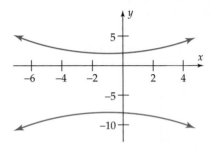

27. Hyperbola, center $(-3, 2)$; vertices $(-4, 2)$, $(-2, 2)$;
foci $(-3 - \sqrt{5}, 2)$, $(-3 + \sqrt{5}, 2)$;
asymptotes $y - 2 = \pm 2(x + 3)$

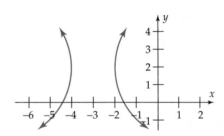

29. Hyperbola, center $(1, -4)$; vertices $(1, -4 - \sqrt{8})$,
$(1, -4 + \sqrt{8})$; foci $(1, -7)$, $(1, -1)$; asymptotes
$y + 4 = \pm\sqrt{8}(x - 1)$

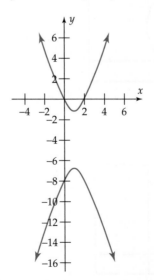

31. Hyperbola, center (3, 3); vertices (3, 1),(3, 5);
foci $(3, 3-\sqrt{20})$, $(3, 3+\sqrt{20})$; asymptotes
$$y - 3 = \pm\frac{1}{2}(x - 3)$$

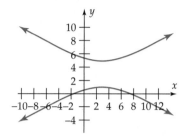

33. circle; center (3, 4)

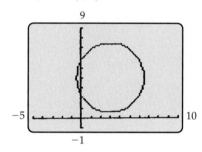

35. ellipse; center (3, 4)

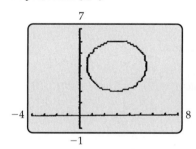

37. hyperbola; center (−2, 2)

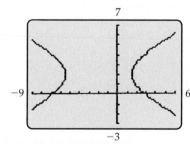

39. $\dfrac{(y - 3)^2}{4} - \dfrac{(x + 2)^2}{6} = 1$

41. $\dfrac{(x - 4)^2}{9} - \dfrac{(y - 2)^2}{16} = 1$

43. $y^2 - 2x^2 = 6$ **45.** $\dfrac{(y - 2)^2}{4} - \dfrac{(x - 3)^2}{9} = 1$

47. $\dfrac{(x + 3)^2}{3} - \dfrac{(y + 3)^2}{4} = 1$

49. The graphs are shown below on one set of coordinate axes.

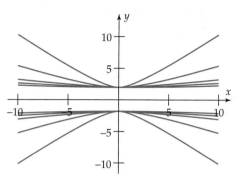

As b increases, the graphs get flatter. Although the hyperbola may look like two horizontal lines for very large b, it is still a hyperbola, with asymptotes $y = \pm 2x/b$ that have nonzero slopes.

51. The asymptotes of $\dfrac{x^2}{a^2} - \dfrac{y^2}{a^2} = 1$ are $y = \pm\dfrac{a}{a}x$ or $y = \pm x$.
Since they have slopes 1 and −1 and $1(-1) = -1$, they are perpendicular.

53. The equation is $\dfrac{x^2}{1,210,000} - \dfrac{y^2}{5,759,600} = 1$ (measurement in feet). The exact location cannot be determined from only the given information.

55. $\dfrac{x^2}{25} - \dfrac{16y^2}{225} = 1$ **57.** $\sqrt{5} \approx 2.24$

59. $\sqrt{3}$ **61.** $3/2$

Section 10.3, page 710

1. $6x = y^2$ **3.** $2x^2 - y^2 = 8$ **5.** $x^2 + 6y^2 = 18$

7. $y^2 = 16x$ **9.** $x^2 = -32y$

11. Focus: (0, 1/12), directrix $y = -1/12$

13. Focus: (0, 1), directrix $y = -1$

15. Focus: (−2, 0), directrix $x = 2$

17. Opens to the right.
Vertex: (2, 0), focus: (9/4, 0), directrix: $x = 7/4$

19. Opens to the left.
Vertex: (2, −1), focus: (7/4, −1), directrix: $x = 9/4$

21. Opens to the right.
Vertex: (2/3, −3), focus: (17/12, −3), directrix: $x = -1/12$

23. Opens to right.
Vertex: (−81/4, 9/2), focus: (−20, 9/2), directrix: $x = -41/2$

25. Opens upward.
Vertex: (−1/6, −49/12), focus: (−1/6, −4), directrix: $y = -25/6$

27. Opens downward.
Vertex: (2/3, 19/3), focus: (2/3, 25/4), directrix: $y = 77/12$

29. Line segment from $(-4, -2)$ to $(4, -2)$

31. Line segment from $(5, -10)$ to $(5, 10)$

33. Line segment from $(-2, 1)$ to $(2, 1)$

35.

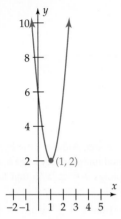

37.

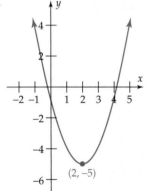

39.

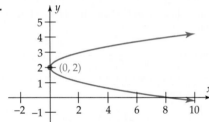

41.

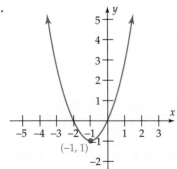

43. $y = 3x^2$

45. $y = 13(x - 1)^2$

47. $x - 2 = 3(y - 1)^2$

49. $x + 3 = 4(y + 2)^2$

51. $y - 1 = 2(x - 1)^2$ **53.** $(y - 3)^2 = x + 1$

55. Parabola, vertex $(3, 4)$

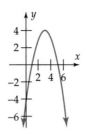

57. Parabola, vertex $(2/3, 1/3)$

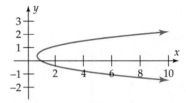

59. Circle, center $(1, 2)$

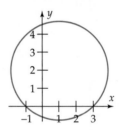

61. Hyperbola, center $(-4, 2)$, vertices $\left(-4 \pm \sqrt{2}, 2\right)$

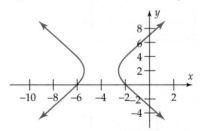

63. $y = (x - 4)^2 + 2$ **65.** $y = -x^2 - 8x - 18$

67. $y = (x + 5)^2 - 3$ **69.** $b = 0$

71. $\left(9, \dfrac{-1 \pm \sqrt{34}}{2}\right)$

73. (a) The focus is $(p, 0)$. To find the points on the parabola directly above and below the focus, let $x = p$ and solve $y^2 = 4px$ for y: $y^2 = 4p^2$, so that $y = \pm 2p$. Hence, the endpoints are $(p, 2p)$ and $(p, -2p)$.

(b) The focus is $(0, p)$. Let $y = p$ and solve $x^2 = 4py$ for x. Then $x = \pm 2p$ and the endpoints are $(-2p, p)$ and $(2p, p)$.

75. The receiver should be placed at the focus, $\frac{2}{3}$ ft or 8 in from the vertex.

77. 43.3 ft   **79.** 5.072 meters   **81.** 27.44 ft

Special Topics 10.3.A, page 717

1. $x = 5 \cos t$ and $y = 5 \sin t$ $(0 \le t \le 2\pi)$

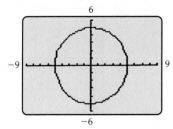

3. $x = 3 \cos t + 4$ and $y = 3 \sin t - 2$ $(0 \le t \le 2\pi)$

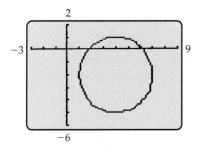

5. $x = \sqrt{5} \cos t - 1$ and $y = \sqrt{5} \sin t$ $(0 \le t \le 2\pi)$

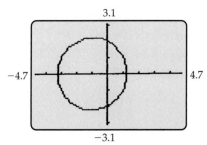

7. $x = \sqrt{10} \cos t$ and $y = 6 \sin t$ $(0 \le t \le 2\pi)$

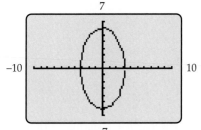

9. $x = \dfrac{1}{2} \cos t$ and $y = \dfrac{1}{2} \sin t$ $(0 \le t \le 2\pi)$

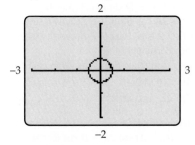

11. $x = 2 \cos t + 1, y = 3 \sin t + 5$ $(0 \le t \le 2\pi)$

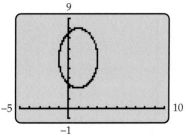

13. $x = 4 \cos t - 1, y = \sqrt{8} \sin t + 4$ $(0 \le t \le 2\pi)$

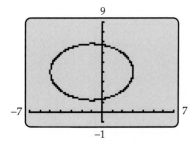

15. $x = \sqrt{10}/\cos t$ and $y = 6 \tan t$ $(0 \le t \le 2\pi)$

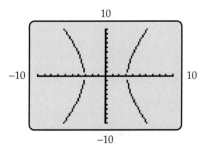

17. $x = 1/\cos t$ and $y = \dfrac{1}{2} \tan t$ $(0 \le t \le 2\pi)$

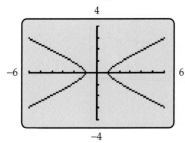

19. $x = 4 \tan t - 1, y = 5/\cos t - 3$ $(0 \le t \le 2\pi)$

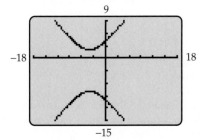

21. $x = 1/\cos t - 3, y = 2 \tan t + 2 \quad (0 \le t \le 2\pi)$

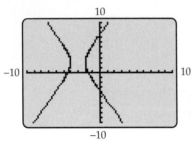

23. $x = t^2/4, y = t$

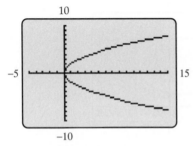

25. $x = t, y = 4(t - 1)^2 + 2$

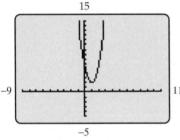

27. $x = 2(t - 2)^2, y = t$

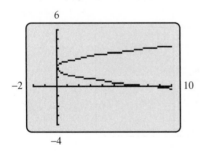

29. Circle; center $(0, -5)$, radius 3

31. Ellipse; center $(4, 0)$ **33.** Hyperbola; center $(2, 4)$

35. Hyperbola; center $(0, -3)$

37. Parabola; vertex $(4, -3)$

39. (a) $x^2 + y^2 = \cos^2(.5t) + \sin^2(.5t) = 1$

(b) $x^2 + y^2 = \cos^2(.5t) + (-\sin(.5t))^2 = \cos^2(.5t) + \sin^2(.5t) = 1$

(c) Since $\cos(.5t)$ and $\sin(.5t)$ each have period 4π, the parametric equations trace out a half-circle when $0 \le t \le 2\pi$.

There are many correct answers to Exercises 40–43, some of which are shown here.

41.

43.

Section 10.4, page 722

1. Ellipse   **3.** Parabola

5. Hyperbola

7. Ellipse. Window $-6 \le x \le 3$ and $-2 \le y \le 4$.

9. Hyperbola. Window $-7 \le x \le 13$ and $-3 \le y \le 9$.

11. Parabola. Window $-1 \le x \le 8$ and $-3 \le y \le 3$.

13. Ellipse. Window $-1.5 \le x \le 1.5$ and $-1 \le y \le 1$.

15. Hyperbola. Window $-15 \le x \le 15$ and $-10 \le y \le 10$.

17. Parabola. Window $-19 \le x \le 2$ and $-1 \le y \le 13$.

19. Hyperbola. Window $-15 \le x \le 15$ and $-15 \le y \le 15$.

21. Ellipse. Window $-6 \le x \le 6$ and $-4 \le y \le 4$.

23. Parabola. Window $-9 \le x \le 4$ and $-2 \le y \le 10$.

Special Topics 10.4.A, page 727

1. $\left(\dfrac{5\sqrt{2}}{2}, -\dfrac{\sqrt{2}}{2} \right)$

3. $\left(\dfrac{\sqrt{3}}{2}, -\dfrac{1}{2} \right)$

5. $\dfrac{u^2}{2} - \dfrac{v^2}{2} = 1$

7. $\dfrac{u^2}{4} + v^2 = 1$

9. $\theta \approx 53.13°; x = \dfrac{3}{5}u - \dfrac{4}{5}v; y = \dfrac{4}{5}u + \dfrac{3}{5}v$

11. $\theta \approx 36.87°; x = \dfrac{4}{5}u - \dfrac{3}{5}v; y = \dfrac{3}{5}u + \dfrac{4}{5}v$

13. (a) $(A\cos^2\theta + B\cos\theta\sin\theta + C\sin^2\theta)u^2 + (B\cos^2\theta - 2A\cos\theta\sin\theta + 2C\cos\theta\sin\theta - B\sin^2\theta)uv + (C\cos^2\theta - B\cos\theta\sin\theta + A\sin^2\theta)v^2 + (D\cos\theta + E\sin\theta)u + (E\cos\theta - D\sin\theta)v + F = 0$

(b) $B' = B\cos^2\theta - 2A\cos\theta\sin\theta + 2C\cos\theta\sin\theta - B\sin^2\theta = 2(C - A)\sin\theta\cos\theta + B(\cos^2\theta - \sin^2\theta)$ since B' is the coefficient of uv

(c) Since $\sin 2\theta = 2\sin\theta\cos\theta$ and $\cos 2\theta = \cos^2\theta - \sin^2\theta$, $B' = (C - A)\sin 2\theta + B\cos 2\theta$.

(d) If $\cot 2\theta = \dfrac{A - C}{B}$, then $\dfrac{\cos 2\theta}{\sin 2\theta} = \dfrac{A - C}{B}$, so that $B\cos 2\theta = (A - C)\sin 2\theta$. Since $-(A - C) = C - A$, we have: $B' = (C - A)\sin 2\theta + B\cos\theta = 0$.

15. (a) From Exercise 13(a) we have $(B')^2 - 4A'C' =$
$(B\cos^2\theta - 2A\cos\theta\sin\theta + 2C\cos\theta\sin\theta -$
$B\sin^2\theta)^2 - 4(A\cos^2\theta + B\cos\theta\sin\theta +$
$C\sin^2\theta)(C\cos^2\theta - B\cos\theta\sin\theta + A\sin^2\theta) =$
$[B(\cos^2\theta - \sin^2\theta) + 2(C - A)\cos\theta\sin\theta]^2 -$
$4(A\cos^2\theta + B\cos\theta\sin\theta + C\sin^2\theta)(C\cos^2\theta -$
$B\cos\theta\sin\theta + A\sin^2\theta) = B^2(\cos^2\theta - \sin^2\theta)^2 +$
$4(C - A)^2\cos^2\theta\sin^2\theta + 4B(C - A)(\cos^2\theta -$
$\sin^2\theta)\cos\theta\sin\theta - [4AC(\cos^4\theta + \sin^4\theta) +$
$4(A^2 + C^2 - B^2)\cos^2\theta\sin^2\theta - 4AB(\cos^3\theta\sin\theta -$
$\cos\theta\sin^3\theta) + 4BC(\cos^3\theta\sin\theta - \cos\theta\sin^3\theta)] =$
$B^2(\cos^4\theta - 2\cos^2\theta\sin^2\theta + \sin^4\theta + 4\cos^2\theta\sin^2\theta) -$
$4AC(2\cos^2\theta\sin^2\theta + \cos^4\theta + \sin^4\theta)$ (everything else
cancels) $= (B^2 - 4AC)(\cos^4\theta + 2\cos^2\theta\sin^2\theta +$
$\sin^4\theta) = (B^2 - 4AC)(\cos^2\theta + \sin^2\theta)^2 = B^2 - 4AC$
(b) If $B^2 - 4AC < 0$, then also $(B')^2 - 4A'C' < 0$. Since
$B' = 0$, $-4A'C' < 0$ and so $A'C' > 0$. By Exercise 14,
the graph is an ellipse. The other two cases are proved
in the same way.

Section 10.5, page 738

1. $-5 \le x \le 6$ and $-2 \le y \le 2$
3. $-3 \le x \le 4$ and $-2 \le y \le 3$
5. $-3 \le x \le 3$ and $-2 \le y \le 2$
7. $-5 \le x \le 4$ and $-4 \le y \le 5$
9. $0 \le x \le 14$ and $-15 \le y \le 0$
11. $-2 \le x \le 20$ and $-11 \le y \le 11$
13. $-12 \le x \le 12$ and $-12 \le y \le 12$
15. $-2 \le x \le 20$ and $-20 \le y \le 4$
17. $-25 \le x \le 22$ and $-25 \le y \le 26$
19. $y = 2x + 7$ **21.** $y = 2x + 5$
23. $x = y^3 - y$ **25.** $y = x^8 - 1$
27. $x = e^y$ (or $y = \ln x$)   **29.** $x^2 + y^2 = 9$
31. Both give a straight line segment between $P = (-4, 7)$ and
$Q = (2, -5)$. The parametric equations in (a) move from P
to Q, and the parametric equations in (b) move from Q to P.
33. $x = 6\cos t + 7$ and $y = 6\sin t - 4$ $(0 \le t \le 2\pi)$
35. $x = 3\cos t + 2$ and $y = 3\sin t - 2$ $(0 \le t \le 2\pi)$
37. $x = 5\cos t$ and $y = \sqrt{18}\sin t$ $(0 \le t \le 2\pi)$
39. $x = \dfrac{3}{\cos t}$ and $y = \sqrt{15}\tan t$ $(0 \le t \le 2\pi)$
41. $x = \sqrt{24}\tan t + 5$ and $y = \dfrac{6}{\cos t} + 2$ $(0 \le t \le 2\pi)$
43. (a) $\dfrac{d - b}{c - a}$ (b) $y - b = \dfrac{d - b}{c - a}(x - a)$
(c) Solving both equations for t, we obtain $t = \dfrac{x - a}{c - a}$ and
$t = \dfrac{y - b}{d - b}$.
Hence, $\dfrac{y - b}{d - b} = \dfrac{x - a}{c - a}$; that is,
$y - b = \dfrac{d - b}{c - a}(x - a)$.

45. $x = 18t - 6$ and $y = 12 - 22t$ $(0 \le t \le 1)$

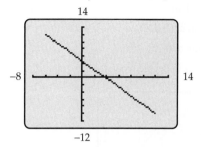

47. $x = 18 - 34t$ and $y = 10t + 4$ $(0 \le t \le 1)$

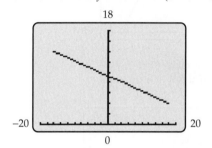

49. (a) $k = 1$ $k = 2$

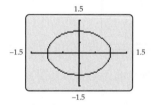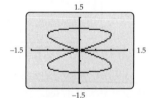

$k = 3$ $k = 4$

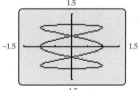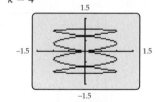

(b) $k = 5$ $k = 6$

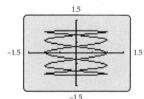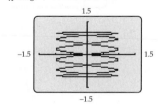

51. (a)–(d)

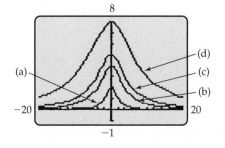

(e) It has the same shape (a bit like a witch's hat), but its peak is at (0, 6).

53. Local minimum at $(-6, 2)$

55. Local maximum at $(4, 5)$

57. (a) Calculator is in degree mode for this graph.

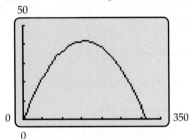

(b) $t = 3.23$ seconds

(c) $y = 41.67$ feet

59. (a)

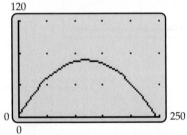

(b) The arrow will go over the wall.

61. $v \approx 105.29$ ft/sec.

63. (a)

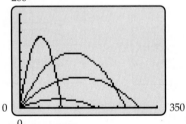

(b) $40°$

(c) It appears that the ball would go farthest for a $45°$ angle.

65. (a) $OT = 3t$ (see page 737). The figure shows that $\cos\left(t - \dfrac{\pi}{2}\right) = \dfrac{CQ}{3}$ and $\sin\left(t - \dfrac{\pi}{2}\right) = \dfrac{PQ}{3}$. Hence, $CQ = 3 \cos\left(t - \dfrac{\pi}{2}\right)$ and $PQ = 3 \sin\left(t - \dfrac{\pi}{2}\right)$. So $x = OT - CQ = 3t - 3 \cos\left(t - \dfrac{\pi}{2}\right)$ and $y = CT + PQ = 3 + 3 \sin\left(t - \dfrac{\pi}{2}\right)$.

(b) $\cos(t - \pi/2) = \cos t \cos \pi/2 + \sin t \sin \pi/2 = (\cos t)(0) + (\sin t)(1) = \sin t$. Therefore, $3t - 3 \cos(t - \pi/2) = 3t - 3 \sin t = 3(t - \sin t)$. $\sin(t - \pi/2) = \sin t \cos \pi/2 - \cos t \sin \pi/2 = (\sin t)(0) - (\cos t)(1) = -\cos t$. Therefore, $3 + 3 \sin(t - \pi/2) = 3 - 3 \cos t = 3(1 - \cos t)$.

67. (a) $OT = 3t$ (see page 737). The figure shows that $\cos\left(t - \dfrac{3\pi}{2}\right) = \dfrac{CQ}{3}$ and $\sin\left(t - \dfrac{3\pi}{2}\right) = \dfrac{PQ}{3}$. Hence, $CQ = 3 \cos\left(t - \dfrac{3\pi}{2}\right)$ and $PQ = 3 \sin\left(t - \dfrac{3\pi}{2}\right)$. So $x = OT + CQ = 3t + 3 \cos\left(t - \dfrac{3\pi}{2}\right)$ and $y = CT - PQ = 3 - 3 \sin\left(t - \dfrac{3\pi}{2}\right)$.

(b) $\cos(t - 3\pi/2) = \cos t \cos 3\pi/2 + \sin t \sin 3\pi/2 = (\cos t)(0) + (\sin t)(-1) = -\sin t$. Therefore, $3t + 3 \cos(t - 3\pi/2) = 3t - 3 \sin t = 3(t - \sin t)$. $\sin(t - 3\pi/2) = \sin t \cos 3\pi/2 - \cos t \sin 3\pi/2 = (\sin t)(0) - (\cos t)(-1) = \cos t$. Therefore, $3 - 3 \sin(t - 3\pi/2) = 3 - 3 \cos t = 3(1 - \cos t)$.

69. (a) Particles A and B do not collide.

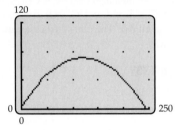

(b) $t \approx 1.10$.

(c) A and C are closest near $t = 1.13$, but do not collide.

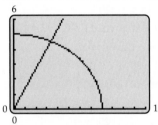

(d) The given function is obtained by applying the distance formula to the points $(8 \cos t, 5 \sin t) = $ A and $(3t, 4t) = $ C. $y = \sqrt{(8 \cos t - 3t)^2 + (5 \sin t - 4t)^2}$ There is a minimum at $t = 1.1322$; however, the minimum is not zero.

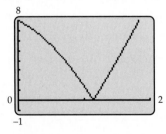

71. (a) As shown on next page, the center Q of the small circle is always at distance $a - b$ from the origin O. Suppose t is the angle that OQ makes with the x-axis. Then the coordinates of Q are $x = (a - b) \cos t$, $y = (a - b) \sin t$. Examining the smaller circle in detail, we see that the change in x-coordinate from Q to P is $b \cos u$, where u is the angle that PQ makes with the positive *x*-axis. Likewise, the change in *y*-coordinate from *Q* to *P* is $-b \sin u$. Therefore, the coordinates of *P* are

$$x = (a - b) \cos t + b \cos u$$
$$y = (a - b) \sin t - b \sin u.$$

The angles *t* and *u* are related by the fact that the inner circle must roll without "slipping." This means the arc length that *P* has moved around the inner circle must equal the arc length that the inner circle has moved along the circumference of the larger circle. In other words, the arc length from *P* to *W* must equal the arc length from *S* to *V*. Since the length of a circular arc is the radius times the angle, this means

$$bu = (a - b)t, \text{ or } u = (a - b)t/b.$$

Substituting this for *u* in the above equations will give the desired parametric equations.

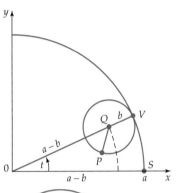

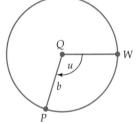

(b)

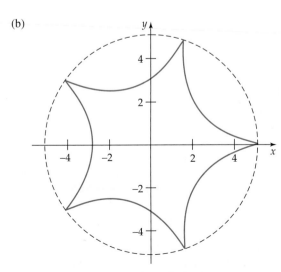

(c)

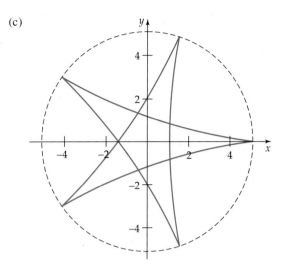

Section 10.6, page 752

1. $P: \left(2, \dfrac{\pi}{4}\right)$, $Q: \left(3, \dfrac{2\pi}{3}\right)$, $R: (5, \pi)$, $S: \left(7, \dfrac{7\pi}{6}\right)$, $T: \left(4, \dfrac{3\pi}{2}\right)$, $U: \left(6, -\dfrac{\pi}{3}\right)$, $V: (7, 0)$

3.

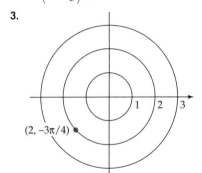

(2, −3π/4)

5.

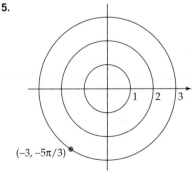

(−3, −5π/3)

Many answers are possible in Exercises 7–12 in addition to those given here.

7. $\left(3, \dfrac{7\pi}{3}\right), \left(3, -\dfrac{5\pi}{3}\right), \left(-3, \dfrac{4\pi}{3}\right), \left(-3, -\dfrac{2\pi}{3}\right)$

9. $\left(2, \dfrac{4\pi}{3}\right), \left(2, -\dfrac{8\pi}{3}\right), \left(-2, \dfrac{\pi}{3}\right), \left(-2, -\dfrac{5\pi}{3}\right)$

11. $\left(\sqrt{3}, \dfrac{11\pi}{4}\right), \left(\sqrt{3}, -\dfrac{5\pi}{4}\right), \left(-\sqrt{3}, \dfrac{7\pi}{4}\right), \left(-\sqrt{3}, -\dfrac{\pi}{4}\right)$

13. $\left(\dfrac{3}{2}, \dfrac{3\sqrt{3}}{2}\right)$ **15.** $\left(\dfrac{\sqrt{3}}{2}, -\dfrac{1}{2}\right)$

17. $(.4255, -1.438)$ **19.** $(-3.604, 1.736)$

21. $\left(6, -\dfrac{\pi}{6}\right)$ **23.** $\left(\sqrt{2}, \dfrac{\pi}{4}\right)$

25. $\left(6, \dfrac{\pi}{3}\right)$ **27.** $(2\sqrt{5}, 1.107)$

29. $(5.59, 2.6679)$ **31.** $\left(2, \dfrac{3\pi}{2}\right)$

33. $(4.47, 2.0344)$ **35.** $r = 5$

37. $r = 12 \sec \theta$ **39.** $r = \dfrac{1}{\sin \theta - 2 \cos \theta}$

41. $x^2 + y^2 = 9$ **43.** $y = \dfrac{x}{\sqrt{3}}$

45. $x = 1$ **47.** $x^3 + xy^2 = y$

49. $x^2 + y^2 = 2y$ **51.** $x^2 = 16 - 8y$

53.

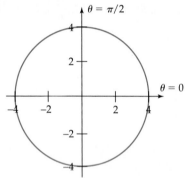

55.

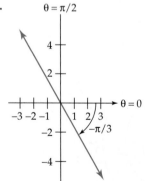

57.

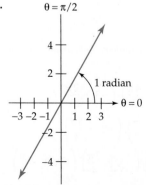

59.

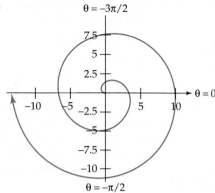

61.

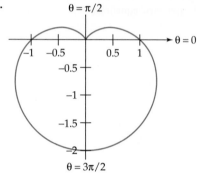

63.

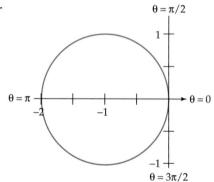

65.

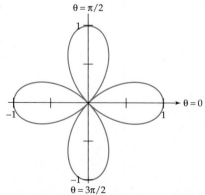

67.

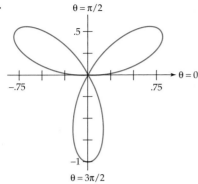

69.

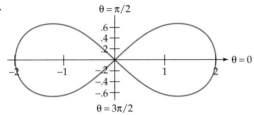

71.

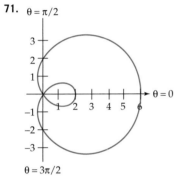

73.

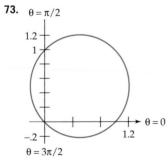

75.

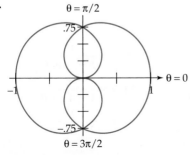

77.

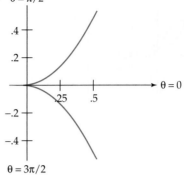

79.

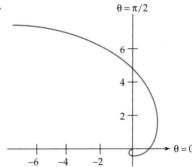

81.

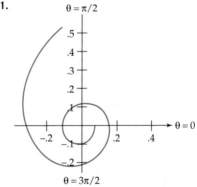

83. (a)

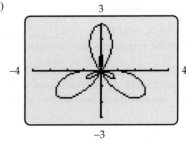

(b)

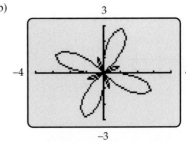

(c)

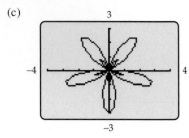

85. $r = a \sin \theta + b \cos \theta$

$r^2 = ar \sin \theta + br \cos \theta$

$x^2 + y^2 = ay + bx$

$x^2 - bx + \dfrac{b^2}{4} + y^2 - ay + \dfrac{a^2}{4} = \dfrac{a^2 + b^2}{4}$

$\left(x - \dfrac{b}{2}\right)^2 + \left(y - \dfrac{a}{2}\right)^2 = \dfrac{a^2 + b^2}{4}$

This is the equation of a circle.

87. The distance from (r, θ) to (s, β) is given by the Law of Cosines.

$$d^2 = r^2 + s^2 - 2rs \cos(\theta - \beta)$$

$$d = \sqrt{r^2 + s^2 - 2rs \cos(\theta - \beta)}$$

Section 10.7, page 761

1. (d) **3.** (c) **5.** (a)

7. Hyperbola, $e = 4/3$ **9.** Parabola, $e = 1$

11. Ellipse, $e = 2/3$ **13.** .1

15. $\sqrt{5}$ **17.** $5/4$

19. (a)

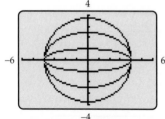

(b) $\sqrt{15}/4, \sqrt{10}/4, \sqrt{2}/4$

(c) The smaller the eccentricity, the closer the shape is to circular.

In the graphs for Exercises 21–32, the x- and y-axes with scales are given for convenience, but coordinates of points are in polar coordinates.

21.

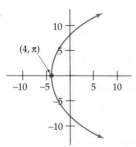

23.

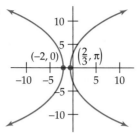

25.

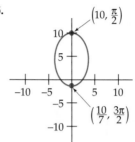

27.

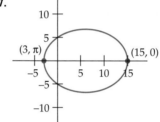

29.

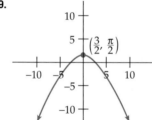

31.

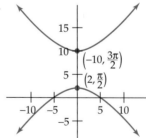

33. $r = \dfrac{6}{1 - \cos \theta}$ **35.** $r = \dfrac{16}{5 + 3 \sin \theta}$

37. $r = \dfrac{3}{1 + 2 \cos \theta}$ **39.** $r = \dfrac{8}{1 - 4 \cos \theta}$

41. $r = \dfrac{3}{1 - \sin \theta}$ **43.** $r = \dfrac{2}{2 + \cos \theta}$

45. $r = \dfrac{2}{1 - 2\cos\theta}$

47. Since $0 < e < 1$, $0 < 1 - e^2 < 1$ as well. The formulas for a^2 and b^2 show that $a^2 = b^2/(1 - e^2)$, so $a^2 > b^2$. Since a and b are both positive, $a > b$.

49. $r = \dfrac{3 \cdot 10^7}{1 - \cos\theta}$

Chapter 10 Review, page 764

1. Ellipse, vertices $\left(0, \pm 2\sqrt{5}\right)$, foci $(0, \pm 2)$

3. Ellipse, vertices $(1, -1)$ and $(1, 7)$, foci $(1, 0)$ and $(1, 6)$

5. Focus: $(0, 5/14)$, directrix: $y = -5/14$

7.

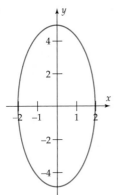

9.

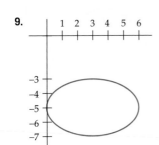

11. Asymptotes: $y + 4 = \pm\dfrac{5}{2}(x - 1)$

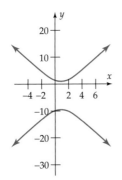

13.

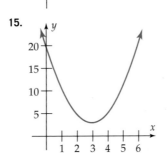

15.

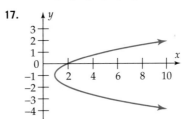

17.

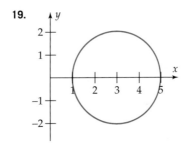

19.

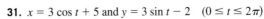

21. Center: $(-4, -5)$, radius: $2\sqrt{2}$

23. $(4, -6)$

25. $\dfrac{(x - 3)^2}{4} + \dfrac{(y - 1)^2}{2} = 1$

27. $\dfrac{y^2}{4} - \dfrac{(x - 3)^2}{16} = 1$

29. $\left(y + \dfrac{1}{2}\right)^2 = -\dfrac{1}{2}\left(x - \dfrac{3}{2}\right)$

31. $x = 3\cos t + 5$ and $y = 3\sin t - 2$ $(0 \le t \le 2\pi)$

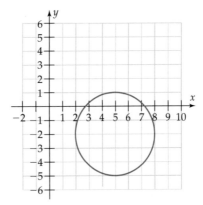

33. $x = 2 \cos t$ and $y = 5 \sin t$ $(0 \le t \le 2\pi)$

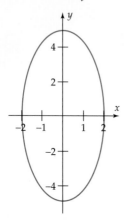

35. $x = 3 \cos t + 3$ and $y = 2 \sin t - 5$ $(0 \le t \le 2\pi)$

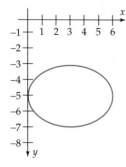

37. $x = 2 \tan t + 1$ and $y = \dfrac{5}{\cos t} - 4$ $(0 \le t \le 2\pi)$

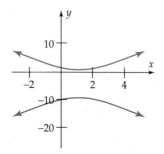

39. $x = (t + 4)^2 - 2$ and $y = t$ (any real number t)

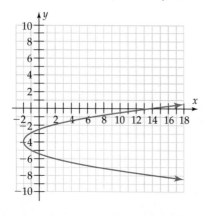

41. 6 feet

43. The receiver should be placed at the focus, $\frac{9}{16}$ feet or 6.25 inches from the vertex.

45. Ellipse

47. Hyperbola

49. $-6 \le x \le 6$ and $-4 \le y \le 4$

51. $-9 \le x \le 9$ and $-6 \le y \le 6$

53. $-15 \le x \le 10$ and $-10 \le y \le 20$

55. $x = \dfrac{1}{2} u - \dfrac{\sqrt{3}}{2} v, y = \dfrac{\sqrt{3}}{2} u + \dfrac{1}{2} v$

57. $45°$

59. $-15 \le x \le 15$ and $-10 \le y \le 10$

61. $-35 \le x \le 32$ and $-2 \le y \le 16$

63. $x = 3 - 2y$

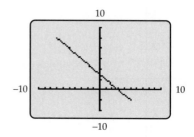

65. $y = 2 - 2x^2$

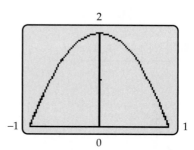

67. (b) and (c)

69.

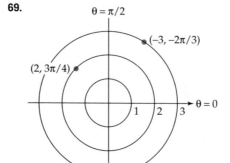

71.

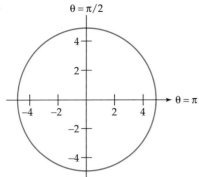

73.

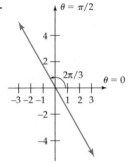

75.

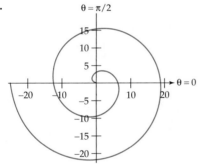

77.

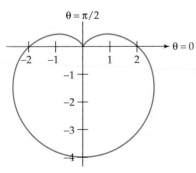

79.

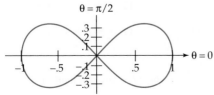

81. $\left(-\dfrac{3}{2}, -\dfrac{3\sqrt{3}}{2}\right)$

83. Eccentricity $= \sqrt{\dfrac{2}{3}} \approx 0.8165$

85. Ellipse

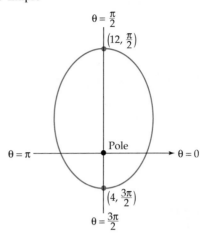

87. Hyperbola

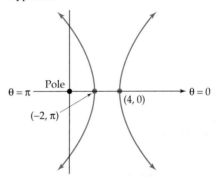

89. $r = \dfrac{24}{5 + \cos\theta}$ **91.** $r = \dfrac{2}{1 + \cos\theta}$

Chapter 10 Test, page 766

1. (a) Focus $(0, 5/4)$; directrix $y = -5/4$
(b) $x^2 = 36y$

2. (a) Hyperbola
(b) Center $(0,0)$; vertices $(0, 3)$, $(0, -3)$; foci $(0, \sqrt{45})$, $(0, -\sqrt{45})$
(c)

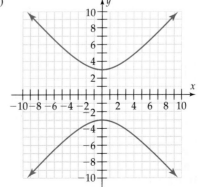

3. Ellipse; center $(-1, 3)$

4. (a) $(2, -2)$

(b)

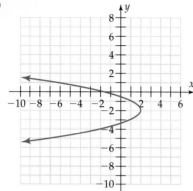

5. $\dfrac{(x + 6)^2}{4} - \dfrac{(y - 1)^2}{16} = 1$

6. $\dfrac{y^2}{4} - \dfrac{x^2}{9} = 1$

7. (a) Parabola
(b) Hyperbola

8. (a) Ellipse
(b) Center $(1, -2)$; vertices $(-3, -2), (5, -2)$

9. 41.42 ft

10. (a) Ellipse
(b) Center $(0, 0)$; vertices $(0, -\sqrt{7}), (0, \sqrt{7})$;
foci $(0, -\sqrt{3}), (0, \sqrt{3})$

(c)

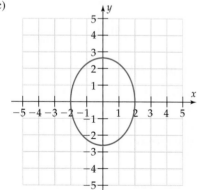

11. Hyperbola; $-15 \leq x \leq 50$ and $-200 \leq y \leq 75$

12. $S = \left(7, \dfrac{7\pi}{6}\right); T = \left(5, \dfrac{3\pi}{2}\right)$

13. $x = 5 \cos t - 3$ and $y = 5 \sin t + 4$ $(0 \leq t \leq 2\pi)$

14. $x = -3(t - 4)^2 - 5$ and $y = t$ (t any real number)

15. (a) $(3\sqrt{2}, 3\sqrt{2})$
(b) $\left(18, -\dfrac{\pi}{6}\right)$

16. $x = y^2 - 3y + 2$

17.

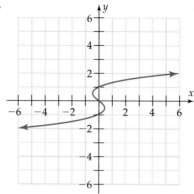

18. (a) $x = 40\sqrt{2}\, t$ and $y = 40\sqrt{2}\, t - 16t^2 + 5$
(b) About 3.62 seconds later

19. $x = \dfrac{3}{\cos t}$ and $y = 4 \tan t$ $(0 \leq t \leq 2\pi)$

20. $x = 2 \cos t - 4$ and $y = 2 \sin t + 2$ $(0 \leq t \leq 2\pi)$

21.

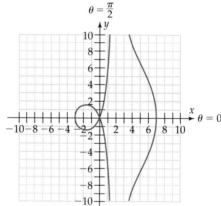

22. $r = \dfrac{20}{9 + 4 \sin \theta}$

Chapter 11

Section 11.1, page 781

1. The values are a solution.

3. The values are a solution.

5. The values are not a solution.

7. $x = 6/5, y = 8/5$ **9.** $x = 2/7, y = -11/7$

11. $r = 5/2, s = -5/2$ **13.** $x = \dfrac{3}{2}c, y = d - \dfrac{1}{2}c$

15. $x = 28, y = 22$ **17.** $x = 2, y = -1$

19. Inconsistent

21. $x = b, y = \dfrac{4b - 2}{5}$, where b is any real number.

23. $x = -6, y = 2$ **25.** $x = 13.2, y = 3.6$

27. $x = 7/11, y = -7$ **29.** $x = 1, y = 1/2$

31. (a) Solve each equation for y, and observe that a positive slope represents an increasing population; a negative slope represents a decreasing population.
(b) 1988

33. 2115

35. $x = \dfrac{rd - sb}{ad - bc}, y = \dfrac{as - rc}{ad - bc}$

37. Since the two lines have slopes $-\frac{1}{2}$ and 2, they are perpendicular and intersect at one point, regardless of c. Therefore, there is exactly one solution for the system. Alternatively, solve the system to find that the only solution is $x = \dfrac{3c + 8}{15}, y = \dfrac{6c - 4}{15}$.

39. $c = -3, d = 1/2$

41. Equilibrium quantity is 20,000; equilibrium price is $17.

43. Equilibrium quantity is 4000; equilibrium price is $180.

45. 145 adults; 55 children

47. Boat: 18 mph; current: 2 mph

49. $1\frac{1}{2}$ lb cashews and $4\frac{1}{2}$ lb peanuts

51. 24 grams of 50% alloy; 16 grams of 75% alloy

53. (a) Electric: $y = 2000 + 960x$, solar: $y = 14,000 + 114x$
(b) Electric: $6800, solar: $14,570
(c) The costs will be the same when $x = 14.2$. Electric heating will be cheaper before that, solar afterwards.

55. (a) $R = 250,500x$
(b) $C = 1,295,000 + 206,500x$
(c) $P = R - C = 44,000x - 1,295,000$
(d) $R = 109,000x$, $C = 440,000 + 82,000x$,
$P = R - C = 27,000x - 440,000$
(e) 50.3 weeks, $918,000
(f) It would be better to open off Broadway if a run of 50 weeks or less is expected.

57. (a) $R = 60x$
(b) 400 hedge trimmers

59. Since the break-even point is above the maximum number that can be sold, the product should not be produced.

61. $14,450 at 9%, $6,450 at 11%

63. (a) $y = .373x + 27.02$
(b) $y = .624x + 23.65$ (c) 2003

65. 80 bowls and 120 plates

Special Topics 11.1.A, page 790

1. $x = 3, y = 9$ or $x = -1, y = 1$

3. $x = \dfrac{-1 + \sqrt{41}}{4}, y = \dfrac{21 - \sqrt{41}}{8}$ or $x = \dfrac{-1 - \sqrt{41}}{4}$, $y = \dfrac{21 + \sqrt{41}}{8}$

5. $x = 7, y = 3$ or $x = 3, y = 7$

7. $x = 0, y = -3$ or $x = 8, y = 1$

9. $x = 2, y = 0$ or $x = 4, y = 2$

11. $x = 4, y = 3$ or $x = -4, y = 3$ or $x = \sqrt{21}, y = -2$ or $x = -\sqrt{21}, y = -2$

13. $x = -1.6237, y = -8.1891$ or $x = 1.3163, y = 1.0826$ or $x = 2.8073, y = 2.4814$

15. $x = -1.9493, y = .4412$ or $x = .3634, y = .9578$ or $x = 1.4184, y = .5986$

17. $x = -.9519, y = -.8145$

19. No solutions

21. $x = 2 + \sqrt{5}, y = -1 + \sqrt{5}$ or $x = 2 - \sqrt{5}, y = -1 - \sqrt{5}$

23. $x = -4.8093, y = 19.3201$ or $x = -3.1434, y = 7.7374$ or $x = 2.1407, y = 7.7230$ or $x = 2.8120, y = 11.7195$

25. $x = -3.8371, y = -2.2596$ or $x = -.9324, y = -7.7796$

27. $x = -1.4873, y = .0480$ or $x = -.0480, y = 1.4873$ or $x = .0480, y = -1.4873$ or $x = 1.4873, y = -.0480$

29. The intersection point nearest the origin is $(2.4, 3.2)$.

31. Two possible boxes: One is 2 by 2 by 4 meters and the other is approximately 3.1231 by 3.1231 by 1.6404 meters.

33. -13 and -4 **35.** 30 and 4.25; 4.25 and 30

37. 16 and -14 **39.** 16.5 feet by 10 feet

41. 32.1 cm and 14.0 cm

Section 11.2, page 803

1. $\begin{pmatrix} 2 & -3 & 4 & 1 \\ 1 & 2 & -6 & 0 \\ 3 & -7 & 4 & -3 \end{pmatrix}$ **3.** $\begin{pmatrix} 2 & -\frac{5}{2} & \frac{2}{3} & 0 \\ 1 & -\frac{1}{4} & 4 & 0 \\ 0 & -3 & \frac{1}{2} & 0 \end{pmatrix}$

5. $3x - 5y = 4$
$9x + 7y = 2$

7. $\begin{aligned} x + \quad\quad z \quad\quad &= 1 \\ x - \ y + 4z - 2w &= 3 \\ 4x + 2y + 5z \quad\quad &= 2 \end{aligned}$

9. $x = 3/2, y = 5, z = -2, w = 0$

11. $x = 3 - 2t, y = 5 - 3t, z = 2, w = t$, where t is any real number.

13. $y = 3/2, x = 3/2, z = -3/2$

15. $z = t, y = -1 + \frac{1}{3}t, x = 2 - \frac{4}{3}t$, where t is any real number.

17. $x = -14, y = -6, z = 2$

19. $x = 100, y = 50, z = 50$

21. $z = t, y = -2t + 1/2, x = t$, where t is any real number.

23. $x = 2, y = 3$

25. No solution

27. $z = t, y = t - 1, x = -t + 2$, where t is any real number.

29. $x = y = z = 0$

31. $x = -1, y = 1, z = -3, w = -2$

33. $x = 7/31, y = 6/31, z = 1/31, w = 29/31$

35. $x = 1/2, y = 1/3, z = -1/4$

37. $x = 3/5, y = 1/5$ **39.** $x = -1, y = 2$

41. $A = 1, B = 3$

43. $A = -3/25, B = 3/25, C = 7/5$

45. $A = 2, B = 3, C = -1$

47. $3000 from friend, $6000 from the bank, $1000 from the insurance company

49. $15,000 in the mutual fund, $30,000 in bonds, $25,000 in the fast-food franchise

51. 6 cups of Progresso, 9 cups of Healthy Choice, 2 cups of Campbell's. The serving size is 1.7 cups.

53. Three possible solutions: 18 bedroom models, 13 living room models, 0 whole-house models, or 16 bedroom models, 8 living room models, 2 whole-house models, or 14 bedroom models, 3 living room models, 4 whole-house models

55. (a) X-ray 1: $a + b$. X-ray 2: $a + c$.
(b) $a = .405, b = .345, c = .195$
(c) A is bone, B is tumorous, C is healthy.

57. 2000 chairs, 1600 chests, and 2500 tables.

59. (a) $x + t = 1000$
$x + y = 1100$
$y + z = 700$
$t + z = 600$
(b) $z = 600 - t, y = 100 + t, x = 1000 - t$
(c) The smallest number of cars leaving A on 4th Avenue is 0; the largest is 600. The smallest number of cars leaving A on Euclid is 400; the largest is 1000. The smallest number of cars leaving C on 5th Avenue is 100; the largest is 700. The smallest number of cars leaving C on Chester is 0; the largest is 600.

Section 11.3, page 816

1. AB is a 2×4 matrix, but BA is not defined.

3. AB is a 3×3 matrix, BA is a 2×2 matrix.

5. AB is a 3×2 matrix, but BA is not defined.

7. $\begin{pmatrix} 13 & 4 & 3 \\ 25 & 5 & -12 \end{pmatrix}$ **9.** $\begin{pmatrix} 1 & -1 \\ 3 & -1 \\ 6 & 2 \end{pmatrix}$

11. $\begin{pmatrix} 1 & -1 & 1 & 2 \\ 4 & 3 & 3 & 2 \\ -1 & -1 & -3 & 2 \\ 5 & 3 & 2 & 5 \end{pmatrix}$

13. $AB = \begin{pmatrix} 7 & 8 \\ 14 & 7 \end{pmatrix}$, $BA = \begin{pmatrix} -1 & 2 \\ 8 & 19 \end{pmatrix}$

15. $AB = \begin{pmatrix} 8 & 24 & -8 \\ 2 & -2 & 6 \\ -3 & -21 & 15 \end{pmatrix}$ $BA = \begin{pmatrix} 19 & 9 & 8 \\ -10 & 2 & 0 \\ 0 & 0 & 0 \end{pmatrix}$

17. $\begin{pmatrix} -2 & 1 \\ 3/2 & -1/2 \end{pmatrix}$

19. The matrix has no inverse.

21. The matrix has no inverse.

23. $\begin{pmatrix} -3 & 2 & -4 \\ -1 & 1 & -1 \\ 8 & -5 & 10 \end{pmatrix}$

25. $x = -1, y = 0, z = -3$ **27.** $x = -8, y = 16, z = 5$

29. $x = -.5, y = -2.1, z = 6.7, w = 2.8$

31. $x = 10.5, y = 5, z = -13, v = 32, w = 2.5$

33. $x = -1149/161, y = 426/161, z = -1124/161,$
$w = 579/161$

35. $x = y = z = v = w = 0$

37. The system is inconsistent and has no solution.

39. $w = t, z = s, y = \dfrac{5}{4} - \dfrac{1}{4}s, x = -2 - 3t$, where s and t are any real numbers.

41. The equation of the parabola is $y = \dfrac{35}{8}x^2 + \dfrac{13}{4}x - \dfrac{117}{8}$.

43. (a) $a = \dfrac{115}{9933}, b = \dfrac{7991}{9933}, c = 315$
(b) 1983: 342.35 ppm; 1993: 357.34 ppm; 2003: 374.65 ppm

45. (a) $y = -\dfrac{31}{32}x^2 + \dfrac{61}{2}x + \dfrac{895}{8}$
(b) 1993: $194.66; 1998: $293.88; 2002: $338.38

47. $a = 1, b = -4, c = 1$

49. One pair of jeans is $34.50, one jacket is $72, one sweater is $44, and one shirt is $21.75.

51. Box A: 15,000, box B: 18,000, box C: 54,000

Chapter 11 Review, page 820

1. $x = -5, y = -7$ **3.** $x = 1, y = 1$

5. 2000; about 69 days

7. 6.25 lbs of the 40% alloy and 43.75 lbs of the 80% alloy

9. $x = 3, y = 9$ or $x = -1, y = 1$

11. $x = 1 + \sqrt{7}, y = 1 - \sqrt{7}$ or $x = 1 - \sqrt{7}, y = 1 + \sqrt{7}$

13. $\begin{pmatrix} 3 & -5 & 2 & -9 \\ 3 & 2 & 0 & 0 \\ 1 & 1 & 3 & 19 \end{pmatrix}$ **15.** $\begin{pmatrix} 1 & 7 & -1 & 2 \\ 3 & -5 & -1 & -1 \\ 2 & -2 & -3 & 0 \end{pmatrix}$

17. $x = -1, y = 1, z = 5$ **19.** $x = 2, y = 4, z = 6$

21. $z = t, y = -2t, x = 1 - t$, where t is any real number.

22. No solutions

23. (c)   **25.** $A = 1, B = 3$

27. $\begin{pmatrix} -2 & 3 \\ -4 & -1 \end{pmatrix}$

29. The product AE does not exist.

31. $\begin{pmatrix} -3 & -2 \\ -7 & -5 \end{pmatrix}$ **33.** $\begin{pmatrix} 1 & 2 & -2 \\ -1 & 3 & 0 \\ 0 & -2 & 1 \end{pmatrix}$

35. $x = -1/85,\ y = -14/85,\ z = -21/34,\ w = 46/85$

37. $y = 5x^2 - 2x + 1$

39. (a) $y = -.1x^2 + 6.1x + 11.2$
    (b) 2000: 62.2 hours, 2015: 101.2 hours

41. 100 faculty members

43. 30 pounds of corn, 15 pounds of soybeans, 40 pounds of by-products

Chapter 11 Test, page 822

1. 9 boxes of Basho Bites, 6 boxes of Health Nuggets

2. $A = 2,\ B = 3,\ C = 4$ **3.** $a = -1,\ b = 2,\ c = 3$

4. $x = 5,\ y = 3$

5. (a) $x = 2,\ y = -\dfrac{1}{2}$
    (b) $x = -1.2,\ y = 3.8$
    (c) $x = -2,\ y = e^{-2}$ and $x = -1,\ y = e^{-1}$

6. (a) One solution ($x = 5,\ y = -3$)
    (b) Infinitely many solutions, all of the form $x = b$,
      $y = \dfrac{5b - 6}{3}$
    (c) No solutions
    (d) 4 solutions

7. 62.5, 37.5

8. (a) $x = \dfrac{1}{3},\ y = -2,\ z = 2$
    (b) $x = 3,\ y = -\dfrac{1}{2},\ z = \dfrac{11}{2}$

9. For the following matrices A, B, compute AB if it is defined, and then compute BA if it is defined.
    (a) $AB = \begin{pmatrix} 6 & 5 & 1 \\ 1 & 3 & 3 \\ -2 & 3 & 3 \end{pmatrix},\ BA = \begin{pmatrix} 3 & 6 & -3 \\ 0 & 8 & 2 \\ -2 & -2 & 1 \end{pmatrix}$

    (b) AB undefined, $BA = \begin{pmatrix} -3 & 9 & 1 \\ -12 & 1 & 3 \end{pmatrix}$

    (c) $AB = BA = \begin{pmatrix} 1 & 0 \\ 0 & 1 \end{pmatrix}$ (the identity matrix)

Chapter 12

Section 12.1, page 835

1. $a_1 = 8,\ a_2 = 10,\ a_3 = 12,\ a_4 = 14,\ a_5 = 16$

3. $a_1 = 1,\ a_2 = \dfrac{1}{8},\ a_3 = \dfrac{1}{27},\ a_4 = \dfrac{1}{64},\ a_5 = \dfrac{1}{125}$

5. $a_1 = \dfrac{1}{2},\ a_2 = \dfrac{1}{2},\ a_3 = \dfrac{3}{8},\ a_4 = \dfrac{1}{4},\ a_5 = \dfrac{5}{32}$

7. $a_1 = -\sqrt{3},\ a_2 = 2,\ a_3 = -\sqrt{5},\ a_4 = \sqrt{6},\ a_5 = -\sqrt{7}$

9. $a_1 = 3.9,\ a_2 = 4.01,\ a_3 = 3.999,\ a_4 = 4.0001,\ a_5 = 3.99999$

11. $a_1 = 2,\ a_2 = 7,\ a_3 = 8,\ a_4 = 13,\ a_5 = 14$

13. $a_1 = 3,\ a_2 = 1,\ a_3 = 4,\ a_4 = 1,\ a_5 = 5$

15. $a_n = (-1)^n$ **17.** $a_n = \dfrac{n}{n + 1}$

19. $a_n = 5n - 3$ **21.** $a_n = 3 \cdot 2^{n-1}$

23. $a_n = 4\sqrt{n}$

25. $a_1 = 4,\ a_2 = 11,\ a_3 = 25,\ a_4 = 53,\ a_5 = 109$

27. $a_1 = -16,\ a_2 = -8,\ a_3 = -4,\ a_4 = -2,\ a_5 = -1$

29. $a_1 = 2,\ a_2 = 0,\ a_3 = 3,\ a_4 = 1,\ a_5 = 4$

31. $a_1 = 1,\ a_2 = -2,\ a_3 = 3,\ a_4 = 2,\ a_5 = 3$

33. $a_0 = 2,\ a_1 = 3,\ a_2 = 3,\ a_3 = 9/2,\ a_4 = 27/4$

35. $\displaystyle\sum_{i=1}^{11} i$ **37.** $\displaystyle\sum_{i=1}^{7} \dfrac{1}{2^{i+6}}$ or $\displaystyle\sum_{i=7}^{13} \dfrac{1}{2^i}$

39. $\displaystyle\sum_{i=1}^{7} \dfrac{2^i}{i}$ **41.** 28

43. 34   **45.** 45

47. 224   **49.** 15,015

51. $a_1 + a_2 + a_3 = -10$
    $a_1 + a_2 + a_3 + a_4 + a_5 + a_6 = -2$

53. $a_1 + a_2 + a_3 = 5$
    $a_1 + a_2 + a_3 + a_4 + a_5 + a_6 = 0$

55. $\displaystyle\sum_{n=1}^{6} \dfrac{1}{2n + 1} \approx .9551$ **57.** $\displaystyle\sum_{n=1}^{5} \dfrac{(-1)^{n+1} n}{n + 7} \approx .2558$

59. 2.613035   **61.** 33.465

63. 1.5759958

65. $a_{12} = \$7299.70,\ a_{36} = \$2555.50$

67. (a) About 1,351,200 in 2004 and 1,427,100 in 2007
    (b) About 8,486,700

69. (a) About \$42.60 in 2005 and \$51.84 in 2007
    (b) About \$298.80

71. (b) $a_{17} = 59,\ a_{18} = 61,\ a_{19} = 67,\ a_{20} = 71$

73. $a_1 = 4,\ a_2 = 9,\ a_3 = 25,\ a_4 = 49,\ a_5 = 121$

75. $a_1 = 3,\ a_2 = 7,\ a_3 = 13,\ a_4 = 19,\ a_5 = 23$

77. (a) 1, 1, 2, 3, 5, 8, 13, 21, 34, 55
    (b) 1, 2, 4, 7, 12, 20, 33, 54, 88, 143
    (c) The nth partial sum is $a_{n+2} - 1$.

79. $5(1)^2 + 4(-1)^1 = 1 = 1^2$
    $5(1)^2 + 4(-1)^2 = 9 = 3^2$
    $5(2)^2 + 4(-1)^3 = 16 = 4^2$
    $5(3)^2 + 4(-1)^4 = 49 = 7^2$
    $5(5)^2 + 4(-1)^5 = 121 = 11^2$
    $5(8)^2 + 4(-1)^6 = 324 = 18^2$
    $5(13)^2 + 4(-1)^7 = 841 = 29^2$
    $5(21)^2 + 4(-1)^8 = 2209 = 47^2$
    $5(34)^2 + 4(-1)^9 = 5776 = 76^2$
    $5(55)^2 + 4(-1)^{10} = 15129 = 123^2$

81. Since $a_n = a_{n-1} + a_{n-2}, a_n - a_{n-1} = a_{n-2}$.
Thus,
$$a_3 - a_2 = a_1$$
$$a_4 - a_3 = a_2$$
$$a_5 - a_4 = a_3$$
$$\cdots\cdots\cdots$$
$$\underline{a_{k+2} - a_{k+1} = a_k}$$
Adding: $\quad a_{k+2} - a_2 = \sum_{n=1}^{k} a_n$

Since $a_2 = 1, a_{k+2} - 1 = \sum_{n=1}^{k} a_n$.

Section 12.2, page 842

1. The sequence is arithmetic, with common difference 2.

3. The sequence is not arithmetic.

5. The sequence is arithmetic, with common difference log 2.

7. The sequence is arithmetic, with common difference -1.

9. $a_1 = 9, a_2 = 13, a_3 = 17, a_4 = 21, a_5 = 25$
The sequence is arithmetic, with common difference 4.

11. $c_1 = -1, c_2 = 1, c_3 = -1, c_4 = 1, c_5 = -1$
The sequence is not arithmetic.

13. $a_1 = 1, a_2 = 4, a_3 = -1, a_4 = 6, a_5 = -3$
The sequence is not arithmetic.

15. $a_1 = 1, a_2 = 2, a_3 = 3, a_4 = 4, a_5 = 5,$
The sequence is arithmetic, with common difference 1.

17. $a_{n+1} - a_n = [3 - 2(n+1)] - (3 - 2n) = -2$
The sequence is arithmetic, with common difference -2.

19. $a_{n+1} - a_n = \left(4 + \dfrac{n+1}{3}\right) - \left(4 + \dfrac{n}{3}\right) = \dfrac{1}{3}$
The sequence is arithmetic, with common difference $\frac{1}{3}$.

21. $a_{n+1} - a_n = \dfrac{5 + 3(n+1)}{2} - \dfrac{5 + 3n}{2}$
$$= \dfrac{5 + 3n + 3 - 5 - 3n}{2} = \dfrac{3}{2}$$
The sequence is arithmetic, with common difference $\frac{3}{2}$.

23. $a_{n+1} - a_n = [c + 2(n+1)] - (c + 2n) = 2$
The sequence is arithmetic, with common difference 2.

25. $a_5 = 13; a_n = 2n + 3$ **27.** $a_5 = 5; a_n = \dfrac{n + 15}{4}$

29. $a_5 = 8; a_n = \dfrac{21 - n}{2}$ **31.** $a_5 = 8.4; a_n = 7.9 + .1n$

33. $a_1 = 6; a_n = 2n + 4$ **35.** $a_1 = -7; a_n = 5n - 12$

37. $a_1 = -3; a_n = 7n - 10$ **39.** $a_1 = -6; a_n = \dfrac{3n - 15}{2}$

41. 87 **43.** $-21/4$ **45.** 30 **47.** -81

49. 710 **51.** $-285/2$ **53.** $470/3$

55. (a) $a_n = 3812 + 342n$ (b) $6206; 7232$

57. 428 seats

59. 23.25, 22.5, 21.75, 21, 20.25, 19.5, 18.75 inches

61. 2550 **63.** 20,100

65. \$77,500 in tenth year; \$437,500 over 10 years

67. (a) $a_n = 127.4 + 9n$
(b) \$235.4 billion
(c) \$1277.4 billion

69. (a) $c_n = 15,100.8 + 728n$ (b) \$614,120

Section 12.3, page 850

1. Arithmetic **3.** Geometric **5.** Arithmetic

7. Geometric **9.** Geometric

11. Both arithmetic ($d = 0$) and geometric ($r = 1$)

13. $a_6 = 160; a_n = 5 \cdot 2^{n-1}$ **15.** $a_6 = 1/256; a_n = 4^{2-n}$

17. $a_6 = -5/16; a_n = \dfrac{5(-1)^{n-1}}{2^{n-2}}$

19. $a_6 = 4/27; a_n = 36(1/3)^{n-1}$

21. $a_6 = -16/125; a_n = -\dfrac{2^{n-2}}{5^{n-3}}$

23. $a_n \div a_{n-1} = (-1/2)^n \div (-1/2)^{n-1} = -1/2$. The sequence is geometric, with common ratio $-1/2$.

25. $a_n \div a_{n-1} = 5^{n+2} \div 5^{n-1+2} = 5$. The sequence is geometric, with common ratio 5.

27. $a_n \div a_{n-1} = (\sqrt{5})^n \div (\sqrt{5})^{n-1} = \sqrt{5}$. The sequence is geometric, with common ratio $\sqrt{5}$.

29. $a_n \div a_{n-1} = 10\,e^{\cdot 4n} \div 10\,e^{\cdot 4(n-1)} = e^{\cdot 4}$. The sequence is geometric with common ratio $e^{\cdot 4}$.

31. $a_5 = 1; a_n = (-1/4)^{n-5}$

33. $a_5 = 5000; a_n = \dfrac{1}{2}(10)^{n-1}$

35. $a_5 = 1/4; a_n = 4^{4-n}$

37. $a_5 = 10\sqrt[3]{2}; a_n = 5(\sqrt[3]{2})^{n-1}$

39. $\dfrac{315}{32}$ **41.** 381 **43.** 254

45. $-\dfrac{4921}{19,683}$ **47.** $\dfrac{665}{8}$

49. (a) $a_n = 5.912(.9619)^{n-1}$
(b) 4.01
(c) About 2015

51. (a) $3.9631(1.0515)^{n-1}$
(b) 2000: \$6.228 billion;
2004: \$7.613 billion;
2008: \$9.307 billion

53. (a) $b_n = 204.74(1.065)^{n-1}$
(b) About \$2401.98

55. 23.75 ft

57. (a) $c_n = 20\left[\dfrac{1 - e^{-.5775n}}{1 - e^{-.5775}}\right]$
(b) About 45.337 mg

59. \$21,474,836.47 **61.** \$1898.44

63. Since $a_n = a_1 r^{n-1}$, $\log a_n = \log a_1 + (n-1)\log r$.
Therefore, $\log a_n - \log a_{n-1} = [\log a_1 + (n-1)\log r] - [\log a_1 + (n-2)\log r] = \log r$.
This shows that the sequence $\{\log a_n\}$ is arithmetic with common difference $\log r$.

65. $a_k = 2^{k-1}$ and $r = 2$. The sum of the preceding terms is
$$\sum_{n=1}^{k-1} a_n = a_1\left(\frac{1 - r^{k-1}}{1 - r}\right) = 1\left(\frac{1 - 2^{k-1}}{1 - 2}\right) = 2^{k-1} - 1$$
Thus, each term $a_k = 2^{k-1}$ equals 1 plus the sum of the preceding terms.

67. 3 years and 2 months

Special Topics 12.3.A, page 856

1. 1   **3.** $3/47$

5. $833\frac{1}{3}$ **7.** $4 + 2\sqrt{2}$

9. $2/9$ **11.** $597/110$

13. $\dfrac{10,702}{4995}$ **15.** $\dfrac{174,067}{99,900}$

17. (a) Since $\displaystyle\sum_{n=1}^{\infty} 2(1.5)^n = 2(1.5)^1 + 2(1.5)^2 + 2(1.5)^3 + \cdots = 3 + 3(1.5)^1 + 3(1.5)^2 + \cdots$, this is a geometric series with $a_1 = 3$ and $r = 1.5$.
(b) $f(x) = 6(1.5^x - 1)$.
(c) The function increases faster and faster as x gets larger; the graph does not approach a horizontal line, and the series does not converge.

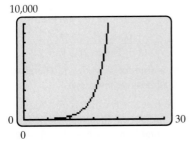

19. We may conjecture that the sum is 1.

21. About 45.58923 mg

Section 12.4, page 862

1. 720   **3.** 56   **5.** 220   **7.** 15

9. 100   **11.** 40   **13.** 0   **15.** 64

17. 3,921,225   **19.** $x^4 + 4x^3 + 6x^2 + 4x + 1$

21. $x^4 + 8x^3 + 24x^2 + 32x + 16$

23. $x^5 + 5x^4 y + 10x^3 y^2 + 10x^2 y^3 + 5xy^4 + y^5$

25. $a^5 - 5a^4 b + 10a^3 b^2 - 10a^2 b^3 + 5ab^4 - b^5$

27. $32x^5 + 80x^4 y^2 + 80x^3 y^4 + 40x^2 y^6 + 10xy^8 + y^{10}$

29. $x^3 + 6x^2\sqrt{x} + 15x^2 + 20x\sqrt{x} + 15x + 6\sqrt{x} + 1$

31. $1 - 10c + 45c^2 - 120c^3 + 210c^4 - 252c^5 + 210c^6 - 120c^7 + 45c^8 - 10c^9 + c^{10}$

33. $x^{-12} + 4x^{-8} + 6x^{-4} + 4 + x^4$

35. 56   **37.** $-8i$

39. $g(x) = x^3 + x^2 - 4$

41. $g(x) = x^4 - 8x^3 + 24x^2 - 29x + 15$

43. $10x^3 y^2$ **45.** $35c^3 d^4$ **47.** $\dfrac{35}{8u^5}$

49. 4032   **51.** 160   **53.** $(x + 1)^5$

55. $(2z + 1)^4$

57. (a) $\dbinom{9}{1} = \dfrac{9!}{1!8!} = \dfrac{9 \cdot 8!}{8!} = 9$ $\dbinom{9}{8} = \dfrac{9!}{8!1!} = 9$

(b) $\dbinom{n}{1} = \dfrac{n!}{1!(n-1)!} = \dfrac{n(n-1)!}{(n-1)!} = n$

$\dbinom{n}{n-1} = \dfrac{n!}{(n-1)!1!} = n$

59. Since $2 = 1 + 1$,
$$2^n = (1 + 1)^n = \binom{n}{0}1^n + \binom{n}{1}1^{n-1} \cdot 1 + \binom{n}{2}1^{n-2} \cdot 1^2 + \cdots + \binom{n}{n}1^n = \binom{n}{0} + \binom{n}{1} + \binom{n}{2} + \cdots + \binom{n}{n}$$

61. $\cos^4\theta + 4i\cos^3\theta\sin\theta - 6\cos^2\theta\sin^2\theta - 4i\cos\theta\sin^3\theta + \sin^4\theta$

63. (a) $f(x + h) - f(x) = \binom{5}{1}x^4 h + \binom{5}{2}x^3 h^2 + \binom{5}{3}x^2 h^3 + \binom{5}{4}xh^4 + \binom{5}{5}h^5$

(b) $\dfrac{f(x + h) - f(x)}{h}$
$$= \frac{h\left[\binom{5}{1}x^4 + \binom{5}{2}x^3 h + \binom{5}{3}x^2 h^2 + \binom{5}{4}xh^3 + \binom{5}{5}h^4\right]}{h}$$
$$= \binom{5}{1}x^4 + \binom{5}{2}x^3 h + \binom{5}{3}x^2 h^2 + \binom{5}{4}xh^3 + \binom{5}{5}h^4$$

(c) $\binom{5}{1}x^4$ or $5x^4$

65. (a) $f(x + h) - f(x)$
$$= \binom{12}{1}x^{11}h + \binom{12}{2}x^{10}h^2 + \cdots + \binom{12}{12}h^{12}$$
(b) $\dfrac{f(x + h) - f(x)}{h}$
$$= \frac{h\left[\binom{12}{1}x^{11} + \binom{12}{2}x^{10}h + \cdots + \binom{12}{12}h^{11}\right]}{h}$$
$$= \binom{12}{1}x^{11} + \binom{12}{2}x^{10}h + \cdots + \binom{12}{12}h^{11}$$
(c) $\binom{12}{1}x^{11}$ or $12x^{11}$

67. (a) $f(x + 10)$ adds 10 to each input; hence, if $x = 0$, $f(x + 10) = f(10)$, that is, if $x = 0$, the value of the function for $1990 + 10 = 2000$ is calculated.

(b) The graph of $g(x) = f(x + 10)$ is the graph of $f(x)$ shifted 10 units to the left.

(c) $g(x) = -.052x^3 - .86x^2 + 40.54x + 652.8$

69. (a) $(n - r)! =$
$(n - r)[(n - r) - 1]! = (n - r)[(n - (r + 1)]!$

(b) $(n - r)! = (n + 1 - r - 1)! = [(n + 1) - (r + 1)]!$

(c) $\dbinom{n}{r + 1} + \dbinom{n}{r} = \dfrac{n!}{(r + 1)![n - (r + 1)]!} +$

$\dfrac{n!}{r!(n - r)!} = \dfrac{(n - r)n!}{(r + 1)![(n - r)[n - (r + 1)]!} +$

$\dfrac{(r + 1)n!}{(r + 1)r!(n - r)!} = \dfrac{(n - r)n!}{(r + 1)!(n - r)!} +$

$\dfrac{(r + 1)n!}{(r + 1)!(n - r)!} = \dfrac{[(n - r) + (r + 1)]n!}{(r + 1)!(n - r)!} =$

$\dfrac{(n + 1)n!}{(r + 1)![(n + 1) - (r + 1)]!} =$

$\dfrac{(n + 1)!}{(r + 1)![(n + 1) - (r + 1)]!} = \dbinom{n + 1}{r + 1}$

(d) Since each entry in row n has form $\dbinom{n}{k}$, the sum of two adjacent entries is $\dbinom{n}{r + 1} + \dbinom{n}{r}$. These are the entries above and to the right and left of $\dbinom{n + 1}{r + 1}$. This explains the assertion.

71. The terms of $(1 + .001)^{1000}$ are all positive and the sum of the first two terms is 2 (why?). Hence, the entire sum is greater than 2.

Section 12.5, page 872

1. When $n = 1$, both sides of the equation are equal to 1; thus the statement is true. Assume that the statement is true for $n = k$. Then $1 + 2 + 2^2 + 2^3 + \cdots + 2^{k-1} = 2^k - 1$. Add 2^k to both sides of the equation to get

$$1 + 2 + 2^2 + 2^3 + \cdots + 2^{k-1} + 2^k = 2^k - 1 + 2^k = 2(2^k) - 1 = 2^{k+1} - 1.$$

Thus, the statement is true for $n = k + 1$. Therefore, by the Principle of Mathematical Induction, the statement is true for all positive integers n.

3. When $n = 1$, both sides of the equation are equal to 1; thus, the statement is true. Assume that the statement is true for $n = k$. Then

$$1 + 3 + 5 + 7 + \cdots + (2k - 1) = k^2.$$

Add the next odd integer, $(2k + 1)$ to both sides:

$$1 + 3 + 5 + 7 + \cdots + (2k - 1) + (2k + 1) = k^2 + (2k + 1) = (k + 1)^2.$$

Thus, the statement is true for $n = k + 1$. Therefore, by the Principle of Mathematical Induction, the statement is true for all positive integers n. That is, the sum of the first n odd integers is n^2.

5. When $n = 1$, both sides of the equation are equal to $\frac{1}{2}$; thus, the statement is true. Assume that the statement is true for $n = k$. Then

$$\frac{1}{2} + \frac{1}{4} + \frac{1}{8} + \cdots + \frac{1}{2^k} = 1 - \frac{1}{2^k}.$$

Add $\dfrac{1}{2^{k+1}}$ to both sides of the equation:

$$\frac{1}{2} + \frac{1}{4} + \frac{1}{8} + \cdots + \frac{1}{2^k} + \frac{1}{2^{k+1}} =$$

$$1 - \frac{1}{2^k} + \frac{1}{2^{k+1}} = 1 + \frac{-2 + 1}{2^{k+1}} = 1 - \frac{1}{2^{k+1}}.$$

Thus, the statement is true for $n = k + 1$. Therefore, the statement is true for all positive integers n.

7. When $n = 1$, both sides of the equation are equal to $\frac{1}{2}$; thus, the statement is true. Assume that the statement is true for $n = k$. Then

$$\frac{1}{1 \cdot 2} + \frac{1}{2 \cdot 3} + \frac{1}{3 \cdot 4} + \cdots + \frac{1}{k(k + 1)} = \frac{k}{k + 1}.$$

Add the next term $\dfrac{1}{(k + 1)(k + 2)}$ to both sides of the equation:

$$\frac{1}{1 \cdot 2} + \frac{1}{2 \cdot 3} + \frac{1}{3 \cdot 4} + \cdots + \frac{1}{k(k + 1)} +$$

$$\frac{1}{(k + 1)(k + 2)} = \frac{k}{k + 1} + \frac{1}{(k + 1)(k + 2)}$$

$$= \frac{(k + 1)^2}{(k + 1)(k + 2)} = \frac{k + 1}{k + 2}.$$

Thus, the statement is true for $n = k + 1$. Therefore, the statement is true for all positive integers n.

9. When $n = 1$, we have $1 + 2 > 1$, which is true. Assume that the statement is true for $n = k$. Then $k + 2 > k$. Add 1 to both sides of the inequality to get $k + 2 + 1 > k + 1$, which is equivalent to $(k + 1) + 2 > (k + 1)$, and the statement is true for $n = k + 1$. Therefore, the statement $n + 2 > n$ is true for all positive integers n.

11. When $n = 1$, we have $3^1 \geq 3(1)$, which is true. Assume that the statement is true for $n = k$. Then $3^k \geq 3(k)$. Multiply both sides of the inequality by 3 to get $3^{k+1} \geq 9k$ or equivalently, $3^{k+1} \geq 3k + 6k$. Since k is a positive integer, $6k > 3$, hence, $3^{k+1} \geq 3k + 6k \geq 3k + 3 = 3(k + 1)$. Thus, the statement is true for $n = k + 1$. Therefore, $3^n \geq 3n$ is true for all positive integers n.

13. When $n = 1$, we have $3(1) > 1 + 1$, which is true. Assume that the statement is true for $n = k$. Then $3k > k + 1$. Add 3 to both sides of the inequality to get $3k + 3 > k + 4$, and since $k + 4 > k + 2$, we have $3k + 3 > k + 2$ which is equivalent to $3(k + 1) > (k + 1) + 1$, and the statement is true for $n = k + 1$. Therefore, the statement $3n > n + 1$ is true for all positive integers n.

15. When $n = 1$, $2^{2n+1} + 1 = 2^3 + 1 = 9$, and the statement is true, since 3 is a factor of 9. Assume that the statement is true for $n = k$. Then 3 is a factor of $2^{2k+1} + 1$. Thus, there is an integer p that we can multiply by 3 to get $2^{2k+1} + 1$. That is, $3p = 2^{2k+1} + 1$, or $3p - 1 = 2^{2k+1}$. Now consider

$$2^{2(k+1)+1} + 1 = 2^{2k+3} + 1 = 2^{2k+1}2^2 + 1 =$$
$$(3p - 1)2^2 + 1 = 12p - 3 = 3(4p - 1).$$

Thus, 3 is a factor of $2^{2(k+1)+1} + 1$, and the statement is true for $n = k + 1$. Thus, 3 is a factor of $2^{2n+1} + 1$ for all positive integers n.

17. When $n = 1$, $3^{2n+2} - 8n - 9 = 64$, and the statement is true, since 64 is a factor of 64. Assume that the statement is true for $n = k$. Then 64 is a factor of $3^{2k+2} - 8k - 9$. Thus, there is an integer p that we can multiply by 64 to get $3^{2k+2} - 8k - 9$. That is, $64p = 3^{2k+2} - 8k - 9$ or $3^{2k+2} = 64p + 8k + 9$. Now consider

$$3^{2(k+1)+2} - 8(k + 1) - 9 = 3^{2k+4} - 8k - 17$$
$$= 9(3^{2k+2}) - 8k - 17$$
$$= 9(64p + 8k + 9) - 8k - 17$$
$$= 9 \cdot 64p + 64k + 64$$
$$= 64(9p + k + 1).$$

Therefore, 64 is a factor of $3^{2(k+1)+2} - 8(k + 1) - 9$, and the statement is true for $n = k + 1$. Thus, 64 is a factor of $3^{2n+2} - 8n - 9$ for all positive integers n.

19. When $n = 1$, both sides of the equation are equal to c, and the statement is true. Assume that the statement is true for $n = k$. Then

$$c + (c + d) + (c + 2d) + (c + 3d) + \cdots +$$
$$(c + (k - 1)d) = \frac{k(2c + (k - 1)d)}{2}.$$

Add $c + kd$ to both sides of the equation to obtain

$$c + (c + d) + (c + 2d) + \cdots + (c + (k - 1)d) +$$
$$(c + kd) = \frac{k(2c + (k - 1)d)}{2} + (c + kd)$$
$$= \frac{2kc + k(k - 1)d + 2c + 2kd}{2}$$
$$= \frac{2c(k + 1) + (k + 1)kd}{2}$$
$$= \frac{(k + 1)(2c + kd)}{2},$$

and the statement is true for $n = k + 1$. Thus, it is true for all positive integers n.

21. (a) $x^2 - y^2 = (x - y)(x + y)$
$x^3 - y^3 = (x - y)(x^2 + xy + y^2)$
$x^4 - y^4 = (x - y)(x^3 + x^2y + xy^2 + y^3)$

(b) *Conjecture:*

$$x^n - y^n = (x - y)(x^{n-1} + x^{n-2}y + x^{n-3}y^2 + \cdots + y^{n-1})$$

From part (a) we see that the statement is true for $n = 2$. Assume that the statement is true for $n = k$. Then $x^k - y^k$
$= (x - y)(x^{k-1} + x^{k-2}y + x^{k-3}y^2 + \cdots + y^{k-1})$.

Consider

$$x^{k+1} - y^{k+1} = xx^k - yy^k = xx^k - xy^k + xy^k - yy^k$$
$$= x(x^k - y^k) + (x - y)y^k$$
$$= x(x - y)(x^{k-1} + x^{k-2}y + x^{k-3}y^2$$
$$+ \cdots + y^{k-1}) + (x - y)y^k$$
$$= (x - y)[x(x^{k-1} + x^{k-2}y + x^{k-3}y^2$$
$$+ \cdots + y^{k-1}) + y^k]$$
$$= (x - y)(x^k + x^{k-1}y + x^{k-2}y^2$$
$$+ \cdots + xy^{k-1} + y^k),$$

and the statement is true for $n = k + 1$. Thus, it is true for all positive integers n.

23. This statement is false. For example, 9 is an odd positive integer, and it is not a prime.

25. When $n = 1$, the statement is true. Assume that the statement is true for $n = k$. Then $(k + 1)^2 > k^2 + 1$. Then

$$(k + 2)^2 = k^2 + 4k + 4 = k^2 + 2k + 1 + 2k + 3$$
$$= (k + 1)^2 + (2k + 3)$$
$$> k^2 + 1 + (2k + 3) = k^2 + 2k + 1 + 3$$
$$= (k + 1)^2 + 1 + 3$$
$$> (k + 1)^2 + 1,$$

and the statement is true for $n = k + 1$. Thus, $(n + 1)^2 > n^2 + 1$ is true for all positive integers n.

27. This statement is false; counterexample: $n = 2$, $n^4 - n + 4 = 18$, and 4 is not a factor of 18.

29. When $n = 5$, we have $2(5) - 4 > 5$, which is true. Assume that the statement is true for $n = k$, where $k \geq 5$. Then $2k - 4 > k$. And $2(k + 1) - 4 = 2k - 4 + 2 > k + 2 > k + 1$. Thus, the statement is true for $n = k + 1$. Therefore, by induction, the statement is true for all $n \geq 5$.

31. When $n = 2$, we have $2^2 > 2$, which is true. Assume that the statement is true for $n = k$, where $k \geq 2$. Then $k^2 > k$. Thus $(k + 1)^2 = k^2 + 2k + 1 > k + 2k + 1 > k + 1$. So the statement is true for $n = k + 1$. Thus, $n^2 > n$ for all $n \geq 2$.

33. When $n = 4$, we have $3^4 > 2^4 + 10(4)$, which is true. Assume that the statement is true for $n = k$, where $k \geq 4$. Then $3^k > 2^k + 10k$. So we have

$$3^{k+1} = 3 \cdot 3^k > 3(2^k + 10k) = 3 \cdot 2^k + 30k >$$
$$2 \cdot 2^k + 30k > 2^{k+1} + 10k + 10$$
$$= 2^{k+1} + 10(k + 1).$$

Therefore, the statement is true for $n = k + 1$. Thus, $3^n > 2^n + 10n$ for all $n \geq 4$.

35. (a) When $n = 2$, you can move the stack in three moves (that is, $2^2 - 1$ moves). When $n = 3$, you can move the stack in seven moves (that is, $2^3 - 1$ moves). When $n = 4$, it takes 15 moves (that is, $2^4 - 1$ moves).

(b) We conjecture that it takes $2^n - 1$ moves to move n rings. Clearly, this holds true for $n = 1$. Assume that the statement is true for $n = k$. That is, it takes $2^k - 1$ moves to move k rings. Now consider $k + 1$ rings. You can move the top k rings in $2^k - 1$ moves, take one move to move the bottom ring, and take $2^k - 1$ moves to move the top k rings onto the bottom. This gives a total of $2^k - 1 + 1 + 2^k - 1 = 2(2^k) - 1 = 2^{k+1} - 1$ moves. Thus the conjecture is true for $n = k + 1$. It will take $2^n - 1$ moves to move n rings for all positive integers n.

37. *De Moivre's Theorem:* For any complex number $z = r(\cos \theta + i \sin \theta)$ and any positive integer n, $z^n = r^n[\cos(n\theta) + i \sin(n\theta)]$. *Proof:* The theorem is obviously true when $n = 1$. Assume that the theorem is true for $n = k$, that is, $z^k = r^k[\cos(k\theta) + i \sin(k\theta)]$. Then
$$z^{k+1} = z \cdot z^k =$$
$$[r(\cos \theta + i \sin \theta)](r^k[\cos(k\theta) + i \sin(k\theta)]).$$

According to the multiplication rule for complex numbers in polar form (multiply the moduli and add the arguments) we have:
$$z^{k+1} = r \cdot r^k[\cos(\theta + k\theta) + i \sin(\theta + k\theta)]$$
$$= r^{k+1}\{\cos[(k + 1)\theta] + i \sin[(k + 1)\theta]\}.$$

This statement says the theorem is true for $n = k + 1$. Therefore, by induction, the theorem is true for every positive integer n.

Chapter 12 Review, page 874

1. $a_1 = -3; a_2 = -1; a_3 = 1; a_4 = 3$

3. $a_1 = 1; a_2 = 1/4; a_3 = 1/9; a_4 = 1/16$

5. $\left(\dfrac{2}{5}\right)^n$

7. (a) About \$4,444,000,000; about \$5,374,000,000; about \$7,150,000,000
(b) 2016

9. $\displaystyle\sum_{n=1}^{10} \dfrac{n}{n+1}$

11. 81 **13.** 40/9

15. (a) About \$25,472 (b) About \$50,431
(c) About \$111,107

17. $a_n = 9 - 6n$ **19.** $a_n = 6n - 11$

21. -55 **23.** $a_n = 2 \cdot 3^{n-1}$

25. $a_n = \dfrac{3}{2^{n-8}}$ **27.** $121/4$

29. $b = 35/4, c = 27/2, d = 73/4$

31. (a) $a_n = 7.898(1.136)^{n-1}$
(b) About \$13.153 billion
(c) About \$66.736 billion

33. (a) $b_n = 100\left(\dfrac{1 - e^{-2.7726n}}{1 - e^{-2.7726}}\right)$
(b) About 106.6665
(c) About 106.6666

35. 2 **37.** 455

39. 1140 **41.** $n + 1$

43. $g(x) = x^3 - 5x^2 + 10x - 5$

45. 80

47. When $n = 1$, both sides of the equation are equal to 1; thus, the statement is true. Assume that the statement is true for $n = k$. Then
$$1 + 5^1 + 5^2 + \cdots + 5^{k-1} = \dfrac{5^k - 1}{4}.$$
Add 5^k to both sides of the equation:
$$1 + 5^1 + 5^2 + \cdots + 5^{k-1} + 5^k = \dfrac{5^k - 1}{4} + 5^k$$
$$= \dfrac{5^k - 1 + 4 \cdot 5^k}{4} = \dfrac{5 \cdot 5^k - 1}{4} = \dfrac{5^{k+1} - 1}{4}.$$
Thus, the statement is true for $n = k + 1$. And by induction, it is true for all positive integers n.

49. Clearly, the statement is true for $x = 0$; hence, assume that $x \neq 0$. Then the statement is true for $n = 1$. Assume that the statement is true for $n = k$; then $|x^k| < 1$. Multiply both sides of this inequality by $|x|$ to obtain $|x^k| \cdot |x| < 1 \cdot |x|$. Thus, $|x^{k+1}| = |x^k| \cdot |x| < |x| < 1$. Thus, the statement is true for $n = k + 1$. Thus, it is true for all positive integers n.

51. When $n = 1$, both sides of the equation are equal to 1; thus, the statement is true. Assume that the statement is true for $n = k$. Then
$$1 + 4^1 + 4^2 + \cdots + 4^{k-1} = \dfrac{4^k - 1}{3}.$$
Add 4^k to both sides:
$$1 + 4^1 + 4^2 + \cdots + 4^{k-1} + 4^k = \dfrac{4^k - 1}{3} + 4^k$$
$$= \dfrac{4^k - 1}{3} + \dfrac{3 \cdot 4^k}{3} = \dfrac{4 \cdot 4^k - 1}{3} = \dfrac{4^{k+1} - 1}{3}.$$
Thus, the statement is true for $n = k + 1$. And by induction, it is true for all positive integers n.

53. When $n = 1$, $9^n - 8n - 1 = 0$, and the statement is true, since 8 is a factor of 0. Assume that the statement is true for $n = k$. Then 8 is a factor of $9^k - 8k - 1$. That is, there is an integer p that we can multiply by 8 to get $9^k - 8k - 1$. Thus, $8p = 9^k - 8k - 1$ or $8p + 8k + 1 = 9^k$. Consider $9^{k+1} - 8(k + 1) - 1$. Rearrange this expression to get $9(9^k) - 8k - 9 = 9(8p + 8k + 1) - 8k - 9 = 72p + 64k = 8(9p + 8k)$, so 8 is a factor of $9^{k+1} - 8(k + 1) - 1$. Thus, the statement is true for $n = k + 1$. Therefore, by induction, the statement is true for all positive integers n.

Chapter 12 Test, page 876

1. $-9, -8, -11, -10, -13$

2. (a) \$3782 (b) \$23,247

3. (a) $-2/9$ (b) $C_n = \dfrac{2(-1)^{n-1}}{3^{n-4}}$

4. 9825

11. $-3x^3 + 15x + 8$

13. $-5xy - x$

15. $15y^3 - 5y$

17. $12a^2x^2 - 6a^3xy + 6a^2xy$

19. $12z^4 + 30z^3$

21. $12a^2b - 18ab^2 + 6a^3b^2$

23. $x^2 - x - 2$

25. $2x^2 + 2x - 12$

27. $y^2 + 7y + 12$

29. $-6x^2 + x + 35$

31. $3y^3 - 9y^2 + 4y - 12$

33. $x^2 - 16$

35. $16a^2 - 25b^2$

37. $y^2 - 22y + 121$

39. $25x^2 - 10bx + b^2$

41. $16x^6 - 8x^3y^4 + y^8$

43. $9x^4 - 12x^2y^4 + 4y^8$

45. $2y^3 + 9y^2 + 7y - 3$

47. $-15w^3 + 2w^2 + 9w - 18$

49. $24x^3 - 4x^2 - 4x$

51. $x^3 - 6x^2 + 11x - 6$

53. $-3x^3 - 5x^2y + 26xy^2 - 8y^3$

55. 3

57. -6

59. 6

61. 1

63. 5

65. $x - 25$

67. $9 + 6\sqrt{y} + y$

69. $\sqrt{3}x^2 + 4x + \sqrt{3}$

71. $3ax^2 + (2a + 3b)x + 2b$

73. $abx^2 + (a^2 + b^2)x + ab$

75. $x^3 - (a + b + c)x^2 + (ab + ac + bc)x - abc$

77. $16x^{n+k}$

79. $y^{r+s} - 4y^r + y^s - 4$

81. $4x^2 + 32x$

83. The statement $3(y + 2) = 3y + 2$ is false when $y = 1$, since $3(1 + 2) = 9$ but $3(1) + 2 = 5$. The mistake is the value 2. The correct statement is $3(y + 2) = 3y + 6$.

85. The statement $(x + y)^2 = x + y^2$ is false when $x = 1$ and $y = 2$, since $(1 + 2)^2 = 9$ but $1 + 2^2 = 5$. The mistake is the first part of the expression $x + y^2$. The correct statement is $(x + y)^2 = x^2 + 2xy + y^2$.

87. The statement $(7x)(7y) = 7xy$ is false when $x = 1$ and $y = 1$, since $(7 \cdot 1)(7 \cdot 1) = 7^2 = 49$, but $7 \cdot 1 \cdot 1 = 7$. The mistake is dropping a 7 on the left side. The correct statement is $(7x)(7y) = 49xy$.

89. The statement $y + y + y = y^3$ is false when $y = 1$, since $1 + 1 + 1 = 3$ but $1^3 = 1$. The mistake is the use of 3 as an exponent. The correct statement is $y + y + y = 3y$.

91. The statement $(x - 3)(x - 2) = x^2 - 5x - 6$ is false when $x = 3$, since $(3 - 3)(3 - 2) = 0$ but $3^2 - 5(3) - 6 = -12$. The mistake is the sign on the 6. The correct statement is $(x - 3)(x - 2) = x^2 - 5x + 6$.

93. If x represents the number chosen, the values at each step are x, $x + 1$, $(x + 1)^2 = x^2 + 2x + 1$, $x - 1$, $(x - 1)^2 = x^2 - 2x + 1$, $(x^2 + 2x + 1) - (x^2 - 2x + 1) = 4x$; $\dfrac{4x}{x} = 4$.

95. Answers will vary.

Section 1.C, page 944

1. $(x + 2)(x - 2)$

3. $(3y + 5)(3y - 5)$

5. $(9x + 2)^2$

7. $(\sqrt{5} + x)(\sqrt{5} - x)$

9. $(7 + 2z)^2$

11. $(x^2 + y^2)(x + y)(x - y)$

13. $(x + 3)(x - 2)$

15. $(z + 3)(z + 1)$

17. $(y + 9)(y - 4)$

19. $(x - 3)^2$

21. $(x + 5)(x + 2)$

23. $(x + 9)(x + 2)$

25. $(3x + 1)(x + 1)$

27. $(2z + 3)(z + 4)$

29. $9x(x - 8)$

31. $2(5x + 1)(x - 1)$

33. $(4u - 3)(2u + 3)$

35. $(2x + 5y)^2$

37. $(x - 5)(x^2 + 5x + 25)$

39. $(x + 2)^3$

41. $(2 + x)(4 - 2x + x^2)$

43. $(-x + 5)^3$

45. $(x + 1)(x^2 - x + 1)$

47. $(2x - y)(4x^2 + 2xy + y^2)$

49. $(x + 2)(x^2 - 2x + 4)(x - 2)(x^2 + 2x + 4)$

51. $(y^2 + 5)(y^2 + 2)$

53. $(9 + y^2)(3 + y)(3 - y)$

55. $(z + 1)(z^2 - z + 1)(z - 1)(z^2 + z + 1)$

57. $(x^2 + 3y)(x^2 - y)$

59. $(x + z)(x - y)$

61. $(a + 2b)(a^2 - b)$

63. $\left(x + \sqrt{8}\right)\left(x - \sqrt{8}\right)(x + 4)$

65. If $x^2 + 1 = (x + c)(x + d) = x^2 + (c + d)x + cd$, then $c + d = 0$ and $cd = 1$. But $c + d = 0$ implies that $c = -d$ and hence that $1 = cd = (-d)d = -d^2$, or equivalently, that $d^2 = -1$. Since there is no real number with this property, $x^2 + 1$ cannot possibly factor in this way.

Section 1.D, page 951

1. $\dfrac{9}{7}$

3. $\dfrac{195}{8}$

5. $\dfrac{x - 2}{x + 1}$

7. $\dfrac{a + b}{a^2 + ab + b^2}$

9. $\dfrac{1}{x}$

11. $\dfrac{29}{35}$

13. $\dfrac{121}{42}$

15. $\dfrac{ce + 3cd}{de}$

17. $\dfrac{b^2 - c^2}{bc}$

19. $\dfrac{-1}{x(x + 1)}$

21. $\dfrac{x + 3}{(x + 4)^2}$

23. $\dfrac{2x - 4}{x(3x - 4)}$

25. $\dfrac{x^2 - xy + y^2 + x + y}{x^3 + y^3}$

27. $\dfrac{-6x^5 - 38x^4 - 84x^3 - 71x^2 - 14x + 1}{4x(x + 1)^3(x + 2)^3}$

29. 2

31. $\dfrac{2}{3c}$

33. $\dfrac{3y}{x^2}$

35. $\dfrac{12x}{x - 3}$

37. $\dfrac{5y^2}{3(y + 5)}$

39. $\dfrac{u + 1}{u}$

41. $\dfrac{(u + v)(4u - 3v)}{(2u - v)(2u - 3v)}$

43. $\dfrac{35}{24}$

45. $\dfrac{u^2}{vw}$ **47.** $\dfrac{x+3}{2x}$ **49.** $\dfrac{x^2y^2}{(x+y)(x+2y)}$

51. $\dfrac{cd(c+d)}{c-d}$ **53.** $\dfrac{y-x}{xy}$ **55.** $\dfrac{-3(y-1)}{y}$

57. $\dfrac{-1}{x(2x+2h)}$ **59.** $\dfrac{xy}{x+y}$ **61.** $\pi/4$

63. $1/25$

65. The statement $\dfrac{1}{a}+\dfrac{1}{b}=\dfrac{1}{a+b}$ is false when $a=1$ and

$b=1$, since $1/1+1/1=2$ but $\dfrac{1}{1+1}=\dfrac{1}{2}$. The correct

statement is $\dfrac{1}{a}+\dfrac{1}{b}=\dfrac{a+b}{ab}$.

67. The statement $\left(\dfrac{1}{\sqrt{a}+\sqrt{b}}\right)^2=\dfrac{1}{a+b}$ is false when $a=1$

and $b=1$, since $\left(\dfrac{1}{\sqrt{1}+\sqrt{1}}\right)^2=\dfrac{1}{4}$ but $\dfrac{1}{1+1}=\dfrac{1}{2}$.

The correct statement is $\left(\dfrac{1}{\sqrt{a}+\sqrt{b}}\right)^2=\dfrac{1}{a+2\sqrt{ab}+b}$.

69. The statement $\dfrac{u}{v}+\dfrac{v}{u}=1$ is false when $u=1$ and $v=2$,

since $\dfrac{u}{v}+\dfrac{v}{u}=\dfrac{1}{2}+\dfrac{2}{1}=2\dfrac{1}{2}$, not 1. The correct statement

is $\dfrac{u}{v}+\dfrac{v}{u}=\dfrac{u^2+v^2}{vu}$.

71. The statement $\left(\sqrt{x}+\sqrt{y}\right)\cdot\dfrac{1}{\left(\sqrt{x}+\sqrt{y}\right)}=x+y$ is false

when $x=1$ and $y=1$, since $(1+1)\cdot\dfrac{1}{(1+1)}=1$ but

$1+1=2$. The correct statement is

$\left(\sqrt{x}+\sqrt{y}\right)\cdot\dfrac{1}{\left(\sqrt{x}+\sqrt{y}\right)}=1.$

Appendix 2, page 961

1. $b=\sqrt{20}=2\sqrt{5}$ **3.** $c=\sqrt{63}=3\sqrt{7}$

5. $b=\sqrt{50}=5\sqrt{2}, c=5$ **7.** $b=\sqrt{75}=5\sqrt{3}, c=5$

9. Angle $A=53.13°$
Angle $R=36.87°$
$r=3, s=4, t=5$

11. Angle $A=65°$ and Angle $R=25°$.
Thus, $\angle A\cong\angle T, \angle C\cong\angle R, AC\cong TR$.
By ASA, the triangles are congruent.

13. $\angle u\cong\angle v$, that is, $\angle POD\cong\angle QOD$. $\angle PDO\cong\angle QDO$, since all right angles are congruent. $OD\cong OD$. By ASA, PDO and QDO are congruent. Hence, corresponding parts are congruent and $PD\cong QD$.

15. $\angle u\cong\angle v$, that is, $\angle BAC\cong\angle DCA$. $AB\cong DC$ and $AC\cong AC$. By SAS, the triangles ABC and CDA are congruent.

17. $\angle x\cong\angle y$, that is, $\angle BAC\cong\angle DAC$. $\angle u\cong\angle v$, that is, $\angle BCA\cong\angle DCA$. $AC\cong AC$. By ASA, the triangles ABC and ADC are congruent.

19. $c=7.5, r=28, t=35$ **21.** $r=3, t=3.5$

23. Since $\dfrac{AB}{RS}=\dfrac{15}{10}=\dfrac{AC}{RT}$ and $\angle A\cong\angle R$, the triangles are similar by S/A/S. $a=15$, $\angle S\cong\angle T$, and both have measure $60°$, since the triangles are equilateral.

25. Since $\dfrac{AC}{RS}=\dfrac{6.885}{4.59}=\dfrac{5.25}{3.5}=\dfrac{BC}{TS}$ and $\angle C\cong\angle S$, the triangles are similar by S/A/S. (S/S/S can also be used.) $\angle T\cong\angle B$, and both have measure $62.46°$. $\angle A\cong\angle R$, and both have measure $42.54°$.

27. Since $\dfrac{a}{r}=\dfrac{5}{17.5}=\dfrac{b}{s}$ and $\angle C\cong\angle T$, the triangles are similar by S/A/S. Since the triangles are isosceles right triangles, $\angle A, \angle B, \angle R, \angle S$ all have measure $45°$. $c=5\sqrt{2}$ and $t=17.5\sqrt{2}$.

29. Since $\angle A\cong\angle R$ and $\angle B\cong\angle T$, the triangles are similar by AA. $\angle C\cong\angle S$ and both have measure $41.5°$; $s=2.4$.

31. 55 ft **33.** 80 ft

35. Distance from Marietta to Columbus: 92.4 mi., distance from Marietta to Cincinnati: 171.2 mi.

37. In the figure, $AD\cong CR, CD\cong AR, AC\cong CA$. Hence, by SSS, triangle $ADC\cong$ triangle CRA, and so $\angle x\cong\angle u$. Similarly, $\angle v\cong\angle y$. Since $\angle x+\angle w+\angle y=180°$, it follows that $\angle u+\angle w+\angle y=180°$.

Index of Applications

Astronomy

black holes, 927–929
brightness of star, 487
comets, 685, 762
earth's orbit around sun, 682–683
periodic orbits of sun and moon,
    571–572, 685
planetary orbits, 349, 412–413
radio telescope, 709–710, 712
satellite orbit, 604, 605, 686
spacecraft, 686
speed of light, 14
sun and moon cycles, 465

Athletics

baseball diamond geometry, 42, 604
baseball thrown, dropped, or hit, 444–445,
    454, 498, 503, 542, 736, 815–816
batting averages, 286–287
car race speed, 72, 114
football field geometry, 50
football thrown, dropped, or kicked, 740
golf ball trajectory, 734–735, 740
gunshot, 740
jogging distances, 175
rowing and running, 110–111,
    114–115, 354
running, 113
stadium seating, 841
swimming distances and time, 652, 665
swimming pool depth, 52
swimming water pressure, 34

Biology and Life Sciences

AIDS cases, number of, 30, 50–51, 100,
    217, 851
Alzheimer's disease, 131
blood alcohol content, 149
blood pressure, 215, 465, 482–483,
    487, 509
computer-aided tomography (CAT)
    scanners, 805
death rates, comparable, 782
disease modeling, 425
drug concentration in bloodstream, 849, 851,
    855–856, 875
FDA drug approval, 129
fungus, growth of, 203
healthy weight range, 15, 66–67
heart disease, death from, 74, 126

illness, temperature and, 216
infant mortality rates, 417
infection, spread of, 384, 392, 408
life expectancy, 130–131, 367, 418, 843
medication, concentration of, 91
multiple births, 416
radioactive decay, 400–401
sizes in scientific notation, 14
trees, size of, 133–134

Business and Manufacturing

advertising spending, 128, 194, 214, 215,
    287, 384
airline passengers, 131
automobile markets, 784
book sales, 836
break-even point, 783–784
car leases, 316
computer software "learning curve," 421
cost functions, 188, 315, 316, 317
costs by unit of production, 120, 132,
    336, 337
hourly earnings, 134, 216–217
input-output analysis, 823–824
ladder building, 843
minimum wage, 214
mixtures, 111, 132, 783, 804–805
net revenue, 875
output by labor and materials, 350, 784,
    806, 821
pensions, 874
prices, 132, 249
production costs, 63–64, 67, 91
production demand, 66
profit margin by unit of production,
    157–158, 160
profits, 121–124, 128, 216, 233, 248,
    282, 843
radio programming, 161
research and development spending, 130
retail expenditures, 875
revenue growth, 44
revenue maximization, 119
salary and wages, 408, 421
sales averages per call, 249
sales growth, 844
sales totals, 791
subdivisions, 316
supply and demand, 137–139
ticket prices, 778–779, 782, 783, 804
time spent discussing budgetary items, 34–35
unemployment statistics, 489
van loads, 805

wireless telephone service, growth in,
    174–175, 207, 408
yield, maximizing, 282

Chemistry and Physics

air quality standards, 820
angle of elevation, sound waves and, 564
balloon filling, 206–207, 213, 221
boiling point of water, 66
cable-hung weight, 652, 653
camera focal length, 303
concentration in solution, 66, 407
daylight hours, 509, 566
density of light, 35–36
exponential decay, 362, 407, 408
flywheels, 512
gas in parts per million, 366, 818
GPS device position, 789, 791
gravitation, 302, 303
gunshot, 566, 569
light rays, 564
mixtures of metals, 780–781
object on a ramp, 310–311, 313, 321–322,
    649–650, 651–652, 659, 662, 664
parabolic satellite dish, 712, 765
particles path, 700
pendulum swings, 39, 488, 490
pistons, 488, 512
pressure of an enclosed gas, 38
puddle, shrinking, 200
pulling an object along ground, 660, 662,
    664, 665
radioactive waste, 303
resistance and resistors, 31, 36–37, 38, 302
rocket travel, 91, 317, 335–336, 554, 570, 595
salt solution, 111, 317
solution, removing items from, 368
sound wave, speed of, 450
spring-hung weight, 38, 484–485, 566, 569
temperature change, 269
temperature of filled container, 209–210
volume of a container, 215
weight supported by beam, 38
whispering room, 685

Construction

arch, 340, 364–365, 367, 765
banner, 619
box, 109–110, 114, 117, 119, 132, 133, 161,
    166–167, 172, 269, 277, 282, 299–300,
    302–303

Subject Index

Trigonometry

If t is a real number and P is the point where the terminal side of an angle of t radians in standard position meets the unit circle, then

$$\cos t = x\text{-coordinate of } P \qquad \sin t = y\text{-coordinate of } P$$

$$\tan t = \frac{\sin t}{\cos t} \qquad \csc t = \frac{1}{\sin t} \qquad \sec t = \frac{1}{\cos t} \qquad \cot t = \frac{\cos t}{\sin t}$$

Point-in-the-Plane Description

For any real number t and point (x, y) on the terminal side of an angle of t radians in standard position:

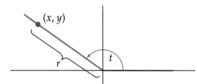

$$\sin t = \frac{y}{r} \qquad \cos t = \frac{x}{r} \qquad \tan t = \frac{y}{x} \quad (x \neq 0)$$

$$\csc t = \frac{r}{y} \quad (y \neq 0) \qquad \sec t = \frac{r}{x} \quad (x \neq 0) \qquad \cot t = \frac{x}{y} \quad (y \neq 0)$$

Periodic Graphs

If $A \neq 0$ and $b > 0$, then each of $f(t) = A\sin(bt + c)$ and $g(t) = A\cos(bt + c)$ has

$$\text{amplitude } |A|, \qquad \text{period } 2\pi/b, \qquad \text{phase shift } -c/b.$$

Right Triangle Trigonometry

$$\sin \theta = \frac{\text{opposite}}{\text{hypotenuse}}$$

$$\cos \theta = \frac{\text{adjacent}}{\text{hypotenuse}}$$

$$\tan \theta = \frac{\text{opposite}}{\text{adjacent}}$$

Special Values

θ		$\sin \theta$	$\cos \theta$	$\tan \theta$
Degrees	Radians			
0°	0	0	1	0
30°	$\dfrac{\pi}{6}$	$\dfrac{1}{2}$	$\dfrac{\sqrt{3}}{2}$	$\dfrac{\sqrt{3}}{3}$
45°	$\dfrac{\pi}{4}$	$\dfrac{\sqrt{2}}{2}$	$\dfrac{\sqrt{2}}{2}$	1
60°	$\dfrac{\pi}{3}$	$\dfrac{\sqrt{3}}{2}$	$\dfrac{1}{2}$	$\sqrt{3}$
90°	$\dfrac{\pi}{2}$	1	0	undefined

Special Right Triangles

Law of Cosines

$$a^2 = b^2 + c^2 - 2bc\cos A$$
$$b^2 = a^2 + c^2 - 2ac\cos B$$
$$c^2 = a^2 + b^2 - 2ab\cos C$$

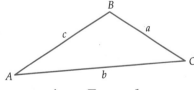

Law of Sines

$$\frac{a}{\sin A} = \frac{b}{\sin B} = \frac{c}{\sin C}$$

Area Formulas

Herron's Formula: Area $= \sqrt{s(s-a)(s-b)(s-c)}$, where $s = \dfrac{1}{2}(a + b + c)$ \qquad Area $= \dfrac{1}{2}ab\sin C$